s in the Environment
g Organisms

sles as a Case Study

metals play key roles in life – all are toxic above a threshold bioavailability, yet many essential to metabolism at lower doses. It is important to appreciate the natural history of an organism in order to understand the interaction between its biology and trace metals. The countryside and indeed the natural history of the British Isles are littered with the effects of metals, mostly via historical mining and subsequent industrial development. This fascinating story encompasses history, economics, geography, geology, chemistry, biochemistry, physiology, ecology, ecotoxicology and, above all, natural history. Examples abound of interactions between organisms and metals in the terrestrial, freshwater, estuarine, coastal and oceanic environments in and around the British Isles. Many of these interactions have nothing to do with metal pollution. All organisms are affected, from bacteria, plants and invertebrates to charismatic species such as seals, dolphins, whales and seabirds. All have a tale to tell.

Philip S. Rainbow has more than 40 years' experience of research into the biology of trace metals. As lecturer, reader and professor at Queen Mary University of London, he taught students at undergraduate, master's and PhD levels, and often ran courses overseas. From 1997 to 2013, he was Keeper of Zoology and subsequently the head of the Department of Life Sciences at the Natural History Museum in London. He has published more than 250 refereed scientific publications, including two co-authored and seven edited books, and also upwards of 30 popular articles. In 2002, he was awarded the Environmental Pollution Kenneth Mellanby Review Award.

Trace Metals in the Environment and Living Organisms

The British Isles as a Case Study

PHILIP S. RAINBOW

Natural History Museum

CAMBRIDGE
UNIVERSITY PRESS

CAMBRIDGE
UNIVERSITY PRESS

University Printing House, Cambridge CB2 8BS, United Kingdom

One Liberty Plaza, 20th Floor, New York, NY 10006, USA

477 Williamstown Road, Port Melbourne, VIC 3207, Australia

314–321, 3rd Floor, Plot 3, Splendor Forum, Jasola District Centre, New Delhi – 110025, India

79 Anson Road, #06–04/06, Singapore 079906

Cambridge University Press is part of the University of Cambridge.

It furthers the University's mission by disseminating knowledge in the pursuit of
education, learning, and research at the highest international levels of excellence.

www.cambridge.org
Information on this title: www.cambridge.org/9781108470933
DOI: 10.1017/9781108658423

First published 2018

Printed in the United Kingdom by TJ International Ltd. Padstow Cornwall

A catalogue record for this publication is available from the British Library.

Library of Congress Cataloging-in-Publication Data
Names: Rainbow, P. S., author.
Title: Trace metals in the environment and living organisms : the British isles as a case study / Philip S. Rainbow,
   Natural History Museum, London.
Description: Cambridge, United Kingdom ; New York, NY : Cambridge University Press, 2018. |
   Includes bibliographical references and index.
Identifiers: LCCN 2018017273| ISBN 9781108470933 (hardback) | ISBN 9781108456869 (paperback)
Subjects: LCSH: Metals–Environmental aspects–British isles.
Classification: LCC QH545.M45 R35 2018 | DDC 572/.51–dc23
   LC record available at https://lccn.loc.gov/2018017273

ISBN 978-1-108-47093-3 Hardback

CONTENTS

PREFACE

This is a book about the natural history of metals. Without metals, there would be no life, and there are habitats in the world where, because of metals, there is no life. This is neither a story of metals in their elemental form, not a discourse on a lump of lead or an ingot of gold. It is a tale of how metals, whether in ionic form in a salt such as zinc chloride, or bound in a molecule as is iron in the protein haemoglobin in our red blood cells, have an all-important role in the biology of life around us.

This is a story of the biology of so-called trace metals, the metals that are to be found in very small 'trace' amounts in the physical environment of the rocks, soils, sediments and waters of the world, and in all forms of life from bacteria to algae and land plants, to fungi, and to invertebrates and vertebrates. The actors in this story are metals such as copper, zinc, lead, iron, manganese, tin, silver, cadmium and arsenic. Pedantic chemists will query whether some of the elements to be considered in this book fall strictly into a precise chemical definition of a metal, and will favour the term metalloid for the likes of arsenic. That argument can be left to another place and another time. There are other metals, the so-called major metals, such as sodium, potassium, calcium and magnesium, that occur in greater abundance on earth, including in living organisms, but again this book is not about them. Trace metals have a biological importance in stark contrast to their tiny concentrations in the physical and biological compartments of the world. All are toxic to life above a threshold concentration, a concentration that may still be staggeringly small to our first impressions, for these toxic concentrations often register only on a parts per million scale. And yet, many of these same metals are actually essential to life at doses below this toxic threshold – in the case of selenium, not far below this toxic threshold. Without enough of an essential metal such as zinc, copper or iron, organisms suffer deficiency symptoms and ultimately just cannot survive without a minimum supply. We have not found (if we ever will) an essential role in metabolism, as part of an enzyme or a key biological molecule such as haemoglobin, for the non-essential metals such as lead or mercury, but these metals are still taken up into organisms with the potential to interfere with life's processes. The division into non-essential or essential is not a division into toxic or non-toxic. All trace metals are toxic if present in high enough concentrations, and indeed some essential metals, for example copper, are amongst the most toxic metals known, if present above that toxic threshold.

The biology of trace metals is not a simple story. Life has evolved in the presence of these toxic metals, and indeed life has evolved because of the presence of these metals. Trace metals have a vast range of potential chemical reactions with the likes of proteins and other key molecules in the metabolism of organisms. The resulting chemical associations have provided a fantastic resource for the action of natural selection on the evolution of the biochemistry of life in all its variants. The biochemical basis of our respiration to deliver energy could not exist without the binding of iron to key molecules called cytochromes, nor could many of our enzymes function without the incorporation of a trace metal to play a key catalytic role. And yet, simultaneously there have evolved the systems within organisms to stop these potentially vital, but potentially toxic, metals binding in the wrong place to the wrong protein at the wrong time. We have evolved in a toxic world but have the systems to cope.

So how have different organisms coped with and exploited trace metals in their biology? The answer is wonderfully and diversely, and this book explores many such examples. Why do some animals contain huge concentrations of some trace metals and not others, while even relatively closely related animals have very different metal concentrations in their bodies? Why do oysters and mussels handle zinc so very differently, so that what is a very low zinc concentration in an oyster would be off the scale in a mussel? Why do some hymenopteran insects concentrate zinc in their ovipositors, and some polychaete worms copper in their jaws? Do other worms concentrate arsenic to deter the feeding efforts of would-be fish predators, and does the high vanadium content of the outer surface of some sea squirts stop them being fouled by barnacles and other sessile marine organisms? Why do some oceanic seabirds have fantastic concentrations of toxic metals in their kidneys, even in the absence of metal pollution? In the sea, how do chitons store potentially toxic amounts of iron before its transport to the radula to harden the developing teeth which will be used to rasp the surface of rocks to scrape off the algae growing there?

All such questions have an effect on the natural history of the organisms concerned, and in turn are affected by the natural history of these organisms. To understand the mechanisms controlling the uptake, accumulation and toxicity of trace metals in organisms, it is necessary to understand the role of natural history in these processes. Many specific aspects of the biology of an animal, such as its method of feeding and respiration or burrowing activity, critically affect how an animal interacts with metals. It is crucial to appreciate the importance of natural history in understanding the interaction of metals and organisms.

There is also an applied aspect to such a discussion, given the ability of metals to play such a significant ecotoxicological role in habitats affected by the output of historical activities such as mining or by other modern industrial effluents. How do some organisms cope better than others in streams or estuaries affected by metal-rich mining waste? How do communities change in such circumstances, and so what?

The British Isles offer an ideal template to explore such questions. The countryside and indeed the natural history of Britain are littered with the effects of metals, mostly via historical mining. The tale to be told encompasses history, geography, geology, chemistry, biochemistry, physiology, ecology, ecotoxicology and, above all, natural history.

The book is a natural history voyage from the ancient mines of Britain, usually situated in upland areas, through the catchments of the surrounding terrestrial regions, to streams, rivers, estuaries, coastal waters and the oceans. It is a story of how metals are found in ores, how they have been mined and consequently enter the local environment with consequences for the growth of plants and terrestrial animals associated with those plants, before affecting the life of upland streams by direct runoff or via the metal-rich sediments that are transported into these streams, either deliberately or accidentally. What changes occur to local freshwater communities, and what are the adaptations of those organisms that survive? From stream to river to estuary. Many of the estuaries in Cornwall, for example, are full of metal-rich sediments washed down in the heyday of local copper mining in the second half of the nineteenth century, lying there still and affecting the local biota. Out to sea, there are fewer examples of the metal-associated effects of humankind, but many examples of how marine organisms handle the metals that inevitably enter their bodies naturally and the physiological uses to which they may be put. Finally, oceanic life is still interacting with metals. While there is some entry of anthropogenically derived metals transported in the atmosphere into the ocean, particularly in polar regions, charismatic marine organisms such as whales still have remarkable physiological methods of handling the trace metals that are taken up quite normally from their often metal-rich diets. Similarly, the kidneys of some oceanic seabirds have no right to function given their naturally accumulated huge loads of the poisonous metal cadmium.

Many of the interactions between trace metals and organisms have nothing to do with metal pollution. All organisms are affected, from bacteria, plants and invertebrates to charismatic species such as seals, dolphins, whales and seabirds. All have a tale to tell.

ACKNOWLEDGEMENTS

I have relied on the help, collaboration and friendship of many people over many years in order to arrive at a position to undertake the writing of this book. First and foremost, my very good friend Sam Luoma, who was my co-author on a tome on metal contamination in aquatic environments, has been an outstanding guide and colleague, as we have worked together to stress the role of biology in the chemistry-dominated world of metal ecotoxicology, and indeed put more 'eco' into 'ecotoxicology'. Claude Amiard-Triquet and Jean-Claude Amiard have similarly influenced my knowledge and thinking considerably, in our happy and productive research collaborations in the worlds of metal ecophysiology and ecotoxicology since the turn of the millennium. Wen-Xiong Wang has been another strong collaborator and friend, generous with his time and resources, as we have exploited the natural laboratory that is the coastal region of Hong Kong as a source of metal-related environmental questions. Over a longer period stretching further back in time, I have learnt so much marine biology from my great friend Geoff Moore. Bill Langston is a continuing source of good sense on the biology of trace metals, and the sadly now late Dave Phillips was a forceful and appreciated collaborator on research projects and a book on biomonitoring in the 1990s. I also owe debts of gratitude to Stephen White, Jason Weeks, Dayanthi Nugegoda, Paul O'Brien, Darren Martin, Michael Depledge, Carmen Casado-Martinez, Wojciech Fialkowski, Farhan Khan, Judit Kalman and Islay Marsden for their productive research collaboration over many years. I would have achieved little without the support of Brian Smith over the last twenty years, and I gratefully acknowledge his considerable contribution.

It is a pleasure to thank Christopher Rainbow for the use of his colour photographs, and for his time and patience in their acquisition. I am also grateful to the Natural History Museum for permission to include colour photographs under their copyright. I thank Harry Taylor for these beautiful illustrations. My thanks too to Kevin Brix and Charlie Arneson for their kind permission to use their photographs. Mike Rumsey, Hölger Thues, Jeff Duckett and Emma Sherlock of the Natural History Museum provided much appreciated guidance through the worlds of ores, lichens, bryophytes and earthworms respectively. William Yeomans is to be thanked for his interest in the black-tailed trout of Leadhills, which he brought to my attention. Paul Henderson kindly read draft chapters for me, but remaining errors are down to me.

Arguably, this book is a culmination of research experience over a career of forty years. All would have been impossible without the love and support of my family, particularly my wife Mary. I am forever in her debt.

1 Introduction

Box 1.1 Definitions

arsenate The arsenate ion AsO_4^{3-} is the common form in which arsenic is dissolved in oxygenated aquatic ecosystems. The arsenic in arsenate is arsenic 5, with an oxidation state of 5.

arsenite The arsenite ion AsO_3^{3-} is a dissolved form of arsenic, less common in aquatic ecosystems than arsenate but more toxic. Arsenic is in the form of arsenic 3, a more reduced oxidation state than As 5.

biomagnification Accumulation via trophic transfer of a metal to a higher tissue concentration in a predator than occurred in its prey occupying a lower trophic level in a food chain. Biomagnification is not a general principle of the food chain transfer of metals in inorganic form, but will occur in the case of the trophic transfer of organometals such as methyl mercury. Biomagnification is a general principle of the food chain transfer of organic contaminants.

biosphere That part of the Earth and its atmosphere which is inhabited by living organisms.

ecotoxicity Toxic effect of a chemical upon populations, communities and ecosystems.

ecotoxicology Study of the toxicological effects of chemicals upon populations, communities and ecosystems, integrating ecology and toxicology.

essential metal A trace metal required in very small doses for the functioning of the metabolism of an organism.

galvanization Process of applying a protective zinc coating to iron or steel.

heavy metal Term often used loosely without strict definition to refer to metals that interact in small doses with organisms to produce toxic effects. Definitions may refer to metals with a specific gravity greater than 4 or even greater than 5. Such definitions, however, would include some metals in Groups 1 and 2 of the Periodic Table which are not generally included in lists of 'heavy metals', implying some expected chemical characteristics of the metal in question.

major (metal) ion Ion of sodium, potassium, calcium or magnesium, present in high concentrations in all biological systems.

metal An element which is shiny and of high density, is malleable and ductile, is a good conductor of heat and electricity and usually enters chemical reactions as a positively charged ion.

metalloid An element with chemical properties between those of a metal and a non-metal, as defined by its position in the Periodic Table, for example arsenic.

microgram (μg) One millionth of a gram (10^{-6}g).

non-essential metal A trace metal for which no requirement for the functioning of the metabolism of an organism has been shown.

organometal A bound combination of a metal and an organic entity. Examples are methyl mercury, tetra-ethyl lead and tributyl tin.

Periodic Table An arrangement of the chemical elements in order of their atomic number, such that elements with similar chemical properties occur at regular intervals and fall into groups of chemically related elements.

ppb Abbreviation for 'parts per billion'. Typically used to express the amount of metal dissolved in water, for example micrograms per litre (μg/L).

ppm Abbreviation for 'parts per million'. Typically used to express the amount of metal per unit (usually dry) weight in organisms, often expressed in micrograms per gram (μg/g) or milligrams per kilogram (mg/kg), or of the amount of metal dissolved in water, for example micrograms per millilitre (μg/ml) or milligrams per litre (mg/L).

rare earth elements (lanthanides) Fifteen elements with atomic numbers 57 (lanthanum) to 71 (lutetium), widely used as industrial catalysts.

trace metal Term often used loosely without strict definition with the implication that the metal is present in only 'trace' concentrations. Where defined, a concentration limit of 0.01% (100 parts per million, ppm) may be quoted, although this limit is often exceeded by the concentrations of many 'trace metals' in different organisms. Some authors will apply the term 'trace metal' only to essential metals. Here the term 'trace metal' is used to include metals and metalloids that have the potential to be toxic to organisms at low doses, whether essential or non-essential.

1.1 Metals

Prologue. The widely used labels 'heavy metals' and 'trace metals' are difficult to define, not least because the variably included elements show a continuum of chemical and biological characteristics. The term 'trace metal' is used here to cover a wide range of metals and metalloids that typically occur in low concentrations in organisms and are potentially toxic, whether or not they are essential or non-essential.

Metals are familiar to all of us. At least most are. We know about iron, tin, zinc and many others, including mercury, the only metal to exist in elemental form as a liquid at room temperature. Metals are elements, and being elements they have the ability to form compounds with other elements. For example, rust basically consists of iron oxide – a compound formed by the reaction of iron with oxygen. Gold and platinum, on the other hand, are relatively inert. They do not form compounds very readily and stay in elemental metal form for eons, the very property that makes them 'precious' and suitable for use in coinage and jewellery. Most metals, however, do not stay in elemental form and exist as compounds.

Many simple inorganic compounds of metals will dissolve in water, separating into ions, for example positive sodium ions and negative chloride ions in the case of salt. These ions can be present in the environment around us, dissolved in the water of soils, sediments, streams, rivers, seas and oceans. Yet, the compounds formed by metals are not always simple salts made up of two inorganic ions. Metals may bind with organic compounds such as proteins that make up much of our bodies, and indeed the organic material of all living organisms. Thus, iron is bound with a relatively complicated organic molecule to make up haemoglobin, which will bind reversibly with oxygen and carbon dioxide in our blood, delivering oxygen to tissues for use to generate energy and removing the resulting waste product, carbon dioxide.

So what defines a metal? This is such an apparently simple question, but one that is almost impossible to answer with objective certainty. What define a metal are its physical and chemical characteristics. There is a definition – metals are elements which are shiny and of high density, are malleable and ductile, are good conductors of heat and electricity and usually enter chemical reactions as positively charged ions. The appearance of 'usually' should already warn us that not all is straightforward. The Periodic Table lists all the elements, classifying them on the basis of their atomic structure, which is inevitably linked to their chemical properties. While different distinct groupings of elements with similar chemical properties can be recognised in the Periodic Table, in some regions of the table chemical properties may show relatively little change between adjacent elements. Thus different elements show the defining characteristics of metals to different extents. There is even a term, metalloid, for elements that show some but not all of the properties of a metal. For example, arsenic and antimony are regularly referred to as metalloids (or even sometimes semimetals), and occasionally aluminium and selenium may be included in this category. Of the 90 elements

occurring naturally on Earth, most (67) might be considered to be metals.

Two labels for metals commonly occur when discussing interactions between metals and organisms. These are the terms 'heavy metals' and 'trace metals'. It is difficult enough to define a metal. It is even more difficult to provide a consistent, objective definition for each of these two collective terms.

The term 'heavy metal' has historically often been used to refer to those metals that are considered environmental pollutants. To be a pollutant, a metal would be toxic to organisms and thus have an ecotoxicological effect on biological communities. To be toxic, a metal would need chemical properties that mean that it binds to the wrong molecules in an organism – in the wrong place, at the wrong time, disrupting normal metabolism. And yet, 'heavy' implies that these metals share a common property of high physical density. Thus definitions of heavy metals refer to metals with a specific gravity greater than 4, or even greater than 5, with no reference to chemical properties. Furthermore, such a definition based on weight would include heavy elements such as radium, the actinides (naturally occurring radioactive elements) and the lanthanides (rare earth elements), elements that are not usually considered to be 'heavy metals'. The term 'heavy metal' is avoided here.

The term 'trace metal' is also difficult to pin down. By implication, it is a metal present in trace amounts in the environment, and some definitions include an upper concentration limit of 0.01% by dry weight (equivalent to 100 parts per million or $\mu g/g$) in an organism. This definition would encompass most of the metals commonly referred to as trace metals, although, as we shall see later in this book, there are spectacular examples of much higher concentrations in particular organisms of metals typically called trace metals. Furthermore, the term 'trace' is also sometimes used with the implication that the element is required in the metabolism of living organisms, in effect is an 'essential' element, a supposition not included in the definition based on environmental concentrations. We shall come to the question of essentiality in more detail shortly, but it

Table 1.1 Selected trace metals and their abbreviations.

Aluminium (Al)	Mercury (Hg)
Antimony (Sb)	Molybdenum (Mo)
Arsenic (As)	Nickel (Ni)
Cadmium (Cd)	Platinum (Pt)
Chromium (Cr)	Selenium (Se)
Cobalt (Co)	Silver (Ag)
Copper (Cu)	Tin (Sn)
Gold (Au)	Titanium (Ti)
Iron (Fe)	Tungsten (W)
Lead (Pb)	Vanadium (V)
Manganese (Mn)	Zinc (Zn)

remains the case that other authors include both essential and non-essential metals as 'trace metals' (Luoma and Rainbow, 2008).

There are chemical definitions of different categories of elements including metals based on their chemical problems that avoid the unsatisfactory terms 'heavy metals' and 'trace metals', but the knowledge of chemistry required to use such definitions is probably more than we need to assume here. This book is concerned with the metals that interact with living organisms in small amounts, all of which are toxic above a given threshold, while many of these same toxic metals are essential to life at lower concentrations. Many of them are the metals that have been mined in the British Isles for centuries, even thousands of years in the cases of copper and tin, often with ecotoxicological consequences on the surrounding vegetation and animal life. In order not to invent another collective term, the term 'trace metals' will be used here, stressing that, while some are essential metals and others non-essential, all are potentially toxic. Table 1.1 lists these trace metals. The list includes the so-called metalloids arsenic and antimony, and their part-time colleagues, aluminium and selenium. Apologies to those seeking better chemical categorisations, but pragmatism prevails here.

Absent from the list are other metals that are also required for life, typically in ionic form and referred to as the major metal ions. These major ions are the ions of sodium, potassium, calcium and magnesium. They are present in organisms in relatively high concentrations and do not exert toxic effects as do trace metals by binding inappropriately to disrupt metabolic pathways.

1.2 Metals and Humans

Prologue. Humankind has mined and smelted metals since the Bronze and Iron Ages for the production of artefacts that include tools, weapons and jewellery. Today, the various physical and chemical properties of the different metals concerned underpin technological development in many industries. Yet there has been a price to pay as particular metals of high ecotoxicity such as mercury and cadmium have been released into the environment with deleterious effects on the local biota, including humans.

From early history, humankind has recognised that metals can be used to advantage, and for eons metals have been mined to serve mankind's purposes. The very names Bronze Age and Iron Age highlight the importance to our ancestors of the development of the technology to enable the smelting of metal-rich ores to release their metals for the production of artefacts.

The Bronze Age is a period of history characterised by the use of copper initially, and then bronze, the alloy formed wholly or chiefly of copper and tin, to make tools, vessels and weapons. The addition of tin to copper yields a product that is more easily melted and better suited for casting. Bronze is also harder than copper. The pure copper stage (the Chalcolithic Age) of the Bronze Age goes back to about 5000 BC. Bronze first came on the scene in about 3500 BC, and the copper mines of Cyprus ('copper island') later contributed significantly to the booming trade of the Bronze Age world of the Mediterranean. In Bronze Age Britain, prehistoric humans were fashioning copper and then bronze tools from about 2500 to 800 BC. The largest Bronze Age copper mine in Northwest

Europe was on the Great Orme in north Wales, with peak activity between 1900 and 1600 BC (MacGregor, 2010). The copper and tin mines of Cornwall were also of worldwide importance. Cornish tin in particular was used in the production of a considerable proportion of the bronze objects made in the second millennium BC. Some of the early bronze objects also contained some arsenic in combination with copper and tin, resulting in a harder alloy. As we shall see in Cornwall, arsenic often occurs as an impurity in copper ores. Its initial incorporation into bronze, therefore, may have occurred by accident, but later arsenic-bearing minerals were intentionally added during smelting.

Bronze continues to be used today, for example in coinage (often with the addition of zinc) and in bearings, typically as phosphor bronze. The addition of phosphorus increases strength and hardness.

The technological availability of another metal, iron, brought the Bronze Age to an end, and the Iron Age began. The Iron Age was the final technological stage in the Stone–Bronze–Iron Age sequence, beginning in about 1200 BC. Iron replaced the more brittle bronze in implements and weapons. Iron had been used as a scarce and precious metal earlier, but the development of metallurgical techniques to smelt iron now allowed the exploitation of iron ore deposits, which are more widespread and plentiful than those of copper ores. During the first millennium BC, the ancient world rapidly converted to using iron for the manufacture of tools and weapons. While the Iron Age, for the first time, put weapons into the hands of many with inevitable consequences, the Iron Age also allowed the development of the iron plough and promoted consequent changes in agricultural procedures and indeed structures of society.

Iron is one of the most abundant elements on the Earth. While the Earth's core is largely metallic iron, in the Earth's crust, iron has reacted with many other elements, not least oxygen, to produce iron-rich minerals, including ores with the potential for technological and commercial exploitation after smelting.

Smelting is the process of the extraction of a metal from its ore. Smelting uses heat and an agent (typically carbon in the form of coke or charcoal) to decompose the ore, the carbon binding with any oxygen in the ore mineral to release elemental metal. Limestone is often added to remove final impurities, a process known as fluxing, leaving behind slag, a glassy mass usually containing a mixture of metal oxides and silicon dioxide. Thus, iron is typically extracted by heating iron ore with coke and limestone in a blast furnace to produce cast iron or pig iron. Steel is an alloy of iron with a small amount of carbon (0.1–2.1%), and this alloy is at the root of the infrastructure of our industrialised society today. Alloy steels contain additional elements such as manganese, nickel, chromium, molybdenum, titanium and vanadium, added to modify the characteristics of the steel.

Before the start of the Bronze Age, defined by the smelting initially of copper, then the production of bronze itself, two other metals, lead and tin, were the first metals to be smelted. The smelting of these two metals is relatively simple, the heat of a wood fire being sufficient to release the elemental metals from some ores. Cast lead beads date from about 6500 BC. Lead, however, is too soft to be used in construction or weapon manufacture, and its technological impact at the time was low. On the other hand, lead is easy to cast and shape, and its day came later when it was used extensively for the piping and storage of water (giving its name to 'plumbing'), and the sheathing of the wooden hulls of boats, by the ancient Greeks and Romans. The Romans had another use for lead, using lead acetate, known as sugar of lead, as a sweetener for wine, with potential toxic effects on heavy consumers. In the eighteenth century in Britain, Devonshire colic affected cider drinkers in the West Country, an affliction identified to be lead poisoning caused again by the addition of sugar of lead to sweeten the cider, and also by the use of lead linings in the cider presses. The toxicity of lead is not a health concern restricted to ancient or recent history. Lead-based paints commonly contain lead in the form of lead chromate (chrome yellow) or lead carbonate (white lead), but they are a health hazard, and sale to the general public in the United Kingdom was banned in 1992. Red lead (lead oxide) is used in the production of batteries, lead glass and rust-proof primer paints. Lead is widely used in roofing. In this elemental form, lead cannot enter biological systems and is not toxic. Lead is also still widely used today in lead–acid batteries, originally invented in 1859. These batteries are particularly popular in the automobile industry, for they are able to supply high surge currents as required by starter motors.

In the twentieth century, an organic compound of lead, tetra-ethyl lead, known as antiknock, was widely added to petrol to prevent pre-ignition and increase a fuel's octane rating. As the fuel was consumed, lead was released into the atmosphere attached to fine particles or aerosols, providing the potential for the uptake of toxic lead into biological systems, including humans, with subsequent neurotoxic effects. Legislation in most (but not all) parts of the world, including Europe and the United States, has now banned the addition of lead to fuel. This banning has had another positive effect. It has allowed the expanded use of catalytic converters on vehicles to reduce the emission of toxic pollutants such as hydrocarbons, carbon monoxide and oxides of nitrogen, the lead in antiknock previously having acted as a catalyst poison, inactivating the converter by forming a coating on the catalyst.

Ores of tin were less common than those of lead, and tin is a soft, weak metal only marginally harder than lead. Therefore, as a pure metal, tin had even less technological impact in prehistory than lead, its significant contribution being its key role in the development of bronze. Tin does, however, have a well-known use today, giving its very name to sealed containers that preserve their food contents. In 1795, Nicholas Appert of Paris discovered that foodstuffs boiled in glass bottles and immediately sealed would keep for several months, thereby providing a solution to Napoleon Bonaparte's publicised problem of keeping food fresh for his troops on their marches (Alexander and Street, 1951). This discovery led to others, and Peter Durand, an Englishman, patented the process of sealing food in containers of iron covered with a thin film of tin, so preventing rust. The coating of iron with tin was not in itself novel, for tinned iron had been used in Bavaria in the thirteenth century for decorative articles and for parts of armour (Alexander and Street, 1951). The product of coating

iron or steel with tin is known as tinplate. Tinplate does not rust and has been used in huge quantities for the production of tin cans to hold preserved foods or other biodegradable products.

In the second half of the twentieth century, an organic compound of tin, tributyl tin (TBT), rose to prominence as a wide spectrum biocide given its very toxic nature. TBT was particularly used as an antifouling agent in the marine environment in the expectation that it would be broken down very quickly in the sea after release from antifouling paint. In the 1970s, however, it was realised that TBT was actually having severe ecotoxicological effects on non-target organisms in the sea, such as oysters and dogwhelks (in the latter case at remarkably low concentrations), and the use of TBT as an antifouling agent is now banned in many countries across the world (Luoma and Rainbow, 2008).

Today, copper makes a very important contribution to our technology in its own right. Copper is very ductile and malleable, is corrosion resistant, and has very high conductivities for electricity and heat. Thus copper wiring has underpinned the electrical industry, and copper piping is used extensively in plumbing. Another alloy of copper is brass, this time with zinc in various proportions according to the ductility and malleability desired. Brass is used for decoration given its bright appearance, and its low frictional properties make it fit for use in doorknobs, locks, gears, bearings, valves and ammunition casings. Brass also has acoustic properties that make it suitable for the production of musical instruments, with a section of an orchestra named after it. Copper has long been used to make coinage, typically in some variation of bronze. The Romans did at times use pure copper in coins, but more usually the coins of lower denomination were made of bronze. In Britain, copper was also used for humbler coinage until 1860, when bronze was substituted (Alexander and Street, 1951). This 'bronze' actually contained zinc as well as tin and copper, until the days of decimalisation. Today, the decimal coins are made of copper-plated steel, nickel-plated steel and alloys of copper and nickel (cupro-nickel) or nickel and brass. The biocidal properties of copper are reflected in its widespread use today in antifouling paints. Before the development of antifouling paints, metallic copper itself was used to sheathe the hulls of boats, beginning in the eighteenth century. The copper sheathing, as in the case of the less effective lead sheathing used by the Romans, was required for protection of the hull, not only against surface-fouling organisms slowing down a vessel by disrupting flow over the hull, but particularly against wood-boring marine animals such as shipworms, destroying wooden hulls by their extensive burrowing.

The use of zinc in brass and coinage has been introduced earlier, but zinc also has other uses. Zinc has been used for roofing but its major use has been in the process of galvanisation – the coating of iron or steel with zinc, primarily to prevent corrosion, particularly in the automobile industry. Zinc is also a key ingredient in batteries, including general-purpose zinc–carbon batteries and heavy-duty zinc chloride batteries, which have been used to power the likes of torches, toys, clocks and transistor radios for a 100 years or so. Zinc–air batteries can be very small, as used in hearing aids, larger for use in cameras or very large to power electric vehicles.

Another metal, nickel, is similar to iron in some of its properties, with a high melting point and greater strength and hardness. On the other hand, nickel is strongly resistant to corrosion and can be used to electroplate iron, brass and silver. Being corrosion resistant and of attractive colour, electroplated nickel silver (EPNS) is common in tableware. Nickel is used today chiefly in alloys, as in the nickel–copper and nickel–brass alloys used in coinage, but particularly in nickel–steels, the addition of nickel providing increased hardness, strength and resistance to wear. Nickel is yet another metal used in batteries, originally in combination with cadmium in rechargeable nickel–cadmium batteries. These, however, have now been replaced by nickel–metal hydride batteries, involving mixtures of many different and often uncommon trace metals and rare earth elements.

Gold is the classic precious metal, being very resistant to corrosion. It is so chemically unreactive that it can be found in elemental form as nuggets, as veins in rock or as grains in alluvial sediments. Since gold is also the most malleable and ductile of metals while being

relatively rare, it has long been used in the production of jewellery and other culturally valuable artefacts. In 1833, near the town of Mold in north Wales, a gold cape was discovered that was eventually dated back to about 2000 BC in the Bronze Age (MacGregor, 2010). The cape had been made with great skill from a single sheet of very thin gold. It may have originated from elsewhere in the Bronze Age world, and its occurrence in north Wales may be related to the presence of a major trading centre associated with the local Great Orme copper mine (MacGregor, 2010).

Pure gold is, however, too soft to be used alone in artefact production, and other metals are added for hardening, thereby defining the 'carat' qualities of gold jewellery today. A carat is a 24th part, and so 22 carat gold has 22 parts of pure gold by weight, and two parts of other alloy metals. In 22 carat gold jewellery, silver and copper together will make up the final two parts of the 24, while in gold coins it is often copper alone. Eighteen carat gold is popular in jewellery, for this alloy mixture is relatively serviceable and robust without too much loss of colour. White golds are also alloys of gold with other metals, usually nickel or palladium in sufficient amount to make the final product white. The corrosion resistance of gold and its rarity have also meant that it has been used in coinage since the first millennium BC. Gold previously played a fundamental role as a 'standard' in monetary systems, predominantly in the nineteenth century, and today central banks continue to keep a portion of their monetary reserves as gold bullion. Gold does also have practical technological uses, for example in electronics, as an alloy with mercury to form an amalgam for dentistry or as a protective coating on other metals (gold plating).

Platinum is even more corrosion resistant than gold and is one of the rarest elements on Earth. It is therefore a desirable precious metal for use in jewellery. In spite of its high cost, platinum does have applications where corrosion resistance is a top priority, and it is used in catalytic converters, electrical systems and medical equipment.

Silver is also used as a precious metal in jewellery in its own right, as well as historically being used in coinage. Silver is relatively resistant to corrosion, but not to the extent of gold. Indeed, the burning of sulphur-containing gas in the home can lead to the blackening of silver objects as a result of the reaction of silver with oxides of sulphur. So-called silver coins usually contain alloys of silver with other metals such as copper, nickel and zinc, and often now contain no silver at all. Even 'sterling silver' contains 7.5% of other metals, usually copper, because pure silver is generally too soft for the production of functional objects such as cutlery and tableware. Nevertheless, the corrosion resistance of silver has led to its wide employment historically in the brewing and milk industries to line vats. Silver also has very high heat and electrical conductivities, hence its use in electrical conductors, and it is also used in mirrors. Large amounts of silver were required by the photographic industry in the form of silver nitrate used to develop films, until so much of the domestic market was replaced by digital photography.

Chromium is another metal used in electroplating, usually being deposited on nickel to increase hardness and durability. While chromium is used to contribute hardness to alloy steels, it is also the key ingredient of stainless steel. Stainless steel by definition resists corrosion and is widely used in cutlery, for example. Stainless steel contains at least 10.5% of chromium. The chromium is vital, for it binds with oxygen in the air to form a passive film of chromium oxide. This film then prevents access of the oxygen to the iron component of the steel to form rust. Salts of chromium have been used as pigments in paints since the 1800s, for example chrome yellow (lead chromate) and chrome red (lead chromate mixed with lead oxide), and also in glass manufacture. Other key industrial applications of chromium include the use of chromium salts to tan leather and as wood preservatives.

Cobalt is mainly used as a component of magnetic, high-strength, wear-resistant alloys, as in turbines. Its compounds, such as cobalt aluminate (cobalt blue), are used as blue pigments in paints and glass. Manganese is present in nearly all steels and in aluminium alloys, improving resistance to corrosion. Manganese is also used in the manufacture of batteries, manganese dioxide being a key component in zinc-carbon batteries. Molybdenum is another trace metal added to alloy steels, particularly in combination with nickel and chromium, and molybdenum disulphide is used as a high-temperature, high-pressure lubricant. Titanium is a low-density metal with a very high strength-to-weight ratio and very good corrosion resistance. It is therefore used widely to produce strong lightweight alloys in military, aerospace, medical, sporting goods and other industries. Titanium dioxide is used as a catalyst and in the production of white pigments. Tungsten is the metal used classically in the filaments of incandescent electric light bulbs. Being very hard and of high density, tungsten is used in military projectiles for increased penetration, and in drill bits. Vanadium is also used in alloy steels, and vanadium pentoxide is a catalyst in the industrial production of sulphuric acid.

Mercury is the only metal to exist as a liquid at room temperature and is very volatile. Mercury is arguably the most toxic of all metals, a very potent neurotoxin, and can enter biological systems in dissolved form, in food or when breathed in. In sediments in the environment, inorganic mercury can be converted by bacteria into the organic compound methyl mercury (an organometal) which readily crosses biological membranes. Methyl mercury can also be biomagnified up aquatic food chains to reach bioaccumulated concentrations of toxicological concern in top predators, such as long-lived tuna. Biomagnification describes the accumulation via trophic transfer of a metal to a higher tissue concentration in a predator than occurred in its prey occupying a lower trophic level in a food chain. Biomagnification is not a general principle of the food chain transfer of metals in inorganic form, but will occur in the case of the trophic transfer of organometals such as methyl mercury.

The neurotoxic nature of mercury has long been recognised, as for example in the depiction of the 'mad hatter' in Lewis Carroll's *Alice in Wonderland*, written in 1865. Beaver fur was used in the nineteenth century in Britain to make good-quality top hats, while rabbit fur was used for cheaper hats. In the case of rabbit fur, however, it was necessary to treat the fur with mercury nitrate to roughen the fibres so that they would mat well. Beaver fur has naturally serrated edges and did not require such mercury treatment. Beaver fur was, however, scarce and therefore expensive, so alternative furs were used. As a result, hatters breathed in toxic mercury fumes with neurotoxic consequences, including aberrant behaviour. Mercury is used in batteries, thermometers, barometers, mercury switches and fluorescent lamps. Other metals, such as tin, silver and gold, will dissolve in liquid mercury to form amalgams widely used in dentistry. In most cases, there is now social and environmental pressure to phase out industrial uses of mercury, simply because of its high toxicity and ecotoxicity.

Cadmium is another notoriously toxic trace metal, with previous industrial uses being phased out for this reason. Cadmium was used in nickel-cadmium batteries, but these are now obsolescent as they are replaced by better performing nickel–metal hydrid batteries. Cadmium was also used for corrosion-resistant plating on steel. Cadmium compounds have been used to stabilise plastics and as pigments, as in the cases of yellow cadmium sulphide and cadmium selenide, known as cadmium red. A new modern application has been the use of cadmium telluride in solar panels. Cadmium is not commonly commercially mined for its own sake, but as a by-product, for example in the exploitation of zinc sulphide ores.

The very word arsenic is synonymous with poison for many of us, and indeed arsenic is one of the more toxic trace metals. In the nineteenth century, arsenical pigments were in demand to colour cotton fabrics and wallpapers, and arsenic was also used in the making of glass, the manufacture of shot and fireworks and the tanning of leather. Its high toxicity also led to the use of arsenic in pesticides and herbicides, but such applications are now declining. Today, arsenic does have industrial uses, as it can be added to copper alloys such as bronze for strengthening, and it is employed in semiconductor manufacture.

Of other trace elements to be considered in this book, selenium is used as a pigment to colour ceramics and glass, and also as a decolouriser in glass-making to remove the green tint resulting from iron impurities. Selenium is also a semiconductor and

is used in photocells, particularly thin-film solar cells of increasing importance for renewable energy production. Antimony is used in alloys with tin and lead and in lead-acid batteries, and antimony trioxide is used as a flame retardant in clothing, seat covers and toys.

While there is debate as to whether aluminium can be called a trace metal, given that it is the third most common element on Earth, after oxygen and silicon, it is included here because aluminium ecotoxicity can be significant. As a metal, aluminium has a remarkably low density, and aluminium alloys are vital to the aerospace, automobile and construction industries. Aluminium is also used in drink cans.

Late in the twentieth century, the industrial use of a new manmade form of metals exploded, with an inevitable subsequent appearance in the environment with potentially ecotoxicological effects. Nanoparticles are particles of between 1 and 100 nanometres in diameter, and many contain one or more trace metals. While they may occur naturally, as in the emissions of volcanoes, many forms of metal-containing nanoparticles are being manufactured for specific industrial and commercial purposes (Yon and Lead, 2008), usually exploiting their high surface area to volume ratios, which make the nanoparticles very reactive or catalytic. Common metal-containing nanoparticles include titanium dioxide nanoparticles used in sunscreens, cosmetics and paints, and silver-rich nanoparticles used in washing machines and clothing, such as socks, to kill bacteria and reduce odours. Gold nanoparticles are used in medicine for drug delivery and are also important in the optical electronic industry. Quantum dots are nanoparticles made of semiconducting materials and are therefore also used in the electronics world. One such quantum dot contains a double load of toxic trace metals, consisting of cadmium and selenium, with potential ecotoxicological concern on release of both into the environment.

While the release of toxic metals by humans into the environment from mining activities has been with us for thousands of years, the release of metal-rich nanoparticles in industrial and domestic effluents is a new and expanding source of toxic trace metals into the environment. We shall see in this book how habitats and biological communities cope with the toxic challenge of trace metals in the environment, however novel their source.

1.3 Sources and Global Cycles of Metals

Prologue. What was the origin of metals on Earth? How do metals enter the biosphere, and how are they subsequently distributed by biogeochemical cycling? Humankind has disturbed such cycling, not least by changing the different rates of flux of particular metals between different environmental components of the Earth to cause the presence of an excess of an ecotoxic metal at a particular location at a particular time.

The early Universe consisted almost entirely of the element hydrogen, with perhaps some helium present. Other elements were subsequently formed by nuclear fusion. A series of nuclear fusion reactions results in the formation of elements up to the atomic mass (56) of iron, but the formation of heavier elements by nuclear fusion requires higher temperatures than are found in stars. These heavier elements are formed by the addition of neutrons into the atomic nucleus, and iron remains one of the most abundant elements in stars. Suns were formed by the gravitational collapse of parts of a molecular cloud called a nebula. Most of the collapsing mass coalesced into a central sun, but the remaining material made up a protoplanetary disc from which planets, moons and asteroids were formed, producing a solar system.

Our solar system, and therefore the Earth, formed 4.55 billion years ago, as the protoplanetary disc cooled and elements reacted with each other to form complex materials, eventually producing lumps of solid matter. Today the Earth has an onionlike structure with a core, mantle, crust and atmosphere, but originally it was probably a homogeneous ball of matter, later forming a series of layers of decreasing density outwards as it cooled. The high-density central core is composed mainly of iron and nickel, solid in the centre but liquid in the outer core. The mantle is mostly solid rock, but is in slow viscous motion over geological time. It contains a wide array of elements, including metals. The crust above the mantle is also solid rock formed by the cooling of the original molten magma. The Earth's crust, together with the outermost mantle, is made up of rigid plates (tectonic plates) which glide over the underlying parts of the mantle, driven by the convection currents therein carrying heat from the Earth's interior to nearer the surface. These tectonic plates move apart, for example along mid-ocean ridges. The emerging hot mantle material forms new rock, often in the form of lava from erupting volcanoes. Along mid-ocean ridges, the new crust formed at the bottom of the sea is actually widening the ocean, as in the case of the Atlantic. The introduction of new mantle material at the bottom of the sea brings with it a supply of metals from below. As mantle material is emitted under the sea, a cycle may be set up where sea water is drawn down through faults or porous sediments to be emitted again at very high temperatures (up to 465°C), but now rich in trace metals such as iron, copper and zinc, together with sulphide. It was late in the twentieth century that it was appreciated that these hydrothermal vents were home to spectacular biological communities, based on the primary production of organic matter by bacteria carrying out chemosynthesis using locally abundant sulphides as an energy source, as opposed to the more familiar primary production of organic matter by photosynthesising plants harnessing light energy in most ecosystems on Earth. While tectonic plates are pushed apart at spreading centres, there are complementary subduction zones where a tectonic plate is forced below its neighbour, typically associated with earthquakes and volcano activity.

During the cooling of the Earth, the molten rocks solidified at the surface to produce the crust. Over time, water emitted from the mantle condensed to form the oceans, leaving thicker parts of crust protruding above as the continents. The continents shared tectonic plates with their adjacent portions of thinner oceanic crust, and therefore were carried along in the several cycles of plate subduction and formation that have occurred during the history of the Earth. Mountain ranges are produced by continental uplift when adjacent tectonic plates collide at subduction zones. The continents as we know them today are therefore not formed of the original rock that first cooled at the surface of the young Earth and are continually moving (continental drift) as the tectonic plates continue their dance across the surface of the planet. The continents are not geologically homogeneous by any means, being made up of different minerals as a result of their specific geological history. Thus continents contain rocks originally formed at the bottom of the sea, and tectonically raised regions that were once hydrothermal vents will contain high concentrations of sulphides of trace metals. These very rocks are some of the ores targeted in mining for metals. As continents erode, they deliver their metal loads into streams, rivers and seas.

Therefore, metals occur naturally on Earth, and they cycle between its different environmental compartments. Metals are a component of the rocks of the Earth's crust and are in the soils and the dust that result from the erosion of these rocks. After the interaction of rainfall and its subsequent runoff from land, metals are present in dissolved form or in particulate matter in lakes, streams, rivers, estuaries, coastal seas and oceans. They are found in the atmosphere in organic and inorganic particulates such as dust, aerosols and gases, as in the case of mercury. These different components together with their constituent biota act as reservoirs for vast amounts of metals which also move between them in the process of biogeochemical cycling. Such cycling occurs naturally, but it is the intervention of human activities that has changed the different rates of flux of metals between the environmental components and ultimately caused the presence of an excess of a toxic metal in a particular place at a particular time. Human activities have brought about metal pollution. The mining of metal-rich ores in particular releases metals into the local catchment. Subsequent industrial processes of metal smelting and refining, manufacture and disposal (whether or not intentional) will also release metals into one or more of the Earth's environmental compartments.

The biogeochemical cycling of metals occurs on a global scale, particularly via the transport of small metalliferous particles in the atmosphere. Volcanoes, for example, blast particulate metals high into the atmosphere with the potential to be transported around the globe. Lower regions of the atmosphere will receive similar particles from forest fires, windblown dust and sea salt sprays (see Table 1.2) (Nriagu, 1989), again with the potential for their transport over significant distances. Thus dust from the Sahara Desert has deposited aluminium, iron, manganese and associated trace metals into the Atlantic and Mediterranean for millions of years, and dust from the deserts of China feeds metals into the Pacific (Luoma and Rainbow, 2008). Table 1.2 summarises the worldwide emissions of many trace metals into the atmosphere, in terms of million kilograms per year, huge totals given the minute 'trace' concentrations of these metals in individual inorganic and organic materials. The best composite global database available is for the 1980s, and the main principles apply today. It is noticeable how the activities of humankind have increased the emission of all the trace metals into the atmosphere, and in most cases anthropogenic sources represent the majority contribution (Table 1.2). Such sources include the burning of coal and oil and the incineration of refuse. Combustion of coal, both in electric power stations and in industrial and residential burners, is the major source of airborne mercury, molybdenum and selenium and a significant source of arsenic, chromium, manganese and antimony (Table 1.2). The burning of oil releases particularly vanadium but also much nickel and tin into the atmosphere (Table 1.2). In fact, the presence of raised accumulated concentrations of vanadium in organisms is often taken as evidence for local contamination by oil. These data compiled in the 1980s show the then huge, but now decreasing, release of lead into the atmosphere as fuel containing tetra-ethyl lead was combusted in vehicle engines (Table 1.2). Chromium and manganese emissions are primarily derived from the iron and steel industry, while other metal industries release lead, arsenic, cadmium, copper and zinc into the atmosphere (Table 1.2) (Nriagu and Pacyna, 1988).

While metals do enter the sea from the atmosphere, it is nevertheless the rivers that move the greatest masses of metals from land to sea, a flux that has again been further hugely increased by anthropogenic

Table 1.2 **Worldwide emissions of selected trace metals into the atmosphere in 1983 (million kg per year).**

	As	Cd	Cr	Cu	Hg	Mn	Mo	Ni	Pb	Sb	Se	Sn	V	Zn
Natural Sources														
Windblown dust	2.6	0.21	27	8.0	0.05	221	1.3	11	3.9	0.78	0.18		16	19
Sea spray	1.7	0.06	0.07	3.6	0.02	0.86	0.22	1.3	1.4	0.56	0.55		3.1	0.44
Volcanoes	3.8	0.82	15	9.4	1.0	42	0.4	14	3.3	0.71	0.95		5.6	9.6
Forest fires	0.19	0.11	0.09	3.8	0.02	23	0.57	2.3	1.9	0.22	0.26		1.8	7.6
Organic particles	3.9	0.24	1.11	3.3	1.4	30	0.54	0.73	1.7	0.29	8.4		1.2	8.1
Total (median)	12	1.4	44	28	2.5	317	3.0	30	12	2.5	10.4		28	45
Anthropogenic Sources														
Coal combustion	3.5	0.88	20	8.1	3.5	19	4.8	24	15	2.3	2.8	1.7	15	20
Oil combustion	1.0	0.25	2.4	3.4	–	2.4	0.94	43	3.9	–	0.83	5.9	124	3.8
Metal production														
Mining	0.08	0.003	–	0.80	–	0.83	–	0.80	3.4	0.18	0.18	–	–	–
Other	17	8.5	28	35	0.22	33	–	15	80	2.2	1.9	1.7	1.5	1.5
Refuse incineration	0.45	1.44	1.4	2.1	2.2	11	–	0.60	3.1	0.90	0.10	1.5	2.0	2.0
Phosphate fertilisers	–	0.27	–	0.69	–	–	–	0.69	0.27	–	0.001	–	–	–
Other	4.0	0.71	1.8	1.2	0.30	–	–	2.7	270*	–	–	–	–	–
Total (max)	26	12	54	51	6.2	66	5.7	87	376	5.5	5.8	11	142	142
Total (median)	19	7.6	30	35	3.6	38	3.3	56	332	3.5	3.8	6.1	86	86
Total emissions (median)	31	9.0	74	63	6.1	355	6.3	86	344	6.0	14.2	11	114	177
% anthropogenic	61	84	41	56	59	12	52	65	97	58	27		75	75

Note: Data are median estimates for natural sources and maximum estimates for anthropogenic sources. *In 1983, an estimated 250 million kg of lead was emitted into the atmosphere annually from the use of lead in vehicle fuel.

Sources: After Nriagu and Pacyna (1988); Nriagu (1989).

Table 1.3 **Anthropogenic inputs of selected trace metals into aquatic ecosystems in 1983 (million kg per year).**

	As	Cd	Cr	Cu	Hg	Mn	Mo	Ni	Pb	Sb	Se	V	Zn
Mining of metals	0.8	0.3	0.7	9	0.2	12	0.6	0.5	2.5	0.4	1.0	–	6
Smelting and refining	13	3.6	20	17	0.04	51	0.4	24	8.8	7.2	20	1.2	44
Manufacturing													
Metals	1.5	1.8	58	38	0.75	20	5.0	7.5	22	15	5.0	0.75	138
Chemicals	7.0	2.5	24	18	1.5	15	3.0	6.0	3.0	0.4	2.5	0.35	5.0
Pulp and paper	4.2	–	1.5	0.39	–	1.5	–	0.12	0.9	0.27	0.9	–	1.5
Petroleum products	0.06	–	0.21	0.06	0.02	–	–	0.06	0.12	0.03	0.09	–	0.24
Atmospheric fallout	7.7	3.6	16	15	1.8	20	1.7	16	113	1.7	1.1	9.1	58
Domestic wastewater	15	3.0	78	48	0.60	171	4.5	102	12	4.5	7.5	4.5	81
Steam electric	14	0.2	8.4	23	3.6	18	1.2	18	1.2	0.36	30	0.6	30
Sewage sludge dumping	6.7	1.3	32	22	0.31	106	4.8	20	16	2.9	3.8	4.3	31
Total (max)	70	16	239	190	8.8	414	21	194	180	33	72	21	395
Total (median)	41	9.4	142	112	4.6	262	11	113	138	18	41	12	226

Note: Data are maximum estimates unless stated.
Source: After Nriagu and Pacyna (1988).

activities (Table 1.3). Although metals released from mining activities have very significant local ecotoxicological effects on adjacent terrestrial and aquatic habitats, as will be discussed in detail in this book, mining is not one of the major sources of trace metal pollution into aquatic ecosystems on a global scale. The major sources of trace metal pollution in aquatic systems (including coastal seas and oceans) are domestic wastewater effluents (As, Cr, Cu, Mn, Ni), coal-burning power stations (As, Hg, Se), iron and steel plants (Cr, Mo, Sb, Zn), smelters for other metals (Cd, Ni, Pb, Se) and the historic marine dumping of sewage sludge (As, Mn, Pb) (Table 1.3) (Nriagu and Pacyna, 1988). Deposition from the atmosphere is the major route of input of lead into natural waters, particularly the oceans (Nriagu and Pacyna, 1988; Luoma and Rainbow, 2008).

Atmospheric fallout will also deliver metals into the soils of terrestrial ecosystems, as in the particular cases of lead (again), cadmium, mercury and vanadium (Table 1.4). More importantly, soils receive large amounts of trace metals from many different industrial wastes, particularly in the form of ash derived from coal burning and the general waste from commercial product disposal (Nriagu and Pacyna, 1988). Urban refuse contributes an important source of Cu, Hg, Pb and Zn to the land. Organic wastes from animal husbandry, logging and agricultural and food production also introduce significant metal loadings into soils (Table 1.4). The application of mined phosphate-based fertilisers onto agricultural land may cause local problems of cadmium contamination, as in the case of the poorly buffered soils of Western Australia. The disposal of sewage sludge on land can also be of high local significance as a source of trace metals. Mining will probably introduce very high quantities of trace metals into the local terrestrial habitat, especially in the form of dumps of mine tailings, which also represent a significant potential source of metals into the local water catchment, particularly via mine tailings ponds. Furthermore, smelting and refining industries may also lead to the dumping of slags and their like on land.

We have been concerned up to now with the global picture, but it is possible to be more specific about the amounts of selected trace metals, in this case, arsenic, cadmium, lead and mercury, entering the UK environment (Hutton and Symon, 1986). These four metals are of the greatest environmental concern and have been the subject of legislation restricting their release into the environment. The data in Tables 1.5 and 1.6 are again for the 1980s, for direct comparison with the global data presented in the previous tables. Today the release of lead into the atmosphere from vehicle fuel combustion is much reduced, and the marine dumping of sewage sludge with its inevitable contaminant load ceased in the UK in 1998, following the earlier 1990 lead of other Northwest European countries. As Table 1.5 shows, fossil fuel (oil and coal) combustion is a major source of emissions of mercury and arsenic into the UK atmosphere, and refuse incineration is a major source of atmospheric cadmium. Large quantities of metals go to landfill, mostly from municipal waste disposal, while the disposal of coal ash in landfill is most important for arsenic. Agricultural land in the UK receives its largest inputs of lead, mercury and arsenic by atmospheric deposition, while the application of phosphate fertilisers delivers the biggest cadmium load to these soils (Table 1.5). The sewage system receives both domestic and industrial effluent waste containing metals. The subsequent direct discharge of sewage and the disposal of sewage sludge also represent significant sources of toxic metals, whether applied to landfill or arable land (Table 1.5) or released into rivers and coastal waters (Table 1.6). Metals applied to land inevitably have the potential to enter the aquatic environment via run off into streams, then rivers, estuaries and coastal waters.

While the biogeochemical cycling of trace metals involves the flux of very large quantities of metals naturally, it is clear that the activities of humankind have become the most important driver in this cycling today. The additional anthropogenic mobilisation of trace metals into the biosphere represents an extra toxic challenge to the biota of the world. This ecotoxicological challenge may not be immediately obvious at a global scale, although atmospheric transport is adding significantly to the input of toxic metals into otherwise uncontaminated polar aquatic

Table 1.4 Anthropogenic inputs of selected trace metals into soils, and total discharge on land, in 1983 (million kg per year).

	As	Cd	Cr	Cu	Hg	Mn	Mo	Ni	Pb	Sb	Se	V	Zn
Agricultural & food wastes	6.0	3.0	90	38	1.5	112	30	45	27	9	7.5	22	150
Animal wastes, manure	4.4	1.2	60	80	0.2	140	24	36	20	0.8	1.4	11	320
Logging, wood waste	3.3	2.2	18	52	2.2	104	3.3	23	8.2	5.5	3.3	9.9	65
Urban refuse	0.7	7.5	33	40	0.26	42	4.4	10	62	1.3	0.62	0.4	97
Sewage sludge	0.24	0.34	11	21	0.8	11	0.32	22	9.7	0.2	0.14	1.5	57
Other organic wastes	0.25	0.01	0.48	0.61	–	0.63	0.4	3.2	1.6	0.11	0.08	0.76	2.1
Metal manufacture waste	0.21	0.08	2.4	7.6	0.8	4.9	0.16	2.5	11	0.16	0.19	0.22	19
Coal fly ash	37	13	446	335	4.8	1,655	74	279	242	22	60	67	484
Fertilizer	0.02	0.25	0.38	0.58	–	–	0.02	0.55	2.3	–	0.10	0.13	1.1
Peat	0.5	0.11	0.19	2.0	0.02	17	0.75	3.5	2.6	0.45	0.41	1.7	3.5
Commercial product waste	41	1.6	610	790	0.82	500	3.2	32	390	4.0	0.2	2.8	620
Atmospheric fallout	18	8.4	38	36	4.3	46	4.0	37	263	3.9	2.6	21	135
Total input soils (max)	112	38	1,309	1,403	15	2,633	145	494	1,039	47	76	138	1,954
Total input soils (median)	82	22	896	954	8.3	1,670	88	325	796	26	41	132	1,372
Mine tailings	11	4.1	–	787	2.8	–	16	64	390	24	0.41	14	620
Smelter slag and wastes	9.0	3.2	–	790	0.28	–	6.5	65	390	16	0.2	6.0	620
Total land discharge (max)	132	45	–	2,980	18	–	168	623	1,819	87	77	158	3,194

Note: Data are maximum estimates unless stated.
Source: After Nriagu and Pacyna (1988).

Table 1.5 **Estimated anthropogenic inputs of selected trace metals to the atmosphere, landfill and arable land in the UK in 1980 (1,000 kg per year).**

	Cd	Pb	Hg	As
Atmospheric emissions				
Nonferrous metal production	3.7	51	5	ND
Production of articles containing this metal	ND*	6,802	10.1	8.8
Iron and steel production	2.3	478	1.8	9
Fossil fuel combustion	1.9	80	25.5	297
Cement manufacture	1	36	2.5	ND
Municipal waste incineration	5	142	5.9	0.5
Sewage sludge incineration	0.2	1.2	0.6	0.1
Total	*14*	*7,590*	*51*	*315*
Input to landfill				
Nonferrous metal production	20.8	2,945	ND*	208
Production of articles containing this metal	217	6,000	46.5	94
Iron and steel production	42	3,050	ND*	179
Fossil fuel combustion	60	1,270	1.8*	838
Cement manufacture	21	ND	ND*	ND
Municipal waste disposal	525	34,914	62.7	204
Sewage sludge and sewage disposal	11.4	182	1.7	6.7
Phosphate fertilisers	1.3	ND	ND	ND
Total	*899*	*48,361*	*113*	*1,530*
Input to arable land				
Sewage sludge	4.9	98.3	1.0	2.5
Phosphate fertilisers	22	2.4	0.1	6.1
Atmospheric deposition	15.4	1,536	5.1	102.4
Pesticides	0	ND	12	ND
Total	*42.3*	*1,637*	*18.2*	*111*

Note: ND = not determined. * Probably negligible.
Source: After Hutton and Symon (1986).

Table 1.6 **Estimated anthropogenic inputs of selected trace metals to UK coastal waters in 1980 (1,000 kg per year).**

	Cd	Pb	Hg	As
Rivers	39	878	14	118
Direct discharge				
Sewage	24.4	146	1.8	15
Sludge via pipeline	0.3	10	0.1	0.1
Industrial discharges	22.2	134	9.5	195
Dumping				
Sewage sludge	5.1	165	1.4	1.5
Industrial wastes	0.4	245	0.4	53
Dredged material	7.0	789	11.2	ND
Total	*98*	*2,367*	*38*	*383*

Note: ND = not determined.
Source: After Hutton and Symon (1986).

ecosystems. The ecotoxicological effects of anthropogenically released trace metals are, however, certainly noticeable at the local level in so many habitats throughout the British Isles, particularly those affected by our long history of mining for trace metals. In this book, we will explore the natural history of metals, initially as manifested in ecotoxicological effects close to mines, but sequentially then via streams, rivers, estuaries and coastal seas, ultimately to the adjacent ocean where metals still are crucial in their effects on the organisms living therein.

1.4 Trace Metals and Living Organisms

Prologue. Trace metals bind to organic compounds such as proteins. Such interaction led to the incorporation of many trace metals (the essential metals) over evolution into biochemical processes to play key metabolic roles. Yet the same affinity of trace metals to bind with organic compounds is the basis of their potential toxicity, binding in the wrong place at

the wrong time. Notorious episodes in the history of the ecotoxicology of trace metals include Minamata disease, caused by mercury, and itai-itai disease, caused by cadmium. Selenium exemplifies an essential trace metal with a very narrow window between doses that bring about either deficiency or toxicity.

As introduced earlier, all trace metals are toxic above a threshold level. Yet, many of these same metals are essential to life at lower doses, for they have been incorporated over evolution into biochemistry to play key metabolic roles (Luoma and Rainbow, 2008). Trace metals will bind readily to particular organic compounds, especially proteins. Proteins are crucial in the biochemistry and physiology of all organisms, particularly for example in their role as enzymes driving all biochemical pathways. Many enzymes involve a trace metal in their structure to act as a catalytic centre, as in the case of zinc in the enzyme carbonic anhydrase or selenium in glutathione peroxidase. Non-enzymatic proteins may also involve metals, as for example in the use of iron to bind oxygen in the respiratory protein haemoglobin. Given the universality of many biochemical pathways in all the different organisms on Earth, trace metals that are essential to one species are essential to most others, particularly across eukaryotic organisms. There a few exceptions, usually between prokaryotes and eukaryotes. Vanadium is essential for the biochemical process of nitrogen fixation, unique to prokaryotes. While selenium is needed for glutathione peroxidise in both eukaryotes and prokaryotes, it is also required for the anaerobic respiration of prokaryotes, an essential role absent from eukaryotes. Cadmium is a classic example of a non-essential metal, but in the 1990s it was discovered that some oceanic phytoplankton species have a carbonic anhydrase that depends on cadmium, and not the usual zinc (Lane et al., 2005).

A list of essential trace metals generally includes As, Co, Cr, Cu, Fe, Mn, Mo, Sb, Se, Sn, Ti, V, W and Zn. Non-essential metals usually include Ag, Al, Au, Cd (but now qualified), Hg, Ni, Pb and Pt (Luoma and Rainbow, 2008).

The down side of the affinity of trace metals to bind with organic compounds such as proteins is that trace metals present in excess will inevitably bind to biological molecules in the wrong place at the wrong time, unless prevented from doing so. They might replace another trace metal playing a key essential role in a protein, thereby inhibiting the catalytic activity of an enzyme, or they might bind elsewhere on the protein, distorting its structure and preventing its biochemical function. Either way, a toxic effect has been caused.

All trace metals, whether essential or non-essential, are toxic above a threshold level, initially causing sublethal effects but ultimately a lethal effect at a higher exposure concentration, or over a longer exposure duration (Figure 1.1). This threshold level may be extremely low or even absent for non-essential metals. In the case of an essential trace metal, a very low metal exposure will cause negative performance or growth effects as a result of deficiency, while toxic effects will still occur at higher exposures above the toxic threshold. It does not follow that non-essential metals are as a rule more toxic than essential metals. Copper, for example, is essential in small doses and

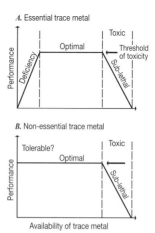

Figure 1.1 Effects of increasing availability of (a) essential and (b) non-essential trace metals on the performance (e.g., growth, production, fecundity, survival, etc.) of an organism. (From Luoma & Rainbow, 2008.)

yet is one of the most toxic metals at higher doses. Nevertheless, it is true that the non-essential metals mercury, silver and cadmium do top most relative toxicity tables of trace metals (Luoma and Rainbow, 2008). If different organisms were to be exposed to a range of dissolved metal concentrations in a standard exposure medium, the dissolved concentrations corresponding to the onset of toxic effects would vary greatly between organisms for the same metal – different species show different sensitivities to the same metal. Similarly, toxic thresholds for one species will vary greatly between metals, and rank orders of metal toxicities will not necessarily be the same between species. Nevertheless, an approximate general (but certainly not universal) order of toxicities of selected trace metals in inorganic form might be

Hg > Ag > Cd > As > Cu > Se > Ni > Pb > Cr > Zn > Sn

reiterating the point that non-essential metals are not necessarily more toxic than essential metals.

Considering the toxicity of trace metals, we are concerned here with two different aspects. The toxic threat of metals to human life is a primary concern. Nevertheless, it will feature less prominently in this book than will the second aspect – the ecotoxicology of trace metals, as exemplified by the effects of metals on the biological communities of different habitats and ecosystems in the British Isles, from the terrestrial through freshwater to the sea. The threat to human health will arise predominantly in the metal contamination of foodstuffs, potentially stemming from the growth of crops in metal-contaminated soils, but more likely from the consumption of long-lived fish at the top of long marine food chains when metals are in organometal form. A classic example is the biomagnification of mercury along oceanic food chains leading to tuna, because most of the bioaccumulated mercury is in the form of methyl mercury, originally derived from inorganic mercury by microbial transformation. Nevertheless, it needs to be remembered that, unlike the case of organic compounds such as organochlorines, biomagnification is not a general principle of the much more common food chain transfer of metals originally absorbed in inorganic form. Other possible threats to human health might be the ingestion of metal-contaminated soils (particularly by infants) and the breathing in of metal-contaminated dusts.

Mercury is infamous in the history of the ecotoxicology of trace metals. The first widely reported instance of a public health impact from aquatic metal contamination occurred in Minamata Bay in Japan in the 1950s (Phillips and Rainbow, 1994). Fish kills and bird mortalities had been noted beforehand, but, in 1953, the local cat population showed severe behavioural aberrations ('dancing disease') and convulsions. In 1956, similar neurotoxic impairment symptoms appeared in the local human population and the name Minamata disease was born. In time, the cause was put down to mercury toxicity, specifically methyl mercury toxicity derived from the diet. A local factory was using mercury salts as catalysts in the production of acetaldehyde, and large amounts of mercury had been emitted in effluent into the local bay. Microbial activity in the sediments transformed the mercury into methyl mercury. Mercury in this form had entered local marine food chains, ultimately severely contaminating the fish and shellfish collected and consumed by the local people. The cats had shown the first symptoms because they were consuming the scraps of fish discarded in food preparation. Alarmingly, the factory continued to discharge effluents contaminated by mercury until 1968. By 2001, nearly 2,000 victims had died and

more than 10,000 had received financial compensation from the factory owners.

Mercury does also provide an example of another principle of metal ecotoxicology, again at variance with the situation for organic contaminants like organochlorines. A high bioaccumulated concentration of a metal in an organism is not necessarily indicative of a toxic effect. Metals can be stored in the tissues of organisms in metabolically inert forms after uptake, in effect locked away out of harm's way. Thus the kidneys and livers of whales and dolphins may contain huge concentrations of mercury, accumulated naturally over time from their diets. These very high bioaccumulated concentrations are not the result of anthropogenic pollution, nor are they indicative of toxicity.

Another very toxic metal, cadmium, figures in a second example of the toxic effects of a trace metal on the local population as a result of anthropogenic activity. In the late 1940s, again in Japan, an usual disease of an overtly 'rheumatic nature' was reported from the Jintsu River basin in Toyama Prefecture (Phillips and Rainbow, 1994). The affliction was known as 'itai-itai' disease, translated literally as 'ouch-ouch' disease, from the cries of pain from the victims, who showed symptoms of softening of bones, skeletal deformation and kidney damage. The cause of itai-itai disease was not clarified until the 1960s, when it was put down to chronic cadmium poisoning. It appears that the primary cause was the discharge of cadmium-rich effluent and sludge from a zinc mine next to the Jintsu River, 50 kilometres upstream. Production from the mine increased in the 1940s and untreated mining effluents were discharged into the river. Downstream, the river water was used to irrigate rice fields, and the resulting rice crop was severely contaminated with cadmium. Cadmium uptake from ingested rice led to cadmium poisoning. A retaining dam at the mine in the 1950s reduced the discharge of contaminated sludge, and the downstream incidence of itai-itai disease diminished rapidly (Phillips and Rainbow, 1994).

Cadmium is still an occupational health hazard today, with potential to cause kidney and bone damage (Järup and Akesson, 2009). The diet is the main source of environmental cadmium exposure to nonsmokers. Atmospheric deposition of airborne cadmium, mining activities and the application of cadmium-containing phosphate fertilisers and sewage sludge on farm land may lead to cadmium contamination of soils, with subsequent increased uptake of cadmium by vegetables and crops destined for human consumption. Smokers will absorb cadmium from the inhalation of cigarette smoke – about 1 µg per day from 20 cigarettes, in comparison with an average cadmium intake from food of between 8 and 25 µg per day (Järup and Akesson, 2009).

While some forms of arsenic are very toxic, others are not. In the waters and sediments of aquatic environments, arsenic usually occurs as arsenate, while it is the less common arsenite which is much more toxic (Neff, 1997). Fortunately for potential consumers, when arsenic is accumulated by marine organisms, it is usually in the form of organoarsenic compounds, particularly arsenobetaine, which is nontoxic and represents no risk to human and other consumers (Neff, 1997). Arsenic derived from geological sources is known to occur in high concentrations in groundwater in certain parts of the world, notoriously in India and Bangladesh, causing chronic health disorders (Bhattacharya et al., 2007). Inorganic arsenic compounds were long used in agriculture as insecticides and herbicides, and as wood preservatives, although this use is now much reduced with increased awareness of arsenic ecotoxicity. In Britain, the industrial revolution has left a legacy of arsenic-contaminated sites, with some local rivers still relatively high in arsenic derived from runoff (Bhattacharya et al., 2007).

Selenium is an excellent example of an essential element for which there is a very narrow window between deficiency and toxicity. While there are no economically significant selenium ores, it is present in high concentrations in high-sulphur coals (up to 34 µg/g) and black shales (600 µg/g), although the selenium content is much lower (less than 2 µg/g) in most coals (Fordyce, 2007; Lenz and Lens, 2009). The ash from high-sulphur coals is further enriched in selenium, with a high associated risk of contaminated leachate entering aquatic systems (Lemly, 2004). Because selenium is an essential micronutrient, it is added to fertilisers for selenium-deficient soils and to animal feed. However, of all essential trace metals,

selenium has one of the narrowest ranges between dietary deficiency (less than 40 µg per day) and dietary toxicity (more than 400 µg per day) in animals, and it is necessary to control carefully the intakes of humans and animals (Fordyce, 2007; Lenz and Lens, 2009). Selenium has an essential metabolic role as part of the enzyme glutathione peroxidise which protects tissues against oxidative damage, and enzymes containing selenium are important in cellular differentiation, growth and development. Selenium deficiency in agricultural animals reared on selenium-deficient soils is common around the world, requiring the use of dietary supplementation, the associated deficiency disease often being described as white muscle disease (Fordyce, 2007; Lenz and Lens, 2009). In the case of humans, selenium deficiency has been implicated in Kreshan disease, a heart disease occurring in China (Fordyce, 2007; Lenz and Lens, 2009). At the other end of the spectrum, ingestion of selenium-rich forage (5 to 40 µg/g) over weeks to months leads to alkali disease and blind staggers in grazing animals (Fordyce, 2007; Lenz and Lens, 2009). The issue of the toxicity of selenium in drinking water has been long debated, and safe upper concentration limits set by government agencies internationally range from 1 to 50 ppb (parts per billion or µg/L), with current opinion favouring the lower end of this range (Gore et al., 2010; Vinceti et al., 2013).

Selenium has also been implicated in ecotoxicological episodes in aquatic environment, notably in the Kesterson National Wildlife Refuge in California in the 1980s (Lemly, 2004; Fordyce, 2007; Lenz and Lens, 2009). Selenium leached from nearby selenium-rich soils was carried into ponds in the refuge in irrigation water used for wetland management, with further concentration by evaporation. Selenium is a rare example of a trace metal that does biomagnify through food chains (Luoma and Rainbow, 2008). It bioaccumulated in local aquatic food chains to give elevated selenium concentrations in every animal group inhabiting the wetlands. This had particularly grave consequences for fish and nesting water bird populations, which showed abnormal development, deformities and reduced survival (Lemly, 2004; Fordyce, 2007; Lenz and Lens, 2009). The drain was closed in 1985, and in 1988 the reservoir and low-lying parts of the refuge were buried under imported topsoil (Luoma and Rainbow, 2008).

1.5 Pollution Hazard and Environmental Regulation

Prologue. In the British Isles, the trace metals of greatest ecotoxicological significance in habitats affected by mining are copper, arsenic, lead and zinc. Since 2000, the safeguarding of Britain's environment from potential metal ecotoxicity has been in line with the Water Framework Directive of the European Union.

The preceding examples have highlighted the potential toxicity, particularly ecotoxicity, of trace metals present in excess, with the apparent paradox in the case of essential metals that insufficient available levels of the metals can also be debilitating. More ecotoxicological examples in the biology of trace metals in the British Isles will be presented in the different habitats explored through this book, counterbalanced, however, by the many nontoxic roles played by trace metals in the natural history of organisms. A further general point does need to be made. The toxicity of a trace metal does not directly reflect its pollution threat or ecological hazard in the real world. A very toxic metal may be extremely rare, and a greater ecotoxicological hazard may be represented by a more abundant metal that is actually less toxic. When considering the mining history of Britain, the subject of Chapter 2, the trace metals of greatest economic importance have been copper, tin, lead and arguably zinc. The list of trace metals of greatest ecotoxicological significance in habitats affected by mining activities is indeed dominated by copper, but also includes arsenic, particularly in Southwest England, and by lead and zinc, for example in mid-Wales, the Northern Pennines and Derbyshire. Even if a high total concentration of metal is present, for example in an aquatic habitat, the chemical form of the metal may mitigate against its ecotoxicological potential, as in the case of arsenic, for which the more common dissolved form arsenate is less toxic than another chemical form, arsenite. The very efficient food chain

transfer of selenium makes it a greater ecotoxicological risk in an aquatic habitat than a relatively low starting dissolved concentration might imply (Luoma and Rainbow, 2008).

In Britain, the recent safeguarding of our aquatic environments from the potential ecotoxicity of contaminants such as trace metals has been in line with the Water Framework Directive (WFD) of the European Union (EU) introduced in 2000. The WFD commits EU member states to achieve good qualitative and quantitative status of all water bodies, including groundwater, rivers, lakes, estuaries and coastal waters. Both good chemical status and good ecological status need to be achieved. As regards chemical contamination, the WFD uses ambient environmental standards to classify water bodies, and these standards aim to prevent ecological damage from contaminants such as trace metals. In particular the WFD establishes maximum acceptable dissolved concentrations of priority hazardous substances, which include the trace metals mercury, cadmium, lead and nickel. While not adopted in the UK, sediment quality guidelines have also been employed elsewhere in the world, for example Australia, Canada and New Zealand, to set supposedly safe concentrations of these contaminants in the sediments of aquatic habitats. It is not enough, however, to meet chemical Environmental Quality Standards, for success under the WFD is also judged by the achievement of ecological goals, using biota to assess the ecological quality of a water body. Currently, the WFD focuses on community-level biological impacts on macroinvertebrates, diatoms, aquatic plants and fish, although a water body will still fail to be classified as good if the average concentration of any metal listed as a priority substance exceeds the relevant Environmental Quality Standard for that specific metal.

It is valuable that the WFD does not rely on chemical guidelines only, because many factors beyond simple total metal concentration affect the interaction of trace metals and organisms and therefore their potential ecotoxicity. It is vital to understand the role of natural history in this interaction, and this aim underlies the approach taken in this book using the British Isles as a case study.

2 Metals and Mining

Box 2.1 Definitions

acid mine drainage (AMD) Outflow of acidic water from active or abandoned coal mines or metal mines.

adit Entrance to an underground mine which is horizontal or nearly horizontal, by which the mine can be entered, drained of water and ventilated and ores extracted at the lowest convenient level.

barytes Mineral consisting of barium sulphate ($BaSO_4$), occurring as a gangue mineral associated with metal ores.

black jack Popular name for the chief ore of zinc, sphalerite, also known as zincblende or blende.

black tin Unrefined tin ore (cassiterite). The term is historically associated with tin mining in Cornwall and Devon, where such tin was smelted to produce metallic tin (white tin).

bloomery Early form of smelter to extract metals, including iron, from their ores. A bloomery consists of a pit or chimney with heat-resistant walls in which pipes near the bottom allow the entry of air by natural ventilation or by bellows. Typically a bloomery is preheated by burning charcoal, before the introduction of metal ore and more charcoal through the top.

calamine Original name given to the two ores of zinc, smithsonite and hemimorphite, assumed until the end of the eighteenth century, to be the same mineral. The name calamine has persisted in the mining industry, however, referring to either smithsonite or hemimorphite indiscriminately.

calcite Mineral composed of calcium carbonate ($CaCO_3$), occurring as a gangue mineral associated with metal ores.

chlorite One of a group of silicate-containing minerals with Mg, Fe, Ni and/or Mn substituted into the silicate mineral lattice in different proportions. It occurs as a gangue mineral associated with metal ores.

fluorspar Mineral composed of calcium fluoride (CaF_2), also known as fluorite. It occurs as a gangue mineral associated with metal ores.

gangue The commercially worthless material that surrounds, or is closely mixed with, a wanted mineral in an ore deposit.

hushing Old mining method which uses a strong flow of water to strip away surface soil to find or exploit veins of metal ore hidden beneath, as in the mining of lead in the Pennines.

lode Deposit of metalliferous ore that fills or is embedded in a crack in a rock formation, or a vein of ore that is deposited or embedded between layers of rock.

ore A mineral or combination of minerals from which a metal can be extracted, with the expectation of yielding a profit after recovery of the costs of mining and processing.

pig iron Brittle form of iron, containing 3 to 5% of carbon and other impurities, solidified from the melted iron produced in a furnace smelting iron ore with a high carbon fuel such as charcoal, coal or coke. Known as cast iron when poured into a mould before cooling.

pig of lead Block of lead solidified from molten lead produced when smelting lead *ore*.

quartz One of the most abundant minerals in the earth's continental crust, made of silicon oxide (SiO_2). It occurs as a gangue mineral associated with metal ores.

slag The glassy mass left as a residue by the smelting of a metallic ore, usually containing a mixture of metal oxides and silicon dioxide.

smelting The extraction of a metal from its ore. Smelting uses heat and often an agent (typically carbon in the form of coke or charcoal) to decompose the ore, the carbon binding with oxygen in the ore mineral to leave behind elemental metal. Limestone is often added to remove final impurities in the metal, a process known as fluxing, leaving behind slag.

stream tin Tin-bearing ore (cassiterite) occurring in the form of rolled fragments or pebbles in alluvial deposits.

tailings Materials left over after separating the valuable fraction from the uneconomic fraction of an ore.

tourmaline Mineral composed of boron silicate compounded with elements such as Al, Fe, Mg, Na, Li or K. It occurs as a gangue mineral associated with metal ores.

white arsenic Almost pure (99.5%) arsenic oxide As_2O_3 refined from arsenic soot (90% As_2O_3) deposited in the chimneys of smelters smelting the arsenic ore mispickel.

white tin Metallic elemental tin. The term is historically associated with tin mining in Cornwall and Devon, where this form of tin was smelted from black tin.

wrought iron Form of iron with a very low carbon content (0.1 to 0.25%) together with inclusions of slag up to 2% by weight. It is tough, malleable and easily welded.

2.1 Introduction

Prologue. This chapter addresses the geology of metals in the British Isles. What metal ores are present and how were they formed? What is the distribution of these ores across Britain and where were they mined? The coming of the industrial revolution brought new sources of trace metals released into the environment.

Britain has a long history of mining for metals, reaching back into the Bronze Age, the subsequent Iron Age and then Roman times. The oldest mines sought copper and tin in particular, the ingredients of bronze, and these were metals well represented in the geology of the British Isles. Indeed, the production of British copper and tin in the ancient world was of great international significance. Over time, other metals were extracted from British mines in significant quantities. Iron, the metal upon which the industrial revolution was based, was highly sought, and lead, silver, zinc, arsenic and even gold were mined. Copper is of particularly high potential ecotoxicity if released into the environment, but ecotoxicological problems will also follow from raised bioavailabilities of other trace metals in terrestrial and aquatic systems. Iron will react with oxygen to form iron oxides and is rarely a chemical danger in the environment. In aquatic systems, however, the deposition of ochre, consisting mainly of iron oxides and hydroxides, can have a physical blanketing effect to cause depletion of the flora and fauna of streams.

It is the mining and other industrial history of locations such as the mining areas of Cornwall, Devon, the Pennines, Wales and Scotland that will set the scene for the following chapters to consider the knock-on effects of trace metals on British natural history, in turn through terrestrial, freshwater, estuarine, coastal and even oceanic ecosystems.

2.2 Geological Origin of Metal Ores

Prologue. Metal ores in the British Isles are typically sequentially deposited by the cooling of hot hydrothermal fluids in faults in sedimentary rocks caused by the intrusion of igneous granite into overlying strata.

In the British Isles, metal ores typically occur in mineral veins in areas where, in the geological past, a mass of igneous granite rock has been intruded into sedimentary rock strata, causing folding and fracturing in these strata (Dines, 1969; Raistrick and Jennings, 1989). The sedimentary rocks into which the granite was intruded have often been affected by the great heat and pressure, and these rocks have been metamorphosed and fractured in a zone around the main granite mass. Hot hydrothermal fluids containing metal salts enter the fractures (faults) from below during the intrusion process, and on cooling these metal salts deposit as veins or lodes of future metal ores (Dines, 1969). There is a recognisable order of deposition of the different metal ores as temperatures cool, and the metal compounds are accompanied by non-metallic compounds that will be considered as waste and referred to as gangue by future miners. Such matrix materials include quartz, calcite, barytes, fluorspar, chlorite and tourmaline (Table 2.1) (Perkins, 1972; Raistrick and Jennings, 1989). The gangue minerals have acted as a flux, helping the metal ores to migrate up into the faults, and they make up the majority of the material filling the fissures containing the metal ores.

The temperature of the surrounding rock is the key factor in the deposition of the minerals, causing the minerals to take up a zoned arrangement with depth – tin first, then copper, lead, zinc (these two often together) and finally iron, in ascending order (Table 2.1) (Perkins, 1972). Ores of secondary metals are associated with these primary metal ores – tungsten in the tin zone, arsenic in the copper zone and silver in the lead/zinc zone. The gangue materials also show a depth distribution, although quartz is found throughout all zones from tin to iron (Table 2.1). Of the other gangue materials, tourmaline is found in the tin zone, chlorite in the copper zone, fluorspar from the copper zone to the lead/zinc zone, barytes from the lead/zinc zone to the lower iron zone and calcite in the iron ore zone (Table 2.1) (Perkins, 1972).

Table 2.1 **Depth distribution of zones of deposition of metal ores and associated secondary ores and matrix (gangue) material, as typified in Southwest England.**

Metal ore zone	Secondary ore	Matrix materials		
Surface				
Iron	Quartz			Calcite
			Barytes	
Lead/zinc	Silver			
		Fluorspar		
Copper	Arsenic			
		Chlorite		
Tin	Tungsten	Tourmaline		
Depth				

Source: Perkins (1972).

The idealised vertical succession of the metal ores and associated gangue material may be displaced laterally, particularly in the case of the shallower metal ore zones, and often the full vertical succession of metal ore zones will not be present. Furthermore, there may be secondary weathering of the metal salts first deposited, for example the oxidation of metal sulphides to oxides and, particularly in the presence of limestone, carbonates.

In Cornwall and Devon, the rock strata into which the granite masses were intruded were sedimentary rocks (sandstones, mudstones {slates} and siltstones), usually of the Devonian period, but with some Carboniferous strata in the north of the region between the latitudes of Launceston and Barnstaple (Dines, 1969). These Palaeozoic sedimentary rock strata were folded and regionally metamorphosed by a major mountain building event, between 350 and 290 million years ago in the Carboniferous and early Permian. In Southwest England, there are five major granite intrusions protruding through the Palaeozoic strata – those of Land's End, Carnmenellis (the area of west Cornwall between Redruth, Helston and Penryn), St Austell Moor, Bodmin Moor and Dartmoor (Figure 2.1) (Dines, 1969). The lodes containing metal ores are usually, but not always, on the margins of the granite intrusions, with tin and copper lodes expected closer to the edge of the granite, and lead and zinc lodes potentially further away (Dines, 1969).

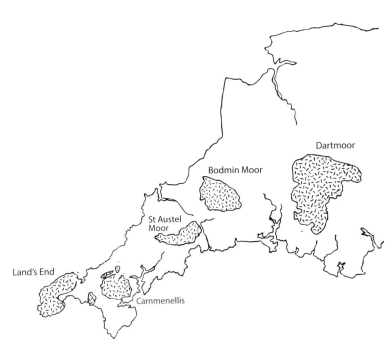

Figure 2.1 Intruded masses of granite in Cornwall and Devon, showing, from west to east, the five major masses of Land's End, Carnmenellis, St Austell Moor, Bodmin Moor and Dartmoor.

In Cornwall and Devon, the granite masses are exposed above the Palaeozoic rocks into which they were intruded, for weathering has stripped off much of any remaining cover, as for example across Dartmoor. In the Pennines, mineral veins are also associated with areas into which a mass of granite rock has been intruded. In contrast to Southwest England, however, the intruded granite is typically still deeply buried, and the mineral veins traverse the usually Carboniferous rocks around and above the granite mass (Raistrick and Jennings, 1989). In the Southern Pennines, Millstone Grit (coarse-grained Carboniferous sandstone) caps the high ground. In the Northern Pennines, in the upper valleys of the rivers Tees, Wear, Derwent, East Allen, West Allen and South Tyne, the surface rocks are mainly Carboniferous limestones (Turnbull, 2006). These northern Pennine limestones are cut by vertical mineral veins containing gangue and metal ores, particularly lead in association with silver (Turnbull, 2006).

In Shropshire, the metal ores, mainly of lead and silver, are again of hydrothermal origin. Hot brines rising from deep in the earth together with gangue materials found their way into fissures, this time in Ordovician rocks, at a time of igneous activity, perhaps in the Devonian period (Brook and Allbutt, 1973). In Wales, the metal-bearing lodes are similarly associated with faulting in the host rocks, in this case Ordovician and Silurian sediments of the lower Palaeozoic, as well as Carboniferous limestones. Lead and zinc lodes with associated gangue materials dominate in Wales, with copper ores less common, except in the case of Parys Mountain on Anglesey (Alloway and Davies, 1971; Davies, 1987). In Scotland, the metal-bearing lodes consist of filled fissures in usually Ordovician and Silurian rocks, commonly of lead and zinc ores (with some silver) together with gangue materials (Wilson, 1921). In Ireland, there are significant metal deposits in the eastern foothills of the Wicklow Mountains, associated with Ordovician rock strata. The metal ores can occur here in veins in siliceous zones, but also as massive metal sulphide deposits a few metres thick, probably of direct volcanic origin (Wright et al., 1999).

2.3 Ores

Prologue. This section reviews the typical dominant ores mined in the British Isles, metal by metal.

2.3.1 Tin

2.3.1.1 Cassiterite

The chief ore of tin mined over the ages in Britain is the oxide of tin (SnO_2) cassiterite, known as black tin in Cornwall and Devon (Plate 1a). Crystals of cassiterite are usually short and square in section, although the ore is often present as a dull crust on a gangue material (Atkinson, 1985). Cassiterite precipitates at high temperature, and so it is found in high-temperature quartz veins, which may also contain tourmaline (Table 2.1) (Perkins, 1972).

2.3.1.2 Stannite

Stannite, a sulphide of tin, copper and iron (Cu_2FeSnS_4), is a minor ore of tin that occurs in small amounts in the tin-bearing veins of Cornwall. The name comes from *stannum*, the Latin for tin. Stannite was first described in 1797 from a sample from Wheal Rock, a mine in the St Agnes District of Cornwall.

2.3.2 Copper

2.3.2.1 Chalcopyrite

Chalcopyrite is the most important copper-bearing mineral in the world (Atkinson, 1987). This importance is reflected in its position as the dominant copper ore in Britain, being present, for example, in Cornwall and Devon (known as yellow copper ore) (Dines, 1969), the Lake District at Coniston (Shaw, 1975), Wales including Parys Mountain on Anglesey (Alloway and Davies, 1971) and Derbyshire in the Southern Pennines (Li and Thornton, 1993a). Chalcopyrite is a sulphide of copper and iron ($CuFeS_2$) (Plate 1b).

2.3.2.2 Chalcocite

Chalcocite, another important copper ore, is also a sulphide of copper, but of copper alone (Cu_2S).

Chalcocite has been found in large quantities in Cornwall and Devon, where it was known as grey copper ore (Dines, 1969).

2.3.2.3 Malachite

Malachite is a green copper carbonate hydroxide material ($Cu_2CO_3(OH)_2$) (Plate 1c), formed secondarily from the weathering and oxidation of the copper sulphides chalcopyrite and chalcocite.

2.3.2.4 Azurite

Azurite ($Cu_3(CO_3)_2(OH)_2$) is the second of the copper carbonate hydroxide minerals, simple copper carbonate not being known to exist in nature. Azurite is blue (Plate 1d). Like malachite, azurite is produced by the secondary weathering of primary copper sulphides (Dines, 1969). Together with malachite, azurite occurs in secondarily altered zones in Cornwall and Devon (Dines, 1969), as well as in the old copper mines of Alderley Edge in Cheshire (Atkinson, 1987).

2.3.2.5 Cuprite

Cuprite, a minor copper ore known as red copper (Dines, 1969), is an oxide mineral composed of copper oxide (Cu_2O) (Plate 1e). Like the copper carbonate hydroxides, malachite and azurite, and another copper oxide mineral, tenorite, cuprite forms in the oxidised zone of copper sulphide deposits.

2.3.2.6 Tenorite

Tenorite is a copper oxide ore (CuO) known as black copper (Dines, 1969) and previously named melaconite. Like cuprite, tenorite can be abundant in South-west England (Dines, 1969).

2.3.3 Arsenic

2.3.3.1 Arsenopyrite

The only arsenic ore of economic value in Britain is arsenopyrite, known historically as mispickel. It is an iron arsenic sulphide ($FeAsS$) (Plate 1f). Arsenopyrite typically occurs in the copper zone with another sulphide, chalcopyrite, but occasionally with cassiterite in upper part of the tin zone (Dines, 1969).

2.3.4 Lead

2.3.4.1 Galena

Galena is the principal ore of lead, being the most abundant and widely distributed, and is easily smelted to release lead. It consists of lead sulphide (PbS) (Plate 1g) and may contain 1 to 2% of silver (argentiferous galena). Galena is widespread in the British Isles, found, for example, in Cornwall and Devon, the Southern and Northern Pennines, Snowdonia in Wales, Scotland and the Mendips in Somerset.

2.3.4.2 Other Lead Ores

While galena is overwhelmingly the most important ore of lead, it can be altered by weathering, such that some lead lodes may contain cerussite (lead carbonate $PbCO_3$, referred to as white lead ore), pyromorphite (lead chlorophosphate $Pb_5(PO_4)_3Cl$), anglesite (lead sulphate $PbSO_4$) and/or mimetite (a lead arsenate chloride $Pb_5(AsO_4)_3Cl$, formed by the oxidation of galena and arsenopyrite) (Wilson, 1921). All these secondary lead ores are of very minor significance in comparison to galena.

2.3.5 Zinc

2.3.5.1 Sphalerite

Sphalerite is the chief ore of zinc, also known as zincblende or blende and popularly as black jack. It is found, for example, in Cornwall and Devon, the Pennines, Wales and Scotland. Sphalerite consists mostly of zinc sulphide, but almost always contains some iron $(Zn,Fe)S$ (Plate 1h). Sphalerite is often found in association with the lead ore galena and the iron ore pyrite.

2.3.5.2 Smithsonite, Hemimorphite and Calamine

Smithsonite consists of zinc carbonate $(ZnCO_3)$ (Plate 1i), occurring in the secondary weathering zone of zinc ore deposits, derived from sphalerite. Often associated with smithsonite is the less common secondary ore of zinc, hemimorphite, zinc silicate $(Zn_4Si_2O_7(OH)_2.H_2O)$, derived from the oxidation of sphalerite.

Until the beginning of the nineteenth century, smithsonite and hemimorphite, being very similar in appearance, were assumed to be the same mineral, and were known under the name calamine. Smithsonite was named after the distinguished English chemist James Smithson, who first identified it as a distinct mineral in 1802. Smithson never married, and on his death in 1829 his estate was left to his nephew, who in turn died without issue in 1835. The bequest then passed to the United States of America to found what became the world famous Smithsonian Institution in Washington. Even once the distinction had been made between smithsonite and hemimorphite, however, the name calamine has persisted in the mining industry, referring to either mineral indiscriminately.

Both smithsonite and hemimorphite are found in Cornwall and Devon (Dines, 1969) and in Scotland (Wilson, 1921), but not in large quantities worth mining. Calamine was mined in the Somerset Mendips (Buckley, 1992; Li and Thornton, 1993a), and in the Northern Pennines from the lead veins of the Allendales and Nentdale (Turnbull, 2006). In the Irish Midlands, there are areas where metal sulphide deposits have undergone oxidation, and both smithsonite and hemimorphite are found at Silvermines (North Tipperary) and smithsonite at Galmoy (County Kilkenny) (Balassone et al., 2008).

2.3.6 Silver

2.3.6.1 Argentiferous Galena

While not a specific ore of silver, argentiferous galena is often the most important mined source of silver. Argentiferous galena may contain 1 to 2% of silver associated with the predominant lead sulphide of galena.

2.3.6.2 Acanthite

Acanthite is the second most important silver ore after argentiferous galena. It consists of the sulphide of silver (Ag_2S) (Plate 1j). It is often confused with argentite, which has the same chemistry but can only exist above 173°C.

2.3.7 Iron

2.3.7.1 Pyrite

Pyrite, or iron pyrite, is an iron sulphide (FeS_2) (Plate 1k). It is the most common of the sulphide ores, and the most common ore of iron. Pyrite has a metallic lustre, and its yellowish hue gives it a superficial resemblance to gold, leading to its common nickname of fool's gold. Pyrite can occur as large deposits, as in the case of Avoca in Wicklow in Ireland and is often mined primarily as a source of sulphur.

When pyrite is exposed to water and air, it will form iron oxides and release sulphate, a process accelerated in freshwater streams by the bacteria *Acidithiobacillus*. The sulphate dissolved in water forms sulphuric acid, leading to acidification of the water and creating acid mine drainage (AMD).

2.3.7.2 Other Iron Ores

Other ores of iron are generally carbonates or oxides. Siderite, also known as chalybite or spathose ore, consists of iron carbonate $FeCO_3$ (Plate 1l). Iron carbonate ores may be argillaceous, containing much clay, as in the Coal Measures (the coal-bearing part of Upper Carboniferous strata) of Cleveland (clay ironstone) and Scotland (blackband). Iron oxide ores include magnetite (iron 2 oxide Fe_3O_4), red haematite (iron 3 oxide Fe_2O_3) and limonite or brown haematite (hydrated iron 3 oxide $Fe_2O_3.xH_2O$).

2.3.8 Manganese

2.3.8.1 Pyrolusite

Pyrolusite is a black mineral of manganese consisting of manganese dioxide (MnO_2). Together with other manganese oxides, pyrolusite may be referred to under the catch-all name of psilomelane (Plate 1m).

2.3.9 Tungsten

2.3.9.1 Wolframite

Tungsten is also called wolfram, and the primary ore of tungsten is wolframite. Wolframite is a mixture of iron and manganese tungstates ((Fe,Mn)WO_4), known as ferberite when it is iron-rich with a low manganese content (Dines, 1969). Wolframite is found associated with cassiterite in quartz veins in the upper part of the tin deposition zone (Table 2.1).

2.3.9.2 Scheelite

Another ore of tungsten that may be found in the tin zone (Table 2.1) is scheelite, a calcium tungstate mineral ($CaWO_4$). While wolframite is the primary ore of tungsten in Southwest England, scheelite does also occur, for example in the Tavistock and St Austell areas (Dines, 1969).

2.3.10 Antimony

2.3.10.1 Antimonite, Jamesonite and Bournonite

Ores of antimony include the sulphide of antimony called antimonite or stibnite (Sb_2S_3), the mixed sulphide of antimony, lead and iron (jamesonite $Pb_4FeSb_6S_{14}$), and the mixed sulphide of antimony, lead and copper (bournonite $PbCuSbS_3$). In Southwest England, antimony ores are usually scattered on the fringes of mineralised areas beyond the lead and zinc deposits, but may be associated with galena, sometimes in workable amounts (Dines, 1969).

2.3.11 Cobalt, Nickel and Molybdenum

Cobalt occurs as arsenides, such as smaltite, and as the sulph-arsenide cobaltite, as well as in oxide form (Dines, 1969). Nickel generally occurs in close association with cobalt and is present as the arsenides niccolite (kupfernickel) and chloanthite and the sulphide millerite (Dines, 1969). Molybdenum occurs as the sulphide molybdenite (Dines, 1969).

2.3.12 Uranium

The major ore of uranium is the oxide uraninite, commonly known as pitchblende because of its black colour. Pitchblende occurs in Southwest England, as may trace amounts of the alteration products of pitchblende – autunite (calcium uranyl phosphate), torbernite (copper uranyl phosphate) and zippeite (hydrous potassium uranium sulphate) (Dines, 1969).

2.3.13 Gold

Unlike other trace metals, gold is chemically stable as the element. It can therefore be found as native gold, or as alloys of gold with silver or copper, in veins of quartz or calcite, perhaps in association with sulphide ores of copper, arsenic, iron or antimony (Herrington et al., 1999). Rather than in veins themselves, gold can be found as grains in alluvial deposits such as river gravels that have resulted from erosion and washing away by water. Gold can be found in Southwest England, but most British gold comes from Wales, particularly from a broad belt north of Dolgellau (Davies, 1987; Herrington et al., 1999).

2.3.14 Aluminium

2.3.14.1 Bauxite

The major ore of aluminium is bauxite, which is a mixture of aluminium oxide/hydroxide minerals such as gibbsite ($Al(OH)_3$), iron oxide ores, the clay mineral kaolinite and small amounts of titanium dioxide. Bauxite is a variety of laterite, weathering to produce laterite soils rich in iron and aluminium. Laterites are considered typical of tropical countries, but they do occur in Northern Ireland, for example at Glenravel, County Antrim, where they once provided a major source of iron and aluminium.

2.4 Geographical Distribution of Ores and Mining Districts

Prologue. The geographical distribution of metal ores, and subsequently of metal mining fields, in the British

Figure 2.2 Principal metalliferous ore fields. (1) West Scotland. (2) Southern Uplands of Scotland. (3) Lake District. (4) Northern Pennines. (5) Yorkshire Dales. (6) Peak District/Derbyshire. (7) north Wales – Halkyn Minera. (8) Snowdonia. (9) central Wales. (10) West Shropshire. (11) Devon–Cornwall. (From Downing et al., 1998, with permission.)

Isles is related to the geological history of granite intrusions into typically Palaeozoic sedimentary rock strata, the concurrent fracturing of these strata and the filling of the faults with hot metal-rich fluids and associated matrix minerals. Figure 2.2 is a simplified map of the principal metalliferous ore fields of Britain, highlighting the important metal mining regions of Cornwall and Devon, the Pennines and central Wales. Ore fields are also present in the Lake District, Shropshire, the Mendips in Somerset, north Wales, including Anglesey, the Southern Uplands of Scotland and the Isle of Man.

2.4.1 Cornwall and Devon

Prologue. The mining fields of Cornwall and Devon have been exploited for the primary metal ores, tin, copper, lead, zinc and iron, as well as the secondary ores of arsenic, silver and tungsten.

In Cornwall and Devon, the Palaeozoic rocks of the Devonian and (in the north) the Carboniferous periods were subjected to earth movements, producing folding, cleavage and jointing with a roughly east-northeast trend, when, in the late Carboniferous or early Permian, granite was intruded to form large bosses. The chief granite bosses are those of Land's End, Carnmenellis, St Austell Moor, Bodmin Moor and Dartmoor (Figure 2.1), in addition to the Scilly Isles off the coast (Dines, 1969). The fissures filled with matrix materials such as quartz, followed by the deposition of tin, copper, lead, zinc and iron as primary ores in sequence as the temperatures cooled (Table 2.1).

The lead, zinc and iron minerals crystallising at lower temperatures could travel further up the faults as temperatures decreased and also further laterally. Thus the ores of lead, zinc and iron occur in some places as peripheral deposits postdating the tin and copper lodes. The vertical ranges of the zones depicted in Table 2.1 vary from place to place, and all zones are not always present. Thus workable bodies of tin are not always present below a well-developed copper zone, as in the cases of some mines north of Camborne and at Devon Great Consols near Gunnislake, both in Cornwall (Dines, 1969). Similarly, lead and zinc ores may be found without an underlying copper zone, as in the region of St Agnes, Cornwall.

While the ore-bearing lodes are concentrated around the edges of the granite bosses (Atkinson, 1985), not all lodes are necessarily on the granite margins (Dines, 1969). Tin and copper lodes can be expected to be closest to a granite boss, as at Camborne, St Ives and St Just, with the lead and zinc lodes potentially further away. However, some tin lodes can be as far away from the granite as the lead and zinc lodes, as at Wheal Jane near Chacewater in Cornwall. Furthermore, iron lodes may occur within the granite, as in the district of St Austell. A notable source of iron (and indeed other metals including zinc, lead and silver) has been the Perran Iron Lode, a linear mineral deposit, 6 kilometres long and between 1 and 30 metres wide, in the Perranporth region of Cornwall. The siderite in the lode has been oxidised to produce limonite as deep as 70 metres (Dines, 1969).

Erosion of overlying sedimentary rock over geological time has exposed the granite bosses of Cornwall and Devon at the surface and made the metal lodes accessible. Over the geological eras, many ores, for example those of copper, lead and zinc, have eroded, dissolved and been lost via streams and rivers to the sea. Cassiterite, the main ore of tin, however, is heavy and also more resistant to dissolution, and so could accumulate locally as alluvial deposits in streams and rivers (Atkinson, 1985).

2.4.2 Pennines

Prologue. The metal mining fields of the Pennines fall into three principal, geologically distinct, regions – the Northern Pennines, the Yorkshire Dales and the Peak District of Derbyshire – and have been exploited particularly for lead and zinc.

In the Pennines, granite masses have been intruded into the beds of limestones, shales and sandstones of the Lower Carboniferous, and the granite is still deeply buried (Raistrick and Jennings, 1989). The fault fractures caused by the intrusions are filled with gangue materials and metal ores (particularly ores of lead and zinc), to become metal lodes above and around the granite.

Figure 2.3 shows the positions of the metal mining fields of the Pennines, which fall into three principal, geologically distinct, regions – the Northern Pennines, the Yorkshire Dales and the Peak District of Derbyshire (Raistrick and Jennings, 1989). These three regions may be referred to as blocks. In the Northern

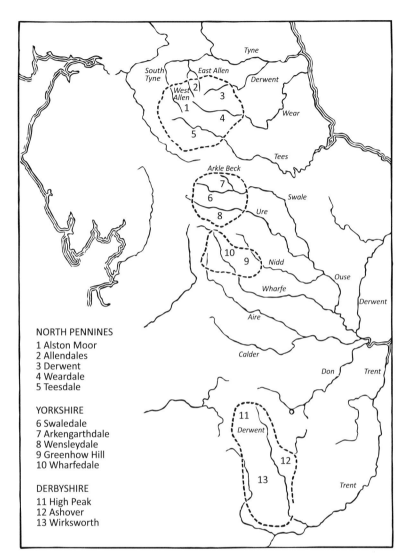

Figure 2.3 The Pennine mining fields. (After Raistrick and Jennings, 1989, with permission.)

Pennines, the Alston Block extends from the South Tyne valley to the upper valley of the Tees. In the Yorkshire Dales, the Askrigg Block reaches from Stainmore to the Aire valley. In Derbyshire, the mining region covers from the central Ashbourne area to the Peak (Raistrick and Jennings, 1989).

In the barren moorland of the Northern Pennines, the Alston Block consists of a succession of Carboniferous sedimentary rocks (limestones, sandstones and shales with a few very thin coals) dipping gently eastwards towards the Durham coalfields and the North Sea, into which a large horizontal body of igneous rock, the Whin Sill, has intruded. The Alston Block can be subdivided into five smaller mining fields (Figure 2.3): Alston Moor, which is in the catchment of the South Tyne and its tributaries; the Allendales, forming the catchments of the East and West Allen Rivers; Weardale, the catchment of the River Wear and its tributaries; the catchment of the upper Derwent, a southern tributary of the Tyne;

and the upper valleys of the River Tees and its tributary the Lune. These dales were destined to become known as the 'lead dales', exploiting the argentiferous galena present (Turnbull, 2006).

In the Yorkshire Dales, there are two important mining fields (Figure 2.3) (Raistrick and Jennings, 1989). To the north, mining has been developed in Swaledale, the Arkle Beck (a tributary of the Swale) and Wensleydale. In the south of the region, the mining field covers upper Wharfedale and the Greenhow area in the Nidd catchment, also extending towards Settle and south to Airedale.

The Derbyshire mining field (Figure 2.3) also divides into two main areas of mining – firstly in the High Peak, and secondly around Wirksworth and Ashover (Raistrick and Jennings, 1989). Metal mineralisation has occurred predominantly in the local Carboniferous limestone. The metal ores present are mainly sulphides including galena (the most common), sphalerite, pyrite and chalcopyrite (Li and Thornton, 1993a). The galena in Derbyshire is usually poor in silver (Edwards et al., 1962). Major gangue materials are calcite, barytes and fluorspar.

Not far from the southern edge of the Peak District, but located in Staffordshire, is Ecton Hill, which has copper-bearing veins but also veins with lead, in sharply folded Lower Carboniferous limestones.

2.4.3 Wales

Prologue. Metal mining regions figure strongly in Wales, not least in central Wales, Snowdonia, the Halkyn Mountain region of northeast Wales and Anglesey (Figure 2.2).

Typically in Wales, metalliferous veins occur in Ordovician and Silurian sediments and in Carboniferous limestone (Alloway and Davies, 1971; Davies, 1987). These lodes contain the lead ore galena and the zinc ore sphalerite, accompanied by quartz, calcite and barytes. The copper ore chalcopyrite does occur, but less commonly than galena and sphalerite, with the striking exception of Parys Mountain in the northeast of Anglesey.

2.4.4 Scotland

Prologue. Mining for lead at Leadhills and Wanlockhead has dominated Scottish metal mining.

Leadhills and Wanlockhead, the highest villages in Scotland, are only a mile apart, although now in the different administrative areas of Strathclyde and of Dumfries and Galloway respectively. Close to the boundary between the old Scottish counties of Lanarkshire and Dumfriesshire, they are the only locations in Scotland where mining for metals has occurred on anything other than a small scale (Wilson, 1921).

In Scottish lodes, it is the lead ore galena and, to a lesser extent, the zinc ore sphalerite that dominate (Wilson, 1921). The other lead ores, cerussite, anglesite, pyromorphite and mimetite, do occur, as well as the mixed lead and antimony ore, jamesonite (Wilson, 1921). Nevertheless, it was galena that was much the major lead ore worked at Leadhills and Wanlockhead, although some cerussite and pyromorphite have been taken in oxidised parts of the Leadhills veins (Wilson, 1921). Of the zinc ores, sphalerite was the only one mined, while smithsonite and hemimorphite also occur. Silver was extracted from argentiferous galena, for example at Coire Buidhe Hill on the south side of Loch Tay (Wilson, 1921). Some copper ores do occur in Scottish lodes, for example along the coast of Dumfries and Galloway, but much less frequently than lead and zinc (Wilson, 1921). Gangue minerals include quartz, calcite and barytes. In addition to Leadhills and Wanlockhead, lead ores also occur at Strontian and Islay in Argyll; at Tyndrum, relatively close by; and at Minnigaf, near Newton Stewart in the south of Dumfries and Galloway (Wilson, 1921). Sphalerite was mined at Wanlockhead and at Newton Stewart (Wilson, 1921).

The metalliferous lodes of the mining fields of Leadhills, Wanlockhead and Newton Stewart lie in a belt of Silurian rock strata, typically shales, intruded into by granite, exemplified by the Cairnsmore of Fleet, an open moorland mountain made of granite about 10 kilometres east of Newton Stewart.

2.4.5 Other Mining Regions

Prologue. Other historical mining regions include areas of Shropshire, Cheshire and Somerset in England.

In Shropshire, the fissures created in Ordovician rocks by probably Devonian igneous activity contain mainly argentiferous galena, a source of lead and silver (Brook and Allbutt, 1973). The mineral veins are confined to one stratum – the Mytton Flags, which are flagstone and gritstone sedimentary rocks more than 300 metres thick. The Snailbeach mine near Shrewsbury was the biggest lead mine in Shropshire, with a reputation for yielding a particularly high volume of lead per unit area, with some small amounts of silver and zinc also taken.

In Cheshire, historic mines at Alderley Edge exploited the secondary copper ores such as malachite and azurite found there in thin veins, copper sulphides being rare (Atkinson, 1987).

In Somerset, for example at Shipham, the Mendips are underlain by Triassic rocks with mineral veins containing smithsonite as the important ore giving up zinc (Li and Thornton, 1993a). Another zinc ore, sphalerite, is also present, as are the two lead ores, galena and cerussite (Li and Thornton, 1993a).

2.5 History of Mining for Metals in the British Isles

Prologue. Mining for metals in Britain began in the Bronze Age and continued with breaks into the twentieth century. Different metals have dominated according to changing demands over centuries. Tin and copper have been particularly important, but significant amounts of silver, lead, zinc, arsenic and iron have also fed national and international markets.

2.5.1 Bronze Age and Iron Age

Prologue. Copper and tin, the ingredients of bronze, were mined in Bronze Age Britain, followed by the addition of iron in the Iron Age.

The Bronze Age can be defined by the smelting first of copper in about 5000 BC, followed by the production of bronze, the alloy between copper and tin, from about 3500 BC. Bronze is harder than copper, and being more easily melted, is more suitable for casting into tools and weapons. Copper was, however, not the first metal to be smelted. That honour fell to lead and tin, with cast lead beads found that date from as early as 6500 BC. These latter two metals are more easily smelted, for they can be released from some ores by the heat of a wood fire. Copper ores typically require higher wood-burning temperatures, for example with extra ventilation provided by bellows. While lead and tin came first, they did not have the advantageous properties of copper, not least its hardness. Lead is too soft to be useful as tools or weapons, and tin is not much harder.

Metals, particularly tin and copper, were certainly being mined in Bronze Age Britain, with copper and then bronze tools being found that date from 2500 to 800 BC (Buckley, 1992). The copper mine on the Great Orme in north Wales was arguably the largest copper mine in Northwest Europe in the first half of the second millennium BC (MacGregor, 2010). The copper and tin mines of Cornwall were also of great international significance in the Bronze Age (Buckley, 1992), with Cornish tin in particular demand for bronze production during that second millennium BC (Atkinson, 1987). Alluvial deposits of cassiterite were the initial source of tin, but outcropping tin lodes also offered an accessible source of tin, as in the sea cliffs at Geevor in the St Just area and at St Agnes in Cornwall. Similarly, copper lodes could also be clearly seen in the cliffs at St Just, Illogan, St Agnes and Perranzabuloe in Cornwall. The metal-bearing lodes of mid- and north Wales were also exploited in the Bronze Age for lead, silver, zinc, copper and even gold (Davies, 1987). Again in the Bronze Age, copper was taken from the Alderley Edge mines in Cheshire (Atkinson, 1987) and possibly from sites in Aberdeenshire in Scotland (Wilson, 1921).

The Iron Age followed the Bronze Age from about 1200 BC, with the development of techniques to smelt iron from its ores. Not only were iron ores more common than those of copper and tin, but iron was less brittle than bronze. Iron soon replaced bronze as the metal of choice for weapons and implements. Southwest England remained of ongoing great importance for the production of tin and copper in the first millennium BC. For example, in the fourth century BC, there was a well-established tin trade between Cornwall and the Mediterranean (Buckley, 1992). In the first century BC, Diodorus Siculus quoted a since lost account of Cornwall's tin trade by Pytheas of Massala (Marseilles), who visited Britain between 325 and 250 BC. Pytheas reported that tin was stockpiled on the island of St Michael's Mount off South Cornwall, before transport across the channel to Gaul, and subsequent transport by horseback to the mouth of the Rhone (Buckley, 1992).

The Iron Age also saw mining for copper in the Lake District (Atkinson, 1987), and exploitation of the lead–zinc ore fields of the Pennines and the Mendip Hills in Somerset (Li and Thornton, 1993a). As befits the age, the Ashdown Forest in Sussex saw tree felling to provide fuel for the production of iron from local iron ore (Christian, 1967).

2.5.2 Roman Britain

Prologue. The Romans in Britain mined lead and copper, in addition to iron, in the Ashdown Forest and gold in Wales.

The Romans put lead to extensive use despite, or because of, its softness, capitalising on its ease of casting and shaping in plumbing and in sheathing wooden boats. The Romans therefore exploited the lead ores of Britain, from the Mendips, Wales, Derbyshire, Yorkshire and the Northern Pennines, but not apparently Southwest England (Dines, 1969; Coyle, 2010). Molten lead emerging from primitive wood-burning hearths into channels solidified into blocks called pigs. Pigs of lead with Roman inscriptions have been found in the Derbyshire Peak District, and there is documentary evidence of the export of Derbyshire pigs of lead by the Romans back to Italy (Edwards et al., 1962). Sites of Roman lead mining in

north Wales included Talargoch at Dyserth and on Halkyn Mountain, both in northeast Wales.

The Romans also mined copper in Britain, but, as for lead, there is no strong evidence for Roman copper (or tin) mining in Cornwall and Devon (Dines, 1969; Buckley, 1992). There were Roman copper mines in Northeast Anglesey and on the Great Orme in north Wales (Atkinson, 1987). Mineral veins were exploited for copper by the Romans near Coniston in the Lake District (Shaw, 1975). The Romans also mined copper at Alderley Edge in Cheshire and in Northwest Shropshire at the Llanymynech mine, southwest of Oswestry (Atkinson, 1987).

From the Roman period forwards, higher smelting temperatures were achieved by the wider use as a fuel of charcoal produced by burning wood with a limited air supply. Charcoal burns at a higher temperature than wood, improving the smelting of tin, lead, copper and particularly iron. The Romans exploited the iron industry of the Ashdown Forest in Sussex. Indeed, a Roman furnace of the first century AD has been found near the River Medway at Ridge Hill Manor, East Grinstead, where the yellow Wadhurst clay is a prime source of iron (Christian, 1967). Local supplies of plum, ash, oak, birch and hazel were used to make charcoal. The Romans also improved access to the iron-producing districts of East Grinstead and Maresfield by opening routes running north to south, in contrast to the existing forest ridgeways, which typically ran east to west (Christian, 1967).

The Romans mined gold in Wales, at the Ogofau mine at Dolaucothi near the village of Pumpsaint between Llanwrda and Lampeter in Dyfed, South Wales (Herington et al., 1999; Coyle, 2010). An estimated 830 kilograms of gold was extracted here in Roman times, from the gold trapped in mixtures of pyrite and arsenopyrite (Coyle, 2010).

2.5.3 The Middle Ages

Prologue. Mining skills were all but lost in Britain in the first centuries of the Middle Ages, which extended from the fifth to the fifteenth century. The mines of the Northern Pennines produced silver at the end of the first millennium, and lead from these mines was in in great demand from Norman times forwards. Tin mining flourished in medieval Cornwall and Devon.

Some mines were active in the early Middle Ages. Silver was gained from the argentiferous galena in the Northern Pennines that was to later yield large quantities of lead. Silver was used by the Anglo-Saxon kings to pay tribute (Danegeld) to the Danes between 990 and 1010, including a payment of about 13 tons of silver in 1007 for two years' relief from attack (Coyle, 2010). In total, about 60 million silver pennies were handed over to the Danes during this period (Coyle, 2010). There was also an increase in the twelfth century in the demand for silver for coinage. The Royal Mint, then based at Carlisle, was supplied with silver from the 'silver mines of Carlisle' located on Alston Moor.

In the ninth century, lead had been sent from Wirksworth in the Derbyshire Peak District to Ely (Coyle, 2010). And in the eleventh century, the Domesday Survey refers to three lead mines at Wirksworth, and one each at Crich, Bakewell, Ashford and Matlock in the Peak District (Edwards et al., 1962). There was probably also lead mining in the nearby Castleton District at this time (Edwards et al., 1962). By the time of the Domesday survey in 1086, however, mining in Wales had virtually ceased (Davies, 1987) and mining had ended in the Alderley Edge copper mines in Cheshire (Atkinson, 1987).

In Scotland, an ancient bronze foundry, possibly dating from the ninth century AD, has been found at the Mote of Mark, near Kipford (Dumfries and Galloway), in an area where thin veins carrying bright green copper ores are fairly abundant (Wilson, 1921). Furthermore, lead mines on Islay are said to have been worked by the Danes (Norsemen) (Wilson, 1921).

From about the twelfth century, there was increased activity in castle, church and cathedral building, with an associated rise in the demand for lead for plumbing, roofing and windows (Edwards et al., 1962; Turnbull, 2006). Indeed, lead would become an important export from medieval Britain, second only to wool (Edwards et al., 1962).

In the Middle Ages, charcoal furnaces, with a stone hearth and bellows for an improved air supply, replaced primitive wood-burning hearths consisting of a shallow hole on the top of a hill facing the prevailing wind for ventilation (Edwards et al., 1962).

2.5.3.1 Pennines

Prologue. The Derbyshire lead mines were particularly active in the later Middle Ages, and the mines of the northern Pennine lead dales also contributed.

King Stephen in the twelfth century granted mineral rights to the Prince Bishops of Durham, who encouraged lead mining in Weardale (Turnbull, 2006). In 1189, 100 cartloads of lead from Derbyshire went to Waltham Abbey in Essex (Coyle, 2010). Also in the late twelfth century, lead was sent from Alston Moor to the palace of King Henry II at Windsor, and lead was exported to France (Turnbull, 2006). Lead exports to Germany went via the River Humber. Lead sent to London used the port of Yarm on Tees, where the cobbles in the High Street are ballast from medieval ships coming to trade for wool and lead (Coyle, 2010). The lead ballasted the wool on the way out.

In Derbyshire, most of the mining area from Wirksworth to Castleton, which covered parts of both High and Low Peak, became known as the King's Field, from which a royalty was levied on the ore raised (Edwards et al., 1962). Mining operations were controlled by a code of laws based on local customs, to be embodied in the practices of the Barmote Courts. The Barmote Courts were set up in 1288 to oversee the Derbyshire lead industry and settle relevant mining disputes, one at Wirksworth to oversee the Low Peak, and the other at Monyash for the High Peak (Edwards et al., 1962). A Barmaster presided over each Barmote Court, with deputy Barmasters at other mining centres such as Matlock, Wensley, Brassington and Tissington. A jury of twelve was selected from twenty-four experienced lead miners, and the Barmote Court met twice a year (Edwards et al., 1962). The courts had jurisdiction over mining on both crown lands and private land. Basically, ancient mining laws entitled the finder of a lead vein to have sufficient land and enough water to work the vein, irrespective of the ownership of the land (Edwards et al., 1962). The ancient rights and customs were jealously guarded by the miners. Indeed, although their powers were strongly regulated by parliamentary acts in the nineteenth century, the Barmote Court, now combined, still meets once a year at Wirksworth.

The reputation of Derbyshire lead miners was high in the thirteenth century. They were sent to open new lead mines in Grassington and Appletreewick in Wharfedale, and in 1294 to exploit the discovery of lead and silver at Combe Martin in Devon (Coyle, 2010). Lead mines were also operating in Weardale, the Derwent valley and Alston Moor in the fifteenth century, according to the 1474 report of a commission to King Edward IV, which gave details of lead mines containing silver in the Northern Pennines (Turnbull, 2006).

2.5.3.2 Cornwall and Devon

Prologue. In medieval Cornwall and Devon, the mining of tin dominated from the thirteenth century. Stannary Parliaments and Stannary Courts were powerful legislative institutions in Cornwall and in the Dartmoor area of Devon, protecting the interests of tin miners. There was little copper sought in Cornwall and Devon again until the middle of the sixteenth century. Lead and silver were also mined in medieval Cornwall and Devon, as at the silver mines on the Bere Alston peninsula north of Plymouth.

As had been the case since the Bronze Age, most tin in Southwest England came from streaming alluvial deposits. The tin ore cassiterite (black tin) had weathered over thousands of years from tin-bearing lodes exposed at the surface. Being relatively insoluble and heavy, it had been washed into river valleys and deposited there in sand and gravel. Tinners would use the river water to wash away lighter waste material and concentrate the black tin before smelting (Buckley, 1992). Typically the tinners would dig trenches up to 15 metres wide in the bottoms of valleys with deep alluvial deposits, and shovel the silt, rich in black tin, into a strong stream of water from a diverted river or brought by a system of ditches across the moor (Coyle, 2010). A team of tinners might work down a valley for some miles, and at convenient points, the ore would be treated in a wooden box to separate the black tin from the silty gangue. The heavier black tin would sink out from the remaining silt, which was washed downstream, often to cause silting up in harbours, to such an extent that stoppages to tin streaming were ordered in 1356 and in 1532 (Coyle, 2010). Occasionally the tinners would recover small grains of gold in the alluvial tin deposits, and so generate a small additional income. Still today, heaps of rubble and waste resulting from medieval tin streaming can be seen in river valleys on Dartmoor, often now overgrown by grass and bilberries (Buckley, 1992).

Tin ore from the tin lodes themselves would increasingly become the major source of tin, with ore taken initially from outcropping lodes and then by mining into cliffs and hillsides. Early tin mines were opencast trenchworks following the lodes down (Buckley, 1992). Even in the thirteenth century, however, some of the first cliff mines had vertical shafts for ventilation, access and the hoisting of tin ore (Buckley, 1992). In the early thirteenth century, Cornwall and Devon produced about 120 tons of metallic tin per year. This total had risen to nearly 600 tons annually by the early fourteenth century before falling to 450 tons per year by the end of that century (Dines, 1969). By the mid-fifteenth century, underground mining for tin had taken over in importance from streaming, being developed, for example, at Gwennap, Kenwyn, Illogan, Gwinear, Breage and St Just in Cornwall (Buckley, 1992).

Tin ore was crushed before smelting, the molten tin then being poured into granite mould stones to form ingots (about 50 to 80 kilograms each). Waterwheels typically powered the crushing of the black tin ore to a fine sand, and the bellows provided a draught for the smelting furnace. The fuel for the smelting was wood, peat or charcoal. The tin ingots produced could be very pure (up to 99.9% tin) according to the purity of the ore (Atkinson, 1985). Mixed ores needed higher temperatures (roasting) to remove impurities and, therefore, more fuel.

The production of tin in Cornwall and Devon was of such importance to the economy of medieval England that special laws for tin miners predate written legal codes in Britain. The Stannary Parliaments and Stannary Courts were separate and powerful legislative institutions in Cornwall and in the Dartmoor area of Devon, protecting the interests of tin mining (Buckley, 1992). Indeed, everyone connected with tin mining in Cornwall and Devon was exempted from any jurisdiction other than the Stannary Courts, in all but the most exceptional circumstances (Buckley, 1992).

In 1201, King John granted a charter to the tin miners of Cornwall and Devon, confirming their 'just and ancient customs and liberties'. The tin miners of both areas originally met together at Hingston Down and referred to themselves as a 'parliament', variously named the Tinners' Parliament or the Stannary Parliament. The Tinners' Charter of 1201 set down considerable rights and privileges for those engaged in the tin industry, giving tinners the right to search for tin on any unenclosed land, as well as exemption from ordinary laws and taxes and from military service, but they were now subject to Stannary Laws (Buckley, 1992). Under the Tinners' Charter, Cornwall was divided into four districts (stannaries), each returning six Stannators to the Stannary Parliament, which had the power to try any cases related to the tin industry under these Stannary Laws. The four stannaries of Cornwall were the following:

- Penwith and Kerrier: including the granite outcrops of Land's End and Carnmenellis
- Tywarnhaile: St Agnes and the Carn Brea area near Camborne
- Blackmore: the Hensbarrow granite upland, now better known as the St Austell moors
- Foweymore: the historic name for Bodmin Moor

While the geographical jurisdiction of each Cornish stannary was more clearly demarcated than would be the case in Devon, for each represented a separate tin-bearing area, the boundaries were not precisely laid down. The relative productivity of the stannaries varied greatly and was in no way related to their size. All tin produced in a stannary had to be weighed and taxed ('coined') before it was sold, with proceeds going to the Duchy of Cornwall or the Crown. This process ('Coinage') took place twice a year, increased later to four times a year, at a Coinage Town. These towns in Cornwall did vary over time, and variously included Penzance, Truro, Helston, St Austell, Bodmin (probably), Liskeard and Lostwithiel (Buckley, 1992). The Coinage was a tin market, with high piles of tin ingots to be sold, and a time of great festivity.

In 1305, the Stannary Charter of Edward I split the stannary institutions between Cornwall and Devon, establishing parliaments and courts for the two counties separately. The jurisdiction of the Cornwall stannary institutions covered the whole of the county, while the Stannary Courts of Devon were based on Dartmoor. Devon stannaries are usually referred to by the names of stannary towns where Coinage took place. The 1305 Stannary Charter established Tavistock, Chagford and Ashburton as Devon's stannary towns, with a monopoly on all tin mining in Devon, a right to representation in the Stannary Parliament and a right to the jurisdiction of the Stannary Courts (Buckley, 1992). Plympton became the fourth Devon stannary town in 1328, after a powerful lobby persuaded the sheriff of Devon that, being nearer the sea, the town had better access for merchants. The Devon stannary towns are all on the fringes of Dartmoor, the granite upland which bore the tin. No definition of the boundaries of the Devon stannaries is known, if indeed one ever existed (Buckley, 1992).

Lead and silver were also mined in medieval Cornwall and Devon. The silver mines at Bere Ferrers on the Bere Alston peninsula between the Rivers Tamar and Tavy were a group of about 12 lead mines mining argentiferous galena of high silver content (Dines, 1969; Dorrington and Pyatt, 1982). They were discovered in 1290, and by 1294 King Edward I was receiving 170 kilograms of silver ore per month (about 2 tons per year), having claimed crown ownership of all metal mines (Dorrington and Pyatt, 1982). The silver mines at Combe Martin in North Devon extended underground with shafts and adits. These mines similarly provided considerable funds to the Crown in the thirteenth and fourteenth centuries. As the silver mines went deeper, they reached the limits of the technology of the time to pump out the water from depth. It was water that closed the ancient Devon silver mines during the 1480s (Buckley, 1992).

2.5.3.3 Lake District

Prologue. Mining for copper at Coniston had continued into medieval times. Lead and silver were also being mined in the Lake District in the fourteenth century (Shaw, 1975).

King Edward III in that century made a grant of mines and appointed Robert de Barton to be keeper of silver/lead mines at Minersdale and Silver Beck and of copper mines at Keswick (Shaw, 1975). Minersdale is the old name for Roughton Gill on the Caldbeck Fells, the valley containing the Roughtongill mine, while Silver Beck refers to the nearby Silvergill mine. The 1474 Commission of Edward IV also reported on the copper mines of Keswick, the most ancient of which was the Goldscope mine, which produced lead in addition to copper (Shaw, 1975).

2.5.3.4 Scotland and Wales

Prologue. In Scotland, lead was mined at Leadhills in the thirteenth century and later. There was some mining of lead in medieval Wales.

In medieval Scotland, there is a record of lead mining at Leadhills in 1239, but there were undoubtedly active lead mines there before then (Wilson, 1921). There are further records in 1466 of the removal of lead from the Leadhills district (Wilson, 1921). There was also mining for lead (and some silver) in Argyll in the fourteenth and fifteenth centuries (Wilson, 1921). In medieval Wales, lead mining that had virtually ceased after Roman times revived somewhat with the demand for lead for castles, churches and abbeys. Thus farmers on Halkyn Mountain supplemented their income with some lead excavation at abandoned Roman mining sites while also seeking and exploiting undiscovered lead lodes near the surface.

2.5.3.5 Iron

Prologue. Medieval Britain needed iron for tools as well as for armour and weapons. Iron production was generally a local industry, exploiting the widespread availability of iron-yielding ores. Bloomeries were used to smelter iron. These employed charcoal as fuel and required large local supplies of wood, as in the Forest of Dean and the Ashdown Forest on the Weald of Sussex and Kent.

Limonite supplied the iron industry of the Forest of Dean, and haematite provided iron in Furness and the Lake District (Shaw, 1975; NAMHO, 2013). Clay ironstones, a mixture of clay and siderite, were, however, the principal sources of iron, being common across much of the British Isles. The clay ironstones were of Lower Cretaceous origin in the Weald and elsewhere in South England, while they dated from the Coal Measures in the Black Country, Derbyshire, and South and West Yorkshire (NAMHO, 2013).

Bloomeries were the earliest forms of smelter capable of smelting iron from its oxide ores. A bloomery consists of a pit or chimney with walls made of earth, clay or stone that can resist heat. Pipes enter near the bottom of the bloomer, allowing the entry of air by natural ventilation or more usually using bellows. The bloomer is preheated by burning charcoal, before the introduction of iron ore and more charcoal through the top. The incomplete combustion of the charcoal produces carbon monoxide, which reacts with the iron oxide ore to produce metallic iron. The product of the bloomery is a mix of slag and iron, which is usually further forged by more heating and beating with a hammer to drive out much of the molten slag from the iron to form wrought iron. This hammering will also shape the iron into a tool or a weapon. Bloomeries were to be later replaced by blast furnaces, again with a source of carbon (typically coke or coal by then) to decompose the ore, but also with limestone added to remove final impurities, to produce cast or pig iron (Alexander and Street, 1951).

There is evidence for the ongoing production of iron from Roman to medieval times in the Forest of Dean, the Weald of Kent and Sussex,

Northamptonshire and southern Lincolnshire, and in parts of West Somerset and East Devon. The Domesday Book of 1086 refers to iron mines near East Grinstead in the Ashdown Forest on the Weald (Christian, 1967), at Lyminge in Kent and at Rhuddlan in northeast Wales (NAMHO, 2013). In medieval times, iron ore was mined in Langdale and Grasmere in the Lake District, and haematite from Furness supplied bloomeries at Coniston (Shaw, 1975).

During medieval times, the demand for iron from agriculture, building and weapon manufacture led to the development of specialist areas for iron production. These areas included the Sussex Weald, the Forest of Dean and Southwest Yorkshire, with iron production also in Somerset and the Blackdown Hills of East Devon (NAMHO, 2013). Particularly key to the suitability of an area for iron production at this time was not only the availability of iron ore, but the availability of wood to make the charcoal needed for smelting in bloomeries. Therefore, in the Middle Ages the iron industry of Britain was concentrated in well-wooded regions, exemplified by the Forest of Dean and the Ashdown Forest. In 1254, for example, iron bars from the Weald provided Henry III with horse shoes and nails (Christian, 1967).

The annual production of iron in medieval England is estimated at about 1,000 tons in 1300, while more iron was imported from Spain and the near continent to meet the full demand (NAMHO, 2013). Thereafter, there was a fall in iron production, perhaps because of the availability of imported iron but also because of the Black Death (1348). In fact, the Weald bucked this trend, possibly because of its proximity to London markets. Demand for iron had increased again by the end of the fifteenth century, by which time water power had been introduced to iron smelting and processing to increase efficiency.

By the late fifteenth century the blast furnace had been introduced into Southeast England (NAMHO, 2013). In a blast furnace, fuel, ore and limestone are continuously introduced into the top while a hot blast of air is blown in at the bottom. The chemical reaction with carbon that turns iron oxide into iron occurs throughout the furnace, as the ore and fuel move down the furnace and the flow of air moves up,

increasing the efficiency of the smelting process. The pig iron produced had fewer impurities than the wrought iron resulting from smelting in a bloomery. Blast furnaces did, however, need more fuel than the bloomeries they replaced. While the Weald continued to have sufficient supplies of wood for charcoal, other areas with good woodland resources also came to prominence for iron production, including the Furness and West Cumbria ore fields using charcoal, and Shropshire exploiting the ironstone of the Coal Measures (NAMHO, 2013).

2.5.4 Sixteenth and Seventeenth Centuries

Prologue. Mines were a source of funds to the Crown which granted licences to search for metals. English monarchs employed Germans with their mining know-how to increase the efficiency of British mining.

With the demise of mining in Britain after the Roman period, Germany had become the centre of mining expertise during the Middle Ages (Turnbull, 2006). So in the sixteenth century, successive English monarchs turned to the Germans to improve mining in England, Wales and Scotland. In 1528, Henry VIII appointed Joachim Hochstetter to be 'principal surveyor and master of all mines in England and Ireland' (Turnbull, 2006). In 1555, in the reign of Mary, a German mining engineer Burchard Cranych was given permission to search for copper, silver and gold (Buckley, 1992). Queen Elizabeth I, needing more bullion at a time of inflation, brought all mines and their products under direct Crown control. She welcomed German miners from the Harz Mountains to develop mines and smelters to increase their productivity (Atkinson, 1985, 1987). In 1568, Elizabeth established the Society of Mines Royal, which was granted a mining monopoly for metals in several English and Welsh counties, including Cumberland and Cornwall.

2.5.4.1 Lake District
Prologue. Mining exploited the exploiting the rich copper veins of the Lake District, such as at Coniston.

Smelters were set up at Keswick before their destruction in the English Civil War. The Greenside Mine on Helvellyn became the largest lead mine in the Lake District.

In the Lake District, the Society of Mines Royal worked copper mines, such as at Coniston, and developed smelters at Keswick, under the guidance of the German mining expert Daniel Hochstetter (Atkinson, 1985, 1987). Hochstetter brought over experienced miners from Germany (often referred to locally as Dutchmen) (Shaw, 1975). He established at Keswick a copper mining and smelting enterprise without equal in Britain, sited at Brigham beside the River Greta, which provided a supply of water to power machinery and bellows (Shaw, 1975). Hochstetter also reopened mines such as the ancient Goldscope Mine in the Newlands valley near Keswick and mines in the Caldbeck Fells (Shaw, 1975). Ores were usually transported to the Keswick smelter by packhorse, and the smelter devoured peat, charcoal and coal from collieries at Caldbeck and Bolton near Wigton. Most charcoal came from Furness, from managed woodland, while peat was cut behind Skiddaw and near Mungrisdale (Shaw, 1975). Initially lead smelted and refined at Keswick was brought from Alston Moor in the Northern Pennines, indicating that none of the local Lake District lead mines was then in operation. Mines at Grasmere and Stoneycroft did start to extract lead but in disappointing amounts, although Roughtongill mine on the Caldbeck Fells was able to keep up a steady supply of lead ore (Shaw, 1975).

The copper mines at Coniston showed notable production of copper in the sixteenth and early seventeenth centuries (Coyle, 2010). The English Civil War, however, brought disaster to the mining and smelting industry of the Lake District. In 1651, Cromwell's men are supposed to have been responsible for the destruction of the copper smelter and mining works in Keswick, an act from which the copper enterprise never really recovered (Shaw, 1975). By then, most of the best ore had been won from Goldscope mine. The other Keswick mines were not nearly so rich, and the Keswick smelters had been relying increasingly on ore

from Coniston and Caldbeck (Shaw, 1975). It was 40 years before there was a revival of Lake District mining under William of Orange (Atkinson, 1985 and 1987). In 1690, another group of 'Dutchmen' rebuilt the copper works at Keswick and built a lead smelter in Stoneycroft Gill. They also either started or reopened the Greenside mine on the northeast side of Helvellyn, served by Swart Brook to provide water both as a power source and to wash the ore extracted. The Greenside mine was to become the largest lead mine in the Lake District, being worked for about 150 years. Coniston copper mine was worked with vigour, reaching 40 metres in depth and producing about three-quarters of all Lake District copper, more than half of it from one vein of chalcopyrite (Shaw, 1975; Atkinson, 1987).

2.5.4.2 Cornwall and Devon
Prologue. Tin mining in Cornwall and Devon continued to flourish in the sixteenth and seventeenth centuries.

Early in Elizabeth's reign, Daniel Hochstetter had come to Cornwall in search of copper, for example at Perranporth near St Agnes. The Society of Mines Royal followed suit later. The establishment of an Elizabethan copper mining and smelting industry in Cornwall, however, failed, and by the early seventeenth century, the Society of Mines Royal had pulled out of Cornwall (Buckley, 1992). A significant factor had been the lack of an available fuel supply in Cornwall, and copper ore was shipped to Neath in South Wales, where the Society built a copper smelter (Atkinson, 1987). Copper prices were also fluctuating, and copper mining virtually ceased in Cornwall in the seventeenth century.

Tin mining in Cornwall and Devon, on the other hand, continued to flourish in the sixteenth and seventeenth centuries, producing about 450 tons of metallic tin annually from the fifteenth to the beginning of the seventeenth century (Dines, 1969). Underground mining for tin had overtaken streaming in importance during the fifteenth century, although many of the early Tudor tin mines were still opencast mines working down the angle of the lode, even down to 100 metres (Buckley, 1992). Nevertheless, some early Tudor underground tin mines employed hundreds of miners. For example, in the 1580s, 300 workers were employed at Godolphin Bal at Breage in the Mount's Bay District of Cornwall (Buckley, 1992).

Drainage remained a major problem in underground mines, and the sixteenth and seventeenth centuries saw many mine pump inventions patented. It was not until 1674, however, that Samuel Moreland's new metal plunger pump brought significant improvement in the pumping out of water, followed by John Coster's waterwheel-powered pumps at the end of the seventeenth century (Buckley, 1992). Another significant development in Cornish tin mines at the end of the seventeenth century was the introduction of blasting with gunpowder. Gunpowder blasting had been first used successfully in Hungary in 1627 and arrived in Southwest England in the 1680s via Germany and the north of England (Buckley, 1992). Both ore extraction and the driving of adits became more efficient. Improved water pumps and the use of explosives now drove underground mines down well over 100 metres. The annual production of tin from Cornwall and Devon increased during the seventeenth century to more than 1,000 tons per year, perhaps with the demand for pewterware in association with the introduction of explosives for blasting (Dines, 1969). Pewter is a malleable metal alloy, with traditionally 85 to 99% tin, the other metals consisting of copper, antimony, lead or even silver. The Poldice tin mine in the Gwennap area near Redruth in Cornwall was legendary in the seventeenth century for productivity, numbers of miners employed and profits made (Buckley, 1992).

2.5.4.3 Pennines

Prologue. Lead mining in the Northern Pennines continued apace in the sixteenth and seventeenth centuries.

Local mining districts for lead mining continued to grow in the Northern Pennines, although the Alston

Moor lead–silver mines were considered exhausted by 1629 (Turnbull, 2006). The Society of Mines Royal, established by Elizabeth I in 1568, had taken ownership of the Northern Pennines mines as in the Lake District (Turnbull, 2006). The new open-hearth furnaces introduced from 1565 improved smelting efficiency, reducing fuel consumption by half. Later, in the seventeenth century, new pumping technology allowed mines to go deeper.

2.5.4.4 Scotland

Prologue. Up to about the sixteenth century, many Scottish lead–silver mines were worked principally for silver. By the end of that century, however, it was lead that was primarily sought, as at Leadhills.

Lead mining had been established in medieval times at Leadhills, and the year 1549 saw the first authentic record of a lead mine on Islay (Wilson, 1921). By the end of the sixteenth century, lead was the primary target of Scottish miners, as at Leadhills, where lead mining commenced again in 1562 (Wilson, 1921). Lead ore was often exported from Scotland in the sixteenth century, at times because of the lack of available fuel for smelting. Between 1585 and 1590, about 100 tons of lead ore was shipped overseas through the Port of Leith (Wilson, 1921). In 1606, a silver–lead mine opened at Hilderstone near Bathgate in West Lothian but was soon exhausted (Wilson, 1921). James VI of Scotland, who reigned as James I of England from 1603 to 1625, used Cornishmen to develop the Scottish lead–silver mines, actually in preference to the German miners who were nearer at Keswick in the English Lake District. In 1614, about 190 tons of lead ore were exported from Scotland (Wilson, 1921). Lead mining was opened up at Wanlockhead from the 1680s, and Scottish lead mining went forward into the following centuries, greatly dominated by mining in the Leadhills and Wanlockhead regions.

The sixteenth century brought the first authentic records of copper mining in Scotland, with copper being worked at Wanlockhead in 1584 (Wilson, 1921). In 1616, there is reference to copper mines on Islay, in addition to the lead mines (Wilson, 1921). Copper mining, however, was very limited in Scotland.

Gold was discovered in the sixteenth century in the area of Leadhills, and the gold mined here provided much of the gold in the gold coinage of James V of Scotland (1513–1542) (Herrington et al., 1999).

2.5.4.5 North Wales, Staffordshire and the Mendips

Prologue. Copper was mined on the Great Orme in north Wales and in Staffordshire, while zinc was taken in Somerset.

Elsewhere, the copper mines on the Great Orme in north Wales had also been revived by German miners in Elizabethan times and were similarly destroyed in the English Civil War (Coyle, 2010). In Staffordshire, copper ore deposits at Ecton Hill had proved difficult to mine from the sharply folded limestones. In 1638, however, Prince Rupert, the nephew of King Charles I, introduced gunpowder and employed German miners with some success (Atkinson, 1987). In the mid-sixteenth century, zinc ore was mined at Shipham in the Somerset Mendips (Li and Thornton, 1993a).

2.5.4.6 Iron

Prologue. Demand for iron continued to increase in the sixteenth and seventeenth centuries. Blast furnaces were now in use, and coal came to the fore as the fuel of choice.

The introduction of the blast furnace into England in the late fifteenth century had accelerated the exploitation of iron ores, but at the cost of the need for larger amounts of fuel. The Weald and the Forest of Dean had sufficient wood available for charcoal production in the sixteenth and seventeenth centuries, but not indefinitely. William Camden (1551–1623) reported that the Weald was full of iron mines 'in sundry places', with furnaces employing the water power of brooks diverted to run in one channel (Christian, 1967). There were important iron furnaces using water power at Hartfield and Newbridge in the heart of the

Ashdown Forest. Forty French specialists were employed in the Weald in 1544, and, in 1549, Sheffield Park ironworks alone employed 26 men (Christian, 1967). Not surprisingly, huge amounts of wood were being used to support this large, nationally important industry. In fact, deforestation had become such a widespread problem threatening the availability of timber for shipbuilding that Elizabeth I passed legislation in 1558 to restrict the cutting of trees to make charcoal. In 1574, Richard Pedley of the Privy Council was ordered to inquire into 'the great spoile and consumption of oakes and other woodes in the countys of Sussex, Surrey and Kent by means of iron milles and furnaces' (Christian, 1967: 18). The size of the Ashdown Forest iron production industry at the time is reflected in the fact that Pedley visited owners of 77 ironworks, including ones at West Hoathly, East Grinstead, Hartfield, Buxted and Maresfield, in 19 days (Christian, 1967).

From the second half of the sixteenth century, there was now increased incentive to use alternative fuel supplies based on coal, and in the early seventeenth century, Dud Dudley used coal as a fuel to make pig iron (Alexander and Street, 1951). Nevertheless, charcoal still remained an important fuel for the iron furnaces of the seventeenth century. Iron making migrated from the traditional centres in the Sussex Weald and the Forest of Dean to the thick woodlands of the Welsh borders, South Yorkshire and Derbyshire valleys and Furness, in the search for an abundant charcoal supply within reasonable reach of iron ore and water power (Raistrick, 1989). There had been some tradition of iron making in these areas, and the bloomery sites of the sixteenth century became small furnace units in the seventeenth century (Raistrick, 1989). In the Lake District, large iron furnaces were in operation at Langdale, in the Coniston District, and at Langstrath near Borrowdale (Shaw, 1975). In the seventeenth century, pig iron was sent from the furnaces of the Forest of Dean to the forges of the Upper Severn and West Midlands, well served by water power and wood supplies. In these forges, the pig iron was converted into bar and rod iron, a process needing more power and fuel than was available back in the Forest of Dean (Raistrick, 1989).

Coal was taking over from charcoal as the fuel of choice for iron smelting during the seventeenth century, and regions such as Shropshire, with access to coalfields, as well as to clay ironstones of the Coal Measures and the water of the River Severn, were well placed to rise to prominence in iron production. As coal mining developed in the English Midlands, there was a rapid decline in the Ashdown Forest iron industry. The number of furnaces there dwindled, as shortages of charcoal and of water power in the drier southeast of the country also conspired against the survival of a historic industry (Christian, 1967).

From coal came coke, and, by the early 1700s, iron production had moved into a new epoch with the use of coke for smelting (Alexander and Street, 1951).

2.5.5 Eighteenth and Nineteenth Centuries

Prologue. The eighteenth and nineteenth centuries saw British mining for metals reach astonishing peaks, serving the needs of the industrial revolution. Improvements in smelting processes enabled the construction of larger and more efficient furnaces. Technological advances in the engineering of pumps allowed mines to combat flooding and go deeper.

In the eighteenth and nineteenth centuries, Britain assumed an international position of prime importance in mining for metals. These two centuries brought the industrial revolution between about 1760 and 1840. Manufacturing processes switched from hand to machine. There were new chemical industries, better iron production processes, improved efficiency of water power and the increasing use of steam power, underpinned by the switch from wood and charcoal to coal as the fuel supply. The improved production of iron and its subsequent uses, not least to make machine tools, were key to the industrial revolution, which started in Britain and subsequently spread to Europe, North America and the rest of the world. The eighteenth and nineteenth centuries in Britain brought expansion in trade, driven by the industrial revolution and enabled by the building of canals, improved roads and ultimately railways. In addition to iron, metals such as copper, lead and zinc

were in high demand and Britain's mines responded accordingly. In 1740, Britain was the world's greatest producer of metals and produced nearly 20,000 tons of iron per year (Alexander and Street, 1951). A century later, Britain made one and a quarter million tons of iron annually, a figure raised to 9 million tons per year by 1900 (Alexander and Street, 1951).

The smelting of metals was improved by the introduction of the reverberatory furnace burning coal at the very end of the seventeenth century. In a reverberatory furnace, the ore being processed does not come into contact with the coal used as fuel, and impurities such as sulphur in the coal could be burnt off separately so as not to affect the final quality of the metal smelted.

Major technological advances to affect mining in the eighteenth century concerned the efficiency of pumps to keep dry the deeper parts of underground mines. In about 1710, the English engineer Thomas Newcomen invented the first practical steam engine to pump water out of mines. The first recorded Newcomen Engine was erected in 1712 at a coalworks near Dudley in Staffordshire. In 1714, John Coster patented a very efficient pumping engine powered by a waterwheel, which could match Newcomen's atmospheric steam engine. For most of the eighteenth century, these two engines were the most effective methods for pumping water out of mines, both engines being improved in efficiency by John Smeaton in Cornwall in the mid-century (Buckley, 1992). In the 1760s, James Watt improved steam engine technology, particularly by the invention of a separate condenser, taking out his first patent in 1769. Patents, however, restricted the application of Watt's steam engines to mines until the end of the century, and mining engineers continued in their search for improved steam-driven pumps circumventing Watt's patents. Right at the end of the eighteenth century, the Cornishman Richard Trevithick built an improved high-pressure noncondensing steam engine, used to pump water from a mine at Wheal Prosper near Porthleven in Cornwall in 1812. As a result of the increased efficiency of pump engines, the depth of mines increased significantly over the eighteenth century. At the start of the century, mines reached

150 metres deep, while at the end of the century mines such as Dolcoath in the Camborne area of Cornwall extended more than 300 metres in depth (Buckley, 1992).

The nineteenth century, however, also brought foreign competition, as increased exploration uncovered the mineral resources of the world. Newly discovered ores were often richer and labour cheaper than in Britain, affecting the price of metals on the world market. British mines were often now uneconomical, and there were regular waves of emigration, particularly of Cornish miners, to the United States, Canada, Australia and South Africa as metal prices fell, especially so in the case of copper in the late nineteenth century (Coyle, 2010). Ironically, when, and if, a metal price recovered, there would often be a subsequent lack of experienced labour back home to take advantage of any new boom. And new overseas mines continued to open up, for example for copper in Chile and for tin in Malaya and Bolivia.

2.5.5.1 Cornwall and Devon

Prologue. The eighteenth century brought a revival of copper mining in Devon and Cornwall, to last until the end of the 1800s. The previously successful tin mining industry took second place to copper mining, as Southwest England dominated world copper production. But copper mines were being exhausted and also suffered under foreign competition in the second half of the nineteenth century. There was a brief revival then for Cornish tin mining, before this too succumbed to overseas competition in the 1890s. Arsenic, often claimed from the spoil heaps of old copper mines, found a temporary market in insecticides in the second half of the 1800s. Cornwall and Devon also produced lead, silver and zinc in the later nineteenth century but essentially lost out to competition from more productive mines elsewhere in Britain or from exhaustion of local lodes.

Copper **Prologue.** Increasing national and international demand heralded a revival of copper mining in Devon and Cornwall. In the middle of the

eighteenth century, Cornwall had become the biggest producer of copper in the world. By the middle of the nineteenth century, however, many Cornish copper mines had been exhausted. Furthermore, the price of copper kept falling during the second half of the nineteenth century as new copper sources across the world came onto the market. Copper mining ceased at the turn of the twentieth century.

By the eighteenth century, demand for copper was increasing. Brass foundries in Birmingham and Bristol needed copper for the manufacture of cylinders, valves and taps (Atkinson, 1987). Copper was also needed to protect the wooden hulls of ships from wood-boring organisms. So, the eighteenth century brought a revival of copper mining in Devon and Cornwall. In the 1680s, Cornwall had produced just a few tons of copper ore per year, but, by the 1720s, Cornwall was raising 6,000 tons of copper ore annually (Buckley, 1992). The annual Cornish production of copper ore reached 7,500 tons in the 1740s, and 26,000 tons in the 1770s (Figure 2.4). In the middle of the eighteenth century, Cornwall had become the biggest producer of copper in the world by a wide margin (Buckley, 1992).

The large tonnages of copper ore extracted, and associated profits made, outstripped the previous achievements of the leading Cornish tin mines, Godolphin Bal in the sixteenth century and Poldice in the seventeenth century (Buckley, 1992). In the first decades of the eighteenth century, copper mines such as Wheal Fortune at St Hilary (south of Hayle) and mines in the Redruth, Camborne and St Day area such as Chacewater (in the Carnon valley and later known as Wheal Busy), Longclose (in Camborne and later incorporated into South Crofty) and Dolcoath (also in Camborne) produced large tonnages of copper ore and enormous profits. In the middle of the eighteenth century, the nearby Poldice mine, in the Gwennap area near Redruth, famous for nearly 200 years as a tin mine, switched priority from mining tin from its principal tin lode to gaining copper from an adjacent copper lode (Dines, 1969). By 1788, the annual output of copper ore from Poldice exceeded that of tin ore. Brass foundries in Bristol, new in 1700 and based

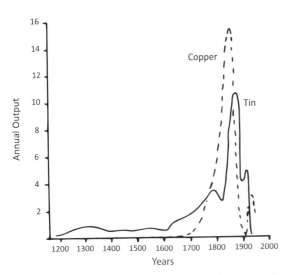

Figure 2.4 The annual output of copper and tin (thousands of tons of metal per year) from Cornwall and Devon from 1200 forwards. (After Dines, 1969.)

Figure 2.5 Camborne Mining District in 1904. (From Dines, 1969.)

firmly on Cornish copper, were the biggest producers of brass in the world by 1750 (Atkinson, 1987).

In the eighteenth and early nineteenth centuries, copper mining changed the face of Cornwall, as Camborne, Pool, St Day, Chacewater and other villages and hamlets became centres of industry (Figure 2.5) (Buckley, 1992). Insignificant tin mines were transformed into large underground copper mines employing hundreds of men. Mines expanded and coalesced into large joint enterprises. What was to become known as the Great County Adit was started in 1748 to drain the Poldice mine, ultimately into the River Carnon (Buckley, 1992). By the 1740s, the Poldice mine, now mining its rich copper resources, was facing water drainage problems at depth. These were addressed by means of the Poldice Deep Adit – the embryo of the Great County Adit. The adit expanded and developed into the most ambitious drainage adit ever driven in Cornwall, with branches to many neighbouring mines in the Redruth and Camborne area and in the Carnon valley by the 1790s. The Great County Adit continued expansion in the nineteenth century, and by 1880 it consisted of a network of drainage channels 40 miles long, serving more than 60 mines (Buckley, 1992). At its peak, the Great

County Adit discharged more than 65 million litres per day into the River Carnon and thence into its estuary, Restronguet Creek. The Great County Adit was a key factor in the success of an area that was for many decades the richest mining district in the world.

Nevertheless, the problems that had affected the smelting of copper in Cornwall and Devon in the seventeenth century had not gone away, particularly the lack of an available fuel supply. It was still necessary to ship copper ore to the copper smelters of Neath and Swansea in South Wales, where coal was abundantly available close to a good harbour. Attempts were made in the eighteenth century to reverse the trade – to import coal from South Wales to smelters in Cornwall. In 1754, copper furnaces were built in Camborne, but the final delivery of coal to these furnaces required the inefficient use of mules with panniers (Atkinson, 1987). In 1756, the Cornish Copper Company bought land (Copperhouse) at Hayle, the best local port near the copper mines of the Redruth and Camborne region. The company built a dock 100 metres long and set up copper furnaces, importing Welsh coal and exporting copper (Atkinson, 1987). By 1819, however, Cornish copper smelting had ended as a result of increasing transport and fuel costs. The copper smelters of South Wales flourished on Cornish and Devonian copper ore, and smelters and foundries in Bristol also benefitted.

Another factor affected the success of Cornish copper smelting, and indeed of the Cornish copper mining industry more widely. In 1768, a huge body of shallow copper ore was discovered on Parys Mountain in Anglesey, in competition with the copper mines of Cornwall, which were going deeper and deeper to win ore at greater economic cost. From 1770 to 1800, Anglesey, not Cornwall, dominated world copper production. Anglesey ore was selling with profit at £50 a ton, while the ore brought up from Cornwall's deep mines needed nearer to £80 a ton to be profitable (Buckley, 1992).

The start of the nineteenth century brought a revival for the copper industry of Cornwall and Devon. Competition from Anglesey fell away, and Watt's patents on steam engines came to an end, releasing the creativity of Cornish mining engineers pumping water out of deep mines. Cornish mine owners moved their money into Welsh copper smelting, bringing under their own control the costs of smelting their ores. In 1805, the price of copper rose to £138 a ton, and Cornish and Devonian copper mines were back in business (Buckley, 1992). Cornwall again became the world's largest producer of copper, and, over the next few decades, the output of copper ore from Cornwall and Devon increased dramatically (Figure 2.4).

In the first half of the nineteenth century, the mines producing the largest amounts of copper ore straddled the Cornwall–Devon border near Gunnislake, in the Callington and Tavistock district (Figure 2.6) (Dines, 1969; Atkinson, 1987). This mining district was about 12 miles long and 4 miles broad, extending westwards from the edge of Dartmoor, into the top of the Tamar valley, and down into the Bere Alston peninsula between the Tamar and Tavy rivers (Dines, 1969), the site of the medieval silver lead mines. The rich copper vein serving the mines in the Gunnislake area was itself nearly 3 kilometres long, up to 13 metres wide and at least 600 metres deep. Of the Gunnislake copper mines, Devon Great Consols, incorporating Wheal Maria, produced 742,400 tons of copper ore between 1845 and 1903. Bedford United produced about 65,950 tons, and Wheal Crebor about 40,000 tons of copper ore, over similar periods (Dines, 1969). Mines in the Gunnislake area shared the advantage of other mines in the deep Tamar valley that they could be drained by adits and not rely to a great extent on pump engines. These mines were also served by

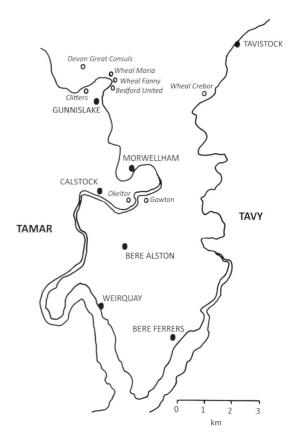

Figure 2.6 Selected mines of the Tavistock and Callington District in the valleys of the Tamar and Tavy.

Morwellham Quay at the top of the navigable part of the River Tamar (Figure 2.6). Morwellham Quay had a long history as a port. Tin ore in the twelfth century, and lead and silver ores in the thirteenth century, were shipped from there. At the end of the eighteenth century, pack horses carried ore from local mines down to the quay, but in 1817 the Tavistock Canal was opened to bring ore to the top of the valley above Morwellham. An inclined plane then delivered the ore to the quay below on the River Tamar. The Devon Great Consols mine was the principal user of Morwellham Quay, which was, for a time, the 'richest copper port in Queen Victoria's Empire'.

In the 1840s and 1850s, Cornwall and Devon represented the most important copper mining district in the world (Atkinson, 1987; Burt et al., 1987). By the 1860s, however, many Cornish copper mines had been exhausted and closed, especially in the Gwennap region near Redruth. Their copper lodes were being mined out more quickly than new lodes could be discovered (Buckley, 1992). Furthermore, the price of copper kept falling during the second half of the nineteenth century as new copper sources across the world came onto the market. The price fell from £87 per ton between 1846 and 1850 to £82 a ton in the early 1870s, and to £56 at the end of the 1880s (Burt et al., 1987). Cornish copper production halved between 1860 and 1870, and halved again during the 1870s (Figure 2.4) (Atkinson, 1987; Burt et al., 1987). The ageing copper mines of Cornwall were simply unable to compete with rapidly increasing production from rich, low-cost copper mining enterprises in Australia, North America, Africa, Chile, Peru and, nearer home, Spain (Atkinson, 1987; Burt et al., 1987).

In the second half of the nineteenth century, the rich copper mines of the Tamar valley fared better than most southwestern copper mines, but even these cut back production (Atkinson, 1987; Burt et al., 1987). Devon Great Consols weathered the recession best. By 1886, Devon produced half of Britain's copper, and, in the 1890s, Devon Great Consols produced 98% of British copper (Atkinson, 1987). But, by 1901 the Devon Great Consols Mine was closed.

Tin Prologue. Tin mining in Cornwall and Devon was overtaken by copper mining in the eighteenth century, before making a comeback in the second half of the nineteenth century. Foreign competition effectively ended tin mining by 1900.

The flourishing tin mining industry of Cornwall initially continued its success into the eighteenth century, with the production of metallic tin increasing to 3,000 tons a year by mid-century, while Devon produced about 50 tons (Dines, 1969). But in the middle of the eighteenth century, tin mining was taking second place to copper mining.

The second half of the nineteenth century did see a reverse in the fortunes of tin and copper mining in the southwest, as Cornwall in particular switched back to tin (Figure 2.4) (Buckley, 1992). As the price of copper

fell during the second half of the nineteenth century, the price of tin, on the other hand, increased from £84 a ton in 1846 to 1850 to £130 a ton the early 1870s, peaking then at £153 a ton (Burt et al., 1987; Buckley, 1992). In Cornwall, copper mines going deeper and deeper had meanwhile discovered hew lodes of tin below the copper lodes.

Correspondingly, Cornish tin production more than doubled by the early 1870s, and held at a high level until the early 1890s (Figure 2.4) (Burt et al., 1987). The output of Cornwall reached 6,000 tons of metallic tin per year in 1850, and peaked at 10,000 tons of tin in 1870 (Figure 2.4), about half the world's yield at the time (Dines, 1969). Cornwall had switched back from a copper county to a tin county, as its mines went deeper in the second half of the nineteenth century. Western Cornwall led the way. The copper and tin Dolcoath mine at Camborne was the largest Cornish producer of tin in every year between 1845 and 1913, with the annual output of black tin ore growing steadily until levelling off at over 2,000 tons a year in the late 1880s and 1890s. (Dines, 1969; Burt et al., 1987). The Dolcoath Main Lode, the major source of tin, was remarkable – up to 900 metres deep, and nearly 20 metres wide in places (Buckley, 1992). Dolcoath had trebled tin production during the 1850s and 1860s, contributing to an all-time peak Cornish tin output in 1871 (Figure 2.4) (Burt et al., 1987). After Dolcoath, the largest tin producing mines were the nearby Camborne mines of Carn Brea, Tincroft, East Pool and Agar, Grenville and Basset, all on the Great Flat Lode discovered in the early 1870s (Dines, 1969; Burt et al., 1987). In addition to these mines in the Camborne Redruth area, other significant nineteenth century tin mines in West Cornwall were working in the St Agnes and Levant districts on the North Cornish coast (Burt et al., 1987).

The tin market, however, in turn slumped in the 1890s for the familiar reason of overseas competition. Cornish tin output halved, and tin production was suspended in many mines (Burt et al., 1987). Alluvial tin had been discovered in Malaya earlier in the nineteenth century, but political instability and associated piracy there in the 1860s and 1870s caused a fall in Malayan tin production. When the price of tin reached the record £153 a ton in the 1870s, Cornwall had cashed in. The tin boom could not last. Ironically, it was the Royal Navy that restored political order in Malaya, and eastern tin production resumed (Buckley, 1992). In the late 1870s, the price of tin dropped to £35 a ton. While there was some subsequent price recovery to above £100 a ton in the 1880s, there was another fall to £64 in 1896 (Buckley, 1992). Tin discoveries in Bolivia, Indonesia and Tasmania and elsewhere in Australia, ironically involving expatriate Cornishmen, had added to the overseas competition. Cornish tin mines in the 1890s amalgamated to become more efficient to meet this competition, as the copper mines had done 100 years earlier in the face of competition from Anglesey. But the Cornish tin boom was over.

Over time, the power of the ancient Stannary Parliaments had declined. The last Stannary Parliament in Devon met in 1748, and the last in Cornwall in 1752 (Coyle, 2010). In 1838, Coinage was replaced by tax paid at the smelting house, following extensive petitioning by the Cornish tin industry for simplification of the taxation rules (Buckley, 1992). The principal purpose for Coinage town status had now ceased, although Coinage towns do still retain certain historic rights, reflecting their past importance.

With the decline of Cornish copper and tin mining, the Great County Adit was being neglected and not maintained adequately. In the winter of 1876, major floods caused huge quantities of silt to be washed down the River Carnon to the top of Restronguet Creek, permanently blocking navigation up the creek to the quays at Devoran (Coyle, 2010).

Arsenic   Prologue. Existing markets for arsenic were supplemented from the 1860s by insecticides in the United States. Arsenic was gained from mispickel in discarded spoil from copper mines and temporarily sustained Cornish mining at the end of the nineteenth century.

During the nineteenth century, a new economic player increasingly emerged onto the mining scene in Cornwall and Devon – arsenic. The predominant arsenic ore in Southwest England is arsenopyrite,

known as mispickel. Arsenopyrite is mainly associated with the important copper ore chalcopyrite in the copper zone, but it is occasionally found with black tin in the top of the tin zone (Table 2.1) (Dines, 1969). Mispickel had long been considered as an undesirable contaminant of a copper (or tin) lode, with arsenic oxide being emitted from smelters or being deposited as a grey powder in smelter chimneys. This 'arsenic soot' contained about 90% arsenic oxide, but could be collected and further refined to 'white arsenic', consisting of almost pure arsenic oxide (Dines, 1969).

In the nineteenth century, there was a market for white arsenic. It was used in the making of glass, the manufacture of shot and fireworks and in the tanning of leather. In particular, arsenical pigments were in demand to colour cotton fabrics and wallpapers, with the newly expanding Lancashire cotton industry a major outlet for Cornish arsenic (Barton, 1971b). In 1812, a former tin smelting house (the Perran Works) near Perranarworthal in the Carnon valley was adapted to produce commercial quantities of arsenic (Barton, 1971b). In the 1830s, a second arsenic works was established at nearby Bissoe, in response to the increasing demand for arsenic from the growing dyestuff industry (Barton, 1971b). Attention had turned to the previously unwanted arsenic in the thousands of tons of discarded mispickel in local spoil heaps, a ready supply for the expanding arsenic industry of the Carnon valley.

In the middle of the nineteenth century, however, the dense fumes of arsenic oxide that were being emitted from the furnace chimneys of the arsenic works of the Carnon valley were having significant ecotoxicological effects on local crops and livestock (Barton, 1971b). The Perran Works, extended in 1849, emitted arsenic fumes day and night, and it was impossible to live downwind of the prevailing southwesterly winds. In 1850, cattle and horses died, bees were killed, fruit trees were blackened and vegetables would not grow on the land of Richard Thomas, which extended to within 50 metres of the Perran Works (Barton, 1971b). The Works was prosecuted as a public nuisance, but, by 1851, further arsenic works had opened in the Carnon valley around Bissoe (Barton, 1971b). For a period in the mid-nineteenth century, sulphur by-products of arsenic production, such as sulphuric acid, were also very marketable. Most sulphur was imported from the volcanic regions of Sicily, particularly after the removal of import duty in 1825. For a period around 1840, however, import of Sicilian sulphur ceased after a monopoly established by the King of Naples increased Sicilian prices, promoting sulphur production in the Carnon valley (Barton, 1971b). This situation was not to last. Sicilian prices were lowered and the Sicilian sulphur trade resumed. Furthermore, sulphur from Spanish pyrite at Huelva came on the market from about 1865, and Cornish sulphur production was hit severely (Barton, 1971b).

In the 1860s, a new market for arsenic developed, a market that would help sustain Cornish mining for the next decades. Arsenic-based insecticides were proving very effective. The international demand for arsenic rose, with a consequent increase in its price. The United States became a strong market for Cornish arsenic. The Colorado beetle was a severe pest of potato crops, and the boll weevil was ravaging cotton crops (Barton, 1971b). While this insecticide-based market did fluctuate from year to year, as the populations of the insect pests varied in abundance, it was a godsend for Cornish arsenic mining. In the 1870s, there was also an increased demand from Germany for arsenic to be used in the manufacture of dyestuffs, a demand that could not be fully met from the local mines in the Harz region (Barton, 1971b). New arsenic enterprises sprang up while the price of arsenic was high in the second half of the 1800s, with works near Gunnislake and the Tamar valley, Plympton, Hayle and Swanpool near Falmouth (Barton, 1971b). In 1868, Devon Great Consols near Gunnislake, the greatest copper mine in Britain, started exploitation of the arsenic ores in its possession (Barton, 1971b). The predominant position in arsenic production switched eastwards from West Cornwall. In the 1870s, Devon Great Consols had been joined by Okeltor, Gawton (Figure 2.6) and Friendship, to the effect that these few mines in the Callington and Tavistock District were now producing half the world's output of arsenic (Dines, 1969). The annual output of white arsenic from Cornwall and Devon at this period peaked at 8,000 tons (Dines, 1969).

The arsenic works were still creating ecotoxicological problems, as in the case of Great Wheal Busy (previously the Chacewater mine), where there were losses of livestock in the surrounding blighted fields (Barton, 1971b). Some progress was made on this front by the increasing installation at arsenic works of a cooling and washing flue to remove arsenic before the chimney stack, the waste water being led away to a settling pond (Barton, 1971b). Atmospheric contamination of nearby farmland was reduced, but terrestrial contamination had potentially been replaced by aquatic contamination if ever the settling pond waters were to enter local watercourses.

As in the cases of copper and tin, the heyday of arsenic production could not last. From the 1880s forwards, foreign competition spelt the end of the Cornish and Devonian arsenic trade. The major foreign competition came from Sweden, and in the 1890s much of the arsenic mining in Cornwall and Devon ceased. The total annual output of white arsenic from Cornwall and Devon fell to 900 tons (Dines, 1969). Two major arsenic works in West Cornwall, at Bissoe and Gwithian, survived the turn of the century, but it was a change in US legislation in 1909 that foretold the end of Cornish arsenic mining. Not unexpectedly, the copper ores of Butte in Montana also contain arsenic. As in Cornwall, the arsenic emitted from smelters at the great Anaconda copper mine was having a similar ecotoxicological effect on cattle, sheep and horses over a wide local area (Luoma and Rainbow, 2008). In 1909, US legislation forbade the discharge of arsenic fumes into the atmosphere, and the Anaconda mine was required to remove arsenic from its smelter emissions (Barton, 1971b). The arsenic previously lost up the chimneys was now available for the local US insecticide industry. Cornwall was no longer the requisite source of arsenic, and the US market for Cornish arsenic to make insecticides increasingly fell away, to finally cease in 1925.

Lead, Silver and Zinc   Prologue. Cornwall and Devon did produce lead, but in smaller amounts than elsewhere in Britain, until the exhaustion of lead lodes by the late nineteenth century. Silver was also produced, particularly in the third quarter of the nineteenth century. Zinc production, often from spoil heaps, peaked between 1850 and 1885, but lost out to competition from mines elsewhere in Britain.

The lead, silver and zinc mines of Cornwall and Devon had not been idle in the eighteenth and nineteenth centuries, although never attaining the dominant position of such mines in the Pennines, Wales and Scotland (Burt et al., 1987). The industrial revolution had brought increasing demand for lead, for example for piping, building materials, castings and lead-based paints. Lead production particularly was strong in the first half of the nineteenth century in Southwest England. The chief areas of lead production were St Agnes, Liskeard and Callington (Dines, 1969), while East Wheal Rose, at St Newlyn East near Newquay in Cornwall, was briefly the country's largest single lead mine in the mid-1840s (Burt et al., 1987).

In the Tamar valley, nineteenth century steam power driving pumps had enabled local mines to go deeper and under the River Tamar, to rich pockets of lead ore containing silver (Dorrington and Pyatt, 1982). By 1809, the Bere Alston mines were again amongst the leading lead and silver mines in Britain, with local smelting at nearby Weirquay on the River Tamar. As in the case of Cornish attempts to smelt copper, coal needed to be imported, and the opportunity was also taken to import ore to be smelted in the Weirquay smelters (Dorrington and Pyatt, 1982). By 1845, there were 18 smelters operating there. By 1865, however, these had ceased functioning, leaving behind spoil heaps of both local and foreign origin (Dorrington and Pyatt, 1982). Once again, the lack of a local supply of coal in Cornwall and Devon had proved decisive.

The large-scale production of lead in Southwest England in the first half of the 1800s collapsed in the third quarter of the century (Dines, 1969; Burt et al., 1987). Between 1845 and 1850, the annual output of lead ore from Cornwall and Devon had averaged over 10,000 tons, but this annual yield had fallen to below 1,000 tons by 1878 (Dines, 1969). In the 1880s, only a few hundred tons of lead per year emerged from Cornwall. East Wheal Rose was still the main local producer of lead, with input also from the Trelawney,

Mary Ann and Herodsfoot mines near Callington (Burt et al., 1987). Production of lead ore ceased in 1887 (Dines, 1969). The cumulative total production of lead ore from Cornwall and Devon between 1845 and 1886 had been 311,000 tons (Dines, 1969).

In the case of lead, the main reason for the fall in production in Cornwall and Devon was not foreign competition and adverse price movements, but the exhaustion of the lead lodes (Burt et al., 1987). Elsewhere in Britain in the early 1870s, there was record lead production in association with high lead prices (Burt et al., 1987). By the end of the nineteenth century, however, the British lead industry was seriously challenged by competition from Spain, Germany and the United States, in the latter case by the development of ore fields in Colorado and Idaho (Turnbull, 2006). In 1883, silver–lead ores were discovered at Broken Hill in New South Wales, Australia, and, by the 1890s, these mines alone produced an annual output of 50,000 tons of lead ore (Turnbull, 2006).

The production of silver was able to offset the demise of lead in Cornwall and Devon to some extent. The chief silver mining regions, exploiting silver-rich lead ore, were St Agnes (the Chivertons, Cargoll and East Wheal Rose mines), St Austell, Liskeard and Callington (the Herodsfoot, Wrey Ludcott, Mary Ann and Trewetha mines) and the Bere mines in the Bere Alston peninsula (Dines, 1969; Burt et al., 1987). At East Wheal Rose in 1852, 48,000 ounces (1,360 kilograms) of silver were obtained from the argentiferous galena mined (Burt et al., 1987). The income from this silver (£12,000) matched the income from the associated lead. The richness in silver of Cornwall's argentiferous galena was second only to that of neighbouring Devon, on the other side of the River Tamar and effectively working the same ores, at nearly 47 ounces per ton of lead (1.35 kilogram per tonne) (Burt et al., 1987). Not all of Cornwall's silver was from lead mines, because some lodes of lead were so rich in silver that this was mined without associated lead production, as, for example, at Great Crinnis, near St Austell (Burt et al., 1987). Specifically, silver ores were mined at Rosewarne and Herland (near Gwinear) and at Wheal Newton, near Callington (Dines, 1969). Silver was also reclaimed from copper ore at Levant, in the St Just area near Wadebridge, and at the Prince of Wales mine at Callington (Dines, 1969). During the third quarter of the nineteenth century, Cornwall became the most important silver mining county in Britain, accounting for 40% of national silver production in the mid-1850s (Burt et al., 1987). Silver production in Cornwall peaked at 300,000 ounces (8,500 kilograms) in 1869 (Burt et al., 1987). After 1875, however, the production of both lead and silver in Cornwall had dropped to be of little significance.

Zinc ores typically occur with lead and silver ores (Table 2.1), but zinc production from Cornish lead mines was less significant than that of silver (Burt et al., 1987). Indeed, until the nineteenth century, the zincblende mined in lead mines was frequently discarded in spoil dumps (Dines, 1969). The nineteenth century, however, brought the development of galvanising techniques whereby a layer of protective zinc is applied to iron or steel, and there was a new demand for zinc. Discarded spoil heaps of zincblende at lead mines were now a convenient source of desirable zinc. Zinc production in Cornwall and Devon was at a maximum between 1850 and 1885, and a cumulative total of 85,000 tons of zinc ore was worked in this period (Dines, 1969). Three quarters of this total came from the St Agnes district in Cornwall, with smaller contributions from the Gwinear, Camborne, St Austell, Callington and Dartmoor districts (Dines, 1969). In Cornwall, the West Chiverton and Cargol mines (both in the Perranzabuloe mining district of St Agnes) were the only important lead mines to achieve a high level of zinc output (Burt et al., 1987). Otherwise, the production of zinc in Cornwall was from mines other than lead mines, the zinc ore being mined alone or in association with copper, tin or iron mining. Such mixed mines included Great Retallack on the Perran Iron Lode, nearby Budnick Consols close to Perranporth and Pencourse Consols near St Austell, which were the leading zinc producers of the 1850s and 1860s (Burt et al., 1987). These mines were responsible during that period for up to a third of the British production of zinc (Burt et al., 1987). The Perran Iron Lode has thick masses of zincblende, apparently deposited before the

iron (see Table 2.1), so mines on this lode have been amongst the major producers of zinc (Dines, 1969). Other British zinc mining regions, however, also responded to the increasing demand for zinc for galvanisation, and, by 1870, Cornish zinc production represented only 5% of national production (Burt et al., 1987). In the late 1870s, there was a resurgence of zinc mining at West Chiverton, and in the early 1880s at Duchy Peru on the Perran Iron Lode, which produced 20,000 tons of zinc ore in total between 1858 and 1886 (Dines, 1969; Burt et al., 1987). But zinc-yielding mines in the Northern Pennines, Wales and the Isle of Man were all expanding production in competition. West Chiverton and Duchy Peru closed in the mid-1880s, and Cornish zinc production collapsed, rarely again to exceed a few hundred tons per year (Burt et al., 1987).

Other Metals Prologue. Small amounts of gold, manganese, tungsten, antimony, cobalt, nickel, molybdenum and uranium were mined in Cornwall and Devon.

Gold was recovered from southwest England in the eighteenth and nineteenth centuries (Herrington et al., 1999). Streams worked for alluvial tin had turned up a few flakes or, more rarely small nuggets, of gold for centuries. In 1808, a gold nugget weighing 62 grams was recovered from the Carnon Stream Tin Works at Perranarworthal, at the head of Restronguet Creek (Herrington et al., 1999). Similarly, the Treore stream at Port Isaac in North Cornwall gave up gold in some quantity to panning (Herrington et al., 1999). Gold was mined in the old copper mines around North Molton in North Devon, particularly the Britannia and Bampfylde mines, in the nineteenth century (Herrington et al., 1999).

Some manganese was mined in the Callington and Tavistock district at the top of the Tamar in the mid-nineteenth century. For example, the Chillaton and Hogstor mine near Tavistock, which continued activity until 1907, produced about 50,000 tons of manganese ore in total (Dines, 1969). Very little manganese was produced in Cornwall as opposed to Devon. Devon was producing more than 8,000 tons of manganese ore per year in the early 1870s, while the Cornish annual output was only a few hundred tons of ore (Burt et al., 1987). Most of the albeit little Cornish production was from iron lodes north of St Austell Moor (Dines, 1969). Much of this was from the Ruthvoes mine near Indian Queens, Newquay, which yielded up almost half of Cornwall's manganese output, mainly in the early 1880s (Dines, 1969; Burt et al., 1987). The remaining Cornish manganese came from the Launceston area, immediately next to the large Devonian manganese producers at Chillaton (Burt et al., 1987).

Wolframite is the main ore of tungsten in Southwest England. It had long been considered as an undesirable gangue in tin lodes, being difficult to separate from cassiterite of similar density (Dines, 1969). Some production of tungsten is reported from 1858, but serious mining for tungsten did not begin until the twentieth century, with the increased use of tungsten to harden steel (Dines, 1969; Burt et al., 1987). Between 1859 and 1913, Cornwall produced nearly 4,900 tons of tungsten ore, virtually the entire national output, with four mines dominating production – East Pool, South Crofty and Carn Brea and Tincroft, all from the Camborne and Redruth region, and Clitters United near Gunnislake (Burt et al., 1987).

A little antimony was mined in the eighteenth and nineteenth centuries from the Wadebridge region of Cornwall (the Bodannon, Pengenna and Trevennick mines) and also southeast of Callington (the Trebullet mine) (Dines, 1969; Burt et al., 1987). The local ores of antimony, such as jamesonite, are sulphides, and antimony can be found with argentiferous galena yielding lead and silver, as at Trevennick. The total Cornish output of antimony ore between 1845 and 1913 was 35 tons (Burt et al., 1987).

Ores of cobalt have been reported from about a dozen mines in Southwest England, but only about six have raised the ore (Dines, 1969). Mines such as the Dolcoath mine and East Pool mine, both in the Redruth and Camborne region, and Dowgas mine near St Austell, each produced only a few tons of cobalt ore in the nineteenth century (Dines, 1969; Burt et al., 1987). Nickel usually occurs in close association with cobalt (Dines, 1969). A chief producer has been the St

Austell Consols mine, producing about 7 tons of nickel ore to go with 250 tons of mixed cobalt and nickel ores (Dines, 1969). East Pool mine produced 4 tons of nickel ore and Fowey Consols 8 tons (Dines, 1969). Molybdenum occurs rarely in Southwest England, but was mined, in unknown amounts, from Dobwalls near Callington (Dines, 1969).

Before the 'discovery' of radioactivity, uranium was used to give a yellowish-green colour to glass, and most of the limited output of uranium from Southwest England in the first half of the nineteenth century went to this purpose (Dines, 1969). Wheal Owles at St Just, East Pool near Camborne and St Austell Consols rarely produced 1 ton of uranium ore per year in combination, before they all stopped production in the 1880s (Dines, 1969; Burt et al., 1987). From the 1890s, uranium was sought for its radioactive properties. The chief Cornish producers were then Wheal Trenwith at St Ives and the South Terras mine near St Austell. Cornwall produced up to 103 tons of uranium ore per year through to 1913 (Dines, 1969; Burt et al., 1987).

2.5.5.2 Pennines

Prologue. By the mid-eighteenth century, the northern Pennine dales represented the most important lead mining district in Britain, the world's largest lead producer, a position which was lost by the end of the nineteenth century as a result of foreign competition. The demand in the nineteenth century for zinc for galvanisation promoted zinc production in Nentdale in the Northern Pennines in the second half of the 1800s.

The lead industry of the Pennines also responded to increased demand for lead in the eighteenth and nineteenth centuries. The end of the seventeenth century had brought increased efficiency to mining in the Pennines, in the form of improved smelting processes and better pumps to drain deep mines (Raistrick and Jennings, 1989). From the early years of the eighteenth century, Newcomen's atmospheric steam engines became available to pump water out of mines going deeper. A Newcomen engine was installed at a lead mine at Winster in the Peak District of Derbyshire in 1717, followed by three at nearby Yatestoop by 1730 (Willies, 1982). The Yatestoop mine had an annual production of almost 3,000 tons of lead at its peak, and was briefly, in the mid-eighteenth century, the most important lead mine in Britain (Willies, 1982).

The end of the seventeenth century had also brought improved business practices, typically involving the formation of large companies. The London Lead Company was established in 1692, operating in the Derwent valley and Teesdale, and the Blackett-Beaumont Company operated in the Allendales and Weardale until the end of the nineteenth century (Raistrick and Jennings, 1989; Turnbull, 2006). These companies leased large areas from landowners to gain control of scores of mines and directed the whole chain of mining, dressing, smelting, refining and merchanting of lead (Raistrick and Jennings, 1989; Turnbull, 2006). By the mid-eighteenth century, Britain was world's largest lead producer, and the Northern Pennine dales represented the most important lead mining district in Britain (Turnbull, 2006).

In the Northern Pennines, the Allendales and upper Weardale (Figure 2.3) were very important lead mining regions in the eighteenth and nineteenth centuries. Revitalised Alston Moor lead mines centred on Nenthead and Killhope and the Teesdale lead mines on Middleton (Turnbull, 2006). Mining settlements expanded, and, with them over time, the associated industries of dressing, smelting, roads, railways, tramways and water supply lines. Indeed, the villages of Allenheads and Nenthead were purpose-built for workers in the mining industry.

Lead mining also flourished in the eighteenth and nineteenth centuries in the Yorkshire Dales of the mid-Pennines, in Nidderdale, Wharfedale, Airedale, Ribblesdale and Bowland (Raistrick, 1973). Particularly important were the Greenhow mining field, straddling the watershed of the rivers Nidd and Wharfe and the site of the productive Greenhow Hill lead mine, and the Grassington Moor mining area in Upper Wharfedale (Figure 2.3) (Raistrick, 1973). The lead mines of Grassington Moor were most prosperous between 1821 and 1861, with output averaging 965 tons of lead per year, but there was a fall in output thereafter as the mines became exhausted (Raistrick, 1973). Lead mines on Malham Moor produced small amounts of lead ore (up to 16 tons per year) at the start of the nineteenth century (Raistrick, 1973).

Further south in the Peak District, lead production during the eighteenth century was considerable, involving mines such as Yatestoop (Willies, 1982). Peak District lead production peaked in the late eighteenth century but did remain fairly high until 1850. The Matlock area became the chief local area for smelting at this time (Willies, 1982). Derbyshire's last surviving lead mine was the Mill Close mine, on the west side of Darley Dale, which struck a deep rich vein of lead in the 1860s (Edwards et al., 1962; Willies, 1982).

By the end of the nineteenth century, however, the dominance of the British lead industry was collapsing in the face of foreign competition. There was a continuing fall in lead prices in the second half of the century, and British lead ore production fell away (Turnbull, 2006). Some lead mining in the Northern Pennines survived into the twentieth century, but the great days of lead mining in the Pennines were over. And the decline of the mining industry of the Northern Pennines was associated with a significant depopulation of areas that had previously supported thriving mining communities.

Ores of zinc are typically associated with ores of lead (Table 2.1), as in the cases of zincblende and calamine in the lead veins of the Allendales and Nentdale in the Northern Pennines, and of calamine at Malham in the Yorkshire Dales (Turnbull, 2006). The market for zinc, however, was limited in the eighteenth century and had been mostly met with zinc produced more cheaply in Germany, until the new demand in the nineteenth century for zinc for galvanisation. Zincblende was then mined from mines in Nentdale, and Pennine spoil heaps were redressed for zinc ores (Turnbull, 2006). A zinc smelter was opened in 1845 at Tindale, a coal mining village northwest of Nenthead, and this smelter served the surrounding area for the next fifty years (Turnbull, 2006). The mining of zinc benefitted some regions, such as the Nenthead district, as the price of lead fell in the second half of the nineteenth century. Zinc mining in Nentdale peaked in 1895, and did survive into the twentieth century (Turnbull, 2006).

The mining industry has left its mark on the face of the Pennines landscape, in the form of old mine buildings with flues and chimneys, spoil heaps, opencast workings, shafts, adits and wheel pits. Hushing is an old mining method which uses a strong flow of water to strip away surface soil to find or exploit veins of metal ore hidden beneath. Typically, water is allowed to collect behind specially constructed dams high on hillsides and then released as a torrent. Surviving signs of this hushing process are long trenches on hillsides excavated by the powerful flow of water, as for example at Monks Moor above Middleton-in-Teesdale, and also *V*-shaped nicks on cliff faces caused similarly. Lead mining has left characteristic scars on the limestone country of Derbyshire (Edwards et al., 1962). Hollows denote the positions of old shafts, and there are often attendant spoil heaps covered in grass. Lead veins (known as rakes in Derbyshire) can be traced for miles across the Peak District countryside by the presence of narrow ridgelike spoil banks, sometimes planted with trees to prevent grazing livestock being subject to lead poisoning (Edwards et al., 1962). Thus, Deep Rake near Calver has been planted with beech and sycamore (Edwards et al., 1962).

2.5.5.3 Wales

Prologue. Mining of the lead–silver–zinc lodes of Wales was extensive in the eighteenth and nineteenth centuries. Parys Mountain on Anglesey dominated the world's production of copper from 1770 to 1800, and other copper mines were active in Snowdonia. Gold mining north of Dolgellau was particularly active in the nineteenth century.

Mining of the lead–silver–zinc lodes of mid- and north Wales had its heyday in the eighteenth and nineteenth centuries following the industrial revolution, peaking in the middle decades of the 1800s

(Davies, 1987). Indeed, the output of the lead–silver–zinc mines of Montgomeryshire (now part of Powys), Cardiganshire (now Ceredigion) and Flintshire (the site of Halkyn Mountain) at this time was second in importance only to that of the Pennines (Coyle, 2010).

There were many small mines in mid-Wales with the potential for growth in the eighteenth century (Coyle, 2010). For example, the Van mine, near Llanidloes in the old county of Montgomeryshire, grew in importance over the period and into the twentieth century. In total, it produced 100,000 tons of lead ore, 30,000 tons of zinc ore and 770,000 ounces (nearly 22,000 kilograms) of silver (Coyle, 2010). In Ceredigion, lead, silver and zinc mining centred on the upper valleys of the rivers Ystwyth and Rheidol, at the large and important Cwm Ystwyth and Cwm Rheidol mines respectively. Ore production there peaked in the eighteenth century. The principal ore extracted was argentiferous galena, with a high silver content, especially at Cwm Ystwyth, while associated sphalerite, bearing zinc, was only occasionally processed and often dumped in spoil heaps around the mines. Halkyn Mountain on the west bank of the Dee estuary in Flintshire was intensively mined for lead (with silver and, later, zinc) in the eighteenth and nineteenth centuries (Davies and Roberts, 1978; Coyle, 2010). Rich lead veins were discovered in the Halkyn Mountain area at intervals through the eighteenth century, and, by 1845, Flintshire was producing 10,000 tons of lead ore per year.

Initially in the eighteenth century, smelting of the lead ore extracted took place locally near the mines, but diminishing supplies of wood for charcoal required the smelting to be removed to specialist smelters with close access to coalfields. The smelters concerned were near Swansea in the Tawe valley in South Wales and near Holywell on the Dee estuary (Davies, 1987).

Copper mining in Wales in the eighteenth and nineteenth centuries has been introduced earlier in this chapter because of its competition with copper mining in Southwest England at the end of the eighteenth century. To be precise, Parys Mountain on Anglesey was the world's greatest producer of copper from 1770 to 1800, after the discovery in 1768 of the

Great Lode, a shallow body of copper ore (chalcopyrite) that was not rich but was abundant and very accessible (Atkinson, 1987; Buckley, 1992). Parys Mountain lies a mile or two inland of the port of Amlwch on the northeast coast of Anglesey (Figure 2.7). Copper had been mined there in Roman times, but little thereafter. Most copper gained locally in the period leading up to the mid-eighteenth century was by precipitation of copper onto scrap iron placed in the copper-rich waters running off the mountain (Coyle, 2010). By the 1760s, demand for copper had increased considerably, partly for the production of brass, but particularly to serve the needs of the navy for copper sheathing of their wooden boats to prevent the settlement of fouling organisms, and especially the ravages of wood-boring shipworms (Rowlands, 1966; Coyle, 2010). Thus, there was a stimulus for copper mining, and an increased incentive for prospectors to seek new lodes of copper. Interest in the copper resources of Parys Mountain was revived, and mining restarted in 1761 (Rowlands, 1966). Mining was not particularly successful, however, because of problems with water flooding the mine shafts. But this all changed on 2 March 1768, when a vein of copper ore was discovered just a few feet below the surface of the soil. Exploitation took the form of a large open pit with blasting by gunpowder, and by 1782 deep shafts were also in operation for underground mining (Rowlands, 1966). Production peaked in the 1780s and early 1790s, when about 40,000 tons of copper ore were being produced per year, albeit of variable quality (Rowlands, 1966). Production declined from 1800, and, although there was a mini-recovery from 1811 to the early 1820s (1,000 to 9,000 tons of ore annually), copper mining on Parys Mountain was essentially exhausted (Rowlands, 1966).

Not surprisingly, there had been an ecotoxicological fallout from the intense copper mining activity. From 1768, Parys Mountain, which had previously been covered with gorse, was quickly made bare, and copper-rich dust and sulphur fumes prevented any regrowth of vegetation (Rowlands, 1966). By the end of the eighteenth century, the vegetation in an area extending about half a mile on every side of the mountain had been destroyed (Rowlands, 1966).

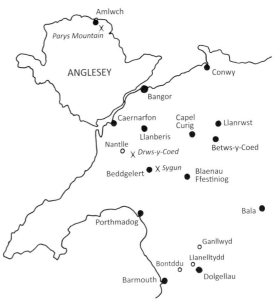

Figure 2.7 Significant copper and gold mining localities of north Wales.

Other copper mines were also active in the eighteenth and nineteenth centuries in north Wales (Bick, 1982). There were copper mines working in nearby Snowdonia across the Menai Strait in the mid-eighteenth century, and the great copper discovery on Parys Mountain in 1768 provided a considerable stimulus to further copper mining and prospecting there (Bick, 1982). Most Snowdonia copper mines were located in a region north of Porthmadog and Blaenau Ffestiniog and south of Llanberis (Figure 2.7) (Bick, 1982). Drws-y-Coed in Nantlle Vale, for example, was a very old copper mine, reputedly dating back to the thirteenth century and employing Cornish miners in 1761 (Bick, 1982). Production of copper ore peaked at Drws-y-Coed between 1820 and 1840, approaching 1,000 tons a year, assisted by miners who had moved in the 1830s from Parys Mountain, by now worked out (Bick, 1982). Local production by the 1880s, however, rarely exceeded 200 tons of ore per year, and Drws-y-Coed staggered into the twentieth century (Bick, 1982). The total recorded output of Drws-y-Coed is 13,000 tons of copper ore, but this is probably an underestimate, with double that amount more likely (Bick, 1982). Another copper mine of local repute was the Sygun copper mine near Beddgelert, active in the nineteenth century (Figure 2.7) (Bick, 1982). Many records of the output of the copper mines of Snowdonia actually come from the Swansea smelters to which the ore was sent for smelting at the beginning of the nineteenth century. An output of nearly 65,000 tons of copper ore is recorded from Snowdonia between 1804 and 1913, but a more realistic estimate may be 160,000 tons (Bick, 1982). Nevertheless, compared to Cornwall or Anglesey, Snowdonia copper mining can be considered to be a backwater, and usually a loss-making one at that (Bick, 1982).

Wales hosts the richest gold mining region of Britain, in the form of a broad belt north of Dolgellau in Merioneth (mostly now part of Gwynedd), where gold mining was particularly active in the nineteenth century (Herrington et al., 1999; Coyle, 2010). The heyday of the Dolgellau mines was in the 1860s, but some gold mining continued well into the twentieth century (Herrington et al., 1999; Coyle, 2010). Of the more than 3,500 kilograms of gold considered to have been recovered in Britain between 1860 and 1909, 90% came from this mining belt. The gold is mainly associated with Cambrian slates, while grains and small nuggets of gold have been found in the gravels of the River Mawddach (Herrington et al., 1999). Two mines stand out. Gold was discovered at Clogau-St Davids in Bontddu, near Barmouth (Figure 2.7), in 1836 and 1837 while prospecting for copper, this mine going on to provide gold for Royal Family wedding rings in the twentieth century (Herrington et al., 1999; Coyle, 2010). There was a big discovery of gold at the Gwynfynydd mine near Ganllwyd, Dolgellau (Figure 2.7) in 1867, followed by others in 1900 and 1904 (Coyle, 2010).

2.5.5.4 Scotland

Prologue. The lead mines of Leadhills and Wanlockhead were worked practically continuously during the eighteenth and nineteenth centuries. Zinc was worked only on a very small scale in Scotland, and mining for copper was very limited. Some silver and gold were produced.

The lead mines of Leadhills and Wanlockhead averaged an output of 1,000 tons of lead per year through the 1800s (Wilson, 1921). Between 1850 and 1900, Leadhills produced 69,354 tons of lead ore with an associated 215,682 ounces (6,120 kilograms) of silver, while Wanlockhead delivered 71,787 tons of lead ore and 312,762 ounces (8,867 kilograms) of silver (Figure 2.8) (Wilson, 1921). Elsewhere in Scotland, lead mines were opened at Strontian in western Lochaber in 1722 and at nearby Tyndrum in North Stirling in 1739, while the Blackcraig lead mine at Newton Stewart was discovered accidentally in 1763 (Wilson, 1921). The lead mines in all three districts lasted into the nineteenth century. The Strontian mines yielded 239 tons of lead per year between 1852 and 1872, after an earlier higher output of about 400 tons per year (Wilson, 1921). The Tyndrum mines raised 5,017 tons of lead between 1741 and 1768, but only a total of 430 tons of lead between 1856 and closure in 1865. Lead production from the East and West Blackcraig mines totalled 2,685 tons of lead (1854–1881) and 975 tons of lead (1853–1872) respectively, but lead mining at Blackcraig had effectively ended by 1882 (Wilson, 1921). The lead mines on Islay were worked for a period in the eighteenth century and were active again in the third quarter of the nineteenth century, until final closure in 1880. These Islay lead mines produced a total of 1,426 tons of lead, and 18,424 ounces (522 kilograms) of associated silver between 1862 and 1880 (Wilson, 1921).

No zinc ore was produced at Leadhills, and at Wanlockhead no zincblende was saved before 1880. However, with the increased demand for zinc needed for galvanisation, 1,518 tons of zincblende were produced at Wanlockhead between 1880 and 1900. Output at Wanlockhead after that increased to about 1,000 tons per year in the first decade of the twentieth century (Figure 2.8) (Wilson, 1921). Zinc was worked only on a very small scale elsewhere in Scotland in the eighteenth and nineteenth centuries, with some small output from Strontian (33 tons of zincblende at Corrantee in 1871) and the Blackcraig mines of Newton Stewart (33 tons of zincblende in 1865) (Wilson, 1921).

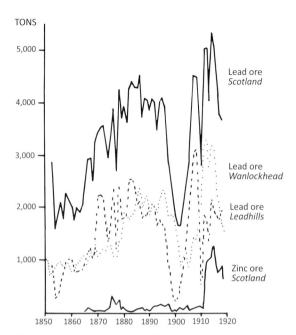

Figure 2.8 Annual production (tons) of lead and zinc ores from Scottish mines between 1850 and 1920. (After Wilson, 1921. Courtesy of British Geological Survey.)

Mining for copper in Scotland was, as previously, very limited in the eighteenth and nineteenth centuries, and usually of only local importance (Wilson, 1921, Atkinson, 1987). The Sandlodge mine in Sandwick in Shetland produced copper ore (chalcopyrite) from 1790 into the late nineteenth century, but without significant economic success. There were small copper mines in Kirkcudbrightshire (now Dumfries and Galloway), the largest of which was the Enrick mine, near Gatehouse of Fleet. The Enrick mine was operational between 1820 and 1857, with the copper ore extracted being sent to Swansea for smelting (Atkinson, 1987).

A very rich deposit of silver ore was discovered in 1711 at Alva, in the county of Clackmannanshire, in the Central Lowlands, west of Stirling (Wilson, 1921). The Alva mine was briefly very profitable, yielding £4,000 of silver per month, but it soon became exhausted.

Collection of alluvial gold in Scotland may date from the thirteenth century, and, in 1828, the Breadalbane nugget (quartz with 62 grams of gold) was found in the headwaters of the River Tay in Perthshire (Herrington et al., 1999). In the late 1860s, there was a mini-gold rush in Scotland after a nugget of gold weighing above 15 grams was found in Helmsdale Water, in Sutherland between Inverness and Wick (Herrington et al., 1999). More alluvial gold was found in the Suisgill and Kildonan Burns, tributaries of the Helmsdale River, and, by 1869, as many as 500 gold prospectors were panning the local alluvial deposits (Herrington et al., 1999). About 93 kilograms of gold was found over a two-year period.

2.5.5.5 Lake District

Prologue. The Lake District saw a good deal of mining in the eighteenth century, with lead rather than copper, being the metal particularly sought. The nineteenth century saw the greatest mining activity in the Lake District since Elizabethan times

In the eighteenth century, the Coniston copper mine continued in operation, but the Keswick copper mines were little worked and the Keswick smelter ceased activity early in the century (Shaw, 1975). The lead mines, on the other hand, were worked vigorously, serving the Stoneycroft smelter continuously, as well as smaller lead smelters at the Roughtongill mine on Caldbeck Fells and at the Hartsop Mine in Patterdale. Lead ore from Greenside was carried on horseback to the Stoneycroft smelter over Sticks Pass.

The nineteenth century delivered improved roads and the development of railways (Shaw, 1975). Subsequently, it brought the end of the use of the packhorse to ferry ore and of the boat transport of ore across lakes such as Coniston, Ullswater and Derwentwater. Early in the century, the Coniston copper mines further expanded, employing up to 600 men over a period lasting 50 years (Shaw, 1975). In the Helvellyn region (Figure 2.9), the Greenside mine in Patterdale expanded to employ 300 men mining lead (Shaw, 1975). In the late 1800s, there was a similar expansion of lead mining at other lead mines in the Lake District, including the Goldscope mine, the Yewthwaite mine and the Brandelhow mine, southwest of Derwentwater, and the Gategill mine at

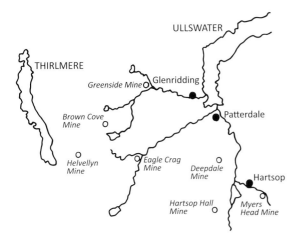

Figure 2.9. Mines of the Helvellyn mining field in the Lake District. (After Shaw, 1975, with permission.)

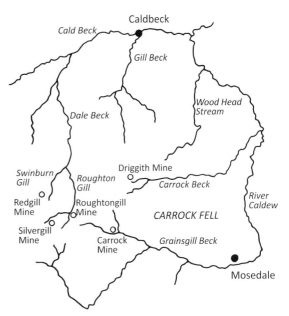

Figure 2.10 Major mines of the Caldbeck Fells in the Lake District. (After Shaw, 1975, with permission.)

Threlkeld, all in the vicinity of Keswick (Shaw, 1975). On Caldbeck Fells (Figure 2.10), the Roughtongill and Driggith lead mines operated at full capacity in the nineteenth century, and both were worked out by the 1870s (Shaw, 1975). In 1874, there were twelve lead mines and three copper mines producing ore in the Lake District. Towards the end of the century, however, large metal ore discoveries were being made worldwide, and foreign competition caused lead and copper prices to fall below the economic limit of most Lake District mines. Of the copper mines, only Coniston lingered on. In the case of lead, the Greenside mine was reorganised and modernised to be kept in production with a smaller workforce. Metal mining had ceased completely on Caldbeck Fells by 1878 with the closure of Roughtongill (Shaw, 1975). The heyday of Lakeland mining was past.

2.5.5.6 Shropshire, Staffordshire and Somerset

Prologue. There was lead mining in Shropshire, copper mining at Ecton Hill in North Staffordshire, and mining for zinc in the Somerset Mendips.

The lead mines of Shropshire were restricted to a small hilly area within a 3 mile radius of the hamlet of Shelve, 12 miles southwest of Shrewsbury (Brook and Allbutt, 1971). The mines were known to the Romans,

but remained mostly unworked until the nineteeth-century when, in Victorian times, they punched above their weight. In 1872, ten percent of all Britain's lead ore came from ten active Shropshire mines, including the Snailbeach mine, the biggest mine in the area (Brook and Allbutt, 1973). The local lead ore was of very high quality, and these productive mines were managed by the finest mining engineers of the time. The hilly nature of the region allowed these engineers to drive adits from low in river valleys to meet the lead veins in the hills, avoiding the need for expensive pumps to drain the deep mines. Nearby coal from the Pontesbury coalfield just to the north provided the fuel for a lead smelter at Snailbeach, although other ore was shipped by rail to be smelted in Wales, Deeside and Bristol (Brook and Allbutt, 1973). Output from the Snailbeach mine peaked in the 1850s at 3,500 tons of lead ore annually. In 1872, Shropshire produced 7,386 tons of lead ore out of a total for England, Wales, Scotland, Ireland and the Isle of Man of 81,564 tons (Brook and Allbutt, 1973). This ore gave up 5,602 tons of lead and 2,960 ounces (172 kilograms) of silver in 1872, and yet was produced by only 10 mines out of a total of 455 British mines (Brook and Allbutt, 1971). By 1875, five Shropshire mines produced 7,600 tons lead ore out of a British total of 77,746 tons. By 1900, British lead ore production was down to 32,000 tons, with 1,200 tons from Shropshire. Between 1845 and 1913, 241,000 tons of lead ore came from the Shropshire mines, more than half from the Snailbeach mine alone (Brook and Allbutt, 1971). As with lead mining elsewhere in Britain, Shropshire lead mining was hit by the fall in lead prices in 1884, and in 1905, the Snailbeach mine produced only 20 tons of lead ore.

Ecton Hill in north Staffordshire was mined for copper by Cornishmen between 1766 and its closure in 1817, producing more than 50,000 tons of copper ore (Atkinson, 1987). Ecton Hill was, therefore, one of the richest copper mines in England in the late eighteenth century. Ore was originally transported to Denby in Derbyshire for smelting, but, in 1869, a nearer smelter was opened at Whiston in the Churnet valley in Staffordshire. Most of the Staffordshire copper went to the manufacture of brass, although 300 tons were supplied in the late 1700s to the Royal Navy for copper sheathing of its vessels for protection against shipworms. In the 1790s, Ecton Hill was producing 4,000 tons of copper ore per year, but by 1800 most of the ore had been worked out.

In the Somerset Mendips, at Shipham and the neighbouring hamlet of Rowberrow, the mining of calamine for zinc was very active in the eighteenth and early nineteenth centuries (Li and Thornton, 1993a). The availability of zinc ore at Shipham and Rowberrow provided a convenient source of zinc for the nearby Bristol brass industry. There were more than a hundred small zinc mines in Shipham alone in the middle of the eighteenth century, with hundreds of local residents involved. In addition to Shipham and Rowberrow, zinc extraction was also carried out at nearby Burrington, Winscombe and East Harptree, while it was the extraction of lead, typically associated with zinc, that was the primary industry at Priddy, Charterhouse, Chewton-on-Mendip and East Harptree (Davies and Ballinger, 1990). By the start of the 1800s, however, the local calamine industry was in decline. By 1853, all mining operations had ceased, leaving a legacy of disturbed contaminated ground at Shipham.

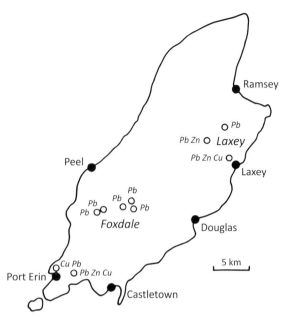

Figure 2.11. The Isle of Man. Larger mines producing lead (Pb), zinc (Zn) and/or copper (Cu), mostly in the major mining regions of Foxdale and Laxey.

2.5.5.7 Isle of Man

Prologue. Ores of copper, and of lead with zinc, occur on the Isle of Man.

It is likely that mining for copper and lead on the Isle of Man goes back at least to the thirteenth century (Southgate et al., 1983). The eighteenth century brought a marked increase in mining on the Isle of Man, reaching a peak in the nineteenth century before a subsequent rapid decline in the first two decades of the 1900s (Southgate et al., 1983). There were two major mining areas on the Isle of Man, the Foxdale region producing lead, and the Laxey region producing zinc, lead and some copper (Figure 2.11). Between 1845 and 1882, mines on the Isle of Man produced 131,127 tons of lead ore, yielding 93,297 tons of lead, and 3,462,701 ounces (98,166 kilograms) of silver (Hunt, 2011).

2.5.5.8 Ireland

Prologue. Copper was mined in considerable quantity in the Vale of Avoca in the eastern foothills of the

Wicklow Mountains in southeastern Ireland. Some lead and gold were also produced in Ireland.

The East and West Avoca Mines, lying on either side of the Avoca River, were major copper-producing mines in the eighteenth and nineteenth centuries (Wright et al., 1999). Copper was first discovered in the early eighteenth century at the East Avoca Cronebane mine, below silver being extracted from an easily worked, weathered vein (Wright et al., 1999). Copper mining began in 1720 at East Avoca, followed in 1826 by the development of West Avoca. Mining of copper continued at Avoca until 1888, with 216,000 tons of copper ore being produced in total over the two centuries (Wright et al., 1999). Furthermore, pyrite was present in abundance at Avoca and was mined as a source of sulphur from 1839 when the Sicilian sulphur trade with Britain was disrupted.

Lead was also mined in County Wicklow from the early nineteenth century. The Ballinafunshoge mine in the Vale of Glenmalure opened in about 1800. Better-known lead mines were present from about 1825 in the Glendalough and nearby Glendasan valleys, separated by Camaderry Mountain bearing the lead mineral vein. The Glendalough and Glendasan lead mines survived into the twentieth century, Glendalough closing at the end of World War I and Glendasan in 1957.

In the eighteenth century, small nuggets of gold were found in the alluvial deposits of the Ballin valley stream in County Wicklow, which proved to be the most productive gold region in Ireland. Indeed, the largest nugget of gold ever found in the British Isles, the Wicklow nugget (682 grams), was found in the Balin valley stream in 1795 (Herrington et al., 1999).

2.5.5.9 Iron

Prologue. With the rise of coke as a fuel, the iron industry moved from wooded districts to the coalfields in Scotland, the Midlands and South Wales, and the Coal Measures ironstones were important sources of iron.

By the early 1700s, iron production had moved into a new epoch with the use of coke for smelting iron by Abraham Darby I in his furnace at Coalbrookdale in Shropshire in 1730 (Alexander and Street, 1951). Coalbrookdale was destined to become the most famous ironworks in the world, located in the Ironbridge Gorge. Iron Bridge itself was constructed of cast iron across the River Severn in 1779 by Abraham Darby III, the grandson of the first Abraham Darby (NAMHO, 2013). With the development of coke-fired blast furnaces during the eighteenth century, the Coal Measures ironstones rose to prominence as a source of iron. And the iron industry moved from wooded districts to the coalfields in Scotland, the Midlands and South Wales. The rapid decline in the seventeenth century of the Ashdown Forest iron industry continued to extinction in the eighteenth century. Today, splendid firebacks in local houses, iron grave slabs in churchyards at Waghurst, West Hoathly, East Grinstead, Hartfield, Maresfield, Rotherfield and Uckfield, and many evocative place names are almost all that remain of a once powerful national iron industry centred on the Ashdown Forest (Christian, 1967). Nevertheless, it still took a long time for the use of charcoal to cease in Britain. The last British charcoal iron furnace was not adapted to the use of coke until 1920, at Blackborrow near Windermere (Turnbull, 2006).

At Coalbrookdale, the first Abraham Darby essentially moved iron smelting from a domestic onto an industrial scale (Raistrick, 1973). Abraham Darby had originally been a partner in brass foundries in Bristol. He switched to iron founding when he set up the Coalbrookdale Company, a business that was to remain in the family for 150 years. Darby produced marketable goods such as pots and kettles directly from iron from his forge. Coalbrookdale was well placed on the Coalbrook, a tributary of the River Severn, for the distribution of these products downriver to Gloucester and Bristol. Furthermore, a local coalfield lay across the deep valley of the Coalbrook and the deeper gorge at Ironbridge, cut by the River Severn. There was therefore significant local exposure of coal, which was relatively free from impurities and well suited for coking. In addition to a plentiful fuel supply of good local coal, there was also abundant water available from the Severn to power the bellows for forges and furnaces. Moreover, there was a good previous local history of charcoal-based iron making from the copious supplies of local Coal Measure ironstones, and the Coalbrookdale Company was set up to prosper. Prosper it did, and Ironbridge Gorge is often referred to as the birthplace of the industrial revolution. Demand for iron just grew and grew over the eighteenth and nineteenth centuries, enhanced by the development in 1856 of the Bessemer process to make steel (Alexander and Street, 1951). Iron and steel were needed more and more with improved engines harnessing the power of steam, the building of bridges and the development of railways using cast-iron rails, iron ships, guns, tools and, ultimately at the end of the 1800s, horseless vehicles. Hand in hand with the huge increase in the use of iron after the industrial revolution were associated increases in demand for the other principal players in this story, metals such as copper, tin, lead and zinc.

With the great development in the eighteenth and nineteenth centuries of iron production based on the Coal Measures, yielding ironstones as well as coal itself, the historical mining regions of Britain, exploiting their veins of iron ore, became less and less significant sources of iron production.

In Cornwall and Devon, the historically chief sources of iron had been lodes near Exmoor and lodes in and around the St Austell granite mass, extending north towards Wadebridge and west to the St Agnes region and the Perran Iron Lode (Dines, 1969). The largest and deepest mine developed on the Perran Iron Lode was the Duchy Peru mine at Perranzabuloe in the St Agnes district, producing significant zinc in addition to iron. Particularly active in the 1880s, Duchy Peru produced 32,400 tons of iron ore between 1858 and 1886 (Dines, 1969). Gravel Hill mine on the cliffs at the north end of Perran Beach was known to have been active before 1728, and produced 7,400 tons of limonite between 1874 and 1882 (Dines, 1969). The Cornish iron lodes, however, were far from coal and far from potential markets, and many of the lodes were vertical and more difficult to work than the ironstones of the Coal Measures. Between 1855 and

1875, Cornwall and Devon did produce between 24,000 and 87,000 tons of iron ore (usually limonite) per year, but only 6,500 tons in the years after 1884 (Dines, 1969). In the second half of the nineteenth century, Cornwall never produced more than 1% of total British iron production (Burt et al., 1987).

In the Northern Pennines, siderite (spathose iron ore) was mined in Weardale and also at Nenthead and Alston (Turnbull, 2006). The Northern Pennines had good access to local coalfields, and iron mining on a large scale commenced in Weardale from the 1840s. The railway reached the Slitt Pasture ironstone mine in Weardale in the late 1850s (Turnbull, 2006). The local iron ore was originally smelted at Stanhope, but, in 1845, furnaces were built at Tow Law and Tudhoe on the edge of the coalfield. Iron mining declined in the Northern Pennines from the 1880s and had practically ceased by the end of the nineteenth century.

In the Lake counties in the nineteenth century, railways spelt the end for the Lakeland iron furnaces (Shaw, 1975). Iron ore was available in West Cumberland and Furness with nearby supplies of coal and access to steam power, and the railways brought improved transport efficiencies. Iron output in the Lake District was limited by the restrictions of charcoal and waterpower, and the Lakeland furnaces were too uncompetitive to survive.

2.5.6 Twentieth Century Forwards

Prologue. The twentieth century saw a continuing decline in British mining with the odd temporary resurgence when needs must, or when the international metal price was high enough for mining to be commercially viable again.

On entering the twentieth century, British mining was at the mercy of fluctuating international metal prices. The most accessible British ores had long been mined out. Meanwhile, new overseas metal ore deposits were relatively unexploited, and usually came hand in hand with cheaper labour costs. Prices would be distorted by the demands of two world wars for the manufacture of armaments, ships, planes and vehicles. War also had an effect on access to overseas mining districts and on the transport of their exports. The First World War in particular created a shortage of manpower in British mines, as the nation fed men in vast numbers into the trenches. A combination of increased demand and lack of availability of imported metal in wartime would create circumstances whereby previously uneconomical British mines needed to be put back into operation to gain vital metal resources. Tungsten, for example, was in demand to harden steel to make armour-piercing weapons. The twentieth century, and indeed the twenty-first century, also brought new uses for metals, such as in semiconductors or other components in the electronics industry, or in new metaliferous nanoparticles with novel characteristics of industrial or domestic value.

2.5.6.1 Cornwall and Devon

Prologue. Southwest England featured prominently in this cycle of feast and famine controlled by international metal prices. Cornish tin mining struggled on into the twentieth century until finally ceasing in 1998. Arsenic production also continued into the 1900s.

Tin was one of two major players at the start of this mining history in the Bronze Age. Tin was the prominent metal, still, in Cornish mining in the twentieth century. The 1890s had brought a fall in the international price of tin, and while many Cornish tin mines closed, others had resorted to amalgamation as a survival route (Burt et al., 1987). The South Crofty mine in Camborne, first formed in 1854 from three previously separate mines (Longclose, Dudnance and Penhilick), expanded rapidly from the beginning of the twentieth century to become established as the leader of the tin industry in the years before World War I (Burt et al., 1987). By the late 1960s, South Crofty covered nearly six square miles, and included many of what had been the most productive mines in Cornwall – all of the Carn Breas, Cooks Kitchens, Croftys, Dolcoaths, Pools, Roskears, Setons and Tincrofts, and many others. Similarly, the twentieth century Geevor mine, between the villages of Pendeen and Trewellard in the St Just mining district on the

Land's End peninsula, expanded greatly from its nineteenth century limits to include the Boscasswells and Levants, in addition to Botallack, Carne, Carnyorth and Pendeen Consols (Burt et al., 1987).

The Cornish output of tin had halved in the 1890s and never recovered again to the production levels of the late nineteenth century (Figure 2.4). Medium-sized tin mining operations had disappeared, a situation exacerbated by mine exhaustion and a lack of capital investment in the Cornish mines, as investors chased more profitable foreign ventures. The price of tin did come back early in the twentieth century from the low of £64 a ton in 1896, to double in 1900, and reach £181 a ton in 1906, and £210 a ton just before World War I (Burt et al., 1987; Buckley et al., 1992). But Cornish tin production could not respond and remained persistently low. The Great War brought a fall in the price of tin to £151 a ton, a loss of mining labour as men went to war, and materials and gear were in short supply. There was a cutback on development work in the Cornish tin mines in order to concentrate the reduced manpower resources into production, no new tin grounds were uncovered, plant was in poor condition and cash reserves dwindled. Fewer than 20 Cornish mines survived the First World War (Buckley et al., 1992).

Foreign competition had put paid to any significant copper, lead and zinc mining in Cornwall and Devon by the end of the nineteenth century, but the production of arsenic did continue into the twentieth century. The output of white arsenic from Devon and Cornwall had fallen to 900 tons annually by 1904, but the years of World War I brought increased demand, and therefore increased production then of 2,000 tons per year. Some old copper mines were reopened during the war to access the associated arsenic, as in the case in 1915 of Wheal Fanny, part of Devon Great Consols near Gunnislake (Dines, 1969). The dumps of many other mines were also reworked to gain the arsenic ore previously thrown away. The arsenic works at Bissoe and Gwithian in West Cornwall were still functioning to process arsenic (Barton, 1971b). There was then a postwar slump in the price of arsenic, albeit with a temporary increased demand in 1923 for arsenic for use again in insecticides for the

US cotton fields. But very effective new insecticides were replacing those based on arsenic. The production of white arsenic in Devon and Cornwall fell to between 100 and 200 tons a year in the 1930s and below 100 tons annually in the 1950s (Dines, 1969). Arsenic was by then a minor by-product of the output of active tin mines such as South Crofty and adjacent East Pool in Camborne (Barton, 1931b).

Cornish tin mining struggled on further into the twentieth century in the shape of South Crofty, East Pool and Geevor (Buckley, 1992). East Pool received government support in the Second World War but closed thereafter. South Crofty and Geevor carried on with difficulty, falling behind in modern technology in spite of ore reserves and an adequate tin price. Nevertheless, a disused tin mine, Wheal Jane, near Baldhu and Chacewater in the Carnon valley, received investment in the 1960s and reopened in 1971 (Buckley, 1992). Wheal Jane had been active since the mid-eighteenth century, and, between 1847 and 1895 had produced 3,832 tons of black tin, 740 tons of copper ore, 302 tons of lead ore, 2,921 ounces (83 kilograms) of silver, 586 tons of zinc ore, 33,340 tons of pyrite, 86 tons of arsenopyrite and 243 tons of white arsenic (Dines, 1969). The arsenic revenue had seen the mine through from 1885 to 1895.

Cornish tin mining in the second half of the twentieth century was buoyed up by an international agreement amongst tin producing and consuming nations to buy up all surplus production (Atkinson, 1985). This agreement caused the price of tin to rise to an artificial high at over £10,000 per ton. The agreement had created an oversupply of tin and was terminated in October 1985. The tin price fell to below £4,000 a ton, and many tin producers throughout the world were forced to close down. The Cornish tin mines were hit badly, and, in receipt of substantial government aid, needed to pursue programmes of modernisation to reduce costs. Nevertheless, Geevor shut in 1990, Wheal Jane closed in February 1991 and mining ceased at South Crofty in 1998 (Atkinson, 1985; Buckley, 1992). Geevor had produced about 50,000 tons of black tin during the twentieth century. South Crofty, the last survivor, had been active for more than 300 years, suspending mining on four

occasions, but never totally closing. South Crofty had incorporated into its set some of the greatest tin mines of the nineteenth century – Dolcoath, Carn Brea and East Pool. Interest is still regularly revived in the potential for further mining in the likes of South Crofty, as metal prices rise on the international market.

There has been an unfortunate aftermath to the closing down of Wheal Jane in 1991, which was soon followed by the switching off of the water pumps draining the mine (Bowen et al., 1998). Like many Cornish mines, Wheal Jane penetrated below the water table, artificially depressed by pumping. If pumping stops, water levels rise, and water will enter any horizontal adits draining the mine to an adjacent valley bottom. So was the case with Wheal Jane. In January 1992, more than 45,000 cubic metres of water, heavily contaminated with arsenic, cadmium and iron amongst other metals, escaped from the Nangiles Adit into the Carnon valley, just to the southeast of the discharge of the Great Adit (Bowen et al., 1998). These discharged waters were also extremely acidic (pH 1.0), for they contained sulphuric acid, oxidised from the sulphur in the pyrites in the mine. From the Carnon River, the acid mine drainage overflow entered Restronguet Creek and eventually Falmouth Bay. Left behind in the Carnon River were large smothering deposits of ochre, oxides and hydroxides of iron precipitating out with neutralisation of the acid (Bowen et al., 1998). There were clear ecotoxicological effects of the metal-contaminated flood on the local freshwater and estuarine flora and fauna, including waterfowl and fish. Remedial solutions have included lime dosing (to neutralise the acid), the installation of flocculation and settling ponds behind tailings dams and the establishment of wetlands as natural filters (Bowen et al., 1998).

Minor metals have not been ignored in Cornwall and Devon in the twentieth century and thereafter. For example, there have been recent new discoveries of gold in the Crediton area of Devon (Herrington et al., 1999). Renewed interest in tungsten led to an increase in the annual yield of tungsten from the two counties in the second decade of the 1900s to 250 tons, but this decreased to an average yearly yield of

41 tons between 1920 and 1934 (Dines, 1969). In 1907, at South Crofty, tungsten generated more than 20% of the receipts from the sales of all ores, producing 40% of total income when combined with arsenic (Burt et al., 1987). In the second decade of the twenty-first century, the Hemerdon tungsten and tin mine on the edge of Dartmoor, near Plympton, has been reopened. The reopened mine can exploit the world's fourth-largest deposit of tungsten and could supposedly contribute up to 4% of the world's tungsten production, presently overwhelmingly dominated by China.

2.5.6.2 Pennines

Prologue. In the Pennines, some lead mining continued into the 1900s.

The rich Mill Close lead mine in Darley Dale in Derbyshire was particularly active in the first half of the twentieth century (Edwards et al., 1962). World War I stimulated an increase in the demand for lead, and the Mill Close Mine increased its output of lead ore, a raised production level that continued into the early 1930s. In 1931, Mill Close, then the largest lead mine in the country, produced 26,000 of the 29,000 tons of lead ore raised in Britain (Edwards et al., 1962). A disastrous flood in 1940, however, put an end to mining operations there. Other twentieth century mines in Derbyshire included a mine near the High Tor at Matlock and the Glebe mine at Eyam (Edwards et al., 1962). Eyam is the village best known for choosing to isolate itself when the plague was discovered there in August 1665 rather than let the infection spread.

2.5.6.3 Lake District

Prologue. Copper and lead mining in the Lake District ended in the twentieth century. Tungsten was mined on the Caldbeck Fells.

In the Lake District, the start of World War I temporarily stimulated some increased mining in the Coniston copper mines, but, in effect, there was little significant production thereafter (Shaw, 1975).

In the case of Lakeland lead mining, the slump in metal prices at the end of World War I was also the beginning of the end for the lead mines near Keswick, as at Thornthwaite at the foot of Sea Howe (Shaw, 1975). The Gategill mine at Threlkeld, at the southern foot of Blencathra to the east of Keswick, which had produced lead and zinc, finally closed in 1928. Force Crag, at the head of the Coledale valley above Braithwaite, had mined lead in the nineteenth century, and zinc into the twentieth century, but it was the production of barytes that kept the mine open until its final abandonment in 1991. The great Greenside mine in Patterdale on the Helvellyn mining field (Figure 2.9) was completely worked out in 1962, and lead mining ended in Patterdale (Shaw, 1975).

On the Caldbeck Fells, tungsten, needed for armaments, was mined during World War I from the Carrock Wolfram mine in the valley of the Grainsgill Beck (Figure 2.10) (Shaw, 1975). This mine, the only British tungsten mine outside Southwest England, employed more than 100 men for most of the war. But, reduced demand and a lower tungsten price followed in peacetime, and mining ended in 1918. In spite of renewed interest in tungsten during World War II and the Korean War, no tungsten ore was actually mined then. In fact, the Carrock mine did reopen in 1977 and produced about 16,000 tons of ore per year until 1981, when the tungsten price had fallen again and the mine shut.

2.5.6.4 Wales

Prologue. Copper mining in Snowdonia scraped into the twentieth century.

In Wales, copper mining in Snowdonia just about continued into the twentieth century, as at Drws-y-Coed in Nantlle Vale and Glasdir at Llanelltyd, near Dolgellau (Figure 2.7) (Bick, 1982). Glasdir shut in 1914 and Drws-y-Coed in 1920. When the copper price rises, so does renewed interest in Parys Mountain on Anglesey, as in the 1980s and 1990s, but with no resumption of mining as yet. Gold mining continued in Merioneth for much of the twentieth century, Clogau shutting in 1995 and Gwynfynydd in 1998 (Coyle, 2010).

2.5.6.5 Scotland

Prologue. Lead mining at Leadhills and Wanlockhead had ceased by the end of the 1930s.

In Scotland, mining at Wanlockhead and Leadhills was yet active in the first two decades of the twentieth century, with both Wanlockhead and Leadhills still returning an average of 1,000 tons of lead ore per year until 1920 (Figure 2.8) (Wilson, 1921). Lead mining, however, then declined steeply. Lead mining closed at Leadhills in 1928, and effectively at Wanlockhead by the end of the 1930s. Gold was discovered in the late twentieth century at Cononish near Tyndrum in North Stirling, and a mine was constructed to await opening when gold prices allow (Herrinton et al., 1999; Coyle, 2010).

2.5.6.6 Ireland

Prologue. The Avoca copper mines closed in the 1980s. A gold mine in Northern Ireland also produces silver and lead.

In Ireland, the Avoca copper mines in County Wicklow remained active, on and off, until 1982 (Wright et al., 1999). West Avoca went as deep as 300 metres, but East Avoca used open-cast pits in the 1970s and 1980s. Unfortunately, acid mine drainage from the Avoca mine workings rendered the Avoca River seriously contaminated down to the estuary at Arklow (Wright et al., 1999).

In Northern Ireland, there is a modern gold mine at Cavanacaw, near Omagh, County Tyrone. After gold was discovered in the 1980s, the Cavanacaw mine opened in 2007. It is an open-pit mine which also produces silver and lead.

2.5.6.7 Aluminium Smelting

Prologue. Bauxite was mined at Glenravel, County Antrim, Northern Ireland, the resulting alumina being smelted in the West Highlands of Scotland.

Northern Ireland also saw attempts to extract bauxite for its aluminium content at the very end of the nineteenth century. Into the twentieth century, British Aluminium mined bauxite from Glenravel, County Antrim, and processed it into alumina at Larne, nearby on the east coast. Alumina consists of aluminium oxide, which is the starting material for smelting into aluminium metal. The alumina was shipped from Larne to aluminium smelters in the West Highlands of Scotland, with the potential for cheap hydroelectric power. The Foyers smelter, on the eastern banks of Loch Ness, opened in 1895, followed by smelting works at Kinlochleven, at the head of Loch Leven, which reached full production in 1909. By 1911, the Scottish Highlands produced one third of the world's output of primary aluminium. By the second half of the twentieth century, however, aluminium smelters sited elsewhere in Europe nearer consumers, and in North America, dominated any Scottish aluminium smelting. Nevertheless, a smelter at Lochaber in the western Highlands, established in 1929, still produced about 1% of the world's aluminium at the start of the twenty-first century.

2.5.6.8 Avonmouth

Prologue. The twentieth century saw the development of the very large zinc and lead smelting works at Avonmouth, just to the west of Bristol.

The Avonmouth smelting works, established in 1917, were fed directly by zinc, lead and mixed ore concentrates unloaded at Avonmouth Docks. Avonmouth was the site of development of the Imperial Smelting Process in 1950, using one of the largest blast furnaces of its kind in the world, to smelt lead and zinc together (Coyle, 2010). This process involves the use of a spray of molten lead droplets to absorb zinc vapour, subsequently separating the zinc and lead again on cooling. The Imperial Smelting Process (ISP) recovers much more lead from starting ores than previous processes, increases lead and zinc production efficiency and decreases labour and maintenance costs. Cadmium is a common contaminant of zinc sulphide ores, and the ISP will also separate out cadmium for production. In 1979, an Imperial Smelting Furnace (ISF) lead–zinc smelter commissioned at Avonmouth in 1967, produced 77,000 tons of zinc and 414 tons of cadmium, the lead bullion produced being refined overseas (Hutton, 1984). An inevitable by-product of smelting, however, is the emission of particulate metals into the atmosphere from smelter chimneys, even with particulate control devices. The ISF at Avonmouth, in the 1970s, was estimated to emit 3 to 4 tons of cadmium, 40 to 60 tons of zinc, and 20 to 30 tons of lead per year, with probably higher emissions before the updating of the smelter in 1971 (Hutton, 1984). Such emissions, not surprisingly, have the potential to have ecotoxicological effects on terrestrial habitats downwind, typically deciduous woodland in the case of Avonmouth. With the cost of smelting being lower elsewhere in the world, smelting ceased at Avonmouth at the end of the 1970s, although the site remained as a stockholding and distribution centre until 2003.

2.5.7 Conclusions

Prologue. The rich mining history of the British Isles inevitably left behind an ecotoxicological legacy.

It is clear then that the British Isles have a rich history of the mining of trace metals, mining that has been of enormous economic significance over centuries and even millennia. It is inevitable, however, that the mining, and indeed subsequent smelting, of these toxic metals would have an environmental legacy. Some of that legacy is visible today in the heaps of spoil left behind in Dartmoor river valleys by medieval tin streamers, or in the trenches and scars on cliff faces in Derbyshire denoting sites of hushing, practised centuries ago. Less immediately obvious, but of greater environmental significance, is the invisible residual toxic metal contamination of soils, streams, rivers, estuaries and even coastal seas. This contamination and its ecotoxicological effects are still with us today. The following chapters explore the details of this metal contamination and consider how life in terrestrial, freshwater, estuarine and marine environments copes with the raised toxic metal bioavailabilities inevitably encountered. The good news is that, for the most part over time, life has adapted and survived. Some of those adaptations are remarkable. Over the next chapters, we will come across familiar names of locations, streams, rivers and estuaries already met in this chapter's brief run through of history. Perhaps it will be impossible to ever look at the beauty of the sailing scene presented today by Restronguet Creek in quite the same way again.

Box 3.1 Definitions

absorption Uptake of a metal into an organism by crossing the cell membrane (e.g., of an epithelial cell on the surface or in an alimentary tract), with the potential to interact with physiological processes within an organism.

accumulation (bioaccumulation) Net outcome of uptake and excretion (if any) of a trace metal by an organism, expressed in terms of metal content (μg) or metal concentration (μg/g).

active transport Movement of molecules across a cell membrane against a concentration gradient. This process requires energy, such as chemical energy from adenosine triphosphate (ATP).

adsorption Passive adhesion of a trace metal onto the outside of a solid surface, such as a sediment particle, a detritus particle, or a living organism with an exoskeleton beyond the external epithelium. Adsorbed metal does not interact with the physiological processes within an organism. *Desorption* is the passive release of adsorbed metals back into solution.

antioxidant A molecule that reacts with highly reactive free radicals to block their potential to cause damage to cells. Antioxidants include glutathione and enzymes such as catalase and superoxide dismutase.

bioaccessibility Relative measure of the release into solution of a trace metal in a potentially ingested item through the action of a defined cocktail of chemicals mimicking a digestive process (typically of humans).

bioaccumulated metal guideline A measure of ecotoxicological exposure based on a bioaccumulated metal concentration. An example is an accumulated trace metal concentration in a biomonitor that can be correlated with the onset of independently measured ecotoxicological effects on local community structure.

bioavailability A relative measure of that fraction of the total ambient contaminant such as a trace metal that an organism actually takes up when encountering or processing environmental media, summated across all possible sources of contaminant, including solution and diet as appropriate.

bioavailable A description of a chemical form of a contaminant such as a trace metal that can be taken up into a living organism, for example from solution or from ingested food.

biodynamic modelling Mathematical modelling of the bioaccumulation of a metal for a species from a balance amongst uptake rate from solution, uptake rate from food, the rates of efflux of metal from the body after uptake from either route and, where appropriate, the rate of growth. Biodynamic modelling predicts the accumulation of a metal by an organism at a location by combining the aforementioned quantified physiological processes with site-specific metal concentrations and geochemical conditions. Also called *biokinetic modelling*.

bioindicator A species, ecotype of a species or group of species, the presence, abundance or absence of which indicates a particular environmental factor or change.

biological monitoring The use of biological systems to assess the structural and functional integrity of terrestrial and aquatic ecosystems.

biomarker A biological response (e.g., a biochemical, cellular, physiological or behavioural variation) that can be measured at the lower levels of biological organisation in tissues or body fluids or at the level of the whole organism.

biomonitor An organism that accumulates chemical contaminants such as trace metals in its tissues, the accumulated concentration of which provides a relative measure of the total amount of contaminant taken up by all routes by that organism, integrated over a preceding time period.

biomonitoring The use of the bioaccumulated concentrations of contaminants such as trace metals in organisms to provide relative measures of the total amount of contaminant taken up by all routes by those organisms, integrated over a preceding time period.

biotic index A number on a scale grading the quality of a site according to the diversity and abundance of the organisms present, with particular reference to key taxa with known pollution insensitivities.

carrier protein A protein that traverses the lipid bilayer of the apical membrane of epithelial cells. It will bind metal ions, pass them across the lipid bilayer and release them on the intracellular side of the membrane. Also termed a *transporter protein*.

chelating agent A typically organic compound that will bind to a trace metal ion to form a dissolved complex. Humic acids, including fulvic acids, are naturally occurring organic compounds that will chelate dissolved metal ions.

complexation The process by which a metal ion is surrounded by a number of organic or inorganic molecules or other inorganic ions (ligands) bonded to it.

detoxification The process by which a metal released intracellularly in an organism is bound by a ligand of such high affinity that it in effect prevents that metal from binding to other molecules, either to play an essential role or to cause toxicity. Trace metal detoxification processes may involve soluble ligands (e.g., metallothioneins, glutathione), insoluble granules or intracellular deposits (collectively termed metal-rich granules).

digestive gland Organ derived from blind ending extensions of the midgut of molluscs, producing digestive enzymes, carrying out absorption of digested food and involved in the detoxification of accumulated trace metals. Sometimes referred to as the hepatopancreas, a term restricted here to decapod crustaceans.

efflux rate constant A constant (K_e) defining the proportion of the tissue concentration of metal excreted per day. Efflux rate constants are both species-specific and metal-specific. They may (but not necessarily), however, vary according to whether the metal has been taken up by different uptake routes (e.g., from solution or diet), if the metal taken up enters different physiological pathways of detoxification. Thus efflux rate constants may also be expressed for metals originally taken up from solution (K_{es}) or from food (K_{ef}).

eukaryote Any organism whose cells contain a membrane-bound cell nucleus and other membrane-bound cell organelles. All multicellular organisms such as animals, fungi and plants are eukaryotes, as are many single-celled organisms, often grouped as protists. Eukaryotes represent a distinct level of cellular organisation from prokaryotes.

facilitated diffusion Diffusion of metal into a tissue as facilitated by a carrier protein to aid passage across the cell membrane down a concentration gradient. Facilitated diffusion does not require energy.

ferritin A ubiquitous intracellular protein that stores iron in detoxified form, releasing it in a controlled fashion.

free metal ion A positive metal ion that, although hydrated, is not complexed by organic or inorganic species to form a complex metal ion in which a central metal ion is surrounded by a number of molecules or other ions bonded to it.

free radical A highly reactive atom, ion or molecule with the potential to react with molecules in cells to cause damage.

functional ecology The branch of ecology that focuses on the roles or functions that species play in an ecological community, emphasising physiological, anatomical and life history characteristics.

genome The whole hereditary information of an organism, including both genes and noncoding regions of the DNA.

genomics Study of an organism's genome.

genotoxic Causing damage to DNA, and thereby capable of causing mutations or cancer.

glutathione A tripeptide molecule consisting of three amino acids (glutamic acid–cysteine–glycine) that is ubiquitous in cells at relatively high concentrations. Glutathione will bind, and potentially detoxify, several trace metals, including copper and iron. Glutathione also acts as an antioxidant and is used as a biomarker of oxidative stress that might indicate exposure of an organism to toxic metals.

growth dilution A reduction in accumulated metal concentration (µg/g) caused by the rate of growth (production of tissue) outpacing the rate of accumulation of metal content (µg).

haemocyanin The copper-based respiratory protein in the blood of many malacostracan crustaceans and gastropod molluscs, in which the copper reversibly binds oxygen for transport between tissues.

haemosiderin An insoluble form of stored iron in cells, produced from the degradation of ferritin in lysosomes.

hepatopancreas Organ derived from blind ending extensions of the midgut of decapod crustaceans, producing digestive enzymes and carrying out absorption of digested food. The organ is involved in the detoxification of accumulated trace metals.

humic acids (humates) The predominant component of humic substances, the major organic constituents of soil, streams, lakes, seas and oceans. The complex mixture of many different acids is produced by the microbial biodegradation of organic matter and resistant to further biodegradation. Humic acids will chelate dissolved metal ions.

ion channel A channel formed from a protein which traverses the lipid bilayer of the membrane of epithelial cells and controls the flow of a major ion across the cell membrane. This is a specific example of a protein channel. The proteins form temporary aqueous pores of a specific diameter; they are specifically suited, for example, for a major metal ion such as calcium, potassium or sodium. Ion channels open and close according to the cell's requirements.

isoforms Slightly different forms of the same protein resulting from co-occurring slightly different genes, typically very closely related gene duplicates.

ligand An ion, molecule or a molecular group that binds with a trace metal to form a larger complex.

lipofuscin Yellow-brown pigment granules found in residual bodies (lysosomes) in eukaryotic cells, resulting from the lysosomal digestion of lipids and proteins. Also called the ageing pigment.

lysosome A membrane-enclosed organelle in a eukaryotic cell that contains enzymes that break down biological molecules such as proteins, as well as particulate material taken up by endocytosis into the cell, or obsolete components of the cell itself. The interior of a lysosome is acidic. A lysosome containing finally indigestible material is called a residual body.

metabolically available metal Intracellular metal that has not been detoxified and is therefore available to bind with molecules in the cell, either to play an essential metabolic role or to cause toxicity.

metabolite A product of metabolism, the chemical processes occurring within a living organism. The term metabolite is usually restricted to small molecules.

metabolome The complete set of metabolites found in a biological sample, which is variable under different contaminant exposure conditions.

metabolomics The study of the metabolome of an organism.

metallothionein-like proteins (MTLP): Proteins which appear to be metallothioneins but await final analytical identification of all characterising features of metallothioneins.

metallothioneins (MT) Low molecular weight cytosolic proteins, with high contents of the sulphur-containing amino acid cysteine, which bind and detoxify certain trace metals. Synthesis of metallothioneins can be induced by the presence of these metals (e.g., Zn, Cu, Cd, Ag, Hg).

metal-rich granules (MRG) Insoluble inclusions in cells that contain bound trace metals. MRG come in a variety of forms, including crystals (e.g., ferritin crystals), spherical crystalline bodies (e.g., pyrophosphate granules based on calcium) and residual bodies of lysosomes.

neutral red retention (NRR) Laboratory test that is used as a biomarker of the exposure of an organism to contaminants, including toxic metals. The NRR test measures the stability of the lysosomal membrane in affected cells. Cells are treated with the dye neutral red, and the time is measured for the leakage of the dye from the lysosomes into the rest of the cell. A short retention time indicates low lysosomal membrane stability caused by contaminant exposure.

ordination A multivariate statistical technique that groups objects (e.g., sites) characterised by many variables (e.g., species abundances, environmental metal concentrations) along axes, so that similar objects cluster near each other, and dissimilar objects are further apart. Ordination techniques include principal components analysis (PCA) and multidimensional scaling (MDS).

physicochemical Relating to both physical and chemical properties, for example of an exposure medium.

phytochelatins A group of trace metal-binding peptides originally isolated from plants, rich in the amino acid cysteine. The major intracellular detoxificatory molecules binding Cd, Zn, Pb and Hg in green plants, macrophytic algae and fungi.

pollution-induced community tolerance (PICT) An approach to measuring the response of a pollution-induced selective pressure on a community. PICT is tested by comparing communities collected from contaminated and reference sites to contaminant exposure under controlled conditions. Any increased community tolerance that results from the elimination of more sensitive species is considered to be strong evidence that community restructuring has been caused by the contaminant.

prokaryote Single-celled microorganism with no membrane-bound cell nucleus nor any other membrane-bound organelles within the cell. Archaea and Bacteria make up the two domains of prokaryotes. Prokaryotes represent a distinct level of cellular organisation from eukaryotes.

proteome The full set of proteins encoded by the genome of an organism. The term has taken on a more specific meaning to refer to all the expressed proteins at a given time under defined conditions.

proteomics The study of the proteome.

pyrophosphate A form $(P_2O_7^{4-})$ of phosphate ion that has a high affinity for many metals, making it an ideal binding partner in detoxified metal-rich granules.

resistance Metal resistance is a characteristic of all members of a species which survive relatively well generally in conditions of high trace metal contamination, without specific population-based physiological acclimation or genetically based adaptation to the high metal bioavailability in a particular local habitat (cf. *tolerance*).

richness The number of taxa (usually species) present in a defined sampling unit.

residual body A lysosome containing indigestible material such as lipofuscin, left over after the enzymatic breakdown of digestible molecules or material. The deposits in residual bodies may be rich in metals.

scope for growth (SFG) An estimate of the surplus energy available to an animal for growth and reproduction, calculated from the difference between energy assimilated from food and the energy used in respiration, used as a physiological biomarker. SFG is interpreted to decrease when energy is required to cope with the extra physiological cost of handling (detoxifying or excreting) high amounts of trace metals taken up in contaminated environments.

sediment quality triad (SQT) A methodology for the ecotoxicological assessment of the chemical contamination of a sediment, involving at least three measures: (i) sediment concentrations of toxic chemicals, (ii) toxicity testing of sediment samples and (iii) investigation of the biological community composition.

siderophore A lipid-soluble compound secreted by usually prokaryotes, but also by the roots of grasses (a *phytosiderophore*), that will specifically bind iron 3, mobilising it from insoluble iron hydroxide, to be taken up in this complexed form across the cell membrane.

species diversity A single number that incorporates information on the number of species and their abundances in a defined sampling unit.

species evenness A measure of how equally the individuals in a community are distributed amongst the species present.

taxon (plural taxa) A taxonomic group of any rank, including all subordinate groups, considered to be sufficiently distinct from other such groups to be treated as a separate unit.

tolerance Metal tolerance is a characteristic of a particular metal-exposed population of a species, able to show higher survival, growth, etc. than other populations of its own species, when exposed to a very high bioavailability of a toxic metal. Metal tolerance may be inheritable after selection over several generations or be derived by physiological acclimation and be restricted to the one exposed generation.

transcriptome The set of all RNA molecules (transcripts) produced by an organism, reflecting the genes actively being expressed at a particular time and varying with environmental conditions, including exposure to raised trace metal bioavailability.

transcriptomics The study of the transcriptome of an organism.

transferrins Proteins in the blood of animals that bind iron strongly yet reversibly, in two binding sites.

trophically available metal (TAM) That fraction of the accumulated metal concentration in a food (prey) organism that is available for uptake by the feeding (predator) animal. This fraction is not a constant characteristic of the prey item, for it will vary with the digestive powers of the predator.

univariate statistic A single variable. Indices of species diversity, species evenness and species richness are univariate.

uptake The entry of a trace metal across the cell membrane into the cell(s) of an organism (typically) via any or all uptake routes available. Uptake is only synonymous with accumulation if there is no excretion of metal that has entered the organism.

uptake rate Rate of uptake of a metal into an organism, typically expressed as micrograms metal per gram tissue per day (μg/g/d).

uptake rate constant A constant (K_u) that describes the uptake rate of a metal per unit metal concentration in a medium. For dissolved metals, the units are typically micrograms metal per gram tissue per day, per microgram per litre of water (μg/g/d per μg/L), the units cancelling out to give litres per gram per day (L/g/d). Uptake rate constants are both species-specific and metal-specific but vary between exposure media if the dissolved bioavailability of the metal varies between the media, for example with change of salinity.

ventral caecum (plural caeca) One of (typically) a pair of blind ending extensions of the midgut of an amphipod or isopod crustacean, carrying out equivalent functions to the hepatopancreas of decapod crustaceans.

3.1 Introduction

Prologue. All trace metals are taken up by all organisms and then typically accumulated, potentially with toxic effects. This chapter explores the general principles controlling the uptake, accumulation, essentiality and toxicity of trace metals, applicable to all organisms irrespective of habitat.

Natural history can be defined as the study of nature, natural objects and natural phenomena. The use of the term is now commonly confined to the study of living organisms in their environment, but, in its wider meaning, natural history also covers the physical and chemical world around us. When considering the natural history of trace metals, it is important to appreciate the importance of the geology and the chemistry of the metals, for they interact with the biology of these metals. The geology of trace metals has been explored in the previous chapter, considering what metals are where in the British Isles in significant amounts, and the history of their exploitation and therefore release into the biosphere. So what? This is a biological question, covering metabolic deficiency and ecotoxicology at opposite ends of a concentration gradient in the case of essential trace metals, but only potential ecotoxicology for non-essential metals.

This chapter sets the biological scene more fully than can be attempted in an introduction. There are many general principles to be explored as organisms take up and typically accumulate trace metals, the accumulation being a net outcome of uptake and excretion (if any) of these metals. Why does uptake happen? Understandably so for essential metals, but why for non-essential metals? When does toxicity occur? What are the physiological mechanisms preventing the toxic effects of metals after uptake into an organism? All trace metals are toxic at some point. A crucial starting point is that all trace metals are taken up by all organisms, usually without direct physiological control on the part of the receiving organism. The rate of uptake depends on the concentration of the metal to which the organism is exposed, and to a large extent on other physicochemical characteristics of the medium from which the metals are taken up, whether it be the surrounding water or the food ingested by an animal. How does toxicity relate to uptake rate? Why is toxicity not related to total bioaccumulated concentration? This chapter will address these questions and, in doing so, it will set the scene for an understanding of the interaction of metals and the different biota occupying the habitats explored in turn in subsequent chapters.

3.2 Evolution and Trace Metals

Prologue. Life has evolved in the presence of trace metals. Trace metals have different chemistries, and therefore different potential uses in the biochemistry that is shared by life on the planet, and in the biochemistry that is restricted to some taxa and not others. Furthermore, metals exist on Earth in different abundances and chemical forms that can be taken up by organisms. It is the interaction between the chemical properties that make them useful in biochemistry, and their availability and ease of access by organisms that have controlled what metals have evolved to be essential today.

The abundance of the elements in the universe is a result of the greater or lesser ease of the formation of their atomic nuclei by nuclear fusion in stars, and the subsequent stability of these nuclei (Williams and Fraústo da Silva, 1996). Particularly stable forms, with their atomic numbers, are helium (4), carbon (12), oxygen (16), magnesium (24), silicon (32) and iron (56), and these include some of the most abundant elements on Earth (Williams and Fraústo da Silva, 1996). The early atmosphere and corresponding early sea on Earth lacked free oxygen, presenting reducing conditions with hydrogen, methane, carbon monoxide and hydrogen sulphide present (Williams and Fraústo da Silva, 2003). Many trace metals form insoluble sulphides, and these precipitated from solution in this sea, as did much aluminium, magnesium and calcium bound with abundant silicon to form relatively insoluble silicates (Williams and Fraústo da Silva, 2003). This left sodium, chloride and potassium ions in solution in this early sea in abundance, together with some calcium and magnesium ions not precipitated out as silicate (Williams and Fraústo da Silva, 2003). Some trace metals are stable in reduced state, as in the cases of iron 2 and manganese 2, and these would have been present in the early sea as dissolved trace metals.

Although such a sea may have lasted for a billion years or so, oxygen began to enter the picture, probably initially produced from the chemical breakdown of water by light energy, with perhaps the iron salts in the sea acting as catalysts (Williams and Fraústo da Silva, 2003). This oxygen was mopped up by the sulphides present, forming sulphates, and in time releasing metals such as Co, Cr, Cu, Mo, Ni, Zn and V into solution from their insoluble sulphides (Williams and Fraústo da Silva, 2003). The balance shifted progressively to a more oxygenating environment, greatly increased from about 3 billion years ago after life had evolved, with the subsequent enhanced large-scale biogenic release of oxygen from water (Williams and Fraústo da Silva, 1996). More sulphides were oxidised to sulphates, and iron 2 was oxidised to iron 3, which came out of solution as insoluble iron oxides and hydroxides. Manganese was similarly oxidised from manganese 2 to manganese 4, to precipitate bound to oxygen in oxides and hydroxides. Most trace metals, however, could now exist in solution in the sea in the absence of sulphides, albeit at low concentrations given their initial low abundances, usually as the positive metal ion such as Zn^{2+} or Cd^{2+}. Vanadium, molybdenum and selenium, on the other hand, could be oxidised further to produce the negative ions vanadate, molybdate and selenate.

Catalysts change the rates of chemical reactions, and it was the catalytic properties of trace metals that made them so useful in the evolution of life, originally prokaryotes and subsequently eukaryotes, including the plants and animals familiar today. It is the chemical nature of trace metals that they readily bind with the elements sulphur and nitrogen. These elements are common in the structure of many of the organic molecules of life, particularly in proteins. As the first cells evolved, metals such as iron and manganese were available to bind with sulphur and nitrogen in the proteins that were the first enzymes, and to increase their catalytic efficiency under the driver of natural selection. Because it could be oxidised and reduced in turn between the two oxidation states iron 2 and iron 3, iron crucially facilitated the transfer of electrons, as in the series of biochemical reactions of anaerobic respiration delivering energy for the cell (Fraústo da Silva and Williams, 1993). Manganese played the key catalytic role in the biogenic release of oxygen from water in photosynthesis. Molybdenum and vanadium were harnessed as

catalysts in the prokaryote oxidation of nitrogen from ammonia (NH_3) to nitrogen (N_2) and ultimately to nitrate (NO_3^-). The latter stage of this oxidation sequence is referred to as the fixation of nitrogen, vital to making nitrogen available for use by most organisms, particularly eukaryotes such as plants. Such trace metals had become essential (Harrison and Hoare,1980).

The very chemical properties that have made trace metals so useful in the evolution of the biochemistry of life are those that make all trace metals toxic in high abundance (Figure 1.1). Because trace metals will bind readily with organic compounds such as proteins, it is inevitable that, unless otherwise prevented from doing so, trace metals taken up in excess will bind to biological molecules in the wrong place at the wrong time to cause a toxic effect. This will happen irrespective of whether the metal is essential in lower quantities or is non-essential. Trace metals present in excess might replace another metal carrying out a vital function in a protein such as an enzyme, or they might bind to another part of the protein, distorting its structure and preventing its biochemical function. Either way, a toxic effect has been caused. We will be concerned in this chapter as to how organisms cope with such a toxic threat.

3.3 Uptake of Trace Metals

Prologue. There are several potential routes for water-soluble trace metals to cross the lipid bilayer of a cell membrane and enter a cell and hence the body of an organism.

To survive, cells need essential trace metals to cross a formidable barrier, the cell membrane, a lipid bilayer barring their passage from the environment into the cell. Thus the first stage of the interaction between trace metals and a living organism is the process of metal uptake, the transport of a metal across the cell membrane into a cell which might, for example, be an epithelial cell on the outside of the organism, such as the gill of an aquatic animal or a cell lining the alimentary tract. This uptake process may also be called absorption. Once a trace metal has been taken up, it may be excreted again or, more usually, accumulated in the organism.

Given the long-shared evolutionary history of life and essential trace metals, it is not surprising that the methods by which metals are transported across the cell membrane are shared across the tree of life, with different uptake routes being prominent in different organisms and under different environmental conditions.

The eukaryotic cell membrane basically consists of a lipid bilayer crossed by many proteins. Because the lipid bilayer acts as a barrier to the passage of water and water-soluble chemicals into the cell, the embedded proteins control the passage of soluble chemicals across the membrane. Dissolved trace metals, which are typically in ionic form in solution, would not cross the apical membrane into the cell without the intercession of such membrane proteins.

There is more than one mechanism by which a trace metal can cross the apical membrane to gain entry into a cell and thence the rest of the organism (Figure 3.1) (Rainbow and Luoma, 2015). One uptake route may dominate, but different routes will have different importance, varying, for example, with the metal and the chemical form of the metal; the species of organism and its physiological state; and the nature of the epithelium concerned and its physiological role. Thus different routes will be more important for the uptake of trace metals across a gut cell compared to a gill cell. Routes of uptake of trace metals (Figure 3.1) include carrier proteins (also called transporter proteins); protein channels forming a temporary aqueous pore of a certain diameter to transport a specific major ion, such as calcium; co-transport with specific carrier systems for molecules, such as amino acids, particularly in the gut; direct diffusion through the lipid bilayer of uncharged metal forms, including some organometal compounds; and the endocytosis of metals bound in particles.

A key principle in the uptake of trace metals is that they will pass down a concentration gradient from a high concentration to a low concentration by diffusion. Diffusion is a passive process that does not require energy, in contrast to active transport, which

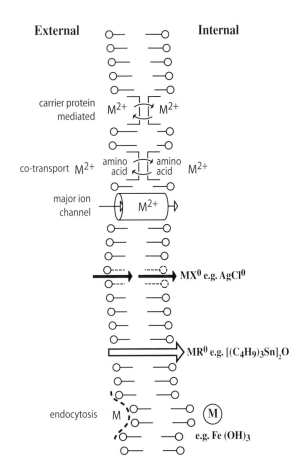

Figure 3.1 A simplified diagram of the possible routes by which a trace metal (M) may cross the apical membrane of a eukaryotic cell. (From Luoma and Rainbow, 2008.)

uses energy to move ions or molecules across a cell membrane against a concentration gradient. So long as there is a concentration gradient from the outside to the inside of a cell, and so long as there is an intermediary transport system such as a carrier protein to cross the membrane, then metals will enter the cell passively. Paradoxically, the total concentrations of trace metals within cells (and whole organisms) are typically far higher than the total dissolved concentrations of the metals in the bathing environment. How then can the metals enter by passive diffusion? The answer lies in the fate of the trace metals upon entry into a cell. Their chemical properties ensure that they are immediately bound to organic molecules such as proteins in the cell. Although the absorbed metals may then be passed on to other cell molecules with stronger binding affinities for the metal, the bound metals from the start are not in a position to diffuse back out of the cell. In short, the concentration gradient of diffusible metal remains, and metals will continue to enter the cell passively, even though the internal total metal concentration in the cell is much higher than the total dissolved concentration outside.

A carrier protein is a protein that traverses the lipid bilayer of the apical membrane of epithelial cells (Figure 3.1). It will bind metal ions on the outside, pass them across the lipid bilayer and release them on the intracellular side of the membrane. Because of the internal binding of the released metal, there is a concentration gradient of diffusible metal, the diffusion being facilitated by the carrier protein. This uptake mechanism is therefore termed passive facilitated diffusion and is often considered to be the most important route of entry of trace metals into organisms. Dissolved metals in media bathing aquatic organisms are usually present in different chemical forms according to the physicochemical makeup of the medium. For the dissolved metals that exist as positive metal ions, such as Zn^{2+} (generalised as M^{2+} in Figure 3.1), experimental results are best explained if it is this free metal ion that binds with the carrier protein for transport across the membrane. No free metal ion remains unbound in the cell, and there is a concentration gradient out to in, even if outside dissolved concentrations are very, very low. It is logical that carrier proteins for essential trace metals would have come under positive selection pressure in evolution. Chemistry, however, will not be ignored, and it is unavoidable that non-essential metals with similar chemical properties to particular essential metals will also bind to their carrier proteins, probably less efficiently but nevertheless with resulting entry into the cell. Thus, it appears that cadmium will hijack zinc transporters. In reality, all trace metals, essential or not, will enter all organisms to some degree.

A major ion channel (Figure 3.1) is a pore-forming transmembrane protein that controls the flow of a major ion, such as calcium, sodium or potassium, across a cell membrane. It is possible for the free metal ions of trace metals to trespass into these channels if of a similar size to the intended major metal ion, as in the case of cadmium and calcium channels. The free Cd^{2+} ion (0.092 nanometres) is of a very similar ionic radius to the Ca^{2+} ion (0.094 nanometres), and some cadmium from the external medium will inevitably enter epithelial cells via calcium channels (Luoma and Rainbow, 2008). The significance of this route of entry for cadmium, as opposed to the use of a trace metal carrier protein (perhaps for zinc), will depend on the specific organism in question and any physiological requirement for high calcium uptake. A mollusc laying down a calcium carbonate shell or a newly moulted crab recalcifying its exoskeleton can be expected to have high rates of calcium uptake from the medium. More cadmium than usual may then enter along with the increased passage of calcium through calcium channels. Similarly, uptake of silver in the gills of freshwater rainbow trout *Oncorhynchus mykiss* includes the entry of Ag^+ ions into Na^+ channels in the apical membranes of gill cells (Bury and Wood, 1999).

Trace metals that exist in solution as negative ions (e.g., arsenate, chromate, molybdate, selenate, vanadate) have the potential to enter cells via ion channels for sulphate or phosphate.

The co-transport of trace metals with specific carrier systems for particular molecules can occur (Figure 3.1), as in the case of amino acids taken up in the alimentary tract of animals for the later synthesis of proteins. Co-transport is a major mechanism for the uptake of methyl mercury, which forms a complex with the amino acid cysteine that mimics another amino acid (methionine). Specific co-transporters for methionine mistake the methyl mercury–cysteine complex for methionine and transport it into the gut cell (Luoma and Rainbow, 2008).

Uncharged chemical forms are more soluble in lipid than their ionic counterparts, and so some trace metals may enter the cell in this form. In particular, many organometal compounds are typically lipid-soluble and therefore able to cross the lipid cell membrane relatively easily, as in the case of tributyl tin.

A trace metal may also enter a cell in particulate form by endocytosis, the process by which a particle or molecule is engulfed by an in-tucking of the cell membrane (Figure 3.1). Particulate iron oxides or hydroxides can be taken up in this way by the gill cells of mussels (Luoma and Rainbow, 2008). The uptake of metals in particulate form is also very likely in the gut epithelial cells of many invertebrates relying on intracellular digestion within cells as opposed to extracellular digestion in the gut lumen. Metalliferous nanoparticles can be taken up by endocytosis, particularly but not exclusively in the guts of animals, although metals may leach from nanoparticles into the surrounding medium and be taken up in dissolved form.

Many prokaryotic microorganisms are capable of secreting lipid-soluble compounds that will specifically bind metal ions, and in some cases facilitate their passage across the cell membrane (Harrison and Hoare, 1980). In many cases, however, the role of these compounds seems to be in chemical warfare against competing microbes, sabotaging the integrity of the opponents' membranes. A siderophore is one such compound secreted by prokaryotes that mobilises iron 3 from insoluble iron hydroxide usually present in oxygenated conditions, to be taken up by in this bound form across the cell membrane (Harrison and Hoare, 1980). The synthesis of siderophores is stimulated under conditions of iron deprivation to fulfil the essential need for iron.

Trace metals can, therefore, enter an epithelial cell and thus the body of an organism by one of a variety of uptake routes, and potentially by several routes simultaneously. All routes of entry are not equally important, and some routes may never be quantitatively significant in the total uptake of a trace metal by particular organisms. The relative importance of different trace metal uptake routes will change under different conditions, whether these conditions are the physicochemical conditions of the medium bathing the cell surface or the physiological conditions within an organism. The free metal ion appears to best model the form of metal to bind with carrier proteins, to enter major ion channels and perhaps to bind to co-transporters, but not the form of metal feeding into the other routes of uptake (Figure 3.1). Arguably the carrier protein route, perhaps with the major ion channel route under some circumstances, accounts for most of the uptake of dissolved metals from solution by aquatic eukaryotic organisms under most environmental conditions. So it is the availability of free metal ions that is a major controller of the uptake of dissolved metals from the media encountered by animals and plants. It is time to explain the concept of bioavailability.

3.4 Bioavailability

Prologue. The physicochemical form of a trace metal in the environment affects whether or not it can be taken up by an organism. A trace metal is bioavailable to a specified organism only if it can be taken up by that organism.

Metals exist in the environment in different chemical forms. Few are present naturally as elemental metal, although precious metals such as gold and platinum are, almost by definition, very resistant to taking part in chemical reactions to form compounds or to dissolve in solution as metal ions. These precious metals will indeed be found as pure elements. Clearly a lump of gold or of lead offers no toxic threat to a living organism, for it cannot be absorbed into the body to cause a toxic effect. A metal needs to be in dissolved form or to be incorporated into a digestible food item to be available for uptake by an organism. To be dissolved may not be enough, because a dissolved metal may form compounds in which it is so tightly bound, while still dissolved, that the metal cannot be taken up by an organism.

A metal is said to be bioavailable only if it is in a form that can be taken up by an organism. The term 'bioavailability' therefore refers to that fraction of the total ambient metal that an organism actually takes up when encountering or processing environmental media, totalled across all possible sources of metal, including water and food as appropriate (Luoma and Rainbow, 2008; Rainbow and Luoma, 2015). For example, the metal bioavailable to a seaweed or a freshwater alga would be a fraction of the dissolved metal in the surrounding medium, while the metal bioavailable to a freshwater or marine invertebrate would include fractions of both dissolved metal and metal derived from ingested food. Necessarily there is a correlation between the bioavailability of a trace metal to an organism, totalled across all sources, and its total uptake rate.

The bioavailability of a metal typically increases with its concentration (be it in water or food), but it remains possible for two media of the same dissolved concentration of a metal to have different bioavailabilities, according to what other dissolved substances are present in the different media.

Bioavailability is strictly dependent on the specific organism being considered. Thus what metal is bioavailable to a particular species in a habitat will depend on what an animal is eating as well as the dissolved metal to which it is exposed. Nevertheless, the term 'bioavailability' does tend to be used without such specific allocation to a particular organism. As we shall see, it probably is valid to extrapolate from the specific to the general, particularly in the case of dissolved bioavailabilities. A medium that is of high metal bioavailability to one aquatic species is likely to be of high metal bioavailability to another aquatic species. High dissolved bioavailabilities do tend to result in raised accumulated concentrations in most biota, and so the diets of many local animals will also have raised metal concentrations. So it is understandable that a habitat may be referred to as presenting high bioavailabilities of particular metals, even though the different organisms therein may actually be tapping into these high bioavailabilities in subtly different ways. We shall return to this point when discussing biomonitoring later.

3.4.1 Dissolved Bioavailability

Prologue. The free metal ion is considered to best model the form of dissolved trace metal that can be taken up across the cell membrane.

We have seen that it is the free metal ion that is considered to best represent the form of many trace metals in solution that binds with carrier proteins or the proteins of major ion channels, accounting predominantly for the uptake of these trace metals from solution. Thus the bioavailable dissolved form of the trace metal is considered to be (or strictly, is best modelled by) the free metal ion, a concept is referred to as the free ion activity model (Campbell, 1995). In most natural waters, only a proportion of the total dissolved metal concentration is present in the form of the free metal ion, for much of the remaining dissolved metal forms complexes with other inorganic ions or with organic compounds. Complexation is the process by which a metal ion is surrounded by a number of organic or inorganic molecules or other inorganic ions bonded to it. Indeed, a free metal ion is itself complexed by water molecules and is strictly a hydrated free metal ion. Complexation, however, by other complexing agents to form a complex metal ion involves stronger bonds than those involved in hydration. These complexation bonds are typically strong enough to outcompete with membrane proteins that might otherwise bind with the metal ion for transport into the cell. Thus, perhaps to oversimplify, free metal ions are bioavailable while complex metals ions are not. Importantly, the complexation of free metal ions in solution is an equilibrium process, so there will always be a percentage of the dissolved metal present as the free metal ion. This percentage may be very small. Furthermore, as free metal ions bind to membrane proteins for transport into cells, the free metal ions in the external medium will be replenished from the equilibria with the different complex metal ions in solution. Uptake of metals will not exhaust the supply of free metal ions, and metal uptake from solution simply continues.

The complexation of dissolved trace metals present as positive ions is best exemplified in seawater. Table 3.1 shows the inorganic complexation of trace metal ions in seawater. This table in fact ignores any complexation by organic compounds that might be present in the sea. Such organic complexation might take precedence over inorganic complexation, particularly in the case of copper, and to a lesser extent zinc, in coastal waters and even in some oceanic waters (Donat and Bruland, 1995). What is clear from the table is the fact that most dissolved metal is not present as the free metal ion, although it must be

Table 3.1 **Inorganic complexation of dissolved metals in oxygenated seawater.**

Metal		Probable main dissolved species
Ag	Silver	$AgCl_2^-$, $AgCl_4^{3-}$, $AgCl_3^{2-}$
Al	Aluminium	$Al(OH)_4^-$, $Al(OH)_3^0$
As	Arsenic	$HAsO_4^{2-}$
Ca	Calcium	Ca^{2+}
Cd	Cadmium	$CdCl_2^0$, $CdCl^+$, $CdCl_3^-$
Co	Cobalt	Co^{2+}, $CoCO_3^0$, $CoCl^+$
Cr	Chromium	CrO_4^{2-}, $NaCrO_4^-$
Cu	Copper	$CuCO_3^0$, $Cu(OH)^+$, Cu^{2+}
Fe	Iron	$Fe(OH)_3^0$, $Fe(OH)_2^+$
Hg	Mercury	$HgCl_4^{2-}$, $HgCl_3Br^{2-}$, $HgCl_3^-$
K	Potassium	K^+
Mg	Magnesium	Mg^{2+}
Mn	Manganese	Mn^{2+}, $MnCl^+$
Mo	Molybdenum	MoO_4^{2-}
Na	Sodium	Na^+
Ni	Nickel	Ni^{2+}, $NiCO_3^0$, $NiCl^+$
Pb	Lead	$PbCO_3^0$, $Pb(CO_3)_2^{2-}$, $PbCl^+$
Se	Selenium	SeO_4^{2-}, SeO_3^{2-}, $HSeO_3^-$
V	Vanadium	HVO_4^{2-}, $H_2VO_4^-$, $NaHVO_4^-$
Zn	Zinc	Zn^{2+}, $Zn(OH)^+$, $ZnCO_3^0$, $ZnCl^+$

Source: Luoma and Rainbow (2008).

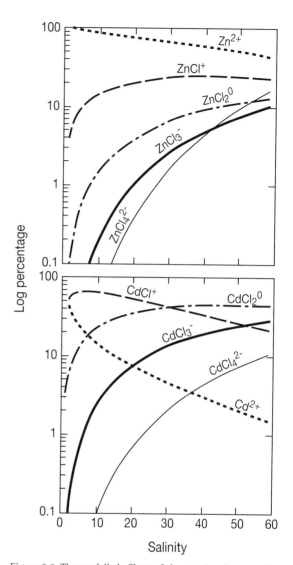

Figure 3.2 The modelled effects of changes in salinity on the inorganic complexation of dissolved zinc and cadmium. (From Luoma and Rainbow, 2008.)

reiterated that complexation is an equilibrium process and some free metal ion will always be present, even in small amounts. The free metal ion Zn^{2+} is the most common form of dissolved zinc in seawater in the absence of organic complexation, but, even in this case, it represents only 40% of the total dissolved zinc present (Luoma and Rainbow, 2008). Chloride complexation is particularly important for cadmium, mercury and silver in seawater, with less than 3% of dissolved cadmium in the form of free Cd^{2+} (Luoma and Rainbow, 2008). It follows that, as salinity is reduced in passage up an estuary into freshwater, the concentrations of inorganic ions, not least chloride ions, decrease and more free metal ion is released from the complexation equilibria. Figure 3.2 illustrates such a reduction in inorganic speciation with reducing salinity for zinc and cadmium, the percentage of free cadmium ion present increasing spectacularly with the drop in chloride.

It follows that two media of the same total dissolved concentrations of a trace metal, for example zinc or cadmium, may have different concentrations of free metal ions contributing to the total concentration according to what else is dissolved in the medium. They will have different bioavailabilities of zinc or cadmium. Thus the intertidal pool-dwelling prawn *Palaemon elegans* (Plate 2a) will take up more zinc from 50% seawater than it will from full seawater, even when the total dissolved zinc concentration is the same (Luoma and Rainbow, 2008). The bioavailability, and correspondingly the uptake rate, of zinc increases in the more dilute medium because it is the availability of the free zinc ions for binding to zinc transporters in the gill cells that controls uptake. The uptake of the trace metal has been controlled by the physicochemical conditions of the exposure medium, not by the prawn itself.

Dissolved organic matter (DOM) is an important source of complexing agents for dissolved trace metals in both seawater and freshwater (Donat and Bruland, 1995). Organic complexation reduces the fraction of dissolved copper in inorganic form in some surface oceanic waters to less than 1%, and the free copper ion itself is only about 0.01% of the total dissolved copper present (Donat and Bruland, 1995). Nearly 99% of dissolved zinc is organically complexed in Central North Pacific surface waters (Donat and Bruland, 1995). In freshwater, dissolved humic acids derived from the microbial breakdown of organic matter are particularly important metal complexing agents. The presence of organic complexing agents in a medium reduces the proportion of dissolved trace metal present as the free metal ion. The bioavailability of the dissolved metal has been reduced in comparison with another medium of the same dissolved metal concentration but lacking dissolved organic matter. Uptake rates of many dissolved trace metals into local aquatic biota will be reduced simply because of the presence of DOM.

Although it is now recognised that the bioavailability of many dissolved trace metals is related to the concentration of the free metal ion, it is very difficult analytically to measure the concentrations of the free metal ion and of all the different forms (chemical species) of most metals dissolved in natural waters. It has, therefore, been tempting for freshwater chemists to use equations to model how much of a dissolved metal might be present as the free metal ion (or in particular complexed forms) if we know the concentrations in the exposure medium of other inorganic ions, DOM and the pH (essentially the hydrogen ion H^+ concentration). Acid conditions can promote the availability of the free metal ion, the hydrogen ions replacing metals in dissolved metal complexes. On the other hand, the extra H^+ ions at low pH may also compete with free metal ions at a membrane protein binding site and decrease the rate of metal uptake. The balance between these two contradictory effects determines bioavailability and uptake rate. The presence of other dissolved metals may also affect the release of the free ion from different forms of organic and inorganic complexation, or compete with the metal at the membrane protein binding site. The concentrations of the free metal ion and other chosen low molecular weight metal complexes can be calculated from first chemical principles, essentially by considering the total concentration of the dissolved metal itself, the concentrations of other metals, the pH and the concentrations of potential complexing agents and their affinities for the dissolved metal. One of the most successful and commonly used such metal speciation models in freshwater is the Windermere Humic Aqueous Model (WHAM) (Tipping, 1994; Tipping et al., 1998, 2011a). The predicted 'bioavailable' dissolved metal concentrations calculated by WHAM in its different updated guises are usually better predictors of the toxicological effects of dissolved metals than are total dissolved metal concentrations. However, it still needs to be remembered that such models concern only dissolved metal concentrations and dissolved metal bioavailabilities. They ignore the ecotoxicologically significant role of metal uptake from the diet in the real contaminated world.

Some trace metals will not follow the Free Ion Activity Model because their commonly dissolved ionic forms in water are negative ions (e.g., arsenate, chromate, molybdate, selenate and vanadate)

(Table 3.1), offering potential for their entry into cells via major ion routes for sulphate or phosphate.

Talk of the effect of complexation, inorganic or organic, on free metal ion concentrations, and hence dissolved bioavailabilities of trace metals, should not obscure the fact that the most important factor remains the dissolved concentration of the relevant metal. The concentration of any free metal ions present will inevitably increase with total dissolved concentration, even if the free metal ion only represents a low percentage of this total dissolved concentration.

3.4.2 Effects of Physiology

Prologue. In certain cases, an organism may make a physiological change that affects the rate of uptake of dissolved trace metals. Nevertheless, as a general rule, the uptake rates of dissolved trace metals are controlled by the physicochemistry of the medium, not the physiology of the organism concerned.

The preceding discussion on bioavailability has stressed the point that the uptake rates of dissolved metals by organisms are controlled by the physicochemical features of the exposure medium, and not by the organism itself. But is this always true? For most organisms, it appears so, but not for all. Furthermore, some animals can make physiological changes that affect metal uptake rates, although those physiological changes may be made in response to environmental factors other than trace metal exposure (Luoma and Rainbow, 2008).

Different organisms, even phylogenetically related organisms, take up trace metals at different rates under identical physicochemical conditions. These interspecific biological differences are often large, perhaps corresponding to the number of membrane uptake sites presented by the organism, reflecting differences in their functional ecology. Functional ecology focuses on the roles or functions that species play in an ecological community, emphasising physiological, anatomical and life history characteristics. Thus the method of respiration of an aquatic animal will affect how much water is passed over permeable external surfaces, affecting the potential

Table 3.2 **Uptake rates of Zn and Cd from solution by three marine crustaceans, a decapod palaemonid prawn (*Palaemon elegans*), an amphipod (*Echinogammarus pirloti*) and a barnacle (*Austrominius modestus*) under identical physicochemical conditions (artificial seawater, 33 salinity, 10°C).**

	Zinc	Cadmium
Dissolved concentration (µg/L)	100	31.6
Uptake rate (µg/g/d)		
Palaemon elegans	1.52	0.074
Echinogammarus pirloti	2.68	0.72
Austrominius modestus	26.9	2.75

Source: Luoma and Rainbow (2008).

uptake of dissolved metals. Table 3.2 compares the uptake rates from solution of zinc and cadmium by three crustaceans under identical physicochemical conditions (Luoma and Rainbow, 2008). The decapod crustacean, the palaemonid prawn, has the lowest dissolved uptake rates of both metals, followed by the amphipod, with the barnacle showing much the highest dissolved uptake rates (Table 3.2). This might be expected from their comparative functional ecology. A decapod could be expected to have a low rate of metal uptake per unit body weight because most of the body is covered by an impermeable cuticle, with permeability restricted to the gills. A barnacle, on the other hand, has a large permeable surface area represented not least by the thoracic legs, and moves large volumes of water across these permeable surfaces during suspension feeding. The amphipod might have a higher proportion of permeable exoskeleton than a decapod but less than a barnacle.

Adaptation to habitat may also lead to differences between closely related species in the uptake rates of

Table 3.3 **Uptake rates of Zn (μg/g/d) from solution by three decapod crustaceans under specified identical physicochemical conditions.**

Conditions	Crustacean	Zn uptake rate
20 μg/L Zn, 33 salinity, 10°C	*Pandalus montagui* (sublittoral marine)	0.93
	Palaemon elegans (littoral pools)	0.58
100 μg/L Zn, 16.5 salinity, 10°C	*Palaemon elegans* (littoral pools)	5.27
	Palaemonetes varians (brackish water)	1.80

Source: Luoma and Rainbow (2008).

dissolved trace metals. For example, differences between marine and estuarine or freshwater species in the permeability of their external surface to water may be the foundation of interspecific differences in trace metal uptake rates. A reduction in permeability to water in comparison to marine relatives is an adaptation of several crustaceans to life in estuaries and freshwater, limiting their uptake of water by osmosis in media of low osmotic pressure (Mantel and Farmer, 1983). Furthermore, the zinc uptake rates of caridean decapod crustaceans (commonly but inconsistently called prawns or shrimps) are lower, all else being equal, in those carideans that live at lower salinity. As Table 3.3 shows, the uptake rate of dissolved zinc by the fully marine *Pandalus montagui* exceeds that of the tide pool-dwelling *Palaemon elegans*, which is greater than that of the brackish water inhabitant *Palaemonetes varians* (Luoma and Rainbow, 2008). This interspecific decrease in dissolved zinc uptake rate follows differences in permeability expected as an adaptation to life in low

salinity (Mantel and Farmer, 1983), probably as a result of selected changes in the morphology of the outside body surface that has additionally resulted in a reduction in the number of trace metal uptake sites.

But what about physiological changes made by an organism that affect metal uptake rates, as opposed to typical physiological adaptations differing between species? These can occur, albeit not in many species, and, in some animal examples, apparently not necessarily as a direct response to changes in trace metal bioavailabilities.

Oceanic species of phytoplankton are able to grow at maximal rates at much lower free metal ion concentrations of essential metals, such as iron, manganese and zinc, than their coastal water counterparts growing in higher dissolved concentrations (Sunda, 1989; Sunda and Huntsman, 1998). Furthermore, species of marine phytoplankton, particularly oceanic phytoplankton, are able to regulate the rates of uptake of iron, manganese and zinc in response to low ambient free metal ion concentrations and thereby maintain nearly constant essential metal concentrations (Sunda, 1989; Sunda and Huntsman, 1998). Fish also appear to have the ability to regulate the uptake of dissolved trace metals such as copper, iron and zinc at different dissolved concentrations, by making responsive changes to the number and nature of the membrane proteins involved in their uptake (Hogstrand and Wood, 1996; Bury and Handy, 2010).

Intraspecific variation in metal uptake rates may also occur as a result of a physiological change made by the organism in response to a change in an environmental factor such as salinity. For example, the zinc and cadmium uptake rates of the crab *Carcinus maenas* actually decrease with decreased salinity, in contradiction to the expected increase associated with physicochemically driven increases in free metal ion concentrations (Chan et al., 1992). The physiological response of this euryhaline crab to decreased salinities reflects a decrease made in the apparent water permeability of the integument of the crab reducing osmotic uptake of water in low salinities. This physiological response is of sufficient magnitude to offset the physicochemical enhancement of trace metal uptake. Another euryhaline crustacean, the littoral amphipod *Orchestia gammarellus*, shows a similar adaptive response to low salinity that counteracts the enhanced availability of free metal ions of zinc and cadmium at low salinities (Rainbow et al., 1993; Rainbow and Kwan, 1995). Interestingly, a third euryhaline crustacean, the intertidal pool-dwelling prawn *Palaemon elegans*, shows no physiological offsetting of the physicochemical enhancement of the uptake of dissolved zinc at low salinities (Nugegoda and Rainbow, 1989a, b).

Thus the physiology of an individual organism can affect trace metal uptake rates from solution. This is an indirect physiological effect in the cases of some, but not all, of those remarkable animals that are euryhaline, being able to survive in a wide range of salinities.

There are examples where exposure to exceedingly high trace metal bioavailabilities can result in physiological acclimation and even the genetic selection of a tolerant population of a species (Rainbow and Luoma, 2011a). These examples are rare, however, and are dealt with in more detail in Section 3.8. Even in these cases, it is usually not metal uptake rates that are affected but other parameters of metal accumulation kinetics to bring about metal tolerance (Rainbow and Luoma, 2011a). It is enough to conclude here that it is typically physicochemistry and not physiology that generally controls metal uptake rates by most organisms in different dissolved metal bioavailabilities.

3.4.3 Bioavailability from Food

Prologue. The bioavailability to an animal of trace metals in ingested food is controlled by ingestion rate, the concentration and chemical binding of the trace metals in the food item and the digestive and absorptive powers of the animal itself.

The use of high dissolved metal exposures in toxicity testing in the second half of the twentieth century provided the basis for a belief that uptake from solution is the most (if not the only) ecotoxicologically significant source of metals to animals. It has, however, become increasingly apparent over the last

decade or so that the diet is an important route, and often the most important route, for the uptake of trace metals by animals in the field, including under metal polluted conditions (Luoma and Rainbow, 2008). Uptake of trace metals from food is a function of how much food an animal ingests (dependent on feeding rate), the metal concentration in the food and how much of that ingested metal can be extracted and assimilated in the alimentary tract of the animal.

The feeding of an animal is clearly a feature of its functional ecology, thus its natural history. Feeding rate varies with many factors, not least the availability of food and the feeding preferences of the animal, with consequences for the different dietary bioavailabilities of metals from different diets. A suspension feeder will ingest large quantities of small particles, often including detritus particles, with a high surface area to volume ratio and high organic content promoting adsorption of trace metals. A deposit feeder will ingest and process large volumes of potentially metal-rich sediment particles. Either feeding process will promote the dietary uptake of metals, in comparison with a macrophagous feeder such as carnivorous crustacean taking in large pieces of animal tissue. Animal tissue, especially muscle, is not particularly metal-rich.

Assimilation efficiency is a measure of the percentage of the metal ingested that is assimilated into (initially) the alimentary tract epithelium of the feeding animal. Thus this metal has been taken up with the potential to interact with the physiology of the animal concerned. The assimilation efficiency of a metal will vary between species but also within a species. The factors affecting assimilation efficiency, and hence the bioavailability of metals from the diet, include the total concentration of the metal in the diet, the chemical form of the metal in the food, the digestive processes releasing trace metals from the ingested meal, the chemical form of metal that is released by digestion and whether that form can be taken up by a gut epithelial cell. Given the variety of alimentary tract structures and digestive processes found in animals, it is to be expected that the assimilation efficiency for a particular metal will vary considerably between species. Intraspecific variation, not least related to the chemical form of the metal bound in ingested food, is additional to this interspecific variation.

As discussed for bioavailability in general, the bioavailability of a metal in a particular diet depends on the feeding animal and is not a generalised feature of the food item. An important cause of the differences in assimilation efficiency, and thus the trophic bioavailability, of the metal in a particular food is variation in the digestive processes of the feeding animals. Ingested metals may be released from a solid food item for transport across the membrane of the animal digestive tract by powerful solubilising forces in the digestive tract – the process of extracellular digestion. Intracellular digestion, on the other hand, is the process by which an animal cell engulfs a food particle by endocytosis, to be digested intracellularly within a food vacuole. Intracellular digestion is the predominant form of digestion in some invertebrate animals, for example bivalve molluscs, but, more commonly, metals are released from food particles by extracellular digestive processes in the alimentary tract. The metals may then cross the apical membrane of the gut epithelial cells by any of the uptake mechanisms described previously, but the proportional contributions of the different uptake routes are likely to be different in the gut than at an external permeable surface. Metals are typically complexed in alimentary tract fluids with high concentrations of small organic molecules. Gut epithelial cells are adapted for the uptake of many of the organic molecules, such as amino acids released by digestion, and trace metals may be taken up by co-transport with amino acids such as histidine and cysteine.

It is now well established that one factor controlling the assimilation of a trace metal from the diet is the chemical form of that trace metal accumulated in the food organism (Rainbow et al., 2011a). A very useful approach to investigate this factor has been to separate the metals accumulated in food organisms into different soluble and insoluble subcellular fractions (Figure 3.3a) (Wallace et al., 2003). The two soluble fractions both consist of proteins. One soluble fraction (Figure 3.3a) is made up of the remarkable proteins called metallothioneins (MT), which are

adapted to bind particular trace metals. In Figure 3.3a, these are labelled metallothionein-like proteins (MTLP). These are proteins which appear to be metallothioneins (and almost certainly are metallothioneins) but which await final analytical identification of all the biochemical features characteristic of metallothioneins. This final confirmation requires biochemical skills and facilities beyond most metal ecophysiologists, and use of the term metallothionein-like proteins is more defensible. Metallothioneins have a unique stability to high temperature and are therefore easily separated from all the other soluble proteins present by exposure to raised temperature which denatures and precipitates all the other proteins. The second protein fraction contains all these other soluble proteins in the cells – fundamentally the enzymes. These are labelled in Figure 3.3a as heat-sensitive proteins, an operational category reflecting how they have been separated from the metallothioneins. The insoluble metal-binding fractions are cell organelles (e.g., mitochondria), cell debris and metal-rich granules (Figure 3.3a). Metal-rich granules (MRG) will figure strongly in this book, for they usually represent the final form of binding of accumulated trace metals, rendering them unavailable to play any metabolic role, toxic or essential. Cellular debris is not surprisingly a catch-all dustbin category without further identification.

In an attempt to seek general principles governing the trophic availability of dietary metals, it was first proposed that only metal bound to the soluble subcellular fraction in planktonic diatoms is available to be assimilated (with 100% efficiency) by copepod crustaceans feeding on them (Figure 3.3b) (Reinfelder and Fisher, 1991). This proposal has now been tested for many filter-feeding invertebrates, but has been shown to be an oversimplification as a general principle (Rainbow et al., 2011a). The operational separation of metals accumulated in food items into the five fractions identified in Figure 3.3a has also been applied to predators feeding on invertebrate prey. The chemical form of cadmium accumulated by a prey item, in this case the oligochaete worm *Limnodrilus hoffmeisteri*, did affect Cd assimilation after ingestion by the prawn *Palaemonetes pugio*

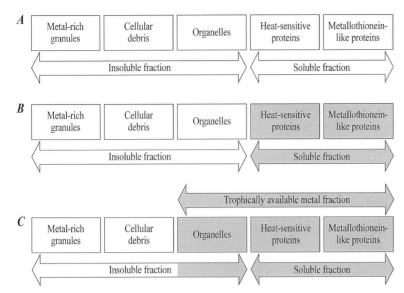

Figure 3.3 (a) Fractionation of metal accumulated in prey into five components (after Wallace et al., 2003). (b) Highlighted soluble fractions considered trophically available to copepods (after Reinfelder and Fisher, 1991). (c) Highlighted fractions represent the proposed trophically available metal (TAM) component of Wallace and Luoma (2003). (From Rainbow et al., 2011a, with permission.)

(Wallace and Lopez, 1996, 1997; Wallace et al., 1998). As in the case of the copepods feeding on diatoms (Reinfelder and Fisher, 1991), cadmium associated with soluble cytosolic proteins in the prey was 100% trophically available to the prawn predator, but in addition cadmium bound to cell organelles was 70% trophically available (Wallace and Lopez, 1996, 1997; Wallace et al., 1998). Cadmium bound to insoluble Cd-rich granules was not assimilated. Thus more than just the soluble component of metals accumulated in prey might be trophically available to a predator (Rainbow et al., 2011a).

The next significant step in the search for a general principle of trophic bioavailability involved a study of how the subcellular fractionation of cadmium and zinc accumulated in the tissues of two bivalve molluscs affected their trophic transfer to another prawn, *Palaemon macrodactylus* (Wallace and Luoma, 2003). In this case, the assimilation efficiency results were best explained if trace metals bound to cell organelles were added to metals bound to the soluble fraction (both soluble protein fractions) to form what was termed the trophically available metal (TAM) fraction

of metals accumulated in the bivalve prey (Figure 3.3c) (Wallace and Luoma, 2003).

This proposed TAM fraction stimulated research to test its generality as an underlying principle to explain the trophic availability to predators of metals accumulated by prey. Subsequent research has in fact shown that different subcellular components of accumulated trace metals appear to be trophically available to different degrees to different predators (Rainbow et al., 2011a). This is particularly true in the case of predatory neogastropod molluscs. These are sea snails, including whelks, with remarkably strong digestive powers which can assimilate metals to some degree from all five subcellular fractions of accumulated metals in prey tissue, including insoluble metal-rich granules (Cheung and Wang, 2005; Rainbow et al., 2007). The TAM fraction as originally defined does not, therefore, account for all trophically available metal in the diet of all predators. The strength of digestion in a specific predator affects exactly what metal-binding fractions release their metals for assimilation and to what extent. For example, metals bound in metal-rich granules in prey tissues appear to

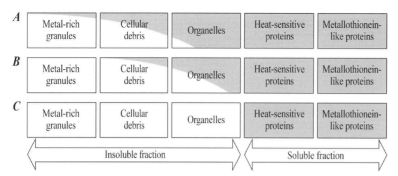

Figure 3.4 Fractionation of metal accumulated in prey into five components (after Wallace et al., 2003). (a) Highlighted areas covering all five fractions to some degree represent metal accumulated in prey that is trophically available to a carnivorous neogastropod mollusc (after Cheung and Wang, 2005; Rainbow et al., 2007). (b) Highlighted areas (from four fractions) represent metal accumulated in prey that is trophically available to a predator with weaker digestive powers than a neogastropod mollusc. (c) Highlighted areas (from two fractions) represent metal accumulated in prey that is trophically available to a planktonic copepod filtering phytoplankton (after Reinfelder and Fisher, 1991). (From Rainbow et al., 2011a, with permission.)

be more susceptible to the assimilatory powers of neogastropod molluscs (Cheung and Wang, 2005; Rainbow et al., 2007) than those of palaemonid decapod crustaceans (Wallace and Lopez, 1996, 1997; Wallace et al., 1998).

Figure 3.4 shows schematically how the trophically available metal fraction in prey can vary with the strength of the digestive and assimilative powers of the predator. In Figure 3.4a, metals in all five fractions of accumulated metal in prey are trophically available to some degree to a neogastropod mollusc (Cheung and Wang, 2005; Rainbow et al., 2007). Figure 3.4b illustrates the case of a predator with weaker digestion which may assimilate all metals from the soluble components in the prey, none from metal-rich granules, but some metal to different degrees from organelles and cellular debris. In fact, as we shall see later in this chapter, metal-rich granules do not fall into a single chemical category, and there will be additional variation in their digestibility, even by a single predator (Rainbow et al., 2011a). Figure 3.4c depicts the subcellular metal components in phytoplankton that may be trophically available to planktonic copepods (Reinfelder and Fisher, 1991), particularly if feeding superfluously. Superfluous feeding is the method of feeding reportedly used by copepods feeding on very dense concentrations of

phytoplankton. Ingestion is so rapid that passage through the gut may simply act as a press to squeeze out soluble material for alimentary uptake, with limited, if any, digestion of solid components of the ingested phytoplankton. It is logical that, under these circumstances, only metal in the soluble component of the ingested diatoms would be trophically available. Other invertebrate herbivores filtering phytoplankton do show stronger assimilation than depicted in Figure 3.4c, better represented by Figure 3.4b.

The uptake of metal from ingested food in the alimentary tract of an animal using external digestion can be arbitrarily divided into two phases: firstly, the release of soluble metals from the ingested meal into the gut medium; and, secondly, the subsequent uptake of these metals into the gut epithelium. Both processes contribute to trophic bioavailability, but the first process alone is considered in the concept of bioaccessibility. Since trophic bioavailability ultimately concerns the ability of an animal to assimilate metals released into solution in the gut (phase two), in addition to phase one, it is difficult to model chemically in the laboratory. The first phase, however, can be modelled by using chemical substitutes for digestion – in short, to measure the relative bioaccessibility of metals in materials that are potentially ingested. The bioaccessibility of a metal in a potential food item is therefore a relative measure of its release into solution from that solid material through the action of a defined cocktail of chemicals mimicking a digestive process (typically of humans). Thus, it is now common to use defined media of particular pH and digestive enzyme and other protein concentrations to measure the relative bioaccessibilities of trace metals in a food item such as an oyster (Bragigand et al., 2004) or in garden soil that might be ingested by an infant (Oomen et al., 2002; Wragg et al., 2011).

3.4.4 Uptake from Sediments

Prologue. Uptake of trace metals from sediments results from a combination of uptake of metals dissolved in pore water and/or in any water current irrigating the burrow of an animal, and uptake from ingested sediment in the case of deposit-feeding animals.

An animal such as a worm living in a burrow in a sediment, be it in freshwater, estuarine or marine conditions, can take up metals from the surrounding water and from its food supply, just as in the case of any other aquatic animal.

The water between the particles of a sediment (the interstitial or pore water) has long been considered to be the major source of metals to such burrowers. An oligochaete or polychaete worm is soft bodied. It therefore presents a potentially large permeable surface area to the water bathing its body, with a corresponding potential to take up large amounts of dissolved metals. Interest has therefore historically been centred on the physicochemical nature of pore waters and their dissolved trace metal contents. In many sediments, particularly organically rich sediments, the organic carbon content of the sediments is used by microbes as an energy source in respiration. Initially this respiration is aerobic and uses up available oxygen. Aerobic respiration is then followed by anaerobic respiration in what is now a reducing environment. Such reducing conditions are easily recognised by the rotten eggs smell of hydrogen sulphide released when digging in such sediments, which are particularly common in the intertidal mudflats of

estuaries. In addition to hydrogen sulphide, metal sulphides have been deposited in these sediments, predominantly black iron sulphide. Furthermore, the pore waters present in such muds will contain high concentrations of iron 2 (and manganese 2), the reduced forms of the metals, which are very soluble in the absence of oxygen. Much research has therefore been carried out in an attempt to predict pore water concentrations of trace metals, trying to predict the ecotoxicology of metal-rich sediments to the local biota, particularly the infauna burrowed therein. Of particular interest has been the relationship between the total amounts of trace metals present in the sediment and the quantities of available sulphide, applying the rule of thumb that an excess of sulphides available will bind all trace metals and not release them into pore waters for potential uptake by local burrowing animals (Luoma and Rainbow, 2008; De Jonge et al., 2009). Unfortunately, this concept is flawed, for burrowing animals do not come into contact with oxygen-deficient pore water as such. A burrowing animal will create a respiratory current (sometimes simultaneously acting as a suspension feeding current) to pass oxygenated water from the overlying water column across its body, including gills or other permeable surfaces. Pore water will make little or potentially no contribution to this bathing water. Consequently, the metal concentrations and other physicochemical characteristics of the pore water have little relevance to the uptake of dissolved trace metals by burrowing animals. As well illustrated in Plate 2b, burrows of the intertidal polychaete worm *Hediste diversicolor* are easily distinguishable in the mud of Restronguet Creek, Cornwall, UK, a sediment rich in trace metals, particularly iron, contributing the black iron sulphide dominating anoxic regions of the mud. The burrows, in contrast, show the local oxidation of iron sulphide to red/brown iron oxides and hydroxides, caused by the flow through of currents of oxygenated water from the overlying water column. The worms are simply not in contact with anoxic pore water. Any uptake of dissolved trace from water will be controlled by the same physicochemical factors as that of any aquatic invertebrate in an oxygenated aquatic habitat.

Interestingly enough, the paradigm that uptake of metals by burrowing organisms is controlled by pore water conditions is further demolished by the increasing number of observations that sediment ingesting infauna obtain much the greater proportion of their accumulated trace metals from the diet anyway (Casado-Martinez et al., 2009a; Rainbow et al., 2009b).

3.5 Accumulation

Prologue. The accumulation of trace metals by an organism is the net outcome of uptake and any excretion. The accumulated metal content of an organism can be conceptually divided into two categories: metabolically available and detoxified metal.

There is remarkable variation in the accumulated trace metal concentrations of invertebrates, between even relatively closely related species for one metal, and within one species for different metals, even in the absence of raised metal bioavailabilities. This huge variation is well illustrated in Table 3.4, which provides examples of accumulated concentrations of three trace metals – zinc, copper and cadmium – within a single taxon, the crustaceans.

Six crustacean taxa are represented in Table 3.4: barnacles and five malacostracan groups – krill, two peracarid taxa (amphipods and isopods), and two decapod taxa – with examples from both metal-contaminated and noncontaminated sites. Barnacles, even from uncontaminated sites, have zinc concentrations an order of magnitude or more above those of the other crustaceans, except the two isopod species (terrestrial woodlice) from zinc-contaminated sites. On the other hand, barnacles from uncontaminated sites typically have body copper concentrations equivalent to, or below, those of malacostracans from uncontaminated sites. Barnacles, however, increase their body copper concentrations well above those of any malacostracan when at a copper-contaminated site, for example Dulas Bay in Anglesey, Wales, receiving copper-rich run off from Parys Mountain. Differences between body cadmium concentrations in

the three crustacean taxa are not so marked (Table 3.4).

These crustacean data illustrate the general point that there is typically no absolute definition of what is a high or low accumulated concentration of one particular metal across invertebrates. A high concentration in one invertebrate may be low for another, and vice versa.

A factor that will change the accumulated body concentration (but not the content) of a trace metal in an organism is a change in weight of that organism. Growth will cause a reduction in accumulated metal concentration if the rate of growth outpaces the rate of accumulation of metal content, a feature termed growth dilution. Growth does not necessarily always cause a reduction in accumulated concentration, for the growth rate may be outpaced by the rate of metal accumulation, thereby resulting in an increased accumulated metal concentration even in the presence of growth (Figure 3.5). Once growth has ceased, ongoing accumulation will again have the potential to cause an increase in accumulated metal concentration (Figure 3.5). Loss of weight, resulting, for example, from the use of stored energy reserves or the emission of gametes, has the potential to increase accumulated metal concentration with or without changes in accumulated metal content. The same effects can take place at organ or tissue level within an animal.

The accumulated concentration of a trace metal in an animal is the net outcome of the uptake of that metal from solution and diet, and any excretion of that metal. In many cases, particularly for non-essential metals accumulated by invertebrates, excretion may be insignificant. When significant excretion does occur, it may be predominantly via one excretion route or by any combination of excretion routes, as in the case of uptake routes. A high accumulated concentration can, therefore, result from different combinations of high and low uptake and excretion rates.

Figure 3.6 is a schematic depiction of the accumulated metal content of an aquatic invertebrate such as a decapod crustacean. The accumulated metal content has been conceptually divided into two categories: metabolically available and detoxified metal. When a metal crosses the apical membrane of

Table 3.4 Variability in trace metal concentrations amongst crustaceans. A selection of body concentrations (µg/g dry weight) of Zn, Cu and Cd in a systematic range of crustaceans from clean and metal-contaminated (C) sites, contaminated with at least one, but not necessarily all three, metals.

Species	Location	Zinc	Copper	Cadmium
Cirripedia (barnacles)				
Semibalanus balanoides	Dulas Bay, Wales (C)	50,280	3,750	–
	Menai Strait, Wales	1,220–19,230	170	–
	Southend, England	27,840	232	28
Austrominius modestus	Southend, England	3,460	–	23
	Southend, England	4,900–11,700	20–169	41–50
Amphibalanus improvisus	Woolwich, Thames estuary, England (C)	33,000–153,000	183–913	23–28
Malacostraca				
Peracarida				
Amphipoda				
Orchestia gammarellus	Restronguet Creek, England (C)	392	139	9.8
	Dulas Bay, Wales (C)	126	105	9.1
	Millport, Scotland	152–186	64–78	1.6
Talitrus saltator	Dulas Bay, Wales (C)	235	203	–
	Millport, Scotland	178–264	27–40	–
Themisto gaudichaudii	NE Atlantic Ocean	66	31	53
Isopoda				
Oniscus asellus	Haw Wood, Avon, England (C)	524	454	154
	Wetmoor Wood, Avon, England	63	93	16

Table 3.4 (**cont.**)

Species	Location	Zinc	Copper	Cadmium
Porcellio scaber	Haw Wood, Avon, England (C)	897	651	49
	Wetmoor Wood, Avon, England	186	171	2.4
Eucarida				
Euphausiacea (krill)				
Meganyctiphanes norvegica	NE Atlantic Ocean	102	58	0.66
Decapoda				
Dendrobranchiata				
Gennadas valens	NE Atlantic Ocean	47	26–70	1.9
Sergia robustus	NE Atlantic Ocean	54	16–24	2.3
Pleocyemata				
Caridea				
Acanthephyra purpurea	NE Atlantic Ocean	44	29	2.2
Systellaspis debilis	NE Atlantic Ocean	47	30–100	8.7
Palaemon elegans	Millport, Scotland	81	110	0.9
Pandalus montagui	Firth of Clyde, Scotland	58	57	–
Brachyura (crabs)				
Carcinus maenas	Firth of Clyde, Scotland	97	30	0.87

Sources: Rainbow, 1987, 1998, 2002; Rainbow et al., 1989; Hopkin, 1990a; Luoma and Rainbow, 2008; Rainbow and Luoma, 2011b.

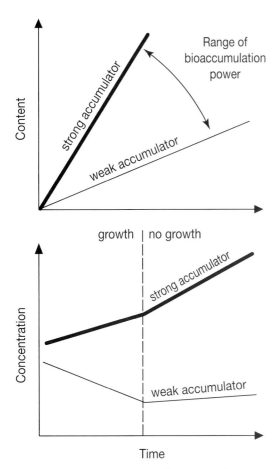

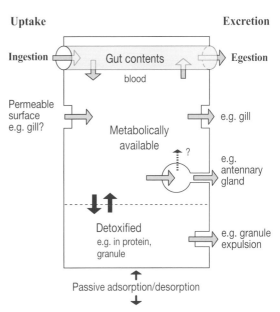

Figure 3.6 A schematic representation of the body metal content of an aquatic invertebrate such as a crustacean. When metal first enters the body, it will initially be metabolically available to interact with the animal's physiology before potentially being stored in detoxified form, probably elsewhere in the body after internal transport via body fluids. Detoxified storage may be permanent or temporary. Trace metals taken up into the body may or may not be excreted, either from the metabolically available component or from the detoxified store. Adsorbed metals may be desorbed again and do not interact with the physiology of the animal. (From Luoma and Rainbow, 2008.)

Figure 3.5 Changes in trace metal content and concentration (content per unit weight) over time in organisms with different powers of accumulation of that trace metal. As a result of relative rates of growth and metal accumulation, content may increase while concentration decreases (growth dilution). (From Luoma and Rainbow, 2008.)

an epithelial cell of the gill or alimentary tract and enters an invertebrate, it has the potential to bind with molecules in the cell, for example proteins, which have a high affinity for that metal. In short, it is metabolically available. This binding may be vital in metabolism, as in the case of essential trace metals being incorporated into enzymes, or it may be the cause of toxic effects, the wrong metal binding in the wrong place to cause a lack of function of a key molecule in the cell. It is crucial, therefore, that the binding of that incoming metal is controlled by

temporary or permanent detoxification, whether the metal is essential or not, because all trace metals will cause toxic effects if left unattended in a cell in metabolically available form above a threshold con-centration. Such detoxification can include binding to soluble detoxificatory proteins such as metal-lothioneins, or ultimately the formation of insoluble metal-rich granules, denying access of the metal to metabolic pathways. Ultimately, this division into metabolically available and detoxified metal com-ponents is for the most part arbitrary, for they involve a network of interconnected physiological pathways handling the metals at different rates of reaction. Attempts can be made to equate these conceptual categories to recognisable subcellular

Table 3.5 **Essential metal requirements in enzymes of metabolising soft tissue (µg/g dry weight).**

	Number of metal-associated enzymes	Percentage of total number of enzymes	Average number of metal atoms per enzyme molecule	Estimated enzyme metal requirement in tissue
Copper	30	1.40	2.95	26.3
Zinc	80	3.74	1.41	34.5

Source: Rainbow and Luoma, 2011b.

forms of bound metal (Wallace and Luoma, 2003; Wallace et al., 2003). This is perhaps easier to achieve in the case of detoxified metal, for example metal bound with metallothionein or in insoluble metal-rich granular form, but is more difficult in the case of metabolically available metal (Rainbow and Luoma, 2011b). For the moment, anyway, we are unable to define with any confidence metabolically available metal in terms of real subcellular fractions that can be isolated biochemically. Thus, at present we cannot measure its concentration in a tissue.

Exchangeable adsorbed metal on the external surface of an aquatic invertebrate, such as a crustacean or an insect larva, may make up a significant proportion of the body content of some metals (e.g., iron, manganese) for some species under some environmental conditions and has been included in Figure 3.6. For most metals, however, the adsorbed fraction is usually quantitatively not significant. Furthermore, it plays no role in the ecophysiology or ecotoxicology of the metal in the invertebrate for it has not been taken up into the body.

3.5.1 Metabolic Requirements for Essential Trace Metals

Prologue. It is possible to make theoretical calculations of the accumulated concentrations of the essential trace metals copper and zinc needed to meet the metabolic requirements of metal-associated enzymes. For certain crustaceans, copper is also required in the respiratory protein haemocyanin.

Some oceanic decapod crustaceans may approach a situation of copper deficiency.

It is potentially useful to attempt to make theoretical estimates of the amounts of particular essential trace metals needed to fulfil metabolic needs in an aquatic invertebrate and thus gain an understanding of the relative size of the metabolically available component of accumulated metal depicted in Figure 3.6 (Rainbow 2002, 2007; Rainbow and Luoma, 2011b). For example, for copper and zinc, theoretical calculations can be based on the number of copper- and zinc-bearing enzymes and their contributions to the total concentrations of enzymes in metabolising tissues, albeit with many assumptions being made (Table 3.5) (Rainbow and Luoma, 2011b). There may be other essential roles for metals beyond their contribution to metal-associated enzymes. For example, malacostracan crustaceans (including amphipods and decapods) contain the copper-based respiratory protein haemocyanin for the transport of oxygen in the blood. So an estimate of the copper requirement of haemocyanin needs to be included in any estimate of the total metabolic requirement of a decapod crustacean for copper (Rainbow and Luoma, 2011b). Table 3.6 shows such an estimate for copper in the caridean decapod *Pandalus montagui*, taking into account the weight of the metabolically inert exoskeleton in the calculations of enzyme requirements. The total theoretical figure estimated is 38.1 µg/g (Table 3.6). A measured mean concentration of copper in *P. montagui* collected in the

Table 3.6 **Estimates of the essential requirements for copper (µg/g dry weight) in the caridean decapod *Pandalus montagui* compared with a measured concentration.**

Pandalus montagui	
% distribution dry weight	
Exoskeleton	40.0
Blood	2.1
Soft tissues	57.9
Haemocyanin	
Blood Cu concentration	44 µg/ml
Blood volume in body	0.52 ml/g dry weight
Blood Cu concentration in body	22.9 µg/g
Enzyme requirement	
Metabolising soft tissue	26.3 µg/g
Whole body	15.2 µg/g
Total body Cu metabolic requirement	
Haemocyanin	22.9 µg/g
Enzymes	15.2 µg/g
Total	38.1 µg/g
Measured body Cu concentration	
Mean ± SD	57.4 ± 18.9 µg/g

Source: Rainbow and Luoma, 2011b.

Firth of Clyde, Scotland, is 57.4 µg/g (Nugegoda and Rainbow, 1988), indicating that such theoretical estimates may have some credibility. Copper concentrations as low as 20 µg/g in barnacles from sites with no copper contamination (Table 3.6) would not be unexpected, for barnacles lack haemocyanin and the metabolic requirements for copper would cover only the needs of copper-associated enzymes.

If there is a minimum physiological requirement for an essential trace metal such as copper or zinc in an invertebrate, it is relevant to ask whether there are any examples where invertebrates may be close to a state of metal deficiency. Arguably, some oceanic decapod crustaceans approach a situation of copper deficiency. Small specimens of the mesopelagic decapod *Systellaspis debilis* have low copper concentrations (about 30 µg/g), only reaching more typical caridean copper concentrations of about 100 µg/g in large adults (Table 3.4) (White and Rainbow, 1987; Rainbow and Abdennour, 1989; Ridout et al., 1989). Following the calculations laid out in Table 3.6, many juvenile *S. debilis* would only have sufficient body copper to meet enzymatic needs, while larger adults have enough copper for haemocyanin as well. Indeed, juvenile *S. debilis* have been shown to contain little or no haemocyanin, while adults contain a more typical caridean decapod quotient (Rainbow and Abdennour, 1989). Furthermore, juvenile *S. debilis* undergo less distinct vertical migrations than the adults (Roe, 1984), perhaps related to a lack of sufficient copper for haemocyanin and consequently decreased respiratory performance.

3.5.2 Detoxification

Prologue. Much, if not most, accumulated trace metal will be detoxified by strong binding, for example in soluble metallothioneins or in insoluble metal-rich granules of different chemical forms.

The second category of an accumulated trace metal depicted in Figure 3.6 is that of detoxified metal. Organisms have evolved physiological mechanisms to prevent a buildup of all potentially toxic metals, both essential and non-essential, in cells in metabolically available form. Therefore, trace metals are typically bound in cells to selected molecules that hold them out of harm's way – either irreversibly, or at times reversibly in the case of essential metals needed later to fulfil a metabolic role. These metals have been detoxified by binding to specific binding sites with a high affinity for that metal. Such binding sites may, for example, be on a small soluble molecule such as glutathione (Mason and Jenkins, 1995), a soluble protein such as a metallothionein (Amiard et al., 2006) or ultimately in a metal-rich granule of one of several different types (Hopkin, 1989; Mason and Jenkins, 1995; Marigomez et al., 2002).

Glutathione is a small molecule, a tripeptide consisting of three amino acids (glutamic acid–cysteine–glycine), that is ubiquitous in cells at relatively high concentrations. Through the presence of the sulphur-containing amino acid cysteine, glutathione will bind, and potentially detoxify, several trace metals, including copper and iron (Mason and Jenkins, 1995).

Metallothioneins (MT) are cellular proteins, widespread in most living organisms, that bind to particular trace metals such as Ag, Cd, Cu, Hg and Zn, their synthesis often being promoted (induced) by exposure to these metals (Amiard et al., 2006). MT are typically of relatively low molecular weight compared to most proteins and have a high content of the sulphur-containing amino acid cysteine, the sulphur atoms of several cysteines providing two multiple binding sites for the trace metals (Amiard et al., 2006). MT also show remarkable heat stability, a characteristic that facilitates their operational separation from other cellular proteins (mostly enzymes) (Wallace et al., 2003). MT play a role in the cellular detoxification of both essential and nonessential metals. MT induction, however, can vary, for example with organism, metal, exposure concentration and different MT isoforms (slightly different forms of the same protein resulting from co-occurring, slightly different genes) that might be present (Amiard et al., 2006). Like other proteins, MT are turned over in the cell, a process involving their breakdown in lysosomes (autophagic vacuoles in cells). Given such turnover, an increased rate of MT synthesis on metal exposure may be associated with an increased rate of MT turnover, but not necessarily an increased concentration of MT (Amiard et al., 2006; Luoma and Rainbow, 2008).

Phytochelatins are a group of trace metal–binding peptides originally isolated from plants (Mason and Jenkins, 1995). They are the major intracellular detoxificatory molecules binding especially Cd, but also Zn, Pb and Hg, in plants, macrophytic algae and fungi. Phytochelatins are also rich in the amino acid cysteine, for they consist of a series of about two to ten glutathione-like peptides (glutamic acid–cysteine) followed by glycine, assembled by the enzyme phytochelatin synthase (Mason and Jenkins, 1995). Originally phytochelatins were thought to be present in only plants and yeasts, but they have now also been found in cyanobacteria (blue green bacteria, once called blue-green algae), all groups of algae and indeed in animals, including nematodes, flatworms and earthworms amongst the invertebrates (Mason and Jenkins, 1995; Bundy et al., 2014).

Two well-known proteins bind iron in organisms. Transferrin transports iron between animal tissues, while ferritin is a ubiquitous intracellular iron-storage protein found in almost all living organisms, including bacteria, algae, plants and animals (Harrison and Hoare, 1980; Ford et al., 1984). Transferrin binds iron in two binding sites in the protein, strongly but reversibly, controlling the level of any free iron in the blood of animals (Harrison and Hoare, 1980). Ferritin acts both as a temporary store, preventing the buildup of metabolically available iron in cells, and as a long-term mobilisable reserve of iron. Ferritin is a complex molecule made up of 24 symmetrical protein subunits that form a near-spherical hollow shell. The central cavity of the shell is accessed through small channels between the subunits and is occupied by an iron oxide–phosphate complex (Harrison and Hoare, 1980; Ford et al., 1984). Iron oxide is typically insoluble, but the protein shell around it means that the large molecule is a soluble store of detoxified iron. Ferritin, like other proteins, will be degraded in lysosomes, the product being haemosiderin, an insoluble form of detoxified storage of iron in animals (Harrison and Hoare, 1980).

Metal-rich granules are trace metal-rich insoluble cellular inclusions that take several different forms, including phosphate granules, crystals and residual bodies of lysosomes (Hopkin, 1989; Mason and Jenkins, 1995; Marigomez et al., 2002).

Figure 3.7 A zinc pyrophosphate granule in the body of the barnacle *Lepas anatifera* from the Northeast Atlantic Ocean. Unstained, scale bar 0.5 μm. (From Walker et al., 1975; Rainbow, 1987.)

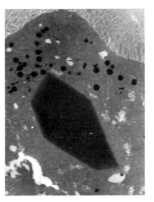

Figure 3.8 A crystal of ferritin in an R/F cell of the ventral caecum of the stegocephalid amphipod *Stegocephaloides christianiensis*. Long axis is 14 μm. (From Moore and Rainbow, 1984, with permission.)

Phosphate granules usually contain concentric layers of calcium and magnesium phosphate, with the additional presence of many other metals often including potassium, manganese, zinc, iron and lead. Phosphates occur in different forms, including pyrophosphate, which binds a catholic array of metals in very insoluble form (Hopkin, 1989; Mason and Jenkins, 1995), as in the zinc pyrophosphate granules present in large quantities in the bodies of barnacles (Figure 3.7) (Walker et al., 1975; Pullen and Rainbow, 1991). In order for a metal-rich granule to function as a metal detoxification granule, it is important that the trace metals bound therein are not released again in metabolically available form. Generally calcium phosphate–based granules offer permanent binding sites for the potentially toxic bound trace metals, which are not then available for subsequent release, Thus these granules can act as safe permanent dustbins for unwanted trace metals, whether or not the granule itself is subsequently expelled from the body or stored for the lifetime of the organism (Figure 3.6).

It is important, however, to distinguish these calcium phosphate–based detoxification granules from another type of calcium-based granule that can occur in animal tissues. These are calcium carbonate granules that are usually extracellular and may be very large (up to 1 millimetre in diameter) (Hopkin, 1989). These granules are usually very pure with no other metals present in significant amounts (Hopkin, 1989). These calcium carbonate granules represent mobilisable stores of calcium needed by an animal such as a snail with a calcium carbonate shell. To fulfil this role, such granules must be easily resolubilised, a characteristic that is incompatible with any role as a trace metal detoxification granule.

Another type of metal-rich granule is that of a pure crystal of a metal-rich compound. Such crystals are best exemplified by a metal-rich protein, usually found in soluble form – the iron-binding protein ferritin. There are examples known where ferritin may be present in a cell in such a high concentration that it forms a highly distinctive crystal. Ferritin crystals are a characteristic feature of the ventral caecum epithelium of a particular family of marine amphipods – the stegocephalids (Figure 3.8) (Moore and Rainbow, 1984) – and of cells of the radula sac of molluscs such as chitons (Towe et al., 1963). In both cases, the ferritin crystals represent extremely concentrated stores of detoxified iron. In the stegocephalids, the ferritin crystals are acting as a permanently detoxified store of iron absorbed from their diet of cnidarians such as sea pens and jellyfish, to be later expelled from the gut on completion of the epithelial cell cycle (Moore and Rainbow, 1984). In chitons, the ferritin crystals are mobilisable stores of iron needed to harden the developing teeth of the radula of these molluscs rasping algae off the surfaces of intertidal and subtidal rocks (Towe et al., 1963).

A common type of intracellular metal-rich granule contains sulphur (apparently in organic form) together with trace metals, particularly copper, but also Ag, Cd, Hg and Zn (Hopkin, 1989; Mason and Jenkins, 1995; Marigomez et al., 2002). These granules appear to be residual bodies of lysosomes in which metallothioneins binding these metals have been broken down (Hopkin, 1989). Some metals, particularly cadmium, may be released back into the cell by this lysosomal process, requiring further detoxificatory binding by newly synthesised MT. Copper, on the other hand, is more likely to stay in the lysosomal residual body, since MT associated with copper are particularly resistant to lysosomal degradation because of the high chemical stability of copper–sulphur bonds (Langston et al.,1998). Crustaceans from the highly copper-contaminated site of Dulas Bay, Anglesey, Wales, provide good examples of such sulphur-rich lysosomal residual bodies, in all cases with copper as the predominant trace metal present. Thus body cells of the barnacle *Semibalanus balanoides*, with extraordinarily high accumulated copper concentrations of 3,750 µg/g (Table 3.4), contain such copper-rich granules (Figure 3.9) (Walker, 1977a), a rare occurrence in barnacles in comparison with the universal presence of zinc pyrophosphate granules in all barnacles (Rainbow, 1987). Similarly, cells in the ventral caecum epithelium of two amphipods, *Corophium volutator* and *Orchestia gammarellus*, in Dulas Bay also contain these copper-rich, sulphur-rich lysosomal residual bodies (Icely and Nott, 1980; Nassiri et al., 2000). The high copper availability in Dulas Bay has induced the synthesis of MT binding copper with subsequent increased breakdown of this MT in lysosomes, leading to the accumulation of these metal-rich granules (residual bodies) in the ventral caecal cells. In both these latter cases, the granules will be expelled from the alimentary tract at the end of a cell cycle, while the same granules in the bodies of the barnacles will remain there permanently without access to an expulsion route (Figure 3.6).

Haemosiderin has a similar lysosomal residual body origin, but being formed from the lysosomal

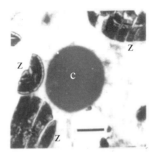

Figure 3.9 A copper-rich granule (C) and zinc pyrophosphate granules (Z) from the body of the barnacle *Semibalanus balanoides* from the highly copper-contaminated site of Dulas Bay, Anglesey, Wales. Unstained, scale bar 0.5 µm. (From Walker, 1977a; Rainbow, 1987.)

breakdown of ferritin, haemosiderin lacks sulphur or trace metals other than iron (Harrison and Hoare, 1980).

3.5.3 Accumulation, Detoxification and Toxicity

Prologue. A trace metal will exert a toxic effect when the total rate of uptake of the metal combined across all the uptake routes into an organism is greater than the combined rates of excretion (if significant) and detoxification (Luoma and Rainbow, 2008; Croteau and Luoma, 2009; Casado-Martinez et al., 2010a).

Uptake rates typically increase with the local bioavailabilities of the metals. Under most circumstances, the combined rate of uptake of a trace metal across all metal uptake routes is easily matched by the rate of excretion of the metal, the rate of metal detoxification or a combination of the two. Only in conditions of extremely high local metal bioavailabilities does toxicity ensue. Above the threshold of toxicity (Figure 1.1), the total rate of metal uptake is now greater than the combined rates of excretion and detoxification (Figure 3.10). Under these circumstances, metals that have been taken up will accumulate intracellularly in metabolically available form. These metals will bind where they are not wanted with

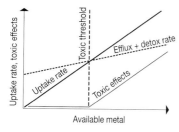

Figure 3.10 Schematic representation of how the uptake rate (combined across all routes of uptake) of a trace metal and (after a threshold) the manifestation of toxic effects will increase with the bioavailability of the trace metal to an organism. Toxic effects occur when the uptake rate exceeds the combined rates of efflux and detoxification. (From Luoma and Rainbow, 2008.)

toxic effects. Above the threshold of toxicity, toxicity is directly related to uptake rate. Since the rate of uptake of a metal increases with increased metal bioavailability, it follows then that, in the toxic range of exposures, toxicity increases with metal bioavailability (Figure 3.10).

Because toxicity results from the net outcome of the comparative rates of uptake, excretion and detoxification, organisms can show big interspecific differences in the rates of these different processes without succumbing to toxicity. Thus a species with a high total rate of metal uptake is not necessarily at greater risk of metal toxicity, so long as it has a high rate of excretion or detoxification, or combination of both. A species with a low intrinsic rate of metal uptake may be at greater risk of toxicity, if the rates of excretion and detoxification are also relatively low. The relative sensitivity of a species to toxic metal exposure will therefore be a result of the net outcome of the three processes and not simply be related to a high metal uptake rate. Different species can be characteristically sensitive or resistant to exposure to a particular trace metal, simply as a result of the interaction of their rates of uptake, excretion and detoxification. As we shall see later, many forms of risk assessment of the ecotoxicology of trace metals call on lists of metal sensitivities of different species in a habitat, in an attempt to set up guidelines to protect a given percentage (often 95%) of the species present.

3.5.4 Accumulation Patterns

Prologue. Interspecific differences in the rates of uptake, excretion and detoxification of particular trace metals result in the occurrence of a range of trace metal accumulation patterns and resulting accumulated metal concentrations, differing both between metals in a single species and between species for one metal.

Excretion rates can differ greatly both interspecifically for one metal and intraspecifically for different metals. For example, the excretion rate of a metal can be insignificant if all incoming metal is stored permanently within the body in detoxified form. On the other hand, incoming metals, notably essential metals such as zinc, can be excreted very rapidly by some species, particularly vertebrates and certain invertebrates. Metal excretion may be in soluble form, for example from gills, and/or in insoluble detoxified form, for example from the alimentary tract or the kidney. Thus amphipod crustaceans excrete copper-rich, sulphur-rich granules from the ventral caeca of the alimentary tract (Icely and Nott, 1980; Nassiri et al., 2000), and mussels excrete zinc-rich granules from the kidney (George and Pirie, 1980). Excretion of a trace metal may be predominantly by one route or by different routes according to the route of uptake. For example, a metal taken up from solution may be translocated to a particular organ for temporary storage to be followed by excretion, perhaps in insoluble form, or may be lost from the gill with or without intermediate translocation to another organ. Similarly, a metal taken up in the alimentary tract may remain in the gut epithelial cell into which it was assimilated to be later lost from that gut cell, or it may be translocated elsewhere in the body to be excreted from another organ such as the gill or kidney.

The accumulation of zinc by different crustaceans (Box 3.2, Table 3.4) is a good example of the possible variation between even related species of the accumulation patterns shown for a single trace metal (Rainbow and White, 1989; Luoma and Rainbow, 2008).

Box 3.2 Accumulation of Zinc by Crustaceans

The accumulation pattern of zinc by the caridean decapod *Palaemon elegans* is a classic example of the regulation of the body concentration of a trace metal by an invertebrate (Rainbow, 2002; Luoma and Rainbow, 2008). When exposed to increasing dissolved zinc concentrations, the decapod maintains a constant body concentration of zinc, until, beyond a threshold dissolved bioavailability, the body zinc concentration accumulated over the period of exposure increases progressively with the zinc exposure concentration until death (Rainbow and White, 1989; Rainbow and Luoma, 2011c). There is significant uptake of zinc from solution by *P. elegans* at all dissolved zinc exposures examined, with the zinc uptake rate increasing in proportion to exposed zinc concentration (Rainbow and White, 1989). Thus the decapod is regulating the body concentration of zinc by balancing the uptake rate of new zinc with an increased rate of zinc excretion, until a threshold dissolved bioavailability of zinc is reached. At this point of regulation breakdown, *P. elegans* can no longer match the high rate of zinc uptake with the rate of zinc excretion, and net accumulation of body zinc begins. At a high enough dissolved zinc exposure, the difference between the rates of zinc uptake and excretion is such that a threshold concentration of metabolically available zinc is exceeded, and toxicity follows (Rainbow and Luoma, 2011c).

The regulated body concentration of zinc in *P. elegans* is about 80 µg/g Zn (Rainbow and Luoma, 2011c). Significant net accumulation of zinc begins at a dissolved zinc exposure above 100 µg/L. At 100 µg/L Zn, about 15% of total body zinc (11.6 µg/g Zn per day) is turned over daily by *P. elegans*, highlighting the importance of excretion in this zinc accumulation pattern (Rainbow and Luoma, 2011c). After regulation breakdown, the maximum total body zinc concentration in the decapod is only about 180 µg/g (Rainbow and Luoma, 2011c), in strong contrast to the much higher body zinc concentrations achieved in many other crustaceans (Table 3.4). Since *P. elegans* suffers lethal zinc toxicity when its body concentration is only 180 µg/g Zn, it is highly unlikely that this decapod uses irreversible detoxification of excess zinc. The decapod is arguably regulating its body concentration of zinc to meet its metabolic needs. Furthermore, some zinc may be stored in reversibly detoxified form as a ready supply of this essential metal, and *P. elegans* does contain metallothionein (Rainbow and Luoma, 2011c). Consideration of theoretical estimates of metabolic needs and inspection of minimum zinc concentrations measured in crustaceans suggest that 50 µg/g Zn might be needed to meet the essential requirements for zinc by crustaceans. The regulated body concentration of 80 µg/g Zn in *P. elegans* can therefore be interpreted to consist of 50 µg/g Zn in metabolically available form to meet metabolic requirements, with a further 30 µg/g Zn in a reversibly detoxified store (Rainbow and Luoma, 2011c).

The same numbers can be used to estimate the maximum concentration of zinc in metabolically available form that can be tolerated by *P. elegans* and by extrapolation other crustaceans. The maximum body concentration of zinc tolerated by *P. elegans* is 180 µg/g. In the absence of irreversible permanent detoxification, most of this accumulated metal will be in metabolically available form, except for the contribution of 30 µg/g Zn reversibly bound to metallothionein. Thus a value of 150 µg/g Zn is an estimate for the maximum tolerable concentration of metal in metabolically available form before lethal toxic effects follow for *P. elegans* (Rainbow and Luoma, 2011c). This concentration is the critical concentration of metabolically available metal and can be extrapolated to other crustaceans.

Therefore, barnacles and amphipods are also considered to contain 50 µg/g Zn in metabolically available form to meet metabolic requirements and potentially a further 30 µg/g Zn in reversibly detoxified form bound to metallothionein. Excess accumulated zinc beyond this 80 µg/g would be irreversibly detoxified and stored (whether permanently or temporarily before excretion in detoxified form) where the physiological mechanisms exist (Figure 3.6). There is no theoretical limit to the concentration of metal that can be stored in irreversibly detoxified form, but species will differ in the extent (and rate) to which they carry out permanent detoxification.

In barnacles, all zinc taken up from solution is accumulated without significant excretion (White and Rainbow, 1989; Rainbow, 2002). Any excretion of zinc taken up from the diet is also extremely limited, such zinc having a half-life of 1,346 days in *Austrominius modestus* (Rainbow and Wang, 2001). Correspondingly, accumulated body concentrations of zinc in barnacles can reach extremely high values (e.g., 50,000 µg/g or more; Table 3.4). The vast majority of this accumulated Zn is necessarily in detoxified form, bound, in this case, in zinc pyrophosphate granules (Figure 3.7) (Rainbow, 1987). Thus the zinc accumulation patterns in the caridean decapod and in barnacles could not be more different. The zinc accumulation pattern of barnacles consists of strong accumulation with accumulated zinc permanently stored in detoxified form without excretion, resulting in some of the highest accumulated concentrations of any trace metal in any animal tissue (Eisler, 1981; Rainbow, 1987). In contrast to the decapod example, it is a minute percentage of the total accumulated zinc concentration in barnacles that is metabolically available (50 µg/g) or in temporary reversibly detoxified form bound to metallothionein (perhaps up to 30 µg/g), a combined sum of 80 µg/g in a total Zn concentration ranging from about 2,000 µg/g to in excess of 100,000 µg/g (Table 3.4).

The zinc accumulation pattern of amphipod crustaceans is also that of net accumulation (Rainbow and White, 1989), with detoxified storage, but in this case with excretion of the detoxified metal-rich granules holding much of the accumulated zinc (Figure 3.11). Zinc taken up from both solution and the diet accumulates in lysosomes in cells of the ventral caeca, to be excreted with the faeces on completion of the cell cycle of the ventral caeca epithelial cells (Figure 3.11) (Nassiri et al., 2000; Rainbow, 2002). This is not a process of regulation, for the body concentration of zinc in the amphipods reaches a new steady-state level as zinc bioavailabilities change. The zinc bioavailability is reflected in the number of zinc-rich granules in the ventral caeca at any one time, and hence the zinc concentration of both the ventral caeca and the whole body to which the ventral caeca make a major contribution (Figure 3.11). As the body zinc concentration increases, the proportion of accumulated zinc in the detoxified component increases, from an estimated starting point at which most zinc in an amphipod from an uncontaminated site will probably be in metabolically available form (50 µg/g Zn) with potentially some bound reversibly to metallothionein (perhaps up to 30 µg/g Zn), to give a combined total of the order of 50 to 80 µg/g Zn (Rainbow and Luoma, 2011c). Body Zn concentrations in amphipods above this estimated range (Table 3.4) will include accumulated zinc permanently detoxified in the ventral caecal cells progressing through their cell cycle, which takes about 30 days or more (Luoma and Rainbow, 2008). The final excretion of this detoxified zinc means that the accumulated zinc concentrations of amphipods will not approach those of barnacles, although they will rise, albeit over a smaller range, in response to raised zinc bioavailabilities (Table 3.4).

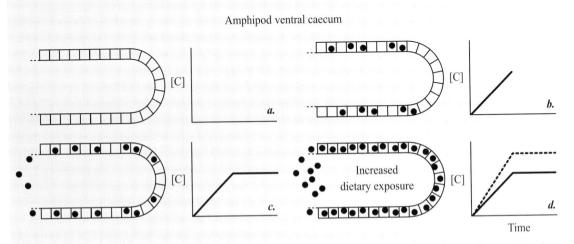

Figure 3.11 (a, b, c) The accumulation of detoxified metal (e.g., Zn or Cu) in the ventral caecal cells of an amphipod crustacean exposed to a trace metal in the diet, and the corresponding changes in total body metal concentration [C] over time. (d) The effect of increased dietary exposure to the metal. (From Luoma and Rainbow, 2008.)

The same crustaceans can be used to exemplify differences between the accumulation patterns of different metals within single species, in this case copper in comparison with zinc.

The decapod *Palaemon elegans* appears to regulate the body concentration of copper (ca. 110 to 130 μg/g Cu) over a wide range of dissolved copper exposures, as in the case of zinc (Rainbow and White, 1989). After regulation breakdown at high dissolved copper bioavailabilities, when the rate of copper uptake exceeds the rate of copper excretion, the accumulated copper concentration in *P. elegans* can, however, reach higher levels (ca. 600 μg/g) than in the case of zinc. Under these circumstances, cells in the hepatopancreas of the decapod contain copper-rich, sulphur-rich residual bodies from the lysosomal breakdown of metallothionein binding copper (Luoma and Rainbow, 2008). These copper-rich granules will be excreted via the gut as the hepatopancreas epithelial cells complete their cell cycle. The accumulation of copper by *P. elegans* therefore shows two excretory routes: excretion from the metabolically available component during copper regulation, and additional excretion from an insoluble detoxified store during the net accumulation phase that follows regulation breakdown at high copper bioavailabilities. Thus, in contrast to the situation for zinc, the decapod will use permanent detoxification to cope with the excess of metal being taken up at the very high metal bioavailabilities associated with regulation breakdown. As for zinc, however, eventually the rate of copper uptake is so great that it exceeds the combined rate of copper excretion (by whatever routes) and copper detoxification, and toxicity ensues.

The regulated body concentration of copper (ca. 110 to 130 μg/g) in *P. elegans* is more than sufficient to meet estimated metabolic requirements for copper in both enzymes and haemocyanin in a caridean decapod (38 μg/g; Table 3.6), the remainder presumably being bound reversibly to the likes of metallothionein.

Copper concentrations in the bodies of barnacles from uncontaminated sites are generally low, even down to 20 μg/g Cu (Table 3.4), but still meeting theoretical estimates of copper-enzyme requirements (15 μg/g Cu, Table 3.6). Barnacles lack haemocyanin (Table 3.6). Barnacles do, however, have the potential to accumulate very high concentrations of copper when exposed to copper contamination in the field (Table 3.4), and show strong net accumulation of copper from solution at all exposures in the laboratory with no suggestion of regulation (Rainbow and White, 1989). Almost all incoming copper is accumulated in detoxified form, probably bound initially with soluble metallothionein (Rainbow, 2002). Barnacles from very copper-contaminated sites such as Dulas Bay (Table 3.4) have many copper-rich, sulphur-rich granules in their body tissues (Figure 3.9) (Walker, 1977a), which are residual bodies resulting from the lysosomal breakdown of metallothioneins binding copper.

The copper accumulation pattern of amphipods is again intermediate between the apparent regulation of the body copper concentration by caridean decapods and the strong net copper accumulation of barnacles (Luoma and Rainbow, 2008). Amphipods do not regulate body copper concentrations, and amphipods exposed to a range of dissolved copper exposures show net accumulation at all exposures (Rainbow and White, 1989). Like zinc, copper is accumulated in the cells of the ventral caeca (Figure 3.11), this time in the form of copper-rich, sulphur-rich granules (Nassiri et al., 2000), again residual bodies from the lysosomal breakdown of copper-binding metallothionein. Copper detoxified in these granules is excreted from the gut on completion of the cell cycle of the ventral caeca epithelial cells (Figure 3.11). As for zinc, this is not a process of regulation of the body copper concentration which reaches a new steady-state level as copper bioavailabilities change (Figure 3.11). The bioavailability of copper is reflected in the number of copper-rich granules, and hence the copper concentration, in the ventral caeca. The proportion of accumulated copper in the detoxified component increases as the body copper concentration increases, from a starting point at which most copper in an amphipod from an uncontaminated site will probably be in metabolically

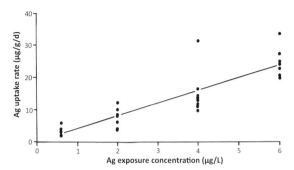

Figure 3.12 *Scrobicularia plana*. Relationship between uptake rates of radiolabelled Ag of individual clams from solution and dissolved radiolabelled metal concentration. The slope of the best-fit line is the uptake rate constant (µg/g/d per µg/L or L/g/d). (After Kalman et al., 2014, with permission.)

available form (Tables 3.4 and 3.6) Amphipods have haemocyanin and their metabolic requirements for copper might therefore usually be of the order of 38 µg/g Cu, as calculated for a caridean decapod (Table 3.6).

In the case of a third trace metal, all three crustaceans accumulate the nonessential metal cadmium. Indeed, no crustacean regulates the body concentrations of non-essential trace metals such as cadmium or lead (Luoma and Rainbow, 2008). Thus, in caridean decapods (*Palaemon elegans*), amphipods (*Echinogammarus pirloti*) and barnacles (*Austrominius modestus*), all cadmium taken up from solution is accumulated with no significant excretion over at least a 28-day period (Rainbow and White, 1989). The accumulated cadmium is necessarily detoxified, typically bound to soluble metallothionein. The cadmium-binding metallothionein will be broken down in lysosomes, but, unlike copper, cadmium is rarely visualised in lysosome residual bodies (Langston et al., 1998).

3.5.5 Biodynamic Modelling

Prologue. Biodynamic modelling uses measurements of the uptake rates of a metal from solution and the diet, the rates of loss of metal from the body after uptake from either route and, where appropriate, the

rate of growth to calculate the predicted accumulated concentration of the trace metal in the body of an aquatic organism in a particular habitat.

A new approach has recently been applied to studies of the bioaccumulation of trace metals by (particularly aquatic) organisms – that of biodynamic modelling (Wang et al., 1996; Luoma and Rainbow, 2005). Because metal uptake rates depend on site-specific metal concentrations in water and food and on local geochemical conditions, it is possible to use biodynamic modelling to predict accumulated concentrations of metals in different species at different locations, with much success (Luoma and Rainbow, 2005, 2008). Furthermore, biodynamic modelling allows the calculation of the relative importance of solution and food as metal sources to an aquatic animal.

The physiological parameters controlling the uptake and loss of a metal by an organism during bioaccumulation are measured in the laboratory, typically using radiolabelled metal to distinguish newly accumulated metal from accumulated metal already in the body before the start of any experiments. Over an environmentally realistic range of dissolved trace metal concentrations, the rate of metal uptake from solution by an aquatic invertebrate is directly proportional to the dissolved concentration (e.g., Figure 3.12). The uptake rate constant is the slope of the line describing this relationship (Figure 3.12). This uptake rate constant for a particular trace metal is measured in the laboratory for a particular species under a given set of physicochemical variables of a medium (for example, different salinities which might affect the availability of the free metal ion). It is then possible to calculate the uptake rate from solution at any dissolved concentration of the trace metal. It is also possible to use the same experiments to follow the excretion of the metal from the body after uptake from solution (e.g., Figure 3.13). Such efflux of metal newly taken up is usually in two phases, the second slower phase representing the excretion (if any) of the metal from solution that has been taken up and accumulated in the body (Figure 3.13). From such a graph can be

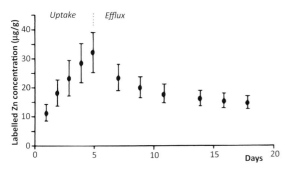

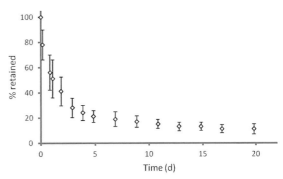

Figure 3.13 *Scrobicularia plana*. A plot against time of the accumulated labelled Zn concentration (mean ± SD, $n = 7$) in clams exposed individually to 25 µg/L radiolabelled Zn for 5 days (uptake phase), followed by 13 days of depuration (efflux phase) in a medium without added dissolved metal. The efflux phase can be divided into an initial fast phase and a second slower phase. The latter represents the excretion of the metal that has been taken up from solution and accumulated in the body. The efflux rate constant can be calculated as the proportion per day of this accumulated metal that is excreted in the second slower phase. (From Kalman et al., 2014, with permission.)

Figure 3.14 *Scrobicularia plana*. Percentage of radiolabelled arsenic retained in the deposit-feeding clams after uptake from ingested radiolabelled sediment (mean ± SD). The first rapid phase of metal loss represents the defaecation of sediment containing radiolabelled metal that has not been assimilated. The second slower phase depicts excretion from the clams of metal that has been assimilated in the gut and accumulated in the soft tissues. Back extrapolation of a line through the points in the second part of the graph to intercept the *y*-axis gives the assimilation efficiency of As from sediment, in this case 25%. The same line can also be used to calculate the efflux rate constant of metal accumulated from the diet (the proportion per day of this accumulated metal that is excreted in the second slower phase of efflux). (From Kalman et al., 2014, with permission.)

calculated the efflux rate constant of this accumulated metal, the proportion that is excreted per day during the second slower phase. With a knowledge of the uptake rate of the trace metal from solution and the efflux rate constant after uptake from solution, it is then possible to calculate how much metal has been accumulated by an organism from solution under steady-state conditions (Luoma and Rainbow, 2005).

In the case of an aquatic animal, we are also interested in how much metal is accumulated from the diet. Similarly, it is possible to calculate this from measurements made in the laboratory, extrapolated to the field with different concentrations of the metal in local food material (Luoma and Rainbow, 2005). The key parameter to be measured in the laboratory is the assimilation efficiency, the efficiency with which an animal extracts metal from its food and absorbs it into its tissues, expressed as a percentage of the total metal ingested. Figure 3.14 depicts the percentage of radiolabelled arsenic retained in the deposit-feeding bivalve *Scrobicularia plana* after uptake from ingested radiolabelled sediment (Kalman et al., 2014). The first rapid phase of metal loss represents the

defaecation of sediment containing radiolabelled metal that has not been assimilated. The second slower phase depicts excretion from the bivalves of metal that has been assimilated in the gut and accumulated in the soft tissues. The assimilation efficiency (25%) in *S. plana* of arsenic in the sediment can be obtained from the graph, as can the efflux rate constant of metal accumulated from the diet (the proportion per day of this accumulated metal that is excreted in the second slower phase of efflux) (Figure 3.14) (Kalman et al., 2014). With a knowledge of the ingestion rate of the animal, the assimilation efficiency, the concentration of metal in the food item and the efflux rate constant of metal accumulated from the diet, it is then possible to calculate how much metal has been accumulated by an organism from food under steady-state conditions.

Knowing the separate rates of net accumulation of metals taken up from water and food, the total metal

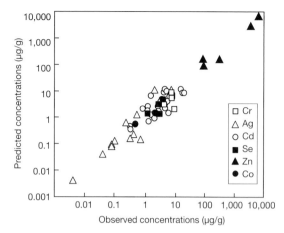

Figure 3.15 Comparison of trace metal bioaccumulation predicted by biodynamic modelling in 15 different studies to independent observations of bioaccumulation in the species of interest from the habitat for which the physiological parameters were developed (Luoma and Rainbow, 2005). The data covered 6 metals, 13 species of animals, 3 phyla and 11 marine, estuarine and freshwater ecosystems. The agreement between predictions and independent observations from nature was strong (From Luoma and Rainbow, 2008.)

concentration accumulated by an organism can be calculated under any environmental circumstances. Growth rate constants can allow for the effect of growth dilution if necessary.

It is important to know how well a theoretical biodynamic model using physiological parameters of bioaccumulation measured in the laboratory can actually predict metal concentrations bioaccumulated by organisms in the field. To this end, a comparison was made of accumulated concentrations of trace metals predicted by biodynamic modelling in 15 different studies against independently measured bioaccumulated concentrations in these species from habitats with known environmental metal concentrations (Luoma and Rainbow, 2005). The data covered six trace metals, thirteen animal species and eleven marine, estuarine and freshwater habitats. Figure 3.15 shows the comparison. The agreement between predicted and independently measured bioaccumulated concentrations was very good, showing that the assumptions underlying biodynamic modelling are valid. Table 3.7 similarly compares measured metal concentrations in coastal

invertebrates against those predicted by biodynamic modelling, further confirming the applicability of biodynamic modelling in studies of trace metal bioaccumulation. Biodynamic modelling, together with an understanding of the trace metal accumulation patterns of different organisms, therefore, allows the prediction and explanation of widely differing accumulated trace metal concentrations in aquatic invertebrates, combining geochemical measurements of environmental metal concentrations with the estimation of key physiological parameters of metal bioaccumulation in a particular species.

3.5.6 Food Chains and Trace Metals

Prologue. Trace metals are taken up by animals from the diet and are accumulated. It follows, therefore, that metals can be transferred along food chains. As a general principle, trace metals are not biomagnified along food chains.

Two major factors control the transfer of trace metals along food chains: the trophic bioavailabilities of the metals accumulated by the food species to the feeding animal at any stage (trophic level) in the food chain, and the accumulation pattern for each metal shown by each animal in the food chain whatever its trophic level.

Not all metal accumulated by a prey species will be trophically bioavailable to the feeding animal, particularly if much of the accumulated metal in the food item is strongly bound in MRG, such as pyrophosphate granules. For example, phosphate-based MRG binding zinc or manganese in the kidney of the queen scallop *Aequipecten opercularis* or the digestive gland of the periwinkle *Littorina littorea* were present undigested in the faeces of the carnivorous netted dogwhelk *Tritia reticulata* fed on these tissues (Nott and Nicolaidou, 1990). Similarly, zinc phosphate MRG in the body tissue of the barnacle *Semibalanus balanoides* passed through the gut of another predatory gastropod, the dogwhelk *Nucella lapillus*, feeding on the barnacles (Nott and Nicolaidou, 1990). MRG in the form of lysosomal residual bodies containing sulphur and copper in the digestive gland of the gastropod *Murex*

Table 3.7 **Comparisons between observed mean metal concentrations (µg/g dry weight, range or standard deviation [SD]) accumulated by aquatic invertebrates in particular field situations and those predicted by biodynamic modelling.**

Species	Location	Metal	Predicted	Range	Observed	Range/SD
Arenicola marina	Thames Estuary	Ag	0.24	0.12–1.19	0.59	0.15–0.94
(lugworm)	Thames Estuary	As	14.0	8.8–41.5	17.1	14.9–20.4
	Thames Estuary	Cd	1.22	0.05–3.47	3.20	1.61–5.17
	Dulas Bay, Wales	Zn	89.1	14.1–218	89.6	68.4–141
Austrominius modestus	Southend, Thames	Cd	9	7.6–10.6	23.3	6.6
(barnacle)	Southend, Thames	Zn	2950	1500–4400	3463	1160
Hediste diversicolor	Blackwater Estuary	Ag	0.96	0.54–1.59	1.38	0.94–1.82
(ragworm)	Blackwater Estuary	As	9.6	–	9.1	1.0
	Tavy Estuary	Cd	4.87	4.85–4.89	2.62	1.87–3.37
	Blackwater Estuary	Zn	297	292–301	237	211–263
Scrobicularia plana	Tavy Estuary	Ag	0.57	–	0.49	0.10–0.99
(bivalve)	Gannel Estuary	As	55.0	–	50.3	30.1–62.7
	Gannel Estuary	Zn	1910	–	1510	734–2550

Sources: Rainbow and Wang, 2001; Casado-Martinez et al., 2009a, 2010b; Rainbow et al., 2009b, 2011b; Kalman et al., 2014.

trunculus were partially digested by the hermit crab *Clibanarius erythropus*, releasing magnesium and chlorine but retaining much of the copper and sulphur (Nott and Nicolaidou, 1994).

It is important to dispel the myth that, as a general principle, trace metals are biomagnified along food chains, with successively higher accumulated metal body concentrations at each ascending trophic level. They are not. Biomagnification is indeed a general characteristic of the transfer of organic contaminants such as organochlorines up food chains, as the contaminants increasingly concentrate in the lipid components of animals at higher trophic levels of food chains. This is not the general case for most trace metals. The accumulated concentrations of a metal in each animal in a food chain depend on the accumulation pattern of a particular metal in a particular species, irrespective of its trophic level. Given the vast number of species making up food chains, the large number of trace metals and the variation in degree of storage accumulation in different accumulation patterns that might be adopted by different animals, it is not surprising that there are examples of food chains that do show an increase in accumulated concentrations in animals at higher trophic levels. There are, however, as many or more that show a successive diminution of accumulated concentrations, or indeed ups and downs, or downs and ups, from bottom to top of a food chain (Wang, 2002). For example, the classic marine planktonic food chain of phytoplankton to copepods to fish shows reduced accumulated concentrations of trace metals such as cadmium, silver and zinc at higher trophic levels, largely as a result of the effective efflux of the metals by copepods and their very low assimilation by marine fish (Wang, 2002).

While trace metals do not follow a general principle of biomagnification up food chains, there are specific examples where biomagnification is the rule. Organometals, such as methyl mercury, behave in this context mostly as organic compounds and do show biomagnification. The trace element selenium, often called a metalloid because its chemical properties fall somewhat between those of a metal and a nonmetal, will also show biomagnification up food chains (Luoma and Rainbow, 2008).

While methyl mercury will have chemical characteristics associated with its organic component, inorganic mercury will behave much more typically as a trace metal with chemical similarities to cadmium and silver in its uptake routes, its detoxification by binding with metallothionein and its lack of biomagnification up food chains. In many aquatic situations, however, methyl mercury is the dominant form of mercury present, and so the physiology of the total mercury content of an aquatic organism will reflect the organic nature of its predominant chemical form. High concentrations of methyl mercury, therefore, can accumulate in fish at the top of food chains, particularly if these fish are long-lived, methyl mercury being excreted from fish much more slowly than inorganic mercury (Luoma and Rainbow, 2008). Biomagnification of methyl mercury is further enhanced when the top predatory fish in the food chain are themselves feeding on predatory fish at the trophic level below, as in the case of pike *Esox lucius* in lakes or tuna in the oceans. Methyl mercury can also be accumulated to high concentrations in piscivorous seabirds and

marine mammals occupying top spot in a marine food chain.

The commonest form of dissolved selenium in natural waters is selenate, rather than its rarer more bioavailable form selenite, and uptake of selenium from solution is so slow that it cannot explain the bioaccumulation of selenium in aquatic animals (Luoma and Rainbow, 2008). Selenium uptake from food, however, is much more efficient, and it is diet that supplies most of the selenium accumulated by animals (Luoma and Rainbow, 2008). Thus selenium is transferred efficiently from one trophic level to the next, and biomagnification does occur.

3.5.7 Significance of Bioaccumulated Trace Metal Concentrations

Prologue. In the usual presence of significant detoxification, there is no relationship between the total accumulated trace metal concentration in an organism and its toxicity to that organism. There will, however, be a threshold concentration of metabolically available accumulated metal that will initiate toxic effects.

A myth that needs dispelling is any general principle that total body trace metal concentrations in organisms in the field can reach a specifiable threshold concentration at which a species experiences toxic effects, including ultimately lethality. Such a principle can hold for organic contaminants, but it should not be extended to trace metals, given the ability of most organisms to accumulate trace metals in a detoxified form. Thus, as a general rule, there is no lethal total body concentration of an accumulated metal that can be recognised as a limiting concentration in field-collected specimens.

The explanation lies in the widespread occurrence of storage detoxification of accumulated trace metals, whether permanent or temporary. For example, as previously discussed, the bodies of barnacles can contain very variable amounts of accumulated zinc, from about 2,000 µg/g to in excess of 100,000 µg/g Zn (Table 3.4), with the metabolically available component representing about 50 µg/g of this total

(Rainbow and Luoma, 2011c). Like any other organism, a barnacle will suffer toxic effects when the local bioavailability of zinc increases so that the total rate of zinc uptake exceeds the rate of detoxification, excretion being effectively insignificant in barnacles. New zinc entering the body then accumulates in the metabolically available fraction, and the toxic threshold concentration of metabolically available body zinc is reached. This point can be reached whatever the concentration of detoxified zinc stored in zinc pyrophosphate granules, and, therefore, whatever the total zinc concentration of the body. Thus, toxicity occurs independently of how much zinc has been detoxified over the previous lifetime of the barnacle, and, therefore, regardless of the total body zinc concentration (Rainbow and Luoma, 2011c).

The same point holds for organisms with metal accumulation patterns that involve temporary storage of detoxified metal with subsequent excretion, as in the case of zinc and copper in amphipods (Figure 3.11). The greater the detoxification of accumulated metal, the more impossible it will be to define any lethal body concentration applicable to field-collected specimens, where bioavailabilities are variable. The talitrid amphipod crustacean *Talitrus saltator* collected in the field typically contain body concentrations of copper in the range 30 to 40 µg/g, but these amphipods in copper-contaminated habitats may have accumulated concentrations of 200 µg/g Cu or more (Table 3.4). If amphipods of different starting body concentrations of copper were to be exposed to very high dissolved copper bioavailabilities in the laboratory, for example in toxicity testing experiments, then so much copper would enter so rapidly that almost all the copper would accumulate in the metabolically available fraction without significant detoxification. This lethal load in the metabolically available component would be the same for all the amphipods exposed to the high copper, but the total body concentration at death would differ because of the different starting concentrations of detoxified, and therefore total, accumulated copper present in the body, a difference of about 160 µg/g Cu (Table 3.4). Thus, there is no reproducible lethal total body

concentration of copper that can be applied to amphipods collected in the field (Luoma and Rainbow, 2008; Rainbow and Luoma, 2011b).

For a regulator of trace metal body content such as the caridean decapod *Palaemon elegans* in the case of zinc, the lethal body concentration concept is more valid (ca. 180 µg/g Zn, as discussed previously). This is because most, if not all, of the body zinc appears to remain in metabolically available form (up to 150 µg/g Zn at high zinc bioavailabilities), with no permanent storage detoxification. These figures were derived from laboratory experiments, but, in practice, the detection of a lethal body concentration in such a species in a field study is highly unlikely (Rainbow and Luoma, 2011b). Under conditions of very high trace metal bioavailability, once regulation has broken down when the uptake rate exceeds the excretion rate, the metabolically available (and therefore body) metal concentration will increase from the typical regulated concentration over a very short time period.

But are there any exceptions at all to the general principle that there is no lethal body concentration of a metal applicable to an organism in the field? As stated previously, if there is no storage detoxification of accumulated metal, there is then theoretically a lethal body concentration that can be applicable in the field. The same principle would apply if any detoxified component is a fixed percentage of total accumulated metal content (Rainbow and Luoma, 2011b).

The answer to the question posed may well be 'Yes' for selenium and certain organometal compounds, already discussed as not behaving as typical trace metals when it comes to their biology. Total accumulated selenium concentrations can be correlated with toxic effects, at least in fish and birds (Skorupa, 1998; Janz et al., 2010), and there is no known accumulated detoxified store of selenium, analogous to the insoluble detoxification of other trace metals (Luoma and Rainbow, 2008). There is net bioaccumulation of selenium, however, and some species are somewhat more resistant to an elevated body burden of selenium than others. Thus selenium concentrations in eggs of birds can be correlated with toxic effects, but with interspecific differences. For example, in the case of nonviability of chicks, the EC50 (the concentration at which 50% of the population shows the toxic effect) in eggs varies from 58 µg/g Se in stilt *Himantopus mexicanus* to 105 µg/g Se in avocet *Recurvirostra americana* (Skorupa et al., 1998).

Bioaccumulated concentrations of tributyl tin (TBT) can also be correlated with observed toxic effects (Meador, 2000). TBT can be metabolised to less toxic forms and there is no bioaccumulation in detoxified form. The behaviour of TBT in this context is somewhere between behaviour as a traditionally accumulated trace metal and behaviour as an organic compound (Luoma and Rainbow, 2008).

While there is generally no valid lethal total body concentration applicable to a particular species that can be recognised as a toxic threshold body concentration in field-collected specimens, the concept has the potential to be adapted to define a lethal body concentration of metabolically available accumulated metal. Such an approach is more biologically valid and has been attempted for zinc in crustaceans (Rainbow and Luoma, 2011c). The value of 150 µg/g Zn is the estimate made for the maximum tolerable concentration of metal in metabolically available form before lethal toxic effects follow for the caridean decapod *Palaemon elegans*. This concentration can be defined as the critical concentration of metabolically available metal and was extrapolated previously to other crustaceans. It is all very well defining this critical concentration of metabolically available metal, but can the concentration of accumulated metal in metabolically available form actually be measured in an organism? One approach that is being increasingly adopted is to use subcellular compartmentalisation of accumulated metal in an attempt to measure this metabolically available fraction (Wang and Rainbow, 2006). This approach assumes that the metabolically available fraction of accumulated metal does correspond to a combination of different subcellular compartments of accumulated metal. The detoxified accumulated metal fraction is arguably easier to identify as the metal in MRG or bound to metallothionein (Wallace and Luoma, 2003; Wallace et al., 2003), although, even in this case, other

forms of detoxificatory binding may be missed. It is debateable, however, whether and which of the remaining non-detoxified subcellular fractions can be identified as metal-sensitive, as has been proposed for metals bound to other proteins and organelles (Wallace et al., 2003), and whether they do represent the metabolically available fraction (Rainbow et al., 2015). Against a background of physiological processes handling metals at different rates, it may ultimately prove impossible to equate a subcellular component, operationally separated by techniques such as centrifugation, with the conceptually important metabolically available component of accumulated metal.

Although, with the exception of selenium and the likes of TBT, it is not possible to define lethal total accumulated body concentrations of particular trace metals for field-collected specimens, it is certainly possible to recognise abnormally high bioaccumulated concentrations of particular metals in different invertebrates. Table 3.4 has already highlighted the very different ranges of naturally accumulated zinc, copper and cadmium concentrations in different taxa of crustaceans, showing how an accumulated metal concentration that may be typical of one taxon would be very high or very low for another taxon, as a result of the different accumulation patterns adopted. Table 3.8 confirms this conclusion for another group of closely related animals, in this case example species of different families of coastal bivalve molluscs. A zinc concentration of 300 µg/g would be high for a mussel, but low for an oyster or a semelid bivalve (Table 3.8). A copper concentration of 250 µg/g would be high for both a mussel and a semelid bivalve, but routine for an oyster (Table 3.8). In the case of silver, high concentrations reached in oysters and semelids are unheard of in mussels (Table 3.8). Thus what is a high or a low accumulated concentration of a trace metal can only be decided with reference to the species (and arguably the taxonomic family) of the organisms concerned.

Furthermore, an understanding of what is a definable high accumulated metal concentration in a particular species allows us to conclude that a specimen with such an atypically high accumulated

Table 3.8 **Accumulated concentrations (µg/g dry weight) of trace metals, categorised as typical or high, in the soft tissues of bivalve molluscs from different families.**

Family	Ag		Cu		Zn	
Species	Typical	High	Typical	High	Typical	High
Mytilidae (mussels)						
Mytilus edulis	0.01–1	3–20	4–20	50–270	50–200	300–600
Ostreidae (oysters)						
Crassostrea gigas	1–3	10–70	30–500	1,000–5,000	1,500–4,500	6,000–14,000
Semelidae						
Scrobicularia plana	0.1–5	10–300	9–60	120–360	300–1,000	2,000–5,000

Sources: Bryan et al., 1980, 1985; RNO, 2006; Luoma and Rainbow, 2008; Rainbow and Luoma, 2011b.

concentration has been exposed to an atypically high local bioavailability of the metal concerned. Thus, a knowledge of the significance of different accumulated metal concentrations is invaluable, for such concentrations provide information on high and potentially ecotoxicologically significant trace metal bioavailabilities in different habitats (Luoma and Rainbow, 2008). This knowledge can, therefore, be used in biomonitoring programmes to measure the relative bioavailabilities of different metals at different sites in a habitat, or at one location at different times – in short, to monitor trace metal pollution.

3.6 Biomonitors

Prologue. Any organism that is a net accumulator of a trace metal has the potential to act as a biomonitor. High accumulated metal concentrations in biomonitors indicate the presence of atypically high bioavailabilities of those metals.

A biomonitor is an organism which accumulates trace metals in its tissues, the accumulated metal

concentration of which provides a relative measure of the total amount of metal taken up by all routes by that organism, integrated over a preceding time period (Luoma and Rainbow, 2008). Thus, the bioaccumulated concentrations are integrated measures of the uptake and accumulation of metals from all sources to that organism, solution and diet in the case of aquatic invertebrates. The use of biomonitors can, therefore, identify localities of high or low metal bioavailability to that chosen biomonitor, and identify changes of metal bioavailability over space and/or time. Such information is a prerequisite to the efficiency of any subsequent search for ecotoxicological effects in metal-contaminated habitats.

It is important to note that the use of the term 'biomonitoring' is more restricted here than might be the case for some authors. For example, the term 'biomonitoring' has been used synonymously with the term 'biological monitoring', to mean the wider use of biological systems to assess the structural and functional integrity of aquatic and terrestrial ecosystems (Clements and Newman, 2002). Here, the term 'biological monitoring' is used to cover the wider context of the use of any biological system in ecosystem assessment, which would include biomonitoring as defined here.

Strong accumulators typically show a wider range of accumulated trace metal concentrations than weak accumulators over a metal bioavailability range. Hence they offer greater discriminatory ability between sites (Fialkowski and Rainbow, 2006) or between sampling occasions at the same site. Nevertheless, weak accumulators can still be used as biomonitors, often introducing the advantage of reflecting a shorter recent period of metal exposure in the accumulated metal concentration at the time of sampling. Often conclusions on local metal bioavailabilities are drawn from data for single biomonitors. Strictly those results only apply to the integrated metal sources available to that biomonitor. A biomonitoring programme should, therefore, employ a suite of biomonitors to cover a range of potential metal sources in a habitat. If a suite of biomonitors is used, comprising a selected group of biomonitors that take up and accumulate metals from a variety of sources (for example, solution, suspended detritus and plankton, deposited sediment, etc., in an aquatic habitat), comparative information on the relative importance of different bioavailable sources of metals in a specific habitat can be provided. Although it is indeed preferable to use a large suite of biomonitors in a biomonitoring programme, it is usual for metal contamination in a habitat, for example an estuary receiving waste from a metal mine via a metal-rich stream, to be manifested in more than one physical component of the aquatic system. Thus, analysis of even just one or a few selected biomonitors will still show that the habitat is contaminated with metal in a form that is bioavailable to at least some resident organisms.

Many algae and plants and most invertebrates are accumulators of trace metals, storing most of the metal taken up in detoxified form, whether temporarily or permanently, and have the potential to act as a biomonitor. There is now enough published information to define high or low accumulated metal concentrations in a variety of biomonitors. These are contenders for use as complementary biomonitors in biomonitoring programmes in many different

habitats, including, for example, terrestrial leaf litter systems, freshwater streams, estuaries or other coastal habitats.

Table 3.9 summarises relevant accumulated metal concentrations for up to nine trace metals in a range of organisms that would be strong candidates to be used in a suite of biomonitors in UK estuaries and adjacent coastal habitats, seeking to identify sites of high or low metal contamination.

The widespread intertidal brown seaweed, the bladder wrack *Fucus vesiculosus*, will take up trace metals only from solution, so long as it is not in contact with sediment.

Several suspension feeding invertebrates will reflect metal bioavailabilities in both suspended matter and solution. Differences between them, in ranking of sites but not in absolute metal concentrations, which will be species-specific (Phillips and Rainbow, 1988), will come about because of differences in the exact nature and size of the suspended plankton and detritus particles selected by each species, and in the relative importance of solution and diet to the subsequent accumulation of a metal in each species, as can be elucidated by biodynamic modelling. Suspension feeders in Table 3.9 include two barnacles, the northwest European estuarine species *Amphibalanus improvisus* and the cosmopolitan invasive fouling barnacle *Amphibalanus amphitrite*. Use of widespread cosmopolitan species as biomonitors is particularly to be encouraged, for they sample a wide international database of sites with different degrees of metal contamination (Rainbow and Phillips, 1993). Barnacles have the advantage of being strong accumulators of many trace metals, particularly zinc (Tables 3.4 and 3.9). Two suspension feeding bivalves are also strong candidates as biomonitors in UK coastal waters, the mussel *Mytilus edulis* and the oyster *Crassostrea gigas* (Table 3.9). *M. edulis* is a widely used biomonitor for trace metals (RNO, 2006). Although originally distributed on either side of the north Atlantic, including therefore northwest Europe, this species is now cosmopolitan after anthropogenic introductions for mariculture across the world. The mussel is a relatively weak accumulator of trace metals such as copper and zinc compared to barnacles and oysters (Table 3.9), but

Table 3.9 Accumulated concentrations of trace metals (μg/g dry weight), categorised as 'typical' or 'high' when possible, in a macrophytic alga *Fucus vesiculosus*, bodies of invertebrates (the polychaete *Hediste diversicolor*, the barnacles *Amphibalanus amphitrite* and *Amphibalanus improvisus*, the talitrid amphipods *Orchestia gammarellus* and *Talitrus saltator*) and soft tissues of bivalves (the mussel *Mytilus edulis*, the oyster *Crassostrea gigas*, the tellin *Macoma balthica* and the semelid *Scrobicularia plana*), all suitable biomonitors in coastal waters. The low end of the typical range is indicative of uncontaminated conditions; the high end of the typical range is indicative of concentrations representative of moderate contamination on a regional scale. The 'high' concentrations are indicative of atypically raised bioavailability of that metal in the local habitat.

	Ag	As	Cd	Co	Cr	Cu	Ni	Pb	Zn
Fucus vesiculosus									
Typical	0.1–1.0	4–30	0.2–2.5	0.6–6.0	0.6–5	7–75	1–10	1–20	20–100
High	2–10	50–500	4–100	16–70	55	200–3,500	30–60	20–610	500–11,000
Hediste diversicolor									
Typical	0.1–1	6–10	0.05–1.0	1–15	0.1–1	10–30	1.5–6	0.2–5	100–300
High	3–40	20–250	2.5–10	25–30	2–10	50–4,000	13–40	10–1,000	400–2,000
Amphibalanus amphitrite									
Typical	0.7–9.0	7–150	0.8–10	0.1–2.1	1–10	52–100	1.3–8.2	0.4–10	2,700–10,000
High	22.5	250–460	50–170	9–11	20–40	500–9,400	100	35–40	12,000–50,000
Amphibalanus improvisus									
Typical	0.1–10		3–10	3–15	3–25	15–150	4–40	2–30	1,500–10,000
High	10–20		10–30			200–920	40–70	30–130	20,000–150,000
Orchestia gammarellus									
Typical			0.5–5			30–100	5.3–9.6		80–200

Table 3.9 **(cont.)**

	Ag	As	Cd	Co	Cr	Cu	Ni	Pb	Zn
High			5–16			120–370			250–500
Talitrus saltator									
Typical			1–5		1–5	30–75			100–250
High			35		180	200–400			300–500
Mytilus edulis									
Typical	0.01–1	3–20	0.2–3	0.02–2	0.2–3	4–20	0.1–5	0.2–20	50–200
High	3–20	50–200	5–90	5–10	10–450	50–270	10–30	30–130	300–600
Crassostrea gigas									
Typical	1–3	3–20	1–5	0.2–1.0	0.2–2.0	30–500	1–5	0.5–5	1,500–4,500
High	10–70	30+	10–50		10+	1,000–5,000		20–140	6.000–14,000
Macoma balthica									
Typical	0.2–5	10–35	0.1–2	0.6–10	0.3–5	20–80	0.3–5	0.5–10	300–600
High	50–300	50–100	5–10	20	10–20	100–350	10–20	30–150	1,000–3,000
Scrobicularia plana									
Typical	0.1–5	10–30	0.2–3	2–20	0.4–7	9–60	1–10	4–40	300–1,000
High	10–300	100–300	5–45	30–110	15–25	120–360	15–25	100–1,500	2,000–5,000

Sources: Bryan et al., 1980, 1985; Bryan and Gibbs, 1983; Rainbow et al., 1989, 2002, 2011c; Moore et al., 1991; McEvoy et al., 2000; Langston et al., 2004; RNO, 2006; Luoma and Rainbow, 2008; Morillo and Usero, 2008; Fialkowski et al., 2009; Rainbow and Luoma, 2011a, b; Nasrolahi et al., 2014; Johnstone et al., 2016.

does accumulate sufficiently different body concentrations under different metal bioavailabilities to allow distinction to be made between sites of different degrees of metal contamination. Oysters, including the Pacific oyster *C. gigas*, are strong accumulators of trace metals, particularly of copper and zinc (Table 3.9). This oyster represents a reverse case of anthropogenic introduction for mariculture to that of the mussel, being introduced into northwest Europe from the Indo-Pacific and now spreading around UK and other Northwest European coasts. Both *M. edulis* and *C. gigas* have extensive comparative databases of accumulated metal concentrations internationally in coastal habitats of different degrees of trace metal contamination (RNO, 2006; Luoma and Rainbow, 2008).

Deposit feeders ingest organic particles that have settled on, or become incorporated into, bottom sediments. Some deposit feeders such as the estuarine polychaete ragworm *Hediste diversicolor* will ingest whole sediment, while bivalves such as the semelid *Scrobicularia plana* and the tellinid *Macoma balthica* selectively feed on organic material newly settled on the top of the sediment. Thus the ragworm feeds on older sediment particles than these two bivalves, and the polychaete and bivalves are therefore sampling sediment-associated metals deposited over different time periods. Furthermore, *S. plana* and *M. balthica* typically sample newly deposited sediment from different regions of an estuary, *S. plana* being more abundant towards the top of an estuary. High and low accumulated metal concentrations of these three common British deposit feeders are presented in Table 3.9.

Classically, trace metal biomonitors have been used to distinguish geographical differences in trace metal bioavailabilities, as across the variably metal contaminated estuaries of Southwest England left as a legacy of the metal mining history of Cornwall and Devon (Bryan et al., 1980, 1985; Bryan and Gibbs, 1983) or down the Thames estuary as a result of industrial and domestic metal-rich effluents (Rainbow et al., 2002). Biomonitors can also record changes in metal bioavailabilities at specified locations over time. For example, a study of the metal-rich estuaries of Southwest England showed no general decrease in the high bioavailabilities of trace metals in their sediments from the 1970s and 1980s to 2003 and 2006 (Rainbow et al., 2011c).

High accumulated metal concentrations in biomonitors indicate the presence of atypically high bioavailabilities of those metals. They do not themselves indicate that these high metal bioavailabilities are having an ecotoxicological effect on the resident community in a habitat. If, however, particular accumulated metal concentrations in selected biomonitors can be correlated with independently measured ecotoxicological effects on a local community, then they can be used as easily measurable bioaccumulated metal guidelines indicative of such ecotoxicological effects (Luoma and Rainbow, 2008). This concept is illustrated by the use of accumulated copper concentrations in larvae of species of the caddisfly *Hydropsyche* as bioaccumulated metal guidelines correlating with copper-induced changes in the local mayfly community assemblages of streams in North America and Southwest England affected by mining (Luoma et al., 2010; Rainbow et al., 2012). Such copper concentrations in the caddisflies can be used as a threshold to correlate with the loss of certain mayflies, especially heptageniid and ephemerellid mayflies, from the local freshwater stream community. This approach is particularly applicable in aquatic habitats in which metals are the dominating factor causing an ecotoxicological effect, as in waters affected by effluent from metal mining. An understanding of the significance of such accumulated metal concentrations in local metal resistant bioaccumulators will also provide information on the relative importance of these metals as ecotoxicological agents in habitats with mixed trace metal contaminants or indeed other causes of ecotoxicity.

3.7 Biomarkers

Prologue. Biomarkers are used to investigate whether raised trace metal bioavailabilities in a habitat have any ecotoxicological significance. Biomarkers are available at all levels of biological organisation from the molecular to the population.

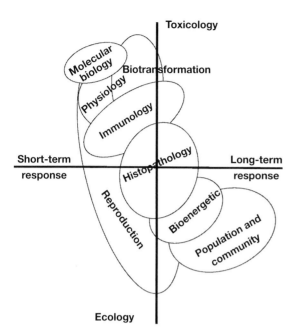

Figure 3.16 Biomarkers: latency between exposure of fish to pollutants and the occurrence of biological effects at different levels of biological organisation. (From Amiard-Triquet et al., 2013, with permission.)

While biomonitors supply important information on when, where and how much of a metal in bioavailable forms is present in a habitat, biomonitoring does not directly provide an answer to the ecotoxicological question, 'So what?' The 'So what?' question asks whether an observed high bioavailability of a metal is actually having an ecotoxicological effect on any of the local biota, and to what extent. What is the ecotoxicological significance of any raised trace metal bioavailability? To answer these questions, we turn to biomarkers (Luoma and Rainbow, 2008; Amiard-Triquet et al., 2013).

A biomarker is a biological response, for example a biochemical, cellular, physiological or behavioural variation that can be measured in tissues or body fluids, or at the level of the whole organism (Luoma and Rainbow, 2008). A biomarker is a measurable biological response to the local presence of a signifi-cant bioavailable quantity of a contaminant, specif-ically a trace metal in the context of this book.

Ecotoxicological effects of toxic metal exposure occur at different levels of biological organisation – sequentially through the molecular, biochemical, cel-lular, physiological, individual organism, population and community levels (Figure 3.16). Any biological response will be first elicited at the lowest level of biological organisation. Thus, a single trace metal ion taken up may bind to a single protein molecule. A singular event like this has no toxic significance, but with increased metal uptake at higher metal exposures, more and more metal will bind to intra-cellular molecules with the potential for toxic effects. The molecular response is the most sensitive, and soonest manifested, followed in turn by later effects progressing up the hierarchy of biological organisa-tion (Figure 3.16). While biomarkers at the lowest levels of biological organisation are the most sensi-tive, it is the toxicological effects reflected by bio-markers at the population and community levels that are the most ecotoxicologically significant. If metal pollution (high metal bioavailability) is producing ecotoxicological effects observable at the community level, then biota are being greatly affected. The task facing environmental scientists is to detect where the ecotoxicological effects of metal contamination are causing damage to biota, hopefully at an early enough stage that the ecological damage is not pro-found and irreversible.

Many biomarkers of trace metal pollution in dif-ferent habitats have been proposed and assessed and are now being widely employed (Amiard-Triquet et al., 2013). An ideal trace metal biomarker in eco-toxicology would be metal-specific, such as the decreased activity of the enzyme d-amino levulinic acid dehydratase (ALAD) in response to lead expos-ure (Amiard and Amiard-Triquet, 2013). Most bio-markers, however, lack specificity, for they typically respond to more than one stressor and are indicative of the general health status of an organism. Never-theless, biomarkers are still of considerable ecotox-icological assessment value. A biomarker is of particular ecotoxicological importance if it is sensi-tive (and therefore detectable at low levels of bio-logical organisation), and if links can be established between its early detection in exposed biota in the field and consequent ecotoxicological effects at higher levels of biological organisation, ultimately up to the population and community levels (Amiard and Amiard-Triquet, 2013; Moore et al., 2013).

At the molecular level, trace metals may bind to DNA and interrupt normal metabolic functioning, with the ultimate potential ecotoxicological effect being carcinogenesis (the development of tumours), heritable mutations and teratogenesis (the malformation of embryos). A first stage in any genotoxic action of a trace metal is often the formation of an adduct, a piece of DNA covalently bonded to the metal, and a highly reactive free radical may be generated. Damage to DNA may be reparable to a point, but ultimately damage to chromosomes and cell division processes can occur. The comet assay is a commonly used biomarker of the early stages of damage to DNA, for it is a laboratory technique that identifies the presence of broken DNA fragments in individual cells of exposed organisms.

Biochemical level biomarkers include the metal-binding proteins metallothioneins previously introduced. MT are widespread, nonenzymatic proteins that are induced by, and bind to, particular trace metals. Metallothioneins are characterised by low molecular weight (12–15 kDa); have a high content of the sulphur-containing amino acid cysteine, which provides binding sites for the trace metals; and are heat stable (Amiard et al., 2006). Since MT are inducible and play a role in the cellular detoxification of both essential and non-essential metals, there was initially a great deal of excitement in their potential for use as specific biomarkers for toxic metal exposure in the field. While there are field examples of the induction of MT in organisms in metal-contaminated field conditions, MT induction can, however, be variable, for example with organism, metal, exposure bioavailability and the different MT isoforms that might be present (Amiard et al., 2006). Moreover, it is now appreciated that MT can also be induced by other stress factors that are not related to metal toxicity, such as anoxia, handling, starvation, freezing and the presence of antibiotics, herbicides or vitamins (Amiard et al., 2006). Furthermore, like other proteins, MT are turned over rapidly in cells. Thus, an increased

rate of MT synthesis on metal exposure may be associated with an increased rate of MT turnover, but not an increased concentration of MT, which is the typical measure used as a biomarker. Thus metallothioneins are no longer considered to be the specific biomarkers of toxic metal exposure once hoped. Nevertheless, any observed induction of MT in organisms at sites under investigation for metal ecotoxicology can contribute to a package of relevant biological field observations indicating metal ecotoxicity.

Other, more general, biochemical stress responses that are used as biomarkers include stress proteins and biochemical defences against free radicals, which can be generated by trace metal exposure (Roméo et al., 2009). Stress proteins belong to a set of protein families originally called heat shock proteins but now known to be induced by exposure to many stressors, including, for example, organic compounds, ultraviolet radiation, salinity change as well as toxic metals. Antioxidant defences against free radicals (which may, for example, damage DNA) include antioxidant enzymes such as superoxide dismutase and catalase (Roméo et al., 2009). Other biomarkers of oxidative stress include concentrations of malondialdehyde, a breakdown product of the oxidation of lipids caused by free radicals, and glutathione, identified previously as able to bind metals in cells but also a scavenger of free radicals (Roméo et al., 2009). A now widely employed biomarker is total oxyradical scavenging capacity (TOSC), an integrated measure of the antioxidant defences of an organism, the stimulation of which may be at least partly promoted by toxic metal exposure (Regoli, 2000). While such more general biochemical stress biomarkers may not be specific to toxic metal exposure, they contribute to a general battery of biomarkers assessing the health status of an organism under investigation for potential ecotoxicological stress.

With modern molecular biological approaches, we are in the era of 'omics' – studies on the genome, the proteome, the transcriptome and the metabolome.

Genomics is the study of the genome, the entirety of an organism's hereditary information including both genes and noncoding regions. Proteomics is the study of the full set of proteins (the proteome) encoded by the genome. Transcriptomics is the study of the transcriptome, the set of all RNA molecules produced, reflecting the genes actively being expressed at a particular time and varying with environmental conditions, including exposure to raised trace metal bioavailability. Metabolomics studies the metabolome, the complete set of small molecule metabolites found in a biological sample, which is again variable under different contaminant exposure conditions. Transcriptomic and metabolomic studies are typically still at the laboratory stage of investigation, but they do have great potential for use in biomarker studies in the field, as we try to understand which genes are promoted or show reduced transcription, and which metabolites are present, under conditions of raised trace metal exposure.

More classic biomarkers have been developed at the cytological level of biological organisation. Genotoxic damage caused by toxic metal exposure can lead to chromosomal changes visible in cells under the microscope. For example, aberrations have been identified in the structure of chromosomes in the salivary glands of larvae of the chironomid midge *Chironomus acidophilus* in Afon Goch, a small river draining the former copper mining area of Parys Mountain in Anglesey (Michailova et al., 2009). Afon Goch is subject to periodic acid runoff and has high water concentrations of iron, manganese and zinc, as well as copper (Michailova et al., 2009).

Other popular cytological biomarkers of toxic metal pollution centre on lysosomes (Moore et al., 2013). Increased functioning of the lysosomal system is a sign of general stress. Thus there are several responses of lysosomes to stress that are used as biomarkers, including changes in lysosomal size and number in particular cells, production of lipofuscin and the destabilisation of the lysosomal membrane (Moore et al., 2006, 2013). Lipofuscin is a pigment that is the end product of the oxidation of cell components which might be caused by the presence of free radicals in a cell (Moore et al., 2006). The increased production of lipofuscin in lysosomes indicates increased lysosomal turnover activity, potentially caused by exposure to toxic metals, but also to organic contaminants (Moore et al., 2006). Lipofuscin granules may also

contribute to the detoxification of excess cellular trace metals, as an insoluble end product of the lysosomal breakdown of metallothioneins. The functional stability of the lysosome membrane, and hence its permeability, changes with degree of exposure to contaminants, including trace metals (Moore et al., 2006). The assessment of lysosomal membrane stability based on the dye neutral red (neutral red retention {NRR}) has become widely used as a biomarker in ecotoxicology because it is both simple and quantitative (Moore et al., 2006, 2013). The NRR test measures the stability of the lysosomal membrane in affected cells. Typically, blood cells from the animal of interest are applied to glass slides, treated with neutral red, and the time is measured for the leakage of the dye from the lysosomes into the rest of the cell. A short retention time indicates low lysosomal membrane stability caused by contaminant exposure. Measurement of lysosomal stability is a sensitive, low organisational-level biomarker, and desirably its quantification has been correlated with other biomarkers at other levels of biological organisation, including TOSC and scope for growth discussed later in this section (Moore et al., 2006, 2013).

Biomarkers at the level of the organism can be morphological or physiological. There are relatively few records of morphological abnormalities of organisms resulting from field exposure to high bioavailabilities of trace metals, and these usually concern diatoms or chironomid midge larvae. There is a significant correlation between valve size reduction in diatoms and trace metal contamination, and abnormal development appears to be more common as diatom valves decrease in size (Falasco et al., 2009). Mouthpart deformities have long been known to occur in chironomid midge larvae living in metal-contaminated sediments, and cause/effect relationships between the presence of such deformities and sediment concentrations of trace metals, particularly copper, have been demonstrated in the laboratory (Martinez et al., 2003). Mouthpart deformities of *Chironomus tentans* larvae include fused teeth, split teeth, missing teeth, extra teeth and abnormally shaped teeth on the mandible, with apparently different effects being associated with different metal exposures (Martinez et al., 2003). Measures of chironomid mouthpart deformities in the field have been used as a contributing index in the biological assessment of the environmental quality of freshwater sediments (Warwick, 1988; De Pauw and Heylen, 2001). It appears, therefore, that mouthpart deformities in chironomid midge larvae may offer relevant contributing evidence in any ecotoxicological investigation of very metal-rich streams.

Physiological biomarkers include measurements of growth rates and feeding rates. Growth rates of larvae of the chironomid midge *Chironomus riparius*, measured as increases in body length, were inhibited in streams contaminated with metals from mining in Portugal (Faria et al., 2008). Feeding rates feature strongly in studies of the effects of trace metal contaminants in freshwater, for example in the case of freshwater amphipod crustaceans such as *Gammarus pulex* (Maltby et al., 2002).

Scope for growth (SFG) is a physiological biomarker that integrates different physiological measures in a calculation of the energy balance of an animal. Thus, SFG is a theoretical estimate of the surplus energy available to an animal for growth and reproduction, calculated from the difference between energy assimilated from food and the energy used in respiration. SFG is interpreted to decrease when energy is required to cope with the extra physiological cost of handling (detoxifying or excreting) high amounts of toxins, such as trace metals taken up in contaminated environments. SFG has been best developed for mussels *Mytilus edulis* in coastal waters, and is considered a robust indicator of their health status in the field, responding to exposure to a range of contaminants that includes trace metals (Widdows et al., 1995, 2002). Furthermore, a correlational link has been established between lysosomal membrane stability and SFG in mussels *M. edulis* in the marine environment, meeting the desired objective of linking biomarkers at different levels of biological organisation (Moore et al., 2006, 2013). In freshwater, the amphipod *Gammarus pulex* shows reduced SFG when exposed to 0.5 mg/L zinc (Maltby et al., 1990a). Furthermore, reduced SFG is associated in *G. pulex* with a decrease in the size of offspring

released in the subsequent brood (Maltby and Naylor, 1990). SFG is, therefore, a very useful integrated measure of how well an animal is coping with a high bioavailability of a contaminant such as a toxic metal. However, it is not easy to measure, requiring rapid access to laboratory facilities for the measurement of such physiological parameters as feeding rate, assimilation efficiency and respiration rate.

Another integrative response at the level of the organism that can be used as a biomarker is the behaviour of an animal, specifically those aspects of behaviour affected by toxic exposure (Amiard-Triquet, 2009). Behavioural changes clearly have the potential to induce knock-on effects at the population level, as in the case of a reduced swimming capacity of a fish to capture prey and avoid predators (Amiard-Triquet and Amiard, 2013). Furthermore, behavioural disturbances originate in biochemical and physiological impairments, such as neurotoxicity and hormone disruption, linking responses at different levels of biological organisation (Amiard-Triquet and Amiard, 2013).

Ecotoxicological effects of trace metals observed at the population level have clear ecological significance for the biota in a metal-contaminated habitat. Such effects, for example, might be on the numbers of individuals present, population age structure, reproductive rates or recruitment rates. The selection of metal-tolerant strains is another population-level effect considered in more detail in Section 3.8. Laboratory-based studies have shown that metal exposure can affect population parameters such as the rate of population increase in *Gammarus pulex* decreased by copper exposure Maund et al., 1992) or life-table parameters such as percentage hatch, juvenile survival, fecundity and time to maturity in the freshwater gastropod mollusc *Biomphalaria glabrata* affected by cadmium exposure (Salice and Miller, 2003). In the field, population densities of the aquatic larvae of many families of insects, and subsequently adult emergence rates, decreased with increased water metal concentrations in metal-contaminated streams in the Rocky Mountains in Colorado, including the Colorado Mineral Belt that has been mined for the past 150 years (Schmidt et al., 2013). Thus the ecotoxicological effects of metal exposure on the populations of the insect larvae were not all manifested until later in the lifecycle, during metamorphosis and emergence.

Recruitment to the adult population of postlarval juveniles, settling and metamorphosing after larval development in the plankton, appears to be an especially susceptible stage to toxic metal exposure in the case of estuarine and marine benthic invertebrates. Restronguet Creek in Cornwall has extraordinarily high sediment concentrations of trace metals, including As, Cu, Fe, Pb, Mn and Zn (Bryan and Gibbs, 1983). The observed impoverishment of bivalve molluscs in Restronguet Creek has been attributed to the ecotoxicity of local copper and zinc dissolved and sediment bioavailabilities to bivalve larvae and juveniles, with consequent deleterious effects on the recruitment of bivalves such as the tellinid bivalve *Macoma balthica*, the cockle *Cerastoderma edule* and the mussel *Mytilus edulis* that might otherwise be expected to be present (Bryan and Gibbs, 1983).

3.7.1 Community Effects

Prologue. Ecotoxicologically significant high trace metal bioavailabilities will ultimately have a deleterious effect on local biological community structures.

Ecotoxicological effects of exposure to toxic metals at the population level have clear implications for the community structure of the local biota, via such effects as loss of species and changes in the abundances of individuals of those species still present. Any resulting changes to community structure are acting at the highest level of biological organisation recognised in the hierarchy of biomarkers. There are many instances in the literature where high metal contamination is associated with an alteration in the nature of the biological community (Luoma and Rainbow, 2008).

Community-level biological monitoring is based on the assumption that the composition and organisation of communities reflect local environmental conditions and respond to any significant anthropogenic alterations to these conditions (Clements and Newman, 2002). A second assumption is that species differ in their sensitivity to anthropogenic stressors such as high trace metal bioavailabilities, which thereby cause structural changes in community composition (Clements and Newman, 2002).

Given the assumption that species differ in their sensitivity to raised bioavailabilities of toxic metals, then an ecotoxicologically significant high bioavailability of a trace metal in a habitat will change the numbers of species present and the abundances of those species remaining. Measures of the numbers of species present and their abundances have typically been first steps in any analysis of community structure, for example of benthic invertebrates in metal-contaminated streams. Species richness is the number of species present in a prescribed sampling unit. Species richness, however, is highly dependent on sampling effort, for example the area sampled. Species richness also increases with the sample size and number of individuals collected, to eventually level off as the point is approached when all the species present have been recorded. A particular disadvantage of measures of species richness is that they do not take into account differences in abundance amongst the species that are present.

Species diversity indices provide a single value that incorporates information on both the number of species and species abundances. Species diversity indices are often used in biological monitoring studies to compare communities in different locations, although they too can be criticised for theoretical and statistical assumptions (Clements and Newman, 2002). Simpson's Index, for example, is very sensitive to abundances of dominant species and relatively insensitive to rare species. The Shannon–Wiener Index is more sensitive to rare species and is considered more relevant to ecotoxicology studies (Clements and Newman, 2002). Another measure of community structure is species evenness, a measure of how equally the individuals in a community are distributed amongst the species present.

While single-number (univariate) indices are simple, they do have the serious deficiency of the loss of information that inevitably occurs when details of community composition are reduced to a single number. Multivariate statistical techniques are used more frequently today, such as principal components analysis (PCA) or multidimensional scaling (MDS) (Sparks, 2000). Many multivariate techniques aim to replace the original large set of variables (for example, species abundances or metal concentrations in sediment or biomonitors at a number of sites) by a much smaller set of derived variables (often referred to as axes) which still retain most of the relevant information. This process is termed ordination. Ordination attempts to present results by plotting graphs of the new variables against each other, allowing us to visualise relationships between sites, identifying those sites which are alike or different from each other. Thus the biological communities at sites with ecotoxicologically significant trace metal contamination should be identifiable from a background of expected communities at control sites.

Another approach taken to overcome the deficiencies of univariate measures of community diversity has been the development of biotic indices, based on the relative abundance of sensitive and resistant species present in a habitat.

Biotic indices are particularly popular in the ecotoxicological assessment of communities of benthic macroinvertebrates in streams and rivers. Most of the research and development of such biotic indices have, however, centred on organic pollution and not on the ecotoxicological effects of trace metals. It follows that different species show different relative sensitivities according to the contaminant of concern, and a biotic index relevant to the assessment of organic pollution is not directly applicable to an assessment of toxic metal pollution, and vice versa.

One such biotic index is the biological monitoring working party (BMWP) score, produced by a working party convened in 1976 by the UK Department of the Environment (Metcalfe, 1989; Jones et al., 2010). Taxa are ranked according to their sensitivity to organic pollution, with corresponding scores from 10 (sensitive) to 1 (resistant), which are summated to give the total score. Since the BMWP score was developed as an index to assess the ecological effects of organic pollution, it cannot, therefore, be expected from first principles to be suitable as a biotic index for metal contamination in freshwater. Nevertheless, the use of the BMWP has been evaluated in this context, specifically with respect to coal mine water discharges (Environment Agency, 2009a) and acid mine drainage (AMD) (Jarvis and Younger, 2000; Gray and Delaney, 2008).

3.8 Tolerance

Prologue. As pointed out by Sam Luoma, the presence of a metal-tolerant population of an organism in a particular habitat is evidence that local bioavailabilities of that toxic metal are of ecotoxicological significance, clearly to that species but potentially also to other members of the local biota (Luoma, 1977).

Exposure to a high enough bioavailability of a toxic metal can act as a selection pressure on an exposed population, selecting for physiological traits that are the most metal tolerant in the gene pool of that population. This selection pressure can ultimately lead to the establishment of a metal-tolerant population.

It needs to be stressed that the term 'metal-tolerant' is used here to refer to a particular metal-exposed population of a species, whether or not that tolerance is inheritable after selection over several generations or has been derived by physiological acclimation and is restricted to the one exposed generation (Amiard-Triquet et al., 2011). The term 'metal-resistant' is used to refer to all members of a species which survives relatively well generally in conditions of metal contamination, without specific physiological acclimation or genetically based adaptation to the high metal bioavailability in a particular local habitat.

The presence of metal-tolerant populations has been long known in UK streams affected by metal mining activities (Kelly, 1988). Populations of the green alga *Stigeoclonium tenue* abundant in zinc-contaminated streams in the Northern Pennines showed increased tolerance to dissolved zinc exposure, in comparison to populations in streams with lower zinc levels (Harding and Whitton, 1976). A population of another green alga, *Chlorella vulgaris*, from the River Hayle draining disused copper mines in Cornwall, has been reported to be copper-tolerant (Foster, 1977). Furthermore, 19 metal-tolerant strains of green algae have been identified from the River Hayle and another Cornish river, the lead-rich River Gannel, with several of the copper-tolerant strains from the Hayle also showing lead cotolerance (tolerance to more than one metal in the same population) (Foster, 1982b). A parallel study of the freshwater isopod crustacean *Proasellus meridianus* in these same two Cornish rivers has also demonstrated the presence of metal tolerance and co-tolerance (Brown, 1977b).

Restronguet Creek in Cornwall receives discharge from the Carnon River, draining a catchment with a long history of mining (Dines, 1969), and Restronguet Creek sediments contain extraordinarily high concentrations of As, Cu, Fe, Mn and Zn (Bryan and Gibbs, 1983). Restronguet Creek was reported in the 1970s and 1980s to house metal-tolerant populations of three sediment-ingesting burrowing invertebrates, the polychaete *Hediste diversicolor* (Cu and Zn tolerant), the bivalve mollusc *Scrobicularia plana* (Cu) and the amphipod crustacean *Corophium volutator* (Cu) (Bryan and Gibbs, 1983). Other metal-tolerant invertebrate populations reported from Restronguet Creek are those of the crab *Carcinus maenas* (Cu and Zn) and the polychaete *Nephtys hombergii* (Cu) (Bryan and Gibbs, 1983). The seaweed *Fucus vesiculosus* (bladder wrack) also showed Cu tolerance (Bryan and Gibbs, 1983). The continuing presence into the twenty-first century of the copper and zinc-tolerant population of *Hediste diversicolor* at the top of Restronguet Creek has been confirmed, substantiating the ongoing ecotoxicological significance of high local copper and zinc bioavailabilities in the local sediments (Grant et al., 1989; Mouneyrac et al., 2003). Copper and zinc tolerances of this upper Restronguet Creek population of *H..diversicolor* are genetically based, inheritable down generations, but bring a selective disadvantage beneath a threshold sediment concentration of each metal (Hateley et al., 1989; Pook et al., 2009). The metabolic costs of the metal tolerance in this worm population result in a scope for growth 46 to 62% less than that of non-tolerant worm populations (Pook et al., 2009). Thus metal tolerance comes at a considerable physiological cost, only worth paying when metal exposure is ecotoxicologically significant.

The argument that the evolution of metal tolerance in a population of a single species shows that the local metal exposure is of ecotoxicological significance can be extended to the whole biological community. Thus the selection pressure associated with an ecotoxicologically significant metal bioavailability will lead to an increased average tolerance to that metal amongst all species in the local biological community (Blanck et al., 1988). Strictly, it should be recognised that this increased average tolerance might well be based on or include changes in the relative abundance of metal-resistant species as defined in this chapter, in addition to any selection of metal-tolerant populations of particular species. Pollution-induced community tolerance (PICT) is a potential ecotoxicological tool to assess the effects of a toxicant on communities (Blanck and Wängberg, 1988; Blanck et al., 1988). PICT is tested by comparing responses of communities collected from metal-contaminated and reference sites to metal exposure under controlled conditions. Any increased community tolerance that results from the elimination of more sensitive species is considered to be strong evidence that community restructuring has been caused by the toxic metal (Clements and Rohr, 2009). The need to carry out experiments to measure the difference in tolerance between communities does constrain the application of PICT as an assessment tool. Nevertheless, PICT has been tested in several different communities beyond the marine periphyton community on which the PICT hypothesis was first developed (Blanck and Wängberg, 1988; Clements and Newman, 2002). Such communities include freshwater periphyton and phytoplankton, marine phytoplankton, estuarine nematode worms and freshwater and coastal macroinvertebrates (Clements and Rohr, 2009). The trace metals concerned include arsenic, cadmium, copper and zinc. Amongst the assumptions behind PICT is that the communities most likely to be suitable for PICT assessment are those that show a large amount of variation in metal sensitivity amongst species (Clements and Newman, 2002). PICT may prove particularly suitable for comparison of microbial populations collected from different field sites (Tlili and Montuelle, 2011; Virsek et al., 2013).

An effect of the presence of metal-tolerant populations in a community is that measures of community structure will not be as severely reduced as might be supposed in a community exposed to very high metal bioavailabilities. Thus Restronguet Creek is home to a benthic coastal invertebrate community that, although impoverished, is surprisingly diverse given the ecotoxic sediments present (Bryan and Gibbs, 1983). There may be further consequences of the development of metal-tolerant populations in a metal-contaminated community, particularly the potential extra trophic transfer of metals accumulated by metal-tolerant invertebrates to predators higher up the food chain. Thus copper accumulated to very high concentrations by Restronguet Creek *Hediste diversicolor* is transferable to predators (Rainbow et al., 2004, 2006b; Rainbow and Smith, 2013). When these predators are flounders (*Platichthys flesus*), visiting the estuary temporarily at high tide, or wading birds stopping off to feed at low tide at particular seasons of the year, the high copper dose ingested may have ecotoxicological knock-ons beyond the confines of Restronguet Creek.

This chapter has explored the biological principles that will apply to the case examples that will be covered in detail in the rest of this book. The case studies chosen will concentrate on habitats and catchments introduced in Chapter 2, as trace metals derived from mining and industry are delivered from land through freshwater systems into their estuaries and beyond.

4 Terrestrial Environment

Box 4.1 Definitions

acid soil Soil containing an excess of hydrogen ions (pH less than 7).

ALAD The enzyme d-amino levulinic acid dehydratase, part of the metabolic pathway for the synthesis of haem needed to make haemoglobin. The activity of this enzyme is inhibited by lead, and this inhibition provides a very sensitive and specific biomarker for lead ecotoxicity.

Arachnida A taxon of terrestrial arthropods that includes spiders, scorpions, pseudoscorpions, harvestmen, mites and ticks. The body of arachnids is organised into two parts, the prosoma and the opisthosoma.

basic soil Soil containing an excess of salts (for example, calcium hydroxide) able to neutralise acids (pH greater than 7).

bioaccumulation factor (BAF) The ratio of the metal concentration at steady state in an organism to the metal concentration in its food or other defined matrix in its environment (e.g., soil). The term is often used without definition, occasionally (and confusingly) as a synonym for BCF. Use of the term BAF often depends on untenable assumptions, for example, that there is a direct and exclusive relationship between the concentrations concerned, and that the metal concentration in the organism is actually in steady state.

bioconcentration factor (BCF) The ratio of the metal concentration at steady state in an organism to the metal concentration in the water in which it lives. This term is often used without definition, occasionally (and confusingly) as a synonym for BAF. Use of the term BCF often depends on untenable assumptions. For example, in the case of aquatic animals, a BCF ignores the fact that a significant proportion of the metal accumulated is derived from the diet. A BCF is more suitable for use in the case of a plant or alga, but, even then, the assumption that the accumulated metal concentration in the organism is in steady state is probably erroneous.

British National Vegetation Classification (NVC) A system of classifying natural habitat types in Great Britain according to the vegetation they contain.

bryophytes Mainly green flowerless plants that reproduce sexually by means of spores. Examples include mosses, liverworts and hornworts.

calaminarian grassland community The community of plants recognised by the British National Vegetation Classification system to occur in upland habitats with soils with bioavailabilities of trace metals that are high enough to be toxic to most plants.

calamine soil Soil enriched in zinc, lead and cadmium.

calcareous Rich in calcium carbonate, as in chalk and other forms of limestone.

calcicole A plant that thrives in a calcareous soil, being intolerant of acid soils.

calcifuge A plant that thrives in acid soils, being intolerant of basic soils such as those containing chalk.

chalk A soft, white, porous sedimentary rock, a form of limestone composed of calcite (calcium carbonate $CaCO_3$). Typically formed as a calcareous coccolithophore ooze on the ocean bottom before geological uplift to underlie terrestrial habitats (e.g., the chalklands of southern England).

chelicera (plural chelicerae) One of the paired first appendages of arachnids, often fanglike and modified as grasping or piercing mouthparts.

chloragogenous tissue A loose assemblage of cells surrounding the gut of polychaete and oligochaete worms, for example earthworms, that is an important centre of metabolism, including the detoxification of trace metals and the synthesis of haemoglobin.

collembolans (springtails) Wingless terrestrial arthropods, living in leaf litter and soil. Primarily detritivores and microbivores, feeding directly on microbial organisms.

community A group of interacting populations of organisms that overlap in time and space. In plant ecology, communities may be formally described by reference to one or more dominant species, as under the British National Vegetation Classification system or in the case of metallophytic lichen communities.

contaminated land exposure model (CLEA) A model used by the UK Department for Environment, Food and Rural Affairs (DEFRA) and the Environment Agency to derive Soil Guideline Values for the concentrations of toxic metals in different soils, below which long-term human health risks from exposure are considered to be tolerable or minimal.

critical soil concentration A threshold trace metal concentration in a soil above which it is considered likely that there will be an ecotoxicological effect on local biota.

detritivorous Feeding on detritus.

detritus Fragmented particulate organic matter derived from the decomposition of plant and animal remains.

enchytraeid An oligochaete worm resembling a small earthworm, for example, potworms living in organically rich soils.

environmental genomics Study of genetic material recovered directly from environmental samples (such as soil, sediment or water samples) to produce largely unbiased samples of all genes from all members of the sampled community.

flowering plants (angiosperms) A taxon of seed plants, characteristically with flowers, enclosed seeds and fruit. It is the most diverse group of green plants.

green plants A taxon that includes some green algae, bryophytes, pteridophytes and the seed plants which consist of the gymnosperms and the flowering plants.

gymnosperms A taxon of seed plants that includes conifers and cycads. The seeds of gymnosperms are not enclosed as in flowering plants, but develop, for example, on the surface of scales or leaves modified to form cones.

habitat The particular ecological area that is occupied by an organism.

haemoglobin The iron-based respiratory protein, widespread amongst invertebrates and vertebrates, in which the iron reversibly binds oxygen.

Hemiptera The order of insects (known as the true bugs) that includes aphids, planthoppers, leafhoppers, cicadas and shield bugs.

hyperaccumulation Accumulation by an organism of an atypically very high concentration of a trace metal. Flowering plants that accumulate atypically high concentrations of trace metals, specifically in their shoots, are known as hyperaccumulators.

insectivore Common name applied to a group of mammals including shrews, moles and hedgehogs that feed primarily on arthropods (including insects) and earthworms. Formerly a recognised order in the classification of mammals, the group is now divided into three different orders. Shrews and moles are together in one, and hedgehogs in another, of these three orders.

isopod crustacean One of an order of dorso-ventrally flattened crustaceans that live in oceans, seas, estuaries and freshwater and even on land, as in the case of woodlice. Isopods have direct development with no free-living planktonic larval stage. Examples include species of the terrestrial genera *Oniscus* and *Porcellio*, the freshwater genera *Asellus* and *Proasellus*, and the sea slater *Ligia oceanica* living in coastal strandlines.

keystone species A species of a plant or animal that plays a unique and critical role in maintaining the structure and function of an ecosystem out of proportion to its abundance.

lichen A symbiosis between at least two organisms: a fungus and a photosynthetic partner, which may be a green alga or a blue-green bacterium.

limestone A sedimentary rock composed largely of the minerals calcite and aragonite (different crystal forms of calcium carbonate). Most limestone is composed of secretions of marine organisms such as coral or (in the case of chalk) planktonic coccolithophores.

Malpighian tubules Blind ending tubules opening into the junction of the midgut and hindgut of some insects, myriapods and arachnids, typically occurring in multiples of two. Malpighian tubules are bathed in blood and are considered to carry out excretion and osmoregulation. Malpighian tubules, in at least the insects and the arachnids, have probably evolved separately without derivation from a common ancestor.

metallophyte A plant that can grow in atypically high bioavailabilities of trace metals. Here the term 'plant' is used loosely without taxonomic restriction, for metallophytes can include, for example, lichens, bryophytes, pteridophytes and flowering plants.

midgut diverticula Blind ending branches of the midgut of (particularly) arachnids, producing digestive enzymes, carrying out absorption of digested food and involved in the physiological detoxification of accumulated trace metals. The physiological equivalent of the digestive gland of molluscs or the hepatopancreas of decapod crustaceans.

mycorrhizal fungi Fungi which establish (usually mutualistic) symbioses with the roots of plants, providing a direct physical link between soil and the plant. The symbiotic association is intracellular in the case of arbuscular mycorrhizal fungi (AMF), or extracellular in the case of ectomycorrhizal fungi.

NOEC The estimated highest exposure concentration (*No Observable Effect Concentration*) of a toxic metal in a particular medium in laboratory toxicity tests that will cause no specified effect on the population of organism being tested. Similar acronyms include the NOAEL, the No Observable Adverse Effect Level.

ooze A type of marine sediment that consists (at least 30%) of the skeletons, shells or other calcareous or siliceous secretions of planktonic organisms that deposit after death of the organism. Calcareous oozes are typically derived from coccolithophores, foraminiferans and pteropods, and siliceous oozes from diatoms and radiolarians.

plant A term that strictly is limited to a single taxon (the green plants) that includes some green algae, bryophytes, pteridophytes and the seed plants which consist of the gymnosperms and the flowering plants. Nevertheless, the term 'plant' is still used less restrictedly to refer also to a variety of unrelated photosynthetic organisms, including red algae, brown algae, some protists, blue-green bacteria and lichens.

pteridophytes Plants with vessels, fronds, roots and occasionally stems or trunks, for example, ferns, horsetails and club mosses. They reproduce sexually via spores, lacking flowers or seeds. They are strictly not a single taxon, for ferns and horsetails are more closely related to gymnosperms and flowering plants (seed plants) than to club mosses.

Red Data List An inventory compiled by the International Union for the Conservation of Nature (IUCN) of the global conservation status of threatened species. Precise criteria are used to evaluate the extinction risk of species, globally and regionally.

seed plants Plants that develop from seeds, consisting of the gymnosperms (e.g., conifers) and flowering plants.

serpentine Referring to a variable group of rock-forming minerals that are hydrous magnesium iron silicates $(Mg,Fe)_3Si_2O_5(OH)_4$ that may contain trace metals such as cobalt, chromium, manganese, nickel and iron.

serpentine soils Soils derived from serpentine minerals found in particular British locations, including the Lizard Peninsula (Cornwall), Shetland and several mainland Scottish sites. Serpentine soils harbour a distinctive flora.

Soil Guideline Value (SGV) A concentration of a toxic metal in different soils, below which long-term human health risks from exposure are considered to be tolerable or minimal, as derived from the contaminated land exposure assessment model. An SGV represents a trigger value to indicate to a risk assessor that soil metal concentrations above this level may pose a significant possibility of significant harm to human health, a situation requiring further investigation.

species sensitivity distribution A ranking of the relative sensitivities of different species to a metal, often expressed in terms of No Observable Effect Concentrations (NOEC). Typically used to identify the NOEC at the 5% point, the metal concentration in a particular medium (e.g., water, soil) below which 95% of the species ranked will show no toxic effect.

woodlouse (plural woodlice) A terrestrial isopod crustacean found in soils and leaf litter, typically detritivorous. Genera include *Oniscus* and *Porcellio*. Other common names include slater, sowbug and pillbug.

4.1 Introduction

Prologue. Soils and dusts may contain raised concentrations of trace metals as a result of anthropogenic contamination from mining, smelting and other industrial processes. Trace metals are taken up by plants and enter terrestrial grazing and decomposition food chains, with potential ecotoxicological effects on the flora and fauna (including livestock and humans) in both natural and managed terrestrial ecosystems. At the other end of the bioavailability spectrum, crops and livestock may show symptoms of essential trace metal deficiency.

Trace metals occur in all the different environmental compartments of the Earth, passing through biogeochemical cycles often accelerated today by the actions of humans. The terrestrial environment is no exception.

Trace metals occur in variable concentrations in soils and dust ultimately derived from the erosion of rocks of the Earth's crust. From soils, trace metals are taken up into plants, with the subsequent potential to be accumulated and passed on to terrestrial herbivores and carnivores through food chains. Thus trace metals are present in both physical and biological components of the terrestrial environment, accumulating or passing through at different rates according to the terrestrial habitat concerned. The terrestrial biological component covers the full range of the tree of life, including microbes, algae, land plants, invertebrate detritivores, herbivores and carnivores, as well as vertebrates such as small mammals and birds, and the top predators that hunt them.

An increased bioavailability of any trace metal, essential or non-essential, has the potential to cause an ecotoxicological effect, initially sublethal but ultimately lethal, above a threshold bioavailability that causes the total uptake rate of that metal to exceed the capacity of the recipient organism to detoxify and/or excrete it. Given the history of metal mining in Britain, it is not surprising that areas of the British Isles show severe metal contamination of local soils, derived from the processes of mining and subsequent treatments such as smelting (Table 4.1). Such toxic metal contamination can be so strong as to cause recognisable ecotoxicological effects on local terrestrial organisms, at all biological levels from the molecular to the individual, and then to the population and the ecological community. These effects might at times be offset to a degree by the evolution of metal tolerance in affected populations. Ecotoxicological effects resulting from high toxic metal bioavailabilities in soils can extend to crops, livestock and even human health (Underwood, 1962; Thornton and Webb, 1979; Thornton, 1993).

The primary cause of raised toxic metal concentrations in British soils is the local mining of the metals. In the particular regions highlighted in Chapter 2, huge quantities of metal ores have been mined over centuries with inevitable local terrestrial contamination. Metal-rich dust particles may spread the contamination from the immediate vicinity of mines. Smelting emits yet more metal-bearing particles into the atmosphere, to be dispersed 10 kilometres or more from the smelter, before deposition onto the soils (Table 1.5) and vegetation of local habitats (Little and Martin, 1972; Hutton and Symon, 1986).

Other anthropogenic inputs of metals into soils may come via the atmosphere or by direct application of specific metal-rich material (Tables 1.4, 1.5, 4.1). In the case of metal deposition from the atmosphere, typically in particulate form, anthropogenic sources include the burning of coal and oil and the incineration of refuse and sewage sludge (Hutton and Symon, 1986). Coal burning in power stations, industrial works and residential houses releases significant amounts of antimony, arsenic, chromium, manganese, mercury, molybdenum and selenium into the atmosphere, while oil combustion provides nickel, tin and particularly vanadium (Tables 1.2, 1.5) (Nriagu and Pacyna, 1988; Nriagu, 1989). Fortunately, in

Britain the days of the addition of lead to petrol are now over, eliminating this source of lead into the atmosphere, and in turn its consequent deposition onto roadside and urban soils and vegetation. The iron and steel industries release chromium and manganese into the atmosphere, and other metal-associated industries emit arsenic, cadmium, copper, lead and zinc (Tables 1.2, 1.5). Another source of metal input into soils is the direct application of materials such as ash from the burning of coal and urban refuse, sewage sludge, phosphate fertilizers which bear cadmium (Taylor, 1997) and organic wastes from animal husbandry and other agricultural production processes (Tables 1.4, 1.5, 4.1).

As discussed in Chapter 3, there are several routes by which trace metals enter organisms, with the subsequent potential to meet essential metal requirements, but also to bring about ecotoxicological effects. Correspondingly, there are different potential routes for the uptake and subsequent accumulation of trace metals by organisms in terrestrial habitats, the metal being derived originally from soils, but also from dust particles. The biological consequences of metal uptake and accumulation in different terrestrial organisms will feature strongly in this chapter.

Land plants with roots will take up metals from the soil solution via these roots, while algae or microbes in or on the soil will absorb metals over their surface in contact with the soil particles or soil solution. From the roots, plants will translocate trace metals to stems and leaves, often with degrees of control, and certainly variability, as to what metals reach what parts, and accumulate there or not. Metals in the different parts of a plant may be consumed by herbivores, be they leaf-eating caterpillars, browsing garden snails, sap-sucking aphids or grazing vertebrates such as rabbits or cows. The metals can then be transferred through food chains that may involve birds or mammals. Uneaten plants may die off, or more often drop their leaves annually, to suffer decomposition through the agency of bacteria and fungi. The metals accumulated in the plants have the potential to be transferred to the microbial tissue, the main source of nutrition of a detritivore. Thus, leaf litter and associated microbes feed the likes of woodlice, and the

Table 4.1 **Concentrations of selected trace metals in (mostly) British uncontaminated and contaminated soils, metal sources added to soils, and in dusts (µg/g dry weight).**

	Ag	As	Cd	Co	Cr	Cu	Hg
Uncontaminated soils							
'Normal' background/ control sites	0.05 (0.01–8)[1]	6 (0.1–40)[1]	0.35 (0.01–2)[1]	8 (0.05–65)[1]	70 (5–1,500)[1]	30 (2–250)[1]	0.06 (0.01–0.5)[1]
	0.6[2]	18 (4–80)[3]	2.5[2]	40 (16–57)[4]		24 (7–47)[3]	0.1[2]
	1.3 (1.0–1.8)[4]	8.3–11[8]	1.9 (1.8–2.0)[4]	24 (4–54)[5]		19	0.10[9]
	0.8 (0.3–1.3)[5]	7–33[11]	0.30[12]			(14–31)[4] 14–26[13]	
			0.96 (0.5–1.2)[14]			13 (9.8–15)[14]	
						16–150[15]	
Soils contaminated by:							
Mining	0.4–13.6[2]	60–2,500[3]	4.4–372[2]			29–2,000[3]	1.9–5.8[2]
		30 (20–39)[8]	0.6–75[13]			116–1,410[4]	
		350 (171–466)[8]	0.4–540[16]			370–2,700[17]	
		228 (23–1,080)[11]	159 (24–440)[19]			98–1,970[22]	
		8,510–26,53[17]	8.8–98[20]			790[24]	
		8,200–180,000[21]	<2–520[22]			26–2,740[25]	
		322 (144–892)[23]	0.1–350[25]				
Smelting, refining, etc.		31 (17–44)[8]	3.6–6.1[2]			9,270 (2,600–18,700)[14]	0.1–0.2[2]
		522 (210–925)[8]	28 (9.6–58)[14]				
Other industry		1,390 (5–5,270)[27]					
Phosphate fertiliser			0.85 (0.20–2.59)[29]				
Atmospheric deposition							
Smelter, refinery			69[12]			494 (129–880)[14]	
			7.4 (5.5–9.7)[14]			1,950[31]	
			33[30]			160[32]	
			218[31]				
			26[32]				

Mn	Mo	Ni	Pb	Sb	Se	Sn	V	Zn
1,000 (20–10,000)[1]	1.2 (0.1–40)[1]	50 (2–750)[1]	35 (2–300)[1]	1 (0.2–10)[1]	0.4 (0.01–2)[1]	4 (1–200)[1]	90 (3–500)[1]	90 (1–900)[1]
100 (5–250)[5]		40 (3–105)[5]	59 (34–144)[3]	0.3–8.6[6]		7.6[7]		95 (51–220)[3]
				1.0–1.4[8]				
			42 (24–56)[4]	0.48[10]				129 (80–171)[4]
			10–150[11]					33–60[13]
			17–50[13]					42–180[15]
			2,590–28,900[2]	4.0–50.0[2]				478–45,900[2]
			90–2,900[4]	23.4 (6.5–51.4)[8]				114–1,020[3]
			85–87,400[13]					107–10,400[13]
			35–48,000[16]					10–49,400[16]
			5,800–65,300[18]					15,000 (2,040–50,000)[19]
			597–2,610[20]					1,100–7,810[20]
			1,140–26,500[22]					180–64,000[22]
			170–24,600[25]					160–45,000[25]
			151–27,200[26]					
			9,770–30,000[2]	103 (50–154)[8]				380–1,380[2]
			28,800–300,000[18]					
			2,090 (39–3,640)[27]					
			635 (300–1,170)[28]					
			1,910[12]					3,190[12]
			190–2,400[18]					1,700[30]
			398[30]					22,400[31]
			13,300[31]					543[32]
			263[32]					

Table 4.1 (cont.)

	Ag	As	Cd	Co	Cr	Cu	Hg
Urban soil		77[7]	1.3 (<1–40)[33]			161[7] 73 (13–2,320)[33]	0.23[9]

Metal sources

	Ag	As	Cd	Co	Cr	Cu	Hg
Coal ash		104 (40–205)[35]	0.30 (0.13–0.82)[35]				
Incinerated refuse ash	55–220[36]	16 (7.6–22)[35] 9–74[36]	473 (239–750)[35] <1–477[36]	25–54[36]	730–1900[36]	69–2,000[36]	0.09–25[36]
Sewage sludge			12 (3–1,500)[37] 1–3[38] 12[39] 30–70[40]	12 (2–260)[37] 7.5[39]	250 (40–8,800)[37] 19–56[38] 660[39] 200–1,070[40]	800 (200–8,000)[37] 349–564[38] 370[39] 505–2,500[40]	4.4 (1.9–5.2)[37] 0.3–0.8[38]
Phosphate fertilisers			42–100[29]				

Dust

	Ag	As	Cd	Co	Cr	Cu	Hg
Urban house dust		1–330[41] 2–122[41]	6.9 (<1–8,040)[22] 193 (12–387)[28] 7.7 (<1–336)[33]			205 (3–48,800)[22] 208 (9–5,300)[33]	
Urban road dust			2.0 (<1–280)[22] 4.0 (<1–180)[33]			81 (7–3,030)[22] 115 (18–2,400)[33]	

Note: Single values represent a summary value, for example a mean, a mean derived from transformation of raw data or a median. Ranges may similarly be values derived from raw data.

Sources: Bowen (1979);[1] Li and Thornton (1993a);[2] Colbourn et al. (1975);[3] Alloway and Davies (1971);[4] Davies (1971);[5] Tschan et al. (2009);[6] Marchant et al. (2011);[7] Li and Thornton (1993b);[8] Tipping et al. (2011b);[9] Shacklette and Boerngen (1984);[10] Thornton and Webb (1979);[11] Martin et al. (1982);[12] Davies and Ballinger (1990);[13] Hunter et al. (1987a);[14] Thornton (1975a);[15] Davies and Roberts (1978);[16] Porter and Peterson (1977);[17]

Mn	Mo	Ni	Pb	Sb	Se	Sn	V	Zn
			432[7]			58[7]		424 (58–13,100)[33]
			647 (1–13,700)[33]					
			950 (80–3,680)[34]					
			59 (17–176)[35]		5.4 (2.1–23)[35]			
2,000–8,500[36]		39–960[36]	9,160 (4,400–17,200)[35]	139–760[36]	4.4 (2.5–66)[35]	1,200–2,600[36]	110–166[36]	800–26,000[36]
			60–5,400[36]		1.4–13[36]			
400 (150–2,500)[37]	5 (2–30)[37]	80 (20–5,300)[37]	700 (120–3,000)[37]		2.3 (1.6–7.0)[37]	120 (40–700)[37]		3,000 (700–49,000)[37]
762[39]		13–31[38]	58–621[38]					253–592[38]
		82[39]	771[39]					1,130[39]
		150–350[40]						2,390–5,860[40]
			561 (5–36,900)[22]					1,090 (81–115,000)[22]
			3,370 (276–8,620)[28]					1,320 (81–115,000)[33]
			1,100 (5–36,900)[33]					
			786 (45–9,660)[22]					388 (46–22,900)[22]
			2,940 (1,040–6,510)[28]					513 (121–5,150)[33]
			1,350 (172–9,660)[33]					

Colbourn and Thornton (1978);[18] Matthews and Thornton (1982);[19] Davies and Ginnever (1979);[20] Rieuwarts et al. (2014);[21] Culbard et al. (1988);[22] Xu and Thornton (1985); [23] Arnold et al. (2008);[24] Morgan and Morgan (1988);[25] Appleton et al. (2013);[26] Hartley et al. (2009);[27] Muskett et al. (1979);[28] Taylor (1997);[29] Read et al. (1998);[30] Spurgeon and Hopkin (1996a);[31] Goodman and Roberts (1971);[32] Thornton et al. (1985);[33] Clark et al. (2008);[34] Wadge et al. (1986);[35] Lisk (1988);[36] Berrow and Burridge (1980);[37] Kilkenny and Good (1998);[38] Mackay et al. (1972);[39] Shelton (1971);[40] Mitchell and Barr (1995).[41]

metals can enter a new food chain. Again, metals derived from the soil can be transmitted through local terrestrial food webs, in this case based on the decomposition cycle. It is worth noting that plants will also receive an input of metals from the atmosphere, typically by the deposition of dust particles. Whether or not these deposited metals enter the plant, they may still be ingested by herbivores or detritivores and be transmitted along local food chains.

Some soil-dwelling invertebrates may take up trace metals from soils directly via the soil solution, the metals being released from binding to soil particles by simple chemical processes, and then entering the invertebrate's body via permeable external surfaces. This process can be described as uptake from solution, a process that is routine in the case of aquatic animals. Indeed, earthworms, prime examples of soil-dwelling invertebrates that are able to take up metals from the soil solution, are, in terms of their evolution, strictly freshwater worms able to survive in the lumpy freshwater medium that is soil (Barrington, 1967).

Trace metals may also pass directly from soils to animals by ingestion, with subsequent digestion and uptake from the alimentary tract. Thus soil ingestion is a very important metal uptake route for earthworms. Birds or small mammals such as moles feeding on earthworms will be exposed not only to the metals accumulated by the worm, but to metals still associated with soil particles in the gut of the consumed earthworm. Grazing vertebrates, particularly cattle, will also inevitably consume soil as they crop grass. Even humans may consume soil particles inadvertently with salad and other vegetables, or less inadvertently in the case of small infants sampling soil in the back garden.

As emphasised in Chapter 3, the total concentration of a metal in a medium, such as a soil, may not be a measure of the fraction of that metal concentration that is bioavailable to an exposed organism by whatever routes of uptake available to that organism. Thus much of a trace metal associated with a soil particle may be too tightly bound chemically and not be released by processes of chemical equilibria into the soil solution for potential uptake by plants or earthworms. Similarly, some of the metal in or on a soil particle may be too tightly bound to be released for subsequent alimentary absorption by the digestive processes of an animal ingesting the soil. The strength of such digestive processes will also vary between the different animals concerned, highlighting that bioavailability, in this case trophic bioavailability, is a concept specific to a particular organism (Rainbow et al., 2011a). Nevertheless, it remains a truism that the higher the total concentration of a trace metal in a soil, the higher the likely metal bioavailability to a resident organism.

There are links between human health and trace metals in the terrestrial environment (Underwood, 1962; Thornton and Webb, 1979; Thornton, 1993). The most obvious, and major, route of uptake of trace metals by humans is by the ingestion of food and drinking water. While many foods may originate in aquatic environments, not least in the case of seafood, crops grown in any number of soils are key components of our food supply, as is the meat derived from livestock fed on such crops (Thornton and Webb, 1979). The transfer of metals to humans from vegetables and arable crops grown in metal-rich soils, directly or indirectly via a food chain, will be considered in this chapter. We also ingest soil occurring on garden vegetables as well as indoor dust, both of which carry metal loads (Underwood, 1962; Thornton and Webb, 1979). Little children will put into their mouths various items dirty with dust and soil, not least their hands, dropped food and toys. We will also inhale dust, with the potential for uptake of metals in the lungs; the higher the metal concentration of the dust, the greater the potential metal uptake (Thornton and Webb, 1979).

It is important to note that essential trace metals can cause biological effects by their absence as well as their presence. The absence of an essential trace metal in a soil, or more typically its presence at a bioavailability below a minimum threshold, can lead to the manifestation of deficiency effects, perhaps in the reduced growth of plants or in the presence of deficiency symptoms in agricultural livestock grazing affected pastures. In the British Isles, there are well-documented disorders in crops and livestock attributable to deficiencies of the essential metals copper, cobalt and selenium (Underwood, 1962; Thornton and Webb, 1979; Thornton, 1993).

So Chapter 4 is a chapter on the natural history of metals in the terrestrial environment.

4.2 Metal Concentrations in Soils

Prologue. British soils contain a range of trace metal concentrations, including very high levels caused by anthropogenic contamination from mining, smelting and other activities. Not all the trace metals present in soils are in bioavailable form. Calcium in soils will interact with the bioavailability of aluminium and iron present and affect plant growth.

All soils contain trace metals, the concentrations therein being primarily dependent on the nature of the rock from which the mineral component of the soil has been derived by erosion. Thus for each trace metal there is a wide range of so-called normal or background concentrations in soils, even before any anthropogenic influence has been brought to bear. Table 4.1 provides such ranges for many trace metals.

The major anthropogenic factor increasing metal concentrations in British soils is mining and its associated industries of smelting and refining. Table 4.1 provides general summaries of metal concentrations in soils affected by British mining, these consisting of concentrations at mining sites, at smelting and refining sites, as well as at sites downwind of smelters and refineries affected by the deposition of the inevitable metal-rich particles emitted into the atmosphere from smelter and refinery chimneys. Given the history of mining in the British Isles, it is not surprising that the major trace metals of mining origin that contaminate British soils are copper and arsenic, the ores of which often occur in association, and the other pairing of lead and zinc (Table 4.1). In addition, silver occurs in variable concentrations with lead ore, as does cadmium with zinc. It is notable in Table 4.1 that local mining can increase typical soil metal concentrations by many orders of magnitude, and local ecotoxicological effects are all but inevitable. Tables 4.2 and 4.3 break down the mining and smelter/refinery data into

more geographical detail for selected trace metals, including arsenic, copper, lead and zinc. Locations familiar from Chapter 2 are, not unexpectedly, encountered again, and will continue to appear throughout this chapter.

4.2.1 Mining Sites

Prologue. Former mining sites with very high soil concentrations of trace metals occur in Southwest England, the Northern and Southern Pennines and northeast and mid-Wales and at Shipham in Somerset.

In Southwest England, many soils in the Tamar valley (Colbourn et al., 1975), and the Hayle and Camborne region of Cornwall (Xu and Thornton, 1985; Li and Thornton, 1993b), for example, show the legacy of the intense mining activities, not least in the nineteenth century (Table 4.2). Arsenic, in particular, reaches huge concentrations of 10,000 µg/g and more in soils and spoil heaps at mining sites in the Tamar valley (Colbourn et al., 1975; Rieuwarts et al., 2014) and at the adjacent Devon Great Consols mine (Langdon et al., 2001) (Table 4.2). In comparison, typical background soil concentrations are less than 50 µg/g As (Table 4.1). Soil concentrations of copper are also raised at mining sites in both southwestern regions (Table 4.2), up to 2,000 µg/g Cu, in comparison to background concentrations of less than 100 µg/g Cu (Table 4.1).

In contrast, in the Peak District of Derbyshire in the Southern Pennines and in Weardale in the Northern Pennines, it is lead that shows greatly raised concentrations in soils affected by mining (Table 4.2). Zinc concentrations may also be raised from a background concentration usually below 200 µg/g but typically not by the same high factor as lead (Tables 4.1, 4.2). Background soil concentrations of lead are usually below 150 µg/g (Table 4.1), while soil concentrations up to 30,000 µg/g Pb can occur at Pennine mining sites (Table 4.2). At Winster near Wirksworth in the Peak District, raised concentrations of antimony (a trace metal relatively unstudied in British soils) have been measured in the topsoil, up to 51 µg/g

Table 4.2 Concentrations of selected trace metals in soils affected by mining in Britain (µg/g dry weight).

	Ag	As	Cd	Cu	Hg	Pb	Sb	Zn
Southwest England								
Tamar valley		60–2,500[1] 8,200–180,000[2]		29–2,000[1]		60–1,010[1]		114–1,020[1]
Hayle, Camborne		389 (334–461)[3] 322 (144–892)[5]		98–1,970[4]				
Devon Great Consols		8,980[6]		1,730[6]				
The Pennines								
Winster, Wirksworth, Derbyshire		30 (20–39)[3]	20 (4.4–38)[7]			18,000 (3,280–28,900)[7]	23 (6.5–51)[3]	1,530 (478–3,180)[7]
Derbyshire						11,000–19,400[8] 1,140–26,500[4] 171–10,500[9]		
Rookhope and Weardale, N. Pennines						151–27,200[9] 160–19,000[10]		
Wales								
Halkyn, Clwyd			6.1 (0.4–540)[11]			886 (35–48,000)[11]		728 (10–49,400)[11]
Ystwyth valley, Ceredigion						1,420 (90–2,900)[12]		455 (95–810)[12]
Near Parys Mountain, Anglesey				642 (116–1,410)[12]		379 (171–1,000)[12]		335 (190–650)[12]

Lake District

Coniston					790[13]			
Carrock Fell					725[6]		10,300[6]	
Shipham, Somerset	9,760–45,900[7] 107–10,400[14] 15,000 (2,040–50,000)[15] 1,100–7,810[16] 180–64,000[4]	38 (31–50)[3]	2,590–3,740[7] 85–87,400[14] 3,830 (475–7,800)[15] 597–2,610[16]	1.9–5.8[7]		66–372[7] 0.6–75[14] 159 (24–440)[15] 8.8–98[16] <2–520[4]	350 (171–466)[3]	11–14[7]

Note: Single values represent a summary value, for example a mean, a mean derived from transformation of raw data or a median. Ranges may similarly be derived from raw data.

Sources: Colbourn et al. (1975);[1] Rieuwarts et al. (2014);[2] Li and Thornton (1993b);[3] Culbard et al. (1988);[4] Xu and Thornton (1985);[5] Langdon et al. (2001);[6] Li and Thornton (1993a);[7] Colbourn and Thornton (1978);[8] Appleton et al. (2013);[9] Williamson and Evans (1973);[10] Davies and Roberts (1978);[11] Alloway and Davies (1971);[12] Arnold et al. (2008);[13] Davies and Ballinger (1990);[14] Matthews and Thornton (1982);[15] Davies and Ginnever (1979).[16]

Table 4.3 **Concentrations of selected trace metals in soils affected by smelting and refining in Britain (µg/g dry weight).**

	As	Cd	Cu	Hg	Pb	Sb	Zn
Smelter/Refinery Sites							
Southwest England							
New Mill As calciner Camborne, Cornwall	522 (210–925)[1]						
The Pennines							
Stone Edge, Derbyshire	31 (17–45)[1]	3.6–6.1[2]		0.1–0.2[2]	9,770–30,000[1]	103 (50.0–154)[1]	380–1,380[2]
Foxlane smelter, Derbyshire					28,800–300,000[3]		
Merseyside							
Cu refinery		28 (9.6–58)[4]	9,270 (2,600–18,700)[4]				
Atmospheric fallout							
Avonmouth smelter		69[5] 33[7] 218[6] 5.9[8]	1,950[6]		1,910[5] 398[7] 13,300[6] 291[8]		3,190[5] 1,700[7] 22,400[6] 406[8]
Foxlane smelter, Derbyshire					190–2,700[9]		
Merseyside Cu refinery		7.4 (5.5–9.7)[4]	494 (129–880)[44]				
Swansea		26[10]	160[10]		263[10]		543[10]

Note: Single values represent a summary value, for example a mean, a mean derived from transformation of raw data or a median. Ranges may similarly be derived from raw data.

Sources: Li and Thornton (1993b);[1] Li and Thornton (1993a);[2] Colbourn and Thornton (1978);[3] Hunter et al. (1987a);[4] Martin et al. (1982);[5] Spurgeon and Hopkin (1996a);[6] Read et al. (1998);[7] Spurgeon and Hopkin (1999a);[8] Colbourn and Thornton (1978);[9] Goodman and Roberts (1971).[10]

(Table 4.2), compared with local background concentrations of less than 2 µg/g Sb (Table 4.1) (Li and Thornton, 1993b).

In the region of Halkyn Mountain in northeast Wales, soils affected by mining can contain raised concentrations of lead, zinc and cadmium that match or exceed the raised concentrations in the Pennines (Table 4.2) (Davies and Roberts, 1978). Thus, measured lead concentrations have reached 48,000 µg/g, zinc concentrations 49,400 µg/g and cadmium concentrations 580 µg/g. This cadmium concentration contrasts strongly with background concentrations of cadmium in soils that usually lie below 2 µg/g Cd (Table 4.1). Raised soil concentrations of lead and zinc have also been measured in soils in the mining region of the Ystwyth valley in Wales (Table 4.2) (Alloway and Davies, 1971).

In the Shipham region of the Mendips in North Somerset, the combination of zinc, lead and cadmium is again prominent when considering soil concentrations raised by local mining activity (Table 4.2). Historically, Shipham is renowned for the mining of zinc, but lead mining also took place nearby (Davies and Ballinger, 1990; Li and Thornton, 1993a). Local contamination of soils (and vegetables) by cadmium in Shipham has occasionally attracted press attention, and indeed local soil concentrations of cadmium are amongst the highest in Britain (Davies and Ginnever, 1979; Davies and Ballinger, 1990).

4.2.2 Smelter and Refinery Sites

Prologue. Examples of smelter and refinery sites with very high soil concentrations of trace metals occur in Cornwall, Derbyshire and Merseyside.

Soils at smelter sites are similarly raised in particular trace metal concentrations (Table 4.1, 4.3). The historical arsenic calcination site at New Mill, in the Camborne region of Cornwall, has raised arsenic soil concentrations (Table 4.3), in line with soil concentrations at the nearby Camborne mining site at Fraddam (Table 4.2) (Li and Thornton, 1993b). The Stone Edge smelter site in Derbyshire received ore mainly from the Winster mining site (Table 4.2)

(Li and Thornton, 1993a). Correspondingly, soils from this site have similarly greatly raised concentrations of lead, zinc, cadmium and antimony (Tables 4.2, 4.3).

Surface soils at a Merseyside copper refinery complex, established in the first half of the twentieth century, contained elevated cadmium and copper concentrations, up to 58 µg/g Cd and 18,700 µg/g Cu respectively, when sampled in the 1980s (Table 4.3) (Hunter et al., 1987a).

4.2.3 Smelter and Refinery Emissions

Prologue. Sites downwind of smelters and refineries also show raised trace metal soil concentrations.

Smelter and refinery chimneys inevitably release metal-rich particles into the atmosphere and these will be deposited downwind onto local habitats (Table 4.3).

The spread and biological effects of metal-rich smelter emissions in the United Kingdom have been studied most extensively in the case of the Avonmouth primary zinc, lead and cadmium smelting works near Bristol (Little and Martin, 1972; Martin et al., 1982; Hopkin, 1989). In the 1970s, some 50 tonnes of zinc, 30 tonnes of lead and 3 tonnes of cadmium were emitted as fine particles each year from its chimneys (Hopkin, 1989). Most of these emitted particles were metal sulphates, sulphides and oxides with diameters below 2.5 µm. Such small particles may be transported considerable distances, and metals derived from the Avonmouth smelting works have been detected in soil, vegetation and invertebrates up to 25 kilometres to the northeast in the direction of the prevailing wind (Hopkin, 1989). Soil concentrations in Haw Wood, 3.2 kilometres northeast of the smelting works, averaged 69 µg/g Cd, 1,910 µg/g Pb and 3,190 µg/g Zn in 1979 (Table 4.3), in contrast to soil concentrations of 0.30 µg/g Cd, 41 µg/g Pb and 118 µg/g Zn 28 kilometres away at Midger Wood (Martin et al., 1982).

In the case of the Merseyside copper refinery complex, average soil concentrations of 7.4 µg/g Cd and 494 µg/g Cu were measured 1 kilometre distant from the complex (Table 4.3), still well above average concentrations of 0.96 μg/g Cd and 13 μg/g Cu at a control site (Hunter et al., 1987a).

4.2.4 Other Anthropogenic Input into Soils

Prologue. Other anthropogenic input of trace metals into British soils has been reduced since the mid-1900s.

Other sources of the anthropogenic input of trace metals into soils in the British Isles are compiled into Table 4.1. Many are now mainly of historic interest as practices (particularly agricultural practices) have changed. Strikingly, the data for lead concentrations in urban soils, house dust and road dust (Table 4.1) predate the banning of the addition of tetra-ethyl lead as an antiknock agent in petrol. There is now also control in Britain and elsewhere in Europe of the nature of sewage sludge that can be added to agricultural soils, in particular limits to the sludge concentrations of arsenic, cadmium, chromium, copper, lead, mercury, molybdenum, nickel, selenium and zinc (Department of Environment, 1989). Fly ash from the burning of coal and refuse can contain high concentrations of trace metals – particularly arsenic in coal fly ash, cadmium and lead in refuse fly ash and selenium in both (Table 4.1) (Wadge et al., 1986). While coal fly ash is used in agriculture in many countries of the world, with the potential to increase crop yields by providing the likes of calcium, iron, potassium, magnesium and sodium, as well as zinc (Basu et al., 2009), the toxic metal contents of both fly ashes usually consign them to landfill in Britain (Hutton and Symon, 1986; Wadge et al., 1986). Otherwise, more pulverised fly ash from coal-fired power stations in Britain is being repurposed, for example by incorporation into concrete.

4.2.5 Bioavailability of Trace Metals in Soils

Prologue. The total concentrations of trace metals in a soil typically exceed the concentrations that are bioavailable to different organisms associated with that soil. Chemical extraction techniques attempt to model such bioavailable metal concentrations.

The soil metal concentrations listed in Tables 4.1 to 4.3 are so-called total or total recoverable metal concentrations. Typically, soils have been dried and then digested in very strong acid to produce an acidic digest containing all the metals that were associated with that soil sample. The metal concentrations are then measured in the digest by standard chemical analytical techniques, back calculation giving the starting metal concentration per unit weight of dried soil. Typical strong acids used in such digestion procedures are concentrated nitric acid on its own, a 4:1 mixture of concentrated nitric and perchloric acids, or occasionally aqua regia, a 1:3 mixture of concentrated nitric and hydrochloric acids. These are seriously strong acids capable of dissolving most soil particles.

The metals associated with a soil particle may be adsorbed onto the surface, binding chemically with either organic components of the soil particle and/or with inorganic constituents such as iron or manganese oxides that might be present. A metal may also be part of the original mineral from which the soil particle has been derived by erosion, in effect part of the very structure of the soil particle, as in the case of a soil particle originating from a metal ore. In this latter case, the chemical binding of the metal to the likes of oxide or carbonate may be very strong indeed. An extremely strong acid, not least aqua regia, is capable of dissolving most soil particles completely, releasing into solution even any metals integrally bound in the structure of the soil particle.

But what does this mean in a biological context? Only metal that is bioavailable to an organism has the capacity to interact with the biology of that organism. And bioavailability varies according to the organism concerned. In the case of plants with roots, it will be (a portion of the) metal that has been released into the soil solution from soil particles that may be taken up by the roots – i.e., be bioavailable to the plant by this route. We are ignoring for the moment atmospheric deposition of metals onto leaves, which may or may not enter the plant (be bioavailable to the plant) by this route. Chemical exchange of trace metals between soil particles and the soil solution will vary with the chemical nature of the soil solution (not least its acidity and the presence or absence therein of organic compounds that might bind metals) and therefore with soil type but will really only concern metals that are bound to the outside of a soil particle. Metals that are part of the chemical structure of the original soil particle will not be so released into the soil solution. Already, then, it is clear that a total metal concentration in a soil, as measured using very strong acid digestion, will not be an absolute measure of the metal in the soil that is bioavailable for uptake by the roots of plants.

Similarly, it will be (a proportion of) metals released into the soil solution that will be available for uptake across the permeable external surface of an earthworm, i.e., be bioavailable for uptake by this route. The bioavailable fraction of metals in the soil solution available for uptake by plant roots and by earthworms respectively may well be very similar, given the near-universal nature of metal uptake mechanisms as described in Chapter 3, but not necessarily so. And the metals released from the same soil particles ingested into the gut of an earthworm may well differ in quantity and chemical nature from those bioavailable to the worm in the external soil solution, given the presence of digestive enzymes and various metal-binding agents in the alimentary tract of the worm. The total bioavailability of a soil-associated metal to an earthworm will therefore consist of both metal bioavailable for external uptake from the soil solution and metal trophically bioavailable in the alimentary tract. The relative contribution of these two uptake routes in earthworms will vary according to the characteristics of the soil concerned. And trophic availability will vary between animals (Rainbow et al, 2011a). It is clear that there is no universal concentration value that can be called the bioavailable metal concentration of a soil.

Total metal concentrations in soils may still give an indication of the relative bioavailabilities of the metals present, but a potentially unreliable one. It is not surprising then that there are various other chemical methods used to measure the metal contents of soils that are relatively easily released, i.e., are exchangeable. It is important to remember that no

chemical extraction technique can measure a bioavailable metal concentration in an environment (bioavailability being a concept specific to a particular organism). Nevertheless, some chemical extraction methods may model metal bioavailabilities (generalised across different organisms) more closely than others. The very strong acids chosen for the extraction of the total metal content of a soil will inevitably provide an overestimate of the metal content that might be bioavailable to different organisms. Weaker chemical agents, therefore, have been proposed to measure the concentrations of the so-called exchangeable metals present in soils, as possible models of the soil metal concentrations that might be bioavailable to particular organisms by particular uptake routes (e.g., plants via root uptake, or earthworms via epidermal uptake). There is, however, no consistency in the choice of such weaker chemicals, and, therefore, in quite what metal fraction in a soil is deemed exchangeable. Nevertheless, it makes sense that the concentration of any such exchangeable metal is more likely to reflect and model what might be bioavailable to a particular organism than does the total metal concentration of a soil.

In a 1971 study of the raised concentrations of cadmium, lead and zinc in soils downwind of the lead and zinc smelting plant at Avonmouth, total metal concentrations in the soils were measured after acid digestion in a 4:1 mixture of concentrated nitric and perchloric acids (Little and Martin, 1972). In addition, soil samples were extracted with a 2.5% acetic acid solution to model the soil concentrations of these metals that might be bioavailable for uptake by elm (*Ulmus glabra*) and hawthorn (*Crataegus monogyna*). In 1979, the same research team, again investigating the atmospheric deposition of these three metals and copper downwind of the Avonmouth smelter, used concentrated nitric acid alone for total metal concentration analysis of soils, and 0.5 molar ammonium nitrate to model the soil metal concentrations that might be available in the soil for uptake by trees, shrubs and herbaceous plants (Martin et al., 1982). Yet another weak chemical agent was used to model the soil availabilities to plants of cadmium, copper, lead and zinc in a study of soils of the southern Peak

District associated with mining and smelting sites, in this case a dilute nitric acid extraction using 0.05 molar (M) nitric acid (Colbourn and Thornton, 1978). The concentration of nitric acid in this diluted form is about 300 times lower than that in undiluted concentrated nitric acid (about 15 M), as used to extract total metal concentrations in soils. Relatively weak chemical agents now used widely to define exchangeable metal concentrations in soil are acetic acid (0.5 M, about 3%) and the metal chelating agent ethylenediaminetetraacetic acid (EDTA) (0.05 M) (Archer and Hodgson, 1987).

The use of sequential extraction, involving the application of a series of chemical agents of increasing strength, has also been common in attempts to model soil concentrations that might be bioavailable, particularly via the roots of plants. A study of arsenic in garden soils and vegetable crops in the historical mining region of Hayle–Camborne–Godolphin in Cornwall used deionised water and then a dilute acid-fluoride solution in series to model the arsenic fraction in the soils that was potentially bioavailable to vegetables such as beetroot, carrots, lettuce, onions and peas (Table 4.4a) (Xu and Thornton, 1985). The soil concentrations of arsenic extracted by these weak agents totalled less than 2.5% of the total arsenic present in the soils, the total arsenic here being extracted with magnesium nitrate. The sequential extraction system applied to model the soil bioavailability of copper and cadmium to grasses (*Agrostis stolonifera*, *Festuca rubra*), coltsfoot (*Tussilago farfara*) and the horsetail (*Equisetum arvense*) in a study of soils affected by emissions from the copper refinery complex in Merseyside involved double-distilled water, acetic acid and EDTA (Hunter et al., 1987a). This sequential extraction system released between 43% and 93% of total copper, and between 32% and 76% of total cadmium, from soils at the refinery site, at a site 1 kilometre downwind, and at a control site (Table 4.4b).

A more complex sequential extraction procedure was used to model potentially available soil copper to earthworms in soils from copper-contaminated mine spoil sites at Carrock Fell in the Lake District and at the Devon Great Consols mine at the top of the Tamar valley (Table 4.4c) (Langdon et al., 2001). The bioavailable copper modelled was the total copper bioavailability integrated across both uptake routes of epidermal uptake from the soil solution and trophic uptake from ingested soil particles. The sequential procedure detailed in Table 4.4c extracted more than 90% of the total copper present in the contaminated soils at Carrock Fell (96%) and Devon Great Consols (99%) and at a control uncontaminated site (93%). A strength of such a sequential extraction system is that it can highlight the differences in the proportional contributions of the different forms of chemical binding of a metal in different soils. This is shown clearly between the two copper-contaminated soils, with a much higher percentage of defined exchangeable copper in the Devon Great Consols soil than the Carrock Fell sample (Table 4.4c).

The concept of a sequential extraction system to separate, in turn, the different forms of chemical binding of a metal in a soil is attractive. Nevertheless, extracted metal concentrations need to be interpreted with caution, for the soil-associated metals are distributed between different metal-binding components of a soil in a series of chemical equilibria. The extraction of a single chemical form of bound metal, as exemplified in Table 4.4, is likely to cause changes in the distribution of the soil metals between the other remaining chemical forms binding the metal, simply as a result of the presence of such chemical equilibria. Interpreted with caution, however, sequential extraction remains a widely used approach to investigate the potential bioavailabilities of soil-associated metals.

Given the variability between soil-inhabiting organisms in the routes used for the uptake of soil-associated trace metals, it is impossible to come up with a single chemical extraction system to model the bioavailability of a soil-associated metal across species. It is not surprising, therefore, that there is an array of different extraction systems used by different authors to obtain a fractional soil metal concentration that is an improvement upon a total metal concentration when attempting to assess the comparative metal bioavailabilities (and hence comparative potential ecotoxicities) of different metal-contaminated soils (van Gestel, 2008). The important point, however, is to remain aware that, even if a particular extracted soil metal concentration correlates well with

Table 4.4 **Examples of sequential extraction procedures used to model fractions of the metal concentrations of soils that might be bioavailable for uptake by the roots of plants and by earthworms.**

(a) *As concentrations (μg/g) in extracts of garden soils of Cornwall*

	Total	Water-soluble	Dilute acid–fluoride extractable	Percentage of total extracted
	(saturated magnesium nitrate)	(deionised water)	(0.03 M ammonium fluoride + 0.025 M hydrochloric acid)	
Mean	322	0.89	4.97	1.8
Range	144–892	0.31–2.78	0.45–17.0	0.5–2.2

Source: Xu and Thornton (1985)

(b) *Mean ($\pm$ SE) Cu and Cd concentrations (μg/g) in extracts of soils affected by emissions from a copper refinery complex in Merseyside*

	Total	Water-soluble	Acetic acid	EDTA	Percentage of total extracted
	(4 conc. nitric acid: 1 conc. perchloric acid, 120°C)	(double-distilled water)	(0.5 M acetic acid)	(0.5 M EDTA, pH 4.0)	
Copper					
Refinery site	9,270 ± 2,200	9.3 ± 2.5	2,850 ± 1,070	5,800 ± 1,670	93
1 km site	494 ± 105	1.7 ± 0.4	4.9 ± 1.2	268 ± 52.5	56
Control site	13.3 ± 1.4	0.17 ± 0.01	0.21 ± 0.03	5.3 ± 0.3	43
Cadmium					
Refinery site	28.3 ± 6.95	0.05 ± 0.01	12.6 ± 1.43	8.9 ± 1.68	76
1 km site	7.4 ± 0.9	<0.01	0.94 ± 0.18	2.34 ± 0.17	44
Control site	0.97 ± 0.1	<0.01	0.07 ± 0.01	0.24 ± 0.02	32

Source: Hunter et al. (1987a)

(c) *Mean Cu concentrations (μg/g) in extracts of soils from copper-contaminated mine spoil sites at Carrock Fell (Cumbria) and Devon Great Consols mine (Devon)*

	Total	Exchangeable	Surface bound to oxide and carbonate	Bound to Fe/Mn oxides	Organically bound	Residue (% total)
		(1 M magnesium chloride, pH 7.0)	(1 M sodium acetate, pH 5.0)	(0.04 M hydroxylamine hydrochloride in 25% acetic acid)	(3:10 0.02 M nitric acid: 30% hydrogen peroxide; then 1.2 M ammonium acetate in 10% nitric acid)	(1:3 conc. nitric acid: conc. hydrochloric acid)
Carrock Fell	1,320	4	68	1,090	108	48 (3.6)
Devon Great Consols	12,930	11,810	327	210	429	156 (1.2)
Control site	14	2	4	0	7	0

Source: Langdon et al. (2001)

ecotoxicological effects observed on an organism, that fractional soil concentration remains a model, not an absolute measure, of the bioavailability of a soil-associated metal to that organism. Only the bioaccumulated metal concentration in the organism concerned can provide an integrated measure of the total bioavailability of a soil-associated metal to that organism.

4.2.6 Bioaccessibility

Prologue. Bioaccessibility is a relative measure of the release of a trace metal from ingested soil or dust particles into solution in the gut of a consumer, modelled by the action of a defined cocktail of chemicals mimicking digestion (typically of humans).

Another measure commonly used in the context of metal concentrations in soils is that of bioaccessibility, particularly when considering soil ingestion as an exposure route for the uptake of toxic metals by

humans, especially infants (Oomen et al., 2002). The uptake of a metal from ingested items involves two stages that in effect define the trophic bioavailability of an ingested metal to the consuming animal. In the specific case of ingested soil, the first stage is the release of metals from the soil particle by the processes of digestion in the gut, and the second stage is the subsequent uptake of these released metals from the gut lumen into the gut epithelium of the consumer. It is the efficiency of the first stage that is described as the bioaccessibility of the ingested metal. Bioaccessibility is therefore not synonymous with trophic bioavailability, which requires the combination of both stages: digestion, then uptake. The concept of bioaccessibility can only be applied to animals using extracellular digestion, as opposed to intracellular digestion whereby gut epithelial cells will engulf small food particles by endocytosis, combining digestion and absorption into a single step. Intracellular digestion is more typical of many invertebrates, for vertebrates, particularly mammals, rely on

Box 4.2 Bioaccessibility: Unified BARGE Method

The UBM uses a two part in vitro simulation of the human gastrointestinal tract (Wragg et al., 2011; Appleton et al., 2013). A gastric extraction simulates the digestive action of the stomach. A gastrointestinal extraction, using sequential extraction of the gastric phase followed by the intestinal phase, simulates the joint action of stomach and intestine digestive processes. Two bioaccessibility measures therefore arise from the UBM, a 'stomach' value (BS-Pb in Table 4.5) and a 'stomach and intestine' value (BSI-Pb in Table 4.5). Typically, it is the stomach value that is the higher value, probably because of re-adsorption of metals to the (albeit now modified) soil particles or to the precipitation of metals in the less acidic conditions of the intestine. In a study of lead bioaccessibility in urban topsoils (Table 4.5), the 'stomach' phase value was chosen as the more suitable (Appleton et al., 2012). A wider study of lead-contaminated soils from Pennine mining areas as well as these urban soils (Table 4.5), compared the two measures of lead bioaccessibility to gain insight into the geochemical controls of lead bioaccessibility in different types of soil (Appleton et al., 2013). As in the case of bioavailability, differences in the chemical makeup of different soils affect the comparative bioaccessibilities of metals in these different soils (Table 4.5).

extracellular digestion with subsequent alimentary uptake of solubilised forms.

Bioaccessibility has the potential to vary between animal consumers because of the interspecific variation of digestive processes. In an attempt, then, to produce a standard measure of bioaccessibility for the comparative investigation of soils (and dusts), the measurement of bioaccessibility typically involves a defined standard digestive medium. The bioaccessibility of a metal in a potential food item can be now defined as a relative measure of its release into solution from that solid material through the action of a defined cocktail of chemicals mimicking a digestive process, typically of humans (Oomen et al., 2002; Wragg et al., 2011). There are several in vitro digestion models simulating the human gastrointestinal tract to assess the bioaccessibility of metals in soils. One of the more widely accepted methods has been developed by the BioAccessibility Research Group of Europe (BARGE), the Unified BARGE Method (UBM) (Box 4.2) (Wragg et al., 2011).

The concept of bioaccessibility has also been extended to metals associated with dust, typically

defined as fine (less than 100 μm) settled or airborne particulate material (Turner, 2011).

4.2.7 Trace Metals and Soil Types

Prologue. Plant ecologists have long been familiar with the concept that different plants grow preferentially in different soil types that might be defined by the presence, absence or relative concentration of different metals in the soil.

4.2.7.1 Calcicoles and Calcifuges

Prologue. A particularly important such metal is calcium. The indirect effects of soil chemistry on the bioavailability of aluminium and iron affect the growth of calcicole and calcifuge plants in soils of different calcium concentrations.

Classic examples of plants growing differentially in different soils are calcicole and calcifuge plants. Calcicole plants are capable of thriving in a calcareous soil. A calcareous soil is mostly or partly composed of calcium carbonate, thus a soil that contains chalk or

Table 4.5 **Estimates of the bioaccessibility of lead in topsoils from mining areas (Derbyshire and Rookhope, Northern Pennines)[1] and urban areas (Glasgow, London, Northampton, Swansea).[2]**

	Total Pb (mean µg/g in <2mm fraction	BS-Pb (mean µg/g in <250µm fraction)	% BS-Pb	BSI-Pb (mean µg/g in <250µm fraction)	% BSI-Pb
Derbyshire	1,710	895	52%	172	10%
Rookhope	1,840	1,010	55%	507	28%
Glasgow	836	418	50%	–	–
London	1,740	1,200	69%	–	–
Northampton	85	37	44%	–	–
Swansea	821	563	69%	–	–

Note: Measures quoted are derived from the UBM,[3] simulating lead bioaccessibility in the 'stomach' (BS-Pb) and the 'stomach and intestine' (BSI-Pb).[1,3,4]

Sources: Appleton et al. (2013),[1] Oomen et al. (2002),[2] Wragg et al. (2011),[3] Appleton et al. (2012).[4]

other forms of limestone. Calcicole plants include the grasses of chalkland grassland communities such as sheep's fescue *Festuca ovina* and meadow oat grass *Avenula pratensis* in heavily grazed chalk grasslands (NVC Community CG2 of the British National Vegetation Classification system). Calcicole species other than grasses in this chalkland community include salad burnet (*Sanguisorba minor*), bird's foot trefoil (*Lotus corniculatus*), fairy flax (*Linum catharticum*), small scabious (*Scabiosa columbaria*) and wild thyme (*Thymus praecox*) (Lousley, 1950). Calcifuge plants, on the other hand, thrive in acid soils, the very word *calcifuge* being derived from the Latin 'to flee chalk'. Calcifuge plants include heather (*Calluna vulgaris*) and bilberry (*Vaccinum myrtillus*).

Life, however, is not so simple that it is the presence of a high or low concentration of calcium in the soil solution to which the plants are responding. In calcareous soils, calcium is bound to carbonate. The chalk has often been derived from the calcium carbonate laid down as plates (coccoliths) on phytoplankton (coccolithophores) living in the prehistoric past. The coccoliths of these phytoplankton, typified by the cosmopolitan species *Emiliania huxleyi* abundant throughout temperate, subtropical and tropical oceans of the world today, sediment out after death to form calcareous oozes that may be kilometres thick on the ocean floor. It is a late Cretaceous calcareous ooze that has been uplifted over geological time to now outcrop widely as chalk in southern England and form the impressive White Cliffs of Dover and the Needles on the Isle of Wight. Still worth reading on this topic is a classic essay *On a Piece of Chalk* by Thomas Henry Huxley, based on a public lecture that he delivered in 1868 to the working men of Norwich at a meeting of the British Association for the Advancement of Science. The essay, a masterpiece in what is now termed the public understanding of science, reconstructs the geological history of Britain from a simple piece of chalk in a wonderfully clear demonstration of the methods of science.

A calcareous soil is a basic or alkaline soil in which any acid has been neutralised. Whether a soil is acid or alkaline will affect the bioavailability of trace metals to plants via the soil solution and root system. Thus, the bioavailabilities of aluminium or the essential metal iron, for example, will be very different in a chalkland soil as opposed to a more acidic sandy or heathland soil. It appears, then, that it is not the presence or absence of calcium per se that defines whether a plant is a calcicole or a calcifuge. Calcicoles grown on acidic soils often develop the symptoms of aluminium toxicity, while calcifuge plants grown on alkaline soils develop symptoms of iron deficiency. In the former case, the bioavailability of aluminium in the soil solution to the roots of the calcicole plants has been increased to a toxic level. In the latter case, there is reduced bioavailability of iron in the soil solution to the roots of the calcifuges, as a result of the chemical form of the iron in an alkaline soil.

4.2.7.2 Aluminium and Laterites

Prologue. Laterites are soil types rich in iron and aluminium. Aluminium toxicity is an important growth-limiting factor for many plants in acid soils.

Aluminium is one of the most abundant elements in soils. In neutral or weakly acidic soils, aluminium exists in the form of stable insoluble aluminosilicates or oxides (Matsumoto, 2000). In more acid soils, however, the aluminium becomes solubilised, ultimately to Al^{3+}. These soluble forms are more bioavailable to the roots of plants and are, therefore, potentially toxic. In fact, aluminium toxicity is an important growth-limiting factor for many plants in acid soils below pH 5.0, when the phytotoxic form Al^{3+} predominates (Rout et al., 2001). In contrast, at neutral or higher pH in soil, aluminium is found as nontoxic hydroxides, including insoluble $Al(OH)_3$ (gibbsite) and the aluminate anion $(Al(OH)_4^-)$ (Matsumoto, 2000).

Laterites are soil types rich in iron and aluminium that are typically formed in tropical countries by the weathering of underlying parent rock, for example Tertiary basalts in (now temperate) Northern Ireland (Hill et al., 2000). According to the nature of the parent rock, laterites can also be high in nickel (Hill et al., 2000). Bauxite, the major ore of aluminium is a variety of laterite, consisting mostly of three aluminium oxide/hydroxide minerals, including gibbsite in a mixture of two iron oxides, the clay mineral kaolinite and some titanium dioxide. Lateritic soils derived from bauxite, therefore, will contain relatively high concentrations of aluminium, in addition to iron.

4.2.7.3 Serpentine Group of Soils

Prologue. Serpentine soils harbour a distinctive flora and are toxic to many plants. This toxicity has been attributed simplistically to raised bioavailabilities of nickel, chromium and/or cobalt, but other physico-chemical factors are almost certainly involved.

Serpentine soils represent another soil type for which their particular chemical compositions have been postulated to influence the vegetation growing on them (Proctor, 1971a,b; Proctor and Woodell, 1971). 'Serpentine' is actually an imprecise term, referring to a variable group of minerals that are derived hydrothermally from the mineral olivine (Proctor and Woodell, 1971), a magnesium iron silicate known as peridot when of gem quality. Serpentine minerals are, therefore, magnesium iron silicates, and they variably contain additional trace metals such as cobalt, chromium, manganese, nickel and iron. Serpentine soils, derived from serpentine minerals, are found in particular British locations, including the Lizard peninsula in Cornwall, Shetland and several mainland Scottish sites (Figure 4.1) (Brooks, 1987). The concentrations of the trace metals cobalt, chromium and nickel in these serpentine soils (Table 4.6) are well above those considered background (Table 4.1), dramatically so in the case of nickel.

Serpentine soils harbour a distinctive flora and are toxic to many plants. The characteristic vegetation of British serpentine soils has been described in terms of up to five different plant communities (Table 4.7) (Proctor and Woodell, 1971; Brooks, 1987). Most of the serpentine vegetation falls into two types: that associated with debris or scree (with less than 25% cover), and that of heath or grassland with complete vegetation cover (Table 4.7) (Proctor and Woodell, 1971). In the latter case, a sedge grassland develops after grazing (Brooks, 1987). Rock crevices bear their own distinctive vegetation, and in poorly drained sites there are mires which may be base-rich or base-poor, again with distinct plant communities (Table 4.7).

The toxicity of serpentine soils has historically been attributed to raised toxic bioavailabilities of nickel, chromium and/or cobalt in serpentine soils, but this simplistic hypothesis is now rejected (Slingsby and

Table 4.6 **Concentrations of selected elements (including trace metals) in serpentine soils in Britain (µg/g dry weight).**

	Ca	Co	Cr	Mg	Ni
England					
The Lizard, Cornwall	13,600	139	2,410	69,000	2,030
Scotland					
Shetlands	12,380	322	10,380	158,000	5,460
Shetlands	4,640	77	1,670	150,000	2,200
Aberdeen	15,900	277	4,880	132,000	4,630

Sources: Slingsby and Brown (1977); Brooks (1987).

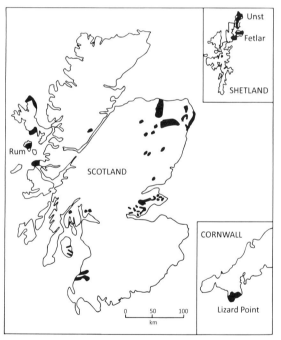

Figure 4.1 Serpentine soils (solid black areas) in the British Isles. (From Brooks, 1987, with permission of Springer Nature.)

(Proctor, 1971a, b; Slingsby and Brown, 1977). While some serpentine plant species may accumulate high concentrations of nickel in roots and foliage, as in the case of species of bentgrass *Agrostis* growing on Unst serpentine debris in the Shetlands, they do not necessarily show symptoms of nickel toxicity (Slingsby and Brown, 1977; Brooks, 1987). Nor do oat plants grown on serpentine soils (Slingsby and Brown, 1977). Other driving factors proposed to explain serpentine soil toxicity include a deficiency of major nutrients, especially phosphorus, and a high magnesium/calcium ratio (Table 4.6), promoting magnesium toxicity (Slingsby and Brown, 1977; Brooks, 1987). There certainly does not seem to be a single chemical explanation of the toxicity of serpentine soils. The answer probably lies in the interaction of 'a symphony of factors', which might also include physical factors such as surface soil instability, drought and wind exposure, and also varies in effect between serpentine soils and sites (Spence, 1970; Brooks, 1987).

Brown, 1977; Brooks, 1987; Slingsby et al., 2010). The high trace metal concentrations in serpentine soils do not consistently translate into high metal bioavailabilities to the roots of plants, with variation between the serpentine soils and between the plants tested

4.3 Microbial Community

Prologue. High trace metal bioavailabilities in soils will impact negatively on the biomass and diversity of soil microbes, and on soil microbial processes,

Table 4.7 **Dominant taxa of plant communities on Scottish serpentine sites with a characteristic serpentine flora.**

Heath–grass–sedge	Scree/debris	Base-rich crevices	Base-rich mire	Base-poor mire
Calluna vulgaris	*Arenaria norvegica*	*Asplenium viride*	*Carex flacca*	*Carex demissa*
Heather	Arctic sandwort	Green spleenwort	Blue sedge	Common yellow sedge
Erica cinerea	*Cardaminopsis petraea*	*Asplenium adiantum-nigrum*	*Agrostis stolonifera*	*Schoenus nigricans*
Bell heather	Northern rock cress	Black spleenwort	Creeping bentgrass	Black bog rush
Carex pulicaris	*Silene maritima*	*Draba incana*	*Ranunculus flammula*	*Armeria maritima*
Flea sedge	Sea campion	Hoary whitlow grass	Lesser spearwort	Thrift
Juniperus communis		*Cystopteris fragilis*	*Carex pulicaris*	
Common juniper		Brittle bladder fern	Flea sedge	
Deschampsia flexuosa		*Silene maritima*	*Festuca rubra*	
Wavy hair grass		Sea campion	Red fescue	
Polygala serpyllifolia		*Avenula pratensis*		
Heath milkwort		Meadow oat grass		
		Cardaminopsis petraea		
		Northern rock cress		

Sources: Spence (1970); Brooks (1987).

although some microbial populations show trace metal tolerance.

The microbial community of soils, essentially consisting of fungi and prokaryotes such as bacteria, is crucial to the ecological functioning of the soil ecosystem. This community drives key processes such as the decomposition of organic matter and the regeneration of nutrients such as nitrate and phosphate, mineralising organic nitrogen and phosphorus compounds to inorganic ions that can be taken up by primary producers, as for example via the roots of plants.

High trace metal bioavailabilities in soils will impact negatively on the biomass and diversity of soil microbes, and thus on soil microbial processes, measured, for example, in terms of rates of microbial respiration, mineralisation or nitrogen fixation (Brookes, 1995; Giller et al., 1998). Such ecotoxicological effects are particularly apparent in the vicinity of smelters, being manifested as the accumulation of deep layers of leaf litter in woodland as microbial decomposition processes have been inhibited (Giller et al., 1998). Another historical example of metal ecotoxicity to soil microbial processes concerns agricultural soils to which sewage sludge has been applied (Giller et al., 1998), prior to the establishment late in the twentieth century of limits to the concentrations of trace metals permissible in sewage sludge destined for agricultural application.

It may well be the case that soil microorganisms are more sensitive to toxic metal stress than soil animals or plants growing on the same soils (Giller et al., 1998). There is, in fact, great variability in the data concerning what soil concentrations of metals are toxic to soil microbial communities (Brookes, 1995; Giller et al., 1998). This variability is caused by the indirect relationship between total metal concentrations and bioavailability (in this case, to different microbes) in soils, as well as by differences in results obtained in necessarily relatively short-term laboratory toxicity experiments (of inevitably varying design) and longer-term field studies. With the caveat required by the presence of such variation in measured toxic concentrations of trace metals in soil, microbial respiration has been reported to be decreased at soil concentrations above 1,000 µg/g copper or zinc (Brookes, 1995). Similarly, nitrogen mineralisation in soil has been reported to be inhibited at around 1,000 µg/g copper, zinc or nickel, at about 100 to 500 µg/g lead or chromium and at around 10 to 100 µg/g cadmium (Brookes, 1995). While treating such toxic concentration data with caution as to their general applicability, it is still clear that many British soils contaminated with trace metals derived from mining and associated smelting and refining processes (Tables 4.1 to 4.3) will be toxic to soil microorganisms, with consequent effects on soil microbial processes.

A particularly important specific microbial community in soils is that of mycorrhizal fungi, which establish (usually) mutualistic symbioses with the majority of plants, providing a direct physical link between the soil and the roots of the plant. The mycorrhizal fungi feed water and nutrients directly into the plant roots and receive in return sugars from the plant. The symbiotic association is intracellular in the case of arbuscular mycorrhizal fungi (AMF), or extracellular in the case of ectomycorrhizal fungi. High concentrations of toxic metals in soils receiving long-term applications of metal-containing sewage sludge do reduce AMF spore numbers, as well as AMF species richness and diversity (Del Val et al., 1999).

A wide range of macrofungi (predominantly ascomycetes or basidiomycetes) concentrate trace metals in their fruiting bodies (e.g., mushrooms), even when growing on soils without atypically raised metal concentrations (Lepp, 1992). Species of *Agaricus*, for example, bioaccumulate cadmium and mercury from such soils, while the related species *Amanita muscaria* (the fly agaric) has raised concentrations of cadmium, selenium and vanadium (Lepp, 1992). Table 4.8 lists genera of fungi that have at least one species known to accumulate atypically raised concentrations of trace metals from uncontaminated soils.

Soil bacteria and fungi do have the ability to adapt to high toxic metal exposure, resulting in the selection of metal-tolerant strains or populations in soils with high metal bioavailabilities (Giller et al., 1998). The distinction made in Chapter 3 between metal resistance and metal tolerance is maintained here. The term 'metal-resistant' is used to refer to all members of a species which survives relatively well generally in conditions of metal contamination, while metal-tolerant refers to a particular metal-exposed population of a species, with increased survival compared to its conspecifics in conditions of high metal bioavailability. Mechanisms of metal tolerance in bacteria and fungi are varied. These mechanisms include the binding of metals to proteins, cell walls or molecules excreted extracellularly; compartmentation of metals within cells; the formation of insoluble metal

Table 4.8 **Genera of fungi that have at least one species known to accumulate atypically raised concentrations of trace metals from uncontaminated soils.**

Genus	Trace metal
Agaricus	Ag, Au, Cd, Co, Cu, Hg
Amanita	Cd, Cu, Hg, Se, V
Boletus	Ag, Cu, Hg, Se
Calvatia	Ag, Cu
Clitocybe	Cu, Hg
Coprinus	Cd, Co, Cu
Cortinarius	Cd, Cu, Hg
Cytoderma	Cd, Hg
Laccaria	Ag, Cd
Lactarius	Hg
Lycopordon	Ag, Cu, Hg, Se
Macrolepiota	Au, Cu
Russula	Cd, Cu, Hg
Sarcodon	Hg
Suillius	Fe

Source: Lepp (1992).

sulphides; decreased metal uptake; and enhanced export of metals from cells, including, in the case of mercury, volatilisation of the metal from the cell (Giller et al., 1998).

Specific examples of the increased abundance of microbial metal tolerance with increased metal contamination of the soil include a higher level of metal tolerance in populations of species of the Gram-negative bacterial genus *Pseudomonas* isolated from soil around industrial sites (Campbell et al., 1995). Similarly, a correlation was found between the proportion of metal-tolerant bacteria and total zinc and cadmium concentrations in highly metal-contaminated soils from mining areas (Olson and Thornton, 1982). An increased proportion of cadmium-tolerant isolates of Gram-negative bacteria was found in the 1980s in soils with high cadmium, copper, lead and zinc concentrations resulting from long-term application of sewage sludge (Barkay et al., 1985), before the regulation of metal concentrations in sewage sludge applied to agricultural soils.

Investigation of the microbial communities of soils is particularly amenable to the use of environmental genomics, in which genetic material is recovered directly from environmental samples (such as soil samples) to produce largely unbiased samples of all genes from all members of the sampled community. Environmental genomics can therefore be used to compare the structure of soil microbial communities under different conditions of metal contamination. The concept of pollution-induced community tolerance (PICT) reflects the fact that the selection pressure associated with an ecotoxicologically significant metal bioavailability will lead to an increased average tolerance to that metal amongst all species in the local biological community (Chapter 3). PICT is tested by comparing communities collected from metal-contaminated and reference sites to metal exposure under controlled conditions. A potential comparative measure of microbial activity under different degrees of soil metal contamination is the community respiration rate, which can be determined in short-term incubations of a whole soil microbial community (Bérard et al., 2014). The use of such a measure could be an important tool to investigate PICT in the context of metal soil contamination. The simultaneous application of environmental genomics has the potential to add valuable information on changes in microbial community structure associated with any observed increase in community metal tolerance.

4.4 Lichens

Prologue. Lichens grow on hard surfaces, including metal-rich rocks such as ore deposits in the region of mines, particularly spoil heaps. Lichens are not single organisms, but consist of a symbiosis between at least two organisms: a fungus and a photosynthetic partner. The latter may be a green alga or a blue-green bacterium.

4.4.1 Metallophytic Lichen Assemblages

Prologue. Distinctive lichen assemblages are associated with different metal-rich mine spoils.

Metallophytes are plants that can grow in atypically high bioavailabilities of trace metals, the term 'plant' here being used so loosely that it includes lichens. Metallophytic lichens can be divided into specialist obligate metallophytes, apparently surviving only in the presence of raised bioavailabilities of metals, and facultative metallophytes, that can grow in metal-rich habitats but are not confined to them. Lichens in this latter category are typically generalist pioneer species, able to grow in a wide range of habitats, including metal-rich ones.

Lichen assemblages occurring on mine spoil were originally considered to belong to a single community, the *Acarospora sinopica* community (Purvis, 1996). It is now appreciated that this metallophyte lichen community, based on the rust-coloured *A. sinopica* (Plate 3a), typifies sites where rocks rich in iron sulphides have been weathered by bacterial action to release both iron and sulphuric acid (Purvis, 1996). Moreover, three distinct assemblages of metallophyte lichens are now recognised, associated in turn with rocks rich in iron sulphide, alkaline rocks rich in copper and rocks rich in lead and zinc (Table 4.9) (Purvis, 1996; Gilbert, 2000; Smith et al., 2009).

The distinctive metallophyte lichens found on rocks rich in ferrous sulphides that have undergone weathering to release oxidised ferric iron and sulphuric acid are *Acarospora sinopica* itself, as well as *Lecanora epanora* and *Rhizocarpon furfurosum*, together with others listed in Table 4.9 (Purvis, 1996; Purvis and Halls, 1996; Gilbert, 2000). This iron sulphide-rich rock lichen community (Table 4.9) can be further subdivided into two associations. *Lecanora epanora* and *Lecanora handelii* (Plate 3b) are amongst the lichens found on vertical, dry, sheltered overhanging rock surfaces, while *A. sinopica*, *Lecanora subaurea* (Plate 3c), *Lecidea silacea* (Plate 3d) and *Miriquidica atrofulva* (Plate 3e) are found on exposed sunny horizontal surfaces (Purvis, 1996; Gilbert, 2000). Both these latter associations are found on silicate-based rocks with iron sulphide at the Coniston copper mines in the Lake District and at Parys Mountain on Anglesey (Purvis, 1996; Gilbert, 2000).

The copper sulphide ores chalcopyrite and chalcocite may be present in the iron sulphide–rich rocks that are weathered to produce the acid- and iron-rich conditions on which the *Acarospora sinopica* lichen community thrives, as at Coniston and Parys Mountain. There is, however, another distinct lichen community that occurs on copper-rich rocks in which copper sulphide ores are associated with significant amounts of calcium carbonate, creating an alkaline environment (Purvis 2010a). Other copper ores are also typically present in these circumstances, including the copper carbonate hydroxide minerals malachite and azurite. The lichen community typical of these higher-pH, copper-rich rocks is typified by *Lecidea inops*, often with *Psilochia leprosa* (Plate 3f), *Stereocaulon leucophaeopsis* and *S. symphycheilum* (Plate 3g) present (Table 4.9) (Purvis, 1996, 2010a; Smith et al., 2009). This lichen community was first discovered at Coniston by William Purvis and Peter James in 1984 (Purvis and James, 1985), and has since been found elsewhere in the Lake District at Dale Head, and in Scandinavia (Purvis, 1996; Gilbert, 2000; Smith et al., 2009). The faithful species of this alkaline copper–rich rock lichen community is *Lecidea inops* (Plate 3h), a species first described in 1874 from a Scandinavian copper mine and not found in other metal-rich lichen communities (Purvis, 2010a). While *L. inops* belongs to a complex of closely related taxa, including *L. auriculata* and *L. diducens*, it is a chemically, ecologically, morphologically and distributionally distinct species (Purvis, 1996). In the United Kingdom, *L. inops* is known only from copper-rich sites in the Lake District, and has been added to the list of plants protected under Schedule 8 of the UK Wildlife and Countryside Act (1981) (Purvis, 1996).

Table 4.9 **A selection of metallophyte lichen species occurring on metal-rich rocks in the British Isles, subdivided into (a) specialist obligate metallophytes and (b) metal-resistant generalist pioneer species.**

Iron sulphide–rich rocks	Alkaline copper–rich rocks	Lead/zinc-rich rocks
(a) Obligate metallophytes	*(a) Obligate metallophytes*	*(a) Obligate metallophytes*
Acarospora sinopica	*Lecidea inops*	*Gyalidea roseola*
Lecanora epanora	*Psilolechia leprosa*	*Gyalidea subscutellaris*
Lecanora handelii	*Stereocaulon leucophaeopsis*	*Placynthiella hyporhoda*
Lecanora subaurea	*Stereocaulon symphycheilum*	*Stereocaulon delisei*
Lecidea silacea		*Stereocaulon leucophaeopsis*
Miriquidica atrofulva		
Rhizocarpon furfurosum		*(b) Metal-resistant generalists*
Stereocaulon leucophaeopsis		
		Bacidia saxenii
		Bacidia viridescens
(b) Metal-resistant generalists		*Cladonia cariosa*
		Peltigera neckeri
Arthonia lapidicola		*Peltigera venosa*
		Sarcosagium campestre
		Steinia geophana
		Stereocaulon condensatum
		Stereocaulon dactylophyllum
		Stereocaulon glareosum
		Stereocaulon nanodes
		Stereocaulon pileatum
		Vezdaea acicularis
		Vezdaea aestivalis
		Vezdaea cobria
		Vezdaea leprosa
		Vezdaea retigera
		Vezdaea rheocarpa

Sources: Purvis (1996); Gilbert (2000); Smith et al. (2009).

Within the same lichen community, the more common and widely distributed *Psilochia leprosa* is mostly restricted to alkaline copper–rich environments, often on dry, steep or overhanging rocks (Purvis, 1996). In fact, *P. leprosa* is also found on walls, as on the north-facing granite walls of derelict mine buildings at the Poldice mine, Redruth, Cornwall, where it is associated with mortar containing copper minerals (Purvis, 2010a). This lichen can also be found on the walls of old smelter flues, and on church walls, growing on the mortar beneath copper lightning conductors, but not at urban sites with significant acidification (Purvis, 2010a). *P. leprosa* has even been found growing on a church wall on mortar beneath copper grilles, and has the potential to grow on copper lettering on tombstones in the absence of acidification. *Stereocaulon leucophaeopsis* and *S. symphycheilum* are to be found on exposed copper-rich boulders, and can tolerate more acid conditions (Gilbert, 2000).

While the *Lecidea inops* lichen community is distinct from the *Acarospora sinopica* community, the two may be found in close proximity, as at Coniston, where regions of alkaline copper ores may be flanked by rocks rich in iron sulphides (Gilbert, 2000).

It was stressed previously that *Lecidea inops* is a true distinct species, but other apparently distinct lichen species on copper-rich rocks may be just ecotypes of well-known species found elsewhere. In 1826, Christian Sommerfeldt described a greenish lichen *Lecidea theiodes* from a copper mine in Saltdalen, northern Norway, as a new species (Purvis, 2010b). More than 150 years later, using modern analytical techniques, it has been realised that what had been described as a new species was in fact an environmental modification of the widespread grey lichen *Lecidea lactea* growing on copper-rich rocks and accumulating green copper compounds (Purvis, 2010b). Other lichen species also have copper ecotypes with the green colour distinctive of accumulated copper compounds. Some of these have inadvertently been described as new species. British lichen species with green copper ecotypes include *Myriospora smaragdula* (Plate 3i), *Buellia aethalea* (Plate 3j) and *Lecanora polytropa*. In Europe, about 12 lichen species are known to turn greenish-yellow when growing on copper-rich rocks (Purvis, 2010b). Apart from the example of *M. smaragdula*, green copper ecotypes of lichens appear to be uncommon in Britain, but are best developed at copper mines in Cornwall (Gilbert, 2000). The presence of atypical green ecotypes of a widespread lichen does have implications as a bioindicator for metal prospectors (Purvis, 2010b).

The third metallophyte lichen assemblage occurs on rocks rich in lead and zinc (Table 4.9), although no characteristic lichen community has been formally described (Gilbert, 2000). Nevertheless, it has long been recognised that lead–zinc mineralised rocks on spoil heaps support a distinctive lichen flora dominated by species of *Stereocaulon* (Purvis, 1996). From an ecological point of view, there are two types of lead–zinc ore deposits according to the nature of the rock in which the lead–zinc mineralisation occurred (Gilbert, 2000). In northeast Wales, the Northern Pennines, Derbyshire and the Mendips, mineralisation occurred in an alkaline, limestone environment. In the Lake District, central Wales, Scotland (Leadhills, Strontian, Tyndrum) and the Isle of Man, mineralisation took place in acid rocks. In Cornwall, both types of lead–zinc ore deposits occur.

Spoil heaps at the alkaline lead–zinc mines of the Northern Pennines and Derbyshire can host *Bacidia saxenii, B. viridescens, Sarcosagium campestre, Steinia geophana, Vezdaea acicularis, V. leprosa* (Plate 3k), *V. retigera* and *V. rheocarpa* (Gilbert, 2000). In the Northern Pennines, colonies of the Red Data List lichen *Peltigera venosa* can be found at mines near the Cumbria–Northumberland border, as at the Black Burn mine near Alston (Purvis and Halls, 1996). The metal-rich shingle beside the rivers draining the mining districts of the Northern Pennines and Derbyshire also hosts lichens of the lead–zinc lichen assemblage. For example, riverside shingle of the Rivers South Tyne and Tyne in Northumberland provides a habitat for many of the lead–zinc metallophyte lichen species shown in Table 4.9, including *P. venosa* (Gilbert, 2000).

The acid lead–zinc mine spoils of the Lake District, central Wales and Scotland have a rich and distinctive lichen vegetation (Gilbert, 2000), including metallophyte species listed in Table 4.9. In central Wales, for example, mine sites frequently host *Stereocaulon condensatum* and *Gyalidea subscutellaris*, while the first modern record of the rare metallophyte lichen *Placynthiella hyporhoda* was on acid ground at the Ceunant mine, near Bwlch in Powys (Purvis and Halls, 1996). In 1962, the first and only UK record of *Gyalidea roseola* was made at the Feish Dhomhnuill mine in Strontian, Scotland (Gilbert, 2000). The shingle alongside the Rivers Rheidol and Ystwyth, draining the mining region of central Wales, supports *Cladonia cariosa*, *Stereocaulon condensatum* and *S. glareosum* from the lead–zinc list in Table 4.9 (Gilbert, 2000).

Lichens associated with lead–zinc rocks can turn up in unexpected places. Mine spoil from lead–zinc mines is recognised as a persistent weed killer. So, in mid-Wales, the Cambrian Railway Company made widespread use of spoil from the Y Fan lead mines at Llanidloes, Powys, to surface the track, even up to 50 kilometres away from the mine (Woods, 1988; Gilbert, 2000). Twenty-five years after the tracks had been closed, they were still relatively free of any weeds. They did, though, harbour assemblages of lichens akin to those of lead–zinc mine spoil tips, with, for example, *Vezdaea leprosa*, *V. retigera* and *Steinia geophana* present (see Table 4.9) (Woods, 1988). The Forestry Commission in mid-Wales, as well as farmers and private companies, also used lead mine spoil as top dressing for tracks, and the metallophyte lichen *Vezdaea acicularis* was first described from such a track in the Towy valley, Carmarthenshire (Woods, 1988; Gilbert, 2000). As in copper-resistant lichens, their lead–zinc counterparts also have the potential to appear in churchyards, growing, for example, in association with lead lettering on tombstones or the lead on church windows (Purvis and James, 1985; Purvis, 2010b). *Vezdaea leprosa* also commonly grows in the drip zone below galvanised (zinc-coated) wire fencing, for example along motorways (Woods, 1988; Purvis, 1996). Similarly, runoff from acid rain falling on the galvanised steel of electricity pylons can promote the presence of lead–zinc metallophytes such as *Bacidia saxenii*, *Sarcosagium campestre*, *Steinia geophana*, *Vezdaea leprosa* and the rarely recorded *Stereocaulon glareosum* beneath the pylon (Purvis and Halls, 1996; Gilbert, 2000).

4.4.2 Detoxification of Trace Metals

Prologue. Any accumulated trace metal in high concentration needs to be detoxified. For example, lichens contain several copper-binding chemicals able to detoxify accumulated copper. Lichens have the potential to act as biomonitors of airborne trace metal contamination.

Copper accumulated by *Lecidea lactea* is bound to a large degree to norstictic acid, giving the lichen a green colour, which is accentuated in its copper ecotypes growing on copper-rich rocks (Purvis, 2010a, b). Norstictic acid is a widespread chemical compound produced in the metabolism of lichen-forming fungi, often being deposited as crystals. And norstictic acid will bind accumulated copper as part of a copper detoxification process in many lichens (Purvis, 2010b), including the green copper ecotypes of *Acarospora smaragdula* and *Buellia aethalea* (Purvis et al., 1987). On the other hand, the characteristic species of the alkaline copper–rich rock lichen community, *Lecidea inops*, binds accumulated copper in the form of crystals of copper oxalate, which may contribute up to 40% of the lichen dry weight (16% copper) (Purvis, 1996). Copper ecotypes of *Lecanora polytropa* also use copper oxalate to bind accumulated copper, and this same copper compound has also been identified in *Acarospora rugulosa* and *Lecidea lactea* (Chisholm et al., 1987). It is interesting then that *L. lactea* uses both norstictic acid (Purvis et al., 1987) and oxalic acid to detoxify copper, accentuating the point that different metal detoxification mechanisms are available to organisms. In fact, other potential copper-binding compounds are also present in lichens, including psoromic acid with a similar chemical structure to norstictic acid (Purvis et al., 1987). Thus, there is no universal copper detoxification mechanism by which all lichens may avoid the toxic effects of copper.

Other trace metals are also bound as oxalates in lichens (Purvis, 1996). *Pertusaria corallina* growing on manganese ore contained manganese oxalate, and ferric oxalate is known from *Caloplaca aurantia* growing on ferruginous dolomite. Magnesium oxalate has been found at the lichen–rock interface of *Tephromela atra* growing on a Scottish serpentine rock (iron magnesium silicate). *Stereocaulon vesuvianum* from a lead smelting mill in West Yorkshire contained lead oxide in the body of the lichen, and lead carbonate at the lichen–rock interface. Accumulated iron in lichens also needs detoxification, and *Acarospora sinopica* accumulates iron in the form of hydrated iron oxides (Purvis, 2010b).

Lichens growing on metal-rich rocks take up and accumulate dissolved metals originating from these rocks (Garty, 2001). Moreover, lichens growing on any surface have the potential to take up metals dissolved in rainwater and can also incorporate airborne particulate metals into their tissue (Garty, 2001). It follows, therefore, that, by measurement of their accumulated metal concentrations, lichens have the potential to act as biomonitors of airborne metal contamination, dissolved or particulate (Purvis, 1996). Thus lichens have been used as biomonitors of the atmospheric fallout of metals around smelters, industrial centres, urban areas, mines and along roadside verges, their accumulated metal concentrations falling off rapidly, then more slowly, with distance from the emission source (Purvis, 1996). Lichens can be used, for example, to monitor atmospheric contamination with vanadium and nickel derived from the combustion of oil in power plants (Purvis, 2010b) or (more historically) the airborne lead contamination caused by the combustion of leaded petrol in vehicles (Garty, 2001). Beyond the British Isles, lichens have been used as biomonitors of the long-range atmospheric transport of metals to regions such as the High Arctic, once considered pristine and beyond the reach of anthropogenic contamination (Garty, 2001).

4.5 Mosses and Ferns

Prologue. Mosses and their relatives, the liverworts and hornworts, make up the bryophytes, mainly green flowerless plants found not only on land but also in freshwater. Bryophytes take up dissolved trace metals and accumulate them to high concentrations.

Terrestrial bryophytes absorb water over their entire surface together with dissolved trace metals (Tyler, 1990). These metals are usually taken up from rainwater but may also be absorbed from water in the soil on which the mosses are growing. Metal ions are mainly inactivated by binding with the negative charges on the cell walls of the mosses, limiting their subsequent access into the interior of the cells, where they can play a biological role, essential or toxic (Tyler, 1990).

4.5.1 Mosses

Prologue. Many moss species are relatively resistant to high availabilities of copper, zinc, lead and manganese. Mosses can be used as biomonitors to assess the rates of atmospheric deposition of trace metals.

There is very little information available on any toxic effects caused by the exposure of mosses to trace metal contamination in the field in Britain (Tyler, 1990). Laboratory exposures, on the other hand, have shown up, not unexpectedly, interspecific differences in the sensitivities of different mosses to high dissolved concentrations of different trace metals. Thus, many moss species are known to be relatively resistant to high availabilities of copper, zinc, lead and manganese (Tyler, 1960). Some mosses are even known by common names that include a metal as an epithet, for example the 'copper mosses'.

The Field Guide to the Mosses and Liverworts of Britain and Ireland includes all such British mosses (Atherton et al., 2010). *Scopelophila cataractae*, tongue-leaf copper moss, is paradoxically present only on the most toxic zinc-rich substrates, being found on very moist zinc-rich mine spoil or on the banks of adjacent streams. *Grimmia atrata*, copper grimmia, shows a preference for outcrops of acidic rocks rich in trace metals. *Mielichhoferia elongata*, elongate copper moss, is found in Yorkshire (near Ingleby Greenhow, Cleveland) and at two sites in Scotland (for example, Corrie Kander in Aberdeenshire) and is restricted to rock crevices in extremely

Table 4.10 **Trace metal concentrations (µg/g dry weight) in the moss *Hypnum cupressiforme* at sites in the Swansea and Neath valleys, 8 or 25 kilometres downwind of an urban-industrial complex in Swansea, and at two control (C) sites, in the 1960s.**

Site	Cd	Cu	Ni	Pb	Zn
Swansea valley (8 km)	9.5	68	193	260	345
Swansea valley (25 km)	3.0	19	25	103	152
Neath valley (8 km)	4.5	62	46	348	264
Neath valley (25 km)	1.0	18	12	79	140
Gwendraeth (C)	1.0	15	6.5	50	108
Gower (C)	1.8	11	10	65	74

Source: Goodman and Roberts (1971).

acidic, metal-rich rocks (Wilkins, 1977; Atherton et al., 2010). In spite of its common name, *M. elongata* is found on exposed ridges of iron-rich Lower Jurassic shales in Yorkshire but is apparently absent from mine workings and areas of mine spoil. In fact, accumulated concentrations of copper in *M. elongata* and in *G. atrata* are not unusually high for mosses, at only 7 to 12 µg/g (Wilkins, 1977). It appears that copper is not an important factor controlling the distribution of these two so-called copper mosses, rendering the common name inappropriate. The key to the distributions of *M. elongata* and *G. atrata* is apparently their ability to tolerate very high acidity in the form of high concentrations of sulphuric acid produced from the oxidative weathering of iron sulphide (Wilkins, 1977; Tyler, 1990). Another species of *Mielichhoferia*, *M. mielichhoferiana*, grows on metal-rich rock in mountains. More logically, lead moss, *Ditrichum plumbicola*, is a tiny moss entirely restricted to waste from lead mines, growing on highly metal-toxic spoil throughout the lead mining regions of Britain. In fact, the most abundant turf-forming moss on metalliferous mine waste is *Weissia controversa var. densifolia*, green-tufted stubble moss, lacking any indication in its common name as to its

habitat preferences. This moss is found typically at lead mines but can also be found in the zinc-rich conditions under galvanised roadside barriers and dripping galvanised roofs.

Away from mining sites, mosses have been widely used as biomonitors to assess the rates of atmospheric deposition of trace metals (Tyler, 1990). The deposition of metals from the atmosphere is in fact the primary factor determining the accumulation by mosses of, for example, cadmium and lead (Harmens et al., 2012). Indeed, correlations have been established between the accumulated concentrations of these metals in mosses, and their total atmospheric deposition rates in many European countries, including the United Kingdom (Harmens et al., 2012). The moss widely employed for such atmospheric metal biomonitoring in Britain is *Hypnum cupressiforme*, cypress-leaved plait moss (Table 4.10). Other suitable biomonitoring species in the British Isles and elsewhere in Europe are red-stemmed feather moss *Pleurozium schreberi*, glittering wood moss *Hylocomium splendens*, and neat feather moss *Pseudoscleropodium purum* (Harmens et al., 2012). Usually the apical part of the moss is used for analysis. It is generally considered that the concentrations therein are integrated measures of atmospheric deposition for the previous three years, the growth time of the part of the moss analysed. As a variation on the analysis of live indigenous moss samples, the moss bag technique uses the deployment for about one month of standardised air-dried moss samples in nylon net bags to monitor metal deposition rates in metal-contaminated regions (Goodman and Roberts, 1971; Tyler, 1990).

4.5.2 Ferns

Prologue. Ferns are pteridophytes, which are green plants with vessels, fronds and roots, but not flowers.

A fern known to be a metallophyte in the British Isles is the forked spleenwort, *Asplenium septentrionale*. The forked spleenwort has a wide native distribution, including Europe, North America and Asia, living in crevices in rocks. It is considered rare in Britain, but it has colonised derelict lead mine workings, for example in Wales, being locally abundant where it does occur (Condry, 1981).

4.6 Flowering Plants

Prologue. The term 'plant' is often used loosely to refer to a variety of unrelated photosynthetic organisms, including blue green bacteria, some protists, lichens, red algae, brown algae, green algae, bryophytes, pteridophytes, conifers and flowering plants. Strictly, biologists use the term to refer to a single taxon (the green plants), which consists of green algae, bryophytes, pteridophytes and the seed plants made up of the gymnosperms (for example, conifers) and the flowering plants. This section concerns the flowering plants.

4.6.1 Uptake, Accumulation and Detoxification

Prologue. Flowering plants take up trace metals from the soil solution via their roots and accumulate them in the shoots. Three different accumulation patterns have been described: those of hyperaccumulators, indicators and excluders. Phytochelatins play the major, but not exclusive, role in the detoxification of trace metals taken up by flowering plants.

Flowering plants take up trace metals from the soil solution via their roots. The physicochemical characteristics of the soil solution interact with the dissolved concentrations of the metals therein to control how much metal enters the root cells, in effect how much of the soil solution metal is bioavailable to the plant. Plant roots may release metal-binding agents that render particular trace metals more bioavailable for uptake, as in the case of grasses and iron-binding siderophores termed phytosiderophores (Crowder, 1991; Callahan et al., 2006). Trace metals can also enter plants through the leaves, mainly through stomata (Crowder, 1991), but uptake through the roots is dominant.

Metals may bind interchangeably on the cell walls of plant roots. Subsequently, the entry of trace metals across the cell membrane into root cells follows the general principles outlined in Chapter 3. Thus, there are many specific carrier proteins for essential trace metals such as zinc or copper. Non-essential metals such as cadmium or lead trespass as opportunistic hitchhikers onto carrier proteins for the likes of zinc or iron or through protein channels for major ions such as calcium (Clemens, 2006; Peralta-Videa et al., 2009). Arsenic is taken up as arsenate via phosphate channels, and selenium as selenate via sulphate channels (Clemens, 2006; Peralta-Videa et al., 2009).

After uptake into the root cells, trace metals are quickly bound to the first of a series of organic molecules with strong metal affinities, preventing them from binding to other molecular sites where they might exert a toxic effect. Such organically bound metals accumulate in the vacuoles of root cells (Clemens, 2006), but a portion of the metal taken up will be translocated to other tissues aboveground, proportions varying between metals and plants (Peralta-Videa et al., 2009).

Flowering plants accumulate trace metals in the shoots, including the leaves (Baker, 1981; Peralta-Videa et al., 2009). These accumulated metal

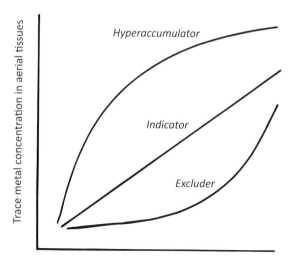

Figure 4.2 Patterns of accumulation of trace metals in the shoots and leaves of flowering plants in response to increasing trace metal bioavailabilities in soils.

concentrations in the aerial parts of the plant may or may not rise in direct proportion to metal concentrations in the soil (Figure 4.2). Hyperaccumulators will accumulate particularly high concentrations of particular metals in foliage, even when soil metal levels are low (Figure 4.2) (Baker, 1981). Where there is a direct relationship between concentrations of a metal in shoots and leaves and in the soil (Figure 4.2), the plants may be referred to as indicators (Baker, 1981). These indicators have clear potential as biomonitors of soil metal bioavailabilities. Some plants maintain accumulated concentrations of particular trace metals in the aerial shoots at low levels, even when growing in metal-contaminated soils, up to a critical metal bioavailability in the soil, after which there is relatively unrestricted root-to-shoot transport (Figure 4.2). The term 'excluders' has been applied to such plants (Baker, 1981), perhaps a little misleadingly, because the metals are not excluded from the plant as a whole but held in the roots after uptake. Furthermore, the same plant may act as an excluder for one metal but accumulate another in aerial parts in proportion to soil bioavailability (Baker, 1981). Ratios of the concentrations of zinc accumulated in leaves and roots indicate that alpine pennycress

Thlaspi caerulescens and spring sandwort *Minuartia verna* are accumulators of zinc, while bladder campion *Silene vulgaris* is a zinc excluder (Baker, 1981).

Trace metals taken up by flowering plants need to be detoxified if they are not to cause toxic effects in a cell. Many compounds have been identified in the cells of flowering plants that bind strongly to particular trace metals and presumably play a role in their detoxification. In Chapter 3, attention was drawn to three important such detoxificatory molecules: metallothioneins, glutathione and phytochelatins. Originally, phytochelatins were thought to be responsible for trace metal detoxification in flowering plants, while metallothioneins were more important in other organisms, particularly animals. It is now known, however, that this is an oversimplification, for metallothioneins are present in plants, and appear to be functional to some degree (Cobbett and Goldsbrough, 2002; Callahan et al., 2006). Furthermore, phytochelatins are not restricted to flowering plants but are present in many other groups of organisms, including animals (Bundy et al., 2014). Nevertheless, phytochelatins are still considered to play the major role in the detoxification of trace metals taken up by flowering plants. Like metallothioneins, phytochelatins are inducible by exposure to raised trace metal bioavailabilities. But, while MTs are directly encoded by genes which increase MT production in response to trace metal exposure, phytochelatins are assembled by the enzyme phytochelatin synthase from cysteine-rich dipeptides originating from glutathione (Cobbett and Goldsbrough, 2002). There are therefore no phytochelatin genes as such. There are, however, genes for the enzyme phytochelatin synthase and for glutathione, and it is the biosynthesis of these that is increased in response to trace metal exposure.

There are also other candidates for trace metal detoxificatory compounds in flowering plants. Several of these are small molecules containing a nitrogen atom able to bind with the metal. Nicotianamine is found in all plant cells where it chelates both Fe^{3+} and Fe^{2+} ions; it also binds nickel, as in alpine pennycress *T. caerulescens* (Callahan et al., 2006). Moreover, nicotianamine is the biochemical precursor of another metal-binding compound, the amino acid, mugineic acid, which is secreted as a siderophore by the roots of grasses to promote the uptake of iron in complexed form (Callahan et al., 2006). A very important compound binding trace metals in plants is the amino acid histidine, which uses not only nitrogen but other parts of the molecule to bind metals, particularly nickel and zinc (Krämer et al., 1996; Callahan et al., 2006). Another amino acid, proline, will also bind trace metals in plants. Also present in high concentrations in the cell vacuoles of flowering plants are organic acids such as citric acid and oxalic acid. These organic acids are capable of binding metal ions, as in the case of zinc and cadmium in alpine pennycress (Callahan et al., 2006). There is also evidence that the concentrations of such organic acids increase in flowering plants in response to an increasing concentration of, for example, zinc in the soil (Callahan et al., 2006). These organic acids may then be part of metal detoxification processes in some plants, particularly for zinc.

4.6.2 Metallophytes

Prologue. Metallophytes are plants that can grow in soils with atypically high bioavailabilities of trace metals. The calaminarian grassland community occurs in upland habitats with soils with bioavailabilities of trace metals that are high enough to be toxic to most plants.

Metallophytes are commonly found on the spoil heaps of mines. Some metallophytes appear to be obligate metallophytes, which are said to survive only in the presence of metals, while facultative metallophytes can grow in metal-rich soils, but are not confined to them (Baker et al., 2010). As previously, a distinction is made here between metal resistance and metal tolerance. The term 'metal-resistant' is used to refer to all members of a species which survives well in conditions of metal contamination. Metal-tolerant refers to a particular metal-exposed population of a species which shows increased survival in conditions of high metal bioavailability, compared to other populations of the same species growing elsewhere. Both metal-resistant species and metal-tolerant populations (or

ecotypes) of more widely distributed species can be considered as metallophytes. The obligate metallophyte *Viola calaminaria*, the zinc violet (Baker et al., 2010), is restricted to calamine soils (soils enriched in zinc, lead and cadmium) in western Europe, and can be considered a metal-resistant species. Many facultative metallophytes, on the other hand, may be specialised ecotypes or metal-tolerant populations of species also occurring in the absence of metal enrichment. Facultative metallophytes include spring sandwort *Minuartia verna* (called lead sandwort in parts of northern England), thrift *Armeria maritima* (Plate 2c), sea campion *Silene uniflora* (Plate 2d) and alpine pennycress *Thlaspi caerulescens* (Baker et al., 2010).

The British National Vegetation Classification system recognises a particular plant community amongst upland habitats that is found on soils with bioavailabilities of trace metals that are high enough to be toxic to most plants – the calaminarian grassland community. Typical situations for this community are mine workings and spoil heaps such as on Halkyn Mountain in north Wales, and stable river gravels rich in lead and zinc as on the South Tyne and Allen river gravels in Northumberland. Calaminarian grassland vegetation is normally species-poor, with limited vegetation cover. Plants characteristic of the community are spring sandwort *M. verna* and alpine pennycress *T. caerulescens*, which occur together with common bentgrass *Agrostis capillaris*, birdsfoot trefoil *Lotus corniculatus*, fairy flax *Linum catharticum*, harebell *Campanula rotundifolia*, sheep's sorrel *Rumex acetosella* and wild thyme *Thymus polytrichus*. Facultative metallophytes present include ecotypes of sheep's fescue *Festuca ovina*, sea campion *S. uniflora* and thrift *A. maritima*. The fern *Asplenium septentrionale*, forked spleenwort, benefits from the lack of competition from faster growing plants.

The British Isles have many sites with the potential to host metallophytes, not least in Wales and the Pennines, although these sites are usually far apart and individually of very limited extent. In the Northern Pennines of Northumberland and Cumbria, metallophytes are to be found at mine sites such as

Table 4.11 **Metallophytes found at mine sites and on metalliferous fine-grained sediment river alluvium on the banks of rivers in the Northern Pennines.**

Spring sandwort	*Minuartia verna*
Alpine pennycress	*Thlaspi caerulescens*
Bladder campion	*Silene vulgaris*
Thrift	*Armeria maritima*
Mountain pansy	*Viola lutea*
Pyrenean scurvygrass	*Cochlearia pyrenaica*
Kidney vetch	*Anthyllis vulneraria*
Common restharrow	*Ononis repens*
Sheep's fescue	*Festuca ovina*
Meadow oatgrass	*Avenula pratensis*
Common bentgrass	*Agrostis capillaris*
Dune helleborine	*Epipactis dunensis*

Source: Lunn (2004).

Whitesike, near Garrigill in Cumbria, and on the metalliferous fine-grained sediment river alluvium on the banks of the South Tyne, Nent and West Allen rivers, draining into the River Tyne (Table 4.11). In fact, metallophytes occur on the South Tyne alluvium well below the orefield, even on the banks of the Tyne itself below Hexham, the metal-rich sediment having washed down over time (Lunn, 2004). Notable river bank sites with metallophytes are at Garrigill, Low Nest, Williamston, Lambley and Beltingham on the South Tyne, below Nenthead on the Nent, at Nine-banks on the West Allen and at Close House on the Tyne (Lunn, 2004). Another Northern Pennine site hosting the calaminarian grassland community is Haggs Bank, situated halfway between Nenthead and Alston in Cumbria, on the south-facing side of the valley of the River Nent. Metallophyte species present here in numbers include spring sandwort and alpine pennycress, which grow in open areas of bare soil with apparently high metal toxicity. Also abundant are Pyrenean scurvygrass *Cochlearia pyrenaica* and

mountain pansy *Viola lutea* growing in areas of closed vegetation with common bentgrass, birdsfoot trefoil and fairy flax.

In the mining areas of Wales, alpine pennycress grows near Llanrwst in the Conwy valley of north Wales and along the River Ystwyth in mid-Wales, just downstream from lead mines near Llanafan (Condry, 1981). Spring sandwort is common at the sites of lead mines at Halkyn Mountain, colonising patches of disturbed ground on spoil sites, and is also plentiful at a copper mine at Hermon in Coed y Brenin, Dolgellau (Condry, 1981). Thrift occurs at this same copper mine. Sea campion is abundant near lead mines in northern and mid-Wales, well away from its usual habitat on sea cliffs.

Alpine pennycress also grew as part of a self-seeded sward on reclaimed land mined for the zinc ore calamine, with cadmium present as a contaminant, near Shipham in Somerset (Matthews and Thornton, 1982).

Metallophytes vary in their metal accumulation patterns, and hence their strategies to cope with the potentially toxic uptake of trace metals from the metal-rich soils that they inhabit. The zinc violet *Viola calaminaria* is an excluder of zinc, lead and cadmium, while alpine pennycress *T. caerulescens*, spring sandwort *M. verna* and thrift *A. maritima* are accumulators of these three metals (Bothe, 2011). In fact, alpine pennycress is a hyperaccumulator of zinc and cadmium, but not lead (Bothe, 2011). Spring sandwort continuously forms new leaves which are initially low in trace metals. Metals are then deposited in the leaves during growth before the metal-rich leaves subsequently drop off (Bothe, 2011). There is some confusion as to the exact systematic resolution of thrift plants, with the metallophyte form growing on mine spoil tips assigned to a different subspecies, *Armeria maritima halleri*, to that growing on the coast, *A. maritima maritima* (Bothe, 2011). The coastal form of thrift has special glands, derived from stomata, on the leaves that apparently excrete excess salt, and it is possible that similar glands may excrete excess metal in the metallophyte (Bothe, 2011). It is ironic that the plant family to which thrift belongs has the Latin name

Plumbaginaceae, after the genus *Plumbago*, the Latin for lead being *plumbum*. Indeed, plants in this genus have the common name leadwort. It appears, however, that the name does not refer to any ability to grow on lead-rich soils and may be a reference to the lead blue colour of the flowers of species in the genus.

4.6.3 Accumulation from Metal-Contaminated Soils

Prologue. Flowering plants that accumulate a trace metal in their leaves in proportion to the bioavailability of the metal in the soil can be used as biomonitors of these soil metal bioavailabilities.

Known as indicators, these plants (or, more usually, their leaves) can provide a direct assessment of soil metal contamination that is potentially of ecotoxicological significance. This potential ecotoxicological effect may not only apply directly to the plant itself but also to animals further up food chains reliant on the plant as a source of energy, including food chains involving detritivores. Such measures of soil metal bioavailability are of particular relevance where there has been anthropogenic input of trace metals into the soil, as in the cases of mining and smelting or refining.

4.6.3.1 Mining Sites

Prologue. There have been several studies of concentrations of trace metals, particularly arsenic, lead, zinc and cadmium, in flowering plants growing on soils affected by mining in the British Isles.

Studies vary as to whether they have analysed samples of mixed vegetation growing on metal-rich soils, selected plant species from the field, or experimentally grown specific vegetables on mining-affected soils. Furthermore, such studies have usually failed to distinguish between indicator and hyperaccumulator species amongst the plants analysed (Figure 4.2).

Concentrations of lead, zinc and cadmium were raised in mixed grass samples collected from a site

Table 4.12 Concentrations (μg/g dry weight) of lead, zinc and cadmium in soil and mixed grass samples from the area of Shipham, Somerset.

	Lead	Zinc	Cadmium
Shipham Mining area			
Site 1			
Soil	3,740	45,900	372
Mixed grasses	9.13	192	2.06
Site 2			
Soil	12,300	13,400	114
Mixed grasses	6.00	92	1.45
Site 3			
Soil	2,590	9,760	66
Mixed grasses	4.75	318	1.80
Old Red Sandstone area			
Site 4			
Soil	88	304	2.50
Mixed grasses	2.13	40	0.71
Site 5			
Soil	73	193	1.10
Mixed grasses	2.13	57	0.88
Site 6			
Soil	79	146	0.80
Mixed grasses	2.00	60	1.01

Source: Li and Thornton (1993a).

with raised topsoil concentrations of these three metals at Shipham in Somerset, in comparison with another site from an Old Red Sandstone area south of Shipham (Table 4.12) (Li and Thornton, 1993a). Although the soil concentrations of these metals at the latter site were low in comparison with the Shipham site, they were still above typical background concentrations for uncontaminated soils (Table 4.1). The grass samples consisted of cropped common

Table 4.13 **Concentrations of arsenic (µg/g dry weight) in selected vegetables growing in garden soils with a range of As concentrations (Table 4.4a) in the Hayle–Camborne–Godolphin region of Cornwall.**

	Lettuce	Onion	Beetroot	Carrot	Pea	Bean
Mean	0.85	0.20	0.17	0.21	0.04	0.04
Range	0.15–3.88	0.10–0.49	0.02–0.93	0.10–0.93	0.01–0.11	0.02–0.09

Source: Xu and Thornton (1985).

pasture grass species, with appreciable quantities of Yorkshire fog (*Holcus lanatus*) and perennial ryegrass (*Lolium perenne*) (Li and Thornton, 1993a). The accumulated metal concentrations in the mixed grasses have thus responded to a degree to the higher soil concentrations of the three metals (Table 4.12).

In another study at Shipham, samples of the aerial parts of mixed grasses showed increasing concentrations of cadmium from less than 0.5 to about 2.5 µg/g Cd dry weight, when growing on soils with a range of cadmium concentrations up to 100 µg/g dry weight (Matthews and Thornton, 1982). Thereafter, the accumulated cadmium concentrations in the leaves of the grasses stayed constant at just below 3 µg/g, even when soil cadmium concentrations approached 1,000 µg/g Cd. The grass leaves accumulated much less cadmium than did the leaves of members of the daisy family (Compositae). The daisy (*Bellis perennis*) accumulated up to about 70 µg/g Cd when growing on the same range of cadmium-contaminated soils, followed by dandelion (*Taraxacum* sp.) with up to 25 µg/g Cd and yarrow (*Achillea millefolium*) with up to 12 µg/g Cd. The grasses are still taking up cadmium from the cadmium-contaminated soil but store it in the roots without passage to the leaves (Matthews and Thornton, 1982). The roots of perennial ryegrass and Yorkshire fog contained the highest proportion of total accumulated cadmium in the grasses, with concentrations of cadmium in the roots of Yorkshire fog reaching 300 µg/g Cd (Matthews and Thornton, 1982).

Arsenic has figured strongly in studies of the concentrations of trace metals in plants growing on mining-contaminated soils in Cornwall (Table 4.2). Arsenic concentrations in the edible parts of selected vegetables growing in garden soils in the historical mining area of Hayle–Camborne–Godolphin have been compared against arsenic concentrations in the soil (Tables 4.4a, 4.13) (Xu and Thornton, 1985). The soil concentrations were not only total arsenic concentrations, but also arsenic concentrations extracted with water and with a weak chemical extractant, in an attempt to model the bioavailable portion of arsenic in the soil (Table 4.4a). The accumulated arsenic concentrations in the garden vegetables varied with species and, except in the case of lettuce, did not exceed 1 µg/g As dry weight (Table 4.13). Even for lettuce, the As concentrations were well below the 1959 UK statutory limit for most foods of 1 µg/g As fresh weight, equivalent to 20 µg/g As dry weight for a salad leaf. Total arsenic concentrations in lettuce, onion, beetroot and pea did correlate with arsenic concentrations in the soil, both for total soil arsenic and for extractable soil arsenic, the vegetable arsenic concentration increasing with the soil concentration (Xu and Thornton, 1985). The lack of a correlation for carrots and beans may reflect a different arsenic accumulation pattern for these two vegetables compared to the others. The use of weak extracting agents to model the bioavailability of the arsenic in the soil to the vegetables gave no improved prediction over simple total arsenic soil concentrations. Arsenic does bind strongly to iron oxide, and it might be necessary to take the iron content of the soil into account when trying to model and predict the bioavailability of soil arsenic to plants growing in that soil (Xu and Thornton, 1985).

Concentrations of lead and zinc have been measured in grasses growing in a mine waste revegetation scheme involving lead and zinc-rich mine tailings from a holding dam in the southern Pennine orefield of Derbyshire, as well as in associated plant litter

Table 4.14 **Concentrations of lead and zinc (µg/g dry weight) in grasses, plant litter and topsoil from a tailings dam revegetation scheme in the Southern Pennine orefield, Derbyshire, and a control grassland site in Northumberland.**

	Lead	Zinc
Tailings dam		
Aerial grass components	48–112	113–207
Plant litter	630–1,550	1,120–1,650
Topsoil	3,410–4,410	$1,925 \pm 643$ *(95% CL)*
Control site		
Aerial grass components	5.5–15	17–35
Plant litter	58–82	73–121
Topsoil	94–120	62 ± 5.4 *(95% CL)*

Sources: Andrews et al. (1989a,b).

(Andrews et al., 1989a, b). Grasses were cropped at ground level before analysis of their different components. Comparisons were made of accumulated lead and zinc concentrations in these grasses and plant litter, in grasses and plant litter from an uncontaminated grassland north of Hexham in Northumberland and in soils at the two sites (Table 4.14). Accumulated concentrations of both metals were significantly increased in the grasses, and particularly in the plant litter, at the mine tailings site (Table 4.14) (Andrews et al., 1989a, b). These raised metal concentrations in the plant material offer potential for increased metal transfer through local food chains involving invertebrates and mammals, with an associated risk of ecotoxicological consequences.

4.6.3.2 Smelter Emissions

Prologue. Metal contamination of soils downwind of metal smelters and refineries can also increase accumulated metal concentrations in plants growing in natural habitats, such as woodland that might initially be considered to be remote from the metal source.

The Avonmouth zinc, lead and cadmium smelting works near Bristol was such a source of metals, emitted in particulate form into the atmosphere to fall out subsequently to contaminate soils up to 25 kilometres downwind (Hopkin, 1989).

Figure 4.3 shows the distributions of the concentrations of zinc and lead in topsoil and leaves of the elm *Ulmus glabra*, along a gradient downwind of the prevailing wind in 1971 (Little and Martin, 1972). The soil concentrations are those that could be extracted by a weak acetic acid solution in an attempt to model the soil metal that might be bioavailable to rooted plants. The leaves were unwashed and, therefore, the leaf metal concentrations include metal present as a surface deposit on the leaves, in addition to metal transported from soil to leaf. Washing showed that an average of 64% of the lead, 28% of the zinc and 20% of the cadmium in the elm leaves was superficial (Little and Martin, 1972). Nevertheless, all the leaf metal, whether absorbed or superficial, has the potential to be ingested by grazers or by detritivores after leaf fall. The highest concentrations of metals in the elm leaves were 8,000 µg/g Zn, 5,000 µg/g Pb and 50 µg/g Cd (all dry weight) in leaves collected 250 metres from the smelter, although raised leaf concentrations were still present 10 kilometres or more from the smelter (Figure 4.3) (Little and Martin, 1972). There was a significant correlation between extracted soil metal concentrations and leaf metal concentrations for zinc and lead in leaves of the elm, and also between soil and leaf concentrations of zinc, lead and cadmium in the hawthorn *Crataegus monogyna* (Little and Martin, 1972). It is clear that the accumulated concentrations of zinc, lead and cadmium found in leaves of elm and hawthorn downwind of the Avonmouth smelter in 1971 represented abnormally high accumulations, setting the scene for atypical transfer of these toxic metals through local woodland food chains.

A further study of the downwind effect of the Avonmouth smelter on accumulated metal concentrations in plants and plant litter of local woodlands

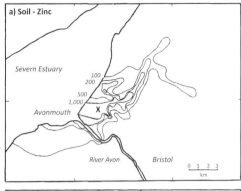

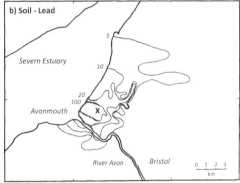

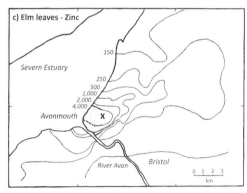

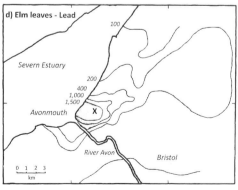

was carried out in 1979 (Martin et al., 1982). Table 4.15 shows that concentrations of zinc, lead, cadmium and also copper were appreciably raised in samples of hazel leaves, dog's mercury, leaf litter and soil at Haw Wood, 3.2 kilometres downwind, in comparison to Midger Wood, 28 kilometres downwind. High accumulated metal concentrations were also present in other plant samples from Haw Wood, including leaves of oak *Quercus robur*, field maple *Acer campestre*, beech *Fagus sylvatica* and ash *Fraxinus excelsior*, as well as in further herbaceous plants growing in the understory (Martin et al., 1982). It was concluded that the vegetation, litter and soil of Haw Wood were heavily contaminated with zinc, lead and cadmium, and to a lesser extent copper, with the pattern of concentrations as anticipated from the order of magnitude of metal emissions from Avonmouth (Martin et al., 1982). Concentrations of the trace metals were generally higher in the litter than in any other component of the woodland ecosystem. Trace metal concentrations in the deposited litter also increased with depth in the soil, as decomposition progressed (Martin et al., 1982).

Table 4.16 shows concentrations of trace metals extracted from surface soil and present in the leaves of the grass red fescue at sites downwind of an urban-industrial complex in Swansea in the 1960s (Goodman and Roberts, 1971). Increases in concentrations of cadmium, copper, nickel, lead and zinc in the grass followed increases in extractable metal concentrations in the surface soil (Table 4.16). The increases in both soil and grass concentrations of nickel and copper between 3 and 8 kilometres down the Swansea valley reflected the presence of a nickel refinery at Clydach, 5 kilometres downwind of the centre of Swansea (Goodman and Roberts, 1971).

Figure 4.3 Avonmouth, Bristol. Distributions of concentrations (μg/g dry weight) of zinc and lead in the top 5 centimetres of soil (extracted with 2.5% acetic acid) and in the unwashed leaves of elm (*Ulmus glabra*) (digested in 4:1 nitric:perchloric acids) downwind of the smelter (X) in October 1971. (a) Zn in soil, (b) Pb in soil, (c) Zn in elm leaves, (d) Pb in elm leaves. (After Little and Martin, 1972, with permission.)

Table 4.15 **Concentrations (µg/g dry weight) of zinc, lead, cadmium and copper in samples of leaves of hazel (*Corylus avellana*), dog's mercury (*Mercurialis perennis*), leaf litter (surface and progressively deeper in horizons F and H), and soil from Haw Wood and Midger Wood (3.2 and 28 kilometres respectively northeast of the Avonmouth smelter) in October 1979.**

	Zinc	Lead	Cadmium	Copper
Haw Wood				
Hazel leaves	478	270	4.6	34
Dog's mercury	662	155	29	18
Surface litter	644	293	6.2	44
Litter (F horizon)	1450	1030	22	67
Litter (H horizon)	2350	1810	52	88
Soil (0–1 cm)	944	203	17	35
Midger Wood				
Hazel leaves	33	11	0.15	6.5
Dog's mercury	135	11	0.76	6.0
Surface litter	68	31	0.79	4.2
Litter (F + H horizons)	182	86	1.43	14
Soil	118	41	0.30	21

Source: Martin et al. (1982).

The copper refinery complex in Merseyside exemplified in Table 4.4b caused raised total and sequentially extractable concentrations of both copper and cadmium in soils at the refinery site and at a site 1 kilometre away (Hunter et al., 1987a). These increased soil concentrations were associated in turn with raised concentrations of both metals in two grasses in local grassland swards: creeping bentgrass *Agrostis stolonifera* and red fescue *Festuca rubra* (Table 4.17). Copper concentrations in red fescue at the Merseyside refinery site (73 µg/g, Table 4.17) were much higher than any of those measured in the same grass in the Swansea study (Table 4.16). On the other hand, more copper was accumulated by the grass at the 8 kilometre Swansea site (3 kilometres downwind of the nickel production centre) than at the site 1 kilometre downwind of the Merseyside refinery (Tables 4.16, 4.17). Cadmium concentrations in red fescue were higher along the first few kilometres of the Swansea valley (Table 4.16) than at the Merseyside refinery site or at the Merseyside site 1 kilometre downwind (Table 4.17).

Even in the absence of woodland, there are still tree species that can be used as biomonitors of the spread of bioavailable metals from metal industry centres (Migeon et al., 2009). Willow and poplar trees of the family *Salicaceae* show particularly high accumulation of zinc and cadmium in their leaves (Migeon et al., 2009; Remon et al., 2013). Concentrations as high as 950 µg/g Zn and 44 µg/g Cd dry weight have been measured in leaves of the hybrid poplar *Populus tremula x Populus tremuloides* near a zinc and lead production plant in northern France (Migeon et al., 2009). Other flowering plant species considered to be good indicators of soil metal contamination are the dandelion *Taraxacum officinale* and shepherd's purse *Capsella bursa-pastoris* (Remon et al., 2013).

The Avonmouth, Swansea and Merseyside studies consistently showed that emissions from metal smelters or refineries can cause raised accumulated concentrations in vegetation several kilometres downwind of the smelter or refinery site.

4.6.3.3 Roadside Lead

Prologue. Lead emitted in the exhaust of vehicles using leaded petrol could be taken up by roadside vegetation.

Although now mainly of historical interest, there was much concern in the late twentieth century as to the fallout into the local environment of lead emitted from the exhaust pipes of vehicles using leaded petrol.

Table 4.18 gives results from a study published in 1967 of concentrations of lead in the leaves of

Table 4.16 **Trace metal concentrations (μg/g dry weight) in surface soil (extracted in 0.5 M acetic acid, or EDTA for Cu) and in leaves of the grass red fescue *Festuca rubra* at sites in the Swansea and Neath valleys, 1.5, 3, 8 or 25 kilometres downwind of an urban-industrial complex in Swansea, and at two control (C) sites, in the 1960s.**

Site	Cd	Cu	Ni	Pb	Zn
Surface soil					
Swansea valley (1.5 km)	26.0	160	20	263	543
Swansea valley (3 km)	8.0	172	7	46	310
Swansea valley (8 km)	3.0	214	234	14	150
Swansea valley (25 km)	0.3	80	6	8	45
Neath valley (8 km)	2.0	210	2	17	105
Neath valley (25 km)	0.5	77	5	9	75
Gwendraeth (C)	0.4	35	3	6	20
Gower (C)	0.5	35	3	16	68
Red fescue					
Swansea valley (1.5 km)	40.0	15	15	814	1,190
Swansea valley (3 km)	9.0	15	15	86	245
Swansea valley (8 km)	2.5	26	82	21	152
Swansea valley (25 km)	1.4	12	7	13	43
Neath valley (8 km)	3.0	13	20	40	136
Neath valley (25 km)	1.3	12	8	16	60
Gwendraeth (C)	0.7	9	6	5.5	25
Gower (C)	0.8	5	8	12.5	25

Source: Goodman and Roberts (1971).

privet *Ligustrum ovalifolium* collected from two categories of sites in various industrial and rural districts in England, Scotland and Wales (Everett et al., 1967). Category I sites were sited along main roads, while Category II sites were located in gardens or parkland remote from main roads. There was usually a higher concentration of lead in the privet leaves collected near main roads, except in more rural locations such as Ambleside, Hamilton, Lossiemouth and Plymouth (Table 4.18). The conclusion here is that lead from petrol had raised the privet leaf lead concentrations near main roads, except where the volume of traffic in more rural sites had been too low to have a significant effect. Lead concentrations in privet leaves from both Category I and Category II sites collected in the most industrial areas, such as Birmingham, Newcastle and Sheffield, were higher than those from other sites (Table 4.18). It can be concluded that the amount of background lead in the air at these sites was high, even away from the immediate proximity to main roads. At Birmingham and Sheffield, there was still further increased lead accumulation by the privet next to the main roads, but, in Newcastle, this effect was masked by the very high lead concentrations in privet leaves at sites of both categories (Table 4.18).

The banning of lead from petrol has now solved any potential ecotoxicological effects of this source of lead in the environment today.

4.6.4 Hyperaccumulation

Prologue. Flowering plants that accumulate atypically high concentrations of trace metals in their shoots are known as hyperaccumulators. Hyperaccumulation is considered to be an adaptive character, but its adaptive role is controversial.

Table 4.17 **Copper and cadmium concentrations (mean ± SE, μg/g dry weight) in soils (total and cumulative sequentially extracted concentrations; see Table 4.4b) and associated grasses (creeping bentgrass *Agrostis stolonifera* and red fescue *Festuca rubra*) growing on those soils, affected by emissions from a copper refinery complex in Merseyside.**

	Total soil concentration	Cumulative sequentially extracted soil concentration	*Agrostis stolonifera*	*Festuca rubra*
Copper				
Refinery site	9,270 ± 2,200	8,660 ± 1,160	122 ± 31	73 ± 13
1 km site	494 ± 105	275 ± 31	25 ± 2.4	23 ± 2.5
Control site	13 ± 1.4	5.7 ± 0.2	9.8 ± 0.7	7.6 ± 0.5
Cadmium				
Refinery site	28 ± 7.0	22 ± 1.3	3.3 ± 0.4	2.9 ± 0.3
1 km site	7.4 ± 0.9	3.3 ± 0.1	1.32 ± 0.07	1.30 ± 0.09
Control site	0.96 ± 0.1	0.31 ± 0.01	0.63 ± 0.06	0.51 ± 0.05

Source: Hunter et al. (1987a).

Originally, the term 'hyperaccumulator' (Baker, 1981) was applied to plants that accumulated more than 1,000 μg/g dry weight of nickel in their leaves (Macnair, 2003). The concept has since been extended to other metals, such as zinc (>10,000 μg/g), manganese (>10,000 μg/g), copper (>1,000 μg/g), selenium (>1,000 μg/g), chromium (>500 μg/g), cadmium (>100 μg/g), cobalt (>100 μg/g), and lead (>100 μg/g) (Baker, 1981; Bothe, 2011). As a rule of thumb, hyperaccumulators have a metal concentration in their aerial parts that is approximately ten times higher than in a non-hyperaccumulator growing on the same metal-contaminated soil (Bothe, 2011).

Hyperaccumulation appears generally to be a species-level phenomenon, although there is also intraspecific variation between populations (Macnair, 2003). There is some taxonomic grouping of the species that are hyperaccumulators, with a substantial number in the family *Brassicaceae*, the cabbage family, particularly in the pennycress genus *Thlaspi* (Macnair, 2003). This genus does, though, include

species that are hyperaccumulators and those that are not. For example, alpine pennycress *Thlaspi caerulescens* is a hyperaccumulator of zinc, cadmium, nickel, manganese and cobalt (Macnair, 2003; Callahan et al., 2006). It is not, on the other hand, a hyperaccumulator of lead, copper, iron, aluminium or chromium, which, while being accumulated strongly in the roots, are not translocated to the shoots (Reeves, 2006; Bothe, 2011). *T. goesingense*, an Austrian species of *Thlaspi*, also hyperaccumulates zinc and nickel, but another British species, field pennycress *T. arvense*, does not (Macnair, 2003). *T. caerulescens* appears to have a greater number of zinc transporters taking up zinc from soil to root cell than does *T. arvense*, followed by a particularly effective zinc transport system from root to shoot (Macnair, 2003; Callahan et al., 2006). In younger tissues of *T. caerulescens*, zinc is bound to histidine but, in older tissues, there is additional binding of zinc to organic acids in leaf cell vacuoles (Callahan et al., 2006). Unlike zinc, hyperaccumulated cadmium in *T. caerulescens* is bound to sulphur in phytochelatins

Table 4.18 **Mean concentrations of lead (µg/g dry weight) in leaves of privet *Ligustrum ovalifolium* collected from sites along main roads (Category I) or from sites remote from main roads (Category II), in selected industrial or rural districts in England, Scotland and Wales, in a study published in 1967. If the ratio is significant, then there is a significant difference between lead concentrations of privet leaves in Categories I and II.**

| | Lead concentration | | | |
District	Category I	Category II	Ratio I/II	Significance of ratio
Ambleside	38	34	1.1	None
Ayr	189	22	8.5	Highly significant
Birmingham	148	73	2.0	Highly significant
Bournemouth	68	4.7	14.6	Highly significant
Brentwood	58	21	2.8	Highly significant
Edinburgh	166	45	3.7	Highly significant
Exeter	62	20	3.1	Highly significant
Hamilton	36	29	1.3	None
Lossiemouth	24	24	1.0	None
Newcastle	211	240	0.9	None
Norwich	43	20	2.2	Significant
Plymouth	35	37	0.9	None
Sheffield	224	71	3.1	Highly significant
Mean of 33 districts	*87*	*45*	*1.9*	

Source: Everett et al. (1967).

in younger tissues, although cadmium will also bind to organic acids in the vacuoles of older leaves (Callahan et al., 2006). There is no evidence that phytochelatins, in spite of their importance in flowering plants, play a role in zinc or nickel hyperaccumulation in general, nickel in *T. caerulescens* being bound to nicotianamine (Callahan et al., 2006).

Hyperaccumulation is clearly a derived character, and its taxonomic distribution strongly suggests that is adaptive (Macnair, 2003). What, then, might be its adaptive role? Five principal hypotheses have been proposed: increased metal tolerance; protection against herbivores or pathogens; inadvertent uptake; drought tolerance; and reduction of competition.

Under the first of these hypotheses, hyperaccumulation is a mechanism of tolerance. Tolerance often involves the enhanced detoxified binding of an accumulated toxic metal. In plants that are not hyperaccumulators, such binding occurs in the roots. In hyperaccumulators, detoxified binding predominantly takes place in the shoots, including the leaves, and involves greater amounts of accumulated metals. It is possible that detoxificatory binding is more efficient in shoots than in roots. It is the case that hyperaccumulators are usually very tolerant of exposure to the metal that they hyperaccumulate, as in the case of *T. caerulescens* and zinc (Macnair, 2003). There is, however, little field evidence that hyperaccumulators are more tolerant than other metallophytes inhabiting the same metal-rich habitat (Macnair, 2003). Furthermore, there is genetic evidence in the case of *T. caerulescens* and zinc that tolerance and hyperaccumulation are genetically independent characters – suggesting no causal link between them (Macnair, 2003).

Another potential connection between tolerance and hyperaccumulation is the possibility that hyperaccumulation is a mechanism to excrete accumulated metal (Macnair, 2003). Transport of accumulated metal to the leaves allows the excretion of these metals when the leaves are shed. However, other plants accumulate metals in the roots, and roots are 'turned over' beneath the ground, as are the leaves above. It is not clear, then, why any loss of metal via the leaves should be more efficient than through the roots (Macnair, 2003).

The second hypothesis considers that the hyperaccumulation of particular toxic metals represents a defence against herbivores and pathogens such as fungi. Plants live in a world where they are continually challenged by herbivores, pathogens and parasites, expending significant energy in the production of secondary metabolites that are toxic to these challengers. It is an attractive hypothesis that it would be less energetically costly to a plant growing in soils containing high concentrations of toxic metals to accumulate high concentrations of metals as alternative toxins to deter herbivores or pathogens (Macnair, 2003). This inorganic defence hypothesis considers that metal hyperaccumulation by plants can affect herbivores in two ways. The first is through toxicity of the metal, ingestion leading to lethal or sublethal effects on the feeding herbivore (Vilas Boas et al., 2014). The second means is by deterrence, in which the plant tissues with the higher accumulated metal contents are ingested to a lesser extent in a choice against tissues with lower metal concentrations (Vilas Boas et al., 2014). A combination of toxicity and deterrence is not excluded.

There is evidence that the hyperaccumulation of nickel by plants can be toxic to a number of generalist herbivores (Macnair, 2003). A comparison of nickel hyperaccumulators grown on normal soil against those grown on nickel-rich soil, or of hyperaccumulators against nonhyperaccumulators, shows that plants with high nickel are always more toxic to herbivores than plants with more normal nickel concentrations (Macnair, 2003). While a demonstration that accumulated nickel in a leaf is toxic does imply that a plant hyperaccumulating nickel will gain protection from herbivores, in practice the plant is still being damaged before the herbivore has ingested sufficient leaf material to suffer a toxic effect. Deterrence of a herbivore would be more effective from the point of view of the plant. Such deterrence has been demonstrated in the case of alpine pennycress growing in either a low or high zinc medium (Pollard and Baker, 1997). The high zinc *T. caerulescens* plants accumulated ten times as much zinc in their leaves as the plants grown in the low zinc medium. When the plants were presented to three herbivores (locusts *Schistocerca gregaria*, slugs *Deroceras caruanae* and caterpillars of the large white butterfly *Pieris brassicae*), all three showed a feeding preference for the low zinc plants. The caterpillars even rejected the high zinc leaves without apparently ingesting them (Pollard and Baker, 1997).

The evidence, however, does not always support the inorganic defence hypothesis. In the case of a Californian nickel hyperaccumulator, the milkwort jewelflower *Streptanthus polygaloides*, there was no real evidence that the accumulated nickel provided protection against herbivorous insects or mammalian herbivores (Boyd, 2004). Moreover, hyperaccumulated nickel or zinc does not appear to offer protection against specialist feeders such as phloem-feeding aphids (Macnair, 2003). Similarly, evidence to support the hypothesis that metal hyperaccumulation can serve as protection against fungal attack is not strong (Macnair, 2003). Furthermore, many hyperaccumulating plants have relatively low concentrations of metal, for example zinc, as seedlings, and it is seedlings that are particularly susceptible to predation by generalist herbivores such as snails (Macnair, 2003).

The inadvertent uptake hypothesis has it that the hyperaccumulation of a metal is not an adaptation for an increased accumulated concentration of the metal itself, but rather that the plant has become adapted for the acquisition of another required nutrient, and the metal in question is hyperaccumulated inadvertently (Macnair, 2003). Thus hyperaccumulation should really be considered as non-adaptive. It is not clear, however, what the required nutrient might be. Iron, an essential metal, is a possibility. Iron is present in aerated soils in the oxidised form iron 3 with low bioavailability, such that iron is frequently a limiting factor for the growth of plants. Thus some grasses secrete phytosiderophores to increase the uptake of iron, and these siderophores can also increase the uptake of zinc (Macnair, 2003). Furthermore, in situations of iron deficiency, plants may show upregulation of iron transporter proteins, increasing the uptake of iron into root cells (Macnair, 2003). Such iron transporters may also transport other metal ions including zinc and cadmium, so the uptake of these metals will also be increased inadvertently. While this hypothesis is attractive, there is, however, no strong evidence that zinc or nickel hyperaccumulators accumulate more iron than closely related species that are not hyperaccumulators. Nor is there evidence that there is competition between the uptake of iron and the uptake of cadmium, zinc or nickel by hyperaccumulators indicating sharing of uptake routes (Macnair, 2003).

A fourth hypothesis considers that hyperaccumulation might be an adaptation for drought tolerance. All nickel hyperaccumulators occur on serpentine soils, some being endemic, although they normally remain a minority component of the serpentine flora (Macnair, 2003). Serpentine soils support a distinct flora, although no single chemical factor explains this feature. Several physical factors might be considered as additional drivers controlling the nature of the unusual serpentine flora, and one such factor may be drought (Macnair, 2003). Serpentine soils weather quickly and have poor water-retaining properties, many of the associated plants showing xeromorphic characteristics to retain water. The synthesis of soluble compounds such as organic acids, and their storage in cell vacuoles, will increase the osmotic pressure therein, promoting the retention of water and

maintaining turgor. The use of metal ions obtained from the soil for such osmotic purposes may require less energy than the synthesis of organic compounds to increase osmotic pressure. However, it would still be necessary to synthesise metal-binding organic compounds to detoxify any trace metal ions held in the cell vacuoles of leaves. Furthermore, the concentrations of trace metals in the leaves, although high in comparison to those in other plants, are still too low to have much osmotic impact (Macnair, 2003). A concentration of 1,000 µg/g Ni (the defined threshold for a hyperaccumulator) is almost negligible in comparison with the total concentrations of ions in a plant cell that contribute to its osmotic pressure. And it has been shown that nickel or zinc hyperaccumulator status does not provide increased drought resistance to the hyperaccumulators yellow tuft alyssum *Alyssum murale* (a southern European species introduced into the United States) and alpine pennycress *Thlaspi caerulescens* (Whiting et al., 2003; Boyd, 2004).

The reduction of competition hypothesis takes an ecological viewpoint. It can be hypothesised that the toxic trace metals released from the decomposition of the leaves shed by hyperaccumulators would be present in high enough concentration to inhibit the local growth of metal-sensitive plant species that would otherwise compete with the hyperaccumulator (Macnair, 2003). This process would not, however, inhibit the growth of metallophytes, and many hyperaccumulators, not least nickel hyperaccumulators on serpentine soils, grow on metal-rich soils in the company of other metallophytes.

While it does appear that hyperaccumulation is of adaptive value, given its recurring convergent evolution, there does not seem to be a generally accepted single adaptive role to explain all cases of hyperaccumulation. Nevertheless, hyperaccumulation of metals does exist, and it does have the potential to cause ecological consequences.

Can the presence of hyperaccumulators have an ecological effect at higher trophic levels of the local ecosystem? The accumulated metal in the leaves of hyperaccumulators can deter and be toxic to invertebrate herbivores. Such an effect could potentially affect the population dynamics of these local herbivores in general, but only if hyperaccumulators were common in any particular habitat. Hyperaccumulators, however, are usually not common in the flora of the habitats in which they are found (Macnair, 2003). Furthermore, metalliferous sites typically have a poor flora with low productivity, a feature that will anyway be important in determining local herbivore population dynamics.

On the other hand, hyperaccumulators might have a significant but restricted ecological effect resulting from the trophic transfer of metals from plant to a specific herbivore, and potentially thence up a particular food chain. If the herbivore can survive the ingestion of metal-rich foliage, there is potential for the trophic transfer of atypically large amounts of toxic metal to the herbivore and thence to higher trophic levels. These large amounts have the potential for sublethal toxic effects, not only in the herbivore but also in predators depending on that herbivore. There is also another aspect of the trophic transfer of metals from a hyperaccumulator to a herbivore able to survive ingestion of its leaves. The herbivore could take up the toxic metal from the plant and in turn use it as a toxic defence against its own predators (Macnair, 2003; Boyd, 2004).

4.6.5 Metal Tolerance

Prologue. Ecotoxicologically significant bioavailabilities of trace metals in soils can lead to the selection of metal-tolerant populations of plants.

A metal-tolerant population or ecotype of a plant species shows increased survival in conditions of high toxic metal bioavailability when compared to other populations of the same species growing in the absence of metal contamination. The metal-tolerant population has been selected under high metal conditions from the background genetic variation present in a normal population. Natural selection has favoured genotypes leading to physiological mechanisms to cope with the extra toxic metal bioavailability. This metal tolerance has a genetic basis and can, therefore, be inherited by the offspring of the

metal-tolerant plants, leading to the establishment of an ongoing metal-tolerant population at a particular metal-contaminated site.

Metal tolerance is not bestowed by a single physiological attribute but appears to be derived usually from a combination of physiological mechanisms (Baker, 1987). Furthermore, different metal-tolerant populations of the same species, and certainly metal-tolerant populations of different species, will use different physiological mechanisms, as a result of the random genetic variation present upon which natural selection can act. Tolerance to one metal does not automatically confer tolerance to another metal in a population. Nevertheless, it is not uncommon for a single population to have independently acquired tolerance to more than one metal simultaneously (co-tolerance). There is a physiological cost to these physiological mechanisms. Consequently, metal-tolerant populations typically are outcompeted by nontolerant populations of the species under non-contaminated conditions.

4.6.5.1 Occurrence of Tolerant Populations

Prologue. British populations of different flowering plants, typically from mine spoil tips, have been found to be tolerant to one or more of arsenic, cadmium, copper, lead, nickel and zinc.

The first record of a metal-tolerant plant population was from Germany in 1934, a copper-tolerant population of red campion *Silene dioica* from a copper mine. Metal-tolerant populations of plants growing on mine spoil heaps in Britain have been known since 1952, when Professor A. D. Bradshaw demonstrated lead tolerance in a mine site population of common bentgrass *Agrostis tenuis*. A population of *A. tenuis* growing at a disused lead mine at Goginan, near Aberystwyth in Wales, could grow in mine soil containing about 10,000 µg/g (1%) of lead, while a population of the same grass from an uncontaminated pasture a few hundred metres away showed no growth and 50% mortality within three months when grown in the mine soil (Bradshaw, 1952). Mine population plants were smaller and slower growing than the pasture population in uncontaminated soil, illustrating the physiological energy cost to be paid for metal tolerance. Two years of growth of the mine population in uncontaminated soil did not cause any loss of lead tolerance upon retesting in mine soil, confirming the genetic nature of the lead tolerance (Bradshaw, 1952). Three plants out of 60 from the uncontaminated pasture did show some tolerance to the high bioavailability of lead, illustrating the natural presence of genetic variation upon which natural selection could act (Bradshaw, 1952). In the late 1950s and 1960s, lead tolerance was reported in another grass, sheep's fescue *Festuca ovina* (Baker et al., 2010).

There have now been many reports (Table 4.19) of British populations of different plants, typically from mine spoil tips, tolerant to one or more of arsenic, cadmium, copper, lead, nickel and zinc (Gregory and Bradshaw, 1965; Jain and Bradshaw, 1966; Antonovics et al., 1971). For example, both lead-tolerant and zinc-tolerant populations of *Agrostis tenuis* have been found on spoil at the Goginan lead mine in Ceredigion; at Frongoch zinc and lead mine near Pontrhydygroes, also in Ceredigion; at Trelogan zinc and lead mine on the Halkyn-Minera orefield in Flintshire; and at a zinc and lead mine at Llanrwst in the Conwy valley in Gwynedd. Lead-tolerant and copper-tolerant populations of *A. tenuis* have also been reported from the Parys Mountain copper mine on Anglesey. Zinc-tolerant and copper-tolerant populations of common bentgrass have been found at the Drws-y-Coed copper mine, near Rhyd Ddu in Snowdonia. Further zinc-tolerant ecotypes of *A. tenuis* have been described from the Aberllyn zinc mine between Betws-y-coed and Llyn Parc in Gwynedd, and from the Lambriggan zinc and lead mine near St Agnes in Cornwall.

Other grasses with metal-tolerant populations include sheep's fescue (*Festuca ovina*), red fescue (*Festuca rubra*) and sweet vernal (*Anthoxanthum odoratum*), each with zinc-tolerant strains at the Trelogan zinc and lead mine in Flintshire (Table 4.19) (Gregory and Bradshaw, 1965; Jain and Bradshaw, 1966). Further metal-tolerant populations of red fescue are known from copper mine sites at Ecton, Staffordshire (co-tolerant to copper, lead and zinc) and the Great Orme in north Wales (co-tolerant to copper and zinc) and from near a copper refinery at Prescot, Lancashire (co-tolerant to copper and zinc) (Wong, 1982).

Arsenic-tolerant populations of both common bentgrass, *Agrostis tenuis*, and its close relative creeping bentgrass, *A. stolonifera*, are known from the site of the Gawton mine in the Tamar valley near Tavistock, Devon (Porter and Peterson, 1977; Benson et al., 1981). An arsenic-tolerant strain of another grass, Yorkshire fog *Holcus lanatus*, has also been collected from the Gawton site (Meharg and Macnair, 1991). Cadmium-tolerant populations of common bentgrass *A. tenuis* and red fescue *F. ovina* are known from mine sites in Belgium and Germany (Simon, 1977). A British example of cadmium tolerance derives from exposure to cadmium emitted from the chimneys of the Avonmouth smelter, in this case a population of Yorkshire fog *H. lanatus* collected from a site about 3 kilometres downwind of the smelter (Coughtrey and Martin, 1977).

4.6.5.2 Evolution of Tolerance

Prologue. The evolution of a metal-tolerant plant population can be extremely rapid.

So, the evolution of metal-tolerant ecotypes is not uncommon (Table 4.19). Co-tolerance of a population to more than one metal occurs quite frequently, but probably by independent evolution. Tolerance to one metal does not automatically confer tolerance to another, with different physiological mechanisms probably being involved. The degree of tolerance is related to the metal concentration of the local soil, determining the strength of the natural selection pressure in action (Antonovics et al., 1971). This natural selection has acted on the genetic variation present in normal populations. Individuals with some metal tolerance occur naturally but are only favoured in conditions of high metal bioavailability in the local soil. Metal-tolerant genotypes appear to be present in normal populations at a frequency of between one or two per thousand (Berry, 1977) and 1 or 2% (Antonovics et al., 1971; Walley et al., 1974).

Table 4.19 **Flowering plant species found in Britain with ecotypes known to be tolerant to trace metals.**

Order	Family	Species	Common name	Metal
Grasses				
Poales	Gramineae	*Agrostis canina*	Velvet bentgrass	Pb, Zn
Poales	Gramineae	*Agrostis stolonifera*	Creeping bentgrass	Pb, Zn, As
Poales	Gramineae	*Agrostis tenuis*	Common bentgrass	As, Cu, Ni, Pb, Zn
Poales	Poaceae	*Anthoxanthum odoratum*	Sweet vernal grass	Zn
Poales	Poaceae	*Festuca ovina*	Sheep's fescue	Pb, Zn
Poales	Poaceae	*Festuca rubra*	Red fescue	Zn
Poales	Poaceae	*Holcus lanatus*	Yorkshire fog	As, Cd, Zn
Flowers				
Asterales	Asteraceae	*Taraxacum officinale*	Dandelion	Cu
Asterales	Asteraceae	*Tussilago farfara*	Coltsfoot	Cu
Asterales	Campanulaceae	*Campanula rotundifolia*	Harebell	Zn
Brassicales	Brassicaceae	*Thlaspi caerulescens*	Alpine pennycress	Zn
Caryophyllales	Caryophyllaceae	*Minuartia verna*	Spring sandwort	Zn
Caryophyllales	Caryophyllaceae	*Silene dioica*	Red campion	Cu
Caryophyllales	Caryophyllaceae	*Silene vulgaris*	Bladder campion	Zn
Caryophyllales	Plumbaginaceae	*Armeria maritima*	Thrift	Zn
Caryophyllales	Polygonaceae	*Rumex acetosa*	Sorrel	Cu, Zn
Lamiales	Plantaginaceae	*Plantago lanceolata*	Ribgrass	Zn
Lamiales	Scrophulariaceae	*Mimulus guttatus*	Monkey-flower	Cu
Malpighiales	Linaceae	*Linum catharticum*	Fairy flax	Zn
Malpighiales	Violaceae	*Viola lutea*	Mountain pansy	Zn

Sources: Antonovics et al. (1971), Coughtrey and Martin (1977), Benson et al. (1981), Meharg and Macnair (1991).

The evolution of a metal-tolerant population can be extremely rapid, even in one or two generations (Antonovics et al., 1971; Berry, 1977). Many of the mine tips shown to host established metal-tolerant populations of plants have only been 50- to 100-years-old. The establishment of such a metal-tolerant population might possibly be derived from the import of seeds from older sites with long-established metal-tolerant populations. This does, however, seem unlikely, for metal-tolerant populations have been found on contaminated slag heaps in Wales resulting from the smelting of imported ore, for example, in the Swansea area, 50 miles from the nearest mine (Gregory and Bradshaw, 1965; Antonovics et al., 1971). Furthermore, zinc-tolerant populations of sheep's fescue *Festuca ovina* and velvet bentgrass *Agrostis canina* have been found growing under galvanised fences several hundred miles from any mining area, within 30 years of fence erection (Antonovics et al., 1971).

Metal-tolerant populations are normally surrounded by non-tolerant populations, and so gene flow is likely between them. Yet, the geographical distribution of the tolerant population usually changes abruptly at a boundary (over a metre or so), suggesting that gene flow is quite low and/or selection pressure is high (Antonovics et al., 1971).

The question arises as to whether metal tolerance is restricted to certain families and genera of flowering plants, for example the family *Caryophyllaceae* or the grass genera *Agrostis* and *Festuca* (Table 4.19). In fact, this appears not to be the case, in spite of the bias introduced by the preference of active research groups for particular plant models.

Nevertheless, the ability to evolve metal tolerance may be associated with the presence of a particular character, such as the ability to grow in low nutrient conditions (Antonovics et al., 1971). Lead and copper-tolerant populations of *Agrostis tenuis*, for example, are better able to grow in culture conditions of low phosphate, in comparison to conspecific populations from pasture soils (Antonovics et al., 1971). Indeed, addition of complete fertiliser at normal agricultural rates enormously improves the growth of plants on mine soils irrespective of their origin. Even metal-tolerant plants sometimes need addition of fertiliser to grow on bare mine soils (Antonovics et al., 1971). Calcium availability is also an important variable. Acid mine soils usually host *Agrostis tenuis* and *Festuca ovina*, while calcareous mine soils support *A. stolonifera* and *F. rubra*. Metal tolerant ecotypes typically show competitive inferiority on

non-contaminated soils, but this inferiority may not always be considerable. It may be differential under different conditions, the difference in competitive ability depending, for example, on phosphate levels (Antonovics et al., 1971). Thus, the evolution of tolerance to metals is only one facet of the evolution of plants on abandoned mine workings. There is also independent natural selection for morphological and physiological characteristics which adapt plants for the harsher mine environment, such as tolerance of low phosphate or low calcium.

4.6.5.3 Mechanisms of Tolerance

Prologue. Metal tolerance results from different physiological mechanisms in different metal-tolerant populations.

Few generalisations can be made as to the specific physiological adaptations that confer metal tolerance on the individuals of a metal-tolerant ecotype. Metal-tolerant plants still take up metals from the soil via the roots, and tolerance is usually effected by a suite of physiological mechanisms developed to varying degrees for different metals in different species and populations.

In zinc- and copper-tolerant strains of *A. tenuis*, zinc and copper are concentrated in the cell walls and are therefore not metabolically available to play a toxic role in metabolism. In the case of zinc, the pectate fraction of the cell wall fraction contained more zinc in zinc-tolerant than in nontolerant strains of this grass (Antonovics et al., 1971).

In zinc- and lead-tolerant strains of sea campion *Silene uniflora*, there was restricted transport of zinc and lead to the shoots, and much lower accumulated concentrations of these metals in the shoots, in tolerant plants compared to non-tolerant plants exposed to high zinc and lead concentrations in soil (Baker, 1981). The zinc-tolerant plants had higher concentrations of zinc in the roots compared to the non-tolerant plants but lower concentrations in the shoots, the so-called exclusion accumulation pattern. Some zinc-tolerant populations also show reduced uptake of zinc from soil. Such reduced uptake was not, however, general across zinc-tolerant strains of sea campion, while all these strains did show exclusion of zinc from the shoots (Baker, 1981).

The physiological mechanism of arsenic tolerance has been investigated in the arsenic-tolerant strain of Yorkshire fog *Holcus lanatus* from the Gawton mine in the Tamar valley in Devon (Meharg and Macnair, 1991). Differences in the uptake, transport and accumulation of arsenic occurred between arsenic-tolerant and non-tolerant ecotypes. The tolerant plants had restricted arsenic uptake, yet these plants were capable of accumulating arsenic to high concentrations over longer time periods. Arsenate is taken up via the phosphate uptake system, and the reduction of arsenic uptake by the arsenic-tolerant plants was put down to an alteration in this phosphate uptake system. After uptake, a greater proportion (75%) of the accumulated arsenic in the plant was transported to the shoots in the tolerant plants, than in the non-tolerant plants (50%). It appears, therefore, that there is efficient detoxification of the accumulated arsenic in the shoots, probably by its chemical conversion to relatively non-toxic organic arsenic compounds (Meharg and Macnair, 1991).

4.6.6 Aluminium

Prologue. Aluminium deserves specific mention, for aluminium toxicity is the major factor limiting crop productivity in acid soils, which comprise up to 40% of the world's arable lands.

In mildly acidic soils or soils of neutral pH, aluminium is in the form of stable insoluble aluminosilicates or oxides (Matsumoto, 2000). These chemical forms of aluminium are not readily bioavailable for uptake by the roots of plants, and are, therefore, non-toxic. As soils become more acidic, aluminium is chemically transformed into more soluble and bioavailable forms, such as the aluminium hydroxide ions $Al(OH)^{2+}$ and $Al(OH)_2^+$, and ultimately the free metal ion Al^{3+} (Matsumoto, 2000). These bioavailable forms, particularly Al^{3+}, are potentially toxic to plants (Kochian, 1995).

The initial and most dramatic symptom of aluminium toxicity in plants is the inhibition of root elongation at the root apex (Kochian, 1995; Matsumoto, 2000) This in turn leads to secondary toxic responses in the plant as a result of the damaged root system.

The physiological mechanisms of aluminium toxicity in plants have not been fully established. The Al^{3+} ion has a similar ionic radius to the magnesium ion Mg^{2+}, and will block magnesium channels, thereby reducing magnesium uptake (Kochian, 1995; Matsumoto, 2000). The Al^{3+} ion will similarly interfere with the metabolic functioning of magnesium in cells. It has been proposed that aluminium may also block calcium channels at the root apex, reducing calcium uptake. There is, however, less strong evidence for aluminium interfering with the metabolism of calcium in plants than for magnesium, for which it shows a better size fit (Kochian, 1995; Matsumoto, 2000). A further possible toxic mechanism proposed for aluminium is that the Al^{3+} ion binds to the iron phytosiderophores secreted by certain grasses, thereby reducing the uptake of essential iron.

Research, primarily on crop plants, has shown that some plant species or varieties show resistance to aluminium toxicity (Kochian, 1995; Rout et al., 2001). This aluminium resistance has been shown to be genetically based, as in the case of particular strains of rye or Chinese spring wheat which carry genes for aluminium tolerance (Kochian, 1995; Matsumoto, 2000). Again, there is a lack of certainty as to the physiological basis of such aluminium resistance. Possible mechanisms include the secretion of aluminium binding agents such as the organic acids citric, malic and/or oxalic acid, reducing the bioavailability of the free aluminium ion for uptake. Alternatively, and more controversially, it has been suggested that aluminium-resistant strains may be able to bring about root-induced increases in the adjacent soil pH, which will in turn affect the local soil chemistry of aluminium and reduce its bioavailability (Kochian, 1995; Matsumoto, 2000). Mechanisms of aluminium resistance could, on the other hand, be internal, for example the enhanced cellular production of detoxificatory binding agents such as the aforementioned organic acids or a metallothionein-like protein.

While the anthropogenic contamination of soils with aluminium is not a significant ecotoxicological factor affecting soils in the British Isles, aluminium toxicity will still be playing a significant ecological role in British soils without the intervention of mankind. We have seen how calcicole plants develop symptoms of aluminium toxicity when growing on acidic soils. Aluminium toxicity will also be a strong limiting factor to plant growth in acidified wetland soils (Crowder, 1991).

4.7 Invertebrate Herbivores

Prologue. Concentrations of trace metals in plants will increase with increased soil metal concentrations, and some plants (hyperaccumulators) will contain very high trace metal concentrations even when growing in soils with relatively low metal concentrations. Very high accumulated metal concentrations may cause toxic effects when shoots are ingested by herbivores. This section explores the natural history of trace metals in herbivorous invertebrates.

4.7.1 Slugs and Snails

Prologue. Slugs and snails are major invertebrate herbivores in terrestrial habitats. These gastropod molluscs will inevitably ingest, and potentially assimilate and accumulate, any trace metals in the plant tissues on which they are feeding. This dietary load of trace metals will be atypically high when the molluscs are feeding on vegetation growing on metal-contaminated soil.

Slugs and snails store accumulated concentrations of cadmium, lead and zinc in the digestive gland, with particularly high concentrations in this organ when they are feeding on metal-contaminated vegetation (Tables 4.20, 4.21). Copper, on the other hand, is more evenly distributed in the body, because it is a component of the respiratory protein haemocyanin and is stored in cells surrounding the blood vessels

(Coughtrey and Martin, 1976a; Hopkin, 1989). Haemocyanin is not as widespread amongst animals as the iron-bearing respiratory protein haemoglobin, but it is characteristic of gastropod molluscs as well as malacostracan crustaceans.

Concentrations of cadmium, lead and zinc increase in the digestive gland, and correspondingly the whole body (soft tissues excluding the shell), of the garden snail *Cornu aspersum* (formerly known as *Helix aspersa*) feeding on vegetation with increased accumulated metal concentrations (Tables 4.20, 4.21). The contaminated site in Table 4.20 was 1 kilometre downwind of the Avonmouth smelter, that in Table 4.21a 3.2 kilometre downwind. In both cases, the cadmium, lead and zinc concentrations in both digestive gland and whole body are raised above the equivalent concentrations in snails from uncontaminated sites. While the Avonmouth snails had higher concentrations of the three metals in the digestive gland at the site nearer the smelter, total metal concentrations in the snails did not show the contamination gradient quite so clearly (Tables 4.20, 4.21). Like the digestive gland concentrations of cadmium, lead and zinc in the snails nearer the Avonmouth smelter, the digestive gland concentrations of copper are also amongst the highest ever recorded in the garden snail (Table 4.20) (Hopkin, 1989). The digestive gland does not dominate a list of accumulated tissue and body copper concentrations as it does for cadmium, lead and zinc, but copper concentrations are still higher in both digestive glands and total bodies of the snail as metal contamination increases (Tables 4.20. 4.21).

The same pattern of increased stored trace metal concentrations in the digestive gland and whole body with increased metal contamination is seen for the black slug *Arion ater* (Table 4.21). Again, in the case of copper, the accumulated concentration in the digestive gland may be lower than the total body concentration (Table 4.21). The zinc concentration in the digestive gland of the slugs collected at the disused Cwm Ystwyth lead and zinc mine is remarkably high (7,130 µg/g Zn), much higher than the already high zinc concentrations reported for the garden snail

Table 4.20 **Concentrations (mean ± SE, μg/g dry weight, *n* = 6) of cadmium, copper, lead and zinc in tissues of the garden snail *Cornu aspersum* from a site 1 kilometre downwind of Avonmouth smelting works and from an uncontaminated site (Kynance Cove, Cornwall).**

	Cadmium	Copper	Lead	Zinc
Smelter site				
Digestive gland	271 ± 17	186 ± 38	490 ± 96	1,780 ± 300
Gut	30 ± 14	181 ± 39	143 ± 46	204 ± 17
Reproductive tissue	7.5 ± 1.8	135 ± 27	16 ± 3.3	58 ± 3.6
Rest	17 ± 2.5	347 ± 60	18 ± 6.4	96 ± 22
Total	50 ± 11	228 ± 43	121 ± 24	418 ± 71
Kynance Cove				
Digestive gland	20 ± 5.2	118 ± 41	31.8 ± 4.8	429 ± 71
Gut	9.5 ± 4.2	95 ± 33	5.6 ± 0.9	67 ± 9.6
Reproductive tissue	5.5 ± 4.2	56 ± 16	0.75 ± 1.04	52 ± 4.5
Rest	1.1 ± 0.2	138 ± 36	0.63 ± 0.66	49 ± 2.1
Total	4.9 ± 0.6	104 ± 28	5.8 ± 1.1	103 ± 14

Source: Hopkin (1989).

(Tables 4.20, 4.21). The ability to accumulate very high zinc concentrations seems to be a feature of slugs in general. For example, mean accumulated total body concentrations in three other slug species collected at Blaise, 3.2 kilometres downwind of the Avonmouth smelting works, in another study, were higher than that in the garden snail at the same site – the garden slug *Arion hortensis* (1,410 μg/g Zn), the orange-banded Arion *Arion fasciatus* (1,110 μg/g Zn) and the grey field slug *Agriolimax reticulatus* (948 μg/g μg/g Zn), in comparison to the garden snail *Cornu aspersum* (400 μg/g Zn) (Coughtrey and Martin, 1976b).

The high accumulated concentrations of cadmium, lead and zinc in the digestive glands of slugs and snails reflect the fact that this organ is the site of the storage of these metals in detoxified form. The digestive gland is the site of assimilation and absorption of nutrients from ingested food in gastropods and is an ideal position to control the absorption of trace metals, both essential and non-essential, from the diet. The metabolic availability of assimilated copper to the rest of the body also requires physiological control exercised by the digestive gland. Much of the absorbed copper, however, is required for the synthesis of haemocyanin, and copper is not stored in detoxified form in the digestive gland to the same degree as the other three trace metals (Table 4.20).

4.7.1.1 Detoxification

Prologue. Snails present good examples of the use of metallothioneins and metal-rich granules for the detoxification of accumulated trace metals.

Metallothioneins (MT) are intracellular proteins that bind to particular trace metals including Cd,

Table 4.21 **Total body and digestive gland concentrations (mean $\pm$ SE, µg/g dry weight, $n = 6$) of cadmium, copper, lead and zinc: (a) in samples of the garden snail *Cornu aspersum* from Blaise, a metal-contaminated site 3.2 kilometres downwind of Avonmouth smelting works, and from a control site (Housel Bay, Lizard, Cornwall); and (b) in samples of the black slug *Arion ater* from the disused Cwm Yystwyth lead and zinc mine in the Ystwyth valley, Wales, and from an uncontaminated site on the campus of Aberystwyth University, Wales.**

	Cadmium	Copper	Lead	Zinc
(a) *Cornu aspersum*[1]				
Blaise, Avonmouth				
Digestive gland	198 $\pm$ 24	60 $\pm$ 12	84 $\pm$ 15	1,570 $\pm$ 171
Total body	65 $\pm$ 7.6	68 $\pm$ 10	29 $\pm$ 3.4	413 $\pm$ 46
Lizard (control)				
Digestive gland	15 $\pm$ 1.9	31 $\pm$ 3.4	45 $\pm$ 29	260 $\pm$ 35
Total body	6.0 $\pm$ 0.6	46 $\pm$ 4.0	19 $\pm$ 8.0	95 $\pm$ 7.7
(b) *Arion ater*[2]				
Cwm Ystwyth mine				
Digestive gland	67 $\pm$ 2.5	72 $\pm$ 18	129 $\pm$ 6	7,130 $\pm$ 800
Total body	20 $\pm$ 4	90 $\pm$ 10	94 $\pm$ 23	1,230 $\pm$ 60
Aberystwyth (control)				
Digestive gland	10 $\pm$ 2.5	44 $\pm$ 5	11 $\pm$ 1	3,200 $\pm$ 190
Total body	2 $\pm$ 0.2	50 $\pm$ 4	5 $\pm$ 0.9	280 $\pm$ 20

Sources: Coughtrey and Martin (1976a);[1] Ireland (1979b).[2]

Cu and Zn, their synthesis typically being induced by exposure to these metals. Different MT isoforms (slightly different forms of the same protein resulting from co-occurring slightly different genes) are often present and play somewhat different roles in the trace metal physiology of the organism concerned. Metal-rich granules are trace metal-rich insoluble cellular inclusions that take several different forms, including phosphate granules and sulphur-containing residual bodies in lysosomes in which the breakdown of metallothioneins has occurred.

Terrestrial snails provide one of the best examples of the use of different MT isoforms for different physiological roles when handling potentially toxic trace metals (Dallinger et al., 1997; Palacios et al., 2011). Terrestrial snails, exemplified by the Roman snail *Helix pomatia*, have two MT isoforms that specifically bind either cadmium or copper (Palacios et al., 2011). The Cd-binding MT occurs in the digestive gland, is induced by cadmium exposure and binds 6 Cd^{2+} ions per molecule of the MT protein. The physiological role of this MT isoform is solely the detoxification of Cd entering the body of the snail from ingested food via the digestive gland. At very high cadmium exposures, residual bodies resulting from the autolysis of this Cd-MT can be seen in cells of the digestive gland of exposed snails (Hödl et al., 2010). The Cu-binding MT is part of the normal physiology of copper, which is needed to make the respiratory protein haemocyanin. This Cu-MT binds 12 ions of Cu^+ and is not restricted to the digestive gland, but found in all major organs of the snail (Palacios et al., 2011; Dallinger et al., 2005). Copper still needs to be detoxified when present in body cells, and the Cu-MT represents a reversible temporary store of this essential trace metal. The Cu-MT isoform is found in pore cells which occur in all the major organs, regulating the metabolic functioning of copper and synthesising haemocyanin (Dallinger et al., 2005). While Cu-MT concentrations and total copper concentrations are usually relatively constant in the different organs, copper concentrations will increase in the pore cells upon copper exposure of the snails (Dallinger et al., 2005), as reflected in the tissue copper concentrations reported in Tables 4.20 and 4.21. Under conditions of high copper exposure, pore cells will show the presence of insoluble inclusions bearing copper (Dallinger et al., 2005), interpretable as residual bodies of lysosomes in which the Cu-MT has been broken down.

Also important in trace metal detoxification in the digestive glands of slugs and snails are phosphate granules. These are typically made up of concentric layers of calcium and magnesium phosphate, with the additional presence of many other metals, often including potassium, manganese, zinc, iron and lead (Hopkin, 1989). The phosphates occur in different forms, including orthophosphate and, most importantly, pyrophosphate. Pyrophosphate binds many metals in very insoluble form, ensuring that the toxic metals bound therein are not released again in metabolically available form. Thus phosphate-based granules can act as safe permanent stores for potentially toxic trace metals, whether or not the granule itself is subsequently expelled from the body. These calcium pyrophosphate granules are, therefore, the sites where the high concentrations of zinc and lead (and also manganese) are bound in detoxified form in the digestive glands of terrestrial gastropods (Tables 4.20, 4.21). While zinc and lead are each bound in such granules in the digestive gland, there may be subtle differences in the granules concerned, their physiology and their subsequent fate. Snails (*Cornu aspersum*) exposed to contaminated zinc and lead conditions accumulated both metals as expected, but the excretion patterns of the two metals differed (Gimbert et al., 2008). The snails appeared able to excrete the granules containing most of the lead accumulated, while accumulated zinc was excreted at a much lower rate. It would appear then that the detoxification physiology of zinc and lead differs somewhat in the digestive glands of snails, possibly because zinc may also be bound in metallothioneins, however temporarily, as well as in phosphate granules.

It is important to distinguish these calcium phosphate-based detoxification granules from another type of calcium-based granule common in snails (Hopkin, 1989). These are calcium carbonate granules that are extracellular, and are usually very pure with no other metals present. These calcium carbonate granules represent temporary stores of calcium needed by the snail to supply its calcium carbonate shell. Therefore, these granules must be easily resolubilised, a feature incompatible with any role in trace metal detoxification.

4.7.1.2 Biomonitoring

Prologue. Slugs and snails can be used as biomonitors of trace metal bioavailabilities in terrestrial habitats.

The accumulated metal concentrations of slugs and snails increase with increased metal contamination, rendering them suitable as trace metal biomonitors. More specifically, the isolated digestive gland of terrestrial gastropods appears to be particularly suitable for trace metal biomonitoring purposes, with very high accumulated concentrations and a greater power to distinguish between metal bioavailabilities at different sites. A caveat needs to be introduced, because larger, older gastropods may have higher accumulated metal concentrations than smaller, younger ones, simply as a function of age. Thus, slugs or snails of a similar size should be analysed, restricted to a single species in order to avoid any effects of interspecific differences.

Suitable biomarkers of the exposure of snails to very high cadmium bioavailabilities would be the degree of saturation of induced Cd-MT (the number of cadmium ions bound per Cd-MT molecule up to a maximum of six), and the presence of lysosomal residual bodies containing cadmium in digestive gland cells of the snail (Hödl et al., 2010).

4.7.2 Insect Herbivores

Prologue. Many insects are herbivorous, ingesting vegetation that might be growing on metal-contaminated soils, with different degrees of specificity of choice of diet. The caterpillars of butterflies and moths, in particular, can be very specific to certain host plants as their source of nutrition. Other insect herbivores in British terrestrial habitats include bees, ants, grasshoppers, aphids and weevils.

Not unexpectedly, herbivorous insects typically show increased accumulation of trace metals when feeding on plants that have accumulated higher than usual metal concentrations when growing on metal-enriched soil. We have seen earlier how the copper refinery complex in Merseyside in the 1980s caused raised concentrations of copper and cadmium in soils at the refinery site itself and at another site 1 kilometre away (Table 4.4b). In turn, raised

copper and cadmium bioavailabilities in the soils at these two sites caused increased accumulated concentrations of the two metals in the two grasses, creeping bentgrass *Agrostis stolonifera* and red fescue *Festuca rubra*, dominating local grassland swards (Table 4.17). Copper and cadmium were moved by trophic transfer to accumulate in local insect herbivores, in the case of this grassland community, in grasshoppers, ants, aphids, caterpillars and weevils (Table 4.22). The local caterpillar was that of a brown butterfly, the grayling *Hipparchia semele*, for which red fescue is a primary food plant. In all these herbivores, copper and cadmium concentrations at the refinery site were greatly raised above background controls, as was also the case, albeit to a lesser extent, at the site 1 kilometre from the refinery (Table 4.22).

Detoxificatory calcium phosphate-based granules binding the likes of manganese, zinc and iron can also be found in herbivorous insects (Ballan-Dufrançais, 2002). Such granules have been found in cells of the midgut of the European red wood ant *Formica polyctena* and in the Malpighian tubules of this ant and of the European mole cricket *Gryllotalpa gryllotalpa*. Not unexpectedly, sulphur and metal-rich lysosomal residual bodies also occur, as in the midgut cells of houseflies *Musca domestica* in which they contain copper (Sohal et al., 1977; Ballan-Dufrançais, 2002).

Insects also provide good examples of the use of trace metals in biology, other than playing an essential biochemical role in metabolism. Many insects use the trace metals zinc and/or manganese to harden the cutting edges of their mandibles (Hillerton and Vincent, 1982; Quicke et al., 1998). Concentrations of these two metals can reach as high as 10% (100,000 µg/g) dry weight in the cuticle of these mandibles. Several orders of insects show examples of such mineralisation to harden the jaws, including orthopterans (grasshoppers and locusts) in addition to beetles; stick insects; caterpillars of butterflies and moths; and ants and parasitoid wasp larvae in the list of British examples in Table 4.23. Zinc is the preferred trace metal in the mandibles of orthopterans, phasmids and lepidopterans. Manganese can be present

Table 4.22 **Copper and cadmium concentrations (mean ± SE, μg/g dry weight, *n* = 30–250) in herbivorous insects collected in pitfall traps in established grassland swards dominated by creeping bentgrass *Agrostis stolonifera* and red fescue *Festuca rubra* at a control site and at sites affected by emissions from a copper refinery complex in Merseyside (see Table 4.17). The samples of ants, aphids and weevil species were mixed.**

Order Family Species	Common name	Copper			Cadmium		
		Control site	1 km site	Refinery	Control site	1 km site	Refinery
Orthoptera							
Acrididae							
Chorthippus brunneus	Common field grasshopper	385 ± 2.1	66 ± 5.9	333 ± 38	0.19 ± 0.01	0.32 ± 0.04	1.96 ± 0.1
Hymenoptera							
Formicidae	Ants	33 ± 3.1	131 ± 4.9	731 ± 166	1.2 ± 0.1	5.4 ± 1.3	38 ± 5.5
Hemiptera	Aphids	30 ± 2.7	62 ± 5.3	265 ± 98	0.8 ± 0.1	3.5 ± 1.2	11 ± 3.2
Lepidoptera							
Nymphalidae							
Hipparchia semele	Grayling caterpillar	14 ± 1.1	69 ± 13	160 ± 22	0.6 ± 0.2	7.1 ± 1.3	22 ± 4.3
Coleoptera							
Curculionidae	Weevils						
Otiorhynchus sulcatus							
Barynotus obscurus		30 ± 1.6	66 ± 3.9	421 ± 91	0.6 ± 0.1	3.6 ± 0.4	15 ± 1.4
Ceutorhynchus sp.							

Source: Hunter et al. (1987b).

together with zinc in hymenopteran and coleopteran mandibles, particularly in wood-boring beetle larvae or by itself in the mandibles of some other beetles (Table 4.23) (Hillerton and Vincent, 1982). The use of trace metals to harden jaws is not a feature unique to insects. The same characteristic shall be seen again later in this book, particularly in the case of ragworms (nereidid polychaetes) living in coastal waters.

Table 4.23 **Occurrence of manganese and zinc in the cutting region of the mandibles of insects found in Britain.**

Order 　Species	Common name	Stage and feeding type	Metal
Coleoptera			
Anobium punctatum	Furniture beetle	Wood-boring larva	Mn, Zn
Hylotrupes bajalus	House longhorn beetle	Wood-boring larva	Mn, Zn
Tenebrio molitor	Mealworm beetle	Adult – stored products	Mn
Phaedon cochleariae	Mustard leaf beetle	Herbivorous adult	Mn
Leptinotarsa decemlineata	Colorado beetle	Herbivorous adult	Mn
Chrysolina menthastri	Mint beetle	Herbivorous adult	Mn
Agriotes lineatus	Lined click beetle	Herbivorous larva	Mn
Phasmida			
Carausius morosus	Laboratory stick insect	Herbivorous adult	Zn
Lepidoptera			
Daphnis nerii	Oleander hawkmoth	Herbivorous larva	Zn
Pieris brassicae	Large white butterfly	Herbivorous larva	Zn
Hymenoptera			
Formica rufa	Red wood ant	Herbivorous worker	Mn, Zn
Formica lugubris	Hairy wood ant	Herbivorous worker	Mn, Zn
Calameuta filiformis	Reed stem borer	Herbivorous larva	Zn
Xiphidria sp.	Alder woodwasp	Herbivorous larva	Zn
Ormyrus nitidulus	A chalcid wasp	Parasitoid larva	Mn, Zn
Sycophila biguttata	A chalcid wasp	Parasitoid larva	Zn
Perithous sp.	An ichneumonid wasp	Parasitoid larva	Zn
Mutilla europaea	Eastern velvet ant	Parasitoid larva	Zn

Sources: Hillerton and Vincent (1982), Quicke et al. (1998).

4.8 Invertebrate Detritivores

Prologue. While herbivores feed on living plant material, a very large amount of plant material passes through detritus-based food chains after the death of a plant or after leaf fall from deciduous shrubs and trees. Thus there is potential for metals accumulated in plant tissue to be taken up and accumulated by invertebrate detritivores. Major detritivores of the leaf litter and soils of terrestrial habitats in the British Isles are earthworms, woodlice, collembolans and millipedes.

As in the case of herbivores feeding on plants growing on metal-contaminated soils, detritivores will also ingest plant detrital material with high accumulated metal concentrations, as well as the associated microbes processing the plant detritus. These detritivores also show increased accumulation of trace metals.

We can again use for illustration the study of grassland swards affected by the copper refinery complex in Merseyside in the 1980s (Hunter et al., 1987a, b). As in insect herbivores (Table 4.22) feeding on the grasses with raised accumulated concentrations of copper and cadmium (Table 4.17), these two trace metals were also accumulated to atypically raised concentrations by local detritivores, consisting of earthworms, woodlice, millipedes, collembolans and fly larvae, gaining their metals from the ingestion of local plant detritus (Table 4.24). For all these invertebrate groups, there were clearly progressive increases in accumulated concentrations of copper and cadmium from control samples, to samples taken at the 1 kilometre site and finally at the refinery site itself (Table 4.24). A slight caveat does, however, need to be introduced here, because most samples consisted of mixed species, although the woodlice samples predominantly consisted of the common shiny woodlouse *Oniscus asellus*, and *Orchesella villosa* dominated the collembolan samples. There is therefore a remaining chance that differences between accumulated concentrations of different composite samples at different sites might have been caused by different contributions of different species with different typical background concentrations. We shall see later in this section how two different common woodlice species, *O. asellus* and *Porcellio scaber*, accumulate different concentrations of trace metals under the same conditions of metal exposure. However, the increases in accumulated concentrations observed here (Table 4.24) are so high that it is extremely unlikely that the cause is any factor other than increased bioavailabilities of the metals to the

detritivores at the metal-contaminated sites. Again, because of the presence of mixed species in these samples, it will also be necessary to be cautious in any comparisons of the accumulated concentration data in Table 4.24 with other published concentration data for specific named species. We can, nevertheless, still conclude with confidence that there were raised bioavailabilities of copper and cadmium to the invertebrate detritivores at the refinery site and at the site at 1 kilometre distant.

4.8.1 Earthworms

Prologue. Earthworms are classic examples of invertebrate detritivores living in soil and will accumulate high concentrations of trace metals in metal-contaminated soils.

Different earthworm species show differences in their ecology, illustrated by different feeding methods, depth of burrow (if any) and soil preferences (Sherlock, 2012).

The largest and most common British earthworm is *Lumbricus terrestris* (Plate 4a), mainly found in grassland, and thus lawns, with a preference for basic soils, especially clay. This earthworm lives deep in the soil in a vertical burrow, coming up to the surface at night to draw leaves and other dead plant material into the burrow. Some soil can be ingested. A related species, *Lumbricus rubellus* (Plate 4b), known, particularly to anglers, as the red worm, is more of a surface dweller, for example in leaf litter, feeding directly on decaying plant material with little or no soil consumed. The red worm appears to have the lowest specificity of all British earthworms. It is recorded from all habitats, being particularly prevalent in organically rich environments as well as in acid grassland and woodland. This acid tolerance equips *L. rubellus* for life in many mining-contaminated soils. An earthworm with a similar method of surface feeding is another red-coloured worm *Dendrodrilus rubidus* (Plate 4c), previously known as *Dendrobaena rubida*. *D. rubidus* is often found in woodland, favouring acidic soils and soils with a high organic content, as is its rarer close

Table 4.24 Copper and cadmium concentrations (mean ± SE, μg/g dry weight, $n = 30 – 60$) in detritivorous invertebrates collected in pitfall traps in established grassland swards dominated by creeping bentgrass *Agrostis stolonifera* and red fescue *Festuca rubra* at a control site and at sites affected by emissions from a copper refinery complex in Merseyside (see Table 4.17). All samples contained mixed species.

Higher taxon Order Family Species	Common name	Copper			Cadmium		
		Control site	1 km site	Refinery	Control site	1 km site	Refinery
Oligochaeta	Earthworms	24 ± 1.3	155 ± 19	1,170 ± 300	4.1 ± 0.3	34 ± 2.8	107 ± 25
Crustacea							
Isopoda	Woodlice						
Oniscus asellus		78 ± 7.7	836 ± 126	2,390 ± 270	15 ± 1.0	130 ± 46	231 ± 131
Armadillidium vulgare							
Diplopoda	Millipedes						
Julidae							
Tachypodoiulus niger		138 ± 6	511 ± 40	780 ± 126	5.6 ± 0.2	14 ± 0.4	19 ± 2.4
Polydesmidae							
Oxidus sp.							
Collembola	Springtails						
Orchesella villosa		50 ± 2.3	175 ± 28	2,370 ± 510	2.1 ± 0.3	12 ± 2.3	52 ± 9.7
Others							
Insecta							
Diptera	Fly larvae	26 ± 2.0	85 ± 15.7	210 ± 27	2.2 ± 0.4	6.9 ± 0.9	25 ± 3.9

Source: Hunter et al. (1987b).

relative *Dendrodrilus octaedra*. *D. rubidus* is often present at old mining sites. Examples of earthworms that live in permanent horizontal burrows in the soil, rarely coming to the surface and consuming the soil itself, are *Allolobophora chlorotica*, *Aporrectodea rosea* (previously known as *Allolobophora rosea*) and *Aporrectodea caliginosa* (previously *Allolobophora caliginosa*). *A. chlorotica* is found in most habitats, but it prefers neutral to basic grassland and arable soils. *A. rosea* is widespread, being found in woodland and grassland, preferring neutral or basic soils. *A. caliginosa*, the grey worm (Plate 4d), is very common, again being found in most habitats but being particularly dominant in grasslands. *A. calaginosa*, however, is not found in the most acidic soils. Earthworms in a further ecological category live in areas of very high organic matter such as under cowpats or in compost heaps, as in the case of the brandling *Eisenia fetida* (Plate 4e).

4.8.1.1 Trace Metal Accumulation

Prologue. Earthworms take up trace metals from the soil solution via their permeable skin and in the alimentary tract from ingested soil or litter particles. Accumulated trace metal concentrations in earthworms increase with trace metal contamination of soils.

Table 4.25 presents the accumulated concentrations of cadmium, copper, lead and zinc in two acid-tolerant earthworm species, *Lumbricus rubellus* and *Dendrodrilus rubidus*, from sites affected by mining in Britain. Both these species are considered to be litter-dwelling, with little, if any, direct soil ingestion (Sherlock, 2012). Nevertheless, it is still necessary to be wary of the contribution of the gut contents to total body metal concentrations, for metals still in the gut have not been assimilated and thus not accumulated by the worm. In Table 4.25, all the worms analysed have been allowed to depurate their gut contents by feeding on damp filter paper before analysis, so that the contribution of these gut contents to any total measured metal concentration is negligible. The need for depuration is even greater when analysing the accumulated metal concentrations of earthworms that are specifically soil ingestors, as in the cases of *Aporrectodea rosea* and *A. caliginosa*. If it is not possible to allow depuration by the earthworms, some researchers will resort to dissection and the flushing out of the gut of the worm with distilled water before metal analysis (Hopkin, 1989).

The data in Table 4.25 mostly concern earthworms collected from the abandoned Cwm Ystwyth lead and zinc mine in Ceredigion in mid-Wales, and from old lead mines in Shropshire. It is clear from Table 4.25 and Figure 4.4 that accumulated concentrations of lead and zinc in both *L. rubellus* and *D. rubidus* reach extremely high levels in response to high metal concentrations in the local soil. Similarly, cadmium concentrations accumulated by each species can increase greatly in response to local soil cadmium concentrations, albeit over a lower concentration range than for lead or zinc (Table 4.25, Figure 4.4). While copper concentrations in soils affected by mining do not reach the extremely high concentrations achieved by lead or zinc (Table 4.2), copper concentrations accumulated by the local earthworms do show a significant, although small, increase with increased soil copper concentration over this more limited soil concentration range (Table 4.25, Figure 4.4). Accumulated copper concentrations in the two earthworm species do not increase to the extent seen for the other three metals (Table 4.25, Figure 4.4), suggesting a different accumulation pattern in these earthworms for copper than the strong accumulation apparent for cadmium, lead and zinc.

Earthworms take up trace metals by two routes: from the soil solution via their permeable skin, and in the alimentary tract from ingested food particles. The latter will vary from recognisable plant detritus such as leaf litter to soil particles, depending on the earthworm species concerned and on the local availability of food materials. Thus both soil solution and diet contribute to the total metal uptake by earthworms. The exact fraction of each that is bioavailable to the worm is controlled by physicochemical factors, such as the chemical characteristics of the soil solution or the ingested food particles, and species-specific

Table 4.25 **Total concentrations (µg/g dry weight) of selected trace metals in earthworms and associated soils at sites affected by mining and at control sites. Earthworms were typically depurated in petri dishes with damp filter paper for between one and four days to empty gut contents.**

Species		Cadmium	Copper	Lead	Zinc
Site					
Lumbricus rubellus[1]					
Cwm Ystwyth, Ceredigion, Wales	Earthworm	*na*	13 ± 5.8	3,590 ± 900	739 ± 181
Control: Keele University, Staffordshire	Earthworm	*na*	15 ± 5.2	25 ± 6.2	646 ± 77
Lumbricus rubellus[2]					
Control: Dinas Powys, Glamorgan, Wales	Soil	0.69 ± 0.06	23 ± 0.5	91 ± 5	416 ± 70
	Earthworm	9.3 ± 1.3	19 ± 2.0	26 ± 5	498 ± 62
Llantrisant, Glamorgan, Wales (A)	Soil	8.31 ± 0.15	22 ± 0.2	6,260 ± 160	904 ± 34
	Earthworm	102 ± 19	24 ± 2.1	1,310 ± 340	2,350 ± 470
Llantrisant, Glamorgan, Wales (B)	Soil	3.56 ± 0.27	61 ± 3.5	11,900 ± 1,050	792 ± 61
	Earthworm	29.5 ± 5.4	30 ± 1.7	2,950 ± 550	788 ± 24
Castell, Ponterwyd, Ceredigion, Wales	Soil	20.8 ± 0.6	154 ± 7	445 ± 22	4,650 ± 360
	Earthworm	205 ± 44	59 ± 11	420 ± 81	3,450 ± 830
Cwm Ystwyth, Ceredigion, Wales (A)	Soil	94.4 ± 5.4	94 ± 8.1	13,900 ± 640	183,000 ± 17,500
	Earthworm	71 ± 12	24 ± 3.7	4,180 ± 830	5,170 ± 670
Cwm Ystwyth, Ceredigion, Wales (B)	Soil	0.46 ± 0.05	26 ± 1.0	667 ± 11	520 ± 94
	Earthworm	10 ± 2.5	19 ± 0.9	2,600 ± 450	719 ± 56

Table 4.25 **(cont.)**

Species		Cadmium	Copper	Lead	Zinc
Site					
Cwm Ystwyth, Ceredigion, Wales (C)	Soil	6.75 ± 0.25	103 ± 6	4,640 ± 440	3,180 ± 140
	Earthworm	61 ± 9.3	26 ± 3.1	1,420 ± 220	2,090 ± 230
Snailbeach, Shropshire	Soil	207 ± 34	31 ± 4.3	7,320 ± 930	79,300 ±14,200
	Earthworm	331 ± 41	28 ± 2.2	1,480 ± 250	2,510 ± 340
Pennerley, Shelve, Shropshire	Soil	124 ± 9	775 ± 225	16,700 ± 2,100	52,500 ± 3,010
	Earthworm	214 ± 12	66 ± 6.0	1,770 ± 140	3,270 ± 700
Roman Gravels, Shelve, Shropshire	Soil	318 ± 18	816 ± 360	7,230 ± 380	96,800 ± 15,600
	Earthworm	487 ± 49	61 ± 17.2	1,490 ± 230	5,270 ± 840
***Lumbricus rubellus*[3]**					
Control site plus 11 mining sites	Earthworm	8–577	8–104	4–10,400	394–3,870
***Dendrodrilus rubidus*[1]**					
Cwm Ystwyth, Ceredigion, Wales	Earthworm	*na*	17 ± 3.1	7,590 ± 1,480	309 ± 38
Control: Keele University, Staffordshire	Earthworm	*na*	13 ± 5.4	37 ± 5.3	251 ± 25
***Dendrodrilus rubidus*[3]**					
Control site plus 13 mining sites	Earthworm	10–1,790	9–34	12–13,000	308–1,690

Note: Data are means (± SE, $n = 3 - 6$)[1,2] or ranges of geometric means.[3] *na: not analysed.*
Sources: Ireland and Richards (1977);[1] Corp and Morgan (1991);[2] Morgan and Morgan (1988).[3]

biological features such as the number and nature of metal-transporting systems in the epithelia of the external surface and the alimentary tract. Additionally, in the case of uptake from food, the choice of food, the strength of digestion of food particles in the gut and the gut passage time will affect the trophic bioavailability of metals in ingested particles. It is being increasingly realised that, even in metal-contaminated soils, not all metal uptake by an earthworm is via the permeable external surface from the soil solution, and that the ingestion route can be very important (Vijver et al., 2003). The proportional contribution of each route varies with earthworm species, not least because of different feeding habits, and with the relative concentrations (and bioavailabilities) of metals in the soil solution and in the ingested food (Hobbelen et al., 2006).

After uptake, trace metals are accumulated to different degrees by earthworms, the accumulated concentrations being dependent on the net outcome of the total uptake rate and the rate of any excretion.

Experimental exposures of the earthworms *Eisenia fetida* and *Lumbricus rubellus* have shown that the two non-essential metals cadmium and lead are continuously accumulated from metal-contaminated soils, over a period as long as 42 days, with no equilibrium plateau metal concentration being reached (Spurgeon and Hopkin, 1999b; Giska et al., 2014). Any excretion of accumulated cadmium or lead, even after return to uncontaminated soil, is extremely slow or absent. Such a strong net accumulation pattern for both cadmium and lead also holds for other earthworms investigated, and is consistent with the accumulated cadmium and lead field concentration data presented in Table 4.25 for *L. rubellus* and *Dendrodrilus rubidus*.

The accumulation pattern of earthworms for the essential metal copper is very different from that for cadmium or lead (Nahmani et al., 2007; Ardestani et al., 2014). Experimental exposures of *E. fetida* and *L. rubellus* to copper-contaminated soils have shown that a high excretion rate of copper maintains a stable body copper concentration in the worms over time at most copper soil concentrations (Spurgeon and Hopkin, 1999b; Giska et al., 2014). This accumulation

pattern is not regulation of body copper concentrations, since steady-state body copper concentrations in the earthworms do increase with copper exposure. This accumulation pattern of strong copper excretion, but with some increase with steady-state body copper concentration with copper exposure, fits the field data (Table 4.25, Figure 4.4) for *L. rubellus* and *D. rubidus* (Morgan and Morgan, 1988). It is not uncommon for invertebrates to show excretion of accumulated

(a) Lumbricus rubellus

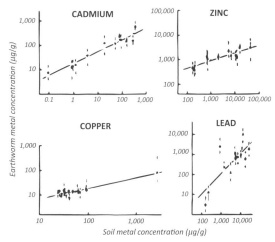

(b) Dendrodrilus rubidus

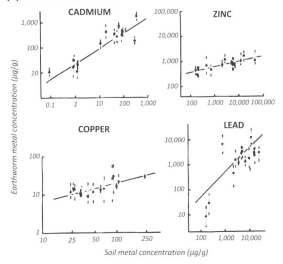

Figure 4.4 Relationships between the Cd, Cu, Pb and Zn concentrations (μg/g dry weight) in British soils and earthworms (means and ranges). (a) *Lumbricus rubellus* from 12 sites and (b) *Dendrodrilus rubidus* from 14 sites. (After Morgan and Morgan, 1988, with permission.)

copper (Luoma and Rainbow, 2008), whether or not this results in the regulation of the body copper concentration to the same constant level over a wide range of copper exposures.

Another essential metal, zinc, can also be excreted at a high rate by *Eisenia fetida* (Spurgeon and Hopkin, 1999b), *Lumbricus rubellus* (Giska et al., 2014) and other earthworms (Nahmani et al., 2007; Ardestani et al., 2014). In spite of this high rate of zinc excretion, earthworms do still show significant net accumulation of zinc, increasing with the zinc concentration of the soil to which they are exposed (Spurgeon and Hopkin, 1999b; Giska et al., 2014). The same conclusion of significant zinc accumulation by *L. rubellus* and *D. rubidus* in proportion to the local soil zinc concentration can be drawn from the field data in Table 4.25 and Figure 4.4. Thus the accumulation pattern of earthworms for zinc is intermediate between that of copper and those of cadmium and lead. In the case of copper, strong excretion restricts copper accumulation to small (but significant) increases with increased copper exposure. There is no significant excretion of cadmium nor lead by earthworms, so all metal taken up is accumulated. For zinc, there is again significant excretion, but the difference between the rates of zinc uptake and zinc excretion is greater than in the case of copper, and there is considerable zinc accumulation in proportion to soil zinc exposure.

4.8.1.2 Detoxification of Trace Metals

Prologue. Earthworms show two types of metal-rich granules involved in the detoxification of accumulated trace metals in the chloragogenous tissue surrounding the gut.

The extra cadmium, lead and zinc accumulated to high concentrations by earthworms living in metal-contaminated soils (Table 4.25) do need to be stored in detoxified form. It is debatable whether the highest copper concentration (ca. 100 µg/g Cu; Table 4.25, Figure 4.4) accumulated by earthworms at high soil copper exposures will need any extra detoxification over and above that supplied by

copper physiological mechanisms based on metallothionein.

The key tissue in earthworms responsible for the physiological handling and detoxification of trace metals is the chloragogenous tissue, a loose assemblage of cells surrounding the gut. Indeed, the respiratory protein haemoglobin is synthesised in specific sites in the chloragogenous tissue. This tissue is bathed by the fluid in the body cavity of the earthworm, and rupture of the chloragogenous cells releases dissolved and particulate material into this fluid. Amongst the particles so released are metal-rich granules (MRG), which may be stored in the body cavity or excreted by paired excretory organs present in many of the segments of the earthworm (Hopkin, 1989).

Earthworms show two types of metal-rich granules involved in the detoxification of accumulated trace metals in the chloragogenous tissue (Hopkin, 1989). Concentrically structured phosphate-based MRG, containing calcium, magnesium and potassium, bind zinc, lead and iron in detoxified form. In comparison to such phosphate-based MRG in other invertebrates, these granules in earthworms happen to contain an atypically large amount of organic matter as a matrix, contributing to an unusually high content of sulphur therein. As expected in a detoxification granule, the phosphate-based MRG in the chloragogenous tissue contain raised levels of both zinc and lead in earthworms collected from the spoil heaps of abandoned zinc and lead mines. Thus unusually high amounts of zinc and lead are present in the phosphate-based MRG of both *Lumbricus rubellus* and *Dendrodrilus rubidus* collected from the abandoned Cwm Ystwyth mine, Ceredigion (Ireland and Richards, 1977), and from a disused lead mine at Draethen near Caerphilly, Glamorgan (Morgan and Morris, 1982; Morgan and Morgan, 1989a). The same increase in both zinc and lead in phosphate-based MRG has also been observed in *L. rubellus* collected from spoil heaps of old mines at Llantrisant, Glamorgan (Morgan and Morgan, 1989a), and at Wemyss, Ceredigion (Morgan and Morgan, 1989b).

The second type of MRG found in earthworm chloragogenous tissue is a lysosomal residual body, derived from the breakdown of MT binding trace metals (Hopkin, 1989). Such lysosomal residual bodies in other invertebrates typically contain copper and some zinc in association with sulphur, all being derived from MT autolysis. In earthworms, however, copper has not been recorded in these MRG, probably because accumulated copper concentrations do not reach high enough levels. Similarly, no zinc is present in these earthworm lysosomal residual bodies, for most (if not all) accumulated zinc is bound to phosphate-based granules. The typical metal present in earthworm lysosomal residual bodies is actually cadmium, in association with sulphur. Cadmium is rarely seen in these MRG in other invertebrates. Not only is Cd-binding MT usually turned over very rapidly to release the cadmium to bind to newly synthesised MT without deposition in insoluble form in the lysosome, but accumulated cadmium concentrations in other invertebrates very rarely, if ever, reach the high accumulated cadmium concentrations seen in earthworms. The remarkably high cadmium concentrations attained in *Lumbricus rubellus* and *Dendrodrilus rubidus* from cadmium-contaminated mining sites (577 and 1,790 µg/g Cd dry weight, respectively; Table 4.25) are unheard of in other invertebrates. So, it is cadmium that is present in association with sulphur in the lysosomal residual bodies derived from metallothionein breakdown in *D. rubidus* collected from the disused lead mine at Draethen, Glamorgan (Morgan and Morris, 1982).

Given the presence in earthworm chloragogenous tissue of lysosomal residual bodies derived from the breakdown of metallothionein, it follows that metallothionein in soluble form is also involved in the detoxification (temporary or permanent) of trace metals such as cadmium, copper and zinc in earthworms. Metallothioneins binding cadmium, copper or zinc have indeed been identified in many earthworm species, including *Lumbricus rubellus*, *L. terrestris*, *Dendrodrilus rubidus*, *D. octaedra* and *Eisenia fetida* (Bengtsson et al., 1992; Stürzenbaum et al., 1998). At least two isoforms of cadmium-binding MT have been found in *L. rubellus*, with the implication of subtly different functions (Stürzenbaum et al., 1998). As an alternative to copper detoxification by binding to metallothionein, copper has also been shown to bind to the amino acid histidine in *Lumbricus rubellus*, an increase in histidine levels being induced by copper exposure (Gibb et al., 1997).

4.8.1.3 Ecotoxicology: Tolerance and Resistance

Prologue. Some earthworm populations living in mining-contaminated soils show tolerance to raised soil bioavailabilities of trace metals. In addition, different earthworm species are differentially resistant to exposure to metal contamination as a species characteristic.

Since earthworm populations are, in effect, in permanent residence in a particular habitat, it is no surprise that some populations living in mining-contaminated soils display tolerance to raised soil bioavailabilities of toxic metals in comparison to populations of the same earthworm species living elsewhere. Populations of both *Lumbricus rubellus* and *Dendrodrilus rubidus* collected from metal contaminated mine spoil heaps at an abandoned tungsten mine at Carrock Fell, Cumbria, and a disused copper and arsenic mine at Devon Great Consols, Devon, show tolerance to both copper and arsenic (Langdon et al., 1999, 2001). Another Cumbrian population of *D. rubidus*, in this case collected from mine spoil at the abandoned Coniston copper mine, also showed tolerance to copper (Arnold et al., 2008). It was not determined in any of these cases whether the tolerance was a result of physiological acclimation or had a genetic basis.

Distinct from the establishment of metal-tolerant populations of a particular earthworm species is the inherent variation between species in their resistance to toxic metal exposure. Thus different earthworm species are differentially resistant to exposure to metal contamination as a species characteristic. This variation has been clearly shown in a study on the population densities and species composition of earthworms at sites in the vicinity of the smelting works at Avonmouth, near Bristol (Spurgeon and Hopkin, 1996a). No earthworms were present at sites within 1 kilometres of the smelting works, and

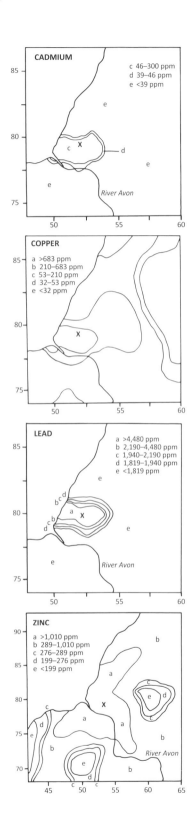

earthworm numbers were still significantly reduced at sites 2 kilometres away. Total earthworm density reduced with increased soil concentrations of cadmium, copper, lead and zinc, the concentrations of these metals covarying. Interspecific differences in metal resistance resulted in the more metal-resistant species *Lumbricus rubellus*, *L. castaneus* and *L. terrestris* being present at sites close to the smelter where the less metal-resistant species *Aporrectodea rosea*, *Aporrectodea caliginosa* and *Allolobophora chlorotica* were absent (Spurgeon and Hopkin, 1996a). Toxicity testing indicated that it was sensitivity to zinc that was the factor limiting the distribution of the different earthworms (Figure 4.5) (Spurgeon et al., 1994; Spurgeon and Hopkin, 1996a).

Thus it is apparent that the degree of trace metal contamination of a soil can have direct ecotoxicological consequences on local earthworm populations, in terms of both the abundance of local earthworms and the species composition of local earthworm assemblages. Such ecotoxicological effects will have knock-on effects on the decomposition of plant detritus and soil formation in the contaminated habitat. There may also, however, be indirect effects on the biology of other animals dependent on earthworms as a food source. Firstly, low numbers, or indeed the absence, of earthworm prey will affect the success of particular predators. Secondly, the presence of high concentrations of metals in the earthworms that are still present as potential prey is likely to lead to the increased ingestion of dietary metals by those predators. Whether more metal is subsequently assimilated from the alimentary tracts of such predators feeding on metal-rich earthworms remains to be seen. It is quite likely, for example, that the zinc and lead bound in phosphate-based granules in the

Figure 4.5 Maps showing the areas in which the estimated toxicity values for cadmium, copper, lead and zinc in the earthworm *Eisenia fetida* are exceeded in soils near the smelting works (X) at Avonmouth. (a) 14-day LC50; (b) 56-day NOEC mortality; (c) 56-day EC50 cocoon production; (d) 56-day NOEC cocoon production; (e) no toxic effect. Axes are Ordnance Survey Grid numbers expressed in kilometres. (After Spurgeon et al., 1994, with permission.)

earthworms may be more digestion-resistant than metals such as cadmium associated with sulphur in lysosomal residual bodies (Rainbow et al., 2011a). Later sections in this chapter consider the trophic transfer of toxic metals from earthworms to animals higher in any food chain, including mammals and birds.

4.8.2 Woodlice

Prologue. Woodlice are important invertebrate detritivores in British terrestrial habitats.

The insects dominate the terrestrial environment, but they are not the only arthropods found on land. They have been joined by their evolutionary relatives, the woodlice. Woodlice are isopod crustaceans, glorifying under a large number of local names, including pill bugs, sow bugs, cheesy bobs and chiggy pigs. Their closest marine relatives are the sea slaters living in the strandline of beaches, illustrating the evolutionary route taken by woodlice to invade land, probably during the Carboniferous (Hopkin, 1991). Important in this evolutionary transition has been the previous evolution by isopods of a development strategy that replaces the free-living planktonic larva of most marine crustaceans by the holding of fertilised eggs, and subsequently hatched juveniles, in a fluid-filled brood pouch. The young woodlice hatch out in a sufficiently well-developed state to cope with life in the terrestrial environment. To different degrees in different species, some woodlice have also developed infolds of the cuticle to form an internal network of branching hollow tubules (pseudotracheae) for gas exchange while restricting water loss by evaporation. Woodlice characteristically feed on decaying plant detritus, thereby exposing them to trace metals accumulated by the plants from which the detritus originated. The digested products of ingested food (including metals) are absorbed in the ventral caeca, blind-ending tubular extensions of the midgut. Woodlice typically have two pairs of ventral caeca, the cells of which are the main players in trace metal physiology, carrying out detoxification and any storage of absorbed metals.

Two particular British woodlouse species concern us here: the common shiny woodlouse *Oniscus asellus* (Plate 5a) and the common rough woodlouse *Porcellio scaber* (Plate 5b), the two most widespread and abundant woodlice in Britain (Hopkin, 1991). *O. asellus* is found in almost every terrestrial habitat where conditions are damp, particularly under rotting wood in deciduous woodland. *P. scaber* is similarly widespread, but prefers slightly drier conditions than *O. asellus*. It is pertinent that *O. asellus* lacks pseudotracheae, while *P. scaber* possesses two pairs of pseudotracheal networks (visible as white patches) on its lower surface.

4.8.2.1 Trace Metal Accumulation

Prologue. Accumulated trace metal concentrations in woodlice increase with increased metal contamination of leaf litter.

Table 4.26 presents total concentrations of cadmium, copper, lead and zinc in these two woodlouse species from metal-contaminated and uncontaminated British sites. Table 4.26 also lists concentration ranges of zinc in each species deemed to be typical of sites with different degrees of zinc contamination in the local plant detritus (Hopkin et al., 1989a). Table 4.27 compares accumulated concentrations of the four metals in the two woodlouse species collected at the same sites. Accumulated concentrations of all four metals increase in both species with proximity to a source of metal contamination and thus the degree of contamination of local leaf litter. Furthermore, the two species differ in the absolute concentrations accumulated when feeding on the same leaf litter at the same site (Table 4.27). *Porcellio scaber* accumulates higher concentrations of copper and zinc than *Oniscus asellus*, as reflected in the different ranges of woodlouse zinc concentrations considered indicative of different degrees of zinc contamination (Tables 4.26, 4.27). On the other hand, *O. asellus* accumulates higher concentrations of cadmium and lead than *P. scaber* (Table 4.27).

A study of accumulated metal concentrations in *Porcellio scaber* in the region of Avonmouth, Bristol

Table 4.26 **Total concentrations (μg/g dry weight) of selected trace metals in two species of woodlice (*Oniscus asellus* and *Porcellio scaber*) at metal-contaminated and control sites.**

Species Site	Cadmium	Copper	Lead	Zinc
Oniscus asellus				
Avonmouth smelter, 2 km[1]	202		297	704
Avonmouth smelter, 2.8 km[1]	94		107	315
Avonmouth smelter, 28 km[1]	12		55	120
Avonmouth smelter, 1 km[2]	152 ± 37	567 ± 138	413 ± 150	488 ± 103
Control: Kynance Cove, Cornwall[2]	5.9 ± 1.0	115 ± 42	4.2 ± 0.9	108 ± 25
Control: Mimram valley, Hertford[3]		143 ± 17		172 ± 28
Uncontaminated[4]				0–150
Low contamination[4]				150–250
Moderate contamination[4]				250–350
High contamination[4]				350–500
Very high contamination[4]				>500
Porcellio scaber				
Avonmouth and Shipham[5] *(Figure 4.6)*				
Severn Beach (1)				961
St Andrews Road (2)	106	1,000	328	1,500
Hallen Wood 3 km (3)	146	991	142	1,440
Westerleigh (4)	5.5			
Swing Bridge (5)			101	
Swing Bridge (6)			32	
Ashton (7)			6.9	
Shipham (8)	70	67	52	900
Charterhouse (9)			113	
Long Ashton (10)				215
Bishop Sutton (11)	1.7			

Table 4.26 **(cont.)**

Species	Cadmium	Copper	Lead	Zinc
Site				
Corston (12)			0.7	
Langridge (13)		124		
Control: Mimram valley, Hertford[3]		218 ± 18		413 ± 91
Uncontaminated[4]				0–350
Low contamination[4]				350–450
Moderate contamination[4]				450–550
High contamination[4]				550–1,000
Very high contamination[4]				>1,000

Note: Data are means (± SE, n = 6–10) or ranges.
Sources: Martin et al. (1976);[1] Hopkin (1989);[2] Rainbow (unpublished);[3] Hopkin et al. (1989a);[4] Hopkin et al. (1986).[5]

and Shipham in 1983 has provided much of the data for the different sites in Table 4.26, summarised in Figures 4.6 and 4.7 (Hopkin et al., 1986). Woodlouse concentrations of all four metals were raised near, and immediately downwind of, the Avonmouth smelter (Figures 4.6 and 4.7), emitting considerable amounts of these metals into the atmosphere at the time (Hopkin et al., 1986). There were additional hotspots for cadmium, lead and zinc at the disused zinc mine at Shipham (Figures 4.6 and 4.7), to be expected given the common co-occurrence of lead and cadmium in zinc ores. Woodlice at site 1 (Severn Beach) showed high concentrations of zinc (Table 4.26, Figure 4.6), probably in response to zinc-contaminated rain drainage from a derelict hut built from galvanised corrugated iron at the site of collection, in addition to fallout from the nearby Avonmouth smelter. Site 6 at the Swing Bridge in Bristol was by the A4 trunk road, and the high lead concentrations in the local woodlice (Table 4.26, Figure 4.6) were likely to have been caused by the presence of lead in vehicle exhaust at the time, contaminating the local plant detritus consisting of *Buddleia* leaves (Hopkin et al., 1986).

4.8.2.2 Detoxification

Prologue. Woodlice use lysosomal residual bodies in the S cells of the ventral caeca as detoxified stores of accumulated trace metals.

Table 4.27 illustrates that the ventral caeca of woodlice accumulate very high concentrations of cadmium, copper, lead and zinc, these concentrations increasing with metal exposure. The ventral caeca contain most of the accumulated trace metals in the woodlouse body and represent the site of the storage accumulation of detoxified metal in MRG. There are two types of cells in the ventral caecum epithelium of woodlice: B cells and S cells (Hopkin, 1989). The B cells contain much lipid and glycogen as stored energy reserves, and are also considered to be involved in the secretion of digestive enzymes. The main function of the S cells, on the other hand, is the detoxification and storage of trace metals in MRG, the number and size of these granules increasing with metal input (Hopkin, 1989). Perhaps surprisingly, given the very high concentrations of lead and zinc accumulated in the ventral caeca of woodlice exposed to high lead and zinc

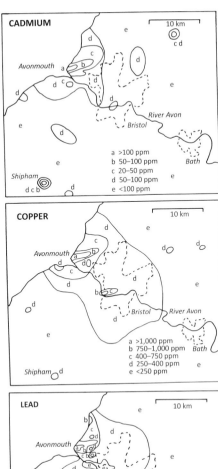

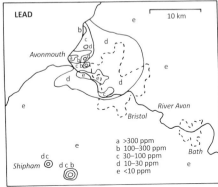

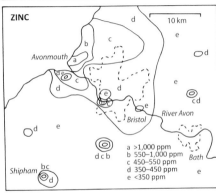

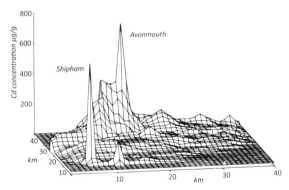

Figure 4.7 Three-dimensional plot of cadmium concentrations (μg/g dry wt) in the ventral caeca of the woodlouse *Porcellio scaber* in the vicinity of Avonmouth, Bristol and Shipham (see Figure 4.6). Highest concentrations were in woodlice close to the smelting works at Avonmouth and the disused zinc mine at Shipham. (After Hopkin et al., 1986, with permission.)

bioavailabilities, woodlice appear to lack the distinct calcium-based phosphate granules, so important in the detoxificatory binding of lead and zinc in snails and earthworms (Hopkin, 1989). Prevalent in the S cells, however, are lysosomal residual bodies, typically containing copper and sulphur (Hopkin, 1989, 1990b). Cadmium is also present in such residual lysosomal bodies in the S cells of woodlice with high accumulated cadmium concentrations. Such sulphur-rich lysosomal bodies with the likes of copper and cadmium are typically derived from the lysosomal breakdown of metallothioneins. There does appear to be detoxificatory binding of lead and zinc to phosphate in woodlice, but this is not manifested in the form of discrete granules. Rather, zinc is deposited as phosphate-rich material on the cell membranes of both B and S cells, and both zinc and lead phosphates will form around lysosomal residual bodies binding copper in S cells (Hopkin, 1989, 1990b; Schill and Köhler, 2004).

Figure 4.6 Concentrations of cadmium, copper, lead and zinc (μg/g dry wt) in the woodlouse *Porcellio scaber* in the vicinity of Avonmouth, Bristol and Shipham. Highest concentrations were typically in woodlice close to the smelting works at Avonmouth and the disused zinc mine at Shipham. (After Hopkin et al., 1986, with permission.)

Table 4.27 **Total concentrations and concentrations of trace metals in the ventral caeca in two species of woodlice (*Oniscus asellus* and *Porcellio scaber*) from three sites, and in plant litter (dead leaves of field maple *Acer campestre*) from two of these sites (μg/g dry weight, mean ± SE, *n* = 5 – 12): (a) near Portishead, Severn Estuary; (b) Wetmoor Wood (uncontaminated) and (c) Haw Wood (contaminated) near Avonmouth**

Site Woodlouse/plant litter	Cadmium	Copper	Lead	Zinc
(a) Portishead, Severn Estuary[1]				
Oniscus asellus				
Ventral caeca	508 ± 96	2,140 ± 250	370 ± 82	963 ± 174
Total	36 ± 6.0	196 ± 18	38 ± 6.2	124 ± 11
Porcellio scaber				
Ventral caeca	128 ± 27	2,570 ± 500	83 ± 16	3,870 ± 590
Total	12 ± 2.1	243 ± 32	19 ± 1.8	367 ± 43
(b) Wetmoor Wood, Avonmouth[2]				
Oniscus asellus				
Ventral caeca	329 ± 44	837 ± 93	275 ± 74	324 ± 52
Total	16 ± 2.5	92 ± 4.3	21 ± 5.6	63 ± 3.2
Porcellio scaber				
Ventral caeca	33 ± 4.2	1,780 ± 380	28 ± 6.1	2,140 ± 460
Total	2.44 ± 0.31	171 ± 13	2.57 ± 0.40	186 ± 21
Field maple dead leaves	0.72 ± 0.14	12 ± 1.3	42 ± 7.2	96 ± 7.6
(c) Haw Wood (Avonmouth)[2]				
Oniscus asellus				
Ventral caeca	2,940 ± 230	7,920 ± 700	8,570 ± 1,240	8,250 ± 1,050
Total	154 ± 10	454 ± 26	508 ± 59	524 ± 42
Porcellio scaber				
Ventral caeca	706 ± 52	9,150 ± 1,340	927 ± 199	12,000 ± 1,600
Total	49 ± 6.3	651 ± 31	116 ± 13	897 ± 111
Field maple dead leaves	26 ± 5.2	52 ± 10	908 ± 200	1,430 ± 323

Sources: Hopkin (1989);[1] Hopkin (1990a).[2]

The B cells of the epithelia of the ventral caeca turn over in a 24-hour digestive cell cycle (Hopkin, 1990a). When these cells disintegrate at the end of the cycle, the cell contents are lost into the lumen of the ventral caecum, to be reabsorbed or to pass into the remainder of the gut to be defaecated. S cells, however, have a much longer residence time. They, therefore, do not disintegrate as often as B cells, if at all, with limited, or no, voiding of their contents, which include MRG (Hopkin, 1990b; Schill and Köhler, 2004).

The fact that the detoxification of zinc by binding with phosphate can occur in both B cells and S cells in woodlice provides a clue as to why *Porcellio scaber* accumulates higher concentrations of zinc than *Oniscus asellus* when exposed to the same trophic supply of zinc (Table 4.27). *O. asellus* stores the majority of assimilated zinc in the B cells of the ventral caecum, while *P. scaber* stores zinc predominantly in the lysosomal residual bodies of the S cells (Schill and Köhler, 2004). Thus, zinc is lost at a faster rate from *O. asellus* as a consequence of the higher rate of turnover of B cells in comparison to S cells, resulting in lower concentrations of zinc in the ventral caeca and thence the whole body.

4.8.2.3 Ecotoxicology

Prologue. The high concentration of zinc in leaf litter was the toxic agent limiting the local distribution of the woodlouse *Porcellio scaber* in woodland near the Avonmouth smelter.

Figure 4.6 shows the very high accumulated concentrations of trace metals in *Porcellio scaber* in the area of the Avonmouth smelter in 1983 (Hopkin et al., 1986). The closest point to the smelter where these woodlice could be found was at the St Andrews Road site (Table 4.26, Figure 4.6). Further research has concluded that, of the four trace metals present in elevated concentrations in the local leaf litter, it was the high concentration of zinc that was the toxic agent limiting the local distribution of *P. scaber* near the smelter (Hopkin and Hames, 1994). No woodlice were found at sites where the zinc concentration in

the leaf litter exceeded 5,000 µg/g, the leaf litter concentration at St Andrews Road being 4,150 µg/g. Trace metal toxicity occurs when the rate of total uptake of the metal is greater than the combination of the rates of excretion and detoxification of the metal. There is, therefore, a rate of metal accumulation (the difference between the rates of uptake and excretion) that is too high for the maximum rate of detoxification to detoxify all accumulating metal. The excess metal accumulated remains in metabolically available form, and toxicity follows, initially sublethal toxicity but ultimately mortality. There are data available for the rate of accumulation of zinc by *P. scaber* that allow the calculation of this critical rate of zinc accumulation (Hopkin, 1990a; van Straalen et al., 2005). *P. scaber* from uncontaminated Wetmoor Wood, well downwind of the Avonmouth smelter, suffered 60% mortality when fed for 20 weeks on leaf litter from metal-contaminated Haw Wood, close to the smelter (Table 4.27). These woodlice, moribund from the toxic effects of ingested zinc, accumulated zinc in the total body at a rate of 8.0 µg/g per day, the rate of the accumulation in the ventral caeca being 105 µg/g per day (Hopkin, 1990a). In a later study of the same woodlouse species, in this case near a zinc smelter in the Netherlands, a zinc accumulation rate in the whole body of 7.5 µg/g per day caused a reduction in growth, while a rate of 8.6 µg/g per day caused 50% mortality (van Straalen et al., 2005). The agreement between the results of these two independent studies is remarkable. In the first study, specimens of the second species *Oniscus asellus* collected from Wetmoor Wood were also fed on Haw Wood leaf litter for 20 weeks (Table 4.27). This exposure resulted in 53% mortality, the equivalent rates of zinc accumulation being 5.7 µg/g per day for the whole body, and 80 µg/g per day for the ventral caeca (Hopkin, 1990a).

So, the very high bioavailabilities of zinc in leaf litter were lethal to *Porcellio scaber* very close to the Avonmouth smelter where litter zinc concentrations exceeded 5,000 µg/g (Hopkin and Hames, 1994). It might be expected, then, that sublethal ecotoxicological effects would be apparent in woodlice present at sites with leaf litter concentrations approaching, but still below, this ecotoxicological threshold. Surprisingly, energy reserves of lipid and glycogen in the ventral caeca of *P. scaber* showed no significant decrease with decreasing distance to the smelter (Schill and Köhler, 2004). It appears, therefore, that these woodlice were coping with the extra metal detoxification necessary without serious impact on their energy reserves, so long as the extra zinc input remained below the ecotoxicological threshold. On the other hand, the other woodlouse species, *Oniscus asellus*, from the same Avonmouth sites did show a significant massive reduction in energy reserves in the ventral caeca with decreasing distance to the smelting works, and with increased metal concentrations in the leaf litter (Schill and Köhler, 2004). Clearly, this species of woodlouse is expending significant energy to cope with the high metal input.

Porcellio scaber did show a significant reduction in the average size of both male and female woodlice with site proximity to the smelter, an effect put down to the combined effects of a reduction in growth rate and increased mortality eliminating the larger (older) woodlice (Jones and Hopkin, 1998). Other ecotoxicological effects of trace metals on woodlice are a reduction of feeding rate (Drobne and Hopkin, 1995), and increased locomotor behaviour (Bayley et al., 1997).

4.8.3 Collembolans

Prologue. Collembolans (springtails) are small detritivores living in moist leaf litter and soil. They accumulate trace metals from their soil and litter environment, storing accumulated metals in the midgut epithelium.

Collembolans, commonly known as springtails, are wingless arthropods, at one time considered to be primitive wingless insects. The common name of springtail results from the fact that most (but not all) collembolans have a forked springing organ folded beneath the abdomen. This organ is held under tension until its release flings the springtail into the air as a defence reaction. Collembolans are small (usually less than 5 millimetres long) and are extremely

common in moist leaf litter and soil. They are detritivores and microbivores, consuming the microorganisms (particularly fungi) breaking down plant detritus. The digestive tract of collembolans lacks caeca or other tubular extensions of the midgut. Peculiarly, collembolans renew the epithelium of the midgut at each moult, the degenerating old epithelium being expelled from the alimentary tract. Collembolans also have a unique thin-walled structure, the ventral tube, situated on the under surface of the first segment of the abdomen. The ventral tube appears to absorb moisture from the environment, helping to maintain the water balance of the collembolan, but also offering a route for the uptake of dissolved metals in the soil solution.

Collembolan species common in Britain include *Orchesella cincta* and its sister species *Orchesella villosa*, which is particularly widespread and common in southern England and Wales, but increasingly scarce moving north. Nevertheless, it was *O. villosa* that was the predominant collembolan in the pitfall traps in grassland swards at sites affected by emissions from the copper refinery complex in Merseyside (Table 4.24). *Folsomia candida* is another very common and widespread collembolan found in soils and has become a 'standard' springtail used in laboratory ecotoxicological testing. *Tomocerus minor* is also extremely common and widespread. In spite of its specific name, it can reach 4.5 millimetres long.

Collembolans accumulate trace metals from their soil and litter environment, storing accumulated metals in the midgut epithelium. The collembolan samples consisting mostly of *Orchesella villosa* at grassland sites affected by the Merseyside copper refinery showed increased accumulated concentrations of copper and cadmium with proximity to the refinery (Table 4.24). Furthermore, *Orchesella cincta* showed increased accumulation of cadmium and lead, but not zinc, from soil from the site of a zinc smelter at Budel in the Netherlands, compared to a reference site (van Straalen et al., 1987). All these three metals were present in raised concentrations in the leaf litter/humus horizon at Budel. The absence of a correspondingly increased zinc concentration in the Budel *O. cincta* suggests an ability on the part of the collembolan to regulate accumulated body zinc concentrations. *O. cincta* can also excrete accumulated cadmium and lead, particularly after removal from a metal-contaminated food source (van Straalen et al., 1987; Janssen et al., 1991). Since net accumulation of cadmium and lead by the collembolan still occurs in the presence of metal contaminated food, it would appear that the excretion rate does not match the uptake rate, leading to raised body concentrations of these two metals. As in other collembolans, excretion of accumulated trace metals by *O. cincta* takes place through exfoliation at every moult of the midgut epithelium storing accumulated trace metals in detoxified form (van Straalen et al., 1987). The midgut cells of *O. cincta* and of *Tomocerus minor* contain typical phosphate-based granules, binding and detoxifying extra lead accumulated (van Straalen et al., 1987; Köhler, 2002). Cadmium in the midgut cells of *O. cincta* is bound by metallothionein (Posthuma et al., 1993).

Collembolans provide a very good example of the evolution by a population of trace metal tolerance, for which the underlying physiological mechanism has been elucidated. The population of *Orchesella cincta* living in litter at an old zinc-lead mine at Plombières in Belgium is cadmium tolerant, as shown by its resistance to cadmium-induced growth reduction when exposed to a cadmium-enhanced diet (van Straalen *et al.*, 1987; Posthuma, 1990). Cadmium is a contaminant of zinc ores, and a high local soil bioavailability of cadmium at a zinc-lead mine site is not surprising. The cadmium tolerance persisted in first-generation laboratory-bred offspring, showing that the tolerance has a genetic basis (Posthuma, 1990). The cadmium-tolerant collembolans showed significantly higher cadmium excretion efficiencies than reference populations (van Straalen et al., 1987; Posthuma et al., 1992). The mechanism of cadmium tolerance is based on increased storage detoxification in the midgut epithelium. This temporary detoxified store of cadmium is subsequently excreted at every moult, thereby leading to enhanced cadmium excretion (Posthuma et al., 1992; Sterenborg and Roelofs, 2003). In the Plombières *O. cincta* population, there has been selection at the gene level for increased

production of metallothionein that binds cadmium (Cd-MT) on exposure to cadmium (Sterenborg and Roelofs, 2003). Cadmium assimilated into the midgut epithelium of the collembolan induces Cd-MT, this induction occurring at an enhanced rate in the Cd-tolerant population at Plombières. This MT and bound cadmium is then broken down in lysosomes in the midgut cells to form Cd-rich lysosomal residual bodies, which are released into the gut lumen at each moult, excreting the cadmium (van Stralen and Roelofs, 2005).

4.8.4 Millipedes

Prologue. Millipedes are detritivorous myriapod arthropods common in soil and leaf litter. They accumulate higher concentrations of trace metals when exposed to contaminated soils and leaf litter.

Millipedes are characterised by the appearance of two pairs of jointed legs per apparent segment. Each apparent segment, however, has been formed by the fusion of two single original segments, each with one pair of legs. No known millipede species actually has a thousand legs, as implied in the name. Typically, millipedes are elongated, slow moving and detritivorous, feeding on decaying leaves, digesting especially the bacterial and fungal component of plant detritus. The gut of millipedes lacks any midgut caeca, but there is a layer of 'hepatic tissue' around the midgut. Millipedes also have a pair of Malpighian tubules leading off the junction of midgut and hindgut.

A mixed selection of millipedes was collected in pitfall traps in the grassland swards affected to varying degrees by emissions of the Merseyside copper refinery complex in the 1980s (Hunter et al., 1987b). The cadmium and copper concentrations accumulated by these millipedes increased with proximity to the refinery (Table 4.24). The typical background concentration of copper in the millipedes from the control site was higher than that of any other invertebrates in the study (Tables 4.22, 4.24, 4.28), and millipedes do contain the copper-bearing respiratory protein haemocyanin. Yet, the extra accumulation of copper

Table 4.28 **Copper and cadmium concentrations (mean $\pm$ SE, µg/g dry weight, $n = 80 - 400$) in predatory beetles, spiders and harvestmen collected in pitfall traps in established grassland swards dominated by creeping bentgrass** *Agrostis stolonifera* **and red fescue** *Festuca rubra* **at a control site and at sites affected by emissions from a copper refinery complex in Merseyside (see Tables 4.17 and 4.24). All samples contained mixed species.**

		Copper			Cadmium		
Higher taxon							
Order							
Family							
Species	Common name	Control site	1 km site	Refinery	Control site	1 km site	Refinery
Arachnida							
Araneida	Spiders						
Lycosidae	Wolf spiders	58 ± 4.2	160 ± 15	887 ± 171	2.6 ± 0.3	35 ± 5.0	102 ± 7.5
Linyphiidae	Money spiders	89 ± 11	200 ± 13	1020 ± 138	2.4 ± 0.3	19 ± 1.3	89 ± 9.8
Opiliones	Harvestmen	42 ± 3.6	100 ± 8.8	1010 ± 108	2.8 ± 0.3	25 ± 1.8	93 ± 10
Insecta							
Coleoptera	Beetles						
Staphylinidae	Rove beetles						
Aleochara sp.							
Philonthus sp.							
Quedius sp.							
Ocypus olens		29 ± 1.8	58 ± 4.1	522 ± 100	0.6 ± 0.1	4.9 ± 0.5	14 ± 1.6
Tachinus sp.							
Xantholinus sp.							
Cholerva sp.							
Carabidae	Ground beetles						
Carabus monilis							
Loricera pilicornis							
Nebria brevicollis		21 ± 0.9	46 ± 2.9	460 ± 96	0.7 ± 0.1	5.6 ± 0.9	15 ± 1.1
Pterostichus diligens							
Pterostichus madidus							
Pterostichus strenuus							

Table 4.28 **(cont.)**

Higher taxon		Copper			Cadmium		
Order							
Family							
Species	Common name	Control site	1 km site	Refinery	Control site	1 km site	Refinery
Predatory beetle larvae		33 ± 3.8	53 ± 14.1	298 ± 34	2.2 ± 0.3	8.1 ± 0.6	21 ± 1.2

Source: Hunter et al. (1987b).

at the refinery site was lower than that of earthworms, woodlice, collembolans, spiders and harvestmen (Tables 4.24, 4.28). Similarly, while the cadmium concentrations of millipedes were raised at the refinery, these accumulated concentrations were much lower than those of earthworms, woodlice and collembolans at the same site (Table 4.24). It would appear, then, that millipedes are not as strong accumulators of copper and cadmium as their detritivorous counterparts. Like collembolans, millipedes replace the midgut epithelium during moulting (Hopkin, 1989), promoting the excretion of any metals accumulated therein. And cadmium and lead, at least, are predominantly stored in the midgut epithelium of millipedes (Köhler et al., 1995).

Zinc in millipedes is mostly stored elsewhere in the body, in this case in a layer of cells beneath the exoskeleton (Hopkin, 1989; Köhler et al., 1995). This stored zinc has no route for excretion. Furthermore, millipedes assimilate more ingested zinc than they do cadmium or lead (Hopkin, 1989; Köhler et al., 1995). Correspondingly, millipedes contain high body concentrations of zinc, of the order of 300 to 500 µg/g, even in the absence of zinc contamination (Hopkin, 1989).

The detoxified storage of absorbed lead, and of the minor component of total body zinc (3 to 7 %), in the midgut cells is in the form of phosphate-based granules (Köhler et al., 1995; Köhler, 2002). These granules are lost when the midgut cells are lost at each moult.

4.9 Invertebrate Predators

Prologue. Invertebrate herbivores and detritivores are prey to predators, both invertebrate and vertebrate. The metals accumulated in high concentrations by the herbivores and detritivores present a potentially ecotoxicologically significant trophic challenge to those predators. The invertebrate predators at greatest risk are those preying on invertebrates living in habitats contaminated by mining or atmospheric emission from refineries and smelters. The invertebrate predators of greatest ecological significance in this context are spiders, centipedes and beetles.

Spiders are arachnid arthropods, taxonomically distinct from the insects, crustaceans and myriapods met earlier in this chapter. Predatory spiders typical of British habitats include the wolf spiders and the money spiders. Another spider, the woodlouse spider *Dysdera crocata*, feeds exclusively on woodlice. Members of two other arachnid groups, harvestmen and pseudoscorpions, are also invertebrate predators in British terrestrial ecosystems. The harvestmen *Phalangium opilio* and *Mitopus morio* are both widespread, common species. The pseudoscorpion *Neobisium muscorum* is widely distributed in litter and under stones, and will feed on collembolans and mites. Centipedes are relatively large, fast moving, terrestrial predators and are found in soil and leaf litter, under stones and in dead wood. Of the insects, it is the beetles that are common predators, particularly carabid ground beetles, as well as staphylinid rove beetles.

A heterogeneous collection of invertebrate predators was collected in the pitfall traps in the grassland swards affected to different extents by the atmospheric emissions of the Merseyside copper refinery complex in the 1980s (Table 4.28). While all these samples suffered from the problem of the mixing of several different species of each group of invertebrate predators, it is probably still safe to conclude that both the copper and cadmium concentrations of wolf spiders, money spiders, harvestmen, rove beetles and ground beetles increased with proximity to the refinery (Table 4.28).

4.9.1 Spiders

Prologue. Spiders are essentially liquid feeders, sucking out digested soft tissues from prey. They are, therefore, exposed to any high concentrations of trace metals accumulated by their invertebrate prey. Spiders accumulate trace metals, particularly cadmium. Some spiders use zinc and/or manganese to harden the cutting edges of their chelicerae.

Spiders are extremely important predators in British terrestrial habitats, catching prey by means of a web or by active hunting. They prey almost exclusively on other arthropods. Spiders are essentially liquid feeders, injecting venom and then digestive enzymes into captured prey. Digested prey tissues are sucked into the stomach and then the midgut of the spider. The midgut gives rise to a large number of branching midgut diverticula, which ramify through the body of the spider, even into the legs. The midgut diverticula are the physiological equivalent of the digestive glands of molluscs or the hepatopancreas of decapod crustaceans. The diverticula produce digestive enzymes, carry out absorption of digested food and are involved in the physiological detoxification of accumulated trace metals. Since spiders remove only soft tissues, in digested form, from the body of the prey, they are exposed to much higher concentrations

of trace metals from the prey than might be implied from the total body metal concentrations of the prey. Spiders, like other arachnids, also have Malpighian tubules, carrying out excretion.

Spiders have the copper-based respiratory protein haemocyanin, contributing no doubt to the higher background concentrations of copper in the spiders (55 to 90 µg/g Cu), than in the beetles (20 to 35 µg/g Cu) shown in Table 4.28. It is often claimed that spiders have relatively high accumulated concentrations of cadmium compared to other terrestrial invertebrates at the same site, including uncontaminated sites (Hopkin, 1989; Hendrickx et al., 2003). This may well be so when some spiders (and indeed harvestmen) are compared against insects (Tables 4.22, 4.28) but not against woodlice or millipedes (Table 4.24).

Wolf spiders are robust, agile, active hunters with good eyesight. They show strong accumulation of cadmium from their relatively high baseline concentration when feeding on cadmium-contaminated prey (Table 4.28) (Hendrickx et al., 2003; Eraly et al., 2010). For example, the pirate wolf spider *Pirata piraticus*, common particularly in marshes hunting the likes of rove beetles, is a strong cadmium accumulator from such prey. This strong accumulation can be put down to a high efficiency (70%) of cadmium assimilation from ingested prey and almost no excretion of the cadmium assimilated ((Hendrickx et al., 2003). Another wolf spider with strong powers of cadmium accumulation from its prey is *Pardosa saltans*, often previously referred to in Britain as *P. lugubris*, the common lugubrious wolf spider (Eraly et al., 2010). This wolf spider is common in litter in woodland, particularly in open spaces therein, or at the edge of a wood or hedge, hunting ground-dwelling prey. The cadmium accumulated in spiders is detoxified, as to be expected, by binding to metallothioneins (Eraly et al., 2010, 2011).

Pardosa saltans have been collected from cadmium and zinc-contaminated litter from old mining sites near Liège in Belgium, including La Rochette à Prayon Pond in Chaudfontaine (Eraly et al., 2011). Average cadmium and zinc concentrations in the wolf spiders from this site were about 18 µg/g Cd and 150 µg/g Zn, in comparison with average concentrations from a control site of under 2 µg/g Cd and just under 100 µg/g Zn. A comparison of the life histories of the wolf spiders across metal-contaminated and control sites showed that adult size and condition decreased with the accumulated cadmium concentrations, indicative of exposure to higher cadmium bioavailabilities. Egg mass, on the other hand, correlated positively with accumulated cadmium concentration. Thus, under the conditions of cadmium contamination, the adult female wolf spiders were smaller and in poorer condition, delayed reproduction and produced heavier eggs. While individual eggs were heavier, these smaller females produced cocoons of lower mass with fewer egg numbers, resulting in reduced fecundity. Concentrations of metallothioneins binding cadmium and zinc were not raised in the wolf spiders from the metal-contaminated sites (Eraly et al., 2011). This does not, however, rule out increased metallothionein induction, for MT synthesis and turnover are often independent of the standing stock MT concentrations that can be measured.

Box 4.3 The Woodlouse Spider

The woodlouse spider *Dysdera crocata* is a specialist predator of woodlice, with the associated risk of the ingestion of the very high contents of trace metals accumulated by these detritivores (Tables 4.24, 4.26, 4.27). Table 4.29 presents the results of an experiment in which *D. crocata* were either starved or fed for 36 days on woodlice *Porcellio scaber* from their collection site in Bristol or from Hallen Wood, 3 kilometres downwind of the Avonmouth smelter (Hopkin and Martin, 1985; Hopkin, 1989). In spite of the different concentrations of cadmium, lead and zinc in the woodlouse prey, there were no significant differences between the concentrations of these metals, nor of copper, between the three groups of spiders after 36 days (Table 4.29). Comparisons of the metal contents of diet and faeces produced by the spider showed that the spiders were assimilating the metals from the food, with net assimilation efficiencies ranging from 35% for copper and zinc to 71% for cadmium (Hopkin and Martin, 1985; Hopkin, 1989). The spiders are, therefore, excreting this assimilated metal very rapidly, and their accumulated concentrations do not increase by the amounts of metals assimilated. *D. crocata* is accordingly well able to cope with the ingestion of woodlice with very high accumulated concentrations of metals.

In fact, the woodlouse spider is detoxifying assimilated metal in cells of the midgut diverticula but is very rapidly excreting the detoxified product as these cells turn over rapidly (Hopkin, 1989). These cells of the midgut diverticula of *Dysdera crocata* correspondingly contain numerous metal rich granules (MRG) – both phosphate-based granules able to bind lead and zinc, and lysosomal residual bodies with copper, cadmium and sulphur derived from metallothionein (Hopkin, 1989, 1990b). The same MRG can be seen expelled in the faeces after cell turnover. The cadmium accumulation pattern of the woodlouse spider, belonging to the family *Dysderidae*, is thus clearly distinct from the strong cadmium accumulation of the wolf spiders of the family *Lycosidae*.

Table 4.29 Accumulation of trace metals by the woodlouse spider *Dysdera crocata* when starved for 36 days or fed on woodlice *Porcellio scaber* (1 woodlouse every 3 days for 36 days) from their collection site in Bristol or from Hallen Wood, 3 kilometres downwind of the Avonmouth smelter, in autumn 1983. Metal concentrations (µg/g dry weight) and dry weights (mg) are means ± SE (n = 4).

	Cadmium	Copper	Lead	Zinc
Woodlice prey				
Bristol				
Concentration	10 ± 1.3	1,220 ± 160	41 ± 5.0	867 ± 108
net assimilated by spider	61	47	60	35
Hallen Wood				
Concentration	181 ± 45	1,420 ± 350	360 ± 90	1,510 ± 380
net assimilated by spider	71	35	57	42
Woodlouse spider				
Concentrations				
Starved	4.49 ± 1.16	100 ± 37	4.90 ± 3.49	115 ± 38
(dw 12.04 ± 4.66 mg)				
Fed *Bristol* woodlice	4.53 ± 1.20	178 ± 63	2.98 ± 1.05	184 ± 51
(dw 10.02 ± 0.68 mg)				
Fed *Hallen Wood* woodlice	4.41 ± 2.37	74 ± 13	2.29 ± 0.68	150 ± 58
(dw 11.80 ± 4.27 mg)				

Sources: After Hopkin & Martin (1985), Hopkin (1989).

While the jaws of their arthropod relatives, the crustaceans, insects and millipedes, consist of paired mandibles on the fourth segment of the head, the equivalent appendages of arachnids are paired chelicerae on the first segment of the prosoma, the first of two body parts of an arachnid. The chelicerae of spiders are typically modified as grasping or piercing jaws. As in the case of the mandibles of some insects (Table 4.23), some spiders have made use of zinc and/or manganese to harden the cutting edges of their chelicerae. For example, the garden spider, *Araneus diadematus*, has high concentrations of zinc (up to

23% dry weight) in the large fang part of each chelicera, with manganese (up to 5% dry weight) present in the marginal teeth adjacent to the fang (Schofield and Lefevre, 1989; Schofield et al., 1989). Manganese, but not zinc, is also present in the claws on other legs of the garden spider (Schofield and Lefevre, 1989). Similar distribution patterns of zinc and manganese also occur in other orbweaving spiders, the ornamental orbweaver *Larinioides patagiatus* and the six-spotted orbweaver *Araniella displicata* (Schofield and Lefevre, 1989).

4.9.2 Pseudoscorpions

Prologue. *Neobisium muscorum* feeds on collembolans and mites and accumulates cadmium.

Like wolf spiders, the pseudoscorpion *Neobisium muscorum* is a strong accumulator of cadmium from ingested prey. The body concentration of cadmium in *N. muscorum* fed on a diet of cadmium-enriched collembolans (22 µg/g Cd) for 30 days, rose from about 2 µg/g Cd to about 20 µg/g Cd (Janssen et al., 1991). The assimilation efficiency of cadmium from the ingested prey was high (nearly 60%), perhaps expectedly because pseudoscorpions also suck out the body contents of their prey. Furthermore, no excretion of the extra accumulated cadmium was observed over a 100-day elimination period (Janssen et al., 1991).

4.9.3 Centipedes

Prologue. Centipedes are predatory myriapods, accumulating higher concentrations of trace metals from their prey in metal-contaminated terrestrial habitats.

Despite their name, centipedes have a varying number of paired legs, from fewer than 20 to more than 300. As in the case of the millipedes, the midgut of centipedes lacks any caeca or other expansions. Haemocyanin, the respiratory protein containing copper, is also present in centipedes (Hopkin, 1989).

Common in Britain is the banded centipede *Lithobius variegatus*, present in woodland, particularly in western Britain, including woods in the Avonmouth

area. It feeds on woodlice, collembolans and millipedes. Table 4.30 presents field concentrations of this centipede from four woods in the region of the Avonmouth smelter in 1982. Two woods, Midger Wood and Wetmoor Wood, are considered uncontaminated, while the other two, Haw Wood and Hallen Wood, each about 3 kilometres from the smelter, are metal-contaminated (Hopkin and Martin, 1983; Hopkin, 1989).

As to be expected, accumulated concentrations of all four metals were higher in the centipedes from the metal-contaminated woods than from the other two uncontaminated woods (Tale 4.30). Background zinc concentrations in the centipedes are high compared to most terrestrial arthropods, substantial amounts of zinc being stored in subcuticular tissue below the exoskeleton (Table 4.30) (Hopkin and Martin, 1983; Hopkin, 1989). Cells of the midgut store cadmium and copper in *L. variegatus*, both in uncontaminated and contaminated habitats (Table 4.30) (Hopkin and Martin, 1983; Hopkin, 1989). In the presence of raised lead bioavailabilities, extra accumulated lead is also stored in the midgut cells (Table 4.30). Accumulated zinc and lead in the midgut cells can be found in calcium phosphate-based detoxification granules, to be later voided into the gut lumen and lost with the faeces (Hopkin, 1989). In the Hallen Wood centipedes, the subcuticular tissue contained 28% of the body zinc, while the midgut contained 16% of body zinc, 70% of body copper, 86 to 95% of body cadmium and 63 to 100% of body lead (Hopkin and Martin, 1983).

4.9.4 Beetles

Prologue. Beetles, particularly carabid ground beetles, are important invertebrate predators in leaf litter habitats in the British Isles. Ground beetles are net accumulators of cadmium and copper, but appear to regulate body concentrations of zinc.

Table 4.28 has already confirmed that the accumulated cadmium and copper concentrations of ground beetles (and also of rove beetles) increased with proximity to a Merseyside copper refinery complex in the 1980s.

Table 4.30 **Accumulation of trace metals by the centipede *Lithobius variegatus* from four sites in the region of the Avonmouth smelter in 1982 – two uncontaminated sites (Midger Wood and Wetmoor Wood) and two contaminated sites (Haw Wood and Hallen Wood), each 3 kilometres downwind of the smelter. Centipedes were depurated for four days. Metal concentrations (µg/g dry weight) of centipede tissues are means (*n* = 7).**

	Cadmium	Copper	Lead	Zinc
Midger Wood				
Remaining tissues	<0.13	15	<0.64	118
Midgut	3.23	61	<3.00	658
Subcuticular tissue	<1.50	61	<11.0	1,840
Exoskeleton	<0.05	8.3	<0.50	439
Total	*0.27–0.38*	*16*	*<1.00*	*367*
Wetmoor Wood				
Remaining tissues	<0.14	17	<0.71	102
Midgut	5.29	45	<3.5	431
Subcuticular tissue	<3.5	49	<16.0	932
Exoskeleton	<0.14	10	<0.7	292
Total	*0.46–0.66*	*17*	*<1.25*	*232*
Haw Wood				
Remaining tissues	<1.82	22	<0.79	137
Midgut	26.8	206	28.1	764
Subcuticular tissue	<0.9	58	<16.0	3,070
Exoskeleton	0.85	11	<0.7	454
Total	*2.6–3.4*	*50*	*2.3–3.7*	*470*
Hallen Wood				
Remaining tissues	<2.26	22	<0.70	147
Midgut	85	398	12	857

Table 4.30 **(cont.)**

	Cadmium	Copper	Lead	Zinc
Subcuticular tissue	<0.8	83	<4.2	2,450
Exoskeleton	1.14	9.4	<0.5	612
Total	*9.6–11*	*61*	*1.3–2.1*	*581*

Source: Recalculated from Hopkin and Martin (1983).

One carabid, the big-eyed bronze beetle *Notiophilus biguttatus*, is widespread and common in open woodland throughout the British Isles. It preys on collembolans and mites, consuming its prey whole. This ground beetle showed high cadmium assimilation efficiency (35%) when fed on cadmium-enriched collembolans over 35 days (Janssen et al., 1991). Their accumulated cadmium concentration climbed rapidly from about 0.75 µg/g Cd to about 3.7 µg/g Cd in five days, then plateaued for the remaining 30 days of the feeding period. When the ground beetles were removed from the cadmium-enriched diet, the accumulated cadmium concentration dropped back to the starting concentration in fewer than ten days, the beetles excreting 38% of accumulated cadmium per day (Janssen et al., 1991).

Cadmium and zinc accumulation have been studied in another British ground beetle, *Poecilus cupreus*, a general predator locally common in the south in open, dry habitats with short grass (Kramarz, 1999). When fed on cadmium-rich prey, *P. cupreus* showed the same cadmium accumulation pattern as *N. biguttatus*. This ground beetle rapidly accumulated cadmium to a plateau concentration (in this case, about 7 µg/g Cd) before a fall back to the starting concentration (0.15 µg/g Cd) on change back to an uncontaminated diet (Kramarz, 1999). In contrast, the body zinc concentration of *P. cupreus* stayed constant (about 58 to 69 µg/g Zn), both during a period of feeding on zinc-enriched prey and after transition back to an uncontaminated diet. *P. cupreus* also showed almost no differences in body zinc concentrations between individuals collected from control sites and sites near a smelter (Kramarz, 1999).

Therefore, in contrast to the net accumulation of cadmium and copper, ground beetles appear to be able to regulate the body concentration of zinc, presumably by matching the rate of excretion of zinc to the rate of zinc uptake, even when this is increased under conditions of high trophic bioavailability of zinc.

4.9.5 Parasitoid Wasps

Prologue. Parasitoid wasps use long ovipositors with tips hardened by zinc and/or manganese to reach insect larvae located deep in wood substrates.

Parasitoid wasps, such as chalcid or ichneumonid wasps, lay their eggs on or in the larvae of other insects. The developing wasp larva then feeds on the body of the host larva, eventually killing it. Host larvae are often to be found living deep in a hard substrate such as wood, and many parasitoid wasps are equipped with a long ovipositor that can drill through the wood to reach its goal. Of significant advantage to the parasitoid wasp is the possession of an ovipositor with a hardened tip, able to withstand wear and tear during drilling. As in the case of certain insect mandibles, zinc and/or manganese are again used by some parasitoid wasps for this purpose, hardening the tip of the ovipositor to increase the effectiveness of drilling (Quicke et al., 1998). As in the mandibles, concentrations of zinc or manganese can again reach 10% dry weight in the cuticle of the ovipositor tips. British examples include the chalcid wasp *Torymus auratus*, which has an ovipositor hardened by zinc (Quicke et al., 1998), used to lay its eggs in the larvae of gall wasps causing plant galls.

On the other hand, the ichneumonid wasp, *Perithous* sp., uses manganese for the mineralisation of the tip of its ovipositor, laying its eggs on the larvae of certain solitary wasps (Quicke et al., 1998).

4.10 Mammals

Prologue. We turn now to the vertebrates in British terrestrial habitats that might be affected by high bioavailabilities of trace metals. This section concerns mammals, particularly small mammals.

When it comes to mammals and life in metal-contaminated terrestrial habitats, most of what is known is about small mammals. Small mammals can be herbivorous, carnivorous or indeed omnivorous, with associated variation in the items in their diet that potentially provide them with a trophic source of trace metals. Examples include rodents such as the wood mouse *Apodemus sylvaticus* (omnivorous), the field vole *Microtus agrestis* (herbivorous) and the bank vole *Myodes glareolus* (herbivorous). Insectivores are represented by the common shrew *Sorex araneus* (carnivorous) and the mole *Talpa europea* (carnivorous). Some information is also available for bats.

4.10.1 Small mammals

Prologue. Different small mammals have different diets and different accumulation patterns for different trace metals, resulting in different accumulated concentrations of trace metals in their different organs under different conditions of trace metal contamination of their habitats.

The wood mouse *Apodemus sylvaticus* is common and is often encountered living close to humans. More widely, wood mice live in woodland, hedgerows, grassland and cultivated fields, giving rise to another common name, the field mouse. Wood mice are omnivorous, although they are primarily seed eaters, feeding particularly on the seeds of trees. They will also eat berries, fruits and roots, as well as small invertebrates such as insects and snails. Wood mice are preyed upon by tawny owls, foxes, weasels and domestic cats.

The field vole *Microtus agrestis* (with other common names, including the short-tailed vole) is one of the most common mammals in Britain and is found in most grassy habitats. It is an herbivore, preferentially feeding on grasses and herbs, including species of bentgrass *Agrostis* and red fescue *Festuca rubra*, met earlier as metallophytes able to grow on soils with atypically high bioavailabilities of trace metals. The field vole will eat other vegetation such as root tubers and even occasionally insect larvae. Nevertheless, the trophic input of metals is essentially from plants. In turn, the field vole is an important dietary item for owls, especially barn owls, and it is also preyed on by kestrels, weasels, stoats and foxes.

The bank vole *Myodes glareolus* (formerly *Clethrionomys glareolus*) is found in woodland, hedgerows, parks and gardens with dense vegetation such as bramble and bracken. It is predominantly herbivorous, eating fruit and nuts, but occasionally it will take insects and other small invertebrates. Unlike the field vole, bank voles are rarely taken by barn owls, but they are a source of food to tawny owls, weasels, stoats and foxes.

A truly carnivorous small mammal is the common shrew *Sorex araneus*, needing to consume two to three times its body weight daily in order to survive. Its diet consists of insects, spiders, worms, slugs and even other vertebrates such as amphibians and small rodents. The common shrew is to be found in woodland, grassland and hedgerows. Its predators include owls, weasels, stoats and foxes, although its taste is unpleasant to domestic cats.

Another carnivore is the mole *Talpa europea*, famous for feeding on earthworms. While moles do prefer earthworms, they will also prey on insects and centipedes, and even mice and shrews. Moles need sufficiently deep soil to construct their burrows, and they inhabit pastures, meadows, arable lands, parks and gardens. Moles become susceptible to predation when they come above ground, as in the case of young dispersing from the mother's nest. The main predators then are birds, including owls, buzzards, ravens and herons.

These small mammals act as important links in the transfer of trace metals along British terrestrial food chains. Their diets vary in the relative contributions of plant and animal material and in the specific plants and invertebrates eaten. Thus, the trophic inputs of metals into these mammals differ interspecifically, with seasonal variation added on top. Furthermore, any trace metals accumulated in the whole bodies, or in particular organs such as the kidney or liver, of these small mammals have the potential to be passed up different food chains to predatory birds and mammals.

Table 4.31 lists accumulated concentrations of arsenic, cadmium, copper, lead and zinc in small mammals collected from sites contaminated by mining and smelter activities and from control sites. Not surprisingly, we have already met many of these sites earlier in the book. While Table 4.31 includes whole body concentrations, kidneys and livers feature strongly, for these are often organs of trace metal accumulation (and storage detoxification) in vertebrates and are easier to analyse. Nowadays, we seek nondestructive methods of trace metal analysis for vertebrates, and a few data for hair samples are also included in Table 4.31.

It is clear from Table 4.31 that accumulated metal concentrations vary between kidneys and livers, and between these organs and whole bodies. Whole body concentrations are usually lower than kidney and liver concentrations, for whole body concentrations are diluted by the inclusion of tissues that lack the metal accumulating properties of kidneys and livers. There are also differences between species in the organ and body metal concentrations accumulated at the same site, and thus what accumulated metal concentrations should be considered as high or low for the different species (Table 4.31). Such differences are not only driven by the differences in diets between species (with correspondingly different trophic inputs of metals) but also by different accumulation patterns shown by the mammals after metal uptake, for example the relative contributions of excretion or storage detoxification of accumulated metal.

Accumulated concentrations of arsenic, cadmium and lead in kidneys, livers and whole bodies typically increase in metal-contaminated habitats as trophic inputs of the metals increase accordingly (Table 4.31). This conclusion, however, does not appear quite as clear-cut for copper and zinc (Table 4.31). Copper concentrations did not increase with copper contamination in kidneys, livers and whole bodies of the wood mouse *Apodemus sylvaticus*, but did in the hair (Table 4.31) (Hunter et al., 1987c, 1989). In the field vole *Microtus agrestis,* copper concentrations increased with copper contamination in kidneys and hair, but not in livers nor whole bodies (Table 4.31) (Hunter et al., 1987c, 1989). Most accumulated zinc concentrations in kidneys, livers and whole bodies do not change in most small mammal species between uncontaminated and zinc-contaminated sites (Table 4.31), implying efficient physiological regulation of accumulated zinc concentrations, probably employing zinc excretion. On the other hand, whole body zinc concentrations do appear raised at contaminated sites in comparison to controls in the specific case of the field vole *Microtus agrestis*, as do the zinc concentrations of both kidney and liver in moles *Talpa europea* (Table 4.31).

The data in Table 4.31 allow the recognition of what are atypically high accumulated concentrations of cadmium, lead and arsenic in kidneys, livers and whole bodies of the different species.

4.10.1.1 Kidneys

Prologue. The kidney is an important site for the detoxified storage of accumulated trace metals in small mammals.

In the case of kidneys, average cadmium concentrations above 8 µg/g Cd are high and indicative of raised cadmium bioavailability in the wood mouse, *Apodemus sylvaticus* (Table 4.31). The average kidney concentrations of wood mice from the Minera mine, near Halkyn Mountain in northeast Wales, and from the Merseyside copper/cadmium refinery site were each raised to about 40 µg/g Cd (Table 4.31). Average kidney cadmium concentrations greater than

Table 4.31 Concentrations (means ± SE (variable n) or ranges, µg/g dry weight) of arsenic, cadmium, copper, lead and zinc in kidneys, livers, hair and whole bodies of small mammals collected from control sites considered uncontaminated (U), or from sites contaminated with metals (As, Cd, Cu, Pb and/or Zn) from mining or smelter emission.

Species / Organ/whole body / Site	As	Cd	Cu	Pb	Zn
Apodemus sylvaticus (wood mouse)					
Kidney					
Minera control, Halkyn, NE Wales (U)[1,2]		1.7 ± 0.4		13 ± 2.2	158 ± 46
Minera mine, Halkyn, NE Wales (Cd, Pb, Zn)[1,2]		40 ± 10		47 ± 11	101 ± 4
Y Fan control, Llanidloes, Wales (U)[1,2]		1.7 ± 0.8		9.4 ± 3.1	
Y Fan mine, Llanidloes, Wales (Cd, Pb, Zn)[1,2]		10 ± 3.0		39 ± 13	
Swansea valley control (U)[1]		2.2 ± 0.4		14 ± 2.0	
Swansea valley smelter (Cd, Pb, Zn)[1]		18 ± 5.5		65 ± 8.0	
Exeter University control (U)[3]	0.64 (0.30–4.2)				
Kolora Park, Exeter, Devon (U)[3]	0.85 (0.30–54)				
Gawton United mine, Devon (As, Cu)[3]	2.9 (0.30–22)				
Devon Great Consols, Devon (As, Cu)[3]	5.1 (1.0–35)				
Morwellham Woods, Devon (As, Cu)[3]	1.5 (0.30–8.2)				
Wheal Exmouth, Devon (Pb, Zn)[3]	1.0 (0.30–3.7)				
Metaleurop Nord smelter control, France (U)[4]		3.3 (0.3–40)		0.6 (nm–2.6)	
Metaleurop Nord smelter, site 097, France (Cd, Pb, Zn)[4]		11 (1.2–76)		18 (3.4–268)	
Metaleurop Nord smelter, site 117, France (Cd, Pb, Zn)[4]		22 (1.3–120)		5.3 (0.6–282)	
Merseyside smelter control (U)[5]		2.0 ± 0.3	22 ± 1.1		
Merseyside smelter 1 km site (Cd, Cu)[5]		8.5 ± 1.6	19 ± 0.7		
Merseyside smelter site (Cd, Cu)[5]		42 ± 17	17 ± 0.3		

Table 4.31 (cont.)

Species			As	Cd	Cu	Pb	Zn
Organ/whole body							
Site							
Liver							
Minera control, Halkyn, NE Wales (U)[1,2]				0.5 ± 0.1		7.9 ± 0.6	133 ± 14
Minera mine, Halkyn, NE Wales (Cd, Pb, Zn)[1,2]				9.8 ± 3.8		12 ± 0.9	85 ± 11
Y Fan control, Llanidloes, Wales (U)[1,2]				0.9 ± 0.4		5.4 ± 0.7	
Y Fan mine, Llanidloes, Wales (Cd, Pb, Zn)[1,2]				2.5 ± 0.4		13 ± 4.1	
Swansea valley control (U)[1]				0.7 ± 0.2		6.6 ± 1.1	
Swansea valley smelter (Cd, Pb, Zn)[1]				4.4 ± 1.6		12 ± 1.5	
Exeter University control (U)[3]			0.77 (0.30–2.0)				
Kolora Park, Exeter, Devon (U)[3]			0.68 (0.30–5.6)				
Gawton United mine, Devon (As, Cu)[3]			3.5 (0.30–30)				
Devon Great Consols, Devon (As, Cu)[3]			5.8 (1.1–33)				
Morwellham Woods, Devon (As, Cu)[3]			1.3 (0.30–18)				
Wheal Exmouth, Devon (Pb, Zn)[3]			0.89 ((0.30–2.8)				
Metaleurop Nord smelter control, France (U)[4]				1.2 (0.21–16)		(nm–5.4)	
Metaleurop Nord smelter site 097, France (Cd, Pb, Zn)[4]				2.8 (0.26–23)		18 (3.4–68)	
Metaleurop Nord smelter site 117, France (Cd, Pb, Zn)[4]				7.4 (0.29–196)		5.3 (0.6–282)	
Merseyside smelter control (U)[5]				0.4 ± 0.1	16 ± 0.6		
Merseyside smelter 1 km site (Cd, Cu)[5]				1.8 ± 0.5	16 ± 0.5		
Merseyside smelter site (Cd, Cu)[5]				18 ± 4.8	16 ± 0.4		
Hair							
Metaleurop Nord smelter control, France (U)[4]				0.05 (nm–10)		0.6 (nm–4.4)	
Metaleurop Nord smelter site 097, France (Cd, Pb, Zn)[4]				0.4 (nm–4.5)		15 (1.9–87)	

Site					
Metaleurop Nord smelter site 117, France (Cd, Pb, Zn)[4]		0.1 (nm–3.6)		2.4 (0.9–7.1)	
Antwerp control 1, Belgium (U)[6]		0.01 ± 0.01	7.3 ± 0.2	0.26 ± 0.08	204 ± 5
Antwerp control 2, Belgium (U)[6]		0.01 ± 0.00	7.8 ± 0.3	0.18 ± 0.02	194 ± 6
Hoboken smelter, Antwerp, Belgium (Cd, Cu, Pb, Zn)[6]		0.12 ± 0.01	8.5 ± 0.4	2.7 ± 0.2	217 ± 7
Hageven smelter, Limburg, Belgium (Cd, Cu, Pb, Zn)[6]		0.07 ± 0.02	7.5 ± 0.4	0.78 ± 0.17	215 ± 6
Merseyside smelter control (U)[5]		0.42 ± 0.06	11 ± 0.7		
Merseyside smelter 1 km site (Cd, Cu)[5]		0.60 ± 0.18	15 ± 1.1		
Merseyside smelter site (Cd, Cu)[5]		1.08 ± 0.12	19 ± 1.6		
Whole body					
Minera control, Halkyn, NE Wales (U)[1,2]		0.29 ± 0.06		0.9 ± 0.2	96 ± 5
Minera mine, Halkyn, NE Wales (Cd, Pb, Zn)[1,2]		2.64 ± 0.71		8.6 ± 3.3	107 ± 5
Y Fan control, Llanidloes, Wales (U)[1,2]		0.34 ± 0.10		1.2 ± 0.2	112 ± 4
Y Fan mine, Llanidloes, Wales (Cd, Pb, Zn)[1,2]		0.96 ± 0.16		14 ± 4.2	115 ± 6
Swansea valley control (U)[1]		0.27 ± 0.08			
Swansea valley smelter (Cd, Pb, Zn)[1]		0.76 ± 0.21			
Exeter University control (U)[3]	0.38 (0.05–0.99)				
Kolora Park, Exeter, Devon (U)[3]	0.84 (0.38–3.6)				
Gawton United mine, Devon (As, Cu)[3]	13 (4.7–56)				
Devon Great Consols, Devon (As, Cu)[3]	17 (3.1–71)				
Morwellham Woods, Devon (As, Cu)[3]	17 (2.1–320)				
Wheal Exmouth, Devon (Pb, Zn)[3]	3.7 (0.32–21)				

Table 4.31 (cont.)

Species Organ/whole body Site	As	Cd	Cu	Pb	Zn
Microtus agrestis (field vole)					
Kidney					
Y Fan control, Llanidloes, Wales (U)[1]		0.8 ± 0.0		8.3 ± 1.8	
Y Fan mine, Llanidloes, Wales (Cd, Pb, Zn)[1]		8.9 ± 2.1		60 ± 20	
Control, Derbyshire (U)[7,8,9]		1.8 ± 0.3		5.9 ± 0.9	121 ± 33
Revegetated mine tailings, Derbyshire (Cd, Pb, Zn)[7,8,9]		5.2 ± 0.9		21 ± 2	123 ± 29
Merseyside smelter control (U)[5]		1.7 ± 0.2	20 ± 0.6		
Merseyside smelter 1 km site (Cd, Cu)[5]		24 ± 6	20 ± 1.0		
Merseyside smelter site (Cd, Cu)[5]		89 ± 23	25 ± 1.9		
Liver					
Y Fan control, Llanidloes, Wales (U)[1]		0.27 ± 0.02		4.7 ± 0.4	
Y Fan mine, Llanidloes, Wales (Cd, Pb, Zn)[1]		1.1 ± 0.2		14 ± 2	
Control, Derbyshire (U)[7,8,9]		1.1 ± 0.2		6.1 ± 1.9	113 ± 22
Revegetated mine tailings, Derbyshire (Cd, Pb, Zn)[7,8,9]		1.9 ± 0.2		13 ± 2	120 ± 27
Merseyside smelter control (U)[5]		0.7 ± 0.1	15 ± 0.9		
Merseyside smelter 1 km site (Cd, Cu)[5]		8.7 ± 1.1	15 ± 0.8		
Merseyside smelter site (Cd, Cu)[5]		23 ± 3	16 ± 0.4		
Hair					
Merseyside smelter control (U)[5]		0.28 ± 0.11	12 ± 0.9		
Merseyside smelter 1 km site (Cd, Cu)[5]		0.61 ± 0.22	18 ± 2.5		
Merseyside smelter site (Cd, Cu)[5]		1.1 ± 0.4	26 ± 5.1		

Whole body

Site				
Minera control, Halkyn, NE Wales (U)[1,2]			2.8 ± 0.5	121 ± 9
Minera mine, Halkyn, NE Wales (Cd, Pb, Zn)[1,2]			45 ± 17	192 ± 25
Y Fan control, Llanidloes, Wales (U)[1,2]		0.13 ± 0.02	2.8 ± 0.5	121 ± 9
Y Fan mine, Llanidloes, Wales (Cd, Pb, Zn)[1,2]		0.62 ± 0.08	43 ± 9.1	169 ± 24
Control, Derbyshire (U)[7,8,9]		0.88 ± 0.05	7.7 ± 2.7	103 ± 13
Re-vegetated mine tailings, Derbyshire (Cd, Pb, Zn)[7,8,9]		1.84 ± 0.12	60 ± 14	153 ± 20
Merseyside smelter control (U)[5]		0.58 ± 0.06	7.4 ± 1.0	

Myodes glareolus (bank vole)

Kidney

Site				
Y Fan control, Llanidloes, Wales (U)[1]		4.6 ± 0.8	8.5 ± 2.2	
Y Fan mine, Llanidloes, Wales (Cd, Pb, Zn)[1]		17 ± 2.6	47 ± 26	
Frongoch control, Ceredigion, Wales (Pb, Zn)[10]		0.3 ± 0.1	0.3 ± 0.1	95 ± 3
Frongoch mine, Ceredigion, Wales (Pb, Zn)[10]		1.9 ± 0.6	16 ± 3.7	72 ± 1
Exeter University control (U)[3]	0.69 (0.30–4.2)			
Kolora Park, Exeter, Devon (U)[3]	0.62 (0.30–4.8)			
Gawton United mine, Devon (As, Cu)[3]	2.5 (0.30–32)			
Devon Great Consols, Devon (As, Cu)[3]	4.1 (3.6–4.6)			
Morwellham Woods, Devon (As, Cu)[3]	1.0 (0.30–9.2)			
Wheal Exmouth, Devon (Pb, Zn)[3]	0.60 (0.30–1.3)			

Table 4.31 (cont.)

Species	As	Cd	Cu	Pb	Zn
Organ/whole body					
Site					
Liver					
Y Fan control, Llanidloes, Wales (U)[1]		1.1 ± 0.6		7.0 ± 0.7	
Y Fan mine, Llanidloes, Wales (Cd, Pb, Zn)[1]		5.1 ± 1.0		10 ± 2	
Frongoch control, Ceredigion, Wales (Pb, Zn)[10]		0.1 ± 0.0		0.6 ± 0.1	117 ± 3.7
Frongoch mine, Ceredigion, Wales (Pb, Zn)[10]		0.3 ± 0.1		8.0 ± 0.8	77.7 ± 2.3
Exeter University control (U)[3]	0.71 (0.30–11)				
Kolora Park, Exeter, Devon (U)[3]	0.61 (0.30–6.1)				
Gawton United mine, Devon (As, Cu)[3]	2.4 (0.30–23)				
Devon Great Consols, Devon (As, Cu)[3]	4.4 (4.3–4.4)				
Morwellham Woods, Devon (As, Cu)[3]	1.20 (0.30–2.5)				
Wheal Exmouth, Devon (Pb, Zn)[3]	0.43 (0.30–2.5)				
Whole body					
Minera control, Halkyn, NE Wales (U)[1,2]				2.4 ± 0.6	
Minera mine, Halkyn, NE Wales (Cd, Pb, Zn)[1,2]				16	
Y Fan control, Llanidloes, Wales (U)[1,2]		0.17 ± 0.05		2.6 ± 0.4	143 ± 34
Y Fan mine, Llanidloes, Wales (Cd, Pb, Zn)[1,2]		0.87 ± 0.16		21 ± 5.3	143 ± 25
Frongoch control, Ceredigion, Wales (Pb, Zn)[10]		0.16 ± 0.02		0.03 ± 0.00	70 ± 4
Frongoch mine, Ceredigion, Wales (Pb, Zn)[10]		0.07 ± 0.01		25.3 ± 3	80 ± 3
Exeter University control (U)[3]	1.7 (0.66–2.5)				
Kolora Park, Exeter, Devon (U)[3]	2.1 (0.89–6.2)				
Gawton United mine, Devon (As, Cu)[3]	192 (4.4–380)				

Devon Great Consols, Devon (As, Cu)[3]	8.8 (0.55–58)			
Morwellham Woods, Devon (As, Cu)[3]	2.8 (0.79–11)			
Wheal Exmouth, Devon (Pb, Zn)[3]	0.45 (0.24–0.66)			

Sorex araneus (common shrew)

Kidney

Control, Derbyshire (U)[7,8,9]	4.1 ± 0.7		21 ± 7.3	204 ± 148
Re-vegetated mine tailings, Derbyshire (Cd, Pb, Zn)[7,8,9]	158 ± 15		90 ± 30	281 ± 153
Merseyside smelter control (U)[5]	21 ± 1.6	31 ± 2		
Merseyside smelter 1 km site (Cd, Cu)[5]	156 ± 25	33 ± 3		
Merseyside smelter site (Cd, Cu)[5]	253 ± 75	51 ± 6		

Liver

Control, Derbyshire (U)[7,8,9]	2.9 ± 0.6		7.1 ± 2.1	169 ± 83
Revegetated mine tailings, Derbyshire (Cd, Pb, Zn)[7,8,9]	236 ± 30		22 ± 5	180 ± 40
Merseyside smelter control (U)[5]	14 ± 1.4	24 ± 2		
Merseyside smelter 1 km site (Cd, Cu)[5]	245 ± 45	42 ± 4		
Merseyside smelter site (Cd, Cu)[5]	578 ± 124	63 ± 14		

Hair

Merseyside smelter control (U)[5]	0.54 ± 0.08	13 ± 1.3		
Merseyside smelter 1 km site (Cd, Cu)[5]	1.11 ± 0.24	21 ± 5.3		
Merseyside smelter site (Cd, Cu)[5]	1.63 ± 0.37	52 ± 11		

Table 4.31 (**cont.**)

Species Organ/whole body Site	As	Cd	Cu	Pb	Zn
Whole body					
Minera control, Halkyn, NE Wales (U)[2]				0.9 ± 0.1	
Minera mine, Halkyn, NE Wales (Cd, Pb, Zn)[2]				11 ± 2.7	
Control, Derbyshire (U)[7,8,9]		1.2 ± 0.1		11 ± 3.9	129 ± 16
Revegetated mine tailings, Derbyshire (Cd, Pb, Zn)[7,8,9]		53 ± 3		99 ± 13	172 ± 18
Merseyside smelter control (U)[5]		3.2 ± 0.3	12 ± 1		
Merseyside smelter site (Cd, Cu)[5]		93 ± 9	38 ± 7		
Talpa europea (mole)					
Kidney					
Arnhem control, Netherlands (U)[11]		59 (30–125)	25 (22–27)	22 (17–31)	131 (105–152)
Budel smelter sites, Netherlands (Cd, Pb, Zn)[11]		223 (88–352)	30 (25–37)	24 (8–38)	349 (215–449)
Liver					
Arnhem control, Netherlands (U)[11]		30 (25–48)	23 (21–26)	9 (6–11)	15 (11–120)
Budel smelter sites, Netherlands (Cd, Pb, Zn)[11]		200 (122–234)	27 (22–30)	10 (5–12)	212 (160–244)

Note: nm = not measurable.

Sources: Johnson et al. (1978);[1] Roberts et al. (1978);[2] Erry et al. (2000);[3] Tête et al. (2014a);[4] Hunter et al. (1989);[5] Beernaert et al. (2008);[6] Andrews et al. (1989a);[7] Andrews et al. (1989b);[8] Andrews et al. (1984);[9] Milton et al. (2003);[10] Ma (1987).[11]

5 µg/g Cd can be considered high in the field vole, *Microtus agrestis* (Table 4.31). A high kidney concentration of nearly 90 µg/g Cd was reached in field voles at the Merseyside copper/cadmium refinery site. The threshold kidney cadmium concentration indicative of raised cadmium bioavailability is probably again 10 µg/g Cd in the bank vole, *Myodes glareolus* (Table 4.31). Average cadmium concentrations get much higher in the common shrew *Sorex araneus* in contaminated conditions, reaching above 250 µg/g Cd, compared to an average control concentration of about 4 µg/g Cd (Table 4.31). Cadmium concentrations in the kidneys of common shrews from the very contaminated Merseyside copper/cadmium refinery site increased with age (measured as weight) of the shrews, attaining nearly 600 µg/g Cd in the oldest individuals, indicating increased storage of accumulated cadmium over time (Hunter et al., 1989). Moles, *Talpa europea*, from a contaminated habitat can also have an extremely high average cadmium concentration in the kidney, exceeding 200 µg/g Cd (Table 4.31).

High average lead concentrations in the kidney fall in the ranges of about 40 to 65 µg/g Pb in wood mice, 20 to 60 µg/g Pb in field voles and 15 to 50 µg/g Pb in bank voles (Table 4.31). The average lead concentration in the kidney of contaminated common shrews can reach 90 µg/g Pb, from a baseline of about 20 µg/g Pb (Table 4.31). Average arsenic kidney concentrations of 3 to 5 µg/g As were registered in both wood mice and bank voles from arsenic-contaminated sites, although arsenic concentrations in individuals of both species from sites considered uncontaminated reached or exceeded these averages (Table 4.31). Copper concentrations in the kidneys of wood mice appear to be regulated at about 20 µg/g Cu (Table 4.31). This same concentration of 20 µg/g Cu is apparently also the baseline copper concentration in the kidneys of field voles, but this can increase to 25 µg/g Cu in copper-contaminated conditions, implying a breakdown of any regulation (Table 4.31). Common shrews have higher average control concentrations of copper in the kidney (30 µg/g Cu), but this can increase under contaminated conditions to above 50 µg/g Cu (Table 4.31). Kidney copper concentrations

in shrews from the contaminated Merseyside copper/cadmium refinery site did not, however, increase with age of the common shrew (Hunter et al., 1989). With the exception of the mole, small mammals appear to regulate zinc concentrations in the kidneys to relatively constant levels, this level being highest in common shrews (Table 4.31).

4.10.1.2 Livers

Prologue. The liver is another important organ for the detoxified storage of accumulated trace metals in small mammals.

Accumulated concentrations of cadmium and lead are lower in the livers of wood mice *A. sylvaticus* than in their kidneys (Table 4.31), indicating that the kidney is the more important site of storage detoxification of both metals. As in the kidney, average cadmium and lead concentrations in the livers of wood mice increased in contaminated conditions, each from a lower base to lower raised concentrations than in the kidney (Table 4.31). Average cadmium concentrations of 2 µg/g Cd or above in the liver are high in wood mice, as are average lead concentrations of 10 µg/g Pb (Table 4.31). Similar patterns are seen in the field vole *M. agrestis*. Liver concentrations of both cadmium and lead are lower in the livers than in the kidneys of field voles, and both increase in contaminated conditions, but not to the extent shown by the kidney concentrations (Table 4.31). Average lead concentrations of 10 µg/g Pb or above in the liver are high in field voles, as in wood mice (Table 4.31). It is difficult to define quite what is a raised liver cadmium concentration in field voles (Table 4.31). Liver concentrations above 2 µg/g Cd are high, but the threshold concentration indicative of raised cadmium bioavailability may be lower than this (Table 4.31). Similarly, in bank voles *M. glareolus*, liver concentrations of both cadmium and lead are lower than kidney concentrations (Table 4.31). Both increase in contaminated conditions, from a lower baseline concentration to lower highs than in the kidneys of the same animals (Table 4.31). Average cadmium concentrations of

5 µg/g Cd or above in the liver are high in bank voles, as are average lead concentrations of 10 µg/g Pb (Table 4.31).

As for kidneys, insectivorous common shrews *S. araneus* again have higher liver concentrations of cadmium and lead than any of the rodents (Table 4.31). The average liver concentration in contaminated common shrews can reach nearly 600 µg/g Cd, compared to a control concentration of about 3 µg/g Cd (Table 4.31). In common shrews from the Merseyside copper/cadmium refinery site, cadmium concentrations in the liver also increased with age of the shrews, reaching more than 1,000 µg/g Cd in the oldest individuals (Hunter et al., 1989). In the common shrew, liver concentrations of cadmium are greater than those in the kidney (Table 4.31), indicating a greater storage detoxification role for the liver than the kidney. This detoxified cadmium store in both organs of the common shrews builds up over time at cadmium-contaminated sites (Hunter et al., 1989). Average liver concentrations of lead in the common shrew can reach 22 µg/g Pb under contaminated conditions, compared to a background 7 µg/g Pb (Table 4.31). The livers of moles *T. europea* living in contaminated soils also have very high cadmium concentrations reaching 200 µg/g Cd (Table 4.31). Mole liver concentrations of lead do not appear to rise with soil contamination, while liver zinc concentrations do increase, contrarily to the other small mammals (Table 4.31).

In contrast to cadmium and lead, the arsenic concentrations in the livers of wood mice and bank voles are not lower than those in the kidney (Table 4.31), suggesting equal importance of the two organs in arsenic detoxification and accumulation. Average arsenic liver concentrations increase in wood mice and bank voles in arsenic-contaminated habitats, although, as in the kidney, liver arsenic concentrations in individuals of both species from uncontaminated sites reached or exceeded these raised averages (Table 4.31).

Copper concentrations in the livers of wood mice and field voles did not increase in the rodents exposed to copper contamination, apparently being regulated at about 20 µg/g Cu and about 15 µg/g Cu respectively (Table 4.31). On the other hand, average copper concentrations in the livers of common shrews did increase with copper contamination, from a baseline of about 24 µg/g Cu to about 63 µg/g Cu (Table 4.31). Liver copper concentrations in individuals at the contaminated Merseyside copper/cadmium refinery site also increased with age, reaching more than 150 µg/g Cu in the oldest individuals (Hunter et al., 1989). Copper concentrations in the livers of moles are apparently regulated, at between 23 and 30 µg/g Cu (Table 4.31). Accumulated zinc concentrations in livers appear to be regulated in wood mice, field voles, bank voles and common shrews, but not in moles (Table 4.31).

4.10.1.3 Whole Bodies

Prologue. Concentrations of arsenic, cadmium, lead and copper, but not zinc, in the whole bodies of small mammals are lower than those in the kidney or liver.

The whole body concentrations of arsenic, cadmium, lead and copper increase in contaminated conditions, cadmium, copper and lead, reaching the highest body concentrations in the common shrew (Table 4.31). In the case of zinc, there does seem to be an increase in total body concentration under contaminated conditions in the case of the field vole, but not for wood mice, bank voles or common shrews (Table 4.31).

The observed increases in the total body concentrations of cadmium in the three rodents under contaminated conditions are relatively modest (Table 4.31). Maximum average body concentrations of cadmium were about 2.6 µg/g Cd, 1.8 µg/g Cd and 0.9 µg/g Cd, compared to background concentrations of about 0.3 µg/g Cd, less than 0.6 µg/g Cd and less than 0.2 µg/g Cd in wood mice, field voles and bank voles respectively (Table 4.31). Common shrews reached the much higher contaminated whole body concentrations of 53 µg/g Cd at a revegetated mine tailings site in Derbyshire, and 93 µg/g Cd at the copper/cadmium refinery in Merseyside, from a background control concentration of about 1 to 3 µg/g Cd (Table 4.31). Whole body cadmium concentrations of wood mice, field voles and common shrews all increased with proximity to the Merseyside copper/cadmium smelter, but to a much greater extent in the shrews (Hunter et al., 1987c, 1989). Whole body copper concentrations of wood mice and field voles did not appear to increase with proximity to the Merseyside smelter, but whole body copper concentrations in the common shrew did increase from about 12 µg/g Cu at the control site to about 39 µg/g Cu at the smelter site (Hunter et al., 1987c, 1989).

Whole body concentrations of lead in the rodents showed large increases under contaminated conditions (Table 4.31). Whole body lead concentrations in wood mice increased from about 1 µg/g Pb to about 10 to 15 µg/g Pb in the presence of lead contamination (Table 4.31). Equivalent data for field voles covered a wider range, from about 3 µg/g Pb to 40 to 60 µg/g Pb, and those for bank voles from also about 3 µg/g Pb to 15 to 25 µg/g Pb (Table 4.31). Common shrews had the highest average whole-body lead concentrations, nearly 100 µg/g Pb in common shrews from the revegetated mine tailings site in Derbyshire (Table 4.31). High lead concentrations in bones contribute to the high body lead concentrations in common shrews at this site, to a greater extent than in field voles from the same site.

The consistently higher concentrations of cadmium, copper and lead accumulated in the kidneys, liver and whole bodies of the common shrew in comparison to the three rodents (Table 4.31) deserve further comment. There may well be differences in the physiological handling after uptake of these metals between the rodents and the insectivore as regards relative degrees of excretion versus detoxified storage. Nevertheless, the nature of the diet combined with the amount consumed will be strong contributory factors to a very high trophic input of the two metals in the shrew. In the Merseyside refinery study, wood mice were feeding mainly on the seeds of bramble (*Rubus fruticosus*), rosebay willow herb (*Epilobium angustifolium*) and gorse (*Ulex europaeus)* (Hunter et al., 1987c). Their mean daily dietary input of cadmium was calculated to be 0.22 µg Cd per gram per day at the control site and 0.94 µg per gram per day at the copper/cadmium refinery (Hunter et al., 1987c). Field voles were feeding

predominantly on creeping bentgrass *Agrostis stolonifera* (Table 4.17), with mean daily dietary inputs of 0.35 µg Cd per gram per day at the control site and 2.11 µg Cd per gram per day at the copper/cadmium refinery site. Common shrews were preying mostly on spiders and harvestmen (Table 4.28) and had mean daily inputs of 0.58 and 25.0 µg Cd per gram per day at the control and copper/cadmium refinery sites respectively (Hunter et al., 1987c). Thus, the trophic input of cadmium into the common shrew at the control site was already the highest of the three small mammals, but very much the highest at the copper/cadmium refinery site, here being between 10 and 30 times that in the other two small mammals. Similar calculations for dietary copper input gave mean daily dietary inputs of 30, 90 and 260 µg Cu per gram per day at the refinery site for wood mice, field voles and common shrews respectively.

4.10.1.4 Hair

Prologue. Trace metal concentrations in the hair of small mammals can be measured non-invasively. Hair concentrations of lead and cadmium do correlate with accumulated concentrations in the kidney and liver of wood mice under contaminated conditions.

There is a clear interest in developing non-invasive measures of the exposure of mammals to trace metal contamination. Table 4.31, therefore, presents trace metal concentrations in hair samples of wood mice *A. sylvaticus* for which accumulated concentration data are also available for kidneys and livers. In the vicinity of the Metaleurop Nord former lead and zinc smelter near Calais, France, which shut in 2003, there was a relatively strong significant correlation between lead concentration in each of the kidney and the liver and that in the hair of the local wood mice (Table 4.31). Thus these hair lead concentrations were relatively good predictors of the lead concentrations in the organs (Tête et al., 2014a). Cadmium concentrations in the hair of wood mice were also significantly correlated with accumulated cadmium concentrations in each of the kidney and liver, but more weakly than in the case of lead (Tête et al.,

2014a). Thus, hair can also be used to predict cadmium concentrations in the two organs, but with more caution than in the case of lead. Similarly, in a Belgian study, the hair of wood mice from smelter sites contained higher cadmium and lead concentrations than at control sites, but no site differences were apparent in hair concentrations of copper nor zinc (Table 4.31) (Beernaert et al., 2008). Cadmium concentrations were also raised in hair samples of wood mice, field voles and common shrews collected near the copper/cadmium refinery in Merseyside, as indeed were copper concentrations (Table 4.31).

4.10.1.5 Ecotoxicology

Prologue. What measurements can be used to determine whether a population of small mammals is suffering any ecotoxicological effect from high trace metal exposure? Criticism is presented of the concept of a critical accumulated concentration in an organ of a small mammal, above which accumulated trace metals are considered to be exerting a toxic effect on that mammal. A more justifiable indicator of incipient toxicity is a threshold level of dietary input of a trace metal – an intake Lowest Observable Adverse Effect Level (LOAEL), measured as micrograms of metal per unit body weight per day.

The question arises as to whether any of the small mammals discussed previously, and listed in Table 4.31, were suffering any ecotoxicological effects from their extra accumulated loads of trace metals, especially in the kidney and liver. Mammal ecologists have been attracted to the concept of a critical tissue concentration of an accumulated trace metal in an organ such as the kidney or liver, above which these accumulated trace metals are considered to be exerting a toxic effect (Cooke, 2011; Ma, 2011). Chapter 3 made the point, however, that, in the presence of any detoxified storage of a trace metal, there could be no toxicologically critical concentration of total metal accumulated in an organ, much of it, by definition, being in a form that is not metabolically available. It is the concentration of metabolically available metal that has not been detoxified

that might reach a critical level in an organ and bring on toxic effects. This threshold is reached more quickly (and at a lower total accumulated concentration) when the rates of metal uptake into the animal, and consequently into the organ, are increased by raised local bioavailability of the metal. It is clear from Table 4.31 that storage detoxification of cadmium and lead, at least, is occurring in the kidney and liver of small mammals. Cadmium, for example, is certainly bound to metallothionein in these organs in small mammals, as in mammals in general. Nevertheless, the concept of a critical total concentration of a metal in an organ remains prevalent for mammals (Cooke, 2011; Ma, 2011). It is not surprising that different studies have identified the onset of toxicological effects at different 'critical' accumulated concentrations, particularly in laboratory studies in which mammals are fed a metal-contaminated diet at a high rate (Cooke, 2011; Ma, 2011).

In mammals, the kidney is usually considered the critical organ in toxicity studies of both cadmium and lead, being the first organ in which damage can be observed or adverse functional changes occur (Hunter et al., 1989; Cooke, 2011; Ma, 2011). Cells of kidneys with very high accumulated concentrations of cadmium or lead show evidence of cell damage, typically at the ultrastructural level. The degree of ultrastructural damage in kidney cells increases with the concentration of accumulated cadmium or lead present in the kidney, and then other symptoms of incipient kidney failure start to occur (Cooke, 2011; Ma, 2011). These symptoms might include the presence of abnormal quantities of protein in the urine; an increase in kidney weight as a percentage of total body weight; and, in the case of lead toxicity, excretion of d-amino levulinic acid dehydratase (ALAD) in the urine. Mammalian livers with highly elevated accumulated cadmium concentrations also show ultrastructural damage. The question arises then as to what might count as a very high accumulated concentration of cadmium or lead in the kidney or liver that would be indicative of the onset of toxic effects.

Mice and rats exposed in the laboratory to cadmium have shown symptoms of protein in the urine and cell damage in kidney tubules, in association with kidney concentrations of 105 to 260 µg/g Cd dry weight (Cooke, 2011). Thus, a kidney concentration of 105 µg/g Cd has been proposed as the Lowest Observable Adverse Effect Level (i.e., the critical total concentration of cadmium in the kidney) for small mammals in the field (Shore and Douben, 1994a; Cooke, 2011). Average kidney concentrations in all the rodents from cadmium-contaminated locations listed in Table 4.31 were below this threshold. On the other hand, bank voles, *Myodes glareolus*, near the Rönnskärsverken sulphide ore smelter in Sweden, had kidney concentrations of 14 to 56 µg/g Cd dry weight and did show symptoms of kidney damage, including protein in the urine (Leffler and Nyholm, 1996). In contrast to the rodents, cadmium-contaminated common shrews have considerably higher average kidney concentrations than the proposed 105 µg/g Cd, as in the case of shrews from the revegetated mine tailings site in Derbyshire and from the Merseyside copper/cadmium refinery site (Table 4.31). In the latter case, the degree of ultrastructural damage in the kidneys of individual common shrews increased with the kidney cadmium concentration over the range 150 to 560 µg/g Cd (Hunter et al., 1987c). Corresponding cadmium concentrations in the livers of these individual common shrews were 300 to 1,000 µg/g Cd, with increasing ultrastructural damage to liver cells. In spite of this observable damage to kidney and liver cells, there was no evidence from urine analysis of any renal dysfunction in these common shrews, which were in seemingly good condition on capture (Hunter et al., 1987c). Thus, the accumulation of cadmium to high concentrations in the kidney or liver of a small mammal is not in itself a symptom of cadmium toxicity. It is a feature of the cadmium physiology of common shrews that they can accumulate much more cadmium in the kidney than can wood mice and voles. As in the case of common shrews, moles will also accumulate cadmium concentrations in the kidney well above the proposed 105 µg/g Cd threshold (Table 4.31).

Criticism of the concept of a critical total accumulated metal concentration is based on the principle that toxicity can occur at any total accumulated metal concentration, if, and when, the total uptake rate exceeds a threshold that cannot be matched by the rates of storage detoxification and excretion. Indeed, the Swedish bank vole example highlights the inherent difficulty in trying to pin down any such total critical metal concentration in an organ or whole body. The implication here is that cadmium was entering these bank voles so quickly that toxicity ensued before there had been time for further buildup of accumulated cadmium in their kidneys to approach the proposed 105 µg/g Cd figure. A conceptually more justifiable approach to understanding the onset of the ecotoxicological effects of a toxic metal that is being taken up and accumulated is to define a threshold uptake rate. This might be known as an intake Lowest Observable Adverse Effect Level, and 3.5 to 7.5 µg Cd per gram per day has been proposed as an intake LOAEL for cadmium in laboratory rodents (Shore and Douben, 1994a). Such a cadmium uptake rate threshold for rodents would explain the absence of ecotoxicological symptoms in the wood mice and field voles at the refinery site at Merseyside, with average daily dietary Cd inputs of 0.94 µg per gram per day and 2.11 µg per gram per day respectively (Hunter et al., 1987c). Common shrews would need a different cadmium intake LOAEL than the rodents. Common shrews at the Merseyside refinery site had an average daily Cd input rate of 25.0 µg per gram per day, and these shrews showed no physiological evidence of kidney failure or other signs of cadmium toxicity (Hunter et al., 1987c). In the light of the evidence in Table 4.31, moles, too, would need a higher cadmium intake LOAEL than rodents.

Similar attempts to define a critical accumulated concentration in the kidneys and livers of mammals have been made for lead. It has been proposed, for example, that for mammals in general, kidney lead concentrations greater than 15 µg/g dry weight are associated with structural and functional kidney damage, and kidney lead concentrations above 80 µg/g Pb are associated with loss of body weight and death (Ma, 2011). An average kidney lead concentration of 70 µg/g dry weight has also been proposed as a threshold concentration above which some clinical manifestation of lead toxicity might be expected in small mammals, the equivalent liver concentration being 15 to 25 µg/g Pb (Shore and Douben, 1994b; Ma, 2011).

Such generalised proposals also inevitably fall foul of interspecific differences in lead accumulation patterns. Field voles *Microtus agrestis* trapped at the Y Fan mine site in Wales had an average kidney lead concentration of 60 µg/g Pb (Table 4.31). These field voles showed evidence of kidney toxicity, expressed as ultrastructural damage in kidney cells, and an increase in kidney weight as a percentage of total body weight (Roberts et al., 1978). Wood mice *Apodemus sylvaticus* from the site of the Monera mine, with an average kidney lead concentration of 47 µg/g Pb (Table 4.31), also showed an increase in kidney weight as a percentage of total body weight as an indicator of lead ecotoxicity (Roberts et al., 1978). Wood mice from the Y Fan mine site, with an average kidney lead concentration of 39 µg/g Pb (Table 48), did not show this kidney weight effect (Roberts et al., 1978). Nor did bank voles from Y Fan with an average kidney lead concentration of 47 µg/g Pb (Table 4.31) (Johnson et al., 1978). As a further example of interspecific variability, common shrews with a higher kidney concentration of 225 µg/g Pb, acquired by laboratory feeding with lead-contaminated earthworms, showed reduced body weight and an increased mortality rate (Ma, 2011). Similarly, it has also been reported that moles are likely to suffer severe lead intoxication when the average kidney concentration exceeds 200 µg/g Pb (Ma, 2011).

The Metaleurop Nord smelter in northern France contaminated the local area with atmospheric fallout of cadmium, lead and zinc until 2003, reflected in ongoing high concentrations of cadmium and lead in the kidneys and livers of local wood mice *Apodemus sylvaticus* (Table 4.31). There was evidence of both kidney and liver damage in wood mice, with high accumulated cadmium and lead concentrations in these organs (Tête et al., 2014b). The kidney concentrations of cadmium and lead in the kidneys of some individual wood mice reached 120 µg/g Cd and 282 µg/g Pb, the average concentrations of their sampled population being 38 µg/g Cd and 16 µg/g Pb respectively. Equivalent liver concentrations reached 32 µg/g Cd and 6 µg/g Pb in individual wood mice

and averaged 11 µg/g Cd and 1.3 µg/g Pb in their parent populations. The degree of cell degeneration in kidney tubules was significantly related to the lead concentration of the kidney (Tête et al., 2014b). Thus, average kidney concentrations of both cadmium and lead in populations of wood mice showing ecotoxicological effects were below the proposed critical total concentrations of 105 µg/g Cd and 70 µg/g Pb, although kidney concentrations of both metals exceeded these thresholds in some individuals.

Another anthropogenic source of lead into terrestrial ecosystems is the deposition of lead pellets from shotgun ammunition, particularly concentrated in the vicinity of clay pigeon shooting ranges (O'Halloran et al., 1988; Ma, 1989). Decomposition of gunshot pellets in most soils may typically take tens or hundreds of years by physical erosion. In acid soils, however, there is dissolution of pellets to release soluble lead in bioavailable form (Ma, 1989). Wood mice, bank voles and common shrews collected from a shooting range in coastal dunes in the Netherlands had strongly elevated average lead concentrations in kidney, liver and bone (Ma, 1989). Kidney concentrations of lead, for example, exceeded any proposed critical renal concentrations (Ma, 1989). Bank voles and shrews from the shooting range had significantly increased kidney weight as a percentage of total body weight, indicating lead poisoning (Ma, 1989).

In conclusion, an accumulated trace metal concentration in an organ such as the kidney cannot directly tell us that the metal is having a significant ecotoxicological effect on that organ and on that individual mammal. Nevertheless, that high accumulated concentration is indicative that the mammal has been exposed to an atypically high toxic metal bioavailability, with increased associated risk of an ecotoxicological outcome.

There is, however, another factor that needs consideration, a factor that will be relevant again later in this book, when we consider the enormously high accumulated concentrations of particular trace metals in the kidneys of long-lived pelagic seabirds in the absence of anthropogenic metal contamination. How much of an avian or mammalian kidney needs to be functional for the individual bird or mammal to survive? Much of the high concentrations of the trace metals accumulated are in detoxified form in the kidney cells, and these cells show considerable evidence of ultrastructural cell damage. Are all kidney cells similarly affected with a consequent decrease in total kidney functionality, or are some kidney cells (newly produced) relatively free of accumulated trace metals and able to function sufficiently to meet all the renal needs of the animal concerned? Does a high accumulated concentration of a trace metal in a mammalian or avian kidney necessarily come with a price to be paid in terms of kidney functionality, or is high detoxified storage accumulation a normal feature of the trace metal physiology of otherwise healthy mammals or birds? Trace metals will present an ecotoxicological challenge when taken up at a faster rate than can be detoxified under this accumulation mechanism, but any such ecotoxicological threshold could occur at any accumulated metal concentration in the kidney.

4.10.2 Bats

Prologue. Bats comprise another group of British insectivorous mammals, with the potential to contain high accumulated concentrations of trace metals in their kidneys, if feeding on insects living in metal-contaminated habitats.

Table 4.32 presents accumulated concentrations of cadmium, lead and mercury in bats found dead or fatally injured in Cornwall and Devon between 1988 and 2003 (Walker et al., 2007). Pipistrelles (63% of the 272 bats collected) were not separated into the more recently recognised cryptic species *Pipistrellus pipistrellus* and *P. pygmaeus*. The other bat species shown are the brown long-eared bat *Plecotus auritus* (22%), the whiskered bat *Myotis mystacinus* (6%) and Natterer's bat *Myotis nattereri* (5%). There were interspecific differences in the kidney metal concentrations (Table 4.32). Whiskered and Natterer's bats had the highest median kidney concentrations of mercury, with the significantly lowest mercury concentrations in the pipistrelles. Natterer's bats again had the highest kidney

concentrations of cadmium. Kidney lead concentrations were, however, lowest in Natterer's bats, but did not differ between the other species. In the pipistrelles, adults had significantly higher kidney cadmium concentrations than juveniles (Walker et al., 2007). Kidney concentrations of lead did not decrease over time in any of the bats, despite concurrent reductions in atmospheric lead with the phasing out of lead in petrol (Walker et al., 2007). For what they are worth, the proposed critical kidney concentrations for cadmium (105 µg/g Cd) and lead (70 µg/g Pb) are not exceeded by any of the bat kidney concentrations in Table 4.32. Nor do the mercury concentrations in the kidneys of these bats (Table 4.32) exceed a proposed critical tissue concentration for non-marine mammals of 70 µg/g Hg dry weight (Thompson, 1996; Walker et al., 2007).

4.11 Birds

Prologue. Birds living in terrestrial habitats affected by mining and smelting inevitably come into contact with raised bioavailabilities of trace metals through their diet.

The diet is not the only form of interaction between terrestrial birds and trace metals. Choughs *Pyrrhocorax pyrrhocorax* nest in caves, and the shafts of the old copper mines of Snowdonia and the lead mines of mid-Wales have provided sites for choughs to breed inland (Condry, 1981). Old mine walls in Wales have also provided homes for the stock dove *Columba oenas*, pied wagtails *Motacilla alba*, grey wagtails *Motacilla cinerea* and occasionally redstarts *Phoenicurus phoenicurus* (Condry, 1981).

Nevertheless, it is the diet and interspecific differences in diets that are particularly relevant to the natural history of trace metals in birds in British terrestrial habitats. Amongst the passerines, great tits *Parus major* and blue tits *Cyanistes caruleus* feed predominantly on insects and other arthropods, as do pied flycatchers *Ficedula hypoleuca*. House sparrows *Passer domesticus* mostly eat the seeds of grains and weeds. Another passerine, the blackbird *Turdus merula*, is more omnivorous, feeding on insects and

Table 4.32 **Concentrations of cadmium, lead and mercury (μg/g dry weight) in kidneys of British bats collected in 1988 to 2003 from Southwest England.**

	Pipistrelles	Brown long-eared bat	Whiskered bat	Natterer's bat
Cadmium				
Median	1.42	0.83	1.61	6.27
IQR	0.34–3.38	0.24–2.75	0.63–4.31	0.29–13.9
Maximum	29	13	16	19
Lead				
Median	2.45	3.38	4.05	1.16
IQR	1.26–4.85	1.83–5.08	2.43–9.61	0.62–1.98
Maximum	70	25	24	5.68
Mercury				
Median	0.93	1.52	3.00	2.65
IQR	0.93–0.93	0.93–2.22	2.16–5.50	0.93–4.74
Maximum	5.08	8.14	8.01	7.87

Note: IQR is the interquartile range – the range of values covered by the central 50% of an ascending dataset.
Source: Walker et al. (2007).

earthworms as well as berries and fruits. Doves and pigeons are herbivorous, feeding on seeds, fruits and other parts of plants. For example, feral pigeons *Columba livia* eat seeds and berries, but will also scavenge and take insects and spiders. Birds of prey are of particular concern, preying on other birds, but also on small mammals which may contain high concentrations of toxic metals such as cadmium and lead in their kidneys and livers. We turn to birds of prey first.

4.11.1 Birds of Prey

Prologue. The trace metal of past historic ecotoxicological concern for British birds of prey was mercury, because of its food chain transfer and biomagnification in the form of the organometal methyl mercury.

Birds of prey are at risk of lead poisoning if feeding on birds or mammals containing lead shot in their bodies or carcasses.

Barn owls *Tyto alba* take field voles and shrews. Tawny owls *Strix aluco* prey on wood mice and bank voles but will take other mammals, birds and even earthworms and beetles. Little owls *Athene noctus* prey on insects, earthworms and amphibians but also small birds and mammals. Kestrels *Falco tinnunculus* mostly eat shrews, mice and voles, especially field voles, but also invertebrates in late summer (Newton et al., 1993; Pain et al., 1995). Buzzards *Buteo buteo* eat mainly small mammals, but will also feed on birds, amphibians, reptiles and insects, as well as scavenging on carrion (Pain et al., 1995). Sparrowhawks *Accipiter nisus* take mainly passerine birds, but will eat pigeons (Newton et al., 1993; Pain et al., 1995). Peregrine falcons *Falco peregrinus* prey on medium-sized birds such as larger passerines, pigeons and doves, as well as waterfowl and waders near water (Pain et al., 1995).

The concentrations of contaminants in birds of prey were attracting considerable attention in the decades subsequent to the publication of Rachel Carson's *Silent Spring* in 1962. This milestone book had compiled evidence to show that organochlorine pesticides were causing significant detrimental effects on the environment, particularly in birds of prey at the top of food chains. Furthermore, the chemical industry in North America at the time, without opposition from public officials, had deliberately spread considerable disinformation on the ecotoxicology of these pesticides. Organochlorines, like other organic contaminants, typically biomagnify up food chains. They then cause significant toxic effects, such as eggshell thinning in birds of prey, with subsequent mortalities. It was not known at the time whether toxic metals might also be biomagnified up food chains leading to birds of prey, again with potential ecotoxicological effects. In fact, we now know that trace metals in their more usual inorganic form are not biomagnified up food chains as a general principle, although biomagnification of organometals such as methyl mercury will occur. So, during the later decades of the twentieth century, collections of birds of prey, found dead from a variety of causes, were built up in freezers for later analysis for trace metals, in addition to organochlorine contaminants (Newton et al., 1993; Pain et al., 1995). Tables 4.33 and 4.34 show the concentrations of mercury and lead in selected organs of several British birds of prey collected from the 1960s to the 1990s.

In the 1960s in Britain, sources of mercury into the environment were for the most part industrial, but there was some agricultural input too. The industrial sources consisted of the combustion of fossil fuels, the smelting of sulphide ores, the production of cement, the incineration of refuse and point sources such as chlor-alkali plants (Newton et al., 1993). In agriculture, mercury was used as a fungicidal dressing for seeds, often in methyl form, with the consequent risk of mercury accumulation by granivorous birds and then biomagnification of methyl mercury up food chains to birds of prey, potentially with ecotoxicological effects (Scheuhammer, 1987). Between the 1960s and 1990, the industrial uses and disposal of mercury were controlled more rigorously in Britain, as was the use of mercury-based seed dressings.

The birds of prey involved in the study of mercury accumulation summarised in Table 4.33 were sparrowhawks *Accipiter nisus* and kestrels *Falco tinnunculus*, and the organ of interest was the liver. Samples from each species were divided into five time blocks between 1963 and 1990 (Table 4.33) to look for downward trends over time, as mercury release into the environment was falling. The livers of sparrowhawks and kestrels had similar accumulated concentrations of mercury, and both species did show a significant downward trend over time (Table 4.33) (Newton et al., 1993). Concentrations of mercury in the livers of birds (other than fish-eating birds) typically fall into the range of 1 to 10 µg/g Hg dry weight (Scheuhammer, 1987). The mercury concentrations of the livers of the sparrowhawks and kestrels fell into this same range, towards the top end before 1975, and at the lower end by 1990 (Table 4.33).

As previously discussed for mammals, attempts have been made to propose accumulated trace metal concentrations that may be indicative of adverse ecotoxicological effects in the organs of birds. A proposed lethal threshold concentration of mercury in the liver of birds is 20 µg/g Hg wet weight (Shore et al., 2011), equivalent to 60 µg/g Hg dry weight, using a wet weight to dry weight ratio of 3 (Pain et al., 1995). A lower threshold mercury concentration of 2 µg/g Hg wet weight (6 µg/g Hg dry weight) in the liver has been proposed as indicative of a sublethal effect depressing reproduction in birds (Shore et al., 2011). Both sparrowhawks and kestrels had liver mercury concentrations well below the proposed lethal concentration over the whole time period of the study shown in Table 4.33. Nevertheless, the early liver concentrations of the 1960s and 1970s did exceed the lower proposed threshold concentration indicative of an adverse sublethal effect on reproduction in birds. Reproduction appears to be one of the most sensitive endpoints of mercury ecotoxicity, for mercury accumulates in the egg-white proteins and acts

Table 4.33 **Accumulated concentrations (µg/g dry weight) of mercury in livers of birds of prey in Britain between 1963 and 1990.**

Species	Mercury	
	Geometric	*1 SD*
Date	*mean*	*range*
Sparrowhawk *Accipiter nisus*		
1963–1970	4.60	2.55–8.28
1971–1975	5.64	3.28–9.72
1976–1980	3.52	1.50–8.29
1981–1985	2.26	0.65–7.87
1986–1990	0.98	0.15–6.28
Kestrel *Falco tinnunculus*		
1963–1970	5.77	2.98–11.2
1971–1975	3.90	1.77–8.62
1976–1980	1.27	0.43–3.74
1981–1985	0.77	0.19–3.11
1986–1990	0.22	0.02–2.21

Source: Newton et al. (1993).

Table 4.34 **Accumulated concentrations (µg/g dry weight) of lead in livers of birds of prey in Britain between 1981 and 1992.**

Species	Lead	
	Median	*Maximum*
Sparrowhawk *Accipiter nisus*	0.55	12
Kestrel *Falco tinnunculus*	0.69	10
Peregrine *Falco peregrinus*	0.48	22
Buzzard *Buteo buteo*	1.34	909
Little Owl *Athene noctua*	0.82	14

Source: Pain et al. (1995).

as an embryo toxicant (Wolfe et al., 1998). The evidence is not, however, available to show whether these birds of prey really were suffering depressed reproduction as a result of mercury accumulation, particularly because the same birds had also accumulated organochlorine pesticides and PCBs (Newton et al., 1993).

It is the kidney that is the major reservoir of inorganic mercury in birds, where it is bound to metallothionein (Scheuhammer, 1987). A proposed lethal threshold mercury concentration in the kidney of birds is 40 µg/g Hg wet weight (120 µg/g Hg dry weight) (Shore et al., 2011).

Elsewhere in the world, for example in California, high bioavailabilities of selenium can occur, with environmental consequences (Luoma and Rainbow, 2008). Interestingly enough, these are not all negative. In mammals, high accumulated concentrations of selenium in the liver can offset the toxicity of accumulated mercury via a process of detoxification to form mercury selenide, so allowing the accumulation of atypically high mercury concentrations without toxic effect (Scheuhammer, 1987). The situation is less clear in the case of bird livers, in which accumulated mercury concentrations are usually below a threshold at which a significant selenium-mercury interaction occurs (Scheuhammer, 1987). Furthermore, a different selenium-mercury interaction appears to be present in the kidney from the liver. So, any consideration of accumulated concentrations of mercury in the organs of birds living in terrestrial (and indeed freshwater) habitats in the British Isles can ignore any effect of selenium, although this will not necessarily be the case out at sea.

Table 4.34 shows concentrations of lead accumulated in the livers of sparrowhawks, kestrels, peregrine falcons, buzzards and little owls found dead between 1981 and 1992. Median lead concentrations in the livers of the birds of prey were below 1.2 µg/g Pb dry weight, and below 2.2 µg/g Pb in a wider range of birds of prey analysed (Pain et al., 1995). On the other hand, very high lead concentrations, greater than 20 µg/g Pb, were found in one peregrine falcon and one buzzard (22 and 909 µg/g Pb respectively) (Pain et al., 1995). It was concluded that these exceptional concentrations of accumulated lead in the liver were from birds that had likely ingested lead gunshot in the flesh of their prey. Buzzards will feed on carrion, including dead game birds and rabbits, so this is not unexpected. Peregrine falcons take live birds, including pigeons, ducks, waders and some game birds (Pain et al., 1995). Not all birds that have been shot will be killed and may live on to become easy prey to a hunter such as a peregrine falcon. Furthermore, the stomachs of birds of prey are very acidic. Ingested lead shot may be dissolved rapidly, allowing lead to enter the blood and be transported to the liver of the bird (Pain et al., 1995). Typical liver concentrations of lead in sparrowhawks were relatively low in comparison to the other birds of prey (Table 4.34), but adult birds, both males and females, had higher lead concentrations in the liver than juveniles (Pain et al., 1995).

Attempts have also been made to define threshold liver concentrations of lead in birds that are indicative of adverse ecotoxicological effects. Thus it has been proposed that 6 µg/g Pb dry weight in the liver would indicate above background exposure, and 20 µg/g Pb dry weight would indicate acute exposure (Pain et al., 1995). In the case of birds of prey in the order *Falconiformes*, which includes sparrowhawks, kestrels, peregrine falcons and buzzards, more specific proposals for threshold concentrations of lead in the liver are 2 µg/g Pb wet weight (6 µg/g Pb dry weight) as indicative of subclinical lead poisoning, 6 µg/g Pb wet weight (18 µg/g Pb dry weight) for clinical lead poisoning and 10 µg/g Pb wet weight (30 µg/g Pb dry weight) for severe clinical lead poisoning (Franson and Pain, 2011). Thus, all median lead concentrations

in the livers of the falcons (and presumably the owls) in Table 4.34 can be considered indicative of background exposure of the birds to lead. Maximum liver concentrations in particular individual birds (Table 4.34), however, may indicate subclinical lead poisoning in some sparrowhawks, kestrels and little owls; clinical lead poisoning in the peregrine falcon highlighted previously; and severe clinical lead poisoning in the buzzard. It is likely then that the peregrine falcon and the buzzard had died from lead poisoning, resulting from exposure to lead shot in their prey. Another British bird of prey at risk of ingesting lead shot when scavenging on carcasses is the red kite *Milvus milvus* (Fisher et al., 2006).

Male and female birds have not yet been separated in this account of the accumulation of lead by birds. In fact, female birds accumulate lead more quickly than male birds (Scheuhammer, 1987). The physiology of lead in organisms has strong links to that of calcium. Indeed, we have seen how accumulated lead is often detoxified by binding to granules based on calcium phosphate. Female birds have an increased need for calcium to form eggshells, and correspondingly the absorption of calcium in the alimentary tract of female birds is typically greater than that of males (Scheuhammer, 1987). The increased uptake of calcium is associated with increased uptake of lead, given the sharing of uptake routes. Thus, female birds have more accumulated lead than their male counterparts. It is the same sharing of physiological processes by calcium and lead that leads to high concentrations of lead in the bones of birds, at typical background concentrations of 2 to 15 µg/g Pb dry weight with a low turnover rate (Scheuhammer, 1987; Franson and Pain, 2011).

Table 4.35 presents a comparison of arsenic concentrations in organs of sparrowhawks, kestrels and barn owls found dead between 1970 and 1996 in an area of Southwest England with a history of arsenic contamination against those concentrations from these birds from Southwest Scotland, in the absence of arsenic enrichment (Erry et al., 1999). Accumulated arsenic concentrations in kidney, liver and muscle tissue of kestrels from Southwest England were about three times higher than in the organs of their counterparts in Southwest Scotland, but there were no such regional differences in the sparrow-hawks nor barn owls (Table 4.35). The arsenic concentrations of the three organs were also significantly higher in the kestrels from Southwest England than in the other two birds of prey also from Southwest England (Table 4.35). Differences between kestrels and sparrowhawks may result from differences in their diets. Sparrowhawks almost exclusively take small birds, while kestrels will feed on small mammals, which will accumulate arsenic to high concentrations in the kidney and liver when living in arsenic-contaminated habitats (Table 4.31). Barn owls similarly to kestrels, however, also take small mammals, so dietary differences are a less likely factor in this comparison. It is possible, then, that barn owls and kestrels simply show different powers of arsenic accumulation linked to different excretion rates of arsenic taken up from the diet. Erry et al. (1999) concluded that none of the arsenic burdens found (even in the kestrels from Southwest England) was likely to be associated with adverse health effects in the birds.

4.11.2 London Pigeons

Prologue. London pigeons in the late 1970s contained high concentrations of lead, probably derived from the vehicle combustion of leaded petrol, with apparent symptoms of lead toxicity. Pigeons at Heathrow Airport had high accumulated cadmium concentrations but without apparent ecotoxicological effects.

Also of environmental concern in the 1970s were the potential ecotoxicological effects of the atmospheric fallout of lead from the exhausts of vehicles running on petrol to which tetra-ethyl lead had been added as an antiknock agent, a practice now banned in Britain. Table 4.36 presents data on the accumulation of lead, and two other trace metals cadmium and zinc, in organs of pigeons *Columba livia* collected along a transect from central London to Heathrow Airport in the late 1970s (Hutton and Goodman, 1980).

In the case of lead, there were clear gradients of lead concentrations in the kidneys, livers, brain, bone

and blood of the pigeons, decreasing from central London, westwards to Heathrow Airport (Table 4.36). The tissue lead gradients observed were associated with an increasing degree of lead exposure with

Table 4.35 **Accumulated concentrations (μg/g dry weight) of arsenic in organs of birds of prey in Southwest England and Southwest Scotland between 1970 and 1996.**

Species	Arsenic	
Location		
Organ	*Geometric mean*	*Range*
Sparrowhawk *Accipiter nisus*		
SW England		
Kidney	0.076	0.049–0.507
Liver	0.062	0.039–0.500
Muscle	0.043	0.030–0.208
SW Scotland		
Kidney	0.067	0.049–0.325
Liver	0.060	0.039–0.402
Muscle	0.057	0.030–0.919
Kestrel *Falco tinnunculus*		
SW England		
Kidney	0.278	0.049–1.14
Liver	0.346	0.039–1.41
Muscle	0.187	0.030–3.86
SW Scotland		
Kidney	0.094	0.049–0.252
Liver	0.121	0.039–0.500

Table 4.35 **(cont.)**

Species	Arsenic	
Location		
Organ	Geometric mean	Range
Muscle	0.057	0.030–0.282
Barn owl *Tyto alba*		
SW England		
1970–1996		
Kidney	0.054	0.049–0.142
Liver	0.061	0.039–0.223
Muscle	0.040	0.030–0.155
SW Scotland		
1970–1996		
Kidney	0.068	0.049–0.265
Liver	0.048	0.039–0.194
Muscle	0.048	0.030–0.174

Source: Erry et al. (1999).

proximity to the centre of London. This lead exposure would have been in the form of atmospheric lead present in the city air, derived from vehicle exhausts at the time, and lead in London street dusts contaminating the food of the pigeons (Hutton and Goodman, 1980). Typical background concentrations of lead in the organs of adult birds are 0.5 to 5 µg/g Pb dry weight in the liver, 1 to 10 µg/g Pb dry weight in the kidney and 2 to 15 µg/g Pb dry weight in bone (Scheuhammer, 1987). A lead concentration of 20 µg/g Pb dry weight would indicate acute exposure to lead (Pain et al., 1995). Pigeons and doves apparently have higher accumulated concentrations of lead than many other birds (Franson and Pain, 2011). Specific threshold concentrations of lead in the liver, kidney

and blood have, therefore, been proposed for birds of the pigeon family as indicative of subclinical, clinical and severe clinical lead poisoning (Table 4.37). The lead concentrations accumulated in the organs of the pigeons in the London area in the late 1970s (Table 4.36) clearly exceeded many of these proposed thresholds (Table 4.37). The kidney data indicated severe clinical lead poisoning in the pigeons from Chelsea, clinical lead poisoning at Mortlake and subclinical lead poisoning at Heathrow Airport. Liver data suggested clinical lead poisoning of the pigeons at Chelsea, and subclinical lead poisoning at Mortlake and Heathrow. Blood data indicated subclinical lead poisoning at Chelsea and Mortlake. And the London pigeons did indeed show evidence of kidney and blood toxicity associated with lead poisoning (Hutton, 1980). The symptoms of lead poisoning were detected mainly in the pigeons with the largest lead burdens, these symptoms including increased kidney weight and ultrastructural damage in kidney and liver cells. Also seen was a depression in the activity of the enzyme ALAD in blood, liver and kidney (Hutton, 1980). This enzyme participates in the synthesis of haem needed for haemoglobin and is notoriously sensitive to lead.

Accumulated cadmium concentrations in the kidneys, livers and brains of pigeons also showed differences along the transect (Table 4.36). For cadmium, however, it was the pigeons at Heathrow that had the highest accumulated concentrations, while concentrations in the Chelsea pigeons reached a second lower peak (Table 4.36). As for lead, therefore, there was raised exposure of the pigeons to cadmium in central London, but, in contrast with lead, there was very much increased exposure at Heathrow Airport. While individual pigeons from Heathrow reached the very high kidney and liver concentrations of 302 and 34 µg/g Cd dry weight respectively, there was great variability at this site, with some pigeons with much lower values (see the wide standard errors in Table 4.36) (Hutton and Goodman, 1980). These results do indicate an appreciable degree of cadmium contamination at Heathrow, possibly associated with the release of cadmium with wear of aircraft engines, bodywork and tyres (Hutton and Goodman, 1980).

Table 4.36 **Accumulated concentrations (mean ± SE, variable n) of cadmium, lead and zinc in kidneys, livers, brains, bones (all µg/g dry weight) and blood (µg per 100ml) in feral pigeons *Columba livia* collected in the London area in the late 1970s. Collection sites were at Chelsea (London SW10, 4 kilometres from the centre of London), Mortlake (London SW14, 8 kilometres), Heathrow Airport (18 kilometres) and a control site in rural Cambridgeshire.**

Organ	Cadmium	Lead	Zinc
Collection site			
Kidney			
Chelsea	12 ± 2.1	321 ± 45	169 ± 6
Mortlake	1.52 ± 0.31	49 ± 18	102 ± 13
Heathrow Airport	51 ± 23	9.9 ± 2.6	183 ± 29
Cambridgeshire control	1.75	4.3 ± 1.3	191 ± 19
Liver			
Chelsea	2.45 ± 0.28	22 ± 2.0	147 ± 8
Mortlake	0.40 ± 0.07	10 ± 2.4	79 ± 6.4
Heathrow Airport	9.48 ± 3.15	6.1 ± 1.1	239 ±36
Cambridgeshire control	0.54 ± 0.05	2.0 ± 0.3	204 ± 32
Brain			
Chelsea	0.19 ±0.02	12 ± 1.0	54 ± 2.7
Mortlake	0.14 ± 0.02	6.3 ± 1.1	53 ± 5.5
Heathrow Airport	0.47 ± 0.03	3.3 ± 0.3	61 ± 1.7
Cambridgeshire control	nd	nd	nd
Bone			
Chelsea		669 ± 46	198 ± 5
Mortlake		282 ± 74	153 ± 10
Heathrow Airport		108 ± 28	166 ± 19
Cambridgeshire control		5.7 ± 1.1	158 ± 6
Blood			
Chelsea		101 ± 13	
Mortlake		45 ± 10	

Table 4.36 **(cont.)**

Organ	Cadmium	Lead	Zinc
Collection site			
Heathrow Airport		16 ± 5.5	
Cambridgeshire control		19 ± 2.6	

Note: nd = not determined.
Source: Hutton and Goodman (1980).

Table 4.37 **Proposed threshold concentrations of lead in livers and kidneys (both µg/g dry weight) and blood (µg per 100 ml) in pigeons and doves (*Columbiformes*) that are indicative of subclinical, clinical and severe clinical lead poisoning.**

	Liver	Kidney	Blood
Subclinical Pb poisoning	6	6	20
Clinical Pb poisoning	18	45	200
Severe clinical poisoning	45	90	300

Note: Liver and kidney concentrations have been converted from fresh weight to dry weight by applying a factor of 3.
Source: Franson and Pain (2011).

Most (67 to 97%) of the body burden of accumulated cadmium in birds is stored in the liver and kidneys, bound to metallothionein with a low turnover rate and, therefore, accumulation over time (Scheuhammer, 1987; Wayland and Scheuhammer, 2011). Cadmium concentrations are higher in the kidney, which is the critical organ in chronic cadmium toxicity (Wayland and Scheuhammer, 2011). At high enough cadmium accumulation rates, the rate of metallothionein production will increasingly fail to detoxify the incoming cadmium with subsequent toxic effects. Assuming a wet weight to dry weight tissue ratio of 3, median background concentrations of cadmium in the kidneys and liver of terrestrial birds are 1.2 (range 0.06 to 172) and 0.3 (0.3 to 28) µg/g Cd dry weight respectively (Wayland and Scheuhammer, 2011). Most terrestrial birds have kidney and liver cadmium concentrations below 6 µg/g Cd dry weight, higher values in the ranges above being provided by the red grouse *Lagopus lagopus*, which feeds on willow, known to be a cadmium-accumulating plant (Wayland and Scheuhammer, 2011). Cadmium concentrations in the kidneys and livers of pigeons from both Heathrow and Chelsea (Table 4.36) exceeded the previously quoted median concentrations. Furthermore, cadmium concentrations in both organs of the Heathrow pigeons and in the kidney of Chelsea pigeons were greater than 6 µg/g Cd dry weight (Table 4.36). These accumulated cadmium concentrations are clearly atypically high and indicative of significant exposure of the pigeons to raised bioavailabilities of cadmium. It is more difficult, though, to conclude whether these accumulated cadmium concentrations are indicative of a cadmium ecotoxicological effect. Nevertheless, it has been proposed that an accumulated cadmium concentration of 65 µg/g Cd wet weight (195 µg/g Cd dry weight) in the kidney of adult birds may be associated with a 50% probability of alterations in energy metabolism or structural and functional damage to the kidneys and other organs ((Wayland and Scheuhammer, 2011). A threshold effect level for liver concentrations in adult birds may lie between 135 and 210 µg/g Cd dry weight (Wayland and Scheuhammer, 2011). On these criteria, not even the Heathrow Airport pigeons would have been suffering from cadmium ecotoxicology, in spite of the raised local cadmium bioavailability.

Accumulated zinc concentrations in the pigeons of the London area were all similar across the locations investigated (Table 4.36), indicating physiological regulation of organ concentrations of this essential metal.

4.11.3 Pheasants, Partridge and Lead Shot

It has been shown earlier how birds of prey might be at risk of lead poisoning after ingesting lead shot in the bodies of carrion or wounded birds, but some birds are at risk from the direct ingestion of lead shot. Game birds such as common pheasants *Phasianus colchicus* and grey partridge *Perdix perdix* can pick up lead shot pellets along with grit for their gizzards and consequently suffer from lead poisoning (Fisher et al., 2006).

4.11.4 Biomonitoring and Ecotoxicology

Prologue. Accumulated concentrations of many trace metals in the kidneys, livers, bones and blood of birds can be used to biomonitor metal bioavailabilities to the birds concerned. Feathers and eggs represent non-invasive sources of biomonitoring information.

We have seen that accumulated concentrations of the trace metals arsenic, cadmium, lead and mercury, but not zinc, in the kidneys and livers of particular birds increase with increased local metal bioavailabilities. Concentrations of lead in the blood and bones also rise with increased exposure to lead. Thus, these organs can be used to biomonitor local bioavailabilities of these metals to birds. The use of kidneys, livers and bone to this end is all very well if the dead bodies of birds are available. There is a great advantage, however, in seeking non-destructive methods to monitor metal bioavailabilities to birds, over time or between different geographical locations. Interest has, therefore, centred on the use of feathers as non-destructive biomonitoring tools. The extraction of blood for metal analysis is another possibility. Eggs too have come under investigation, particularly because effects on reproductive success, manifested as

lethal or sublethal effects on the embryo, are often early symptoms of trace metal ecotoxicity in birds.

As for all biomonitors of trace metal bioavailability, the ideal avian biomonitor should have particular preferable attributes (Dmowski, 1999; Hollamby et al., 2006). A suitable avian biomonitor should be abundant, easily identified and sampled, long-lived and large enough for tissue analysis. Its biology should be well known, particularly with respect to its feeding, moulting and reproductive biology, with an understanding of the effects of age and sex on metal accumulation in different tissues. The bird should preferably be non-nomadic and non-migratory at least for the part of the lifecycle when sampling occurs, so that results can be associated with a particular geographical area. Any territorial behaviour also needs appreciation. The species should be widespread in order to allow metal accumulation data to be compared over different geographical areas. The proposed biomonitor should be hardy, and not sensitive to high bioavailabilities of the toxic metals of interest. Of particular importance in the case of birds is that there should be public and regulatory acceptance of the proposed species as a biomonitor and of the methods of sampling (Hollamby et al., 2006). Widespread common species that meet many of these criteria and have been used as biomonitors include feral pigeons *Columba livia*, magpies *Pica pica*, great tits *Parus major*, blue tits *Cyanistes caruleus*, starlings *Sturnus vulgaris*, blackbirds *Turdus merula* and house sparrows *Passer domesticus*. Tissues of birds of prey such as sparrowhawks *Accipiter nisus*, kestrels *Falco tinnunculus*, buzzards *Buteo buteo* and barn owls *Tyto alba* have also been analysed in metal biomonitoring programmes, but these tissues are collected on an adventitious basis when bodies of recently dead birds are found (Tables 4.33 to 4.35).

4.11.4.1 Feathers

Prologue. Feathers represent very useful non-destructive tools for trace metal biomonitoring, their value increased by a knowledge of moulting patterns of the bird species concerned. Museum specimens can provide historical data, particularly for mercury.

Most birds moult annually, usually after the breeding season (Furness, 2015). While the new feathers are growing, they have an active blood supply, and trace metals are commonly transferred from other organs in the body to the newly growing feathers (Burger, 1993). In the new feathers, the trace metals bind with keratin, the major protein constituent of feathers (as indeed of hair in the case of mammals). Keratin has a high content of amino acids that contain sulphur, a favoured binding site of many trace metals. The accumulated metal content of a feather reflects the amount of metal accumulated by the bird in the period between moults, typically the last year. The correlation between concentrations of trace metals in the feathers and in other organs is usually high for most trace metals, although most data are from seabirds, discussed later in this book (Burger, 1993). The proportion of body metal that is contained in the feathers does vary with the metal, from a high of 90% for mercury, through 40 to 60% for lead and nickel, 30 to 50% for manganese to less than 30% for cadmium, copper, iron and zinc (Burger, 1993).

The dynamics of the movement of trace metals from internal organs such as the liver and kidney of birds via the blood to the new feathers is best known for mercury (Burger, 1993; Furness, 2015). Even if mercury is released into the environment as inorganic mercury, it can be transformed by microbial action to the organic form methyl mercury, typically in sediments with low oxygen, low pH and high concentrations of dissolved organic matter promoting the growth of the sulphate-reducing bacteria that are largely responsible for mercury methylation. Methyl mercury biomagnifies up food chains, and almost all mercury in vertebrates, especially in top predators, is in the form of methyl mercury (Furness, 2015). Birds and mammals can excrete mercury because it binds strongly to keratin, in hair in the case of mammals, but in feathers in the case of birds. In addition, female birds can transfer mercury to developing eggs, which are also rich in proteins with sulphur-containing amino acids, potentially with ecotoxicological results in the egg (Furness, 2015). Birds have mercury concentrations in the blood in equilibrium with mercury

levels in the diet, and, between moults, mercury builds up in the bird's internal organs, such as the liver, kidneys and even muscle. At the moult, much of the accumulated mercury is transported to the newly growing feathers, and, by the end of the moult, most of the mercury is now in the feathers (Furness, 2015). It is also the case that the first feathers to be regrown (logically the first to be moulted) contain the highest concentration of mercury, with decreasing mercury concentrations in later feathers, in the order of formation during the moult (Burger, 1993; Furness, 2015).

Feathers, therefore, represent very useful non-destructive biomonitoring tools. The amount of metals transferred to the feathers reflect the amount accumulated previously over a known period by the bird. This amount in turn is affected by the local metal bioavailability, particularly in the diet of the bird concerned. Nevertheless, it remains necessary to take into account the biology of metal accumulation in the feathers, not least the moult pattern and the effects of age and gender, and sample accordingly. It is also the case that different parts of feathers may contain different accumulated metal concentrations, not least because the mass of a feather may vary along the feather axis (Burger, 1993; Bortolotti, 2010).

Another factor to be considered is that metals may be deposited onto the outside of feathers after their growth, the amount of such external metal not necessarily being in direct correlation with the amount of a metal previously accumulated by the bird (Dmowski, 1999). The strength of the cleaning agent used, with or without the application of ultrasound, does range from simple double-distilled water, through relatively mild sodium hydroxide and detergents, to organic solvents such as acetone, hexane and ether (Dmowski, 1999; Scheifler et al., 2006). The importance of the contribution of external metal on a feather to its total metal concentration will vary between metals, not least according to the strength of its accumulation within the feather. Thus any contribution of external mercury on a feather to the total mercury concentration is usually low (Dmowski, 1999). Yet, a study of lead in tail feathers of blackbirds *Turdus merula* from both rural and urban

Table 4.38 **Mean ($\pm$ SD) concentrations of lead in breast feathers, unwashed and washed (0.25 M sodium hydroxide) outermost tail feathers (μg/g dry weight) and blood (μg per 100 ml) in blackbirds *Turdus merula* from rural and urban sites in eastern France in 1999–2000.**

| | Breast feathers | Outermost tail feathers | | Blood |
		Unwashed	Washed	
Rural	1.37 ± 1.07	1.82 ± 1.14	0.73 ± 0.65	5 ± 10
Urban	3.15 ± 1.77	7.75 ± 4.50	4.89 ± 3.18	15 ± 9

Source: Scheifler et al. (2006).

sites in eastern France used washing with a relatively mild solution of sodium hydroxide to show that 37% of the lead in these feathers from urban birds was from external contamination (Table 4.38) (Scheifler et al., 2006). Another complicating factor is that some of the metal found externally on a feather may have an internal origin, for metals may be present in the secretions of the preening gland, which produces oil used by birds for waterproofing.

The first use of feathers for the biomonitoring of trace metal bioavailabilities involved mercury, specifically in response to a need to monitor changes in mercury pollution deriving from the use of methyl mercury as a seed dressing in the 1950s and 1960s (Furness, 2015). Methyl mercury from this source was being accumulated by granivorous birds, such as house sparrows and pigeons, before trophic transfer to birds of prey. Another advantage of the use of feathers as biomonitoring tools soon became apparent. The concentrations of mercury in recently collected feathers could be compared against concentrations in feathers of historic taxidermy specimens in museums, providing a time context (Furness, 2015). A small proviso, however, might need to be taken into

account, because sometimes mercury, in the form of inorganic mercuric chloride, was used as a preservative of taxidermy specimens. This potential problem is easily overcome by analysing inorganic and organic mercury separately in the feathers, all methyl mercury being of internal origin after accumulation by the living bird.

Table 4.39 illustrates two biomonitoring studies using the outermost tail feathers of great tits *Parus major* to monitor the bioavailabilities of several trace metals along a metal pollution gradient of up to 8.5 kilometres away from a metal industrial plant (Union Minière, now called Umicore) near Antwerp in Belgium. This metal plant is involved in the refining and recycling of many trace metals and in the manufacture of specialised products from these metals. The concentrations of all the metals analysed in the feathers increased significantly (usually many-fold) with proximity to the plant in both studies (Table 4.39). There was no significant effect of age nor gender of the birds on the accumulated concentrations in the feathers (Janssens et al., 2001). Accumulated concentrations were also comparable between the two studies. Feather concentrations of arsenic, cadmium, copper, lead, mercury, nickel and zinc in these great tits (Table 4.39) are amongst the highest recorded for birds, indicating extremely high toxic metal exposure of the birds near the metal plant (Janssens et al., 2001). Indeed, these great tits did show symptoms of metal ecotoxicity, particularly as regards blood characteristics and egg hatching success (Janssens et al., 2003; Geens et al., 2010).

The study of lead in blackbirds *Turdus merula* from rural and urban sites in eastern France showed that lead concentrations in the outermost tail feathers and breast feathers were significantly higher in the urban birds (Table 4.38). The lead concentrations of the washed outermost tail feathers of the urban blackbirds were raised to a mean of nearly 5 μg/g Pb, compared to less than 1 μg/g Pb in rural birds (Table 4.38). These increased lead concentrations accumulated in the urban blackbirds resulted from increased lead concentrations in their earthworm prey (Scheifler et al., 2006). These feather concentrations are, nevertheless, far below lead concentrations (up to 231 μg/g) in

Table 4.39 **Mean ($\pm$ SE, n = 10 – 69) concentrations of selected trace metals (µg/g dry weight) in the outermost tail feathers of great tits** *Parus major*, **along a metal pollution gradient up to 8.5 kilometres away from a metal industrial plant (Union Minière) near Antwerp, Belgium, and at a control site (Brasschaat), in 1998 to 1999[1] and in 2006.[2]**

Metal	Union Minière	F8	F7	University Antwerp	F4	Brasschaat
Dates	*(0–0.35 km)*	*(0.4–0.6 km)*	*(2.5 km)*	*(4 km)*	*(8.5 km)*	*(control 20 km)*
Arsenic						
1998–1999	23 ± 2.8	12 ± 1.1	7.3 ± 2.9	0.88 ± 0.16	nm	0.96 ± 0.24
2006	31 ± 3	21 ± 3	2.6 ± 0.2	nm	0.51 ± 0.09	nm
Cadmium						
1998–1999	9.3 ± 1.0	5.3 ± 0.6	4.6 ± 1.7	1.5 ± 0.2	nm	0.6 ± 0.2
2006	11 ± 3	17 ± 6	0.39 ± 0.04	nm	0.19 ± 0.08	nm
Cobalt						
1998–1999	0.66 ± 0.17	0.13 ± 0.03	0.27 ± 0.08	0.02 ± 0.03	nm	0.04 ± 0.02
2006	0.27 ± 0.05	0.25 ± 0.07	0.02 ± 0.01	nm	0.08 ± 0.08	nm
Copper						
1998–1999	55 ± 8.3	29 ± 3.4	21 ± 8.2	7.0 ± 0.5	nm	6.5 ± 1.4
2006	37 ± 4	37 ± 5.0	9.8 ± 0.9	nm	9.4 ± 0.4	nm
Lead						
1998–1999	231 ± 31	122 ± 11	55 ± 14.7	15 ± 1.6	nm	8.1 ± 1.3
2006	140 ± 18	110 ± 19	8.1 ± 1.3	nm	4.1 ± 0.4	nm
Mercury						
1998–1999	3.13 ± 0.64	1.82 ± 0.31	2.88 ± 0.48	0.24 ± 0.11	nm	0.84 ± 0.34
Nickel						
1998–1999	2.78 ± 0.34	1.28 ± 0.09	1.50 ± 0.36	0.84 ± 0.05	nm	0.31 ± 0.17
2006	13 ± 3.0	14 ± 6.0	7.2 ± 2.2	nm	5.7 ± 1.1	nm

Table 4.39 **(cont.)**

Metal	Union Minière	F8	F7	University Antwerp	F4	Brasschaat
Dates	*(0–0.35 km)*	*(0.4–0.6 km)*	*(2.5 km)*	*(4 km)*	*(8.5 km)*	*(control 20 km)*
Silver						
1998–1999	3.6 ± 0.6	1.7 ± 0.3	1.0 ± 0.7	0.09 ± 0.04	nm	0.13 ± 0.07
2006	3.9 ± 0.5	2.6 ± 0.5	0.11 ± 0.02	nm	0.16 ± 0.04	nm
Zinc						
1998–1999	264 ± 64	122 ± 10	182 ± 29	126 ± 9	nm	120 ± 11
2006	240 ± 26	216 ± 25	142 ± 13	nm	184 ± 24	nm

Note: Feathers were washed twice in deionised water and 1 M acetone. nm = not measured.
Sources: Janssens et al. (2001)[1]; Geens et al. (2010).[2]

equivalent feathers in the great tits affected by atmospheric lead fallout near the metal industrial plant at Antwerp (Table 4.39). Correspondingly, the accumulated lead concentrations in the earthworm prey of the urban blackbirds (up to a mean of 2.6 µg/g Pb dry weight) were only about double those in the rural earthworms (Scheifler et al., 2006) and orders of magnitude short of the accumulated lead concentrations of earthworms affected by lead mining (up to nearly 8,000 µg/g Pb) (Table 4.25).

Trace metal concentrations have also been measured in the feathers of European birds of prey, not least as a result of interest in the biomagnification of methyl mercury up food chains (Lodenius and Solonen, 2013). After allowance for any external deposition, the trace metals in feathers are ultimately derived from the diet. It is not surprising, then, that feather metal concentrations in the birds of prey that hunt small mammals reach higher values than those in raptors taking other birds, even in unpolluted situations (Table 4.40). Of the birds of prey listed in Table 4.40, it is the barn owl that has the highest feather concentrations of cadmium, copper, lead, mercury and zinc.

The mercury data in Table 4.40 include feather concentrations in museum specimens from the nineteenth century and concentrations from most decades of the twentieth century (Lodenius and Solonen, 2013). The data available for barn owls, tawny owls and buzzards confirm that the highest feather mercury concentrations are from Swedish birds collected in 1944 to 1965 (32 µg/g), 1965 to 1966 (6.1 µg/g) and 1965 to 1966 (16 µg/g) respectively. Lower feather concentrations of mercury were present in barn owls collected in Belgium in 2001 (0.8 µg/g) and in Sweden between 1921 and 1943 (1 µg/g). In tawny owls, the lowest mercury feather concentrations (1.7 µg/g) were from Swedish birds collected in 1862. In buzzards, Swedish birds collected in in 1861 had mercury feather concentrations of only 4 µg/g. Thus the concentrations of mercury in the feathers of these birds of prey show the now expected raised bioavailability of mercury to birds in the middle of the twentieth century, when both industrial and agricultural uses of mercury were at a high (Newton et al., 1993). Feather concentrations of mercury between 5 and 40 µg/g have been linked to impaired reproductive ability (Wolfe et al., 1998), suggesting that at least the barn owls and buzzards collected in the 1950s and 1960s might have been suffering ecotoxicological effects from the accumulated mercury.

Table 4.40 **Ranges of average concentrations (μg/g dry weight) of cadmium, copper, lead, mercury and zinc in feathers of adult birds of prey (categorised as mammal-eaters or bird-eaters) from unpolluted areas in Europe, including Belgium, Finland, Germany, Italy, the Netherlands, Spain and Sweden.**

Diet Species	Cadmium	Copper	Lead	Mercury	Zinc
Mammal-eaters					
Kestrel *Falco tinnunculus*				1.9	
Buzzard *Buteo buteo*	0.03–0.15	6.1–9.2	1.0–1.5	2.4–16	140
Barn owl *Tyto alba*	0.08–2.3	5–27	7–76	0.8–32	62–363
Tawny owl *Strix aluco*	0.05	4.4	0.2	1.7–6.1	120
Little owl *Athene noctua*	0.05–0.09	7.5	2–4.6	0.23	31
Bird-eaters					
Peregrine falcon *Falco peregrinus*	0.01	8.1	0.1	3.5–49	47
Sparrowhawk *Accipiter nisus*	0.01–0.17	4.8–12	0.3–4.3	0.7–8.4	35–130

Source: Lodenius and Solonen (2013).

The dark pigment melanin in feathers can bind zinc (Châtelain et al., 2014). Correspondingly, feral pigeons that are more melanic can bind more zinc in their feathers than their less dark counterparts. Dark pigeons do well in urbanised environments, and it has been suggested that an increased ability to store zinc in the feathers may be of selective advantage in urban environments (Châtelain et al., 2014).

4.11.4.2 Blood

Prologue. Concentrations of non-essential trace metals in the blood of birds reflect recent dietary inputs of these trace metals.

Trace metal concentrations in the blood of birds are affected by the bioavailabilities of the metals in the diet, the metals taken up in the alimentary tract being passed in the blood to internal organs such as the liver, kidney and bone for storage. Any subsequent transfer to feathers will occur at the next moult. Trace metal concentrations in the blood therefore have the potential to reflect recent, short-term bioavailabilities of the metals in the diet, while concentrations in organs and feathers reflect bioavailabilities over longer periods. Analysis of trace metal concentrations in the blood of birds is, therefore, a useful potential tool to biomonitor short-term changes in metal bioavailabilities. It does, though, have the drawback of the difficulty of measuring what are usually very low metal levels in contrast to those in organs and feathers.

A dataset of the concentrations of trace metals in the blood of birds is offered by the great tits *Parus major* collected along the metal pollution gradient away from the metal industrial plant near Antwerp in Belgium (Table 4.41) (Geens et al., 2010). Blood concentrations of cadmium and lead differed significantly between sites, being raised closest to the Union Minière metal plant (Table 4.41). On the other hand, the blood concentrations of cobalt, copper, nickel and zinc did not vary along the metal pollution gradient (Table 4.41). These conclusions contrast with those for the feather concentrations (Table 4.39), which differed significantly along the gradient for all metals (Geens et al., 2010). There was also a significant correlation between the blood and feather concentrations of cadmium and lead, reflecting the fact that they increase together on approach to the metal plant. The probable reason for the different conclusions for blood and feathers is that there may be physiological control of some metal concentrations in the blood, particularly for essential metals (Geens et al., 2010). While these blood metal concentrations are held relatively constant, the rate of turnover from blood to storage organ (and ultimately to feathers at the time of moult) will be increased to deliver more metal for accumulated storage. Another contributory factor may be the external metal contamination of feathers, adding a source of metal independent of the metal taken up by the birds from the diet. As in the case of the feather metal concentrations, the blood metal concentrations were unaffected by age or gender of the birds.

For the moment, anyway, it may be best to conclude that the use of blood concentrations for biomonitoring should be restricted to non-essential metals such as cadmium, lead or mercury, until physiological regulation of blood concentrations of the metal of interest can be ruled out.

The great tits did show ecotoxicological symptoms, presumed to result from the extremely high toxic metal exposure of the birds near the metal plant (Janssens et al., 2001; Geens et al., 2010). In the blood, the haemoglobin concentration, the volume percentage of red blood cells, the mean volume of red blood cells and the mean amount of haemoglobin in each red blood cell were all lower in the great tits near the metal plant (Geens et al., 2010). Such a weakened haematological status can have serious implications for the fitness and breeding capacity of these great tits. All these blood parameters were significantly negatively correlated with the lead concentrations in the blood (Geens et al., 2010). It must be remembered, though, that such a correlation does not necessarily prove that it is the lead that is the primary or only negative ecotoxicological agent. Nevertheless, the great tits nearest the metal plant did have average

Table 4.41 **Mean ($\pm$ SE, *n* = 10 – 69) concentrations of selected trace metals (µg per 100 ml) in the blood of great tits *Parus major* along a metal pollution gradient up to 8.5 kilometres from a metal industrial plant (Union Minière) near Antwerp, Belgium in 2006.**

Metal	Union Minière	F8	F7	F4
	(0–0.35 km)	(0.4–0.6km)	(2.5 km)	(8.5 km)
Cadmium	1.6 ± 0.3	1.1 ± 0.2	0.7 ± 0.1	0.7 ± 0.1
Cobalt	35 ± 8	25 ± 5	25 ± 4	23 ± 7
Copper	13 ± 2	9 ± 2	20 ± 3	13 ± 3
Lead	28 ± 2	17 ± 2	3 ± 1	2 ± 2
Nickel	11 ± 3	14 ± 3	13 ± 5	18 ± 3
Zinc	5.5 ± 0.5	6.9 ± 0.5	7.9 ± 0.7	6.9 ± 0.7

Source: Geens et al. (2010).

blood lead concentrations (28 µg per 100 ml) (Table 4.41), above the threshold of 20 µg Pb per 100 ml considered to be associated with symptoms of subclinical lead poisoning in pigeons (Franson and Pain, 2011). Furthermore, lead exposure is known to affect the activity of the enzyme ALAD, and this enzyme is involved in the synthesis of haem needed for haemoglobin. It is not unlikely, therefore, that the high bioavailability of lead near the plant is causing the observed negative blood symptoms in the local great tits (Geens et al., 2010).

Mean lead concentrations in the blood of urban blackbirds *Turdus merula* in eastern France were raised to 15 µg Pb per 100 ml, in comparison to their rural counterparts (5 µg Pb per 100 ml) (Table 4.38). This raised blood lead concentration is still below the threshold of 20 µg Pb per 100 ml considered to be associated with symptoms of subclinical lead poisoning in pigeons. Understandably, there was no correlation between lead concentrations in the blood and the body condition indices of the blackbirds studied (Scheifler et al., 2006).

4.11.4.3 Eggs

Prologue. Eggs are easily sampled biomonitoring tools, of particular interest because reduced reproductive success is an early symptom of trace metal ecotoxicology in birds.

The eggs of birds are attractive, non-destructive biomonitoring agents because they are easily sampled without the stress of handling a bird and they are formed in a specific period (Dauwe et al., 1999). It is easy to check whether there is variation in the trace metal concentration of an egg with order of laying, and the removal of one egg from a nest usually has only a minor effect on population parameters (Dauwe et al., 1999, 2005a). Furthermore, accumulated metal concentrations in eggs are of special interest, because reduced reproductive success, as a decrease in success of hatching, is often an early symptom of trace metal ecotoxicology in birds.

For data, we can return to the great tits *Parus major* along the metal pollution gradient away from the metal industrial plant (Union Minière) near Antwerp, supplemented by data for blue tits *Cyanistes caruleus* (Table 4.42). Both great tits and blue tits are insectivorous during the breeding season and are at a high trophic level in local food chains through which trace metals are being transferred. Table 4.42 presents metal concentrations in both egg contents and eggshells for the two species near the metal plant, and for great tits 4 kilometres away at the University of Antwerp. In great tits, metal concentrations in the egg contents differed significantly between the two sites only for lead, with a 15-fold increase in concentration at the metal plant site (Table 4.42). Metal concentrations in the eggshells of the great tits were raised significantly at the plant site for arsenic and cadmium, as well as lead (Table 4.42). The egg contents of the great tits had higher copper and zinc

concentrations than the eggshells at both sites, but, at the metal-polluted site, the eggshells had significantly higher arsenic and lead concentrations than the egg contents (Table 4.42). Similarly, in the blue tits at the metal-polluted site, it was again the egg contents that had the higher concentrations of copper and zinc, and the eggshells the higher concentrations of arsenic and lead (Table 4.42). There were no interspecific differences in the concentrations of any of the metals in either egg contents or eggshells at the polluted site where both species occurred (Table 4.42).

It can be concluded that eggs (of at least great tits and blue tits) can be used for the biomonitoring of arsenic, cadmium and lead bioavailabilities to terrestrial birds. Eggshells in particular had the advantage over egg contents, because they showed up the geographical variation in arsenic and cadmium bioavailabilities, in addition to that in lead bioavailability in this study. Furthermore, eggshells can be collected after hatching. The concentrations of copper and zinc in the egg contents appear to be physiologically regulated (Table 4.42), ruling out any role in biomonitoring differences in copper or zinc bioavailabilities between sites. Nevertheless, the measurement of the concentrations of some trace metals in the egg contents can be informative. It is in the egg contents that any ecotoxicological effect of an accumulated trace metal on embryo survival will occur. Indeed, the 15-fold increase in the lead concentration of the egg contents of the great tits at the metal plant site (Table 4.42) does draw attention to the likelihood of lead being implicated in any ecotoxicological effects on reproductive success in this great tit population, as was concluded previously for its possible role as a causative factor in the detrimental blood symptoms in these great tits.

Measures of reproductive success have been made on the populations of great tits and blue tits along this metal pollution gradient in Belgium (Janssens et al., 2003; Dauwe et al., 2005b). The breeding performance of great tits was studied in 1998, 1999 and 2000 at four study sites up to 4 kilometres along the metal pollution gradient (sites Union Minière, F8, F7 and University of Antwerp in Table 4.39) (Janssens et al.,

Table 4.42 **Mean ($\pm$ SE, n = 5 – 12) concentrations (μg/g dry weight) of selected trace metals in the eggs (divided into egg contents and eggshells) of great tits *Parus major* and blue tits *Cyanistes caruleus* collected from the region of a metal industrial plant (Union Minière) near Antwerp, Belgium, and (great tits only) a control site 4 kilometres away (University of Antwerp) in 1998.**

Species	Arsenic	Cadmium	Copper	Lead	Zinc
Egg part					
Site					
Great tit					
Egg contents					
Union Minière	0.45 $\pm$ 0.25	0.8 $\pm$ 0.6	5.5 $\pm$ 0.7	2.0 $\pm$ 0.4	62 $\pm$ 3
University of Antwerp	0.22 $\pm$ 0.14	0.05 $\pm$ 0.01	4.8 $\pm$ 0.8	0.13 $\pm$ 0.05	69 $\pm$ 13
Eggshell					
Union Minière	4.2 $\pm$ 0.8	0.31 $\pm$ 0.08	3.2 $\pm$ 0.5	15 $\pm$ 4	28 $\pm$ 5
University of Antwerp	1.2 $\pm$ 0.6	0.08 $\pm$ 0.02	1.7 $\pm$ 0.2	0.37 $\pm$ 0.16	19 $\pm$ 6
Blue tit					
Egg contents					
Union Minière	0.33 $\pm$ 0.13	0.17 $\pm$ 0.14	5.4 $\pm$ 0.7	2.2 $\pm$ 0.6	60 $\pm$ 8
Eggshell					
Union Minière	3.7 $\pm$ 1.2	0.15 $\pm$ 0.02	2.8 $\pm$ 1.1	7.4 $\pm$ 1.1	32 $\pm$ 8

Source: Dauwe et al. (1999).

2003). Hatching success of the great tits was significantly reduced at the two most polluted sites. Similarly, the success of fledging (the rearing of a young bird until it is ready to fly) also significantly decreased at the two sites nearest the metal plant. Together, the reductions in hatching success and fledging success caused a reduction in overall breeding success at the two most polluted sites. Furthermore, there was also a significantly higher incidence of laying interruptions near the pollution source. Other reproductive parameters were unaffected by the metal pollution. The onset of laying, clutch size, length of incubation period and the proportion of nests with low quality eggs all showed no significant difference between the sites.

A parallel study of reproductive success in blue tits from the same four study sites, this time in 1999, 2000 and 2001, however, showed contrasting results (Dauwe et al., 2005b). The only effect of the metal pollution was a higher proportion of females with egg laying interruptions at the two sites closest to the metal plant. There were differences in the hatching success of the blue tits between sites, but these could not be related to the metal pollution gradient. The start of egg laying, clutch size and, importantly, hatching and fledging success did not differ significantly amongst sites. Thus, there was no significant effect of trace metal pollution on the overall breeding success of the blue tits. The reproductive success of the blue tits was better in all study sites than that of the great tits, The great tits were clearly more sensitive to trace metal exposure than the blue tits at the same sites, highlighting the difficulty of generalising between species when it comes to predicting any ecotoxicological effects of trace metal exposure (Dauwe et al., 2005b).

In the case of the eggs of birds of prey, mercury concentrations have been measured in the egg contents of addled or deserted eggs of peregrine falcons *Falco peregrinus* collected between 1963 and 1986 in Britain (Newton et al., 1989). The eggs were subdivided into the two categories of inland sites and coastal sites. Inland site data were also separated out for the two time periods 1971 to 1974 and 1981 to 1986. In the eggs from the inland sites, the mean mercury concentrations decreased significantly between these two periods, from 0.59 µg/g Hg dry weight in 1971 to 1974 to 0.08 µg/g Hg in 1981 to 1986. Such a fall is in line with the previously highlighted decreases in both industrial and agricultural uses of mercury in Britain from the middle of the twentieth century. The peregrine falcon eggs from coastal sites had significantly higher mercury concentrations than those from inland – 1.27 µg/g Hg as opposed to 0.21 µg/g Hg in eggs pooled over both periods. The brood sizes of a subsample of the peregrine falcons decreased in correlation with increasing mercury concentrations in the eggs. It could not be concluded, however, that this brood size reduction was a direct effect of the mercury, because the egg mercury concentrations correlated with the concentrations of the organochlorines dichlorodiphenyldichloroethylene (DDE, a breakdown product of dichlorodiphenyltrichloroethane {DDT}) and polychlorinated biphenyls (PCBs) in the eggs. DDE causes eggshell thinning in birds of prey, and the egg concentrations of DDE in this study correlated negatively with shell indices representing measures of shell thickness. Nevertheless, the egg concentrations of mercury and DDE together best explained the variance in brood sizes of the peregrines, suggesting an ecotoxicological role for mercury in affecting the reproduction of the peregrine falcons before the 1980s.

We can again ask the question whether the concentration of a trace metal in an organ of a bird, in this case mercury in eggs, can be correlated with an ecotoxicological effect. It has been proposed that reproduction in nonmarine birds may be impaired if the mercury concentration in the egg exceeds 0.6 µg/g Hg wet weight (Shore et al., 2011), equivalent to 3 µg/g Hg dry weight assuming a wet weight to dry weight ratio of five. Of 106 inland peregrine falcon eggs in the study, 19 had mercury concentrations above 1 µg/g Hg dry weight, 3 above 2 µg/g Hg dry weight, but none above 3 µg/g Hg dry weight (Newton et al., 1989). Of the 22 coastal peregrine falcon eggs, 12 eggs had mercury concentrations above 1 µg/g Hg dry weight, 7 above 2 µg/g Hg dry weight and 6 above 3 µg/g Hg dry weight. The highest of these concentrations was 6.4 µg/g Hg dry weight in a coastal egg from Wester Ross in Scotland in 1973 (Newton et al., 1989). It would appear, therefore, that any ecotoxicological effect of mercury in the eggs would have occurred in the coastal peregrines, before mercury bioavailabilities dropped later in the twentieth century.

The higher mercury concentrations in the eggs of the coastal peregrine falcons might be attributable to greater mercury contamination of estuarine and marine environments compared to inland environments. It is more likely, however, to be an effect of the different natural diets of coastal and inland peregrine falcons (Newton et al., 1989). Coastal peregrines take waders and seabirds (particularly auks) rather than the pigeons, grouse and passerines of their inland counterparts. And, as we shall see in later chapters, seabirds typically have higher accumulated mercury concentrations than inland birds.

4.12 Environmental Effects: Community Ecotoxicology

Prologue. Several terrestrial habitats in the British Isles have been affected by increased bioavailabilities of trace metals derived from mining activities, or from the atmospheric fallout of metals emitted from smelting works. Organisms at all trophic levels in local food webs have taken up and accumulated atypically high concentrations of these potentially toxic metals. Particular ecotoxicological effects have been recognised above in specific organisms in these metal-contaminated habitats. This section will consider such ecotoxicological effects at the community level, attempting to address the 'so what?' question. What has been the effect, if any, of such high bioavailabilities of toxic metals on the structure and function of the ecological communities affected?

Sites in the immediate vicinity of the mines themselves, typically with attendant spoil heaps, or of smelters are almost inevitably industrial barrens with high phytotoxic bioavailabilities of trace metals in the local soil. Such sites, in effect, lack the necessary topsoil to support any vegetation and a consequent ecological community. This section, therefore concentrates on metal-affected habitats with vegetation, however limited in biodiversity, to answer the question posed in the prologue.

4.12.1 Avonmouth Smelter

Prologue. There were clear ecotoxicological effects of high trace metal bioavailabilities on the biological communities in woodland habitats downwind of the Avonmouth smelting works in the 1970s and 1980s.

The region of Avonmouth, near Bristol, has been used for smelting since Roman times, but it was the 1920s and thereafter that saw the development there of very large zinc and lead smelting works. By the 1970s, it was estimated that about 50 tonnes of zinc, 30 tonnes of lead and 3 tonnes of cadmium were emitted each year in fine particulate form from the Avonmouth chimneys (Hopkin, 1989). Raised metal concentrations in local plants and animals were detectable in woodlands up to 25 kilometres downwind (Hopkin, 1989). Metal production from the Avonmouth smelting works ceased in the 1970s. Much of the relevant research on the effects of the toxic metals derived from Avonmouth was carried out in the 1970s and 1980s. Since trace metals can remain in soils and leaf litter for many years after deposition, passing then into the plants and animals of local food webs, it is by no means certain that any ecotoxicological effects on local organisms have by now passed into history.

Many details of the atypically high accumulation by local biota of the trace metals derived from the Avonmouth smelting works have populated earlier sections of this chapter. Furthermore, several ecotoxicological effects or indicators have been recognised at biological levels below the community. Thus a population of the grass Yorkshire fog *Holcus lanatus* from a site 3 kilometres downwind of the smelter site showed tolerance to cadmium (Coughtrey and Martin, 1977), indicating the presence there of an ecotoxicologically significant bioavailability of cadmium to this plant. The very high bioavailabilities of zinc in local leaf litter were considered to be lethal to the woodlouse *Porcellio scaber* very close to the Avonmouth smelter, where litter zinc concentrations exceeded 5,000 $\mu g/g$ (Hopkin and Hames, 1994). Surprisingly, though, the energy reserves in this woodlouse showed no significant decrease with decreasing distance to the smelter (Schill and Köhler, 2004). On the other hand, another woodlouse, *Oniscus asellus*, from the same Avonmouth sites, did show a significant massive reduction in energy reserves with decreasing distance to the smelting works and with increased metal concentrations in the leaf litter (Schill and Köhler, 2004). In contrast to *P. scaber*, *O. asellus* was, therefore, expending significant energy to cope with the high metal input approaching the smelting works. *Porcellio scaber* was not completely unaffected, however, for this woodlouse did show a significant reduction in average size with site proximity to the smelter, probably as a combined result of a reduction in growth rate and increased mortality eliminating older woodlice (Jones and Hopkin, 1998).

At the community level, the most obvious ecotoxicological effect of the metal fallout from Avonmouth was the presence of a considerable layer of undecomposed leaf material on the surface of the soil at sites such as Blaise and Hallen Woods, each about 3 kilometres downwind (Table 4.43) (Hutton, 1984). Indeed, the standing crop of leaf litter in local woods decreased with distance away from Avonmouth, in association with falling leaf litter metal concentrations and decreasing acidification (Table 4.43) (Coughtrey et al., 1979). Seventy-seven per cent of the variation in the amount of leaf litter present could be accounted for by the litter concentrations of zinc and

Table 4.43 **Mean ($\pm$ SE, n = 5 – 12) standing crops of leaf litter (g per square metre), litter concentrations of selected metals (µg/g dry weight), and pH in deciduous woodlands at increasing distances (km) from the Avonmouth smelter works in 1977.**

Wood	Distance from smelter	Litter mass	Cadmium	Copper	Lead	Zinc	pH
Moorgrove	2.5	8,345 $\pm$ 2,131	23 $\pm$ 2.3	73 $\pm$ 5	1,052 $\pm$ 123	764 $\pm$ 187	4.0 $\pm$ 0.1
Blaise	2.9	13,160 $\pm$ 1,411	32 $\pm$ 2.5	47 $\pm$ 2	721 $\pm$ 12	1,844 $\pm$ 116	6.3 $\pm$ 0.1
Hallen	2.9	8,343 $\pm$ 598	62 $\pm$ 6.1	135 $\pm$ 12	2,179 $\pm$ 272	2,469 $\pm$ 221	3.9 $\pm$ 0.1
Haw	3.1	7,910 $\pm$ 640	98 $\pm$ 8.2	100 $\pm$ 6	1,545 $\pm$ 141	2,814 $\pm$ 730	5.0 $\pm$ 0.1
Leigh	6.8	1,784 $\pm$ 157	7.2 $\pm$ 1.0	16 $\pm$ 1	191 $\pm$ 17	169 $\pm$ 8	5.8 $\pm$ 0.1
Wetmoor	23.0	913 $\pm$ 100	1.5 $\pm$ 0.1	20 $\pm$ 8	44 $\pm$ 3	80 $\pm$ 9	5.5 $\pm$ 0.2
Midger	28.5	3,104 $\pm$ 268	5.7 $\pm$ 0.7	15 $\pm$ 1	103 $\pm$ 13	202 $\pm$ 15	5.7 $\pm$ 0.3

Source: Coughtrey et al. (1979).

cadmium, together with increased acidification. Increased litter accumulation is known from other non-British woodlands affected by trace metal contamination, and results from a decrease in the rate of litter decomposition, caused by reduced microbial activity in the fallen leaves, by the absence of certain litter-dwelling detritivorous invertebrates or by a combination of the two (Coughtrey et al., 1979; Hutton, 1984). Hallen Wood, only about 3 kilometres downwind of Avonmouth, did have an almost complete absence of earthworms and, apparently, millipedes in the early 1980s (Table 4.44). The reduced numbers of millipedes were put down to reduced survival rates of the juveniles (Hopkin, 1989). Changes in the population densities of soil and litter invertebrates in Hallen Wood are selective, for other invertebrate populations appeared unaffected, while

collembolans (and also mites) had higher population densities than in uncontaminated woodlands (Table 4.44).

In the case of earthworms, no species was present at sites within 1 kilometre of the smelting works, and earthworm numbers were still significantly reduced at sites up to 3 kilometres away (Spurgeon and Hopkin, 1996a). Total earthworm density reduced with increased soil and litter concentrations of cadmium, copper, lead and zinc, the concentrations of these metals covarying (Spurgeon and Hopkin, 1996a). The more metal-resistant species *Lumbricus rubellus*, *L. castaneus* and *L. terrestris* were present at sites close to the smelter, as in the case of *L. rubellus* alone in Hallen Wood (Table 4.44). Less metal-resistant earthworm species such as *Aporrectodea rosea*, *Aporrectodea calaginosa* and *Allolobophora chlorotica*

Table 4.44 **Population densities of major groups of invertebrates (numbers of individuals per square metre) in leaf litter and soil from a contaminated deciduous woodland about 3 kilometres downwind of the Avonmouth smelting works (Hallen Wood) and a similar but uncontaminated woodland (Wetmoor Wood) about 23 kilometres downwind, in the early 1980s**

		Hallen Wood	Wetmoor Wood
Higher taxon			
Order			
Family			
Species	Common name		
Oligochaeta	Earthworms		
Lumbricus rubellus		17	3
Lumbricus terrestris		0	29
Aporrectodea longa		0	4
Aporrectodea caliginosa		0	30
Octolasion cyaneum		0	9
Crustacea			
Isopoda	Woodlice		
Oniscus asellus		56	20
Trichoniscus pusillus		0	151
Diplopoda	Millipedes		
Julidae		0	11
Polydesmidae		8	79
Glomeridae		0	21

		Hallen Wood	Wetmoor Wood
Chilopoda	Centipedes		
Lithobiomorpha			
Lithobiidae		112	116
Geophilomorpha		328	263
Collembola	Springtails	20,800	8,690
Insecta			
Coleoptera (adults)	Beetles	902	48
Coleoptera (larvae)	Beetle larvae	120	4
Diptera (larvae)	Fly larvae	4,590	291

Source: Hopkin (1989).

were absent near the smelting works. It was concluded previously that it was sensitivity to zinc that was the factor limiting the distribution of the different earthworms (Figure 4.5) (Spurgeon et al., 1994; Spurgeon and Hopkin, 1996a).

A further study in 1885 to 1986 investigated the invertebrate community sampled by pitfall traps at six woodland sites between 3 and 23 kilometres downwind of the Avonmouth smelting works, with simultaneous measurements of trace metal (Cd, Cu, Pb, Zn) concentrations in soil and leaf litter (Read et al., 1998). Pitfall traps will sample ground-running invertebrates, including millipedes, harvestmen, spiders and carabid ground beetles, but few, if any, earthworms. Perhaps surprisingly, no results were presented for woodlice. Beetle larvae, hymenopterans, pseudoscorpions and gastropod molluscs were more numerous at metal-contaminated sites, while earwigs were more prevalent at the uncontaminated sites. In fact, the numbers of beetle larvae and hymenopterans showed a positive trace metal effect. Most of the hymenopterans were bees, wasps and flying parasitoids rather than ants.

The earlier study had indicated a reduced number of millipedes at the metal-contaminated Hallen Wood (Table 4.44), but the more comprehensive later study did not bear out this conclusion (Read et al., 1998). The millipede *Chordeuma proximum* was present in large numbers at the most metal-contaminated site, and *Glomeris marginata* (the pill millipede) was present at most sites, its population numbers also correlating with litter metal concentrations. The occurrence of the flat-backed millipede *Polydesmus angustus* was also closely related to high metal levels. On the other hand, *Tachypodoiulus niger* (the black millipede or white-legged snake millipede) and *Ophyiulus pilosus* (a widespread and common millipede) occurred at sites away from the effects of high trace metal concentrations (Read et al., 1998).

Amongst the harvestmen collected, there were again species preferring metal-contaminated sites and others found at the non-contaminated sites (Read et al., 1998). Present at the most metal-contaminated sites were *Leiobunum blackwalli* (the numbers of which showed a positive correlation with soil metal concentrations), the usually more common *Leiobunum rotundum* and *Nemastoma bimaculatum*. The main harvestmen species at noncontaminated sites were *Mitostoma chrysomelas* and *Analasmocephalus cambridgii*.

In the case of spiders, there was a smaller number of species at the high metal sites, and a larger potential number to be found at the uncontaminated sites (Read et al., 1998). Box 4.4 provides details of the spiders associated with the woodlands affected by the Avonmouth smelter.

Box 4.4 Spiders and the Avonmouth Smelter

Of the spider species at the metal-contaminated sites downwind of Avonmouth, many were present in only low numbers (Read et al., 1998). Some species, however, were common at the contaminated sites, their numbers showing a positive correlation with metal concentrations. Amongst these species were *Coelotes atropos*, *Anyphaena accentuata*, *Robertus lividus* and two linyphyiid money spiders *Micragus herbigradus* and *Lepthyphantes zimmermanni*.

A large number of lycosid wolf spider species of the genus *Pardosa*, including *P. lugubris*, were to be found at the low metal end of the contamination gradient of sites (Read et al., 1998). Money spiders (*Linyphyiidae*), which use webs to capture their prey, were common across all sites. Wolf spiders (*Lycosidae*) are ground runners which hunt their prey by sight, and they were conspicuous by their absence at the high metal sites. The hunting niche of the wolf spiders may therefore be vacant in the woods highly contaminated by trace metals. There were two species of large ground-running agelenid spiders found that might occupy this niche, *Coelotes atropos* and, in smaller numbers, *Coelotes terrestris*. Indeed, the distribution of *C. atropos* was almost a mirror image of those of the lycosids. *C. atropos* does have a retreat but will forage away from it, in the manner of a wolf spider. The absence of wolf spiders from metal-contaminated habitats may be restricted only to woodland habitats, because lycosids were caught in pitfall traps in metal-contaminated grassland swards close to the Merseyside copper refinery met earlier in this chapter (Hunter et al., 1987b). We also saw earlier that lycosid spiders take up and accumulate cadmium strongly in comparison to another spider (*Dysdera crocata*, the woodlouse spider), and lycosids generally have higher cadmium, copper, lead and zinc body concentrations than web-building spiders such as linyphyiids (Eraly et al., 2011). Furthermore, the adult size and condition of the wolf spider *Pardosa lugubris* (reported as *P. saltans*) from zinc and lead-rich sites in Belgium decreased with accumulated body concentrations of cadmium, indicating an ecotoxicological effect (Eraly et al., 2011). In the wolf spider population that had the highest accumulated concentrations of cadmium and zinc, reproductive output was down and reproduction postponed. It is possible, but by no means proven, that wolf spiders have atypically high trace metal uptake rates in comparison to other spiders, enhancing any risk of trace metal ecotoxicity in metal-contaminated woodlands with appropriate metal-contaminated prey for them to feed on.

Data for carabid ground beetles also identified species found in high metal conditions and those better represented in the uncontaminated woodlands downwind of Avonmouth (Read et al., 1998). The ground beetles particularly associated with high metal concentrations were *Cychrus caraboides*, *Carabus nemoralis*, *Carabus violaceus*, *Metabletus foveatus*, *Paranchus albipes* and *Pterostichus nigrita*. Those more prevalent at the uncontaminated sites were *Carabus granulatus*, *Leistus rufomarginatus*, *Bembidion tetracolum*, *Agonum viduum*, *Agonum obscurum* and *Pterostichus nigrita*. Thus ground beetle species even in the same genus, for example *Carabus*, may have different distributions with respect to metal contamination (Read *et al.*, 1998).

When considering the combined data for all the different taxa collected in this study, it was concluded that the uncontaminated sites had higher species diversities than the metal-contaminated ones (Read et al., 1998). Furthermore, the species diversity of spiders was negatively correlated with the soil concentrations of copper, lead and zinc, but not simply as a result of the loss of the wolf spiders from the most contaminated sites (Read et al., 1998). While it was clear that metal contamination from the Avonmouth smelting works was affecting the woodland invertebrate communities immediately downwind, the ecotoxicological effect on these communities was subtler than a wholesale reduction in the occurrence and numbers of particular macroinvertebrates. As for functional groups, the study also concluded that the metal-contaminated sites were not deficient in

macrodecomposer species. Nevertheless, it must be remembered that data for earthworms were few, and absent in the case of woodlice. It is also the case that neither microarthropods nor microorganisms were surveyed in the study.

As a general conclusion for the woodlands immediately downwind of the Avonmouth smelting works in the 1970s and 1980s, there was a clear ecotoxicological effect of metal deposition on the decomposition rate of leaf litter in the woods. This effect may be based on the negative effects of the trace metal exposure on the microorganisms decomposing the leaf litter, but it has not been ruled out that a reduction of detritivorous activity on the part of earthworms (and woodlice?) may also be significant.

4.12.2 Merseyside Smelter

Prologue. The site of the copper refinery complex on Merseyside was sufficiently contaminated with copper in the 1980s to cause a significant ecotoxicological effect of reduced diversity in the local grassland flora.

Copper smelting took place at St Helens on Merseyside in the late eighteenth century, using ore from Parys Mountain on Anglesey, but it was in the first half of the twentieth century that a large copper refinery complex was developed there. By the 1980s, this complex included a copper fire refinery, a copper electrolytic refinery, a brass foundry and a plant for the production of high-grade copper rod and copper-cadmium alloys (Hunter et al., 1987a). Nevertheless, the second half of the twentieth century brought industrial closures, and electrolytic copper refining ceased on Merseyside in 1991. The copper refinery complex was situated centrally in an urban-industrial conurbation, with nearby amenity grassland and recreational open spaces. It was the grassland sward habitats present up to 1 kilometre away from the refinery perimeter that were chosen in the 1980s for a study of the ecotoxicology of copper and cadmium emitted from the complex. The study sites, therefore, consisted of a copper refinery site, a site 1 kilometre away and a control site (Hunter et al., 1987a).

Results have already been presented in this chapter on copper and cadmium concentrations in soils (Tables 4.3, 4.4), accumulated in grasses (Table 4.17), as well as in herbivorous (Table 4.22), detritivorous (Table 4.24) and predatory (Table 4.28) invertebrates and small mammals (Table 4.31) at these sites. Copper and cadmium concentrations were all well above background concentrations in soil, grasses and invertebrates at the refinery and at the site 1 kilometre away, indicating the presence of atypically raised copper and cadmium bioavailabilities at these sites (Hunter et al., 1987a, b).

Similarly, there were raised accumulated cadmium concentrations in the kidneys, livers and hair of wood mice *Apodemus sylvaticus*, field voles *Microtus agrestis* and common shrews *Sorex araneus* at each of these sites (Table 4.31), again indicative of raised cadmium bioavailabilities to these small mammals (Hunter et al., 1987c). Accumulated copper concentrations in the kidneys and livers of wood mice, and in the livers of field voles, from the two sites appeared to be regulated to an approximately constant level (Table 4.31), but raised copper concentrations in the kidneys of field voles, and the hair of both rodent species, do indicate raised copper bioavailabilities too (Table 4.31). Raised accumulated copper concentrations in livers, kidneys and hair of common shrews from the two sites indicate raised copper bioavailabilities to this small mammal species as well (Table 4.31). In spite of high accumulated cadmium concentrations in their kidneys with associated evidence of apparent ultrastructural damage to kidney cells, and a very high average daily input rate of cadmium of 25 µg per gram per day, the common

shrews from the refinery site appeared to be in good condition, with no apparent symptoms of kidney failure or other signs of cadmium toxicity (Hunter et al., 1987c, 1989). These same common shrews were also apparently coping well with a mean daily dietary input of copper of 260 µg per gram per day at the refinery site.

We now ask the question whether any ecotoxicological effects of cadmium and/or copper at the community level were detected in the grassland habitats in the Merseyside study.

The diversity of the grassland flora was certainly reduced at the refinery site, in this case to only four species: creeping bentgrass *Agrostis stolonifera* and red fescue *Festuca rubra*, which dominated, together with *Equisetum arvense* (field horsetail or common horsetail) and coltsfoot *Tussilago farfara* (Hunter et al., 1987a, c). The control grassland site had a more mixed species composition, but still with high proportions of creeping bentgrass and red fescue. The populations of both creeping bentgrass and red fescue at the refinery site were tolerant to copper but not cadmium (Hunter et al., 1987c; Wu and Bradshaw, 1972), indicating the presence at the refinery site of an ecotoxicologically significant bioavailability of copper to both these grasses. The implication is that it is the raised bioavailability of copper, not of cadmium, that has restricted floral diversity here.

All expected major groups of invertebrates were present at all three Merseyside sites (see Tables 4.22, 4.24, 4.28), although the earthworms and woodlice in particular were less abundant at the refinery site (Hunter et al., 1987b). The abundance and species composition of the other invertebrate groups were remarkably similar between the sites. Detritivorous invertebrates were still abundant at all sites, including earthworms and woodlice in addition to collembolans and millipedes (Hunter et al., 1987b). Considering the previous discussion of the Avonmouth woodland sites, it is noteworthy that millipedes were present at all three Merseyside sites, in the form of the julid *Tachypodoiulus niger* and a polydesmid species of the genus *Oxidus* (Table 4.24). In contrast, at Avonmouth, *T. niger* occurred at sites away from the effects of high trace metal concentrations (Read et al., 1998). Distribution data for spiders at the Merseyside sites also contrasted with the Avonmouth conclusions, particularly in the case of wolf spiders. The most abundant spiders at the Merseyside sites were indeed lycosid wolf spiders, as well as linyphyiid money spiders (Hunter et al., 1987c). Wolf spiders had been absent from metal-contaminated woodland habitats near Avonmouth, but they were abundant and the undoubted single greatest food item for common shrews at the Merseyside refinery site.

In conclusion, the refinery site on Merseyside was sufficiently contaminated with copper in the 1980s to cause a significant ecotoxicological effect on the local grassland flora, reducing floral diversity and selecting for copper tolerance in two grasses. Earthworms and woodlice were lower in abundance at the refinery site, but macroinvertebrate detritivores were still well represented there (Hunter et al., 1987b). Any ecotoxicological effect on the local decomposition of detritus is unknown. The abundance of wolf spiders at the Merseyside refinery site, but their absence from metal-contaminated sites at Avonmouth, is intriguing. For what they are worth, soil concentrations of cadmium were lower on Merseyside, but soil concentrations of copper higher on Merseyside than at Avonmouth (Table 4.3). Concentrations of cadmium in the grasses at the Merseyside refinery site (Table 4.17) were lower than those in the leaf litter in the most contaminated sites near Avonmouth (Table 4.43), and copper concentrations were similar (Tables 4.17, 4.43). Of course, in contrast to the Merseyside grasses, the Avonmouth woodland leaf litter also had very high concentrations of zinc (concluded to be ecotoxic to woodlice) and of lead (Tables 4.17, 4.43). And comparative accumulated concentrations in plant material may not translate directly into comparative trophic bioavailabilities. Perhaps it was a high bioavailability of zinc along the food chains leading to wolf spiders in the likes of Hallen Wood near Avonmouth that was ultimately responsible for the local absence of these spiders. Indeed, it may be different trophic bioavailabilities of potentially toxic metals in the different invertebrates constituting the diets of wolf spiders in grassland and woodlands that contributed to any raised ecotoxicological threat to the wolf spiders in the woods near Avonmouth.

A final point might be made in the case of the potential ecotoxicology of the metals emitted from the Merseyside smelter chimneys. While the Avonmouth smelters were primarily smelting zinc and lead ores, with some cadmium contamination, the Merseyside smelters were smelting copper ores. Many copper ores also contain significant amounts of arsenic ores. In the 1970s and 1980s, it was relatively routine to use the analytical technique of atomic absorption spectrophotometry to measure concentrations of trace metals such as cadmium, copper, lead and zinc in biological material. Measurement of arsenic was, however, much less straightforward when using atomic absorption spectrophotometry. Arsenic analysis necessitated a preliminary chemical step to generate the hydride of arsenic before spectrophotometric measurements could be taken. Thus, it was rare then for arsenic to be measured in biological samples destined for cadmium and copper analysis, and accumulated arsenic concentrations were not measured in the Merseyside study. It remains possible, then, that arsenic originating in copper ores may also have been emitted in significant quantities from the Merseyside chimneys, contributing to any potential ecotoxicological metal effects on local habitats. Today, much analysis of trace metals in biological material is carried out by inductively coupled plasma (ICP) spectrometry, measuring concentrations of arsenic simultaneously with those of many other metals. Correspondingly, in recent years there has been an increase in the number of metal ecotoxicology studies involving arsenic.

4.12.3 Spoil Tips

Prologue. Spoil tips in the Northern Pennines and the Tamar valley have been surveyed for soil fauna and flora respectively.

4.12.3.1 Northern Pennines Lead Mines
Prologue. High soil lead concentrations in mine spoil tips in Weardale were not correlated with decreased abundance of soil fauna.

The soil fauna of spoil tips at disused lead mines at Westgate in Weardale, County Durham, were investigated in 1971 (Williamson and Evans, 1973). The spoil heaps, with a gradient of soil lead concentrations (Table 4.45), had been colonised by a poor ground vegetation, chiefly species of grasses of the genera *Agrostis* and *Festuca*. All sites contained diverse soil faunas, even in soils with total lead concentrations of 19,000 µg/g (Table 4.45). There were no negative correlations between the positions of sites in rank orders of abundances of different taxa and soil lead concentrations, indicating no ecotoxicological effect of the lead on the position of any taxon in this rank order. There was a positive correlation between the rank order of mites in the ranking of abundances and soil lead concentrations, perhaps indicating that mites are benefitting competitively from high lead concentrations. Further statistical analysis indicated that site 17 (9,000 µg/g Pb) was the most favourable site for the soil fauna, and site 4 (800 µg/g Pb) the least favourable. Clearly factors other than the soil lead concentration were affecting the abundance of the soil fauna in these lead mine spoil tips, even though a soil lead concentration of 19,000 µg/g is amongst the highest recorded at British mining sites (Table 4.2).

4.12.3.2 Tamar Valley

Prologue. Sparse vegetation on two spoil tips in the Tamar valley does not appear to be a result of trace metal ecotoxicology.

The comparative floral ecology of two spoil tips in the Tamar valley, one at Drakewalls mine on the Cornish bank of the River Tamar close to Gunnislake and the other at the smelter site at Weirquay on the Devon side near Bere Alston, was investigated at the end of the 1970s (Dorrington and Pyatt, 1982).

Drakewalls mine was one of 80 or more mines within the parishes of Calstock, Callington and Stoke Climsland in the Tamar valley, an area of copper, tin, silver, lead and arsenic production between 1815 and 1910. Between 1875 and 1883, Drakewalls mine was the principal producer of tin in the area, and some

copper and a little arsenic were also gained from the mine before closure in 1910 (Dorrington and Pyatt, 1982). Weirquay was the site of smelters primarily serving the lead and silver mines of the Bere Alston peninsula, undergoing a revival from the beginning of the nineteenth century. By 1845, there were 18 furnaces smelting ore from the Bere Alston mines, together with ores imported from Wales, France, Spain and Newfoundland (Dorrington and Pyatt, 1982). Within 20 years, the lack of local coal had sounded the death knell for the Weirquay smelters. The spoil heaps that remained at Weirquay therefore had a mixed origin, both locally from Devon but also from further afield in Britain and abroad.

Soil concentrations of trace metals in the Drakewalls and Weirquay spoil tips are shown in Table 4.46. Soil concentrations of copper, lead and zinc were higher at Drakewalls, with more tin at Weirquay. No soil concentration from either site can really be considered high in comparison to other mining or smelting sites (Table 4.2). Only the soil concentrations of copper and lead at Drakewalls (Table 4.46) even exceed background concentrations (Table 4.1), barely so in the case of copper.

Tips at both sites were only sparsely vegetated, even though attempts had been made in the past to introduce some additional cultivated plants to each (Dorrington and Pyatt, 1982). The Drakewalls tip had sparse vegetation at the top, but a dense growth of herbs and shrubs at its base. The dominant plant species were blackberry, gorse, ivy, red campion and wavy hair grass. Also present on the Drakewalls tip were the lichens *Cladonia pyxidata* and *Parmelia saxatilis*. None of these particular plants or lichens has been identified as a metallophyte, although this is the case for one plant species found on this tip, sheep's sorrel *Rumex acetosella*. At Weirquay, blackberry and gorse were also common, as were goosegrass, small-flowered cranesbill, broad-leaved dock, bladder campion, wood sage and brittle bladder fern *Cystopteris fragilis*. Also numerous on the Weirquay tip were the fungus *Clavaria argillacea* (moor club) and the lichens *Graphis scripta* and *Lecidella elaeochroma* (as *Lecidea limitata*). Again, none of these species at Weirquay is regarded as a metallophyte.

Table 4.45 Numbers of individuals of different taxa of soil fauna extracted from each of 18 soil cores from spoil heaps at disused lead mines at Westgate, Weardale, County Durham, in 1971, with total lead concentrations (μg/g Pb dry weight after 1:2:4 perchloric: hydrochloric: nitric acids extraction) in the soil at each site.

Higher taxon	Taxon Stage	1	2	3	4	5	6	7	8	9	10	11	12	13	14	15	16	17	18
Oligochaeta	Earthworms	1	1	2	1	–	2	–	–	–	–	6	–	–	–	–	2	–	–
Chilopoda	Centipedes	1	–	4	–	6	1	12	–	–	3	3	7	7	2	2	–	23	1
Collembola	Springtails	159	392	324	252	187	147	280	334	227	180	102	275	265	152	62	243	358	268
Insecta																			
Thysanoptera	Thrips	2	5	2	3	2	3	5	3	2	2	–	1	16	1	–	2	5	8
Coleoptera	Beetles																		
	Adults	6	1	2	2	1	1	4	2	13	8	5	2	19	1	4	3	3	2
	Larvae	6	13	2	1	5	13	15	1	16	4	2	3	20	2	4	9	11	7

Order	Group																		
Diptera	Flies																		
	Larvae	–	7	3	1	1	2	3	3	5	1	5	–	–	2	–	8	4	3
Hemiptera	Bugs	16	3	11	1	5	5	13	1	13	5	–	6	5	19	20	6	7	19
Arachnida																			
Acarina	Mites	340	236	239	167	248	221	232	385	250	457	113	284	352	312	416	402	277	374
Araneida	Spiders	–	2	–	2	–	–	–	1	1	1	7	–	–	–	2	–	–	–
Soil lead concentration		160	660	680	800	900	1,700	1,800	2,100	2,550	2,620	2,700	3,050	3,850	4,500	7,100	7,750	9,000	19,000

Source: Williamson and Evans (1973).

Table 4.46 **Concentrations of selected trace metals, potassium and phosphate (µg/g dry weight) and pH in soil from spoil tips from two sites in the Tamar valley, the Drakewalls mine site and the Weirquay smelter site.**

	Drakewalls	Weirquay
Copper	348	6.0
Lead	1,060	2.6
Tin	35	–
Zinc	483	4.9
Potassium	16	14
Phosphate	2.4	3.9
pH	4.6	3.1

Source: Dorrington and Pyatt (1982).

Thus, although the vegetation on both tips is considered sparse, soil metal concentrations are not very high and all but no metallophytes are present. It is unlikely, therefore, that either spoil tip is an example of trace metal ecotoxicology in action. The cause of the reduced floral biodiversity lies elsewhere. The soils of both tips had phosphorus and potassium concentrations (Table 4.46) that are very low in comparison with a balanced fertile soil sample, such as John Innes no.3 potting compost with 18 µg/g of phosphate and 804 µg/g of potassium (Dorrington and Pyatt, 1982). Severe phosphate and potassium deficiencies may, therefore, be more to blame for poor vegetation growth on both tips than any raised soil bioavailability of a toxic metal.

It is worth remembering that the interaction of the several factors affecting plant growth and biodiversity in metal-rich habitats is not simple, as indeed was the case for naturally occurring serpentine soils. While serpentine soils may be high in cobalt, chromium and especially nickel, metal toxicity alone does not appear to be the single driving factor controlling the nature of serpentine floras. Other possible driving factors include a deficiency of major nutrients, especially phosphorus, as might be the case here.

4.12.4 Sewage Sludge Application

Prologue. Long-term application of sewage sludge to agricultural ground may cause elevated concentrations of trace metals in the soil. There are now regulatory limits on the concentrations of trace metals permissible in sewage sludge destined for agricultural application, limiting any potential ecotoxicological risk (Department of Environment, 1989).

Statutory limits for soil metal concentrations in the United Kingdom following sewage sludge application, for example, are 300 µg/g dry weight for zinc, 135 µg/g dry weight for copper and 75 µg/g dry weight for nickel, with a lower advisory limit of 200 µg/g dry weight for zinc (Creamer et al., 2008).

A far-sighted long-term experiment was set up at ADAS Gleadthorpe in Nottinghamshire in 1982, to assess long-term responses in the biodiversity of soil invertebrates to elevated levels of copper, nickel and zinc brought about by the application of artificially spiked sewage sludge (Creamer et al., 2008). Initial application took place in 1982, with a further application in 1986, in an attempt to produce low and high experimental soil concentrations of each metal, the high concentrations to exceed the aforementioned statutory limits. Years later, between 2002 and 2004, the biodiversity and activity of soil invertebrates were measured in the different experimental soils. Also measured in 2002 were total metal concentrations of the soils, as well as concentrations of metals extracted with diethylene triaminepentaacetate (DTP) in an attempt to model bioavailable metal concentrations (Table 4.47). In practice, the mean total zinc concentration (378 µg/g) of the soil in the high Zn experiment did exceed the statutory value of 300 µg/g),

Table 4.47 **Mean ($\pm$ SE, n = 6) total and DTP-extracted concentrations (μg/g dry weight) of copper, nickel and zinc in control (addition of no sludge or of sludge with no metal spike) and experimental (metal-spiked sludge added) soils in 2002, after the application in 1982 and 1986 of artificially spiked sewage sludge.**

Soil	Copper		Nickel		Zinc	
	Total	DTP	Total	DTP	Total	DTP
No sludge control	14 ± 1.4	2.0 ± 0.5	11 ± 0.9	0.4 ± 0.0	53 ± 2.8	3.7 ± 0.2
No metal sludge	19. ± 2.1	2.9 ± 0.4	12 ± 0.8	0.6 ± 0.1	68 ± 4.3	7.1 ± 0.5
Low Cu	176 ± 11	79 ± 3.7	6.4 ± 0.4	0.5 ± 0.0	51 ± 0.2	8.4 ± 0.1
High Cu	509 ± 21	235 ± 4.0	9.1 ± 1.2	0.7 ± 0.0	79 ± 3.4	15 ± 0.8
Low Ni	14 ± 1.5	1.5 ± 0.1	25 ± 3.4	3.6 ± 0.3	57 ± 4.1	4.1 ± 0.1
High Ni	11 ± 1.1	1.3 ± 0.1	38 ± 1.4	8.4 ± 0.4	52 ± 2.1	3.6 ± 0.4
Low Zn	14 ± 0.5	2.9 ± 0.3	10 ± 0.7	0.8 ± 0.0	200 ± 10	77 ± 7.2
High Zn	22 ± 1.2	5.8 ± 0.8	9.9 ± 0.6	1.0 ± 0.1	378 ± 17	176 ± 20

Note: Total metal concentrations followed extraction with aqua regia. Metal concentrations after extraction with DTP attempt to model bioavailable concentrations.
Source: Creamer et al. (2008).

while the mean soil copper concentrations in both low and high Cu experiments (176 and 509 μg/g respectively) did likewise, both being greater than 135 μg/g Cu. The soil concentration of nickel in either the low Ni or high Ni experiment, however, failed to match the statutory limit of 75 μg/g Ni.

Data on the mean abundances of soil invertebrates and subsequent univariate measures of community structure are presented in Table 4.48. There were no adverse effects on the abundance or overall community diversity of the soil invertebrates in soils which had metal concentrations below the statutory limits (Creamer et al., 2008). Some individual taxa of soil invertebrates did, however, show sensitivity to one or both of copper and zinc in the soils with metal concentrations above the statutory limits (Tables 4.47,

4.48). Earthworm abundance was significantly reduced in the two soils which had received copper-spiked sewage sludge, and the abundances of nematodes and enchytraeids were sensitive to the high copper and zinc concentrations (Table 4.48). While taxon richness and evenness were unaffected, the Shannon–Wiener diversity index did decrease significantly in both copper treatments and the high zinc treatment (Table 4.48). Paradoxically, the abundance of collembolans was highest in the high copper treatment, and mite abundance showed an overall trend for greater abundance in all the metal treatments (Table 4.48). The collembolans and mites may be taking advantage of changes in abundance of other soil invertebrate groups, in particular enchytraeids and nematodes, indicating some intermediate

Table 4.48 Mean ($\pm$ SE, n = 16) abundances (thousands per square) and univariate measures of community structure (taxon richness, Shannon–Weiner diversity index, evenness) of soil invertebrate taxa in control (addition of no sludge or of sludge with no metal spike) and experimental (metal-spiked sludge added) soils in 2002 to 2004, after the application in 1982 and 1986 of artificially spiked sewage sludge.

	Controls		Copper-spiked		Nickel-spiked		Zinc-spiked	
	No sludge	No metal sludge	Low Cu	High Cu	Low Ni	High Ni	Low Zn	High Zn
Nematodes	18.6 ± 6.3	32.4 ± 5.3	30.3 ± 9.9	16.6 ± 6.4	18.7 ± 4.9	30.8 ± 8.0	16.8 ± 5.3	4.8 ± 1.9
Oligochaetes								
Enchytraeids	23.1 ± 6.0	30.6 ± 4.0	10.0 ± 4.2	8.4 ± 4.2	21.4 ± 4.4	26.6 ± 4.7	12.1 ± 2.4	3.7 ± 1.9
Earthworms	0.08 ± 0.0	0.11 ± 0.0	0.01 ± 0.0	0.01 ± 0.0	0.13 ± 0.1	0.07 ± 0.0	0.08 ± 0.0	0.05 ± 0.0
Collembolans	17.7 ± 4.0	22.6 ± 4.2	15.9 ± 3.2	30.9 ± 12.6	10.6 ± 1.4	21.0 ± 5.8	23.0 ± 3.9	12.3 ± 2.3
Proturans	0.14 ± 0.1	0.02 ± 0.0	0.29 ± 0.2	0.54 ± 0.5	0.0	0.10 ± 0.1	0.27 ± 1.4	2.2 ± 1.4
Insects								
Thrips	0.00	0.08 ± 0.1	0.15 ± 0.1	0.03 ± 0.0	0.15 ± 0.1	0.03 ± 0.0	0.07 ± 0.1	0.13 ± 0.0
Beetles	0.4 ± 0.2	0.8 ± 0.2	0.3 ± 0.1	1.7 ± 0.5	2.1 ± 1.6	1.5 ± 0.8	0.9 ± 0.3	0.6 ± 0.1
Fly larvae	0.4 ± 0.2	0.5 ± 0.2	0.3 ± 0.1	0.4 ± 0.2	1.2 ± 0.8	0.2 ± 0.1	0.2 ± 0.1	0.2 ± 0.1
Hemipterans	1.5 ± 0.6	1.5 ± 0.6	0.8 ± 0.3	0.5 ± 0.3	5.1 ± 4.6	0.6 ± 0.3	1.3 ± 0.6	0.3 ± 0.2
Hymenopterans	0.2 ± 0.1	0.2 ± 0.1	0.2 ± 0.1	0.1 ± 0.0	0.04 ± 0.0	0.9 ± 0.8	0.2 ± 0.1	0.2 ± 0.1

Arachnids

Mites	69.4 ± 10.8	57.0 ± 5.1	95.9 ± 12.0	124 ± 45	69.8 ± 12.3	97.6 ± 18.1	139 ± 51	103 ± 23
Spiders	0.07 ± 0.0	0.09 ± 0.0	0.22 ± 0.1	0.07 ± 0.0	0.15 ± 0.1	0.10 ± 0.1	0.27 ± 0.1	0.06 ± 0.0
Taxon Richness	5.3 ± 0.3	5.0 ± 0.3	4.4 ± 0.4	4.2 ± 0.4	4.9 ± 0.4	4.6 ± 0.3	5.6 ± 0.3	4.6 ± 0.4
Diversity Index	0.64 ± 0.0	0.70 ± 0.0	0.52 ± 0.0	0.52 ± 0.1	0.69 ± 0.1	0.60 ± 0.0	0.64 ± 0.1	0.56 ± 0.1
Evenness	0.82 ± 0.0	0.84 ± 0.0	0.82 ± 0.0	0.86 ± 0.0	0.82 ± 0.0	0.83 ± 0.0	0.83 ± 0.0	0.84 ± 0.0

Source: Creamer et al. (2008).

disturbance of the soil invertebrate community. It may not be a coincidence that it was in the case of mites that there was a positive correlation between the rank order of abundances and that of soil lead concentrations in the Weardale spoil tips considered earlier in this chapter (Williamson and Evans, 1973).

Litter decomposition was reduced in the soils treated with copper and zinc, although there was no direct relationship between litter decomposition rates and soil invertebrate abundance or diversity (Creamer et al., 2008). So, long-term exposure to sufficiently raised soil copper and zinc concentrations derived from the application of sewage sludge can affect the ecological functioning of the affected soils.

4.12.5 Hyperaccumulation: Ecological Effects

Prologue. Accumulated metal concentrations in the leaves of hyperaccumulators are potentially of ecotoxicological significance to invertebrate herbivores feeding on these leaves, with potential knock-on effects in associated food chains.

Not all raised bioaccumulated concentrations of trace metals that may have ecotoxicological consequences are the result of anthropogenic contamination. Certain plants are hyperaccumulators of particular trace metals, even in the absence of metal contamination, although admittedly particularly enhanced in the presence of raised metal bioavailabilities. Examples of hyperaccumulators include certain, but not all, species of the genus *Thlaspi* in the cabbage family *Brassicaceae*, for example, alpine pennycress *Thlaspi caerulescens* (Macnair, 2003).

Hyperaccumulators have an accumulated concentration of nickel or other trace metals in their leaves which is about ten times higher or more than in a non-hyperaccumulator growing on the same soil. By definition, hyperaccumulators have a concentration of more than 1,000 µg/g dry weight nickel; or >10,000 µg/g zinc or manganese; or >1,000 µg/g copper or selenium; or >500 µg/g chromium; or >100 µg/g cadmium, cobalt or lead in their leaves.

Such accumulated metal concentrations are potentially of ecotoxicological significance to herbivores

feeding on these leaves, particularly invertebrate herbivores. Indeed, one of the hypotheses for an adaptive role for hyperaccumulation by plants proposes that protection against herbivory is significant enough to lead to the evolutionary selection of hyperaccumulation Macnair, 2003). If a herbivore does survive the ingestion of metal-rich foliage, there is then potential for the trophic transfer of atypically large amounts of toxic metal to the herbivore, and thence to animals occupying higher trophic levels in relevant food chains. Thus the presence of the nickel hyperaccumulator *Alyssum pintodasilvae* (5,600 µg/g dry weight Ni in the leaves) on serpentine soils in northeastern Portugal led to increased nickel accumulation in local grasshoppers (herbivores) and spiders (predators) (Peterson et al., 2003). Such trophic transfer brings with it the potential for sublethal toxic effects, not only in the herbivore, but also in predators feeding on that herbivore in local food webs.

Furthermore, the high toxic metal concentrations in the leaves of hyperaccumulators can potentially have an ecotoxicological effect on the fungi and other microbes involved in leaf litter decomposition and on the invertebrate detritivores ingesting the dead leaves. For example, again in Portugal, detritivorous woodlice *Porcellio dilatatus* feeding on leaf litter derived from the nickel hyperaccumulator *Alyssum pintodasilvae* showed significantly greater mortality (83%) than woodlice fed litter from a non-hyperaccumulator plant (8% mortality) (Gonçalves et al., 2007).

The community structure of the local insect herbivores can also be affected by the presence of hyperaccumulating plants if a particular herbivore possesses characteristics that allow it to cope with a high metal diet, so enabling it to prosper disproportionately when that hyperaccumulator is present (Macnair, 2003; Boyd, 2004). The milkwort jewelflower *Streptanthus polygaloides* of California, a hyperaccumulator of nickel, is the specialist food plant for the herbivorous hemipteran insect *Melanotrichus boydi*. *M. boydi* is not affected by the high nickel concentrations of *S. polygaloides*, and populations of this plant bug are promoted by the presence of the milkwort jewelflower.

Another aspect of the trophic transfer of toxic metals from a hyperaccumulator to an herbivore able to survive ingestion of its leaves can be of ecological significance. An herbivore taking up the toxic metal from the hyperaccumulating plant might in turn use the metal as a toxic defence against its own predators (Macnair, 2003; Boyd, 2004). The nickel accumulated by the bug *M. boydi* feeding on the milkwort jewel-flower *S. polygaloides* may protect the bug against one of its invertebrate predators, the goldenrod crab spider *Misumena vatia* (Boyd and Wall, 2001). This spider had lower survivorship when fed exclusively on *M. boydi* in laboratory experiments. Killing a predator after it has been eaten is not a very effective form of antipredator defence for the bug, but it does remain possible that predators in the field may learn to avoid *M. boydi* with its high accumulated load of nickel (Macnair, 2003).

4.12.6 Ecotoxicological Risk Assessment

Prologue. Exposure to high trace metal bioavailabilities can have significant ecotoxicological effects in terrestrial habitats at levels of biological organisation from the molecular to the community. This section addresses methods for the detection and assessment of ecotoxicological effects at the community level.

4.12.6.1 Chemical Criteria

Prologue. It would be very useful in any ecotoxicological management programme to have an indication of a likely ecotoxicological threat from a simple chemical measurement, for example of a soil metal concentration.

Critical Soil Concentrations Prologue. Is it possible to define a particular trace metal concentration in the soil that can be considered as critical, indicating a likely ecotoxicological effect on the local biota?

Attempts have been made to define soil quality criteria, or critical metal concentrations in soils, which might be associated with ecotoxicological outcomes in associated soil organisms and communities (van Straalen and Denneman, 1989; van Straalen, 1993). It is not difficult to think of conceptual problems with this approach. Soil concentrations of trace metals do not always relate directly to soil bioavailabilities, not least because of the different metal binding properties of the variable geochemical components of different soils. And, bioavailabilities to what? To different species of plants, or earthworms, or other soil ingestors? Nevertheless, while we must usually be careful to define a metal bioavailability in terms of a particular organism of interest, it does remain generally true that a high metal bioavailability of a metal to one species in a habitat can often be extrapolated to other species of similar biology, particularly feeding method. So, let us be pragmatic and consider soil quality criteria.

Inevitably, we start with toxicity testing (Box 4.5).

Box 4.5 LC50s, EC50s and NOECs

While exposure of biota to raised toxic metal bioavailabilities will cause sublethal toxic effects before any lethal point is reached, most toxicological data for invertebrates are in the form of laboratory results on exposure concentrations of single metals that cause mortality. The classic toxicity test is the 96-hour acute toxicity test, with death as the endpoint. This provides an estimate of the concentration of metal in a particular medium (in this case, soil) that will kill 50% of the population of invertebrate tested – the LC_{50}, where LC refers to Lethal Concentration. If the endpoint is a sublethal toxic effect, the test will provide an EC_{50}, EC signifying an Effect Concentration. Such testing will also provide an estimate of the highest exposure concentration to cause no toxic effect – the No Observable Effect Concentration (NOEC).

These toxicity tests do have the advantage of providing single numbers to define toxicity, but there are severe limitations in extrapolating the results to real field situations. Many metal exposures in field-contaminated situations involve more than one toxic metal, the toxicities of which may interact, while laboratory data still often concern single metal exposures. Laboratory toxicity testing with soils often involves the addition of extra metal (spiking) to a starting soil. Such newly added metal is unlikely to be bound in the same chemical form (and therefore be of the same bioavailability) as the preexisting metal, even if left to stand for several weeks. Furthermore, the probably different physicochemical conditions in the soils at different contaminated sites will have different consequent effects on the bioavailabilities (and therefore ecotoxicities) of the metals therein. LC_{50} concentrations, even for the same species, will therefore differ between these soils. To be fair, such imitations are well recognised now, and systematic adjustments (using application factors or other mathematical equations) are typically made to laboratory-derived toxicity measures to try to take into account such variation in geochemistry between soils.

Species Sensitivity Distributions and HC5 Prologue. A species sensitivity distribution is a ranking of the relative sensitivities of different species to a metal, often expressed in terms of No Observable Effect Concentrations (NOEC), enabling the identification of a Hazardous Concentration (HC5) below which 95% of the species tested will show no toxic effect.

In an ecotoxicological risk assessment of metal exposure, a common approach is to rank the NOECs of a metal for relevant organisms, available from laboratory toxicity tests, to produce a species sensitivity distribution. For example, laboratory-derived NOECs for reproduction in soil invertebrates, including earthworms, snails, slugs, woodlice, mites and springtails, have been compiled for different trace metals (van Straalen, 1993). These data may then be modified by appropriate mathematical models to allow for geochemical differences between soils, for example differences in the clay or humus contents of different soils that might affect comparative bioavailabilities. The resulting standardised NOECs are then ranked with the aim of identifying the standardised NOEC, below which 95% of the species tested will show no toxic effect (van Straalen and Denneman, 1989; van Straalen, 1993). With appropriate

statistical modification to allow for factors such as the number of species tested, a Hazardous Concentration (HC) can be calculated. This is typically an HCp, where p is the percentage of species with an NOEC lower than the HC. Thus, an HC5 would theoretically cause no toxic effect to 95% of the species tested.

An early attempt to come up with soil HC5 values for soil invertebrates produced numbers of 0.20 µg/g cadmium, 2.7 µg/g copper and 77 µg/g lead (van Straalen, 1993). While there is a logical basis to such a calculation, many assumptions are involved, particularly the statistical distribution of the NOEC data assembled. Furthermore, the extent and quality of the original laboratory toxicity data are crucial factors, even before consideration of the problems associated with the extrapolation of the laboratory data to the real world of different soils. It is to be stressed that these quoted HC5 values were an early example of how an HC5 may be calculated. So, not too much credence should be applied to the absolute values. Indeed, these HC5 values are much lower than soil critical metal concentrations proposed previously for forest soil invertebrates based on a review of the literature – 10 to 50 µg/g Cd, <100 µg/g Cu, 100 to 200 µg/g Pb and <500 µg/g Zn (Bengtsson and Tranvik, 1989).

A major problem with the HC5 approach is the assumption that the NOEC toxicity data are distributed in a particular mathematical fashion, an assumption that is often not met (Hopkin, 1993). Predictions of the 5% point on such data distributions are usually theoretical, with no actual experimental data so low in the distribution of data. In fact, the calculated HC5 of 0.20 µg/g for cadmium and that of 77 µg/g for lead (van Straalen, 1993) are in fact exceeded by most soils considered uncontaminated in industrialised countries (Hopkin, 1993). Table 4.1, for example, lists background cadmium concentrations in soils up to 2.5 µg/g, and background lead concentrations usually below 150 µg/g, but as high as 300 µg/g Pb. Similarly, background copper concentrations in Table 4.1, while mostly below 100 µg/g, do extend to 250 µg/g Cu, in comparison to the HC5 value of 2.7 µg/g Cu. Paradoxically, copper is essential, and agricultural soils with less than about 5 µg/g Cu have been regarded as being copper deficient (Scheinberg, 1991; Hopkin, 1993). Under this scenario, an HC5 of 2.7 µg/g Cu would suggest that 5% of soil animals are affected adversely by a soil concentration of copper that may not provide the minimum copper requirements of at least some of the remaining 95% (Hopkin, 1993). Figure 1.1 showed how an increasing bioavailability of an essential trace metal will cause changes in the performance of an organism from positions of deficiency through optimal performance to a toxic situation. Different organisms will have different positions of these curves along the bioavailability axis, replaceable here by the concentration of an essential metal in a given soil (Figure 4.8). So, different soil concentrations of an essential (but still toxic) metal will cause different effects on different soil invertebrates, as is indeed the basis of a species sensitivity distribution. Figure 4.8 shows such a situation for two different species of soil invertebrates and copper, in a given soil of defined geochemical characteristics. One species is an invertebrate whose performance is reduced by a soil copper concentration of only 2 µg/g Cu. This same soil concentration, on the other hand, is insufficient to supply the minimum essential copper requirement of the second species. While it is certainly the case that the

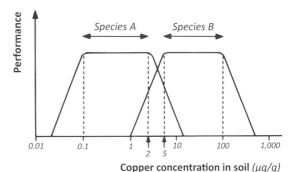

Figure 4.8 Schematic diagram of the relationships between performances of two different soil invertebrates and the concentration (µg/g dry wt) of the essential metal copper in a given soil. Species A is an invertebrate whose performance is reduced by a soil copper concentration of only 2 µg/g Cu. On the other hand, this same soil concentration is insufficient to supply the minimum essential copper requirements of Species B. (After Hopkin, 1993, with permission.)

performance curves of different soil invertebrates will occupy different positions on the horizontal axis in Figure 4.8, it would be remarkable indeed if the spread of these curves would be as great as implied by the adoption of an HC5 of 2.7 µg/g Cu in the light of a 5 µg/g Cu deficiency threshold.

There are also conceptual problems associated with an HC5 approach, irrespective of the values chosen. Acceptance of the HC5 approach implies that 5% of soil species are not worthy of environmental protection (Hopkin, 1993). What if one such species is a Red Data List species that already has legal protection but has a high metal sensitivity? Again, what if another of these 5% species is a keystone species, playing a critical role in maintaining the structure and function of the soil ecosystem. For example, the loss of a large earthworm from a woodland soil may result in a decrease in the litter decomposition rate and an accumulation of undecomposed litter on the surface of the soil. HC5 concentrations might better be set for a defined suite of indicator soil organisms, including, for example, earthworms, woodlice and snails (Hopkin, 1993).

Attempts to address some of the problems associated with the adoption of a single value for a critical trace metal concentration in soil have necessarily led to more complicated procedures – usually mathematically more complicated procedures – to generate more meaningful critical values. The advantage for environmental management of the establishment of what might be a critical metal concentration criterion for soils has not gone away. Nevertheless, considerable allowance does need to be made for the inevitably different geochemistries of different soils with consequently different effects on the bioavailabilities and ecotoxicities of their associated trace metals.

Soil Solution Free Metal Ion Concentrations Prologue. One approach is to identify the concentration of the free metal ion in the soil solution as the best model for the bioavailable form of metal taken up from the soil solution (Sauvé et al., 1998).

While this approach addresses the uptake of trace metals from the soil solution, for example by plants or earthworms, it does not allow directly for ingested soil as a route for uptake by deposit-feeding soil invertebrates such as earthworms.

There is, nevertheless, logic in considering it as a valid attempt to model the bioavailability of metals in soils to plants.

A species sensitivity distribution is still drawn up, but in this case consisting of NOECs in terms of the free metal concentration in the soil solution (Sauvé et al., 1998). The NOEC could, for example, refer to a toxic endpoint such as crop yield, or a measure of soil microbial activity such as a microbial enzyme activity. The NOEC at the 5% point (or indeed at any other percentage point) can be defined as the critical free metal ion concentration in the soil solution. Use of the free metal ion has already allowed for the differential effects on metal bioavailabilities of soil pH and organic matter content between soils. It does, therefore, eliminate much of the variability associated with the use of total metal concentrations in soils in the prediction of toxic effects on biota. In fact, the free metal ion concentration is not itself measured but estimated mathematically from measurements of total metal concentration, organic matter concentration and pH in the soil (Sauvé et al., 1998). Tiny soil solution concentrations of the free

metal ions of copper and lead are toxic in terms of crop yield and soil microbial activity. They are perhaps best understood if translated back into total soil metal concentrations in an agricultural soil of typical organic content at a defined pH. Thus, at pH 7, a total concentration of lead of 2,276 µg/g dry weight in such a soil will have a toxic effect on average on 50% of the organisms tested in a range of bioassays, including crop plant yields and soil microbial activities (Sauvé et al., 1998). The comparable figure for copper is lower at 1,766 µg/g dry weight under similar soil conditions. Both critical metal soil concentrations fall considerably with increased acidification of the soil, to 177 µg/g Pb and 103 µg/g Cu at pH 5.5 (Sauvé et al., 1998). These latter estimated critical soil concentrations still seem low in the ranges of soil metal concentrations listed in Table 4.1, but those at pH 7 appear to be more defensible than the first attempts introduced earlier (van Straalen, 1993).

The use of the free metal ion concentrations in the soil solution as the most appropriate model for the toxicity of trace metals in soils has gained much support, as for example in a later study involving cadmium and zinc, as well as copper and lead (Lofts et al., 2004). This study again used mathematical equations to predict soil solution free metal ion concentrations from total metal concentrations in the soil, taking into account the organic content of the soil and the soil pH. In this case, the soil pH was also considered to take into account any effects of other cations such as calcium, as well as hydrogen ions, on free metal ion concentrations, simplifying the mathematics involved. Chronic toxicity data from the literature for plants, invertebrates, microbial processes and fungi were assembled into species sensitivity distributions ranked against estimated free metal ion concentrations in the soil solution of many soils (Lofts et al., 2004). From these distributions, it was then possible to define critical soil solution concentrations of the free metal ions of cadmium, copper, lead and zinc that corresponded to the 5% toxicity point, thereby aiming to protect 95% of the species involved from soil metal toxicity. From these critical free metal ion concentrations, it is possible by back calculation to come up with the different total metal concentrations considered critical in soils of different pH and organic matter contents.

This approach has been used in an assessment of the risk to soil fauna of seven trace metals (cadmium, chromium, copper, lead, nickel, vanadium and zinc) in a study of their distribution in more than 1,000 soil samples across Britain (Spurgeon et al., 2008). The samples were taken in 1998 to 1999 from predominantly rural sites, classified into one of 17 broad habitat classifications. The total metal concentrations in the soil samples fell out into two clusters, consisting of Cr, Cu, Ni, V and Zn on the one hand (with Cu and Zn as one identifiable subcluster and Cr, Ni and V as another), and Cd and Pb on the other. Because the sites sampled did not include mining and smelting sites, there was no evidence for a distinct Cd, Pb, Zn cluster as might be attributable to their common co-occurrence in ore bodies. The associations between Cu and Zn, and between Cd and Pb, may be down to widespread industrial usage and the effects of urbanisation (Spurgeon et al., 2008). An association of Cr, Ni and V probably results from their common occurrence in particular rock types from which some of the soils would have originated. Cr and Ni were identified earlier as co-occurring in serpentine minerals.

In the risk assessment, free metal ion concentrations in the soil solution were estimated using the total metal concentration, the soil organic matter content and the soil pH (Lofts et al., 2004; Spurgeon et al., 2008). A critical free metal ion concentration in the soil solution was also calculated for each site using the toxicity data compiled previously (Lofts et al., 2004). The free metal ion concentration calculated for each site was expressed as a fraction of the critical free metal ion concentration appropriate for the site. Of the 1,083 samples, the percentages with trace metal concentrations above the critical limit were 1% for Cd, 27% for Cu (with 20% more than three times higher), 24% for Pb (16% more than three times higher) and 44% for Zn (22% more than three times higher) (Spurgeon et al., 2008). This situation would be surprising, and probably does not reflect the real situation in British soils.

It would seem, therefore, that the calculation of critical free metal ion concentrations from laboratory toxicity tests, often with metal spiking, is still suffering from the artificiality of laboratory testing in comparison to what is happening in the real world. In natural soils, there is probably a lower risk of metal ecotoxicity than is inferred from the aforementioned critical concentrations based on laboratory testing data (Spurgeon et al., 2008). Nevertheless, it can probably be concluded that the greatest risk of any ecotoxicological effect from trace metals in British soils is from zinc. Furthermore, the risk appears greatest in habitats such as arable and horticultural, improved grassland and built-up areas, where soil metal concentrations are more frequently elevated (Spurgeon et al., 2008).

Up the Food Chain   Prologue. It is necessary also to consider the toxicity of soil-derived metals to organisms one or more steps removed from direct exposure in the soil, as in the case of animals higher up a food chain.

The NOEC data used in the preceding examples have been data on the toxicity of metal exposures directly to particular soil organisms. It is also possible to consider the toxicity of soil-derived metals to animals higher up a food chain, taking in metals by trophic transfer. Moles, shrews and badgers preying on earthworms would fit into this category.

A study in the vicinity of the Avonmouth smelting works in the 1990s attempted to assess the ecotoxicological risk to such predators of high accumulated concentrations of cadmium, copper, lead and zinc in earthworms, the metals being derived from the local soils contaminated by emissions from the smelter (Spurgeon and Hopkin, 1996b). A first stage in the risk assessment concerned the prediction of accumulated metal concentrations in the earthworms from metal concentrations in the soil. Given the discussion earlier in this chapter on the accumulation of trace metals by different earthworm species from soils of different bioavailabilities, it is a false assumption that any relationship between soil metal concentrations and accumulated concentrations in different species of earthworms can be represented by a single factor.

Nevertheless, in order to keep matters simple, this was the approach taken (Spurgeon and Hopkin, 1996b).

At this point, it is necessary to make a brief diversion to consider the meaning and application of such single factors (see Box 4.6).

Let's return to the risk assessment of the ecotoxicological risk to predators of the trophic transfer of trace metals from earthworms living in metal-contaminated soils (Spurgeon and Hopkin, 1996b). Relationships were investigated between the datasets of accumulated

Box 4.6 Bioconcentration Factors and Bioaccumulation Factors

This box essentially represents a plea against the use of single factors, variously termed bioconcentration factors (BCF) or bioaccumulation factors (BAF), in studies of trace metal bioaccumulation.

Both terms are often used without definition, and occasionally (and confusingly) as synonyms for each other. Where definitions are offered, they may vary between authors. As defined here, a BCF is the ratio of the metal concentration at steady state in an organism to the concentration in the water in which it lives. A BAF is the ratio of metal concentration in steady state in an organism to its food or another defined matrix in its environment (e.g., soil). The use of either term is often on the basis of untenable assumptions. For example, in the case of aquatic animals, use of a BCF ignores the fact that a significant proportion of the metal accumulated is derived from the diet. A BCF is more suitable for use in the case of a plant or alga. Even then, the assumption that the accumulated metal concentration in the organism is in steady state is probably erroneous, except perhaps in an animal that regulates its accumulated concentration to an approximately constant level. And, in this case, there would be no constant direct relationship between an unchanging, regulated, accumulated metal concentration and the variable ambient water concentration, so eliminating the possibility of the existence of a constant BCF of any validity (McGeer et al., 2003).

The use of a BAF implies that there is a direct and exclusive relationship between the concentrations concerned. In a reverse of the preceding argument for a BCF, a BAF between metal concentrations in an aquatic animal and its diet ignores the contribution to the total accumulated metal of any metal taken up from solution. An earthworm bathed by a soil solution would, for example, potentially be taking up metal from this dissolved source as well as from any ingested soil. And as discussed earlier, not all earthworms ingest soil. Like a BCF, a BAF assumes that the metal concentration in the organism is actually in steady state, an assumption that may well be invalid. Nevertheless, it does need to be stated that the use of such single factors is more defensible when considering the accumulation of organic contaminants, where more straight-line relationships occur consistently between environmental concentrations and bioaccumulated concentrations. Still, the transgression of such factors into simplistic interpretations of metal bioaccumulation should be avoided.

metal concentrations of Cd, Cu, Pb and Zn in up to six species of earthworms (*Lumbricus terrestris, Lumbricus rubellus, Lumbricus castaneus, Allolobophora chlorotica, Aporrectodea rosea* and *Aporrectodea caliginosa*) and soil concentrations of these metals. In spite of the combination of data for earthworms of different feeding habits and metal accumulation patterns, and in spite of observed site differences in these generalised relationships, the authors took a pragmatic (if conceptually untenable) approach to define a mean BAF for each metal across all earthworm species and all sites. These values were then carried forward in the risk assessment. The next stage involved the estimation of predator sensitivity by using literature NOEC values for dietary metal concentrations toxic to birds and mammals, to come up with HC5 values – the concentrations of the metals in the diets of birds and mammals together that would protect 95% of the species investigated. If the diets were earthworms, these HC5 values could then be related by the previously calculated BAFs to come up with maximum permissible concentrations of the metals in the soils from which the earthworms had obtained the metals. These HC5 values would theoretically ensure that 95% of the potential predator species would not suffer dietary toxic effects. These maximum possible soil concentrations were calculated to be 0.017 µg/g Cd, 18.9 µg/g Cu, 30.4 µg/g Pb and 36.1 µg/g Zn (Spurgeon and Hopkin, 1996b). Comparisons of these theoretical critical concentrations with soil concentrations in Table 4.1 show that, as for some of the earlier estimates of critical soil concentrations, they fall into the background range, being exceeded by metal concentrations even in uncontaminated agricultural soils. Time to think again.

What can we conclude about critical soil concentrations for metals? However they are calculated, we are moving away from the concept of a simple one critical soil concentration of a metal that fits all soils to an array of different critical toxic metal concentrations for different soils. Their calculation may still involve assumptions on the relationships between the geochemical characteristics of soils and the bioavailabilities (and potential ecotoxicities) of the metals concerned, and assumptions on what is meant by

bioavailability when generalised across biota. Nevertheless, the calculation of such critical concentrations is now on an increasingly justifiable conceptual basis, and potentially might allow the provision of logically defensible information to environmental managers attempting to assess the ecotoxicological risk of particular toxic metal concentrations in metal-contaminated soils. For the moment, these critical concentrations would be better used as relative indicators of the likelihood of adverse effects rather than absolute values for direct prediction (Spurgeon et al., 2008).

4.12.6.2 Biological Criteria
Prologue. An alternative approach in ecotoxicological risk assessments of trace metal exposures in terrestrial habitats is to use biological criteria as opposed to chemical concentrations in soils.

Underlying problems associated with the use of physical measurements such as critical soil concentrations have encouraged terrestrial ecotoxicologists to turn to biological criteria to identify the presence (actual or potential) of ecotoxicological effects caused by raised trace metal bioavailabilities in terrestrial ecosystems. Biological criteria do have the inbuilt advantage of responding directly to the local bioavailabilities of metals in soils, bypassing the vexing question of the variable relationships between soil metal concentrations and their bioavailabilities (to whatever organisms).

Critical Tissue Concentrations and Acceptable Daily Inputs Prologue. In the real world, the concept of a critical tissue concentration, above which accumulated trace metals are considered to be exerting a toxic effect on the animal concerned, is still widely used.

In the search for simple numbers in ecotoxicological risk assessments, it would be an attractive option to define a particular concentration in a tissue, organ or whole body of an organism living in a terrestrial habitat that can be considered critical, being indicative of a significant toxicological effect on that organism. We have, however, already discussed in this chapter the problems associated with the definition of such a critical tissue concentration of a metal in, for example, an organ of a small mammal or bird that would indicate a toxicological threat to that species. Problems arise because, in the presence of any detoxified storage of a trace metal, there can be no toxicologically critical concentration of total metal accumulated in an organ. Much of the total metal accumulated is detoxified and therefore in a safe form that is not metabolically available. It is the concentration of metabolically available metal that has, by definition, not been detoxified that can reach a critical threshold level in a tissue to cause toxic effects. This threshold is reached more quickly, and at a lower total accumulated concentration, when the rates of metal uptake into the organism, and consequently into an organ, are increased by increased local bioavailability of the metal. Toxicological effects can, and will, occur at different accumulated tissue concentrations, thereby contradicting the concept of a 'critical' total tissue concentration.

Nevertheless, the concept of a critical tissue concentration has remained attractive to ecotoxicologists. Such tissue critical concentrations – in this case, in the kidneys of black-tailed godwits *Limosa limosa* and badgers *Meles meles* – were used in a risk assessment study for the back calculation of critical soil concentrations of cadmium and lead that would indicate an ecotoxicological threat to wildlife (de Vries et al., 2007). The study also used earthworms as the food chain link in the trophic transfer of trace metals from soil to predators, represented by a vermivore, the black-tailed godwit and the more omnivorous badger. This study did not use a BAF to link soil and earthworm concentrations, but relied on equations that recognised the nonlinear relationship between these concentrations. Furthermore, these equations also allowed for soil pH and other geochemical properties of the soil that would affect the bioavailability of soil metals. Instead of calling upon NOEC toxicity data, the next step in the risk assessment used a critical tissue concentration, in this case of cadmium

or lead in the kidney of a black-tailed godwit or badger, to calculate an acceptable daily input (ADI) of each metal that would not cause exceedance of this critical tissue concentration. The kidney was chosen for each predator because it was considered the most sensitive organ for the accumulation of cadmium and lead. Table 4.49 shows the critical tissue concentrations of cadmium and lead in the kidneys of each predator, chosen from the literature, and the calculated ADI of each metal from the diet that would lead to those accumulated metal concentrations in the kidneys over the exposure period chosen. In this particular scenario, badgers were assumed to be feeding on earthworms only. Replacement by some of these earthworms by plant material such as grass would have reduced the dietary metal intake of the metals by the badger, so this example (Table 4.49) can be seen as a worst case scenario. Table 4.50 subsequently shows the estimated critical soil concentrations of cadmium and lead in different soils that would have led to high enough accumulated metal concentrations in earthworms (and also grass) to have produced these threshold maximum ADI at defined feeding rates. As it happens, the critical soil concentrations for cadmium (Table 4.50) implying an ecotoxicological risk to a vermivore and an omnivore are again very low in comparison to soil cadmium concentrations in the

Table 4.49 **Calculated ADI (mg per day) of cadmium and lead by black-tailed godwits *Limosa limosa* and badgers *Meles meles* that would result in proposed critical tissue concentrations (µg/g dry weight) in their kidneys over a defined exposure period (days).**

	Critical tissue concentration		Exposure period	ADI	
	Cd	Pb		Cd	Pb
Godwit	200	90	122	0.253	0.114
Badger	200	90	365	1.781	0.801

Source: de Vries et al. (2007).

Table 4.50 **Estimated critical soil concentrations (µg/g dry weight) of cadmium and lead in different soils that would have led to high enough accumulated metal concentrations in earthworms and grass to produce maximum acceptable daily inputs of the metals to black-tailed godwits *Limosa limosa* and badgers *Meles meles* (see Table 4.49), at feeding rates of 0.1 kilogram per day fresh weight of earthworms by godwits, and 0.5 kilogram per day fresh weight of earthworms together with 0.5 kilogram per day fresh weight of grass by badgers.**

Soil use	Soil type	pH	Critical soil Cd concentration		Critical soil Pb concentration	
			Godwit	Badger	Godwit	Badger
Agriculture	Sand	5.5	0.14	0.26	123	157
Agriculture	Clay	6.5	0.66	1.2	534	668
Agriculture	Peat	6.0	1.0	1.9	1,024	1,297
Nature	Sand	4.5	0.07	0.12	69	88
Nature	Clay	6.0	0.47	0.82	412	514
Nature	Peat	4.5	0.33	0.60	426	539

Source: de Vries et al. (2007).

real world (Table 4.1). Estimated critical soil concentrations for lead (Table 4.50) may be more defensible when compared to soil lead concentrations listed in Table 4.1.

In spite of the use of critical tissue concentrations criticised previously, this study does show a way forward because it used the concept of an ADI of a trace metal (de Vries et al., 2007). A major tenet of this book has been that toxicity in an individual organism occurs when the rate of uptake of a trace metal exceeds the maximum combined rate of detoxification and excretion. Thus, toxicity is related to uptake rate and not accumulated concentration (in the presence of any detoxification). And the ADI is a measure of uptake rate in animals such as mammals and birds relying only on the diet for trace metal input. Thus,

unconnected to any concept of a critical tissue concentration, there will be a maximum daily input of a trace metal that can be tolerated before the onset of toxic symptoms – in short, an ADI.

Biomarkers and Tolerance Prologue. There are now suitable biomarkers of the potentially ecotoxicologically significant exposure of terrestrial organisms to trace metal exposure in the field. Metal-tolerant populations of plants and earthworms indicate the presence of local bioavailabilities of that toxic metal of ecotoxicological significance in British terrestrial habitats.

Biomarkers are biological responses to trace metal exposure that can be measured at the lower levels of biological organisation in tissues or body fluids or at the level of the whole organism. Progress has been made in recent years on the identification of suitable biomarkers in suitable terrestrial organisms, not least in earthworms, that signify that the organism has been exposed to a raised trace metal bioavailability, and, in some cases, an ecotoxicologically significant trace metal challenge (Kammenga et al., 2000; Spurgeon et al., 2003). In earthworms, biomarkers include the measurement of lysosomal membrane stability using the neutral red assay, the expression of metallothioneins and metabolomic profiling (Spurgeon et al., 2003). At a higher biological level of organisation, the reproduction of earthworms has been shown to be affected by elevated metal concentrations in soils close to the Avonmouth smelter (Spurgeon and Hopkin, 1995; Spurgeon et al., 2003).

At the population level, the presence of a metal-tolerant population of an organism in a particular habitat is evidence that local bioavailabilities of that toxic metal are of ecotoxicological significance, clearly to that species, but potentially also to other members of the local biota. And there are indeed examples of metal-tolerant plants and also invertebrates in metal-contaminated terrestrial habitats in Britain. Species of bentgrass (genus *Agrostis)* and fescue (genus *Festuca*) are well represented in a list of metal-tolerant populations at British mine sites, along with another grass, Yorkshire fog (*Holcus lanatus*) (Table 4.19). The trace metals with the ecologically significant soil bioavailabilities promoting the selection of tolerant strains in these grasses include arsenic, copper, lead, nickel and zinc (Table 4.19). Other flowering plants with metal-tolerant populations in Britain belong to many different plant families and include alpine pennycress *Thlaspi caerulescens*, red campion *Silene dioica*, bladder campion *Silene vulgaris* and thrift *Armeria maritima* (Table 4.19). Populations of earthworms (*Lumbricus rubellus* and *Dendrodrilus rubidus*) with tolerance to both copper and arsenic have been collected from spoil heaps at mines at Carrock Fell, Cumbria and Devon Great Consols, Devon (Langdon et al., 1999, 2001). Another Cumbrian population of *D. rubidus*, this time from the Coniston copper mine, showed

tolerance to copper (Arnold et al., 2008). The population of the collembolan *Orchesella cincta* living in litter at the old zinc-lead mine at Plombières in Belgium is cadmium tolerant (van Straalen et al., 1987).

Thus, we have biomarkers of the ecotoxicological effects of trace metals in terrestrial environments, at levels of biological organisation from the molecular, biochemical, cellular, physiological, individual organism to population levels. While biomarkers at the lowest levels of biological organisation are the most sensitive, it is the toxicological effects at the community level that are the most ecotoxicologically significant. If a high metal bioavailability is producing ecotoxicological effects observable at the community level, then biota are being greatly affected. It is all very well detecting lower-level biomarkers. We need to link these biomarkers to effects at the community level for such biomarkers to be of greater value to community ecotoxicologists. This is a problem that is not restricted to the terrestrial environment. This same need to link biomarkers at different levels of biological organisation also applies to freshwater and marine ecotoxicology. As we shall see in later chapters, considerable progress is being made in this context in coastal waters, particularly in the case of biomarkers of trace metal ecotoxicology in the mussel *Mytilus edulis*.

At the community level of biological organisation, we particularly need information that will allow us to detect (and ultimately predict) ecotoxicological effects on the structure and function of the ecological communities affected. We have seen in this chapter that raised soil bioavailabilities of, probably, zinc were associated with reduced species diversity of invertebrates and increased leaf litter accumulation at woodland sites affected by atmospheric fallout from smelters at Avonmouth (Read et al., 1998). Raised copper bioavailabilities close to the Merseyside smelters reduced floral biodiversity in affected grassland swards (Hunter et al., 1987b). So, how should we go about assessing any risk of significant ecotoxicological effects at the community level caused by trace metals in terrestrial habitats affected by trace metal contamination?

Bioindicators Prologue. A bioindicator species can be defined as a species whose presence, abundance or absence indicates a particular environmental factor or change.

Life is rarely so simple that the presence of a single species can indicate the presence of an ecotoxicologically significant bioavailability of a trace metal in a soil. Nevertheless, the presence of the metallophyte *Viola calaminaria*, the zinc violet, comes close (van Straalen, 1998). The zinc violet appears to be an obligate metallophyte, specifically restricted to calamine soils which are enriched in zinc, lead and cadmium (Bizoux et al., 2004; Baker et al., 2010). Unfortunately, from a British point of view, *V. calaminaria* is endemic to a small region in eastern Belgium, the southern Netherlands and western Germany (Bizoux et al., 2004). Indeed, the zinc violet is present at the Plombières mine site in Belgium that is home to the cadmium-tolerant population of the collembolan *Orchesella cincta* (Posthuma, 1990). Rather than consider a single bioindicator species, more is to be gained from a consideration of a bioindicator role for a whole community, for example the soil arthropod community (van Straalen, 1998).

Statistical Measures of Community Structure Prologue. Both univariate and multivariate statistical measures of community structure are available to terrestrial ecotoxicologists.

Statistical measures of community structure range from univariate measures such as species abundance, richness, diversity indices and evenness to the multivariate measures now more widely employed. The study on the effects on the biodiversity of soil invertebrates of the addition of metal-enriched sewage sludge to agricultural soils used univariate statistics (Creamer et al., 2008). The pitfall trap study of invertebrates in woods downwind of Avonmouth employed multivariate statistics (Read et al., 1998).

There is no doubt that the use of multivariate statistics is more likely to detect more subtle differences in ecological community structure between uncontaminated and metal-contaminated habitats than is achievable with simpler univariate measures.

However, the use of multivariate statistics typically requires the collection of a large amount of data on the abundances of component taxa (identified either to species or, if justified, a higher taxon level) in the different habitats. This is a procedure that is labour intensive and requires a considerable degree of systematic expertise. It is little wonder that terrestrial ecotoxicologists pine for simple comparative numbers.

Bioaccumulated Metal Guidelines Prologue. A bioaccumulated metal guideline is a measure of ecotoxicological exposure based on a bioaccumulated metal concentration, for example in a biomonitor that can be correlated with the onset of independently measured ecotoxicological effects on local community structure.

There is another possible way forward that returns us to consideration of bioaccumulated metal concentrations, but not in terms of critical tissue concentrations supposed to indicate incipient toxicity of the metal concerned to the organism in which the concentration has been measured. This approach has already received much criticism in this chapter.

High accumulated metal concentrations in biomonitors indicate the presence of atypically high bioavailabilities of those metals, strictly to that biomonitor. The high accumulated concentrations do not themselves indicate that these high metal bioavailabilities are having a toxicological effect on the biomonitor itself, or an ecotoxicological effect on the resident community in a habitat. On the other hand, if particular accumulated metal concentrations in selected biomonitors can be correlated with independently measured ecotoxicological effects on a local community, then they can be used as bioaccumulated metal guidelines indicative of these ecotoxicological effects (Luoma and Rainbow, 2008). For example, in the freshwater habitat, accumulated copper concentrations in larvae of species of the caddisfly *Hydropsyche* have been validated as suitable bioaccumulated metal guidelines, correlating with copper-induced changes in the local mayfly community assemblages of streams in North America and Southwest England affected by mining (Luoma et al.,

2010; Rainbow et al., 2012). Such use of bioaccumulated metal guidelines can be extended to the terrestrial environment. Suitable biomonitors would include earthworms and woodlice. If accumulated metal concentrations in such biomonitors are calibrated against negative changes in associated communities, for example of soil invertebrate communities, measured by either univariate or multivariate statistical techniques, then a bioaccumulated concentration in the biomonitor becomes a surrogate measure of a significant ecotoxicological effect at the community level. After calibration, a simple analytical measure bypasses the need for a great deal of counting and identification of a chosen local biological community.

4.13 Livestock

Prologue. In any discussion of the natural history of metals in the British Isles, it is impossible to ignore the fact that agricultural land is a very significant component of the terrestrial ecosystem. Indeed, the natural history of agricultural land, whether pasture or arable, is part of the natural history of the British Isles. Crops and farm livestock are similarly part of our natural history, and part of the natural history of metals in these isles.

4.13.1 Ecotoxicology

Prologue. Ecotoxicological effects resulting from high trace metal bioavailabilities in soils can extend to crops and livestock, while low bioavailabilities in some agricultural soils can lead to essential metal deficiencies.

4.13.1.1 History
Prologue. The ecotoxicological consequences for crops and livestock of high bioavailabilities of trace metals in soils affected by mining activities have been known for centuries.

In the first decade of the twentieth century, farmers in Cardiganshire (now Ceredigion) in Wales were complaining that both their soils and their livestock were suffering from lead poisoning, as an undesirable effect of local mining (Griffith, 1919). This was by no means a new situation. In 1815, Walter Davies (a Welsh poet, editor and antiquarian, commonly known by his chosen bardic name of Gwallter Mechain) produced a report for the Board of Agriculture on the agriculture and economy of south Wales, in collaboration with Edward Williams (Iolo Morganwg). In this report, Walter Davies stated: 'During conversation with a Cardiganshire gentleman he observed how niggardly Nature had bestowed her blessings on his native county. We endeavoured to frame an apology for Nature; and among other instances of her liberality, mentioned the silver and lead mines. "O," he exclaimed, "that is a curse, and not a blessing; the mines enrich a person or two in an age, and entail poverty on hundreds of generations to come." "The waters from the mines" he added, "spread sterility over the adjacent fields, and kill all the fish in the rivers."' Furthermore, in 1873, the Rivers Pollution Commissioners reported that of all the lead mining districts in Great Britain, Montgomeryshire (now part of Powys) and particularly Cardiganshire were the only two areas where land and livestock suffered injury to any marked extent (Frankland and Morton, 1873; Griffith, 1919). The valleys of the Ystwyth, Rheidol, Clarach and Dyfi were highlighted in this regard. Finally, the *Report of the Royal Commission on Land in Wales 1896* quoted a witness who referred to a meadow which 45 years previously had been flooded by the adjacent river, then strongly impregnated with lead from local mines (Griffith, 1919). The flooding had left a mineral deposit. Prior to the flooding, he had kept ten cows and a bull on the meadow. Subsequent to the flooding, the mineral deposit made the meadow dangerous for horses, and only suitable for three or four cows, although butter from these cows was quite unsaleable (Griffith, 1919).

At the start of the twentieth century, the lead mines of North Ceredigion were typically surrounded by enormous heaps of mine spoil, containing lead ore in the form of lead sulphide, galena and lesser amounts of zinc ore, again as the sulphide, blende (Griffith, 1919). Initially, drainage water from these heaps would be low in lead and zinc. Over time, however, the sulphides would be oxidised by the action of air and water to sulphuric acid, so producing acid mine drainage, particularly rich in lead. The drainage water would also carry fine metal-rich material that settled out on stream and river beds, to depths of centimetres to a metre or more (Griffith, 1919). Regular floods caused metal-rich water to cover local farmland, but more importantly would resuspend the metal-rich sediment and dump it on the fields. The river valleys suffering most were those of the Ystwyth, Rheidol and Clarach and the upper reaches of the Teifi. Livestock drinking stream or river water suffered poisoning, as in the case of sheep near Cwm Ystwyth. Otherwise, the danger to livestock came from the lead-rich sediment dumped on the fields, in addition to lead adsorbed from floodwater onto the external surfaces of plants. Back at the mines, farmers learnt not to let poultry have any direct access to the toxic spoil heaps. And spoil heaps remained, even after the closure of the mines that had produced them.

Back on the fields, the Ceredigion grassland pastures presented a now expected characteristic flora (Griffith, 1919). Common bentgrass *Agrostis capillaris* dominated and was the only plant present in the very toxic regions, albeit scantily so. Yorkshire fog *Holcus lanatus* would occur in damper areas. As contamination decreased, sheep's fescue *Festuca ovina*, mountain pansy *Viola lutea* and sea campion *Silene uniflora* made an appearance. Clovers were conspicuous by their absence. On affected arable land, seeds of cereals might germinate, and the seedlings appeared to flourish for a time, until becoming stunted or dying.

All farm stock suffered when grazing on the contaminated pasture. Sheep, horses and poultry suffered most, cattle less so (Griffith, 1919). Sheep often aborted, and any lambs born did not thrive. Ewes would waste away and would be disposed of after three years, instead of the usual four years. Horses reared on the contaminated fields suffered symptoms of diarrhoea, limb stiffness and paralysis and ultimately (and characteristically) became 'broken winded' (a chronic respiratory condition), surviving at most four years. Cattle fared much better, although they were said not to have thrived as well as those grazing on uncontaminated pasture. Poultry were particularly sensitive if given access to contaminated land or brooks, laying badly, producing shell-less eggs and perishing early. We saw earlier how the physiology of lead is linked to the physiology of calcium, and female birds transport lead to the eggs with the calcium destined for eggshell production (Scheuhammer, 1987). The excess lead in this case clearly disrupted the formation of eggshells by the affected chickens. One is reminded of the words of the Cardiganshire gentleman reported in 1815, that 'the waters from the mines spread sterility over the adjacent fields', a situation still evident a century later.

In other areas of lead mining, lead poisoning in calves and lameness and poor thriving in lambs were noted in the lead mining areas of the Northern Pennines in the 1950s and 1960s, as were losses of lambs in north Derbyshire (Thornton and Webb, 1979).

The arsenic industry of the Carnon valley in Cornwall in the middle of the nineteenth century was notorious for the unwanted effects of the toxic emissions from its chimneys on local livestock. In 1850, cattle and horses died on farmland downwind of the Perran Works emitting dense fumes of arsenic oxide (Barton, 1971b). There were also losses of livestock nearby in the fields surrounding Great Wheal Busy (previously known as the Chacewater mine) (Barton, 1971b). Cattle grazing on arsenic-contaminated pasture in Cornwall have shown symptoms of arsenic toxicity in the form of dysentery and respiratory distress (Mitchell and Barr, 1995). Ingestion of arsenic by these cattle may have been as high as 50 milligrams of arsenic per day through grazing, with 60 to 70% attributable to the arsenic contained in the soil accidentally ingested with the grass (Mitchell and Barr, 1995).

We saw earlier how the emissions of an urban-industrial complex in Swansea in the 1960s caused the accumulation of very high concentrations of trace metals, including lead, in the grass red fescue *Festuca rubra* at nearby sites in the Swansea valley (Table 4.16). In October 1969, a farmer living within 2 kilometres west of Swansea centre purchased a horse which he kept in stalls over winter, feeding it on hay cut from the surrounding fields in summer 1969 (Goodman and Roberts, 1971). The horse was put out to grass in spring 1970 but was soon confirmed clinically to be suffering from lead poisoning. It died in June that year. Analysis of the 1969 hay and 1970 grass confirmed the presence of very high accumulated lead concentrations of about 100 µg/g dry weight in both (cf. Table 4.16) (Goodman and Roberts, 1971). A concentration of 100 µg/g lead was recognised as a lethal value for fodder for cattle, and horses are considered even more susceptible to lead. Analysis of kidney tissue from the horse showed a lead concentration of 40 µg/g wet weight, also considered to be lethal (Goodman and Roberts, 1971).

High soil concentrations of molybdenum lead to toxicity in cattle, as in soils in Somerset derived from the Lower Lias formation, which contain 20 µg/g Mo or more (Thornton and Webb, 1979). Molybdenum toxicity is actually caused indirectly, because the raised dietary concentrations of molybdenum reduce the absorption of the essential metal copper and cause symptoms of copper deficiency (Thornton and Webb, 1979). Deficiencies of essential metals in livestock are addressed in more detail later in this section.

By the end of the twentieth century, clinical problems of metal poisoning in livestock were only rarely reported in Britain, although subclinical effects of metals on the health of livestock could not be completely ruled out (Thornton and Webb, 1979).

4.13.1.2 Critical Soil Concentrations

Prologue. A critical soil concentration is a threshold trace metal concentration in a soil above which it is considered likely that there will be an ecotoxicological effect on local biota.

We discussed previously how community ecotoxicologists have been attracted to the concept of a single value to define critical soil concentrations of trace metals, below which there should be a negligible threat of a significant ecotoxicological effect on associated ecological communities. The same attraction exists for toxicologists concerned with an ecotoxicological threat to livestock from dietary metal inputs.

The aforementioned study that attempted to calculate critical soil concentrations of cadmium and lead for wildlife represented by black-tailed godwit and badgers took a similar approach to come up with critical soil concentrations of cadmium, lead and mercury for livestock (de Vries et al., 2007). The starting point again was a consideration of presumed critical tissue concentrations of these toxic metals in the kidneys and livers of cows and sheep considered to be associated with the manifestation of toxic effects of the metals in the livestock (Table 4.51). These critical tissue concentrations were then used to calculate the maximum ADI of the three metals in cows and sheep that would result in these accumulated concentrations in kidneys and livers (Table 4.51). Uptake of metals from air or water was ignored as negligible in these calculations. Using specified ingestion rates of fodder (grass) and associated accidental intake of soil while grazing, and assuming fixed ratios (bioaccumulation factors) between fodder metal concentrations and accumulated tissue concentrations in the livestock, calculations could be made of the critical metal concentrations in the fodder to bring about the maximum ADI shown in Table 4.51. Finally, mathematical equations linking concentrations of metals in soils and in plants grown on that soil, allowing where possible for factors such as soil pH, clay and organic matter contents, provided estimates of critical soil concentrations of each of the three metals in different soils that would cause critical metal concentrations in the fodder to bring about these maximum ADI in Table 4.51. Table 4.52 illustrates one such set of critical soil concentrations in different soil types, in this case for the kidneys of cattle as the most sensitive animal organ (de Vries et al., 2007).

It is necessary, of course, to be aware of assumptions made in these model calculations and of the lack

Table 4.51 **Calculated maximum ADI (mg per day) of cadmium, lead and mercury by cattle and sheep that would result in proposed critical tissue concentrations (μg/g fresh weight) in their kidneys or livers. Ingestion rates of grass were assumed to be 16.9 and 2.5 kilograms per day, and of soil 0.41 and 0.10 kilograms per day, by cows and sheep respectively.**

	Organ	Critical tissue concentration			Acceptable daily intake		
		Cd	Pb	Hg	Cd	Pb	Hg
Cow	Kidney	5	3	1.4	29	604	38
	Liver	1.4	2	2	44	857	219
Sheep	Kidney	4	5	1	5	–	5.6
	Liver	2	5	4	2.8	–	182

Source: de Vries et al. (2007).

of availability of particular information (for example, for lead in sheep for Table 4.51). Similarly, an equation for a mathematical relationship between soil and grass lead concentrations was also lacking (de Vries et al., 2007). The study also used critical tissue concentrations and bioaccumulation factors, which are both concepts criticised earlier in this chapter. Nevertheless, values for critical soil concentrations considered to represent thresholds of risk for cattle have emerged (Table 4.52). At these critical soil concentrations (Table 4.52), it is accidental soil ingestion that is the dominant uptake pathway, highlighting the importance of chosen assumed ingestion rates to the model. The estimated critical soil concentrations for cadmium in the different soils (Table 4.52) are strikingly higher than those calculated for wildlife safety (Table 4.49), and perhaps more credible in the light of concentrations of cadmium in contaminated and uncontaminated soils, as listed in Table 4.1. Similarly, critical soil concentrations estimated for lead and mercury (Table 4.52) would fall well into the contaminated ranges given for these two metals in Table 4.1.

4.13.2 Deficiency

Prologue. At the other end of the gradient of trace metal bioavailabilities in soils, a very low local bioavailability of an essential trace metal in the soil may be as important as a high trace metal bioavailability for crops and in grass pastures supporting grazing livestock.

Copper deficiency in cereal crops, with a consequent fall in the yield of grain, is associated with some soils derived from peaty, sandy and calcareous parent materials (Thornton and Webb, 1979). A total soil concentration of 2 μg/g Cu (cf. Table 4.1) has proved a useful indication of soils that give rise to symptoms of deficiency in cereals, although soils with a little more copper may also prove copper deficient, if much of the soil copper is not in a form that is bioavailable to the plant roots (Thornton and Webb, 1979). Indeed, the slightly higher copper concentration of 5 μg/g is sometimes used to identify a copper deficiency threshold in soils (Scheinberg, 1991; Hopkin, 1993). Copper deficiency in cattle has also occasionally been reported on sandy soils in Britain (Thornton and

Table 4.52 **Estimated critical soil concentrations (µg/g dry weight) of cadmium, lead and mercury in different soils that would have led to high enough accumulated metal concentrations in grass (and ingested soil) to produce maximum acceptable daily inputs of the metals in cattle to cause proposed critical tissue concentrations of these metals in the kidney (see Table 4.51).**

Metal	Soil type	Critical soil concentration
Cadmium	Sand	67
	Clay	62
	Peat	66
Lead	Sand	1,382
	Clay	1,382
	Peat	1,386
Mercury	Sand	68
	Clay	68
	Peat	67

Source: de Vries et al. (2007).

Webb, 1979). Symptoms of copper deficiency in cattle and sheep include the loss of full control of body movements, defective hair or wool growth, anaemia and poor bone mineralisation leading to fractures. In cattle in particular, symptoms of copper deficiency (hypocuprosis) also include a failure to grow or put on weight as expected and diarrhoea. The defective production of melanin in hair leads to a pale coat. In sheep, a sensitive sign of copper deficiency is 'swayback' in lambs, a description of the lack of control of body movements which varies along a gradient of severity. Poor growth follows, and the quality of any wool produced is poor, a condition called 'steely wool'.

Copper deficiency in livestock can still occur even if there is apparently enough copper present in the soil. The presence of excess molybdenum in the diet of cattle and sheep, while still not enough to cause molybdenum toxicity, can interfere with the absorption of copper to cause hypocuprosis in cattle and swayback in sheep (Thornton and Webb, 1979; Thornton, 1993). Soils with more than 5 µg/g Mo may support vegetation with greater than 2 µg/g dry weight Mo, and the consumption of such vegetation can lead to diarrhoea, loss of production and growth retardation in cattle (Thornton and Webb, 1979). The majority of British soils contain less than 2 µg/g Mo, but soils derived from marine black shales can contain up to 100 µg/g Mo. Such soils are actually widespread across England and Wales (Thornton and Webb, 1979). A geochemical survey based on stream sediments showed up an area of Derbyshire near Ashbourne with atypically high molybdenum concentrations in soil, derived from underlying black Namurian shale from the Carboniferous (Thornton, 1975a; Thornton and Webb, 1979). Clinical symptoms of copper deficiency in cattle were limited to about 25 square kilometres of this 150 square kilometre area, but subsequent blood tests showed that more than 75% of cattle in the wider region were deficient in copper but not displaying clinical signs (Thornton, 1975a; Thornton and Webb, 1979). Subsequent copper supplementation trials showed significant live weight gain responses by young cattle in the area.

Cobalt is important biologically as a constituent of vitamin B12. Cobalt deficiency (pine) in sheep is known on soils derived from granite in Southwest England; on soils derived from sandstones, limestones, shales and acid igneous rocks in northern England; and on soils derived from acid igneous and sandstone rocks in Scotland (Thornton and Webb, 1979). Weaned lambs at pasture in late summer or autumn are particularly susceptible. Symptoms include growth retardation, small size, poor body and wool condition, anaemia, lethargy and watery discharge from the eyes. Cobalt-deficient herbage is usually found on soils containing 10 µg/g Co or less (cf. Table 4.1), preferably measured after extraction with acetic acid as a model for bioavailability (Thornton and Webb, 1979).

Selenium deficiency (and/or vitamin E deficiency) leading to degenerative white muscle disease in sheep and calves is known from several parts of England, Wales and Scotland, again in apparent association with sandy soils (Thornton and Webb, 1979). In Britain, white muscle disease typically affects rapidly growing lambs, particularly male lambs, of two- to six- weeks-old, causing them to be reluctant to move and then unable to rise.

4.14 Human Health

Prologue. There are clear links between human health and trace metals in the terrestrial environment.

A major route for the uptake of trace metals by humans is by the ingestion of food and, usually to a lesser extent, drinking water. Much of this food is derived from vegetables and other crops grown on soils with different bioavailabilities of trace metals. Much also comes from livestock which have grazed directly on grass growing on these different soils or which have been fed on fodder grown on soils of different trace metal bioavailabilities. To a limited extent as adults, we may also ingest soil with garden vegetables, as well as indoor dust, both of which carry trace metal burdens. Little children certainly will put into their mouths various items contaminated with dust and soil. We will also inhale dust, with the potential for uptake of metals in the lungs.

4.14.1 Risk to Human Health

Prologue. An understanding of the significance and relative importance of the different routes of the uptake of potentially toxic trace metals into humans is vital in any assessment of the potential risks to human health.

4.14.1.1 Environmental Geochemistry and Human Health
Prologue. Geochemical maps can highlight potential areas of high trace metal bioavailabilities with implications for human health.

A useful initial stage in assessing the human health risks of trace metals on a national scale has come from combining environmental geochemistry and health studies (Thornton and Webb, 1979; Thornton, 1993). Geochemical maps, for example the Wolfson Geochemical Atlas of England and Wales (Webb et al., 1978), are based on the systematic sampling and analysis of metal concentrations of the likes of rocks, soils, stream sediments, lake sediments, surface waters and vegetation. Such geochemical maps can be used for mineral exploitation, but also have great relevance for agriculture, pollution studies and, here, for human health, allowing geochemical hotspots for particular trace metals to be identified and followed up. Not surprisingly, such metal anomalies are mostly related to metal contamination arising from historical mining and smelting activities.

A comprehensive environmental health study on the effects of cadmium was carried out in about 1980 near the village of Shipham in Somerset, identified as a geochemical hotspot for zinc and cadmium (Thornton, 1993). Shipham had been a site for the mining of zinc ore between 1700 and 1850, the zinc ore also containing appreciable amounts of cadmium. The study showed that more than 60% of household garden soils in Shipham had cadmium concentrations above 60 µg/g Cd, and household dust metal concentrations averaged 26 µg/g Cd and 2,300 µg/g Zn. Local Shipham residents had an increased dietary input of cadmium compared to the UK norm, but this rarely exceeded the UN's Food and Agriculture Organisation (FAO)/World Health Organisation (WHO) provisional tolerable weekly input of 420 micrograms of cadmium for a 60 kilogram adult (Table 4.53). In a comparison with a nearby control population, the Shipham residents showed only slight health and biochemical differences attributable to cadmium (Thornton, 1993). It appeared that the high

Table 4.53 **Provisional tolerable weekly inputs (PTWI, micrograms per week per 60 kilogram adult) of toxic metals considered of human health concern, as recommended by the Joint FAO and WHO Expert Committee on Food Additives (JECFA).**

	PTWI µg per week per 60 kg adult
Inorganic Arsenic	900
Cadmium	420
Lead	1,500
Mercury	96

Source: Morais et al. (2012).

local bioavailabilities of zinc and calcium had provided a degree of health protection against any adverse effects of the raised cadmium bioavailability to the Shipham population.

4.14.1.2 Food

Prologue. Concentrations of trace metals in foods need to be considered in any risk assessment of human exposure to potentially toxic inputs.

Cadmium can also be used as an example of a typical risk assessment of the transfer of a toxic metal from soils to the human food chain (Jackson and Alloway, 1992). Incidence of acute cadmium toxicity is rare, and any health risk from cadmium exposure would be through chronic dietary exposure. The human kidney is the main target organ, with an average cadmium residence time of 30 years or more. A concentration of 200 µg/g Cd fresh weight in the cortex of the human

kidney has been associated with kidney dysfunction, with symptoms such the presence of abnormal quantities of protein in the urine. The typical method of assessment of exposure of the population to cadmium would use government data on the average consumption of foods of different types, together with the cadmium concentrations analysed in these foods, to come up with an estimate of the national average dietary input of cadmium – in this case, 140 (range 90 to 180) µg Cd per person per week chain (Jackson and Alloway, 1992). This national average is safely below the FAO/WHO JECFA provisional tolerable weekly input (PTWI) of 420 µg Cd per 60 kilogram person per week (Table 4.53). To make an assessment for a particular subpopulation, as in the previously discussed case of Shipham, local data would be needed on diets and cadmium concentrations in local foods. In order to restrict the dietary input of cadmium, the European Union has proposed maximum levels of cadmium in different foodstuffs (Table 4.54). A risk assessment may go further and mathematically relate cadmium concentrations in specific vegetables or other crops to concentrations in soils, allowing for different physicochemical attributes of the soil as previously described for livestock (de Vries et al., 2007). Such relationships can then be used to set other limits, such as the maximum concentrations of particular metals such as cadmium in sewage sludge that can be applied to arable land.

The combined use of JECFA provisional tolerable weekly inputs (Table 4.53) and EU-regulated maximum levels in various foodstuffs (Table 4.54) is the basis of the control of input via the human diet for cadmium, lead and mercury, trace metals of particular concern as regards human health.

In the case of mercury, it is the organic form methyl mercury that is of toxicological concern (Morais et al., 2012). In practice, methyl mercury enters the human diet in large, long-lived predatory marine fish such as tuna, with some input from shellfish. EU statutory limits for mercury in different foodstuffs (Tables 4.54, 7.16) reflect the importance of particular long-lived fish with high fat content (methyl mercury is fat soluble) as a dietary source of methyl mercury, the form of effectively all the mercury in the fish.

Table 4.54 **Maximum concentrations (μg/g fresh weight) of cadmium, lead and mercury in different selected foodstuffs under EU Commission Regulations Nos 1881/2006, 629/2008 and 1005/2015.**

Foodstuff	Cadmium	Lead	Mercury
Meat (not offal) of cattle, sheep, pigs, poultry	0.05	0.1	
Edible offal of cattle, sheep, pigs, poultry		0.5	
Liver of cattle, sheep, pigs, poultry	0.5		
Kidney of cattle, sheep, pigs, poultry	1.0		
Muscle meat of fish*	0.05–0.3	0.3	0.5–1.0
Crustaceans – muscle (white meat)	0.5	0.5	0.5
Bivalve molluscs	1.0	1.5	0.5
Cephalopod molluscs (not viscera)	1.0	0.3	0.5
Dried seaweed products	3.0		
Cereals, legumes and pulses		0.2	
Cereals, excl. bran, germ, wheat and rice	0.1		
Bran, germ, wheat and rice	0.2		
Vegetables, excl. brassicas, leaf vegetables		0.1	
Brassicas, leaf vegetables, fresh herbs, fungi		0.3	
Vegetables, excl. leaf, root and stem vegetables	0.05		
Stem vegetables, root vegetables and potatoes	0.1		
Leaf vegetables, fresh herbs and selected fungi	0.2		
Remaining fungi	1.0		
Fruit	0.05		
Fruit, excl. berries and small fruit		0.1	
Berries and small fruits		0.2	
Food supplements	1.0	3.0	0.1

* Maximum concentrations vary with species of fish (see Table 7.16).

There is a JECFA PTWI for inorganic arsenic (Table 4.53), but no EU statutory limits for arsenic levels in food. There is, however, a 1959 UK statutory limit of 1 µg/g fresh weight of arsenic in most foods, exceptions including seafood. Fish and other seafood in fact account for more than 90% of the arsenic in the British diet, but the large majority of this arsenic is in non-toxic organic form, as opposed to inorganic arsenic, which is indeed toxic. We saw earlier how vegetables grown in arsenic-contaminated garden soils in the historical mining area of Hayle–Camborne–Godolphin in Cornwall (Table 4.4a) still had accumulated arsenic concentrations (Table 4.13) that were well below the 1959 UK statutory limit of 1 µg/g As fresh weight (Xu and Thornton, 1985). Although the total arsenic concentrations in the soils were high, at an average of 322 µg/g As (Table 4.4a) in comparison with more normal soil concentrations (Table 4.1), extractable arsenic concentrations in the soils were very much lower, at about 2% of total arsenic (Table 4.4a). It would appear, then, that soil concentrations of arsenic in a form bioavailable to the growing vegetables were not greatly elevated.

High exposure to lead can have neurological effects on young children, affecting neurobehavioral and intellectual development. Fortunately, lead is now widely banned as an additive to petrol, including in Britain. For the general public, it is the consumption of food that is typically the major source of lead exposure. A recent study has modelled the exposure of one- to five-year-old children across Europe to lead through the consumption of locally produced food in non-polluted rural areas in the first two decades of the twenty-first century (Bierkens et al., 2012). Typically, across Europe, the model predicted that these children obtained about 70% of total lead uptake from the diet, 20% from drinking water, 7% from soil and dust ingestion and inhalation, and 1 % from air inhalation. Further models predicted the concentrations of lead in the children's blood. Increased blood lead concentrations are known to correlate with impaired performance on IQ tests in young children. For example, an increase in blood lead from 10 to 20 µg of lead per 100 milligram is associated with a drop of 2.6 IQ points (Bierkens et al., 2012). Furthermore,

there does not seem to be a lower threshold of blood lead, below which no IQ impairment occurs. The predicted average blood lead concentration of one- to five-year-old children in the United Kingdom was only about 0.5 µg of lead per 100 milligram, in a range of 0.4 to 1.6 µg across Europe. Thus, the average modelled decrease in IQ would be 0.22 IQ points per child in the United Kingdom attributable to consumption of local food. This is reassuring, although the modelling is as yet by no means perfect (Bierkens et al., 2012).

In a Derbyshire study of a historic lead–zinc mining area in the 1970s, households were grouped according to the lead concentrations of the soil in their gardens (Thornton, 1993). The amounts of lead in the hair and blood of children aged two to four increased with the concentrations of lead in the garden soil and in household dust. Reassuringly, none of the lead concentrations in the children was high enough to be considered hazardous at the time, even though lead concentrations in soil and dust were very high, peaking at 28,000 µg/g and 25,000 µg/g. The major pathway of lead uptake into these local children was through the inhalation and involuntary ingestion of dust particles (Barltrop et al., 1975; Thornton, 1993).

The villages of Leadhills and Wanlockhead in Scotland were the centre of lead mining in Scotland from medieval times until the start of the twentieth century. A survey was carried out in 1984 of blood lead concentrations in adults and children living in the area (Moffat, 1989). Blood lead concentrations averaged 15.9, 12.4 and 17.6 µg Pb per 100 ml in adult men, adult women and children respectively from Leadhills and Wanlockhead. Equivalent blood lead concentrations were 11.0, 8.3 and 10.4 µg Pb per 100 ml at a control village, Moniaive, lying 26 miles to the southwest in the same general belt of Silurian rock strata. Although raised at Leadhills and Wanlockhead, the blood lead concentrations there were very similar to, and in some cases lower than, those found in British cities at the time, and lower than the 1982 recommended UK action level of 25 µg Pb per 100 ml (Moffat, 1989). There was no evidence to suggest that the consumption of locally grown vegetables was associated with the raised blood lead concentrations at Leadhills and Wanlockhead. The most important contributor to raised blood lead levels in these two villages was lead in drinking water, explaining 11% of the variance in blood lead concentrations.

4.14.1.3 Drinking Water

Prologue. Historically, drinking water had the potential to be a significant route of uptake of a trace metal where water supplies had been affected by local mining for the metal.

Historically, poisoning from lead in water caused health problems in lead mining communities in upland Ceredigion in the nineteenth century. And, as we have just seen, drinking water was the major contributor to raised blood levels to residents of Leadhills and Wanlockhead in 1984.

Today, dissolved concentrations of hazardous trace metals in drinking water are monitored to ensure that they are below the levels specified in the UK Drinking Water Standards (DWS) listed in Table 4.55. Metal concentrations in local waters may exceed these guidelines, but such contaminated waters would not be used for drinking water abstraction. For example, the UK DWS for lead in drinking water is now 10 μg/L (Table 4.55), while surface water concentrations of lead in old Welsh mining areas measured in the 1980s varied from 1 to 25 μg/L Pb (Abdullah and Royle, 1972). Lead in drinking water also has the potential to exceed guideline concentrations in household taps in soft and acid water regions, if old lead pipes are still in use.

4.14.1.4 Soil

Prologue. The uptake of metals directly from soils needs to be taken into account in any risk assessment of trace metal exposure to human health.

Soil may be ingested directly by humans, not least by young children. A Swedish study reported that children aged one to six years in urban Uppsala ingested 100 milligrams of soil per day by hand-to-mouth

Table 4.55 **UK Drinking Water Standards (DWS). Maximum dissolved concentrations (μg/L) allowable at the point of compliance, consumers' taps.**

	UK DWS (μg/L)
Arsenic	10
Cadmium	5
Chromium	50
Copper	2,000
Lead	10
Mercury	1
Nickel	20
Selenium	10

activity (Ljung et al., 2006). Similarly, soil ingestion rates in the United States have been estimated to be 50 milligrams of soil per day by one- to five-year-old children and 10 milligrams of soil per day by older children and adults (Murphy et al., 1989).

Garden soils in the historical lead mining village of Winster in the Derbyshire Peak District were analysed in 1988 (Cotter-Howells and Thornton, 1991). The soil samples contained very high concentrations of lead (average 7,140 μg/g, range 2,400 to 22,800 μg/g), compared to a more typical background soil concentration range of 10 to 300 μg/g Pb (Table 4.1). In fact, the lead in the soil grains consists of the very stable mineral form pyromorphite, formed from the weathering of galena (lead sulphide). Pyromorphite has extremely low solubility, and (fortunately) apparently very low bioavailability to humans when ingested.

4.14.1.5 Dust

Prologue. Dust from either or both dietary and inhalation routes can represent a significant source of toxic metals to humans.

Dust is typically defined as fine settled or airborne particulate material, less than a tenth of a millimetre (100 µm) in size (Turner, 2011). Any original mineral component of dust particles may well contain trace metals, and all dust particles, whether of physical or biological origin, will also adsorb trace metals. There is subsequent potential for the dust particles to settle out and contaminate food preparation surfaces and food itself or other articles put into the mouth or to be inhaled into the lungs.

Dust from powdered rocks containing trace metals, notably ores, will clearly contain trace metals in high concentrations that are bioaccessible and bioavailable to different degrees. Such dust was a particular health hazard for miners. 'Belland' is a dialect term in the Peak District for finely powdered lead ore associated with local mining and smelting. The inhalation of belland was the greatest risk to the health of a Derbyshire lead miner, leading to severe symptoms of breathing difficulties and swelling of joints and limbs. In Ceredigion, the equivalent lung disease of lead miners was known as 'y belen' (the ball).

Amounts of metals in dusts can be related to the amounts of metals in local soils (Thornton et al., 1985), the dusts having at least part of their origin in the soils, for example by wind suspension of soil particles on drying out. In the case of metal-contaminated soils and dusts, there is a distinction between mining and smelting communities in these soil-to-household dust metal concentration relationships (Cotter-Howells and Thornton, 1991). This distinction results, at least in part, to the different physical and chemical properties of the dust particles originating from the two different operations. The surface properties and moisture content of smelter particles may render them more likely than mining-

derived particles to adhere to clothes, shoes and pets for transport indoors to contribute to household dust (Murphy et al., 1989; Mitchell and Barr, 1995).

In the case of lead in soils and associated local house dust in the mining communities of Derbyshire and Shipham, the concentration of lead in the household dust was made up of two components: a component consisting of 15% of the soil lead concentration, to be added to the second component, a background constant of 500 µg/g Pb attributable to indoor sources (Barltrop et al., 1975; Ljung et al., 2006). Given that household dusts in the Derbyshire mining village of Winster in 1988 contained an average of 1,560 µg/g Pb (Cotter-Howells and Thornton, 1991), more than two-thirds (68%) of the lead in the household dust on average was derived from local soil. The high indoor contribution of 500 µg/g Pb to lead concentrations in household dusts in these mining villages, in comparison to total concentrations as low as 5 µg/g Pb in household dust elsewhere (Table 4.1), may have been attributable to household paint containing lead (Murphy et al., 1989).

A small secondary metal recovery factory in north London in 1976 caused a considerable increase in the concentration of cadmium in dust collected within 30 metres of the factory boundary, compared to control dusts collected from a minimum of 1.5 kilometres away (Muskett et al., 1979). The factory melted bulk metal scrap to recover lead, as well as occasionally processing jewellers' waste for silver and gold. The dust near the factory contained an average of 193 µg/g Cd (range 12 to 387 µg/g Cd), as opposed to a control average of 16 µg/g Cd (range 11 to 26 µg/g Cd). The peak level of cadmium measured in the air (including airborne dust) near the factory was 11 µg per cubic metre, a very high figure compared to typical urban levels of 1 to 3 µg Cd per cubic metre. The present workplace exposure limit for cadmium in air is 25 µg per cubic metre, if exposed for eight hours continuously. There was no significant increase in the lead concentration of the dust near the factory compared to the controls, with averages (ranges) of 3,370 (276 to 8,620) µg/g Pb, as opposed to 1,540 (368 to 1,940) µg/g Pb. It is possible that the added effect of lead released into the atmosphere from petrol combustion at the time was obscuring any effect of the factory emissions on local dust lead concentrations.

At Winster in the Peak District in 1988, was there any effect of the high concentrations of lead measured in household dust and garden soils on blood levels of lead in children? Even though these environmental lead concentrations translated to a relatively large amount of lead on the hands of Winster children, the short answer is 'no' (Cotter-Howells and Thornton, 1991). The average lead levels in the blood of Winster children were 9.4 µg Pb per 100 ml for one- to eight-year-olds, and 6.9 µg Pb per 100 ml in one- to three-year-olds. Both measures were within the normal UK range at the time, and well below the 1982 recommended UK action level of 25 µg Pb per 100 ml (Cotter-Howells and Thornton, 1991). Contemporary analyses gave average blood lead concentrations (all µg Pb per 100 ml) of 11.7 in Birmingham one- to three-year-olds, 10.7 in Edinburgh six- to nine-year-olds, between 7.4 and 8.3 in London five- to seven-year-olds and 6.4 to 6.8 in rural Suffolk five- to seven-year-olds. Thus, the blood lead concentrations in the Winster children might have been considered elevated for a rural habitat, but were similar to, or only slightly elevated above, comparative urban values. The results for Winster were not in line with other studies of the expected effect of increased dust and soil lead concentrations on the blood lead levels of local children. Usually, a concentration average of 500 µg/g Pb in household dust would be expected to be associated with a blood level of lead in children of 15.0 µg Pb per 100 ml (Cotter-Howells and Thornton, 1991), Thus the high concentrations of lead in the garden soils and household dust of Winster were not reflected in high bioavailabilities of the lead to children or adults, a probable reflection of its presence in the form of pyromorphite, a lead compound of low chemical extractability and therefore bioavailability.

The phasing out of lead in petrol from the mid-1980s has resulted in a welcome considerable fall in blood lead concentrations in children. A study in the early 1990s of blood levels in London schoolchildren between five- and seven-years-old already showed up a significant downward trend in average blood lead

levels from 1986 to 1992, with maximum blood levels of 17, 21, 15 and 9 µg Pb per 100 ml in 1986, 1987, 1991 and 1992 respectively (O'Donohoe et al., 1998). Indeed, the 1982 recommended UK action level of 25 µg Pb per 100 ml has now been replaced by the lower 1995 WHO recommended clinical threshold for concern of 10 µg Pb per 100 ml. Few if any UK children now exceed this blood level, except under specific circumstances of lead poisoning. Nevertheless, even the 10 µg Pb per 100 ml figure might be seen as a risk management tool rather than a threshold for lead toxicity (Bellinger, 2004). As previously stated, there seems to be no lower threshold of blood lead in young children below which there is no IQ impairment. Indeed, a blood level as low as 5 µg Pb per 100 ml may cause some neurological effect in young children (Bierkens et al., 2012).

It is not enough simply to measure total trace metal concentrations when including dusts and soils in risk assessments of the toxic metal exposures of humans. The total bioavailability of a trace metal to humans would be a relative measure of that fraction of the total metal concentrations in different media (food, water, etc.) actually taken up into the body, summated across all possible uptake routes. These uptake routes in our case would be food, drinking water, ingested soil, ingested dust and inhaled dust. Each of these metal sources would have its own bioavailability, contributing to the total bioavailability and thence total uptake of the metal into the body. The bioavailability of a metal in a particle of ingested soil or dust would be the fraction of its total metal content that is absorbed by the body after ingestion. This absorption involves two processes: firstly, the release of the metal from the particle by digestion in the gut; and secondly, the subsequent uptake of this released metal from the gut lumen into the gut epithelium. The efficiency of the first stage is described as the bioaccessibility of the metal in the ingested soil or dust particle (see Section 4.2.6, Table 4.5).

The incorporation of bioaccessibility information into risk assessments of toxic metal exposures will improve their relevance. A study of the UBM 'stomach' bioaccessibility of chromium in Glasgow soil samples collected in 2001/2002 and 2005 showed that bioaccessible metal (and therefore the resultant bioavailable metal) in these soils was considerably less than 100%, averaging only 5% (Broadway et al., 2010). Any risk assessment based on the total metal concentrations of the soils would inevitably overestimate the amount of chromium available for absorption from ingested soil and dust particles. The bioaccessibility of arsenic has been measured in samples of garden soil and household dust from a residential area in east Cornwall in the vicinity of a historical mining site (Rieuwerts et al., 2006). Similarly to the case of chromium in Glasgow, bioaccessibility concentrations were only 10 to 20% of the total arsenic concentrations. Again, bioaccessible concentrations can be used to improve subsequent risk assessments of local human exposure to arsenic.

Trace metals can also enter the human body after their release from inhaled dust particles in the lung. Not all inhaled particles enter the lungs, larger particles being stopped in the upper respiratory tract, but fine particles (less than 10 µm) will do so.

An extreme example is the case of smoking, the smoke particles offering a source of trace metals, particularly cadmium, which can be taken up after inhalation (Morais et al., 2012). Tobacco smoke is one of the largest single sources of cadmium in humans, and the absorption of cadmium from the lungs is much greater than from the alimentary tract (Morais et al., 2012). Smokers will absorb about 1 µg Cd per day from inhalation of the smoke of 20 cigarettes, compared to an average cadmium intake from food of between 8 and 25 µg per day (Järup and Åkesson, 2009). Such total intake rates are still below the provisional tolerable daily input of 60 µg Cd per day for a 60 kilogram adult recommended by the FAO and WHO (Table 4.53). Some arsenic is also absorbed from cigarette smoking. About 20 µg As will be released from 20 cigarettes, in comparison to about 90 µg ingested per day by an average UK adult in 1982 (Mitchell and Barr, 1995). Most of the dietary arsenic is in non-toxic organic form, so uptake rates of arsenic, even for smokers, do not approach the FAO/WHO provisional tolerable daily input of about 130 µg of (inorganic) arsenic per day for a 60 kilogram adult (Table 4.53) (Morais et al., 2012).

In a risk assessment, it is necessary to calculate the exposure to a trace metal that results from inhalation, to be added to any exposure from ingestion. In the case of arsenic, indoor air arsenic can be considered to originate from two sources: the ingress of outdoor dust and the resuspension of household dust Mitchell and Barr, 1995; Murphy et al., 1989). In the case of the first source, an average of 30% of the outdoor dust transported indoors is estimated to remain suspended. The second source, the resuspension of household dust, is considered to account for 50% of the indoor suspended dust. Arsenic concentrations in dust in parts of Cornwall are often four or more times those of control dusts, and so the inhalation of dust does need inclusion in local risk assessments of arsenic exposure (Mitchell and Barr, 1995). Nevertheless, risk assessments of chronic arsenic exposure in Southwest England have failed to identify a significant general health risk (Mitchell and Barr, 1995). It always remains the case, however, that more detailed risk assessments of the exposure to arsenic of certain population subgroups based on location and/or age may indicate some need for remedial action (Mitchell and Barr, 1995).

The risk assessment study of exposure to chromium in Glasgow soils did consider inhalation as well as ingestion (Broadway et al., 2010). Furthermore, this risk assessment used a chemical model of chromium bioaccessibility for uptake in lung fluid in addition to the UBM 'stomach' bioaccessibility model. A lung simulation fluid (Gamble's Solution) was employed, extracting chromium from the fraction of dust of less than 10 µm, the size fraction that can reach the lungs (Broadway et al., 2010). The chemical form of chromium is very important in risk assessments of chromium toxicity, chromium 6 being considered toxic, and chromium 3 non-toxic. Total concentrations of chromium measured in environmental media will contain both forms of chromium. Many environmental guidelines will make the assumption for safety purposes that all the chromium present is in the form of chromium 6. This is often unlikely but does provide a considerable safety factor. The less than 10 µm size fraction of the two most contaminated Glasgow dust samples (out of 27) did contain sufficient lung-bioaccessible chromium 6 (in addition to a larger major chromium 3 component), to make dust inhalation the most potentially harmful uptake route (Broadway et al., 2010).

4.14.1.6 Soil Guideline Values

Prologue. Soil Guideline Values (SGV) are used as indicators of chemical contamination in soil, below which long-term human health risks from exposure are considered tolerable or minimal. SGV are considered to indicate whether there is a need to assess further whether remedial action is required.

The approach of the Department for Environment, Food and Rural Affairs (DEFRA) and the Environment Agency (EA) in the United Kingdom is to use the contaminated land exposure assessment (CLEA) model to derive SGV (Environment Agency, 2009b). SGV do not represent thresholds for an unacceptable intake of the contaminant, in this case a toxic metal, but they do represent trigger values to indicate to a risk assessor that soil concentrations above this level may pose a significant possibility of significant harm to human health. This possibility will be linked to the margin of exceedance, the duration and frequency of exposure and other site-specific factors. In short, SGV are screening tools for risk assessors. SGV do not take into account other sources of metal uptake by humans that are not soil based, for example toxic metal contamination of drinking water.

The CLEA model uses generic assumptions about the biogeochemistry of different metals in the soil environment, and about human behaviour, to estimate child and adult exposures to trace metals for those living, working and/or playing on metal-contaminated sites over long-term periods (DEFRA and Environment Agency, 2002). SGV are derived using the CLEA model by comparing these estimates of exposure with Health Criterion Values that represent a tolerable or minimal risk to health from chronic exposure. Such a Health Criterion Value might be a tolerable weekly or daily input of a metal, as shown in Table 4.53. The CLEA model estimates the average daily exposure (per body weight) to a contaminant in

a soil via the following routes: (a) ingestion of homegrown produce, (b) ingestion of soil, (c) ingestion of dust, (d) inhalation of dust, (e) inhalation of vapour (e.g., elemental mercury) and (f) absorption through the skin (e.g., in the case of methyl mercury) (Environment Agency, 2009b, c). We have discussed previously how the concentrations of metals in vegetables and crops can be linked by mathematical relationships of different complexities to concentrations in soils, allowing for factors such as pH and soil organic content that affect soil bioavailability. The CLEA model takes such soil factors into account to calculate simple soil-to-plant concentration factors applicable to different green vegetables, root vegetables, fruit, etc. The CLEA does not, however, consider bioaccessibility percentages for ingested or inhaled soils or dusts, using total metal concentrations of soils and dusts in estimating exposure rates via these routes. Use of total metal concentrations in this way introduces a safety factor, given that such bioaccessibility percentages are usually well short of 100% (see Table 4.5).

Table 4.56 lists SGV for selected trace metals. Arsenic, cadmium, chromium, lead and mercury have figured in earlier discussions, but two further toxic metals, nickel and selenium, also appear in the table. Nickel does occur in laterite rocks and soils, particularly in Scotland (Table 4.6), but not in sufficient quantities to be mined in the British Isles. Nickel is, however, released from the burning of oil and some coals, and it has an important industrial role in the production of stainless steel, nickel plating, batteries and alloys for the manufacture of coinage. Sewage sludge and some agricultural fertilisers applied to arable land may add nickel to the soil. The most significant nickel contamination of soil in Britain is localised to specific smelter or plating works. Selenium is also not present in sufficient concentrations in British rocks to be mined, but it is present in high-sulphur coals and is especially enriched in ash from the burning of these coals.

How do the SGV in Table 4.56 compare to metal concentrations measured in particular contaminated soils in Britain? Arsenic and chromium provide examples.

Concentrations of arsenic in garden soils in 1984 in the historical mining area of Hayle–Camborne–Godolphin in Cornwall averaged 322 µg/g As, with a range of 144 to 892 (Table 4.4a). These soil concentrations are well above the SGV for arsenic in residential and allotment soils (Table 4.56) and are therefore worthy of further consideration in any risk assessment. In fact, extractable concentrations of arsenic in these garden soils were very much lower, at about 2% of total As concentrations (Table 4.4a). And arsenic concentrations in the edible parts of selected vegetables growing in these soils (Table 4.13) were well below the 1959 UK statutory limit for most foods of 1 µg/g As fresh weight (Xu and Thornton, 1985). It is clear, then, that the bioavailabilities of arsenic to vegetables growing in these Cornish garden soils were not elevated to the point of any concern for human health, even though total soil concentrations of arsenic were high. There would be no concern here, then, in any risk assessment as regards the uptake of arsenic by the local population from the soil to vegetable transfer route.

What about soil and dust ingestion and inhalation routes for arsenic in contaminated Cornish soils? The SGV refer only to soil concentrations, for the CLEA model incorporates the contribution of local soil to local dust in its calculations (Environment Agency, 2009b, c). Total concentrations of arsenic in garden soils in a residential area in East Cornwall near a historical mining site averaged 262 µg/g As (range 23 to 471 µg/g) in the first decade of the twenty-first century (Rieuwerts et al., 2006). These were considerably elevated compared to control garden soil concentrations from a village in Southeast Cornwall with no history of mining, averaging 22 µg/g As (range 8 to 40 µg/g). Concentrations of total arsenic in the mining community garden soils are again well above SGV for arsenic in residential and allotment soils, while the control Cornish soil concentrations fall below these SGV (Table 4.56). At first sight, therefore, the comparison of the SGV for arsenic and the measured total arsenic concentrations in the garden soils of the mining community indicates a need for further investigation in a risk assessment, concentrating now on the soil and dust ingestion and inhalation routes of

Table 4.56 **Environment Agency SGV (µg/g dry weight) of toxic metals derived from the CLEA model to assess risk to human health from long-term exposure to metal contamination in soils under different land use situations. SGV are 2009 figures, except for chromium and lead, which are 2002 values.**

	Residential	Allotment	Commercial
Inorganic arsenic	32	43	640
Cadmium	10	1.8	230
Chromium	130*	130	5,000
Lead	450*	450	750
Mercury			
Inorganic mercury	170	30	3,600
Methyl mercury	11	8	410
Nickel	130	230	1,800
Selenium	350	120	1,300

Notes: Figures are based on a sandy loam soil with 6% soil organic matter.
* SGV for residential land without plant uptake are also available for chromium (200 µg/g Cr) and lead (450 µg/g Pb).

arsenic uptake. The risk assessment study did consider the bioaccessibility of arsenic in ingested local dust, generally recording percentages of 10 to 20 % for the stomach phase (Rieuwerts et al., 2006). Even with this percentage markdown, however, the CLEA model came up with estimated daily intakes by local young children up to six-years-old of arsenic from ingested soil and dust that often exceeded the dose (0.3 µg per kilogram body weight per day) used for the derivation of the SGV in Table 4.56 (Rieuwerts et al., 2006).

The health criterion for arsenic refers to the risk of contracting cancer caused by arsenic exposure. Further consideration of risk does need to take into account that cancer is caused by long-term exposure, and intake of arsenic from dust and soil ingestion does decrease as a child gets older (Rieuwerts et al., 2006). Furthermore, many of the homes sampled were not occupied by children. Nevertheless, it remains possible that some infants living in historical mining areas of Cornwall are exposed to potentially unsafe doses of arsenic originating in soil and dust through ingestion and inhalation, and there does appear to be a need for local health care protection programmes (Rieuwerts et al., 2006).

The SGV for chromium assume that all chromium in the soil is in the more toxic form chromium 6. The SGV are 130 µg/g Cr for residential and allotment soils, and 200 µg/g Cr for residential soils not being used to grow plants (Table 4.56). In the risk assessment study of exposure to chromium in Glasgow soils quoted previously, total chromium concentrations in 25 out of 27 samples averaged 117 µg/g Cr, with a range from 65 to 229 µg/g (Broadway et al., 2010). Although 7 of these 25 samples exceeded the 130 µg/g Cr SGV, concentrations of separately measured chromium 6 (cr 6) in the 25 samples only averaged 6 µg/g Cr 6, with a range from less than 2 to 23 µg/g. The percentages of orally bioaccessible chromium in a subsample of these 25 soil samples actually matched the percentages of Cr 6 present. Even if all the bioaccessible chromium were to be chromium 6, bioaccessible chromium 6 concentrations in the soils would still be far below the chromium SGV of 130 µg/g. There is no toxicological concern, therefore, over the chromium in these 25 Glasgow soil samples. The two remaining soil samples, however, had total chromium

concentrations of 3,680 and 658 µg/g, of which high percentages were in the form of chromium 6: 40% (1,485 µg/g Cr 6) and 26% (171 µg/g Cr 6) respectively. Thirty-one per cent (1,160 µg/g) and 18% (116 µg/g) of the total chromium in these two soil samples were orally bioaccessible. However, none of this orally bioaccessible chromium in either of the two most chromium contaminated soil samples was in the form of chromium 6.

Furthermore, the total concentration of lung bioaccessible chromium in each sample was 350 and 63 µg/g Cr respectively, of which only 18 and 13 µg/g was in the form of chromium 6. The concentrations of orally bioaccessible and lung bioaccessible chromium 6 in soil samples probably overlap and should not be summated, when measurable. In this example, even the two most chromium-contaminated soil samples had chromium 6 concentrations that were bioaccessible either in the gut or lungs, which offered no toxic threat when considered against the relevant chromium SGV (Table 4.56).

It is clear then that Soil Guideline Values for toxic metals in soils in different circumstances represent very good starting points in assessments of risk to local human health when used as intended (Environment Agency, 2009b). SGV indicate to risk assessors when a follow-up study of local factors is needed. Such factors might include local population age distributions, durations and levels of activity indoors and out, relative consumption of local vegetables and the oral and lung bioaccessibilities of metals in local soils and dusts. A detailed risk assessment might then lead, as necessary, to the establishment of a local health care protection programme.

We have come to the end of this account of trace metals in the British terrestrial environment. Terrestrial habitats are, however, not isolated ecosystems. Runoff from terrestrial sites with raised trace metal bioavailabilities is likely to contain enhanced bioavailabilities of toxic metals as it drains into our streams and rivers. It is to the freshwater environment, then, that we turn next.

5 Freshwater

Box 5.1 Definitions

acid volatile sulphide (AVS) Sulphide that is released when a sediment is extracted with dilute hydrochloric acid. Predominantly iron and manganese sulphides.

alga (plural algae) Any of various chiefly aquatic, eukaryotic, photosynthetic organisms, ranging in size from single-celled forms (e.g., phytoplanktonic protists such as diatoms and dinoflagellates) to macrophytic green algae, brown algae and red algae. Algae were once considered to be plants but are now classified separately because they lack true roots, stems and leaves.

anadromous An animal, such as a salmonid fish, that is born in freshwater, spends most of life at sea and migrates back to freshwater to spawn.

benthic Pertaining to the bottom of an aquatic habitat.

benthos Organisms living in or on the bottom of an aquatic habitat such as a river bed, sea bed or lake floor.

blue green bacteria A taxon of photosynthetic bacteria (*Cyanobacteria*) (therefore prokaryotes) carrying out photosynthesis. Previously known as blue green algae.

caddisfly A member of the insect order *Trichoptera* with larvae that live in freshwater. Examples include species of the genera *Hydropsyche* and *Plectrocnemia*.

Charophyta (charophytes) Green algae closest related to land plants, including the genera *Spirogyra* and *Mougeotia*.

chironomid midges Non-biting flies of the family *Chironomidae*, widespread globally. Their aquatic larvae are common inhabitants of freshwater systems, particularly sediments rich in organic carbon. Examples of chironomids include species of the genera *Chironomus*, *Chaetocladius* and *Eukiefferiella*.

Chlorophyta (chlorophytes) The largest division of green algae, including the genera *Chlorella*, *Tetraselmis*, *Stigeoclonium*, *Microspora* and *Ulothrix*.

Chrysophyta (chrysophytes) Golden-brown algae mostly found in freshwater, including the genera *Hydrurus* and *Chrysonebula*.

colloid A dispersion of particles, ranging between 1 and 1,000 nanometres (0.001 to 1 μm) in diameter, that are able to remain evenly dispersed in a water medium without settling to the bottom.

cumulative criterion unit (CCU) A combined unit summarising the summated effects of different toxic metals in a mixture. Typically the individual dissolved concentration of each metal is expressed as a fraction of the dissolved concentration of the metal considered by the US Environmental Protection Agency to be toxic to aquatic organisms. These fractional values are then summated for all metals to come up with a CCU score, for example at each site in a survey.

detritus Fragmented particulate organic matter derived from the decomposition of plant and animal remains.

diatoms Photosynthetic unicellular protists abundant in freshwater and marine habitats, either benthic or planktonic. Diatoms are dominant primary producers in the marine plankton of temperate and polar regions. Benthic diatoms are common inhabitants of shallow streams.

Diptera (dipterans) The order of insects containing the 'true flies' with two wings, the second pair typical of insects being much reduced. Examples are houseflies, craneflies, midges, gnats and mosquitoes. Larvae typically feed on decaying plant or animal matter in many habitats including freshwater.

DOM (Dissolved Organic Matter) Matter present in any aquatic system, including soil or sediment pore waters. DOM often consists of humic acids, but is usually difficult to define chemically in its totality.

eutrophication Excessive richness of nutrients in a body of water, encouraging algal growth. Dying or dead algae will sink to deeper water or to the bottom, where their organic matter is broken down by microbial action, depleting the amount of oxygen present with knock-on ecotoxic effects.

flagellate Description of a cell or organism with one or more whiplike organelles called flagella (singular flagellum). Many protistans, for example *Euglena*, are described as flagellate and are united in a single subphylum.

fulvic acids (fulvates) Humic acids of low molecular weight and high oxygen content.

green algae Historically, members of a large group of unrelated green algae, but the term is often now restricted to the algae in the taxon that includes the green plants. Genera of green algae include *Stigeoclonium* and *Spirogyra*.

hardness Feature of a freshwater sample that is determined by the dissolved concentrations of cations with a charge greater than 1. In practice, the ions usually concerned are those of calcium Ca^{2+} and magnesium Mg^{2+}, derived by leaching from local calcium-containing minerals, such as calcite and gypsum, and/or from dolomite, which contains both magnesium and calcium.

instar Developmental stage between moults of an arthropod such as an insect.

interstitial Pertaining to, or occurring within, the pore spaces between sediment particles.

lentic Relating to standing or relatively still water, such as freshwater ponds and lakes.

littoral Pertaining to the shore, commonly the seashore but also the shore of a lake.

lotic Relating to, or designating, natural communities living in rapidly flowing water, such as streams and rivers.

macroinvertebrates Invertebrates that are large enough to be seen without the aid of a microscope.

mayfly A member of the insect order *Ephemeroptera* with larvae that live in freshwater, while the winged adult stage is notoriously short-lived. Examples include species of the genera *Baetis, Ephemerella* and *Heptagenia*, members of the baetid, ephemerellid and heptageniid families respectively.

meiofauna (meiobenthos) A loose grouping of benthic organisms, larger than microorganisms but smaller than macrofauna, that will pass through a 1 millimetre mesh but not a 45 μm mesh. Meiofauna live in freshwater, estuarine and marine environments.

microphyte (adjective microphytic) Microscopic alga in freshwater or marine systems, usually less than 1 millimetre long. Typically they are unicellular but may be in chains or groups.

midge The common name given to many different families of flies, including chironomid midges.

oligochaete worm Worm of the subclass *Oligochaeta* of the phylum *Annelida*, made up of many aquatic (e.g., *Tubifex*) and terrestrial worms, including all earthworms (e.g., *Lumbricus, Dendrodrilus*) and enchytraeids. Oligochaetes typically burrow in sediment, which they ingest.

ochre Naturally occurring pigment consisting mainly of iron oxides and hydroxides, ranging in colour from yellow to deep orange or brown. Ochre typically precipitates in freshwater streams receiving coal or other mine acid water discharge.

periphyton Complex mixture of algae, cyanobacteria and other microbes, and detritus that is attached to submerged surfaces in most aquatic habitats.

pore water The water occurring within the pore spaces between sediment particles.

protists (protistans) Typically unicellular members of a large and diverse group of eukaryotic microorganisms, including diatoms and dinoflagellates, which are important primary producers in aquatic habitats, and non-photosynthetic protistans once collectively referred to as protozoans.

protozoans Former collective form for unicellular protistans that were considered animal-like, being non-photosynthetic.

red algae Members of a large group of mostly marine multicellular algae, including many small seaweeds of British coastal waters, such as species of the genera *Chondrus* and *Corallina*. Freshwater genera of red algae include *Lemanea*.

stonefly A member of the insect order *Plecoptera* with larvae that live in freshwater, particularly fast-flowing streams with stony beds. Examples include species of the genera *Leuctra* and *Nemoura*.

tubificid An oligochaete worm (e.g., *Tubifex*) of the family *Tubificidae* (now renamed *Naididae*), common in freshwater and estuarine sediments which are organically rich.

Water Framework Directive (WFD) An EU directive committing EU member states to achieving good qualitative and quantitative status of all water bodies, including rivers, lakes, estuaries, coastal waters and groundwater. Both ecological status and chemical status need to be at least good to achieve what is termed Good Water Status.

Weight of Evidence (WOE) An approach to ecological risk assessment that combines information from multiple lines of evidence to determine the ecotoxicological status of a contaminated habitat.

5.1 Introduction

Prologue. This chapter on the natural history of metals in freshwater will explore what trace metals occur where and in what concentrations in British freshwater systems, and the biology of these trace metals in the different organisms that make up the freshwater biota.

Trace metals occur naturally on Earth, as components of the rocks of the Earth's crust. Over time, the rocks have eroded, often with the assistance of rainfall and water runoff, and the metals have been released into the atmosphere, the terrestrial environment and the parts of the hydrosphere that make up the freshwater, estuarine and marine environments. In the British Isles, freshwater systems are predominantly made up of streams and rivers, although significant lake systems are present, not least in the Lake District.

Although particulate trace metals may be deposited directly from the atmosphere onto water bodies, trace metals mainly enter freshwater systems in runoff from land. The actions of mankind have accelerated the cycling of metals into freshwater, in effect increasing the rate of entry of dissolved and particle-associated metals into the streams draining river and lake catchments. The mining of metal-rich ores inevitably releases metals into the local catchment, but often greater quantities of toxic metals are introduced into freshwater systems in runoff and effluents from smelters, iron and steel plants and other industrial

works, and from domestic and urban wastewater effluents (Table 1.3).

Such industrial and urban emissions of trace metals have significant effects on the ecologies of the rivers and estuaries into which they are introduced. Nevertheless, in the British Isles, the wider geographical effects of trace metals on the natural history of freshwater systems have still been predominantly caused by historical mining. Mining activities have led to raised concentrations of dissolved metals in the streams draining mining sites, but also, perhaps more significantly and persistently, to much increased loadings of metal-rich sediment particles carried down into streams, particularly during periods of heavy rain and storms. These metal-rich particles fall out from suspension onto the bottom of the stream, to a degree depending on their size and on the rate of water flow.

The process of the mining of metal-rich ores inevitably produces large amounts of solid waste or tailings, consisting of gangue as well as metal-bearing minerals of insufficient metal content to be economically viable. Such tailings may be stored in heaps at the mine site, and rainwater draining from these heaps will contain raised concentrations of metals on a drainage path to a local stream. Typically, such mine drainage is very acidic, further increasing its ecotoxicological potential. Alternatively, such solid waste may be impounded in tailings ponds, the covering of water preventing the export of metal-rich dust into the atmosphere. Such tailings ponds can, however, be the source of environmental problems, if metal-rich water seeps through embankments or even the base of the tailings pond. Much more dramatically, tailings dams themselves may fail, leading to a flood of usually acidic, metal-rich water and sediment slurry to explode down local water courses. An example of such a tailings dam failure, fortunately not in the British Isles, occurred on 25 April 1998 in Southwest Spain. A tailings dam at the Los-Frailes lead–zinc mine at Aznalcóllar, near Seville, burst and released nearly 5 million cubic metres of tailings and acid metal-contaminated water into the Rio Agrio, a tributary of the Rio Guadiamar. The tailings were carried about 40 kilometres down these rivers; covered thousands of hectares of farmland; and only narrowly avoided the Donana National Park, a UNESCO World Heritage Site and one of Europe's most important wetlands for birdlife.

5.1.1 Biology of Trace Metals in Freshwater

Prologue. The interaction of trace metals and life in freshwater follows the general principles governing the biology of trace metals outlined in Chapter 3.

Freshwater organisms will take up trace metals and typically accumulate them, with or without some excretion. A proportion of the essential trace metals taken up will be maintained in metabolically available form for biochemical use in essential metabolic processes. The remaining accumulated essential trace metals, and all accumulated non-essential trace metals, will need to be converted into a detoxified form to prevent toxic interaction with the organism's metabolism, binding in the wrong place at the wrong time. Toxic effects will occur when toxic metals are taken up into the organism, by one or several routes of uptake, at a faster rate than can be coped with by the processes of detoxification and excretion combined.

Dissolved metal is an obvious source of trace metals for organisms living in freshwater but is not the only source. Animals, whether invertebrate or vertebrate, will also take up metals from the diet. Animals can be herbivores, carnivores or omnivores, being generalist or specialist feeders to different degrees. Dietary sources of metals will also include sediments in the case of deposit feeders or indeed suspension feeders if sediments are resuspended by water flow.

Rooted aquatic plants have the potential to take up metals from the pore water of the sediments in which they are rooted. They may also take up dissolved metals from the water surrounding their stems and leaves. Macrophytic algae, or the phytoplankton more commonly encountered in lakes, will take up metals only from solution in the water column. Typically, freshwater animals will take up dissolved trace metals through their permeable external surfaces, such permeable surfaces often being restricted to specialist areas (the gills) in larger invertebrates and in fish. Freshwater bivalve molluscs have very extensive areas of gills for suspension feeding, these gills allowing the entry of trace metals as well as oxygen and water. In the absence of specialist gills, the likes of small oligochaete worms, flatworms and small crustaceans present large areas of permeable body surface to the surrounding water and will take up dissolved metals through these areas.

The predominant invertebrate life in British streams and rivers consists of insect larvae. Typically, the external covering of an insect larva is a chitinous cuticle, rendered impermeable by such processes as tanning or the secretion of a lipid epicuticle. Insects rely on tracheal tubes, opening on the surface of the body through spiracles, to convey oxygen in air directly to the tissues. Thus aquatic insects typically lack a permeable respiratory epithelium in direct contact with the surrounding water with the potential for trace metal uptake from solution. There are, however, some insect larvae that do have a permeable epithelium in contact with the aquatic medium for osmoregulatory purposes, as in the case of the anal papillae of some mosquito larvae, which provide a possible site for dissolved metal uptake. There are also subtle variations as to how larvae of different insect orders obtain oxygen. Hanging down from the surface film, mosquito larvae have a respiratory siphon to provide a direct air supply. Other insect larvae, however, access dissolved oxygen in the water by using tracheal gills, as in caddisflies, mayflies and stoneflies. Tracheal gills are typically thin plates or filaments on the external surface of the larva that have a rich supply of tracheal tubes and a thin permeable cuticle, so that oxygen can diffuse from the water into the tracheal system. There is no route for dissolved metal uptake here.

Like their freshwater insect counterparts, many British freshwater gastropod molluscs also have a terrestrial evolutionary ancestry, in this case as land snails. So-called pulmonate gastropods have a lung for air breathing. Many freshwater gastropods, therefore, return to the water surface to breathe. Yet, even pulmonate gastropods still have large areas of permeable external surface, for example on the foot, with potential for the uptake of dissolved metals.

Fish have gills where they will take up dissolved trace metals, but the diet of fish is a very significant metal source. Most of the food items of freshwater fish will have accumulated trace metals in the local freshwater habitat, either directly from the water or through trophic transfer up food chains. Certain freshwater fish, and trout and salmon are obvious examples, will take adult insects from the water surface, not least caddisflies and mayflies. Much of the metal contents of these flying insects will have been accumulated during the longer-lived aquatic larval stage, possibly in the same stream or river, but more likely in a different but still relatively local water body. Ironically, then, these fish will be ingesting trace metals accumulated from a different freshwater system than the one in which they live. Migratory fish will be taking up metals, both from solution and diet when feeding, in different aquatic environments which may include both marine and freshwater systems, for example in the case of salmon.

Birds and mammals associated with freshwater systems are air breathing, and so all trace metal input will be from the diet. Ducks, geese and swans are herbivorous, while kingfishers prey on fish. Otters are classic carnivores, feeding predominantly on fish.

The uptake of trace metals by freshwater biota depends on their bioavailabilities – their bioavailability from solution, and, in the case of animals, their trophic bioavailability. In both cases, bioavailability still ultimately depends on the biological characteristics of the individual species of relevance, but general

chemical principles affecting bioavailabilities do still apply.

The bioavailability of a dissolved trace metal, typically in the water column but also in sediment pore water for rooted plants or burrowing invertebrates, will depend on the organism concerned, but also on the chemical form of the dissolved metal. Dissolved metals in natural waters, including freshwaters, may not all be in the form of the dissolved ion standing alone, the free metal ion. They may well be complexed by other inorganic ions, such as hydroxides, or by organic molecules, such as humic acids derived from the microbial breakdown of organic matter. The free metal ion is considered to represent the form of many trace metals in solution that binds with the cell membrane proteins that account predominantly for the uptake of these trace metals from solution. Complexation involves chemical bonds that are typically strong enough to outcompete the membrane proteins that might otherwise bind the metal ion for transport into the cell. Importantly, the complexation of free metal ions in solution is an equilibrium process. There will always be a percentage of the dissolved metal present as the free metal ion available for uptake. Nevertheless, it can still be concluded that the greater the complexation (inorganic or organic) of dissolved trace metal ions in an aquatic medium, the lower will be the dissolved bioavailability of the metal. In freshwater systems, including pore waters of freshwater sediments, it is dissolved organic matter (DOM), a component that is rarely well defined chemically, that is responsible for most dissolved trace metal complexation.

In freshwater, a further complicating factor affecting dissolved bioavailabilities of trace metals is the acidity of the medium (Luoma and Rainbow, 2008). Acid conditions (low pH) are caused by high concentrations of hydrogen ions H^+. Hydrogen ions can replace trace metals in dissolved metal complexes, releasing free metal ions, and thereby promoting dissolved trace metal bioavailability. In contradictory fashion, extra hydrogen ions may also compete with free metal ions at a membrane protein binding site and decrease the uptake rate of the dissolved trace metal. The balance between these two opposing effects determines the dissolved bioavailability and uptake rate of a trace metal in different freshwater systems. As a general conclusion, however, acidic conditions in a metal-contaminated freshwater system represent a scenario of high ecotoxicity to the resident biota.

The trophic bioavailability of a trace metal in the diet of a freshwater animal will depend both on the feeding and digestive biology of the animal concerned, and on the chemical binding of the metal in the food item consumed (Rainbow et al., 2011a). The uptake of a trace metals from the diet depends on how much food an animal ingests, the metal concentration in the food and how much of that ingested metal can be digested and assimilated in the gut of the animal. The feeding of an animal is a key feature of its natural history. Feeding rates vary considerably, for example with the feeding preferences of the animal and the availability of food, with consequences for the different dietary bioavailabilities of metals from different diets. A suspension feeder, such as a net-feeding caddisfly larva, will ingest large quantities of small particles, often detritus particles, with a large surface area to volume ratio and high organic content, promoting their accumulation of high concentrations of trace metals. A deposit feeder, such as a tubificid oligochaete worm, will ingest and process large volumes of potentially metal-rich sediment particles. Either feeding process will promote a high dietary input of metals, in comparison with a carnivore feeding on animal tissues. Assimilation efficiency, the percentage of the metal ingested that is assimilated, will vary between species, but also within a species, according to the chemical form of the trace metal in the food and the rate of passage of the food through the gut.

As in the terrestrial environment, organisms that are net accumulators of trace metals can be used as biomonitors of trace metal bioavailabilities in the habitat in which they live. As we shall see in this chapter, trace metal biomonitors, for example hydropsychid caddisflies (Rainbow et al., 2012), are playing an increasing part in identifying British freshwater systems at ecotoxicological risk from high bioavailabilities of toxic metals. Considerable advances have been made in recent years to assess and safeguard our freshwater systems from metal pollution, under the Water Framework Directive of the European Union. The final section of this chapter will review progress and the techniques available to us in the achievement of good qualitative and quantitative status of our water bodies under this directive.

5.2 Metal Concentrations in Freshwater

Prologue. An obvious first question to ask is, 'How much trace metal is present in the water?'

This is a deceptively simple question, but it immediately opens up a can of worms. In a freshwater body such as a stream or a river, the water can be transporting a wide size range of materials, from dissolved ions and molecules, through colloids, to suspended silt, sand and even pebbles and rocks under extreme water flows. Trace metals may be in any of these forms, from free metal ions, to dissolved metal complexes of an array of molecular sizes, to colloids, and in association with suspended particles of all sizes. Small, organically rich particles such as detritus particles or resuspended muds or silts may be particularly metal-rich. Under calmer flows, these suspended particles will precipitate to contribute to the sediments accumulating at the bottom of streams and rivers.

What is needed is a definition of a dissolved metal. Typically, a dissolved metal is operationally defined as metal in a form that will pass through a filter of 0.45 μm (Bowe, 1979; Luoma and Rainbow, 2008). Perversely, filters of 0.2, 0.3 or 0.4 μm are sometimes used, but more often for oceanic waters. The pore size of 0.45 μm was first chosen because it excludes nearly all bacteria. This pore size does, however, let through most colloids, which are dispersions of particles ranging between 1 and 1,000 nanometres (0.001 to 1 μm) in diameter, that are able to remain evenly dispersed in a water medium without settling to the bottom. Trace metals can be found in all these forms in freshwater – as free metal ions, as small and large dissolved metal complexes and associated with colloids. The complexes may be small, as when metal ions are complexed with inorganic anions such as hydroxides and chloride, or of increasing molecular size, as in the case of organic complexation with DOM such as humates including fulvates. The distributions of dissolved trace metals between these forms will vary with the chemistry of the trace metals themselves, with pH and with the nature and concentration of the DOM (Florence, 1977; Turner et al., 1981). Differences in these distributions will affect the relative bioavailabilities of the dissolved metals. Above the 0.45 μm cutoff, the trace metals in the water can be considered to be in suspended particulate form, with a greater likelihood of deposition with increasing particle size and decreasing water flow.

When water concentrations of trace metals are quoted for particular water systems, they may well be broken down into dissolved and particulate metal concentrations, the two together giving a total concentration of the metal in the water.

Table 5.1 attempts to set the scene for expected background concentrations of dissolved trace metals in freshwater systems, both internationally and in the British Isles. Table 5.2 presents dissolved concentrations of selected trace metals in British streams and rivers affected by mining contamination, with data for local streams presumed unaffected, where available. It is clear from Table 5.2 that streams and rivers draining mining areas in the British Isles have a continuing history of greatly raised dissolved concentrations of the trace metals with which we have become familiar. Thus high dissolved concentrations of copper are present in particular Cornish rivers (perhaps in association with arsenic) and in Afon Goch draining Parys Mountain in Anglesey (Table 5.2). Lead and zinc have been reported in high dissolved concentrations in the lead and zinc mining regions of, for example, mid-Wales and the Northern Pennines (Table 5.2).

In addition to these dissolved concentrations in the streams affected by mining, there will also be trace metals associated with the suspended material above 0.45 μm in diameter. Table 5.3 presents examples of the relative contributions of dissolved and suspended particulate metal to total water concentrations of

metals in two mining-affected freshwater systems, the River Carnon in Cornwall and Afon Goch in Anglesey. In the presence of high sediment loadings in the water column of streams and rivers affected by mining contamination, it is the suspended particulate matter that may well hold the majority of the total water content of particular trace metals (for example, iron; Table 5.3). This metal moiety will contribute to the metal content of the sediments when water flows are low enough to allow the suspended matter to deposit out.

We shall return to consider in detail some of the streams and rivers listed in Table 5.2 after an introduction to the concentrations of trace metals found in their sediments. These sediments have ultimately been derived predominantly from mine waste.

5.3 Metal Concentrations in Freshwater Sediments

Prologue. Streams and rivers draining historic mining areas in the British Isles contain sediments with extremely high concentrations of particular trace metals.

Table 5.4 presents sediment concentrations of selected trace metals in British streams and rivers affected by mining contamination. When available, sediment concentrations of trace metals are also shown for local streams presumed to be unaffected by mining.

Streams and rivers draining mining areas contain sediments with exorbitantly high concentrations of particular trace metals compared to local control levels (Table 5.4). There are expected associations between high sediment concentrations of lead and zinc, occasionally with cadmium, in rivers draining the lead–zinc mining regions of the Northern Pennines, Derbyshire, and mid-Wales, and associations between high copper and arsenic sediment concentrations in rivers such as the Tamar in southwestern England (Table 5.4). Where copper is present in high concentrations together with lead and/or zinc in river sediments, then that river, for example the River Carnon in Cornwall, is draining a catchment containing a heterogeneous collection of different metal-bearing ores (Table 5.4). Most mining affected rivers typically have high concentrations of

Table 5.1 **Dissolved concentrations (µg/L) of selected trace metals in freshwater systems, including concentrations that are considered to be low.**

	Freshwater[1]		Freshwater background[2]	Remote rivers[3]	Control British streams
	Range	Median		Range	
Ag	0.01–3.5	0.3	0.3		
Al	8–3,500	300	<30		
As	0.2–230	0.5	2		<3–11
Cd	0.01–3	0.1	0.07	0.0006–0.0018	<0.1–0.7
Co	0.04–8	0.2	0.05		<4–20
Cr	0.1–6	1	0.5		
Cu	0.2–30	3	1.8	0.2–2.4	0.2–14
Fe	10–1,400	500	<30		2–400
Hg	0.0001–2.8	0.1	0.01		
Mn	0.02–130	8	<5		1–100
Mo	0.03–10	0.5	1		
Ni	0.02–27	0.5	0.3	0.5–1.3	<2–26
Pb	0.06–120	3	0.2	0.006–0.017	2–10
Sb	0.01–5	0.2	0.1		
Se	0.02–1	0.2	0.1		

Table 5.1 **(cont.)**

	Freshwater[1]		Freshwater background[2]	Remote rivers[3]	Control British streams
	Range	Median		Range	
Sn	0.004–0.09	0.009	0.03		
V	0.01–20	0.5	0.9		
Zn	0.2–100	15	0.5	0.02–0.25	11–130

Note: Control data for British streams are from Table 5.2.
Sources: Bowen (1979);[1] Förstner and Wittmann(1983);[2] Luoma and Rainbow (2008).[3]

iron (and often manganese) (Table 5.4) derived from iron-rich ores associated with the ores bearing the trace metals of interest.

Table 5.4 also quotes an attempt, based on Cornish river sediments, to define what sediment concentrations can be considered to indicate moderate and high contamination by arsenic, copper, lead or zinc (Abrahams and Thornton, 1987). These definitions provide a degree of context for the ranges of very high river sediment concentrations quoted throughout Table 5.4.

Given that active mining ceased in many mining districts in Britain more than 100 years ago, and that sediments will move downstream under strong water flows, however intermittent, more recently deposited river sediments are usually lower in metal concentrations than their historical counterparts (Table 5.4). Table 5.4 includes data for metal concentrations in sediments deposited by rivers overflowing onto their flood plains on occasions of high rainfall. Flood sediment particles will on average be smaller, lighter and more organically rich than the sediments that remain on the river bottom. Such small, organically rich mud or silt particles offer large surface areas with many metal binding sites. It is not surprising, therefore, that floodplain sediments often have higher metal concentrations than adjacent sediments that remain on the river bottom (Table 5.4).

In the next section, we shall explore in more detail some of the streams and rivers highlighted in Tables 5.2 and 5.4. At this point, we can generally assume that a high dissolved concentration will translate into a high dissolved bioavailability of the metal to the local biota, even though the relationship between dissolved concentration and dissolved bioavailability is variable, depending on such factors as the presence of DOM or the pH. Similarly, for the moment, we will assume that a high concentration of a trace metal in a sediment, as listed in Table 5.4, will also translate into a high bioavailability of the sediment metal to the likes of rooted macrophytes, to soft-bodied burrowing invertebrates and to deposit-feeding invertebrates ingesting the sediment. As will have become increasingly apparent, there is not necessarily a direct relationship between the metal concentration in a sediment and its bioavailability, which will be variably affected by how strongly any sediment metal is bound chemically in that sediment.

5.4 Metal-Contaminated Freshwater Systems

Prologue. Several of the streams and rivers highlighted in Tables 5.2 and 5.4 deserve further attention. Usually they are the streams and rivers associated

Table 5.2 Dissolved concentrations (μg/L) of selected trace metals in British streams and rivers affected by mining contamination, and local control streams and rivers where data are available.

	Ag	As	Cd	Co	Cu	Fe	Mn	Ni	Pb	Zn
SW England										
River Carnon										
1982[1]	<1	14–58	24.8 (6.4–106)	37–79	593 (370–973)	5,020 (2,670–10,400)	1,780 (955–3,320)	66–172	2–31	12,500 (3,630–57,200)
1992[2]		0–249								
1992–94[3]			1–22	0–100	20–1,300	0–49,000	250–1,800		0–22	120–23,000
River Hayle										
1973[4]					20–400	50–400				40–1,800
1997–2003[5]					5–150					
River Gannel										
1970–73[6]										150
1975–76[7]					0–20	100–3,000			20–700	20–2,000
Cornwall control rivers[1]		<3–11	<0.1–0.7	<4–20	2–14	76–1,650	2–139	<2–26	2–33	27–409
Wales										
Afon Goch, Anglesey										
1977[8]					2,850–10,600	11,000–110,000	2,150–7,750			5,550–21,700
1990–92[9]					0–59,900	0–260,000	10–49,000			170–41,900

Site								
River Ystwyth								
1970-71[10]		1.1-1.3			7-15		2-6	200-270
1982[11]	0.00-0.01		1.5-36				10-2,100	20-15,400
River Rheidol								
1970-71[10]		1.0-3.4			9-28		1.3-2.4	50-130
1982[11]	0.01-0.23		0.9-103				4-770	110-25,000
River Conwy system								
1969[12]		0-10.7	0.8-5.6			0-9.8	2.7-31.8	94-3,260
2000-2[13]							17-50	300-342
1969 local controls[12]		0-1.2	0.8-4.3			0.3-7.9	0.5-16.5	2-75
Wales control rivers[10]		0.4	0.7	1.7	0.8	0.5	0.7	11
Northern Pennines								
River Nent system								
Gillgill Burn 1973-74[14]		1-60	<5-14	<2-7,400	<5-2,010		7-3,700	50-30,200
1977[15]		2.8-18.7	1.2-6.9	35-6,370	53-1,070	1.2-63.2	1-518	1,090-15,800
Caplecleugh 1981[16]		1.7-5.3		<20-80	120-180	1-13	1-13	5,250-8,650
Nenthead stream 2006[17]		0.18-0.33	1.4-1.9	2.7-12.1	18-190		12-36	590-2,810

Table 5.2 (cont.)

	Ag	As	Cd	Co	Cu	Fe	Mn	Ni	Pb	Zn
West Allen River										
1979–80[18]										40–4,150
River Derwent										
1978–79[19]			0–1.2						3–35	7–580
Mining site streams										
Means 2006[17]		0.16–0.55	0.04–5.32		0.5–2.5	30–90	5–190	1–12	1–36	9–2,810
Lake District										
Mining site streams										
Threlkeld stream 2006[17]		0.39	19.1		5.2	9.0	1,100	76	156	11,000
Means 2006[17]		0.10–278	0.02–19.1		0.2–7.5	2.5–9.0	2–1,100	0.1–76	0.1–156	2–11,000
Scotland										
Leadhills										
Glengonnar Water[20]										
3 yr means 2001–10			0.4–1.0						14–32	40–125
Maximum 1999–2011			1.8						130	170

Note: Single values represent a mean value, and ranges are also included where available.

Sources: Bryan and Gibbs (1983);[1] Hunt and Howard (1994);[2] Neal et al. (2005);[3] Brown (1977a);[4] Durrant et al. (2011);[5] Bryan and Hummerstone (1973b);[6] Foster (1982a);[7] Foster et al. (1978);[8] Boult et al. (1994);[9] Abdullah and Royle (1972);[10] Jones et al. (1985);[11] Elderfield et al. (1971);[12] Brydie and Polya (2003);[13] Say and Whitton (1980);[14] Nuttall and Younger (1999);[15] Patterson and Whitton (1981);[16] Environment Agency (2008a);[17] Abel and Green (1981);[18] Burrows and Whitton (1983);[19] Scottish Environment Protection Agency (2011).[20]

Table 5.3 **Example contributions of suspended particulate matter (>0.45 μm) to average total water concentrations (μg/L) of trace metals in mining contaminated streams or rivers.**

Suspended particulate matter	Cd	Cu	Fe	Mn	Pb	Zn
Carnon River, Cornwall[1,2]						
1980–81 concentration (μg/L)	0.8	96–110	2,060–4450	10–15	11–27	30–70
%age contribution to total water concentration	3.1	14.0–16.7	47.0–51.8	0.8–1.5	28.0–64.3	0.0–0.7
Afon Goch, Anglesey[3]						
1990–92 concentration (μg/L)		0–520	50–7,710	0–60		10–120
%age contribution to total water concentration		0–100	0.1–100	0–3.5		0.1–5.6

Sources: Bryan and Gibbs (1983);[1] Bryan and Hummerstone (1973a);[2] Boult et al. (1994).[3]

with historical mining sites, and their adjacent terrestrial habitats have often already been encountered in Chapter 4.

5.4.1 Southwest England

Prologue. It is clear from Tables 5.2 and 5.4 that rivers in Southwest England (Figure 5.1) are amongst the most metal-contaminated in the British Isles.

5.4.1.1 River Carnon

Prologue. The dissolved and sediment trace metal concentrations in the River Carnon are extremely high for arsenic, copper, iron, lead, manganese and zinc.

The River Carnon in Cornwall drains one of the most heavily mined river catchments in Britain (Chapter 2). This catchment extends from the historical copper and tin mining districts of Chacewater, St Day, Carharrack, Gwennap and Baldhu to the east of Redruth, past Twelveheads and Bissoe, down to Devoran, at the head of the river estuary Restronguet Creek (Figure 5.2). Restronguet Creek is part of the extensive Fal Estuary system on the south coast of Cornwall.

There was a postscript to the story of mining in the Carnon valley, a postscript that was to have a significant effect on metal concentrations in the Carnon River at the end of the twentieth century. The closure of the Wheal Jane mine meant that water was no longer being pumped out of the mine workings. Underground water levels rose, causing the local adit, the Nangiles Adit, to burst in January 1992 (Hunt and Howard, 1994). This released more than 45,000 cubic metres of acidic metal-contaminated water into the River Carnon, just downstream of the discharge of the Great County Adit.

Not surprisingly given the history of local mining, the dissolved metal concentrations in the River Carnon (Table 5.2) are extremely high for arsenic, copper, iron, lead, manganese and zinc, both before and after the bursting of the Nangiles Adit in 1992 (Bryan and Gibbs, 1983; Hunt and Howard, 1994; Neal et al., 2005). Even prior to 1992, the River Carnon water was quite acidic (pH 3.8), helping keep the high metal concentrations in solution. Additionally, the river contained significant amounts of trace metals in particulate form in the water column, for example copper, iron and lead (Table 5.3) and also arsenic (71% total arsenic at Devoran in 1982) (Bryan and Gibbs, 1983). The Great County Adit was the

Table 5.4 **Sediment concentrations (μg/g dry weight) of selected trace metals in British streams and rivers affected by mining contamination, and local control streams where data are available.**

	As	Cd	Co	Cu	Fe	Mn	Ni	Pb	Zn
SW England									
River Carnon									
1972[1]		5 (2–18)		1,650 (42–6,500)	57,000 (28,000–97,000)	1,010 (260–3,990)		470 (48–1,440)	1,080 (150–7,000)
River Hayle									
1972–75[2]				2,400					
1973[3]				100–5,000	500–40,000				100–1,500
2004–8[4]				133–1,720					390–884
Red River									
1972[1]		4 (2–7)		590 (56–1,680)	43,000 (15,000–99,000)	930 (220–3,920)		295 (10–1,030)	605 (69–800)
River Gannel									
1972[1]		7 (3–35)		100 (22–520)	45,000 (29,000–65,000)	1,440 (230–6,100)		605 (43–4,410)	1,000 (150–7,000)
1972–75[2]								6,610	
River Tamar									
2000s[5]	73 (5–11,000)	1.1 (0.3–22)		85 (11–8,000)				43 (13–450)	194 (48–1,900)
Cornwall control streams[1]		3–4		46–563	39,000–48,000	657–1,000		89–402	255–391
Cornwall moderately contaminated[6]	100–500			100–500				100–500	300–700

Cornwall highly contaminated[6]	>500	>500			>500	>700
Wales						
River Ystwyth						
1917[7]					1,100,000	1,800,000
1980s channel bed[8]	0.1–1.5	10–50			100–2,500	100–750
2012 flood sediment[8]	2–14	20–117			700–54,000	150–4,000
River Rheidol						
1917[7]					1,000,000	1,200,000
1970s channel bed[8]	0.1–1.5	25–75			100–500	300–400
2012 flood sediment[8]	2–7	40–75			500–1,400	400–8,000
2000s flood sediment[5]	0.4–4.4	28–76			440–2,520	120–680
River Conwy system						
1969 River Conwy[9]	50 (20–130)	43 (30–60)	3,260 (1,000–8,500)	54 (50–85)	165 (60–400)	530 (200–1,300)
1969 tributaries[9]	81 (20–130)	63 (30–200)	4,980 (1,600–>10,000)	82 (40–300)	2,370 (85–>10,000)	3,700 (1,000–>10,000)
2000–2[10]					212–523	1,090–1,190
1969 local controls[9]	70 (10–1,300)	21 (5–300)	3,680 (300–>10,000)	55 (20–160)	160 (20–400)	460 (<50–850)
Northern Pennines						
River Nent system						
1980s flood sediment[5]					5,260 (224–15,800)	16,300 (4,360–38,000)

Table 5.4 (**cont.**)

	As	Cd	Co	Cu	Fe	Mn	Ni	Pb	Zn
1993 flood sediment[11]		91		179	49,000	3,060		3,130	28,600
West Allen River									
1980s flood sediment[5]		17 (5–33)		32 (22–40)				1,300 (98–3,170)	763 (74–1,130)
South Tyne									
1993 flood sediment[11]		<4–14		20–64	24,600–53,200	1,100–2,940		561–2,860	935–5,300
River Derwent									
1978–79[12]		0.6–13.8						96–3,120	82–2,760
Derbyshire									
River Ecclesbourne 1976[13]		3–117						794–25,700	447–10,700
Scotland									
Leadhills									
Glengonnar Water									
1982–84[14]		up to 4						up to 16,400	up to 1,680
1990s[15]		0.14 (0.04–0.2)		170 (80–450)				15,300 (3,000–35,900)	830 (510–2,050)
Tyndrum									
River Fillan system 2000s[16]								25–11,000	750–30,000

Isle of Man

River Neb 1978–79[17]	<5–>20	200–>500	250–>5,000	250–>5,000
Glen Maye River 1978–79[17]	<5–>20	<50–100	<100– >5,000	100–>5,000
Laxey River 1978–79[17]	<5–>20	<50–500	100–5,000	100–>5,000

Note: Single values represent a mean value, and ranges are also included where available.

Sources: Aston et al. (1974);[1] Brown (1976);[2] Brown (1977a);[3] Khan et al. (2011);[4] Environment Agency (2008b);[5] Abrahams and Thornton (1987);[6] Griffith (1919);[7] Foulds et al. (2014);[8] Elderfield et al. (1971);[9] Brydie and Polya (2003);[10] Hudson-Edwards et al. (1996);[11] Burrows and Whitton (1983);[12] Moriarty et al. (1982);[13] Scottish Environment Protection Agency (2011);[14] Rowan et al. (1995);[15] Pulford et al. (2009);[16] Southgate et al. (1983).[17]

Figure 5.1 Selected rivers contaminated with trace metals in Southwest England.

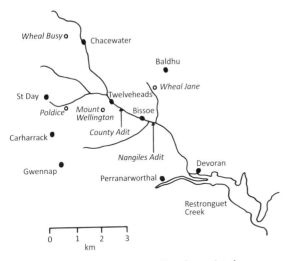

Figure 5.2 River Carnon, Cornwall and associated mining areas. (After Bryan and Gibbs, 1983. Reproduced by permission of the Marine Biological Association, UK.)

major source of dissolved copper into the Carnon River in the early 1980s, while the then-active Wheal Jane was contributing significant amounts of lead, zinc and also cadmium (a contaminant of zinc ore) to the river (Bryan and Gibbs, 1983).

In February 1992, immediately after the Nangiles Adit episode, dissolved arsenic concentrations in the River Carnon rose from an upstream concentration of 1 µg/L to 249 µg/L, 240 µg/L of this arsenic being in the toxic form of arsenite. By July 1992, dissolved concentrations of arsenite had fortunately fallen back to less than 1 µg/L As in the river (Hunt and Howard, 1994).

In line with the high dissolved trace metal concentrations in the River Carnon, its sediments also show extremely high metal concentrations, notably of copper, iron, lead, manganese and zinc (Table 5.4). In the sediment classification presented in Table 5.4 (Abrahams and Thornton, 1987), the sediments of the River Carnon can be considered highly contaminated in copper, lead and zinc. Data are lacking for concentrations of arsenic in River Carnon sediments. Arsenic concentrations, however, are extremely high in the estuarine sediments of Restronguet Creek (Bryan and Gibbs, 1983), and so it can be safely concluded that the river sediments are similarly highly contaminated with arsenic.

5.4.1.2 River Hayle

Prologue. The River Hayle has high dissolved concentrations of copper, and high sediment concentrations of copper and arsenic.

The River Hayle (Figure 5.3) has its source near the village of Crowan to the south of Camborne, and flows for nearly 20 kilometres to reach the sea at the town of Hayle, at the southern end of St Ives Bay on the north Cornish coast. The slates and sandstones of the catchment of the Hayle are characterised by high-temperature hydrothermal veins cross-cut by younger, lower-temperature veins. The former veins contain cassiterite, arsenopyrite and chalcopyrite with tin, arsenic and copper respectively, and the latter are characterised by chalcopyrite, sphalerite and galena, providing copper, zinc and lead (Rollinson et al., 2007).

The catchment of the Hayle River has had a long association with firstly tin and subsequently copper mining (Chapter 2). Tin was mined at Godolphin in the sixteenth century. In the case of copper, Wheal Fortune at St Hilary was particularly productive in the

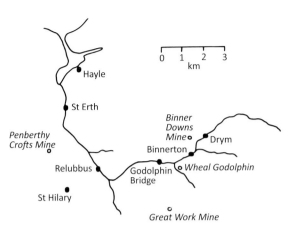

Figure 5.3 River Hayle, Cornwall and associated major mines.

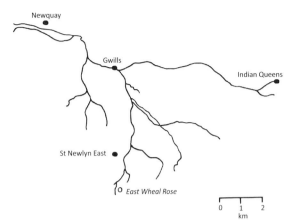

Figure 5.4 River Gannel, Cornwall.

first half of the eighteenth century, and the Godolphin mine was very active a century later up to the 1840s.

High copper (and to a lesser extent zinc) concentrations persist in dissolved form and in the sediments in the River Hayle (Tables 5.2 and 5.4), and in the suspended particulate material in the river water (Brown, 1978). Stream sediments in the Hayle catchment also contain very high concentrations of arsenic, up to 14,000 µg/g (Abrahams and Thornton, 1987), as might be expected given the common association of copper and arsenic in ores. Again not unexpectedly, the Hayle also contains high concentrations of iron, in solution, in particulate form in the water and in sediments (Tables 5.2 and 5.4).

5.4.1.3 Red River
Prologue. The sediments of the Red River have raised concentrations of copper, lead and zinc.

The Red River rises by Bolenowe, near Troon just south of Camborne, and flows for about 13 kilometres before entering the north end of St Ives Bay at Godrevy (Figure 5.1). The catchment area of the Red River includes the important historical mining regions of Camborne, Pool and Tuckingmill. The ancient practice of tin streaming took place in the Red River, before the advent in the catchment of underground mining for copper, tin and some zinc,

particularly in the eighteenth and nineteenth centuries (Chapter 2).

Extensive mining activities provided the suspended sediment which coloured the river red, although today the Red River has lost its previously distinctive colour. Over its history, the Red River has been much modified, having been diverted and canalised, to become in effect an industrial drain. Concentrations of trace metals in its sediments are significantly raised (Table 5.4), for example in the cases of copper, lead and zinc, in addition to associated iron and manganese (Table 5.4).

5.4.1.4 River Gannel
Prologue. The River Gannel near Newquay contains very high sediment concentrations of lead and zinc.

The River Gannel (Figure 5.4), rising in the village of Indian Queens, drains the old lead mining region of Newlyn Downs, to the southeast of Newquay, the home of East Wheal Rose (Chapter 2).

Expectedly, concentrations of lead, and its common associate zinc, are very high in the sediments of the River Gannel (Table 5.4). Dissolved concentrations of lead and zinc in the Gannel are also high, but not as high as in rivers draining historical lead mining regions elsewhere in the British Isles (Table 5.2). Total concentrations of lead in the river water, on the other hand, are extremely high, reaching 35,000 µg/L Pb in the 1970s (Brown, 1977a). Lead in the River Gannel, therefore, appears to be mostly associated with sediment particles, whether deposited or in the water column.

5.4.1.5 River Tamar

Prologue. The sediments of the River Tamar are very high in arsenic, copper and zinc, together with raised concentrations of lead.

The River Tamar forms the southern part of the county boundary between Cornwall and Devon. The river rises about 6 kilometres from the north Cornish coast near Bude before it flows for nearly 100 kilometres southwards across the peninsula to enter Plymouth Sound. On its way, the Tamar passes Launceston and then flows through the historic mining region of Gunnislake, just north of the head of its estuary (Figure 5.5). Amongst the tributaries

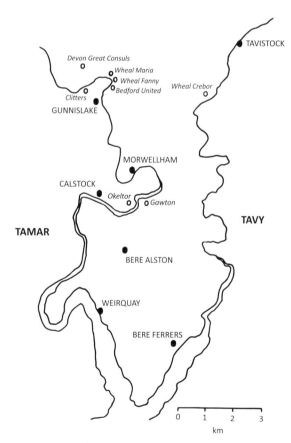

Figure 5.5 Southern part of the River Tamar and adjacent River Tavy, Cornwall and Devon.

of the River Tamar are the Lynher (or St Germans River) in Cornwall, and the Tavy in Devon, flowing through Tavistock.

The most important ore deposits in the Tamar catchment lie in a belt 18 kilometres by 6 kilometres in the Gunnislake to Calstock area, extensively mined for copper, arsenic and lead during the nineteenth century (Chapter 2). The Gunnislake mines were drained by adits into the River Tamar and were served by Morwellham Quay at the top of the Tamar Estuary. Ore deposits extend down into the Bere Alston peninsula and into the adjacent catchment of the Tavy (Figure 5.5) and yielded silver at Bere Alston.

Against this historical mining background, it is not surprising that sediments of the River Tamar are highly contaminated with arsenic, copper and zinc, with moderate contamination of lead (Table 5.4). The highest concentrations of arsenic in River Tamar sediments in the 1970s (up to 5,000 µg/g As) were in the Gunnislake and Callington areas, with total water arsenic concentrations exceeding 250 µg/L As at Gunnislake (Aston et al., 1974).

5.4.2 Wales

Prologue. Rivers in Wales contaminated with trace metals derived from mining include Afon Goch on Anglesey, the Rivers Ystwyth and Rheidol in Ceredigion, and the River Conwy in north Wales.

5.4.2.1 Afon Goch

Prologue. Afon Goch contains very high concentrations of copper, zinc, iron and manganese.

Afon Goch rises on the south side of Parys Mountain on Anglesey and flows south and then northeast for 11 kilometres before entering its estuary Dulas Bay (Figure 5.6). The first kilometre of the stream drains a stark region of mine spoil (Boult et al., 1994), a manifestation of the copper mining on Parys Mountain after the discovery of the large shallow body of copper ore in 1768 (Chapter 2).

The legacy of intense copper mining activity between 1770 and 1800 can be seen in the

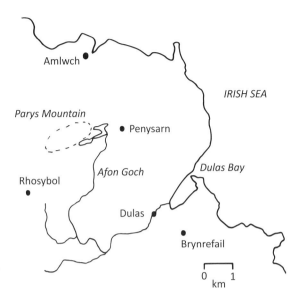

Figure 5.6 Afon Goch, Anglesey. (After Boult et al., 1994, with permission.)

concentrations of metals in Afon Goch. The sulphide ore of Parys Mountain oxidises to give sulphuric acid. So the waters of the Afon Goch are acidic, with very high concentrations of copper, zinc, iron and manganese. (Foster et al., 1978; Boult et al., 1994). The dissolved concentrations of these trace metals are indeed amongst the highest in any stream or river in Britain (Table 5.2). In addition to the dissolved metal, the contributions of metal associated with suspended particles to total water concentrations of copper and zinc in the Afon Goch may approach 100% (Table 5.3).

5.4.2.2 Rivers Ystwyth and Rheidol

Prologue. The Rivers Ystwyth and Rheidol were historically contaminated with lead and zinc from mining.

The historical lead, zinc and silver mining region of mid-Wales is centred on Ceredigion and Powys, particularly that part of Powys that was Montgomeryshire. We shall concentrate here on the Rivers Ystwyth and Rheidol, which both reach the sea at Aberystwyth (Figure 5.7). The trace metal status of these two rivers has been the most studied (Tables 5.2 and 5.4), while the River Clarach to the north of Aberystwyth (Figure 5.7) drains the same

mining region with a particular concentration of historical silver mines (Carpenter, 1924; Jones, 1940, 1949, 1958; Newton, 1944).

Mining of ores containing lead, zinc and silver in mid-Wales peaked in the middle of the nineteenth century (Chapter 2). In Ceredigion, mining was particularly important in the upper valleys of the Rivers Ystwyth and Rheidol. The large mines at Cwm Ystwyth and Frongoch drained into the Ystwyth, and the Cwm Rheidol mine and the Goginan mine at Melindwr drained into the Rheidol (Figure 5.7). Lead and silver were the two main metals extracted, while the less sought after zinc ore was often left in spoil heaps. Given the relatively low abundance of silver, it was lead and zinc that drained into the rivers in particularly high concentrations, whether dissolved (Table 5.2) or associated with mineral particles that were originally in suspension before deposition as bottom sediments (Table 5.4). The Cwm Ystwyth mine closed in 1893 and the Frongoch mine in 1903, although there was some minor activity at these mines at the time of the First World War to extract lead and zinc from spoil heaps (Jones, 1940). In the Rheidol valley, mining did continue spasmodically in the first half of the twentieth century (Newton, 1944). Cwm Rheidol, for example, reopened briefly in 1938, when a large quantity of heavily metal-polluted water was discharged into the River Rheidol (Newton, 1944).

The Rivers Ystwyth and Rheidol, as well as other rivers of Ceredigion, were already known to be severely contaminated with lead and zinc by the time of the Rivers Pollution Act of 1876 (Griffith, 1919; Davies, 1987). There had been a general improvement in the water quality of the River Ystwyth by the time of a survey in 1919, but the beds of the Ystwyth and Rheidol were still then covered with a layer of sediment of mining origin that ranged in depth from a few centimetres to more than a metre (Carpenter, 1924). In 1919, the total lead concentration in the water of the lowest reaches of the Ystwyth was 400 µg/L Pb, decreasing to below 100 µg/L Pb by 1922 (Davies, 1987). Dissolved lead concentrations in the Rheidol decreased from a range of 200 to 500 µg/L in 1919 to 1921 (Carpenter, 1924) to less than 40 µg/L in 1971 to 1972 (Davies, 1987).

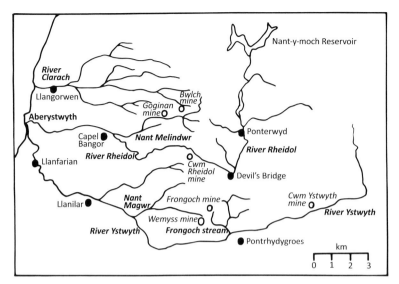

Figure 5.7 Rivers Ystwyth, Rheidol and Clarach, mid-Wales, and associated major mines.

While lead and zinc concentrations in the waters and sediments generally of the rivers Ystwyth and Rheidol are lower than in the past, extremely high concentrations of these metals (Tables 5.2 and 5.4) can still occur at the river entry points of adits from old mines such as Cwm Ystwyth and Frongoch on the Ystwyth and Cwm Rheidol and Goginan on the Rheidol (Figure 5.7) (Jones et al., 1985).

5.4.2.3 River Conwy

Prologue. The water and sediments of the River Conwy are high in lead and zinc from mining near Llanrwst.

The Conwy catchment in north Wales contains lead and zinc ores that outcrop on the western side of the river valley, particularly in the historical mining district of Llanrwst (Figure 5.8). Although mining there goes back to Roman times, most mining took place in the second half of the nineteenth century (Brydie and Polya, 2003), with lead being more sought after than zinc, which was often left in spoil tips (Elderfield et al., 1979).

As a legacy of mining in the catchment, the River Conwy and some of its tributaries contain high concentrations of lead and zinc, both in solution

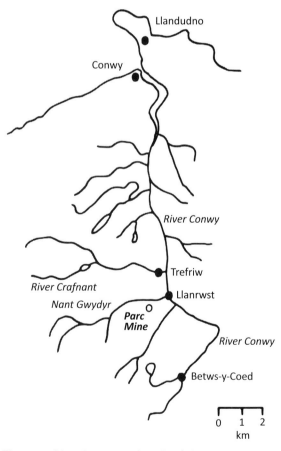

Figure 5.8 River Conwy, north Wales. (After Brydie and Polya, 2003, with permission.)

(Table 5.2), and in the sediments (Table 5.4). Two tributaries draining the Llanrwst district, the River Crafnant and Nant Gwydyr (Figure 5.8), are particularly contaminated by lead and zinc (Johnson and Eaton, 1980; Gao and Bradshaw, 1995; Brydie and Polya, 2003).

The former lead and zinc mine Parc mine, now derelict, lies in a clearing of the Gwydyr Forest in the Conwy valley, 2 kilometres southwest of Llanrwst (Figure 5.8). The heyday of the Parc mine was in the second half of the nineteenth century, but mining there did continue in the twentieth century before final closure in the 1960s. This closure left behind a large mine tailings heap, and this heap was a significant source of lead and zinc into the local stream (Nant Gwydyr), both via leaching of dissolved metals and by bulk erosion (Gao and Bradshaw, 1995). In the late 1970s, a reclamation programme was put in place, involving drainage engineering, recontouring and grassing of the heap after addition of coarse quarry waste material. This stabilisation work prevented the further release of particulate material from the tailings heap and improved the water quality of the Nant Gwydyr. Nevertheless, the Nant Gwydyr can still be considered metal-contaminated. In the 1990s, it was contributing about 1 tonne of zinc and 0.2 tonne of lead, as well as 0.05 tonne of cadmium, each year to the River Conwy, mostly now from the old mine adit (Gao and Bradshaw, 1995).

5.4.3 Northern Pennines

Prologue. Ores containing lead, zinc and often silver are found in the Northern Pennines, for example at the heads of catchments that drain northwards into the southern Tyne via the Rivers Nent, West Allen and East Allen, and north into the Tyne itself via the River Derwent, but also southwards into the headwaters of the Rivers Wear and Tees.

The exploitation of the lead and silver ores of the Northern Pennines has a long history (Chapter 2). In the eighteenth and nineteenth centuries, lead mining boomed on Alston Moor, based at Nenthead and Killhope and drained by the South Tyne (Figure 5.9). Lead mining in the Allendales (drained by the West and East Allen Rivers) and in the Derwent valley (Figure 5.9) also peaked in these two centuries, as it did in Weardale and in Teesdale.

The zinc ore associated with the lead ore in Nentdale and the Allendales was in demand for galvanization in the nineteenth century. Zinc was, therefore, sought then from the spoil heaps of the local lead mines, as well as being mined specifically, particularly in Nentdale at the end of the century. Zinc production ended in the Nent valley in the early 1900s, although during the Second World War some of the old spoil material was reworked to obtain zinc for the war effort (Nuttall and Younger, 1999).

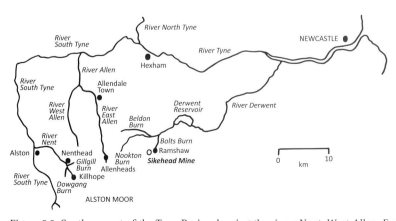

Figure 5.9 Southern part of the Tyne Basin, showing the rivers Nent, West Allen, East Allen, South Tyne and Derwent.

With such a mining history, it is inevitable that the local rivers such as the Nent, West Allen and Derwent, together with some of their tributaries, should have high concentrations of lead and zinc (together in some cases with cadmium), both in dissolved form and in deposited sediment (Tables 5.2 and 5.4).

5.4.3.1 River Nent

Prologue. The River Nent drains the peat-covered fells of Alston Moor, flowing north through the mining districts of Killhope and Nenthead to join the South Tyne by Alston (Figure 5.9).

Amongst the small metal-rich tributaries of the River Nent is the narrow Gillgill Burn, high in both lead and zinc (Table 5.2). The Gillgill Burn rises approximately 1.5 kilometres to the east of Nenthead and drains a large complex of disused mine workings (Armitage, 1980; Say and Whitton, 1980, 1981). Also just east of Nenthead is a small tributary stream, only 14 metres long, leading from the Caplecleugh Low Level adit draining the old Caplecleugh lead–zinc mine, and high in zinc (Table 5.2) (Pearson and Whitton, 1981; Say and Whitton, 1981). A particularly important Nentdale adit still carrying out a major mine dewatering function is the Nent Force Level (Younger, 1998; Nuttall and Younger, 1999). This adit was driven at the end of the eighteenth century, totalling nearly 8 kilometres long and taking 66 years to complete. Nent Force Level closely follows the course of the Nent valley, draining many old lead–zinc mines and flowing into the River Nent near Alston. The discharge of the Nent Force Adit is still particularly rich in zinc, but not lead (Younger, 1998; Nuttall and Younger, 1999).

5.4.3.2 River West Allen

Prologue. The River West Allen rises in the old lead-mining district of Coalcleugh. The river flows north for about 15 kilometres before joining the adjacent East Allen River to form the short Allen River to enter the South Tyne (Figure 5.9).

The West Allen catchment contains more zinc-bearing lodes than the East Allen catchment (Abel and Green, 1981). Furthermore, much of the old zinc-bearing mine waste material in the East Allen catchment has been removed from the surface, while there remained many old mine heaps in the West Allen valley, at least until the 1980s (Abel and Green, 1981). The upshot is that in the 1980s, the River West Allen contained considerably higher levels of dissolved zinc than the East Allen, dissolved concentrations that are high on a national scale (Table 5.2). Concentrations of both zinc and lead are also high in the sediments of the River West Allen (Table 5.4).

5.4.3.3 River Derwent

Prologue. The River Derwent in Northeast England drains moorland on the edge of the Northern Pennine orefield, on the border between County Durham and Northumberland.

The River Derwent (Figure 5.9) is formed by the confluence of the Beldon Burn and the Nookton Burn, but it is the southern tributary Bolts Burn that introduces lead, zinc and cadmium into the river (Burrows and Whitton, 1983). Bolts Burn drains a district of old lead mines such as the Sikehead mine at Ramshaw. The River Derwent, correspondingly, has raised concentrations of dissolved lead and zinc and also cadmium (Table 5.2), as well as high sediment concentrations of these three metals (Table 5.4). Although these metal levels are not as high as those in the River Nent, for example, they still have the potential to be of ecotoxicological significance. The River Derwent then flows northeast via the Derwent Reservoir, west of Consett, to join the Tyne near the MetroCentre in Newcastle (Figure 5.9).

A word of caution is needed here because there is more than one River Derwent in England. The name Derwent is said to refer to oak trees, so a River Derwent would lie in a valley originally thick with oak trees. The other Rivers Derwent are in Derbyshire, Yorkshire and the Lake District. In

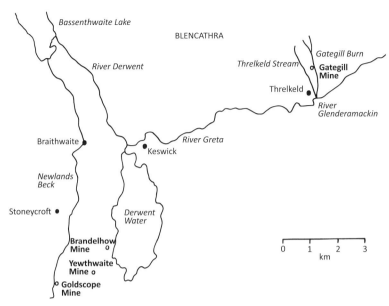

Figure 5.10 The rivers Glenderamackin, Greta, Derwent and Newlands Beck, with associated major mines near Keswick in the Lake District.

Derbyshire, the River Derwent rises in the Peak District and flows for more than 100 kilometres to join the River Trent south of Derby. There is another Derwent Reservoir along its course. The River Derwent in Yorkshire flows from the North York Moors, through the Vales of Pickering and York to join the Yorkshire Ouse at Barmby on the Marsh. The River Derwent in the Lake District rises at Styhead Tarn under Scafell Pike, flows north through the valley of Borrowdale and through Derwentwater before being joined by the River Greta by Keswick (Figure 5.10). This River Derwent then flows through Bassenthwaite Lake and west via Cockermouth to reach the Irish Sea at Workington.

5.4.4 Derbyshire

Prologue. The Peak District of Derbyshire has a long history of lead mining (Chapter 2).

5.4.4.1 River Ecclesbourne

Prologue. The River Ecclesbourne drains the historical lead mining district of Wirksworth in the Peak District.

The River Ecclesbourne in Derbyshire is a tributary of the Derbyshire River Derwent, which it joins near Duffield. It rises near Wirksworth before it flows south for about 16 kilometres to Duffield (Moriarty et al., 1982). The Romans mined lead at Wirksworth, and the Wirksworth District was also important in lead mining in the Middle Ages, particularly in the thirteenth century, when the Barmote Courts were established (Chapter 2). Lead production in the Peak District was considerable during the eighteenth century and remained high until the middle of the nineteenth century.

Not unexpectedly, in the 1970s, the sediments of the River Ecclesbourne were shown to be high in lead, zinc and cadmium (Table 5.4).

5.4.5 Lake District

Prologue. Keswick represents an example of an historical mining area in the Lake District.

5.4.5.1 Threlkeld, Keswick

Prologue. The Threlkeld stream draining into the River Glenderamackin contains high dissolved concentrations of cadmium, lead and zinc.

The area around the historical Gategill lead and zinc mine at Threlkeld near Keswick drains via several small streams into the River Glenderamackin and then the River Greta (Figure 5.10). The River Greta was adjacent to the sixteenth century copper smelter at Brigham (Chapter 2) and joins the River Derwent by Keswick.

Threlkeld stream contained the highest 2006 - dissolved cadmium, lead and zinc concentrations of 11 Lake District mining site streams summarised in Table 5.2.

5.4.6 Scotland

Prologue. Leadhills and nearby Wanlockhead were the most important mining districts in Scotland, while some mining also took place in Tyndrum in North Stirling.

5.4.6.1 Leadhills

Prologue. Glengonnar Water drains the Leadhills District.

Glengonnar Water, also called Elvan Water, is a small river rising in the Lowther Hills just to the east of Leadhills village, and flows north to join the River Clyde near Abington (Figure 5.11). The Leadhills and Wanlockhead districts contained ores of lead, zinc and silver, with lead the predominant target metal.

Dissolved concentrations of lead, zinc and cadmium in Glengonnar Water today are somewhat raised above background (Table 5.2), while sediment concentrations of lead and zinc in the river are still highly contaminated (Table 5.4).

5.4.6.2 Tyndrum

Prologue. Tyndrum was a lead mining district of some significance in the eighteenth and nineteenth centuries.

Most of the mine waste from the disused Tyndrum lead–zinc mine has remained either in dumps at the mine site, with runoff to the nearby River Fillan, or

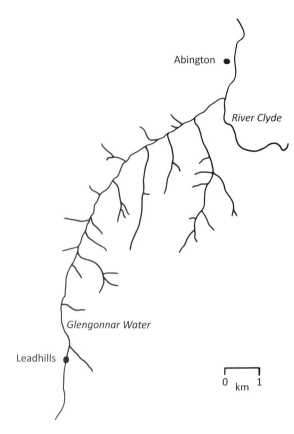

Figure 5.11 Glengonnar Water, Leadhills, Scotland. (After Rowan et al., 1995, with permission.)

retained behind a dam above the River Fillan (Figure 5.12) (Pulford et al., 2009). Analyses of sediments in the River Fillan show them still to be very high in lead and zinc (Table 5.4).

5.4.7 Isle of Man

Prologue. Ores of copper and of lead, in association with zinc and some silver, are present on the Isle of Man.

There were two dominant mining districts on the Isle of Man (Chapter 2). The Foxdale district, drained by the Rivers Neb and Glen Maye, produced lead, while the Laxey district, drained by the Laxey River on the east of the island (Figure 5.13), produced lead, zinc and some copper (Southgate et al., 1983).

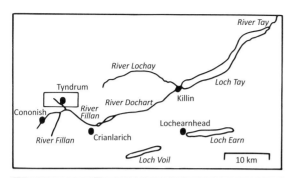

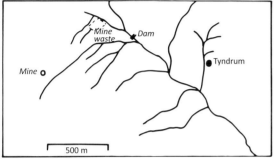

Figure 5.12 Tyndrum and the River Fillan, Scotland. (After Pulford et al., 2009, with permission.)

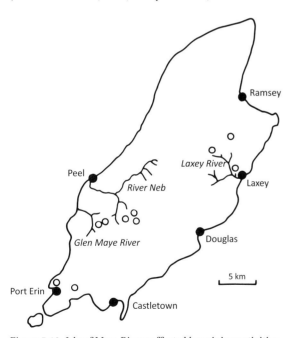

Figure 5.13 Isle of Man. Rivers affected by mining activities.

At the end of the 1970s, the sediments of the rivers Neb, Glen Maye and Laxey still contained raised concentrations of cadmium, copper, lead and zinc (Table 5.4).

5.5 Concentration and Bioavailability

Prologue. The bioavailabilities of trace metals in freshwaters and their associated sediments depend not only on total concentrations but also on other physicochemical factors.

The preceding section has identified particular streams or rivers draining mining districts that have very high concentrations of trace metals, in solution and/or in their sediments. The assumption is generally made that such high concentrations reflect high bioavailabilities of the metals concerned. On balance, this is a valid assumption as a generality. Nevertheless, it is appropriate to comment further on the relationships between concentrations and bioavailabilities of trace metals in freshwater systems.

5.5.1 Dissolved Bioavailability

Prologue. Ultimately, the dissolved bioavailability of a trace metal in a freshwater habitat depends not only on its total concentration but also on the degree of its complexation by binding agents that reduce the equilibrium concentration of the free metal ion.

The key staring point here is that the same dissolved concentration will not necessarily translate into the same dissolved bioavailability to a particular organism in different freshwater systems. The availability of the free metal ion is usually considered the best model for the local dissolved bioavailability of a trace metal (Chapter 3). In estuaries and seas, inorganic complexation of dissolved trace metals by the likes of chloride ions is very important, often exceeding organic complexation by any dissolved organic molecules (Chapter 6). In freshwater systems, however, it is organic complexation of dissolved trace metals that is usually more important, reducing the dissolved bioavailabilities of metals in the presence of dissolved organic matter DOM, for example in lakes or downstream of sewage effluents.

The dissolved bioavailability of a trace metal may also be affected by the presence of other dissolved metal ions, including major ions such as calcium ions (Chapter 3). Calcium ions can compete with trace metal ions, such as cadmium or zinc, for uptake sites on the surface of an organism. Thus, differences in calcium concentrations (hardness) of freshwaters will cause differences in the bioavailabilities of such dissolved trace metals in the different waters, even if total dissolved trace metal concentrations are the same.

Differences in the pH (hydrogen ion concentration or acidity) of different freshwaters will affect dissolved trace metal bioavailabilities. This is effected both by the displacement by hydrogen ions of trace metals from any complexation by DOM (increasing bioavailability) and by competition by hydrogen ions with free trace metal ions for binding at biological uptake sites (decreasing bioavailability).

The upshot, then, is that while it is valid to assume generally that a high dissolved trace metal concentration is indicative of a high dissolved bioavailability, there will be subtle differences between freshwater systems. The same organism exposed to the same dissolved metal concentration will not necessarily take up (and usually subsequently accumulate) the same amount of trace metal in two different freshwater systems. We will see shortly, therefore, that particular plants and animals may have different trace metal uptake rates from the same dissolved metal concentration in different habitats, and therefore different accumulated trace metal concentrations. Differences in trace metal uptake rates will also have different consequences for the potential ecotoxicology of any dissolved metal exposure to an organism, the ecotoxicity of a trace metal depending on its uptake rate (see Chapter 3).

5.5.2 Sediment Bioavailability

Prologue. The total bioavailability of a trace metal associated with a freshwater sediment can be subdivided into two separate components: trophic bioavailability and dissolved bioavailability in the pore water of the sediment.

Animals that ingest sediment particles, whether deposited on the bottom of a freshwater body or suspended in the water column, have the potential to take up any trace metals associated with those sediment particles after digestion in the alimentary tract. Such metals are described as trophically bioavailable. Secondly, metals associated with deposited sediment particles may be released by equilibrium into the surrounding pore water, with the potential then to be taken up into the roots of a submerged plant or across the permeable epithelium of a burrowing invertebrate.

For either route of metal uptake, internally in the gut or externally from pore water, the bioavailability of the sediment-associated trace metal will be affected by the strength of chemical binding of the metal with the sediment particles. Trace metals will bind strongly with the organic components of muds, and if these organic components differ in quantity and quality between sediments, then so will the total bioavailability of the trace metals. Furthermore, any oxides of iron in a sediment will bind particularly strongly with certain trace metals, including arsenic, copper and lead (Bryan and Gibbs, 1983). These metals will, therefore, be less bioavailable in a river sediment with a high iron content than in a low iron sediment, even when total trace metal concentrations in the sediment are the same.

Similarly, many trace metals will bind strongly to sulphides in sediments with potential consequences for their bioavailabilities. This particular form of binding of trace metals in sediments is the mechanistic basis of a proposed technique to assess trace metal bioavailabilities in freshwater sediments, and hence the potential ecotoxicities of trace metals in these sediments (Luoma and Rainbow, 2008).

Sulphides, particularly hydrogen sulphide with its characteristic smell of bad eggs and black iron sulphide, are common in anoxic sediments that lack the oxygen that would oxidise the sulphides to sulphates. Sulphide binds strongly with most trace metals, and will, therefore, inhibit any exchange of the trace metals into the sediment pore water. Thus the pore waters of anoxic sediments have greatly reduced dissolved concentrations of trace metals. This observation led to the proposal that these low dissolved concentrations would correlate with low pore water bioavailabilities of trace metals (Di Toro et al., 1991, 1992). Furthermore, the ecotoxicological potential of a sediment resulting from its toxic metal content could be assessed by comparative measures of the trace metal and sulphide concentrations of a sediment. If the sediment concentration of sulphide (strictly the acid-extractable sulphide called the acid volatile sulphide {AVS}) exceeds the equivalent concentrations of trace metals, then the trace metals would effectively not be released into the pore water to be bioavailable and cause toxicity (Di Toro et al., 1991, 1992; Chapman et al., 1998). This proposal was tested and apparently substantiated by exposing test invertebrates to homogenised sediments with different AVS and trace metal concentrations (Di Toro et al., 1992). There is, however, a fundamental flaw in this approach to predicting the metal ecotoxicity of sediments, because in practice burrowing invertebrates are typically not bathed in anoxic pore water (Luoma and Rainbow, 2008). If invertebrates do burrow in anoxic sediments, they do so by pumping oxygenated water from the water column above into their burrows, simply so that they can respire. Their burrows therefore are oxygenated. Indeed, it is often possible to see the rust-coloured lings of their burrows, caused by the presence of iron oxides oxidised from the surrounding iron sulphides present in the otherwise black milieu of the surrounding anoxic sediment (Plate 2b).

Thus, well-designed field experiments have now confirmed that AVS is not a significant variable in describing variation in trace metal uptake and accumulation in freshwater invertebrates burrowing in sediments (De Jonge et al., 2009, 2010). Such invertebrates can accumulate metals from field sediments even when there is an excess of AVS present in the sediments (De Jonge et al., 2009, 2010).

While it is tempting to chemists to seek chemical surrogates with which to assess the bioavailabilities

of toxic metals in sediments, it remains the case that bioavailabilities depend on the organism concerned. Only the concentration of a trace metal accumulated in an organism can provide a valid assessment of the total bioavailability of a metal in a sediment to that organism. Thus we should rely on biomonitors to provide assessments of trace metal bioavailabilities, and thus ecotoxicological potentials, in field situations. Suitable biomonitors to be used for this purpose in British streams affected by mining will be identified later in this chapter, as we consider the bioaccumulation of trace metals by different freshwater organisms.

5.6 Acid Mine Drainage

Prologue. Much of the drainage into freshwater systems from mining sites is very acidic, a generality that has led to the coining of the term 'acid mine drainage' (AMD).

The origin of AMD lies in the sulphides, the common form of many ores of trace metals, particularly pyrite (iron sulphide), but also chalcopyrite (copper iron sulphide), chalcocite (copper sulphide), arsenopyrite (arsenic iron sulphide), galena (lead sulphide) and sphalerite (zinc iron sulphide) (Chapter 2). When these sulphides are exposed to air in the presence of water at the mine surface, they are oxidised to sulphate with the release of hydrogen ions, so creating acid conditions, in effect producing sulphuric acid (Kelly, 1988). In the case of pyrite, the iron 2 ions released are oxidized to iron 3 ions, and sulphide to sulphate, through the catalytic agency of acidophilic iron and sulphur-oxidising bacteria such as *Leptospirillium ferrooxidans* and *Acidithiobacillus ferrooxidans* (Kelly, 1988; Luoma and Rainbow, 2008). Near mines, at a low pH of between 2 and 4, colonies of these bacteria can appear as streamers in the drainage water, or as slimes on the bottom of the drainage stream. AMD is also typically associated with the classic orange-coloured precipitate of ochre, formed of hydroxides and oxides of iron 3, particularly above a pH of 4 as the AMD is increasingly neutralised on passage downstream. The iron in ochre does not represent a toxic chemical challenge to the local biota, but ochre can have a physical blanketing effect that causes depletion of the stream flora and fauna. Furthermore, some trace metals, for example arsenic, copper and lead, may come out of solution and adsorb onto the freshly precipitated iron oxides and hydroxides (Kelly, 1988; Luoma and Rainbow, 2008). There is potential for these toxic metals to be ingested by invertebrate grazers in the stream or to be ingested by deposit feeders if the iron oxide particles with their trace metal loadings are incorporated into the bottom sediments.

One example of a dramatic discharge of AMD into a British freshwater system was the catastrophic release in January 1992 of heavily metal-contaminated AMD from the Nangiles Adit draining the disused Wheal Jane mine into the River Carnon in Cornwall (Hunt and Howard, 1994; Bowen et al., 1998). This AMD had ultimately been derived from the sulphide ores (not least pyrite) in the mine and had an initial pH as low as 1. On the way down the Carnon River to Restronguet Creek, neutralisation of the acid discharge led to the considerable deposition of ochre along the riverbed. The metal-contaminated flood had clear ecotoxicological consequences on the river's flora and fauna, including fish, and a remediation programme was required (Bowen et al., 1998). The ensuing remediation has included dosing with lime to neutralise the acid, the establishment of settling ponds behind tailings dams and the use of constructed wetlands as natural filters (Whitehead and Neal, 2005; Whitehead and Prior, 2005).

Another example of the impact of AMD on a freshwater system comes from the abandoned copper and sulphur mines at Avoca in County Wicklow in Southeast Ireland (Gray, 1997; Wright et al., 1999). Copper was mined at the East and West Avoca mines on either side of the Avoca River from the eighteenth century until 1982, together with sulphur in the form of pyrite, particularly in the nineteenth century (Chapter 2).

The mining of pyrite (in addition to chalcopyrite and some sphalerite) set the scene for the production of AMD (Gray, 1996, 1997), not least because open-cast pits were used at East Avoca in the 1970s and 1980s. As a result, AMD has severely contaminated the Avoca River right down to its estuary at Arklow, with no fish life being present for more than 200 years (Wright et al., 1999).

The ecotoxicological effects of AMD in freshwater systems are complex, so that it is sometimes difficult to characterise, quantify or predict these effects (Gray, 1997). The detrimental effects of AMD consist of a mixture of trace metal toxicity, sedimentation processes, acidity and salinisation. There are no specific indicator species for AMD in streams and rivers, although oligochaete worms and dipteran fly larvae (particularly chironomid midge larvae) often dominate macroinvertebrate assemblages downstream of AMD discharges. Mayflies are particularly sensitive to AMD and are amongst the last invertebrate groups to recolonise previously affected streams and rivers (Gray, 1997).

An attempt has also been made to develop an Acid Mine Drainage Index (AMDI) as a chemical index of impact assessment, based solely on the concentrations of key chemical contaminants typical of mine drainage (Gray, 1996). These characteristic contaminants are pH, sulphate, iron, zinc, aluminium, copper and cadmium, with pH and sulphate given the greatest weighting in the calculation of the AMDI. The AMDI varies from 0 to 100. AMDI values between 0 and 20 indicate raw AMD with little or no dilution, as in the case of seepage from mine spoil heaps. AMDI values between 90 and 100 indicate water uncontaminated by AMD.

In a study of the Avoca mines in the early 1990s, raw AMD draining from spoil heaps had AMDI values as low as 0.6 (Gray, 1996). A spring near the mine site with an AMDI of 3.1 had dissolved trace metal concentrations of 433,000 µg/L Zn, 165,000 µg/L Cu and 1,030 µg/L Cd, at a pH of 2.6 (Gray, 1996). The Avoca mines are drained by two deep adits, Ballymurtagh Adit on the west side, and the Deep Adit on the east side. These two adits had a mean AMDI of 31 and 25 respectively, indicating AMD that has been diluted, from groundwater in the case of an adit. The Ballymurtagh Adit had dissolved trace metal concentrations of 73,000 μg/L Zn, 4,000 μg/L Cu and 255 μg/L Cd, at a pH of 3.6,

still very high on a national scale (Table 5.2). Downstream close to Arklow at the head of the Avoca estuary, the AMDI had recovered to 94; the pH to 8.2; and the dissolved trace metal concentrations had fallen to 160 μg/L Zn, 7 μg/L Cu and effectively 0 μg/L Cd (Gray, 1996).

Box 5.2 Coal Mining

Acid mine drainage can also emanate from coal mines when the coal mined has a high sulphur content. The major form of sulphur in coal is pyrite (iron sulphide) (Environment Agency, 2008a). So the prerequisites for the production of AMD are in place. The added load of trace metals derived from their sulphide ores will be missing, but streams receiving coal mine AMD will still show the presence of acidophilic iron and sulphur-oxidising bacteria and then ochre. The deposition of ochre, even in the absence of further trace metals, will cause depletion of the biota, both by physical blanketing and by causing reduced light penetration for plant growth (Environment Agency, 2009a).

AMD from coal mines can have significant ecotoxicological effects on the fauna of affected streams, as in the case of invertebrates in streams in the Durham coalfield (Jarvis and Younger, 1997) and invertebrates and brown trout (*Salmo trutta*) in the south Wales coalfield (Scullion and Edwards, 1980a,b).

Another product of coal mining has the potential to have an ecotoxicological effect on freshwater life, and that is the ash derived from the burning of coal. Much of this ash goes to landfill, but there is still a risk of leachate draining into local freshwater systems. High sulphur coal may be enriched in selenium or arsenic, which would then be a contaminant of any leachate.

5.7 Iron-Rich Streams

Prologue. The Ashdown Forest in the centre of the Weald in Southeast England has a long history of iron production. It straddles the watersheds of two small lowland rivers: the Medway, flowing north to the Thames Estuary; and the Sussex Ouse, flowing south to the English Channel.

The hills of the Weald consist of soft Ashdown Sandstone, with pockets of Wadhurst Clay, a source of iron ore, which is often present as small nodules. The small streams of the Ashdown Forest,

whether flowing north or south (Figure 5.14), are both iron-rich and acid-rich (Townsend et al., 1983; Hildrew, 2009).

One such iron- and acid-rich stream is Broadstone Stream (Figure 5.14), the bottom of which is bright orange with ochre (Hildrew, 2009). Broadstone Stream, like some of its neighbours, has an impoverished fauna and lacks fish (Hildrew, 2009). In fact, the structure of the invertebrate communities in the Ashdown Forest streams is strongly related to variation in stream pH, and not to either total or soluble iron concentrations (Townsend et al., 1983; Hildrew, 2009). More acidic sites have low numbers of

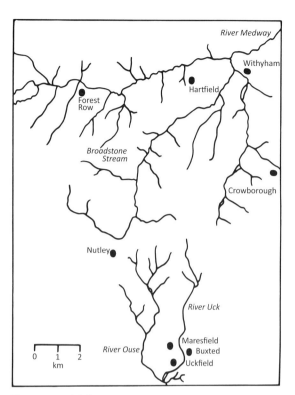

Figure 5.14 Ashdown Forest iron-rich streams. (After Hildrew, 2009, with permission.)

with the silken nets of a web-spinning predatory caddis larva that seemed just impossibly abundant' (Hildrew, 2009: 181). The caddisfly larva in question is *Plectrocnemia conspersa*, and another common large predatory insect larva in Broadstone Stream is the alderfly *Sialis fuliginosa* (Hildrew, 2009). The presence in abundance of these relatively large predatory insect larvae is correlated with the absence of fish, which might otherwise control their populations. With a reduction in stream acidity, the numbers of invertebrate species in the collector and predator categories increase, and these are now joined by grazer/scrapers and filter feeders (Townsend et al., 1983). This reduction in stream acidity also correlates with the appearance downstream of sparse populations of brown trout (*Salmo trutta*), then bullheads (*Cottus gobio*) and finally low numbers of other fish species (Hildrew, 2009).

While the streams of the Ashdown Forest are often referred to as iron-rich streams, it is perhaps ironic that it is not the concentration of iron but the low pH that is the likely cause of faunal impoverishment.

individual invertebrates and low species richness. At these acid sites, the only functional groups of invertebrates present are collectors, shredders and predators (Townsend et al., 1983). In the words of Alan Hildrew, the bed of Broadstone Stream is 'festooned

5.8 Aluminium

Prologue. A metal that can have a significant ecotoxicological effect in freshwater systems is the common metal aluminium.

Box 5.3 Aluminium in Freshwater

Aluminium becomes more soluble at acidic pHs. Correspondingly, the likelihood for aluminium to be more bioavailable and potentially toxic to biota is greater at low pH (Gensemer and Playle, 1999). It is now appreciated that aluminium can be a major factor affecting freshwater organisms in acidic habitats.

The key to understanding the solubility, bioavailability and potential toxicity of aluminium in freshwater is an understanding of its chemical speciation. Aluminium is relatively insoluble at pH 6 to 8, covering the range of most circumneutral freshwaters. Aluminium at these pHs forms the mineral gibbsite $Al(OH)_3$, which precipitates from solution (Gensemer and Playle, 1999). At lower pHs, dissolved aluminium can be in the form of inorganic hydroxy species $Al(OH)_2^+$ and $Al(OH)^{2+}$, but mostly as the uncomplexed free-metal ion Al^{3+}, which increases in importance below pH 5 (Gensemer and Playle, 1999). In effect, Al^{3+} is the only form of dissolved aluminium below pH 4. Paradoxically, aluminium is also soluble again at pHs above 8, with increasing domination of the soluble hydroxy species $Al(OH)_4^-$, the only form of aluminium above pH 9. In practice, freshwater systems at these high pHs are rare and need not concern us here.

When considering the interaction between freshwater organisms and aluminium, it is the free metal ion Al^{3+} that is all important. There remains the potential for this ion to be complexed by both inorganic ions and by organic molecules, reducing its availability to be taken up by biota. Potential inorganic complexing agents are the anions fluoride and (less strongly) sulphate (Gensemer and Playle, 1999). Dissolved aluminium can also be complexed by small organic molecules such as citrate, oxylate and acetate produced by microorganisms and plants, but additionally by large organic molecules such as humic and fulvic acids (Gensemer and Playle, 1999). In short, Al^{3+} ions can be complexed by DOM, increasing the solubility of aluminium, but decreasing its bioavailability and potential toxicity. DOM is probably more important than fluoride in complexing dissolved aluminium in most freshwater systems (Gensemer and Playle, 1999). A complicating factor, however, in any direct relationship between toxicity and the availability of the free aluminium metal ion, is the potential at low pH for abundant hydrogen ions to displace aluminium ions from uptake sites on the surface of organisms (Gensemer and Playle, 1999).

Lake acidification to a pH below 5 causes profound changes in the community structure of phytoplankton and other pelagic communities, an ecotoxicological effect ascribed to aluminium ecotoxicity (Gensemer and Playle, 1999). Similarly, studies of acidification-induced negative changes in stream periphyton community structure have implicated aluminium toxicity as the responsible factor (Gensemer and Playle, 1999). Diatoms appear more sensitive to aluminium exposure than green algae, while, in the animal world, aquatic invertebrates are generally considered less sensitive to aluminium exposure than are fish (Gensemer and Playle, 1999). The toxic effects of aluminium on invertebrates at low pH with high Al^{3+} bioavailability appear to affect ionic regulatory systems, such as major ion channels, as an additive effect on top of the effects of H^+ ions (Gensemer and

Playle, 1999). Aluminium is a gill toxicant to fish, causing effects on ionic regulation, respiration or a mixture of both. At low pH with high Al^{3+}, the ionic regulation effects predominate, presumably by interference with major ion channels in the gill epithelia of the fish (Gensemer and Playle, 1999). Respiratory effects are more important in moderately acidic waters, the toxic effects being explained by the precipitation of aluminium hydroxide onto the gill surface causing smothering and interfering with gas exchange (Gensemer and Playle, 1999).

5.9 Microscopic Life

Prologue. In considering in detail the interaction between the biota of freshwater streams and rivers contaminated by trace metals and the raised bio-availabilities of these toxic metals, we start with life under the microscope.

Bacteria are common in freshwater sediments. Bacteria will also be present in the surface film attached to submerged surfaces in streams and rivers, particularly so in the case of the photosynthetic blue-green bacteria. Also present in the surface film will be small eukaryotic organisms. Much of the biota living in this film are photosynthetic and make up the periphyton, a heterogeneous mixture of blue green bacteria, protistans and other microphytic algae. Small animals may be present too, including protistans that are not photosynthetic (historically termed protozoans), rotifers, tardigrades, small crustaceans, oligochaetes and insect larvae.

5.9.1 Bacteria

Prologue. The presence of high bioavailabilities of toxic metals in the sediments and surface films of streams contaminated with trace metals can be expected to have a strong selective effect on their bacterial communities.

Such selective effects might be changes in the structure of the bacterial communities, including, potentially, an increased average tolerance to a toxic metal amongst all species in that bacterial community, a concept termed pollution-induced community tolerance (PICT) (see Chapter 3). In fact, the bacterial communities of British mining streams and rivers remain relatively unexplored. Recent developments in molecular techniques to analyse the taxonomic composition of microbial communities offer great potential to analyse any community differences in response to trace metal exposure. Such molecular methods have been used, for example, to analyse biofilm bacterial communities at 23 urbanized stream sites in Auckland, New Zealand, variously impacted by copper, lead and zinc (Ancion et al., 2013). Concentrations of the three metals in biofilms explained 7% of the variation in the bacterial biofilm communities.

5.9.2 Periphyton

Prologue. The periphyton of metal-contaminated mining streams and rivers has been well studied, with most research conducted on benthic diatoms. Other photosynthetic protistans do occur in the periphyton, as well as the developing stages of what will become macrophytic algae, visible to the naked eye.

Species of *Euglena*, a flagellate, represent protistans other than diatoms in the periphyton of metal-contaminated streams. *Euglena mutabilis* was found in 1974 in regions of the Gillgill Burn near Nenthead in the Northern Pennines that were particularly high in zinc (Say and Whitton, 1980). Indeed, *E. mutabilis* is to be found widely in Europe and North America in highly acidic streams, including those with high concentrations of trace metals (Hargreaves et al., 1975; Nakatsu and Hutchinson, 1988).

5.9.2.1 Diatoms

Prologue. The diatoms of freshwater streams are clearly affected by trace metal contamination, which causes changes in diatom community composition, if not their elimination entirely. Diatoms also show structural abnormalities in the presence of high trace metal bioavailabilities.

A survey in summer 1953 showed that the uncontaminated upper regions of the River Ystwyth in mid-Wales had a luxuriant flora, including diatoms (Jones, 1958). Twenty species of diatoms were present. Two species dominated, *Tabellaria flocculosa* and *Hannaea* (as *Ceratoneis*) *arcus*, with three other species sub-dominant, six species common and nine species rare (Jones, 1958). The diatoms disappeared downstream, immediately below the Cwm Ystwyth mine, releasing lead and zinc into the river (Table 5.2). It was not until Llanfarian, well past the Cwm Ystwyth, Frongoch and Wemyss mine discharges and approaching Aberystwyth (Figure 5.7), that there was any evidence for the return of diatoms to the River Ystwyth (Jones, 1958). At Llanfarian, diatom remains were found in the gut contents of larval stoneflies, mayflies, black-flies and beetles (Jones, 1958).

In the Northern Pennines, the River Nent (Figure 5.9) receives inputs from the Gillgill Burn, high in both lead and zinc, and Caplecleugh Low Level with high zinc concentrations (Table 5.2). A 1974 survey of the Gillgill Burn showed a remarkable species richness of diatoms, with 16 diatom species still being present at dissolved zinc concentrations above 3,000 µg/L (Say and Whitton, 1980). Four particular diatom species, *Caloneis bacillum* (as *C. lagerstedtii*), *Pinnularia appendiculata*, *Pinnularia borealis* and *Eunotia tenella*, were found at sites with the highest dissolved zinc concentrations. Four diatom species were recorded in the nearby Caplecleugh Low Level in 1981: *Halamphora* (as *Amphora*) *veneta*, *Pinnularia subcapitata*, *Surirella minuta* (as *S. ovata*) and *Achnanthes minutissima* (Patterson and Whitton, 1981). Interestingly, *A. minutissima* was one of the subdominant diatom species in the uncontaminated upper reaches of the River Ystwyth in 1953, to be eliminated below the Cwm Ystwyth mine (Jones, 1958).

More widely in metal-contaminated streams in the Northern Pennines, species of diatoms frequent at high zinc sites are *Caloneis bacillum* and *Pinnularia subcapitata*, already met previously, together with *Pinnularia sudetica* and *Neidium perminutum* (Say and Whitton, 1981).

A broad survey in 1997 of 51 metal-contaminated streams in Cornwall (including the Rivers Carnon, Hayle, Gannel and the Red River) and in mid-Wales (including the rivers Ystwyth and Rheidol) investigated correlations between diatom assemblage compositions and environmental characteristics of the sites sampled (Hirst et al., 2002). Amongst these site characteristics were dissolved concentrations of aluminium, cadmium, copper, iron, lead, manganese and zinc, in addition to pH, altitude, conductivity and stream width. To measure the combined toxic effect of a mixture of trace metals, their individual dissolved concentrations were each expressed as a fraction of the dissolved concentration of the metal considered by the US Environmental Protection Agency to be harmful to aquatic organisms (Hirst et al., 2002). These fractional values were then summated at each site for all metals to come up with a cumulative

criterion unit (CCU) score. The most important metals to contribute to CCU scores were lead, copper and zinc (Hirst et al., 2002).

There were observed variations in diatom assemblage composition across the sites, with pH and conductivity explaining the major variations (Hirst et al., 2002). The strongest predictor of diatom assemblages, however, was the CCU score, although there were no significant variations in diatom species diversity, richness or evenness with metal concentration (Hirst et al., 2002). Diatom species apparently insensitive to high metal concentrations included *Achnanthidium* (as *Psammothidium*) *helveticum*, *Eunotia subarcuatoides*, *Pinnularia subcapitata* and *Sellaphora seminulum* (Hirst et al., 2002). *P. subcapitata* we have met previously as being present in the zinc-rich Caplecleugh Low Level in 1981 and more widely in other zinc-rich Northern Pennine streams. Metal-sensitive diatom species included *Fragilaria capucina* var. *rumpens*, *Achnanthes oblongella* and *Tabellaria focculosa* (Hirst et al., 2002). *T. flocculosa* was a dominant diatom species in the uncontaminated upper region of the River Ystwyth in 1953, to be eliminated below the lead–zinc mines (Jones, 1958).

The ranges of dissolved concentrations of cadmium, copper, lead and zinc across the sites were 0.03 to 6.6 µg/L Cd, 0.7 to 838 µg/L Cu, 0.04 to 302 µg/L Pb and 10 to 3,330 µg/L Zn (Hirst et al., 2002). All maximum concentrations are raised above background but do not reach the extremes possible in British streams contaminated by mining (Table 5.2).

Another large study of diatom communities at 34 sites in metal-contaminated streams was carried out in 2006 in the Northern Pennines (including Nenthead stream) and the Lake District (including Threlkeld stream) (Table 5.2) (Environment Agency, 2008a). One hundred and four diatom taxa were found across the 34 sites, with between 4 and 35 taxa at each site. The study explored the relationship between diatom community composition and dissolved concentrations of cadmium, copper, lead, nickel and zinc, having first eliminated any effects of the 'natural' characteristics of the sites (pH, altitude, alkalinity, stream width). There were significant relationships between diatom community structure and the dissolved concentration of each trace metal, with dissolved zinc showing the most significant such relationship. Dissolved cadmium concentrations covaried with dissolved zinc concentrations, so the conclusion of the study was that dissolved zinc (and/or cadmium) had the most effect on diatom numbers, explaining about 30% of the variability (Environment Agency, 2008a). The only diatom species that appeared to be more common in the metal-rich streams of the Northern Pennines was *Halamphora oligotraphenta* (as *Amphora veneta* var. *capitata*). Otherwise, relatively trace metal–insensitive diatom species were *Achnanthidium minutissimum, Meridion circulare* and *Fragilaria capucina*.

This study of the metal-contaminated streams in the Northern Pennines and the Lake District, in fact, concluded that the approach of using diatom community analysis is not especially sensitive in seeking ecotoxicological effects of trace metal contamination in streams affected by mining (Environment Agency, 2008a). The ranges of dissolved trace metal concentrations in the study were <0.01 to 5.3 µg/L Cd, 0.2 to 7.5 µg/L Cu, 0.12 to 75.9 µg/L Ni, 0.1 to 156 µg/L Pb and 1.1 to 11,000 µg/L Zn. A comparison of these dissolved concentration ranges with the data in Table 5.2 suggests that it is not surprising that zinc was showing the strongest ecotoxicological effect on diatom community structure.

A study of diatoms in the metal-contaminated River Dommel in Belgium in 2006 to 2007 identified *Sellaphora seminulum* and *Eolimna minima* as insensitive to high zinc concentrations, and *Tabellaria flocculosa* and *Fragilaria capucina* var. *rumpens* as being associated with low metal concentrations (De Jonge et al., 2008). Similarly, *S. seminulum* was considered to be metal-insensitive by the study of metal-contaminated streams in Cornwall and Wales (Hirst et al., 2002). Furthermore, *T. flocculosa* and *F. capucina* var. *rumpens* were also considered metal-sensitive in this latter study (Hirst et al., 2002).

Diatoms show structural developmental abnormalities resulting from field exposure to very high bioavailabilities of toxic metals (Falasco et al., 2009). Abnormalities in the development of diatoms include a reduction in the size of the two valves of the diatom in correlation with trace metal contamination (Falasco et al., 2009). Particular diatom species, including *Achnanthidium minutissimum, Fragilaria gracilis, F. capucina* var. *rumpens, F. crotonensis* and *F. tenera*, show enhanced abnormal morphologies, such as valve distortion and abnormal valve striation, in response to increased trace metal (Cd, Cu, Hg, Zn) bioavailabilities at metal-contaminated sites (Falasco et al., 2009). In streams in the Northern Pennines, for example, there is a good relationship between dissolved concentrations of zinc and the numbers of 'twisted' cells of *Fragilaria capucina*, the presence of distorted cells in more than one-tenth of the diatom population indicating metal enrichment (Environment Agency, 2009a). Percentage distortion increased further as dissolved zinc concentrations approached and exceeded 1,000 µg/L (Environment Agency, 2009a).

5.9.3 Meiofauna

Prologue. Meiofauna are defined by size, being small animals larger than 45 µm but smaller than 1 millimetres. Meiofauna are present in both sediments and surface films of freshwater systems, including those contaminated by trace metals.

A study has been made of the meiofauna of the sediments of two metal-contaminated freshwater systems in Cornwall, the Rivers Lynher and Seaton, draining the southeast corner of Bodmin Moor (Burton et al., 2001). The compositions of the meiofaunal communities were analysed at a series of sites which represented a gradient in dissolved concentrations of copper, zinc, aluminium and iron. The meiofaunal communities were very different at sites with high metal levels from those at sites with low metal levels. Copper, either singly or in combination with zinc, aluminium or dissolved organic carbon, was the most important correlate with meiofaunal community composition (Burton et al., 2001). Not all meiofaunal species were adversely affected by the increased presence of metals, and some, most notably certain cyclopoid copepods such as *Diacyclops languidodes*, *D. bisetosus* and *Eucyclops serrulatus*, were abundant at high metal concentrations. Moreover, meiofaunal communities at sites with high metal concentrations were not significantly less diverse than at sites with low metal concentrations. Nevertheless, the meiofaunal communities had still been changed, in correlation with the raised dissolved metal concentrations (Burton et al., 2001).

The potential exposures to copper and zinc in this study, though, do need to be put into context. Dissolved copper concentrations across the studied sites ranged from 10 to 766 µg/L, and dissolved zinc from 52 to 874 µg/L (Burton et al., 2001). While these dissolved concentrations are raised above background concentrations for Cornish rivers, they do not approach the extremely high dissolved copper and zinc concentrations in severely contaminated Cornish streams and rivers (Table 5.2). It remains likely that meiofaunal sediment communities at severely contaminated sites will not only be changed but could also be expected to show decreased diversities.

Meiofauna are also to be found in surface films in metal-contaminated streams. In the New Zealand study of urbanised stream sites in the Auckland region, concentrations of copper, lead and zinc in biofilms explained 9% of the variation in their ciliate protozoan communities (Hargreaves et al., 1975).

5.10 Algae

Prologue. The term 'algae' refers to a heterogeneous group of aquatic photosynthetic organisms across a multitude of different taxonomic groups. Algae are eukaryotic and range in size from microphytic single-celled forms such as phytoplanktonic or benthic diatoms to macrophytic green algae, brown algae and red algae.

Many of the large macrophytic algae live on the seashore (seaweeds) but do also occur in freshwater, particularly so in the case of green algae. The term 'alga' has also been applied in the past to a group of prokaryotic photosynthetic organisms, the blue green bacteria, under the epithet 'blue green algae'.

Blue-green bacteria are capable of existing in macrophytic form, not least as filaments in some of the streams considered here. It is not, therefore, surprising that they have historically been called algae, and blue-green bacteria appear first in this section.

5.10.1 Blue Green Bacteria

Prologue. Blue green bacteria are found in trace metal-contaminated streams.

A survey in the 1930s of the River Rheidol in mid-Wales showed the presence in some abundance of the macrophytic blue green bacteria *Chamaesiphon incrustans*, *Chamaesiphon britannicus* and *Leptolyngbya tenuis*, even downstream of the discharge from the Cwm Rheidol mine (Figure 5.7) (Reese, 1937; Jones, 1949). In the River Ystwyth in 1953, patches of the '*Chamaesiphon* type' were present on stones in the relatively barren stretch of river below the entry of the Frongoch stream (Figure 5.7) (Jones, 1958).

In the catchment of the River Nent in the Northern Pennines (Figure 5.9), the Gillgill Burn in 1974 contained 17 species of blue green bacteria, with species of *Synechococcus* largely restricted to regions of the stream with higher zinc levels (Say and Whitton, 1980). In nearby Caplecleugh Low Level in 1981, there were three species of blue green bacteria reported: *Pseudanabaena catenata*, and unnamed species in each of the genera *Lyngbya* and *Phormidium* (Patterson and Whitton, 1981). In metal-rich Northern Pennine streams in general, narrow filamentous forms of blue green bacteria of the family Oscillatoriaceae (which includes the genera *Lyngbya*, *Phormidium* and *Plectonema*) are widespread (Say and Whitton, 1981). Sheets of species of the morphologically similar genera *Plectonema* and *Schizothrix* are also to be found in shallow zinc-rich Northern Pennine streams (Say and Whitton, 1981).

5.10.2 Green Algae

Prologue. While brown algae and red algae are groups of taxonomically related algae, the term 'green algae' has been widely used to refer to a collection of algae with no taxonomic connection. More strictly today, however, the term 'green algae' is used to refer to algae that are placed in a taxonomically defined group, containing the green algae and the land plants. The green algae in this group are subdivided into two divisions, each containing taxonomically related green algae. One division, the Chlorophyta, or chlorophytes, contains most of the green algae and figures in the streams discussed here. The other division contains one group of green algae, the Charophyta or charophytes, in addition to their relatives the land plants. Charophytes are also represented here.

5.10.2.1 Mid-Wales

Prologue. The trace metal-contaminated Rivers Ystwyth and Rheidol have been surveyed for green algae.

In surveys in 1920, 1940 and 1953, the River Ystwyth from the Cwm Ystwyth mine down to Aberystwyth (Figure 5.7) lacked the presence of green algae (Carpenter, 1924; Jones, 1940, 1958). The upper uncontaminated part of the river, on the other hand, contained species of the chlorophyte genus *Stigeoclonium* and the charophyte genera *Zygnema*, *Mougeotia* and *Spirogyra* (Jones, 1958). By 1973, green algae had returned to the River Ystwyth. The filamentous chlorophyte *Klebsormidium rivulare* (as *Hormidium rivulare*) was widespread along the river from above the old mines to below (McLean and Jones, 1975). Another chlorophyte *Ulothrix* sp. was next most dominant, while species of the charophyte genera *Zygnema* and *Spirogyra* could exist in areas of some moderate metal contamination (McLean and Jones, 1975).

In the contaminated River Rheidol nearby, there were no green algae in 1920 (Carpenter, 1924), but a survey in the 1930s recorded the presence there of the chlorophytes *Stigeoclonium falklandicum* and *Ulothrix* sp., as well as species of the charophyte *Mougeotia* (Reese, 1937; Jones, 1949). In a 1947 study, the chlorophytes *Stigeoclonium* sp. and *Draparnaldia glomerata* were growing well in the lower reaches of the Rheidol between Capel Bangor and Aberystwyth (Figure 5.7) (Jones, 1949).

5.10.2.2 Northern Pennines

Prologue. Green algae occur in lead- and zinc-contaminated streams in the River Nent system, with some local chlorophyte populations developing tolerance to zinc.

In the Northern Pennines, a 1974 survey of the lead and zinc-rich Gillgill Burn feeding into the River Nent (Figure 5.9), revealed 11 chlorophyte species still present at dissolved zinc concentrations above 3,000 µg/L (Say and Whitton, 1980).

Particularly zinc-resistant was *Klebsormidium rivulare* (as *Hormidium rivulare*), met above in the River Ystwyth and considered to occur frequently in acid mine drainages (Say and Whitton, 1980). Populations of this chlorophyte from the Gillgill Burn showed a genetically based tolerance to zinc, indicating that the zinc exposure is ecotoxicologically significant (Say et al., 1977). Six different Gillgill Burn populations of *K. rivulare* were zinc-tolerant, the higher the level of zinc exposure in the field, the higher the tolerance to zinc subsequently exhibited in laboratory assays. Field selection for zinc tolerance in *K. rivulare* appeared to begin at dissolved zinc concentrations above 2,000 µg/L Zn (Say et al., 1977; Say and Whitton, 1980). Another chlorophyte collected from sites in the Gillgill Burn with high zinc levels was *Stigeoclonium tenue* (Say and Whitton, 1980). Gillgill Burn populations of this green alga also showed zinc tolerance (Harding and Whitton, 1976). Collection sites of zinc-tolerant *S. tenue* in the Gillgill Burn had average dissolved zinc concentrations of 7,090 µg/L Zn (Whitton, 1970). Similarly to *K. rivulare*, the threshold dissolved zinc concentration for field selection of tolerance in *S. tenue* was considered to be about 2,000 µg/L Zn (Whitton, 1970). *Stigeoclonium tenue* was also found in abundance at sites lower down the River Nent below Nenthead in 1978 (Armitage, 1980). In contrast, another chlorophyte, *Microspora amoena*, was only found at downstream sites in the Gillgill Burn in 1974, in relatively low dissolved zinc concentrations (Say and Whitton, 1980).

Three green algae were recorded from the zinc-enriched Caplecleugh Low Level at Nenthead in 1981 (Patterson and Whitton, 1981). Two species of the charophyte genus *Mougeotia* were abundant, and the now familiar chlorophyte *Klebsormidium rivulare* was also present.

Conspicuous by their absence from metal-enriched streams of the Northern Pennines are species of the chlorophyte genus *Cladophora* (Environment Agency, 2009a). Indeed, most reports in the literature indicate that species of *Cladophora* in freshwater are exceptionally sensitive to trace metal exposure (Whitton, 1970; Kelly, 1988).

5.10.2.3 Cornwall

Prologue. The algal flora of the Rivers Hayle and Gannel in Cornwall have also come under investigation. The River Hayle is heavily contaminated by copper, with also arsenic and some zinc present, while the River Gannel is lead and zinc-rich.

In a survey carried out between 1975 and 1976, the algae that were dominant, or otherwise common, at metal-contaminated sites in both rivers were all green algae, both chlorophytes and charophytes (Foster, 1982a). Both total algal abundance and the number of algal species were depressed at high metal sites. There were associations of green algal species in the field samples, and these correlated with dissolved trace metal concentrations. In fact, there was a close similarity between the green algal flora of metal-contaminated sites on the copper-polluted River Hayle and the lead-polluted River Gannel. This similarity indicates that the degree of trace metal contamination, rather than the dominant trace metal itself, determined the species association present. Mine sites high in dissolved trace metal concentrations were characterised by a community based on the filamentous chlorophyte genus *Microspora*: *M. stagnorum, M. pachyderma* and *M. willeana*. Sites with a low level of metal contamination were characterised by a charophyte green algal community of species of *Spirogyra* and *Mougeotia* (Foster, 1982a). At least one population of *Microspora stagnorum* from high metal sites on the rivers Hayle and Gannel showed the presence of copper tolerance, over and above a general high copper resistance shown by this chlorophyte (Foster, 1982b).

A population of another chlorophyte from the River Hayle, the unicellular *Chlorella vulgaris*, has been reported to show copper tolerance (Foster, 1977). A strain of *C. vulgaris* collected from a site with a total water concentration of 120 μg/L Cu downstream of a disused copper mine, had much higher growth rates in laboratory-dissolved copper exposures than a control strain collected above the mine site (<2 μg/L Cu) (Foster, 1977). The growth rate of the tolerant strain was unaffected by exposure to 100 μg/L dissolved copper, and this strain could still grow at 1,000 μg/L Cu. The control strain, on the other hand, had a decreased growth rate at an exposure of 50 μg/L dissolved Cu, and growth was completely inhibited at 300 μg/L Cu. The copper tolerance was maintained after 30-fold subculturing in the laboratory, confirming its genetic basis. The copper-tolerant strain accumulated five to ten times less copper than the control strain in laboratory copper exposures, suggesting copper exclusion to be the mechanism underlying the tolerance (Foster, 1977).

5.10.3 Red Algae

Prologue. Red algae are relatively uncommon in freshwater but have been reported from the Rivers Ystwyth and Rheidol.

In regions of the River Ystwyth barren of green algae in 1920, 1940 and 1953, there were often flourishing growths of red algae in the genera *Batrachospermum* and (particularly) *Lemanea* (Carpenter, 1924; Jones, 1958). To belie their classification as red algae, species of both these genera are in fact greenish in colour. Similarly, species of *Batrachospermum* and *Lemanea mamillosa* were also found in metal-contaminated parts of the nearby River Rheidol in the 1920s, 1930s and 1940s (Carpenter, 1924; Reese, 1937; Jones, 1949).

5.10.4 Golden-Brown Algae

Prologue. The golden-brown algae belong to the class *Chrysophyceae*, mostly to be found in freshwater. Chrysophytes do occur as macrophytes in metal-contaminated streams.

In the Gillgill Burn in 1974, there were seven chrysophyte species present at sites with dissolved zinc concentrations above 3,000 µg/L (Say and Whitton, 1980). At sites considered intermediate in zinc exposure levels in Gillgill Burn, there were conspicuous growths of the chrysophyte macrophytes *Hydrurus foetidus* and *Chrysonebula holmesii*. These algae are gelatinous, and their presence in intermediate zinc exposures may be down to reduced algal competition or decreased grazing (Say and Whitton, 1980).

5.10.5 Accumulation and Detoxification in Algae

Prologue. Algae are typically net accumulators of trace metals, reaching higher bioaccumulated concentrations as dissolved exposures increase (Whitton et al., 1981).

Mixed filamentous green algae from the Rivers Hayle and Gannel had raised accumulated concentrations of copper, lead and zinc at corresponding high metal sites on either river (Foster, 1982a). Accumulated concentrations of trace metals in the algae from low to high metal sites ranged from less than 100 to 7,000 µg/g Cu, from about 20 to 23,000 µg/g Pb and from about 50 to 2,000 µg/g Zn respectively. Species of the charophyte *Mougeotia* accumulated 219,000 µg/g Zn, 1,930 µg/g Pb and 83 µg/g Cd in Caplecleugh Low Level near Nenthead in the Northern Pennines (Patterson and Whitton, 1981).

Such high bioaccumulated concentrations of trace metals need to be detoxified in order to avoid toxic effects on the organisms concerned. A major mechanism for the detoxification of accumulated trace metals in both microalgae and macroalgae is their binding by phytochelatins (Perales-Vela et al., 2006).

5.11 Bryophytes

Prologue. Bryophytes, both liverworts and mosses, can represent the dominant flora in British streams contaminated with trace metals. The bryophyte flora of a stream is reduced both in biomass and number of species as trace metal contamination increases.

5.11.1 Mining-Affected Streams and Rivers

Prologue. Metal-contaminated freshwater systems surveyed for bryophytes include those of the Ystwyth and Rheidol in mid-Wales, the River Nent in Rivers the Northern Pennines and several streams in the Lake District.

5.11.1.1 Mid-Wales
Prologue. Some bryophyte species show resistance to lead and zinc exposure in the Rivers Ystwyth and Rheidol, especially the liverwort *Scapania undulata*.

In 1939, only a few scanty tufts of the liverwort *Scapania undulata* could be found in the River Ystwyth between the Cwm Ystwyth mine and Aberystwyth (Figure 5.7). Yet, in the uncontaminated part of the river, above any mine effluents, the moss *Fontinalis antipyretica* was common (Jones, 1958).

In surveys undertaken in the River Ystwyth between 1972 and 1973 (McLean and Jones, 1975), *S. undulata* was again the most metal-resistant bryophyte present (as it was in the River Clarach), being the only bryophyte found at the most metal-contaminated sites, including immediately below entry points of mine drainage streams. This liverwort was less common in cleaner areas of the River Ystwyth. At less contaminated sites in the Ystwyth could be found the mosses *Fontinalis squamosa*, *Platyhypnidium* (as *Eurynchium*) *riparioides*, *Racomitrium aciculare* and *Scorpidium scorpioides*. *Brachythecium rivulare* occurred in streams near the River Ystwyth that were relatively contamination-free. Much the same collection of bryophyte species was recorded from the River Ystwyth and associated mine drainage streams (the Cwm Ystwyth, Frongoch and Wemyss mines) in 1974, with the addition of the mosses *Philonotis fontana* growing at the sides of streams, and *Hygrohypnum ochraceum* (Burton and Peterson, 1979). *P. fontana* was found in all three of these mine drainage streams; *S. undulata* and *P. riparioides* at Cwm Ystwyth and Wemyss; and *F. squamosa, H. ochraceum* and *R. aciculare* only at Cwm Ystwyth (Figure 5.7) (Burton and Peterson, 1979).

The liverwort *Scapania undulata* and the moss *Hygrohypnum luridum* were collected along the River Ystwyth in 1982 (Jones et al., 1985). *S.undulata* was found in streams at the Cwm Ystwyth, Frongoch and Wemyss mines, as well as in the main river itself (Figure 5.7). *H. luridum* was present in the Frongoch stream and extended further down the main river than *S. undulata*.

In the River Rheidol in 1948, the moss *Fontinalis antipyretica* was abundant in the upper reaches above any mine effluents, and the liverwort *Jungermannia gracillima* was also sometimes present (Jones, 1949). Below the Cwm Rheidol mine and further downstream (Figure 5.7), there were only occasional clumps of *F. antipyretica*, but no other bryophytes (Jones, 1949). Yet, by 1974, *Scapania undulata, Philonotis fontana* and *Fontinalis squamosa* were recorded from the stream draining the Cwm Rheidol mine (Burton and Peterson, 1979), indicating reduced, but still significant, metal contamination. In 1982, *S. undulata* was again reported from the River Rheidol below the Cwm Rheidol mine (Jones et al., 1985). This liverwort and the moss *Hygrohypnum luridum* were also to be found below the Goginan mine, on the Nant Melindwr, a tributary of the Rheidol (Figure 5.7) (Jones et al., 1985).

In the lead- and zinc-rich streams of the historical mining region of mid-Wales, it is, therefore, the liverwort *Scapania undulata* that shows the greatest resistance to the legacy of metal contamination. Then in streams with less, but still moderate, metal contamination can variably be found the mosses *Fontinalis squamosa, Platyhypnidium riparioides, Racomitrium aciculare, Philonotis fontana* and *Hygrohypnum ochraceum*.

5.11.1.2 Northern Pennines
Prologue. Bryophytes do occur in the zinc-contaminated streams of the Northern Pennines, with the liverwort *Scapania undulata* particularly zinc-resistant.

In the Gillgill Burn on the River Nent, South Tyneside (Figure 5.9), in 1974, the only bryophyte present at

dissolved zinc concentrations above 3,000 µg/L was the now familiar liverwort *Scapania undulata* (Say and Whitton, 1980).

A wider survey of the metal-contaminated streams of the Northern Pennine Orefield during 1974 and 1975, confirmed that the distribution of bryophytes in the streams could be related to zinc levels in the streams (Say and Whitton, 1981). *S. undulata*, as expected, was found growing at high zinc levels, at times together with the liverwort *Jungermannia atrovirens* (as *Solenostoma triste*) or the moss *Bryum pallens*. *J. atrovirens* characterised small seepages. *B. pallens* was found only in calcareous adit waters, often in the presence of a thick coating of ochre, frequently with the moss *Dicranella varia* as an associated bank species. In zinc-rich streams of surface origin containing *S. undulata*, the moss *Philonotis fontana* often occurred as a bank species. To pass down a gradient of zinc contamination in the Northern Pennine streams, adit waters with lower levels of zinc contamination were characterised by *Brachythecium rivulare*, while such surface water streams were often dominated by *Platyhypnidium riparioides* or *Hygrohypnum ochraceum* (Say and Whitton, 1981).

In line with these generalisations, which agree well with the findings for bryophytes in mid-Wales, a 1981 survey of the Caplecleugh Low Level adit at Nenthead (Figure 5.9), recorded the presence only of *Bryum pallens* and *Dicranella varia* (Patterson and Whitton, 1981). In 2006, *Platyhypnidium riparioides* was collected from the lead- and zinc-rich Nentsberry stream flowing into the River Nent (Environment Agency, 2008a).

5.11.1.3 Lake District
Prologue. Several now familiar bryophytes were also collected from metal-contaminated streams of the Lake District in 2006 (Figure 5.10) (Environment Agency, 2008a).

Scapania undulata, for example, was present in the copper-contaminated Lever's Water Beck at Coniston. The remaining examples are bryophytes present in streams contaminated by lead and zinc in 2006. *Hygrohypnum ochraceum* was present in Newlands Beck, west of Derwentwater , below the added effect of local spoil heap runoff. *Fontinalis squamosa* occurred in Eller Gill near Keswick. On the Caldbeck Fells (Figure 2.10), *Brachythecium rivulare* was found in Roughton Gill, and *Platyhypnidium riparioides* was present in Wood Head stream, which contains copper and arsenic in addition to lead and zinc (Environment Agency, 2008a).

5.11.2 Accumulation and Biomonitoring

Prologue. Bryophytes living in freshwater streams will take up and accumulate trace metals from solution. Bryophytes are good biomonitors of dissolved trace metal bioavailabilities in freshwater systems.

Uptake and subsequent accumulation of a trace metal by a bryophyte will depend on the dissolved bioavailability of the metal, in turn particularly driven by the dissolved concentration of the metal. The higher the dissolved concentration, the higher the accumulated concentration in the bryophyte, with the different factors affecting dissolved bioavailability modulating the degree of increased uptake in different streams of different physicochemistries.

The trace metals accumulated by bryophytes will consist partly of metal passively adsorbed onto the cell walls, and partly of metal taken up into the cell. It is this latter absorbed metal that can interact with the physiology of the bryophyte and, in high concentration, cause ecotoxicological effects.

Table 5.5 provides examples of the raised concentrations of trace metals accumulated by bryophytes in metal-rich streams in Britain. The bryophytes concerned have already been introduced, as have many of the mining-contaminated streams. Most of the freshwater systems included in Table 5.5 are in historical lead and zinc mining regions in mid-Wales, the Northern Pennines and the Lake District. Table 5.5 additionally contains bioaccumulation data for three of the bryophytes, *Fontinalis antipyretica*, *F. squamosa* and *Platyhypnidium riparioides*, from

the River Etherow in the Manchester area affected by industrial pollution in 1979 to 1980 (Say et al., 1981).

The metal-resistant liverwort *Scapania undulata* reached phenomenally high accumulated concentrations of lead (up to 49,400 µg/g) in the River Clarach system in 1982, and also very high lead concentrations in the River Ystwyth system (up to 13,400 µg/g) that year (Jones et al., 1985). In 1972 to 1973, the moss *Fontinalis squamosa* accumulated up to 10,800 µg/g Pb in the Clarach system (McLean and Jones, 1975). In Wood Head stream on the Caldbeck Fells in the Lake District (Figure 2.10) in 2006, the moss *Platyhypnidium riparioides* had accumulated 26,400 µg/g Pb (Environment Agency, 2008a). These elevated accumulated lead concentrations are well above control lead concentrations up to about 100 µg/g Pb in these species (Table 5.5).

Accumulated zinc concentrations in bryophytes in zinc-contaminated streams can also be greatly raised above control concentrations of about 100 µg/g Zn (Table 5.5). *S. undulata* in the River Ystwyth system in 1982 reached 6,000 µg/g Zn (Jones et al., 1985), *Hygrohypnum* sp. in Brown Gill in the South Tyne system in 2006 accumulated up to 4,620 µg/g Zn (Environment Agency, 2008a), and *P. riparioides* in Wood Head stream on the Caldbeck Fells in 2006 contained up to 4,830 µg/g Zn (Environment Agency, 2008a).

Cadmium can be associated with zinc ores, and, correspondingly, cadmium bioavailabilities can be raised in local mining-contaminated streams. Thus *S. undulata* from the River Ystwyth and Rheidol systems in 1982 presented accumulated cadmium concentrations up to 21 and 16 µg/g Cd respectively (Table 5.5). Cadmium concentrations in *P. riparioides* in Wood Head stream on the Caldbeck Fells in 2006 had somewhat higher accumulated concentrations (28 µg/g Cd) (Environment Agency, 2008a). Although less spectacular than the highest accumulated concentrations of lead and zinc, these high cadmium concentrations in the bryophytes are still greatly elevated above control cadmium concentrations below 1 µg/g Cd (Table 5.5).

In streams draining old copper mining districts, the local bryophytes also contain raised copper and

Table 5.5 Ranges of concentrations (μg/g dry weight) of selected trace metals in bryophytes from British freshwater systems affected by mining or industrial contamination, including sites above the entry of metal contamination.

Bryophyte / Location	Date	Ag	As	Cd	Cu	Pb	Zn
Liverwort							
Scapania undulata Water Earwort							
Wales							
River Ystwyth system							
Ystwyth and mine streams[1]	1974				10–80	61–8,900	68–1,930
Ystwyth and tributaries[2]	1982	0.01–8.0		0.20–21	10–140	34–13,400	64–6,000
River Rheidol system							
Mine streams[1]	1974				7–61	765–7,810	777–3,560
Rheidol and tributaries[2]	1982	0.62–1.8		6.3–16	33–53	570–3,100	212–4,250
River Clarach system							
Clarach and tributaries[3]	1972–73				64–203	3,880–14,800	245–1,950
Clarach and tributaries[2]	1982	6.2–7.7		2.2–3.5	108–160	18,300–49,400	530–1,480
Lake District							
Coniston							
Lever's Water Beck[4]	2006		163	0.43	2,020	160	111

Table 5.5 (**cont.**)

Bryophyte	Location	Date	Ag	As	Cd	Cu	Pb	Zn
Mosses								
Brachythecium rivulare River Feather-moss								
	Wales							
	River Clarach system							
	Clarach and tributaries[3]	1972–73				36–90	1,330–8,210	381–1,290
	Lake District							
	Caldbeck Fells							
	Roughton Gill[4]	2006		110	4.47	199	4,490	877
Fontinalis antipyretica Greater Water-moss								
	Manchester industrial							
	River Etherow[5]	1979–80			1.0–354	14–103		
Fontinalis squamosa Alpine Water-moss								
	Wales							
	River Ystwyth system							
	Ystwyth and mine streams[1]	1974				15	133	102
	River Rheidol system							
	Mine streams[1]	1974				34–37	1,300–1,390	2,630–2,840

River Clarach system						
Clarach and tributaries[3]	1972–73			43–105	1,170–10,800	125–1,280
Lake District						
Keswick						
Eller Gill[4]	2006	162	13.6	51	1,060	594
Manchester industrial						
River Etherow[5]	1979–80				53–99	64.1–5,430
Hygrohypnum ochraceum Claw Brook-moss						
Wales						
River Ystwyth system						
Ystwyth and mine streams[1]	1974			13–15	91–2,450	98–780
Lake District						
Keswick						
Newlands Beck[4]	2006	45	0.76	104	5,510	159
***Hygrohypnum* sp. 1** Brook-moss						
Wales						
River Ystwyth system						
Ystwyth and tributaries[2]	1982	0.06–0.60	0.51–11	24–42	208–3,650	100–1,990
River Rheidol system						
Rheidol and tributaries[2]	1982	2.2	4.3	45	4,770	690

Table 5.5 (**cont.**)

Bryophyte / Location	Date	Ag	As	Cd	Cu	Pb	Zn
Hygrohypnum sp. 2 Brook-moss							
Northern Pennines							
South Tyne system							
Brown Gill[4]	2006		10	10	36	12,500	4,620
Platyhypnidium riparioides (formerly *Rhynchostegium riparioides*) Long-beaked Feather-moss							
Wales							
River Rheidol system							
Mine streams[1]	1974				10	1,640	3,270
River Clarach system							
Clarach and tributaries[3]	1972–73				378	9,650	378
Lake District							
Caldbeck Fells							
Wood Head stream[4]	2006		1,910	28.4	741	26,400	4,830
Northern Pennines							
South Tyne system							
Nentsberry Stream, Nent[4]	2006		5	3.8	25	2,820	1,250
River Wear system							
Rookhope Burn[6]	1986			10.4		560	2,850

Manchester industrial					
River Etherow[5]	1979–80	2.7–433	23–189	47–110	94–6,710
Racomitrium aciculare Yellow Fringe-moss					
Wales					
River Ystwyth system					
Ystwyth and mine streams[1]	1974		21	4,650	764
River Rheidol system					
Mine streams[1]	1974		19–43	897–1,220	564–1,080

Sources: Burton and Peterson (1979);[1] Jones et al. (1985);[2] McLean and Jones (1975);[3] Environmental Agency (2008a);[4] Say et al. (1981);[5] Kelly et al. (1987).[6]

arsenic concentrations (Table 5.5). *S. undulata* in Lever's Water Beck above Coppermine valley near Coniston in the Lake District contained up to 2,020 µg/g Cu in 2006 (Table 5.5). *P. riparioides* in Wood Head stream the same year had 741 µg/g Cu (Table 5.5). While again not as high as the accumulated concentrations of lead and zinc in bryophytes in metal-contaminated streams, these high copper concentrations are still well raised above background bryophyte concentrations of about 50 µg/g Cu or less (Table 5.5). Arsenic concentrations in the same *P. riparioides* in Wood Head stream in 2006 reached 1,910 µg/g As, well above any other bryophyte As concentration in Table 5.5. Background As concentrations in bryophytes can be as low as 5 to 10 µg/g As (Table 5.5). Therefore, the concentrations of about 160 µg/g As in *S. undulata* from Lever's Water Beck, Coniston, and in *F. squamosa* from Eller Gill, Keswick, should also probably be considered high and indicative of raised arsenic bioavailabilities (Table 5.5).

Thus bryophytes make good biomonitors of dissolved trace metal bioavailabilities in freshwater systems. Indeed, the results from mining-contaminated streams enable us to interpret the bryophyte data for the industrially affected River Etherow near Manchester in 1979 to 1980 (Say et al., 1981). Bryophyte accumulated concentrations were clearly very high indeed for cadmium and zinc in the River Etherow (Table 5.5), indicating greatly raised bioavailabilities of these trace metals. For example *Platyhypnidium riparioides* in the River Etherow contained up to 433 µg/g Cd, and *Fontinalis antipyretica* up to 354 µg/g Cd, astoundingly high concentrations (Table 5.5). Zinc concentrations in *P. riparioides* and *F. squamosa* respectively in the River Etherow reached up to 6,710 and 5,430 µg/g Zn, again extremely high and matching the highest zinc concentrations in bryophytes from zinc-contaminated mining streams (Table 5.5). There is a suggestion too of somewhat raised copper bioavailability in the River Etherow, *P. riparioides* reaching 189 µg/g Cu (Table 5.5) (Say et al., 1981). Lead concentrations in the River Etherow bryophytes cannot be considered as raised (Table 5.5).

All the bryophytes listed in Table 5.5 can be considered good biomonitors of trace metal bioavailabilities in metal-contaminated streams and rivers, and many of them appear on a proposed list of potential plant biomonitors (Whitton et al., 1981).

A variation on the theme of using in situ bryophytes to biomonitor trace metal bioavailabilities in streams or rivers is to collect moss samples from a clean submerged site, place them in mesh bags and transplant them for up to two days to the sites chosen to be monitored (Kelly et al., 1987). The transplanted moss reaches a new equilibrium bioaccumulated trace metal concentration over this period, reflecting the metal bioavailability of the new site. Both *Platyhypnidium riparioides* and *Fontinalis antipyretica* have been used to this end in a biomonitoring study of cadmium, lead and zinc bioavailabilities in Northern Pennine streams (Kelly et al., 1987).

5.12 Flowering Plants

Prologue. Flowering plants are conspicuous by their absence from metal-contaminated streams and rivers.

In the first half of the twentieth century, there were no aquatic flowering plants in metal-contaminated regions of the Rivers Ystwyth and Rheidol. Flowering plants only occurred in the River Ystwyth in the final stages of recovery of the river from metal contamination (Newton, 1944) and only well downstream in the River Rheidol (Jones, 1949). The plants concerned in both rivers were a species of water starwort of the genus *Callitriche* and water crowfoot *Ranunculus aquatilis*, the latter being very sensitive to any lead or zinc contamination entering the river (Newton, 1944).

Similarly, the metal-rich streams of the Northern Pennines were devoid of aquatic flowering plants at the times of study (Say and Whitton, 1980, 1981; Patterson and Whitton, 1981).

5.13 Invertebrates

Prologue. While small in number in comparison to their marine counterparts, there are still many invertebrate taxa in British freshwater systems. These range in size from the microscopic invertebrates in the meiofauna to large predatory dragonfly larvae and very large bivalve molluscs such as swan mussels.

Inevitably there is much to say about the ecologically important insect larvae that occur in mining-affected freshwater systems, namely stoneflies, mayflies, caddisflies and chironomid midges in particular. Crustaceans will also be represented, specifically amphipods and isopods. Two more taxa will be discussed in some detail: oligochaete worms living in sediments; and freshwater molluscs, including both bivalves and gastropods.

Firstly, we need to mention those many other invertebrate taxa that appear on fauna lists in streams and rivers or, rather, are often missing from fauna lists of metal-contaminated habitats. Relatively little is known about the biology of trace metals in these invertebrates, although their presence on, or their disappearance from, such fauna lists does provide information on the ecotoxicology of the trace metals contaminating a particular freshwater habitat. Brief reference has already been made to rotifers and tardigrades in surface films and to cyclopoid copepod crustaceans in sediments. Harpacticoid copepods are also to be found in freshwater sediments. Additional invertebrate taxa that may be listed in freshwater fauna lists are hydras, flatworms, nematode worms, nematomorph worms, leeches, collembolans, water mites and water spiders.

Between 1919 and 1922, Kathleen Carpenter studied the fauna of the river systems of mid-Wales, particularly the Rivers Ystwyth, Rheidol and Clarach (Figure 5.7), polluted with lead and zinc from historical lead mining (Carpenter, 1924). She showed the complete absence then from the most contaminated streams of flatworms, oligochaetes, leeches, molluscs, larger crustaceans such as amphipods and isopods, caddisflies as well as fish. The fauna of the most contaminated streams was limited to a few insect larvae: the chironomid *Tanypus nebulosus* (the clouded-winged midge), the blackfly *Simulium latipes* and the baetid mayfly *Cloeon simile*. The fauna of the Lower Ystwyth before 1922 consisted of only nine species, which included two of the preceding:

T. nebulosus and *C. simile.* The other seven species were a cyclopoid copepod; a water flea; a water mite; the water boatman or common backswimmer *Notonecta glauca* (a hemipteran bug); two water beetles, *Dytiscus marginalis* (the great diving beetle) and *Oreodytes sanmarkii*; and the stonefly *Leuctra nigra* (Carpenter, 1924). By 1953, many invertebrate taxa had returned to the mainstream of the River Ystwyth, downstream of the entry points of old mine effluents (Jones, 1940, 1958). The fauna, by then, consisted of one species of flatworm (*Phagocata vitta*) and several species each of water mites, stoneflies, mayflies, caddisflies, hemipterans, beetles and flies, amongst which were a blackfly, a cranefly and several species of midges, including chironomids. Oligochaetes, leeches, crustaceans and molluscs were conspicuous by their absence.

The invertebrate fauna of the nearby River Rheidol (Figure 5.7) in 1947 to 1948, downstream of historical mines, was much richer than that of the contemporary River Ystwyth. In addition to the taxa identified previously for the Ystwyth, the Rheidol also hosted a hydra; oligochaetes; two gastropods; one bivalve; leeches; small crustaceans; a collembolan; and, further amongst the insect larvae, dragonflies and an alderfly (Jones, 1949). The numbers of species of stoneflies, mayflies, caddisflies, hemipteran bugs, beetles and flies were also all much higher than in the Ystwyth (Jones, 1949).

In the Northern Pennines, the River Nent (Figure 5.9) draining into the South Tyne River is rich in zinc. Samples were taken between 1976 and 1977 of the benthic invertebrates living at sites along the Nent system characterised by a gradient of zinc levels from high to low (Table 5.6) (Armitage, 1980; Armitage and Blackburn, 1985). Across the range of sites, there was a significant negative correlation between the total water concentration of zinc and the numbers of invertebrate taxa per site (Armitage, 1980). There were three high zinc sites with total water concentrations above 2,000 µg/L zinc, a site close by at the mouth of the Dowgang Level draining an old lead mine and two sites on the Gillgill Burn tributary (Table 5.6). These three high zinc sites had reduced numbers of both individual invertebrates and

Table 5.6 **Selected insect larvae present at zinc-contaminated sites in the Nent River system in the Northern Pennines in 1976 to 1977, affected to different degrees by zinc contamination from historical mining activities.**

Total water Zn concentration (µg/L)	20–360	770–1,100	1,210–1,680	2,080–2,900	7,600
(number of sites)	(4)	(4)	(6)	(2)	(1)
Plecoptera (stoneflies)					
Leuctridae					
Leuctra		L. hippopus	L. hippopus	L. hippopus	
Leuctra	L. inermis	L. inermis	L. inermis	L. inermis	
Leuctra	L. moselyi	L. moselyi	L. moselyi	L. moselyi	
Leuctra	L. fusca	L. fusca	L. fusca		
Chloroperlidae					
Chloroperla	C. torrentium	C. torrentium	C. torrentium		
Perlodidae					
Isoperla	I. grammatica	I. grammatica	I. grammatica		
Perlodes	P. microcephala		P. microcephala		
Nemouridae					
Nemoura				N. cinerea	
Nemoura	N. cambrica	N. cambrica	N. cambrica		
Protonemura	P. meyeri		P. meyeri		
Protonemura	P. praecox	P. praecox	P. praecox		
Amphinemura	A. sulcicollis	A. sulcicollis	A. sulcicollis		
Taeniopterygidae					
Brachyptera	B. risi	B. risi	B. risi		

Ephemeroptera (mayflies)

Baetidae

Baetis	*B. rhodani*	*B. rhodani*	*B. rhodani*
Baetis	*B. scambus*	*B. scambus*	*B. scambus*
Centroptilum	*C. luteolum*		

Heptageniidae

Ecdyonurus	*E. dispar*	*E. dispar*	
Electrogena	*E. lateralis*	*E. lateralis*	
Rhithrogena	*R. semicolorata*	*R. semicolorata*	

Ephemerellidae

Serratella	*S. ignita*	*S. ignita*	*S. ignita*

Trichoptera (caddisflies)

Limnephilidae

Potamophylax	*P. cingulatus*		*P. cingulatus*
Stenophylax		*S. sequax/lateralis*	*S. sequax/lateralis*

Hydroptilidae

Hydroptila	Hydroptila sp.	Hydroptila sp.	Hydroptila sp.

Hydropsychidae

Hydropsyche	*H. instabilis*	*H. instabilis*	*H. instabilis*
Hydropsyche	*H. pellucidula*	*H. pellucidula*	*H. pellucidula*

Rhyacophilidae

Rhyacophila	*R. dorsalis*	*R. dorsalis*	*R. dorsalis*

Polycentropidae

Plectrocnemia	*P. conspersa*	*P. conspersa*	*P. conspersa*

Table 5.6 (cont.)

Total water Zn concentration (µg/L)	20–360	770–1,100	1,210–1,680	2,080–2,900	7,600
(number of sites)	*(4)*	*(4)*	*(6)*	*(2)*	*(1)*
Plectrocnemia	*P. geniculata*	*P. geniculata*	*P. geniculata*	*P. geniculate*	
Polycentropus	*P. flavomaculatus*	*P. flavomaculatus*	*P. flavomaculatus*		
Chironomidae (chironomid midges)					
Orthocladiinae					
Chaetocladius	*C. melaleucus*	*C. melaleucus*	*C. melaleucus*	*C. melaleucus*	*C. melaleucus*
Cricotopus	*C. pulchripes*	*C. pulchripes*	*C. pulchripes*	*C. pulchripes*	
Cricotopus		*C. trifascia*			
Eukiefferiella	*E. claripennis*	*E. claripennis*	*E. claripennis*		
Eukiefferiella	*E. minor*	*E. minor*	*E. minor*	*E. minor*	
Limnophyes			*Limnophyes* sp.	*Limnophyes* sp.	
Psectrocladius			*P. psilopterus*	*P. psilopterus*	
Brillia	*B. bifida*	*B. bifida*	*B. bifida*		
Heleniella	*H. ornaticollis*		*H. ornaticollis*		
Metriocnemus	*Metriocnemus* spp.	*Metriocnemus* spp.	*Metriocnemus* spp.		
Orthocladius	*O. frigidus*	*O. frigidus*	*O. frigidus*		
Parametriocnemus			*P. stylatus*		
Rheocricotopus	*R. effusus*	*R. effusus*	*R. effusus*		
Chironominae					
Polypedilum	*Polypedilum* sp.	*Polypedilum* sp.	*Polypedilum* sp.		
Micropsectra	*Micropsectra* sp.	*Micropsectra* sp.			
Prodiamesinae					
Prodiamesa	*P. olivacea*	*P. olivacea*			

Diamesinae			
Diamesa	*Diamesa* sp.	*Diamesa* sp.	*Diamesa* sp.
Potthastia		*P. gaedii*	*P. gaedii*
Pseudodiamesa			*P. branickii*
Pseudokiefferiella	*P. parva*	*P. parva*	*P. parva*
Tanypodinae			
Macropelopia	*M. notata*	*M. notata*	*M. notata*
Thienemannimyia	*Thienemannimyia* spp.	*Thienemannimyia* spp.	*Thienemannimyia* spp.

Note: Zinc contamination is expressed in terms of total water concentrations (µg/L) of zinc.

Sources: Armitage (1980); Armitage and Blackburn (1985).

invertebrate taxa. The Dowgang Level site (7,600 µg/L Zn) supported only two taxa: a chironomid midge, *Chaetocladius melaleucus* (family *Orthocladiinae*); and a lumbricid oligochaete, not specified further. The two high zinc Gillgill Burn sites (2,080 µg/L and 2,900 µg/L Zn) contained 18 and 19 macroinvertebrate taxa respectively, in comparison with at least 57 taxa at low zinc sites along the Nent (Armitage, 1980; Armitage and Blackburn, 1985). Across these two high zinc sites in the Gillgill Burn were one species of flatworm, two tubificid oligochaete species and one species of water mite, all unidentified. All the other invertebrate taxa at the high zinc sites in the Burn were insects, usually larvae: four stoneflies (*Nemoura cinerea* and three species of *Leuctra*), three caddisflies (*Plectrocnemia conspersa, P. geniculata* and a species of *Stenophylax*), one water beetle (*Helophorus brevipalpis*) and 11 dipteran fly species (Table 5.6). These flies included two species of blackfly, two cranefly species and seven chironomid species. Amongst these chironomids were *C. melaleucus* found at the Dowgang Level site, four other orthocladiines and two tanypodine chironomids (Armitage, 1980; Armitage and Blackburn, 1985). There were no mayflies present at the two high zinc Gillgill Burn sites. As in the mining-contaminated river systems of mid-Wales, leeches, crustaceans and molluscs were absent from sites of high metal contamination, although oligochaetes were present at the equivalent Nent sites (Armitage, 1980).

By the end of the twentieth century, it was routine for freshwater ecologists to employ multi-variate statistical techniques to detect patterns in datasets of invertebrate abundance across a series of sites. Ordination techniques, for example, allow the recognition of sites with different degrees of similarity of their invertebrate populations. Thus it is possible to identify sites with reduced macroinvertebrate assemblages caused by the ecotoxicological action of significant trace metal contamination, from a background of expected communities at uncontaminated sites. It is also possible to identify which invertebrate species occur in which assemblages along a gradient of metal-contaminated sites.

Anthony Gower and colleagues used ordination techniques to analyse the macroinvertebrate communities, present in 1991 to 1992, at 46 sites on 12 Cornish river systems, affected to different degrees by trace metal contamination from historical mining activities (Gower et al., 1994). These river systems included the Rivers Carnon, Hayle and Gannel. From a group of 39 physicochemical variables measured at each site, it was dissolved copper concentration that gave the strongest correlation with the invertebrate data, suggesting it might be a major driving factor in determining macroinvertebrate community structure. The statistical analysis distinguished four groups of sites along a gradient of copper concentrations on the basis of their different macroinvertebrate communities (Table 5.7). In the group of sites with the highest dissolved copper concentration (average 832 µg/L Cu), mayflies were absent, but flatworms, chironomids (especially of the subfamily *Orthocladiinae*) and a caddisfly were abundant. Dominant species were the flatworm *Phagocata vitta*, the orthocladiine chironomids *Chaetocladius melaleucus* and *Eukiefferiella claripennis* and the caddisfly *Plectrocnemia conspersa*. This group of sites also had the highest average dissolved concentrations of zinc (572 µg/L Zn) and aluminium (414 µg/L Al), and the lowest average pH (5.6) (Gower et al., 1994). We have already met the chironomid *C. melaleucus* as the most zinc-resistant insect in the River Nent system in the Northern Pennines (Table 5.6). Furthermore, it was the net-spinning caddis *P. conspersa* that dominated low-pH sites in the iron-rich streams of the Ashdown Forest (Hildrew, 2009).

Table 5.7 shows the presence and absence of the species of larvae of stoneflies, mayflies, caddisflies and chironomid midges across the four groups of sites distinguished in the statistical analysis. Species richness of all macroinvertebrates fell along the series from Group 1 to Group 4, although species numbers of the selected insect taxa in Table 5.7 were similar in groups 2 and 3 (Gower et al., 1994). There were changes in the percentage contribution of major taxa to the macroinvertebrate community across the four groups (Gower et al., 1994). Stoneflies contributed the largest proportion (19%) of individual

Table 5.7 **Selected insect larvae present in each of four groups of sites on 12 Cornish river systems in 1991 to 1992, affected to different degrees by trace metal contamination from historical mining activities. The four groups are characterised by a gradient of increasing average dissolved copper concentration.**

	Group 1	Group 2	Group 3	Group 4
Dissolved metal concentration (µg/L)				
Copper	15.9	42.1	165	832
Zinc	262	51.7	506	572
Plecoptera (stoneflies)				
Leuctridae				
Leuctra	*L. nigra*			
Leuctra	*L. fusca*			
Leuctra	*L. hippopus*		*L. hippopus*	
Chloroperlidae				
Chloroperla	*C. torrentium*		*C. torrentium*	
Perlodidae				
Isoperla	*I. grammatica*	*I. grammatica*	*I. grammatica*	
Nemouridae				
Protonemura	*P. meyeri*	*P. meyeri*	*P. meyeri*	
Amphinemura	*A. sulcicollis*	*A. sulcicollis*	*A. sulcicollis*	*A. sulcicollis*
Ephemeroptera (mayflies)				
Baetidae				
Alainites	*A. muticus*			
Baetis	*B. rhodani*	*B. rhodani*	*B. rhodani*	
Baetis	*B. vernus*	*B. vernus*	*B. vernus*	
Heptageniidae				
Rhithrogena	*R. semicolorata*	*R. semicolorata*		
Trichoptera (caddisflies)				
Limnephilidae				
Chaetopteryx	*C. villosa*			
Potamophylax	*Potamophylax* spp.	*Potamophylax* spp.	*Potamophylax* spp.	

Table 5.7 **(cont.)**

	Group 1	Group 2	Group 3	Group 4
Sericostomatidae				
Sericostoma	*S. personatum*	*S. personatum*		
Hydropsychidae				
Hydropsyche	*H. pellucidula*	*H. pellucidula*		
Hydropsyche	*H. siltalai*	*H. siltalai*	*H. siltalai*	
Rhyacophilidae				
Rhyacophila	*R. dorsalis*	*R. dorsalis*	*R. dorsalis*	*R. dorsalis*
Polycentropidae				
Plectrocnemia	*P. conspersa*	*P. conspersa*	*P. conspersa*	*P. conspersa*
Chironomidae (chironomid midges)				
Orthocladiinae				
Rheocricotopus	*R. effusus*			
Tvetenia	*T. calvescens*			
Thienemanniella	*Thienemanniella* spp.	*Thienemanniella* spp.	*Thienemanniella* spp.	
Brillia	*B. bifida*	*B. bifida*	*B. bifida*	*B. bifida*
Eukiefferiella	*E. claripennis*	*E. claripennis*	*E. claripennis*	*E. claripennis*
Chaetocladius	*C. melaleucus*	*C. melaleucus*	*C. melaleucus*	*C. melaleucus*
Corynoneura	*C. lobata*		*C. lobata*	*C. lobata*
Chironominae				
Micropsectra	*Micropsectra* sp.		*Micropsectra* sp.	
Polypedilum	*Polypedilum* sp.	*Polypedilum* sp.	*Polypedilum* sp.	
Prodiamesinae				
Prodiamesa	*P. olivacea*	*P. olivacea*	*P. olivacea*	
Diamesinae				
Diamesa	*Diamesa* sp.	*Diamesa* sp.	*Diamesa* sp.	*Diamesa* sp.
Tanypodinae				
Trissopelopia	*T. longimana*	*T. longimana*	*T. longimana*	*T. longimana*

Source: Gower et al. (1994).

macroinvertebrates in Group 1, followed by mayflies, chironomid midges and oligochaetes. Several of the stonefly species were confined to Group 1, and the abundance of those stoneflies that did appear in other groups was usually highest at the Group 1 sites. In Group 2, the mayflies contributed more than one-third of the macroinvertebrates present, the most frequent and abundant being *Baetis rhodani* and *Rhithrogena semicolorata* (Table 5.7). Caddisfly larvae were at their highest abundance in Group 2. *Hydropsyche pellucidula* was the most widespread and abundant caddisfly species in Group 2, but did not follow its congeneric relative *H. siltalai* into Group 3. In Group 4, the orthocladiine chironomids constituted 55% of the macroinvertebrates present, with *Chaetocladius melaleucus* and *Eukiefferiella claripennis* particularly prominent. Oligochaetes made their strongest contribution to macroinvertebrate numbers in Group 3, but were much reduced in Group 4. The flatworm *Phagocata vitta*, already met in the Rivers Ystwyth and Rheidol (Jones, 1949, 1958), was present in Group 4, where it constituted as many as 30% of the individual macroinvertebrates found (Gower et al., 1994). There were no mayflies in Group 4 and the stoneflies, present only in the guise of *Amphinemura sulcicollis* (Table 5.7), were much reduced in number. Of the caddisflies, it was *Plectrocnemia conspersa* that became increasingly prominent in groups 3 and 4.

This study highlighted the need to be wary about general statements concerning the metal resistance of particular insect taxa at the taxonomic level of order, for example the Ephemeroptera (mayflies) or Trichoptera (caddisflies). There were distinct differences between the metal resistances of different mayfly families within the Ephemeroptera. The baetids were more copper resistant than the heptageniid *Rhithrogena semicolorata*, and there were no ephemerellid mayflies found (Gower et al., 1994). Other authors have noted the metal sensitivity of ephemerellid and heptageniid mayflies, and the relative insensitivity of baetid mayflies (Rainbow et al., 2012), for example in North American mining-contaminated river systems such as Colorado mountain streams (Clements et al., 2000) and the Clark Fork River in Montana (Luoma et al., 2010). Furthermore, there was an identifiable order of metal sensitivities within the Trichoptera in the Cornish rivers (Table 5.7) (Gower et al., 1994). For example, amongst the net-spinning caddisfly larvae, *H. pellicidula* was more copper sensitive than

H. siltalai, and both were more copper sensitive than *Plectrocnemia conspersa* (Table 5.7).

The 1997 survey of 51 metal-contaminated streams in Cornwall and mid-Wales that investigated correlations between diatom assemblage compositions and the environmental characteristics of the sites sampled, similarly considered the composition of macroinvertebrate assemblages (Hirst et al., 2002). The species richness, species diversity and total abundance of the macroinvertebrates declined with increasing copper concentrations at the sites (Hirst et al., 2002), in line with the 1991/1992 data for the Cornish sites (Gower et al., 1994). For the large mixed Cornish and Welsh dataset, however, overall macroinvertebrate assemblage composition did not change with copper or zinc dissolved concentrations at sites (Hirst et al., 2002). The absence of any such clear correlations of single metal exposure with assemblage compositiion may well have been the result of combining so many sites dominated on the one hand by copper (in Cornwall), and on the other by lead and zinc (in mid-Wales). Nevertheless, the abundance of some individual macroinvertebrate taxa in the mixed dataset did vary with copper or zinc concentration or with CCU score (Hirst et al., 2002). For example, the abundance of stoneflies in the families *Leuctridae and Nemouridae* declined significantly with increasing dissolved copper concentrations, as in the earlier Cornish study. The abundances of chironomid midge larvae and of beetles also fell significantly with increasing copper concentration in the joint study. Heptageniid mayflies, however, were not scarcer at metal-rich sites (Hirst et al., 2002), despite this family of mayflies usually being regarded as metal sensitive (Clements, 2000; Clements et al., 2000; Rainbow et al., 2012). Baetid mayflies similarly showed no decrease in abundance at metal-rich sites (Hirst et al., 2002), more expectedly given their perceived relative metal insensitivity compared to heptageniids and ephemerellids (Gower et al., 1994; Rainbow et al., 2012). The observation that heptageniid mayflies were still present at the metal-rich sites in the joint Cornish-Welsh study does not necessarily contradict the more general conclusion that heptageniids (together with ephemerellids) are more metal-sensitive than baetids.

It has already been noted that the most metal-contaminated sites in this early 1990s Cornish and Welsh study do not fall into the extreme contamination categories of other comparative studies of metal-contaminated British river systems (Armitage, 1980; Gower et al., 1994). Thus heptageniids were still present at the most contaminated sites, and baetids showed no decrease in abundance there (Hirst et al., 2002).

It is time to explore in more detail the biology of trace metals in particular insect larvae. To be highlighted are stoneflies (*Plecoptera*), mayflies (*Ephemeroptera*), caddisflies (*Trichoptera*) and chironomid midges (*Diptera, Chironomidae*), all prominent in the previous, more general discussion on the structure of macroinvertebrate communities in metal-affected streams and rivers.

5.13.1 Stoneflies

Prologue. Stoneflies belong to the insect order *Plecoptera*, with larvae that live in freshwater, reaching their greatest abundance in fast-flowing, well-oxygenated streams and rivers with stony and gravel bottoms. Their preference for small upland streams has brought stonefly larvae into a position of strong likelihood for exposure to metal-contaminated effluents from British mining activities, not least in the Pennines or upland regions of mid-Wales. Stoneflies are affected by high trace metal contamination by mining effluent, to the point of selective elimination.

Stoneflies also occur in lowland streams and rivers, but in smaller numbers. They are considered intolerant of waters of poor quality with high organic and low oxygen contents. Stonefly larvae are typically herbivorous, grazing the periphyton microflora of surface films on submerged rocks or feeding as shredders on benthic algae or dead leaves. Thus the acidification-tolerant larvae of *Leuctra nigra* feed on iron bacteria. Larger species, particularly in the families *Perlodidae* and *Perlidae*, may be carnivorous on other smaller insect larvae.

No stoneflies were present in the mine-affected regions of the Ystwyth in 1919, with a single stonefly, *Leuctra nigra*, present further downstream in the

lower Ystwyth (Carpenter, 1924). In the River Nent system in the Northern Pennines in the mid-1970s (Armitage, 1980), there were no stoneflies at the most zinc-contaminated site (7,600 µg/L Zn) at the mouth of the Dowgang Level (Table 5.6). The next two most zinc-contaminated Nent system sites, both in the Gillgill Burn with total water zinc concentrations of 2,080 and 2,900 µg/L Zn, had four stoneflies between them (Table 5.6). *Nemoura cinerea* was the most abundant stonefly at the more zinc-contaminated site of the two, which also contained *Leuctra hippopus*. Two other species of *Leuctra*, *L. inermis* and *L. moselyi*, joined these two stoneflies at the 2,080 µg/L Zn site (Table 5.6). Six Nent system sites, including the third site on the Gillgill Burn, had total zinc concentrations in the water between 1,210 and 1,680 µg/L Zn. While these six sites lacked *N. cinerea*, which was only found in the two high zinc Gillgill Burn sites, they did between them contain the three aforementioned *Leuctra* species and a further nine stoneflies (Table 5.6). No new stonefly species appeared at lower zinc exposures. The threshold exposure corresponding to a major elimination of most stoneflies lay between total zinc concentrations in the water of 1,680 and 2,080 µg/L (Table 5.6).

Stoneflies were well represented in the 1991 to 1992 study of 46 sites on 12 Cornish river systems, covering a gradient of exposures to trace metal contamination from historic mining (Gower et al., 1994). Dissolved copper concentration was highlighted as the probable major driver affecting the community structure of macroinvertebrates across the sites. Four groups of sites along a gradient of dissolved copper exposure were recognised on the basis of their different macroinvertebrate communities (Table 5.7). Stoneflies contributed the largest proportion (19%) of individual macroinvertebrates in Group 1, the group of sites with the lowest dissolved copper concentration. While five of the seven stonefly species in Group 1 were also to be found in Group 3 (Table 5.7), their abundance was typically highest at the sites in Group 1 (Gower et al., 1994). The nemourid *Amphinemura sulcicollis* was the only stonefly to be found at Group 4 sites with the highest copper exposure (Table 5.7). *A. sulcicollis* was also found in the Nent system, where it occupied a position in the groups of sites exposed to 1,210 to 1,680 µg/L total zinc water concentrations or lower (Table 5.6). Another nemourid stonefly, *Protonemura meyeri*, occupied sites of Group 3 copper exposure in Cornwall (Table 5.7) and also sites of some considerable zinc exposure (1,210 to 1,680 µg/L total Zn) in the Nent (Table 5.6). These two nemourid stoneflies seem to be showing resistance to trace metal exposure, although the nemourid stonefly *Nemoura cinerea*, which was most zinc resistant in the Northern Pennines (Table 5.6), was absent from the Cornish streams and rivers (Gower et al., 1994). Three further stoneflies in other families, *Leuctra hippopus* (*Leuctridae*), *Chloroperla torrentium* (*Chloroperlidae*) and *Isoperla grammatica* (*Perlodidae*), also showed relatively high resistance to either copper or zinc exposure across the two datasets (Tables 5.6 and 5.7).

Table 5.8 lists accumulated concentrations of selected trace metals in stoneflies collected from British streams and rivers, including many affected by historical mining. There may be subtle interspecific differences between trace metal bioaccumulation patterns even between closely related species, and ideally it remains preferable to identify organisms for analysis to species level. Nevertheless, in the field this is not always pragmatically feasible. So Table 5.8 includes data on specific species (*Perla bipunctata*), on mixed species in a single genus (*Leuctra* spp.) and in mixed species grouped at family level (*Perlodidae*). When using organisms as biomonitors of summated local trace metal bioavailabilities to that organism, the aspiration is that intersite differences in bioaccumulated trace metal concentrations will be much greater than any differences caused by physiological variation, or in this case by a lack of complete species separation. Given the huge ranges in bioaccumulated concentrations of stoneflies from different sites, particularly so in the cases of arsenic, lead and zinc (Table 5.8), it would appear pragmatically acceptable to identify stoneflies only to the level of genus in a biomonitoring study. It might even be possible to draw conclusions from bioaccumulation data collected at the family level (De Jonge et al., 2014).

The data in Table 5.8 show that stoneflies are good biomonitors of local bioavailabilities in streams and rivers of arsenic, cadmium, copper, lead and zinc.

Atypically high bioaccumulated concentrations of each of these trace metals are found in streams and rivers identified earlier in the chapter as containing high dissolved (Table 5.2) and sediment (Table 5.4) concentrations of the trace metal concerned. Similarly, the stonefly biomonitoring data (Table 5.8) correlate well with bryophyte biomonitoring data for metal-contaminated British freshwater systems (Table 5.5).

While concentrations of cadmium accumulated by the stoneflies do not show the wider ranges observed for arsenic, copper, lead and zinc, it is still easy to recognise atypically high bioaccumulated concentrations, above those identified as control or background (Table 5.8). Thus, bioaccumulated cadmium concentrations might typically lie between 0.4 and 2 µg/g in stoneflies from uncontaminated systems, but reached as high as 66 µg/g Cd in species of *Leuctra* collected from the South Tyne in 2014 (Table 5.8). Arsenic concentrations in stoneflies typically lie in the range 15 to 30 µg/g in uncontaminated situations, but can reach in excess of 3,000 or 4,000 µg/g in arsenic-contaminated Cornish streams and rivers (Table 5.8). Control copper concentrations in stoneflies are of the order of 10 to 60 µg/g with bioaccumulated concentrations up to and exceeding 1,000 µg/g Cu, for example in the River Hayle in Cornwall with high copper bioavailabilities (Table 5.8). Background lead concentrations in stoneflies are probably up to 30 µg/g, with bioaccumulated lead concentrations reaching as high as 2,500 µg/g under conditions of lead contamination (Table 5.8). Zinc concentrations in stoneflies have a higher baseline between about 140 and 300 µg/g, but can similarly reach concentrations above 5,000 µg/g Zn, for example in the upper tributaries of the River Wye in mid-Wales or in the South Tyne in the Northern Pennines (Table 5.8).

A striking feature of the data in Table 5.8 is the recent date of the collections quoted, many of which come from a study between 2013 and 2015 (Jones et al., 2016). It is a daunting conclusion that many of the streams and rivers listed in Table 5.8, still have very high trace metal bioavailabilities in spite of local mining having ceased as much as a century or more ago.

Table 5.8 **Mean accumulated concentrations (μg/g dry weight) of selected trace metals in stoneflies in British streams and rivers affected by mining contamination and in uncontaminated control streams and rivers.**

Family		As	Cd	Cu	Pb	Zn
Genus or species						
Region						
Catchment						
River/stream	Date					
Leuctridae						
Leuctra spp.						
SW England						
Bolingey stream, Perranzabuloe[1]	2014	45–53	17–25	75–147	88–296	843–2,270
Porthleven stream, Porthleven[1]	2014	3,570–4,610	4.2–9.9	81–531	7.7–50	203–353
Hayle[1]	2014	2,880–3,330	3.4–5.5	769–1,120	22–30	348–544
Wales						
Rheidol						
Afon Melindwr[1]	2013		3.1–8.1	55–90	164–280	268–340
Ystwyth						
Nant Magwr[1]	2013		15	41–42	27–399	259–1,320
Wye[1]	2014	84–170	8.6–15	15–113	18–104	308–5,450
Northern Pennines						
South Tyne[99]	2014	392–638	1.9–66	26–110	42–796	370–4,290
Derwent[2]	1978/79		1.2–21		27–391	212–1,290
England and Wales						
Controls[1]	2013/14	15–30	0.4–2.0	10–60	5–30	140–300
Leuctra sp.						
Northern Pennines and Lake District[3]	2006		0.2–24	22–578	4.1–2,490	182–5,600
Perlidae						
Perla bipunctata						
Northern Pennines						
Derwent[2]	1978/79		0.8–12		4–128	232–522

Table 5.8 **(cont.)**

Family	As	Cd	Cu	Pb	Zn
Genus or species					
Region					
Catchment					
River/stream — Date					
Derwent[4] — 1992			18–33	33–90	194–509
East Allen[4] — 1992			32–35	39–107	319–577
West Allen[4] — 1992			23–29	38–51	384
Chloroperlidae					
Chloroperla sp.					
Northern Pennines					
Derwent[2] — 1978/79		0.9–3.7		101–295	226–750
Perlodidae					
Northern Pennines and Lake District[3] — 2006		0.1–15	11–399	0.2–584	163–3,000
Isoperla grammatica					
Northern Pennines					
Derwent[2] — 1978/79		2.4–3.6		22–271	304–689
Perlodes microcephala					
Northern Pennines					
Derwent[2] — 1978/79		1.2–50		11–161	316–962
Nemouridae					
Amphinemura sulcicollis					
Northern Pennines					
Derwent[2] — 1978/79		3.7–5.1		68–596	203–1,060

Sources: Jones et al. (2016);[1] Burrows and Whitton (1983);[2] De Jonge et al. (2014);[3] Rainbow, Gibson and Smith (unpublished).[4]

5.13.2 Mayflies

Prologue. Mayfly larvae are widespread in British streams and rivers, including those contaminated by trace metals. Different mayflies show different trace metal sensitivities, baetids being considered relatively metal-resistant while heptageniid and ephemerellid larvae are considered metal-sensitive. As another generalisation, mayflies appear more trace metal-sensitive than stoneflies or caddisflies. Species of *Baetis* are good biomonitors of trace metal bioavailabilities in streams moderately affected by mining contamination.

Mayflies belong to the insect order *Ephemeroptera*, a name derived from the ancient Greek meaning 'lasting for a day'. Indeed, adult mayflies are famous for their short lives, although they may not last exactly one day. Many adult mayflies live for only a few hours, some for a few days and some even for less than one hour (Lancaster and Downes, 2013). Adult mayflies will aggregate in huge mating swarms near water where they lay their eggs. Most of a mayfly's life is spent as an aquatic larva. This larval stage generally lasts in the order of weeks to months, although aquatic development may be as long as three years (Lancaster and Downes, 2013). Most mayfly larvae are detritivorous or herbivorous, feeding on detritus or periphyton in the surface film, while a few are carnivorous (Brittain, 1982).

Different mayfly larvae prefer different freshwater habitats. Species of the heptageniid mayfly genera *Ecdyonurus* and *Rhithrogena* cope with rapid flowing water over rocks (Harris, 1952) and are typically grazers or scrapers of the surface film. On the other hand, the larvae of *Ephemera* (unsurprisingly an ephemerid) are confined to fine sand and mud (Harris, 1952) and are detritivorous. Larvae of species of *Baetis* (a baetid) are distributed more widely, for they are able to live both on rocks and amongst weeds in flowing water, and correspondingly are absent from ponds or lakes (Harris, 1952). Larvae of another baetid, *Cloeon simile*, the Lake Olive, prefers slow flowing or still water (Harris, 1952). Baetid larvae include both fine particle detritivores and periphyton grazers (Brittain, 1982).

Reference has been made earlier to the generalisation that there are differences between mayfly families as to their sensitivities to exposure to trace metal contamination. Research in the United States has provided strong evidence that baetid larvae are relatively trace metal-insensitive as mayflies go, while heptageniid and ephemerellid larvae are considered metal-sensitive (Clements, 2000; Clements et al., 2000).

What about evidence from mining-contaminated streams in the British Isles? The single mayfly species present in the most lead- and zinc-contaminated streams of the mid-Wales catchments of the Rivers Ystwyth, Rheidol and Clarach in the early 1920s was a baetid, in this case *Cloeon simile* (Carpenter, 1924). In the zinc-contaminated system of the River Nent in 1976, two baetids, *Baetis rhodani* (the large dark olive) and *B. vernus* (the medium olive), were found at sites exposed to higher total water concentrations of zinc than any of the sites occupied by the heptageniids *Ecdyonurus dispar, Electrogena lateralis* and *Rhithrogena semicolorata* (Table 5.6). The ephemerellid *Serratella ignita* was, however, also found at the same relatively high zinc sites as the baetids (Table 5.6). Another perceived generalisation, that mayflies are more metal-sensitive than stoneflies or caddisflies, does seem to hold for the Nent River data (Table 5.6).

Data for the Cornish river systems predominantly affected by copper in the early 1990s also confirm that the two baetids *B. rhodani* and *B. vernus* are present in a group of sites of higher dissolved copper exposure than the sites occupied by the heptageniid *R. semicolorata* (Table 5.7). As in the Nent system (Table 5.6), no mayfly larvae were present in the most metal-exposed Cornish sites, which did still contain stoneflies and caddisflies (Table 5.7).

A later 2009 study of three copper-contaminated Cornish river systems (Rainbow et al., 2012) also showed that baetids (again represented by *B. rhodani* and *B. vernus*) were present at sites with high copper bioavailabilities, in the absence from these sites of the heptageniid *R. semicolorata*, which occurred only at the lowest copper exposure site (Table 5.9a). Baetids also showed higher copper resistance than the ephemerellid *S. ignita* (Table 5.9a). The same two baetid species also appeared to show higher resistance to lead and/or zinc exposure in the lead- and

Table 5.9 **Mayfly larvae (Ephemeroptera) present in trace metal-contaminated Cornish river systems in 2009: (a) at ten sites from three catchments (River Carnon, River Hayle, Red River) contaminated by copper, variably associated with arsenic; (b) at five sites in the River Gannel system contaminated by lead, variably associated with zinc. Measures of trace metal bioavailabilities are mean accumulated concentrations (μg/g dry weight) in caddisflies** *Hydropsyche siltalai* **at the site.**

(a) Copper/arsenic sites

Hydropsyche siltalai Cu (μg/g)	41	102–315	433–575	831–1,200
Hydropsyche siltalai As (μg/g)	42	<35–172	65–67	110 –51
Number of sites	1	4	2	3
Baetidae				
Alainites		*A. muticus*		
Baetis	*B. rhodani*	*B. rhodani*	*B. rhodani*	
Baetis		*B. vernus*	*B. vernus*	*B. vernus*
Heptageniidae				
Rhithrogena	*R. semicolorata*			
Ephemerellidae				
Serratella		*S. ignita*		

(b) Lead/zinc sites

Hydropsyche siltalai Pb (μg/g)	<26–90	209	441
Hydropsyche siltalai Zn (μg/g)	200–636	562	600
Number of sites	3	1	1
Baetidae			
Alainites	*A. muticus*		
Baetis	*B. rhodani*	*B. rhodani*	*B. rhodani*
Baetis	*B. scambus/fuscatus*	*B. scambus/fuscatus*	
Baetis	*B. vernus*		*B. vernus*
Heptageniidae			
Ecdyonurus		*E. torrentis*	
Rhithrogena	*R. semicolorata*	*R. semicolorata*	
Ephemerellidae			
Serratella	*S. ignita*	*S. ignita*	

Table 5.9 **(cont.)**

Ephemeridae		
Ephemera	E. danica	E. danica
Caenidae		
Caenis	C. rivulorum	C. rivulorum
Leptophlebiidae		
Habrophlebia	H. fusca	
Paraleptophlebia	P. cincta	

Source: Rainbow et al. (2012).

zinc-contaminated River Gannel system than any heptageniid or ephemerellid (Table 5.9b). Mayflies as a whole were also absent from several metal-contaminated sites where a caddisfly could survive, in this case *Plectrocnemia conspersa* (Rainbow et al., 2012). In each of the copper-contaminated Lynher and Seaton River systems, the lead-contaminated Gannel River system, and the multicontaminated Carnon River system, *P. conspersa* was present at highly contaminated sites unaccompanied by any mayfly (Rainbow et al., 2012).

It can be concluded, therefore, that British streams and rivers contaminated by trace metals derived from mining do hold to two generalisations concerning mayfly metal sensitivities: baetid mayflies are more metal-resistant than heptageniid and ephemerellid mayflies, and particular stonefly and caddisfly species are more metal-resistant than any mayfly.

Table 5.10 presents bioaccumulated concentrations of five mining-associated trace metals in species of the mayfly genus *Baetis* in British streams and rivers, affected by mining contamination. As we have seen, it is baetid mayflies that are the most likely mayflies to occur in the more contaminated streams and rivers. Correspondingly, species of *Baetis* itself are well represented in the accumulated metal concentration literature for catchments across different geographical regions of Britain associated with mining (Table 5.10).

Mayfly larvae show wide ranges of bioaccumulated trace metal concentrations, especially for zinc (Table 5.10). The higher bioaccumulated

concentrations of metals in the mayflies are found in larvae from water systems containing atypically high dissolved and sediment concentrations of trace metals (Tables 5.2 and 5.4). A water system included in Table 5.10, but not yet met in this chapter, is that of the Rea Brook in Shropshire. The Rea Brook, via its tributary the Ministerley Brook, drains the region of the historically important Snailbeach lead mine, near Shrewsbury, before flowing into the River Severn. Local Shropshire mines excavated zinc ore in addition to lead ore, and the Rea Brook system is thus contaminated with zinc as well as lead. The mayfly biomonitoring data (Table 5.10) agree well with biomonitoring datasets presented earlier for bryophytes (Table 5.5) and stoneflies (Table 5.8).

As discussed for stoneflies, an idealised biomonitoring programme would use data for a single species. The data in Table 5.10 are at least restricted to data for species of a single genus, and there is excellent agreement across the data as to what is a high or low accumulated trace metal concentration. In fact, a study of two species of *Baetis*, *B. rhodani* and *B. vernus*, from the catchment of the River Biala Przemsza draining a zinc and lead mining area of Upper Silesia in Poland has specifically addressed the potential consequences of not separating data for these two species on any conclusions from a biomonitoring programme. Reassuringly, this Polish study concluded that any failure to distinguish between the two species, *B. rhodani* and *B. vernus*, would not affect conclusions to be drawn on relative differences in trace

Table 5.10 **Mean accumulated concentrations (µg/g dry weight) of selected trace metals in baetid mayflies (species of *Baetis*) in British streams and rivers affected by mining contamination, and in uncontaminated control streams and rivers.**

Taxon	Date	As	Cd	Cu	Pb	Zn
Region						
Catchment						
River/stream						
Baetis rhodani						
Yorkshire Dales						
Wharfedale						
Hebden Beck[1]	1993			46–83	61–884	6,020–17,500
Shropshire						
Rea Brook[1]	1993			71–98	427–1,320	3,820–13,500
Baetis spp.						
SW England						
Bolingey stream, Perranzabuloe[2]	2014	29–91	2.7–36	39–166	2.9–211	484–8,720
Porthleven stream, Porthleven[2]	2014	340–6,050	12–19	116–796	5.5–43	1,150–2,340
Hayle[2]	2014	51–936	1.0–7.3	15–555	2.0–28	150–1,040
Wales						
Rheidol						
Afon Melindwr[2]	2013		2.5–12	29–59	33–131	232–2,430
Ystwyth						
Nant Cwmnewydion[2]	2013		1.2–25	32–71	617–632	9,480–10,900
Shropshire						
Rea Brook[2]	2014		1.7–32	28–48	5.3–167	328–8,360
Northern Pennines						
Derwent[3]	1978/79		7.5–72		20–320	1,800–10,500
South Tyne[2]	2014	8.7–31	0.9–2.4	13–20	2.1–80	159–395
Teesdale						
Eggleston Burn[2]	2015		4.2–6.2	18–22	81–136	1,900–2,190
Hudeshope Beck[2]	2015		4.2–10	19–26	24–177	824–2,320
England and Wales						
Controls[2]	2013/14	6.5–35	0.4–2.0	10–50	0.5–30	45–300

Table 5.10 **(cont.)**

Taxon		As	Cd	Cu	Pb	Zn
Region						
Catchment						
River/stream	Date					
Baetis sp. A						
Northern Pennines						
Derwent[4]	1992			13–36	68–114	347–658
East Allen[4]	1992			13–25	50–76	306–333
West Allen[4]	1992			21–52	66–78	410–860
Baetis sp. B						
Northern Pennines						
Derwent[4]	1992			15–36	67–87	306–586
East Allen[4]	1992			25	74	133

Sources: Rainbow, Hurley and Smith (unpublished);[1] Jones et al. (2016);[2] Burrows and Whitton (1983);[3] Rainbow, Gibson and Smith (unpublished).[4]

metal bioavailabilities between sites (Fialkowski et al., 2003; Fialkowski and Rainbow, 2006).

From Table 5.10, it is possible then to distinguish between background trace metal concentrations accumulated by species of *Baetis* and atypically high bioaccumulated concentrations reached under conditions of high local metal bioavailabilities. Background bioaccumulated concentrations of arsenic would lie between 6 and 35 µg/g, with raised concentrations apparently able to exceed 6,000 µg/g As. A background concentration of cadmium in species of *Baetis* would fall between approximately 0.4 and 2 µg/g, but this might rise above 70 µg/g Cd under conditions of atypically raised cadmium bioavailability. The bioaccumulated concentration of copper in these mayflies from uncontaminated sites would be expected to be between 10 and 50 µg/g, while an atypically raised accumulated copper concentration might approach 800 µg/g Cu. Background lead concentrations are of the order of 0.5 to 30 µg/g but

might rise as high as 1,330 µg/g Pb under conditions of very high lead contamination. Bioaccumulated zinc concentrations have a higher baseline of 45 to 300 µg/g and can attain concentrations well into the thousands, reaching as high as 17,500 µg/g Zn (Table 5.10).

5.13.3 Caddisflies

Prologue. Caddisfly larvae are an ecologically important component of invertebrate benthic life in British streams and rivers, including those contaminated by trace metals. Caddisflies are generally considered more resistant to trace metal exposure than mayflies. Species of *Hydropsyche* are good biomonitors of trace metal bioavailabilities in streams contaminated by trace metals.

Caddisflies, sometimes called sedges, make up the insect order *Trichoptera*. Many caddisfly larvae construct transportable cases in which they live, the cases being made up of sand particles, small stones or plant fragments (Edington and Hildrew, 1981). Indeed, it is possible that the word 'caddis' may derive from an old custom of travelling salesmen of pinning samples of ribbon or 'cadace' onto their coats (Edington and Hildrew, 1981). Such cased caddisflies include members of the families *Limnephilidae, Hydroptilidae* and *Sericostomatidae* (Tables 5.6 and 5.7).

Of more relevance here, however, are caseless caddis larvae, which make up about 50 of the approximately 200 species of British caddisflies (Harris, 1952; Edington and Hildrew, 1981). Caseless caddisflies in two families, the *Hydropsychidae* and the *Polycentropidae* (Tables 5.6 and 5.7), spin silken nets which they use to collect food from the flowing water in which they live (Edington and Hildrew, 1981). Species of *Hydropsyche* (grey flags) occur in the streams of interest to us here. *H. siltalai* (Plate 5c) and to a lesser extent *H. pellucidula* are particularly well represented in the metal-contaminated systems of Southwest England (Table 5.7). Another net-spinner, the polycentropid *Plectrocnemia conspersa*, extends further upstream than the *Hydropsyche* species and can be found living in very narrow upper reaches of streams. Larvae of the species of the genus *Rhyacophila* (the sand sedge), such as *R. dorsalis*, are also caseless and are free-roving predators.

After allowing for interspecies variation, caddisflies are generally considered more resistant to trace metal exposure than mayflies (Clements et al., 2000; Rainbow et al., 2012). For example, at the two high zinc sites in the Gillgill Burn, a tributary of the River Nent, there were no mayflies present while caddisflies and stoneflies were represented (Table 5.6). Similarly, in the metal-contaminated river systems of Cornwall in 1991 to 1992, caddisflies and stoneflies appeared in the most copper-exposed group of sites, in the absence of mayflies (Table 5.7).

It remains the case, however, that caddisflies show species differences in their trace metal sensitivities (Tables 5.6 and 5.7). Of the three caddisflies present in the zinc-contaminated Gillgill Burn in 1976 to 1977, two were species of the polycentropid genus *Plectrocnemia, P. conspersa* and *P. geniculata* (Table 5.6). The third was a species of the cased

limnephilid genus *Stenophylax*, either *S. sequax* or *S. lateralis* but not identified further (Table 5.6). In the metal-contaminated river systems of Cornwall in 1991 to 1992, *P. conspersa* again showed as amongst the most trace metal-resistant caddisflies (in this case, copper), accompanied then by the carnivorous *Rhyacophila dorsalis* (Group 4 in Table 5.7). *Hydropsyche siltalai* and an unidentified species of the limnephilid *Potamophylax* joined these two species in the group of sites with the second-highest copper exposure (Group 3, Table 5.7). The second *Hydropsyche* species, *H. pellucidula*, did not appear until the next group of sites down the gradient of copper exposure (Group 2, Table 5.7).

Table 5.11 lists the bioaccumulated concentrations of selected trace metals in species of two net-spinning genera of caseless caddisflies, *Hydropsyche* and *Plectrocnemia*, in British streams and rivers affected by mining contamination.

As for both stoneflies such as species of *Leuctra* (Table 5.8) and mayflies such as species of *Baetis* (Table 5.10), species of both *Hydropsyche* and *Plectrocnemia* (Table 5.11) show a wide range of accumulated trace metal concentrations according to their site of collection and can be used for bio-monitoring. Possessing tracheal gills, caddisfly larvae would have limited potential to take up trace metals dissolved in the surrounding medium. On the other hand, a net-feeding caddisfly larva does have the potential to ingest many small detritus particles filtered from suspension, including perhaps resus-pended fine sediment. Small suspended particles have a large surface area to volume ratio and high organic content, and so are likely to adsorb much metal from solution. They thereby represent a metal-rich food source reflecting metal levels at a site. Indeed, bio-dynamic modelling studies on two *Hydropsyche* species, *H. siltalai* and *H. pellucidula*, have confirmed that both species obtain more than 95% of their accumulated arsenic, and almost 100% of their accumulated silver, from ingested food rather than solution (Awrahman et al., 2015).

The net-spinning caddisfly larvae in Table 5.11 contain ranges of accumulated trace metal concentrations across different sites, enabling the recognition of background and atypically raised

Table 5.11 **Mean (or estimated for standard sized 10 milligram caddisfly larva)* accumulated concentrations (µg/g dry weight) of selected trace metals in net-spinning caseless caddisfly larvae in British streams and rivers affected by mining contamination, and in uncontaminated control streams and rivers.**

Taxon		As	Cd	Cu	Pb	Zn
Region						
Catchment						
River/stream	Date					
Hydropsyche pellucidula						
Northern Pennines						
Derwent[1]	1978/79		1.0–51		73–261	359–659
Hydropsyche siltalai						
SW England						
Hayle[2]	2009	42–151		41–844	11–18	218–508
Hayle[3]*	2013	105–360				
Red River[2]	2009	50–120		102–1,200	<30	214–246
Red River[3]*	2013	51–886				
Porthtowan[2]	2009	172		169	<30	313
Carnon[2]	2009	<35–82		247–315	<30–48	353–414
Gannel[2]	2009	<27–467		19–45	32–441	200–636
Gannel[3]*	2013	85–258				
Seaton[2]	2009	21		835	<25	203
Cornwall controls[2]	2009	25–50		19–45	10–35	200–300
Hydropsyche spp.						
SW England						
Bolingey stream, Perranzabuloe[4]	2014	25–63	0.5–5.1	21–114	3.8–205	146–1,460
Porthleven stream, Porthleven[4]	2014	279–12,300	1.3–15	59–792	5.2–96	154–400
Hayle[4]	2014	106–280	4.0–22	131–334	14–36	324–3,810
Wales						
Rheidol						
Afon Melindwr[4]	2013		1.1–1.3	23–30	90–217	177–285

Table 5.11 **(cont.)**

Taxon	Date	As	Cd	Cu	Pb	Zn
Region						
Catchment						
River/stream						
Ystwyth						
Nant Cwmnewydion[4]	2013		3.5–4.6	19–20	627–632	688–1,080
Shropshire						
Rea Brook[4]	2014	14	5.7–6.8	15–34	19–270	369–1,270
Northern Pennines						
East Allen[4]	2014	28–71	3.6–5.3	24–51	2.1–3.3	355–717
Teesdale						
Eggleston Burn[4]	2015			9.2	514	321
Hudeshope Beck[4]	2015			14–16	32–423	150–386
England and Wales						
Controls[4]	2013/14	7–50	0.5–2.5	10–55	2–35	100–200
Plectrocnemia conspersa						
SW England						
Lynher[2]	2009	<28–38		259–425	<25–99.8	178–232
Darley Brook[5]	1985			229		
Porthtowan[2]	2009	<32		98	<28	426
Carnon[2]	2009	<43–181		95–291	<43	259–452
Gannel[2]	2009	<48		131	204	180
Seaton[2]	2009	<50		684	<44	191
Cornwall controls[2]	2009	<50		90–150	<50	170–200

Sources: Burrows and Whitton (1983);[1] Rainbow et al. (2012);[2] Awrahman et al. (2015);[3] Jones et al. (2016);[4] Gower and Darlington (1990).[5]

bioaccumulated concentrations. A background arsenic concentration in *Hydropsyche siltalai* would be between about 7 and 50 µg/g, but concentrations can exceed 10,000 µg/g As, as in caddisflies from the Porthleven stream in North Cornwall (Table 5.11). Cadmium concentrations fall in a lower range.

Background concentrations expected in uncontaminated conditions would be between about 0.5 and 2.5 µg/g Cd in *Hydropsyche* species (Table 5.11). Accumulated cadmium concentrations in these caddisflies can reach about 70 µg/g under conditions of high cadmium bioavailability, as in the Rivers Derwent and East Allen in the Northern Pennines (Table 5.11). Copper concentrations in *Hydropsyche* species are in the approximate range 10 to 55 µg/g in the absence of copper contamination, but can reach 1,200 µg/g Cu, as in the River Hayle in Cornwall (Table 5.11). Background lead concentrations in species of *Hydropsyche* are of the order of 2 to 35 µg/g, but accumulated concentrations of about 630 µg/g Pb have been recorded in these caddisflies from the River Ystwyth system in mid-Wales (Table 5.11). As for stoneflies and mayflies, background zinc concentrations in *Hydropsyche* species are higher than for the other trace metals, falling between 100 and 300 µg/g (Table 5.11). Atypically raised zinc concentrations in these caddisflies can exceed 1,000 µg/g and approach 4,000 µg/g (Table 5.11).

The bioaccumulated trace metal concentrations presented in Table 5.11, and indeed the equivalent data in Tables 5.8 and 5.10, are typically for late instar larvae. The choice of these larger larvae provides more material for analysis, but also reduces the consequences of size effects on accumulated metal concentrations when making comparisons between sites. Many biomonitoring studies make allowance statistically for such size effects (Rainbow and Moore, 1986; Luoma and Rainbow, 2008). Indeed, some of the data in Table 5.11 are concentrations estimated for standard-sized larvae. Because of the higher surface area to volume ratios of smaller invertebrates, the presence of adsorbed metal on the surface of a very small aquatic arthropod can make a significant contribution to the total bioaccumulated metal concentration. The remainder is the metal that has been taken up and accumulated internally. This latter component represents the metal of greater biological interest, because only this absorbed metal can interact physiologically with the organism and be of ecotoxicological significance. The accumulated concentration of a trace metal in an aquatic arthropod will often follow what is known mathematically as a power function curve, with a considerable rise in total bioaccumulated concentration below a threshold individual dry weight, estimated to be about 2 milligrams (0.002 g) (Rainbow and Moore, 1986). This effect of size is well shown in a study of copper concentrations in

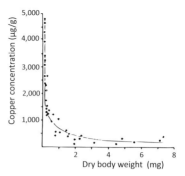

Figure 5.15 *Plectrocnemia conspersa*: relationship between dry body weight (mg) and Cu concentration (µg/g dry weight) in individual caddisfly larvae (instars II–V) collected from the copper-contaminated Darley Brook tributary of the River Lynher, Cornwall, in September 1984. The curve drawn is a best-fit power function curve. (After Darlington et al., 1987, with permission of Springer Nature.)

different larval instars of *Plectrocnemia conspersa* from Darley Brook, a copper-contaminated tributary of the River Lynher in Cornwall (Figure 5.15). In September 1984, larval instar stages II, III, IV and V *P. conspersa* in Darley Brook had respective mean copper concentrations of 3,770, 1,810, 740 and 258 µg/g Cu (Darlington et al., 1987). The effect of dry weight on accumulated copper concentrations in these larvae followed a power function relationship, with accumulated copper concentrations rising steeply in caddisfly larvae below 2 milligram dry weight (Figure 5.15).

The bioaccumulation of atypically high concentrations of trace metals implies the existence of a physiological mechanism to store the metals in detoxified form, whether permanently or for a period before their excretion. Such is the case for the copper accumulated by *P. conspersa* larvae collected from Darley Brook (Darlington and Gower, 1990). Granules containing copper and sulphur are common in the cells of the Malpighian tubules and in the epithelial cells underlying the cuticle over the body (Darlington and Gower, 1990). These granules can be interpreted as products of the lysosomal breakdown of metallothionein binding copper, induced as a mechanism of detoxification of the considerable amounts of copper entering these larvae. Equivalent caddisfly larvae from an uncontaminated nearby stream also contained these granules in the Malpighian tubules and beneath the body cuticle, but in noticeably fewer numbers than in the Darley Brook larvae (Darlington and Gower, 1990). It is likely that a number, at least, of the subcuticular granules are shed with the old cuticle at the time of a moult (Darlington and Gower, 1990), partly reducing the extremely high total copper concentration of the larva, at least for a time at the beginning of the next intermoult stage.

The potential of species of *Hydropsyche* to be used as biomonitors of trace metal bioavailabilities, coupled with their relative insensitivity to metals in comparison with mayflies, has led to their use as ecological indicators of metal ecotoxicity at the community level. High accumulated trace metal concentrations within a particular species are not in themselves indicators of any ecotoxicological effect of the metals on that species. A high accumulated concentration may simply reflect an efficient storage detoxification system, and a high total bioaccumulated concentration does not necessarily indicate that the accumulating organism is under any metal ecotoxicological stress. An atypically high accumulated metal concentration in a biomonitor does indicate a high local bioavailability of the trace metal. While not affecting the biomonitoring species itself, that high bioavailability may be sufficient to have an ecotoxicological effect on other more metal-sensitive members of the local biological community. Hence, a bioaccumulated trace metal concentration in a metal-insensitive

biomonitor can be calibrated for use as an indicator of an ecotoxicological effect on other local organisms (Luoma et al., 2010; Rainbow et al., 2012). This has been shown to be the case for bioaccumulated concentrations of arsenic, copper, lead and zinc in *Hydropsyche siltalai* in metal-contaminated Cornish river systems and ecotoxicological effects on the presence and abundance of mayflies, especially heptageniid and ephemerellid mayflies (Rainbow et al., 2012). Thus it has proved possible to plot the accumulated copper concentrations in *H. siltalai* from these rivers against the combined abundance of heptageniid and ephemerellid mayflies at each site (Figure 5.16). Furthermore, the study was able to define the threshold of accumulated copper concentration (170 µg/g) in the caddisflies that indicated a local copper bioavailability high enough to have a deleterious ecotoxicological effect on the abundance of these two families of mayflies. Similar ecotoxicological thresholds for arsenic, lead and zinc were 85 µg/g As, 300 µg/g Pb and 300 µg/g Zn (Rainbow et al., 2012). It is clear, therefore, at many of the sites listed in Table 5.11, the local bioavailability of one or more of these trace metals would easily be high enough to reduce the abundance of heptageniid and ephemerellid mayflies, or indeed eliminate them completely.

It is a feature of Figure 5.16 that, at some sites, the abundance of the mayflies was still low, even though the bioavailability of copper, indicated by the bioaccumulated copper concentration in

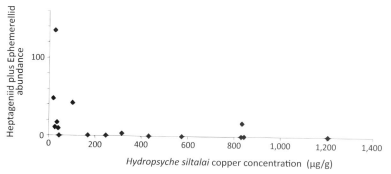

Figure 5.16 A plot of mean accumulated Cu concentrations (µg/g dry weight) in larvae of the caddisfly *Hydropsyche siltalai* against the combined abundance of heptageniid and ephemerellid mayflies (mean numbers in one-minute kick samples) at the site in metal-contaminated Cornish river systems. (After Rainbow et al., 2012, with permission.)

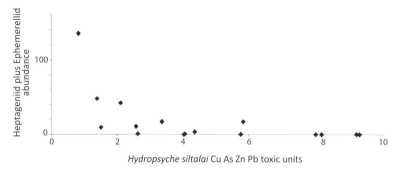

Figure 5.17 A plot of combined mean accumulated trace metal concentrations expressed as toxic units in larvae of the caddisfly *Hydropsyche siltalai* against the combined abundance of heptageniid and ephemerellid mayflies (mean numbers in one-minute kick samples) at the site in metal-contaminated Cornish river systems. (After Rainbow et al., 2012, with permission.)

H. siltalai, was low. The reason for this observation is that the dataset comes from a variety of contaminated Cornish river systems, including the River Gannel, contaminated by lead and zinc, but not copper. Thus the apparently anomalous sites in Figure 5.16 are sites on the Gannel, where it is the lead (and/or zinc) bioavailability, but not copper bioavailability, that is having an ecotoxicological effect on the mayflies (Rainbow et al., 2012). One way to address the question of the effects of a mixture of toxic metals present at a site is to use a CCU of toxicity (Clements, 2004; Rainbow et al., 2012). In this case, the mean accumulated concentration of each metal in *H. siltalai* at each site is divided by the threshold concentration defined previously for the metal that indicates an ecotoxicological effect on the heptageniid and ephemerellid mayflies. These toxic units for the different metals are summated for each site. Figure 5.17 shows the outcome of this process for the Cornish river systems. The new combined toxic unit value gives an excellent explanation and indication of the onset of ecotoxicological effects of combined raised toxic metal bioavailabilities on the heptageniid and ephemerellid mayflies (Figure 5.17).

5.13.4 Chironomid Midges

Prologue. Chironomid midge larvae can be relatively insensitive to high exposure to toxic metals in comparison to other freshwater insect larvae. The larvae accumulate high concentrations of trace metals under conditions of raised bioavailabilities. Chironomid larvae show morphological deformities when exposed to high bioavailabilities of trace metals in sediments.

Midges are an informal assortment of many different kinds of small flies. Flies belong to the insect order *Diptera*. This order includes houseflies, horseflies and hoverflies, together with flies with aquatic larvae such as blackflies, mosquitoes and the non-biting midges, the *Chironomidae*. Because the term 'midge' has no strict taxonomic meaning, so-called midges appear in many dipteran families, including gall midges (*Cecidomyiidae*); phantom midges (*Chaoboridae*); biting midges (*Ceratopogonidae*); and, of interest to us here, the *Chironomidae*.

Chironomids are very diverse, but their biodiversity often goes unnoticed because individual chironomid species are notoriously difficult to identify, other than by taxonomic experts. Indeed, we have already come across this relative difficulty of species identification in Tables 5.6 and 5.7. There are 11 subfamilies of chironomids, five of which we have met in Tables 5.6 and 5.7, the largest subfamilies being the *Orthocladiinae*, the *Chironominae* and the *Tanypodinae*.

Chironomid larvae are common inhabitants of muds or silts at the bottom of British streams or rivers, although they are also associated with all sorts of other inorganic or organic substrates, including gravel, pebbles, cobbles, leaves, wood and plants. While many chironomid larvae burrow, others will live on the surface of the substratum. Chironomid larvae tend to be pale olive or blood red in colour (Harris, 1952). Bright red chironomid larvae are often called blood worms, although this name is also commonly applied to a variety of red worms, including oligochaetes. The red colour is down to the presence of strong concentrations of the respiratory protein haemoglobin, an adaptation to life in low oxygen conditions, particularly organically rich muds. The distribution of high haemoglobin concentrations amongst chironomids does vary between subfamilies. All chironomine chironomids, for example, contain haemoglobin (Harris, 1952).

As already seen in Tables 5.6 and 5.7, chironomids can be relatively insensitive to high exposure to toxic metals in comparison to other aquatic insect larvae. In Kathleen Carpenter's 1919 to 1922 study of the lead- and zinc-contaminated river systems of mid-Wales, the tanypodine chironomid *Tanypus nebulosus* (the clouded-winged midge) was one of only three invertebrates found in the most contaminated streams.

In the system of the River Nent in the Northern Pennines in the 1970s, the orthocladiine chironomid midge, *Chaetocladius melaleucus*, was the only insect present at the most zinc-contaminated site in the Dowgang Level (Table 5.6). The next two high zinc sites (2,080 and 2,900 µg/g Zn in the Gillgill Burn) contained up to seven chironomid species – four more orthocladiines in addition to *C. melaleucus*, and two tanypodines (Table 5.6). Although the number of chironomid species was low at the three most zinc-contaminated sites, the abundance of the seven species, when present, was high (Armitage and Blackburn, 1985). More chironomid species were found on passage down the gradient of zinc contamination over the remaining sites in the system (Table 5.6). There were 18 species present at sites with total water concentrations of zinc between 1,210 and 1,680 µg/L, 16 species at sites with 770 to 1,100, and 30 species across the lowest zinc sites (Table 5.6). The 15 chironomid species occurring only at sites in the lowest zinc contamination category are not shown in Table 5.6.

In metal-contaminated Cornish river systems in the early 1990s (Gower et al., 1994), chironomid midges showed up particularly strongly as relatively copper-insensitive insect larvae (Table 5.7). While the species richness of chironomids did fall along the gradient of sites of increasing copper exposure, chironomids, especially the orthocladiine chironomids *Chaetocladius melaleucus* and *Eukiefferiella claripennis*, were abundant at the most copper-exposed sites (Table 5.7). In these Group 4 sites, orthocladiine chironomids made up 55% of all benthic macroinvertebrates present. Indeed, chironomid larvae as a whole made their highest percentage contribution to the number of individual macroinvertebrates in groups 3 (30%) and 4 (60%), yet, in the latter group, they were only represented by half the number of species found in Group 1 (Table 5.7).

Thus the orthocladiine *C. melaleucus* has come out as both the most zinc-resistant chironomid larva in the River Nent system (Table 5.6) and as one of the five most copper-resistant chironomids in Cornish rivers (Table 5.7). While none of the other six chironomids present at sites in the second-highest category of zinc contamination in the Nent system (Table 5.6) appeared at sites in either of the two highest copper-exposed category in Cornwall (Table 5.7), two chironomid species in the third zinc exposure category (1,210 to 1,680 µg/L Zn) in the Nent system (Table 5.6) are represented in the highest copper-exposed group of Cornish sites (Table 5.7). These are two further orthocladiine species, *Eukiefferiella claripennis* and *Brillia bifida* (Tables 5.6 and 5.7). An unidentified diamesine chironomid species, *Diamesa* sp., also appears in both these categories (Tables 5.6 and 5.7), but we cannot be sure that is the same species in the two studies.

Chironomid larvae, like their stonefly, mayfly and caddisfly larval counterparts, accumulate high concentrations of trace metals under conditions of raised bioavailability. The available database, however, is much more restricted. Unidentified species of *Chironomus* accumulated 59 µg/g Cd, 305 µg/g Pb and 1,500 µg/g Zn at a downstream site on the River Derwent in the Northern Pennines site in the late 1970s (Burrows and Whitton, 1983). To put these concentrations into context, in the zinc- and cadmium-contaminated Riou Mort and its tributaries in France in the late 1970s, chironomid larvae (mainly *Chironomus* gr. *thummi*) accumulated up to 189 µg/g Cd and 3,600 µg/g Zn (Say and Giani, 1981). The same chironomid species collected from industrially contaminated water courses in Flanders, Belgium, in 1991 contained up to 38 µg/g Cd, 489 µg/g Cu, 2,320 µg/g Pb and 1,320 µg/g Zn (Bervoets et al., 1994). Background concentrations of trace metals in these midge larvae fell in the range 0.1 to 0.6 µg/g Cd, 12 to 35 µg/g Cu, 0.5 to 10 µg/g Pb and 128 to 240 µg/g Zn (Bervoets et al., 1994).

Chironomid larvae are examples of invertebrates that show morphological deformities when exposed to high bioavailabilities of toxic contaminants including trace metals in sediments. Deformities of the mouthparts occur in chironomid larvae living in trace metal-contaminated sediments (Di Veroli et al., 2014). Furthermore, cause/effect relationships between the presence of such deformities and sediment metal concentrations have been demonstrated in the laboratory (Martinez et al., 2003). Mouthpart deformities of *Chironomus tentans* larvae include fused teeth, split teeth, missing teeth, extra teeth and abnormally shaped teeth on the mandible, with different effects apparently being associated with different metal exposures (Martinez et al., 2003). With the ever-present caveat that high sediment concentrations of trace metals may not translate directly into high metal bioavailabilities, it would appear that the sediment metal concentrations associated with mouthpart deformities (Janssens de Bisthoven et al., 1992; Martinez et al., 2003) fall within the ranges of high concentrations of metals in freshwater sediments affected by mining contamination in the British Isles (Table 5.4).

As can be seen in Table 5.2, the Afon Goch draining the former copper mining region of Parys Mountain in Anglesey, north Wales, contains very high dissolved concentrations of copper, iron, manganese and zinc at low pH. Specimens of the chironomid larva *Chironomus acidophilus* collected from the Afon Goch in 2004, showed evidence of a genotoxic effect of exposure to high bioavailabilities of toxic metals (Michailova et al., 2009). These chironomids presented genotoxic rearrangements of chromosomes in the salivary gland, possibly associated with a reduction of population fitness. It was proposed that such changes in salivary gland chromosomes of chironomid larvae may provide a very sensitive endpoint for detecting ecotoxicologically significant contamination by trace metals (Michailova et al., 2009).

5.13.5 Further Insects

Prologue. Other freshwater insects deserve mention in less detail.

5.13.5.1 Blackflies

Prologue. Like the chironomid midges, blackflies fall in a family of the *Diptera*, in this case the *Simuliidae*. They too have aquatic larvae.

Blackflies may also be difficult to identify to the level of species, and the relevant taxonomic unit may be a species complex. Blackflies are to be found in freshwater systems contaminated by high bioavailabilities of trace metals derived from mining.

The very limited fauna of the most zinc- and lead-contaminated streams of mid-Wales, studied by Kathleen Carpenter between 1919 and 1922, included a blackfly – *Simulium latipes* (Carpenter, 1924). In the zinc-contaminated River Nent system in the Northern Pennines in the 1970s, the two high zinc sites in the Gillgill Burn (2,080 and 2,900 µg/L Zn – Table 80) contained *Simulium cryophilum* (as *S. brevicaule*) and a blackfly of the *Simulium vernum* complex. Three further blackfly species, *S. aureum*, *S. monticola* and *S. variegatum*, were present at the sites with total water concentrations between 1,210 and 1,680 µg/L Zn (Armitage, 1980).

Blackfly larvae also accumulate high concentrations of trace metals when exposed to raised bioavailabilities. Thus species of *Simulium* in the River Derwent in the Northern Pennines in 1978 to 1979 contained up to 29 µg/g Cd, 617 µg/g Pb and 1,190 µg/g Zn (Burrows and Whitton, 1983).

5.13.5.2 Craneflies

Prologue. Craneflies also belong to the insect order *Diptera*. They fall across several families, including the *Tipulidae*, the *Limoniidae* and the hairy-eyed craneflies, the *Pediciidae*. Cranefly larvae, too, occur in freshwater systems affected by trace metals derived from mining activities.

In the Nent in the 1970s, tipulid cranefly species of the genus *Tipula* itself occurred at the high zinc sites in the Gillgill Burn, together with a limoniid cranefly of the genus *Molophilus* (Armitage, 1980). Hairy-eyed craneflies of the genus *Dicranota* were restricted to the two groups of sites in Table 5.6 with the lowest zinc exposure (Armitage, 1980). On the other hand, down in Cornwall in the early 1990s, one hairy-eyed cranefly *Pedicia rivosa* occurred in groups of sites across the full range of copper exposures, including sites in Group 4, the most copper-contaminated (Table 5.7) (Gower et al., 1994). Another hairy-eyed cranefly, in this case an unspecified species of *Dicranota*, extended its distribution up to Group 3, as did species of the limoniid cranefly family in the genus *Eleophila* (Table 5.7) (Gower et al., 1994).

5.13.5.3 Dagger Flies

Prologue. Other dipteran flies found in metal-contaminated river systems are the dagger flies or balloon flies of the family *Empididae*.

Dagger flies were found at sites of medium zinc exposure in the Nent (1,201 to 1,680 µg/L Zn) (Armitage, 1980), but extended through all copper-contaminated rivers in Cornwall (Gower et al., 1994). Species of the *Widemannia* species group were widespread and well-represented in Group 4 sites with the highest copper exposure (Gower et al., 1994).

5.13.5.4 Alderflies

Prologue. To move away from dipteran flies, alderflies belong to the order *Megaloptera*. Their large aquatic larvae are typically carnivorous.

In the zinc-contaminated Nent system in the 1970s, the alderfly *Sialis fuliginosa* was present but only at sites of relatively low zinc exposure (770–1,100 µg/L Zn, Table 5.6) (Armitage, 1980). We have already met this

species of alderfly, abundant in the iron-rich and acid-rich Broadstone Stream in the Ashdown Forest (Hildrew, 2009). Back in Cornwall, though, another alderfly larva, in this case *Sialis lutraria*, was present in 1973 in the adit of the Wheal Godolphin mine, emptying into the River Hayle, with very high copper and zinc levels (Brown, 1977a).

5.13.5.5 Dragonflies

Prologue. Dragonflies belong to the insect order *Odonata* and typically have very large carnivorous aquatic larvae, also to be found in the presence of raised trace metal levels.

Larvae of the golden-ringed dragonfly *Cordulegaster boltonii* were present across groups 1 to 3 of the range of copper exposures in Cornish rivers in the 1990s (Gower et al., 1994). In one of these Cornish rivers, the River Hayle, larvae of a species of skimmer of the genus *Libelulla* occurred in 1973 at a site at Carbis Mill with high copper exposure Brown, 1977a).

5.13.5.6 Beetles

Prologue. The final group of insects considered here are the beetles of the order *Coleoptera*. Beetles occur in freshwater, both as larvae and as adults, when they are usually referred to as water beetles.

The macroinvertebrate fauna of the Lower Ystwyth River in mid-Wales before 1922 consisted of only nine species, but included were two water beetles *Dytiscus marginalis* (the great diving beetle) and *Oreodytes sanmarkii*, both of the family *Dytiscidae* (Carpenter, 1924). In the Nent system in the 1970s, the small water beetle *Helophorus brevipalpis* (family *Helophoridae*) was found at the high zinc sites in the Gillgill Burn (2,080–2,900 µg/g Zn) (Armitage, 1980). In the copper-contaminated rivers of Cornwall in the 1990s, another dytiscid water beetle, *Agabus guttatus*, occurred at sites in the two highest copper exposure groups (groups 3 and 4, Table 5.7) (Gower et al., 1994). The riffle beetle *Limnius volckmari* (family *Elmidae*) occurred at sites across groups 1 to 3 (Gower et al., 1994). A water beetle of the genus *Hygrobia* was also found in the copper and zinc-rich Wheal Godolphin Adit of the Hayle system in 1973 (Brown, 1977a).

Thus, water beetles do indeed show up at freshwater sites in Britain that can be considered to be very contaminated by trace metals from mining.

5.13.6 Crustaceans

Prologue. Above the size range of the meiofauna, the most common crustaceans in British freshwaters are amphipods and isopods.

5.13.6.1 Amphipods

Prologue. The amphipod crustacean *Gammarus pulex*, the so-called freshwater shrimp, is very common in freshwater systems on mainland Britain not contaminated by trace metals.

Gammarus pulex feeds on detritus and is an important part of the detritus food chain in the streams and rivers of Great Britain, bringing about the reduction of decaying leaves into smaller particles. *G. pulex*, however, was absent from Ireland until introduced into Lough Neagh in the 1950s. The equivalent freshwater amphipod in Ireland is a related species, *Gammarus duebeni*, which lives in brackish or estuarine habitats in England, Wales and Scotland.

Gammarus pulex is not considered to be metal-resistant, and it is typically absent from streams strongly affected by trace metal contamination from mining activity. Thus in the zinc-contaminated Nent system in the Northern Pennines investigated in the mid-1970s, these amphipods were present only in the sites falling below a total water zinc concentration of 1,000 µg/L Zn, at the bottom end of the zinc exposure range of the sites examined (Table 5.6) (Armitage, 1980). Similarly, in 1979, *G. pulex* was present in the River East Allen in the Northern Pennines, but absent from the nearby River West Allen, which had considerably higher levels of zinc while still supporting stoneflies, baetid mayflies, caddisflies and chironomids (Abel and Green, 1981). In the copper-contaminated river systems of Cornwall in the 1990s,

G. pulex was not present at any site outside Group 1, the group of lowest copper exposure (Table 5.7) (Gower et al., 1994). Its absence was attributed to the effects of low water hardness, alkalinity and pH and/or elevated trace metal bioavailabilities.

As is typical of amphipod crustaceans, *Gammarus pulex* is a net accumulator of trace metals, but usually a relatively weak one in comparison with the insect larvae presented in Tables 5.8, 5.10 and 5.11. And indeed, the amphipod will not survive in the highest metal bioavailabilities survived by stoneflies, baetid mayflies and caddisflies listed in these tables. Nevertheless, as can be seen in Table 5.12, accumulated trace metal concentrations in *G. pulex* will be lifted above expected baselines when the local metal bioavailabilities are raised. Notable are the high lead concentrations accumulated by amphipods in Glengonnar Water draining Leadhills in Scotland, and very high zinc concentrations in amphipods in Arkle Beck in Swaledale in the Northern Pennines (Table 5.12).

The typical baseline accumulated copper concentration of 19 to 40 µg/g (Table 5.12) deserves comment. Such a concentration would approximately meet estimated essential copper requirements for enzymes and the respiratory protein haemocyanin in a malacostracan crustacean such as an amphipod (Chapter 3). The lower copper concentrations in some *G. pulex* in Table 5.12, for example in Red Tarn Beck at Greenside in the Lake District (Figure 2.9), may reflect a physiological sacrifice of haemocyanin as a protein energy reserve to an amphipod under stress, at the expense of its respiratory oxygen-carrying function requiring copper. Such a use of haemocyanin is not unknown in amphipods, causing a significant fall in body copper concentrations, for example in non-active littoral amphipods *Talitrus saltator* hibernating deep in burrows at the top of sandy shores in winter (Moore and Rainbow, 1987).

5.13.6.2 Isopods

Prologue. Isopod crustaceans in British streams and rivers include the relatively common species *Asellus aquaticus* and *Proasellus meridianus*.

Table 5.12 **Mean (or estimated for standard-sized 5 milligram amphipod)* accumulated concentrations (µg/g dry weight) of selected trace metals in the amphipod crustacean *Gammarus pulex* in British streams and rivers affected by mining contamination and in uncontaminated control streams and rivers.**

Taxon			Date	As	Cd	Cu	Pb	Zn
Region								
	Catchment							
		River/stream						
Gammarus pulex								
SW England								
	St Lawrence stream, Lanivet, Bodmin[1]		2014	42	3.8	109	3.6	100
	Tavy							
		River Burn[1]	2014	12–41	0.5–3.1	11–22	5.2–54	103–431
Shropshire								
	Rea Brook[1]		2014	7.7–9.4	0.3–7.6	68–77	3.5–92	80–292
Northern Pennines								
	Swaledale							
		Arkle Beck[1]	2014	22–54	1.2–4.7	24–54	3.8–21	463–1,900
Lake District								
	Greenside							
		Red Tarn Beck[1]	2014			8.4–12	17–38	328–679
Scotland								
	Leadhills							
		Glengonnar Water[2*]	1983			55–83	298–322	105–114
	Cumbrae							
		Control stream[2*]	1983			77–100		62–101
England and Wales								
	Controls[1]		2013/14	7–30	0.3–1.5	19–40	1.5–5	60–110

Sources: Jones et al. (2016);[1] Rainbow and Moore (1986).[2]

We met isopods, in the form of woodlice, living in the terrestrial environment. Although of marine origin, isopod crustaceans are also to be found in freshwater, where they go under the common name of water slaters. Water slaters tend to crawl over the bottom of streams, rivers and ponds and over weeds, rather than swim. They are detritivorous, feeding on decaying organic matter. The most widespread water slater in Britain is the common water slater *Asellus aquaticus*, which can tolerate low oxygen levels and live in stagnant water. Of more relevance here, however, is its close relative, the one-spotted water slater *Proasellus meridianus*, an inhabitant of the streams and rivers of, for example, Cornwall. *P. meridianus* appears to be less sensitive to trace metal exposure than *A. aquaticus*, somewhat ironically because the tolerance by the latter of the low oxygen conditions of eutrophic waters has given it a reputation for pollution tolerance. As ever, there is pollution and pollution, and *P. meridianus* appears the better able to cope with trace metal pollution.

That said, even *Proasellus meridianus* is not as metal-insensitive as many of the insect larvae met earlier (Tables 5.6 and 5.7). It was absent from any part of the zinc-contaminated Nent system in the 1970s, even when the amphipod *Gammarus pulex* was present (Armitage, 1980). In reverse, in contrast to *G. pulex*, *P. meridianus* was present at sites other than those in Group 1, the lowest copper exposure category, in copper-contaminated Cornish river systems in the 1990s (Gower et al., 1994).

A clue to the ability of *Proasellus meridianus* to survive in such metal-contaminated river systems in Cornwall may lie in the evolution of metal-tolerant populations in different rivers. One population, in the River Hayle, was found to be tolerant to both copper and lead (Brown, 1977b, 1978), although the Hayle contains very high concentrations of only copper and not lead (Tables 5.2 and 5.4). A second population of *P. meridianus*, tolerant only to lead, was found in the lead-rich River Gannel, which lacks high copper concentrations (Brown, 1977b, 1978). It would appear, therefore, that the mechanism for copper tolerance in the River Hayle population has simultaneously achieved lead tolerance, even in the absence of raised lead bioavailability (Brown, 1977b, 1978). On the other hand, the mechanism for lead tolerance in the River Gannel population is not associated with simultaneous copper tolerance in the absence of raised copper bioavailability.

Increased detoxification of extra accumulated copper appears to be the basis of copper tolerance in the copper-tolerant Hayle population of *P. meridianus* (Brown, 1977b; Rainbow and Luoma, 2011a). These isopods contained many copper- and sulphur-rich intracellular deposits in the epithelial cells of their ventral caeca (Brown, 1977b). These intracellular deposits are presumably derived from the lysosomal breakdown of copper-rich metallothionein induced by the high copper exposure.

A possible basis of lead tolerance in each of the two lead-tolerant populations lies in their reduced accumulation of lead from solution in comparison with controls (Brown, 1977b). On the other hand, when lead is delivered via the diet, both lead-tolerant populations showed strong accumulation of lead (Brown, 1977b), necessarily associated with storage detoxification. In contrast to the situation for copper accumulated in the copper-tolerant Hayle population (1,030 µg/g Cu body concentration), the ventral caeca did not appear to be important storage sites for lead in the lead-tolerant Gannel population of *P. meridianus* (220 µg/g Pb body concentration) (Brown, 1977b). The relatively low accumulated body concentrations of lead in these isopods from the Gannel suggest a limited role for diet as an accumulation route for lead in these isopods, and that their lead tolerance is based on reduced accumulation of lead from solution. Such reduced accumulation of lead from solution could be down to reduced uptake from solution and/or enhanced excretion after uptake.

There is an important feature of the reproductive biology of isopod crustaceans that promotes the potential of independent populations to develop a genetically based tolerance to high local trace metal bioavailabilities. Isopods brood their young in brood pouches. So juveniles, miniature versions of the adults, hatch out to live in the same habitat as their parents. This enables the local selection of particular genetic traits that are adaptive to life in high metal conditions. On the other hand, insects with aquatic larvae still have winged adults that disperse aerially. Even if the adults do not fly far to breed and lay their eggs in a convenient body of water, there is still likely to be dispersal away from any water system in which their larvae developed, even if this is only an adjacent stream. Moreover, adult stoneflies, mayflies and caddisflies laying their eggs in a metal-contaminated stream may well have developed in a local uncontaminated stream. No potential exists here for the genetic isolation of populations in metal-contaminated habitats to develop local metal tolerance. Amphipod crustaceans also brood their young, so the evolution of local metal-tolerant populations might be expected too in *Gammarus pulex*.

5.13.7 Oligochaetes

Prologue. Some sediment-dwelling freshwater oligochaete worms can be found in freshwater systems severely contaminated by trace metals.

Some oligochaete worms have been met earlier in this book, and reference has been made to three oligochaete families: the lumbricids, the lumbriculids and the tubificids (recently renamed as naidids). The first of these families, the lumbricids, also contains the terrestrial earthworms discussed in the previous chapter. Indeed, the genus *Eiseniella* to be encountered here includes both species living in soil, albeit waterlogged soil, and species living in freshwater sediments.

Four oligochaete species were found in the River Rheidol in 1947 to 1948 (Jones, 1949). One, *Lumbriculus variegatus*, a lumbriculid variably called a blood worm or the blackworm, occurred immediately downstream of the old Cwm Rheidol mine, as well as further downstream together with unnamed species of the lumbricid oligochaete *Eiseniella* and two tubificid oligochaetes, *Nais* and *Chaetogaster* (Jones, 1949). In the Northern Pennines, an unidentified lumbricid oligochaete was present in the mid-1970s at the mouth of the zinc-rich Dowgang Level on the River Nent system (Armitage, 1980). Sites on the nearby Gillgill Burn, also highly zinc-contaminated, held two unidentified tubificid oligochaete worms (Armitage, 1980). In Cornwall in 1991 to 1992, the lumbriculid oligochaete *Stylodrilus heringianus* occurred in all four groups of sites, from low to high copper exposures, while the tubificid *Rhyacodrilus coccineus* was present in groups 1 to 3 (Gower et al., 1994). In this Cornish study, oligochaetes made their strongest contribution to macroinvertebrate abundance in Group 3, being much reduced in number in Group 4.

Oligochaetes, therefore, will occur in freshwater systems of high trace metal bioavailabilities.

Incomplete specific identifications, however, prevent generalisations being made as to what species might be considered particularly likely to occur in such contaminated habitats.

Oligochaetes live in sediments and are typically deposit feeders, ingesting considerable volumes of sediment particles from which they extract nutrients. Thus oligochaetes take up trace metals from metal-rich sediments, in addition to any dissolved metal taken up from the surrounding medium, whether in a defined burrow or not. It is becoming increasingly apparent that ingested sediment represents the major source of trace metals to oligochaetes, even when dissolved metal concentrations can be considered high. Thus biodynamic modelling has shown that sediment ingestion is the main route of uptake of arsenic, cadmium, copper and zinc in *Lumbriculus variegatus* (Camusso et al., 2012).

Tubificid oligochaetes (strictly now naidid oligochaetes) have come in for particular attention as regards their trace metal biology, for they can dominate the sediment fauna of both freshwater and estuarine systems. Tubificids, such as species of *Tubifex*, live in organically rich fine muds and hence organically polluted freshwater systems. Tubificids contain haemoglobin as an adaptation to the low oxygen conditions in these muds, and, as a result, may vary in colour pink to red. Species of *Tubifex*, too, accumulate most trace metals from ingested sediments (De Jonge et al., 2009, 2010). Like their terrestrial counterparts, the earthworms, freshwater oligochaetes can accumulate high concentrations of trace metals. Species of *Tubifex* collected between 1990 and 2010 from metal-contaminated lowland rivers in Flanders, Belgium, had accumulated trace metal concentrations (after gut depuration) up to 316 µg/g As, 2,600 µg/g Cd, 1,490 µg/g Cu, 1,260 µg/g Pb and 11,600 µg/g Zn (De Jonge et al., 2010; Bervoets et al., 2016).

We saw earlier how a threshold accumulated metal concentration in a relatively metal hardy biomonitor can be used as an ecological indicator of an ecotoxicological effect on the presence and abundance of other more metal-sensitive members of the freshwater community. That example concerned metal concentrations in the caddisfly *Hydropsyche siltalai* and the combined abundance of heptageniid and ephemerellid mayflies in Cornish streams and rivers (Figures 5.16 and 5.17) (Rainbow et al., 2012). That particular relationship is suitable for application in streams with running water and stony bottoms. Tubificids (and incidentally chironomids) offer a suitable alternative further downstream in slow-flowing water courses with soft bottoms (Bervoets et al., 2016). Metal-contaminated lowland rivers in Flanders, Belgium, offered a suitable test arena. One of the ecotoxicological endpoints chosen to describe the macroinvertebrate community present at each site was the Multimetric Macroinvertebrate Index Flanders (MMIF) (Gabriels et al., 2010; Bervoets et al., 2016). The MMIF has values between 0 and 1, with 1 representing the presence of a community unaffected by any ecotoxicological impact. The MMIF endpoint indicating good ecological quality, as required under the EU Water Framework Directive, was defined as 0.6 (Bervoets et al., 2016). The corresponding threshold accumulated concentrations in tubificids were 85 µg/g As, 28 µg/g Cd, 71 µg/g Cu, 79 µg/g Pb and 930 µg/g Zn (Bervoets et al., 2016).

5.13.8 Molluscs

Prologue. Both gastropod and bivalve molluscs are found in British freshwater systems.

5.13.8.1 Gastropods

Prologue. Most gastropod molluscs in British freshwater systems are pulmonate, taking up and accumulating trace metals mostly from the diet.

Pulmonate gastropods have evolved from terrestrial gastropods and breathe air through a lung that has replaced the gills of an original ancestral marine gastropod. Thus species of the freshwater gastropod *Lymnaea* come to the surface to breathe. Nevertheless, even pulmonate gastropods have large areas of permeable surfaces, for example on the foot, with potential for the uptake of dissolved metals, including both trace metals and major metal ions. A smaller number of British freshwater gastropods, such as the

river snail *Viviparus viviparus*, have an evolutionary history of the colonisation of freshwater by migration upstream through estuaries, without deviation onto land. Such gastropods have kept the gill of their marine ancestors, allowing uptake of dissolved metals here, as well as through any other permeable body surfaces. Whether pulmonate or not, freshwater gastropods typically graze on the surface films on rocks or plants and will take up trace metals from this trophic source.

Indeed, the pulmonate gastropod *Lymnaea stagnalis*, the great pond snail, derives most of its accumulated cadmium, copper and nickel from the diet (Croteau and Luoma, 2008). Increased dietary concentrations of these metals lead to increased accumulation by *L. stagnalis*, associated with increased detoxified storage of the accumulated metal (Croteau and Luoma, 2009). Excess accumulated copper, for example, is detoxified by binding to metallothionein induced by the copper exposure (Ng et al., 2011). Thus specimens of *Radix balthica* (formerly *Lymnaea peregra*) collected in 1980 to 1981 from Ullswater in Cumbria, receiving input of lead from Patterdale and the Glenridding Beck draining the Greenside mine, contained up to 1,860 µg/g Pb in the digestive gland (Everard and Denny, 1984). Even a pulmonate is not isolated from trace metals in solution. *L. stagnalis* has a high rate of uptake of calcium from solution, necessary to serve the calcium needs of shell formation. Since dissolved lead can compete with dissolved calcium for uptake at sites responsible for calcium uptake and thereby inhibit calcium uptake, snails of the genus *Lymnaea* are well known to be very sensitive to chronic lead exposure (Grosell and Brix, 2009).

Another pulmonate, the river limpet *Ancylus fluviatilis* is one of the first macroinvertebrate species to disappear from streams or rivers affected by lead or zinc pollution (Newton, 1944). Thus this mollusc was one of the invertebrates to be eliminated by the effects of mine effluents in the Ystwyth and Rheidol river systems in mid-Wales and one of the last to reappear after the closure of the mines early in the twentieth century (Newton, 1944). The river limpet was back in the River Rheidol downstream of the old Cwm Rheidol effluent stream in 1947 to 1948 (Jones, 1949). In the Nent river system in 1976 to 1977, *A. fluviatilis* occurred only in a downstream tributary, the Foreshield Burn, uncontaminated by lead or zinc, and at a site immediately below the Foreshield Burn with the weakest level of zinc contamination (Armitage, 1980). The river limpet was absent from all upstream Nent sites with metal contamination (Armitage, 1980). Similarly, in the late 1970s, *A. fluviatilis* occurred in the River East Allen in the Northern Pennines, but not in the adjacent River West Allen with considerably higher levels of zinc (Abel and Green, 1981). In Cornwall, *Ancylus fluviatilis* did not occur at any site not in Group 1, the lowest copper exposure group (Gower et al., 1994).

5.13.8.2 Bivalves

Prologue. Lamellibranch bivalve molluscs have very extensive areas of gills for suspension feeding. These gills are permeable and so allow the entry into the body of oxygen, water and indeed trace metals. Their filtered food may consist of resuspended sediment particles, as well as phytoplankton or light detritus particles in suspension, and represents a significant trophic source of trace metal uptake.

There are numerous species of bivalves in British freshwater systems. Bivalves tend to inhabit the lower reaches of rivers, as opposed to the upland streams more likely to be contaminated with trace metals from mining activity. Downstream, however, they may come into contact with the metalliferous effluents of industry, particularly towards estuaries.

Larger bivalves include swan mussels, for example species of *Anodonta*, which require calcium-rich hard water, living in river margins where the sediment is muddy and compact. The common name 'mussel' is applied to several unrelated freshwater, estuarine and marine bivalves. In freshwater, in addition to swan mussels, there are also pea mussels and zebra mussels. Pea mussels, such as species of *Pisidium*, are usually small, roundish bivalves, living in the detritus and mud of standing and slow-flowing waters. The zebra mussel *Dreissena polymorpha*, as its common name suggests, has a shell with distinctive stripes. It is an introduced species, originally native to the Ponto-Caspian region of southern Russia, and has been present in British river systems since the 1820s. In contrast to most bivalves, the zebra mussel is not a burrower but attaches to hard substrates, and can occur in vast numbers, potentially clogging the intake pipes of industrial installations. Another introduced freshwater bivalve, the Asian clam *Corbicula fluminea*, is spreading in Southeast England after being reported for the first time in 1998, in the River Chet in the Norfolk Broads. The Asian clam inhabits lakes, canals and slow-flowing rivers and can also be extremely abundant.

Life in freshwater for a lamellibranch bivalve brings with it considerable physiological problems of water balance. The possession of a permeable gill, expanded considerably to increase the efficiency of suspension feeding, means that a vast surface area is available not only for respiratory exchange but also water uptake. The osmotic pressure of the body fluids of, for example the swan mussel *Anodonta cygnaea*, although maintained at a level lower than the body fluids of its marine relatives, is unavoidably higher than that of the surrounding freshwater medium (Barrington, 1967). Water will inevitably enter down this osmotic gradient and needs to be pumped out again via the kidneys. Even though the urine produced is more dilute than the body fluids originally filtered by the swan mussel's kidney (Barrington, 1967), some major metal ions (for example, sodium and calcium) will be lost from the body. These need to be replaced by active uptake across membrane pumps on the external epithelium on the outside of the body. Dissolved trace metals may enter via these major ion pumps, not least cadmium and lead via calcium pumps. Furthermore, the very presence of a huge gill area will already mean that gill epithelial sites for the passive uptake of trace metals by carrier proteins will be present in potentially large numbers, increasing dissolved trace metal uptake by this route too. Thus a freshwater lamellibranch bivalve is likely to have high uptake rates of dissolved trace metals compared, for example, to insects or crustaceans with external cuticles that can be rendered impermeable, other than in a restricted area of gill.

A swan mussel also needs to build a large shell containing calcium while living in a dilute medium. Swan mussels, therefore, have a remarkably efficient mechanism of internal calcium storage. Swan mussels, including *Anodonta cygnaea*, contain very large numbers of calcium phosphate concretions, particularly in the gills but also in mantle tissues. These calcium phosphate granules can contribute up to 55% of the dry weight of the gill in *A. cygnaea* (Pynnönen et al., 1987; Hinzmann et al., 2014). These calcium granules are not trace metal detoxification granules but are reusable stores of calcium. Because they are redissolved to serve this purpose, it is important that there are no significant contents of toxic metals that might be released on the unsuspecting metabolism. In fact, calcium-based trace metal detoxification granules are rare, if even present, in other organs of *A. cygnaea*, such as the digestive gland or kidney, which are the typical invertebrate organs for the storage of detoxified metals. Trace metals are accumulated and stored by swan mussels in the digestive gland and/or the kidney, include copper and cadmium, which are bound to metallothionein or present in lysosomes autolysing this metallothionein (Bonneris et al., 2005; Cooper et al., 2010).

Swan mussels (family *Unionidae*) have been used as biomonitors of trace metal bioavailabilities in large lowland rivers affected by domestic and industrial discharges. In 1974 to 1975, samples of the duck mussel *Anodonta anatina* were collected from seven sites along the River Thames through the old counties of Oxfordshire, Berkshire and Middlesex (Manly and George, 1977). Sites were defined as rural or urban, not least by the comparative daily discharge of total sewage within 5 kilometres upstream of each site. Trace metal concentrations were then measured in the soft tissues of the duck mussels from these sites (Table 5.13). Accumulated concentrations of lead, nickel and zinc were significantly raised in the duck mussels collected from the urban localities of Reading, Lower Sunbury and Teddington (Table 5.13). Sewage effluent contains all the trace metals analysed.

Furthermore, at the urban sites, there would have also been industrial effluents not discharged through public sewers and additional lead from motor exhaust at this period before the banning of lead in petrol. The lack of increased copper and mercury accumulation in the duck mussels at the urban sites may be down to the presence of copper or mercury-bearing pesticides (or their derivatives) in runoff from agricultural land at the rural sites.

The zebra mussel *D. polymorpha* has been widely used as a biomonitor of trace metal bioavailabilities in the large lowland rivers of Western Europe affected by industrial discharges (Kraak et al., 1991; Mersch et al., 1992). The zebra mussel has the potential to play a similar biomonitoring role here in Britain.

Background concentrations (dry weight) of trace metals in *D. polymorpha* are approximately 1 μg/g Cd, 1 μg/g Cr, 12 μg/g Cu, 0.5 μg/g Pb and 110 μg/g Zn, raised trace metal concentrations being reported from the Rhine, Meuse and Mosel (Kraak et al., 1991; Mersch et al., 1992). Elevated accumulated concentrations of cadmium have reached 90 μg/g in the Rhine, 9 μg/g in the Meuse and 3 μg/g in the Mosel. Chromium concentrations of 4 μg/g have been recorded in *D. polymorpha* from the Mosel. Copper concentrations reached 30 μg/g in zebra mussels from the Rhine and Meuse and 50 μg/g in the Mosel. Raised bioaccumulated lead concentrations have attained 20 μg/g in the Mosel and 25 μg/g in the Rhine. In the case of zinc, elevated accumulated concentrations in zebra mussels have been found in the Mosel (240 μg/g) and the Rhine (600 μg/g).

Zebra mussels originating from an uncontaminated location have also been translocated for known periods of time to contaminated sites in order to follow any increased accumulation of trace metals. Clearly it is important that the translocated species was already present in the river system receiving translocated specimens so that invasive species are not spread further. Zebra mussels were translocated in cages in 2006 to 2008 to sites in the metal-contaminated River Dommel in Flanders, Belgium, and left on each occasion for a period of 42 days (De Jonge et al., 2012). The translocation in 2006 preceded a programme of dredging of the Dommel.

(a)

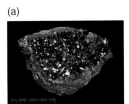

Plate 1a Cassiterite. Total scale is 10 cm. (Photo by H. Taylor, Natural History Museum, with permission.) (A black and white version of this figure will appear in some formats.)

(b)

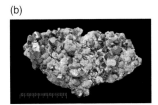

Plate 1b Chalcopyrite. Total scale is 10 cm. (Photo by H. Taylor, Natural History Museum, with permission.) (A black and white version of this figure will appear in some formats.)

(c)

Plate 1c Malachite. Total scale is 20 cm. (Photo by H. Taylor, Natural History Museum, with permission.) (A black and white version of this figure will appear in some formats.)

(d)

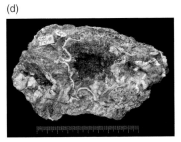

Plate 1d Azurite. Total scale is 20 cm. (Photo by H. Taylor, Natural History Museum, with permission.) (A black and white version of this figure will appear in some formats.)

(e)

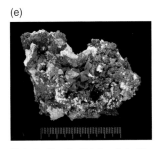

Plate 1e Cuprite. Total scale is 10 cm. (Photo by H. Taylor, Natural History Museum, with permission.) (A black and white version of this figure will appear in some formats.)

(f)

Plate 1f Arsenopyrite. Total scale is 10 cm. (Photo by H. Taylor, Natural History Museum, with permission.) (A black and white version of this figure will appear in some formats.)

(g)

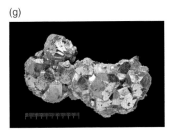

Plate 1g Galena. Total scale is 10 cm. (Photo by H. Taylor, Natural History Museum, with permission.) (A black and white version of this figure will appear in some formats.)

(h)

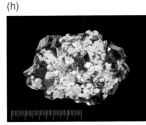

Plate 1h Sphalerite. Total scale is 10 cm. (Photo by H. Taylor, Natural History Museum, with permission.) (A black and white version of this figure will appear in some formats.)

(i)

Plate 1i Smithsonite. Total scale is 10 cm. (Photo by H. Taylor, Natural History Museum, with permission.) (A black and white version of this figure will appear in some formats.)

(j)

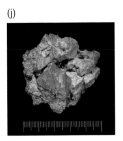

Plate 1j Acanthite. Total scale is 10 cm. (Photo by H. Taylor, Natural History Museum, with permission.) (A black and white version of this figure will appear in some formats.)

(k)

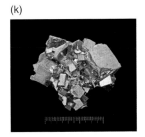

Plate 1k Pyrite. Total scale is 10 cm. (Photo by H. Taylor, Natural History Museum, with permission.) (A black and white version of this figure will appear in some formats.)

(l)

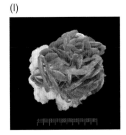

Plate 1l Siderite. Total scale is 10 cm. (Photo by H. Taylor, Natural History Museum, with permission.) (A black and white version of this figure will appear in some formats.)

(m)

Plate 1m Psilomelane. Total scale is 10 cm. (Photo by H. Taylor, Natural History Museum, with permission.) (A black and white version of this figure will appear in some formats.)

(a)

Plate 2a *Palaemon elegans*. A decapod crustacean, commonly referred to as a prawn, that typically lives in intertidal rock pools. The prawn is 4 cm long. (Photograph P S Rainbow). (A black and white version of this figure will appear in some formats.)

(b)

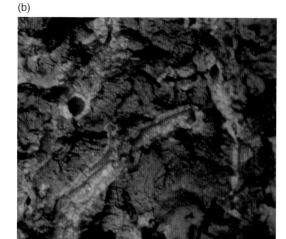

Plate 2b Burrows of the intertidal polychaete worm *Hediste diversicolor* in the mud of Restronguet Creek, Cornwall, UK, a sediment rich in trace metals, particularly iron, contributing the black iron sulphide dominating anoxic regions of the mud. The burrows, in contrast, show the local oxidation of iron sulphide to red/brown iron oxides and hydroxides, caused by the flow through of currents of oxygenated water from the overlying water column. The worms are not in contact with anoxic pore water. (Photograph by Kevin Brix, reproduced from Luoma and Rainbow, 2008.) (A black and white version of this figure will appear in some formats.)

(c)

Plate 2c *Armeria maritima*. Thrift on the coast in the Levant district of Cornwall. (Photo by C.J. Rainbow, with permission.) (A black and white version of this figure will appear in some formats.)

(d)

Plate 2d *Silene uniflora*. Sea campion on the north Cornish coast near the Botallack mine. (Photo by C.J. Rainbow, with permission.) (A black and white version of this figure will appear in some formats.)

(a)

Plate 3a *Acarospora sinopica*. Total scale is 1 cm. (Photo by H. Taylor, Natural History Museum, with permission.) (A black and white version of this figure will appear in some formats.)

(b)

Plate 3b *Lecanora handelii*. Total scale is 1 cm. (Photo by H. Taylor, Natural History Museum, with permission.) (A black and white version of this figure will appear in some formats.)

(c)

Plate 3c *Lecanora subaurea*. Total scale is 10 cm. (Photo by H. Taylor, Natural History Museum, with permission.) (A black and white version of this figure will appear in some formats.)

(d)

Plate 3d *Lecidea silacea*. Total scale is 1 cm. (Photo by H. Taylor, Natural History Museum, with permission.) (A black and white version of this figure will appear in some formats.)

(e)

Plate 3e *Miriquidica atrofulva*. Total scale is 10 cm. (Photo by H. Taylor, Natural History Museum, with permission.) (A black and white version of this figure will appear in some formats.)

(f)

Plate 3f *Psilochia leprosa*. Total scale is 1 cm. (Photo by H. Taylor, Natural History Museum, with permission.) (A black and white version of this figure will appear in some formats.)

(g)

Plate 3g *Stereocaulon symphycheilum*. Total scale is 5 cm. (Photo by H. Taylor, Natural History Museum, with permission.) (A black and white version of this figure will appear in some formats.)

(h)

Plate 3h *Lecidea inops*. Total scale is 5 cm. (Photo by H. Taylor, Natural History Museum, with permission.) (A black and white version of this figure will appear in some formats.)

(i)

Plate 3i *Myriospora smaragdula*. Total scale is 1 cm. (Photo by H. Taylor, Natural History Museum, with permission.) (A black and white version of this figure will appear in some formats.)

(j)

Plate 3j *Buellia aethalea*. Total scale is 1 cm. (Photo by H. Taylor, Natural History Museum, with permission.) (A black and white version of this figure will appear in some formats.)

(k)

Plate 3k *Vezdaea leprosa*. Total scale is 1 cm. (Photo by H. Taylor, Natural History Museum, with permission.) (A black and white version of this figure will appear in some formats.)

(a)

Plate 4a *Lumbricus terrestris*, the common earthworm. (Photo by H. Taylor, Natural History Museum, with permission.) (A black and white version of this figure will appear in some formats.)

(b)

Plate 4b *Lumbricus rubellus*, the red worm. (Photo by H. Taylor, Natural History Museum, with permission.) (A black and white version of this figure will appear in some formats.)

(c)

Plate 4c *Dendrodrilus rubidus*, the red wiggler. (Photo by H. Taylor, Natural History Museum, with permission.) (A black and white version of this figure will appear in some formats.)

(d)

Plate 4d *Aporrectodea caliginosa*, the grey worm. (Photo by H. Taylor, Natural History Museum, with permission.) (A black and white version of this figure will appear in some formats.)

(e)

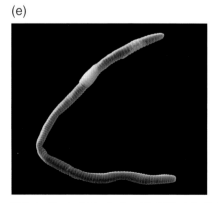

Plate 4e *Eisenia fetida*, the brandling. (Photo by H. Taylor, Natural History Museum, with permission.) (A black and white version of this figure will appear in some formats.)

Plate 5a The woodlouse *Oniscus asellus*. The woodlouse is about 1 cm long. (Photo by H. Taylor, Natural History Museum, with permission.) (A black and white version of this figure will appear in some formats.)

(b)

Plate 5b The woodlouse *Porcellio scaber*. The woodlouse is about 1 cm long. (Photo by H. Taylor, Natural History Museum, with permission.) (A black and white version of this figure will appear in some formats.)

(c)

Plate 5c The final larval stage of the caddisfly *Hydropsyche siltalai*. The larva is about 1.5 cm long. (Photo by H. Taylor, Natural History Museum, with permission.) (A black and white version of this figure will appear in some formats.)

(a)

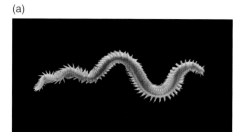

Plate 6a Ragworm *Hediste diversicolor*. The ragworm is about 4 cm long. (Photo by H. Taylor, Natural History Museum, with permission.) (A black and white version of this figure will appear in some formats.)

(b)

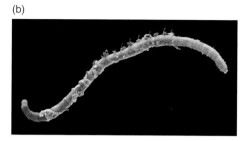

Plate 6b Lugworm *Arenicola marina*. The lugworm is about 15 cm log. (Photo by H. Taylor, Natural History Museum, with permission.) (A black and white version of this figure will appear in some formats.)

(c)

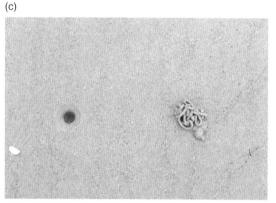

Plate 6c *Arenicola marina*. Lugworm burrow at low tide, with typical cast and depression (5 cm across). (Photo by C.J. Rainbow, with permission.) (A black and white version of this figure will appear in some formats.)

(d)

Plate 6d Beachhopper *Orchestia gammarellus*. The beachhopper is 1 cm long. (Photo by H. Taylor, Natural History Museum, with permission.) (A black and white version of this figure will appear in some formats.)

(e)

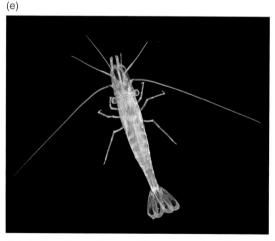

Plate 6e The palaemonid prawn *Palaemonetes varians*. The prawn is 3 cm long. (Photo by H. Taylor, Natural History Museum, with permission.) (A black and white version of this figure will appear in some formats.)

(a)

Plate 7a Peppery Furrow shell *Scrobicularia plana*. The shell is 4 cm long. (Photo by H. Taylor, Natural History Museum, with permission.) (A black and white version of this figure will appear in some formats.)

(b)

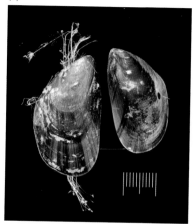

Plate 7b Common Mussel *Mytilus edulis*. Total scale is 1 cm. (Photo by H. Taylor, Natural History Museum, with permission.) (A black and white version of this figure will appear in some formats.)

(c)

Plate 7c Pacific Oyster *Crassostrea gigas* from Southend-on-Sea with attached barnacles *Austrominius modestus*. Total scale is 20 cm. (Photo by H. Taylor, Natural History Museum, with permission.) (A black and white version of this figure will appear in some formats.)

(d)

Plate 7d Queen scallop *Aequipecten opercularis*. The queen scallop is 7 cm long. (Photo by H. Taylor, Natural History Museum, with permission.) (A black and white version of this figure will appear in some formats.)

(e)

Plate 7e The netted dogwhelk *Tritia reticulata*. The shell is 1 cm long. (Photo by H. Taylor, Natural History Museum, with permission.) (A black and white version of this figure will appear in some formats.)

Plate 8a *Systellaspis debilis*. The decapod is 5 cm long. (Photo by C. Arneson, with permission.) (A black and white version of this figure will appear in some formats.)

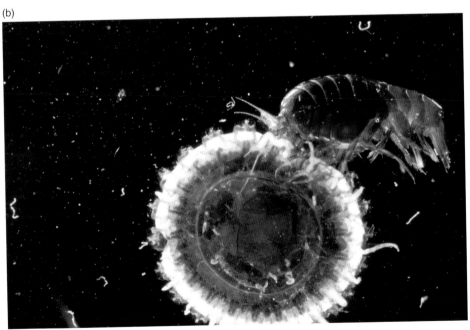

Plate 8b Mesopelagic stegocephalid amphipod *Parandania boecki* (1.6 cm long) feeding on the medusa *Atolla parva*. (Photo by C. Arneson, with permission.) (A black and white version of this figure will appear in some formats.)

Table 5.13 **Mean concentrations of trace metals (± 1 SD, µg/g dry weight) in soft tissues of the duck mussel *Anodonta anatina* collected from rural and urban sites on the River Thames in 1974 to 1975. Also shown are total sewage discharges (thousand litres per day) within 5 kilometres upstream of each sampling site.**

	Sewage discharge	Cd	Cu	Hg	Ni	Pb	Zn
Rural							
Appleton, Oxfordshire	327	3.2 ± 2.1	21 ± 8	7.0 ± 2.9	4.2 ± 2.5	9.8 ± 4.5	403 ± 154
Pangbourne, Berkshire	182	3.9 ± 2.1	74 ± 25	5.4 ± 3.1	1.3 ± 1.1	14 ± 6.4	536 ± 259
Hurley, Berkshire	218	2.0 ± 1.2	99 ± 70	8.2 ± 5.9	4.5 ± 3.9	12 ± 10	498 ± 180
Bray, Berkshire	0	0.4 ± 0.5	92 ± 59	0.4 ± 0.3	5.0 ± 3.9	13 ± 6.7	426 ± 160
Urban							
Reading, Berkshire	477	5.9 ± 2.3	103 ± 34	5.7 ± 2.1	25 ± 9.7	43 ± 20	$1,740 \pm 872$
Lower Sunbury, Middlesex	855	3.9 ± 0.6	65 ± 48	0.9 ± 0.8	16 ± 3.4	26 ± 4.7	$1,390 \pm 512$
Teddington, Middlesex	455	4.6 ± 1.4	66 ± 27	0.7 ± 0.3	13 ± 6.9	23 ± 13	$1,820 \pm 886$

Source: Manly and George (1977).

The highest mean accumulated concentrations of trace metals in the translocated zebra mussels were recorded in 2006, before the dredging of the riverbed, namely 12 µg/g Cd, 37 µg/g Pb and 464 µg/g Zn. All these bioaccumulated concentrations are well above the background concentrations reported previously, implying the presence of atypically raised bioavailabilities of these trace metals to the zebra mussels. Furthermore, this study also noted significant relationships between accumulated trace metal concentrations in *D. polymorpha* (and incidentally resident chironomid larvae) and measured responses of the local biological communities (De Jonge et al., 2012).

While another bivalve invader, Asian clam *Corbicula fluminea*, is much newer to British freshwater systems, it too has a record of use as a trace metal biomonitor, in rivers in both in Western Europe and the United States (Peltier et al., 2008). If *C. fluminea* continues to spread in Britain, it will also have potential as a trace metal biomonitor in urban rivers.

5.14 Fish

Prologue. Many of the British freshwater systems affected by trace metal contamination from mining are relatively small upland streams. The fish characteristic of such habitats are bullheads *Cottus gobio* (also called the Miller's Thumb) and brown trout *Salmo trutta*.

While bullheads belong to the predominantly marine family, the *Cottidae* or sculpins, brown trout belong to the family *Salmonidae* together with their close relatives, salmon. All salmonids spawn in freshwater, but life history strategies thereafter do vary, even within species. Salmonids are classically anadromous, being born in freshwater, spending most of their life at sea, before migrating back to freshwater to spawn. Yet, single salmonid species may consist of both anadromous and wholly freshwater populations, and *S. trutta* is one such species (Maitland and Campbell, 1992). Amongst a plethora of different common names for *S. trutta*, anadromous forms are usually called sea trout (sewin in Wales), and fish permanently resident in freshwater are called brown trout (Maitland and Campbell, 1992). This distinction, however, is by no means consistent. All native trout populations in the British Isles are descended from colonisation by anadromous trout soon after the last glaciation, 13,000 to 15,000 years ago. These populations may still be largely anadromous, completely non-anadromous or intermediate between the two. Furthermore, there has been considerable human redistribution of brown trout between water systems since any original establishment of British populations (Maitland and Campbell, 1992).

When toxic metals from mining enter upland headwaters, there are various possible scenarios affecting the survival of fish populations. If metal entry is right at the top of the system, it is likely that local populations of fish (probably bullheads and/or brown tout) will be eliminated. If metal entry is a little way down from the head of the stream system, fish may survive above the region affected by mine effluent. An anadromous population of brown trout will be eliminated because its survival requires passage through a metal-polluted stretch of river downstream, but any non-anadromous forms may survive above the mining region. After the cessation of mining and

associated reduction in bioavailabilies of toxic metals in the stream or river, recolonisation of the upper reaches of a system could occur by upward migration of downstream fish.

With the strong caveat that is dangerous to extrapolate directly from the world of laboratory toxicity testing based on dissolved exposures, to the real world where trophic intake of metals is so important, it is generally considered that copper is the dissolved trace metal that is most toxic to fish, followed by lead and zinc (Kelly, 1988). Furthermore, amongst fish, salmonids appear particular sensitive to copper, and to a lesser extent zinc (Kelly, 1988).

5.14.1 History of Fish in Metal-Contaminated River Systems

Prologue. Records of the presence and absence of fish are available for the metal-contaminated river systems of the Ystwyth, Rheidol and Conwy in Wales, Glengonnar Water in Scotland and the Hayle in Cornwall.

In Kathleen Carpenter's surveys of the rivers of northern Ceredigion around 1920, fish were completely absent from mining-contaminated rivers and streams. Fish had reappeared in the River Ystwyth by 1939 (Jones, 1940; Newton, 1944). Sea trout made it upstream past the estuary at Aberystwyth, but not past the nearby bridge at Llanfarian (Figure 5.7), some being caught by anglers, but many being seen floating dead or dying. Brown trout, however, were surviving in the mainstream and tributaries of the Ystwyth above the Cwm Ystwyth mine, the first metal entry point (Jones, 1940). If carried down into the mainstream below the mine during floods, these resident brown trout did not survive long, however, and again could be seen floating, dead or dying (Jones, 1940). By 1953, though, brown trout could survive in small numbers downstream in the lower reaches of the River Ystwyth, and the anadromous sea lamprey *Petromyzon marinus* was an occasional visitor (Jones, 1958). Fish such as the bullhead *C. gobio*, the minnow *Phoxinus phoxinus* and the three-spined stickleback *Gasterosteus aculeatus* were, however, still absent from the River Ystwyth then (Jones, 1958). By 1975, fish had returned to most of the River Ystwyth, but were still absent from a 3 kilometre stretch of river immediately below the Cwm Ystwyth mine (Davies, 1987). Salmon, *Salmo salar*, were reported from the River Ystwyth in 1970 (Davies, 1987).

The nearby River Rheidol recovered much more quickly from its mining legacy than the River Ystwyth (Davies, 1987). Three-spined stickleback were recorded downstream in the river in the 1920s, and at the beginning of the 1930s, this stickleback had been joined by the eel *Anguilla anguilla* and by sea trout (Davies, 1987). By the 1950s, the stickleback and eels were well established downstream, and trout were common throughout the system of the Rheidol (Davies, 1987). The first definite record of salmon in the Rheidol was in 1952. In 1969, an entrance plug in an adit of the old Cwm Rheidol mine was accidentally breached, and the resulting discharge caused a major fish kill in the River Rheidol (Davies, 1987).

In the River Conwy system in north Wales, draining the old lead–zinc Parc mine near Llanrwst, trout (*S. trutta*) and salmon (*S. salar*) were to be found in the late 1970s only in uncontaminated streams or in the extreme lowland reaches of the system diluted by uncontaminated tributary drainage (Johnson and Eaton, 1980). They were absent from the Nant Gwydyr draining the old Parc mine, and from the also metal-contaminated River Crafnant (Figure 5.8).

In Scotland, Glengonnar Water, draining the Leadhills lead mining region at the top of the Clyde catchment, does have a resident brown trout population in spite of the high lead and zinc concentrations in its sediments (Table 5.14). These trout are distinctive, for they show blackening of the tails and spinal abnormalities that have been considered indicative of chronic exposure to lead (Clyde River Protection Board, 1984; Scottish Environment Protection Agency, 2011). It is inspiring that the black-tailed trout of Glengonnar Water have recently been captured for posterity in a stained-glass window at Leadhills Primary School (William Yeomans, private communication).

The River Hayle in Cornwall (Figure 5.3) has a long history of mining contamination and is known to be very contaminated by copper and to a lesser extent by zinc (Tables 5.2 and 5.4). The river can arbitrarily be divided into three regions on the basis of dissolved copper and zinc concentrations as measured at the turn of this century (Durrant et al., 2011). An upper region including Drym and Binnerton (Figure 5.3) is low in metal contamination. A middle region, in the region of drainage from mines in the Godolphin area (Figure 5.3), is highly contaminated with copper and zinc, with dissolved concentrations known to be extremely toxic to metal-naïve trout *Salmo trutta*, and far in excess of concentrations known to inhibit trout movement (Durrant et al., 2011). The lower section of the river, down from Relubbus through St Erth to the estuary at Hayle (Figure 5.3), is considered moderately contaminated. The trout *S. trutta* was found in the upper and lower sections of the river but was absent from the highly contaminated middle section (Durrant et al., 2011). A population genetics study showed that this contaminated middle section was not a complete chemical barrier to gene flow between trout populations above and below it (Durrant et al., 2011). There was some gene exchange between trout at Binnerton, just above the contaminated zone, and trout below the middle zone. In this low contamination upper zone, there was, however, a genetic distinction between trout collected at Binnerton and those collected about 1 kilometre further upstream at Drym, which appeared to form a genetically distinct population. There thus appears to be one population of Hale trout capable to a degree of resisting high copper and zinc bioavailabilities and of crossing the very contaminated middle zone. Higher in the river, however, a second population of trout, less able to withstand high copper and zinc bioavailabilities, is able to outcompete the lower river population in uncontaminated conditions.

A complicating factor that may need to be added to this story is the high bioavailability of arsenic, expected but not measured, in the middle section of the Hayle River.

A more recent wider population genetics study of trout populations in Cornish rivers affected by trace metal contamination from historical mining, including the Hayle, has concluded that metal-contaminated

Table 5.14 **Mean concentrations of cadmium and lead (μg/g dry weight with 95% confidence limits) in whole bodies of female bullheads *Cottus gobio* of age groups 0 and 1 from three sites along the River Ecclesbourne (total length 16 kilometres) upstream from its confluence with the River Derwent in the Peak District, in August 1981. Also shown are mean cadmium and lead concentrations (μg/g dry weight) in sediments (<2 millimetres) at each site.**

Site	Distance upstream of River Derwent	Sediment		Age group	Fish body concentration			
		Cadmium	Lead		Cadmium		Lead	
	Km				Mean	95% CL	Mean	95% CL
10	7.6	4.0	676	0	0.13	0.02, 0.79	0.49	0.32, 0.77
				1	0.23	0.15, 0.33	1.16	0.69, 1.94
14	9.5	21	2,190	1	0.40	0.26, 0.62	2.25	1.35, 3.78
17	12.6	60	9,770	0	0.95	0.52, 1.74	2.53	0.67, 9.55

Source: Moriarty et al. (1984).

rivers have trout populations that are genetically distinct from populations in uncontaminated rivers (Paris et al., 2015). Metal-impacted trout populations have low genetic diversity and have experienced severe population declines. Furthermore, the metal-affected populations are also genetically distinct from each other, even when close geographically. It was possible to date the origins of these genetic patterns to periods of intense mining activity, with splits from clean-river fish populations dating back to about 1050, perhaps in association with the development of tin streaming releasing tin and associated trace metals into Cornish river systems (Paris et al., 2015). The results of this wider recent study also confirmed the earlier results that there were two genetically distinct trout populations in the upper and lower Hayle River either side of the Godolphin region (Durrant et al., 2011) and dated this split to about 1860 (Paris et al., 2015). The peak exploitation of mines in the Godolphin region was from 1815 to 1850, and Hayle trout clearly experienced population declines associated with this period of mining activity.

5.14.2 Accumulation

Prologue. Freshwater fish take up and accumulate race metals such as cadmium, lead and mercury.

The reporting of accumulated trace metal concentrations in freshwater fish is not quite as straightforward as for the bryophytes and invertebrates. Firstly, there are usually age effects on accumulated accumulations. These effects may be recognised by dividing populations into different age groups; by plotting accumulated concentrations against total body weight; or by a combination of the two, plotting weight effects within age groups. A clear example of how to allow for such effects is available for bullhead *Cottus gobio* (Moriarty et al., 1984), collected from the lead and cadmium-contaminated River Ecclesbourne in the Peak District (Table 5.4). The power function statistical model is often an appropriate mathematical model of observed relationships between body weight and accumulated trace metal concentration (Moriarty et al., 1984). Secondly, there are also usually clear differences between the different organs of a fish when it comes to accumulated trace metal concentrations. Muscle tissues are usually relatively low in accumulated trace metal concentrations, and indeed muscle concentrations may be regulated to approximately constant levels in the case of essential metals such as zinc. On the other hand, kidneys and livers are potential detoxified storage organs, and often have relatively high accumulated trace metal concentrations (Badsha and Goldspink, 1982; Mason, 1987). And different trace metals may yet be stored in different organs. Furthermore, some unexpected organs may have unexpectedly high concentrations of particular trace metals, for example zinc in eyes and heart (Badsha and Goldspink, 1982), even if these organs do not make a high contribution to total body metal content.

A key point in the presentation of accumulated trace metal data in fish relates to the purpose for which the data were collected. In a comparative biomonitoring study of relative trace metal bioavailabilities to which a particular fish species has been exposed, over space or time, concentration data are usually expressed in terms of dry weight. These concentrations may relate to whole bodies or to particular identified storage organs, preferably after allowance for size effects. If, on the other hand, the aim is to carry out a risk assessment on the potential trophic transfer of a toxic metal from the fish to a consumer, be it a human (Mason, 1987) or a particular piscivorous predator such as an otter *Lutra lutra* (Yamaguchi et al., 2003), then the data may be presented in terms of wet weight. Such risk assessment data may involve just muscle tissue (human consumption) or whole fish according to the feeding method of the consumer. The flesh (i.e., muscle tissue) of eels *Anguilla anguilla* is often of concern for the human food chain. Measured wet weight accumulated metal concentrations may then be compared against recommended regulatory limits, typically for muscle tissue and for particular trace metals of concern, such as mercury, cadmium or lead (Mason, 1987; Jürgens et al., 2013). Fish tissue that has been archived also allows the investigation of changes over time in the exposure of freshwater fish to particular contaminants (Jürgens et al., 2013).

The upshot is that the literature database for accumulated metal concentrations in fish is very variable. Nevertheless, Tables 5.14 to 5.16 summarise some such data for discussion. Because the data were collected for different purposes, accumulated concentrations in Tables 5.14 and 5.15 are expressed in terms of dry weight, those in Table 5.16 in terms of wet weight. Wet weight to dry weight ratios do vary between tissues. Factors of 0.329 and 0.205, for example, were used for conversion of dry weight concentrations to wet weight concentrations in muscle tissue of roach and eel respectively (Yamaguchi et al., 2003). Thus the water contents of these muscle tissues were about 67% and nearly 80% respectively.

Table 5.14 presents cadmium and lead concentration data for young bullheads *Cottus gobio* collected in 1981 at three sites along the River Ecclesbourne draining the old lead-mining district of Wirksworth in the Peak District (Moriarty et al., 1984). The river contains sediments contaminated with cadmium, lead and zinc (Table 5.4), and the three fish collection sites showed a gradient of sediment concentrations of cadmium and lead (Table 5.14). The whole body concentrations of cadmium and lead in the bullheads, however, were not obviously high compared to those in other species of fish from water systems not considered to be metal-contaminated (Badsha and Goldspink, 1982). This is possibly a species effect but may well be a combination of the young age of the bullheads (age groups 0 and 1) and the likelihood that the high sediment concentrations of trace metals in the Ecclesbourne represent tightly adsorbed metal. Thus these high sediment concentrations may not be necessarily reflected in very high dissolved metal bioavailabilities affecting the food chain delivering local prey to the bullheads. Nevertheless, the accumulated cadmium and lead concentrations of the bullheads did decrease downstream in line with falling sediment concentrations (Table 5.14).

The whole fish concentrations of cadmium, lead and zinc in Table 5.15 refer to a late 1970s study based on four sites in the Manchester area and one in Shropshire remote from industrial activity (Badsha and Goldspink, 1982). Of the Manchester sites, Compstall Lake is fed by water from the River

Table 5.15 **Geometric mean concentrations of cadmium, lead and zinc (μg/g dry weight with 95% confidence limits {CL}) in whole bodies of roach** *Rutilus rutilus*, **common bream** *Abramis brama*, **perch** *Perca* **fluviatilis and pike** *Esox lucius* **of different age groups (years) collected from four Manchester sites and one site (Ellesmere Mere) in Shropshire in the late 1970s.**

Species Site	Age group	Cadmium		Lead		Zinc	
		Mean	95% CL	Mean	95% CL	Mean	95% CL
Roach *Rutilus rutilus*							
Rostherne Mere	3	22	21, 23	6.5	5.5, 7.7	356	344, 369
	5	25	23, 27	14	13, 15	374	366, 382
	9	29	27, 31	18	17, 37	387	370, 405
Lyme Park	7	9.3	7.0, 13	11	9.2, 13	326	302, 351
Common bream *Abramis brama*							
Compstall Lake	2	11	9.7, 13	14	12, 16	196	162, 236
Ellesmere Mere	11	nd	nd	nd	nd	168	165, 170
Perch *Perca fluviatilis*							
Rostherne Mere	4+	29	26, 32	6.6	5.7, 7.7	170	164, 177
Ellesmere Mere	2+	13	11, 15	15	12, 19	143	133, 153
Tatton Mere	2+	12	11, 14	26	24, 29	150	145, 155
Pike *Esox lucius*							
Tatton Mere	2+	23	21, 24	10	9.3, 12	367	362, 371

Note: nd = not detectable.
Source: Badsha and Goldspink (1982).

Etherow. The River Etherow was shown earlier to be very high, at the end of the 1970s at least, in levels of cadmium and zinc, but not lead, bioavailable to aquatic bryophytes (Table 5.5). Given the nature of bryophytes, these high bioavailabilities would be dissolved bioavailabilities. The fish collected from Compstall Lake were young (age group 2) common bream *Abramis brama*. The only other site from which this species was collected was Ellesmere Mere in Shropshire, considered to be uncontaminated, and the

Table 5.16 Concentrations of cadmium, lead and mercury (μg/g wet weight) in tissues of eels *Anguilla anguilla*, roach *Rutilus rutilus*, bleak *Alburnus alburnus*, perch *Perca fluviatilis*, pike *Esox lucius*, gudgeon *Gobio gobio* and trout *Salmo trutta* collected from different regions of Britain.

Species / Region	Date	Tissue	Cadmium		Lead		Mercury	
			Mean or range of means	Total range	Mean or range of means	Total range	Mean or range of means	Total range or ± SE
Eel *Anguilla anguilla*								
East Anglia[1]	1980s	Liver	0.06–0.47	0.01–1.95	0.26–0.80	0.03–3.37	0.07–0.59	0.01–3.36
East Anglia[1]	1980s	Muscle	0.02–0.08	0.01–0.41	0.03–0.08	0.01–0.48	0.13–0.39	0.08–1.30
East Anglia[2]	1981–84	Muscle	<0.01	<0.01	0.70	0.40–1.00	0.31	0.09–0.66
SW England[2]	1981–84	Muscle	0.07	<0.01–0.68	0.44	<0.02–2.20	0.13	0.03–0.35
Wales[2]	1981–84	Muscle	0.22	<0.01–1.88	0.92	<0.02–7.40	0.18	0.02–0.38
NE Scotland[2]	1981–84	Muscle	0.69	0.10–1.60	2.06	0.02–4.80	0.36	0.20–0.53
Upper Thames[3]								
R. Windrush	1996	Muscle					0.31	–
R. Thames	1995	Muscle					0.15	± 0.07
Roach *Rutilus rutilus*								
East Anglia[2]	1981–84	Muscle	0.09	<0.01–1.07	1.11	<0.02–8.70	0.05	0.01–0.19
SW England[2]	1981–84	Muscle	0.12	<0.01–0.76	1.57	<0.02–8.70	0.07	0.01–0.20
Wales[2]	1981–84	Muscle	0.10	0.04–0.18	0.32	0.20–0.60	0.07	0.03–0.15
NE England[2]	1981–84	Muscle	0.05	<0.01–0.25	0.89	<0.02–1.90	0.07	0.01–0.18
Upper Thames[3]								
R. Ray	1996	Muscle					0.09	± 0.01
R. Windrush	1996	Muscle					0.09	± 0.02
R. Thames	1996	Muscle					0.06	± 0.01

Table 5.16 (cont.)

Species Region	Date	Tissue	Cadmium Mean or range of means	Total range	Lead Mean or range of means	Total range	Mercury Mean or range of means	Total range or ± SE
River Thames[4]	2007–11	Whole fish					0.02–0.04	–
River Nene[4]	2008	Whole fish					0.02–0.04	–
River Kennet[4]	2011	Whole fish					0.02	–
Bleak *Alburnus alburnus*								
River Thames[4]	2007–08	Whole fish					0.03–0.05	–
Perch *Perca fluviatilis*								
Britain[2]	1981–84	Muscle	0.04	<0.01–0.35	0.92	0.40–1.40	0.12	0.04–0.32
Upper Thames[3]								
R. Ray	1996	Muscle					0.07	± 0.02
Pike *Esox lucius*								
Britain[2]	1981–84	Muscle	<0.01	<0.01	0.58	<0.02–2.00	0.09	0.01–0.28
Upper Thames[3]								
R. Thames	1995	Muscle					0.34	± 0.09
Gudgeon *Gobio gobio*								
Britain[2]	1981–84	Muscle	0.08	<0.01–0.20	0.57	<0.02–2.00	0.12	0.06–0.22
Trout *Salmo trutta*								
Britain[2]	1981–84	Muscle	0.06	<0.01–0.27	0.61	<0.02–4.30	0.07	0.02–0.27

Sources: Barak and Mason (1990);[1] Mason (1987);[1] Yamaguchi et al. (2003);[2] Jürgens et al. (2013).[4]

common bream here were much older (age group 11) (Table 5.15). It is not possible to conclude with any confidence that the fish in Compstall Lake have raised accumulated concentrations of zinc or cadmium (Table 5.15). Accumulated zinc concentrations are usually regulated by fish to approximately constant levels, over a wide range of raised bioavailabilities and hence uptake rates. Indeed, the accumulated zinc concentrations of the young common bream in Compstall Lake and their older conspecifics in Ellesmere Mere are much the same (Table 5.15). The young Compstall Lake common bream did contain measurable accumulated cadmium concentrations, in contrast to the older Ellesmere Mere bream (Table 5.15). It is possible that this indicates raised accumulation of cadmium by *A. brama* in Compstall Lake, but the accumulated concentrations measured do not appear that high in comparison to other fish species, including those of equivalent age (Table 5.15). These data appear to confirm that fish are poor biomonitors of raised trace metal bioavailabilities in comparison to the likes of bryophytes and aquatic insect larvae. Nevertheless, some conclusions can still be drawn from Table 5.15. The accumulated concentrations of cadmium and lead increased with age in roach in Rostherne Mere, and the apparently regulated body concentrations of zinc are higher in roach than in either common bream or perch (Badsha and Goldspink, 1982).

5.14.2.1 Metals in the Flesh of Freshwater Fish

Prologue. What are the concentrations of trace metals of potential ecotoxicological concern in the flesh (muscle tissue) of freshwater fish possibly destined for human consumption?

During the 1980s, concerns were being raised over the human intake of toxic contaminants, particularly organochlorine contaminants, from the flesh (muscle tissue) of fish in the human diet. The trace metals cadmium, lead and mercury came in for attention. The major freshwater fish regularly eaten in Britain were salmonids (salmon and trout). Because these were reputed to be fish of clean waters, they were not considered to be of any major significance as a source of toxic metals to humans (Mason, 1987). There was, however, a growing fishery for eels, *Anguilla anguilla*, not least for export to continental Europe, and eels do frequent industrially contaminated freshwaters and estuaries (Mason, 1987). Surveys were, therefore, undertaken in the 1980s of muscle concentrations of cadmium, lead and mercury in eels in British rivers, either in specific regions such as East Anglia (Essex and Suffolk) (Barak and Mason, 1990) or in wider regional studies across Britain (Mason, 1987) (Table 5.16). Other fish, not least roach, were sometimes included (Table 5.16).

The ranges of mean trace metal concentrations in livers and muscle tissue of eels from 12 river systems in the East Anglian study are presented in Table 5.16. Two rivers, the Rivers Brain and Chelmer, stood out as containing eels with accumulated metal concentrations at the top end of the ranges of liver concentrations for all the rivers (Barak and Mason, 1990). Thus eels from the River Brain had mean liver concentrations of 0.47 µg/g ww Cd, 0.80 µg/g ww Pb and 0.59 µg/g ww Hg, while equivalent concentrations in eels from the Chelmer were 0.39 µg/g ww Cd, 0.61 µg/g ww Pb and 0.16 µg/g ww Hg. The Rivers Brain and Chelmer were certainly receiving metal pollution, especially of mercury. The high mean liver concentrations of mercury in Brain and Chelmer eels were also reflected in very high mean muscle concentrations of mercury in these fish, namely 0.39 µg/g ww in the River Brain and 0.28 µg/g ww in the Chelmer. In contrast, the mean cadmium and lead concentrations in the muscle tissue of the Brain and Chelmer eels were not atypically raised in comparison with the other East Anglian rivers, confirming that fish muscle tissue is not a good biomonitor for cadmium or lead bioavailability.

The wider regional dataset (Mason, 1987) considered eels and roach from up to five arbitrary regions of Britain, with some combined data across regions for perch, pike, gudgeon and trout (Table 5.16). There were significant regional differences in the mercury concentrations in the muscle tissue of eels (but not roach), with levels from Southwest England being below those of the other regions

(Table 5.16). Muscle concentrations of lead in eels (but not roach) also varied regionally, with the southwestern England eels again having the lowest concentrations (Table 5.16). Finally, muscle concentrations of cadmium, too, varied regionally in eels but not roach, with again eels from Southwest England having the lowest concentrations of those that could be compared statistically (Table 5.16). The numbers of samples restricted the interspecific comparisons that could be made. Nevertheless, mercury concentrations in muscles were significantly higher in eels than in roach from Southwest England and East Anglia (Table 5.16). On the other hand, the mean lead concentration in the muscle tissue of roach was significantly higher than that in eels in Southwest England (Table 5.16). There were no significant differences in muscle cadmium concentrations between eels and roach (Table 5.16). Given the wide areas of the regions chosen, it was difficult to subscribe causes to the regional differences seen in the eel data.

Sampling of freshwater fish for mercury analysis continued into the 1990s and the twenty-first century (Table 5.16), typically as part of risk assessments for mammalian predators such as otter (Yamaguchi et al., 2003) and for humans (Jürgens et al., 2013). The 1990s study quoted in Table 5.16 concentrated on the Upper Thames and its tributaries, the Ray and the Windrush, upstream of Oxford (Yamaguchi et al., 2003). As in the wider regional study of the earlier 1980s (Mason, 1987), mercury concentrations in muscle tissues of fish at the same site (Windrush) were higher in eels than in roach (Table 5.16). Perch and roach at the same location (River Ray below Swindon) did not differ in muscle tissue mercury concentrations (Table 5.16). Pike had the highest muscle tissue mercury concentration, higher than eels at the same location (River Thames at Northmoor, just upstream of Oxford) (Table 5.16). High mercury concentrations in pike, a top predator, probably reflect biomagnification of methyl mercury up food chains and accumulation of mercury with age of the fish.

The twenty-first century data in Table 5.16 are mercury concentrations for whole fish (roach *Rutilus rutilus* and bleak *Alburnus alburnus*) collected from the Thames in the London area and from the River Nene, which flows northeast into the Wash (Jürgens et al., 2013). In whole roach and bleak, mercury concentrations increased with size or age. Mercury concentrations are higher in muscle tissue than in whole bodies, although this difference is usually less than a factor of two (Jürgens et al., 2013). In the light of this factor, and given the 1980s data for mercury concentrations in roach muscle tissues (Table 5.16), it was concluded that mercury concentrations in fish were much lower in the first decade of the twenty-first century than 20 to 30 years previously (Jürgens et al., 2013), corroborating earlier indications of a reduction in the mercury contamination of freshwater fish in the United Kingdom (Downs et al., 1999). As noted previously, the industrial uses and disposal of mercury were controlled more rigorously or deliberately reduced in Britain between the 1960s and 1990. A reduction in the mercury contamination of freshwater fish since the 1980s is, therefore, to be expected.

5.15 Birds

Prologue. Birds associated with freshwater systems will take up trace metals from the diet. For example, ducks, geese and swans are typically herbivorous, while herons, egrets, grebes and kingfishers prey on fish.

The general principles of the biology of trace metals in birds laid out in Chapter 4 apply to birds associated with freshwater, even though the specific dietary sources of the trace metals necessarily differ between systems. Trace metals taken up in the alimentary tract of birds are passed into the blood and hence on to other organs, such as the liver, kidney and/or bone for detoxified storage. The accumulated trace metal concentrations in these storage organs typically increase with increased local bioavailabilities of the trace metals. Trace metals will also be transferred from these internal organs to the feathers, but not until the time of the next moult, when the feathers are newly growing and have an active blood supply. The amount of trace metal transferred into the feathers reflects the amount accumulated by the bird since the last moult, the proportion of body metal transferred varying with the trace metal concerned. Trace metals will also be transferred into developing eggs. These trace metals are of particular interest to ecotoxicologists because reduced reproductive success, such as a decrease in the success of hatching of the eggs, is often an early symptom of trace metal ecotoxicology in birds.

5.15.1 Mercury

Prologue. It is appropriate to begin with mercury, the trace metal causing the most environmental concern in the 1980s.

Mercury taken up from the alimentary tract of birds will be transferred to organs throughout the body, including muscles, liver and kidney (Scheuhammer, 1987; Wolfe et al., 1998). Birds will then transfer most of this mercury into the feathers at the next moult, and as much as 90% of total body mercury can then be contained in the feathers (Burger, 1993). Birds will subsequently excrete this mercury when the feathers are shed at the following moult (Furness, 2015). Female birds also transfer mercury to developing eggs, with potential ecotoxicological consequences.

Mercury released into the freshwater environment, typically in inorganic form, can be transformed by microbial action in sediments to methyl mercury (Wolfe et al., 1998; Furness, 2015). Methyl mercury is particularly toxic, producing reproductive and neurological impairment, and biomagnifies up food chains to accumulate in top predators, such as piscivorous birds. Methyl mercury is easily absorbed in the alimentary tract (Scheuhammer, 1991). Predators in aquatic food chains accumulate more mercury than those in terrestrial food chains because of differences in concentrations of methyl mercury in the different food organisms lower down the food chain, confounded by biomagnification up the food chain. An average of more than 50% of the mercury present in fish muscle is in the form of methyl mercury, and this often may approach 100% (Scheuhammer, 1991). In comparison, aquatic invertebrates such as insect larvae and molluscs typically have less than 50% of their mercury content in the form of methyl mercury,

and any methyl mercury in freshwater plants is barely detectable (Scheuhammer, 1991).

As discussed in Chapter 4, bird ecotoxicologists have attempted to define critical concentrations of mercury in the liver, kidney, blood, feathers and eggs of birds which might be considered as thresholds above which particular symptoms of mercury ecotoxicology become apparent (Shore et al., 2011). This concept of a critical threshold accumulated metal concentration has been criticised because toxicity can occur at any accumulated metal concentration, if and when the total uptake rate exceeds a threshold that cannot be matched by the rates of storage detoxification and/or excretion. So-called critical accumulated concentrations have often been derived from laboratory exposures when birds (including the mallard duck *Anas platyrhynchos*) have been fed on high mercury diets (Heinz, 1976). Given the high rates of mercury uptake under these relatively acute circumstances, it is likely that mercury uptake will exceed combined detoxification and excretion at a lower accumulated body or organ mercury concentration than might occur under chronic field exposures of birds to mercury bioavailabilities that are raised but not lethal.

Nevertheless, critical mercury concentrations are to be found in the literature for different organs of birds (Scheuhammer, 1991; Shore et al., 2011). It is piscivorous birds that are the most likely to achieve these concentrations under field conditions of mercury exposure. Because reproductive impairment is the most sensitive indication of mercury ecotoxicity, proposed critical concentrations are lower for the onset of negative effects on reproductive success than for adult bird lethality. Proposed critical wet weight concentrations for non-marine birds do vary between authors, but a recent review suggests threshold liver concentrations of 2 µg/g and 20 µg/g Hg for reproductive impairment and lethality respectively, a kidney concentration of 40 µg/g Hg for lethality, and an egg concentration of 0.6 µg/g Hg for reduced reproductive success (Shore et al., 2011). Experimental studies on mallard ducks specifically similarly concluded that liver concentrations of 2 to 12 µg/g Hg wet weight, and egg concentrations of 1 µg/g Hg wet weight, could be linked to decreased hatchability of eggs and increased hatchling mortality, even in the absence of toxic signs in the adults (Heinz, 1976; Scheuhammer, 1991). Feather concentrations associated with impaired reproduction in mallards fed mercury-rich diets in these experimental studies were 9 to 70 µg/g Hg (Heinz, 1976; Shore et al., 2011), a variability that is hardly useful when attempting to define a theoretical critical concentration. These feather concentrations lie on either side of a proposed critical concentration of 20 µg/g Hg in feathers of wetland birds considered indicative of a mercury ecotoxicological threat (Scheuhammer, 1991).

A more justifiable approach to the use of accumulated concentrations to recognise any ecotoxicological threat from mercury is to consider concentrations of mercury in the diet, or preferably methyl mercury concentrations in the diet. It is dietary concentrations that will determine the uptake rates of mercury by birds. Feather concentrations of greater than 20 µg/g Hg in piscivorous birds can result from dietary concentrations above 1 µg/g Hg dry weight (Scheuhammer, 1991). Similarly, proposed critical concentrations of mercury in the diet are 0.25 µg/g Hg wet weight for reproductive impairment of feeding birds and 6 µg/g methyl Hg wet weight for death of the feeding bird (Shore et al., 2011).

Let us then consider piscivorous bird such as herons and egrets feeding on fish in British waters, remembering that most of the mercury in fish is in the form of methyl mercury (Scheuhammer, 1991; Yamaguchi et al., 2003). From the fish data presented in Table 5.16, it can be concluded that any 0.25 µg/g Hg wet weight threshold was exceeded in muscle tissues of eels, perch and pike from several British rivers in the 1980s and 1990s, with a potential mercury ecotoxicological threat to piscivorous birds. By this century, mercury concentrations in British fish had fallen (Table 5.16), and any widespread ecotoxicological risk from mercury to piscivorous birds is probably now negligible.

5.15.2 Cadmium

Prologue. Cadmium is a second trace metal that can be considered of potential ecotoxicological significance in Europe. The biology of cadmium in freshwater birds follows the general principles for trace metals in birds summarised earlier in this chapter.

Most (70 to 100%) of the body burden of accumulated cadmium in birds is stored in the liver and kidneys, bound to metallothionein with a low turnover rate (Scheuhammer, 1987; Wayland and Scheuhammer, 2011). Correspondingly, a lower percentage (below 30%) of body cadmium is transferred to the feathers at the moult than in the case of mercury. Cadmium concentrations in kidneys and livers increase with increased local cadmium bioavailability (e.g., Table 4.36). Cadmium concentrations are higher in the kidney, which is considered to be the critical organ in chronic cadmium toxicity (Scheuhammer, 1991; Wayland and Scheuhammer, 2011). As for mercury, critical accumulated concentrations of cadmium in different bird organs have been proposed (Wayland and Scheuhammer, 2011). Thus, an accumulated cadmium concentration of 65 µg/g Cd wet weight in the kidney of adult birds may be associated with a 50% probability of alterations in energy metabolism or damage to the kidneys. A threshold toxic effect level for liver concentrations in adult birds may lie between 45 and 70 µg/g Cd wet weight (Wayland and Scheuhammer, 2011).

Cadmium concentrations in tissues of birds vary with habitat, diet, age and physiological status. The highest average cadmium concentrations in bird livers and kidneys are to be found in seabirds, the lowest in terrestrial and freshwater birds (Wayland and Scheuhammer, 2011). Freshwater birds in general have median kidney concentrations of 0.4 µg/g Cd wet weight (range 0.1 to 4.9 µg/g) and median liver concentrations of 0.1 µg/g Cd wet weight (range 0 to 7.6 µg/g) (Wayland and Scheuhammer, 2011). Most wild freshwater ducks have kidney cadmium concentrations between 2 and 8 µg/g dry weight and liver cadmium concentrations of less than 3 µg/g dry weight (Scheuhammer, 1987). These accumulated concentrations are below the aforementioned proposed critical concentrations, even assuming a factor of 5 between wet and dry weights. Unlike the case for mercury, piscivorous birds do not risk increased exposure to dietary cadmium in comparison with their herbivorous counterparts, because cadmium is not biomagnified up food chains.

As for mercury, it is more justifiable to move away from critical tissue concentrations and consider potentially toxic cadmium concentrations in the diet. Adult mallard ducks fed a diet with 200 µg/g Cd dry weight developed kidney lesions within 9 weeks (White et al., 1978). Mallard ducklings fed a diet with 20 µg/g Cd dry weight developed kidney lesions and anaemia over a similar period (Cain et al., 1983). The mallard ducklings accumulated more cadmium than adult ducks on the same diet and were more susceptible to cadmium toxicity (Cain et al., 1983; Scheuhammer, 1987).

How do we translate these experimental findings into the real world? Atypically high cadmium concentrations in freshwater biota in Britain do occur in streams and rivers affected by historical mining (Tables 5.5, 5.8, 5.10 and 5.11). Aquatic flowering plants are absent from the most metal-contaminated habitats. Correspondingly, (mainly) herbivorous birds such as ducks, geese and swans are also absent, and would not, therefore, be exposed to very high dietary concentrations of cadmium. Fish, too, are absent from the most contaminated freshwater habitats, and so piscivorous birds are not at risk from cadmium eco-toxicity. On the other hand, metal-contaminated streams do have an invertebrate fauna, particularly of benthic insect larvae, and these do accumulate very high cadmium concentrations (Tables 5.8, 5.10 and 5.11) above the dietary cadmium concentrations judged to cause cadmium ecotoxicity in birds, particularly immature birds.

One British bird that may come across these cadmium-rich insect larvae is the dipper *Cinclus cinclus*, which inhabits fast-flowing mountain streams and feeds on benthic aquatic invertebrates. A study in British Columbia, Canada, on the related American dipper *Cinclus mexicanus*, involved the measurement of trace element concentrations in diet items, and an estimation of the bird's daily intake of dietary metal for comparison against a proposed tolerable daily intake (TDI) of each trace metal by birds (Morrissey et al., 2005). The TDI uses avian chronic

toxicity tests to calculate the geometric mean of the no observable adverse effect level (NOAEL) and the lowest observable adverse effect level (LOAEL). This mean intake rate (µg metal per gram body weight per day) is then arbitrarily divided by an uncertainty factor of 10 (in this study) to account for differences in sensitivity between bird species. The TDI used for cadmium for the American dipper was 0.54 µg/g/d. This study also showed that concentrations of trace metals in plucked breast feathers from the American dippers could be used to biomonitor differences in trace metal bioavailabilities (Morrissey et al., 2005). An equivalent study on dippers in upland stream systems contaminated by historical metal mining has the potential to identify whether there is any ecotoxicological risk from cadmium (or indeed other trace metals) to this passerine bird in the British Isles.

5.15.3 Lead

Prologue. A third and final trace metal of potential ecotoxicological significance to be considered here is lead.

Lead accumulates in the kidneys and livers of birds, but also in their bones, accumulated concentrations rising with increased local bioavailability of lead. Female birds accumulate lead more quickly than males (Scheuhammer, 1987). The uptake of lead is linked to that of calcium, and the extra uptake by female birds of calcium to make eggshells leads to greater absorption of lead. It is the same physiological link between calcium and lead that leads to high background concentrations of lead in the bones of birds (Scheuhammer, 1987; Franson and Pain, 2011). As for other trace metals, at the time of moult, lead is transferred to the feathers, which may then contain about 50% of the lead content of the bird's body (Burger, 1993). As in the case of cadmium, piscivorous birds are not at an increased risk of lead ecotoxicity, because lead is not biomagnified up freshwater food chains.

The toxic effects of lead in birds include immuno-suppression, anaemia, lethargy, ultrastructural damage in kidney and liver cells, lack of muscular

Table 5.17 **Proposed threshold concentrations of lead in livers and kidneys (both μg/g wet weight) and blood (μg per 100 ml) in ducks, geese and swans (*Anseriformes*) indicative of subclinical, clinical and severe clinical lead poisoning.**

	Liver	Kidney	Blood
Subclinical Pb poisoning	2	2	20
Clinical Pb poisoning	6	6	50
Severe clinical Pb poisoning	10	15	100

Source: Franson and Pain (2011).

coordination and loss of weight (Franson and Pain, 2011). The most sensitive effect, however, is the inhibition of activity of the enzyme d-amino levulinic acid dehydratase (ALAD), an enzyme necessary for the synthesis of haem needed to make haemoglobin (Franson and Pain, 2011).

Again, ornithological ecotoxicologists have sought the holy grail of threshold tissue concentrations of lead indicative of the onset of toxic effects in particular birds, going so far as to recognise differences in lead accumulated between birds of different orders (Franson and Pain, 2011). Table 4.47 listed such proposed critical concentrations (in terms of dry weight) for pigeons, according to severity of toxic effect, while Table 5.17 does the same, now expressed as wet weight concentrations, for ducks, geese and swans.

The benthic invertebrate fauna of mining-contaminated streams may contain very high accumulated lead concentrations (Tables 5.8, 5.10 and 5.11), and these may represent an ecotoxicologically significant source of lead to dippers *Cinclus cinclus*. The proposed TDI of lead for the American dipper *C. mexicanus* in the British Columbian study was 0.64 μg/g/d (Morrissey et al., 2005). In addition, mallard ducks, particularly young mallards, may

ingest benthic macroinvertebrates when feeding. The ducks may, therefore, be at potential risk of lead poisoning if they spend time feeding on invertebrates in effluent channels from disused lead mines.

5.15.3.1 Lead Shot

Prologue. Other potential sources of lead to birds in the freshwater environment are lead shot used when hunting with shotguns and lead fishing weights used historically by anglers.

Shotgun hunting can lead to freshwater birds such as ducks suffering from lead poisoning if shot but not lethally wounded (Franson and Pain, 2011). Furthermore, ducks, geese and swans will also take up lead shot pellets into their gizzards during feeding, whether these be derived from shotgun hunting or from discarded lead fishing weights (Franson and Pain, 2011). Ingested lead shot will be ground down in the gizzards of birds and be subject to dissolution by acids in the stomach. The resulting solubilised lead is absorbed into the bloodstream, to be treated physiologically as any other lead derived from the diet. In fact, the ingestion of lead shot is the primary source of elevated lead exposure and poisoning in waterfowl and most other bird species (Scheuhammer and Norris, 1996). Typically, the ingestion of ten or more lead shot pellets by a bird causes acute lead poisoning, and the bird will be dead within a few days (Scheuhammer and Norris, 1996). Regular ingestion of smaller numbers of shot pellets results in chronic lead poisoning, leading to the exhibition of the symptoms of lead poisoning and ultimately death (Scheuhammer and Norris, 1996).

The ingestion of lead shot pellets and lead fishing weights has been documented for many birds associated with freshwater, including diving ducks, swans and geese, putting these birds at risk of lead poisoning (Franson and Pain, 2011; Pain et al., 2015). A survey of the gizzards of nearly 2,500 wildfowl shot by wildfowlers in the early 1980s in Great Britain found relatively high incidences of lead pellet ingestion in greylag geese *Anser anser* (7%) and ducks, including gadwalls *Anas strepera* (12%), tufted ducks *Aythya fuligula* (12%), pochards *Aythia ferina* (11%), goldeneyes *Bucephala clangula* (7%) and mallards *Anas platyrhychos* (4%) shot by wildfowlers (Mudge, 1983). Of nearly 250 birds found dead, incidences of pellet ingestion were 22% in mute swans *Cygnus olor*, 10% in Bewick's swans *Cygnus columbianus*, 17% in mallards and 7% in each of pochards and tufted ducks (Mudge, 1983). Most ingested lead pellets originated from shotguns, while the split shot of anglers were found in one pochard and four mute swans. Pellet ingestion by mallards was widespread, particularly by ducks shot at inland freshwater bodies. Lead pellet ingestion by British mallards was estimated to cause the death of at least 8,000 mallard ducks each winter (Mudge, 1983).

Seventy per cent of 101 dead mute swans *Cygnus olor* collected throughout Ireland between 1984 and 1987 had died of lead poisoning derived from the ingestion of lead pellets, either shotgun pellets or discarded fishing weights (O'Halloran et al., 1991). Blood samples were also taken from 971 live mute swans (O'Halloran et al., 1991). Between 31% (summer) and 66% (winter) of mute swans at Cork Lough had blood lead concentrations above 40 µg per 100 milligram, considered the maximum tolerable limit of lead in the blood, in line with proposals in Table 5.17. The concentration of lead in the blood of the mute swans depended on the number of lead pellets ingested as detected by X-ray. Birds with one pellet had blood lead concentrations between 52 and 197 µg per 100 milligrams, while a mute swan with 11 ingested pellets had 1,670 µg Pb per 100 milligrams in the blood (O'Halloran et al., 1991). The latter mute swan was certainly doomed.

Measures have been taken. Since 1986 in the United Kingdom, lead fishing weights of between more than 0.06 grams (size 14) and 28.35 grams (1 ounce) have been banned from sale and from use to weight fishing lines. Shot of lower weight (sizes 14 to 8) can be used by anglers but should be kept in spill-proof containers. This legislation was enacted specifically to protect mute swans from lead poisoning through the ingestion of accidentally discarded lead fishing weights. The population of the mute swan *C. olor* in Great Britain had declined sharply during the 1960s, followed by relatively little change through the 1970s

and early 1980s (Kirby et al., 1994). This trend was reversed with large increases in the mute swan population from 1986/1987, when the sale was banned of fishing weights of a size likely to be ingested by the swans. Furthermore, the incidence of lead poisoning in mute swans also became considerably reduced (Kirby et al., 1994).

But what of lead shot pellets derived from hunting with shotguns? The United Kingdom signed the African–Eurasian Waterbird Agreement in November 1999, concerned with the restrictions on the use of lead shot in hunting. While the agreement is binding throughout the United Kingdom, its implementation has differed slightly between England, Wales, Scotland and Northern Ireland. In England, regulations and amendments were introduced between 1999 and 2002. The upshot is that, in England, use of lead shot in hunting is banned over all foreshore; over specified Sites of Special Scientific Interest; and for the shooting of all ducks, geese, coot and moorhen wherever they occur. Similar regulations are in place for the other three devolved governments, coming into force in 2002, 2005 and 2009 for Wales, Scotland and Northern Ireland respectively. The effectiveness of these regulatory measures to protect wildfowl from lead poisoning from shot pellets will, of course, depend on the degree of compliance observed. A report published as recently as 2015 concluded that in the United Kingdom 50,000 to 100,000 wildfowl (up to 3% of the wintering population) are still likely to die in each winter shooting season as a direct result of ammunition-derived lead poisoning (Pain et al., 2015). For migratory swans, this represents a quarter of all recorded deaths (Pain et al., 2015).

The mallard duck *A. platyryrhynchos* omnivorously feeds on plant and animal material and will ingest lead shot pellets. These lead pellets will have been completely eroded in the gizzard or passed through the digestive tract within three weeks in more than 90% of affected birds (Scheuhammer, 1989). Given the number of mallards in Britain that have the potential to suffer lead poisoning from the ingestion of lead shot pellets, there is inevitable interest from ecotoxicologists to monitor lead exposure in these ducks in a non-destructive fashion. The use of blood samples is attractive, both for the measurement of blood lead concentrations but also for a measure of the degree of ensuing lead toxicity. A particularly sensitive indicator of lead toxicity is a depression of the activity of the enzyme ALAD in the blood, specifically in the red blood cells. The activity of this enzyme can be restored (activated) by removal of lead in the laboratory, and a comparison of restored enzyme activity to the (unrestored) inactivated enzyme activity gives a measure of the degree of lead inhibition of the enzyme in a blood sample (Scheuhammer, 1989). A ratio of activated to non-activated enzyme activity of ALAD in the range 1.0 to 1.3 can be considered background, while higher ratios indicate lead exposure. This ALAD activity ratio is a suitable and sensitive indicator of blood lead concentrations over the range 5 to 150 µg Pb per 100 ml in the blood of birds (see Table 5.17) (Scheuhammer, 1989).

Thus blood samples can yield measurements of lead concentrations and ALAD activity ratios, both non-destructive measures providing information on lead exposure and ensuing toxicity to the birds. A study in 1987 of 510 mallards in Canada considered blood lead concentrations below 15 µg Pb per 100 ml to represent an absence of abnormal lead exposure (Scheuhammer, 1989). On the other hand, blood concentrations above 80 µg Pb per 100 ml were considered in this study to represent highly elevated exposure to lead, in practice lead poisoning (see also Table 5.17), most probably as a result of lead shot ingestion. The average ALAD activity ratio of mallards with blood lead concentrations below 15 µg Pb per 100 ml (in fact, an average of 4.8 µg Pb per 100 ml) was 1.19, considered above to be normal. On the other hand, in the case of the mallards with more than 80 µg Pb per 100 ml (average 251 µg Pb per 100 ml), the average ALAD activity was 17.7. Blood lead concentrations in mallards were substantially elevated for two to three months after the ingestion of lead shot (Scheuhammer, 1989). For the purposes of monitoring lead exposure, such as in mallards, the ALAD assay requires less than 100 microlitres of blood and is effective at monitoring low blood lead concentrations that are difficult to measure even with sensitive specialist analytical equipment.

5.16 Mammals

Prologue. The classic British mammal associated with freshwater systems is the otter *Lutra lutra*, a predominantly fish-eating carnivore with a preference for eels and salmonid fish. The only trace metal of potential ecotoxicological concern for the otter in British freshwater systems today is mercury. Mercury is the one trace metal that is biomagnified along freshwater food chains. Furthermore, methyl mercury, the most toxic form of mercury, represents the great majority of the mercury accumulated by fish.

Between 90 and 95% of any methyl mercury in the diet is assimilated by mammals, compared with 7 to 15% for inorganic mercury (Wolfe et al., 1998). Mercury stays in mammals for a long time, with a biological half-life of about 70 days for methyl mercury and 40 days for inorganic mercury (Wolfe et al., 1998). Most methyl mercury assimilated is eventually excreted by mammals in inorganic form in the faeces, with a little in the urine, also as inorganic mercury. Mammals will transfer some mercury into the hair, to be lost when moulting (Wolfe et al., 1998). Remembering all the caveats about the lack of biological justification for the use of threshold tissue concentrations, proposed wet weight tissue concentrations of mercury for lethality in mammals are 25 µg/g Hg in both the liver and kidney and 10 µg/g Hg in the brain (Shore et al., 2011).

A question that needs to be addressed is the concentration of methyl mercury, essentially all the mercury, in a fish diet that is of potential ecotoxicological significance to a mammalian piscivore such as an otter. Controlled feeding experiments have been carried out in the United States on a closely related otter, the North American river otter *Lutra canadensis*, fed diets with various additions of methyl mercury (O'Connor and Nielsen, 1981; Wolfe et al., 1998). The experimental diets contained 2, 4 or 6 µg/g wet weight of methyl mercury, resulting in mean ingestion rates of 0.09, 0.17 and 0.37 µg/g/d of methyl mercury by the otters. At the lowest dose, the otters developed loss of appetite and loss of full control of body movements between days 168 and 199, and later examination showed histological damage to the central nervous system (CNS). When fed the middle dose, otters showed these symptoms between days 101 and 120 and had histological damage to both the CNS and kidneys. The highest dose of dietary methyl mercury was lethal in an average time of 54 days, and again there was damage to both the CNS and kidneys.

Attempts have been made to come up with proposed dietary concentrations of methyl mercury that cause mercury toxicity in mammals in general. An older proposal concluded that the lowest dietary mercury concentration then known (in 1989) to produce signs of methyl mercury intoxication and mortality in a variety of mammals was approximately 5 µg/g Hg dry weight (1 to 1.6 µg/g Hg wet weight) (Scheuhammer, 1991). This figure is lower than the 2 µg/g wet weight diet previously mentioned, but in the same approximate range. More recently, it has been proposed that the dietary threshold concentration for mercury associated with lethality in mammals is 4.3 µg/g Hg wet weight (range 1.1 to 14 µg/g Hg wet weight), when most of the dietary mercury is in the form of methyl mercury (Shore et al., 2011). This is not a lowest observable effect concentration (LOEC), but a concentration at which 50 to 100% of test animals died. The comparable figure for birds is 20 µg/g Hg wet weight, suggesting that mammals are less sensitive to mercury poisoning than birds (Shore et al., 2011). The European Commission (European Food Safety Authority, 2008) geometric mean dietary NOEC for mammals is 0.9 µg/g Hg wet weight, this being a dietary exposure that would trigger a more detailed risk assessment (Shore et al., 2011).

The study in the 1990s of mercury in fish in the Upper Thames and its tributaries, the rivers Ray and Windrush (Table 5.16), considered the risk from mercury in these fish to a mammalian piscivore (Yamaguchi et al., 2013). The mammal chosen was the American mink Neovison vison, an invasive species in Britain having escaped from fur farms in the 1950s and 1960s. There was a thriving mink population in the Upper Thames region at the time, while otters had not then yet returned to the area. The risk assessment involved the calculation of a hazard index, the ratio of the mercury concentration in the local fish in the mink's diet to an independently defined dietary no observable adverse effect concentration (NOAEC). The NOAEC value chosen was 0.05 µg/g Hg wet weight, the mercury being assumed to be in the form of methyl mercury (Giesy et al., 1994; Yamaguchi et al., 2013). This apparently low NOAEC was calculated by using a biomagnification factor of 11 from methyl mercury in the fish diet to a proposed no effect threshold concentration of 1 µg/g Hg wet weight in the liver of the feeding mink. This value had been derived from the liver concentration range of 1 to 2 µg/g Hg wet weight in the livers of healthy wild mink (Giesy et al., 1994). The resulting NOAEC of 0.05 µg/g Hg wet weight in the diet of mink is therefore conservative. A hazard index greater than 1 indicates that the concentration of mercury in the diet is higher than the threshold that might elicit a toxic response from individual mink. In the 1990s Upper Thames dataset (Table 5.16), mercury concentrations in roach, perch, pike and particularly eels all exceeded the proposed dietary NOAEC, to give hazard indices greater than 1 for the mink (Yamaguchi et al., 2013). If otters had been present in the area, the risk of mercury poisoning from a fish diet would have been even greater, because the diet of otters is completely dominated (more than 95%) by aquatic prey, with a preference for eels (Yamaguchi et al., 2013). The primary prey of mink, on the other hand, is rabbit, which makes up 45% of the mink's ingested food in the Upper Thames region, compared to a contribution of 25% from all fish combined.

The more recent study of mercury in roach and bleak from four English rivers between 2007 and 2011 (Table 5.16) also considered safe dietary levels of mercury in fish in the diets of 'wildlife' (Jürgens et al., 2013). A slightly higher safe level for mercury in fish in the diet of the otter Lutra lutra specifically has been proposed to be 0.1 µg/g Hg wet weight (Boscher et al., 2010). None of the mean mercury concentrations in roach and bleak in the twenty-first century dataset for four English rivers exceeded this concentration (Table 5.16). The more general Environmental Quality Standard (EQS) set by the EU (Priority Substances

Directive 2008/105/EC) to protect wildlife is the lower figure of 0.02 µg/g Hg wet weight in prey tissue (Jürgens et al., 2013). Such prey tissue may be fish, molluscs, crustaceans, etc., as appropriate. When using this lower EQS figure for comparison, most (79%) of the whole fish mercury concentrations in roach and bleak collected between 2007 and 2011 were above this EQS (Table 5.16). Mean mercury concentrations below this EQS were only recorded in samples collected from the two most upstream sites in the Upper Thames itself and in the Kennet, a tributary of the Thames (Jürgens et al., 2013).

Where are we today? While there are no comparative data for eels, concentrations of mercury in roach in the River Thames have certainly dropped considerably in the decade or more since the mid-1990s (Table 5.16). The (hopefully continuing) decrease in mercury concentrations in fish in British rivers, coupled with the conservative nature of the NOAEC and EQS used for comparisons, gives us hope that there might no longer be a significant risk of mercury ecotoxicity to otters feeding on eels in British rivers.

Analyses of otters found dead in Southwest England between 1990 and 2004 have provided an interesting record of the decline of lead in the environment over this period (Chadwick et al., 2011). In the United Kingdom, the sale of leaded petrol was phased out at the end of 1999, resulting in a large decline in lead emissions from road transport. In mammals, lead is accumulated primarily in bone. Between 1992 and 2004, concentrations of lead in the bones of the dead otters fell by more than 70% from a mean of about 450 µg/g Pb, in correlation with declines in environmental lead emissions and associated stream sediment concentrations (Chadwick et al., 2011). There is no ecotoxicological significance in this fall, for even the higher concentrations of bone lead would not be considered to be associated with any expected detriment to the health of the otters when alive.

5.17 Community Ecotoxicology

Prologue. Severe contamination of freshwater systems with trace metals such as lead, zinc, copper

and arsenic from mining has obvious significant deleterious effects on their biodiversity. How can we measure the subtler negative changes to freshwater communities along a gradient of trace metal contamination?

'Early in the nineteenth century the river Ystwyth was a fine clear stream abounding in salmon, sewin and trout' (Jones, 1940: 369). With the development of the lead and zinc mines in the Ystwyth valley after 1820, and the introduction of fine-grinding machinery to crush the ores before washing, the river silted up and all the fish disappeared. Kathleen Carpenter surveyed the fauna of the lower Ystwyth between 1919 and 1921. She found it extremely scanty, being limited to nine invertebrate species, a result of the ecotoxicity of lead and zinc in the river system. There was a gradual recovery of freshwater life in the Ystwyth and in neighbouring Ceredigion rivers in mid-Wales in the twentieth century after the cessation of mining. Essentially, the same story has been repeated in the streams and rivers of other historical mining regions of Britain, caused, for example, by zinc ecotoxicity in the River Nent in the Northern Pennines or by the ecotoxicity of copper, amongst other trace metals, in many rivers in Cornwall.

The conclusion is clear. Severe contamination of freshwater systems with trace metals such as lead, zinc, copper and arsenic from mining can devastate the expected flora and fauna present. The most severe contamination will eliminate freshwater life entirely. Thereafter, down the trace metal contamination gradient, microbes, algae, bryophytes and invertebrates (and even eventually fish) can survive, albeit in biological communities initially much changed from what might be expected in the absence of the metal contamination (Beltman et al., 1999; Clements, 2004). Can these differences in community structure be recognised? In the most severe conditions, depletions in the expected fauna and flora are obvious, but detrimental changes in the structure of local biological communities, though still real, become more and more difficult to discern at the bottom end of the contamination gradient. The techniques available to measure such changes to freshwater fauna and flora are the topic of this section (Clements and Rohr, 2009).

5.17.1 Metal-Resistant Species

Prologue. Some species occurring in British freshwater systems can be recognised as relatively resistant to trace metal exposure.

In an ideal world, it would be possible to conclude that the occurrence of such and such a species in a stream indicates the presence of trace metal contamination of ecotoxicological significance. Life, however, is not so simple. There are species that can be identified as being particularly trace metal-resistant, but these species are not necessarily restricted to trace metal-rich habitats. They might well also occur in habitats lower down the contamination gradient, although again they might there be outcompeted by less metal-resistant relatives. While we cannot identify such ideal indicators of high trace metal bioavailabilities by their presence alone in a contaminated stream, it is, nevertheless, possible to identify some of the more trace metal-resistant organisms met in this chapter.

At the prokaryotic microbial level, blue green bacteria of the genus *Synechococcus* and of the family Oscillatoriaceae are to be found in trace metal-rich streams. A freshwater diatom that is relatively insensitive to high trace metal bioavailabilities is *Pinnularia subcapitata*. Another protistan, the flagellate *Euglena mutabilis*, also resists high concentrations of trace metals. A zinc-resistant green alga is *Klebsormidium rivulare* (previously known as *Hormidium rivulare*). The aquatic bryophyte *Scapania undulata*, a liverwort, is zinc-, lead- and copper-resistant.

Amongst invertebrates, the flatworm *Phagocata vitta* is resistant to zinc and lead in mid-Wales and to copper in Cornwall. Certain freshwater oligochaete worms also occur in waters of high trace metal bioavailabilities, including the lumbriculids *Lumbriculus variegatus*, in the presence of lead and zinc, and *Stylodrilus heringianus*, in the presence of copper. In the case of insect larvae, chironomids (primarily) and caddisflies are generally more trace metal-resistant than stoneflies and mayflies. Nevertheless, there is variation in metal sensitivity within these taxa, and some stoneflies may be more metal-resistant than some caddisflies.

Chironomid midge larvae are well known for being resistant to the organic pollution of freshwater systems, but some are also resistant to high levels of trace metal pollutants. The tanypodine chironomid *Tanypus nebulosus* was one of a few insect larvae in the most metal-contaminated streams of the Ystwyth river system in the 1920s (Carpenter, 1924). The orthocladiine chironomids *Chaetocladius melaleucus* and *Eukiefferiella claripennis* dominated the most copper-contaminated sites in the rivers of Cornwall in the 1990s (Table 5.7). *C. melaleucus* was also one of only two invertebrates in the heavily zinc-contaminated Dowgang Level in the River Nent system in the Northern Pennines in 1976, being the most zinc-resistant insect (Table 5.6).

Of caddisfly larvae encountered in British metal-contaminated streams, it is *Plectrocnemia conspersa* that is particularly trace metal-resistant. *Rhyacophila dorsalis* matches the copper resistance of *P. conspersa* in Cornwall (Table 5.7) but is not quite as zinc-resistant in the Nent (Table 5.6). Amongst the stoneflies, the nemourid *Amphinemura sulcicollis* was the most copper-resistant stonefly in Cornish rivers, equalling the distributions of the two aforementioned caddisflies (Table 5.7). In the zinc-rich Nent river system, another nemourid stonefly, *Nemoura cinerea*, and three species of leuctrids, *Leuctra hippopus*, *L. inermis* and *L. moselyi*, were the most zinc-resistant stoneflies, all four matching the local distribution of *P. conspersa* (Table 5.6).

The trace metal resistances of mayflies are generally low in comparison to the other insect larvae mentioned. Within the mayflies, however, baetids, particularly *Baetis rhodani* and *B. vernus* in Britain, are conspicuously more resistant to trace metal exposure than mayflies in other families.

At the bottom end of the scale of trace metal resistance amongst the flora and fauna of freshwater streams lie aquatic flowering plants, amphipods, isopods, molluscs and fish.

5.17.2 Community Analysis and Biotic Indices

Prologue. We address the subtler changes in the biological communities of freshwater systems affected, but not eliminated, by raised trace metal bioavailabilities.

Organisms, such as insect larvae and other benthic invertebrates are still present, some in abundance, but has the presence of trace metal contamination actually caused changes in the resident communities? Initially, ecologists would have used parameters such as species richness or diversity indices (so-called univariate parameters) to describe and compare the structures of ecological communities. Now, the availability of more comprehensive and powerful multivariate statistical techniques, such as principal components analysis (PCA), has led to the recognition of less immediately obvious patterns of organism abundance across a series of sites of different trace metal exposures. Furthermore, these patterns can be related to particular physicochemical variables, such as trace metal concentrations in sediment or water. It is the benthic invertebrate communities of streams and rivers that are usually the subject of such analyses, as in the cases of the trace metal-rich river systems of Cornwall and mid-Wales (Gower et al., 1994; Hirst et al., 2002). Benthic diatom communities of contaminated streams have also been analysed in this way (Hirst et al., 2002).

Such multivariate analyses of biological communities, whether of benthic invertebrates or diatoms, are enormously informative. However, the sorting of the material collected does need a large commitment of man-hours. Furthermore, the required level of taxonomic expertise to identify the sorted taxa to the appropriate taxonomic level is significant, and hard to come by, whether for diatoms or for the wide systematic range of benthic invertebrates present. Difficulties in identifying species of chironomid midge larvae or oligochaete worms readily spring to mind.

Can life be made easier? A new kid on the block with an exciting future is the use of environmental DNA (eDNA) (Ficetola et al., 2008; Taberlet et al., 2012). In any ecosystem, including therefore freshwater systems, extracellular DNA fragments can be found, for example in water or sediment samples, that were once part of the genomes of micro- and macroorganisms living in that habitat (Hoffmann et al., 2016). Once a database has been compiled of the DNA barcodes of potential resident organisms, it would be relatively easy to use automatic molecular techniques to identify and ultimately to estimate relative abundances of chosen taxa in the environmental samples taken. The effort-intensive part is the building up of a barcode database. But this is a task that is a one-off, the database subsequently being available to all future researchers. The potential of using eDNA is extremely powerful. It has already proved possible, for example, to use eDNA in water samples to detect the presence of fish and amphibians in 39 temperate water bodies, with results that at least matched, and usually exceeded, the number of taxa detected using traditional methods (Valentini et al., 2016).

An attractive alternative to the resource-intensive counting and identification of, for example, all the benthic invertebrates collected for PCA analysis is the calculation of a biotic index, reflecting the ecotoxicological effect of trace metals on a biological community. A biotic index is a number on a scale grading the quality of an environment, for example a site on a river, according to the diversity and abundance of the organisms present, with particular reference to key taxa with known pollution insensitivities (Clements and Newman, 2002). This last point is key. Standard community-based indices such as species richness or a diversity index give all species equal weight regardless of their sensitivity to the anthropogenic contamination being investigated. Therefore, such indices may not show up a response such as a trace metal-sensitive species being simply replaced by a trace metal-insensitive species. A particular advantage of the use of a biotic index is that the level of taxonomic identification involved is often at a high level, such as that of the family, requiring relatively few sorting and identification skills (Metcalfe, 1989). Biotic indices assign scores to these individual taxa based on their relative sensitivity to a specific type of pollution. Such decisions, however, are usually based on expert opinion, and therefore do inevitably introduce some undesirable subjectivity into the assignment of scores (Clements and Newman, 2002).

Nevertheless, the biotic index approach has been enthusiastically adopted by ecologists assessing the effects of organic pollution on benthic macroinvertebrates in freshwater habitats. One such index used in the assessment of organic pollution in freshwater systems is the EPT score (Clements and Newman, 2002). Mayflies (*Ephemoptera*), stoneflies (*Plecoptera*) and caddisflies (*Trichoptera*) are relatively sensitive to organic enrichment, while chironomids are generally insensitive to organic pollution. The EPT score makes direct use of these relative sensitivities at these high taxonomic levels to create a biotic index of organic pollution. Another biotic index for the assessment of organic pollution on benthic macroinvertebrates in freshwaters is the Trent Biotic Index (TBI) (Metcalfe, 1989). The TBI is again based on the presence or absence of mayflies, stoneflies and caddisflies but also of amphipod and isopod crustaceans, tubificid oligochaetes and chironomid larvae red with haemoglobin. Like the EPT, the TBI does not take into account abundances, a desirable modification that was left to the next biotic index developed, the Chandler score (Metcalfe, 1989). Subsequently superseding the Chandler score was the biological monitoring working party (BMWP) score, developed in the 1970s (Metcalfe, 1989). The BMWP score does not try to include abundance data. Benthic macroinvertebrate taxa are still ranked according to their sensitivity to organic pollution with corresponding scores from 10 (sensitive) to 1 (insensitive), which are totalled to give a total BMWP score (Jeffries and Mills, 1990). An average score per taxon (ASPT) is then derived by dividing the BMWP score by the number of taxa used in its generation. The ASPT is correspondingly less sensitive to variation in sampling effort than the BMWP.

The BMWP score was developed as a biotic index to assess the ecotoxicological effects of organic pollution on benthic macroinvertebrates in freshwater habitats. It cannot therefore be expected from first principles to be suitable as a biotic index for trace metal pollution caused by mining effluent.

Nevertheless, the use of the BMWP has been evaluated in this context, in the mining-affected streams of mid-Wales (Environment Agency, 2009a), and in the River Avoca in Ireland affected by acid mine drainage (AMD) discharge (Gray and Delaney, 2008). In the mid-Wales streams, there was a correlation between BMWP score (but not ASPT) and total Zn water concentration, albeit with considerable scatter in the data. There was, however, no clear relationship between any other trace metal water concentration (As, Cd, Cu, Fe, Mn, Ni and Pb, as total or dissolved fraction) and BMWP score, ASPT or number of taxa present (Environment Agency, 2009a). On the other hand, the study of AMD in the Avoco River, draining a region of abandoned copper sulphide mines in Ireland, did find the BMWP to be a good index of AMD ecotoxicology (Gray and Delaney, 2008). The BMWP identified significant reductions in both abundance and taxon richness of benthic macroinvertebrates in response to AMD. The effects of AMD are, however, driven by a number of factors, particularly water acidity and the physical effects of iron precipitation as ochre, confounding any direct effects of trace metal toxicity. Furthermore, the ASPT derived from the BMWP was not correlated with any measured parameter such as pH, sulphate, zinc or iron concentrations, unlike the BMWP score. As might, therefore, be expected given its derivation as an index of organic pollution, the BMWP score is not in any way ideal as a biotic index for the ecotoxicological effect of trace metals in metal-contaminated streams.

So, specific biotic indices, based on relative sensitivities to trace metals, will need to be developed if biotic indices are to be used for the assessment of the ecotoxicological effects in freshwaters of trace metals derived from mining. To follow further the biotic index approach developed for the ecotoxicological assessment of organic pollution in freshwaters, a possible model might be the River Invertebrate Prediction and Classification System (RIVPACS) (Wright et al., 1984). RIVPACS represented a major step forward in pollution assessment because it adopted a 'reference condition' approach. The benthic macroinvertebrate fauna at a site under investigation is compared with fauna at similar 'reference sites' that are not subject to any apparent environmental stress (Jones et al., 2010). The site quality is then measured as a ratio (the observed/expected score), where the expected score has been predicted by RIVPACS based on the fauna at similar sites with matching physical, chemical and geographical characteristics. The methodology is, therefore, dependent on the compilation of an extensive dataset of benthic macroinvertebrate assemblages (identified to species level) from many representative sites across the United Kingdom not subject to pollution or other environmental stress. RIVPACS became the main tool used by regulatory authorities in the United Kingdom for the biological monitoring of rivers. In practice, an expected score is derived for the average BMWP score per taxon (ASPT) and for the number of BMWP scoring families (NTAXA) and compared to the observed scores derived from a family-level identification of samples from the test site. Taxonomic identification at the family level makes the sorting and identification of collected benthic macroinvertebrate samples much less resource intensive than species-level identification. There is inevitably, however, a very large one-off cost in compiling the appropriate dataset of the fauna of reference sites.

RIVPACS was developed for the assessment of organic pollution (based on BMWP, ASPT and NTAXA) and not specifically for the assessment of the ecotoxicological effects of trace metals. Nevertheless, a recent study has demonstrated a negative relationship between the observed/expected ratios of the indices NTAXA and BMWP and invertebrate metal body burdens in rivers in northern England, reflecting raised trace metal bioavailabilities (De Jonge et al., 2013). Furthermore, in Australia, a diagnostic index for acid mine drainage has been tested with success against distance downstream of mine discharges, using an Australian predictive model based on RIVPACS (Sloane and Norris, 2003). It appears, therefore, that the predictive modelling approach (RIVPACS) has potential to be adapted for the assessment of the ecotoxicological impairment of freshwater benthic macroinvertebrate communities by trace metal contamination (Jones et al., 2016).

5.17.3 Biomonitors, Bioaccumulated Metal Guidelines and Ecotoxicological Effects

Prologue. High accumulated trace metal concentrations in certain hardy biomonitors can be correlated with the onset of ecotoxicological effects of trace metal exposure on more sensitive species in the community. These threshold accumulated concentrations in the biomonitors can be used as bioaccumulated metal guidelines of trace metal ecotoxicology in particular freshwater systems.

5.17.3.1 Is There an Easier Way?

We have seen throughout this book the value of accumulated trace metal concentrations in selected biomonitors in providing the best evidence of local bioavailabilities of trace metals in a particular habitat. Strictly, these are the local trace metal bioavailabilities to the specific biomonitor. Nevertheless, with the use of a suite of biomonitors and knowledge of their natural histories, general conclusions can be drawn as to the relative trace metal bioavailabilities present, such as in the water column or sediments of sites in freshwater streams or rivers. Biomonitors provide evidence of the distributions over space and time of high trace metal bioavailabilities. But, even high accumulated concentrations of trace metals in biomonitors do not give any direct information as to whether the biomonitor itself is under any ecotoxicological stress from the accumulated toxic metal. In other words, there is no critical body or tissue concentration of a trace metal in an organism in which storage detoxification of accumulated trace metal is present. Nevertheless, an accumulated concentration in a relatively metal-hardy biomonitor can be used as a bioaccumulated metal guideline to indicate the presence of a significant ecotoxicological effect on other more metal-sensitive members of the local biological community.

Thus, it proved possible to define threshold accumulated trace metal concentrations in the caddisfly larva *Hydropsyche siltalai* inhabiting relatively fast-flowing (often upland) streams contaminated with trace metals that are associated with significant decreases in the abundance of ephemerellid and heptageniid mayflies (Figure 5.16). And it is simpler to measure the accumulated trace metal concentrations in 15 late instar *H. siltalai* larvae than to collect, sort, identify and count all benthic macroinvertebrates for PCA analysis, or even a subgroup of them to calculate a biotic index, even if one suitable for trace metal contamination were to be available.

The *Hydropsyche*/mayfly model works well for particular ranges of raised trace metal bioavailabilities in the metal-contaminated streams and short rivers in Cornwall (Rainbow et al., 2012; Awrahman et al., 2016). The principle that a bioaccumulated trace metal concentration in a metal-insensitive biomonitor can be calibrated for use as an indicator of an ecotoxicological effect on other local organisms can be extended to other biomonitors, other ranges of raised trace metal bioavailabilities and also other freshwater habitats. Indeed, accumulated trace metal concentrations in another metal-hardy caddisfly, the polycentropid *Plectrocnemia conspersa*, can be used upstream of the distribution of *H. siltalai* in metal-contaminated Cornish streams, again as predictors of mayfly abundance (Rainbow et al., 2012).

As further examples, the accumulated trace metal concentrations in combined species of the stonefly genus *Leuctra*, combined stoneflies of the family *Perlodidae*, an unidentified species of the heptageniid mayfly genus *Rhithrogena* and combined blackflies of the family *Simuliidae* have all been used to this end in a study of metal-impacted headwaters of 36 streams in Northwest England (De Jonge et al., 2013). These headwater streams were located in the Lake District, in the Howgill Fells between the Lake District and the Yorkshire Dales, in Ribblesdale and in Swaledale. Threshold accumulated concentrations in these biomonitors could be recognised that correlated with decreases in the expected taxonomic completeness of invertebrate communities as predicted by RIVPACS, even though this methodology was developed to monitor changes in freshwater benthic macroinvertebrate communities caused by organic pollution. The use of the heptageniid *Rhithrogena* species as a biomonitor, as opposed to a metal-sensitive target of any ecotoxicological effect of the trace metal present,

indicates that, in this case, a lower range of raised trace metal bioavailabilities is under investigation than in the *Hydropsyche* study.

In a sediment-rich, lowland river system in Flanders in which caddisflies and mayflies are not abundant, threshold accumulated trace metal concentrations have also now been identified in a chironomid midge (*Chironomus* sp. gr. *thummi*) and in combined oligochaete worms of the family *Tubificidae* that correlate with decreases in biotic indices of biological water quality (Bervoets et al., 2016). Again, the two biotic indices chosen, the Belgian Biotic Index and the Multimetric Macroinvertebrate Index Flanders (MMIF), were designed to assess the effects of organic pollution, but good correlations of the biological quality measured were nevertheless obtained against the biomonitor measures of trace metal bioavailabilities (Bervoets et al., 2016). Chironomids and tubificid oligochaetes could be used in this way in the more sediment-rich parts of British streams and rivers affected by trace metal pollution, especially downstream industrial pollution. Similarly downstream in metal-contaminated rivers, it is possible to correlate accumulated trace metal concentrations in transplanted zebra mussels *Dreissena polymorpha* to changes in macroinvertebrate communities, as shown in the River Dommel in Belgium and the Netherlands (De Jonge et al., 2012).

5.17.4 Biomarkers

Prologue. While the accumulated trace metal concentrations in biomonitors do not indicate the onset of ecotoxicological effects resulting from the trace metals in the organism itself, biomarkers do.

A biomarker is a biological response, for example a molecular, biochemical, cellular, physiological, morphological or behavioural variation, that can be measured in tissues or body fluids or at the level of the whole organism. A biomarker responds to the local presence of a significant bioavailable quantity of a contaminant, specifically here a trace metal. Any biological response to toxic metal exposure will be first elicited at the lowest level of biological organisation, the molecular level. With increased trace metal exposure, and therefore increased trace metal uptake by an exposed organism, responses will then occur sequentially at biochemical, cellular, physiological, individual organism, population and, finally, community levels. While biomarkers at the lowest levels of biological organisation are the most sensitive, it is the toxicological effects reflected by biomarkers at the higher levels, particularly the community level, which are the most ecotoxicologically significant.

In this chapter, we have already met biomarkers of high trace metal bioavailabilities causing ecotoxicological effects in freshwater habitats.

At the biochemical level, decreased activity of the enzyme ALAD in the blood of freshwater birds in response to lead exposure is particularly useful given its sensitivity and specificity to lead. Most important is the link made between ALAD activity ratios and blood lead concentrations indicating lead poisoning, for example in mallard ducks (Scheuhammer, 1989).

Metallothioneins (MT) regularly come under scrutiny as biomarkers of raised bioavailabilities of specific trace metals including cadmium, copper, mercury, silver and zinc (Amiard et al., 2006). There are field examples of the induction of MT in organisms in metal-contaminated field conditions, but MT induction can be variable, not least between organisms and between their tissues. MTs can also be induced by other stress factors that are not related to raised bioavailabilities of toxic metals. Furthermore, it is the MT turnover rate, not standing stock MT concentration in tissues, which is the more appropriate measure of MT induction. While MT are no longer considered to be the specific biomarkers of toxic trace metal exposure once hoped, any observed induction of MT in organisms at metal-contaminated sites can contribute to a package of relevant biological field observations indicating the presence of significant metal ecotoxicity. For all its inherent faults, a relatively common freshwater biomarker is the concentration of MT in the liver of freshwater fish collected along suspected trace metal pollution gradients (Bervoets et al., 2013).

Antioxidant metabolic defences need to be put in place against free radicals which can be generated by high trace metal exposure and may damage DNA (Roméo et al., 2009). Such defences include antioxidant enzymes such as superoxide dismutase and catalase, which can be used as biomarkers, as can concentrations of malondialdehyde, a breakdown product of the oxidation of lipids caused by free radicals. An integrated measure of the antioxidant defences of an organism, increasingly used as a biomarker, is the total oxyradical scavenging capacity (TOSC) (Regoli, 2000). While these more general biochemical stress biomarkers are not specific to toxic metal exposure, they do contribute to a general battery of biomarkers available to assess the health status of an organism suspected to be under ecotoxicological stress from raised trace metal exposure.

Another non-specific, but sensitive and useful, biomarker expressed at a relatively low level of biological organisation is the stability of the membrane surrounding lysosomes in cells, as assessed by the neutral red retention (NRR) method. Importantly, lysosomal membrane stability has now been correlated with other biomarkers at higher levels of biological organisation, considered more ecologically relevant, such as TOSC (Moore et al., 2006, 2013). Another cytological biomarker concerns chromosome structure. The salivary gland chromosomes of the chironomid midge larva *Chironomus acidophilus* in the Afon Goch, Anglesey, draining the copper-rich Parys Mountain, showed genotoxic rearrangement as an effect of exposure to high bioavailabilities of toxic metals (Michailova et al., 2009).

Organisms exposed to ecotoxicologically significant raised trace metal bioavailabilities incur a cost in the physiological handling of the extra incoming toxic metals, which need to be detoxified and/or excreted. This involves the use of energy that cannot then be used for growth and reproduction. Scope for growth (SFG) is a physiological biomarker integrating different physiological measures to estimate the surplus energy available to an animal for growth and reproduction. It is calculated from the difference between energy assimilated from food and the energy used in respiration. SFG will decrease when energy is required to cope with the extra physiological cost of detoxifying or excreting the large quantities of trace metals entering under conditions of raised trace metal bioavailabilities in metal-contaminated environments. SFG has proved to be a popular biomarker in freshwater systems downstream of metal-contaminated industrial effluent discharges, using the amphipod crustacean *Gammarus pulex* as the test animal of choice (Maltby et al., 1990b). Again, SFG is not a biomarker specific to trace metal exposure, but it is a further biomarker that can be used when investigating whether organisms at a particular metal-contaminated freshwater site are suffering any ecotoxicological effects. Furthermore, SFG is measuring a physiological cost that clearly has an ecological relevance to the well-being, and ultimately the survival, of an exposed animal. SFG is, however, not easy to measure, needing ready access to laboratory facilities to measure physiological parameters such as feeding rate, assimilation efficiency and respiration rate. Since, however, a correlational link has been established between lysosomal membrane stability and SFG, at least in mussels *M. edulis* in the coastal environment (Moore et al., 2006, 2013), it may prove to be enough to use the NRR lysosome biomarker as a proxy for SFG, which is perceived to be more ecologically relevant.

Two morphological biomarkers have also been introduced in this chapter. Freshwater benthic diatoms show structural developmental abnormalities, such as reduced valve size, when exposed in the field high bioavailabilities of toxic metals (Falasco et al., 2009). In Northern Pennine streams, the presence of distorted or twisted cells in more than one-tenth of the population of the diatom *Fragilaria capucina* is considered indicative of ecotoxicologically significant zinc enrichment (Environment Agency, 2009a). Chironomid midge larvae show mouthpart deformities when exposed to high bioavailabilities of trace metals in sediments (Martinez et al., 2003).

To summarise, therefore, biomarkers can provide information on the effects of trace metal exposure on affected organisms. Many biomarkers are now routinely and simply measured. They offer tools not to be ignored when trying to assess whether a raised trace metal bioavailability is actually of any real ecotoxicological significance in a freshwater habitat.

5.17.5 Tolerance

Prologue. The presence of a metal-tolerant population of a particular species in a particular habitat is evidence that local bioavailabilities of that trace metal are high enough to be of ecotoxicological significance, clearly to that species, but potentially also to other members of the local biological community (Luoma, 1977).

Tolerance may be inheritable after selection over several generations or may have been derived by physiological acclimation and be restricted to the one exposed generation. The occurrence of trace-metal tolerant populations of particular freshwater species in specified habitats has been highlighted regularly throughout this chapter.

Amongst the algae, populations of the green alga *Klebsormidium rivulare* in the Gillgill Burn, draining into the River Nent in the Northern Pennines, showed a genetically based tolerance to zinc (Harding and Whitton, 1976; Say and Whitton, 1980). Populations of another chlorophyte, *Stigeoclonium tenue*, in the Gillgill Burn, were also zinc-tolerant (Whitton, 1970; Say and Whitton, 1980). At the other end of the country, in Cornwall, there are also reports of populations of freshwater green algae in the Rivers Hayle and Gannel which have developed metal tolerance, particularly to copper, but also to lead and zinc (Foster, 1977, 1982b). Thus, populations of the green algae *Chlorella vulgaris* and *Microspora stagnorum*, amongst others, from copper-rich sites on the River Hayle had developed copper tolerance, some strains also showing a co-tolerance to lead (Foster, 1982b).

Copper and lead tolerances have also been shown to occur in populations of the isopod crustacean *Proasellus meridianus* in the Rivers Hayle and Gannel (Brown, 1977b, 1978). The Hayle isopod population was tolerant to both copper and lead, although most parts of the River Hayle contain low lead and high copper levels. The *P. meridianus* population in the lead-rich River Gannel, on the other hand, was tolerant only to lead, the lead-rich River Gannel lacking high copper concentrations.

The demonstration of the development of trace metal tolerance in a population of a species in a suspected metal-contaminated habitat is very powerful evidence for the presence of an ecotoxicologically significant bioavailability of a trace metal. Furthermore, the principle that the presence of metal tolerance shows that the local metal exposure is of ecotoxicological significance can be extended to the structure of the biological community. Thus, it is assumed that the selection pressure resulting from an ecotoxicologically significant bioavailability of a trace metal will lead to the elimination of metal-sensitive species and, therefore, an increased average tolerance to that toxic metal amongst remaining species in the resident biological community (Blanck, 2002; Clements and Rohr, 2009). This PICT approach does also make the assumption that differences in tolerance between reference and apparently affected communities can be detected using short-term experiments. This latter requirement correspondingly limits the types of freshwater communities that can be investigated with the PICT approach, usually to fast-growing small organisms such as bacteria and periphyton or other meiofaunal eukaryotes. The application of an eDNA approach would be particularly suitable in any PICT investigation.

5.17.6 Water Framework Directive

Prologue. The Water Framework Directive (WFD) commits European Union member states to achieving good qualitative and quantitative status of all water bodies, including rivers, lakes, estuaries, coastal waters and groundwater. Under the WFD, both ecological status and chemical status need to be at least good to achieve what is termed 'Good Water Status'.

The WFD requires that the current status of water bodies be determined using a formal classification scheme that characterises both the ecological status of each water body based on biological quality elements and the physicochemical elements that support these biological elements in a water body. The WFD identifies priority hazardous substances, considered to present a particularly strong environmental risk. Amongst these priority hazardous substances are the trace metals cadmium, lead, mercury, nickel and tin (as tributyl tin).

Thus, the WFD places a legal obligation on EU nations to use biota to assess the ecological quality of a water body (Jones et al., 2010). Ecological objectives are designed to protect and, where necessary, restore the structure and function of aquatic ecosystems, and thereby safeguard the sustainable use of water resources. Currently, the WFD focuses on community-level biological impacts on macroinvertebrates, phytoplankton, macrophytic plants and benthic algae and fish, depending on the type of water body.

In 2008, the Water Framework Directive–United Kingdom Advisory Group (WFD-UKTAG) published a method statement for the monitoring, assessment and classification of the condition of benthic macroinvertebrate communities in rivers in England, Northern Ireland, Scotland and Wales in accordance with the WFD. This method, based on RIVPACS methodology, is a software tool known as the River Invertebrate Classification Tool (RICT) (WFD-UKTAG, 2008a). RICT assesses the condition of the benthic invertebrate communities using the indices Number of Taxa (NTAXA) and Average Score per Taxon (ASPT) and provides a list of benthic invertebrate taxa (typically but not exclusively insect families) to be recorded. These measured indices are compared against reference values for each index derived for reference sites (identified by environmental variables) producing an ecological equality ratio (EQR) for each of the two indices. As for RIVPACS, RICT too monitors the ecological impact in rivers of organic enrichment on the condition of benthic macroinvertebrate communities.

The WFD has also stimulated the development of other ecological assessment tools, appropriate for other ecological communities identified in the Directive. Diatoms for Assessing River and Lake Ecological Quality (DARLEQ) uses diatoms as proxies for benthic algae Kelly et al., 2008), LEAFPACS assesses macrophytic plants (WFD-UKTAG, 2009), and Fisheries Classification Scheme 2 (FCS2) concerns fish (SNIFFER, 2011). Both DARLEQ and LEAFPACS are again calibrated against gradients of the effects of nutrient/organic enrichment, while FCS2 is not specific to any stressor.

Tools to monitor and predict the specific effects of trace metal contamination on the ecological communities identified in the WFD still await development.

5.17.6.1 Environmental Quality Standards

Prologue. The WFD uses ambient EQS to classify water bodies on the basis of chemical contamination.

Under the WFD, in addition to meeting ecological objectives, any water body also needs to have good chemical status before it can achieve overall Good Water Status. EQS are used to classify water bodies in terms of numerical dissolved chemical concentrations which may be legally enforceable environmental regulations. These standards have been set in an attempt to prevent ecological damage from identified contaminants, including trace metals. Under the WFD, even if ecological objectives are met, the environmental status of a water body will still fail to be classified as good if the average concentration of any trace metal listed as a priority substance exceeds the EQS for that specific metal.

The measurement of water concentrations of selected trace metals, therefore, constitutes a major tool for regulatory agencies in their efforts to achieve the requirements of the WFD. Table 5.18 lists Environmental Quality Standards for trace metals in UK Inland Surface Waters, as defined by various EU directives over the years. These EQS are not all statutory and have the potential to change further over time. The establishment of these EQS for dissolved concentrations of trace metals that might be of eco-toxicological significance recognises that dissolved trace metal concentrations are not necessarily directly transferable into dissolved trace metal bioavailabilities. The presence of a significant concentration of dissolved calcium ions, for example, reduces the bioavailabilities of several dissolved trace metals, as

Table 5.18 Environmental Quality Standards (EQS) for trace metals in UK Inland Surface Waters. After EU Directive 2013/39/EU (Amending Directives 2000/60/EC and 2008/105/EC) and Directive 76/464/EEC.

	Hardness (mg/L $CaCO_3$)	Maximum allowable concentration (µg/L)
Arsenic		50*
Cadmium	0–50	0.45
	50–100	0.6
	100–200	0.9
	>200	1.5
Chromium 3		32
Cobalt		100*
Copper	0–50	1*
	50–100	6*
	100–250	10*
	>250	28*
Iron		1,000*
Lead		14
Manganese		300*
Mercury		0.07
Nickel		34
Silver		0.1*

Table 5.18 **(cont.)**

	Hardness (mg/L CaCO$_3$)	Maximum allowable concentration (µg/L)
Zinc	0–50	8*
	50–100	50*
	100–250	75*
	>250	125*

Note: * Nonstatutory.
Source: WFD-UKTAG (2008b).

does the presence of dissolved organic matter. Therefore, as the hardness of a freshwater sample increases, a higher total dissolved concentration of such a trace metal is needed to bring about a specific threshold quantity of that dissolved metal in bioavailable form. Therefore, EQS (statutory or not) for cadmium, copper and zinc are higher in freshwater bodies with harder water (Table 5.18).

It is interesting to compare the values in Table 5.18 with measured dissolved trace metal concentrations in streams and rivers contaminated with trace metals from previous mining activity (Table 5.1). Not surprisingly given the discussion in this chapter of observed ecotoxicological effects on the resident biological communities in many of these streams and rivers, many of the dissolved concentrations listed in Table 5.1 comfortably exceed the EQS in Table 5.18. Furthermore, many of the streams affected by such trace metal contamination have relatively soft water, and the relevant EQS would be the lower values given in Table 5.18.

Another feature of this comparison is that several of the so-called background or control dissolved trace metal concentrations in British freshwater systems (Tables 5.1 and 5.2) also exceed the EQS in Table 5.18, particularly in the case of zinc. EQS are intended to be protective of water bodies, but they do need to be environmentally realistic so that regulatory agencies can put in place procedures to meet them.

There is also potential for the further refinement of EQS for dissolved trace metals beyond a crude classification on the basis of water hardness (Table 5.18). The dissolved bioavailabilities of trace metals depend on the chemical speciation of the dissolved metals. The free metal ion is often taken as the model for the dissolved form of a trace metal that is taken up by an organism, at least for the many trace metals that exist in solution as positively charged ions.

In the absence of sufficiently sensitive analytical techniques to measure the concentrations of all the different forms (chemical species) of most trace metals dissolved in natural waters, speciation modelling has proved a very popular alternative. Thus the concentrations of the free metal ion and of other chosen low molecular weight metal complexes present in solution in a particular water body can be calculated from chemical first principles. This calculation involves the total concentration of the dissolved trace metal, the concentrations of potential ligands (ions or molecules that form a complex with a dissolved trace metal ion) and the chemical stability constants defining the affinity of each ligand to the metal. One of the most successful and commonly applied trace metal speciation models in freshwaters is the Windermere Humic Aqueous Model (WHAM), which continues to be updated regularly (Tipping, 1994; Tipping et al., 1998, 2011a). The predicted 'bioavailable' concentrations calculated by WHAM are usually better predictors of the ecotoxicological effects of dissolved metals than are total dissolved metal concentrations. Nevertheless, a speciation model such as WHAM concerns only dissolved metal concentrations and dissolved metal bioavailabilities. This is only part of the ecotoxicology story when it comes to freshwater animals, ignoring the ecotoxicologically significant role of metal uptake from the diet in the real contaminated world, including metal uptake from sediments in the case of deposit ingesting invertebrates. It must be said, however, that the trophic bioavailability of trace metals via food chains might initially still indirectly depend on dissolved trace metals entering the food chain, particularly via the primary producers at the base of food webs. If any assessment of the ecotoxicological potential of trace metals in a freshwater body is to include EQS for dissolved metal concentrations, then there is a case to replace these with 'bioavailable' dissolved concentrations predicted by a model such as WHAM.

The same concept of environmental quality guidelines can be extended to freshwater sediments, although this is not required under the Water Framework Directive. Sediment Quality Guidelines for trace metal contaminants are employed, for example, in Canada and in Australia and New Zealand (Table 5.19). Sediment Quality Guidelines are not place in the United Kingdom, although the Environment Agency has explored their potential use in England and Wales, using the Environment Canada approach of Threshold Effect Level (TEL) (Environment Agency, 2008b). The TEL is the concentration below which sediment-associated contaminants are not considered to represent a significant hazard to aquatic organisms. The TEL is used as a trigger for further investigation rather than a strict environmental quality standard.

It is understandable that there are no statutory sediment EQS in the EU, particularly because of the many factors that affect the bioavailability, and ultimately ecotoxicity, of sediment-associated trace metals, other than total trace metal concentrations. Such factors include sediment particle size and organic matter content. Also important is the choice of the sediment sample to be analysed, in the light of preceding discussion on the flaws of the AVS approach to measuring the ecotoxicity of sediment-associated metals. It is important that what is actually analysed is the oxygenated part of the sediment occupied by burrowing animals, as opposed to metal-rich anoxic regions of sediment that are avoided by the infauna.

Again with the caveat that a high total concentration of a trace metal in a freshwater sediment does not necessarily translate directly into a high trace metal bioavailability, it is still instructive to compare the threshold sediment concentrations listed in Table 5.19 and real sediment concentrations in British

Table 5.19 **Threshold EQS (µg/g dry weight) for trace metals in freshwater sediments for Canada (Interim Sediment Quality Guidelines ISQG 1999),[1] and for Australia and New Zealand (Low Trigger Value, 10% probability of effect 2000).[2]**

	Canada ISQG (µg/g)	Australia and New Zealand Low trigger value (µg/g)
Arsenic	5.9	20
Cadmium	0.6	1.5
Chromium	37.3	80
Copper	35.7	65
Lead	35.0	50
Mercury	0.17	0.15
Nickel		21
Silver		1
Tributyl Tin (µg/L Sn)		5
Zinc	123	200

Sources: Canadian Council of Ministers of the Environment (1999a);[1] ANZECC and ARMCANZ (2000).[2]

freshwater bodies affected by mining (Table 5.4). It is striking, but not surprising, how so many of the mining-affected streams and rivers in Britain have sediments with trace metal concentrations way in excess any EQS in Canada or in Australia and New

Zealand (Table 5.19). As in the case of dissolved trace metal concentrations, it is also apparent that many sediment concentrations of trace metals that are considered to be background, for example for Cornwall streams (Table 5.4), are above the EQS quoted in Table 5.19. This apparent contradiction is another reason, perhaps, why sediment EQS have not been embraced enthusiastically by environmental regulatory authorities in Britain and elsewhere in Europe.

5.17.7 Weight of Evidence (WOE)

Prologue. A weight of evidence (WOE) approach combines evidence from different lines of evidence to determine the ecotoxicological status of a metal-contaminated habitat.

So where are we when faced with the question of whether the biota of a freshwater habitat are suffering from the more subtle ecotoxicological effects of trace metal contamination?

Understandably, the reflex first step is to measure trace metal concentrations, but in what? In countries in the European Union, the WFD requires measurements of dissolved trace metal concentrations to be matched against EQS. Such analyses are not simple and not cheap, given the very low concentrations typically involved (even in contaminated waters). The dissolved concentrations will also be very variable, not least with local rainfall patterns, and will need to be measured often. Dissolved trace metal concentrations also suffer from the flaw that they are measures of total, not bioavailable, concentrations, although chemical speciation modelling will overcome this deficiency to a degree. Unavoidably, though, dissolved concentrations essentially ignore the often major uptake route of toxic metal entry into freshwater animals, which is the diet.

What about sediment concentrations of trace metals? These are certainly informative to a point, but the difficulty of relating total concentrations to bioavailabilities of sediment-associated metals has been discussed earlier in this book. And bioavailabilities to what? To burrowing animals, to rooted plants? At the

bottom line, Sediment Quality Guidelines have not been introduced under the WFD.

The most informative metal concentration measurements to take are those in biomonitors, with the huge advantage of being relative measures of only the bioavailable fractions of local trace metal levels. The same question can be asked: bioavailable to what? This time, however, there is an answer: bioavailable to that organism chosen as the biomonitor, integrated across all metal uptake routes. These uptake routes will be dissolved metal for a bryophyte, dissolved and dietary metal for a freshwater invertebrate, and dietary metal for a bird or mammal living in association with the freshwater body under investigation. The diet will be the sediment for sediment-ingesting invertebrates. Dissolved metal sources will include sediment pore water to some degree in animals burrowing in the sediment, mixed to differing extents by an irrigation current of oxygenated water from the water column above. The careful selection of a suite of biomonitors will cover all potential metal sources to local freshwater plants and animals.

The use of a selected suite of biomonitors therefore answers the question of whether there is an atypically high level of potentially ecotoxicologically significant metal contamination in a freshwater habitat. But is that atypically high metal bioavailability having an ecotoxicologically significant effect? How should we now go about answering the 'So what' question? What techniques should we apply?

There is no single answer. A WOE approach is now favoured, combining evidence from different lines of evidence to determine the ecotoxicological status of a metal-contaminated habitat (Chapman, 2007; Benedetti et al., 2011). WOE determinations may include both chemical and biological measurements, including field components which may be observational or involve experimental manipulation.

What might such a WOE approach look like in the case of a stream suspected to be metal-contaminated to a significant extent? Classically, lines of evidence would include dissolved metal concentrations, sediment metal concentrations and analyses of benthic macroinvertebrate communities, perhaps involving the calculation of biotic indices. Can this be simplified? The short answer is, 'Yes'.

Initial non-quantitative sampling of the biota present might show up the presence of one or more of the more trace metal-resistant species identified at the beginning of this section. It will also provide the information needed to choose the most appropriate suite of trace metal biomonitors. The key step is to identify a biomonitor for which there is information linking its accumulated metal concentrations with an independent ecotoxicological effect on other more sensitive members of the expected biological communities. Are caddisflies of the genus *Hydropsyche* or the species *Plectrocnemia conspersa* present? If so, the measurement of accumulated trace metal concentrations in only about 15 specimens will provide information on local trace metal bioavailabilities and especially on potential ecological effects on local mayfly larval communities. When more than one of the trace metals present may be an ecotoxicological driver, the use of CCU of toxicity can integrate and partition the ecotoxicological effect of each trace metal (Clements et al., 2000; Clements, 2004). Depending on the degree of metal contamination, other possible biomonitors in this context include stoneflies, blackflies and species of the mayfly genus *Rhithrogena* (De Jonge et al., 2013). In the absence of any of these caddisflies, stoneflies or mayflies, for example at sites further downstream or with more sediment, the accumulated trace metal concentrations of chironomid midges or tubificid oligochaetes may serve to identify thresholds of trace metal bioavailabilities affecting local biological communities (Bervoets et al., 2016). Far downstream in industrially contaminated rivers, the zebra mussel *Dreissena polymorpha* might be used in the same way (De Jonge et al., 2012).

Another potential line of evidence in the WOE approach, as yet not specified in the WFD, is the use of biomarkers. Biomarker methodologies are now well established. Biomarkers at the lower levels of biological organisation are the most sensitive and usually most easily measured. Excitingly, these biomarkers are increasingly being linked to biomarkers at higher levels of biological organisation with clear ecotoxicological relevance. Biomarkers should certainly be part of any WOE approach in this context. If there is any evidence of the local development of tolerance, be it of populations or of communities (PICT), this is very relevant evidence to be included in a WOE approach, although the search for such tolerance might not be a high priority in the initial design of the WOE package.

The way forward in the ecotoxicological assessment of mining-affected streams and rivers is, therefore, via an integrated approach involving a multiplicity of tools from different scientific disciplines, reflected in a carefully designed weight of evidence approach.

5.18 Human Health

Prologue. There are two possible routes by which trace metals in freshwater systems might interact with human health. The first is through drinking water, and the second through the human diet.

By law, drinking water in the United Kingdom must be 'wholesome' at the point of supply. Wholesomeness is defined by reference to drinking water standards (DWS), which have been listed in Table 4.55. There is good agreement worldwide on the setting of these standards, which are typically based on WHO guidelines. The toxic metals derived from British mining activities that have been concerning us in this chapter have been arsenic, cadmium, copper, lead and zinc. Comparison of UK DWS (Table 4.55) against measured dissolved concentrations in British streams and rivers affected by mining (Table 5.1) shows that the latter may be well in excess of the former. There are no UK DWS for zinc, indicating a lack of significant concern about the toxicity of zinc in drinking water to human consumers. In practice, these exceedances are not points of concern, because water from such contaminated water courses would not be extracted for public drinking water supplies.

In Britain, few freshwater organisms enter the human food chain, in contrast to their terrestrial or coastal counterparts. We do eat the flesh (muscle tissue) of salmonids (trout and salmon) and of eels, and even these fish spend significant parts of their lives out at sea. Furthermore, fish muscle tissue is generally low in accumulated trace metal concentrations, compared, for example, to storage organs such as the liver or kidney. The exception, of course, is mercury, which does accumulate in fish muscle. Furthermore, as we have seen in this chapter in risk analyses for bird and other mammalian piscivores, the vast majority of the mercury accumulated by fish is in the form of methyl mercury, the most toxic chemical form of mercury. The origin of mercury in British fish does not lie in historic mining activity. Rather, the source of mercury pollution has historically been industrial, and dissemination of mercury into the environment has been predominantly via the atmosphere.

Given that salmonids are generally considered to be fish of clean, unpolluted freshwaters, it is eels that are of potential concern here (Mason, 1987). The maximum concentration of mercury in the muscle meat of eels considered safe by the EU is 1.0 µg/g wet weight (Table 7.16). Data on average mercury concentrations in the muscle tissue of eels collected from British rivers have been compiled in Table 5.16. None of these average mercury concentrations in eel muscle reached 1.0 µg/g wet weight, the highest average concentration being 0.39 µg/g wet weight in a dataset for East Anglian rivers in the 1980s (Barak and Mason, 1990). Several of the mercury concentrations contributing to the average concentrations quoted in Table 5.16 did, however, exceed 1.0 µg/g wet weight, in one case reaching as high as 1.30 µg/g in eels from the River Brain (Barak and Mason, 1990). As discussed previously, mercury concentrations in British freshwater fish have dropped greatly since the 1980s (Table 5.16). It is unlikely, therefore, that eels collected in British rivers today would have mercury concentrations in their muscles that would be of any concern for human health.

So, while trace metals derived from historic mining activities have left an ongoing legacy of ecotoxicological impact on the biology of some of our rivers and streams, the resulting high trace metal bioavailabilities in these metal-contaminated water courses do not appear to present any significant toxicological threat to humans.

In the next chapter, we turn our attention downstream to the estuaries that are the interfaces between our rivers and our coastal seas.

6 Estuaries

Box 6.1 Definitions

antifouling Use of a chemical agent, such as a paint bearing copper or tributyl tin, to prevent encrustation of manmade objects by aquatic fouling organisms such as algae, barnacles or mussels.

beachhopper A talitrid amphipod crustacean to be found in the strandline of many British shores, feeding on decaying seaweed and able to jump. For example, *Orchestia gammarellus*.

chlor-alkali process An industrial process for the electrolysis of a solution of salt to produce chlorine and sodium hydroxide for industrial use. The process uses a mercury cell in which the salt solution floats on a layer of mercury and results in the release of mercury into the environment.

conservative behaviour A description of the distribution of dissolved concentrations of trace metals in a water body. Along an estuary, dissolved concentrations would change in line with seawater dilution and thus typically fall with increasing salinity. In an ocean, the dissolved trace metal concentration would show a constant concentration relative to salinity and would not, therefore, vary with depth.

deposit feeder An aquatic animal that feeds by ingesting organic particles that have settled on, or become incorporated into, bottom sediments. Deposit feeders, a subgroup of *detritivores*, include whole sediment ingestors such as the lugworm *Arenicola marina*, as well as animals selectively feeding on organic material newly settled on the top of the sediment, as in the case of bivalves such as *Scrobicularia plana*.

imposex The imposition of male sexual characters onto females, as exemplified by the effect of tributyl tin on marine gastropod molluscs.

invasive species A non-indigenous or nonnative species. The term is often restricted to species that have been introduced and adversely affect a habitat ecologically and/or economically.

lugworm A large coastal polychaete worm, *Arenicola marina*, which lives in burrows in intertidal sediments, producing coiled casts of defaecated material observable at low tide.

osmosis The net movement of solvent (typically water) molecules across a partially permeable (cell) membrane into a region of higher concentration of dissolved substances (e.g., ions, organic compounds) from a region of lower concentration.

osmotic pressure A measure of the tendency of a solution (for example, in a cell or body fluid) to take in water by osmosis. Strictly the pressure needed to stop the inward flow of water across a semipermeable membrane by osmosis.

prawn The common name for many small decapod crustaceans that can walk and swim. The term 'prawn' is applied variously and inconsistently (particularly in different countries) and does not apply to a single defined taxon of decapods. Nevertheless, the term 'prawn' is often applied in Britain to species of the palaemonid genera *Palaemon* and *Palaemonetes*.

ragworms Coastal polychaete worms, usually of the genera *Nereis, Hediste* or their relatives in the family *Nereididae*, that may burrow in sediments but typically also move using an *S*-shaped wave motion, whether swimming or in contact with the substratum. For example, the estuarine *Hediste diversicolor*, previously called *Nereis diversicolor*.

Ramsar site A wetland site designated to be of international importance under the Ramsar Convention, an intergovernmental environmental treaty on the conservation of wetlands and their resources.

salinity A measure of the total concentration of dissolved salts in seawater, most of which consist of sodium chloride. Salinity is now quoted without units, although historically it was defined in terms of the mass in grams of all dissolved salts in 1 kilogram of evaporated seawater, or parts per thousand (ppt).

sandhopper A talitrid amphipod crustacean to be found burrowed in the sand at the top of sandy shores, feeding on decaying seaweed in the strandline and able to jump. For example, *Talitrus saltator*.

Sediment Quality Triad (SQT) A methodology for the ecotoxicological assessment of the chemical contamination of a sediment, involving at least three measures: (i) sediment concentrations of toxic chemicals, (ii) toxicity testing of sediment samples, and (iii) investigation of the biological community composition.

shipworm A marine bivalve mollusc, typically of the genus *Teredo*, that bores into submerged wood. Its elongated body in the burrow is unprotected by the reduced shell valves, giving the appearance of a worm.

shrimp The common name for many small decapod crustaceans that can walk and swim. The term 'shrimp' is applied variously and inconsistently (particularly in different countries) and does not apply to a single defined taxon of decapods. Nevertheless, the term 'shrimp' is often applied in Britain to species of the genera *Crangon* and *Pandalus*.

suspension feeder An aquatic animal that feeds by ingesting particulate organic suspended in water, such as plankton and/or detritus particles.

6.1 Introduction

Prologue. The streams and rivers that were the subject of the previous chapter discharge into the sea through estuaries. It is the natural history of trace metals in estuaries that occupies this chapter, arranged in a now familiar pattern – presentation of trace metal concentration data, case studies of the estuaries of some of the metal-contaminated rivers met in the previous chapter and a review of the natural history of metals in important estuarine organisms. There is then an exploration of the effects of trace metal exposure on estuarine communities, calling on case studies in British estuaries. Finally, an applied aspect is considered – the use of trace metals as antifouling agents and the ecotoxicological consequences, particularly of the now notorious antifouling agent tributyl tin (TBT).

Estuaries take various geomorphological forms (Barnes, 1974). In the British Isles, for example, we find drowned river valleys, formed by a rise in sea level after the melting of the ice formed in the last glaciation. The Fal estuary system in Cornwall is one such example. We shall delve in some detail into the natural history of trace metals in one branch of the Fal Estuary, Restronguet Creek, receiving water from the River Carnon, already encountered in Chapter 5. Bar-built estuaries develop behind shingle and sand bars formed parallel to the coast, limiting water exchange with the sea, as in the case of the Humber Estuary with a history of industrial metal contamination. Another type of estuary falls into the general category of a fjord, a glacially overdeepened valley into which seawater now penetrates. The sea lochs of Scotland fit here. A definition of an estuary that would accommodate all variants would be 'a region containing a volume of water of mixed origin, derived partly from a discharging river system and partly from the adjacent sea; the region usually being partially enclosed by a land mass' (Barnes, 1974: 2).

It is the mixing of river water and seawater that makes life in estuaries stressful for organisms. Salinity varies from essentially zero at the top freshwater end of an estuary to about 35 at the marine end. Crudely, salinity is a measure of the total concentration of dissolved salts in seawater, the vast majority of which consist of sodium chloride. At a salinity of 35, sodium contributes about 10.8 parts per thousand (ppt), and chloride about 19.4 ppt, an impressive total of 30.2 in the 35 (Bruland, 1983). Furthermore, the gradient of salinities does not stand still. It will move up and down the estuary with the incoming and ebbing tide. Such fluctuation causes osmotic problems for the biota that live there.

The effect on the fauna and flora is that estuaries contain far fewer species than do the freshwater and marine environments on either side (Barnes, 1974). On the other hand, if a species can cope with the variation in external osmotic pressure, then life in the estuary is good. Estuarine faunas and floras show low biodiversity, but high production and abundances of the organisms present (Barnes, 1974).

In the context of this book, estuaries are essentially contaminant traps, and these contaminants include trace metals. In practice, estuaries often contain sediment concentrations of trace metals that are higher than those in the freshwater and marine environments on either side. One cause is the oscillating nature of the waters of an estuary, slowing down the passage through the estuary of dissolved metals and metals associated with suspended particles. This promotes the deposition of suspended sediment particles with their metal loads onto the bottom of the estuary. Of greater significance, perhaps, is the effect of the physicochemical environment of the estuary. The pH of seawater is generally limited to a range between 7.5 and 8.4, averaging about 8.1 on the slightly alkaline side of neutral. Many freshwaters will have a pH on the acidic side of neutral, as for example in the cases of the River Carnon (Bryan and Gibbs, 1983) and Afon Goch (Foster et al., 1978) met previously. An acidic pH increases the solubility of many trace metals, as in the case of iron. However, when the dissolved iron ions encounter the more alkaline pH in an estuary, caused by the admixture of seawater, they precipitate as iron oxides and hydroxides (Bryan and Gibbs, 1983). This precipitation correspondingly increases the iron concentration in the now estuarine sediment. Furthermore, the formation of particulate iron oxides and hydroxides provides surfaces for the adsorption of other particular trace metals, including arsenic and copper. These trace metals come out of solution and are deposited in the sediments in association with the particulate iron oxides and hydroxides (Bryan and Gibbs, 1983; Langston, 1983; Bryan and Langston, 1992). Complexation with humic acids in freshwater will also promote the solubility of, for example, copper. Humic acids will also precipitate in alkaline estuarine waters and take their complexed trace metals down with them into the sediment (Bryan and Langston, 1992).

The biology of trace metals in estuarine organisms again follows the general principles already met in the terrestrial and freshwater environments. The organisms will be different, but the mechanisms of trace metal uptake and accumulation with associated detoxification will be the same. We will be considering the trophic transfer of the metals along different food chains, culminating in different suites of fish and birds. Many of these fish and birds, and also invertebrates such as crabs, will be temporary residents of the estuarine habitat. Flounder will come in and out with the tide to feed on burrowing estuarine invertebrates when the estuary is inundated. On the other hand, waders will feed on the same estuarine mudflats when the tide is out. Many such birds are migratory, and so will only be present at particular times of the year as they migrate to and fro. British estuaries are key to the survival of many of these migratory waders on a world scale. Estuarine organisms also enter the human food chain, in the form, for example, of cockles, mussels and eels. How do high bioavailabilities of trace metals affect such food chains?

In this chapter, we shall, not unexpectedly, cover the estuaries of some of the mining-contaminated rivers met in some detail in the previous chapter. In Southwest England, the River Carnon flows into Restronguet Creek, perhaps the best studied of British metal-contaminated estuaries. Other southwestern estuaries to be considered include those of the Rivers

Hayle and Gannel on the north coast of Cornwall. The Rivers East Looe and West Looe share a common estuary at Looe on the south Cornish coast. The River Tamar, marking part of the border between Cornwall and Devon, is joined along its estuary by the adjacent River Tavy draining Dartmoor. It is a remarkable feature of the estuaries of these mining-affected Cornish rivers just how much sediment is deposited in their estuaries, considering the actual size of the rivers. As you drive into Newquay, look left and consider the very large areas of salt marsh occupying the extensive sediment banks associated with the estuary. How could such a small river deposit so much sediment? The answer, of course, lies in the old mining practices of washing large amounts of waste ore material into rivers, in effect filling their estuaries with sediments with raised trace metal contents. Another British estuary with a mining legacy is Dulas Bay on Anglesey, the estuary of the Afon Goch draining Parys Mountain.

Britain also has estuaries contaminated with trace metal-rich discharges from industry as opposed to mining. Prominent amongst these are the Thames, Severn, Mersey and Humber Estuaries, but also affected by industrial contamination have been the estuaries of the Rivers Tawe, Neath and Loughor in south Wales and Poole Harbour on the Dorset coast.

6.2 Metal Concentrations in the Waters of Estuaries

Prologue. This section is concerned with the concentrations of trace metals in the water columns (in dissolved {<0.45 µm} or particulate form) of British estuaries subdivided into those affected by upstream mining, those affected by industrial discharges and those considered uncontaminated.

As in freshwater, trace metals in the water column are considered to be dissolved if they pass through a filter of 0.45 µm. Another component of the trace metals in the water will be associated with particles, many temporarily before deposition in the sediments, but some in smaller colloidal forms that stay in suspension. Together, the dissolved and particulate components make up the total trace metal concentration in the water of an estuary.

Table 6.1 quotes dissolved trace metal concentrations in British estuaries, subdivided into three categories: those affected by historical mining upstream, those receiving industrial discharges and those considered relatively uncontaminated by trace metals.

As to be expected from discussion in earlier chapters, the waters of Restronguet Creek, the estuary of the River Carnon, contain remarkably high dissolved concentrations of a range of trace metals, particularly zinc, iron and manganese, but also copper, arsenic, cadmium and lead (Table 6.1). Many of these dissolved trace metal concentrations in Restronguet Creek far exceed those in estuaries considered contaminated by industrial effluent discharges (Table 6.1). But Restronguet Creek is atypical, even for mining-affected estuaries, and the dissolved trace metal concentrations in many industrially affected estuaries match those in other mining-contaminated estuaries (Table 6.1).

In the case of the estuaries contaminated with trace metals derived from historical mining, the source of metals into the estuary is upstream. In industrially contaminated estuaries, for example the Thames or Severn Estuaries, on the other hand, discharge of metal-contaminated effluent may occur lower down the estuary. Any mid-estuary entry of trace metals would disrupt any expected downstream gradient in dissolved metal concentrations from top to bottom of the estuary with increasing seawater dilution. Thus in the Thames Estuary, regions of higher salinity may have higher dissolved trace metal concentrations than regions of lower salinity further upstream (Table 6.1) (McEvoy et al., 2000).

One might expect there to be a linear gradient of dissolved trace metal concentrations down an estuary receiving metal input from mining activity upstream in the river. Dissolved concentrations would fall in line with seawater dilution and thus with increased salinity. Such a distribution of dissolved trace metal concentrations is described as conservative (Holliday and Liss, 1976; Bruland, 1983; Howard et al., 1984). Conservative distributions of dissolved trace metals

down estuaries do occur. Thus dissolved concentrations of cadmium, manganese and zinc down Restronguet Creek are almost linearly related to salinity (Bryan and Gibbs, 1983). Similarly, dissolved manganese and zinc behave conservatively down the uncontaminated Beaulieu Estuary in Hampshire in southern England (Holliday and Liss, 1976). Such conservative behaviour is not, however, the rule. As mentioned previously, dissolved iron typically precipitates from solution in the forms of oxides and hydroxides on encountering raised alkalinity in an estuary. So the behaviour of iron is certainly not conservative, dissolved iron concentrations showing a sharp drop in the upper regions of estuaries at salinities below 10 (Bruland, 1983; Holliday and Liss, 1976). And the precipitating iron oxides and hydroxides will offer a surface for the adsorption of the likes of arsenic, copper, lead and silver, which, therefore, come out of solution in a similar non-linear fashion to the dissolved iron (Bryan and Gibbs, 1983; Langston, 1983; Bryan and Langston, 1992). Thus in Restronguet Creek, dissolved iron, arsenic and copper behave non-conservatively (Bryan and Gibbs, 1983), as do iron and arsenic in the Beaulieu Estuary (Holliday and Liss, 1976; Howard et al., 1984).

Restronguet Creek also provides an example of another change in the physicochemical form of dissolved arsenic down an estuary (Langston, 1983; Klumpp and Peterson, 1979). Forty per cent of the dissolved inorganic arsenic ions in the acidic waters of the River Carnon is in the form of arsenite, the more toxic reduced (As3) form of arsenic (Langston, 1983). On entry into the upper estuary, the proportion of arsenite drops to 10% (Langston, 1983). The proportion of dissolved inorganic arsenic as the oxidised form arsenate (As5) increases down the estuary, until it is the only dissolved inorganic form of arsenic in full seawater at the bottom of the estuary (Klumpp and Peterson, 1979).

The relative contributions of the dissolved and particulate fractions to the total concentrations of trace metals in the water of an estuary vary both between metals and between estuaries. As would be expected, the proportion of iron in the water column that is in particulate form is very high at the point

Table 6.1 Mean concentrations (μg/L) of dissolved (0.45 μm filtered) trace metals in British estuaries affected by (a) mining contamination, (b) industrial contamination or (c) considered uncontaminated.

	Salinity	Ag	As	Cd	Co	Cu	Fe	Mn	Ni	Pb	Zn
(a) Mining											
SW England											
Restronguet Creek											
1971[1]				13		65	107	149		250	570
1972–73[2]	15–34			<0.1–1.2	<1–6	5–26	9–60	5–113	3–18	<2–3	17–300
1975–77[3]			3–42	2.8		16					285
1980–81[4,5]	18–34			0.7–38		3–176	3–2,490	15–1,510			22–20,500
ca 1981[6]	0–34		2–110								
2003[7]	0–34	0.002–0.02									
Tamar											
1976–81[8]	0–32					1–6					2–14
ca 1981[6]	0–30		2–6								
1984[9]	0–5					2–13			5–12		2–8
2003[7]	0–34	<0.0005–0.002									
Wales											
Conwy											
1971[4]	0–32			1		4		15		7	16
1976[10]	0–32										5–55
1998[11]	0–33					0.1–2.5	1–80	1–12	0.5–4		2–32

(b) Industrial

Thames										
1986–87[12]	1–25			0.43		31.3		17.3	16.3	92
1995[12]	1–25			0.32		10.7		6.3	9.9	29
1997[13]	0.1–2	0.12	3.6	0.05	0.31	12	14	8	1.55	53
1997[13]	5–12	0.23	5.4	0.16	0.24	11	5	8	1.28	64
1997[13]	14–19	0.15	5.3	0.23	0.11	12	6	9	0.64	58
1997[13]	21–32	0.16	4.4	0.21	0.07	7	2	7	0.33	42
2001[14]	0–34		0.9–2.6	0.04–0.30	<0.01–1.4	0.9–4.8	<2–15	0.3–4.4	0.01–1.06	9–36
Severn										
1970[15]				1–6					0.8–2.5	21–52
1975–79[16]	0–32			0.5–3		1.6–4.4	1–40	1.5–3.6	1.2–2.9	10–23
1988[17]	0–31	1.5–4.5		0.1–0.5		1–5		0.2–8	<0.3–0.6	1–18
Mersey										
1991–92[18]				0.03–0.05		1.3–3.3	1–11	0.9–9.4	0.11–0.88	2–16
Humber										
ca. 1981[6]										
1991–92[18]				0.05–0.22		0.8–3.6	1–59	0.9–6.3	0.02–0.61	4–15
Swansea Bay										
1978–79[19]				0.32–0.63		4–30	31–853	4–61	1.0–4.3	63–1,370
Tyne										
1991–92[18]				0.01–0.13		0.3–1.6	1–120	0.3–2.8	0.06–1.10	1–22
Tees										
1980[20]	0–33			0.1–4		0.5–10	0.6–127	3–305		1–117
1991–92[18]				0.02–0.06		0.5–10	4–120	0.2–1.0	0.05–0.82	1–14
Poole Harbour										
1971[1]				1		6	28	102	47	26
1973[21]				0.16–0.44		2.0–2.5	6–11		5–13	7–9

Table 6.1 (cont.)

	Salinity	Ag	As	Cd	Co	Cu	Fe	Mn	Ni	Pb	Zn
(c) Uncontaminated											
SW & S England											
Helford											
1971[1]				<1		11	217	16		15	28
1972–73[2]	26–34			0.1–0.4	<1–2	2–6	4–34	3–11	1–11	<2–4	7–27
Avon (Devon)											
ca. 1981[6]	0–34		1.0–1.5								
Beaulieu											
1974[22]	0–30						1–420	10–130			
1980–82[23]	11–34		0.2–1.2								5–65

Note: Single values represent a mean value, and ranges of means are included where available.

Sources: Thornton et al. (1975);[1] Boyden et al. (1979);[2] Klumpp and Peterson (1979);[3] Bryan and Gibbs (1983);[4] Bryan et al. (1985);[5] Langston (1983);[6] Tappin et al. (2010);[7] Ackroyd et al. (1986);[8] Morris (1986);[9] Elderfield et al. (1979);[10] Zhou et al. (2003);[11] Power et al. (1999);[12] McEvoy et al. (2000);[13] Langston et al. (2004);[14] Butterworth et al. (1972);[15] Morris (1984);[16] Apte et al. (1990);[17] Law et al. (1994);[18] Chubb et al. (1980);[19] Taylor (1982);[20] Holliday and Liss (1976);[21] Boyden (1975);[20] Howard et al. (1984).[23]

when dissolved iron is converted to particulate iron oxides and hydroxides. This proportion remains high in that region of the estuary until the precipitating particulate iron has reached the bottom to be incorporated into the sediments. Similarly, the other trace metals such as arsenic and copper that adsorb onto the particulate iron will similarly exist predominantly as particulate metal in the water column in that part of the estuary until deposited into the sediment. Thus in Restronguet Creek, about 95% of the total iron in the water is in particulate suspended form (Boyden et al., 1979). Between 40 and 57% of the total water concentration of copper is correspondingly in particle form, as is about 58% of the water concentration of lead at low tide (Boyden et al., 1979). In the case of two metals that do not adsorb onto the precipitating iron in Restronguet Creek, all of the manganese and most of the zinc in the water column is in dissolved form (Boyden et al., 1979).

Table 6.2 provides examples of the relative concentrations of trace metals in dissolved and particulate form in the waters of selected British estuaries contaminated by industrial discharge. It is striking how the particulate form of all the trace metals listed dominates the trace metal in the water column in the industrialised estuaries sampled between 1991 and 1992 (Table 6.2) (Laslett, 1995). The dissolved trace metal component is of increased significance in the cases of As, Cd, Ni and Zn in the Thames Estuary in 2001 (Table 6.2) (Langston et al., 2004).

6.2.1 Speciation and Bioavailability

Prologue. The complexation of dissolved trace metals in estuarine waters by both inorganic anions and dissolved organic matter reduces the availability of the free metal ion and hence the bioavailability of dissolved trace metal.

Chapter 3 discussed how the complexation of dissolved trace metal ions by inorganic anions such as chloride will decrease the availabilities of dissolved free trace metal cations, with knock-on effects on the bioavailabilities of dissolved trace metals to aquatic organisms. A major inorganic complexing anion of many trace metal cations is indeed the chloride ion, the very anion that increases hugely in concentration down an estuary to reach dissolved concentrations of 19.4 ppt in seawater of salinity 35. Thus, the chloride ion is present in estuaries at dissolved concentrations orders of magnitudes greater than the dissolved concentrations of any trace metal, with the increasing potential to complex many dissolved metal cations down the estuary from freshwater to full seawater.

Table 6.3 repeats Table 3.1 for convenience of reference here. It shows the dissolved trace metals that are complexed by dissolved chloride ions (Bruland, 1983; Laslett, 1995). Chloride complexation is particularly important for the dissolved cations of cadmium, mercury and silver (Table 6.3), with less than 3% of dissolved cadmium in the form of the free Cd^{2+} ion in full seawater (Figure 3.2) (Luoma and Rainbow, 2008). Other dissolved anions present in seawater will also increase in dissolved concentration down an estuary and may also play a part in the inorganic complexation of trace metal ions (Table 6.3).

In the Thames Estuary, salinity varies from 1 to 12 at Woolwich, 13 to 27 at Gravesend to above 30 at Canvey Island (Anon, 1988). These intersite differences in salinity are certainly strong enough to be correlated with changes in dissolved trace metal bioavailabilities to estuarine organisms (Luoma and Rainbow, 2008). This physicochemical effect decreasing the dissolved bioavailabilities of many of the trace metals with increasing salinity would enhance any expected effect of sequential dilution of upstream metal contamination, leading to decreased dissolved bioavailabilities downstream.

In addition to the complexation of dissolved trace metal ions by inorganic complexing agents such as chloride anions, the ions of a few dissolved trace metals, particularly copper, are also susceptible to complexation by dissolved organic molecules in seawater (Förstner and Wittmann, 1983). Many dissolved trace metal ions in freshwaters may be complexed by humates, a form of dissolved organic matter. With the exception of copper, and to a lesser extent mercury, there is a rapid decrease in the proportion of dissolved trace metals bound with humates as the salinity

Table 6.2 Mean concentrations (μg/L) of trace metals in dissolved (<0.45 μm) and particulate (>0.45 μm) forms in the water columns of British estuaries contaminated by industrial discharge.

	As	Cd	Co	Cu	Fe	Mn	Ni	Pb	Zn
Thames									
2001[1]									
Dissolved	0.93–2.2	0.04–0.30	0–0.20	1.7–4.8	17–242	0–13	0.1–4.4	0.01–1.1	9.2–25
Particulate	0.17–1.9	0–0.13	0.03–1.3	0.6–17	72–2,310	6–175	1.3–6.3	1.4–32	0–26
Mersey									
1991–92[2]									
Dissolved		0.01–0.06		1.3–3.3		0.1–20	0.8–9.4	0.03–0.88	1.5–17
Particulate		0.49–1.2		24–56		280–970	19–55	33–140	310–450
Humber									
1991–92[2]									
Dissolved		0.05–0.22		0.8–3.6		0.8–59	0.9–6.3	0.02–0.62	3.6–15
Particulate		0.11–0.40		36–55		820–1,400	42–51	140–190	310–500
Tyne									
1991–92[2]									
Dissolved		0.01–0.13		0.3–1.6		1.3–130	0.3–2.8	0.06–1.10	0.5–25
Particulate		0.27–6.1		27–200		230–900	27–50	44–650	80–1,400
Tees									
1991–92[2]									
Dissolved		0.02–0.10		0.5–10		4.0–120	0.2–1.0	0.05–0.82	0.7–14
Particulate		0.64–5.4		61–140		270–680	19–28	110–320	190–1,100

Sources: Langston et al. (2004);[1] Laslett (1995).[2]

Table 6.3 **Inorganic complexation of dissolved metals in oxygenated seawater.**

Metal		Probable main dissolved species
Ag	Silver	$AgCl_2^-$, $AgCl_4^{3-}$, $AgCl_3^{2-}$
Al	Aluminium	$Al(OH)_4^-$, $Al(OH)_3^0$
As	Arsenic	$HAsO_4^{2-}$
Ca	Calcium	Ca^{2+}
Cd	Cadmium	$CdCl_2^0$, $CdCl^+$, $CdCl_3^-$
Co	Cobalt	Co^{2+}, $CoCO_3^0$, $CoCl^+$
Cr	Chromium	CrO_4^{2-}, $NaCrO_4^-$
Cu	Copper	$CuCO_3^0$, $Cu(OH)^+$, Cu^{2+}
Fe	Iron	$Fe(OH)_3^0$, $Fe(OH)_2^+$
Hg	Mercury	$HgCl_4^{2-}$, $HgCl_3Br^{2-}$, $HgCl_3^-$
K	Potassium	K^+
Mg	Magnesium	Mg^{2+}
Mn	Manganese	Mn^{2+}, $MnCl^+$
Mo	Molybdenum	MoO_4^{2-}
Na	Sodium	Na^+
Ni	Nickel	Ni^{2+}, $NiCO_3^0$, $NiCl^+$
Pb	Lead	$PbCO_3^0$, $Pb(CO_3)_2^{2-}$, $PbCl^+$
Se	Selenium	SeO_4^{2-}, SeO_3^{2-}, $HSeO_3^-$
V	Vanadium	HVO_4^{2-}, $H_2VO_4^-$, $NaHVO_4^-$
Zn	Zinc	Zn^{2+}, $Zn(OH)^+$, $ZnCO_3^0$, $ZnCl^+$

Source: Bruland (1983).

increases when river waters enter estuaries (Mantoura et al., 1978). This drop is principally caused by competition for the humates by the vast excess of dissolved calcium and magnesium ions present in the seawater mixing with the freshwater, taking up almost all the complexing capacity of the humates (Mantoura et al., 1978). Dissolved humates will also precipitate in the estuary (Bryan and Langston, 1992). Only copper ions remain complexed with dissolved humates as the salinity increases down an estuary, and typically 10% of dissolved copper ions may still be complexed by humates at the full seawater end of the estuary (Mantoura et al., 1978). As in the case of inorganic complexation, any organic complexation of dissolved trace metal ions also reduces the dissolved bioavailabilities of trace metals to aquatic organisms.

6.3 Metal Concentrations in Estuarine Sediments

Prologue. Presented here are the concentrations of trace metals in the sediments of British estuaries subdivided into those affected by upstream mining, those affected by industrial discharges and those considered uncontaminated. Sediment-associated trace metals have typically originated from the water column, and sediment concentrations reflect the quantities of these inputs. Factors affecting sediment concentrations of trace metals include sediment grain size, organic content and the presence of iron oxides. Different chemical extractants are used to release sediment-associated metals into solution for analysis.

Many estuaries are typified by the extensive deposition of muddy sediments, derived mostly, but not always completely, from upstream. Rivers have usually deposited larger coarser sediments onto the river bottom before they reach their estuaries, but river waters still carry finer silt particles (between 4 and 63 μm in diameter) into the estuary (Barnes, 1974). The slower passage time through the estuary caused by counteracting ebbing and flooding tidal action allows these small particles to deposit into the sediments. Furthermore, the increased concentrations of ions in the waters of the estuary will cause small silt particles to flocculate, clumping together and sinking more rapidly (Barnes, 1974). Further deposition of fine particles into the sediment is caused by the precipitation of iron oxides and hydroxides discussed earlier. Tidal action can cause incoming seawater to move upstream along the bottom of an estuary, and these upstream tidal currents may transport marine sediments into the estuary. The exact nature of sediment deposition in an estuary will vary with river flow, the suspended particle load in the river, the geomorphological profile of the estuary, its degree of exposure to wave action and the strengths of incoming and outgoing tidal currents. Strong currents will resuspend sediments already deposited, and some estuaries show considerable bulk movement of sediments, varying both with the tidal cycle and with river flow rates.

It is not unusual, therefore, for estuarine sediments and the particulate matter suspended in the water column above them to have the same physical and chemical compositions, at least in part. Thus there were no significant differences in the concentrations of As, Cu, Mn, Ni, Pb and Zn between suspended particles and bed sediments in the Thames Estuary in 2001 (Langston et al., 2004).

Table 6.4 lists the trace metal concentrations of estuarine sediments found in the three categories of British estuaries distinguished in the preceding – those affected by historical mining upstream, those receiving industrial discharges and those considered relatively uncontaminated by trace metals.

At first sight, it would appear simple to compare such sediment concentrations between estuaries. But sediments vary in many of their physical and chemical characteristics between estuaries, not least in grain size and organic content. These two characteristics are themselves linked. Muds consist of smaller particles than sands, and these mud particles have a higher organic content than sand particles. Under the same dissolved trace metal exposures, muds will bind and accumulate higher trace metal concentrations than sands, as a result of the higher surface area to volume ratios of the smaller particles and of the high affinity of trace metals for organic matter.

Grain size is a fundamental physical characteristic of any sediment. Sediments will usually consist of a mix of grain sizes, most easily defined by sieving through a series of sieves of decreasing mesh size. At the bottom end of the size scale, particles below 3.9 μm diameter represent clay, while silt particles have sizes between 3.9 and 64 μm. Muds generally consist of silt and clay. Next in size are sand particles with diameters between 63 and 2,000 μm (0.063 to 2 millimetres), perhaps further subdivided into very

fine sand (0.063 to 0.125 millimetres), fine sand (0.125 to 0.25 millimetre), medium sand (0.25 to 0.5 millimetre), coarse sand (0.5 to 1 millimetre) and very coarse sand (1 to 2 millimetres). To ensure reproducibility, many researchers recommend analysis for trace metals of only the fraction of sediment passing through the 63 μm sieve, in effect the silt and clay (mud) (Luoma and Rainbow, 2008). In practice, this fraction contains most if not all of the accumulated trace metals in a sediment. As it happens most of the concentrations quoted in Table 6.4 involve a size limit of 100 micrometres (0.1 millimetres), including very fine sand in with the silt and clay.

Many geochemists actually go further to allow for other variations in the physicochemical nature of sediments under comparison that would affect the trace metal binding affinities of the different sediments. This compensation is referred to as normalisation (Luoma and Rainbow, 2008). In addition to grain size, normalisation may also attempt to allow for organic content, and for the content of iron oxides and hydroxides with high affinities for the likes of copper and arsenic. Furthermore, the concentration of aluminium in a sediment is considered an unambiguous indicator of the amount of clay in a sediment sample, a higher clay content providing a larger surface area for trace metal adsorption. Normalisation of a trace metal concentration to its aluminium concentration is, therefore, common when comparing the trace metal concentrations of different sediments.

We do not need to go quite this far here when comparing the large differences in sediment concentrations of trace metals between contaminated and uncontaminated estuaries. Nevertheless, similar principles will arise later in this chapter when considering the relationships between sediment concentrations of trace metals and their bioavailabilities to the resident organisms. The presence of iron oxide will reduce the bioavailability of lead in ingested sediment to particular deposit-feeding invertebrates simply as a result of the strong binding of lead to the iron oxide in the sediment (Luoma and Bryan, 1978).

Another choice to be made when analysing the trace metal concentration of a sediment is that of acid to digest the sample, or other chemical extractant to

Table 6.4 Mean or typical concentrations (μg/g dry weight) of trace metals in the oxic surface sediments of British estuaries affected by (a) mining contamination, (b) industrial contamination or (c) considered uncontaminated.

	<μm	Digest	Ag	As	Cd	Co	Cr	Cu	Fe	Hg	Mn	Ni	Pb	Sn	Zn
(a) Mining															
Restronguet Creek															
1974–80[1]	100	HNO3	4.1	2,520	1.3	22	37	2,540	63,000	0.45	559	37	396	1,350*	3,520
Pre-1992[2]	100	HNO3	3.8	1,740	1.5	21	32	2,400	49,100	0.46	485	58	341	56	2,820
2003[3]	100	HNO3	4.3					4,210					336		3,620
2003[3]	100	0.6M HCl	1.2					2,890					168		2,200
Hayle															
1974–80[1]	100	HNO3	1.3	550	1.0	28	36	782	51,200	0.06	742	32	218	1,750*	942
ca. 2005[4]		X-ray fluorescence		57–2,290				88–4,140					30–522	632–5,460	264–1,780
Gannel															
1974–80[1]	100	HNO3	2.9	233	3.0	40	35	217	26,800	0.09	1,160	49	2,175	550*	1,215
pre-1992[2]	100	HNO3	4.1	174	1.4	26	24	150	25,400	0.08	649	38	2,750	9	940
2003[3]	100	HNO3	2.0					141					1,230		870
2003[3]	100	0.6M HCl	0.8					98					781		701
East Looe															
1974–80[1]	100	HNO3	1.9	16	0.4	12	25	36	22,300	0.23	464	36	88	67*	151
2003[3]	100	HNO3	2.4					38					110		141
2003[3]	100	0.6M HCl	1.1					23					87		64
West Looe															
1974–80[1]	100	HNO3	1.3	12	0.3	12	25	57	20,100	0.30	477	36	256	54*	145
2003[3]	100	HNO3	2.7					71					485		157
2003[3]	100	0.6M HCl	1.2					42					416		69
Tamar															
1974–80[1]	100	HNO3	0.9	85	1.5	23	44	305	28,100	0.90	758	49	156	106*	392
Pre-1992[2]	100	HNO3	1.2	93	1.0	21	47	330	35,100	0.83	590	44	235	8	452

Table 6.4 (cont.)

	<μm	Digest	Ag	As	Cd	Co	Cr	Cu	Fe	Hg	Mn	Ni	Pb	Sn	Zn
Tavy															
1974–80[1]	100	HNO$_3$	0.8	131	0.6	19	48	290	37,500	0.84	660	42	176	151*	339
2003[3]	100	HNO$_3$	1.2					261					176		351
2003[3]	100	0.6M HCl	0.5					187					129		218
Conwy															
1971[5]	200	HNO$_3$+	<1	6	<1	14	65	7	13,000		397		28	5	99
Dulas Bay															
1993[6]		Sequential extract						6,600	46,000						3,800
ca. 2000[7]		Sequential extract		19	0.6			1,490	58,200		728				1,600
(b) Industrial															
Thames															
1989[8]	63	HNO$_3$/H$_2$O$_2$	4.7	15	1.3		59	61	28,200	0.6	552	157	179	16	219
2001[9]	100	HNO$_3$	0.2–19	5–14	0.1–1.5	6–21	25–108	10–148	13,700–42,400	0.09–6.5	235–891	9–43	16–506	2–66	47–529
2001[9]	100	1M HCl	0.01–11	2–7	0.04–0.8	2.3–19	3–25	5–99	3,550–12,500		139–676	2–19	15–274	1–55	25–352
Severn															
Avonmouth 1970[10]	–	HF/HClO$_4$			4.3								190		590
1974–80[1]	100	HNO$_3$	0.7	10	1.1	16	48	55	34,200	0.56	652	38	105	58*	314
1988[11]	63	HF/HClO$_4$					85	37	48,400			49	70		226
Pre-1992[2]	100	HNO$_3$	0.4	9	0.6	15	55	38	28,300	0.51	686	33	89	8	259
Lydney 2000[12]	63	HNO$_3$			0.1		43	24				19	50		148
Portishead 2000[12]	63	HNO$_3$			1.0		81	46				27	69		237

Mersey															
1980–84[2,13]	100	HNO$_3$	0.7	42	1.3	13	88	94	29,000	3.0	1,250	32	138	8.3	422
Humber															
Pre-1992[2]	100	HNO$_3$	0.4	50	0.5	16	77	54	35,200	0.55	1,020	39	113	5.1	252
Taw															
1972[14]	2000	HNO$_3$/HClO$_4$			21			75			165		533		1,020
1972[14]	2000	CH$_3$COOH			17			21				48	99		641
Neath															
1974–80[1]	100	HNO$_3$	0.7	20	2.1	18	42	89	31,100	1.2	650	40	98	54*	327
Swansea Bay															
1974–80[1]	100	HNO$_3$	0.3	18	0.4	17	60	37	21,500	0.56	836	36	95	45*	256
Loughor															
1974–80[1]	100	HNO$_3$	0.2	22	1.4	13	799	47	38,100	0.17	791	32	77	320*	220
Pre-1992[2]	100	HNO$_3$	0.2	18	0.5	10	207	27	19,300	0.16	597	21	48	161	146
Tyne															
Pre-1992[2]	100	HNO$_3$	1.6	25	2.2	11	46	92	28,200	0.92	395	34	187	5.4	421
Poole Harbour															
1971[5]	200	HNO$_3$/HClO$_4$		11	1	16	65	12	11,000		340		104	20	60
Holes Bay 1973[215]	200	HNO$_3$/HClO$_4$			7	17	44	44	25,700		132	53	114		158
Pre-1992[2]	100	HNO$_3$	0.8	14	1.9	11	49	50	29,300	0.81	185	26	96	7.4	165
(c) Uncontaminated															
Avon (Devon)															
1974–80[1]	100	HNO$_3$	0.1	13	0.3	10	37	19	19,400	0.12	417	28	39	28*	98
Pre-1992[2]	100	HNO$_3$	0.1	13	0.1	10	28	18	18,400	0.12	326	23	68	4	82

Table 6.4 (cont.)

	<μm	Digest	Ag	As	Cd	Co	Cr	Cu	Fe	Hg	Mn	Ni	Pb	Sn	Zn
Blackwater (Esser)															
1994-95[16]	–	HNO₃/HCl			<0.1-1.7		32-65	16-77	24,200–37,400	0.11-0.51	169-400	19-33	27-64		63-190
Solway Firth															
Pre-1992[2]	100	HNO₃	0.1	6	0.2	6	30	7	14,800	0.03	577	17	25	0.4	59

Note: *Fusion technique to include assessment of cassiterite which is not dissolved in HNO₃.

Sources: Bryan et al. (1980);[1] Bryan and Langston (1992);[2] Rainbow et al. (2011c);[3] Rollinson et al. (2007);[4] Thornton et al. (1975);[5] Parkman et al. (1996);[6] Whiteley and Pearce (2003);[7] Attrill and Thomas (1995);[8] Langston et al. (2004);[9] Butterworth et al. (1972);[10] Little and Smith (1994);[11] Duquesne et al. (2006);[12] Langston (1986);[13] Vivian (1980);[14] Boyden (1975);[15] Emmerson et al. (1997).[16]

release associated trace metals without completely digesting the sediment particles. A common choice of acid is concentrated nitric acid, often at 100°C, to give a so-called total trace metal concentration. Some mineral constituents of a sediment may be resistant to solubilisation even by concentrated nitric acid at this high temperature, and other variants for acid digestion are given in Table 6.4. These include mixtures of concentrated nitric and perchloric acids, concentrated hydrofluoric acid (alone or with perchloric acid) and mixtures of acids with hydrogen peroxide. Some minerals are extraordinarily resistant to acid digestion, as in the case of the tin-bearing mineral cassiterite. A fusion technique is required to solubilise and measure the contribution of tin by cassiterite to the total tin concentration of a sediment (Table 6.4).

We are now approaching the question of the ecotoxicological relevance of a total trace metal concentration in a sediment that requires such a chemical assault to measure all the trace metal present. Surely the tin concentration in a sediment in the form of cassiterite is simply not going to be in any way bioavailable to any organism, however encountered. We are distinguishing now between the concentration and the bioavailability of trace metals in a sediment. Thus, some authors will use a weaker extractant, such as 0.6 or 1 molar hydrochloric acid (see Table 6.4), in an attempt to model a bioavailable concentration and to come up with a better comparative measurement of trace metal concentrations in sediments that might have an effect on resident biota. Nevertheless, it does remain the case that total metal concentrations (typically after nitric acid digestion) are the most common comparative concentrations of trace metals measured in sediments. They should not be dismissed as completely biologically irrelevant, for it is likely that there is some relationship (although variable) between such total concentrations of trace metals in sediments and their bioavailabilities. Total trace metal concentrations in sediments are, therefore, a useful starting point when investigating the comparative ecotoxicologies of different metals in different estuaries.

A variation on the use of a particular chemical extractant to model trace metal bioavailability is to use a series of sequential extractions in turn to progressively remove and analyse the trace metals associated with different chemical components of a sediment (Tessier et al., 1979; Luoma and Rainbow, 2008).

6.3.1 Bioavailability of Sediment-Associated Metals

Prologue. The bioavailability of sediment-associated trace metals to estuarine organisms may involve uptake from pore water or alimentary absorption after ingestion and digestion. Bioavailability depends on the strength of chemical binding of a trace metal in a sediment, controlling its release into solution in different media, and on the trace metal uptake routes of the organism concerned, related inevitably to its natural history.

The total bioavailability of a trace metal associated with sediment can be subdivided into two components: the dissolved bioavailability in the pore water of the sediment, and the trophic bioavailability to animals ingesting the sediment. Trace metals associated with sediment particles may be released by equilibrium into the surrounding pore water, with the potential then to be taken up by the roots of a plant or across an external epithelium of a burrowing invertebrate. Animals ingesting sediment particles can take up trace metals associated with those sediment particles after digestion in the alimentary tract. Such digestion may be intracellular after uptake of small particles into digestive cells for subsequent breakdown, or extracellular with complete solubilisation in the gut lumen before absorption.

Whether a trace metal is taken up by an organism from solution at an external surface or internally in the gut by after digestion, the bioavailability of the sediment-associated trace metal to that organism will be greatly affected by the strength of chemical binding of the metal within the sediment particles. Trace metals bind strongly with the organic components of sediments. If these organic components differ in quantity and type between sediments, then so will the total bioavailabilities of the sediment-bound trace metals. Furthermore, oxides of iron in a sediment will bind particularly strongly with certain trace metals, including arsenic, copper and lead (Bryan and Langston, 1992). These trace metals will, therefore, be less bioavailable in an estuarine sediment with a high iron content than in a sediment less rich in iron, even when total trace metal concentrations in the sediment are equal.

Organisms that take up and accumulate trace metals from a sediment, whether from the dissolved pore water route and/or by assimilation after ingestion and digestion, will accumulate more metals from sediments of higher metal bioavailabilities. There will, therefore, be a good mathematical relationship between the accumulated trace metal concentration in the sediment-dwelling organism and some measure of metal concentration in the sediment that best models the metal bioavailability in the sediment. For example, there might be a better linear relationship between accumulated metal concentrations in the organism and the trace metal concentrations in different sediments as extracted by 1 molar hydrochloric (1MHCl) acid (a relatively weak extractant) than with total metal concentrations in the sediment measured after concentrated nitric acid digestion. In this case, the sediment trace metal concentrations measured after the weaker extraction represent better models of the bioavailable trace metal concentrations than do total metal concentrations. The latter might well include measures of some chemically bound metal in the sediment that is not bioavailable to the organism. One example of such a relationship is the good fit of accumulated body concentrations of silver in the estuarine ragworm *Hediste* (formerly *Nereis*) *diversicolor* from various British estuaries to sediment concentrations of silver extracted with 1 M HCl (Figure 6.1) (Bryan, 1985; Bryan and Langston, 1992).

Other expressions of trace metal concentrations in sediments may be more appropriate models of the bioavailable concentrations therein. Mercury binds predominantly to organic matter in estuarine sediments (Bryan and Langston, 1992). Increasing organic contents in sediments reduce the accumulation of mercury by many estuarine organisms, including the polychaete *H. diversicolor* and the deposit-feeding bivalve *Scrobicularia plana*, the peppery furrow

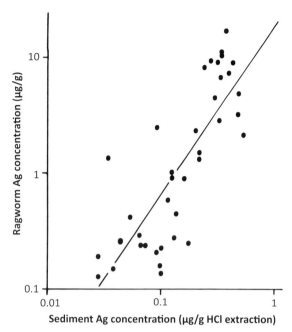

Figure 6.1 Relationship between concentrations of silver accumulated in the ragworm *Hediste diversicolor* and those in the associated surface sediments (1 M HCl extract) of British estuaries, using a log scale. (After Bryan and Langston, 1992, with permission.)

shell (Langston, 1982; Bryan and Langston, 1992). Thus there is a good mathematical relationship between accumulated mercury concentrations in the soft tissues of *S. plana* from British estuaries and the concentrations of mercury in their sediments expressed per unit organic matter (Figure 6.2) (Langston, 1982; Bryan and Langston, 1992).

The effect of the affinities of arsenic and lead for iron oxides on the biovailabilities of these two elements in sediments can also be shown up by the appropriate choice of measure of their concentration in the sediment. Thus there are very good relationships between the concentrations of accumulated arsenic (Figure 6.3) (Langston, 1980) and lead (Figure 6.4) (Luoma and Bryan, 1978) in the soft tissues of *S. plana* from a range of British estuaries and the sediment concentrations of these two trace metals expressed as the ratio of the concentrations of either element to the concentrations of iron, both after extraction with 1 M HCl.

Such relationships cast considerable light on the factors affecting the bioavailabilities of different trace

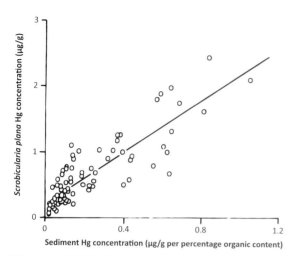

Figure 6.2 Relationship between concentrations of mercury accumulated in the soft tissues of the bivalve *Scrobicularia plana* and those in the associated surface sediments (expressed with respect to organic matter content) of British estuaries. (After Bryan and Langston, 1992, with permission.)

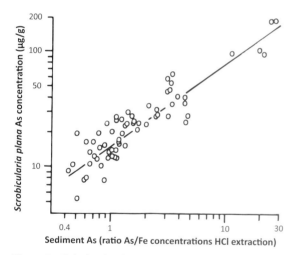

Figure 6.3 Relationship between concentrations of arsenic accumulated in the soft tissues of the bivalve *Scrobicularia plana* and those in the associated surface sediments (expressed as the ratio of the concentration of As (μg/g) to the concentration of iron (mg/g) after extraction with 1 M HCl of British estuaries. (After Langston, 1980, with permission.)

metals in estuarine sediments, and have been used as models of sediment bioavailabilities (Luoma, 1989; Bryan and Langston, 1992). Nevertheless, it is still the case that bioavailabilities depend on the organism

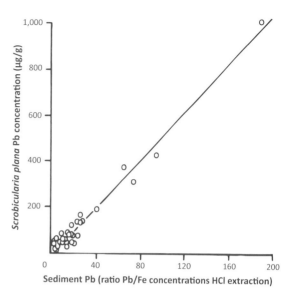

Figure 6.4 Relationship between concentrations of lead accumulated in the soft tissues of the bivalve *Scrobicularia plana* and those in the associated surface sediments (expressed as the ratio of the concentration of Pb (μg/g) to the concentration of iron (mg/g) after extraction with 1 M HCl) of British estuaries. (After Luoma and Bryan, 1978, with permission.)

concerned, and the routes of trace metal uptake available to it. Only the accumulated concentration of a trace metal in an organism provides a valid measure of the total bioavailability of a metal in a sediment to that organism. Again, we should rely on biomonitors to provide assessments of trace metal bioavailabilities, and thus ecotoxicological potentials, in estuaries.

6.3.2 Organometals in Estuarine Sediments

Prologue. Organometal compounds may be formed by microbial action in estuarine sediments, particularly in the cases of mercury, tin and lead, with potential ecotoxicological consequences.

The extensive presence of organically rich muddy sediments in estuaries provides a habitat for considerable microbial activity, typically resulting in anoxic sulphide-rich sediments as oxygen is used up by microbial aerobic respiration before the microbes switch to anaerobic respiration.

Estuarine sediments offer an opportunity for the methylation of particular trace metals, particularly mercury and tin, by microbial action to produce the likes of methyl mercury and methyl tin compounds from inorganic metal precursors. These organometals can diffuse into the pore waters of the sediments and beyond to come into contact with estuarine organisms. Such organometals bypass usual routes of trace metal uptake across biological membranes and are taken up at high rates by aquatic organisms. Thus organometals have high dissolved bioavailabilities. The high uptake rates are associated with the atypically high toxicities of organometals in comparison with their inorganic counterparts. Organometals are also subsequently transferred through estuarine food chains, typically with biomagnification across trophic levels, potentially with ecotoxicological consequences.

Methylation of mercury appears to be greatest in moderately anoxic sediments (Bryan and Langston, 1992). In addition to the microbial formation of methyl mercury, microbial action will also cause demethylation. Furthermore, some forms of methyl mercury, such as dimethyl mercury, are volatile and may be lost from the estuarine system (Bryan and Langston, 1992). The outcome of the concomitant microbial production and degradation of methyl mercury in estuarine sediments is a dynamic equilibrium, producing usually relatively low levels of methyl mercury (<0.02 μg/g), even at mercury-contaminated sites (Bryan and Langston, 1992). And the proportion of total mercury in a sediment that is in methyl form represents less than 0.2% of the total metal in the sediment (Bryan and Langston, 1992). Nevertheless, even though inorganic mercury is the dominant form of mercury in the sediment, it needs to be remembered that methyl mercury is very toxic, has a very high bioavailability and will biomagnify up food chains. Methyl mercury, therefore, still represents a greater ecotoxicological threat than inorganic mercury in estuarine systems.

Inorganic tin in estuarine sediments can also be methylated by microbial action to produce mono-, di- and trimethyl tin (Bryan and Langston, 1992). The biology of methyl tins follows the same principles as the biology of methyl mercury, with enhanced bioavailability in comparison to inorganic forms of the metal and potential biomagnification up estuarine food chains. Nevertheless, methyl tins produced naturally by microbial action in estuarine sediments do not appear to occur in high enough concentrations to be considered potentially detrimental to estuarine organisms (Bryan and Langston, 1992). This conclusion is in strong contrast to the ecotoxicological effects of a different man-made form of organotin, in this case TBT and its mono- and dibutyl tin derivatives, deliberately introduced into estuarine and marine systems as an antifouling agent (Bryan and Langston, 1992). The use and ecotoxicology of TBT are considered in greater detail later in this chapter.

Lead is the third trace metal that can be converted into organic forms, for example ethyl lead compounds, naturally by microbial action in estuarine sediments (Bryan and Langston, 1992). As for naturally produced methyl tins, these organolead compounds are usually present in low concentrations in estuarine sediments. They do not appear to represent an ecotoxicological threat, even when total lead concentrations in the sediment are very high (Bryan and Langston, 1992). The direct introduction of man-made organolead compounds into estuaries can raise organolead concentrations in the local sediments. Such was the case in the sediments of the Mersey Estuary in the late 1970s, with the addition of organolead compounds originating from a factory then manufacturing tetra-ethyl lead additives for petrol (Bryan and Langston, 1992). Local waders such as dunlin *Calidris alpina* were affected by lead ecotoxicity, although the organolead compounds responsible may have entered estuarine food chains via the tellinid bivalve *Macoma balthica* in dissolved form in the water column as opposed to the sediment (Bryan and Langston, 1992).

6.4 Trace Metals in British Estuaries

Prologue. Particular British estuaries rich in trace metals deserve further comment and are discussed here in a series of case studies. These estuaries are

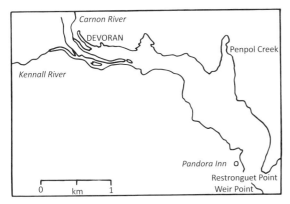

Figure 6.5 Restronguet Creek, Cornwall. (After Bryan and Gibbs, 1983. Reproduced by permission of the Marine Biological Association, UK.)

again divided into those affected by a history of mining, and those contaminated by industrial effluents.

It is to West Cornwall that we turn first. The regions of Redruth and Camborne, and the nearby valley of the River Carnon, have long histories of mining, particularly for tin and copper, but also for arsenic when market conditions were appropriate (Chapter 2). Lead, zinc and iron have also been mined locally. It is comparatively difficult to extract tin from its most common ore, its oxide cassiterite, known locally as black tin. Correspondingly, tin in this form is not very bioavailable, and tin does not feature strongly on the list of trace metals of ecotoxicological concern. The Carnon River flows southeast to drain into its estuary Restronguet Creek (Figure 6.5), a western branch of the Fal estuary system (Figure 6.6) on the south coast of Cornwall. The nearby River Hayle, on the other hand, flows west and north to its estuary on the north Cornish coast in St Ives Bay (Figure 6.7).

6.4.1 Restronguet Creek

Prologue. Restronguet Creek in the Fal estuary system has received atypically huge amounts of dissolved and particulate trace metals (As, Cd, Cu, Fe, Mn, Pb, Sn and Zn) from the River Carnon, resulting from extensive mining in the Carnon valley for centuries.

Restronguet Creek still contains sediment concentrations of some of these trace metals that are unmatched elsewhere in Britain, with consequent very high trace metal bioavailabilities to the resident flora and fauna.

The River Carnon has been a remarkable source of trace metals into Restronguet Creek for centuries, draining the mines in its valley, which switched from relatively minor tin production to huge copper production in the eighteenth and early nineteenth centuries. Arsenic production followed in the late nineteenth century. There was also some output of zinc and lead, according to the local availability of their ores, with cadmium as a waste product. Furthermore, the River Carnon is the recipient of the drainage waters of the Great County Adit, expanded between 1748 and the 1880s to serve more than 60 local mines. At its peak, the Great County Adit discharged more than 65 million litres per day into the Carnon, and thence into Restronguet Creek. The River Carnon, therefore, has historically contained high dissolved and suspended particulate concentrations of tin, copper, arsenic, zinc, lead, cadmium, iron and manganese.

The sediments of Restronguet Creek have built up as mining has developed in the Carnon valley. It is reputed that in medieval times, the creek was navigable as far upstream as Bissoe (Barton, 1971a; Bryan and Gibbs, 1983). And in the nineteenth century, there were quays for ore ships at Devoran (Figure 6.6) (Bryan and Gibbs, 1983), where there is now mud and salt marsh. Restronguet Creek is also remarkable in that the streaming of tin, carried out in the Carnon River over centuries or more, actually extended into the upper reaches of the creek until the nineteenth century (Bryan and Gibbs, 1983). By then, it was becoming increasingly difficult to exclude the tide and remove the silt to get to the tin ore beneath, and tin was then extracted by mining beneath the creek itself (Barton, 1971a; Bryan and Gibbs, 1983).

At the beginning of the 1980s, when the Wheal Jane mine was again active after reopening and the Mount Wellington had only recently shut for the last time (Figure 6.6), it is estimated that the River Carnon was even then discharging annually into Restronguet Creek

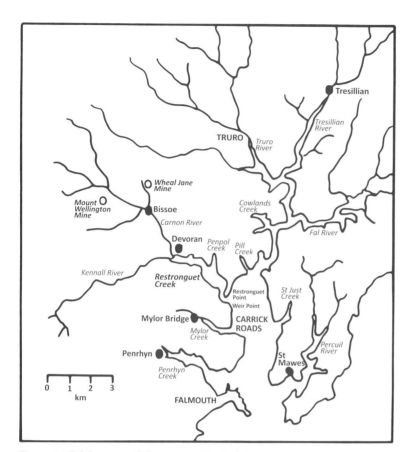

Figure 6.6 Fal Estuary with Restronguet Creek. (After Bryan and Gibbs, 1983. Reproduced by permission of the Marine Biological Association, UK.)

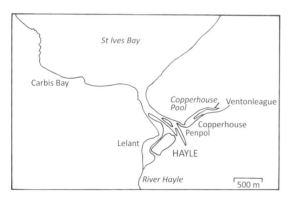

Figure 6.7 Hayle Estuary, Cornwall.

262 tonnes of zinc, 253 tonnes of iron, 33.4 tonnes of manganese, 21.3 tonnes of copper, 5.5 tonnes of arsenic, 1.4 tonnes of lead and 0.7 tonnes of cadmium (Bryan and Gibbs, 1983). In 1992, long after the closure of most of the mines of the Carnon valley and soon after the final closure of Wheal Jane, there was an accidental release into the River Carnon of more than 45,000 cubic metres of acidic drainage water heavily contaminated with trace metals, from the Nangiles Adit just downstream from the Great County Adit (see Chapter 2). This deluge flowed into Restronguet Creek with its contaminant metal load. There are now remedial management procedures in place along the River Carnon.

Dissolved concentrations of trace metals in the river and the estuary (Tables 5.2, 6.1) may not now be as high as they once were historically, but the metal-rich sediments of Restronguet Creek, derived from mining activities upstream, still bear testament to a long history of extremely high mining contamination (Table 6.4). The strong affinities for many trace metals of these sediments, rich in organic carbon and iron oxides, mean that few metals are released back into the overflowing water over time (Rainbow et al., 2011c). Trace metal concentrations in the sediments of Restronguet Creek remain enormously high to this day (Table 6.4). Indeed, there were no significant systematic declines in sediment concentrations of Ag, Cu, Pb and Zn in Restronguet Creek, whether digested in nitric acid or extracted with weaker hydrochloric acid, over the 26-year period between 1977 and 2003 (Rainbow et al., 2011c).

Later in this chapter, we shall explore in more detail the bioavailabilities of these trace metals to members of the estuarine flora and fauna of Restronguet Creek. Suffice to say now that the bioavailabilities of all but one of these trace metals to these resident organisms in the creek are still extraordinarily high (Rainbow et al., 2011c), with very significant ecotoxicological potential. Restronguet Creek is nothing short of a remarkable natural laboratory for the trace metal biologists of today, seeking to understand the ecophysiology and ecotoxicology of trace metals in the real world (Rainbow and Luoma, 2005).

The trace metal that appears to be the one exception is tin. The very high concentrations (>1,000 μg/g) of tin in the sediments of Restronguet Creek are an order of magnitude or more greater than tin concentrations in most estuaries (Table 6.4) (Bryan and Langston, 1992). And yet these high concentrations are poorly reflected in the resident flora and fauna (Bryan and Langston, 1992). The explanation presumably lies in the fact that most of the sediment tin is in the form of particles of cassiterite, so tightly bound that it is not released into pore waters, nor can it be digested and assimilated after ingestion by deposit-feeding invertebrates (Bryan and Langston, 1992).

6.4.2 Hayle

Prologue. The Hayle Estuary on the north Cornish coast has very high sediment concentrations of copper and arsenic in particular.

The Hayle Estuary (Figure 6.7) lies at the southern end of St Ives Bay, divided by Hayle Harbour into north-eastern and southwestern sections. The southwestern

division receives the waters of the River Hayle, draining a catchment with a long history of mining for tin, then copper. The dock at Copperhouse is in the northeastern section. Copper smelters were erected at firstly Penpol at the western end of the Copperhouse Pool in 1721, and then at Ventonleague at the eastern end in 1758 (Pirrie et al., 1999). Tin smelters were also operational at Copperhouse in the nineteenth century (Pirrie et al., 1999).

The sediments of the estuary are particularly high in tin, copper and arsenic, with somewhat raised concentrations also of zinc, and to a lesser extent lead (Table 6.4). As in the case of the Restronguet Creek sediments, the very high levels of tin in the sediments of the Hayle Estuary (Table 6.4) do not translate into high bioaccumulated tin concentrations in the resident flora and fauna, indicating the dominant presence in the sediment of cassiterite particles with low bioavailability (Bryan and Langston, 1992).

6.4.3 Gannel

Prologue. The Gannel Estuary by Newquay has sandy sediments that are rich in lead and zinc.

The Gannel Estuary (Figure 6.8), on the southern outskirts of Newquay, receives the River Gannel draining a catchment with a long history of lead mining. Ores of zinc are associated with the ores of lead in the region, and the sediments of the Gannel Estuary are atypically high in both these trace metals (Table 6.4). Silver can also be found in association with zinc and lead, and silver concentrations in Gannel sediments are also somewhat raised (Table 6.4).

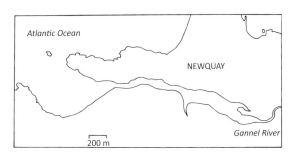

Figure 6.8 Gannel Estuary, Cornwall.

The mouth of the Gannel Estuary faces the Atlantic Ocean and receives the same impressive waves as the adjacent surfing beaches of Newquay. As a result, the middle region of the Gannel Estuary has sediments that are sandier, and less muddy, than their counterparts in other regions of Southwest England. Even in these sandier sediments, there was no evidence for any systematic decline in sediment concentrations of Ag, Cu, Pb and Zn in the Gannel Estuary over the 26-year period between 1977 and 2003 (Rainbow et al., 2011c).

6.4.4 Looe

Prologue. The Looe Estuary has received high quantities of lead derived from mining in the catchment of the West Looe River, with consequences for estuarine sediment concentrations. The East Looe River does not have a history of mining input but did experience input of silver of presumed industrial origin, reflected in sediments of the eastern branch of the Looe Estuary.

The Looe Estuary (Figure 6.9) has two branches, receiving the discharges of the East and West Looe rivers respectively (Bryan and Hummerstone, 1977). The West Looe River drains part of the old lead mining area of Herodsfoot, which produced both lead and silver in the second half of the nineteenth century (Bryan and Hummerstone, 1977). Correspondingly, the sediments of the West Looe branch of the estuary are high in lead and, to some degree, silver (Table 6.4). The East Looe River, on the other hand, does not pass through a mining region, and yet high concentrations of silver have also been found in the sediments of its estuary (Table 6.4) (Bryan and Hummerstone, 1977). As expected, lead concentrations in these East Looe sediments are not particularly raised (Table 6.4). It has been presumed that the source of silver in the East Looe River and Estuary has been an industrial source upstream, independent of any mining activity (Bryan and Hummerstone, 1977).

As in the case of the sediments of Restronguet Creek and the Hayle Estuary, there was no evidence for any systematic decline in sediment concentrations

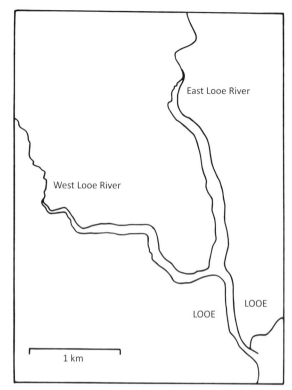

Figure 6.9 Looe Estuary, Cornwall. (After Bryan & Hummerstone, 1977, with permission.)

of Ag, Cu, Pb and Zn in either the West Looe or the East Looe Estuary over the 26-year period between 1977 and 2003 (Rainbow et al., 2011c).

6.4.5 Tamar and Tavy

Prologue. The Tamar Estuary and the associated Tavy Estuary have high sedimentary loadings of arsenic, copper, lead and zinc derived from mining in the valleys of both rivers, notably in the region of Gunnislake on the Tamar.

Before it enters its estuary with Plymouth at its seaward end, the River Tamar passes through the historically important mining area of Gunnislake, the home of the famous Devon Great Consols mine (Figure 6.10). The Gunnislake mines were mined extensively for copper, arsenic and lead in the nineteenth century and produced internationally very

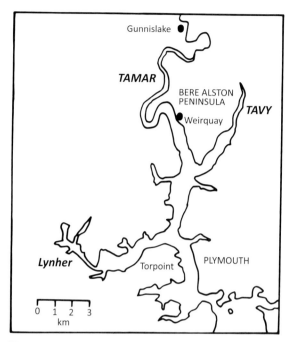

Figure 6.10 Tamar and Tavy Estuary.

significant amounts of copper and arsenic at the time (see Chapter 2). These Gunnislake mines were drained by adits into the River Tamar, the drainage waters carrying their contaminant metal loads into the river and thence into the estuary. Downstream, the Bere Alston peninsula lies on the eastern bank of the upper Tamar Estuary. This peninsula has a long history of silver and lead mining (see Chapter 2), and the Bere Alston lead and silver mines were particularly active in the first decade of the nineteenth century. Mine drainage rich in these two metals entered the adjacent Tamar Estuary. Lead and silver smelting at Weirquay in the nineteenth century was another source of metal input locally into the Tamar Estuary.

Also draining into the Tamar Estuary is the River Tavy, its own smaller estuary lying along the eastern side of the Bere Alston peninsula (Figure 6.10). The upper region of the Tavy drains a mining region of Dartmoor just north of Tavistock. Thus it receives, for example, the waters of the Cholwell Brook draining the Wheal Betsy mine near Mary Tavy, a historic source of tin, lead, zinc and silver (Beane et al., 2016).

After passing through Tavistock, the Tavy drains the eastern end of the rich ore region of Gunnislake and the eastern side of the Bere Alston peninsula. Like the Tamar, the Tavy, too, receives a significant input of copper, arsenic, lead and iron.

Given this history of mining, the sediments of the Tamar and Tavy estuaries are relatively high in copper, arsenic, lead and zinc, though not matching specific sediment concentrations of copper and arsenic in Restronguet Creek and the Hayle Estuary, or lead concentrations in the Gannel Estuary (Table 6.4). As in the case of the sediments of Restronguet Creek and the Hayle Estuary, there was no evidence for any systematic decline in sediment concentrations of Ag, Cu, Pb and Zn in the Tavy Estuary over the 26-year period between 1977 and 2003 (Rainbow et al., 2011c).

This lack of significant decline over decades in the sediment concentrations of trace metals in southwestern estuaries affected by mining shows how tightly the sediments bind to these metals (Rainbow et al., 2011c). These sediments do not readily give up their metals to the overflowing water. We see later in this chapter whether the same metals in the sediments are bioavailable, for example to sediment-ingesting estuarine invertebrates.

6.4.6 Dulas Bay

Prologue. Dulas Bay (Figure 6.11) on the northeast coast of Anglesey has high sediment concentrations of copper, zinc, iron and manganese derived from the mining of Parys Mountain.

Dulas Bay (Figure 6.11) is the estuary of the Afon Goch, the stream that drains the historic copper mining region of Parys Mountain. Copper mining on Parys Mountain escalated when a large, shallow body of copper ore was found there in 1768. Parys Mountain then dominated world copper production for the next three decades. Inevitably, the acidic waters of the Afon Goch transported large amounts of copper, together with zinc, iron and manganese, into Dulas Bay Foster et al., 1978; Boult et al., 1994). It would appear that, at the peak of copper mining,

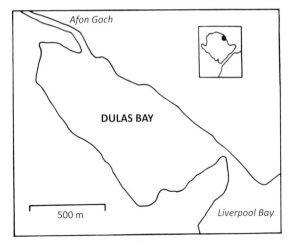

Figure 6.11 Dulas Bay, Anglesey.

considerable copper in dissolved form also issued forth from the mouth of Dulas Bay into the adjacent waters of Liverpool Bay. Local folklore has it that, at this time, local sea captains used to moor their ships off Dulas Bay to defoul their hulls in the ecotoxic copper-rich waters (Foster et al., 1978).

As we saw in Restronguet Creek, iron dissolved in the acid medium of the Afon Goch precipitates as oxides and hydroxides on meeting the raised pH of the estuarine waters of Dulas Bay (Foster et al., 1978). Copper then comes out of solution to adsorb onto this particulate iron, before both metals are deposited in the sediments (Foster et al., 1978). The sediments of Dulas Bay are still enormously high in copper and zinc (Table 6.4). Even in the 1970s, the Afon Goch was transporting annually into Dulas Bay 20 tonnes of manganese, 28 tonnes of copper, 62 tonnes of zinc and nearly 300 tonnes of iron (58% in particulate form) (Foster et al., 1978).

6.4.7 Thames

Prologue. The Thames Estuary has a long history of pollution that includes contamination by trace metals of industrial origin. Cleaner now than during its history up to the second half of the twentieth century, the estuary still has somewhat elevated, although falling, levels of trace metals.

Other, typically larger, British estuaries have been contaminated with toxic metals derived not directly from mining activity, but via the effluents of local industry. We start with the Thames.

The catchment of the Thames comprises nearly 50 main tributaries, draining an area of about 13,000 square kilometres, including the densely populated region of Greater London (Pope and Langston, 2011). The Thames Estuary extends 110 kilometres downstream from Teddington Lock, the upstream limit of tidal influence, to the North Sea beyond Foulness (Figure 6.12). The Thames is one of the largest estuarine systems in the United Kingdom and provides a major input into the southern North Sea (Pope and Langston, 2011).

The history of pollution in the Thames Estuary is mainly a story of severe organic pollution from sewage, with an associated fall in dissolved oxygen concentrations and an increase in ammonia levels, with dramatic deleterious effects on the local estuarine biota and the passage of migratory fish (Wheeler, 1979; Andrews and Rickard, 1980). From the early 1960s through to the late 1970s and thereafter, the estuary recovered from an 'open sewer' to a waterway allowing the passage of migratory fish (Wheeler, 1979; Andrews and Rickard, 1980), mainly as a result of the completion or extension of major sewage plants discharging into the estuary (Power et al., 1999).

The Thames Estuary has also been a recipient of other pollutants, including trace metals and organic contaminants such as organochlorines, in industrial and domestic discharges (Andrews, 1984; Power et al., 1999). Principal sources of contaminants into the estuary are the River Thames itself and the fourteen tributaries feeding into the tidal section (Pope and Langston, 2011). Other major inputs come from the main sewage plants at Mogden, Kew, Crossness, Beckton, Riverside, Longreach, Northfleet, Tilbury, Gravesend, Canvey Island and Southend, all of which discharge into the estuary, together with periodic inputs via storm overflows (McEvoy et al., 2000; Pope and Langston, 2011). Storm water runoff will also enter the estuary directly, particularly in central London (McEvoy et al., 2000). In addition, there are various industrial effluents entering along the

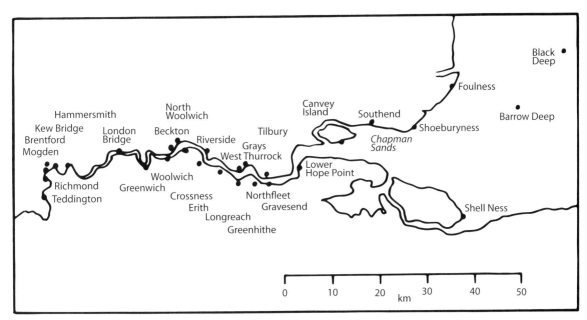

Figure 6.12 Thames Estuary. (After Langston et al., 2004, with permission of the Marine Biological Association, UK.)

estuary, from metal refineries, docks and marinas (Pope and Langston, 2011). Before the cessation of sewage sludge dumping in 1998, the Barrow Deep and South Falls dumping grounds in the outer Thames Estuary received large quantities of sewage sludge contaminated with trace metals (McEvoy et al., 2000). This sewage sludge is now incinerated. Industrial waste and dredge spoil have also been dumped at these sites, increasing sediment concentrations of, for example, lead at Barrow Deep and cadmium at South Falls (McEvoy et al., 2000).

The 1970s saw the introduction of more stringent effluent controls to reduce the toxic metal contents of industrial discharges (Power et al., 1999). Correspondingly, dissolved concentrations of cadmium, copper, mercury, nickel and zinc in the Thames Estuary showed exponential declines between 1980 and 1997 (Power et al., 1999). Dissolved lead concentrations in the estuary also showed a decline over the initial years of this period, but this decline was reversed in the 1990s by the onset of draught, together with the then continuing effect of the use of leaded petrol by Londoners before its eventual phase-out in 1999 (Power et al., 1999). Both total and dissolved concentrations of arsenic, cobalt, copper, iron,

nickel and zinc in Thames estuarine waters also fell significantly over the later shorter period between 1997 and 2001 (Pope and Langston, 2011). Similarly, there were significant falls over this period in total water concentrations of cadmium and manganese and in dissolved concentrations of lead (Pope and Langston, 2011).

To take 2001 as a snapshot, sediment concentrations of most trace metals in the Thames Estuary were still high in comparison to those of uncontaminated estuaries (Table 6.4) (Langston et al., 2004). In comparison with other data from the Marine Biological Association UK, the sediment concentrations of silver in the Thames Estuary were particularly exceptional, falling in the 93rd percentile of the range in estuaries in England and Wales (Langston et al., 2004). Thames Estuary sediments were also very high in cobalt, chromium, lead, manganese, mercury and tin (averages in the 50th to 75th percentile range), while average concentrations of copper, iron, nickel and zinc ranged between the 24th and 50th percentiles (Langston et al., 2004).

Sediment trace metal concentrations in 2001 did generally decrease in a downstream direction (Langston et al., 2004). The concentration gradients were steepest for cadmium, copper, lead, silver and zinc, which were present in highest concentration at Kew, and for mercury and tin, which peaked at Hammersmith (Langston et al., 2004). The implication is that the inputs of these metals into the estuarine system were greater upstream (Langston et al., 2004). A comparison of total metal concentrations in the sediments with those extractable in 1 M HCl showed much of the lead (86%) to be extractable, suggesting an anthropogenic origin, such as urban runoff (Langston et al., 2004). Other trace metals in readily extractable form (>50%) in the sediments were cadmium, copper, manganese, silver, tin and zinc, again suggesting an anthropogenic origin (Langston et al., 2004).

Thames Estuary sediments, particularly those from the inner estuary, were also significantly enriched in comparison with sediments from the central North Sea (Langston et al., 2004). This enrichment was clearest for silver with sediment concentrations in the Thames Estuary sediments being 38-fold higher, on average, than those at Dogger Bank (Langston et al., 2004). Average sediment concentrations of mercury, copper, tin, lead, cobalt and zinc were between two- and sevenfold higher in the estuary than offshore, while chromium, nickel, iron, manganese and cadmium concentrations were still raised, but by less than twofold (Langston et al., 2004). Arsenic, on the other hand, showed higher sediment concentrations offshore than in the estuary (Langston et al., 2004). Arsenic concentrations in the Thames Estuary sediments were also low in the range of average concentrations for estuaries in England and Wales, falling at the 16th percentile (Langston et al., 2004).

There was no clear evidence of any systematic fall in the metal concentrations of sediments in the Thames Estuary as a whole over the short period between 1997 and 2001 (Langston et al., 2004). The only significant fall in total sediment concentration was for cadmium (Langston et al., 2004). Total sediment concentrations of copper, mercury, manganese, nickel, lead, tin and zinc showed no significant change between 1999 and 2001, and there were actually small increases in total sediment concentrations of arsenic, cobalt, chromium, iron and silver

(Langston et al., 2004). In fact, both 1999 and 2001 sediment concentrations of trace metals in the Thames Estuary are not very different from those measured in 1989 (Table 6.4) (Attrill and Thomas, 1995), in contrast to the situation for water concentrations (Power et al., 1999). So again, the sediments appear to hold their metal concentrations over more than a decade, as we saw in the mining-affected estuaries of South-west England. Changes in the trace metal contamination levels of the sediments in the Thames Estuary may need to await bulk movement of the sediments over an extended period.

6.4.8 Severn

Prologue. The Severn Estuary had high dissolved and sediment concentrations of trace metals of industrial origin during the twentieth century, before most of these have fallen to lower present-day values. The Avonmouth smelter was a particular source of large quantities of cadmium, lead and zinc. The Severn Estuary has high hydrodynamic energy associated with its high tidal range, and its sediments are prone to movement by bulk transport.

The Severn Estuary is the largest estuary in the United Kingdom (Apte et al., 1990). In addition to the River Severn itself, which contributes about a quarter of the total freshwater input, there are another nine major rivers which drain into the estuarine system (Figure 6.13) (Morris, 1984; Apte et al., 1990). The Severn Estuary has the highest tidal range in the United Kingdom, causing powerful tidal currents with associated sediment erosion and resuspension (Langston et al., 2010).

The area surrounding the estuary includes Cardiff, Newport and Bristol, and is, correspondingly, an important centre for urban activity with associated sewage disposal (Apte et al., 1990). The region also has a history of considerable industrial activity, not least in south Wales, with coal and steel industries (now much reduced), paper mills and chemical manufacturers (Langston et al., 2010). Metal refining has historically been important, both in south Wales and, until 1980, at Avonmouth on the south bank,

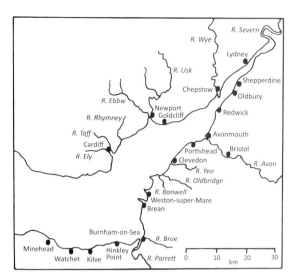

Figure 6.13 Severn Estuary and Inner Bristol Channel.

inputting considerable amounts of trace metals into the estuary (Apte et al., 1990; Langston et al., 2010). The Avonmouth area also hosts incinerators and fertiliser and other chemical plants (Langston et al., 2010).

In the 1970s and 1980s, concentrations of trace metals in the waters (Table 6.1) and sediments (Table 6.4) of the Severn Estuary were certainly high, and indicative of industrial contamination (Morris, 1984; Apte et al., 1990). In the late 1970s, rivers delivered the largest proportions of copper, iron, manganese and nickel into the estuary, together with large amounts of cadmium, lead, mercury and zinc (Owens, 1984; Langston et al., 2010). Industrial discharges emitted major loadings of chromium and mercury, and significant proportions of the inputs of cadmium, copper, iron, lead, manganese and zinc (Owens, 1984; Langston et al., 2010). Domestic sewage was an important source of cadmium and chromium (Owens, 1984; Langston et al., 2010). Atmospheric deposition was particularly important in the outer Severn Estuary, through the agency of the active smelting works at Avonmouth. This deposition contributed about half of the inputs of lead and zinc into the Severn Estuary and also large proportions of cadmium, copper and nickel inputs (Owens, 1984; Langston et al., 2010).

In the case of dissolved trace metal concentrations at that time, nickel typified a trace metal of predominantly riverine origin, decreasing in dissolved concentration with increasing salinity (Morris, 1984). Dissolved concentrations of zinc and cadmium, controlled by input from anthropogenic outfalls and atmospheric deposition, increased with increasing salinity through the Severn Estuary but then decreased with increasing salinity through the Bristol Channel (Morris, 1984). The major input of cadmium was then from the Avonmouth smelter (Radford et al., 1981; Apte et al., 1990). Lead and copper entered the Severn Estuary from several sources, and dissolved concentrations were poorly correlated with salinity in the estuary (Morris, 1984).

The decline in heavy industry locally and the introduction of pollution controls in the later part of the twentieth century have seen a general reduction in the input of trace metals into the Severn Estuary (Langston et al., 2010). Since the late 1970s, trace metal inputs into the estuary from consented discharges and other freshwater sources are estimated to have declined for several trace metals, including mercury (89%), lead (83%), zinc (79%), copper (71%) and cadmium (53%) (Jonas and Millward, 2010; Langston et al., 2010). Consequently, dissolved concentrations of cadmium and lead in the estuary fell about twofold in the 25 years up to 2010 (Langston et al., 2010).

Measurements in 2000 of trace metal concentrations in the sediments of the Severn Estuary also provided clear evidence of substantial declines over the preceding 30-year period (Duquesne et al., 2006). Thus sediment concentrations of lead and zinc between Redwick and Brean on the southern shore fell two- to sixfold, while sediment cadmium concentrations declined nearly tenfold at many sites in the estuary (Duquesne et al., 2006). Sediment contamination by copper, chromium, nickel and lead also declined (Duquesne et al., 2006). Analyses of sediments in 2005 told the same story (Langston et al., 2010). There were reductions across the estuary of 40 to 50% in sediment concentrations of cadmium and mercury in the 25 years leading to 2005, and an 18% reduction in sediment lead concentrations

(Langston et al., 2010). These clear declines in sediment concentrations of contaminant trace metal in the Severn Estuary over a period of a few decades contrast with results discussed earlier for the sediments of the Thames and southwestern estuaries affected by mining. A key point here is probably the high hydrodynamic energy in the Severn Estuary causing considerable turnover of the fine sediments (Langston et al., 2010), including movement downstream. Thus sediments sampled in later years are more recent, and more likely to reflect present or recent inputs of metals from the water column.

This same hydrodynamic energy in the Severn Estuary will tend to distribute sediments over a wide area, but occasionally it is possible to pick out sites of local metal enrichment of the sediment (Langston et al., 2010). Thus, in 2000, the highest sediment trace metal concentrations were found along the Welsh coast at Goldcliff and along the English coast at Portishead (Table 6.4), sites close to the important industrial areas of Cardiff and Avonmouth respectively (Duquesne et al., 2006). Similarly, surveys in 2005 showed sediments close to industrial discharges at Avonmouth still to be high in cadmium, copper, mercury and zinc (Langston et al., 2010). Similarly high sediment concentrations of arsenic, cadmium, chromium, lead, mercury and zinc were found in 2005 at Cardiff (Langston et al., 2010).

Chromium bucks the trend of declining sediment concentrations of trace metals over recent time in the Severn Estuary (Jonas and Millward, 2010; Langston et al., 2010). Chromium sediment concentrations in fine sediments in the estuary appear to have doubled between 1978 and 2005 (Langston et al., 2010). And consented discharges of chromium into the estuary increased over this period, albeit by only 16% (Langston et al., 2010).

6.4.9 Mersey

Prologue. The Mersey Estuary has a long history of industrial metal contamination, not least by mercury emitted by chlor-alkali plants and by organo-lead compounds discharged into the Manchester Ship Canal.

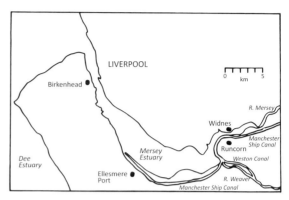

Figure 6.14 Mersey Estuary. (After Langston, 1986, with permission.)

The Mersey Estuary (Figure 6.14) is one of the largest in the United Kingdom, draining a catchment area of about 5,000 square kilometres, including the conurbations of Manchester and Liverpool (Langston et al., 2006). The estuary is 46 kilometres long and its width varies considerably along its length. The main freshwater input into the upper estuary is the River Mersey, while the principal freshwater source into the lower estuary is the Manchester Ship Canal (Langston et al., 2006).

The Mersey Estuary has been the recipient of effluent from large industrial complexes, around Ellesmere Port, Runcorn, Widnes and Manchester (via the Ship Canal). There were also formerly important docks and centres of ship building at Liverpool and Birkenhead (Langston et al., 2006). Development around the estuary has led to the reclamation of large areas of what was once salt marsh, and the shoreline of the estuary is now considerably modified. The Mersey Estuary has a considerable tidal range of about 10 metres, and the resulting strong tidal currents have created deep channels and sandbanks through the estuary (Langston et al., 2006). In spite of the industrial history of the Mersey Estuary, the remaining salt marsh and the intertidal mudflats form an internationally important habitat for wildfowl, including migratory species, with the status of a Ramsar site (Langston et al., 2006).

Industrial pollution of the Mersey Estuary started in the eighteenth century and continued into the second half of the twentieth century, caused by the increasing discharge of liquid waste from textile, tanning, metal processing, chemical and, ultimately, petrochemical industries (Langston et al., 2006). Added to the industrial waste was domestic waste water, sewage and surface runoff from what became a very large populated area. Pollution of the Mersey Estuary was probably at its worst in the 1960s (Langston et al., 2006). In parallel to the situation in the Thames Estuary at the time, the excess load of organic material from sewage, and organically rich industrial effluents, caused the microbial depletion of dissolved oxygen in the estuary to the detriment of the estuarine biota, including migratory fish (Langston et al., 2006). The situation was exacerbated by the mixture of toxic metals and organic chemicals from the large number of local factories. In the 1960s, the estuary was effectively dead, with no fish present.

In the 1970s, North West Water established the Mersey Clean Up Scheme with the aim to reduce discharges of untreated raw sewage and to reoxygenate the Mersey. As in London, the establishment of new sewage plants and the extension of others have gone a long way to restoring oxygen levels and breathing life back into the estuary. Since the 1970s, there have also been schemes to reduce inputs into the Mersey Estuary of many toxic substances, not least mercury, lead and cadmium (Langston et al., 2006).

Mercury played a significant role in the ecotoxicology of the Mersey Estuary in the second half of the twentieth century, via the agency of chlor-alkali plants at Runcorn and Ellesmere Port. The chlor-alkali process is an industrial process for the electrolysis of a solution of salt to produce chlorine and sodium hydroxide (caustic soda) for industrial use. For example, chlorine is used to make PVC plastic and pharmaceuticals. It is present in household bleach and is used on an industrial scale to disinfect drinking water. The chlor-alkali process, however, uses a mercury cell in which the salt solution floats on top of a thin layer of mercury. Unfortunately, the process does result in the release of significant amounts of mercury into the environment. Two other electrolytic methods are available to produce chlorine and caustic soda, and mercury cells are being phased out. In Europe, mercury cell technology still produced just over 50%

of capacity in 2000, but less than 25% in 2013, with the aim of complete phase-out by 2017. Back in the Mersey Estuary, there has been a corresponding replacement of the mercury cell method, with an associated reduction in mercury contamination (Langston et al., 2006). The plant at Runcorn discharged more than 70 tonnes of mercury per year into the estuary in the mid-1970s, but less than 1 tonne per year in 1995, and 298 kilograms in 1998 (Langston et al., 2006). Nevertheless, in the middle of the first decade of the twenty-first century, the Runcorn plant still represented the largest point source discharge of mercury in the United Kingdom. The Ellesmere Port plant used to emit about 0.3 tonnes of mercury per year into the Manchester Ship Canal, but had changed to a mercury-free process by 2005 (Langston et al., 2006).

The Associated Octel plant at Ellesmere Port used to produce tetra-ethyl lead compounds (Langston et al., 2006), forms of organolead, for use as antiknock agents in petrol, a procedure discontinued at the very end of the last century. These organolead compounds were discharged into the Manchester Ship Canal and thence into the Mersey Estuary (Langston et al., 2006). This production has now ceased, but not before these organolead compounds were considered to have had an ecotoxicological effect on birds overwintering in the estuary between 1979 and 1982 (Bryan and Langston, 1992; Langston et al., 2006), as we shall discuss in more detail later.

In 1985, there were 10 electroplating industrial plants discharging about 0.282 tonnes of cadmium per year into the Mersey Estuary (Langston et al., 2006). An European Community directive in that year required regulation of cadmium discharges, and input had dropped to 0.052 tonnes of cadmium per year in 1991 (Langston et al., 2006).

Table 6.4 shows concentrations of trace metals in sediments of the Mersey Estuary in the early 1980s. As would be expected from the preceding discussion, mercury sediment concentrations are the highest listed for all British estuaries. Total concentrations of lead in the sediments of the Mersey Estuary are fairly typical for the industrial estuaries listed, below the sediment concentrations of estuaries affected by lead mining (Table 6.4). However, it is the much lower sediment concentrations of the organolead component of the total lead in the sediments that are of potential ecotoxicological concern. Sediment concentrations of cadmium in the Mersey Estuary in the early 1980s are on the high side in comparison with other estuaries, but still fall beneath the very high cadmium concentrations of Severn Estuary sediments near Avonmouth in 1970 (Table 6.4).

6.4.10 Humber

Prologue. The Humber Estuary has long received input of industrial wastes, including metals from the Rivers Trent and Yorkshire Ouse, as well as local input of waste effluent from industrial plants on the shores of the estuary itself.

The Humber Estuary has the largest catchment area of any UK estuary, at about 25,000 square kilometres, representing more than 20% of the land area of England (Cave et al., 2003). The main rivers draining into the Humber Estuary are the Trent from the south and the Yorkshire Ouse from the west and north. The River Trent and its tributaries drain a large industrial and urban area, encompassing Stoke-on-Trent, Birmingham (via the River Tame), Leicester, Derby and Nottingham. Derby is on the Derbyshire Derwent, a river met in the last chapter, serving the lead-mining region of the Peak District. Tributaries of the Yorkshire Ouse include the Don, draining Sheffield and Doncaster, and the Aire, receiving input from Leeds and Bradford. Tributaries of the Yorkshire Ouse that drain historical mining regions of the Pennines include the rivers Wharfe, Nidd and Swale. Any metal input from mining in these rivers has been well diluted by the time the Ouse flows into the Humber Estuary. The resulting average flow of freshwater through the Humber Estuary into the North Sea is 250 cubic metres per second, much of this water being derived from industrial and domestic effluents (Lee and Cundy, 2001; Cave et al., 2003).

The lower regions of both the Rivers Trent and Yorkshire Ouse are tidal. The River Trent is tidal for about 85 kilometres upstream of its confluence with

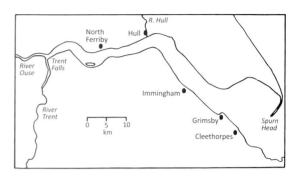

Figure 6.15 Humber Estuary.

the Ouse at Trent Falls, with seawater intrusion reaching Gainsborough, some 40 kilometres inland of Trent Falls (Cave et al., 2003). The tide is excluded from the tributaries of the River Trent by tidal gates and sluices. The tidal limit in the River Ouse is near Acaster, 62 kilometres upstream of Trent Falls, with seawater intrusion reaching Boothferry, about 20 kilometres inland of Trent Falls (Cave et al., 2003). As in the case of the River Trent, the tributaries of the Ouse have weirs restricting tidal incursion.

The Humber Estuary (Figure 6.15) is 62 kilometres long from Trent Falls to Spurn Head, where it enters the North Sea (Cave et al., 2003). On the north bank of the Humber Estuary is the city of Kingston upon Hull, usually abbreviated to Hull, situated on the small river Hull. On the south bank of the estuary, towards the seaward end, are the towns and ports of Immingham and Grimsby and the resort of Cleethorpes.

The Humber Estuary is a shallow, well-mixed estuary with a large tidal range of 7.2 metres and strong tidal currents (Cave et al., 2003). Mean depths range from 3 metres in the inner estuary to 8 metres at the mouth, although continuous dredging maintains shipping channels as deep as 16 metres in the outer estuary. The waters of the Humber Estuary are characteristically turbid, with very high loadings of suspended sediment (Lee and Cundy, 2001). An effect of the considerable redistribution of the sediments by tidal action means that any local enrichment of sediment concentrations by a single industrial source is quickly reduced, and any trace metal enrichment of sediments is spread widely through the estuary (Grant and Middleton, 1990; Cave et al., 2003).

The Rivers Trent and Yorkshire Ouse have long been a combined source of trace metals into the Humber Estuary (Lee and Cundy, 2001; Cave et al., 2003). In the first decade of this century, these rivers still delivered the greatest loads of arsenic, cadmium, copper, mercury, nickel and lead into the estuary (Cave et al., 2003). In the period 1985 to 1992, up to 420 kilograms of copper, 540 kilograms of lead and 2,670 kilograms of zinc per day were input into the Humber Estuary, with much of these metals being trapped in the estuary, bound to fine sediment particles (Lee and Cundy, 2001). These inputs can be expected to be lower today, with the increasing regulation of industrial effluent control and industrial decline since the late twentieth century. Analysis of sediment cores has shown that upstream mining for lead and zinc was the most important source of trace metals into the Humber from the thirteenth to the eighteenth centuries (Grant and Middleton, 1990; Cave et al., 2003). Thereafter, industrial development was the greatest cause of trace metal contamination in the Humber Estuary, with the local development in the eighteenth, nineteenth and twentieth centuries of coal mining, textile and chemical industries (Cave et al., 2003). Added to this industrial development was the effect of considerable urbanisation in the Leeds–Bradford area (Cave et al., 2003). Metal mining has now ceased in the Yorkshire Dales and the Peak District, and coal mining in the Humber catchment all but eliminated. Coal-burning power stations (using imported coal) in the Humber catchment include Ferrybridge, Eggborough and Drax on the tidal River Ouse, the largest coal-burning power station in Europe. Each of the three coal-fired power stations can release up to 1 tonne per year of copper into the Ouse system (Cave et al., 2003). In 1997, the Drax power station released more than 72 tonnes of trace metals into the River Ouse (Cave et al., 2003). While there is some emission of trace metals in effluent cooling water from these power stations, a greater potential source of metals is pulverised fly ash (PFA) resulting from the burning of the coal. PFA contains high concentrations of selected trace metals, such as arsenic and selenium. Disposal of this PFA in local landfill sites in the Selby area does have an inherent risk of potential loss of the trace metals into local water systems.

In addition to the trace metals entering the Humber Estuary via the rivers draining its catchment, there are local sources of sewage in the estuary itself, not least from the city of Hull, essentially untreated until the start of this century (Cave et al., 2003). Furthermore, until 1998, approximately 100,000 tonnes of wet sewage sludge were dumped annually just outside the mouth of the estuary (Cave et al., 2003). In addition, there have been, and are, direct metal discharges from industry into the Humber Estuary (Lee and Cundy, 2001; Cave et al., 2003). Local industrial discharges provide the major sources of chromium and zinc into the estuary, and, until the early 1990s, they were also the main source of copper (Cave et al., 2003). Of two titanium dioxide plants on the south coast of the estuary between Immingham and Grimsby, one shut in 2009, while the other (temporarily?) ceased production of titanium dioxide in the same year in response to recession. These two plants had been producing effluents since the 1950s that contained substantial amounts of aluminium, iron and titanium, staining the local shore red with the iron oxide (Grant and Middleton, 1990). The estuarine sediments associated with this industrial area on the south bank are enriched with these three metals, as well as chromium and vanadium (Grant and Middleton, 1990). On the north bank, the Capper Pass tin smelter at North Ferriby, 13 kilometres to the west of Hull, was one of the largest tin smelters in Europe until its closure in 1991 (Cave et al., 2003). Analyses of estuarine sediments in 1988 also showed very high levels of lead in the sediment on the north bank just east of Hull (Grant and Middleton, 1990). The source of this lead is unknown, and may be down to local sewage discharge, dock or industrial discharges (Grant and Middleton, 1990).

As befits an estuary draining an industrialised catchment, the sediments of the Humber Estuary are enriched with trace metals, comparably to many other industrialised estuaries, and below those of the mining estuaries of the southwest (Table 6.4). The macrotidal nature of the estuary has caused its sediments to become well mixed, making it difficult to identify local point sources of trace metals in the estuary (Grant and Middleton, 1990). An exception to this generality is represented by the emissions of the titanium dioxide processing plants (Grant and Middleton, 1990).

6.4.11 South Wales

Prologue. The region of south Wales around Swansea, Neath and Llanelli has a long history of industry driven by local mining for coal, not least a history of smelting copper and zinc delivering high metal loadings into the Tawe Estuary and Swansea Bay. The tinplate industry of Llanelli has caused very high sediment concentrations of tin and chromium in the Loughor Estuary.

The historical industrial region of south Wales around Swansea, Neath and Llanelli is drained by the Rivers Tawe, Neath and Loughor respectively. The Rivers Tawe and Neath drain into Swansea Bay, while the River Loughor drains into the Bristol Channel on the northern side of the Gower peninsula (Figure 6.16).

Coal has been mined in the Lower Swansea valley since the fourteenth century, providing the fuel for industrial development, including metal smelting, particularly of copper and zinc (Vivian and Massie, 1977). Copper smelting started in 1717 and zinc smelting in the 1850s, with metal smelting peaking in the latter part of the nineteenth century (Vivian and Massie, 1977). The last copper smelter closed in 1920 and the last zinc smelter in 1971 (Vivian and Massie, 1977), leaving only the nickel refinery at Clydach, upstream of Swansea on the River Tawe, still active in reduced form today. Effluent from the nickel refinery enters the River Tawe for transport down to its estuary, but now in greatly lowered amounts compared with the 1960s. The Swansea area has also been the site of large dumps of industrial waste, now removed (Chubb et al., 1980; Vivian, 1980), with the potential for metal-rich drainage runoff, ultimately into the River Tawe. Very high levels of cadmium, lead and zinc were reported in the 1960s in the Nant-y-Fendrod, a small tributary joining the Tawe just above Landore (Figure 6.16), the metals being derived

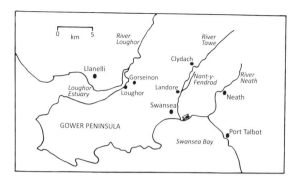

Figure 6.16 South Wales: Swansea Bay receiving the rivers Tawe and Neath, and the estuary of the River Loughor.

both from runoff from tips and via effluent from a local zinc smelter (Vivian and Massie, 1977; Chubb et al., 1980).

Nearby Neath was a market town that, like Swansea, boomed with the coming of the industrial revolution in the eighteenth century, exploiting local coal to supply iron, steel and tinplate industries alongside the River Neath and its estuary.

By the 1970s, the River Tawe (with its tributary the Nant-y-Fendrod) and the River Neath were discharging large amounts of toxic metals derived from local industry through their estuaries into Swansea Bay (Chubb et al., 1980; Vivian, 1980). Sediments at the top of the Tawe Estuary then contained exceedingly high total concentrations of cadmium, nickel, lead and zinc, large proportions of which were extractable with 25% acetic acid (Table 6.4) (Vivian, 1980). Since the 1970s, increasing regulatory control of industrial effluents and the demise of local industry have much reduced the supply of toxic metals into the estuaries of the Rivers Tawe and Neath. Nevertheless, an industrial legacy might reside as continuing high sediment concentrations of trace metals in the sediments, unless the sediments themselves are moved on by strong tidal action or by the intervention of dredging.

Llanelli, on the other side of the Gower peninsula, has been a major centre of tinplate production since the eighteenth century. Indeed, Llanelli was referred to as 'Tinopolis' in the second half of the 1800s. The tinplate industry continued to flourish in the twentieth century and into this millennium, tinplate being mainly used to make tin cans. Tinplate production is closely linked to the steel industry, now declining locally as at Port Talbot, for tinplating involves the addition of a coating of tin by electrolysis onto steel. Indeed, tinplate is a shorthand term for tinplated steel.

Effluents from the tinplate industry into the River Loughor and its estuary have historically been high both in tin, as expected, but also in chromium (Bryan and Langston, 1992). Thus, sediments in the Loughor Estuary in the 1970s and 1980s were exorbitantly high in both tin and chromium (Table 6.4). While higher sediment concentrations of tin have been found in the mining-affected Restronguet Creek and Hayle Estuary (Table 6.4), that tin is in the form of particles of cassiterite of low bioavailability. Much of the tin in the sediments of the Loughor Estuary, on the other hand, did not require application of the fusion technique for analysis (Table 6.4) (Bryan et al., 1980; Bryan and Langston, 1992), suggesting that this industrially derived tin would be more bioavailable to estuarine organisms.

6.4.12 Poole Harbour

Prologue. Holes Bay in the north part of Poole Harbour in Dorset received industrial effluent rich in trace metals up to the late 1970s.

The last specific example of industrial contamination of an estuary to be considered here concerns Poole Harbour on the Dorset coast (Boyden, 1975; Thornton et al., 1975). Poole Harbour extends for about 10 kilometres from Wareham in the west to the suburbs of Bournemouth in the east, with the town of Poole on its northwest shore (Figure 6.17). The main freshwater input into Poole Harbour is from the River Frome in the west.

Holes Bay, a shallow embayment on the northern side of the harbour on the western edge of Poole, was the location in the twentieth century of both metal plating and chemical manufacture industries, emitting effluent rich in trace metals into the bay to add to loads from domestic sewage discharges (Wardlaw, 2005). Alarm bells had begun to ring in 1972 when a new oyster hatchery opened in Poole Harbour at the entrance of Holes Bay. The hatchery immediately encountered ecotoxicological problems in the rearing of oyster larvae in the local estuarine waters (Boyden, 1975). Local estuarine sediments were shown to be high in cadmium, copper, lead and zinc (Table 6.4), and it was concluded that occasional peaks in the dissolved concentration of zinc were sufficiently high, without the addition of the other toxic metals, to have influenced hatchery larval survival (Boyden, 1975). By the late 1970s, the local

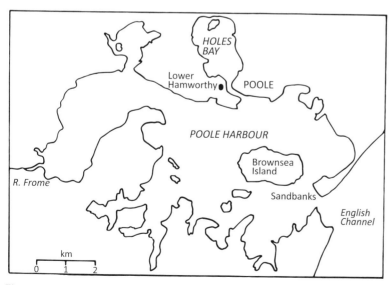

Figure 6.17 Poole Harbour, Dorset.

industrial trade effluents were treated sufficiently to meet the more stringent regulation standards then introduced (Boyden, 1975).

6.5 Salt Marshes and Seagrasses

Prologue. We turn our attention to the biology of trace metals in the flora and fauna of estuaries. In the case of flora, flowering plants are represented in estuaries by the plants of salt marshes and by seagrasses.

Although flowering plants are generally rare in marine systems, several salt-tolerant species can live intertidally, populating salt marshes and mangrove swamps in the upper regions of sheltered shores in temperate and tropical regions respectively. Seagrasses can also grow on the sediments of shores and shallow sublittoral regions throughout the world. In the British Isles, salt marshes are common at the top of estuaries, and beds of seagrasses (often species of the eelgrass genus *Zostera* and its relatives) occur lower down on many shores.

The muds and, to a lesser extent, the sands of estuaries act as a trap for trace metals. Flowering plants, of course, have roots and are, therefore, able to take up trace metals from these sediments, as well as dissolved metals from the overlying water.

The biology of metals in intertidal flowering plants remains relatively unexplored, although these salt-tolerant plants do accumulate trace metals in both roots and shoots in proportion to local bioavailabilities (Bryan and Gibbs, 1983). Thus the salt marsh plant glasswort (*Salicornia europea*) from Restronguet Creek had higher copper and zinc concentrations in shoots (169 µg/g Cu, 289 µg/g Zn) and roots (611 µg/g Cu, 359 µg/g Zn) than glasswort collected from the uncontaminated nearby Avon Estuary in Devon (shoots, 12 µg/g Cu, 62 µg/g Zn; roots, 16 µg/g Cu, 34 µg/g Zn) (Bryan and Gibbs, 1983).

Trace metal concentrations in the seagrass *Zostera noltei* have recently been measured at intertidal sites around Ireland in order to establish a baseline for ongoing monitoring and assessment (Wilkes et al., 2017).

6.6 Seaweeds

Prologue. Seaweeds are the dominant flora of British coasts including estuaries. A common seaweed found in British estuaries is the brown alga *Fucus vesiculosus*, the bladder wrack, while green algae of the genus *Ulva* may also be present. Bladder wrack is a good biomonitor of dissolved bioavailabilities of trace metals in estuaries.

Seaweeds are macrophytic algae, lacking root systems. So, unless the frond of the seaweed is in contact with the local sediment, all the trace metals taken up and accumulated by this frond are derived from dissolved trace metal in the water column. Seaweeds accumulate trace metals in proportion to trace metal exposure, and bladder wrack is considered an excellent biomonitor of dissolved trace metal bioavailabilities in estuaries (Bryan et al., 1985; Luoma and Rainbow, 2008). Correspondingly, there is a good database of accumulated trace metal concentrations in *F. vesiculosus* from many British estuaries (Table 6.5). Furthermore, there is a good appreciation of which of these accumulated concentrations can be considered atypically raised and, therefore, indicative of unusually high local dissolved bioavailabilities of the trace metals concerned (Luoma and Rainbow, 2008; Johnstone et al., 2016).

Table 6.5 presents a selection from this database. The tip of the frond of *F. vesiculosus* consists of younger tissues in which accumulated trace metal concentrations are still increasing (Bryan et al., 1985). It is usual, then, to select for analysis a region of frond about 10 centimetres from the tip, ensuring that it is free from any epiphytes (Bryan et al., 1985; Rainbow et al., 2002). The trace metal concentrations in this part of the frond will provide information on dissolved bioavailabilities integrated over several months, bladder wrack usually growing at 2 to 3 centimetres per month (Knight and Parke, 1950; Bryan et al., 1985).

Certain trace metal concentrations accumulated by *F. vesiculosus* in Table 6.5 stand out as being remarkably high in comparison to what might be

Table 6.5 **Mean, range or range of mean concentrations (μg/g dry weight) of trace metals in bladder wrack *Fucus vesiculosus* from British estuaries affected by (a) mining contamination, (b) industrial contamination or (c) considered uncontaminated.**

	Ag	As	Cd	Co	Cr	Cu	Fe	Hg	Mn	Ni	Pb	Zn
(a) Mining												
Restronguet Creek												
1975–77[1]		38–184				86–91						710–738
1981[2]	0.6–2.2		0.8–1.4	3.8–20	1.6–4.7	717–1,450	2,030–13,500		61–533	4–10	16–68	2,440–4,200
1981[3]		69–382										
2006[4]	0.4	128				166					10	449
Hayle												
1977[2]	0.8		2.3	3.5	4.6	436	4,810		149	22	32	1,860
Gannel												
2006[4]	0.3	31				15					33	206
East Looe												
2006[4]	1.6	28				15					15	186
West Looe												
2006[4]	0.7	22				16					34	135
Looe												
1975–1976[5]	0.8		1.3	5.3	2.2	17	1,450		363	10	38	198
Mouth 1976[2]	0.4		1.4	0.6	0.6	8	121		94	6	3	104
2006[4]	0.5	29				21					31	125
Tamar												
1981[3]		38–59										
1982[2]	0.2		0.8	2.3	7.7	27	631		78	2	11	113
Tavy												
2006[3,4]	0.5	86				88					45	382

Table 6.5 (cont.)

	Ag	As	Cd	Co	Cr	Cu	Fe	Hg	Mn	Ni	Pb	Zn
Dulas Bay												
Mouth 1972[6]						49–97	41–168		52–97	6		228–398
Ystuyth												
1992[7]						5–12			57–251		86–103	521–4,940
Rheidol												
1992[7]						14 28			85–487		46–73	500–3,000
(b) Industrial												
Thames												
1997[8]	0.6–4.4	4–19	0.3–3.0	1.3–4.7	0.6–2.0	10–27	223–1,650	0.02–0.06	67–367	9–31	1–8	53–701
2001[9]	2.2–8.9	6–23	0.3–3.6	1.3–5.4	0.6–2.2	7–36	289–837	0.02–0.08	113–505	6–29	4–18	33–564
2001[10,11]	1.3–1.9		1.9–3.3			13–21	559–2,690		187–353		10–15	230–575
2014[11]	0.2–0.8		0.5–1.0	1.9–2.9	1.9–3.7	12–18	873–1,550		154–228	13–18	1–4	110–301
Severn												
1970[12]			20–220								0.2–9	240–800
Bristol Channel												
1971[13]			2–75								5–19	32–560
Mersey												
1980–84[14]	0.1–0.5	12–49	0.5–2.5		0.7–6.6	10–42	227–1,530	0.07–0.42	69–264	4–23	1–16	209–1,960
Humber												
1981–83[15]			1.6–11			15–97	471–9,130		154–627	6–86	3–22	242–1,040
Forth												
1982[16]								0.1–2.3				

(c) Uncontaminated

Torridge (Devon)

1980[2]	0.3		1.2	2.3	2.2	10	746	193	11	5	99
1981[3]		21–30									
Typical[11]	0.1–1.0	4–30	0.2–2.5	0.6–6.0	0.6–5	7–75	35–250	35–100	1–10	1–20	20–100
High[11]	2–10	50–500	4–100	16–70	55	200–1,500	500–14,000	140–700	30–60	20–610	500–11,000

Sources: Klumpp and Peterson (1979);[1] Bryan and Gibbs (1983);[2] Langston (1984);[3] Rainbow et al. (2011c);[4] Bryan and Hummerstone (1977);[5] Foster (1976);[6] Rainbow and Ward (unpublished);[7] McEvoy et al. (2000);[8] Langston et al. (2004);[9] Rainbow et al. (2002);[10] Johnstone et al. (2016);[11] Butterworth et al. (1972);[12] Nickless et al. (1972);[13] Langston (1986);[14] Barnett and Ashcroft (1985);[15] Elliott and Griffiths (1986).[16]

considered typical in a British estuary. It is difficult to justify the use of terms such as pristine or background in this context, given the fact that estuaries around the British Isles are unlikely to be free of any anthropogenic contamination at all.

As to be expected, Restronguet Creek is one site with very high concentrations of particular trace metals in bladder wrack, indicative of high dissolved bioavailabilities, particularly historically (Table 6.5). The trace metals concerned are copper and zinc, especially, and also arsenic, lead, iron and manganese (Table 6.5). The highest accumulated concentrations date from the 1970s and 1980s, when there was still some mining being carried out in the Carnon valley. By the first decade of this century, accumulated concentrations of copper and zinc in bladder wrack had certainly fallen (Table 6.5), indicating a decline in their dissolved bioavailabilities in Restronguet Creek (Rainbow et al., 2011c).

As also to be expected, accumulated concentrations of lead and zinc in *F. vesiculosus* were also very high in the estuaries of the Rivers Ystwyth and Rheidol, in this case in 1992 (Table 6.5).

In the industrially contaminated estuaries, phenomenally high concentrations of cadmium were accumulated by bladder wrack at sites on the south bank of the Severn Estuary in 1970 (Table 6.5, Figure 6.13) (Butterworth et al., 1972). At Portishead, immediately downstream of the then active Avonmouth smelter, cadmium concentrations in *F. vesiculosus* reached a staggering 220 µg/g (Butterworth et al., 1972). Cadmium concentrations in the seaweed attained 200 µg/g at Clevedon slightly further downstream, with a gradient of accumulated concentrations down to 20 µg/g at Minehead (Butterworth et al., 1972), a concentration that is still very high on any scale (Table 6.5). Bioaccumulated zinc concentrations showed a similar gradient along the southern shore of the Severn Estuary, from the very high concentration of 800 µg/g at Portishead to 240 µg/g at Minehead (Table 6.5) (Butterworth et al., 1972). These bioaccumulated concentrations can be expected now to be long gone after the introduction of tighter environmental regulation of industrial effluents and ultimately the closure of the Avonmouth smelter. Concentrations of cadmium and zinc in bladder wrack were still high at sites in the Bristol Channel beyond the Severn Estuary in 1971 (Table 6.5) (Nickless et al., 1972).

Concentrations of mercury accumulated by *F. vesiculosus* were high (up to 0.42 µg/g) in the early 1980s in the Mersey Estuary, not unexpectedly given the presence of chlor-alkali factories (Langston, 1986). Yet, a more remarkable mercury concentration of 2.3 µg/g was recorded from bladder wrack in the upper regions of the Forth Estuary at Longannet in 1982 (Table 6.5) (Elliott and Griffiths, 1986). Longannet is near Grangemouth and immediately downstream of a site of industrial discharge of inorganic mercury compounds in the mid-1970s (Elliott and Griffiths, 1986). In spite of a reduction in the amount of mercury in this discharge in the 1970s, there was still clearly a high level of bioavailable dissolved mercury at Longannet in 1982 (Elliott and Griffiths, 1986).

The 1980s and 1990s had seen exponential decreases in the dissolved concentrations of many trace metals in the Thames Estuary (Power et al., 1999), and there were additional declines between 1997 and 2001 (Pope and Langston, 2011). Furthermore, there were falls in the dissolved bioavailabilities of these trace metals in the Thames Estuary between 2001 and 2014, reflected in significant declines in bladder wrack concentrations of cadmium, copper, lead, manganese, silver and zinc (Johnstone et al., 2016).

Other seaweeds are also common in British estuaries and may also be used as trace metal biomonitors. These seaweeds include brown seaweeds that are close relatives of bladder wrack. *Fucus ceranoides*, horned wrack, penetrates further upstream in estuaries than *F. vesiculosus*, while *F. spiralis*, spiral wrack, and *Ascophyllum nodosum*, referred to as knotted wrack or egg wrack, occur further downstream. *F. serratus*, serrated or toothed wrack, may be found at the mouth of an estuary.

Green algae are also common, as in the cases of sea lettuce *Ulva lactuca*, and its relative *Ulva* (formerly

Enteromorpha) *intestinalis*, sometimes called gutweed (Bryan et al., 1985). A study of the contamination by selected trace metals of six estuaries on the North Sea coast of Britain in 1984 identified accumulated concentration ranges in *Ulva intestinalis* indicative of moderate and high contamination (Say et al., 1990). These moderate (high) accumulated concentrations are 0.5 to 1.5 (>1.5) µg/g for cadmium, 0.05 to 0.15 (>0.15) µg/g for mercury, 20 to 60 (>60) µg/g for lead and 50 to 150 (>150) µg/g for zinc (Say et al., 1990). High contamination of cadmium was found in the upper regions of the Humber and Tyne Estuaries, of mercury in the upper Forth Estuary, of lead in the Tyne and Thames Estuaries and of zinc in the upper Tyne Estuary (Say et al., 1990).

The trace metals accumulated in estuaries by seaweeds such as bladder wrack have the potential to be transferred along food chains. Grazing gastropods such as limpets, for example *Patella vulgata*, or periwinkles such as *Littorina littorea*, the edible periwinkle, feed directly on seaweeds. Alternatively, decaying seaweed in estuarine strandlines forms the food of detritivores, such as the talitrid amphipod *Orchestia gammarellus*, a beachhopper. The trophic transfer of trace metals to the next trophic level of a food chain does to a degree depend on the chemical binding of the trace metal in the seaweed. Part of the accumulated trace metal content of a seaweed will be exchangeable metal adsorbed onto the algal cell wall, probably bound to alginates in brown algae and carrageenan in red algae (Philips, 1994). Most of the accumulated trace metals, however, will be accumulated within the cells of the seaweed, associated with polyphenols and phytochelatins (Phillips, 1994). Accumulated arsenic in algae is transformed from inorganic arsenate, the form in which it is usually taken up, to less toxic organoarsenic compounds such as arsenosugars (Bryan and Langston, 1992). High accumulated concentrations of arsenic in seaweeds from arsenic-contaminated estuaries, do not, therefore, represent a toxic challenge to animals higher up estuarine food chains.

We have seen that the presence of a metal-tolerant population of an organism at a particular site is evidence that the total local bioavailability of that toxic metal is of ecotoxicological significance, clearly to that species, but potentially also to other organisms present (Luoma, 1977). Experiments in the early 1980s showed that the growth rates of *Fucus vesiculosus* from Restronguet Creek were unaffected by dissolved concentrations of copper that reduced the growth rate of a population of the same species from the Tamar Estuary (Bryan and Gibbs, 1983). High dissolved concentrations of zinc affected the growth rates of both populations similarly. The Restronguet Creek population of bladder wrack was, therefore, tolerant to copper but not zinc (Bryan and Gibbs, 1983).

6.7 Meiofauna

Prologue. Sediment communities of meiofauna, particularly nematodes, respond to elevations in trace metal bioavailabilities in estuarine sediments.

Just as in freshwater, the sediments of estuaries are home to meiofauna, those very small animals in the size range between 45 µm and 1 millimetre. Dominant members of the meiofauna of estuarine sediments are nematode worms and harpacticoid copepods.

A 1991 survey of the sediment meiofaunal communities in five creeks of the Fal Estuary system (Figure 6.6), including Restronguet Creek, with different sediment concentrations of trace metals, particularly copper, showed that there was a strong relationship between meiofaunal nematode community structure and sediment trace metal concentrations (Somerfield et al., 1994). The nematodes were more responsive than the harpacticoid copepods to sediment trace metal concentrations, showing different community compositions, including lower generic richness and diversity, across the gradient of these metal concentrations in the different creeks (Somerfield et al., 1994). The harpacticoid copepods, on the other hand, only showed a difference in community composition in the highly contaminated sediment of Restronguet Creek (Somerfield et al., 1994).

A random selection of nematodes from Restronguet Creek were more tolerant of copper exposure than nematodes from Percuil Creek (Figure 6.6) (Millward and Grant, 1995; Grant and Millward, 1997), the least trace metal contaminated of the Fal Estuary creeks in the previous study (Somerfield et al., 1994). This finding is a demonstration of the usefulness of PICT as an ecotoxicological tool (Blanck et al., 1988). The increase in community tolerance of Restronguet Creek nematodes is a result of the loss of sensitive species and increased abundance of more copper-resistant nematode species, together, apparently, with selection for a copper-tolerant population in at least one nematode species (Grant and Millward, 1997; Millward and Grant, 2000). In fact, more detailed PICT observations showed that a successively weaker impact of copper exposure was also apparent in nematode communities from some of the moderately copper-contaminated sediments of other creeks in the Fal Estuary system, in the order Restronguet, Mylor, Pill and St Just (Figure 6.6) (Grant and Millward, 1997; Millward and Grant, 2000). An expanded PICT study of copper tolerance in meiofaunal nematode communities in estuaries of Southwest England implied that a threshold sediment concentration of 200 μg/g copper (cf. Table 6.4) is needed before selection for nematode community copper tolerance (Millward and Grant, 2000).

A species of nematode that appears relatively trace metal–resistant is *Tripyloides marinus*, abundant in Restronguet Creek but also present in other creeks of the Fal Estuary and throughout northern Europe (Millward, 1996). In fact, the Restronguet Creek population of *T. marinus* has also developed tolerance to copper (Millward, 1996). The physiological basis of the copper resistance of this species in general appears to be based on the efficient detoxification of accumulated copper (and probably also zinc) in intracellular phosphate-based granules in the intestinal epithelium, with their subsequent exclusion into the gut lumen (Millward, 1996). The copper tolerance of the Restronguet Creek population of this nematode can be interpreted to be a result of an increased detoxification rate based on this preexisting detoxification mechanism (Millward, 1996).

6.8 Deposit-Feeding Polychaetes: Ragworms and Lugworms

Prologue. Two ecologically important deposit-feeding polychaetes living in the sediments of British estuaries are the ragworm *Hediste diversicolor* and the lugworm *Arenicola marina*. These polychaetes take up trace metals from solution and from ingested sediment. Both can be used as biomonitors of trace metal bioavailabilities in estuaries.

We have seen in this chapter how the sediments of sheltered estuaries will still retain high concentrations of contaminant trace metals long after the input of dissolved metals into the estuary has been reduced. These sediment-associated metals may be released into the pore water of the sediment with the potential then for dissolved uptake by soft-bodied burrowing invertebrates. Often more importantly, however, these sediment metals are a major source of trace metals to those invertebrates that ingest the sediment. We start here with two such deposit-feeding invertebrates, both polychaete worms – the ragworm *Hediste diversicolor* and the lugworm *Arenicola marina*.

6.8.1 *Hediste diversicolor*: a Ragworm

Prologue. The ragworm *Hediste diversicolor* is a very common burrowing inhabitant of the intertidal muds and sands of British estuaries, being remarkably tolerant of low and varying salinities. This ragworm ingests sediment but can also feed by other mechanisms.

Hediste diversicolor (Plate 6a) will be well known to many natural historians as *Nereis diversicolor*, the specific name referring to the variability between individuals of its colour, which also tends to change from red or brown to green with the start of the breeding season.

Box 6.2 What's in a Name?

The change of the name of the genus of *Hediste diversicolor* from *Nereis* is a good illustration of the activity of taxonomists. When proposing a new species name for a specimen, a taxonomist is actually presenting a hypothesis. Under the Linnean binomial system, the first of the two Latin names refers to the genus. By putting a proposed new species into a particular genus, the taxonomist is hypothesising that the specimen (to be called the type specimen) is sufficiently closely related to other species in that genus to be linked to them under that generic name. On the other hand, it is also sufficiently distinct from those other species in the genus to be designated as a new species in its own right. In the case of *H. diversicolor*, it was previously included in the genus *Nereis*, an increasingly large genus with many species. And yet, within the genus *Nereis* could be recognised groupings of species, more closely related to each other than to other species in the genus. These groupings could be called subgenera, and *Hediste* was one such subgenus. Here we come to another taxonomic hypothesis, for it was later proposed that species of the subgenus *Hediste* were sufficiently distinct from other species of *Nereis* that *Hediste* should be distinguished as a separate genus. Hence the new hypothesis is that *Nereis diversicolor* should become *Hediste diversicolor*. Such taxonomic reevaluations almost become inevitable as a genus becomes large and unwieldly. We shall see the same for the barnacle genus *Balanus* later. Indeed, a similar new generic assignment has happened to another British intertidal ragworm. The impressive large green polychaete, the king ragworm, has also suffered changes of designation of genus, being described in sequence as *Nereis virens*, *Neanthes virens* and presently *Alitta virens*. But watch this space. Other polychaete specialists may yet refute this latest hypothesis.

It was stated previously that *Hediste diversicolor* ingests sediment. Indeed it does, but this statement oversimplifies the range of feeding mechanisms available to this remarkable polychaete (Scaps, 2002). In addition to deposit feeding, *H. diversicolor* can feed as a predator, as a scavenger, as a detritivore and even as a suspension feeder. In the latter role, it can spin a net of mucus at the entrance to its burrow and filter phytoplankton and zooplankton from its respiratory irrigation current (Harley, 1950). Periodically, the net with its nutritious load is swallowed, and a new net secreted.

6.8.1.1 Bioaccumulation

Prologue. *Hediste diversicolor* is a net accumulator of trace metals from the sediment. Accumulated zinc concentrations typically vary only over a small range.

In spite of its variety of feeding methods, it is valid from the point of view of its accumulation of trace metals to consider *Hediste diversicolor* primarily as an ingestor of sediments. Field evidence shows significant relationships between its bioaccumulated trace metal concentrations and those in sediments (Bryan et al., 1980, 1985; Bryan and Langston, 1992). Furthermore, it has been shown that arsenic, copper and lead are mostly bioaccumulated by *H. diversicolor* from ingested sediment (Bryan, 1984). Biodynamic modelling has demonstrated that between 52 and 96% of arsenic, more than 99% of cadmium and more than 98% of zinc accumulated by this ragworm is derived from sediment ingestion (Rainbow et al., 2009b, 2011b).

As indicated by the presence of significant relationships between bioaccumulated and sediment trace metal concentrations, *Hediste diversicolor* can be considered as a net accumulator of trace metals. Indeed, in the cases of copper and cadmium, this net accumulation results in a wide range of accumulated concentrations of copper and cadmium in the ragworm in estuaries affected by mining contamination (Bryan and Hummerstone, 1971, 1973b; Geffard et al., 2005). In contrast to copper and cadmium, the concentration of zinc accumulated by *H. diversicolor* typically ranges by a much smaller factor across metal-contaminated estuaries (Bryan and Hummerstone, 1973b). While sediment zinc concentrations varied by a factor of 30 from about 100 to 3,000 µg/g in a selection of Southwest English estuaries, accumulated zinc concentrations in *H. diversicolor* usually varied less than threefold from 130 to 350 µg/g (Bryan and Hummerstone, 1973b). This situation approaches the regulation of body concentrations of zinc described for some decapod crustaceans in Chapter 3, but is better described as weak net accumulation. Under very high zinc bioavailabilities, and therefore very high zinc uptake rates, this accumulation pattern may break down to bring about much higher accumulated zinc concentrations, as occasionally in Restronguet Creek worms (Rainbow et al., 2009a, b).

Net accumulation of high concentrations of trace metals is necessarily associated with their detoxification for permanent or temporary storage. Indeed, this is the case for *Hediste diversicolor*. Accumulated copper is stored in copper and sulphur-rich lysosomal residual bodies in the epidermis and subsequently in copper and sulphur-rich extracellular granules in the epicuticle (Mouneyrac et al., 2003; Geffard et al., 2005). These lysosomal residual bodies would have been derived from the breakdown of metallothionein induced to bind soluble copper on entry. Some copper is excreted in the long term by *H. diversicolor*, probably in association with renewal of the cuticle (Geffard et al., 2005). When *H. diversicolor* contain very high concentrations of accumulated zinc, as in particular Restronguet Creek worms, much of this

zinc is detoxified and stored in the gut epithelium as spherocrystals based on phosphate (Mouneyrac et al., 2003). At lower, more typical, accumulated zinc concentrations, more of the body zinc in *H. diversicolor* is bound to metallothionein than to insoluble metal-rich granules such as these spherocrystals or lysosomal residual bodies (Rainbow et al., 2006a). Metallothioneins in the ragworm also bind accumulated cadmium and silver (Berthet et al., 2003; Rainbow et al., 2006a). As is typical for marine animals, *H. diversicolor* accumulates arsenic in the the non-toxic organic form arsenobetaine (Geiszinger et al., 2002).

6.8.1.2 Biomonitoring

Prologue. There is a good biomonitoring database of accumulated trace metal concentrations in *Hediste diversicolor* across estuaries with different trace metal bioavailabilities. Bioaccumulated trace metal concentrations in the ragworm identify British estuaries of high trace metal bioavailabilities and show how these bioavailabilities have changed over time.

Table 6.6 presents accumulated concentrations of trace metals in *H. diversicolor* from a selection of mining-affected estuaries, industrially contaminated estuaries and estuaries considered relatively uncontaminated. These data all refer to worms that have been allowed to depurate their gut contents after collection. The sediment in the gut would otherwise contribute a significant proportion of the total measured trace metal concentration of the worm. Since these metals in the gut have not been absorbed by the worm, they have not been bioaccumulated.

The trace metals that have been taken up and accumulated by *H. diversicolor* reflect the metals in the local environment that are bioavailable to the worm. As well as trophically bioavailable trace metals in ingested sediment, the total amount of trace metals bioavailable to the worms would include bioavailable metals dissolved in the water column. These dissolved bioavailable metals come into contact with the permeable external surface of the worm via the respiratory irrigation current that the worm drives through

Table 6.6 Mean, weight-standardised estimate (0.03 g worm), range of concentrations, or range of mean concentrations (µg/g dry weight) of trace metals in the ragworm *Hediste diversicolor* (after depuration of gut contents) from British estuaries affected by (a) mining contamination, (b) industrial contamination or (c) considered uncontaminated.

	Ag	As	Cd	Co	Cr	Cu	Fe	Hg	Mn	Ni	Pb	Sn	Zn
(a) Mining													
Restronguet Creek													
ca. 1970-72[1,2]			0.33-1.1			1,140	425		6.0		3.5		194-225
1977-78[3,4]	3.0-6.2	23-87	0.53-1.8	4.5-14	0.3-0.7	832-1,430	268-554	0.05-0.23	8.3-14	2.3-6.5	1.9-9.8	0.2-0.5	262-405
2002[5,6]			1.3			2,630							278
0.03 g 2003[7]	1.4	90				614					2.5		194
0.03 g 2006[7]	2.0	56				457							179
2007[8,9]	4.8		<1.0-2.3			3,850							1,930
Hayle													
1974[3]	5.3	84	0.47	10	<0.3	1,210	734	0.22	5.7	9.1	4.2	0.08	260
1974-80[4]	8.0		0.17	8.4		947							
Gannel													
ca. 1972[2]			1.7-2.4										153-312
1974-80[4,10]	1.9-5.3	21	0.37-1.5	7.0-14	0.5-0.7	67-257	349-431	0.08	11-16	3.7-13	13-685	1.3	222-466
0.03 g 2003[7]	1.5	20				67					29		197
0.03 g 2006[7]	0.7	23				61							236
East Looe													
1974-80[4,10]	1.2-7.6	15	0.12-0.70	2.2-6.2	0.9-2.2	16-44	385-596	0.08-0.19	9.6-9.7	2.1-3.3	4.4-14	0.5	177-219
0.03 g 2003[7]	0.6	11				21					5		118
0.03 g 2006[7]	1.6	15				12							168
2007[8,9]	2.5					46							294

Table 6.6 (cont.)

	Ag	As	Cd	Co	Cr	Cu	Fe	Hg	Mn	Ni	Pb	Sn	Zn
West Looe													
ca. 1972[2]			0.27										155
1974–80[4,10]	1.0–1.1	20	0.20–0.41	3.5–6.6	4.0	29–55	399	0.17	12	3.3	26	0.5	257
0.03 g 2003[7]	0.5	13				27					37		140
0.03 g 2006[7]	1.0	13				30							177
East & West Looe													
1975–76[11]	0.7–30										2.1–261		
Tamar													
ca. 1970–72[1,2]			0.85			106	458		11		5.8		160–66
1974[3,10]	0.6–0.8	13	0.27–0.53	4.4–8.0	<0.3	46–130	377–591	0.14–0.22	6.9–13	3.1–4.3	5.2–7.0	0.1–0.3	163–179
1974–80[4]	0.8–1.2		0.27–0.65	4.3–11		133–236							
Tavy													
1974–80[4,10]	1.7–2.6	17	1.3–2.7	9.4–25	<0.3	204–218	484	0.47	12	4.5	31	0.2	207
0.03 g 2003[7]	1.2	10	1.2			43					2		157
0.03 g 2006[7]	0.7	13				40							170
2007[8,9]	1.3		2.6			214							472
Plym													
ca. 1970–72[1,2]			1.2–3.6			28	350		7.4		5.9		160–99
1974–80[10]	0.3–1.0	16	0.3–2.9	1.3–4.1	3.9–5.0	28–34	351–524	0.35	8.1–11	5.2–6.2	5.4–9.4	1.06	147–201
(b) Industrial													
Thames													
1997[12]	0.9–7.8	6–18	0.10–0.41	5.3–26	0.04–0.5	17–94	302–1,690	0.06–0.24	8.7–35	1.2–4.6	0.9–5.3	0.04–2.9	144–243
2001[13]	0.6–15	7–27	0.04–0.23	3.9–8.4	0.2–5.7	17–107	383–9,660	0.09–0.25	8.8–132	1.5–12	0.6–10	0.1–1.3	109–259
Severn													
North	8.5–12	14–18	2.5–4.3	4.3–6.1	0.2	16–53	370–448	1.8–2.2	11–12	3.6–5.3	2.1–3.0	<0.1	224–249
1974–80[10]													

	6.3–18	11–16	4.4–5.0	4.1–5.3	0.2–0.3	41–81	384–615	1.4–2.5	9.6–13	5.4	3.3–5.0	<0.1–1.0	203–292
South 1974–80[10]	6.3–18	11–16	4.4–5.0	4.1–5.3	0.2–0.3	41–81	384–615	1.4–2.5	9.6–13	5.4	3.3–5.0	<0.1–1.0	203–292
1978[14]	7.9	13	3.5		0.2	47		1.5		5.2	2.9		250
2005[14]	2.1		0.43		1.8	60		0.61		2.2	1.8		192
Neath													
1974–80[10]	1.5	13	0.85	8.9	<0.3	36	359	1.7	12	6.4	2.5	<0.1	175
Loughor													
1974–80[10]	0.3–0.6	9–17	0.32–0.54	2.6–4.8	0.5–3.1	14–41	402–589	0.03–0.29	10–25	1.9–4.8	0.9–2.3	0.1–0.8	173–204
Mersey													
1980–84[15]	1.5	20	0.7		0.6	46	448	0.91	26	4.8	9.5	0.6	196
	0.4–3.1	14–30	0.06–3.8		0.07–1.6	19–97	265–966	0.26–2.8	9–123	1.8–9.0	3.2–21	0.1–1.8	130–294
1997[16]	1.4	15	0.38		0.5	29	477	0.37	28	3.1	6.0	0.3	175
Tees													
1974–76[17]											4–8		200
(c) Uncontaminated													
Avon (Devon)													
ca 1970–72[1,2]			0.15			33	391		9.8		3.4		150–176
1977[3,10]	0.1	8	0.14	5.1	0.5	19	564	0.07	12	3.3	5.4	0.09	197
Dart (Devon)													
ca 1970–72[1,2]			0.58			22	366		9.5	4.4	4.4		163–185
1974–80[10]	0.1	16	0.73	0.7	<0.3	15	227	0.13	5.6	1.5	1.3	0.3	91
Torridge (Devon)													
ca 1972[2]			0.08										155
1980[3,10]	0.3	13–14	0.22–0.44	7.6–8.2	<0.3	20–21	532–604	0.26–0.28	14–16	4.4–5.4	0.2–2.0	0.09–0.1	163–183

Table 6.6 (cont.)

	Ag	As	Cd	Co	Cr	Cu	Fe	Hg	Mn	Ni	Pb	Sn	Zn
Blackwater (Esser)													
2002[5,6]			3.7			19							258
2003[9]	1.6		0.36										176
2007[8,9]	1.4					25							237
Typical	0.1–1.0	6–10	0.05–1.0	1–15	0.1–1.0	10–30		0.05–0.3		1.5–6	0.2–5	0.04–0.3	100–300
High	3–40	20–250	2.5–10	25–30	2–10	50–4,000		0.5–3		13–40	10–1,000	0.5–3	400–2,000

Sources: Bryan and Hummerstone (1971);[1] Bryan and Hummerstone (1973b);[2] Bryan and Gibbs (1983);[3] Luoma and Bryan (1982);[4] Rainbow et al. (2004);[5] Geffard et al. (2005);[6] Rainbow et al. (2011c);[7] Rainbow et al. (2009a);[8] Rainbow et al. (2009b);[9] Bryan et al. (1980);[10] Bryan and Hummerstone (1977);[11] McEvoy et al. (2000);[12] Langston et al. (2004);[13] Langston et al. (2010);[14] Langston (1986);[15] Langston et al. (2006);[16] Evans and Moon (1981).[17]

6.8 Deposit Feeding Polychaetes: Ragworms and Lugworms 449

its burrow. As shown in Plate 2b, burrows of
H. diversicolor in the black iron sulphide–rich intertidal mud of Restronguet Creek are lined by red/brown
iron oxides and hydroxides. Their formation from
black iron sulphide has been caused by the flow-
through of currents of oxygenated water from the
overlying water column. It follows, then, that the
burrowing ragworms are not in contact with anoxic
pore water that might be rich in dissolved trace metals
in equilibrium with the metal sulphides in the anoxic
sediment. It would seem, therefore, that trace metals
dissolved in the sediment pore water would make little
contribution to the total trace metals taken up by
H. diversicolor, unless they diffuse into the water in
the burrow flushed by the irrigation current.

As shown earlier, it does appear that *H. diversicolor*
derives the bulk of its bioaccumulated concentration of
many trace metals by sediment ingestion, making the
ragworm a biomonitor principally of bioavailable
metals in the sediment. The trace metal silver bucks this
trend to a degree, because between 46 and 80% of the
silver accumulated by *H. diversicolor* is derived from
dissolved silver as opposed to sediment-associated silver
(Rainbow et al., 2009b). This percentage increases with
increased dissolved silver concentrations in estuarine
waters, from levels considered low to those that would
be considered atypically high (Rainbow et al., 2009b).

Table 6.6 confirms that the changes in bioavailabilities of sediment-associated trace metals to
Hediste diversicolor, as reflected in their bioaccumulated concentrations, are typically related to the
changes in total sediment concentrations listed in
Table 6.4. An attempt is also made in Table 6.6 to
define what bioaccumulated concentrations in
H. diversicolor might be considered as typical for UK
estuaries and which are atypically high and indicative
of raised local sediment bioavailabilities. These suggested ranges should be considered as hypotheses and
may change with more data collected.

Thus, in Restronguet Creek, sediment bioavailabilities to *H. diversicolor* have been, and still are, particularly high for copper, zinc and arsenic, and also
raised to a lesser degree for silver and possibly lead
(Table 6.6). Furthermore, there is no consistent evidence for a general decrease in these sediment

bioavailabilities to the ragworm over decades
(Table 6.6) (Rainbow et al., 2011c). Similarly, in line
with total concentrations in the sediments (Table 6.4),
sediment bioavailabilities of copper, arsenic and
probably zinc to *H. diversicolor* were high in the Hayle
Estuary in the 1970s and 1980s (Table 6.6). Sediment
bioavailabilities of lead and zinc to *H. diversicolor*
were, and are, high in the Gannel Estuary (Table 6.6).
In the East Looe branch of the Looe Estuary, silver
bioavailabilities to *H. diversicolor* were high in the
1970s and 1980s but had fallen by the first decade of
this century (Table 6.6). As signified earlier, the presumed source of this silver in the East Looe Estuary
was considered to be industrial and independent of
mining (Bryan and Hummerstone, 1977). The silver
source was, therefore, more likely to be in dissolved
form as opposed to being sediment-associated (Bryan
and Hummerstone, 1977). Given that dissolved silver
is the relatively more bioavailable form of silver to
H. diversicolor (Rainbow et al., 2009b), this input
appears now to have stopped. Bioaccumulated concentrations of silver in East Looe *H. diversicolor*
(Table 6.6) have fallen to a greater extent than would
be expected in the absence of change in the local
sediment concentrations of silver between 1980 and
2003 (Table 6.4). The high concentrations of lead in
the sediments of the West Looe Estuary, historically in
receipt of lead-rich input from the Herodsfoot mining
region, are reflected in high local sediment bioavailabilities of lead to *H. diversicolor* (Table 6.6). The
Tamar and Tavy Estuaries have somewhat raised
sediment bioavailabilities of copper, arsenic and lead
to the ragworm (Table 6.6). In the Plym Estuary,
running along the east side of Plymouth, there were
high bioavailabilities of cadmium, chromium and lead
to *Hediste diversicolor* in the 1970s (Table 6.6).

As in the case of Restronguet Creek, there is no
consistent evidence for any general decreases in
the sediment bioavailabilities of trace metals to
H. diversicolor over decades in the mining-affected
estuaries of the Gannel, West Looe and Tavy
(Table 6.6) (Rainbow et al., 2011c).

In the Thames Estuary, data for 1997 and 2001,
either side of the turn of the century, indicate still
high bioavailabilities to *Hediste diversicolor* of silver,

chromium, copper, lead, tin and zinc (Table 6.6), even after the improved regulation of effluents towards the last decades of the twentieth century. This is not unexpected given the aforementioned conclusion that the total concentrations of metals in the sediments of the relatively sheltered Thames Estuary show little, if any, change over more than a decade (Table 6.4). Looking beyond the summarised data in Table 6.6, accumulated concentrations in *H. diversicolor*, and hence sediment bioavailabilities, in 2001 were highest near the sewage outfall of Crossness for chromium, while those for mercury and lead peaked at West Thurrock (Langston et al., 2004).

In the Severn Estuary in the 1970s and 1980s, bioaccumulated concentrations of cadmium, mercury and silver in particular, but also of lead and zinc, were high in *H. diversicolor* from both south and north banks (Table 6.6). By 2005, all these bioaccumulated concentrations had dropped (Table 6.6), indicating reductions in the local sedimentary bioavailabilities of these trace metals to the ragworm (Langston et al., 2010). Thus, body burdens of cadmium, silver, mercury, nickel and lead in Severn *H. diversicolor* declined by 82%, 73%, 60%, 58% and 39% respectively in the 25-year period up to 2005 (Langston et al., 2010). Such reductions are consistent with falls in most total trace metal concentrations in the sediments of the Severn Estuary, coinciding with increased regulation of local industrial effluents from the 1980s, the closure of Avonmouth smelting works and the

considerable downstream displacement of sediments in such an energy-rich estuarine system (Langston et al., 2010). In contrast, total concentrations of chromium increased in Severn sediments over this period (as mentioned previously), and indeed sedimentary bioavailabilities of chromium to *H. diversicolor* similarly increased (Table 6.6) (Langston et al., 2010).

In the Loughor Estuary by Llanelli in the 1970s, sedimentary bioavailabilities to *H. diversicolor* were atypically high for both chromium and tin (Table 6.6). In spite of the close proximity of the Loughor Estuary to tinplating works, it is noteworthy that chromium and tin bioaccumulated concentrations in local ragworms were equivalent or higher in the Thames Estuary in 2001 (Table 6.6). Highest concentrations in the Thames *H. diversicolor* in 2001 occurred at Crossness and Lower Hope Point for chromium and Gravesend and Grays for tin (Langston et al., 2004).

As to be expected from preceding discussion, mercury concentrations in *Hediste diversicolor* were also high in the Mersey Estuary in the 1980s (Langston, 1986), although these concentrations were matched by bioaccumulated concentrations in the Severn in the 1970s (Table 6.6). By the late 1990s, the accumulated concentrations in the Mersey ragworms had dropped considerably, although still atypically raised (Table 6.6), prior to the complete closure of local chlor-alkali plants (Langston et al., 2006).

Box 6.3 Zinc in the Jaws of Ragworms

Polychaete worms in the genus *Hediste* belong to the family *Nereididae*, which includes the genus *Nereis* itself, as well as the genus *Alitta*, the new home for the king ragworm *A. virens*. All nereidid polychaetes possess an impressive pair of eversible jaws, typically with dark brown to black tips. These jaws contain a remarkably high concentration of zinc, particularly towards the tip (Bryan and Gibbs, 1979, 1980). In *Hediste diversicolor*, the concentration of zinc in the jaws is between 11,100 and 24,300 µg/g dry weight, representing about 20 to 50% of the total body zinc content of worms from estuaries free of zinc contamination (Bryan and Gibbs, 1980). The concentrations of zinc in the jaws of *H. diversicolor* do not increase with zinc bioavailabilities in the sediments in which they live, so the contribution of the jaws to total zinc body concentrations drops to below 10% in ragworms from zinc-contaminated sediments (Bryan and Gibbs, 1980). It appears, therefore, that the presence of zinc in the jaws is not an example of the detoxified storage of excess accumulated zinc in the worms, but an independent structural feature of the jaws themselves (Bryan and Gibbs, 1979, 1980). In fact, the zinc is considered to add hardness and stiffness to the jaws (Bryan and Gibbs, 1980; Lichtenegger et al., 2003), in parallel to the situation observed in Chapter 4 in the mandibles of orthopteran, phasmid and lepidopteran insects (Hillerton and Vincent, 1982).

6.8.1.3 Tolerance and Ecotoxicology

Prologue. Copper and/or zinc tolerant populations of *Hediste diversicolor* have been identified, for example in Restronguet Creek, signifying a high local trace metal bioavailability of ecotoxicological significance.

This book has stressed that the presence of a metal-tolerant population of an organism at a particular site is evidence that the total local bioavailability of that toxic metal is of ecotoxicological significance, clearly to that species, but potentially also to other species present (Luoma, 1977). The population of *Hediste diversicolor* at the top of Restronguet Creek was shown in the 1970s to have developed tolerance for both copper and zinc (Bryan and Hummerstone, 1971, 1973b; Bryan and Gibbs, 1983). It has since been confirmed that this copper- and zinc-tolerant population is still present in the twenty-first century, substantiating the ongoing ecotoxicological significance of the very high copper and zinc bioavailabilities in the local sediments (Mouneyrac et al., 2003). The copper and zinc tolerances of this population of *H. diversicolor* have been shown to be genetically based and, therefore, inheritable down generations (Grant et al., 1989; Hateley et al., 1989). The metal tolerance does, however, come at a metabolic cost, bringing a competitive selective disadvantage beneath a threshold sediment concentration, estimated to be 1,000 µg/g copper and 3,500 µg/g zinc (Grant et al., 1989; Hateley et al., 1989). The metabolic costs of the metal tolerance in this worm population, based on increased rates of storage detoxification of copper and zinc (Berthet et al., 2003; Mouneyrac et al., 2003),

result in a scope for growth 46 to 62% lower than that of non-tolerant worm populations (Pook et al., 2009). The considerable physiological cost of metal tolerance to this ragworm population is, correspondingly, only worth paying when copper and/or zinc exposure is ecotoxicologically significant (Pook et al., 2009).

Another population of *H. diversicolor* shown to exhibit copper tolerance in the 1970s was that in the Hayle Estuary (Bryan, 1976).

The demonstration of the presence of trace metal tolerance in a local population of *Hediste diversicolor* is one piece of evidence to contribute to a weight of evidence approach to investigate the potential eco-toxicology of trace metals in an estuarine habitat (Chapman, 2007: Benedetti et al., 2011). Given its abundance and wide distribution in British and other European estuaries, *H. diversicolor* has proved to be a popular model animal of choice in such investigations of contaminated estuaries (Durou et al., 2007a; Mouneyrac et al., 2010). As explained in Chapter 3, bio-markers at the lower levels of biological organisation are the most sensitive to contaminants and often the easiest to measure. At the higher levels of biological organisation, for example the population and community levels, contaminant-derived changes are those that clearly have the most ecotoxicological significance in a habitat. Biomarkers at all levels of biological organisation have now been developed for *H. diversicolor*, and links made between them (Durou et al., 2007a, b; Gillet et al., 2008; Fossi-Tankoua et al., 2012). At the biochemical level, the antioxidant enzyme catalase is a biomarker of a defence response by the ragworm against the free radicals which can be generated by exposure to trace metals (Durou et al., 2007b; Fossi-Tankoua et al., 2012). Another bio-chemical biomarker of oxidative stress in *H. diversicolor* is the concentration of malondialdehyde (MDA), a breakdown product of the oxidation of lipids caused by free radicals (Durou et al., 2007b). The measure-ment of levels of glycogen, proteins and lipids in *H. diversicolor* provides information on the amounts of energy reserves stored by the worms (Durou et al., 2007b; Fossi-Tankoua et al., 2012). These reserves can be expected to be lower in worms under ecotoxicolo-gical stress from trace metal exposure, with knock-on effects to bring about reduced scope for growth, a physiological biomarker (Pook et al., 2009). Other physiological biomarkers include measures of fecundity, such as the total number of eggs per female worm and sexual maturation patterns (Durou et al., 2007a; Mouneyrac et al., 2010). At the whole organ-ism level, behaviour patterns used as biomarkers in *H. diversicolor* include reduced feeding and burrowing behaviour (Mouneyrac et al., 2010; Fossi-Tankoua et al., 2012), the former with clear implications for the acquisition of energy reserves. Reduced reproductive rates and growth rates will have an effect at the population level, with different population structures of *H. diversicolor* in contaminated and uncontamin-ated estuaries (Durou et al., 2007a; Gillet et al., 2008). Also at the population level, abundance and total biomass of ragworm populations are reduced in sig-nificantly contaminated estuaries (Durou et al., 2007a; Gillet et al., 2008).

6.8.2 *Arenicola marina*: The Lugworm

Prologue. *Arenicola marina* is a net accumulator of trace metals derived from the ingestion of sediment and can be used as a biomonitor of trace metal bioavailabilities.

The lugworm *Arenicola marina* (Plate 6b) is a large polychaete that is widespread on the shores of British estuaries and open coasts, burrowing into muddy sand. The lugworm lives in a burrow which it may occupy for several months, and feeds by ingesting sand and digesting the organic matter present (Green,

1968). The burrow consists of a sand-filled head shaft, a horizontal gallery and a tail shaft. The lugworm in the horizontal gallery ingests sand at the bottom of the head shaft, thereby creating a recognisable depression on the surface of the sand, at times with a small hole (Plate 6c) (Green, 1968). At intervals of about 45 minutes, the lugworm backs up the tail shaft and defaecates on the surface of the sand, leaving a characteristic coiled cylindrical cast on the beach during low tide (Plate 6c) (Green, 1968). When in the burrow, the lugworm will intermittently drive a respiratory irrigation current from the tail to the head end, the current exiting through the head shaft, thereby softening the sand therein (Green, 1968).

Thus, the lugworm is exposed to trace metals dis-solved in the irrigation current, originating from the water column, but potentially receiving some input from diffusing porewater in the sediment. Addition-ally, the lugworm is exposed to the trace metals in the ingested sediment, particularly the metals associated with the sediment organic matter which it digests. Biodynamic modelling studies have confirmed that, under typical environmental conditions, lugworms obtain from ingested sediment more than 90% of their accumulated cadmium and zinc, more than 70% of accumulated silver and between 30 and 60% of accumulated arsenic, (Casado-Martinez et al., 2009a, 2010b). Field and experimental data have shown that increases in the sediment concentrations of cadmium, copper and zinc cause increases in the accumulated concentrations of these trace metals in *A. marina* (Packer et al., 1980; Bat, 1998). Biodynamic modelling has also confirmed that the lugworm is a net accu-mulator of arsenic, cadmium, silver and zinc from sediments (Casado-Martinez et al., 2009a, b, 2010b).

Table 6.7 lists the accumulated concentrations of certain trace metals in *Arenicola marina* collected from several British estuaries, a small database than that for *Hediste diversicolor* (Table 6.6). Nevertheless, several points emerge from Table 6.7. Lugworms live lower down an estuary than *H. diversicolor* and are not to be found in mud with its high affinities for trace metals. While *A. marina* has similar rates of uptake of trace metals from solution and food as *H. diversicolor* (Casado-Martinez et al., 2009a;

Table 6.7 **Means, ranges or ranges of means of concentrations (μg/g dry weight) of trace metals in the lugworm *Arenicola marina* (after depuration of gut contents) from British estuaries affected by (a) mining contamination, (b) industrial contamination or (c) considered uncontaminated.**

	Ag	As	Cd	Cu	Mn	Pb	Zn
(a) Mining							
Gannel							
2008[1,2]	0.21 (0.17–0.25)	83 (47–123)	1.60 (1.29–2.05)				76 (70–82)
Lynher (Tamar)							
2008[1,2]	0.84 (0.81–0.87)	34 (21–44)	0.85 (0.60–1.27)				63 (57–70)
Dulas Bay							
1976[2]			2.1	45	3	22	160
2008[1,2]		16 (13–19)					90 (68–141)
Conwy							
1976[3]			0.8	4	2.5	10	150
(b) Industrial							
Outer Thames (Two Tree Island)							
2008[1,2]	0.59 (0.15–0.94)	17 (15–20)	3.20 (1.61–5.17)				70 (56–105)
Severn (North Bank)							
1976[3]			16–35	12–30	3–10	17–40	170–320
Poole Harbour							
2008[1,2]		23 (13–32)	1.26 (0.66–1.98)				56 (43–90)
(c) Uncontaminated							
Caernarvon (Foryd Bay)							
1976[3]			0.6	8	2.5	10	45

Note: Concentrations from Packer et al. (1980) have been taken from figures and are approximations.
Sources: Casado-Martinez et al. (2009b);[1] Casado-Martinez et al. (2010b);[2] Packer et al. (1980).[3]

Rainbow et al., 2009a), the lugworms would miss the very high trace metal exposures of the ragworms in mud higher up a contaminated estuary. Nevertheless, accumulated concentrations in *A. marina* reflect a raised bioavailability of arsenic in the Gannel Estuary (Table 6.7), as is the case for *H. diversicolor* (Table 6.6). As we have come now to expect, accumulated cadmium concentrations show that cadmium bioavailabilities to the lugworm were extremely high in the Severn Estuary in 1976 (Table 6.7). Copper concentrations in *A. marina* are typically low, at or below about 10 µg/g dry weight (Table 6.7). Yet in the copper-contaminated sediments of Dulas Bay, this bioaccumulated concentration had risen only to 45 µg/g (Table 6.7), an increase far below what would be expected in *H. diversicolor* in similar copper-contaminated sediments (Tables 6.4, 6.6). Similarly to the case for the ragworm (Table 6.6), accumulated zinc concentrations in the lugworm usually fall in a relatively narrow range, in this case between 45 and 80 µg/g Zn (Table 6.7), with accumulated concentrations above 150 µg/g and up to 320 µg/g, indicating very high zinc bioavailabilities (Table 6.7). Such high bioaccumulated concentrations have been found in lugworms from Dulas Bay, Conwy and the Severn Estuary in the 1970s (Table 6.7).

Box 6.4 Arsenic and *Aphelochaeta marioni*

The cirratulid polychaete *Aphelochaeta marioni*, previously known as *Tharyx marioni*, lives buried in soft sediments, intertidally and sublittorally, both within estuaries and beyond. It is widespread around the British Isles and elsewhere in Europe. *A. marioni* is a deposit feeder. The worm remains in its burrow but extends two long palps across the surface of the sediment. These palps each have a canal lined with cilia which transport sediment and other particles back to the mouth of the worm for ingestion.

The remarkable feature of *Aphelochaeta marioni* is that this worm contains extremely high concentrations of arsenic, even in the absence of arsenic contamination of the sediment in which it lives (Bryan and Gibbs, 1983; Gibbs et al., 1983). While *A. marioni* from Restronguet Creek contained 2,340 µg/g As and those from the Tamar 2,280 µg/g As, worms from the uncontaminated Yealm Estuary and Place Cove in the Fal Estuary still contained 2,200 µg/g As and 1,510 µg/g As respectively (Bryan and Gibbs, 1983; Gibbs et al., 1983). In contrast, other cirratulid polychaetes from the same sites contained far lower arsenic concentrations (Bryan and Gibbs, 1983; Gibbs et al., 1983). Thus *Chaetozone caputesocis* from the Tamar and Place Cove sites contained only 24 and 95 µg/g As respectively (Gibbs et al., 1983). *Cirriformia tentaculata* from Place Cove contained 84 µg/g As (Gibbs et al., 1983). The palps of *A. marioni* contained even higher arsenic concentrations than the body of the worm as a whole – 13,000 µg/g in Tamar worms and 3,240 µg/g in Place Cove worms (Gibbs et al., 1983).

The fact that arsenic concentrations in the one species, *Aphelochaeta marioni*, are so different from those in their close cirratulid relatives suggests that these high concentrations are unlikely to be simple reflections of a particular arsenic detoxification mechanism. Do they have a function in some way connected to the specific ecology of *A. marioni*? The distribution of the highest concentrations of arsenic in the palps, the parts of the body most often beyond the protection of the burrow, does raise the possibility that the high arsenic concentrations might confer some protection against predators, through reduced palatability (Gibbs et al., 1983). There is an attraction in this proposal. However, feeding experiments with a fish, the two-spotted goby *Gobiosculus flavescens*, have proved equivocal (Gibbs et al., 1983). Some fish repeatedly rejected *A. marioni* as food, while other individual fish accepted them without hesitation. Similarly, the same fish either rejected or accepted *Chaetozone caputesocis* (with low As concentrations) when presented as food. Thus both high-arsenic and low-arsenic worms were both accepted and rejected as food by the goby (Gibbs et al., 1983). The jury is still out.

6.9 Deposit-Feeding Bivalves

Prologue. Most lamellibranch bivalves are suspension feeders, but a few are surface deposit feeders using the inhalant siphon to take in surface particles. Important bivalve deposit feeders in British estuaries are the peppery furrow shell *Scrobicularia plana* and the Baltic tellin *Macoma balthica*.

The distinctive form of bivalve molluscs is usually attributed to be an adaptation to a burrowing habit. The shell of two hinged valves, the large space between the shells (the mantle cavity), and a flexible foot that can be extended, expanded and then withdrawn as a digging tool all aid in burrowing. The basic respiratory organs of molluscs are paired gills lined with cilia driving a respiratory current through the mantle cavity and over the gills. Most bivalves are lamellibranch bivalves, with greatly expanded gills, producing a strong enough current for it to be a source of suspended food, as well as a source of oxygen. The gills filter the phytoplankton and suspended fine detritus, which are then transferred to the mouth. Many burrowing bivalves seal the edge of the mantle cavity with tissue, leaving only an entrance and an exit for the respiratory/feeding current. Furthermore, the tissue around each of these entry and exit points is often extended into a retractable siphon, reaching up to, and beyond, the surface of the sediment in which the bivalve is buried.

While suspension feeding is typical for lamellibranch bivalves, some of these bivalves, for example members of the families *Semelidae* and *Tellinidae* (the tellins), are actually predominantly surface deposit feeders. They use the inhalant siphon to hoover up surface sediment and detritus particles. And semelid and tellinid bivalves are ecologically very important in British estuaries. In the mud or muddy sand at the top of estuaries, often in the drainage channels of salt marshes on the upper shore, can be found the semelid *Scrobicularia plana*, the so-called peppery furrow shell (Plate 7a). The origin of its common name may lie in the fact that this bivalve was historically a food source in Britain and is said to have a sharp or bitter taste. Indeed, this bivalve can still be bought in France under the name of *lavignon*. *S plana* can be buried 10 centimetres or so beneath the sediment surface. Nevertheless, its presence can be recognised by a star-shaped pattern on the surface of the mud, created by the excursions of the inhalant siphon collecting food particles.

The Baltic tellin *Macoma balthica* can occur in large numbers lower down the estuary, and lower down the shore. It is again a surface deposit feeder using its inhalant siphon to suck up sediment particles, although some populations at sandy sites with strong currents may carry out suspension feeding (Hayward, 2004).

6.9.1 *Scrobicularia plana*: The Peppery Furrow Shell

Prologue. *Scrobicularia plana* accumulates trace metals both from solution and from ingested sediment and is a useful biomonitor of trace metal bioavail-abilities at the top of British estuaries affected by mining or industry.

Scrobicularia plana is a net accumulator of trace metals. Table 6.8 lists concentrations of trace metals in the soft tissues of this bivalve from a selection of metal-contaminated and uncontaminated estuaries around the British Isles. Also included in this table are estimates of what accumulated concentrations in *S. plana* can be considered either typical or high and indicative of raised total local biovail-abilities of trace metals to the bivalve (Luoma and Rainbow, 2008; Rainbow et al., 2011c). As previously, these concentrations should be considered as hypotheses with the potential to change as more data accrue.

S. plana accumulates trace metals both from solu-tion and from ingested sediments (Amiard et al., 1985; Kalman et al., 2014). The rate of accumulation by the bivalve from solution over a four-to-six-day period increases linearly with exposure concentration for arsenic, cadmium, copper, lead, silver and zinc (Amiard et al., 1985; Kalman et al., 2014). Similarly, the accumulated concentrations of arsenic, cadmium, cobalt, copper, lead, silver and zinc in *S. plana* increase linearly with increased concentrations of these trace metals in the sediment (Langston, 1980; Luoma and Bryan, 1982; Kalman et al., 2014). The bioaccumulated concentrations in the bivalve do, therefore, reflect local sediment bioavailabilities of trace metals. *S. plana* is, thus, a biomonitor of these sediment bioavailabilities. In contrast to the situation for many trace metals in the polychaete *Hediste diversicolor*, *S. plana* attains a bigger proportion of its accumulated content of several trace metals from solution, as opposed to ingested sediment (Kalman et al., 2014). Thus, dissolved metal is the predominant source of accumulated arsenic, silver and zinc in *S. plana*, accounting for 50 to 97%, 66 to 99% and 52 to 98% respectively of accumulated metal content under different field conditions (Kalman et al., 2014). This differential dependence between the bivalve and the polychaete on water and food as sources of trace metals can probably be attributed to the fact that

lamellibranch bivalves have relatively huge areas of gill available for metal uptake from solution and, in order to feed, pass large quantities of water across these gills.

Thus *S. plana* is a biomonitor of trace metal bioa-vailabilities in both solution and ingested sediment, and it can fulfil this role in any suite of biomonitors chosen to assess trace metal bioavailabilities in an estuary.

While there are indeed direct linear relationships between the accumulated concentrations of many trace metals in *S. plana* and total sediment concen-trations of these metals (Luoma and Bryan, 1982; Langston, 1980), it is also the case that these linear relationships can be improved by using other measures of the concentrations of trace metals in the sediment, better reflecting their sediment bioavailabilities (Bryan and Langston, 1992). One such measure is the sediment concentration extracted with weak hydrochloric acid, modelling the trace metal bound to the sediment organic matter (Bryan and Langston, 1992). Thus there are improved linear relationships between accumulated concentrations in the tissues of *S. plana* and con-centrations in the sediment when these weakly extracted concentrations are used for cobalt, silver and mercury (e.g., Figure 6.2) (Langston, 1982; Luoma and Bryan, 1982; Bryan and Langston, 1992). Arsenic and lead have strong affinities for iron oxides, affecting their trophic bioavailabilities to a sediment ingestor such as *S. plana* (Bryan and Langston, 1992). This is reflected in the fact that there are improved linear relationships between the concentrations of accumulated arsenic (Figure 6.3) (Langston, 1980) and lead (Figure 6.4) (Luoma and Bryan, 1978) in *S. plana* and sediment concentrations of these two trace metals when expressed as the ratio of the concentrations of either element to the concentrations of iron, both after extraction with 1 M HCl (Bryan and Langston, 1992).

Table 6.8 shows particular features that we have come to expect for trace metal bioavailabilities in British estuaries, both historically and more recently. Thus Restronguet Creek has ongoing very high

Table 6.8 Mean, weight-standardised estimate (0.2 g soft tissues), range of concentrations or range of mean concentrations (μg/g dry weight) of trace metals in soft tissues of the semelid bivalve *Scrobicularia plana* (after depuration of gut contents) from British estuaries affected by (a) mining contamination, (b) industrial contamination or (c) considered uncontaminated.

	Ag	As	Cd	Co	Cr	Cu	Fe	Hg	Mn	Ni	Pb	Sn	Zn
Mining													
Restronguet Creek													
1978-79[1,2,3]	0.2-0.3	160-191	2.3-4.5	4.9-12	1.2-1.3	89-156	2,700-2,860	0.04-0.17	17-19	2.4-4.2	50-72	2.3-2.7	2,580-3,160
0.2 g 2003[4]	1.1					163							1,410
0.2 g 2006[4]	0.21	146				113					27		999
2013[5]	0.5	188			3.1	232	4,670		44		50		2,450
Hayle													
1974-79[1,3,4]	0.1-0.5	97-106	0.7-3.4	6.2-11	0.6-1.3	27-122	962-1,930	0.11-0.17	13-38	2.0-4.7	21-51	0.8-2.9	978-1,560
Gannel													
1974-75[6]	0.2-1.2		1.4-15	19-66	1.5-2.2	25-86	1,120-1,240		57-87	9.8-12	234-828		1,470-2,940
1977-79[1,2,3]	0.7-2.0	98	1.9-14	27-97	2.3-3.8	61-332	1,060-2,730	0.90	59-333	8.8-15	309-1,020	0.9	2,270-4,920
0.2 g 2003[4]	0.8					113					593		2,190
0.2 g 2006[4]		75				39					256		1,190
2013[5]	0.4	50			7.0	50	5,100		300		263		1,510
East Looe													
1977[2]	11-259	31-50	0.4-3.6	15-20	6.6-7.1	95-277	1,620-1,750	0.46-0.79	21-107	16-21	149-246	0.7-1.0	569-2,060
0.2 g 2003[4]	4.6					279					97		1,250
0.2 g 2006[4]	3.8	25				35					22		516
2013[5]	3.1	37			12	113	6,800		176		44		572
West Looe													
1977[2]	6.8-10	28-53	1.3-2.1	8.1-20	4.1-5.9	315-356	580-1,310	0.47-1.3	49-70	7.7-12	225-489	1.2-1.6	1,280-2,430
0.2 g 2003[4]	2.5					235					354		804
0.2 g 2006[4]	2.1	33				121					207		579
2013[5]	2.5	33			5.7	131	3,500		155		244		733

Table 6.8 (cont.)

	Ag	As	Cd	Co	Cr	Cu	Fe	Hg	Mn	Ni	Pb	Sn	Zn
Tamar													
1974[2,7]	0.03–0.4	29–47	0.5–9.8	3.0–37	0.5–4.2	13–52	441–7,280	0.19–0.20	18–97	1.6–9.1	26–296	0.8–1.0	1,450–3,990
Tavy													
1974[2]	0.4	40	7.5	28	2.8	103	1,100	0.75	32	9.3	81	0.8	2,800
0.2 g 2003[4]	1.2					69					40		561
0.2 g 2006[4]	0.4	30				80					23		520
2013[5]	0.5	39			6.5	113	4,600		114		71		2,370
Plym													
1977–78[1,2,3]	0.3–0.9	26–28	6–32	3.4–9.3	2.8–5.9	22–104	597–1,050	0.23	16–23	2.1–3.6	25–40	2.0	1,200–2,170
Industrial													
Thames													
1997[8]	9.3–53	10–25	0.4–1.9	1.9–7.3	0.4–3.2	25–121	332–1,340	0.19–0.74	14–97	1.7–6.2	4–23	0.1–2.6	368–1,310
2001[9]	14–130	10–29	0.3–3.0	3.2–16	1.0–9.9	30–192	716–4,110	0.26–1.2	25–76	3.4–16	10–39	0.2–1.6	464–1,030
Outer 2013[5]	12	17				32	756		16		16		674
Severn													
1976–78[2,3]	2.8–12	13–26	6.3–9.9	5.8–8.7	3.6	41–75	1,220	0.39	52	4.4–9.1	21–77	0.4–0.5	555–1,080
Neath													
1978–79[2]	0.5–0.8	13–25	2.9–3.3	4.3–14	1.9	19–32	973–1,100	0.55–0.70	36–42	5.6	20–49	<0.1–0.6	349–1,010
Loughor													
1979[2]	0.1–0.5	11–20	0.8–1.4	3.9–5.0	2.7–24	9–16	367–1,550	0.02–0.15	30–112	2.7–3.7	10–24	1.0–14	367–556
Mersey													
1980–84[10,11]	0.7 / 0.3–1.6	26 / 18–37	1.8 / 0.7–3.6	6.2	4.7 / 1.5–8.3	47 / 18–100	705 / 392–1,600	1.4 / 0.72–2.5	60 / 32–127	3.6 / 2.4–5.3	43 / 14–65	0.8 / 0.3–1.6	1,870 / 622–3,390
1997[12]	3.2		0.6		6.5	30	1,140	0.59	104	3.6	39	0.6	780
Tyne													
Before 1985[10]	4.6		3.2	5.2	3.6	33	805		20	3.6	109		1,180
Tees													
Before 1985[10]	0.3	17	0.6	4.2	6.9	27	348	0.6	24	1.0	14	0.6	652

Uncontaminated													
Camel (Cornwall)													
1974–80[1,2,6]	0.2–0.6	27	0.3–0.6	4.3–5.4	0.9–1.7	19–77	699–989	0.24	18 – 27	2.7–4.5	8–22	0.7	309–457
Avon (Devon)													
1977[2]	1.0	23	0.5	7.8	4.1	84	541	0.90	36	9.6	23	0.3	683
Dart (Devon)													
1980[2]	0.5	14	0.5	2.8	1.5	19	714	0.29	26	2.9	17	0.7	504
Torridge (Devon)													
1980[1,2]	0.7–1.5	18–37	1.1–2.4	8.0–19	1.7–3.5	17–41	1,360–1,980	0.42–0.72	43–54	4.5–14	13–40	0.5–0.8	403–1,210
Typical[4,13]	0.1–5	10–30	0.2–3	2–20	0.4–7	9–60		0.02–0.5		1–10	4–40	0.1–1	300–1,000
High[4,13]	10–300	100–300	5–45	30–110	15–25	120–360		1–2.5		15–25	100–1,500	5–15	2,000–5,000

Sources: Luoma and Bryan (1978);[1] Bryan et al. (1980);[2] Langston (1980);[3] Rainbow et al. (2011c);[4] Kalman et al. (2015);[5] Bryan and Hummerstone (1978);[6] Bryan and Uysal (1978);[7] McEvoy et al. (2000);[8] Langston et al. (2004);[9] Bryan et al. (1985);[10] Langston (1986);[11] Langston et al. (2006);[12] Luoma and Rainbow (2008).[13]

bioavailabilities to *S. plana* of arsenic, copper and zinc, and, to some extent, a raised bioavailability of lead. The more historical data for the Hayle Estuary confirm high bioavailabilities there of copper and arsenic in the 1970s (Table 6.8). The Gannel Estuary had, and still has, high bioavailabilities of lead and zinc to *S. plana* (Table 6.8). The data highlight the high bioavailability of silver in the East Looe Estuary in the 1970s, a bioavailability that has since dropped considerably (Table 6.8). High bioavailabilities of copper and lead to the bivalve in the West Looe Estuary have been maintained since the 1970s (Table 6.8). Data for the Tamar Estuary in the 1970s show very high bioavailabilities of lead and silver to *S. plana* (Table 6.8), the highest bioaccumulated concentrations occurring at the point of entry of effluent from the old South Tamar mine, near Bere Alston (Bryan and Uysal, 1978). This same location also yielded the highest bioaccumulated concentrations of chromium and silver in *S. plana* (Bryan and Uysal, 1978), although these accumulated concentrations were not themselves remarkably high for this bivalve (Table 6.8). The South Tamar mine was abandoned in 1865 when flooded by water from the estuary breaking into the mine workings (Dines, 1969). In the period from 1849 until closure, this mine produced 7,140 tons of lead ore and 262,470 ounces of silver (Dines, 1969). As for *Hediste diversicolor*, there was a high bioavailability of cadmium to *S. plana* in the Plym Estuary in the late 1970s (Table 6.8).

There were, in fact, significant decreases between 2003 and 2006 in the accumulated concentrations of some trace metals in *S. plana* in some mining-affected estuaries, indicating falls in their bioavailabilities to this bivalve (Rainbow et al., 2011c). Thus, bioavailabilities of silver fell in Restronguet Creek and Tavy; of copper in the Gannel, East Looe and West Looe; of lead in Restronguet Creek, Gannel and East Looe; and of zinc in Restronguet Creek and East Looe (Table 6.8) (Rainbow et al., 2011c). These decreases presumably reflect falls in dissolved bioavailabilities, given the importance of this source of metals to *S. plana*, and the little change observed for sediment bioavailabilities indicated previously by *Hediste diversicolor*.

In the estuaries contaminated by industrial activity, silver and copper bioavailabilities to *S. plana* were high in the Thames Estuary at the turn of this century (Table 6.8). Bioavailabilities of cadmium (not unexpectedly) and silver were high in the Severn Estuary in the 1970s (Table 6.8). The Loughor Estuary in 1979 had high bioavailabilities of chromium and tin to the bivalve (Table 6.8), in line with discussion earlier in this chapter. The Mersey Estuary in the early 1980s had the now expected high bioavailability of mercury (Table 6.8). The bioavailability of lead to *S. plana* was high in the Tyne Estuary prior to 1985 (Table 6.8).

The preceding biomonitoring data all concern trace metal concentrations in the total soft tissue of *Scrobicularia plana*. With sufficient malacological knowledge, it is also possible to dissect out the digestive gland from these soft tissues and analyse this organ as a trace metal biomonitoring tool (Bryan et al., 1980). This may be worth doing if samples of the bivalve have been taken at different seasons, and it is advantageous to remove any effect of the dilution of the total soft tissues with newly developed gonad tissue (Bryan et al., 1980). Furthermore, the digestive gland, as a site of trace metal detoxification and storage, usually contains higher concentrations of trace metals than do the general soft tissues (Bryan et al., 1980). For example, in *S. plana* from the Tamar Estuary in the 1970s, the digestive gland contained 87% of the total zinc in the soft tissues, with an accumulated zinc concentration between 4,800 and 7,140 µg/g, in comparison to a total soft tissue concentration of between 1,450 and 2,790 µg/g Zn (Bryan et al., 1980). Comparable percentages for other trace metals were 95% cadmium, 94% cobalt, 85% lead, 83% nickel, 73% manganese, 65% mercury and 58% tin, but only 36% arsenic, 15% silver and 10% iron (Bryan et al., 1980). Detoxification of high concentrations of accumulated copper and silver by *S. plana* involves binding with sulphur (Truchet et al., 1990), resulting metal-rich granules being interpretable as the residual bodies of lysosomes after autolysis of metallothioneins binding these metals (Kalman et al., 2015). On the other hand, high concentrations of accumulated zinc and lead are bound with phosphate granules (Truchet et al., 1990; Kalman et al., 2015).

In the 1970s, the population of *Scrobicularia plana* in Restronguet Creek was tolerant to copper, but not zinc (Bryan and Gibbs, 1983), indicating that the local bioavailability of copper was of ecotoxicological significance. While the demonstration of such tolerance is a useful tool in any a weight of evidence approach to examining potential trace metal ecotoxicology in estuaries, *S. plana* is also now being routinely used as a model for the measurement of particular biomarkers of trace metal ecotoxicity at various levels of biological organisation (Romero-Ruiz et al., 2008; Boldina-Cosqueric et al., 2010). These biomarkers include concentrations of metallothioneins and MDA, activities of catalase and digestive enzymes, levels of energy reserves and condition indices and burrowing behaviour (Romero-Ruiz et al., 2008; Boldina-Cosqueric et al., 2010).

6.9.2 *Macoma balthica*: The Baltic Tellin

Prologue. As a net accumulator of trace metals, *Macoma balthica* can be used as a biomonitor of trace metal bioavailabilities towards the bottoms of estuaries.

Macoma balthica can be common further down British estuaries, sampling a somewhat different environmental regime than *Scrobicularia plana*. *M. balthica* is predominantly a surface-deposit feeder (Hayward, 2004), extending its inhalant siphon up through the mud or sand from its shallow burrow. Even though it is a burrower, *M. balthica* does change position fairly frequently to sample new areas of surface sediment (Green, 1968). As befits its name, the Baltic tellin is more cold-tolerant than *S. plana*, being found as far north as Finland in the Baltic (Green, 1968). Correspondingly, *M. balthica* survived the very cold winter in Britain in 1963 better than did *S. plana*. The newly settled spat of *M. balthica* are a significant food source for the brown shrimp *Crangon crangon*, while the adult bivalves are preyed upon by flounder

Platichthys flesus, by wading birds such as knot *Calidris canutus* and oystercatchers *Haematopus ostralegu*, and by eider ducks *Somateria molissima*. (Hayward, 2004).

Their position lower down an estuary means that *M. balthica* are exposed to bioavailable sources of trace metals that may be lower than those which *S. plana* experiences in the same estuary. Dissolved metals are more likely to have been diluted by seawater lower in the estuary, and the dissolved bioavailabilities of several trace metals would have been reduced by increased complexation by chloride at higher salinities downstream. Furthermore, the sedimentation of trace metals originating in the incoming river may have been completed in the upper region of the estuary. Nevertheless, *M. balthica* can be used as biomonitors of trace metal bioavailabilities in estuaries, for example in large estuaries contaminated by industry, where the upper reaches of the estuary have been lost to industrialisation.

Table 6.9 lists the concentrations of trace metals accumulated by *M. balthica* in different British estuaries, again with proposed interpretations of what are typical or high bioaccumulated concentrations signifying raised metal bioavailabilities. While this database is smaller than that for *S. plana* (Table 6.8), similar general features can be recognised. Thus, there was a very high bioavailability of silver, as well as copper, to Baltic tellins in the East Looe Estuary in the 1970s (Table 6.9). Silver and copper bioavailabilities to *M. balthica* were also high in the Thames Estuary at the turn of this century (Table 6.9). The Severn Estuary in the 1970s had very high bioavailabilities of cadmium and silver (Table 6.9). Perhaps surprisingly, and contrary to the case for *S. plana*, the Loughor Estuary in the 1970s did not have high bioavailabilities of chromium and tin to *M. balthica* (Table 6.9). Perhaps the high bioavailabilities of these two trace metals of industrial origin were restricted to the sediments of the upper estuary. As now expected, the bioavailability of mercury to *M. balthica* was very high in the Mersey Estuary in the early 1980s (Table 6.9).

Table 6.9 Mean concentrations, ranges of concentrations, or ranges of mean concentrations (µg/g dry weight) of trace metals in soft tissues of the tellinid bivalve *Macoma balthica* (after depuration of gut contents) from British estuaries affected by (a) mining contamination, (b) industrial contamination or (c) considered uncontaminated.

	Ag	As	Cd	Co	Cr	Cu	Fe	Hg	Mn	Ni	Pb	Sn	Zn
Mining													
East Looe													
1976–78[1,2,3]	122	46	0.7	6.8	16	208–338	788–1,540	0.97	21–25	7.7	36–61		1,010–1,160
Tavy													
1981[2]			1.2			105	650		30		25		691
Plym													
1978–80[1,2]	1.0	33	1.0	2.3	1.8	74	626	0.50	9	1.6	9.1	1.5	559
Industrial													
Thames													
1997[4]	5.7–76	9–14	0.2–2.1	0.6–2.4	0.5–4.9	26–255	196–363	0.14–0.45	7–38	0.5–3.3	0.7–16	0.2–1.0	486–840
2001[5]	15–86	13	0.1–0.2	2.3–3.6	0.3–0.9	25–118	361–531	0.17–0.40	12–14	1.6–11	2.0–7.2	0.4–0.9	626–884
Severn													
1976–78[1,3]	6.8–100	22	2.2–9.4	3.1–4.2	3.6–4.2	54–224	1,300–1,810	0.44–1.3	73–356	3.8–13	8.7–19	0.5–1.7	899–1,790
Loughor													
1978[1,3]	0.3	11	0.2	1.1	2.4	32	210	0.12	8	1.6	2.3	1.2	396
Mersey													
1980–84[3,6]	2.3	32	0.3	1.9	2.6	100	906	1.54	36	1.6	11	0.5	721
	0.6–4.3	15–66	0.1–0.5		0.9–4.8	46–190	185–2,640	0.54–4.0	15–70	0.7–3.9	2.1–21	0.2–0.9	320–1,420
Tees													
Before 1985[3]	1.5	18	0.2	1.2	3.0	152	241	1.0	18	0.3	5	0.5	414

Uncontaminated											
Torridge (Devon)											
1980[2]	337		6.2	33		1,130			50			
Typical[5,7]	300–600	0.1–1	0.5–10	8–50	0.3–5	20–80	0.1–0.5	0.3–5	0.6–10	0.1–2	10–35	0.2–5
High[5,7]	1,000–3,000	30–150		100–400	10–20	100–350	1–4	10–20	20	5–10	50–100	50–300

Sources: Bryan et al. (1980);[1] Bryan and Gibbs (1983);[2] Bryan et al. (1985);[3] McEvoy et al. (2000);[4] Langston et al. (2004);[5] Langston (1986);[6] Luoma and Rainbow (2008).[7]

6.10 Deposit-Feeding Amphipod Crustaceans

Prologue. The amphipod crustacean *Corophium volutator* is a surface deposit feeder living in British estuaries.

A final ecologically important estuarine, deposit-feeding invertebrate to be considered here is the amphipod crustacean *Corophium volutator*. There are in fact several species of *Corophium* on European shores (Green, 1968). While *C. volutator* burrows in mud and fine sand, often with a reduced oxygen content, its relative *C. arenarium* lives in better draining sediments with a lower silt and clay content. *C. volutator* also copes better with very low salinities (Green, 1968).

6.10.1 *Corophium volutator*

Prologue. *Corophium volutator* accumulates trace metals in the epithelial cells of the ventral caeca. The cells pass through a cell cycle and detoxified stores of accumulated trace metals are lost through the alimentary tract. Under constant bioavailability conditions, the accumulated trace metal concentrations of the amphipods reach steady state, underpinning the use of these amphipods as biomonitors.

Corophium volutator commonly occurs in the muddy intertidal regions high up estuaries, and it can be found in mining-contaminated estuaries such as Restronguet Creek and Dulas Bay (Bryan and Gibbs, 1983; Icely and Nott, 1980). *C. volutator* is essentially a surface deposit feeder, using its clawed appendages (gnathopods) to sort organic detritus from the surface sediment for ingestion. When under water in its burrow, this amphipod can also filter small suspended sediment particles from an irrigation current, using fine setae on the edge of the gnathopods (Green, 1968).

Table 6.10 presents accumulated concentrations of copper, zinc and iron in *C. volutator* from a small selection of British habitats, including Restronguet Creek and Dulas Bay with sediments extremely high in these trace metals (Table 6.4). Amphipods

Table 6.10 **Mean concentrations (µg/g dry weight) of trace metals in the deposit-feeding amphipod *Corophium volutator* (after depuration of gut contents) from British estuaries and coasts (a) affected by mining contamination or (b) considered uncontaminated.**

	Cu	Fe	Zn
(a) Mining			
Restronguet Creek			
1971[1]	499	700	254
Dulas Bay			
Late 1970s[2]	259	325	109
1984[3]	249		114
(b) Uncontaminated			
Avon (Devon)			
Pre-1983[1]	113	732	149
Menai Strait (North Wales)			
Late 1970s[2]	77	494	104

Sources: Bryan and Gibbs (1983);[1] Icely and Nott (1980);[2] Rainbow and White (unpublshed).[3]

from both these sites show raised accumulated concentrations of copper, but only the Restronguet Creek amphipods have raised body concentrations of zinc (Table 6.10). Amphipod crustaceans, including *C. volutator*, detoxify and accumulate high concentrations of trace metals in the epithelial cells of the ventral caeca, blind-ending tubules of the midgut (Figure 3.11) (Galay Burgos and Rainbow, 1998; Luoma and Rainbow, 2008). In the Dulas Bay amphipods, copper, for example, is detoxified in the form of metal-rich granules containing sulphur (Icely and Nott, 1980). These are the residual bodies of lysosomes breaking down copper metallothionein induced to bind the extra copper entering the amphipod (Luoma and Rainbow, 2008).

The metal-rich granules accumulate in the epithelial cells as they pass along the epithelium of the ventral caecum, prior to expulsion into the gut lumen and eventual defaecation (Figure 3.11) (Galay Burgos and Rainbow, 1998). The upshot is that a steady-state caecum (and hence body) trace metal concentration is reached. This concentration is higher at raised trace metal bioavailabilities that have caused more uptake and more accumulation of the metal in the metal-rich granules. This is clearly seen for copper in both Restronguet Creek and Dulas Bay amphipods, accumulated body concentrations being higher than the uncontaminated baseline of 77 to 113 µg/g copper (Table 6.10). The implication is that there is more bioavailable copper in the Restronguet Creek sediments than at Dulas Bay. The question remains open whether this is down to small regional differences in total copper concentrations in the sediments (Table 6.4), or whether there is another factor affecting the trophic bioavailability of sedimentary copper that differs between the two sites.

It is interesting that accumulated zinc concentrations are not raised in Dulas Bay C. volutator (Table 6.10), given the high concentrations of zinc in the sediments there (Table 6.4). Again, it is difficult to say whether this is the consequence of a difference in local chemical factors affecting the bioavailability of sediment-associated zinc in Dulas Bay, or whether the local population of amphipods in Dulas Bay is showing evidence of physiological adaptation to high zinc bioavailabilities there. Amphipods brood live young, in contrast to many coastal invertebrates that broadcast planktonic larvae far and wide. It is possible, then, for local populations of C. volutator to develop specific local tolerance to a high trace metal bioavailability. Indeed, this is the case for the Restronguet Creek population, which is tolerant to copper but not zinc (Bryan and Gibbs, 1983). The copper tolerance of this population, however, is not associated with the maintenance of what would be considered as typical background body concentration of copper (Table 6.10), but, more likely, an increased rate of copper detoxification (Rainbow and Luoma, 2011a).

The presence of the copper-tolerant population of *Corophium volutator* at the top of Restronguet Creek in 1971 is testament that the local total bioavailability of copper to the amphipod was of ecotoxicological significance. Even with the development of this copper tolerance, this amphipod population was probably still under severe physiological stress. Later surveys, in November 1991 and in March 1992, before and after the overflow from the Wheal Jane mine documented earlier, did not find any *C. volutator* at all in Restronguet Creek (Warwick, 2001).

6.11 Suspension Feeders

Prologue. Suspension-feeding invertebrates filter plankton and suspended detritus, and in some cases resuspended sediment, from the overlying water column. Thus their routes of trace metal uptake and absorption are from solution and from suspended material in the water. Lamellibranch bivalves are the major suspension feeders in British estuaries. Ecologically important suspension-feeding lamellibranchs in estuaries and on adjacent shores are the mytilid mussels and the ostreid oysters, in addition to cockles, clams and scallops.

6.1.1 Mussels

Prologue. The common mussel *Mytilus edulis* is particularly prevalent in British estuaries, as well as on open coastlines.

The mytilid mussels have abandoned any evolutionary history of burrowing. They live as epifauna attached by byssal threads to hard substrata such as rocks, piers or stones and shells, even when these are partially embedded in soft sediments. The more southerly Mediterranean mussel, *Mytilus galloprovincialis*, common on French coasts, is now spreading into Southwest England, south Wales and South and West Ireland, presumably as a result of climate change.

6.11.1.1 Biomonitoring

Prologue. *Mytilus edulis* has a long history of use as a biomonitor of trace metal bioavailabilities in British estuaries and coastal waters, with a good comparative database available.

The common mussel *Mytilus edulis* (Plate 7b) has been widely used as a trace metal biomonitor in British coastal waters (Table 6.11), and there are equivalent mussel watch programmes across the world (Phillips and Rainbow, 1994; Luoma and Rainbow, 2008). Strictly, the mussel is a biomonitor of the combined bioavailabilities of trace metals in solution and in suspended matter in the water column in its local habitat. In comparison with the likes of the polychaete *Hediste diversicolor* and the bivalve *Scrobicularia plana* considered earlier, *M. edulis* samples the lower estuary and beyond. Its presence on open coasts means that it is often the biomonitor of choice when assessing trace metal bioavailabilities in coastal waters beyond the mouth of a specific estuary, for example in the Irish and North seas (Widdows et al., 1995, 2002).

Table 6.11 lists accumulated trace metal concentrations of trace metals in *M. edulis* from a selection of mining-affected estuaries, as well as from industrially contaminated estuaries, an estuary considered relatively uncontaminated and from wider coastal waters. As previously, Table 6.11 includes proposed interpretations of what are either typical or high bioaccumulated concentrations of trace metals in the mussel indicating raised local trace metal bioavailabilities.

In the case of the mining-affected estuaries, Restronguet Creek is conspicuous by its absence from Table 6.11. While occasional specimens may be found at Restronguet Point at the bottom of Restronguet Creek (Figure 6.5), mussels are extremely rare or effectively absent from the creek itself (Bryan and Gibbs, 1983). Mussels transplanted into the middle of Restronguet Creek do survive for several months (Boyden, 1977; Bryan and Gibbs, 1983), suggesting that any ecotoxicological effects preventing establishment of mussel populations in the creek act early in life history, such as at the settling larval or spat stages. *M. edulis* transplanted into Restronguet Creek from the Tamar in 1980 for six months reached a soft tissue zinc concentration of 1,550 µg/g, falling back to 1,000 µg/g after seven months (Bryan and Gibbs, 1983). These accumulated zinc concentrations are phenomenally high for the mussel, amongst the highest ever recorded (Table 6.11). Copper concentrations in these transplanted mussels reached 121 µg/g after seven months (Bryan and Gibbs, 1983), again extremely high for a mussel (Table 6.11). Similarly, bioaccumulated cadmium concentrations reached 7.1 µg/g after seven months (Bryan and Gibbs, 1983), indicative of high cadmium bioavailabilities also (Table 6.11).

In other mining-affected estuaries in Cornwall in the 1970s and 1980s, mussels accumulated high lead concentrations in Penryn Creek, also in the Fal Estuary system (Figure 6.6), and in the West Looe Estuary (Table 6.11). At the same time period, mussels collected from the mouth of the Red River contained 262 µg/g copper and 579 µg/g zinc (Bryan et al., 1985), very high concentrations indeed for a mussel (Table 6.11). Mussels from the Plym Estuary in 1974 confirmed the high bioavailability of cadmium (Table 6.11), seen previously for *H. diversicolor* (Table 6.6) and *Scrobicularia plana* at the time (Table 6.8). These mussels were also exposed to a high bioavailability of zinc in the Plym Estuary in 1974 (Table 6.11).

Mussels from the lower regions of the Thames Estuary below Canvey Island had high concentrations of accumulated silver between 1984 and 2001 (Table 6.11). The 1980s data in Table 6.11 were from sites near Southend at the mouth of the estuary, remarkable in their distance from any upstream river or higher estuary source of bioavailable silver. Both the 1997 and 2001 datasets for the Thames Estuary showed a decrease in silver concentrations accumulated by mussels in a downstream direction from Canvey Island to Shell Ness, beyond the mouth of the estuary (Figure 6.12) (McEvoy et al., 2000; Langston et al., 2004), indicative of the presence of an upstream source of silver. Bioavailabilities of cadmium to *M. edulis* were also high in the Thames Estuary from the 1980s to 2001, with an indication of a decrease

Table 6.11 Mean, range of mean or weight-standardised estimate (0.5g soft tissues) concentrations (µg/g dry weight) of trace metals in soft tissues of the mussel *Mytilus edulis* from British estuaries affected by mining contamination, industrial contamination or considered uncontaminated, and from wider coastal waters.

	Ag	As	Cd	Co	Cr	Cu	Fe	Hg	Mn	Ni	Pb	Sn	Zn
Mining													
Carrick Roads (Fal)													
1975–77[1]		17				15							167
Penryn Creek (Fal)													
1976[2]			2.2			7	274		4.7		85		323
Red River													
Pre-1985[3]	0.24		2.2	6.3	1.7	262	669		16	1.4	9.5		579
East Looe													
Pre-1980[4]	0.2	21	2.3	0.6	2.5	9	401	0.39	6	2.6	45	0.8	113
West Looe													
Pre-1985[3]	0.15		2.4	1.1	1.9	11	328		35	1.9	105		115
Looe													
1975–76[5]	0.55		0.8–2.6	0.02–1.1	0.9–2.7	14	152		0.9–3.5		30		199
Plym													
1974[2]			23			13	175		22		37		330
Industrial													
Thames													
Southend													
1984–85[6]	3.5		8.4	1.0	2.5	11	453		15	6.2	13		300
Pre-1985[3]	1.1												
1997[7]	0.67–8.8	8.2–16	1.1–3.9	0.6–1.7	1.0–2.1	8–13	272–803	0.17–0.58	6–37	0.7–3.1	4.5–10	0.3–0.9	123–288
2001[8]	0.08–8.8	6.8–23	0.7–3.6	0.03–1.4	0.5–1.7	5–13	154–737	0.17–0.56	4–57	2.2–8.1	8.3–18	0.2–0.6	54–231

Table 6.11 (cont.)

	Ag	As	Cd	Co	Cr	Cu	Fe	Hg	Mn	Ni	Pb	Sn	Zn
Bristol Channel													
North 1971[9]			4–60								1–30		62–250
Swansea Bay													
Pre 1985[3]	0.09		12	1.9	7.2	10	289		12	3.6	18		198
Mersey													
1980–84[10]	0.12	20	5.6		3.5	11	450	1.88	17	2.7	21	0.2	355
Tyne													
1977–78[11]													336 65–1,340
Poole Harbour													
1974[12]			3.7–65			7–11	87–154		3 – 5	5–12	7–19		94–162
1974 0.5 g[13]			64–89			10–11	126–199		4	10	17		135–206
Southampton													
Pre-1972[14]		9						1.86					
Solent													
Pre -971[15]	0.03		5.1	1.6	1.5	10	1,700		4	3.7	9		91
Pre-1972[14]		15						0.43					
Forth Estuary													
1982[16]								0.2 – 4					
Leith Docks, Forth													
1982[17]			7.0–16										
Clyde Estuary													
1984[18]			1.7–3.8		3.3–32	7–12		0.22–0.86		1.4–6.0	10–30		114–236

Coastal waters										
Northern Ireland										
1980–81[19]			0.7–7.0		1.7–443	5–15	0.1–3.0	1.3–7.1	1.5–19	51–312
Irish Sea										
1996–97[20]		0.01–3.2	6.2–31	0.2–2.6	0.6–4.2	4–14	0.10–0.32	0.2–3.0	0.6–26	50–202
North Sea										
1990–91[21]		<0.3–5.2				5–10	0.06–0.38		<3–29	58–303
Uncontaminated										
Helford (Cornwall)										
1974[2]		1.1			180	11		4.0	8	103
Typical[8,22]	0.01–1	3–20	0.2–3	0.02–2	0.2–3	4–20	0.05–1	0.1–5	0.2–20	50–200
High[8,22]	3–20	50–200	5–90	5–10	10–450	50–270	1–10	10–30	30–130	300–600

Sources: Klumpp and Peterson (1979);[1] Bryan and Gibbs (1983);[2] Bryan et al., (1985);[3] Bryan et al. (1980);[4] Bryan and Hummerstone (1977);[5] Rainbow and White (unpublished);[6] McEvoy et al. (2000);[7] Langston et al. (2004);[8] Nickless et al. (1972);[9] Langston (1986);[10] Lobel and Wright (1983);[11] Boyden (1975);[12] Boyden (1977);[13] Raymont (1972);[14] Segar et al. (1971);[15] Elliott and Griffiths (1986);[16] Roberts et al. (1986);[17] Miller (1986);[18] Gault et al. (1983);[19] Widdows et al. (2002);[20] Widdows et al. (1995);[21] Luoma and Rainbow (2008).[22]

over this period (Table 6.11) (McEvoy et al., 2000; Langston et al., 2004).

As we have come to expect in the Severn Estuary and the Bristol Channel in the 1970s, bioavailabilities of cadmium (and to a lesser extent zinc) to mussels on the north shore were extremely high (Table 6.11), particularly at Fontygary Bay, near Cardiff (Nickless et al., 1972). Mussels also provided evidence for the expected high bioavailability of mercury in the Mersey Estuary in the early 1980s, together with somewhat raised bioavailabilities then of cadmium and zinc (Table 6.11) (Langston, 1986). A later study confirmed that by the late 1990s, accumulated mercury concentrations in *M. edulis* in the Mersey Estuary had dropped to about 0.5 µg/g from a high in excess of 3 µg/g in 1980 (Langston et al., 2006). Mussels collected from Southampton Water before 1972 also had a high mercury concentration, in this case 1.86 µg/g (Table 6.11) (Raymont, 1972). The source of this mercury might have been resuspended sediment contaminated by local industrial activity (Raymont, 1972).

Mussels from Tynemouth at the mouth of the Tyne Estuary in the late 1970s had a high mean zinc concentration of 336 µg/g, with an individual range from 65 µg/g to the phenomenal concentration of 1,340 µg/g (Lobel and Wright, 1983), matching the zinc concentrations of the transplanted mussels in Restronguet Creek.

In Poole Harbour in 1974, mussels growing at the mouth of the industrialised Holes Bay attained some of the highest cadmium concentrations ever measured in mussel soft tissues (Table 6.11) (Boyden, 1975, 1977). Mussels with soft tissues of 0.5 g dry weight had accumulated cadmium concentrations between 64 and 89 µg/g (Boyden, 1977). Nickel concentrations were also on the high side in these mussels, this trace metal like cadmium also originating from industrial sources in Holes Bay at that time (Boyden, 1975, 1977).

Earlier in this chapter, we noted the remarkably high concentrations of mercury in bladder wrack in the upper Forth Estuary in 1982, at Longannet near the industrial complex of Grangemouth (Table 6.5). Local mussels there at the time also had very high

bioaccumulated concentrations of mercury, between 3.5 and 4 µg/g (Table 6.11) (Elliott and Griffiths, 1986), indicative of high mercury bioavailabilities. Nearby, but further out in the Firth of Forth, very high cadmium concentrations were found in mussels near Leith Docks in 1982 (Table 6.11) (Roberts et al., 1986). This cadmium originated from a sea outfall, discharging effluent from an industrial plant using imported cadmium-rich rock phosphate in the production of fertiliser (Roberts et al., 1986).

In 1984, the Clyde River Protection Board carried out a Mussel Watch programme, surveying 24 sites in the Clyde Estuary (Table 6.11) (Miller, 1986). Atypically high bioavailabilities were identifiable for chromium, and arguably for nickel and cadmium, at the more upstream sites such as Cardross on the northern shore and Woodhall and Port Glasgow on the southern shore (Miller, 1986). This region of the southern shore was highly industrialised and densely populated, and a metal-plating operation there was the potential source of the chromium and nickel (Miller, 1986).

Across the Irish Sea at the beginning of the 1980s, a wider biomonitoring survey of mussels in the coastal waters of Northern Ireland threw up a few hot spots of trace metal bioavailabilities to the mussels (Table 6.11) (Gault et al., 1983). Strangford Lough was the most trace metal-contaminated region in the study, with two separate mussel populations with exceptionally high concentrations of mercury and chromium respectively (Gault et al., 1983). Mercury concentrations in mussels from Whiterock in the Lough at times exceeded 3 µg/g (Table 6.11), matching those in the mussels from the Inner Forth Estuary in the early 1980s (Elliott and Griffiths, 1986). This mercury contamination was attributed to mercury-based antifouling paints, commonly used at the time on the many yachts and pleasure craft moored at Whiterock (Gault et al., 1983). Mussels with the staggeringly high chromium concentrations exceeding 400 µg/g (Table 6.11) were collected from close to the sewage outfall at Killyleagh, emitting not only sewage but also waste from a local tannery (Gault et al., 1983). Otherwise, in Northern Ireland, relatively high cadmium concentrations were found in mussels at

Newtonards at the top of Strangford Lough, with mean concentrations in excess of 4 µg/g (Table 6.11). These mussels also contained the highest nickel and copper concentrations in the study (Table 6.11). The trace metals here possibly originated from local tip leachate and sewage discharge, in the absence of an obvious industrial source (Gault et al., 1983).

It was also possible to identify a small number of coastal sites with somewhat raised trace metal bioavailabilities to mussels, in a wider survey of the Irish Sea in the mid-1990s (Widdows et al., 2002). For example, mussels from sites in Cardigan Bay at Abersoch, Borth and Aberporth, either side of Aberystwyth, had mean lead concentrations of 26, 24 and 15 µg/g lead respectively (Table 6.11) (Widdows et al., 2002). Similarly, mussels from the northwest coast of the Isle of Man had a mean lead concentration of 22 µg/g (Widdows et al., 2002). These concentrations were well above the next highest mean concentration of 6 µg/g lead in the survey (Widdows et al., 2002). They are also above the proposed typical range of mussel lead concentrations (Table 6.11), indicative, therefore, of quite high lead bioavailabilities to the mussels. It is tempting to attribute the raised lead bioavailabilities in Cardigan Bay to lead originating in the old lead mining rivers, the Ystwyth, Rheidol and Clarach (Widdows et al., 2002). Similarly, historical lead mines on the Isle of Man, perhaps in the Foxdale District (Chapter 5), are the probable source of the raised bioavailable lead levels in the case of the Isle of Man mussels (Widdows et al., 2002).

In an equivalent survey in the North Sea in the early 1990s, there were also several coastal sites with recognisably high trace metal bioavailabilities to the local mussels (Widdows et al., 1995). Mussels at Musselburgh, near Edinburgh, had the highest mean concentrations of cadmium at 5.2 µg/g, indicating an atypically high cadmium bioavailability (Table 6.11) (Widdows et al., 1995). Trow Rocks at South Shields, on the south bank of the Tyne opposite Tynemouth, was the collection site of mussels with the highest mean concentrations of lead (29 µg/g) and zinc (303 µg/g) (Table 6.11) (Widdows et al., 1995). Both mean concentrations are indicative of atypically

raised trace metal bioavailabilities to the mussels (Table 6.11). Given that Tynemouth was the location of mussels with even higher accumulated zinc concentrations in the late 1970s (Table 6.11) (Lobel and Wright, 1983), it would appear that the River Tyne was still metal-contaminated to a considerable degree at the start of the 1990s.

6.11.1.2 Accumulation and Detoxification

Prologue. The mussel is a net accumulator of trace metals, with accumulated concentrations increasing with increased local bioavailabilities. Comparative mechanisms of accumulation and detoxification of different trace metals by mussels are discussed.

Trace metals are taken up *Mytilus edulis* both from solution and from the diet of suspended particles (Wang et al., 1996). As befits a lamellibranch bivalve with a large gill area, the contribution of dissolved trace metals to the total uptake of trace metals is significant (Wang et al., 1996). This contribution varies with environmental parameters that affect the rate of dissolved trace metal uptake, the ingestion rate and those factors controlling the binding of the metals in the suspended food particles, differing, for example, between phytoplankton and resuspended sediment particles (Wang et al., 1995, 1996). Thus the contribution of dissolved trace metal to the accumulated trace metal content of a mussel varies between 33 and 67% for silver, 47 and 82% for cadmium and 17 and 51% for zinc (Wang et al., 1996).

Net accumulation of zinc by the mussel does not result in the very high bioaccumulated zinc concentrations encountered in the case of *Scrobicularia plana* (Table 6.8). Although accumulated zinc concentrations in *M. edulis* do increase with rising zinc bioavailability, this increase is usually over a relatively restricted range of bioaccumulated concentrations (Table 6.11). Thus a bioaccumulated zinc concentration of 300 µg/g in the soft tissues of a mussel is atypically high (Table 6.11), whereas the same concentration is low in *S. plana* (Table 6.8). The mussel is, therefore, a weak net accumulator of zinc, an accumulation pattern occasionally (and misleadingly) referred to as partial regulation. The weak net accumulation pattern for zinc by the mussel is based on the temporary storage of detoxified zinc in granules in the kidney, before these are then excreted from the body in the urine (George and Pirie, 1980). After uptake from solution in the gills or from food in the digestive gland, the absorbed zinc is passed rapidly via the blood to the kidney (George and Pirie, 1980). The blood contains only about 1% of the total soft tissue content of zinc in a mussel, while the kidney contains 30% or more (George and Pirie, 1980; Lobel, 1987a, b). The zinc-containing granules in the kidney have a half-life of about two months before being excreted (George and Pirie, 1980; Lobel, 1987a). If the bioavailable supply of zinc is constant, there is a constant standing stock of these granules turning over in the kidney. Consequently, the kidney can contain extremely high accumulated zinc concentrations, with knock-on effects on the total soft tissue concentration (George and Pirie, 1980; Lobel, 1987a, b). Typical mean accumulated zinc concentrations in the kidney of a mussel are about 1,000 µg/g (George and Pirie, 1980), increasing to nearly 8,000 µg/g in mussels from zinc-contaminated habitats, with an individual range from 1,500 to 24,000 µg/g (Lobel, 1987a, b). In mussels transplanted for seven months into Restronguet Creek (to a site somewhat further down the creek than considered previously), mussels with a total soft tissue concentration of 670 µg/g Zn had a kidney concentration of 29,200 µg/g Zn (Bryan and Gibbs, 1983). In contrast, the digestive gland had an accumulated zinc concentration of only 146 µg/g (Bryan and Gibbs, 1983), highlighting the choice of the kidney as the site of zinc detoxification and (temporary) storage of detoxified zinc.

In the detoxification granules in the kidney of mussels, the zinc ultimately is in the form of phosphate with some calcium present (George and Pirie, 1980). The granules of 1 to 3 µm diameter form in membrane-limited vesicles derived from lysosomes, and may coalesce into larger (8 to 10 µm diameter) granules before expulsion from the kidney epithelial cells into the urine (George and Pirie, 1980). It appears that that metallothioneins play a temporary role in the handling of zinc, as well as longer-term roles in the detoxification of accumulated copper and cadmium in mussels (George et al., 1979; Viarengo et al., 1981). Metallothioneins are turned over in cells in lysosomes. Such lysosomes in various stages of development, ending in the formation of lysosomal residual bodies, can be found in cells of the kidneys and digestive gland of mussels (George, 1983; Viarengo et al., 1981, 1984). In fact, the formation of the zinc-rich granules in the mussel kidney cells appears to involve the initial breakdown of zinc-binding metallothionein in lysosomes, for many of the detoxified zinc granules also contain sulphur (George and Pirie, 1980), characteristic of the initial presence of metallothionein.

Copper is also preferentially accumulated in the kidney rather than the digestive gland in mussels under conditions of high copper bioavailability. The mussels transported into the lower end of Restronguet Creek for seven months accumulated a total soft tissue concentration of 50 µg/g, with tissue concentrations of 561 µg/g Cu in the kidney and 53 µg/g Cu in the digestive gland (Bryan and Gibbs, 1983). It is clear that mussels excrete accumulated copper, as they do zinc, but still show net accumulation as copper bioavailabilities rise (Table 6.11). Incoming copper is bound by metallothionein (Viarengo et al., 1981, 1984). In the digestive gland, this copper is excreted into the alimentary tract after autolysis of the copper-bearing metallothioneins in lysosomes (Viarengo et al., 1984).

Incoming cadmium is also bound by inducible metallothionein in the digestive gland and kidney of mussels, the metallothionein destined to be autolysed in lysosomes (George et al., 1979; George, 1983). Copper bound to metallothioneins typically remains in residual bodies after lysosomal breakdown, testament to the strength of the copper-to-sulphur chemical bond. On the other hand, the sulphur-to-cadmium bond in metallothionein is less strong, and cadmium can be released from the lysosome to begin another cycle of metallothionein induction and binding (Bebianno and Langston, 1993). Thus, cadmium is not excreted when lysosomal residual bodies are expelled from kidney or digestive gland epithelial cells. Indeed, accumulated cadmium has a half-life of 300 days in mussel soft tissues, while the metallothionein molecules sequentially binding the cadmium have half-lives of only 25 days (Bebianno and Langston, 1993).

Like zinc, lead accumulated by mussels is also temporarily stored as a phosphate- and sulphur-rich insoluble complex in membrane bound vesicles in kidney cells before excretion in the urine (Schulz-Baldes, 1978). The same granules will also bind chromium in mussels exposed to high chromium bioavailabilities (Chassard-Bouchaud et al., 1989; Walsh and O'Halloran, 1998).

6.11.1.3 Ecotoxicology

Prologue. *Mytilus edulis* has long been used as a model organism in coastal ecotoxicology, playing a key role in the development of suitable biomarkers at different levels of biological organisation.

The common mussel is arguably the laboratory rat of British coastal ecotoxicology. Indeed, the study of biomarkers at the cytological and physiological levels of biological organisation in coastal invertebrates has been pioneered and developed in the mussel *Mytilus edulis* (Widdows et al., 1995; Moore et al., 2006, 2013).

Trace metal exposure can generate free radicals in organisms (see Chapter 3). An appropriate biomarker is the total oxyradical scavenging capacity (TOSC), an integrated measure of the antioxidant defences of an organism, including mussels (Regoli, 2000). Cytological biomarkers centre on lysosomes (see Chapter 3) (Moore et al., 2006). There are several responses of lysosomes to stressors such as high trace metal exposure, including the destabilisation of the lysosomal membrane in cells (Moore et al., 2006). The functional stability of the lysosome membrane, and hence its permeability, changes with degree of exposure. The relatively easy measurement of lysosomal stability in cells using the neutral red retention technique is a sensitive, low organisational level biomarker, well practised in mussels (Moore et al., 2006). Furthermore, and desirably, the measurement of lysosomal stability has been correlated with other biomarkers at other levels of biological organisation

in mussels, including TOSC and scope for growth (see Chapter 3) (Moore et al., 2006, 2013).

Scope for growth (SFG) is a physiological biomarker that integrates different physiological measures to provide an estimate of the surplus energy available to an animal for growth and reproduction (Luoma and Rainbow, 2008). It is calculated from the difference between energy assimilated from food and the energy used in respiration and was pioneered in mussels *Mytilus edulis* in coastal waters (Widdows et al., 1995, 2002). SFG decreases when energy is required to cope with the added metabolic cost of handling (detoxifying and/or excreting) the extra amounts of trace metals taken up in metal-contaminated environments. SFG is considered a robust indicator of the health status of mussels in the field, although it will respond in an integrated manner to exposure of the mussels to a range of other contaminants as well as trace metals (Widdows et al., 1995, 2002). The studies of mussel contamination in the Irish Sea and the North Sea reported in Table 6.11 also measured the SFG of the mussel populations investigated. In these real-world examples, there was concomitant exposure of many of the mussels to other contaminants in addition to trace metals, such as hydrocarbons and organochlorine contaminants (Widdows et al., 1995, 2002). There were clear differences in the SFG of mussel populations from contaminated and non-contaminated sites in both studies, highlighting the value of SFG as a physiological biomarker of the general health status of mussel populations (Widdows et al., 1995, 2002). In neither study was trace metal exposure considered to have had any significant effect on the SFG of any mussel population, even at Trow Rocks at the mouth of the Tyne (Widdows et al., 1995, 2002). In accordance with this conclusion, with the exception of the zinc accumulated by this northeast mussel population, none of the accumulated trace metal concentrations in the mussels in these two studies was really high enough to be considered indicative of severe trace metal exposure (Table 6.11). SFG is, therefore, a valuable biomarker to be used in a suite of biomarkers assessing the integrated ecotoxicological effects of contaminants on field mussel populations. These ecotoxicological effects might be caused solely by trace metal exposures in, for example, mining-affected estuaries in Cornwall lacking significant industrial or domestic input of other contaminants.

6.11.2 Oysters

Prologue. On the shores of British estuaries, the native oyster *Ostrea edulis* is being replaced by the larger introduced oyster *Crassostrea gigas.*

Like the mytilid mussels, ostreid oysters have also abandoned any evolutionary history of burrowing. Oysters live as epifauna, typically cementing one shell valve onto a hard surface, as they use their gills to filter suspended food particles from the water column.

The native British oyster *Ostrea edulis*, the common oyster, was formerly abundant on the shores of British estuaries, but during the twentieth century numbers fell quite considerably (Yonge, 1960). Various causes interacted, including infection by protistan parasites, and the direct and indirect effects of the introduction of foreign oysters for oyster farming (Yonge, 1960). The importation of stocks of the American oyster, *Crassostrea virginica*, introduced two gastropod predators to *O. edulis* as a new food source, the sting winkle *Ocenebra erinaceus* and the oyster drill *Urosalpinx cinerea* (Yonge, 1960). Also introduced from North America was the slipper limpet *Crepidula fornicata*, a remarkable and atypical gastropod because, like a bivalve, it uses its gills to create a feeding current from which it filters suspended food (Yonge, 1960). The slipper limpet became a strong direct competitor to the common oyster for both food and space. While the American oyster is no longer a significant competitor of *O. edulis* in British estuaries, its place has been taken by a much stronger direct competitor, the Pacific oyster, *Crassostrea gigas*. *C. gigas*, as indicated by its specific name, grows quickly to a large size and was again introduced for oyster farming. In 2005, two million individual spat of the Pacific oyster were laid per year in Poole Harbour alone (Jensen et al., 2005b). As our coastal waters warm up, it is really no great surprise that the Pacific oyster has spread, and it is now distributed throughout the British Isles.

6.11.2.1 Accumulation and Detoxification
Prologue. Oysters accumulate very, very high concentrations of copper and zinc, detoxifying these trace metals in blood cells without a route of excretion.

There has been a long history of fishing for the native oyster *Ostrea edulis* in British estuaries. For example, this oyster has been common in the Fal Estuary system for centuries, although it is now suffering from competition with the newly introduced Pacific oyster *Crassostrea gigas* (Plate 7c). *O. edulis* would grow at the bottom end of Restronguet Creek (Bryan and Gibbs, 1983), although it was more abundant further out in Carrick Roads in the main Fal Estuary (Figure 6.6). Many of the Fal oysters, particularly those growing at the bottom of Restronguet Creek, had soft tissues coloured green and were called green-sick oysters (Bryan and Gibbs, 1983). It was known, even in the middle of the nineteenth century, that the green colour was caused by the accumulation of high amounts of copper (O'Shaughnessy, 1866; Bryan and Gibbs, 1983). The colour did not put off potential diners, although oyster farmers might relay oysters from the bottom of Restronguet Creek elsewhere in the Fal Estuary for some decontamination (Bryan and Gibbs, 1983). A study in the 1920s confirmed that Restronguet Creek oysters contained very high concentrations of copper and zinc, especially those oysters that were particularly green (Orton, 1923; Bryan and Gibbs, 1983).

Table 6.12 presents accumulated trace metal concentrations in the two species of oyster, *Ostrea edulis* and *Crassostrea gigas*, from British estuaries, again with proposed typical and high concentration ranges. The oysters accumulate remarkably high concentrations of copper and zinc, even under uncontaminated conditions (Table 6.12). It is likely that the lower concentration ranges of these two trace metals in *C. gigas* (Table 6.12) is attributable to growth dilution, given the very fast growth rate and ultimately larger size of the Pacific oyster.

As we have come to expect, the bioavailabilities of copper and zinc to oysters growing towards the bottom of Restronguet Creek were exorbitantly high through the twentieth century, with high

Table 6.12 Mean, range of mean or weight-standardised estimate* (1 g soft tissues) concentrations (µg/g dry weight) of trace metals in soft tissues of the oysters *Ostrea edulis* and *Crassostrea gigas* from British estuaries affected by mining contamination, industrial contamination or considered uncontaminated.

	Ag	As	Cd	Co	Cr	Cu	Fe	Hg	Mn	Ni	Pb	Zn
Ostrea edulis												
Mining												
Restronguet Creek												
1921[1+]		11–19				1,450–16,500						4,150–10,500
1971[2]		25–39				769–3,900						6,000–14,900
*T (5 month) 1974[2,3]			6.4			ca. 3,000	216		11	6.1	7.8	4,360
1975–77[4]		17	3.0			2,160						7,170
1980[2,5]	0.2–0.9	17–39	3.2–7.9	0.2–0.7	0.2–0.6	552–2,610	150–331		8–20	0.1–0.4	1.1–2.5	4,700–17,100
Carrick Roads (Fal)												
1979[2]			3.5–3.9			609–957	246–386		9–16		2.2–2.8	3,770–5,080
St Just (Fal)												
1921[1+]						1,450						3,000
Lynher (Tamar)												
1980[6+]		9.1				610		0.23			0.6	3,280
Industrial												
Poole Harbour												
1973[7]			5.9–54			86–451	172–394		6–9	2–5	5–8	1,970–3,400
1983[5]	2.8–17		1.6–30	0.3–0.6		40–278	140–219		8–10		0.3–1.2	1,500–3,740

Table 6.12 (cont.)

	Ag	As	Cd	Co	Cr	Cu	Fe	Hg	Mn	Ni	Pb	Zn
Uncontaminated												
Menai Strait (Anglesey)												
*1974[3]			5.2			392	223		18	6.0	6.2	3,440
Blackwater (Esser)												
1970–73[8†]			1.5–7.0			240–375		0.15–0.45			<2.5–13	300–3,000
Typical	1–3	5–10	1–7	0.2–1.0	0.2–2.0	40–500		0.1–0.5		0.1–6	0.3–15	300–4,000
High	20+	20–50	10–60			1,000–17,000						6,000–20,000
Crassostrea gigas												
Mining												
Restronguet Creek												
*T (5 month) 1974[2,3]			12			ca. 2,500	365		17	5.4	9.1	7,110
Industrial												
Thames												
2001[9]	11–64	7.4–18	1.0–4.7	0.2–0.8	0.3–1.6	171–537	267–2,300	0.19–0.42	30–37	1.1–2.6	2.9–4.8	1,690–3,400
Poole Harbour												
T (8 month) 1970[10]			27			396						2,670
Outer 1973[7]			4.6			200	228		18	3	6	1,770
*1974[3]			26			261	313		22		21	2,570
Outer 1983[5]	1.6		4.8	0.5		61	341		27		2.5	1,930

Uncontaminated

Helford (Cornwall)										
T (8 month) 1970[10]	2,590					643		2.2		1–3
Menai Strait										
Pre 1984[11+]	3,680					475				
*1974[3]	4,290		33		237	446		6.3		
Colne (Essex)										
T (8 month) 1970[10]	1,630					207		4.8		
Typical[12]	1,500–4,500	0.5–5	1–5	0.1–0.5		30–500	0.2–1.0	1–5	3–20	1–3
High[12]	6,000–14,000	20–140	10–50	10+		1,000–5,000	0.2–2.0	10–50	30+	10–70

Notes: T = transplanted. + Wet weight concentration data have been multiplied by 5 to give dry weight concentrations.

Sources: Orton (1923);[1] Boyden (1977);[2] Bryan and Gibbs (1983);[3] Klumpp and Peterson (1979);[4] Bryan et al. (1985);[5] Bland et al. (1982);[6] Boyden (1975);[7] Portmann (1979);[8] Langston et al. (2004);[9] Thornton et al. (1975);[10] Pirie et al. (1984);[11] Luoma and Rainbow (2008).[12]

bioavailabilities also of arsenic (Table 6.12). High copper bioavailabilities to *O. edulis* extended into Carrick Roads (Figure 6.6) in the late 1970s (Table 6.12). Oysters of both species highlighted the extremely high bioavailability of cadmium at the entrance to Holes Bay in Poole Harbour in the 1970s and early 1980s (Table 6.12) (Boyden, 1975, 1977), already shown up in the mussel data (Table 6.11). Again, as for mussels (Table 6.11), oysters, in this case *C. gigas*, confirm the relatively high bioavailability of silver in the lower Thames at the turn of the millennium (Table 6.12).

Within oysters, the sites of detoxified storage of the huge amounts of copper and zinc accumulated in contaminated habitats are blood cells (Bryan and Gibbs, 1983; George et al., 1978; Pirie et al., 1984). Table 6.13 shows how the blood cells of green oysters from Restronguet Creek contain 68% of the total soft tissue copper and 47% of the soft tissue zinc (Bryan and Gibbs, 1983). Correspondingly, the blood cells have phenomenally high concentrations of these two metals, way above the already high total soft tissue concentrations (Table 6.13). In contrast, only 3% of soft tissue cadmium is to be found in the blood cells (Table 6.13). We saw previously that extra accumulated zinc and copper in mussels are temporarily stored in the kidney before excretion in granular form into the urine. The blood cells of oysters have no such exit route. So the stored zinc and copper rise in

concentration to impressive heights as uptake and accumulation of these two trace metals continue.

Storage of such high concentrations of zinc and copper in the blood cells of oysters demands detoxification (George et al., 1978; Pirie et al., 1984). In green-sick *Ostrea edulis* from Restronguet Creek, there are three different types of blood cells involved in the detoxified storage of copper and zinc in membrane-limited vesicles within the cells (Pirie et al., 1984). The first cell type stores only zinc, in this case in dense granules (1 μm) containing zinc and phosphorus, presumably as phosphate.; the second cell type contains only copper-rich granules, the copper being associated with sulphur (Pirie et al., 1984). The presumption here is that the copper and sulphur are derived from the breakdown of large quantities of metallothionein induced by copper exposure. A third blood cell type contains both zinc and copper (Pirie et al., 1984). The individual granules in these cells may contain copper only, zinc only or a combination of copper and zinc in varying proportions (Pirie et al., 1984). Specimens of *O. edulis* from Conwy with much lower accumulated copper and zinc soft tissue concentrations than the green-sick Restronguet Creek oysters (175 vs. 1,650 μg/g Cu, and 2,000 vs. 11,700 μg/g Zn) had only the third type of blood cell, storing both copper and zinc (Pirie et al., 1984). Similarly, specimens of *Crassostrea gigas*, from the uncontaminated waters of the Menai Strait in

Table 6.13 **Concentrations (μg/g dry weight) and percentage soft tissue contents of cadmium, copper and zinc in blood cells of green oysters *Ostrea edulis* from Restronguet Creek.**

		Cadmium		Copper		Zinc	
	% dry weight	Concentration	% total content	Concentration	% total content	Concentration	% total content
Total soft tissues		9.0		3,870		14,900	
Blood cells	11%	2.7	3%	23,500	68%	61,800	47%

Source: Bryan and Gibbs (1983).

north Wales, with 475 µg/g Cu and 3,680 µg/g Zn in the soft tissues, also only contained the third type of detoxificatory blood cell, storing both copper and zinc within a mixed granule population (Pirie et al., 1984).

As suggested previously, cadmium, on the other hand, is not immobilised in membrane-limited vesicles in blood cells of oysters, accumulating instead in the digestive gland and kidney (George et al., 1978). In the digestive gland, cadmium is bound by inducible metallothionein (Geffard et al., 2001), this metallothionein being present in the cytosol of the cells before breakdown in lysosomes (Boisson et al., 2003).

6.11.2.2 Ecotoxicology

Prologue. Oysters are generally relatively resistant to trace metal ecotoxicity. They are, however, vulnerable at larval and settlement stages of the lifecycle, for example in Restronguet Creek and at the mouth of Holes Bay, Poole Harbour, in the early 1970s. Studies on the proteomes of oysters make them good candidates for the development of molecular biomarkers of high trace metal exposures.

Oysters as a family of bivalves appear generally resistant to trace metal ecotoxicity in British estuaries. For example, oysters, be they *Ostrea edulis* or *Crassostrea gigas*, will grow in their expected position at the seaward end of Restronguet Creek (Bryan and Gibbs, 1983). Oysters, again of either species, also survive transplanting into the creek from uncontaminated habitats (Boyden, 1977; Bryan and Gibbs, 1983). The remarkable copper and zinc detoxification physiologies of all oysters, whatever their geographical origin, are the clear bases for such trace metal insensitivities. The establishment of mussel populations in Restronguet Creek was inhibited by the ecotoxicological effect of high trace metal bioavailabilities on settling larvae or newly established spat (Bryan and Gibbs, 1983). Oysters, however, settle and live lower down the estuary, with somewhat lower bioavailabilities of copper and zinc, and they do survive this sensitive stage of the lifecycle.

That is not to say that there is no evidence of any ecotoxicological effects of trace metals on oysters in

British estuarine waters. In the early 1970s, the newly established oyster hatchery at the entrance to Holes Bay in Poole Harbour encountered considerable problems in rearing oyster larvae, in this case *O. edulis* (Boyden, 1975). It was subsequently discovered that the hatchery intake water, under the influence of Holes Bay waters contaminated by industrial effluents, occasionally contained exorbitant dissolved levels of zinc, reaching more than 700 µg/L Zn (cf. Table 6.1) (Boyden, 1975). These dissolved zinc concentrations on their own would severely affect the survival of oyster larvae (Boyden, 1975). For example, a dissolved zinc concentration of 125 µg/L will reduce the number of larvae of *C. gigas* able to settle, and spat growth of the oyster is suppressed at 250 µg/L Zn (Boyden et al., 1975). Furthermore, an oyster hatchery at Conwy in north Wales in the early 1970s also suffered intermittent failure of *O. edulis* larvae to develop and settle (Elderfield et al., 1971). Again it was found that, at certain times, dissolved zinc concentrations in the waters of the hatchery fell within the range of 100 to 500 µg/L, toxic to these larvae (Elderfield et al., 1971). The Conwy Estuary receives the waters of the Conwy River draining a zinc-contaminated catchment (Elderfield et al., 1971).

Again in the early 1970s, oysters (*Crassostrea gigas*) being grown from spat in a pond receiving the cooling water of Hinkley Point Power station at the junction of the Severn Estuary and the Bristol Channel showed the characteristic soft tissue colouration of green-sick oysters (Boyden and Romeril, 1974). Analyses confirmed that these oysters contained very high accumulated concentrations of cadmium (17 to 43 µg/g), copper (1,760 to 6,480 µg/g) and zinc (9,860 to 35,100 µg/g) (cf. Table 6.12) (Boyden and Romeril, 1974). While the oysters were indeed surviving and growing, such very high accumulated trace metal concentrations would have implications for selling the oysters for human consumption. We have got used in this chapter to appreciating the high bioavailabilities of cadmium, in particular, in the waters and (potentially resuspended) sediment of the Severn Estuary in the early 1970s. It was not, however, the local estuarine waters or sediments that were the cause of the increased accumulation of cadmium, copper and zinc in these cultivated oysters (Boyden and Romeril, 1974). Rather, the raised bioavailabilities of the metals were caused by the corrosion of galvanised trays used to hold the oysters releasing zinc, and the corrosion of the phosphor bronze blades of a recirculating pump releasing copper (Boyden and Romeril, 1974). It was remarkable that such apparently small sources of metal contamination could cause increased metal accumulation of such a magnitude in the oysters (Boyden and Romeril, 1974).

Oysters have also been investigated as model organisms for the measurement of biomarkers in contaminated estuaries. Previous interest in the applicability of concentrations of metallothioneins in oysters as biomarkers of high trace metal bioavailabilities is now waning, given that metallothionein concentrations vary between organs and season (Mouneyrac et al., 1998; Geffard et al., 2001). And standing MT concentrations are often independent of the ecotoxicologically more significant rates of turnover of induced metallothioneins in cells (see Chapter 3).

Recently interest has switched to the genome, proteome and transcriptome of oysters. The genome is the entirety of an organism's hereditary information, including both genes and non-coding regions, while the proteome is the full set of proteins encoded by the genome. The transcriptome is the full set of RNA molecules produced at a particular time, reflecting the genes actively being expressed then, and varying with environmental conditions, including exposure to raised trace metal bioavailability. As sessile animals living in estuarine and intertidal regions, oysters are faced with dynamically changing environments, and in some estuaries to high toxic metal challenges (Zhang et al., 2012). Indeed, studies of the transcriptomes of oysters reveal an extensive set of genes responding to environmental stress (Zhang et al., 2012), including trace metal exposure (Liu and Wang, 2012). A study of oysters, in this case *Crassostrea hongkongensis*, from habitats along a gradient of trace metal (Cd, Cu, Zn) contamination, identified a proteome pattern of 13 commonly altered proteins in the gills of the oysters (Liu and Wang, 2012). This proteome pattern completely segregated oysters that were metal-contaminated or uncontaminated (Liu and Wang, 2012). Furthermore, the changes in the proteome were linearly related to the integrated metal contamination of the oyster tissues, as represented by the increases in accumulated metal concentrations above those of control oysters (Liu and Wang, 2012). One of the proteins with increased expression in the presence of raised trace metal bioavailabilities was superoxide dismutase (Liu and Wang, 2012), an antioxidant enzyme, itself a biomarker of oxidative stress caused by raised trace metal exposure. In the new 'omics' era, such proteomic and transcriptomic studies in oysters have great potential for the assessment of the ecotoxicological effects of raised trace metal bioavailabilities in estuaries.

6.11.3 Cockles

Prologue. *Cerastoderma edule* accumulates trace metals from solution and from suspended particles in the water column and can be used as a biomonitor.

The next estuarine suspension-feeding lamellibranch bivalve of relevance here is the common cockle *Cerastoderma edule*. Cockles can occur in vast numbers at the bottom of muddy and sandy shores, within estuaries and without, and cockle beds can support large fisheries. For example, 100 tonnes of small seed cockles are laid in Poole Harbour each year, to be harvested later as adults (Jensen et al., 2005b). Cockle spat are eaten by brown shrimp *Crangon crangon* and shore crabs *Carcinus maenas*, while adult cockles are taken by flatfish (flounders *Platichthys flesus*) and wading birds such as knot (*Calidris canutus*), oystercatchers (*Haematopus ostralegus*) and eider ducks (*Somateria mollissima*) (Hayward, 2004).

The cockle is an active burrower. Its burrows are very shallow, the short siphons protruding just above the surface of the sediment. The otherwise sealed mantle cavity is isolated from any pore water in the sediment, and any trace metals taken up originate from the overlying water and the suspended matter therein.

Table 6.14 lists accumulated trace metal concentrations in the common cockle *Cerastoderma edule* from British estuaries and coastal waters.

Table 6.14 Mean or range of mean concentrations (µg/g dry weight) of trace metals in soft tissues of the common cockle *Cerastoderma edule* from British estuaries and coastal waters.

	Ag	As	Cd	Co	Cr	Cu	Fe	Hg	Mn	Ni	Pb	Zn
Mining												
Restronguet Creek												
1981[1,2]	0.6		3.0	5.1		174	1,690		14		7.6	303
Mylor Creek (Fal)												
1980[1]			1.3–1.4			59–91	1,080–1,280		7–15		4.4–8.5	106–233
Carrick Roads (Fal)												
1979[1]			1.2			89	822		39		13	787
St Just (Fal)												
1980[1]			0.5			7.8	760		4.3		2.1	79
Penrhyn Creek (Fal)												
1976[1]			0.5			4.7	328		3.3		6.5	92
Gannel												
1976[1]			3.5			12	1,420		210		69	180
Pre-1985[2]	0.8		3.3	12	1.9	26	1,310		317		120	175
East Looe												
1976[1,3,4]	1.5–2.4	21	0.7	1.6	1.9	9.7	565	0.26	6.2	54	5.3–8.1	54
Tamar												
1980[1]			2.7			7.3	1,050		4.1		4.5	71
Plym												
1980[1]			0.7			3.7	412		1.6		1.8	38
Industrial												
Outer Thames												
Southend pre-1985[2]	11		1.6	3.5	3.2	9	819		11	165	3.5	83
Shore 2001-2[5]	2.3–47	7.6–9.8	0.5–1.4	4.3–10	0.7–4.0	3–19	422–1,430	0.17–0.23	5–18	51–103	1.5–7.8	55–146
Offshore 2001-2[5]	1.6–42	6.4–11	0.3–0.7	2.8–9.4	0.5–2.9	3.5–11	533–2,200	0.12–0.21	15–33	34–85	0.9–5.3	46–87

Table 6.14 (cont.)

	Ag	As	Cd	Co	Cr	Cu	Fe	Hg	Mn	Ni	Pb	Zn
Loughor												
Pre 1985[2]	0.1		0.8		20	14	5,520		152	50	16	117
Swansea Bay												
Pre 1985[2]	0.2		2.4		6.3	9	3,680		104	107	13	100
Mersey												
1980–84[6]	0.2	11	0.8		2.0	9	696	1.86	39	69	7.1	168
Pre 1985[2]	0.5		1.2	1.9		17	1,610		53		18	270
Tees												
Pre 1985[2]	0.2	7.8	0.9	2.0	4.6	15	268	0.43	19	22	3.6	132
Poole Harbour												
1973[7]			1.5–17			4–9	172–502		3–5	35–174	5–14	111–271
Southampton Water												
1970–72[8,9]		6.3					735–1,330	0.80				88–140
1985–86[10]			0.1–2.7			3–30	233–3,090			18–117	5–27	54–171
Uncontaminated												
Kingsbridge (Devon)												
1974[1]			0.5			6.0	305		3.0		4.3	40
Torridge (Devon)												
1980[1,2]	0.01	12	0.4	2.4	0.5	4.2	431	0.26	19	27	0.4	46
Solent												
1970–72[8,9,11]	0.04	1.5–4.5		7.1		11	485–590	0.16	6.3	7.9	0.8	63–130
Typical	0.1–2.0	1–25	0.1–2.0	1.0–2.0	0.5–2.0	3–50		0.1–0.5		5–50	1–20	35–120
High	5–45	5–45	4–20	5–15	5–20	80–250		1–2		100–200	50–120	150–800

Sources: Bryan and Gibbs (1983);[1] Bryan et al. (1985);[2] Bryan and Hummerstone (1977);[3] Bryan et al. (1980);[4] Langston et al. (2004);[5] Langston (1986);[6] Boyden (1975);[7] Raymont (1972);[8] Romeril (1974);[9] Savari et al. (1991);[10] Segar et al. (1971).[11]

Although the common cockle is widely distributed elsewhere in the Fal Estuary system (Figure 6.6), it is not usually found in Restronguet Creek (Bryan and Gibbs, 1983; Bryan et al., 1987a). No spat nor juveniles have been found in the creek, and the occasional cockle found on the surface of the mud at the bottom of the creek is presumed to have been washed in by the tide in rough weather (Bryan and Gibbs, 1983; Bryan et al., 1987a). The data for Restronguet Creek in Table 6.14 refer to a single individual collected on the sediment surface in April 1981 (Bryan and Gibbs, 1983; Bryan et al., 1987a). Accumulated concentrations of copper and zinc in this cockle were very high (Table 6.14). Cockles collected from Carrick Roads, opposite the mouth of Restronguet Creek, also had high copper and zinc concentrations (Table 6.14), presumably as a result of the export of bioavailable copper and zinc from Restronguet Creek. Furthermore, cockles from Mylor Creek, just downstream from Restronguet Creek (Figure 6.6), also had raised accumulated concentrations of copper and zinc in 1980 (Table 6.14).

Cockles from the Gannel Estuary in the 1970s and 1980s had raised concentrations of lead and zinc, now to be expected of biomonitors here (Table 6.14). Cockles at the bottom of the East Looe Estuary in the 1970s showed some evidence of raised silver concentrations (Table 6.14), particular individuals having accumulated silver concentrations up to 6.5 µg/g (Bryan and Hummerstone, 1977). This is again to be expected, given the results for other biomonitors, including *Hediste diversicolor* (Table 6.6), *Scrobicularia plana* (Table 6.8) and *Macoma balthica* (Table 6.12).

In the Thames Estuary, cockles are to be found downstream of Canvey Island and can be dredged from Chapman Sands in midchannel (Figure 6.12) (Langston et al., 2004). In the 1980s, cockles from Southend showed evidence of raised silver bioavailabilities, and high silver bioavailabilities to cockles in the Outer Thames Estuary were still being registered at the turn of this century (Table 6.14). Again, we have seen this high silver bioavailability in the Thames Estuary, for ragworms (Table 6.6), peppery furrow shells (Table 6.8), Baltic tellins (Table 6.9), mussels (Table 6.11) and oysters (Table 6.12). Silver concentrations in the dredged Chapman Sands cockles in 2002 even reached the remarkable average concentration of 42 µg/g (Table 6.14), presumably as a result of some form of anthropogenic silver contamination (Langston et al., 2004).

In other estuarine systems contaminated with trace metals by industrial activity, the Mersey Estuary in the 1980s showed the now familiar raised bioavailability of mercury, in this case to cockles (Table 6.14) (Langston, 1986). There were also raised bioavailabilities to local cockles of cadmium, nickel and zinc near the entrance to the industrialised Holes Bay in Poole Harbour in the early 1970s (Table 6.14) (Boyden, 1975).

A survey of seven cockle populations in Southampton Water (Figure 6.18) in the 1980s identified local sources of raised bioavailabilities of several trace metals (Table 6.14) (Savari et al., 1991). Much of the western side of Southampton Water is industrialised, especially at Marchwood and Fawley in the 1980s (Savari et al., 1991). Cockles collected offshore of Fawley contained the highest accumulated concentrations of cadmium, copper, iron and lead, and concentrations of nickel and zinc were closely second to cockles off Marchwood (Table 6.14) (Savari et al., 1991). The highest accumulated concentrations of cadmium (2.7 µg/g), lead (27 µg/g), nickel (117 µg/g)

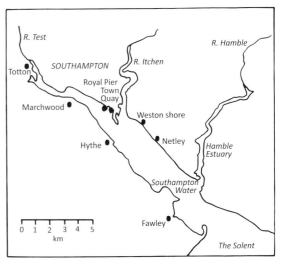

Figure 6.18 Southampton Water.

and zinc (171 µg/g) in either the Fawley or the Marchwood cockles do indicate somewhat raised trace metal bioavailabilities, probably resulting from local industrial effluents (Table 6.14) (Savari et al., 1991).

6.11.4 Clams

Prologue. There is information available on the accumulation of trace metals by the several species of venerid clams found in British estuaries.

The term 'clam' is used to refer to many unrelated bivalves, but particularly in Britain to bivalves of the family *Veneridae*. The genus giving its name to the family is *Venus* herself, typified on British shores by the warty venus *Venus verrucosa*. A few years ago, I would have used the striped venus, *Venus striatula*, as my example, but taxonomists have been at work again, and the striped venus is now *Chamelea gallina*. But why Venus? If you pick up a venus shell, be it a species of *Venus*, *Chamelea* or another venerid genus, look at it side-on. If you are looking at the front of the clam, you will see a usually well-demarcated area shaped as a heart – hence Venus, the goddess of love. If this is not clear, try turning the clam around.

The venerid clam of relevance here is the American hard-shell clam, *Mercenaria mercenaria*, also known as the quahog, imported for mariculture from America in the 1920s and 1930s. Hard-shelled clams were the basis of a fishery in Southampton Water in the 1970s and 1980s (Jensen et al., 2005b). Table 6.15 presents data on the accumulated trace metal concentrations in populations of this clam from Southampton Water and the adjacent Solent at the time. In order to provide a context for low and high accumulated trace metals in venerid bivalves, Table 6.15 also includes trace metal concentrations for two indigenous venerids, the pullet carpet shell *Venerupis corrugata* (once *V. pullastra*) and the cross-cut carpet shell *Ruditapes decussatus*, previously known as *Venerupis decussata*. Incidentally, another species of *Ruditapes* has been imported into a British estuary for mariculture, in this case the Manila clam *Ruditapes philippinarum*. The Manila clam was introduced into Poole Harbour

in 1989, with the expectation that the waters would be too cold to support the production of larvae and recruitment (Jensen et al., 2005a; Humphreys et al., 2007). But such was not the case – the population of Manila clams is flourishing in Poole Harbour, where it now supports a fishery (Jensen et al., 2005a).

In the case of the populations of *Mercenaria mercenaria* in Southampton Water in the early 1970s, there was some evidence for locally raised trace metal bioavailabilities (Table 6.15). Accumulated zinc concentrations above 100 µg/g appear to be atypically high in *M. mercenaria* (Romeril, 1974). Such concentrations were found in clams collected at the top of Southampton Water, near Southampton itself, specifically by Marchwood and the Royal Pier (Figure 6.18) (Romeril, 1974). Accumulated mercury concentrations in this clam ranged from 0.18 to 0.50 µg/g Hg, the higher concentration being found in clams collected by the Town Quay (Raymont, 1972).

The carpet shells *Ruditapes decussatus* from Poole Harbour in 1974 (Table 6.15) were collected by the now disused power station at Lower Hamworthy at the entrance to Holes Bay (Figure 6.17) (Boyden, 1977). As we have come to expect in the case of bivalves in this region of Poole Harbour in the 1970s, accumulated cadmium concentrations in these venerids are very high compared to cadmium concentrations in their close relatives from uncontaminated sites (Table 6.15), indicating raised cadmium bioavailabilities (Boyden, 1977).

6.11.5 Scallops

Prologue. Scallops accumulate trace metals and typically have extremely high concentrations of zinc, silver and cadmium in comparison to other bivalve molluscs.

Members of another lamellibranch bivalve family, the scallops (family *Pectinidae*), are usually to be found offshore living on the bottom. Two commercially important such scallops, the queen scallop *Aequipecten opercularis* (formerly *Chlamys opercularis*) (Plate 7d) and the great scallop *Pecten maximus*, are renowned for their ability to leave the bottom and swim several

Table 6.15 Mean, range of mean or weight-standardised estimate* (1 g soft tissues) concentrations (μg/g dry weight) of trace metals in soft tissues of clams (family *Veneridae*): one introduced species (the American hard-shell clam *Mercenaria mercenaria*), two indigenous species (the pullet carpet shell *Venerupis corrugata* and the cross-cut carpet shell *Ruditapes decussatus*), from British estuaries and coastal waters.

	Ag	As	Cd	Co	Cr	Cu	Fe	Hg	Mn	Ni	Pb	Zn
Mercenaria mercenaria												
Southampton Water												
1970–72[1,2]		3.2–7.1						0.18–0.50				62–138
*Weston 1974[3]			1.4			23	109			9.4	7.8	177
Solent												
1970[4]	1.3		2.1	4.3	0.8	25	5,400		18	11	18	94
Venerupis corrugata												
Carrick Roads (Fal)												
1979[5]			2.5			25	966		8.0		4.9	81
Wembury (Devon)												
1979[5]			0.5			3.0	62		3.0		2.7	25
Ruditapes decussatus												
Poole Harbour												
*1974[3]			8.9			12	368			23	7.8	92

Sources: Raymont, (1972);[1] Romeril (1974);[2] Boyden (1977);[3] Segar et al. (1971);[4] Bryan and Gibbs (1983).[5]

metres through the water like an animated pair of false teeth. Members of the scallop family can, however, be found intertidally at the very bottom of shores, as in the case of the variegated scallop *Chlamys varia* typically to be found under boulders.

Table 6.16 lists accumulated trace metal concentrations in these three scallop species, including a more complete dataset from the French coast at La Rochelle (Bustamente and Miramand, 2005). An immediate feature of Table 6.16 is that scallops accumulate high concentrations of zinc, even in uncontaminated sublittoral habitats. The accumulated background zinc concentrations in scallops exceed those of many of the bivalves considered previously. Nevertheless, the accumulated zinc concentration in *Chlamys varia* from Carrick Roads (2,070 µg/g Zn; Table 6.16) in the Fal Estuary (Figure 6.6) can be considered atypically high, even for a scallop, and indicative of a raised zinc bioavailability (Bryan and Gibbs, 1983). Copper concentrations accumulated by scallops (Table 6.16), on the other hand, are comparable to those of many of the other bivalves. Nevertheless, it is likely that the apparently unremarkable accumulated copper concentration of 54 µg/g in variegated scallops from Carrick Roads (Table 6.16) is also indicative of a high local bioavailability (Bryan and Gibbs, 1983). It is also apparent in Table 6.16 that uncontaminated scallops contain concentrations of both silver and cadmium that would be considered as atypically high in all other bivalves – 10 µg/g for silver, and concentrations of cadmium up to, and exceeding, as much as 100 µg/g (Table 6.16).

High accumulated trace metal concentrations require detoxified storage. A first stage in any consideration of where and how is offered by a consideration of accumulated trace metal concentrations in specified organs. Such data for scallops are offered in Table 6.17.

Very, very high concentrations of zinc are accumulated in the kidney of all three scallop species, even in those scallops not collected from the apparently zinc-contaminated Carrick Roads (Table 6.17). Thus queen scallops from Looe Bay, off South Cornwall, still contained 40,800 µg/g Zn in the kidney, constituting 75% of the total soft tissue load of zinc (Bryan, 1973). A similar situation is seen for manganese in all scallops from all collection sites, with accumulated

kidney concentrations up to 17,300 µg/g, and as much as 93% of soft tissue manganese there accumulated (Table 6.17). Both zinc and manganese are to be found detoxified in many phosphate-based granules in the kidney (George et al., 1980). The kidney of scallops also contains accumulated concentrations of lead (up to 827 µg/g) well above those of other organs or all soft tissues combined, with a majority percentage of total soft tissue lead content (Table 6.17). Although total soft tissue concentrations of copper in scallops are relatively low, greatly raised concentrations are also to be found in the kidney (up to 1,290 µg/g; Table 6.17). Nevertheless, it is the much bigger digestive gland that usually (but not always) contains the majority percentage of soft tissue copper (Table 6.17). The same applies to cadmium in scallops. Highest accumulated concentrations may be found in the kidney, but it is the digestive gland that usually holds most of the body cadmium (Table 6.17), bound to metallothionein (Liu and Wang, 2011). In the case of silver, it is the digestive gland, and not the kidney, that contains both the highest accumulated concentrations and the majority percentage of body silver content (Table 6.17).

6.11.6 Importance of Family

Prologue. Accumulated concentrations of trace metals, whether or not under conditions of high trace metal bioavailabilities, differ greatly between bivalve molluscs of different families, as a result of familial differences in mechanisms of accumulation and detoxification of particular trace metals.

Emerging from much of the preceding discussion on what is a low or a high accumulated trace metal concentration in a bivalve is that the answer does depend greatly on the bivalve itself. For example, the answers for copper or zinc differ greatly between mussels and oysters. What are high bioaccumulated concentrations of either metal in a mussel are low or typical in an oyster.

Table 6.18 explores this concept in more detail, listing typical and high ranges of accumulated concentrations of silver, cadmium, copper, lead and zinc in seven genera of bivalves. These bivalves fall into six

Table 6.16 Mean, range of mean or weight-standardised estimate* (1 g soft tissues) concentrations (μg/g dry weight) of trace metals in soft tissues of three scallops (family Pectinidae), the variegated scallop *Chlamys varia*, the queen scallop *Aequipecten opercularis* and the great scallop *Pecten maximus*, from British and French estuaries and coastal waters.

	Ag	As	Cd	Co	Cr	Cu	Fe	Mn	Ni	Pb	Zn
Chlamys varia											
Carrick Roads (Fal)											
1979[1]			4.0			54	214	51		6.6	2,070
La Rochelle (France)											
1996[2]	1.1–11	14–25	5–9	0.5–1.5	2.6–3.7	9–30		64–339	2.4–3.7	1.3–3.1	313–1,060
Aequipecten opercularis											
Looe Bay											
Pre-1973[1,3]	10		5.5	0.33	2.2	15	113	158	1.6	12	462
*1974[4]			7.3			35			5.5	32	980
Pecten maximus											
Looe Bay											
Pre-1973[3]	2.7		33	0.25		8.9	196	107	0.7	2.0	273
*1974[4]						22		24		13	673
Isle of Man											
1970[5]			13	8.5		3.3	170	140	49	8.3	230
Scottish coastal waters											
†Pre-1973[6]			26–115			2.0–4.5					96–228

Note: *Wet weight concentrations multiplied by 5 to give dry weight concentrations.
Sources: Bryan and Gibbs (1983);[1] Bustamente and Miramand (2005);[2] Bryan (1973);[3] Boyden (1977);[4] Segar et al. (1971);[5] Topping (1973).[6]

Table 6.17 Mean, range of mean or weight-standardised estimate* (1 g soft tissues) concentrations (µg/g dry weight) of trace metals in soft tissues of three scallops (family *Pectinidae*), the variegated scallop *Chlamys varia*, the queen scallop *Aequipecten opercularis* and the great scallop *Pecten maximus*, from British and French estuaries and coastal waters.

	Ag	As	Cd	Co	Cr	Cu	Fe	Mn	Ni	Pb	Zn
Chlamys varia											
Carrick Roads 1979[1]											
Digestive gland			29			164	601	9.2		7.2	251
Kidney			106			704	325	1,360		190	76,300
All soft tissues			4.0			54	214	51		6.6	2,070
La Rochelle 1996[2]											
Digestive gland	3.2–61	11–24	22–40	0.8–2.6	3.1–8.1	22–135		49–174	3.3–6.9	1.1–4.2	110–518
% tissue content	*53–85*	*15–21*	*62–84*	*24–41*	*32–45*	*44–71*		*7–19*	*26–40*	*16–28*	*6–12*
Kidney	0.9–8.3	19–44	40–52	5.8–28	4.3–8.8	33–288		3,590–11,900	11–21	20–59	12,600–38,200
% tissue content	*1–4*	*2–7*	*6–19*	*16–44*	*3–5*	*5–28*		*32–78*	*7–14*	*21–39*	*57–79*
All soft tissues	1.1–11	14–25	4.9–8.5	0.5–1.5	2.6–3.7	9–30		64–339	2.4–3.7	1.3–3.1	313–1,060
Aequipecten opercularis											
Looe Bay pre-1973[3]											
Digestive gland	77		27	1.0	4.7	37	853	30	4.3	10	132
% tissue content	*62*		*42*	*26*	*18*	*17*	*61*	*2*	*25*	*10*	*3*
Kidney	35		41	15	6.6	1,290	330	17,300	78	827	40,800
% tissue content	*3*		*8*	*36*	*3*	*58*	*2*	*89*	*43*	*79*	*75*
All soft tissues	10		5.5	0.3	2.2	15	113	158	1.6	12	462

Pecten maximus

Looe Bay pre-1973[3]

Digestive gland	14	321	1.3	8.1	58	1,300	16	3.6	3.9	407
% tissue content	*63*	*90*	*50*	*47*	*61*	*72*	*1*	*46*	*19*	*16*
Kidney	4.3	79	9.1	3.9	21	149	15,300	23	159	4,800
% tissue content	*2*	*2*	*24*	*2*	*2*	*1*	*93*	*20*	*52*	*52*
All soft tissues	2.7	33	0.3	1.3	8.9	196	107	0.7	2.0	273

Sources: Bryan and Gibbs (1983);[1] Bustamente and Miramand (2005);[2] Bryan (1973).[3]

Table 6.18 Proposed ranges of selected trace metal concentrations (μg/g dry weight) in the total soft tissues of bivalves of seven different families (in six superfamilies in four orders in two subclasses) that are considered typical or high and therefore indicative of raised local trace metal bioavailabilities to the bivalve.

	Ag		Cd		Cu		Pb		Zn	
	Typical	High	Typical	High	Typical	High	Typical	High	Typical	High
Heterodonta										
Veneroida										
Veneroidea										
Veneridae										
Mercenaria	1–2		0.5–3	9+	3–40		3–18		25–100	100–180
Mactroidea										
Cardiidae										
Cerastoderma	0.1–2	5–45	0.1–2	4–20	3–50	80–250	1–20	50–120	35–120	150–800
Tellinoidea										
Tellinidae										
Macoma	0.2–5	50–300	0.1–2	5–10	20–80	100–350	0.5–10	30–150	300–600	1,000–3,000
Semelidae										
Scrobicularia	0.1–5	10–300	0.2–3	5–45	9–60	120–360	4–40	100–1,500	300–1,000	2,000–5,000
Pteriomorpha										
Mytiloida										
Mytiloidea										
Mytilidae										
Mytilus	0.01–1	3–20	0.2–3	5–90	4–20	50–270	0.2–20	30–130	50–200	300–600

Ostreoida										
Ostreoidea										
Ostreidae										
Crassostrea	1–3	10–70	1–5	10–50	30–500	1,000–5,000	0.5–5	20–140	1,500–4,500	6,000–14,000
Pectinoida										
Pectinoidea										
Pectinidae										
Chlamys	1–11		4–115		2–30	54+	1–32		96–1,000	2,000

superfamilies in four orders in two subclasses (Table 6.18). Thus how closely they are related to each other in their evolution varies between pairs of genera being compared. It is clear that there are differences between genera in different families, let alone super-families or orders (Table 6.18). Having stated that, it is also true that the genera in the two most closely related families – the *Tellinidae* and the *Semelidae* in the superfamily *Tellinoidea* – show the closest similarity between the concentration ranges quoted (Table 6.18).

Ultimately, these accumulation differences are down to the different patterns of accumulation for each trace metal in the bivalves, not least the relative proportions of excretion and storage detoxification of the trace metal concerned. Furthermore, the choice of organ as the major site of detoxification is crucial, particularly as regards the presence of a potential excretory route from that organ to the outside of the body. Thus trace metals detoxified and stored in the kidney of a mussel can be excreted in the urine, while the same metals detoxified and stored in the blood cells of an oyster lack a route to the external environment. Even if the same organ is the major site of detoxification in two different bivalves, there may then be differences in the rates of cell turnover in the organ and thus trace metal excretion rates between the two bivalves, allowing different accumulated concentrations to build up or not. Whole soft tissue concentrations may then be affected differentially by growth dilution if the bivalves compared grow at different rates to different sizes.

What is also clear is that differences in trace metal accumulated concentrations between bivalves are not caused by differences in their functional ecology. The bivalves with the highest accumulated trace metal concentrations are semelids and oysters (Table 6.18). The former are deposit feeders and the latter suspension feeders. And oysters are very different from their suspension-feeding relatives in the subclass *Pterio-morpha* (Table 6.18).

6.11.7 Slipper limpets

Prologue. The gastropod *Crepidula fornicata* is a suspension feeder to be found in southern British estuaries.

The slipper limpet *Crepidula fornicata* is that rare animal – a suspension-feeding gastropod mollusc. As in lamellibranch bivalves, the greatly expanded gill of this gastropod creates a powerful incoming current and filters out the phytoplankton borne therein, with the assistance of a great deal of mucus (Yonge and Thompson, 1976). A North American immigrant in the late nineteenth century (Yonge and Thompson, 1976), the slipper limpet is now widespread on estuarine and other shores in in southern Britain. There is, for example, a large population in Poole Harbour, where it is a serious pest of both natural and farmed oyster beds (Underhill-Day and Dyrynda, 2005).

Box 6.5 How the Slipper Limpet Got Its Name

Both the Latin and common names of the slipper limpet deserve comment. *Crepidula* is the Latin word for a little boot or sandal, and hence slipper. And the individual shells of the slipper limpet certainly resemble a diminutive slipper. The term 'limpet' has no precise zoological meaning, but is a name commonly given to many gastropod molluscs that are broadly conical in shape, with no obvious coiling of the shell. The species name *fornicata* refers to the remarkable sex life of the slipper limpet. As in the case of many marine molluscs, the eggs hatch to release planktonic larvae. The planktonic larva of the slipper limpet is short-lived before it settles onto a hard surface on the bottom, particularly onto the shell of another slipper limpet. These first settlers develop into males, but will turn into females over time. Another larva settling on the shell of this female is initially male and able to fertilise the female below. Thereafter, long chains of up to ten or more slipper limpets can be constructed, with females at the bottom, below slipper limpets turning from male to female, and ultimately topped by more recently arrived males (Yonge and Thompson, 1976). These smaller males at the top have a long penis with which they fertilise the females below (Yonge and Thompson, 1976). The species name is indeed appropriate.

Table 6.19 addresses accumulated trace metal concentrations in slipper limpets.

Even the lowest (background) bioaccumulated copper concentrations in slipper limpets are higher than those in most bivalves (Table 6.19). This is to be expected because gastropods typically contain the copper-bearing respiratory protein haemocyanin in the blood. Bivalves on the other hand lack haemocyanin. The copper concentration of slipper limpets in Carrick Roads in 1979 can be considered

Table 6.19 **Mean, range, range of mean or weight-standardised estimate* (0.5 g soft tissues) concentrations (µg/g dry weight) of trace metals in whole soft tissues and digestive glands (DG) of slipper limpets *Crepidula fornicata* from British estuaries and coasts.**

	Ag	Cd	Co	Cr	Cu	Fe	Mn	Ni	Pb	Zn
Carrick Roads (Fal)										
1979[1]		0.5			433	223	22		2.8	84
South Devon										
1979[1]		1.1			132	376	130		2.3	111
Thames										
Whitstable 1984[2]					200	376	116			82
					72–591	107–1,020	54–234			52–143
Whitstable DG 1984[2]					1,990	516	1,080			651
					762–3,840	224–1,100	452–1,820			193–814
Southend 1984[2]	16									
Poole Harbour										
Holes Bay *1974[3]		11			182		26	10	11	85
*1993[4]		1.0–2.7			42–148	214–439	25–37			80–141
Solent										
Lee-on-Solent 1970[5]	1.0	3.9	17	2.0	270	2,000	17	850	3.9	940

Sources: Bryan and Gibbs (1983);[1] Rainbow and White (unpublished);[2] Boyden (1977);[3] Rainbow, Smith and Harris (unpublished);[4] Segar et al. (1971).[5]

atypically high, indicating a raised bioavailability (Table 6.19). This is not surprising, given the position of Carrick Roads at the bottom of Restronguet Creek (Figure 6.6). Copper concentrations also appear high in slipper limpets at Whitstable in the Thames Estuary in 1984 (Table 6.19), indicating raised copper bioavailability. Silver concentrations were also high in Thames slipper limpets in 1984, even as far out as Southend (Table 6.19), a feature that we have seen for other biomonitors.

Mean zinc concentrations in slipper limpets typically fall in a relatively narrow range between 82 and 111 µg/g, individual concentrations varying a little more widely (52 to 143 µg/g) (Table 6.19). This accumulation pattern approaches that of zinc regulation (see Chapter 3). The very high zinc concentration (and indeed nickel and copper concentrations) in slipper limpets collected off Lee-on-Solent in 1970 (Table 6.19) appear anomalously elevated. While there may have been some raising of local trace metal bioavailabilities there at the time, much of these raised accumulated concentrations may be attributable to the presence of somewhat contaminated sediment in the gut and/or gills.

The slipper limpet data for Poole Harbour confirm the high bioavailabilities of cadmium resulting from the industrialised state of Holes Bay

in the 1970s (Table 6.19) (Boyden, 1975, 1977). The higher values in the cadmium concentration range in slipper limpets collected from Poole Harbour in 1993 represent sites in Holes Bay (Table 6.19). Although still a little higher than the cadmium concentrations in slipper limpets from elsewhere in Poole Harbour at the time, it is evident that cadmium bioavailabilities had dropped considerably in Holes Bay since 1974 (Table 6.19).

The Whitstable data show that the digestive glands of slipper limpets contain much higher bioaccumulated concentrations of copper, manganese and zinc than the total soft tissues (Table 6.19). It would appear, then, that the digestive gland is the site of detoxified storage of these trace metals. Any detoxified storage of zinc would presumably be relatively temporary before excretion, given the relative lack of increase of total soft tissue zinc concentrations under different conditions (Table 6.19).

6.11.8 Barnacles

Prologue. Molluscs are not the only ecologically important suspension feeders in British estuaries. The crustaceans are represented by barnacles.

Box 6.6 What Is a Barnacle?

How can a barnacle be described? The most succinct description is attributed to Louis Agassiz, the Swiss biologist who emigrated to the United States in 1847 and was to found the Museum of Comparative Zoology at Harvard University. A barnacle is 'nothing more than a little shrimp-like animal standing on its head in a limestone house and kicking food into its mouth' (Barnes, 1980: 695). And so it is. But how can a crustacean produce a calcium carbonate shell, giving it the superficial appearance of a mollusc? Think of the carapace of a crab and how hard it can be. The crab, another crustacean, lays down calcium carbonate in its exoskeleton for protection. Imagine this calcification process going into overdrive, so that the chitinous cuticle basis of the exoskeleton is swamped by the secreted calcium carbonate. The barnacle now has a shell. But it is only the first two segments of the head of the barnacle that produce this overcalcified exoskeleton, this protective shell. As we have seen to be the case for most marine invertebrates, barnacles produce a planktonic larva from each fertilised egg. This larva develops through six swimming and feeding (nauplius) stages, before it metamorphoses into the cypris larva, the specialist stage that settles onto an appropriate hard surface on the bottom. The first pair of head appendages (the antennules) is crucial in the adhesion of the barnacle onto the substratum, and a dramatic metamorphosis now occurs. The front of the head expands greatly to surround the rest of the body and secretes shell plates. There are usually four further small shell parts (the opercular plates) forming a trapdoor system to serve the rest of the body sitting in its protective box (Figure 6.19). The elongated thoracic legs are furnished with many filter feeding setae and are able to extend out into the water column to filter the suspended plankton and detritus passing by.

A barnacle occupying the intertidal region of British estuaries is *Amphibalanus improvisus*, a barnacle named by Charles Darwin in 1854. Darwin was in fact the world's expert on barnacles, having decided that he would only have a right to expound on the question of the origin of species if he had established his credentials by describing species (Rainbow, 2011). And Darwin chose barnacles. In fact, Darwin named this species as *Balanus improvisus*. The genus *Balanus* over time incorporated a multitude of species, many divisible into several subgenera. As in the case of the polychaete genus *Nereis*, the old genus *Balanus* has now undergone extensive systematic reorganisation, and *Amphibalanus improvisus* has emerged. There is a suggestion that *A. improvisus* is itself an immigrant to Europe, introduced by shipping from the United States early in the nineteenth century (Southward, 2008).

The more common British intertidal barnacle *Semibalanus balanoides*, similarly previously known as *Balanus balanoides*, is present at the mouths of estuaries, but is not typically estuarine. Another barnacle that does occur widely in British estuaries is a further immigrant species, *Austrominius modestus* (Plate 7c). Previously known as *Elminius modestus*, this Australasian barnacle was first noted in Chichester Harbour in 1945, having probably arrived on ships during the World War II. *A. modestus* has now spread right around the British Isles and into continental Europe, common on sheltered shores and in estuaries.

The most easily collected and reproducible part of a barnacle is the discrete body sitting in the cavity beneath the opercular plates (Figure 6.19). This is not strictly the whole body, for it does not include the front part of the head forming the shell plates and the tissue underlying these shell plates. Although really a misnomer, the term 'body' has stuck. The body then is usually that part of the barnacle that is analysed for accumulated trace metal content.

6.11.8.1 Accumulation and Detoxification

Prologue. Barnacles are strong net accumulators of trace metals from solution and suspended particles in the water column. Barnacles are particularly impressive accumulators of zinc, which is detoxified in zinc pyrophosphate granules stored in the body without excretion. The strong accumulating properties of barnacles make them excellent trace metal biomonitors.

Table 6.20 summarises relevant trace metal accumulation data for the bodies of *Amphibalanus improvisus* and *Semibalanus balanoides* in British estuaries and on adjacent coasts.

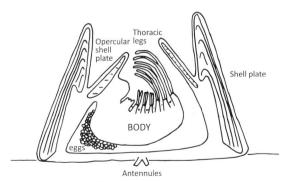

Figure 6.19 Barnacle structure.

Table 6.20 Mean, range of mean or weight-standardised estimate* (0.002 g body) concentrations (µg/g dry weight) of trace metals in bodies of the barnacles *Amphibalanus improvisus* and *Semibalanus balanoides* from British estuaries and other coasts.

	Ag	Cd	Co	Cr	Cu	Fe	Mn	Ni	Pb	Zn
Amphibalanus improvisus										
Thames Estuary										
1984–85[1]	8.0–57	13–28			183–913	700–5,900	38–300		9.8	33,000–153,000
*2001[2]	7.9–11	7.4–9.1			143–239	4,870–6,650	81–96		27–58	19,000–27,800
*2014[3]	8.6–19	3.6–5.2	7.1–15	11–21	139–222	3,790–6,370	79–107	22–33	7.7–8.9	16,800–25,600
Typical[3]	0.1–10	3–10	3–15	3–25	15–150	200–2,000	10–250	4–40	2–30	1,500–10,000
High[3]	10–20	10–30			200–920	2,000–11,000	300–2,500	40–70	30–80	20,000–153,000
Semibalanus balanoides										
Dulas Bay (Anglesey)										
1975–76[4,5]					2,470–3,750					50,300–82,000
1978[6]		60			3,232					113,000
Aberystwyth (Ceredigion)										
Pre-1973[7,8]					100–700		100–300		25–125	3,000–30,000
*1992[9]					148–378		35–94		47–185	1,310–3,530

Menai Strait (N Wales)			
1975–76[4,5]	170–290		1,220–19,200
Thames (Southend)			
1984–85[1]	25	13	
1978[6]	10–28	232	27,800

Sources: Rainbow and White (unpublished);[1] Rainbow et al. (2002);[2] Johnstone et al. (2016);[3] Walker (1977a);[4] Walker (1977b));[5] Rainbow et al. (1980);[6] Ireland (1973);[7] Ireland (1974);[8] Rainbow and Ward (unpublished).[9]

Leaping out from this table are the fantastically high accumulated zinc concentrations in the bodies of barnacles (Table 6.20). Body zinc concentrations remarkably exceeded 100,000 μg/g in *A. improvisus* from Woolwich in the Thames Estuary in the early 1980s (Rainbow, 1987). These accumulated zinc concentrations are greater even than soft tissue concentrations of zinc in oysters (Table 6.12) and body concentrations of zinc in baetid mayfly larvae in zinc-contaminated streams (Table 5.10). Furthermore, they also exceed the zinc concentrations in specialist zinc detoxification organs of invertebrates, such as the ventral caeca of woodlice (Table 4.27) and the kidney of scallops (Table 6.17). Inevitably, this accumulated zinc must be detoxified, and barnacles carry the use of phosphate detoxification granules to an extreme. The incoming zinc is bound in phosphate granules, specifically to pyrophosphate with its catholic taste and very high affinity for many trace metals (see Chapter 3) (Walker et al., 1975; Pullen and Rainbow, 1991). In the case of the Woolwich barnacles, a zinc concentration of 153,000 μg/g would be bound to 204,000 μg/g of pyrophosphate, meaning that more than 35% of the dry body weight of the bodies of these barnacles consists of zinc pyrophosphate granules, even before counting in other metals such as calcium and iron present in the granules. The granules are held in the body tissues below the midgut and cannot be excreted in granular form to the outside (Figure 6.20). They, therefore, accumulate to bring about the very high body zinc concentrations achieved in barnacles in the presence of high zinc bioavailabilities.

Barnacles have very high dissolved uptake rates of trace metals (Luoma and Rainbow, 2008), probably inevitably given the large volumes of water passing over their permeable filtering appendages. Nevertheless, barnacles still obtain most of their accumulated trace metals from their diet of filtered food (Wang et al., 1999; Rainbow and Luoma, 2011c). They ingest large amounts of small suspended particles with high contents of absorbed and adsorbed trace metals. In the case of balanid barnacles, including species of *Amphibalanus* and *Semibalanus*, this food source includes fine suspended detritus and resuspended sediment particles, and a high dietary input of trace metals follows. Their oceanic relatives, the goose barnacles, usually to be found hanging down from floating logs or the hulls of ships, have not evolved the fine filtering apparatus to collect phytoplankton and detritus. They rely on larger suspended particles such as zooplankton as a food source. Goose barnacles, for example species of *Lepas*, do not have the high trace metal uptake rates of their balanid counterparts in estuaries, but, nevertheless, still accumulate zinc in phosphate granules (Figure 3.7) and reach body zinc concentrations above 500 μg/g, even in the open Atlantic (Walker et al., 1975).

The fantastic body zinc concentrations in the bodies of *A. improvisus* in the Thames Estuary in the early 1980s had dropped by 2001 (Table 6.20), with increasing environmental control of industrial effluents. No further drop was recorded in the Thames Estuary between 2001 and 2014 (Table 6.20). *Semibalanus balanoides* collected from the mouth of Dulas Bay in the 1970s also had remarkably high body concentrations of accumulated zinc (Table 6.20). The zinc was derived from the input into Dulas Bay of the Afon Goch, which drains Parys Mountain. The same species at Aberystwyth, by the Ystwyth and Rheidol Estuaries, also showed raised zinc body concentrations in the 1970s and 1980s (Table 6.20), in response to the zinc input from these two rivers (Ireland, 1973, 1974).

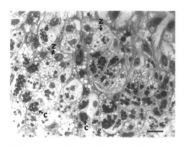

Figure 6.20 Barnacle detoxification granules in the body of *Semibalanus balanoides* from the mouth of Dulas Bay, Anglesey, Wales. Non-staining granules (z) are zinc pyrophosphate granules; dark-staining granules (c) are copper-rich. Scale bar is 10 μm. (After Rainbow, 1987, with permission.)

It is the specimens of *Semibalanus balanoides* collected at the mouth of Dulas Bay in the 1970s that show phenomenally high accumulated concentrations of copper, up to 3,750 µg/g (Table 6.20) (Walker, 1977a). This copper has also been input into Dulas Bay by the Afon Goch, after originating in the copper-rich deposits of Parys Mountain. Typically, copper is detoxified in invertebrates by binding to metallothionein, in turn then broken down in lysosomes. In the Dulas Bay barnacles, so much copper has been accumulated and detoxified that the body is packed with tertiary lysosomes containing copper- and sulphur-rich deposits (Figures 3.9, 6.20) (Walker, 1977a), left as residual bodies after metallothionein degradation. Such copper-rich granules are rarely seen in barnacles, but accumulated copper concentrations above 3,000 µg/g are simply exceptional (Table 6.20).

Table 6.20 also shows that there was a high copper bioavailability to *Amphibalanus improvisus* in the Thames Estuary in the 1980s, and that this had dropped by 2001, to level out to 2014 Johnstone et al., 2016). *S. balanoides* showed the presence of raised copper bioavailability in the vicinity of Aberystwyth in the 1970s and 1990s (Table 6.20).

Lead bioavailabilities to *A. improvisus* in the Thames Estuary were still high in 2001 (Table 6.20), awaiting the effects of the ban on lead in petrol at the end of the 1990s. These lead bioavailabilities had indeed subsequently fallen by 2014 (Table 6.20). As to be expected given the proximity of the estuaries of the lead-contaminated Ystwyth and Rheidol, bioaccumulated lead concentrations were very high in the bodies of *S. balanoides* at Aberystwyth in the 1970s and 1990s (Table 6.20).

Accumulated concentrations of silver and cadmium were high in both *A. improvisus* and *S. balanoides* in the Thames Estuary in the 1980s (Table 6.20). Silver concentrations in *A. improvisus* had dropped by the twenty-first century, but were still high enough in 2014 to be indicative of atypically raised silver bioavailabilities in the Thames Estuary (Table 6.20). Cadmium bioavailabilities to *A. improvisus* in the Thames Estuary had dropped to what are probably typical levels by 2001 and thereafter (Table 6.20).

6.12 Grazing Gastropods

Prologue. The typical way of life of marine gastropods is to graze on the seaweeds of the littoral or immediately sublittoral zones or to be a predator of other invertebrates. In this section, we consider the first of these professions. Grazing gastropods are well represented on the shores of British estuaries and adjacent coasts. Particularly obvious are the periwinkles and limpets.

6.12.1 Periwinkles

Prologue. Periwinkles take up trace metals both from the surrounding water and from the seaweeds on which they graze, the trace metal concentrations in the seaweeds themselves reflecting dissolved trace metal bioavailabilities. Thus, the accumulated trace metal concentrations of periwinkles ultimately depend on dissolved trace metal bioavailabilities, for which they can act as biomonitors.

The common name periwinkle (or winkle) usually refers to species of the genus *Littorina* and its relatives in the family *Littorinidae*. The separation of British periwinkles into different species has been the subject of much soul searching over the last few decades. The edible periwinkle *Littorina littorea*, however, has survived unscathed through this period of systematic confusion and will be the main example here. *L. littorea* grazes preferentially on ephemeral green algae such as *Ulva lactuca* and *Ulva* (formerly *Enteromorpha*) *intestinalis*, softer than the brown fucoid seaweeds which may dominate rocky shores (Hayward, 2004). The two closely related flat periwinkles, *Littorina obtusata* and *Littorina fabalis* (previously *L. mariae*), historically and erroneously lumped as *L. littoralis*, usually live on knotted wrack *Ascophyllum nodosum* and serrated wrack *Fucus serratus* respectively (Hayward, 2004). *L. obtusata* grazes on *A. nodosum* itself, while *L. fabalis* grazes the epiflora growing on *F. serratus* (Hayward, 2004). The common name rough periwinkle is now also known to have been applied to several different presumed species of periwinkle. Remaining today are firstly *L. saxatilis*; secondly *L. compressa*, the black-lined periwinkle once (understandably) called *L. nigrolineata*; and thirdly *L. arcana*, which prefers wave-exposed shores and therefore is not present in sheltered estuaries. A final periwinkle, formerly placed in the genus *Littorina*, is the tiny, black periwinkle *Melaraphe neritoides*. This species is again found on wave-exposed shores and can also be ignored in this chapter on estuaries.

Table 6.21 presents trace metal concentrations accumulated by the body (excluding the shell) of the edible periwinkle *Littorina littorea* on British shores within and without selected estuaries.

Periwinkles only occur towards the bottom of Restronguet Creek, for example at Restronguet Point and Weir Point (Figures 6.5, 6.6) (Bryan and Gibbs, 1983). Edible periwinkles collected there in the 1970s had accumulated concentrations of copper, zinc and arsenic higher than those in this species from any other British estuary (Table 6.21) (Bryan and Gibbs, 1983; Bryan et al., 1983). Periwinkles take up trace metals both from the surrounding water and from the seaweeds on which they graze. The trace metal concentrations in seaweeds reflect dissolved trace metal bioavailabilities. So, the accumulated concentrations of copper, zinc and arsenic in the Restronguet Creek edible periwinkles indirectly reflect the high dissolved bioavailabilities of these three trace metals in the creek and are to be expected (Bryan et al., 1983). Accumulated cadmium and cobalt concentrations in *L. littorea* from Restronguet Creek were also high (Table 6.21), suggesting high local bioavailabilities of these two trace metals too. Accumulated trace metal concentrations from edible periwinkles from other parts of the Fal Estuary system (Figure 6.6) indicate some export of high copper and zinc bioavailabilities beyond the confines of Restronguet Creek (Table 6.21).

Elsewhere in Cornwall, the East Looe Estuary again showed a high bioavailability of silver in the 1970s, in this case to *L. littorea*, and also to some extent raised bioavailabilities of copper, lead and zinc (Table 6.21). The West Looe Estuary in 1976 showed the expected very high bioavailability of lead derived from lead

mining in the catchment, and raised copper bioavailability, to *L. littorea* (Table 6.21). It is possible that the raised silver bioavailability to the periwinkles in the West Looe Estuary, and the raised lead bioavailability in the East Looe Estuary (Table 6.21), might each be a result of dissolved metals originating in the other branch being transported up each branch on a rising tide, after mixing in the combined region of the lower Looe Estuary (Figure 6.9).

The Tamar Estuary in 1978 showed high bioavailabilities of both copper and lead to *L. littorea*. In the same year, the Plym Estuary had high bioavailabilities of cadmium (seen previously for other biomonitors) and, in this case, also lead (Table 6.21).

In contrast to the situation for the barnacle *Semibalanus balanoides* (Table 6.20), edible periwinkles collected in the region of Aberystwyth did not have raised accumulated concentrations of copper or (more expectedly) lead (Table 6.21) (Ireland and Wootton, 1977). It might be the case, then, that the barnacles were taking up high quantities of copper and lead trophically from resuspended fine sediments originating from the Rivers Ystwyth and Rheidol, an uptake route not accessible to the periwinkles.

In the Thames Estuary, edible periwinkles can be found downstream of midestuary at Grays (Figure 6.12) (McEvoy et al., 2000; Langston et al., 2004). In this estuary, there was a very high bioavailability of silver to periwinkles from the 1980s to the turn of the century, decreasing with passage downstream (Table 6.21). Similar results for silver have been seen earlier for other biomonitors. High bioavailabilities of copper, zinc and perhaps mercury in the Thames Estuary in the 1980s had fallen by 2001 (Table 6.21) as local industrial effluents were cleaned up.

The exorbitantly high bioavailabilities of cadmium to many estuarine biomonitors in the Severn Estuary in the 1970s is confirmed for *L. littorea*, in conjunction with raised bioavailabilities of zinc and, to some extent, lead and arguably silver (Table 6.21). These high bioavailabilities are attributable to outputs from the Avonmouth smelter operating at the time. Raised bioavailabilities, particularly of cadmium, to the

Table 6.21 Mean, range of mean or weight-standardised estimate* (1 g body dry weight) concentrations (µg/g dry weight) of trace metals in bodies of the edible periwinkle *Littorina littorea* from British estuaries and coastlines affected by (a) mining contamination, (b) industrial contamination or (c) considered uncontaminated.

	Ag	As	Cd	Co	Cr	Cu	Fe	Hg	Mn	Ni	Pb	Sn	Zn
Mining													
Restronguet Creek													
*1974[1]			2.6			642	364		34		6.5		185
1972–78[2,3]	0.9	71	1.3–4.5	19		261–1,380	275–1,290	0.22	54–158	7.0	0.3–27		197–956
Pre-1983[4]			3.2	14		445	597		76	8.2	6.5		323
Fal Estuary													
1979[2,3]			1.0–1.3			93–464	243–377		10–79	1.3–2.1	0.3–3.3		91–248
East Looe													
1976[3,5]	30–64	29–37	1.6–2.0	1.1–1.2	0.5	154–194	361–400	0.14–0.21	25–53	2.8–3.3	16–17	0.1	114–232
West Looe													
1976[3]	17	36	2.6	3.0		161	458	0.33	133	3.1	70		83
Tamar													
1978[3]	2.4	32	3.2	2.9		207	490	0.66	125		23		123
Plym													
1978[3]	2.6		6.5	1.7		151	658		111		40		106
Aberystwyth													
1974[6]						55			20		6.9		123
Industrial													
Thames													
Grays 1980[3]	101	11	13	1.9	0.7	417	322	0.88	55	5.5	4.0		141
Southend 1984–85[7]	15		3.0										
1997[8]	14–102	8–18	0.5–3.4	0.6–6.5	0.3–1.2	111–238	208–411	0.19–0.60	27–96	2.9–14	1.8–6.0	0.2–1.5	82–140
2001[9]	10–67	6–17	0.7–2.3	0.8–6.7	0.4–1.5	96–208	243–576	0.09–0.33	30–126	2.1–8.1	2.2–6.1	0.3–0.6	52–97

Table 6.21 (cont.)

	Ag	As	Cd	Co	Cr	Cu	Fe	Hg	Mn	Ni	Pb	Sn	Zn
Severn													
1970[10]			25–210								0.1–3.0		140–520
Penarth 1974[6]						249			60		15		186
Shepperdine 1978[3]	12		146	1.5		143	278	0.29	55	5.8	0.9		703
Bristol Channel													
1970[10]			15–40								0.1–0.2		100–150
1971[11]			17–75								3–19		75–210
1974[6]						107			33		9.6		120
1978[3]	9.3–14	16	11–19	1.1–1.4	0.3	140–163	408–430	0.49–0.50	34–81	3.9	3.4–4.0		114–135
Swansea Bay													
1976–79[3]	0.4–2.9	14	5.5–6.3	3.4–15	0.6	133–193	145–282	0.57	85–147	3.3–3.6	4.9–29		101–153
Loughor													
1979[3]	2.4	26	6.8	1.6	1.4	147	394	0.34	84	7.7	4.1		117
Mersey													
1981[3]	2.0	22	3.1	2.5	1.6	113	802	1.48	98	7.8	12		189
Humber													
1978[3]	9.5	29	15	2.8		149	926	0.37	309		4.5		155
Tyne													
1980–81[12]						147							108
Tees													
1978[3]	4.7	24	1.0	2.2		292	343	0.93	53		4.1		95
Poole Harbour													
Holes Bay 1982[3]	26	10	13	5.2		141	574	1.90	60		4.7		120

Uncontaminated

Menai Strait (N Wales)													
Pre 1983[4]			1.6	1.1		77	1,100	98		6.4	6.5		114
Criccieth (W Wales)													
1974[6]						51		28			5.4		92
Torridge (Devon)													
1980[3]	4.2	19	2.0	1.1	0.4	129	399	42	0.42	3.4	2.2		75
Hamble (Hants)													
1979[3]	0.7	21	2.1	1.8	0.1	90	422	42	1.13	2.4	4.1		95
Typical[3]	0.4–5	6–40	0.5–3	0.6–3	0.3–1	50–150			0.10–0.60	1–10	0.1–7	0.1–1.5	50–120
High[3]	5–10	50–75	5–210	5–20	1–2	150–1,400			0.80–2.0	10+	10–70		140–1,000

Sources: Boyden (1977);[1] Bryan and Gibbs (1983);[2] Bryan et al. (1983);[3] Mason and Simkiss (1983);[4] Bryan et al. (1980);[5] Ireland and Wootton (1977);[6] Rainbow and White (unpublished);[7] McEvoy et al. (2000);[8] Langston et al. (2004);[9] Butterworth et al. (1972);[10] Nickless et al. (1972);[11] Lobel et al. (1982).[12]

periwinkles could also be seen downstream in the Bristol Channel, though diminishing with distance (Table 6.21). Cadmium concentrations in both Severn Estuary periwinkles (up to 210 µg/g) and Bristol Channel periwinkles (up to 75 µg/g) in the 1970s are simply astounding in comparison with typical accumulated concentrations (Table 6.21). Table 6.5 confirms that concentrations of cadmium and zinc in 1970 in Severn Estuary seaweeds (in this case, *Fucus vesiculosus*) were similarly very high, particularly for cadmium. It would appear then that the edible periwinkles were responding to raised dissolved bioavailabilities of the trace metals in the Severn Estuary, indirectly via seaweed consumption.

In Swansea Bay on the south Wales coast in the 1970s, bioavailabilities of cadmium, cobalt, copper, lead and zinc were all high (Table 6.21), probably as a result of raised concentrations of dissolved metals in industrial effluents discharging into the Rivers Tawe and Neath. Bioavailabilities of cobalt and (expectedly) chromium were high to periwinkles in the nearby Loughor Estuary receiving effluent from local industrial tinplate works. The Mersey Estuary in 1981 also showed a high bioavailability of chromium to *L. littorea*, together with raised bioavailabilities of lead and zinc, and of mercury from chlor-alkali works (Table 6.21). The Tees Estuary in 1978 also offered a high bioavailability of mercury to edible periwinkles, together with a raised bioavailability of copper (Table 6.21). As again we have come to expect, Holes Bay in Poole Harbour in 1982 had raised bioavailabilities to periwinkles of trace metals of industrial origin, in this case silver, cadmium and cobalt (Table 6.21). Holes Bay also showed a high bioavailability of mercury to *L. littorea*, as did the Hamble Estuary on Southampton Water (Table 6.21). While there might have been some emission of industrial mercury into Holes Bay, another potential mercury source, particularly likely in the Hamble Estuary, is antifouling paint containing mercury, popular at the time to protect the bottoms of sailing boats from fouling organisms. We saw this potential source of environmental mercury earlier in the case of mussels growing at Whiterock in Strangford Lough in Northern Ireland in the early 1980s (Gault et al., 1983).

6.12.1.1 Detoxification of Accumulated Trace Metals

Prologue. Detoxification of accumulated copper is by binding to metallothionein, and is incorporated into a physiological cycle of the synthesis and breakdown of haemocyanin. Calcium phosphate granules detoxify zinc and manganese.

The detoxification of raised accumulated contents of trace metals in *Littorina littorea* takes place against a background physiology handling the essential trace metal copper and the major metal ion calcium.

Like other gastropods, *L. littorea* uses the copper-bearing respiratory protein haemocyanin in the blood to bind and transport oxygen and carbon dioxide. Correspondingly, typical or background accumulated copper concentrations in the periwinkle are above 50 µg/g (Table 6.21), adequate to meet the essential demands of both haemocyanin and copper-bearing enzymes (see Chapter 3). Pore cells in the connective tissue in the body of the periwinkle are involved in the synthesis and metabolic turnover of haemocyanin and contain intracellular deposits storing copper (Mason et al., 1984; Simkiss and Mason, 1984). The copper is associated with sulphur (Mason et al., 1984; Simkiss and Mason, 1984), and the deposits can be interpreted as part of a cycle turning over metallothionein binding the copper needed for haemocyanin synthesis (Martoja et al., 1980). Other accumulated trace metals such as cadmium or zinc that might bind with this metallothionein are conspicuous by their absence from these deposits, suggesting the presence in the pore cells of a protected physiological pathway dedicated to the synthesis of haemocyanin (Simkiss and Mason, 1984).

Also in the body connective tissue in close association with the blood system are so-called calcium cells, containing spherical cytoplasmic granules rich in calcium and magnesium carbonates, but no additional trace metals (Mason and Nott, 1981; Mason et al., 1984). These granules are easily soluble. They can act as temporary stores of calcium needed for physiological purposes, not least shell formation. Furthermore, the carbonate anions released when these granules are solubilised can function to buffer the acid-base balance in the blood of the periwinkles and play a role in ion regulation (Mason and Nott, 1981; Mason et al., 1984). Such calcium-based granules would be inappropriate as detoxification stores of accumulated trace metals, for which release back into circulation would likely have toxic consequences.

Where and how, therefore, is the detoxification of accumulated trace metals achieved in periwinkles? The digestive gland, as to be expected, contains many digestive cells involved with the intracellular digestion of food (Mason et al., 1984). Interspersed between these digestive cells are basophil cells that contain another type of calcium-based granule (Mason et al., 1984). These granules contain pyrophosphate and, in addition to calcium, magnesium and potassium, they bind a wide range of trace metals, but particularly zinc and manganese, as a detoxification mechanism (Mason et al., 1984; Langston and Zhou, 1987). The numbers of these granules increase in the basophil cells of the digestive glands of *L. littorea* from trace metal-contaminated habitats, as do their contents of contaminant trace metals (Mason et al., 1984). Thus these granules in periwinkles from Restronguet Creek contained zinc and manganese, while these two trace metals were absent from the fewer same granules present in *L. littorea* from the Menai Strait (Table 6.21) (Mason and Simkiss, 1983; Mason et al., 1984).

Trace metals are also detoxified in the kidney of edible periwinkles (Mason et al., 1984; Simkiss and Mason, 1984). Kidney cells of *L. littorea* from Restronguet Creek contained granules rich in iron, zinc and manganese (Simkiss and Mason, 1984). A further effect of the exposure of the periwinkles to the trace metal-contaminated conditions of Restronguet Creek was the increased secretion of these granules from the kidney cells into the urine (Mason et al., 1984).

Thus both zinc and manganese can be detoxified in both basophil cells of the digestive gland and in cells in the kidney of *L. littorea* (Mason et al., 1984; Simkiss and Mason, 1984). Newly accumulated zinc accumulates in the kidney, reaching high concentrations in this organ which then stabilise after about a week, as the new zinc accumulation is matched by excretion (Bebianno and Langston, 1995). This new equilibrium concentration in the kidney represents an increased standing stock of zinc-rich granules, now being excreted at an increased rate. Under conditions of high zinc bioavailability, zinc is also being accumulated in the digestive gland (Bebianno and Langston, 1995). The zinc-rich phosphate granules in the basophil cells here have a slower excretion rate than their counterparts in the kidney, and the digestive gland still represents a significant site of zinc detoxification and accumulation (Bebianno and Langston, 1995). In *L. littorea* from Restronguet Creek with a mean total body zinc concentration of 323 µg/g (Table 6.21), the mean zinc concentration in the kidney was 2,880 µg/g and that in the visceral complex containing the digestive gland was 482 µg/g (Mason and Simkiss, 1983). Even under uncontaminated conditions, the kidney of *L. littorea* from the Menai Strait contained 363 µg/g Zn, in a body containing 114 µg/g Zn (Table 6.21) (Mason and Simkiss, 1983).

Extra copper is also accumulated both in the kidney and digestive glands of edible periwinkles from Restronguet Creek, reaching mean concentrations of 647 µg/g Cu in the kidney and 794 µg/g Cu in the visceral complex of animals averaging 445 µg/g Cu in the whole body (Table 6.21) (Mason and Simkiss, 1983). No discrete copper-rich granules have been identified in these organs (Mason et al., 1984; Simkiss and Mason, 1984), in contrast to the pore cells acting independently in the background physiology of haemocyanin. Copper-rich granules have, however, been described, in association with sulphur, in the gill cells of *L. littorea* from Restronguet Creek (Mason et al., 1984). These granules are apparently accumulating as products of the breakdown of metallothionein binding the extra copper taken up under conditions of very high copper bioavailability. And the gills of these same periwinkles contained 1,210 µg/g Cu, higher than the accumulated copper concentrations in associated tissues previously stated (Mason and Simkiss, 1983). Copper bound to soluble metallothionein in cells will still be detoxified but will not appear as insoluble metal-rich granules under electron microscopy.

Cadmium is accumulated to the highest concentrations in the kidney of *Littorina littorea* and to lesser concentrations in the digestive glands and gills. Thus in periwinkles from Restronguet Creek with total body concentrations of 3.2 µg/g Cd (Table 6.21), the kidney contained 20 µg/g Cd, the visceral complex 4.8 µg/g Cd and the gills 3.2 µg/g Cd (Mason and Simkiss, 1983). In uncontaminated *L. littorea* from Penzance in Cornwall, the kidney had the highest accumulated cadmium concentration (13 µg/g) and contained 13% of the total body content of cadmium; the digestive gland had 2.3 µg/g Cd but 54% of body cadmium content; and the gills contained 2.9 µg/g Cd (4%) (Mason and Nott, 1981). Dissolved cadmium taken up in the gills is transported in the blood (bound reversibly to haemocyanin) to the kidney, which reaches a saturated cadmium concentration (Mason and Nott, 1981). Further accumulated cadmium is redistributed to the digestive gland, which stores about half of the total body content of cadmium (Mason and Nott, 1981). The accumulated cadmium is detoxified by binding to newly synthesised metallothionein, in the gills, kidney and digestive gland (Mason and Nott, 1981; Bebianno and Langston, 1995).

6.12.1.2 Tolerance and Ecotoxicology

Prologue. A population of the rough periwinkle *Littorina saxatilis* on the Isle of Man showed tolerance to lead and zinc exposure.

The tolerance of particular populations of a species to a high bioavailability of a trace metal is evidence that the local bioavailability of that trace metal is of ecotoxicological significance (Luoma, 1977). No such tolerant populations have been identified in the case of *Littorina littorea*, perhaps not surprisingly given the presence in the lifecycle of a long-lived planktonic phase effecting the wide distribution of any offspring of a particular population. The rough periwinkle, *Littorina saxatilis*, on the other hand, shows direct development without a planktonic larval phase. Offspring are therefore released into the vicinity of their parents. A population of *L. saxatilis* from the Laxey Estuary on the Isle of Man (Figure 5.13), draining a catchment with a history of predominantly lead and zinc mining, has been shown to be tolerant to exposure to high bioavailabilities of both lead and zinc in comparison to populations of the same species from local control estuaries (Daka and Hawkins,

2004). These high bioavailabilities are illustrated by the higher accumulated body concentrations of lead and zinc in Laxey *L. saxatilis* (471 µg/g Zn, 17 µg/g Pb), in comparison with equivalent accumulated concentrations in this species from Castletown (11 µg/g Zn, 12 µg/g Pb) (Daka and Hawkins, 2004). Accumulated copper concentrations were not raised in Laxey rough periwinkles (80 µg/g Cu), in comparison with Castletown specimens (92 µg/g Cu). The physiological basis of the tolerance to zinc appeared to be a reduced zinc accumulation rate under exposure to raised zinc bioavailabilities (Daka and Hawkins, 2004). Tolerance to lead, on the other hand, was not associated with reduced accumulation, suggesting an increased rate of detoxification of the lead accumulated (Daka and Hawkins, 2004).

Even higher accumulated concentrations of zinc (552 µg/g) and lead (32 µg/g), as well as copper (695 µg/g), were measured in *L. saxatilis* collected in 1978 to 1979 from the mouth of the Avoca River on the east coast of Ireland, indicating very high local bioavailabilities of at least copper and zinc at the time (Wilson, 1982). It is likely that these bioavailabilities would have been of ecotoxicological significance.

6.12.2 Limpets

Prologue. The other littoral grazing gastropods to be considered here are limpets, obvious at low tide on rocky shores around the British Isles.

What is not so obvious is that more than one limpet species may be present. The word *Patella* means kneecap and honours the low conical shape of limpets. In fact, there are three species of *Patella* to be found on British shores (Hayward, 2004). The common limpet *Patella vulgata* lives up to its name. It is widespread around the British Isles, occupying the midshore from open coasts to the mouths of estuaries, grazing on seaweeds. It will be the subject of consideration here. The other two species of *Patella* occur below mean tide level on wave-exposed southwestern and western British coasts (Hayward, 2004). The China limpet *Patella ulyssiponensis* was previously known as *P. aspera*. The black footed limpet *Patella depressa* also used to have a different species name, in this case *P. intermedia*.

6.12.2.1 Limpets as Biomonitors

Prologue. Limpets accumulate trace metals from solution and from their algal food source, accumulated concentrations reflecting local trace metal bioavailabilities.

Table 6.22 summarises accumulated concentrations of trace metals in the common limpet *P. vulgata* from British shores.

An initially surprising feature of Table 6.22 is the very low accumulated copper concentration in common limpets from uncontaminated habitats, often only 10 µg/g or less. Such a copper concentration would supply the essential copper needs of copper-containing enzymes but would not leave any copper available for the copper-bearing respiratory protein haemocyanin (Chapter 3) (Rainbow and Luoma, 2011b). As in the periwinkle *Littorina littorea*, haemocyanin is common in gastropod molluscs, in association with a higher background accumulated body copper concentration of 50 µg/g or more (Table 6.21). Background copper concentrations in *P. vulgata* confirm that it lacks haemocyanin (Table 6.22). At first sight, this would appear to be a surprising omission in its physiology, but the limpet does use another respiratory protein to store oxygen. This is the iron-bearing protein haemoglobin, present, for example, in muscles of the limpet.

Limpets do not occur within Restronguet Creek itself. Nevertheless, specimens of *P. vulgata* collected from the outside of Restronguet Point (Figure 6.5) and elsewhere in the Fal Estuary system (Figure 6.6) in the 1970s and 1980s confirmed the export of raised bioavailabilities of copper and zinc from Restronguet Creek (Table 6.22). There was also evidence for a raised bioavailability of lead to limpets collected from Falmouth Harbour in 1972 (Table 6.22) (Bryan and Gibbs, 1983).

Other raised bioavailabilities of trace metals in Cornish estuaries agree with patterns that we have seen earlier for other biomonitors. Limpets from the mouth of the estuary of the Red River, on the north coast of Cornwall, in 1972 had raised accumulated concentrations of copper, zinc and cadmium (Table 6.22). Limpets from St Ives Bay, affected by the effluent of the River Hayle, contained the highest accumulated concentration of copper recorded (225 µg/g) (Table 6.22). Limpets from the Looe Estuary in the 1970s had high concentrations of lead (originating in the catchment of the West Looe River) and of silver from an industrial source on the East Looe River (Table 6.22).

P. vulgata collected from sites in the vicinity of Aberystwyth in 1992 showed atypically high accumulated concentrations of lead and zinc (Table 6.22). While this might be expected given the outflows of the Rivers Ystwyth and Rheidol at Aberystwyth, it is in contrast to the results for edible periwinkles *Littorina littorea* at the same sites (Table 6.21).

Common limpets from Port Erin on the southwest coast of the Isle of Man in about 1970 had raised concentrations of lead (Table 6.22). It is possible that the raised bioavailability of lead may have originated in the outflow up the west coast of the Glen Maye River (Figure 5.13), draining the Foxdale lead mining district.

Limpets, like the biomonitors before them, have identified the phenomenally high bioavailabilities of cadmium in the Severn Estuary in the 1970s (Table 6.22), resulting from emissions from the Avonmouth smelter complex. Astoundingly high accumulated concentrations in Severn Estuary *P. vulgata* decreased downstream but were still very high in limpets collected from the Bristol Channel (Table 6.22) (Butterworth et al., 1972; Nickless et al., 1972). The huge accumulated cadmium contents of these limpets were detoxified by binding to metallothionein (Howard and Nickless, 1977; Noël-Lambot et al., 1980). There is an indication that these high accumulated loads of cadmium may have been depressing glucose metabolism in Severn Estuary and Bristol Channel limpets (Shore et al., 1975).

Exorbitant cadmium bioavailabilities in the Severn Estuary were accompanied by very high zinc bioavailabilities and high silver bioavailabilities in the 1970s, with the same pattern of decreasing downstream to the Bristol Channel (Table 6.22). Copper bioavailabilities to limpets in the Severn Estuary were also somewhat raised in the 1970s (Table 6.22).

Table 6.22 Mean, range of mean or weight-standardised estimate* (1 g body dry weight) concentrations (μg/g dry weight) of trace metals in bodies of the common limpet *Patella vulgata* from British estuaries and coastlines affected by (a) mining contamination, (b) industrial contamination or (c) considered uncontaminated.

	Ag	As	Cd	Co	Cr	Cu	Fe	Hg	Mn	Ni	Pb	Zn
Mining												
Fal Estuary System												
Restronguet Point												
1975–77[1]		37	16			45						243
Mylor Harbour												
1972–85[2,3]	0.8		7.4–13	0.8	1.5	64–90	669–952		6–12	1.9	8.8–8.9	206–326
Falmouth												
1972–85[2,3]			3.0–9.4			16–46	1,070–2,660		3–24		10–32	177–401
Red River												
1972[2]			24			147	750		15		19	323
Hayle (St Ives Bay)												
Pre–195[3]	0.3		3.3	4.1	1.1	225	2,820		25	1.2	4.9	155
Looe Estuary												
Pre–1980[4]	5.6	33	5.6	0.6	0.5	18	1,160	0.26	6	2.3	30	145
Aberystwyth												
1992[5]			2.2–13			7.8–17			11–41		5.1–26	63–381
Port Erin (Isle of Man)												
Pre–1971[6]			31	0.4		7.7	150		13	2.5	32	84
Industrial												
Severn												
Portishead												
1970–72[7,8]	15		550–593			60					2.8–8.5	458–580

Site / Date												
*1973-74[9]			401–717			24–37	1,260–2,590				14–20	355–434
Pre-1985[3]	12		289	0.93	1.7	35	1,300		102	1.1	6.2	312
Weston												
1972[8]	8.5		535			40					1.8	468
Pre-1980[4]	2.5	15	239	2.4	3.6	41	1,740	0.12	47	4.5	10	279
Watchet												
1970-72[7,8]	19		110–425			32					3.2–4.0	250–340
*1974[9]			191			25	1,370					303
Bristol Channel												
1970[7]			30–70								3.0–4.0	100–150
1971[10]			26–500								2–27	120–375
Tyne												
1980-81[11]							971					150
Uncontaminated												
South Devon Coast												
Pre-1983[2]	12					10	973		6		9	107
Duckpool (Cornwall)												
Pre-1985[3]	1.9		7.8	0.3	0.6	7.6	1,570		14	1.1	0.4	102
Isles of Scilly												
Pre-1985[3]	0.8		5.6	0.04	1.1	5.1	2,360		7	3.6	2.3	81
Typical	0.3–2		2–12			5–20					0.4–10	60–120
High	5–20		20–720			50–225					20–35	200–600

Note: Wet weight concentrations from Peden et al. (1973) have been multiplied by 5.

Sources: Klumpp and Peterson (1979);[1] Bryan and Gibbs (1983);[2] Bryan et al. (1985);[3] Bryan et al. (1980);[4] Rainbow and Ward (unpublished);[5] Segar et al. (1971);[6] Butterworth et al. (1972);[7] Peden et al. (1973);[8] Boyden (1977);[9] Nickless et al. (1972);[10] Lobel et al. (1982).[11]

Box 6.7 Radulae and Iron

Grazing gastropods use the radula, a toothed ribbon, to crop algae growing on the substratum on which they live. The radula of the periwinkle *Littorina littorea* is suited to cutting fine green algae and young sporelings of brown fucoid seaweeds (Hayward, 2004). The teeth of the radula of the limpet *Patella vulgata* are altogether harder and capable of rasping the toughest algae (Hayward, 2004).

The basis of this hardness is the impregnation of the teeth with iron and silicon (Runham, 1961; Runham et al., 1969). The iron is in the form of goethite, a mineral based on iron hydroxide, and the silicon is present as opal, a form of silica (silicon dioxide) (Runham et al., 1969; Van der Wal, 1989). Mature teeth of the limpet radula consists of two parts. The posterior part is iron-rich, consisting of between 44 and 51% iron and 1 to 6% silicon, while the anterior part has 22 to 30% iron and 27 to 32% silicon (Runham et al., 1969).

Iron mineralization of radular teeth is also important in another class of molluscs, the chitons (Towe and Lowenstam, 1967), which similarly graze algae on hard rocks in the littoral and sublittoral zones. There are several species of chitons on British coasts, the most frequently encountered being *Lepidochitona cinerea*.

The cells of the radula sac that secrete the radula in limpets and chitons temporarily store the iron needed for mineralization detoxified by binding in the protein ferritin, both in soluble and crystalline form (Towe et al., 1963; Towe and Lowenstam, 1967).

6.13 Detritivores in the Strandline

Prologue. We move now from grazers to detritivores. The strandlines of British shores, both inside and outside estuaries, are typified by a buildup of decaying brown seaweeds. These brown algae include both wracks, fucoid algae such as species of *Fucus* growing on the midshore, and kelps, larger brown seaweeds such as species of *Laminaria*, from the bottom of shores and below.

6.13.1 Talitrid Amphipod Crustaceans

Prologue. Major detritivorous consumers of the decaying vegetable matter of strandlines are two species of talitrid amphipod crustaceans, the beachhopper *Orchestia gammarellus* and the sandhopper *Talitrus saltator*.

Orchestia gammarellus (Plate 6d) can be found on a wide range of shores other than open sandy beaches and is the talitrid common on the shores of British estuaries. In the British Isles, the sandhopper *T. saltator* typifies fully marine sandy beaches but can tolerate low salinities. It is to be found, for example, on sandy beaches across much of the Baltic Sea, washed by waves of low salinity water (Fialkowski et al., 2009). While *Orchestia gammarellus* and *Talitrus saltator* may be the most familiar talitrids on British shores, they are by no means the only ones (Lincoln, 1979; Wildish, 1987). *Talorchestia deshayesii* is also to be found in strandlines on sandy beaches, while *Orchestia mediterranea* occurs a little lower

Table 6.23 **Copper and zinc body concentrations (μg/g dry weight with 95% confidence limits {CL}) in four species of talitrid amphipod (*Orchestia gammarellus, Orchestia mediterranea, Talorchestia deshayesii* and *Talitrus saltator*) at a standardised body dry weight (0.01g), as estimated from double log regressions of dry weight against trace metal concentration. All amphipods were collected at the same time of year (13–28 September 1986) near Millport on Great Cumbrae Island, Firth of Clyde, Scotland.**

| | COPPER | | | ZINC | |
Species	Concn	CL	Species	Concn	CL
O. gammarellus	76.1	65.2, 88.7	*T. deshayesii*	213	196, 233
O. mediterranea	56.4	50.8, 62.6	*T. saltator*	178	159, 199
T. deshayesii	50.8	46.8, 55.1	*O. gammarellus*	167	147, 190
T. saltator	26.8	24.3, 29.6	*O. mediterranea*	120	108, 134

Source: Moore and Rainbow (1987).

down on rocky shores. *Orchestia cavimana* has invaded freshwater. The alien talitrid *Platorchestia platensis*, with a worldwide distribution, has made it to the shores along British and other European coasts.

6.13.1.1 Body Concentrations of Trace Metals

Prologue. Closely related talitrid amphipods show subtle species differences in body concentrations of copper and zinc. Copper and zinc concentrations do not vary between sexes or with the moult cycle, but body copper concentrations do vary seasonally in association with seasonal differences in body concentrations of haemocyanin.

There are subtle differences in the concentrations of copper and zinc in the bodies of even closely related coastal talitrid species (Moore and Rainbow, 1987). Table 6.23 lists copper and zinc body concentrations in the four talitrids – *Orchestia gammarellus, Orchestia mediterranea, Talorchestia deshayesii* and *Talitrus saltator* – all collected in the close vicinity of Millport on Great Cumbrae Island in the Firth of Clyde, Scotland, at the same time in September 1986 (Moore and Rainbow, 1987). This site was considered to be uncontaminated (Moore and Rainbow, 1987). There

were clear interspecific differences in the accumulated concentrations of both copper and zinc, when comparing standard sized amphipods to overcome any effect of size on accumulated concentrations (Table 6.23) (Moore and Rainbow, 1987).

In such an interspecific comparison, it is necessary to beware physiological effects that might affect these body concentrations, potentially differently between the species. Firstly, therefore, it was confirmed that there were no differences in the body concentrations of copper and zinc between the sexes in three of these species (*O. gammarellus, O. mediterranea, T. deshayesii*), which are sexually dimorphic and readily distinguishable into males and females (Moore and Rainbow, 1987). Another potential source of physiological variation is the moult cycle (Weeks et al., 1992). In fact, it has also been established that body concentrations of neither copper nor zinc vary through the moult cycle in *O. gammarellus, O. mediterranea* and *T. saltator* (Weeks et al., 1992). A third potential source of physiological variation is seasonal change, presumed to have been allowed for by the choice of a single limited period of collection between 13 and 28 September 1986 (Moore and Rainbow, 1987). While the body zinc concentration showed no change with season from January to

November 1987 in any of the three species *O. gammarellus, O. mediterranea* and *T. saltator*, the body copper concentration did vary seasonally in all three species (Rainbow and Moore, 1990). Body copper concentrations of these three species dipped significantly in November but were raised again by March to stay high through the spring and summer (Rainbow and Moore, 1990). In the case of the sandhopper *T. saltator*, the estimated body copper concentration of a 0.01 g amphipod remained between 33 and 39 µg/g Cu between March and September, but dropped to 26 µg/g Cu in November (Rainbow and Moore, 1990). This raises the question as to whether the amphipods in the winter had enough body copper to meet the needs of haemocyanin as well as copper-based enzymes (Rainbow and Luoma, 2011b; Rainbow and Moore, 1990). In practice, *T. saltator* may gain no benefit from the possession of haemocyanin to support respiratory efficiency in the winter, for the amphipod enters a period of winter inactivity, in effect hibernating deep in burrows in the sand (Weeks et al., 1992; Taylor and Spicer, 1986). Haemocyanin may even be used as a protein source of energy in this period, with the loss of the previously associated copper. An advantage to possess haemocyanin returns with increased activity in the spring and summer. The higher body copper concentrations then are more capable of meeting the copper requirements of the respiratory protein (Rainbow and Moore, 1990).

To return to the interspecific comparison made in September 1986 (Table 6.23), it is quite likely that body copper concentrations in *T. saltator* at that time (27 µg/g Cu) were already falling from their spring and summer levels with the onset of the Scottish winter, contributing to the interspecific difference seen for copper. In fact, body copper concentrations in *T. saltator* during spring and summer more typically reach 40 or 50 µg/g (Rainbow and Moore, 1990; Weeks et al., 1992). Nevertheless, when interspecific comparisons were carried out on spring and summer body copper concentrations, there is still a significant interspecific variation in the order *O. gammarellus* > *O. mediterranea* > *T. saltator* (Rainbow and Moore, 1990).

6.13.1.2 Talitrids as Trace Metal Biomonitors

Prologue. There is a good comparative biomonitoring database available for the interpretation of accumulated trace metal concentrations in *Orchestia gammarellus* as indicative of high or baseline trace metal bioavailabilities.

A talitrid amphipod such as *Orchestia gammarellus*, feeding as a detritivore in the strandline, will take up and accumulate trace metals both from solution when covered by the tide and trophically from ingested dead seaweed (Weeks and Rainbow, 1993). Since the trace metal content of the seaweed depends on the local dissolved bioavailability of the metal, *O. gammarellus* can be regarded as being both a direct and an indirect biomonitor of dissolved trace metal bioavailabilities. In practice, *O. gammarellus* in an uncontaminated habitat takes up and accumulates both copper and zinc more from food than from solution (Weeks and Rainbow, 1993).

Amphipods such as *O. gammarellus* are net accumulators of trace metals. Trace metals taken up from diet or solution are accumulated in the cells of the paired ventral caeca arising from the midgut (Figure 3.11). These cells pass through a cell cycle, as they progress from the blind end of the caecum towards its junction with the midgut (Figure 3.11). The cells then disintegrate and their trace metal contents are expelled from the gut with the faeces. Storage of the accumulated trace metals in the ventral caecum cells is in detoxified form. Thus, for example, copper is accumulated in the form of copper-rich, sulphur-rich granules (Weeks, 1992a; Nassiri et al., 2000), which are residual bodies from the lysosomal breakdown of copper-binding metallothionein. If trace metal exposure of the amphipod increases, more metal-rich detoxificatory granules are accumulated in the cells passing through the epithelial cell cycle of the ventral caecum (Weeks, 1992a; Nassiri et al., 2000). Thus the trace metal concentration of each caecum, and correspondingly the whole body of the amphipod, increases upon increased exposure to reach a higher equilibrium concentration (Figure 3.11) (Weeks, 1992a). Therefore, whole body concentrations of trace metals increase in relation to local trace metal bioavailabilities. Talitrid amphipods can, therefore, be used as trace metal biomonitors.

There is a good biomonitoring database available of accumulated trace metals in *Orchestia gammarellus* (Rainbow et al., 1989; Moore et al., 1991), with similar information in the literature for *Talitrus saltator* (Fialkowski et al., 2009; Morrison et al., 2017), and the cosmopolitan *Platorchestia platensis* (Weeks, 1992b).

Table 6.24 lists copper and zinc body concentrations in standard-sized beachhoppers *O. gammarellus* from a variety of metal-contaminated collection sites on British shores. All amphipods in this list were collected between April and August, when body copper concentrations of *O. gammarellus* had levelled off somewhat after a winter low and a March peak (Rainbow and Moore, 1990). Table 6.24 also presents proposed estimates of what are typical and atypically high accumulated body concentrations of these two trace metals in the beachhoppers. Typical summer body concentrations of copper (50 to 80 µg/g) are enough to service essential metabolic requirements, including those of haemocyanin (Rainbow et al., 1989). These body copper concentrations do increase (100 to 275 µg/g) upon exposure to raised copper bioavailabilities. As a result, however, of the continuing turnover of ventral caecal cells, albeit now possessing more copper-rich detoxification granules (Figure 3.11), raised body concentrations of copper, while identifiable, do not reach the very high body concentrations seen, for example, in barnacles (Table 6.20). Typical body concentrations of zinc in *O. gammarellus* (135 to 200 µg/g; Table 6.24) are relatively high in comparison with typical background zinc concentrations of many of the bivalves met earlier. These relatively high background concentrations of zinc are higher than would be needed to meet essential metabolic needs (Rainbow and Luoma, 2011,b, c), and also occur in other talitrids, including *Talitrus saltator* (Fialkowski et al., 2009). It would seem, therefore, that there is quite a high standing stock of detoxified zinc (bound in metallothionein or derived tertiary lysosomes) in ventral caecal cells of talitrids, even in the absence of atypically raised zinc bioavailabilities.

Not unexpectedly, beachhopper body concentrations of copper presented in Table 6.24 are led by data for Restronguet Creek. *O. gammarellus* from other mining-affected estuaries, including the Tamar, Tavy and Gannel in Cornwall and Devon and at Dulas Bay in Anglesey, are also high on the copper list (Table 6.24).

Much of the data in Table 6.24 is taken from two studies in Scotland (Rainbow et al., 1989; Moore et al., 1991). On the eastern side of Scotland, copper concentrations in beachhoppers were high at North Queensferry in the Firth of Forth and upstream in the Forth Estuary at Grangemouth in 1989 (Table 6.24) (Moore et al., 1991). North Queensferry lies between the rail and road Forth bridges on the northern shore of the Firth. There was, however, no obvious local source of high trace metal bioavailabilities at North Queensferry at the time. It is possible, then, that strandline seaweed deposited at North Queensferry had grown at a contaminated site elsewhere in the estuary or firth. One possible source might have been upstream in the Forth Estuary at Grangemouth. That said, however, talitrids collected from Grangemouth itself, had lower (although still atypically raised) copper concentrations than those at North Queensferry (Moore et al., 1991).

The other Scottish study included an investigation of shores on the island of Islay in western Scotland (Rainbow et al., 1989). Islay is famous for its whisky distilleries. These use copper stills and have effluent outfalls onto local shores (Rainbow et al., 1989). In 1987, *O. gammarellus* from shores receiving effluent from three high output distilleries, Laphroaig, Bowmore and Lagavulin, all had atypically raised accumulated copper concentrations (Table 6.24). While local copper bioavailabilities are raised, it is not known whether they are significant enough to have an ecotoxicological effect on local shore communities. In the light of the quality of the products from these distilleries, it is so tempting to conclude that, perhaps, some small environmental prices are worth paying,

In the case of zinc, very high bioavailabilities to *O. gammarellus* could be identified at Restronguet Creek, and also elsewhere in the Fal Estuary system, at

Table 6.24 **Concentrations of copper and zinc (μg/g dry wt) in the talitrid amphipod *Orchestia gammarellus* in 0.01 g dry weight amphipods as estimated from best-fit double log regressions of accumulated concentration against dry weight. All collections were made between April and August. Sites are listed in descending order of copper concentrations.**

Date	Site	Copper Concentration	Zinc concentration
1993[1]	Restronguet Creek, Cornwall	117–271	380–550
1984[2]	Restronguet Creek, Cornwall	139	392
1997[3]	Restronguet Creek, Cornwall	136	169
1989[4]	N Queensferry, Firth of Forth	130	340
1984[2]	Tamar (Torpoint), Cornwall	120	212
1984[2]	Gannel, Cornwall	118	228
1987[2]	Laphroaig, Islay	117	147
1992[5]	Tamar, Cornwall/Devon	43.0–110	162–292
1989[4]	Grangemouth, Firth of Forth	109	153
1997[3]	Dulas Bay, Anglesey	105	126
1992[5]	Tavy, Devon	55.5–104	135–238
1987[2]	Bowmore, Islay	101	160
1987[2]	Lagavulin, Islay	93.2	196
2014[6]	Thames Estuary	49.2–88.4	87–174
1993[1]	St Just, Fal Estuary, Cornwall	79.4	372
1989[4]	Kennetpans, Firth of Forth	78.8	193
1987[2]	Millport, Firth of Clyde	78.1	152
1993[1]	Flushing, Fal Estuary, Cornwall	76.6	360
1989[4]	Dunimarle Castle, Firth of Forth	75.4	209
1984[2]	Kilve, Bristol Channel	75.4	167
1987[2]	Dower House, Islay	74.9	145
2001[7]	Thames Estuary	50.9–72.4	145–259
1987[2]	Loch Indaal, Islay	71.7	143
1989[4]	Torryburn, Firth of Forth	69.1	215
1984[2]	Looe, Cornwall	67.7	140
1995[8]	Blackwater, Essex	51.3–65.0	114–206
1987[2]	Bruichladdich, Islay	64.7	193
1997[3]	Millport, Firth of Clyde	63.8	186
1987[2]	Bunnahabhainn, Islay	63.3	182

Table 6.24 **(cont.)**

Date	Site	Copper Concentration	Zinc concentration
1989[4]	Inverkeithing, Firth of Forth	55.7	192
	Typical	50–80	135–200
	High	100–275	220–550

Sources: Rainbow and Bright (unpublished);[1] Rainbow et al. (1989);[2] Rainbow et al. (1999);[3] Moore et al. (1991);[4] Rainbow and Bhattachary (unpublished);[5] Johnstone et al. (2016),[6] Rainbow et al. (2002);[7] Rainbow and Smith (unpublished).[8]

St Just and Flushing, north of Falmouth (Table 6.24). These latter examples may have been affected by downstream export and deposition of seaweeds originating in Restronguet Creek. Raised zinc bio-availabilities were also present in the 1980s and 1990s in the Tamar, Tavy and Gannel Estuaries (Table 6.24). Amphipods at North Queensferry in 1989 had atypically raised zinc concentrations to go with their high copper concentrations, while zinc concentrations were not raised in talitrids from Grangemouth upstream (Table 6.24). Zinc concentrations in *O. gammarellus* from five sites in the Thames Estuary in both 2001 and 2014 (Table 6.24) can be considered typical, but with one exception. Beachhoppers from Erith, just downstream of the Beckton and Crossness sewage works, in 2001 had raised zinc concentrations (259 µg/g), suggesting a local raised zinc bioavailability at the time, since decreased (Rainbow et al., 2002; Johnstone et al., 2016).

Limited data are available for accumulated cadmium concentrations in *Orchestia gammarellus*. Nevertheless, beachhoppers collected from Kilve on the Somerset shore of the Bristol Channel in 1984 contained 7.4 µg/g Cd, in comparison to concentrations of 1.6 µg/g in Millport amphipods (February 1988) and 1.4 µg/g Cd in specimens from Whithorn in Dumfries and Galloway in Southwest Scotland in March 1988 (Rainbow et al., 1989). It would appear that this high local cadmium bioavailability at Kilve is a (presumably decreasing) legacy from the cadmium originating from the Avonmouth Refinery a few years earlier.

Concentrations of trace metals in other British talitrid amphipods generally reflect the conclusions drawn for *Orchestia gammarellus*. Specimens of *Talitrus saltator* collected from the strandline of the beach by the entrance to Dulas Bay have atypically high body concentrations of both copper and zinc (Rainbow et al., 1989; Fialkowski et al., 2009). *Orchestia mediterranea* from Restronguet Creek and Halton Quay in the Tamar Estuary in 1984 similarly have unusually high copper and zinc body concentrations (Rainbow et al., 1989).

6.14 Predators

Prologue. Predators in British estuarine ecosystems include certain polychaete worms, whelks (neogastropod molluscs), decapod crustaceans (prawns, shrimps and crabs) and fish such as flounders, eelpout and eels.

6.14.1 Polychaete Worms

Prologue. Two families of exclusively predatory polychaetes contain species to be found in British estuaries – the nephtyid catworms and the glycerid bloodworms.

We have already met the omnivorous polychaete *Hediste diversicolor* (Plate 6a), capable of feeding as a predator and scavenger, and its nereidid relative

Alitta virens, the king ragworm, an awesome intertidal predator.

Polychaetes of two further families are specialist predators on British shores, including those of estuaries, burrowing in clean medium grade sand (Hayward, 2004). The catworms belong to the family *Nephtyidae*, of which two species, *Nephtys caeca* and *N. hombergii*, are common. Polychaetes of the family *Glyceridae*, sometimes called bloodworms because of their red colour, are represented by three species: *Glycera tridactyla* (formerly *G. convoluta*), *G. gigantea* (up to 35 centimetres long) and *G. alba*, which is ironically milky white in colour.

6.14.1.1 Catworms

Prologue. *Nephtys hombergii* accumulates trace metals in reflection of local bioavailabilities. Restronguet Creek catworms had high concentrations of copper detoxified by binding with sulphur.

Nephtys hombergii is the more abundant of the two catworm species and also occurs sublittorally. It can reach 20 centimetres long (Hayward, 2004). The head is small. Nevertheless, the catworm has a very large eversible pharynx, lined with a fringe of small, toothlike structures, which it rapidly retracts with its surprised prey enclosed. Prey items include mollusc spat, small crustaceans and other polychaetes (Hayward, 2004).

Table 6.25 lists accumulated trace metal concentrations in *N. hombergii* from British estuaries and from offshore deposits assumed to be uncontaminated by metals. This catworm will take up trace metals from its animal diet and also from any water to which it is exposed in its burrow.

As expected, catworms from Restronguet Creek have accumulated copper concentrations (up to 2,230 µg/g) much raised above a typical range of 12 to 40 µg/g (Table 6.25). Accumulated copper concentrations were also raised in Mylor Creek in the Fal Estuary system, and in the Severn Estuary at Weston-super-Mare in the early 1980s (Table 6.25). Raised bioavailabilities at that time of particular trace metals emitted by the Avonmouth Refinery have become familiar in this chapter. Thus the bioavailability of cadmium to catworms at Weston was expectedly high, as was that of silver (Table 6.25). In addition, Restronguet Creek catworms had raised accumulated concentrations of particularly zinc, and also of lead and silver, in the early 1980s, reflecting raised local bioavailabilities (Table 6.25).

The specimens of *Nephtys hombergii* from Restronguet Creek with the very high accumulated copper concentrations were blackened at the head end, apparently as a result of the deposition of copper sulphide in the body wall (Bryan and Gibbs, 1983). It would appear, therefore, that the detoxification of copper by *N. hombergii* is like that of *Hediste diversicolor* in the high copper exposure conditions of Restronguet Creek (Mouneyrac et al., 2003; Geffard et al., 2005). The accumulated copper would be bound to metallothionein, and then autolysed in lysosomes of the epidermal cells to produce many copper and sulphur-rich extracellular granules (Mouneyrac et al., 2003; Geffard et al., 2005). Moreover, the population of *N. hombergii* in Restronguet Creek has become tolerant to copper exposure (Bryan and Gibbs, 1983).

Box 6.8 Calcium Granules Are Skeletal in Catworms

We have encountered the physiological use of calcium phosphate granules for trace metal detoxification by many invertebrates, the phosphate typically being in the form of pyrophosphate with its high affinities for many trace metals. We have also seen the use of calcium-rich granules (usually as carbonate) as physiological stores of calcium. Polychaetes of the family *Nephtyidae* illustrate another biological use of calcium-rich granules, for they use calcium phosphate granules for skeletal purposes (Gibbs and

Bryan, 1984). Numerous granules of calcium phosphate, as hydroxyapatite with orthophosphate but not pyrophosphate, are disposed in bundles in the main longitudinal blocks of muscle in the body of the catworm (Gibbs and Bryan, 1984). These granules are relatively pure, consisting mostly of calcium and phosphorus, with no trace metals present (Gibbs and Bryan, 1984). Catworms are very active burrowers and are more muscular and less compressible than, for example, the ragworm *Hediste diversicolor*. The robust nature of the catworms is derived from the quantity and arrangement of the calcium phosphate granules, forming ribbons that interleave with the contractile fibres of the muscles (Gibbs and Bryan, 1984). Most of these ribbons of granules are flattened in the vertical plane, permitting strong lateral sinusoidal movement, so typical of *Nephtys* species. This arrangement also resists flexion dorsoventrally, further assisting rapid penetration into sediments as catworms burrow (Gibbs and Bryan, 1984).

6.14.1.2 Glycerids

Prologue. Glycerid polychaetes have jaws that are hardened by very high accumulation of copper, even at uncontaminated sites.

Species of *Glycera* have bodies which taper at both ends. Like catworms, they have a large eversible pharynx. This pharynx, however, has four large black jaws for the grasping of prey (Hayward, 2004).

These jaws concern us here. While nereidid polychaetes harden their jaws with zinc, glycerid polychaetes use another trace metal, this time copper, for the same purpose. Each of the four jaws contains about 1.5% (15,000 µg/g) of copper in dry weight terms (Gibbs and Bryan, 1980). Together, the four jaws contain more than 50% of the body copper in species of *Glycera* from uncontaminated sites. The copper is concentrated in the distal tip of each jaw, where the copper concentration can reach 130,000 µg/g in *G. gigantea* (Gibbs and Bryan, 1980). Zinc is also present throughout each jaw (2,210 to 4,070 µg/g in *G. gigantea*), but is swamped by the copper at the business end of the tooth (Gibbs and Bryan, 1980). The copper in the jaws of glycerid polychaetes is in the form of the copper-based mineral atacamite [$Cu_2(OH)_3Cl$] (Lichtenegger et al., 2002).

As Table 6.25 shows, the accumulated body concentrations of trace metals in glycerids, in this case *Glycera tridactyla*, increase under exposure to high bioavailabilities. Accumulated concentrations of both copper and zinc are highly elevated in *G. tridactyla* from Restronguet Creek (Table 6.25).

6.14.2 Whelks

Prologue. Carnivorous gastropod molluscs are often lumped under the common name of whelks.

Predatory gastropods known as whelks belong to the systematic assemblage known as neogastropods. We met two intertidal neogastropods earlier, the sting winkle *Ocenebra erinaceus* and the oyster drill *Urosalpinx cinerea*, accidentally imported from North America and predators of oysters. Occurring at the bottom of British shores may be young specimens of the otherwise large common whelk *Buccinum undatum*, resident sublittorally and harvested commercially. Nassariid neogastropods are mainly scavengers, represented on British shores by the netted dogwhelk *Tritia reticulata* (Plate 7e), formerly known as *Nassarius reticulatus*. Yet the most common and ecologically dominant neogastropod predator on British shores is the dogwhelk *Nucella lapillus*.

Table 6.25 **Concentrations (μg/g dry weight) of trace metals in two predatory polychaetes,** *Nephtys hombergii* **and** *Glycera tridactyla*, **from British estuaries and sublittoral coastal waters.**

	Ag	Cd	Cu	Fe	Mn	Pb	Zn
Nephtys hombergii							
Restronguet Creek							
Pre-1983[1,2]	1.7–3.3	0.3–0.57	646–2,230	1,720–3,260	7.5–13	44.7	353–518
1994[3]			135–281	444–1,140	6.2–12	6.2	155–173
St Just (Fal Estuary)							
1994[3]			28–33	149–281	6.7		72–88
Place Cove (Fal Estuary)							
Pre-1983[4]	0.31–0.69	0.29–0.54	23–30	353–459	2.2–5.6	3.4–8.4	223–355
Mylor (Fal Estuary)							
1994[3]			117–199	148–320	2.6–3.4		112–146
Torre Sands (Torbay)							
Pre-1985[4]	0.02	0.98	13	400	3.8	3.5	260
Cawsand Bay (Plymouth)							
Pre-1985[4]	0.23	0.22	39	616	8.6	4.6	178
Salcombe (Devon)							
Pre-1985[4]	0.54	1.9	37	364	5.4	3.7	268
Severn Estuary (Weston)							
Pre-1985[4]	2.3	3.6	117	410	12	4.6	259
Glycera tridactyla							
Restronguet Creek							
Pre 1980[1,5]			440–828	1,370	17		298–483
St Just (Fal Estuary)							
Pre-1983[1]			45	954	8.8		169
Place Cove (Fal Estuary)							
Pre-1983[1]	0.5		64	1,060	9.4		240
Torbay (Devon)							
Pre-1983[1]			42	934	7.9		299
Whitsand Bay(Cornwall)							
Pre-1983[1,5]			70	783	4.2		292

Sources: Bryan and Gibbs (1983);[1] Bryan et al. (1985);[2] Williams et al. (1998);[3] Bryan and Gibbs (1987);[4] Gibbs and Bryan (1980.[5]

6.14.2.1 Dogwhelks

Prologue. Dogwhelks take up and strongly accumulate trace metals from their prey such as mussels and barnacles. Dogwhelks, like other neogastropods, have strong digestive powers and can absorb trace metals from some forms of detoxified metal-rich granules in their prey.

The dogwhelk *Nucella lapillus* is common on rocky shores on open coasts and near the mouths of estuaries. It is a predator, feeding on mussels and barnacles and occasionally small limpets (Hayward, 2004). The dogwhelk can be a key factor in determining the structure of ecological communities on British rocky shores.

As predators, dogwhelks, like other neogastropods, will take up trace metals from their animal prey, as well as from the surrounding water via the gills. The trophic route appears to be much the more important of the two, as verified in the cases of iron and zinc for *N. lapillus* feeding on the barnacle *Semibalanus balanoides* (Young, 1977). The trophic input of metals will depend on the quantities of trace metals accumulated by the prey, by the nature of the chemical binding (detoxification) of the trace metals in the prey and by the strength of the digestive and assimilative processes of the neogastropod predator (Rainbow et al., 2011a). The digestive powers of neogastropods appear to be amongst the strongest of invertebrate predators (Rainbow et al., 2011a). Neogastropod digestive processes, for example, can release for alimentary uptake trace metals bound in metal-rich granules in prey tissues, to a greater extent than can the digestive processes of palaemonid decapods (Rainbow and Smith, 2010; Rainbow et al., 2011a). There will still be differences in the trophic release of trace metals from different types of metal-rich detoxification granules in prey tissues. Thus cadmium and silver in cadmium-rich and silver-rich granules (presumably sulphur-rich lysosomal residual bodies) in bivalve prey tissues can be digested and assimilated by the nassariid neogastropod *Tritia reticulata* (Rainbow and Smith, 2010). On the other hand, metal-rich granules based on calcium pyrophosphate can pass unaltered through the digestive tracts of *Tritia*

reticulata feeding on kidney tissue of the scallop *Aequipecten opercularis* (zinc-rich granules) or on the digestive gland of the edible periwinkle *Littorina littorea* (zinc- and manganese-rich granules) (Nott and Nicolaidou, 1990). Similarly, zinc pyrophosphate granules in the body of the barnacle *Semibalanus balanoides* may pass undigested through the alimentary tract of *Nucella lapillus* (Nott and Nicolaidou, 1990). Nevertheless, dogwhelks do take up zinc from barnacle prey (Young, 1977), presumably mostly from zinc bound in other chemical forms (e.g., metallothionein) in barnacle bodies.

Once assimilated, trace metals taken up by neogastropods will themselves need to be detoxified for accumulation. In fact, many neogastropods, not least those belonging to the family *Muricidae*, accumulate high concentrations of trace metals, even in the absence of trace metal contamination. Members of the *Muricidae* on British shores include *Nucella lapillus*, *Ocenebra erinaceus* and *Urosalpinx cinerea* (Table 6.26). The high accumulated body concentrations of trace metals in neogastropods are attributable to their high dietary assimilation efficiencies and low efflux rates (Luoma and Rainbow, 2008). And the site of the detoxified storage of accumulated trace metals is the digestive gland. The digestive glands of muricid gastropods such as *N. lapillus* contain many detoxificatory metal-rich granules (MRG), increased in number under conditions of trace metal contamination (Ireland, 1979a; Nott and Nicolaidou, 1989). Calcium phosphate-based MRG particularly bind zinc, while copper is detoxified predominantly in lysosomal residual bodies associated with sulphur (Nott and Nicolaidou, 1989).

Table 6.26 confirms that neogastropods, particularly of the families *Muricidae* and *Nassariidae*, contain high accumulated concentrations of trace metals, even in the absence of contamination. The most extensive database is that for *Nucella lapillus* (Table 6.26). Even that dataset, however, is too limited to allow the confident proposal of bioaccumulated concentration ranges of trace metals that can be considered as typical or as raised and indicative of raised local bioavailabilities. Suffice to say that body

Table 6.26 **Concentrations (μg/g dry weight) of trace metals in bodies of predatory neogastropod molluscs, three muricids (the dogwhelk *Nucella lapillus*, the sting winkle *Ocenebra erinaceus* and the oyster drill *Urosalpinx cinerea*), one nassariid (the netted dogwhelk *Tritia reticulata*) and one buccinid (the edible whelk *Buccinum undatum*), from the shores of estuaries and shores and below of other coasts of Britain.**

	Ag	As	Cd	Cu	Fe	Mn	Pb	Zn
Muricidae								
Nucella lapillus								
Fal Estuary System								
Restronguet Point								
1975–79[1,2,3]	1.7	48	23	1,000	383	37	5.0	3,350
Weir Point								
1972[2]			14	433	334	19	3.9	1,620
Mylor Harbour								
Pre-1983[2]			12	400	537	18	3.1	1,960
Falmouth								
Pre-1985[2,3]	2.2		6.5–17	177–305	184–327	13–15	1.5–34	503–822
Hayle								
Carbis Bay								
1994[4]			5.3	222	197	9.5		307
Red River								
1972[2]			4.7	809	185	12	1.0	323
Looe								
Pre-1985[3,5]	1.3		5.5–13	51–110	214–270	11–17	1.9–5.1	235–416
Aberystwyth								
1974[6]				246		9.6	5.3	780
1982–83[7]			10					
Port Erin (Isle of Man)								
Pre-1971[8]			73	150	65	12	4.9	860
Severn Estuary								
Brean								
1970[9]			425				27	3,100
Weston								
Pre-1985[3]	8.6		114	114	474	25	19	1,840
Penarth								
1974[6]				458		17	20	2,350

Table 6.26 (**cont.**)

	Ag	As	Cd	Cu	Fe	Mn	Pb	Zn
Watchet								
1970[9]			330				14	1,050
Bristol Channel								
1970–71[9,10]			31–725				1–38	175–4,200
Swansea Bay (Mumbles)								
1982–83[7]			55					
Tyne								
1980–81[11]				282	642			1,120
Criccieth (W Wales)								
1974[6]				254		16	4.8	708
1982–83[7]			9.9					
Maenporth (Cornwall)								
1994[4]			8.7	148	187	11		521
Wembury (Devon)								
Pre-1985[3]	1.6		26	115	274	13	5.0	394
Isles of Scilly								
Pre-1985[3]	1.5		46	28	413	10	2.6	442
Ocenebra erinaceus								
Fal Estuary System								
Carrick Roads								
1979[2]			24	3,770	762	12	8.3	2,850
Urosalpinx cinerea								
Thames Estuary								
Whitstable								
1994[4]			38	178	320	20		1,220
Nassariidae								
Tritia reticulata								
Fal Estuary System								
Carrick Roads								
1979[2]			1.6–2.0	4,080–7,290	3,210–5,350	27–31	7.4–11	906–1,390
Hayle Estuary								
Carbis Bay								
1994[4]			4.5	1,650	260	7.6		787

Table 6.26 **(cont.)**

	Ag	As	Cd	Cu	Fe	Mn	Pb	Zn
Cawsand Bay/Plymouth Sound								
1979[2]			2.1–2.5	1,160–2,600	506–655	66–73	5.6–6.5	617–811
Buccinidae								
Buccinum undatum								
Fal Estuary System								
Carrick Roads								
1979[2]			4.0	510	338	4.8	2.8	3,020
Looe Bay								
*1974[12]			6.5	123	65	6.0	9.1	508
Solent								
Pre 1971[8]	0.6		2.2	180	110	1.7	5.4	620

Note: * = 1 g dry weight body.
Sources: Klumpp and Peterson (1979);[1] Bryan and Gibbs (1983);[2] Bryan et al. (1985);[3] Khan, Taylor and Rainbow (unpublished);[4] Bryan and Hummerstone (1977);[5] Ireland and Wootton (1977);[6] Abdullah and Ireland (1986);[7] Segar et al. (1971);[8] Butterworth et al. (1972);[9] Nickless et al. (1972);[10] Lobel et al. (1982);[11] Boyden (1977).[12]

concentrations above 5 µg/g silver, 100 µg/g cadmium, 300 µg/g copper, 15 µg/g lead and 900 µg/g zinc in *N. lapillus* can confidently be described as atypically high. More accurate thresholds are probably lower.

While absent from Restronguet Creek itself, the dogwhelk *N. lapillus* does occur at the bottom of the creek on Restronguet Point and Weir Point on opposite sides (Figures 6.5 and 6.6) (Bryan and Gibbs, 1983; Bryan et al., 1985). Copper and zinc concentrations in dogwhelks from Restronguet Point in the 1970s were higher than in those from elsewhere in the Fal Estuary system, which were nevertheless still somewhat raised (Table 6.26). Dogwhelks collected near the mouth of the Red River at that time also had very high copper concentrations (Table 6.26). *N. lapillus* collected in 1994 from Carbis Bay, between St Ives and the mouth of Hayle Estuary (Figure 6.7), on the other hand, did not have the raised copper

concentrations that might be expected from the close proximity of output from the Hayle (Table 6.26).

Data for the Severn Estuary in the 1970s and early 1980s tell the same story of raised bioavailabilities of cadmium, and also zinc, silver, lead and copper, presumably related to the Avonmouth smelting works (Table 6.26). The threshold of 100 µg/g indicated previously for cadmium in *N. lapillus* is phenomenal in the context of most bioaccumulated concentrations of cadmium seen in invertebrates in this chapter. Yet, it was still comfortably exceeded in dogwhelks from Brean (425 µg/g) and Watchet (330 µg/g) in 1970 (Table 6.26). High cadmium bioavailabilities to dogwhelks, as well as those of zinc and perhaps lead, were also exported downstream into the Bristol Channel in the early 1970s (Table 6.26).

The high zinc concentration accumulated by dogwhelks at the mouth of the Tyne at the start of the 1980s (Table 6.26) is consistent with the high zinc concentrations in mussels, a favourite prey item, there at the time (Table 6.11) (Lobel and Wright, 1983).

6.14.2.2 Other Neogastropods

It is difficult to place the high accumulated trace metal concentrations of other British neogastropod species on scales from typical to atypically raised (Table 6.26). Nevertheless, copper and zinc concentrations in *Ocenebra erinaceus*, the copper concentration in *Tritia reticulata* and the zinc concentration in *Buccinum undatum*, from Carrick Roads in 1979, all appear to be atypically raised (Table 6.26) (Bryan and Gibbs, 1983).

6.14.3 Prawns and Shrimps

Prologue. Predatory and scavenging decapod crustaceans include those that are commonly, but usually inconsistently, called prawns and shrimps. These decapods can swim as well as walk along the bottom. Caridean decapods, exemplified by palaemonids, are classic regulators of the body concentrations of the essential trace metals zinc and copper over wide ranges of their bioavailabilities, while being net accumulators of non-essential trace metals such as cadmium.

In British estuaries, we are concerned mainly with two families of caridean decapods, the palaemonids (often called prawns) and the crangonids (often called shrimps). On British shores, palaemonid species include *Palaemon elegans* (Plate 2a), common inhabitants of intertidal rock pools; *Palaemon serratus*, larger prawns towards the bottom of rocky shores in southern and western Britain; *Palaemon longirostris*, forming dense shoals in the upper reaches of large estuaries in southern and western Britain including the Thames Estuary; and *Palaemonetes varians* (Plate 6e), brackish water prawns living in salt marsh pools and brackish lagoons (Smaldon et al., 1993). Of the crangonids, *Crangon crangon* is common offshore on sandy and muddy grounds around the British Isles, penetrating the lower reaches of estuaries (Smaldon et al., 1993).

Here we concentrate on the palaemonid prawns *Palaemonetes varians* and *Palaemon longirostris* and

on the brown shrimp *Crangon crangon*. Brown shrimps eat the spat of bivalves such as *Macoma balthica* and *Cerastoderma edule*, polychaetes, smaller crustaceans and even juvenile plaice *Pleuronectes platessa* (Hayward, 2004). In turn, they are eaten by fish that include whiting (*Merlangius merlangus*), sand gobies (*Pomatoschistus minutus*), five-bearded rockling (*Ciliata mustela*), young bass (*Dicentrarchus labrax*) and sea snails (*Liparis liparis*) (Hayward, 2004). The diets of the palaemonids are less well detailed, but *P. varians* will be preyed upon by grey herons (*Arda cinerea*) and mallard ducks (*Anas platyrhynchos*) (Green, 1968).

6.14.3.1 Zinc

When *Palaemonetes varians* was exposed to increasing dissolved zinc concentrations at salinities that varied from 5% to 100% seawater, this decapod maintained a constant body concentration of zinc (90 to 99 µg/g) (Nugegoda and Rainbow, 1989a). This regulated body concentration of zinc was controlled by the prawn increasing its rate of zinc excretion to match the rate of dissolved zinc uptake, itself increasing as the dissolved zinc concentration was raised (Nugegoda and Rainbow, 1989a). Beyond a threshold external dissolved concentration of zinc, regulation broke down as the maximum rate of zinc excretion was exceeded, and net accumulation of body zinc ensued (Nugegoda and Rainbow, 1989a). The threshold dissolved zinc concentration causing regulation breakdown decreased with salinity decrease, as the dissolved bioavailability of zinc at a given dissolved concentration increased as dissolved free zinc ions were released from chloride complexation (see Chapter 3, Figure 3.2) (Nugegoda and Rainbow, 1989a). Thus the threshold dissolved concentration of zinc at regulation breakdown of *P. varians* was 191 µg/L in 100% seawater, 146 µg/L in 50% seawater, and only 19 µg/L in 5% seawater (Nugegoda and Rainbow, 1989a). Incidentally, these dissolved zinc concentration thresholds in 100% and 50% seawater are higher than those for another palaemonid *Palaemon elegans* (92 and 27 µg/L Zn respectively) (Nugegoda and Rainbow, 1989a). This is explicable by the fact that *P. varians* has a lower

dissolved zinc uptake rate at any salinity than
P. elegans (Table 3.3), in line with a lower permeability,
considered to be an adaptation for improved osmotic
water balance in the low salinity conditions of salt
marsh pools. Both littoral palaemonids, in turn,
have higher threshold zinc dissolved concentrations
corresponding to regulation breakdown than the
sublittoral caridean decapod *Pandalus montagui*,
living in fully saline seawater (Nugegoda and
Rainbow, 1989b).

After regulation breakdown, the body zinc concentration of a palaemonid prawn rises as the uptake
of dissolved zinc continues to exceed the maximum
rate of zinc excretion (Nugegoda and Rainbow,
1989a; Luoma and Rainbow, 2008). Much of this
rising accumulated concentration of body zinc is not
in detoxified form and cannot increase too greatly
before this excess of metabolically available zinc
causes sublethal and then lethal toxicity (Luoma and
Rainbow, 2008; Rainbow and Luoma, 2011c). There
have been constant caveats in this book about the
irrelevance of threshold accumulated body concentrations in invertebrates as indicators of the onset of
toxic effects. This is still the case for the vast
majority of invertebrates, which use storage
detoxification (temporary or permanent) of accumulated trace metals. In the absence, however, of
significant storage detoxification, the great majority
of accumulated metal in the body is in metabolically
available form to which a toxic threshold concentration is applicable. This is the case for zinc in
palaemonid decapods (Luoma and Rainbow, 2008;
Rainbow and Luoma, 2011c). In *Palaemon elegans*,
the maximum body concentration of zinc that can
be withstood by the decapod has been estimated to
be 180 µg/g, of which 150 µg/g is in metabolically
available form and 30 µg/g reversibly bound
to metallothionein for safe temporary storage
(Rainbow and Luoma, 2011c). Similar numbers
probably apply to *Palaemonetes varians*. Indeed,
mean total zinc concentrations in *P. varians* after
regulation breakdown on 21-day exposure to high
zinc bioavailabilities did not exceed 200 µg/g
(Nugegoda and Rainbow, 1989a, b).

In *P. varians* from uncontaminated salt marsh
pools at Tollesbury at the head of the Blackwater
Estuary in Essex, the hepatopancreas contained about
7% of the total body zinc content, equivalent to about
14% of the body zinc not in the cuticular exoskeleton
(Nugegoda and Rainbow, 1989a; Rainbow and Smith,
2013). Approximately 58% of this zinc accumulated
in the hepatopancreas was in detoxified form, bound
to metallothionein (Rainbow and Smith, 2013).
Thus, about 5 µg/g of the total body concentration of
84 µg/g zinc was bound to metallothionein in the
hepatopancreas (Rainbow and Smith, 2013).

Table 6.27 lists zinc concentrations in
Palaemonetes varians, *Palaemon longirostris* and
Crangon crangon from British estuaries. As in the case
of the palaemonids, *C. crangon* has also been confirmed to regulate body concentrations of zinc over a
wide range of dissolved bioavailabilities (Amiard
et al., 1985, 1987).

In the case of the zinc concentrations in *P. varians*
given in Table 6.27, those reported for Tollesbury
decapods fit the expected regulated body concentrations described earlier. It may just be the case that zinc
bioavailabilities to this prawn in the salt marsh pools
at Gorseinon, near the tinplate industrial works at
Llanelli and in Dulas Bay, both in 1985, were sufficiently high that zinc regulation was beginning to
break down (Table 6.27). Body zinc concentrations in
Palaemon longirostris in the Thames and Adur
estuaries are consistent with zinc regulation to about
80 to 95 µg/g zinc (Table 6.27). The brown shrimp,
C. crangon, also appears to have expected regulated
mean zinc concentrations (84 to 104 µg/g) when collected from the East and West Looe estuaries, the
Thames Estuary at North Woolwich and the Loughor
Estuary near Gorseinon (Table 6.27). In contrast to
P. varians at this last site, the shrimps were collected
from the presumably more mixed and less zinc-
contaminated waters of the Loughor Estuary itself.
Body zinc concentrations in brown shrimps from the
screens of Hinkley Point power station in the Bristol
Channel in 1984 (Table 6.27) may be raised above
typical regulated levels, as a result of the ongoing
raised zinc bioavailability in the Bristol Channel at the

Table 6.27 **Concentrations (μg/g dry weight) of trace metals in bodies of two palaemonid prawn species (*Palaemonetes varians* and *Palaemon longirostris*) and one crangonid shrimp species (*Crangon crangon*) in British estuaries. Confidence limits (CL) of 95% are provided for most Cu and Zn mean concentrations.**

	Cd	Cr	Cu		Zn	
			Mean	CL	Mean	CL
Palaemonidae						
Palaemonetes varians						
Blackwater Estuary						
Tollesbury						
1983[1,2]			92	81–104	99	95–103
2010[3]			94	87–102	84	78–89
Loughor Estuary						
Gorseinon						
1985[4]	0.83	3.0	224	211–237	130	118–143
Dulas Bay						
1985[4]	0.92		226	216–235	113	109–117
Palaemon longirostris						
Thames Estuary						
North Woolwich						
1983[4]	0.61		103	97–109	85	83–87
Battersea						
2001[5]	0.06	0.19	124		80	
Adur Estuary (Sussex)						
1986[4]			111	103–119	92	89–94
Crangonidae						
Crangon crangon						
East Looe Estuary						
1984[4]	0.47		67	61–73	88	80–97
West Looe Estuary						
1984[4]	0.5		72	68–77	104	88–120
Restronguet Creek						
1984[4]	1.5		210	184–236	241	213–269

Table 6.27 **(cont.)**

	Cd	Cr	Cu		Zn	
			Mean	CL	Mean	CL
Thames Estuary						
North Woolwich						
1983[4]	1.6	1.4	84	79–89	84	75–92
Bristol Channel						
Hinkley Point						
1984[4]	3.0		103	88–118	126	105–47
Loughor Estuary						
Gorseinon						
1986[4]	0.94	4.2	70	67–73	93	86–101

Sources: Nugegoda and Rainbow (1988);[1] Nugegoda and Rainbow (1989);[2] Rainbow and Smith (2013);[3] Rainbow and White (unpublished);[4] Langston et al. (2004).[5]

time. Body concentrations of zinc in *C. crangon* collected in Restronguet Creek in 1984 were certainly very high (Table 6.27). These body zinc concentrations are probably indicative of brown shrimps in which zinc regulation has broken down, and the shrimps may well be suffering ecotoxicological effects.

6.14.3.2 Copper

Decapod crustaceans, including palaemonids such as *Palaemon elegans* (White and Rainbow, 1982; Rainbow and White, 1989), and crangonids such as *Crangon crangon* (Amiard et al., 1985, 1987), also regulate body concentrations of copper over a wide range of dissolved copper bioavailabilities (Luoma and Rainbow, 2008). *P. elegans* regulates the body copper concentration to between 110 and 130 µg/g (Rainbow and White, 1989). Regulation breakdown occurs at high dissolved copper bioavailabilities as in the case of zinc (Rainbow and White, 1989; Luoma and Rainbow, 2008). In contrast to zinc, however, the accumulation of copper after regulation breakdown continues in *P. elegans* to higher body concentrations (about 600 µg/g) before death, as the rate of copper

uptake exceeds the combined rates of copper detoxification and excretion (Chapter 3) (Rainbow and White, 1989; Luoma and Rainbow, 2008). The cells of the hepatopancreas then contain copper- and sulphur-rich residual bodies derived from the lysosomal breakdown of metallothionein binding copper, these being lost into the gut and then the faeces at the end of the hepatopancreatic cell cycle (Luoma and Rainbow, 2008). *P. varians* from the uncontaminated conditions of the Tollesbury salt marshes contained only between 1 and 5% of the total body copper concentration of 94 µg/g in the hepatopancreas, about 10% of the body copper not in the exoskeleton (Nugegoda and Rainbow, 1989b; Rainbow and Smith, 2013). All hepatopancreas copper in these prawns was detoxified by binding with metallothionein (Rainbow and Smith, 2013). It needs to be remembered in this context that decapod crustaceans contain haemocyanin, and a large proportion of total body copper (about 23 µg/g; Table 3.6) will be present in the blood as part of this respiratory protein.

We can extrapolate from *Palaemon elegans* to its two palaemonid relatives, *Palaemonetes varians* and *Palaemon longirostris*, to consider that these two species also regulate body concentrations of copper. In Table 6.27, copper concentrations listed for *P. varians* from Tollesbury appear to be regulated, as do those for *P. longirostris* from the Thames and Adur estuaries. On the other hand, body copper concentrations in *P. varians* from Gorseinon near Llanelli in 1985, and from Dulas Bay in the same year, are both indicating regulation breakdown under conditions of very high local copper bioavailabilities.

In the case of brown shrimps, *Crangon crangon*, mean copper concentrations of shrimps from the East and West Looe Estuaries (1984), from North Woolwich on the Thames Estuary (1983), from Hinkley Point in the Bristol Channel (1984) and from the Loughor Estuary near Llanelli (1986) all appear to fall in the regulated range of 66 to 103 μg/g (Table 6.27). As to be expected, however, the mean copper concentration of *C. crangon* collected from Restronguet Creek in 1984 is well above any regulated level (Table 6.27). As for zinc, the copper concentration accumulated indicates that the regulation of body copper has broken down, and these brown shrimps may also be suffering ecotoxicological effects of copper poisoning.

6.14.3.3 Cadmium and Lead

The body concentration of the non-essential metal cadmium is not regulated by *Palaemon elegans* (White and Rainbow, 1982; Rainbow and White, 1989), nor by *Crangon crangon* (Amiard et al., 1985, 1987). Nor incidentally is the body concentration of lead (another non-essential trace metal) regulated by *C. crangon* (Amiard et al., 1987). Accumulated cadmium concentrations of caridean decapods can, therefore, be expected, to reflect local cadmium bioavailabilities over the full range of local contamination. From Table 6.27, it would appear that brown shrimps in Restronguet Creek (1984), the Thames Estuary at North Woolwich (1983) and the Bristol Channel at Hinkley Point (1984) had all been exposed to raised cadmium bioavailabilities. The increased body cadmium concentration was highest in the Bristol Channel in the 1984 sample, unsurprisingly in the light of evidence presented through this chapter.

6.14.3.4 Chromium

Accumulated chromium concentrations were raised in both *P. varians* and *C. crangon* collected at Gorseinon in the Loughor Estuary in 1985 and 1986 (Table 6.27). The high local bioavailabilities of chromium to these decapods in this south Wales estuary in the 1980s reflect those reported for the ragworm *Hediste diversicolor* (Table 6.6) and the bivalve *Scrobicularia plana* (Table 6.8).

6.14.4 Crabs

Prologue. The classic crab found in British estuaries is the shore crab *Carcinus maenas*. This section explores the biology of copper, zinc, cadmium and lead in the shore crab, interpreting accumulated tissue concentrations in terms of the moult cycle, haemocyanin turnover and accumulation patterns, including regulation, detoxification and variation with local bioavailabilities.

The shore crab *Carcinus maenas* is remarkable. It can withstand the salinity variations of estuaries, in which its distribution extends all the way up into muddy salt marsh creeks. Moreover, it is the most abundant crab on all British shores, adored by children turning over stones at the bottom of rocky shores at low tide. On the other hand, it can also live buried in sand and mud. Adult crabs migrate offshore to live sublittorally (Crothers, 1967). The range of littoral and sublittoral habitats occupied by this common crustacean is extraordinary, and the crab shows exceptional physiological flexibility. The shore crab may be common, but 'common' here should have none of the negative implications often implied by the use of this adjective. The common shore crab is to be admired for its ecological flexibility. On the other hand, it is difficult to convince Australians of this positive view, for this crab is indeed an unwanted alien invader down under, as well as in the western United States.

Carcinus maenas is a predator and scavenger, feeding on a wide range of invertebrates including mussels, periwinkles, limpets and dogwhelks (Hayward, 2004). In estuaries, it will also prey upon both the spat and older individuals of cockles (Hayward, 2004). Like other crabs, *C. maenas* passes through planktonic larval stages (zoea larvae) during development, settling out from the water column at the megalopa stage to take up life on the bottom. An implication of this planktonic larval life of several weeks is that newly settled juvenile crabs will have originated from parents often living well beyond the shore now occupied by their offspring.

Table 6.28 presents concentrations of copper, zinc, cadmium and lead in *Carcinus maenas* from a range of British sites. In many cases, whole body trace metal concentrations have also been broken down into accumulated concentrations in the cuticle, the hepatopancreas and all soft tissues other than the cuticle.

It is a feature of crabs such as the shore crab that more of the weight of the crab is present in the cuticular exoskeleton with increase in size of the crab. The exoskeleton contains trace metals from two sources. Trace metals, particularly iron and manganese, may come out of solution and be adsorbed onto the outside surface of the exoskeleton. More importantly, accumulated trace metals may be passed from the soft tissues to be deposited in the exoskeleton. This transfer may be reversed to some extent upon the moult of the crab. For example, calcium used in calcium carbonate to harden the exoskeleton is partially resorbed premoult for conservation. Cadmium and lead previously deposited with calcium in the exoskeleton may similarly be partially resorbed into the soft tissues then. Anyway, in an intermoult crab, the exoskeleton, while holding a significant and often majority percentage of the total body content of a trace metal, usually has a lower trace metal concentration than the soft tissues (Table 6.28). The upshot is that larger crabs typically have lower total body concentrations of trace metals than smaller crabs at any one location. Thus in Table 6.28, the higher and lower copper and zinc concentrations in whole crabs from Restronguet Creek and the Tamar Estuary were measured in small (about 1 g wet weight) and large (about 50 g wet weight) crabs respectively (Bryan and Gibbs, 1983).

While associated with the redistribution of trace metals between exoskeleton and soft tissues, the

moult cycle also causes changes in total body and other tissue concentrations of trace metals in *C. maenas*, particularly of copper and zinc (Scott-Fordsmand and Depledge, 1997). Furthermore, in early postmoult crabs, calcium is rapidly taken up from solution to replace that lost with the cast moult (Scott-Fordsmand and Depledge, 1997; Nørum et al., 2005). This increased active uptake of calcium via calcium channels in the gills causes increased uptake of cadmium by this route, resulting in increased cadmium accumulation then (Nørum et al., 2005).

6.14.4.1 Copper

Another physiological cycle affecting trace metal concentrations in different tissues, particularly the blood and the hepatopancreas, is that of haemocyanin synthesis and breakdown (Martin and Rainbow, 1998). Haemocyanin uses copper to function as a respiratory protein dissolved in the blood. Haemocyanin in decapod crustaceans has a turnover time of 15 to 30 days (Martin and Rainbow, 1998). In the hepatopancreas, the copper released when haemocyanin is broken down is temporarily stored bound to metallothionein, awaiting recycling to newly synthesised haemocyanin. Copper concentrations in the blood of *C. maenas* are typically of the order of 50 µg/ml, contributing about 10% of the total body content of copper (Rainbow, unpublished; Martin and Rainbow, 1998).

As we have seen in the case of prawns and shrimps, body concentrations of copper appear to be regulated over a wide range of bioavailabilities to approximately constant levels in all decapod crustaceans (Bryan, 1968).

The crab *Carcinus maenas* is no exception (Bryan, 1968; Rainbow, 1985). Shore crabs collected near Millport in the Firth of Clyde in Scotland regulated total body concentrations of copper to about 40 µg/g (with 95% confidence limits of 34 to 46 µg/g) at dissolved copper concentrations up to about 170 µg/L under specified physicochemical conditions (Rainbow, 1985). In Table 6.28, crabs collected from Lynemouth, Millport and Newhaven all have body concentrations within this regulated range. The

Table 6.28 **Concentrations (μg/g dry weight) of trace metals in selected tissues (with percentages of whole crab contents) and whole bodies of the shore crab *Carcinus maenas* from British estuaries and coasts.**

	Cuticle		Hepatopancreas		All soft tissues		Whole crab
	Concn	% content	Concn	% content	Concn	% content	Concn
Copper							
Restronguet Creek							
1975–77[1]							221
1976[2]	97	39	854	22	450	61	185
Pre-1983[3] (large, small)							191, 527
Gannel Estuary							
1976[2]	38	41	301	18	220	59	74
Plym Estuary							
1976[2]	33	45	124	10	149	55	57
Plymouth							
*Pre-1968[4]							77
Tamar Estuary							
Pre-1983[3] (large, small)							77, 181
Dulas Bay							
1976[2]	154	58	1,880	16	741	42	231
Lynemouth, NE England							
*1972–73[5]	22		115				38
Millport, Firth of Clyde							
1984[6]	22	55	77	8	59	45	30
Newhaven, Sussex							
1976[2]	24	48	92	16	109	52	41
Zinc							
Restronguet Creek							
1975–77[1]							198
1976[2]	164	55	469	10	409	46	226
Pre-1983[3] (large, small)							149, 282
Gannel Estuary							
1976[2]	86	50	470	14	351	50	139

Table 6.28 **(cont.)**

	Cuticle		Hepatopancreas		All soft tissues		Whole crab
	Concn	% content	Concn	% content	Concn	% content	Concn
Plym Estuary							
1976[2]	69	53	166	13	226	47	103
Plymouth							
*Pre-1968[4]			280				77
Tamar Estuary							
Pre-1983[3] (large, small)							77, 167
Dulas Bay							
1976[2]	168	65	1,010	9	595	35	225
Lynemouth, NE England							
*1972–73[5]	54		204				83
Millport, Firth of Clyde							
1984[6]	76	59	135	4	159	41	97
Newhaven, Sussex							
1976[2]	60	47	262	18	283	54	104
Cadmium							
Restronguet Creek							
1976[2]	0.7	48	4.2	19	2.2	52	1.1
Gannel Estuary							
1976[2]	1.3	55	4.6	11	3.8	45	1.8
Plym Estuary							
1976[2]	1.6	68	2.6	9	3.0	32	1.9
Dulas Bay							
1976[2]	5.3	49	24	5	36	51	9.3
Lynemouth, NE England							
*1972–73[5]	6.6		5.1				3.4
Millport, Firth of Clyde							
1984[6]	0.6	54	7.8	22	1.8	46	0.9
Newhaven, Sussex							
1976[2]	0.5	39	3.1	22	3.1	61	1.0
Lead							

Table 6.28 (**cont.**)

| | Cuticle | | Hepatopancreas | | All soft tissues | | Whole crab |
	Concn	% content	Concn	% content	Concn	% content	Concn
Restronguet Creek							
1976[2]	16	80	2.1	1	10	20	13
Gannel Estuary							
1976[2]	30	81	39	18	29	19	30
Plym Estuary							
1976[2]	11	74	5.0	3	16	27	12
Dulas Bay							
1976[2]	5.5						4.8
Newhaven, Sussex							
1976[2]	3.8						3.1

Note: *Wet weight data multiplied by 2.5 (cuticle), 5 (hepatopancreas) and 3.5 (whole crab) to obtain dry weight concentrations. In data from Bryan and Gibbs (1983), large crabs weigh about 50 g wet weight, small crabs weigh about 1 g wet weight.

Sources: Klumpp and Peterson (1979);[1] Rainbow (unpublished);[2] Bryan and Gibbs (1983);[3] Bryan (1968);[4] Wright (1976);[5] Rainbow (1985).[6]

equivalent 95% confidence range of regulated soft tissue concentrations of copper in the Millport crabs was 96 to 142 µg/g (Rainbow, 1985). The crabs with total body and soft tissue concentrations of copper clearly well above these estimated regulation ranges were collected from Restronguet Creek and Dulas Bay, as now to be expected given the very high local bioavailabilities of copper at these sites (Table 6.28).

6.14.4.2 Zinc

Body concentrations of zinc also appear to be regulated over a wide range of bioavailabilities to approximately constant levels in all decapod crustaceans (Bryan, 1968).

Carcinus maenas regulates whole body and total soft tissue concentrations of zinc to approximately constant levels when exposed to increasing dissolved zinc bioavailabilities (Rainbow, 1985). Unlike the palaemonid prawns previously discussed, the shore crab does not increase its excretion of zinc over the range of dissolved bioavailabilities investigated to match significant increases in zinc uptake rate (Chan and Rainbow, 1993). For example, at 100 µg/L zinc at 10°C, *Palaemon elegans* has a dissolved zinc uptake rate of 1.52 µg/g Zn per day (Rainbow and White, 1989) and turns over nearly 13% of its body load of zinc per day (White and Rainbow, 1984). *C. maenas*, on the other hand, does not match zinc excretion to dissolved zinc uptake until much higher dissolved exposures (Chan and Rainbow, 1993). At 100 µg/L Zn under the same physicochemical conditions, the shore crab has a dissolved zinc uptake rate of 0.55 µg/g Zn per day, with no significant rate of zinc excretion (Chan and Rainbow, 1993). Under these circumstances, a shore crab would take 200 days to double its total body zinc load (Chan and Rainbow, 1993). Therefore, for the shore crab, growth rate becomes a significant factor in concert with a low zinc uptake rate to maintain relatively constant body zinc concentrations.

Table 6.28 shows that accumulated zinc in shore crabs is stored in the exoskeleton (46 to 65% total body content) and the hepatopancreas (4 to 18% total body content). A further 6 to 12% of body zinc is to be found in the blood (Chan and Rainbow, 1993). In the blood, zinc binds to haemocyanin to give a typical concentration of about 35 µg Zn per ml in *C. maenas* (Martin and Rainbow, 1998). Zinc newly taken up under conditions of raised dissolved zinc bioavailability is added sequentially over time to this zinc bound to haemocyanin, which has a maximum theoretical carrying capacity of zinc of about 120 µg/ml (Martin and Rainbow, 1998). New zinc accumulated in the blood has a very slow rate of turnover with a half-life of 21 days, presumably related to the turnover of haemocyanin itself (Martin and Rainbow, 1998).

Shore crabs from Millport regulated total body concentrations of zinc to about 83 µg/g (with 95% confidence limits of 73 to 94 µg/g) at dissolved zinc concentrations up to about 400 µg/L under specified physicochemical conditions (Rainbow, 1985). The equivalent 95% confidence range of regulated soft tissue concentrations of zinc in the Millport crabs was 120 to 186 µg/g (Rainbow, 1985). Of the crabs listed in Table 6.28, those collected from Plymouth, Lynemouth and Millport all have whole body and/or soft tissue zinc concentrations consistent with these regulated concentration ranges. On the other hand, crabs from Restronguet Creek, the Gannel Estuary and Dulas Bay all had accumulated zinc concentrations clearly above these regulated ranges (Table 6.28), consistent with the expected raised zinc bioavailabilities in these estuaries.

6.14.4.3 Cadmium and Lead

Accumulated concentrations of cadmium are not regulated by the shore crab *Carcinus maenas* (Rainbow, 1985), as is typical of other decapod crustaceans. In shore crabs, cadmium often follows some of the same physiological pathways as calcium (Nørum et al., 2005). Between 39 and 68% of body cadmium can be found in the exoskeleton of intermoult crabs, with a further 5 to 22% stored in the hepatopancreas (Table 6.28), where it is bound by metallothionein (Pedersen et al., 1994, 2014). Cadmium also binds to haemocyanin in the blood, but has a much shorter half-life there than zinc (Martin and Rainbow, 1998). Cadmium newly taken up by *C. maenas* rapidly reaches an equilibrium concentration in the blood, as the rate of removal of cadmium from the blood matches the rate of cadmium uptake into it (Martin and Rainbow, 1998). Thus, in contrast to the long-term storage of zinc bound to haemocyanin in the blood, the blood represents a transient store of cadmium as it is passed, for example, from gills to hepatopancreas (Martin and Rainbow, 1998). The blood of Millport crabs had a cadmium concentration of 0.93 µg/ml (Martin and Rainbow, 1998). Blood concentrations of cadmium in crabs from several of the British estuaries listed in Table 6.28, varied from 0.32 to 5.52 µg/ml, the highest figure being recorded from Dulas Bay (Rainbow, unpublished; Martin and Rainbow, 1998). Of all the crabs listed in Table 6.28, it was again the crabs from Dulas Bay that had the highest soft tissue and whole body concentrations of cadmium.

In the case of lead, the crabs with the highest accumulated concentrations in whole crabs and in the soft tissues were those from the Gannel Estuary (Table 6.28), well known here to have very high lead bioavailability.

6.14.4.4 Detoxification

Accumulated trace metals need to be detoxified. *Carcinus maenas* uses metallothioneins to bind cadmium, copper and zinc in detoxified form, for example in the hepatopancreas (Pedersen et al., 1994, 2014). Different specific metallothioneins play different physiological roles (Pedersen et al., 2014). Thus the specific form of metallothionein induced to bind cadmium in the hepatopancreas is different from that temporarily holding the copper to be delivered to newly synthesised haemocyanin (Pedersen et al., 1994, 2014).

Granules based on calcium are also present in the hepatopancreas of shore crabs. The distinction has already been made between calcium granules containing pyrophosphate that are ideal detoxificatory bins for accumulated trace metals, and much purer, usually calcium carbonate, granules that act as temporary physiological stores. These latter granules usually contain calcium and magnesium, but no trace metals that may otherwise be released back into circulation on the dissolution of these granules on physiological demand. Cells of the hepatopancreas of *C. maenas* do contain calcium granules (Hopkin and Nott, 1979; Simkiss, 1990). These are actually based on orthophosphate, the much more soluble form of phosphate than pyrophosphate (Simkiss, 1990). These granules also contain magnesium and do appear to be suitable to fulfil a storage function for calcium, needed, for example, to begin the recalcification of the exoskeleton after a moult (Hopkin and Nott, 1979). Counterintuitively, however, these calcium phosphate granules do sometimes contain lead, apparently for detoxification (Hopkin and Nott, 1979). Perhaps there is more here than meets the eye, and there is a difference in the subtle composition and occurrence of the calcium phosphate granules containing lead and those that do not.

6.14.4.5 Ecotoxicology

The shore crab *Carcinus maenas* has become a model invertebrate of choice, together with the mussel *Mytilus edulis*, in studies of the trace metal ecotoxicology of estuaries (Luoma and Rainbow, 2008), particularly those involving biomarkers, including metallothioneins (Martin-Diaz et al., 2009; Rodrigues and Pardal, 2014).

In spite of the infamous presence of high bioavailabilities of trace metals, not least copper and zinc, in Restronguet Creek, shore crabs do survive there (Bryan and Gibbs, 1983). Small crabs below 3 centimetres carapace width are not uncommon in the lower and middle reaches of the creek, and larger crabs can be found further upstream during the summer (Bryan and Gibbs, 1983). Small crabs below 1 gram wet weight in Restronguet Creek have particularly high

total concentrations of both copper and zinc, while larger crabs (about 50 g wet weight) have total concentrations of these trace metals that are lower but still raised on a national scale (Table 6.28) (Bryan and Gibbs, 1983). Studies in the 1970s showed that the population of shore crabs in Restronguet Creek was tolerant to both copper and zinc (Bryan and Gibbs, 1983). These studies suggested that decreased uptake rate may be part of the mechanisms of tolerance by Restronguet Creek crabs for each of these two trace metals (Bryan and Gibbs, 1983).

The individual crabs making up the Restronguet Creek population would have originated from parents almost certainly living outside the creek and, therefore, not under any selective pressure to evolve either copper or zinc tolerance. Thus any copper or zinc tolerance would have been selected for in individual crabs settling in, or migrating into, Restronguet Creek, be it brought about by decreased uptake from water and/or diet, by increased excretion, by increased storage detoxification or by any combination of these processes. Thus there is likely to be no single populationwide physiological mechanism for tolerance in these shore crabs, as might be the case, for example, for the in situ copper-tolerant population of the amphipod *Corophium volutator*, which broods its young and lacks a planktonic dispersal phase in the life cycle. In fact, a study in 1996 and 1997 found that the shore crabs then present in Restronguet Creek did not have significantly lower uptake rates from solution of zinc (nor of cadmium or silver) than crabs from Millport (Rainbow et al., 1999). Nor, in fact, did shore crabs from Dulas Bay (Rainbow et al., 1999).

6.14.5 Fish

Prologue. Three species of British estuarine fish in particular have been investigated for their trace metal biology: the flounder, the eelpout and the European eel.

6.14.5.1 Flounder

Prologue. Muscle concentrations of copper and zinc in the flounder *Platichthys flesus* are regulated to relatively constant levels, but those of mercury are of biomonitoring and human ecotoxicological significance. Mercury levels in the muscles of flounder in Liverpool Bay in the 1970s were of concern for human consumption but have now dropped to safe levels.

The flounder, *Platichthys flesus*, is a flatfish up to about 60 centimetres in length, although few exceed 30 centimetres (Henderson, 2014). Typically associated with estuaries, the flounder can extend down to about 100 metres depth on the continental shelf (Henderson, 2014). The youngest flounders are in the shallowest water. Flounders will occasionally enter freshwater, but, during winter, adults migrate offshore (Muus and Dahlstrøm, 1964; Henderson, 2014). They spawn at sea. The eggs and larvae are planktonic, and the juveniles move inshore into estuaries. Flounders may live for 15 years, but are sexually mature at 2- to 3- years old (males) and 3- to 4- years old (females) (Henderson, 2014). Flounders prey upon crustaceans such as amphipods, small prawns and polychaete worms and, when older, on gastropods, on bivalves such as cockles and tellins and on small fish.

While fish are able to move considerable distances, flounders are relatively stationary. It is not unreasonable, therefore, to regard a flounder caught in a particular estuary as a resident of that estuary. Fish are generally considered to regulate the tissue concentrations of essential trace metals such as copper and zinc in many tissues across a range of bioavailabilities, while accumulating non-essential metals such as cadmium and lead in proportion to local bioavailabilities (Luoma and Rainbow, 2008). The flounder is no exception (Bryan et al., 1985; Amiard et al., 1987). So, concentrations of copper and zinc are usually regulated to relatively constant levels in fish muscle tissue, irrespectively of local differences in habitat metal contamination. Typical concentrations of copper and zinc in muscle tissues of flounders fall in the narrow ranges of only about 1 to 2 μg/g dry weight of copper and 24 to 40 μg/g dw of zinc (Franklin, 1987).

Muscle concentrations of non-essential metals such as cadmium and lead, on the other hand, have some biomonitoring significance, although another organ, such as the liver or kidney, would be more involved

with storage detoxification of accumulated metal and be a better bet as a biomonitoring tissue. In fact, whole body trace metal concentrations were used in 1973 in a comparative study of flounders collected from Oldbury-on-Severn, upstream of the Avonmouth Smelter (Figure 6.13), and Barnstaple Bay in North Devon, considered to be a control site (Hardisty et al., 1974). Flounders of five-years-plus at Oldbury had mean accumulated body concentrations (dry weight) of 140 μg/g zinc, 5.2 μg/g cadmium and 28 μg/g lead. Flounders from Barnstaple had mean whole body concentrations of 195 μg/g zinc, 1.7 μg/g cadmium and 19 μg/g lead (Hardisty et al., 1974). The significantly higher cadmium and lead concentrations at Oldbury are explained by the proximity of the Avonmouth smelter, while the higher zinc concentration in the Barnstaple flounders was an unexplained surprise (Hardisty et al., 1974).

We have been concerned so far through much of this book with the significance of accumulated trace metal concentrations in different organisms in providing information on local bioavailabilities of these toxic metals, with the intention of interpreting the ecotoxicological risk of any raised metal bioavailabilities in a particular habitat. A different reason to measure accumulated trace metal concentrations in a particular organism is to assess any potential toxicological danger in the human food chain. This concern has been the driver of many measures of accumulated trace metal concentrations in coastal fish, including flounders, harvested commercially for human consumption (Franklin, 1987). Thus, many analyses have been made of trace metal concentrations in the dorsal muscles of fish, the part of a fish most commonly consumed. Most concentrations of trace metals in fish muscle tissue in the literature are quoted as concentrations per wet weight (ww) (see Table 5.16) for their increased relevance to human dietary considerations.

In the second half of the twentieth century, it was mercury that was the major trace metal in fish of toxicological potential in the human diet. Concentrations of mercury are also not regulated in fish tissues. Ironically, muscle tissue is a good tissue of choice to detect geographical and/or temporal differences in mercury bioavailabilities to fish, as seen in Chapter 5. The British coastal areas of concern in the 1970s were

Table 6.29 **Average concentrations of mercury (μg/g wet weight) in muscle tissues of flounder** *Platichthys* fle*sus* **from four British coastal areas considered to be contaminated to above average levels in the 1970s and 1980s.**

	1970–72	1977–78	1979–80	1982	1984	1985
Thames Estuary	0.72	0.29	0.25	0.29	0.18	0.17
Swansea Bay	0.15	0.17	0.15	0.07	0.19	0.14
Liverpool Bay	0.64	0.33	0.32	0.33	0.19	0.20
Morecambe Bay	0.78	0.34	0.53	0.27	0.27	0.31

Source: Franklin (1987).

Liverpool Bay, Morecambe Bay and Swansea Bay, all receiving mercury-rich discharges from the chlor-alkali industry, and the Outer Thames Estuary, receiving significant mercury input not least via the disposal of sewage sludge (CEFAS, 2001). Annual monitoring of mercury levels in muscle tissue of commercial fish species, including flounder, commenced in the early 1980s, informed by the adoption in 1980 of an Environmental Quality Standard (EQS) of 0.30 μg/g wet weight of mercury in fish flesh (CEFAS, 2001). Table 6.29 lists concentrations of mercury in muscle tissue of flounders from these four areas during the 1970s and 1980s, highlighting falls over time from initially very high concentrations in the early 1970s in Liverpool Bay, Morecambe Bay and the Thames Estuary (Franklin, 1987). By 1985, it was deemed that mercury concentrations in local fish muscle tissues were sufficiently low that it was no longer necessary to continue monitoring in Swansea Bay and the Thames Estuary (CEFAS, 2001). Regular monitoring in Liverpool Bay and Morecambe Bay continued until 1994, with occasional follow-up monitoring thereafter (CEFAS, 2001). Comparable figures in 1998 for flounders from Liverpool Bay and Morecambe Bay were 0.10 and 0.20 μg/g ww of mercury in muscle tissue (CEFAS, 2001).

We saw earlier in this chapter that a very high bioavailability of mercury had been caused in the upper Forth Estuary in Scotland by industrial discharge of mercury-rich effluent at Longannet near Grangemouth in the late 1970s and early 1980s (Elliott and Griffiths, 1986). Flounders collected from Longannet in 1982 correspondingly had average mercury concentrations in muscle tissue of 0.53 μg/g ww, peaking at about 1.3 μg/g ww in large flounders of 30 centimetres length (Elliott and Griffiths, 1986). Equivalent mercury concentrations had dropped a little downstream at Port Edgar to an average concentration of about 0.35 μg/g ww (Elliott and Griffiths, 1986).

6.14.5.2 Eelpout

Prologue. Eelpout *Zoarces viviparus* collected from the Forth Estuary in the early 1980s had high accumulated concentrations of mercury.

Also collected at these two sites along the Forth Estuary in 1982 were specimens of *Zoarces viviparus*, variously referred to as the eelpout or the viviparous blenny (Elliott and Griffiths, 1986). The eelpout is a relatively cold water species found in the North Sea, north of the Thames Estuary up to all of eastern Scotland (Henderson, 2014). The eelpout occurs down to 40 metres depth but also inhabits estuarine and other coastal rocky shores living under stones, amongst seaweed and in rock pools (Muus and Dahlstrøm, 1964; Henderson, 2014). It can survive emersion at low tide and can also burrow in mud. Eelpout prey on small fish and fish eggs and on invertebrates, including molluscs, polychaetes and crustaceans such as amphipods, isopods and shrimps (Muus and Dahlstrøm, 1964; Henderson, 2014).

In 1982, female eelpout at Longannet also had higher average accumulated mercury concentrations in muscle tissue (ca. 0.41 µg/g ww) than those collected at Port Edgar (ca. 0.25 µg/g ww) (Elliott and Griffiths, 1986). There was, however, no difference in the fecundity of the eelpout populations (measured as brood size) between the two sites that might have been attributable to mercury ecotoxicology (Elliott and Griffiths, 1986).

6.14.5.3 Eels

Prologue. The third estuarine fish species considered here is the European eel *Anguilla anguilla*, well known for its remarkable lifecycle (Box 6.9).

Box 6.9 Lifecycle of the European Eel

The European eel *Anguilla anguilla* spawns in the southern Sargasso Sea, about 5,000 kilometres from Europe. Their transparent leptocephalus larvae are then transported for nearly three months across the Atlantic Ocean by the Gulf Stream to Europe, to arrive in estuaries as colourless elvers about 65 millimetres long (Muus and Dahlstrøm, 1964). Elvers transform into glass eels, some of which travel upstream into rivers while others remain in the estuary (Henderson, 2014). The eels may live in freshwaters and estuaries for up to 25 years or more and can reach a length of 1 metre (Henderson, 2014). The eels develop a golden pigmentation and are then called yellow eels. Yellow eels feed on a wide variety of invertebrates and fish during this growth phase (Muus and Dahlstrøm, 1964). Usually, after about seven years in the case of males and at eight- to ten- years old in the case of females, yellow eels stop feeding and change morphologically into silver eels with large eyes and more muscular bodies (Muus and Dahlstrøm, 1964). Silver eels migrate into the sea and disappear into the depths of the Atlantic Ocean. Relying on their stored fat reserves, the eels make their way to the Sargasso Sea to spawn. Eels have long been fished commercially in Europe, but catches have declined considerably in the last 50 years or so, bringing concern for their conservation (Henderson, 2014).

Trace metal concentrations have been measured in eels, both for biomonitoring purposes and to provide information on trace metal input into the human diet. Like other fish, elvers regulate body and tissue concentrations of copper and zinc when exposed to a range of bioavailabilities of these two metals (Amiard et al., 1987). Cadmium and lead are accumulated in proportion to dissolved bioavailabilities (Amiard et al., 1987).

A popular biomarker in eels of raised trace metal bioavailabilities has been the concentration of metallothionein (MT) in the liver (Langston et al., 2002, 2006). In spite of caveats about measuring standing concentrations of MT as opposed to rates of MT turnover, liver concentrations of MT in eels do appear to be correlated with raised bioavailabilities of those trace metals able to induce MT synthesis, such as cadmium, copper, silver and zinc (Langston et al.,

2002). In 1998 in the Thames Estuary, eel liver MT concentrations were raised above a basal level of 2 mg/g dry weight (dw) to as high as 11 mg/g dw in eels at more contaminated upper (Brentford, Kew) and middle estuarine sites (Langston et al., 2002). Lowest liver MT concentrations were found in eels from the mouth of the estuary (Langston et al., 2002). Liver metallothionein concentrations were highly correlated with accumulated liver concentrations of cadmium, copper, silver and zinc, in reflection of raised copper and silver bioavailabilities in the upper and midestuary (Langston et al., 2002). Of these trace metals bound within the MT molecules themselves, the proportions of cadmium, copper and silver increased as the liver concentration of each metal increased (Langston et al., 2002). There was no indication that the binding of these trace metals in the MT reached saturation, implying their successful detoxification over the range of bioavailabilities encountered (Langston et al., 2002).

A comparative study of liver metallothionein in eels from the Weston Canal in the Mersey Estuary in 1999 confirmed that MT (specifically Cd-MT) concentrations were also higher than MT concentrations measured in the livers of eels in the Outer Thames Estuary (Langston et al., 2006). Cd-MT concentrations averaged nearly 5 mg/g dw in liver, indicating raised local cadmium bioavailability in the Weston Canal (Langston et al., 2006).

6.15 Wading Birds

Prologue. Waders are major predators of benthic intertidal invertebrates, feeding, for example, by probing with their bills into mud for burrowing polychaetes or bivalves. Waders therefore have the potential to take up considerable quantities of trace metals in their diets, particularly when feeding in metal-contaminated estuaries containing invertebrates with very high accumulated trace metal concentrations.

Waders consume large proportions of the annual production of the intertidal invertebrates of temperate estuaries, including British estuaries (Evans et al., 1987). Most waders are associated with wetland or coastal environments at some time of the year. Many wader species of Arctic and northern temperate regions are strongly migratory, moving south to overwinter.

Waders belong to the bird order *Charadriiformes*, making up two suborders, the *Charadrii* and the *Scolopaci*. Other suborders in this order include gulls, terns and auks. The *Charadrii* consist of plovers, lapwings, avocets, stilts and oystercatchers. It is the waders of the suborder *Scolopaci* that will mainly concern us here. These waders include dunlin, knot, sanderling, redshank, greenshank, sandpipers, curlew, snipe, godwits and turnstones.

Dunlin and knot are closely related species of the genus *Calidris*. Dunlin *Calidris alpina* are the most common small waders on British coasts, where they can be seen all year round. Their summer breeding grounds, however, are in the uplands of Britain. In the winter, flocks of dunlin can number thousands, moving off the shore at high tide onto neighbouring fields and salt marshes. On the breeding grounds, dunlin feed mainly on insects. On coastal mudflats and beaches, they feed on small polychaete worms and small gastropods, as well as small crustaceans and bivalves. Knot, *Calidris canutus*, also form huge flocks in the winter, feeding in large muddy estuaries that include the Thames, Humber, Morecambe Bay and Strangford Lough. Knot are migratory, visiting Britain in winter from their summer breeding grounds in the Arctic. On mudflats, knot probe for unseen, shallow-buried prey with their bills. Their preferred prey are Baltic tellins, which they swallow whole to be broken up in their muscular gizzard. Knot will also take small cockles, mussels and gastropods.

Redshank and greenshank are both species of the genus *Tringa*. Redshank *Tringa totanus* breed in northern England and Scotland, but about half of the British winter population has migrated from Iceland. Redshank breed in salt marshes and coastal meadows, but in winter they can be seen in numbers on estuaries and coastal lagoons. Redshank hunt on intertidal mudflats for molluscs, crustaceans and polychaetes such as the ragworm *Hediste diversicolor*. They often follow the ebbing tide down the beach. The greenshank *Tringa nebularia* is a migratory species, usually heading south for the winter in Africa. Some greenshank breed in northern Scotland, but many of the greenshank seen in Britain are on migratory passage in spring or autumn. A small proportion will overwinter in western Britain, feeding around freshwater lakes but also in estuaries. On the coast, greenshank take small fish such as gobies, as well as shrimps, crabs and polychaetes.

The curlew *Numenius arquata* is Europe's largest wading bird, with a distinctive long, down-curved bill. Curlew breed on grassland and moorlands across Europe, including Wales, northern England and Scotland. Most populations of curlew migrate south to overwinter on the coast, foraging particularly on the mudflats of estuaries. Curlew take larger estuarine prey items than many other waders. They feed on large polychaete worms, as well as crustaceans, gastropods and bivalves such as *M. balthica*. To satisfy its daily energy needs, a curlew requires each day about 1,200 ragworms (*Hediste diversicolor*) of an average dry weight of 0.05 grams (Evans et al., 1987).

It is clear, then, that waders are very important predators of the invertebrates of British estuaries, not least when overwintering. When foraging in metal-contaminated estuaries, waders will inevitably ingest invertebrates with very high accumulated loads of potentially toxic trace metals. Given the migratory nature of so many waders, occupying particular estuaries for variable time periods of the year, it is difficult to interpret whether the usually temporary high trophic input of toxic metals is of ecotoxicological significance for the birds.

6.15.1 Trace Metals in Waders

Prologue. Trace metals are taken up by waders from the diet, but most absorbed metal is excreted with relatively little percentage retention in the body. Routes of excretion include the feathers lost at each moult. Of the trace metals retained, these may be accumulated in the liver, kidney or bones as well as the feathers, proportions differing between metals.

Organolead compounds were of ecotoxicological significance to waders in the Mersey Estuary between 1979 and 1982.

We have discussed aspects of the biology of trace metals in both terrestrial and freshwater birds. The same general principles continue to apply, as we turn to waders feeding in British estuaries. Waders take up trace metals in the diet. Trace metals will then be excreted regularly through the urine and possibly through the secretions of nasal salt glands (Evans et al., 1987). Seasonal excretion will also occur through loss of feathers at moulting, which typically takes place twice a year in waders, in early spring before breeding and in late summer after breeding (Evans et al., 1987). Non-essential trace metals, particularly mercury, are lost to the feathers premoult, except in the case of cadmium, for which accumulated body concentrations increase with the age of the bird (Evans et al., 1987). Female birds will also eliminate trace metals during egg laying (Evans et al., 1987). Trace metals not excreted directly in the urine or passed to the feathers will be accumulated, often in the liver and/or kidney (Evans and Moon, 1981; Evans et al., 1987). Such accumulated concentrations in waders typically increase over the winter as waders feed in coastal habitats, but decrease again by the following winter, after summer moulting and egg laying (Evans et al., 1987).

The percentages of daily intakes of trace metals that are retained and accumulated in waders are very small (Evans et al., 1987). Curlews *Numenius arquata* feeding on ragworms do not digest the nereidid worm jaws with their high zinc contents, but otherwise appear to be efficient assimilators of other trace metals detoxified and accumulated in the bodies of the worms (Evans et al., 1987). As little as 1% of the trace metals ingested by curlews feeding on *Hediste diversicolor* with high accumulated zinc (200 µg/g) and lead (4 to 8 µg/g) concentrations at Teesmouth in the 1970s (Table 6.6) (Evans and Moon, 1981) was retained by the curlews (Evans et al., 1987). While it is possible that some detoxified forms of zinc and lead in the worms had not been assimilated, it does appear that much of this ingested metal had been excreted after assimilation by the curlews (Evans et al., 1987). Similarly, much less than 1% of the daily zinc ingested by dunlin feeding on *H. diversicolor* at Teesmouth was stored in the kidneys and liver (Evans et al., 1987). Again, less than 1% of the zinc ingested daily by knot feeding on mussels at Teesmouth was added to the accumulated zinc loads in the kidneys and liver (Evans et al., 1987).

Redshank *Tringa totanus* in Restronguet Creek feed on the ragworm *Hediste diversicolor* with their huge accumulated concentrations of copper (Table 6.6) (Bryan and Langston, 1992). While copper concentrations in this polychaete diet are one to two orders of magnitude above normal (Table 6.6), the livers of the Restronguet Creek redshank contained approximately 30 µg/g Cu, only about 1.5 times the control level (Bryan and Langston, 1992). Thus, as for zinc and lead, the waders are retaining only a very small proportion of the copper ingested. The excretion of the excess copper assimilated may or may not have a significant energetic cost, with potential ecotoxicological consequences (Bryan and Langston, 1992).

Much has also been written in this chapter about the extremely high bioavailability of cadmium in the Severn Estuary in the 1970s, resulting from the activities of the Avonmouth smelter. This bioavailability also applied through local estuarine food chains to dunlin *Calidris alpina*. Dunlin sampled in February 1981 had significantly higher concentrations of cadmium in both kidneys and liver than did dunlin collected at the same time from Bangor in north Wales (Evans et al., 1987: Bryan and Langston, 1992). Thus cadmium concentrations in the kidneys were between 15 and 20 µg/g dw in the Severn Estuary dunlin, well above those (less than 5 µg/g) in the dunlin from Bangor, and indeed from other control sites including the Wash and the Firth of Forth (Evans et al., 1987). Again, it is not known whether such high accumulated cadmium concentrations in the dunlin were associated with any deleterious ecotoxic effect (Bryan and Langston, 1992).

Dunlin, knot and redshank overwintering in the Wash in the early 1970s all showed similar patterns of increasing mercury concentrations in the liver as the winter progressed (Parslow, 1973). In dunlin, dry weight mercury concentrations in the liver increased from about 2 µg/g in September to November to 7.2 µg/g in January to March (Parslow, 1973). Comparable data for knot were 1.2 µg/g Hg in August to October to 12 µg/g in January to March, and for redshank 1.6 µg/g in August to 12 µg/g in February to March (Parslow, 1973). During the summer, mercury is transferred from the liver to the feathers premoult to be lost on moulting (Parslow, 1973). The same pattern of winter increase of mercury concentration in the liver (and not the kidneys) of dunlin and knot was observed in populations at Teesmouth in the 1970s (Evans et al., 1987). Liver concentrations of mercury in dunlin rose from an autumn low of about 2 µg/g to nearly 5 µg/g ww in February, and those in knot from about 5 µg/g to nearly 30 µg/g ww over this time period ((Evans et al., 1987; Bryan and Langston, 1992). The much higher concentrations of mercury in the livers of the overwintering Teesmouth waders, particularly after allowance for a factor of about 5 for conversion to dry weight concentrations, suggest considerable mercury contamination at Teesmouth at the time (Bryan and Langston, 1992).

Comparisons of absolute concentrations between autumn and winter samples do need to be treated with some caution, because populations of the waders at a site may have changed as a result of migratory movements. Nevertheless, there is much less migratory movement of the waders during the winter itself, and the general pattern of a winter increase followed by a summer fall in mercury concentrations in the livers of waders is well established (Parslow, 1973; Goede, 1985).

Mercury also occurs in some concentration in the muscles of birds. Waders feeding intertidally lie towards the top of aquatic food chains, characterised by the biomagnification of methyl mercury. Thus in a Swedish study in the 1970s, the pectoral muscles of greenshank *Tringa nebularia* had one of the highest average concentrations of mercury (above 0.4 µg/g ww) of a large selection of birds hunted by peregrine falcons *Falco peregrinus* (Lindberg and Odsjö, 1983). Most equivalent muscle concentrations of mercury in the terrestrial birds in the peregrine's diet were 0.1 µg/g ww or lower (Lindberg and Odsjö, 1983).

Lead will be taken up by waders along with other trace metals during feeding. Assimilated lead will then enter the expected physiological pathways of excretion; detoxified storage accumulation (however temporary) in liver, kidneys and bone; and premoult transfer to the feathers.

Organolead, in the form of tetra-ethyl lead, was added to petrol in the twentieth century to prevent preignition, a practice now phased out. Between 1979 and 1982, there were significant mortalities of dunlin *Calidris alpina*, redshank *Tringa totanus* and curlew *Numenius arquata* (as well as black-headed gulls *Larus rudibundus*, several duck species and other birds) in the Mersey Estuary (Bull et al., 1983; Wilson et al., 1986). These mortalities were attributed to the discharge into the Manchester Ship Canal from a local factory manufacturing tetra-ethyl lead additives for petrol (Bull et al., 1983; Wilson et al., 1986). Average alkyl lead concentrations in the livers and kidneys of dead dunlin from the Mersey Estuary were about 10 and 15 µg/g ww as lead, between 50 and 100% of the total lead in these organs (Bull et al., 1983). Dead casualties of waders in general from the Mersey Estuary had liver lead concentrations extending from about 5 to nearly 50 µg/g ww, while live waders from elsewhere had much lower lead concentrations in the liver, from less than 0.1 µg/g to about 8 µg/g ww (Bull et al., 1983). The waders were apparently accumulating the toxic alkyl lead from their invertebrate prey (Wilson et al., 1986). The local Mersey Estuary Baltic tellin *Macoma balthica* and the ragworm *Hediste diversicolor*, for example, had abnormally high concentrations of 2 µg/g ww of lead as alkyl lead (Bryan and Langston, 1992). Furthermore, the toxic symptoms of the affected birds closely resembled those produced by alkyl lead in experimental birds (Bull et al., 1983; Bryan and Langston, 1992). Unusual circumstances may also have contributed to the wader mortalities between 1979 and 1982 (Wilson et al., 1986). A dry summer and low water flows may have increased local dissolved concentrations of emitted alkyl lead compounds (Wilson et al., 1986). An unusually high biomass of estuarine invertebrates and an early arrival of overwintering migrant waders may also have contributed to the progressive ingestion of

toxic loads by the birds (Wilson et al., 1986). As previously noted, there was improved control of industrial discharges towards the end of the twentieth century, and, by the turn of the millennium, the demand for alkyl lead production as a petrol additive had ceased.

Lead pellets from shotguns can cause lead poisoning of birds, either by their presence in the body of the bird after wounding or by accidental ingestion during feeding. This second source is significant for the likes of ducks and geese (Mudge, 1983) but not apparently for waders with their selective methods of feeding on invertebrate prey.

6.16 Community Effects

Prologue. The most ecologically significant effects of the ecotoxicity of high bioavailabilities of trace metals in estuaries are those at the community level.

As in terrestrial and freshwater habitats, many organisms can survive and live in estuaries that contain greatly enhanced bioavailabilities of potentially toxic trace metals. The hugely increased accumulated trace metal concentrations within many of the inhabitants of such metal-contaminated estuaries are testament to these atypically high trace metal challenges.

But, so what? Are these survivors paying a physiological cost of dealing with the toxic challenge? Are they expending so much energy detoxifying and/ or excreting trace metals that they have less to spare for growth and reproduction? From a natural history point of view, has the ecology of the habitat been affected? Are the same organisms present in the same abundance, carrying out to the same degree the ecosystem functions of production, decomposition, nutrient recycling, etc.? Has the community composition changed, and, if so, so what?

The biological changes most likely to be of ecotoxicological significance are those at the highest level of biological organisation, the community. These must inevitably be preceded by ecotoxicological effects in sequence through lower levels of biological organisation, from molecular, to biochemical, to cellular, to physiological, to individual, to population, to community. Following in the same sequence are the minimum size of the toxic metal challenge needed to elicit a response and the ecotoxicological significance of that response. A few trace metal atoms binding in the wrong place on a protein carrying out a metabolic role are of no significance. On the other hand, a host of trace metal atoms interrupting metabolic processes at the molecular, biochemical, cellular and then physiological levels in a metal-exposed individual will have metabolic and energetic costs that may well affect the reproductive potential of that organism. Is there then an effect or not on the population of that organism? The biological responses referred to here can be used as biomarkers. Such biomarkers are most useful if they are sensitive (and thus by definition at a low level of biological organisation), and can be correlated with (and preferably be used to predict) changes at higher levels of biological organisation. It is these higher-level ecotoxic changes that are of particular concern to natural historians.

6.16.1 Case Studies

Prologue. The classic case study of a British estuary under a significant ecotoxicological challenge from high trace metal bioavailabilities is Restronguet Creek, affected by centuries of mining in the Carnon valley. Community effects have been mitigated by the evolution of local copper- and zinc-tolerant populations of estuarine organisms. The major community change apparent in Restronguet Creek is a diminution in the presence and/or abundance of expected bivalves, affected at larval and settlement stages of the life-cycle. Estuaries exposed to industrial effluents have also shown ecotoxicological effects driven by high trace metal bioavailabilities.

Back in the real world of British estuaries, are there case studies that illustrate the presence of significant ecotoxicological effects of trace metal contamination, particularly at higher levels of biological organisation, not least the community level? In estuaries

contaminated by industrial effluents with high toxic metal loads, the picture is often obscured by the almost inevitable copresence of other toxic contaminants, for example hydrocarbon and organochlorine compounds, in both industrial and domestic effluents, before even considering the effects of sewage. It is not surprising, therefore, that we turn to the mining-contaminated estuaries of Southwest England in our quest for trace metal case studies, and inevitably to Restronguet Creek.

Restronguet Creek still contains exorbitantly high concentrations of many trace metals, notably copper, zinc, arsenic, manganese and iron, in its sediments. The atypically very high concentrations of trace metals in the physical components of this estuary have been translated into very high bioavailabilities of these trace metals to a wide range of local organisms (Bryan and Gibbs, 1983). Furthermore, many of the local populations of flora and fauna in Restronguet Creek show tolerance to one or both of copper and zinc, providing evidence that the local bioavailability of each trace metal is of ecotoxicological significance (Luoma, 1977). Restronguet Creek populations of the brown seaweed *Fucus vesiculosus*, the polychaete worms *Hediste diversicolor* and *Nephtys hombergii*, the bivalve *Scrobicularia plana*, the amphipod crustacean *Corophium volutator* and the crab *Carcinus maenas* have all been shown to be tolerant of raised bioavailabilities of copper (Bryan and Gibbs, 1983). The populations of *H. diversicolor* and *C. maenas* have also been shown to be tolerant of raised zinc bioavailabilities (Bryan and Gibbs, 1983). Moreover, the community of nematode worms in Restronguet Creek is also copper tolerant, constituting an example of PICT (Millward and Grant, 1995; Grant and Millward, 1997). And tolerance does come at an energetic cost.

What of changes to the composition of the biological community in Restronguet Creek? It is clear that the presence of metal tolerant populations of many local estuarine organisms (albeit at a physiological cost) will reduce the negative changes in the presence and population abundance of these organisms to be expected at the community level as a direct ecotoxicological response. It is still likely that the energetic costs of tolerance will have an effect on the reproductive output of local trace metal–tolerant populations, with potential effects at the population level. This may affect the Restronguet Creek population of a brooder and local recruiter such as the amphipod *C. volutator*, with consequences for community structure. The same argument may also apply to other invertebrates lacking planktonic larval dispersal, as in the case of *H. diversicolor* (Bryan et al., 1987a). The crab *C. maenas*, on the other hand, has a long period of pelagic larval dispersal, and the parents of the Restronguet Creek crabs live elsewhere. It is unlikely, then, that any reduction in reproductive output of the Restronguet Creek population of crabs will have any effect on future recruitment and therefore crab population levels in the creek.

So, does Restronguet Creek show changes in community composition that might otherwise be expected in the absence of severe trace metal contamination? In fact, yes (Bryan and Gibbs, 1983). A major difference is seen in the absence or low abundance of bivalve molluscs in Restronguet Creek (Bryan and Gibbs, 1983). Bivalves have a planktonic larval dispersal phase, the veliger larva, ultimately succeeded by a pediveliger with a well-developed foot used in settlement. Restronguet Creek does lack the large populations of mussels *Mytilus edulis*, cockles *Cerastoderma edule*, Baltic tellins *Macoma balthica* and (to a smaller extent) peppery furrow shells *Scrobicularia plana*, otherwise to be expected in a non-contaminated estuary (Bryan and Gibbs, 1983; Bryan et al., 1987a). It is possibly during the act of settlement by the pediveliger in contact with the local metal-contaminated sediment that copper and/or zinc toxicity intervenes, bringing larval recruitment and the establishment of an adult population to a halt (Bryan and Gibbs, 1983). The small gastropod *Peringia* (formerly *Hydrobia*) *ulvae* is also conspicuous by its absence from Restronguet Creek (Bryan and Gibbs, 1983). This snail may be particularly affected by high concentrations of dissolved trace metals in the surface waters of the creek (Bryan and Gibbs, 1983). In spite of the presence in Restronguet Creek in the 1970s and early 1980s of a copper-tolerant

population of the burrowing amphipod *Corophium volutator* (Bryan and Gibbs, 1983), sampling in the early 1990s found this amphipod to be all but absent from the creek (Warwick, 2001). Also absent was the carnivorous isopod crustacean *Cyathura carinata*. These two crustaceans would be expected to be present in some abundance in the absence of trace metal contamination (Warwick, 2001).

There is no doubt that the diversity and abundance of the biota living in the Thames Estuary increased greatly over the last three decades of the twentieth century, as a result of improvements in water quality associated with industrial effluent control and improved sewage treatment (Andrews and Rickard, 1980; Andrews, 1984). Trace metal pollution contributed to the desperate state of the fauna and flora of the Thames Estuary in the middle of the last century, but it is impossible to pick out the specific ecotoxicological effects of high trace metal bioavailabilities on the local community composition. Trace metal bioavailabilities in the Thames Estuary fell greatly over the three decades leading to the turn of this century (Langston et al., 2004) and have continued to fall over the first decade or so of the twenty-first century (Johnstone et al., 2016). Nevertheless, levels of several trace metals, specifically silver, copper and zinc, in the estuary in 2001 did raise the possibility of sublethal impacts on the reproduction, recruitment and development of some local estuarine organisms (Langston et al., 2004).

High bioavailabilities of toxic metals in the 1970s and 1980s can be considered to have been major ecotoxicological drivers of the poor diversity and low abundance of the fauna of the Severn Estuary in the 1970s and 1980s (Langston et al., 2010). By the turn of the millennium, major industries such as the Avonmouth smelter had closed, and stricter effluent contaminant controls had been introduced (Langston et al., 2010). The bioaccumulation of trace metals (except chromium) by local estuarine biomonitors had correspondingly decreased by the first decade of the twenty-first century as bioavailabilities fell (Langston et al., 2010). The high dynamic energy of the Severn Estuary system makes it difficult, however, to link directly high trace metal bioavailabilities, as reflected in bioaccumulated concentrations in biomonitors, with associated ecotoxicological consequences. There is good evidence, nevertheless, of the recovery of diversity and abundance of the benthic macroinvertebrate fauna in the middle part of the Severn Estuary over recent years, as trace metal bioavailabilities have fallen (Langston et al., 2010).

We have seen how high dissolved bioavailabilities of zinc, originating in local industrial effluent, were the probable cause of the mortality of oyster larvae in the early 1970s at the mouth of Holes Bay in Poole Harbour (Boyden, 1975). Surveys were made of the macroinvertebrates living in the intertidal mudflats of Holes Bay in 1972, 1987, 1991 and 2002 (Caldow et al., 2005). Benthic invertebrate numbers remained fairly consistent over three of the decades, but numbers of all invertebrate groups were much reduced in the 1980s (Caldow et al., 2005). Holes Bay received many different forms of contaminants of industrial origin in the 1970s and 1980s. It remains very difficult, therefore, to conclude with any certainty that the fall in faunal diversity in the 1980s can be attributed to the high bioavailabilities of trace metals in Holes Bay at the time. But, they may well have been a factor.

6.16.2 Weight of Evidence

Prologue. A weight of evidence approach should be taken when attempting to identify high trace metal contamination as a cause of ecotoxicity in an estuary.

How, therefore, can we identify high trace metal contamination as a cause of an observed or suspected impoverishment of the diversity and abundance of the local flora and fauna in a specific estuary? Almost inevitably, in industrially contaminated estuaries, other toxic contaminants are likely to be present, including hydrocarbons (oil) and other organic compounds such as organochlorine pesticides or some of the newly emerging persistent organic pollutants (Phillips and Rainbow, 1994).

An important step forward in the development of the methodology to be used in the assessment of the potential ecotoxicological significance of the chemical contamination of a coastal ecosystem was the Sediment Quality Triad (SQT), proposed in the early 1980s (Chapman and Long, 1983; Long and Chapman, 1985). The basis of the SQT is that the assessment of sediment quality in, for example, an estuary must involve at least three measures: (i) sediment concentrations of toxic chemicals (in our case, trace metals); (ii) testing of the toxicity of sediment samples; and (iii) investigation of the community composition of the resident biota, preferably the infauna (Chapman and Long, 1983; Long and Chapman, 1985). The SQT survives today as part of a usually wider framework of measures (Chapman and Hollert, 2006), albeit often after some modification. As we have seen repeatedly in this book, the analysis of trace metal accumulated concentrations in a suite of biomonitors is more biologically informative than trace metal concentrations in sediments or water samples (Phillips and Rainbow, 1994; Luoma and Rainbow, 2008). Furthermore, toxicity testing is more environmentally relevant if moved from the laboratory into the field, for example in mesocosms (Luoma and Rainbow, 2008; O'Brien and Keough, 2013).

The Sediment Quality Triad can be seen as the beginnings of a weight of evidence framework in an ecological risk assessment (Chapman and Hollert, 2006). The weight of evidence (WOE) approach to ecological risk assessment can be defined as the process of combining information from multiple lines of evidence (LOE) to determine the ecotoxicological status of a contaminated habitat (Luoma and Rainbow, 2008; Chapman, 2007). LOE will usually consist of both chemical and biological measurements, including field evidence which may be observational or involve experimental manipulation (Chapman, 2007). Crucial to any assessment will be the choice of appropriate control sites for comparison (Amiard-Triquet and Rainbow, 2009). Recommended LOE for an assessment of the ecotoxicological status of an estuary contaminated by trace metals are summarised in the following subsections (Chapman, 2007; Amiard-Triquet and Rainbow, 2009).

6.16.2.1 Biomonitors

Prologue. A suite of biomonitors provides information as to whether a particular estuary has high or low trace metal bioavailabilities. Accumulated concentrations of trace metals in hardy biomonitors have potential to be used as ecological indicators of metal ecotoxicity at the community level.

The accumulated trace metal concentrations in appropriate biomonitors at different locations, or at the same location at different times, are direct relative measures of the local total bioavailabilities of each trace metal at those locations. Strictly, these measures refer to the bioavailabilities to that particular biomonitor. Nevertheless, the use of a suite of biomonitors allows the investigation of local trace metal bioavailabilities from different sources, such as solution, suspended material or deposited sediments (Phillips and Rainbow, 1994; Luoma and Rainbow, 2008). Good comparative databases from the literature, expanded in this chapter, allow us to interpret what is a low or high local bioavailability of a potentially toxic trace metal. Biomonitors, therefore, inform us whether a particular site has a high or low trace metal bioavailability and in what component of the habitat. It needs to be stressed again that biomonitors do not provide evidence of an ecotoxicological effect in action but are a valuable first step in the assessment of the presence of high trace metal bioavailabilities in an estuary that might potentially be ecotoxic.

What might constitute an appropriate suite of biomonitors in assessing trace metal bioavailabilities in a British estuary suspected to be contaminated by trace metals to an ecotoxicologically significant degree? For dissolved trace metal bioavailabilities, use a seaweed such as bladder wrack *Fucus vesiculosus*. Suspension feeders, such as the mussel *Mytilus edulis*, the oyster *Crassostrea gigas* and the barnacle *Amphibalanus improvisus*, will provide data on the bioavailabilities of trace metals in both solution and in suspended material, including plankton, detritus and resuspended sediment. Different biomonitors will sample different size ranges of suspended material, with different relative dependences on water and suspended food as trace metal sources. To turn to the sediment, the surface deposit feeding bivalves *Scrobicularia plana* and *Macoma balthica* will provide information on the trace metal bioavailabilities of recently deposited surficial sediments. The ragworm *Hediste diversicolor* and the lugworm *Arenicola marina* will sample deeper sediments, of different ages of deposition in a sheltered estuary with limited redistribution by tidal or wave action. Like the suspension feeders, these deposit feeders will rely to different extents between species and trace metals on different trace metal sources, such as solution (water column and/or pore water) and ingested sediment. Analyses of accumulated trace metal concentrations in other estuarine organisms can provide further information, for example on the local trophic transfer of toxic metals to herbivores (periwinkles), detritivores (strandline talitrid amphipods) or predators (neogastropod molluscs, decapod crustaceans or fish).

It is important not to be seduced by the concept of a critical body concentration of a trace metal in an organism indicating incipient lethality (Adams et al., 2010; Rainbow and Luoma, 2011b), a popular concept that can be applied with success to accumulated concentrations of organic contaminants. In the presence of any detoxified storage of a trace metal, there is no toxicologically critical concentration of total metal accumulated in an organism. Much of the accumulated trace metal is in detoxified form and is not metabolically available (Rainbow and Luoma, 2011b). It is the concentration of metabolically available metal that has not been detoxified that can reach a critical threshold level and cause toxic effects (Rainbow and Luoma, 2011b). This critical threshold of metabolically available metal is reached more quickly (and consequently at a lower total accumulated concentration) when the rates of metal uptake into the organism are increased by increased local bioavailability of the trace metal (Adams et al., 2010; Rainbow and Luoma, 2011b). Presently, as yet, it is not possible to link with any confidence the theoretical concept of a pool of metabolically available metal with a biochemically identifiable fraction of accumulated trace metal in an organism (Rainbow and Luoma, 2011b; Rainbow et al., 2015). When this is achieved, it will then be possible to measure the concentrations of this accumulated fraction in field-collected specimens and draw conclusions on their ecotoxicological significance for the species concerned.

On the other hand, accumulated concentrations of trace metals in hardy biomonitors have the potential to be used as ecological indicators of metal eco-toxicity on other species in the expected estuarine community. It was shown in the preceding chapter that a threshold accumulated trace metal concentration in a biomonitor can be used as a bioaccu-mulated metal guideline, an indicator of a trace metal bioavailability sufficiently high to cause an ecotoxic effect at the community level on other more metal-sensitive members of the fauna and flora. The same principle awaits extension to an estuary. Suitable metal-hardy biomonitors might include the ragworm *Hediste diversicolor* and the peppery furrow shell *Scrobicularia plana*, towards the top of an estuary, and the likes of the seaweed *Fucus vesiculosus*, the talitrid amphipod *Orchestia gammarellus* and the mussel *Mytilus edulis* lower down.

6.16.2.2 Environmental Quality Standards

Prologue. Environmental Quality Standards (EQS) are numerical concentrations which may be guidelines or legally enforceable environmental regulations. EQS have limitations as valid arbiters of the ecotoxico-logical significance of trace metal bioavailabilities in estuaries. Nevertheless, a comparison of local meas-ured trace metal concentrations against EQS, where available, may be a valid first step in any environ-mental assessment process.

The original Sediment Quality Triad recommended the measurement of sediment concentrations of trace metals (Chapman and Long, 1983; Long and Chapman, 1985). Such total concentrations do not measure bioavailable concentrations, even after vari-ation in the use of different chemical extractants, such as weak hydrochloric acid, in an attempt to model bioavailable sediment concentrations. The ecotoxico-logical information to be gleaned from trace metal

sediment concentrations, total or extracted by some chemical agent, is not, therefore, as strong as that from an appropriate biomonitor. The same conclusion can be drawn for information to be derived from a dissolved concentration measured in the water column of an estuary.

As we saw in the freshwater environment (Chapter 5), environmental regulators do pine for single number comparisons. We are back to EQS, numerical concentrations which may be guidelines or legally enforceable environmental regulations. EQS relevant here may be either Water Quality Standards or, if ever introduced in the United Kingdom, Sedi-ment Quality Guidelines. In EU countries, under the Water Framework Directive (WFD), any water body needs to have good chemical status, in addition to meeting ecological objectives, before it can achieve overall Good Water Status. And the WFD uses ambi-ent EQS to classify water bodies on the basis of chemical contamination. So, estuarine dissolved and sediment trace metal concentrations are often routinely measured and available. Should we ignore them com-pletely? It would be perverse to do so. So, while being well aware of the limitations of EQS as valid arbiters of the ecotoxicological significance of trace metal bioa-vailabilities, a comparison of local measured trace metal concentrations against EQS, where available, will still be a valid first step in any screening process. Such comparisons will, for example, allow the identification of habitats with very low total concentrations, and, therefore, low maximum potential bioavailabilities. Exceedance of an EQS will highlight the need for further in depth biological assessment.

Table 6.30 lists Water Quality Standards for dis-solved trace metals in UK estuaries, as defined by various EU directives (WFD-UKTAG, 2008b). Since 2008, there has been an increasing realisation amongst government regulators of the lack of direct correlation between total dissolved trace metal con-centrations and their dissolved bioavailabilities. Hence, later proposed WQS for copper have con-sidered the amount of dissolved organic carbon (DOC) present that will complex dissolved copper and reduce its dissolved bioavailabilities (WFD-UKTAG, 2013). Copper WQS have been altered accordingly for high

Table 6.30 **Water Quality Standards (WQS) for dissolved (<0.45 μm) trace metals (μg/L) in UK estuaries, defined as an annual average, for the protection of saltwater life (after EU Directive 2013/39/EU {amending Directives 2000/60/EC and 2008/105/EC}; Directive 76/464/EEC). Also presented are possible marine Sediment Quality Guidelines for trace metal concentrations (μg/g dry weight), as Interim Marine Sediment Quality Guidelines (ISQG) of Environment Canada (1999) and associated Probable Effects Levels (PEL).**

	Water Quality Standard	Sediment Quality Guideline	
	(μg/L)	(μg/g)	
		ISQG	PEL
Arsenic	25	7.24	41.6
Cadmium	5	0.7	4.2
Chromium	15	52.3	160
Copper	5	18.7	108
Lead	25	30.2	112
Mercury*	0.5	0.13	0.70
Nickel	30	15.9	42.8
Zinc	40	124	271

Note: *inorganic mercury.
Sources: WFD-UKTAG (2008b), Cole et al. (1999); Canadian Council of Ministers of the Environment (1999b).

and low DOC conditions in saltwater environments (WFD-UKTAG, 2013). Further changes have also introduced an ambient dissolved background concentration of zinc into the calculations of a local zinc WQS (WFD-UKTAG, 2013). Such changes recognise the inadequacy of dissolved concentrations of trace metals as consistent measures of their dissolved bioavailabilities, reinforcing the case for the use of biomonitors. As to be expected, many of the dissolved trace metal concentrations of the estuaries highlighted in this chapter (Table 6.1) exceed the WQS listed in Table 6.30, often by huge margins.

There are no equivalent Sediment Quality Standards in the United Kingdom, although these often come under discussion. A possible approach would be to follow United States and Canadian examples, which attempt to derive Threshold Effect Levels (TEL) and Probable Effect Levels (PEL) from published databases (Cole et al., 1999; Canadian Council of Ministers of the Environment, 1999b). Table 6.30, therefore, also lists Interim Marine Sediment Quality Guidelines (ISQG), issued by Environment Canada in 1999, and corresponding PEL. Again expectedly, sediment concentrations of many trace metals in the estuaries discussed in this chapter (Table 6.4) far exceed both the ISQG and the PEL listed in Table 6.30.

In the case of the Thames Estuary, trace metal concentrations measured in sediments sampled at 27 sites along the estuary in 2001 can be compared against the values given in Table 6.30 (Langston et al., 2004). The numbers of these 27 sites at which the trace metal sediment concentration exceeded the ISQG were 21 (As), 1 (Cd), 15 (Cr), 22 (Cu), 25 (Hg), 18 (Ni), 22 (Pb) and 18 (Zn). The equivalent numbers of sites exceeding the PEL were 0 (As, Cd, Cr), 2 (Cu), 9 (Hg), 1 (Ni), 8 (Pb) and 2 (Zn). It is quite likely that several of these sediments do contain trace metal concentrations that are ecotoxic, but probably not to the extent indicated. Nevertheless, these brief comparisons have not been standardised for factors, such as sediment organic content or iron oxide content, which will affect the bioavailabilities of the sediment-associated trace metals. It needs to be reinforced, then, that comparison of a measured sediment concentrations of a trace metal at a site for comparison against any agreed SQG can only be a first step to recognise a situation worthy of further ecotoxicological investigation.

6.16.2.3 Biomarkers

Prologue. Biomarkers are biological response (e.g., a biochemical, cellular, physiological or behavioural variation) that can be measured at the lower levels of biological organisation in tissues or body fluids or at the level of the whole organism. Links have been established in estuarine organisms between sensitive low-level biomarkers and biomarkers at higher levels of biological organisation.

What measures then can provide information on the 'so what' question? What are the measures that tell us that a raised trace metal bioavailability is causing a biological response, and whether that response might be of ecotoxicological significance? The answer lies with biomarkers (Amiard-Triquet and Rainbow, 2009; Amiard-Triquet et al., 2013).

Many biomarkers of trace metal pollution are now being employed routinely in estuaries. Many such studies address a wider range of ecotoxic contaminants than just trace metals, and most biomarkers in use typically respond to more than one stressor, being indicative of the general health status of an organism. An ideal biomarker of trace metal ecotoxicology would be trace metal specific, as in the case of a decreased activity of the enzyme ALAD in the blood of vertebrates in response to lead exposure. Such specific biomarkers of trace metal effects, however, are all but absent. Even the induction of the trace metal detoxifying protein metallothionein can be caused by factors other than high trace metal exposure (Amiard et al., 2006). Nevertheless, less specific biomarkers are still of considerable value in a trace metal ecotoxicological assessment, especially when used in combination with other LOE, such as biomonitoring data, in a WOE study. A biomarker is of particular importance in an ecotoxicological risk assessment if it is sensitive (and therefore detectable at low levels of biological organisation) and, if links can be established between this sensitive biomarker and consequent ecotoxicological effects at higher levels of biological organisation, ultimately up to the population and community levels.

The estuarine organisms commonly used for the measurement of biomarkers are the ragworm *Hediste diversicolor*, the mussel *Mytilus edulis*, the peppery furrow shell *Scrobicularia plana* and the common shore crab *Carcinus maenas* (Amiard-Triquet and Rainbow, 2009; Amiard-Triquet et al., 2013).

At the biochemical level, a biomarker of relevance to estuarine trace metal contamination is the trace metal-binding protein metallothionein. The point has been made earlier that MT are turned over rapidly in cells. Thus, an increased rate of MT synthesis on trace metal exposure may be associated with an increased rate of MT turnover but not necessarily with an increased standing stock concentration of MT (Amiard et al., 2006). It is, however, easier to measure MT concentrations than turnover rates in organisms collected in the field. A lack of an increase in MT concentration in an organ or a whole organism should not automatically be interpreted as a lack of MT induction. Nevertheless, the MT concentrations in some organs of some organisms do increase with trace metal exposure and are used as biomarkers of high trace metal exposure. Such organs include the gills of the oyster *Crassostrea gigas* (Geffard et al., 2002) and the liver of the eel *Anguilla anguilla* (Langston et al., 2002).

A particular valuable trace metal biomarker at the biochemical level, specifically for high lead bioavailability, is the activity of ALAD in the blood of fish such as flounder *Platichthys flesus* (OSPAR Commission, 2007).

Trace metals, amongst other contaminants, can catalyse the generation of oxyradicals, also referred to as reactive oxygen species (ROS) (see Chapter 3). So, measures of antioxidant defences in estuarine organisms are relevant biomarkers at the biochemical level, although not specific to trace metal exposure (Amiard-Triquet et al., 2013). Two primary antioxidant enzymes measured as biomarkers are catalase and superoxide dismutase. Glutathione is an oxyradical scavenger, while it is also able to bind trace metals in cells. The concentration of glutathione is, therefore, often included in a battery of biomarkers, measured, for example, in mussels. Another measure of antioxidant defence is the TOSC, again often measured in mussels. If not controlled, ROS can cause lipid peroxidation, and another biomarker is the concentration of malondialdehyde, a breakdown product of lipid peroxidation (Amiard-Triquet et al., 2013).

At the cellular level, several biomarkers of toxic metal exposure involve lysosomes, often studied in the digestive gland cells of mussels (Moore et al., 2006, 2013). Several responses of lysosomes to stress in these cells are used as biomarkers. These include changes in lysosomal size and number (Izagirre and Marigómez, 2009), production of lipofuscin (an end product of the oxidation of cell components) and the destabilisation of the lysosomal membrane through the action of ROS (Moore et al., 2006, 2013). The functional stability of the lysosome membrane, and hence its permeability, changes with degree of exposure to trace metals (Moore et al., 2006). Lysosomal membrane stability is easily measured by the neutral red retention method. Lysosomal stability is a sensitive, low organisational-level biomarker, and, desirably, its quantification has been correlated with biomarkers at other levels of biological organisation, including TOSC, in mussels (Figure 6.21) (Moore et al., 2006, 2013).

ROS generated by high trace metal exposure can damage DNA, and trace metals may also bind directly to DNA. Both cause genotoxic effects (see Chapter 3) (Amiard-Triquet et al., 2013). A segment of DNA bound to a trace metal represents a piece of damaged DNA. The comet assay is a common biomarker of the early stages of damage to DNA, identifying the presence of broken DNA fragments in individual cells of contaminant-exposed organisms. Another manifestation of genotoxic damage is the presence of micronuclei displaced from the main nucleus of the cell. An increased frequency of micronuclei has also been correlated with decreased lysosomal stability in mussels (Moore et al., 2013). Blood cells of bivalve molluscs such as mussels and cockles, or of crustaceans such as crabs, are suitable cells in which to measure these genotoxic biomarkers.

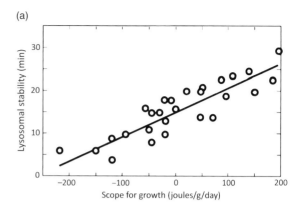

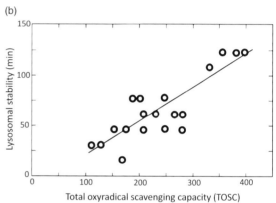

Figure 6.21 Significant correlations of biomarkers from low to higher levels of biological organisation. (a) Between scope for growth and lysosomal stability in the mussel *Mytilus edulis*. (After Allen and Moore, 2004, with permission.) (b) Between lysosomal stability and TOSC in mussels. (After Moore et al., 2006, with permission.)

While some damage to DNA is reparable, ultimately genotoxic effects extend to significant damage to chromosomes. This causes the interruption of cell division processes, perhaps leading to histopathological changes and ultimately to tumour formation. Histopathological changes in tissues can, therefore, also be used as biomarkers (Au, 2004; Amiard-Triquet et al., 2013). Appropriate tissues for inspection are the gills and digestive glands of bivalves, including mussels, oysters and clams, and the livers and gills of flatfish such as flounder (Au, 2004).

Long-term trace metal exposure can affect the genetic profiles of populations of organisms living in contaminated estuaries (Phillips and Rainbow, 1994). In years gone by, protein separation techniques, such as electrophoresis, would have been used to show that the proteins (usually enzymes) produced by the same gene in different populations of an organism were different, indicating that one population had been subjected to a selection pressure such as trace metal exposure. Not today, when the study of the genome, and its derivatives the proteome, the transcriptome and the metabolome, are all but routine. In the ecotoxicological assessment of estuaries, oysters have many advantages as the model organisms of choice in this modern world, not least as a result of the description of the oyster genome (Zhang et al., 2012). The transcriptome of oysters reveals a large set of genes responding to environmental stress (Zhang et al., 2012), including trace metal exposure (Liu and Wang, 2012). Furthermore, trace metal exposure causes changes to the proteomes of oysters in a dose-related pattern along a gradient of trace metal contamination (Liu and Wang, 2012). 'Omics' certainly have parts to play as LOE in a WOE monitoring programme.

Without resort to a study of the whole genome of organisms, molecular biology can still provide other valuable LOE. It is important not to ignore the microbial world in any WOE programme. A bacterial gene commonly assessed in sediments potentially contaminated by trace metals is one conferring resistance to mercury exposure. The mercuric reductase (merA) gene encodes an enzyme that reduces the toxic Hg^{2+} ion into the volatile and relatively inert form of elemental mercury. Often associated with merA is merB, a gene encoding an additional enzyme that degrades toxic organomercury compounds.

Contaminant exposure can cause negative changes to the physiological processes of an organism living in a trace metal contaminated estuary, and these will clearly affect the amount of energy available for growth and reproduction (Amiard-Triquet et al., 2013). Very simple measures such as condition indices may be informative (Luoma and Rainbow, 2008). In bivalves, the condition index is the ratio of soft tissue weight to shell weight, shell volume or shell length. In the estuarine bivalve *Macoma balthica*, there is a negative correlation between the condition index and accumulated copper concentration and hence local copper bioavailabilities (Hummel et al., 1997).

Similarly, there were negative correlations between dry matter content and accumulated concentrations of several trace metals, including cadmium, chromium, lead, nickel and zinc, in mussels *Mytilus edulis* from coastal sites in Northern Ireland (Gault et al., 1983). Condition indices of *Scrobicularia plana* in the Thames Estuary in 2001 showed negative correlations with accumulated concentrations of cadmium, copper, mercury, lead, silver, tin and zinc (Langston et al., 2004).

A more comprehensive biomarker of the physiological health of an organism is that of scope for growth (SFG). SFG integrates measures of several physiological processes to estimate the surplus energy available to an animal for growth and reproduction. The use of SFG was pioneered in the mussel *Mytilus edulis*, making it a most suitable model organism in coastal waters (Widdows et al., 1995, 2002). As in the case of many other biomarkers, SFG will respond to raised bioavailabilities of all toxic contaminants present. Trace metals will only be part of this cocktail in any industrialised estuary. Nevertheless, in estuaries affected only by mining contamination, SFG can provide valuable information on depressions in the physiological status of the local mussels caused by trace metal exposure. Furthermore, a correlational link has been established between lysosomal membrane stability and SFG in mussels, achieving the desirable objective of linking biomarkers at lower and higher levels of biological organisation (Figure 6.21) (Moore et al., 2006, 2013).

Other physiological measures are also available as biomarkers in estuarine organisms. These include levels of energy reserves (glycogen, lipid, protein) and length–weight relationships in the ragworm *Hediste diversicolor* (Durou et al., 2007a, b; Gillet et al., 2008; Mouneyrac et al., 2010; Fossi-Tankoua et al., 2012). Higher levels of energy reserves indicate a better physiological status of the ragworm and can be correlated with the number of eggs per female as a measure of fecundity (Durou et al., 2007a; Mouneyrac et al., 2010). Thus impaired physiological status can be linked directly to reduced reproductive output in *H. diversicolor* (Durou et al., 2007a;

Mouneyrac et al., 2010). There are corresponding negative effects of estuarine sediment contamination on the biomass, population structure and population density of the ragworms, linking biomarkers across these higher levels of biological organisation (Durou et al., 2007a; Gillet et al., 2008; Mouneyrac et al., 2010).

Changes in behaviour are integrative responses at the level of the individual that have links both to biochemical (neurotoxicity) and physiological impairment and effects at the population level (Amiard-Triquet and Amiard, 2013). For example, changes in feeding rate affect energy balance and potentially fecundity, while a reduced locomotion rate may affect predator avoidance or, contrarily, prey capture (Amiard-Triquet and Amiard, 2013). Burrowing speed and feeding rates in *H. diversicolor* both decrease with sediment contamination (Fossi-Tankoua et al., 2012). Similarly, the burrowing speed of the peppery furrow shell *Scrobicularia plana* also falls with sediment contamination (Boldina-Cosqueric et al., 2010). Bivalve molluscs will close their valves in the presence of high dissolved concentrations of copper, with consequences for their feeding rate.

6.16.2.4 Tolerance

Prologue. The demonstration that the local population of an estuarine organism has increased tolerance to exposure to a particular trace metal is evidence that the local bioavailability of that trace metal is of ecotoxicological significance, at least to that species and by extrapolation to other species.

The identification of tolerant populations of one or more species in a particular estuary to a trace metal is an important LOE in a WOE approach to the assessment of ecological risk in that estuary.

As we have seen, local populations of several species, including *Hediste diversicolor*, living in Restronguet Creek show tolerance to one or both of copper and zinc (Bryan and Gibbs, 1983), providing evidence that the local bioavailability of each trace metal is of ecotoxicological significance. Another

population of *H. diversicolor* shown to exhibit copper tolerance in the 1970s was that in the Hayle Estuary (Bryan, 1976).

The same argument can be extended to the community level. It can, therefore, be proposed that the selection pressure associated with an ecotoxicologically significant metal bioavailability will lead to an increased average tolerance to that metal amongst all species in the local biological community. PICT is therefore another LOE in assessing the ecotoxicological effects of a trace metal on estuarine biological communities. Indeed, the community of nematode worms in Restronguet Creek is copper tolerant (Millward and Grant, 1995; Grant and Millward, 1997). PICT studies are particularly appropriate for microbiological communities, including those in estuarine sediments. A study of the microbial community in the sediments of the Hayle Estuary, the Fal estuarine system including Restronguet Creek and a control site in the Kingsbridge Estuary, Devon, was carried out in 2002 to investigate the relationship between copper concentrations in sediment pore water and microbial activity, measured by the microbial incorporation of the amino acid leucine and of thymidine, a constituent of DNA (Ogilvie and Grant, 2008). There was a clear dose response relationship between the sediment concentration of copper and microbial community tolerance to copper exposure (Ogilvie and Grant, 2008).

6.16.2.5 Toxicity Tests

Prologue. Laboratory-based acute toxicity tests using model organisms provide limited information on the potential ecotoxicity of high bioavailabilities of trace metals in contaminated estuaries. Non-acute toxicity testing carried out in the field has more ecotoxicological relevance.

It seems obvious that the toxicity of a sediment under investigation should be tested and confirmed. In practice, this is not as easy as it sounds. Classically, sediment samples would be brought back to the laboratory. Then selected model species would be exposed to these sediments, or often elutriates of these sediments, for standardised periods of time under standardised exposure conditions (Benedetti et al., 2011).

But how suitable are these model organisms? Indeed, several of them do not even live in estuaries. The quest for standardisation has often led to the use of model organisms that are simply ecologically inappropriate. For example, the bioluminescent bacterium *Aliivibrio fischeri* is an attractive test organism for use in whole sediment toxicity tests (Benedetti et al., 2011), because changes in its luminescence are progressive toxic responses and easily measured. Yet *A. fischeri* is widespread at low abundance in marine (often oceanic) planktonic environments, and it is found predominantly in association with the light organs of marine animals such as deep sea squid.

Furthermore, many estuarine and other coastal sediment toxicity tests are carried out using sediment elutriates and planktonic animals (Chapman et al., 1992; Benedetti et al., 2011). The action of a sediment elutriate on a model organism living in the water column bears little resemblance to the real world of sediment ecotoxicology, in which routes of trace metal uptake (and therefore potential toxicity) may be dominated by trophic uptake after sediment ingestion. Classic models for such sediment elutriate toxicity tests include the planktonic veliger larvae of the oyster *Crassostrea gigas* and the nauplius larvae of copepods *Acartia tonsa* and species of *Tigriopus* (Chapman et al., 1992; Matthiessen et al., 1998; Benedetti et al., 2011). These model organisms are at least able to live in estuaries, but not in association with sediment. *A. tonsa* is a planktonic calanoid copepod that is found in coastal waters, including estuaries. Species of the harpacticoid copepod genus *Tigriopus* are common inhabitants of high intertidal rockpools. While not being estuarine, they are able to withstand a wide range of salinities and temperatures and are easily reared over generations in the laboratory. Another test of sediment toxicity based on sediment elutriate measures the sperm fertilisation capability of marine sea urchins (Benedetti et al., 2011), such as *Echinus esculentus* or *Psammechinus miliaris*, rarely encountered even in the outer regions of estuaries.

More environmentally relevant sediment toxicity tests involve testing the survival or burrowing behaviour of invertebrates that do at least live in marine sediments, including intertidal estuarine sediments (Chapman et al., 1992; Benedetti et al., 2011). Such tests are 10-day survival and reburial tests using the North American amphipod crustacean *Rhepoxynius abronius* or, in Britain, the amphipod *Corophium volutator* (Chapman et al., 1992). Another standardised North American test is a 20-day survival and growth test using the nereidid polychaete *Neanthes arenaceodentata* (Chapman et al., 1992). Substitute British test polychaetes would include the ragworm *Hediste diversicolor*, the king ragworm *Alitta virens* and the lugworm *Arenicola marina* (Matthiessen et al., 1998; Casado-Martinez et al., 2010a). A 20-day test of direct sediment toxicity is much closer to the real world of sediment ecotoxicology than many of the other standardised tests listed previously.

Some toxicity tests involve manipulation of the sediments collected from the field, be it from the site under investigation or from a chosen control site. For example, control reference sediments may be spiked by the addition of increasing concentrations of the trace metal under suspicion, in an attempt to match field concentrations at the contaminated site and reproduce their toxic effect, if any, on test organisms (Chapman, 2007). Spiking immediately introduces complications. Spiking times of more than a month (at least) will be needed to bring the added trace metal load into equilibrium with all the trace metal binding sites in the sediment. Even then, there is no guarantee that the spiked sediment will bind the added trace metals in the same physicochemical forms as the contaminated sediments, with consequences for the bioavailability, and hence the toxicity, of the high concentrations in the spiked sediment. A better approach would be to mix sediments from the top and bottom of a contaminated estuary, expectedly with relatively higher and lower concentrations of contaminant metals. This will create a concentration gradient of contaminant metal concentrations in the sediment, with the minimum possible change in the physicochemistry of the sediment metal binding sites along this gradient (Casado-Martinez et al., 2010a).

Another variation of sediment manipulation is to be found in toxicity identification evaluations (TIE), which attempt to determine the chemical causing toxicity (Chapman, 2007). Specific contaminants are chemically removed from a sediment sample and then added back. Sediment toxicity tests are carried out after removal, and after replacement, of the suspected contaminant (Chapman, 2007).

So what to do? What is a protocol of sediment toxicity tests that has environmental relevance and would provide ecotoxicological data of value as an LOE in a WOE study? One approach would be to make up a series of experimental sediments, consisting of (a) the most contaminated sediment samples from the suspect estuary, (b) the least contaminated sediments of the same physicochemical characteristics (e.g., grain size, organic content, etc.) in the same estuary, (c) variable mixtures of these two to create a sediment concentration gradient and (d) sediment from a control estuary of similar physicochemical characteristics. One or more estuarine sediment-dwelling test invertebrates should be used, for example *Hediste diversicolor* and *Corophium volutator*, collected from the control site. This protocol could even be extended to include members of the population of the test invertebrate from the contaminated estuary to seek evidence for local tolerance. Time of exposure should be at least 20 days.

A laboratory microcosm system is appropriate for sediment toxicity testing at the microbial or meiofaunal scale (Austen and Somerfield, 1997). Sediment samples from sites in the Fal Estuary system covering a gradient of sediment trace metal concentrations were defaunated by freezing and then incubated with sediment meiobenthic communities from a control site (Austen and Somerfield, 1997). Changes in the community structure of the meiofaunal nematodes were related to sediment trace metal concentrations (Austen and Somerfield, 1997). Moreover, the changes in community structure observed in the laboratory microcosms were similar to those observed in the creeks of the Fal Estuary system with different sediment trace metal concentrations (Austen and Somerfield, 1997).

While the laboratory offers more easily controlled experimental conditions, the use of field mesocosms in sediment toxicity testing would further increase environmental relevance (O'Brien and Keough, 2013).

Such an approach might involve the transplantation of mesocosms containing defaunated sediments from the suspect estuary and from a control site, back into the site of origin and into the other estuary (O'Brien and Keough, 2013).

6.16.2.6 Community Structure

Prologue. Multivariate statistical techniques can be used to analyse the structure of biological communities in estuaries affected or not by high trace metal bioavailabilities. Such studies are labour-intensive and require considerable taxonomic expertise that is often not available. The use of environmental DNA has potential for efficient studies of the community structure of estuarine habitats.

Ultimately, the biological changes caused in an estuary by exposure to toxic metals are most ecotoxicologically significant if they occur at the community level, resulting in a depleted biota. Such community-level changes are reflected in changed abundances of the organisms present, in the loss of species and, therefore, biodiversity. Affected biota may be keystone species in the local biological community, with major knock-on effects for community structure and ecological functioning.

The statistical analysis of community structures is not easy, as discussed in Chapter 3, and illustrated further in Chapter 5 in the freshwater environment. The same communities of benthic macroinvertebrates, benthic diatoms and meiofauna are often those addressed in surveys of estuaries. Classically, such analyses involve many man-hours of collecting and sorting, and considerable taxonomic expertise and time subsequently to come up with a database suitable for multivariate statistical analysis. The use of environmental DNA for estuarine sediments to identify and estimate the relative abundance of taxa of interest, whether bacteria, benthic diatoms, meiofauna or infaunal macroinvertebrates, will save time and resources (Taberlet et al., 2012). Any such routine use, however, still lies in the future

When analysing the structures of biological communities, we have seen a change from the use of single number (univariate) indices to multivariate statistical techniques such as principal components analysis (PCA) or multidimensional scaling (MDS) (see Chapter 3). Univariate parameters, such as species richness or a diversity index, reduce details of community composition to a single number with an inevitable loss of information. Multivariate techniques still reduce a large original dataset to a small set of derived variables but retain more of the relevant information (Sparks, 2000). Multivariate statistical techniques are better at identifying those sites which are alike or different from each other. Thus, even subtly changed biological communities at sites with ecotoxicologically significant trace metal contamination are separable from a background of expected communities at control sites. For example, the aforementioned study of different meiofaunal nematode communities in sediments of different trace metal concentrations used MDS to determine differences in the structures of these different communities (Austen and Somerfield, 1997).

A classic study of the effects of trace metal contamination of coastal sediments on the community structure of benthic macroinvertebrates was carried out between 1977 and 1983 in the fjords of Norway, some of which were affected by mining pollution (Rygg, 1985). As it happens, the measure of community structure used was univariate, in this case a diversity index. Correlations were sought between this diversity index and sediment concentrations of the trace metals copper, lead and zinc (Rygg, 1985). Correlations were strong for copper, moderate for lead and weak for zinc (Figure 6.22). From a list of the 50 most frequently occurring species, the study identified those most and least resistant to copper. Twenty of these 50 species were absent from sites where the copper concentration of the sediment exceeded 200 µg/g (Rygg, 1985). This chosen concentration of sediment copper is well in excess of proposed Sediment Quality Guidelines (SQG) in Table 6.30. From an inspection of Figure 6.22, it can be seen that depletions in the diversity of benthic invertebrate communities begin in correlation with lower sediment concentrations than 200 µg/g. Application of piecewise regression analysis could split the regressions shown into two regions (a broken-stick pattern), before and after a break point where depletion (a line of negative slope) can be estimated to start. In the case of copper (Figure 6.22), this break point, corresponding to a threshold of ecotoxic action, would be nearer to 100 µg/g than 200 µg/g, a value not out of line with the proposed SQG in Table 6.30.

In summary, implicit in any weight of evidence approach to the assessment of ecological risk created by the trace metal contamination of an estuary is the collection of several lines of evidence. These LOE need to be assessed and integrated into a final conclusion as to whether trace metal contamination is of ecotoxicological significance in the estuary under examination.

6.17 Fouling and Antifouling

Prologue. Copper has been used widely as the toxin in antifouling paints preventing the development of a fouling community on man-made structures in coastal waters. A critical copper leaching rate of 10 µg/cm^2/day is required to prevent settlement and growth of most fouling organisms.

Any man-made surface placed in the sea will be subject to fouling, be it the bottom of a ship, a pier or a pipe for the intake of cooling water for an onshore power station. The surface will firstly adsorb organic molecules. Within a few hours, a bacterial film will develop, soon followed by the growth of other microorganisms such as unicellular sessile diatoms and protozoans. This layer promotes the settlement

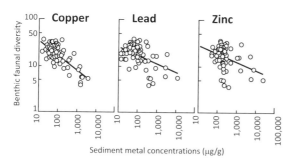

Figure 6.22 Negative correlations between faunal diversity in benthic invertebrate communities in Norwegian fjords and the concentrations of copper (strong), lead (moderate) and zinc (weak) in local sediments. (After Luoma and Rainbow, 2008.)

of fouling invertebrates, their larvae encouraged to settle by chemical cues. Major fouling organisms include barnacles, ascidian tunicates, bryozoans, hydroids and serpulid polychaetes. In shallow waters there may be enough light for macrophytic algae to flourish. Mussels will grow well on piers and in intake pipes.

The outcome is clear. The fouling growth on a ship's hull will increase drag and put up the fuel costs of maintaining speed. Buoys and cables will have increased weight and resistance to currents. There will also be increased resistance to water movement on the legs of the likes of oil platforms. Mussels may block intake pipes. Fouling organisms can also promote corrosion.

Not surprisingly, there has been a long history of attempts to prevent fouling, particularly of ships. Trace metals with their toxic properties have been in the forefront of this effort. Actually, it was the physical properties of particular metals, as opposed to their toxic chemical properties, that first led to their use to protect the hulls of boats. The first ships of any size were made from wood. Not only would fouling organisms grow on the surface of the wood, but wood-boring invertebrates offered an even greater threat by destroying the structure of the wooden hull. Classic amongst these wood borers were shipworms, usually species of *Teredo* and close relatives. Shipworms are not worms at all, but are bivalve molluscs, typically using their shell valves as drill bits to bore into the wood. Within the safety of the burrow, the shell valves are not needed for protection. Correspondingly, these bivalves have soft elongated bodies, giving them a wormlike appearance.

One answer was to use lead sheathing to provide a physical barrier to shipworm attack, making the most of the soft, flexible properties of this metal. It appears that both the Phoenicians in 1,000 BC and the Carthaginians in the fourth century BC used lead sheathing on their wooden ships, often in association with pitch. The ancient Greeks followed suit. But lead is not ideal. A physical barrier of lead might stop shipworm attack, but it offers poor protection against the fouling community which can grow on the surface of the lead. Furthermore, there was an electrochemical reaction between the lead and any iron nails used to

attach it to the hull, promoting the rapid corrosion of these nails.

The use of copper to sheathe boats also has a long history but did not really become widespread until the eighteenth century. In contrast to lead, copper will gradually dissolve in seawater, in effect releasing a constant supply of high local dissolved copper concentrations close to a ship's hull. This chemical agent is toxic to the settling larvae and newly settled juveniles of fouling invertebrates. It was this same toxicity of high concentrations of dissolved copper that had encouraged the sea captains of Liverpool to moor their ships off Dulas Bay to defoul their hulls in the late eighteenth century (Foster et al., 1978). By the 1780s, copper sheathing was widespread in the British navy, allowing the British ships to overtake the fouled French ships in the sea battles at the turn of the nineteenth century. As with lead, however, copper sheathing could not be used with iron nails. Copper nails were fine, but these, like the copper sheathing itself, would dissolve at quite a high rate.

A knowledge of electrochemistry allowed Sir Humphrey Davy in the early nineteenth century to show that the attachment of small pieces of zinc, tin or iron to the copper would prevent its dissolution, the other metals dissolving first (Woods Hole Oceanographic Institution, 1952). This, however, was partly self-defeating because it is the dissolution of the copper that offers the chemical antifouling protection.

The nineteenth century saw the introduction of ships made of iron, now commonly available after the industrial revolution. Wood was becoming in short supply, and iron offered greater strength to counter the power of new steam engines under production. Iron was also more resistant to the explosive shells under development in the world of armaments. But iron hulls without copper would be fouled, and the apposition of copper and iron caused corrosion of the iron hull. The answer lay in the development of paints, firstly an anticorrosion primer in contact with the iron hull, and then a copper-based paint releasing copper continuously to prevent fouling. A new era of antifouling had begun. To be fair, many other toxins were tried in antifouling paints in the nineteenth century, including mercury, arsenic, oils and tars

(Woods Hole Oceanographic Institution, 1952), and the search for other biocides still continues today (Dafforn et al., 2011). Nevertheless, it is copper that generally prevailed into the twentieth century and beyond. There are several types of copper-based antifouling paints used today, many with added extras (booster biocides), such as zinc and toxic organic compounds (Dafforn et al., 2011). These antifouling paints vary in the nature of the matrix that holds and subsequently releases the copper, but all are essentially making use of the local toxicity of dissolved copper (Dafforn et al., 2011).

The key factor here is the leaching rate of the copper from the antifouling paint (Woods Hole Oceanographic Institution, 1952; Barnes, 1948). Figure 6.23 shows the effect of the leaching rate of copper from an antifouling paint on the settlement of larvae of a species of the bryozoan *Bugula* (Woods Hole Oceanographic Institution, 1952). A leaching rate of 10 micrograms of copper per square centimetre per day ($\mu g/cm^2/day$) significantly reduced the percentage successful attachment of these bryozoan larvae (Figure 6.23). Different fouling organisms have different sensitivities to copper, as shown in Figure 6.24 (Barnes, 1948; Woods Hole Oceanographic Institution, 1952). Microbial films are the most resistant to copper, but macrophytic brown algae (species of *Ectocarpus*) and invertebrates are generally prevented from settlement and attachment by this threshold copper leaching rate of 10 $\mu g/cm^2/day$ (Barnes, 1948; Woods Hole Oceanographic Institution, 1952). It is of course fouling by macrophytic seaweeds and invertebrates, in contrast to that by microbial films, that causes problems. Field trials around the world have repeatedly shown that a copper leaching rate of 10 $\mu g/cm^2/day$ is effective to prevent fouling by these larger organisms (Woods Hole Oceanographic Institution, 1952).

The most efficient scenario for an antifouling paint containing copper would be a continual release of copper at this leaching rate for an extended period. In practice, copper release rates do drop over time. In the real world, conventional copper-based antifouling paints have an effective lifetime of up to two years

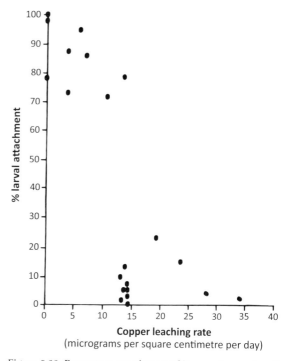

Figure 6.23 Percentage attachment of larvae of a species of the bryozoan *Bugula* in relation to the leaching rate of copper from an antifouling paint. (After Woods Hole Oceanographic Institution, 1952, courtesy of Woods Hole Oceanographic Institution.)

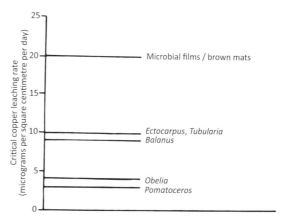

Figure 6.24 Critical leaching rates of copper (micrograms per square centimetre per day) from an antifouling paint that are sufficient to prevent the attachment of marine fouling organisms, including microbial films and brown mats of microscopic algae, and species of macrophytic brown algae (*Ectocarpus*), hydroid cnidarians (*Tubularia* and *Obelia*), barnacles (*Balanus*) and serpulid polychaetes (*Pomatoceros*). (After Barnes, 1948, with permission.)

before needing replacement (Dafforn et al., 2011). Another point to be borne in mind when considering the occurrence of fouling is that the speed of movement of a ship is often enough to prevent the settlement of algal spores and invertebrate larvae. The threshold speeds preventing settlement will vary with the invertebrate concerned, but it remains a truism that ships are at the greatest risk of fouling when in port. This risk is accentuated if the particular nature of the antifouling paint requires the movement of the ship to reach an adequate leaching rate of copper from the paint.

6.17.1 Resistance, Invasive Fouling Species and Tolerance

Prologue. Copper-based antifouling paints select for copper-resistant fouling species and potentially for copper-tolerant populations of these species. Fouling of ships introduces invasive species with these characteristics into British coastal waters.

When the copper leaching rate of an ageing antifouling paint drops below the critical level, fouling organisms begin to settle – firstly the more resistant species, and then the others (Figure 6.24). Thus, a selection process has already been in play – the selection of the more copper-resistant species themselves, in comparison with the community composition of fouling communities on hard surfaces without antifouling paint.

A bryozoan that fouls the hulls of ships is the now cosmopolitan colonial species *Watersipora subtorquata*, the red-rust bryozoan (McKenzie et al., 2012a). As a species, it is remarkably resistant to copper exposure, and its larvae will settle on panels treated with copper-based antifouling paints before many other fouling species (McKenzie et al., 2012a). This bryozoan has been spread around the world from its probably Caribbean origin by international shipping. It has moved from the ships' hulls on which it arrived onto local man-made structures in harbours and marinas in foreign coastal waters. The fouling faunas of marinas in particular are affected by the increased presence of copper-resistant fouling organisms such as *W. subtorquata*, in comparison to the local fauna attached to nearby surfaces lacking antifouling paints. Furthermore, by the very nature of their means of arrival, these copper-resistant fouling organisms are usually invasive species, a further change from the otherwise expected faunal community structure.

Do British marinas now contain copper-resistant and invasive species? New invasive species certainly. These include alien bryozoans, ascidian tunicates (Box 6.10), barnacles, serpulids and seaweeds.

Invasive species of the bryozoan genus *Watersipora* are to be found in British marinas and harbours. There is some confusion whether one or both of *W. subtorquata* and *W. subatra* (the red ripple bryozoan) is here, or even whether these are two names for the same species. The ruby bryozoan *Bugula neritina*, a cosmopolitan fouling species, is widespread in Britain. Other invasive bryozoans introduced into British marinas include the tufty-buff bryozoan *Tricellaria inopinata* and the orange ripple bryozoan *Schizoporella japonica*. *T. inopinata* is native to the Pacific but is now present in southern England. *S. japonica* originates from the Northwest Pacific. It was first recorded from a marina at Holyhead in Anglesey in 2010, but it is now also found in northern Scotland.

Box 6.10 Fouling Sea Squirts

Amongst ascidians, the stalked sea squirt *Styela clava*, native to the Korean region of the West Pacific, was first recorded in Britain at Plymouth in 1953, probably introduced on the hulls of warships returning from the Korean War. *S. clava* has spread rapidly and is now common on piers in UK harbours and marinas, particularly around the south of the British Isles. Amongst its fellow invasive ascidians are the orange-tipped sea squirt *Corella eumyota*, the compass sea squirt *Asterocarpa humilis*, the orange cloak sea squirt *Botrylloides violaceus*, the San Diego sea squirt *Botrylloides diegensis*, the carpet sea squirt *Didemnum vexillum* and the creeping sea squirt *Perophora japonica*. *C. eumyota* is a native of southern circumpolar cold waters across the globe, but is now found along the south and southwest coasts of England and Ireland. Also of southern hemisphere origin, *A. humilis* is to be found in marinas and harbours along the coasts of the English Channel and Irish Sea. *Botrylloides violaceus* is also now present along the southern coasts of Britain, although a native of the western Pacific from Siberia to southern China. In spite of its common name of the San Diego sea squirt, *B. diegensis* is probably exotic to California, as it is to the British Isles, where it occurs in marinas and harbours along southern coasts of England and Wales. Believed to be a native of Japan, *D. vexillum* has also acquired the unflattering common name of 'marine vomit'. It is now present in marinas in Wales and southern England. As indicated by its specific name, *P. japonica* is native to Japan but now to be found along the coasts of the English Channel. Many of these introduced sea squirts appear to be recent arrivals, often first reported only in this century.

An invasive species with a longer history of residency in the British Isles is the Australasian barnacle *Austrominius modestus* (previously *Elminius modestus*), recorded first from Chichester Harbour in 1945. *A. modestus* is now found on rocky shores all round the British Isles as well as in marinas and harbours. The cosmopolitan invasive fouling barnacle *Amphibalanus amphitrite* favours warmer waters. It is often to be found on man-made structures in waters associated with the effluent of British coastal power stations.

A common fouling serpulid polychaete is *Hydroides norvegica*. It is cosmopolitan across the world, including the British Isles, so probably should not be labelled as invasive. In context here, *H. norvegica* is relatively resistant to copper-based antifouling paints, secreting its calcareous tube attached to ships' hulls and man-made structures in marinas and harbours. A correctly labelled invasive serpulid is *Ficopomatus enigmaticus*, the Australian or trumpet tubeworm, also now with a cosmopolitan distribution. It has now been reported from several locations around the British Isles.

Invasive species of both brown and red seaweeds have been introduced into British coastal waters. Wakame or Japanese kelp, *Undaria pinnatifida*, is a large brown seaweed that will grow on the pontoons of marinas. The more infamous invasive brown seaweed from Japan, Japanese wireweed *Sargassum muticum*, grows on natural rocky surfaces in rock pools and shallow water and is not a fouling species. Red algae have joined the list of invasive species growing in British marinas. They include devil's tongue weed *Grateloupia turuturu* from eastern Asia, and siphoned Japan weed *Heterosiphonia japonica* from the Northwest Pacific Ocean.

So it is certainly the case that the fouling communities growing on the man-made structures in British marinas and harbours previously treated with antifouling paint differ from those on other local hard surfaces lacking antifouling paint. The presence of copper-based antifouling paint has led to the selection of a new hard substrate fouling community. This has been achieved firstly by the selection of local species with greater copper-resistance capabilities. Secondly, this same selection process, now acting initially on ships' hulls in foreign waters, has promoted the introduction of invasive species from elsewhere in the world that might have a similar enhanced resistance to copper exposure.

There is evidence from Australia that copper exposure can change fouling communities directly and indirectly (Johnston and Keough, 2002, 2003). Pulsed copper doses had a dramatic direct negative effect on the densities of large space-occupying ascidians on settlement panels. A consequential secondary effect was the positive response of fouling species of many other taxonomic groups as an indirect effect of the copper exposure (Johnston and Keough, 2002). Serpulid polychaetes, considered to be poor competitors for space although good initial colonisers, particularly were present in high densities on the plates affected by copper exposure (Johnston and Keough, 2002, 2003).

What about the potential promotion of dominance by non-indigenous species (NIS)? Another study in Australia investigated the role of copper exposure in facilitating the establishment and spread of NIS in coastal environments (Piola and Johnston, 2008). Forty-seven species surveyed to comprise the local fouling communities were identified as native (32%), NIS (28%) or cryptogenic whose origins were disputed or unknown (28%). Increasing copper exposure of settlement panels decreased the species diversity of native species at all study sites by between 33 and 50%. There was increased dominance of NIS with copper exposure. No single species was responsible for the observed effects, which were driven by a diverse range of species responses. So it is the case that copper exposure from a copper-based antifouling paint does promote the dominance of local fouling communities by introduced species.

Aside from the changes to local fouling communities growing on surfaces covered with ageing antifouling paint, does the use of copper-based antifouling paints in a marina or harbour cause any displaced changes to other local biological communities? Other unpainted hard substrates locally may, for example, become populated by the newly introduced invasive fouling species, or the release of copper from the paints may increase local dissolved bioavailabilities of copper that affect nearby communities growing on unpainted surfaces. It is relevant that newly applied copper-based antifouling paints can have initial copper leaching rates of between 25 and 65 $\mu g/cm^2/day$, well capable of raising local dissolved copper bioavailabilities (McKenzie et al., 2012a).

A study in Southwest England investigated the epifauna, not of nearby unpainted hard surfaces, but of a native large brown seaweed, the kelp *Saccharina latissima*, in four marinas and four reference locations (Johnston et al., 2011). The antifouling paints in the marinas made use of tin and zinc in addition to copper. Accumulated concentrations of copper, tin and zinc in the kelp samples confirmed the much higher dissolved bioavailabilities of these three trace metals in the marinas. And there was an indirect effect of the antifouling paint on the epifaunal community on the kelp. The fouling of the surface of the kelp was light in the reference locations (less than 2% cover), being more extensive in marinas (25 to 80% cover). Expectedly in the light of preceding discussion, the kelp from the marinas also carried the greatest cover and diversity of non-indigenous species. It would appear then that marinas offer a concentrated source of larvae of fouling species, particularly those of invasive species.

Leaching of copper from antifouling paint can also affect local soft bottom communities. In an American example, the infauna of sediments in two San Diego marinas were reduced in species richness in association with concentrations of sediment copper originating from antifouling paint (Neira et al., 2013).

We have seen that the use of antifouling paint can select for the presence of copper-resistant species in the fouling community. Moreover, the scene has also been set for the subsequent selection of copper-tolerant strains within a single fouling species. For example, this has occurred in the fouling filamentous brown alga *Ectocarpus siliculosus* (Russell and Morris, 1970). Cultures of this alga sampled in 1970 from two oceangoing freighters treated with copper-based antifouling paints were ten times more tolerant of dissolved copper than a control shore-based population of the alga (Russell and Morris, 1970). Similarly, a strain of the green alga *Ulva compressa* (as *Enteromorpha compressa*), collected from a fishing vessel based at Anstruther in Fife in Scotland in the early 1980s, was more tolerant to exposure to dissolved copper than a population of the same species collected nearby from the East Rocks at St Andrews (Reed and Moffat, 1983). Furthermore, this copper tolerance was shown to be genetically determined, since the progeny of the original ship-fouling algae were also tolerant of raised dissolved copper concentrations (Reed and Moffat, 1983).

The potential selection of copper-tolerant strains of a species after exposure to copper-based antifouling paint is not necessarily restricted to algae but may well extend to populations of fouling invertebrates settling on surfaces on which the antifouling paint is ageing. The potential is certainly there.

A consequence of the release of copper from anti-fouling paints is that sheltered marinas housing craft treated with copper-based paints can have raised local dissolved bioavailabilities of copper (McKenzie et al., 2011, 2012a, b). Tolerance to dissolved copper was tested for populations of the introduced bryozoan *Watersipora subtorquata* from four sites, including marinas, across Port Hacking, south of Sydney, Australia. These sites differed in their local copper bioavailabilities, as determined by copper accumulation in deployed oysters. In fact, there were no significant differences in the copper sensitivities of the bryozoans from the four sites, indicating no effective local selection for copper tolerance. There was, however, a significant interaction between genotype and environment, signifying the presence of different genotypes of the bryozoan at the different sites. There is, therefore, considerable genetic variation between populations that offers the potential for the selection for copper-tolerant strains (McKenzie et al., 2011, 2012b). Watch this space.

Not all the copper released from antifouling paints is actually in the form of dissolved copper. Particles of antifouling paint are generated during boat maintenance and cleaning, for example during sanding or blasting of hulls (Turner, 2010). Anti-fouling paint particles may also simply flake off poorly maintained or abandoned shore structures or boats (Turner, 2010). These particles reflect the trace metal makeup of the original antifouling paint. Such particles collected from recreational boatyards in the United Kingdom contained 35% dry weight copper and 15% dry weight zinc, in addition to cadmium, chromium, nickel, lead and zinc (Turner, 2010). In turn, these particles may be efficient sources of dissolved copper and other trace metals, given their high surface area to volume ratios. The antifouling paint particles may initially be suspended in the water column before becoming incorporated into local bottom sediments. The trace metals not leached out from the particles may still be bioaccessible and bioavailable to local coastal organisms after ingestion. For example, the mussel *Mytilus edulis* will filter out and ingest small suspended paint particles, and the unselective deposit feeding lugworm *Arenicola marina* will ingest particles present in sediment (Turner, 2010). The trace metals, not least copper, in the paint particles have been shown to be trophically bioavailable to both these invertebrates and therefore have the potential to cause ecotoxicological effects (Turner, 2010).

6.17.2 Tributyl Tin

Prologue. Tributyl tin was used in antifouling paints from the 1960s, but its use is now banned in Britain. TBT is ecotoxic at very low dissolved concentrations, and, for example, had deleterious effects on oyster mariculture in the 1970s and 1980s.

The elephant in the room, yet to be addressed in this account of antifouling, is the story of TBT.

In the 1960s, attention had turned to the use of organometal compounds as biocides. One such compound was tributyl tin, which was then incorporated into antifouling paints. Initially, these were conventional antifouling paints relying on the leaching of the toxic TBT from a soluble matrix. By the 1970s, TBT was being used in antifouling paints in the form of a self-polishing TBT copolymer, this copolymer providing both the biocide and the paint matrix (Dafforn et al., 2011). It is a feature of self-polishing antifouling paints that they have an effective lifetime of about five years, in comparison to the two years or less of conventional antifouling paints. This extended life-time of the paint and the high toxicity of TBT to bacteria, fungi and invertebrates, but not mammals, made TBT copolymer antifouling paints a very attractive proposition. Furthermore, TBT breaks down in the environment to the much less toxic form of inorganic tin. This environmental breakdown of TBT was considered to occur relatively quickly, such that TBT was believed unlikely to present any serious long-term pollution hazard.

The use of TBT in antifouling paints spread in the 1960s and grew rapidly in the 1970s, as it was taken up both by the shipping industry and by owners of small boats. By the late 1970s, TBT was being used on 80% of the world's shipping fleet, in addition to its extensive use on pleasure craft, piers, buoys and mariculture nets and in seawater cooling systems. World production of TBT further increased through the 1980s and thereafter, reaching 32,000 tonnes in 1994 (Langston, 1995). It has been estimated that about 50,000 ships per year were using the port of Finisterre in Northwest Spain at the turn of the mil-lennium and that these ships were releasing about 1.64 tonnes of TBT each year into local coastal waters (Ruiz, 2004).

It is of course the leaching of TBT from paint that brings about its antifouling action. The bad news was that TBT released from antifouling paint did not break down into inorganic tin quite as quickly as expected. Furthermore, dissolved TBT proved to have toxic effects on non-target marine organisms at far lower dissolved concentrations than were ever envisaged (Gibbs et al., 1988; Bryan and Langston, 1992).

The first alarms over TBT came from the oyster industry in France (Alzieu, 1986, 2000). In the late 1970s, repeated failure of larval settlement, reduced growth and abnormal shell development described as balling (a proliferation of chambers in the shells) of oysters *Crassostrea gigas* led to the near collapse of the commercial oyster fishery in the Baie d'Arcachon on the Atlantic coast west of Bordeaux (Alzieu, 1986, 2000). These ecotoxicological effects were attribut-able to the (albeit apparently low) concentrations of TBT in the water of the bay, derived from the use of antifouling paints containing TBT on local small

pleasure craft (Alzieu, 1986). The very low dissolved concentration of 1 nanogram (ng) per litre of tin in the form of TBT was enough to initiate shell thickening in spat of *C. gigas*, 8 ng/L caused sig-nificant shell thickening and 20 ng/L of tin as TBT reduced growth and viability of the oyster spat (Bryan and Langston, 1992). To put these concen-trations into context, total dissolved concentrations of tin (all in inorganic form) are about 1 to 2 ng/L (0.001–0.002 µg/L) in the deep and surface waters of the Atlantic Ocean (Donat and Bruland, 1995). The concentrations of dissolved TBT with a toxic effect on oysters are very low indeed on any scale of dis-solved concentrations of any trace metals causing ecotoxic effects. With the finger pointing at TBT, in 1982 France banned its use on vessels smaller than 25 metres (essentially pleasure craft). Follow-up work in the United Kingdom between 1982 and 1984 confirmed that some British estuaries (for example, the Crouch and Blackwater Estuaries in Essex) also contained high enough dissolved concentrations of TBT to cause reduced meat yields and shell thickening of local stocks of the Pacific oyster *C. gigas* (Thain and Waldock, 1986). The native common oyster *Ostrea edulis* was similarly affected by low concentrations of TBT. In fact, it is likely that the deterioration of common oyster stocks in the Crouch Estuary at the time was caused by the effect of TBT on oyster reproduction and larval settlement (Thain and Waldock, 1986). Molluscs appeared to be particularly sensitive to TBT. In 1985, the United Kingdom set an EQS of 20 ng/L TBT (8 ng/L of tin as TBT) for the protection of marine organisms (Bryan and Langston, 1992). In 1989, this EQS was lowered to 2 ng/L TBT (0.8 ng/L of tin as TBT) (Bryan and Langston, 1992).

6.17.2.1 Imposex and Dogwhelks

Prologue. TBT at low concentrations causes imposex in dogwhelks, ultimately sterilising female dogwhelks with the subsequent reduction and elimination of dogwhelk populations in British harbours and adja-cent coastal waters. These dogwhelk populations are now recovering.

In the 1980s, it had become apparent that a common member of the fauna of British rocky shores was especially sensitive to the ecotoxicological effects of TBT – the dogwhelk *Nucella lapillus* (Bryan et al., 1986). A new word also entered the vocabulary of British marine biologists: imposex (Smith, 1980; Bryan et al., 1986). Imposex is defined as the imposition of male sexual characters onto females, a phenomenon exemplified strongly by the effect of TBT on marine gastropod molluscs (Bryan et al., 1986). The term 'imposex' had been coined in 1980 after observations of reproductive abnormalities in the mud snail *Tritia obsoleta* (as *Nassarius obsoletus*), living near marinas housing yachts in the New York area (Smith, 1980, 1981). It was then suggested that these abnormalities of the reproductive system of the gastropods were linked to the effects of antifouling paints containing TBT (Smith, 1981). In fact, the presence of penis-like outgrowths in females of the dogwhelk *N. lapillus* had already been noted in Plymouth Sound in Devon by 1970, without an understanding of its cause (Blaber, 1970; Gibbs et al., 1987). By the early 1980s, dogwhelks with imposex were found to be common in many marinas, harbours and other rocky shores around Southwest England (Bryan et al., 1986, 1987b; Gibbs and Bryan, 1986; Gibbs et al., 1987). Furthermore, Geoff Bryan, Bill Langston and Peter Gibbs at the Marine Biological Association UK laboratories on Citadel Hill in Plymouth showed unequivocally through a series of laboratory and field studies that it was indeed TBT that was causing imposex in the dogwhelks and other gastropod molluscs (Bryan et al., 1986, 1987b; Gibbs and Bryan, 1986; Gibbs et al., 1987, 1988). And, as in the case of oyster settlement, growth and shell development, these imposex effects in dogwhelks were caused by extremely low dissolved concentrations of TBT (Table 6.31) (Gibbs et al., 1988). The development in female dogwhelks of male sexual characteristics, essentially the formation of a penis and a vas deferens transporting sperm, could be assessed in stages, monitoring the size of the penis in comparison to those of local males (RPS index) and the degree of development of a vas deferens (VDS index) (Table 6.31) (Gibbs and Bryan, 1986; Gibbs et al.,

1988). Negative effects on dogwhelk breeding were already apparent at exposures as low as 1 ng/L of tin as TBT (Table 6.31). The outcome of imposex is that the females become sterilised by the blockage of the oviduct by the developing male reproductive system (Gibbs and Bryan, 1986). There is complete sterilisation of all female dogwhelks by an exposure of 10 ng/L of tin as TBT (Table 6.31). Mortality may even occur in severely affected females by rupture of the capsule gland unable to emit egg capsules (Table 6.31). Furthermore, subsequent reduction of TBT levels did not result in any appreciable reversal of the effects of imposex in individual dogwhelks (Gibbs and Bryan, 1986; Gibbs et al., 1988).

It is also the case that exposure to TBT causes imposex in other gastropod molluscs, particularly neogastropod relatives of the dogwhelk, such as the netted dogwhelk *Tritia reticulata*, the sting winkle *Ocenebra erinaceus* and the oyster drill *Urosalpinx cinerea*. The action of TBT in causing imposex in gastropods can be replicated by the injection of male sex hormones such as testosterone into females. The implication, then, is that TBT is acting as an endocrine disruptor upsetting the balance of male and female sex hormones in the female gastropods.

Unlike many of the marine invertebrates that we have previously encountered, dogwhelks and other neogastropods do not have a planktonic larval stage able to disperse the species over long distances. Dogwhelks lay egg capsules the size of grains of rice in groups on rocks on the shore. Small juvenile dogwhelks hatch directly from these egg capsules to live in the same area as their parents. Adults have limited mobility, but there may be some rafting in of juvenile dogwhelks from elsewhere on drifting pieces of seaweed. Nevertheless, the sterilisation of all local female dogwhelks by TBT exposure means an all but complete stop to local population recruitment. That population of dogwhelks then dies out. Dogwhelks are keystone predators on rocky shores, preying on barnacles and mussels (Hayward, 2004). Their removal, therefore, has consequences for the ecology of rocky shores exposed to TBT, causing significant changes to local community structure (Spence et al., 1990; Hayward, 2004). Loss of dogwhelks from semiexposed

Table 6.31 **Effects of TBT exposure on the reproductive system of female dogwhelks *Nucella lapillus*. RPS: Relative Penis Size; VDS: Vas Deferens Sequence.**

Dissolved TBT (ng Sn per litre)	RPS index (%)	VDS index (%)	Effect on reproductive system in females
<0.5	<5	<4	Normal breeding; development of penis and vas deferens
1–2	40+	4–5(+)	Some females can breed; others sterilised by oviduct blockage; aborted egg capsules in capsule gland
3–5	90+	5(+)	Virtually all females sterilised; initial egg formation apparently normal
10+	90+	5	All females sterile; egg formation repressed; sperm formation initiated
20	90+	5	All females sterile; development of testis; ripe sperm in many
100	90+	5	All females sterile; sperm-ingesting gland undeveloped in some

Source: Gibbs et al. (1988).

rocky shores in the British Isles permits the buildup of dense barnacle cover. This increased abundance of barnacles in turn will reduce the effectiveness of limpet grazing of brown seaweed sporelings (Spence et al., 1990). These seaweeds can then grow over the barnacles to produce a seaweed-dominated shore, where otherwise limpets and barnacles would have flourished (Spence et al., 1990).

In 1970, nearly all female dogwhelks in Plymouth Sound were affected by imposex (Spence et al., 1990; Hayward, 2004). By 1985, imposex had been found in populations of dogwhelks from all around the south-western peninsula of England, from Chesil Beach to the Severn Estuary (Hayward, 2004). As to be expected, the incidence of imposex was particularly high in the sheltered estuaries of Cornwall and Devon, such as those of the Dart, Salcombe, Fal and Helford,

housing large numbers of pleasure craft (Hayward, 2004). Subsequently, imposex was confirmed to be present in dogwhelk populations all around the British Isles, particularly near ports and marinas (Spence et al., 1990; Hayward, 2004). Measurements of dissolved TBT concentrations in British estuaries reflected the link to the local abundance of pleasure craft (Bryan and Langston, 1992). TBT concentrations ranged from less than 1 ng/L of tin as TBT in boat-free estuaries to more than 600 ng/L of tin as TBT near marinas (Bryan and Langston, 1992). Estuaries with TBT levels more than tenfold higher than the 1985 EQS of 8 ng/L of tin as TBT included the popular yachting locations of Salcombe, the Dart, Poole, the Beaulieu, the Crouch and the Blackwater (Bryan and Langston, 1992). Furthermore, TBT concentrations in these estuaries peaked in the spring, when newly painted boats were put back in the water.

With the causal link to TBT now established, the UK government followed the example of France and in 1987 banned the use of TBT on small vessels (less than 25 metres in length) (Dafforn et al., 2011). Similar restrictions were subsequently introduced in the United States (1988), Canada (1989), Australia (1989) and the EU in general (1989) (Dafforn et al., 2011). Yet commercial vessels were still allowed to use TBT antifouling paints. The ecotoxicological effects of TBT were not eliminated by these restrictions only on small boats. TBT antifouling paints on large vessels were still impacting on marine organisms, particularly molluscs, in the vicinity of harbours. So, in November 2001, the International Maritime Organisation (IMO) adopted the International Convention on the Control of Harmful Anti-fouling Systems in Ships (the AFS Convention) to ban the application of TBT antifouling paint on all vessels from 1 January 2003 and to require its absence from all vessels after 1 January 2008 (Dafforn et al., 2011). The AFS Convention did not become internationally binding until September 2008, after its ratification by the necessary 25 states representing at least 25% of the tonnage of world shipping (Dafforn et al., 2011). The EU had already acted, banning the application of TBT paints to any EU-flagged vessel from 1 January 2003 (Dafforn et al., 2011). The scene had now been fully set for the removal of sources of TBT into the marine environment. It would be naïve, however, to consider that TBT is not still being used on domestic vessels in non-signatory countries (Dafforn et al., 2011).

The 1987 UK ban on the use of TBT on small vessels could not be expected to have an immediate effect on levels of TBT in estuaries housing large marinas with their populations of yachts. It is only human nature that stocks of TBT antifouling paints held by individual yacht owners might still be used in the next year or so until exhausted. So, for three or four years after the 1987 ban, sites with boating activity in Essex and Suffolk still had dissolved concentrations of TBT above the EQS of 2 ng/L TBT set in 1987 (Dowson et al., 1992, 1993). These TBT concentrations in 1990 ranged from below 3 to as high as 71 ng/L

(Dowson et al., 1992). But after 1991, these dissolved TBT concentrations declined significantly at most of 22 sites in six east coast English estuarine systems to below the analytical detection limit of 3 ng/L TBT (Dowson et al., 1993). Sediments adsorb TBT, and disturbance of sediments by dredging does have the potential for TBT desorption back into solution (Dowson et al., 1992). Nevertheless, sediment concentrations of TBT in British estuaries have also generally decreased since 1990 (Dowson et al., 1993).

By 1991, there was also evidence of a reduction in the incidence of imposex in British populations of dogwhelks previously strongly affected by TBT exposure (Evans et al., 1991, 1996; Hawkins et al., 2002). Dogwhelk populations sampled in 1986 in areas of high boating activity in Northumberland had shown the presence of severe imposex and poor recruitment (Evans et al., 1991). By 1991, RPS indices in these populations had fallen, there was better survivorship of female dogwhelks and recruitment had improved (Evans et al., 1991). Similar recoveries from the mid-1980s were apparent in 1994 in dogwhelk populations in Devon and Cornwall, in Northeast England and in the Clyde Sea (Evans et al., 1996). Yet imposex in dogwhelk populations near the large Sullom Voe Oil Terminal in the Shetland Islands was still severe in 1994 (Evans et al., 1996). TBT from antifouling paints, then allowed on large vessels such as oil tankers, was still causing ecotoxicological effects in harbours used for international shipping. By the turn of the century, it was clear that the incidence of imposex in dogwhelk populations at sites in Britain associated with pleasure craft had fallen substantially (Hawkins et al., 2002). Recovery was initially rapid, particularly at the worst-affected sites, but then levelled out (Hawkins et al., 2002). It was apparent, therefore, that dissolved TBT was still present at several sites, whether after remobilisation from sediments or originating from large ships (Hawkins et al., 2002). Recovery of gastropod populations from imposex was taking up to ten years (Evans et al., 1991; Hawkins et al., 2002). A complete ban on the use of TBT in antifouling paints would be needed to achieve the complete elimination of observable imposex in dogwhelks (Hawkins et al.,

2002). The EU delivered this ban from the start of 2003, and the IMO followed suit in 2008 (Dafforn et al., 2011). Recovery of dogwhelk populations has again proceeded. It will, however, inevitably take more time to eliminate the hotspots of imposex in dogwhelk populations near docks and ship maintenance facilities, near offshore anchorages and at sites where TBT-contaminated sediments are resuspended by dredging activities (Langston et al., 2015).

So the banning of TBT, initially on small boats and then on larger ships, in Britain is well on the way to solving the ecotoxicological problems caused by this toxic organometal compound. These ecotoxicological effects have affected not just dogwhelks and related gastropods, but also other molluscs, including oysters and other bivalves of both commercial and ecological significance. But there has been another side effect of the banning of TBT. In spite of,

or rather because of, its unexpectedly strong ecotoxic properties, TBT was a remarkably effective antifouling agent, arguably more effective than its successors. It is now therefore easier for fouling organisms to grow on ships' hulls and be transported as invasive species into new habitats. Indeed, many of the invasive species of bryozoans, ascidian tunicates and red algae described in this chapter have colonised our waters only in the last ten years or so. There has been an ecological price to pay for the banning of TBT. Nevertheless, this price might be considered small in comparison to the beneficial elimination of the widespread deleterious effects of TBT on the local marine fauna, particularly on the molluscan fauna.

Estuaries are but a transition stage in the passage of trace metals from land to sea. In the next chapter, we enter the realm of coastal seas and then oceans.

7 Coastal Seas and Oceans

Box 7.1 Definitions

arrow worms Members of a small phylum of mostly planktonic, transparent marine worms, the chaetognaths. Arrow worms are typically abundant, carnivorous members of the zooplankton, ranging from 2 to 120 millimetres long.

baleen plate One of a series of platelike structures made up of keratin, growing down from the upper jaws of baleen whales and used to filter food such as krill from the surrounding seawater.

bathypelagic Pertaining to the pelagic zone of the ocean between 1,000 and 6,000 metres deep.

brown meat A commercial term usually referring to the hepatopancreas, occasionally infiltrated by gonad tissue, of the edible crab *Cancer pagurus*.

cetaceans A group of carnivorous, finned, aquatic (predominantly marine) mammals, commonly referred to as whales. Divisible into two living groups, odontocetes or toothed whales (including porpoises and dolphins), and mysticetes, the baleen whales.

cnidarian Member of the phylum *Cnidaria*, which includes hydras and other hydroids, sea anemones, sea pens, siphonophores and jellyfish. Together with comb jellies, cnidarians may be termed coelenterates.

coccolithophore A unicellular eukaryotic protist member of the marine phytoplankton, which typically secretes calcium carbonate plates (coccoliths) onto its surface. For example, *Emiliania huxleyi*.

copepod crustaceans Small crustaceans that are overwhelmingly abundant in the sea and also occur in freshwater. Copepods may be planktonic, benthic or parasitic. Free living taxa of copepods include calanoids, cyclopoids and harpacticoids.

demersal fish Fish that live and feed on or near the bottom of seas or lakes.

dinoflagellates Photosynthetic unicellular protists abundant in the marine plankton of tropical and subtropical oceans, where they are important primary producers.

epibenthic Living on the bottom of an aquatic habitat.

epipelagic zone The upper zone of an ocean, between the surface and 200 metres deep.

krill Pelagic crustaceans of the taxon *Euphausiacea*, including the genus *Meganyctiphanes*, also called euphausiids. Krill are well known for being the major food of many baleen whales.

medusa (plural medusae) The pelagic stage of the lifecycle of many cnidarians. Medusae are carnivorous and bear gonads. Large medusae are commonly called jellyfish.

mesopelagic Pertaining to the pelagic zone of intermediate depth (200 to 1000 metres) in the ocean.

mysticetes Cetaceans which lack teeth, feeding by filtering krill or small fish with baleen plates in an expanded mouth. Mysticetes (also called baleen whales) include blue whales, fin whales, minke whales and other rorquals.

neuston Members of a community that live on or under the surface film of a water body, typically the ocean.

odontocetes Toothed whales which include porpoises, dolphins, pilot whales, sperm whales and beaked whales. The typical presence of teeth allows these cetaceans to prey on fish and/or squid.

pelagic Pertaining to the water column, as in a sea or lake. The term is used to refer to organisms inhabiting the water column of an ocean.

photic zone The upper zone of a water body such as an ocean in which there is sufficient light for phytoplankton photosynthesis to exceed respiration, allowing the phytoplankton to grow.

phytoplankton Planktonic primary producers, consisting of unicellular protists and photosynthetic prokaryotes.

plankton (adjective *planktonic*) Drifting organisms that are unable to maintain their position or distribution independently of the movement of a body of water. They are often divided into phytoplankton and zooplankton.

pleuston Community of (usually oceanic) organisms that remain permanently at the water surface by their own buoyancy, positioned partly in the water and partly in the air.

protoporphyrin A precursor compound in the metabolic synthesis of a porphyrin, before the incorporation of any metal atom.

porphyrin A water-soluble, coloured organic compound made up of a flat ring of subunits, often containing a central metal atom, such as iron in the case of the haem group of haemoglobin.

primary production Production of organic matter from a simple inorganic substrate, typically carbon dioxide. Plants and autotrophic protists use light energy in photosynthesis, but some prokaryotes use chemical energy in chemosynthesis.

production The production of organic matter. Primary producers convert inorganic carbon into organic matter using light energy (photosynthesis) or chemical energy (chemosynthesis). Secondary producers gain organic matter from primary producers, as in the case of an herbivore feeding on a plant. Tertiary producers eat secondary producers, and so on, along a food chain. Gross production describes all organic matter acquired, while net production describes the organic matter converted into biomass after allowance for that used in respiration.

rorqual A baleen whale of the family *Balaenopteridae*, which includes blue whales, fin whales and minke whales.

scampi A term usually referring in a culinary context to the abdominal (tail) muscles of *Nephrops norvegicus*, the Norway lobster or Dublin Bay prawn, or alternatively to the whole animal itself.

sea pen A colonial marine cnidarian, so-called because of their featherlike appearance reminiscent of an old quill pen. Sea pens typically stick up from a muddy or sandy bottom in which they are anchored, filtering plankton. An example is *Pennatula phosphorea*.

siphonophore A gelatinous, pelagic, marine cnidarian. Each siphonophore may appear to be a single organism but is actually a colony of specialised individuals. Many siphonophores are mesopelagic oceanic predators, unlike the most well-known siphonophore, the Portuguese Man o' War *Physalia physalis*, floating on the ocean surface.

stegocephalid amphipod A member of a family of marine amphipod crustaceans that characteristically contain ferritin crystals in the pair of ventral caeca, detoxifying iron absorbed from cnidarian prey. The name literally means 'horn-headed'. Examples include *Stegocephaloides christianiensis* in British coastal waters, and the pelagic oceanic *Parandania boecki*.

superfluous feeding Method of feeding reportedly used by planktonic copepods feeding on very dense concentrations of phytoplankton. Ingestion is so rapid that passage through the gut may simply act as a press to squeeze out soluble material for alimentary uptake, with limited, if any, digestion of solid components of the ingested phytoplankton.

thermocline Interface in a water body, such as a lake or ocean, between a shallow layer of warm, less dense, surface water and cooler, more dense, deep water, with a very rapid change of temperature over a short distance.

tunicate A marine invertebrate that is part of the phylum *Chordata*, which includes the vertebrates, for many tunicates have a planktonic tadpolelike larva with the chordate characteristics of a dorsal nerve cord and a notochord. Common on rocky shores and on the sea bottom are sea squirts (ascidians) with a tough or gelatinous tunic around the body. Pelagic oceanic tunicates, such as salps and doliolids, are typically gelatinous and barrel-shaped. Tunicates feed by filtering plankton through a perforated pharynx, another chordate feature.

vanadocyte A particular type of blood cell in certain sea squirts that contains a very high concentration of vanadium.

whelk The common name applied to many carnivorous neogastropod sea snails, including the dogwhelk *Nucella lapillus* and the large, edible common whelk *Buccinum undatum.*

white meat A commercial term usually referring to the muscle tissue extracted from the claws and legs of the edible crab *Cancer pagurus.*

zooplankton Animals of the plankton.

7.1 Introduction

Prologue. Beyond estuaries are the coastal seas around the British Isles and then the North Atlantic and Arctic Oceans. There have been significant anthropogenic increases in the bioavailabilities of trace metals historically in these nearshore coastal habitats, such as in Liverpool Bay. Any influence of man decreases on passage away from land and has reduced over time. The natural history of trace metals in these marine environments is not dominated by contamination causing ecotoxicological consequences for the local marine biodiversity or for human consumers of seafood collected in British coastal waters and beyond. There is, however, an exception in the case of methyl mercury in long-lived oceanic fish such as tuna. A new phenomenon appears, that of essential metal deficiency in the oceans, a challenge met and overcome, particularly by oceanic phytoplankton species. We shall meet novel examples of the remarkable use to which accumulated trace metals are put by marine animals, for example in the deterrence of fish predators by specific benthic invertebrates. The chapter finishes with sections on the natural history of trace metals in two charismatic groups of marine animals – the many species of sea-birds around our coasts and the marine mammals now returning in increasing numbers to our seas. Both provide further examples of the accumulation of extremely high trace metal concentrations quite naturally.

The British Isles are surrounded by coastal seas overlying the continental shelf. Beyond to the west lies the North Atlantic Ocean and, to the north, the Arctic Ocean (Figure 7.1). The continental shelf slopes very gently from the bottom of the shore to a depth of about 200 metres, where it gives way to the rapidly descending continental slope. The continental slope marks the start of the ocean, descending steeply to a depth of about 4,000 to 5,000 metres. The continental shelf is wide around the British Isles, extending to the west of Ireland and the north of Scotland. The Irish Sea, the English Channel and almost all of the North Sea lie above this continental shelf (Figure 7.1).

Coastal seas become increasingly remote from any anthropogenic activities that would increase the concentrations, bioavailabilities and ultimately potential ecotoxicological effects of trace metals. Even without man's activities, coastal waters contain dissolved trace metal concentrations above those in the oceans, as trace metals are mobilised naturally from terrestrial deposits to pass through river systems and their estuaries into the sea. Such natural raised concentrations do not, though, approach the ecotoxicologically significant bioavailabilities in some estuaries. Estuaries contaminated with trace metals will increase trace metal concentrations in the waters and sediments of the immediately adjacent coastal sea area, but only over a very limited geographical range. Another way in which anthropogenic activities can raise trace metal bioavailabilities in coastal waters to potentially significant ecotoxicological levels is by the marine dumping of waste. In the twentieth century, it was common to dump trace metal–contaminated sewage sludge and dredged material in the coastal waters around the British Isles – out of sight, out of mind. Such dumping, however, led to considerable

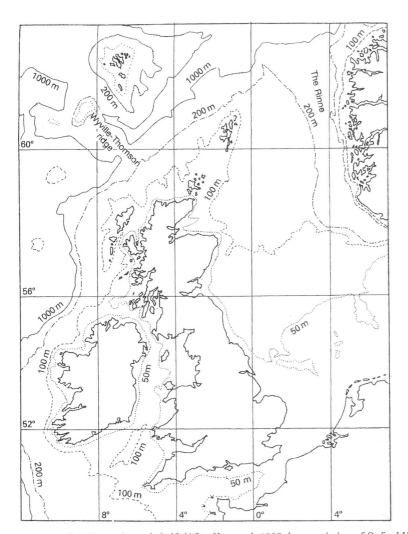

Figure 7.1 Britain continental shelf. (After Hayward, 1995, by permission of Oxford University Press.)

ecotoxicological effects on the benthic marine fauna at dump sites, and sewage sludge dumping has now been phased out.

7.2 Metal Concentrations in Coastal and Oceanic Waters

Prologue. This section presents data on the dissolved concentrations of trace metals in the marine environment, discussing their variation over time and space in coastal waters and their variation with depth in the oceans.

7.2.1 Coastal Waters

Prologue. Coastal waters typically contain higher dissolved concentrations than oceanic waters, a feature that occurs even without anthropogenic intervention.

Table 7.1 lists the dissolved concentrations of trace metals in British coastal waters, together with dissolved concentrations in the surface and deep waters of the North Atlantic Ocean.

There is variability over space and time in the dissolved trace metal concentrations around the

Table 7.1 Concentrations (μg/L) of dissolved (0.2 to 0.4 μm filtered) trace metals in British coastal waters and the North Atlantic Ocean. Also shown are the oceanic distributions of these dissolved metals: R = Recycled or Nutrient-type; S = Scavenged.

	Date	Cd	Co	Cu	Fe	Hg	Mn	Ni	Pb	Zn
Coastal										
North Sea										
Central North Sea[1]	1988–89	0.009–0.03	<0.003–0.008	0.12–0.20	0.04–0.24		<0.19–1.8	0.18–0.27	0.02–0.05	0.15–2.0
off Forth[2]	1983	0.02		0.20					0.03	
off Tyne[3]	1991–92	0.01		0.41		0.005	1.3	0.49	0.04	0.66
off Humber[3]	1991–92	0.02		0.71		0.002	0.73	0.29	0.05	2.2
off The Wash[3]	1991–92	0.03		0.74		0.0003	1.3	1.0	0.19	1.0
Outer Thames[3]	1991–92	0.03		0.83		0.005	0.90	0.90	0.07	0.92
Strait of Dover[4]	1990–91	0.009–0.06	0.003–0.06	0.11–1.1	0.02–4.3		0.10–3.7	0.18–1.4	0.01–0.69	0.09–1.5
English Channel[3]	1991–92	0.02		0.21		<0.0002	0.90	0.24	0.03	0.57
Bristol Channel[3]	1991–92	0.08		1.0		0.0003	0.90	0.71	0.10	2.1
Irish Sea[3]	1991–92	0.03		0.61		0.0006	1.7	0.42	0.07	0.72
Cardigan Bay[3]	1991–92	0.07		0.85		0.0009	0.30	0.56	0.06	2.1
Liverpool Bay[3]	1991–92	0.04		1.5			7.8	0.87	0.17	1.8
North Atlantic Ocean										
Surface[5]		0.0001–0.001	0.001–0.02	0.06–0.08	0.003–0.06	0.0002–0.001	0.05–0.16	0.12	0.02–0.03	0.007–0.01
Deep[5]		0.04	0.001–0.002	0.13	0.03–0.06	0.0002	0.01–0.03	0.83	0.004	0.10
Distribution[5]		R	S	R + S	R + S	S	S	R	S	R

Sources: Tappin et al. (1995);[1] Balls and Topping (1987);[2] Law et al. (1994);[3] Statham et al. (1993);[4] Donat and Bruland (1995).[5]

British Isles, any increases usually resulting from significant local anthropogenic sources. In the light of discussion in Chapter 6, it should be no surprise that dissolved concentrations of cadmium (0.4 to 1.9 µg/L) and zinc (2.8 to 44 µg/L) in the Bristol Channel were very high indeed in 1971 (Abdullah et al., 1972; Abdullah and Royle, 1974). These concentrations were still raised appreciably in the Bristol Channel in the 1990s (Table 7.1). Liverpool Bay, on the eastern edge of the Irish Sea, also had relatively high dissolved concentrations of several trace metals in the 1990s (Table 7.1), reflecting the influence of Liverpool and Manchester via the River Mersey. Again, dissolved concentrations of several of these trace metals in Liverpool Bay had fallen since the beginning of the 1970s, from means of 0.3, 1.7 and 12 µg/L of cadmium, lead and zinc (Abdullah et al., 1972), to 0.04, 0.17 and 1.8 µg/L respectively at the start of the 1990s (Table 7.1). The falls in dissolved trace metal contamination in these coastal water hotspots can be attributed to the increased regulation and limitation of metal-contaminated industrial and domestic discharges since the 1970s.

Anthropogenic input of trace metals into coastal waters is not only via the principal route of contaminated river waters via estuaries. After the phasing out of sewage sludge dumping, there remain dumping of dredged spoil, direct discharges and terrestrial runoff, release from the antifouling paints of shipping, as well as atmospheric fallout via rainfall and dry dust deposition. An indirect source may be release from bottom sediments, particularly if these are resuspended by tidal, wave or storm action.

7.2.2 Ocean Profiles

Prologue. Dissolved trace metal concentrations are not typically constant with oceanic depth but show one of three types of vertical profile.

Table 7.1 also provides data on the dissolved trace metal concentrations in the oceanic waters of the North Atlantic. Concentrations are shown separately for surface waters and deep waters. It is striking how

much these differ (often more than tenfold) for several of the trace metals listed.

Dissolved trace metals show different vertical profiles with ocean depth, dependent upon their biogeochemical interactions with suspended particles in the water column. These vertical distributions fall into three main categories: conservative, recycled and scavenged (Bruland, 1983; Donat and Bruland, 1995).

Trace metals with a conservative distribution interact weakly with particles and show a constant concentration with depth. Concentrations will only change with salinity and mixing of different water masses (Donat and Bruland, 1995). The major metal ions of magnesium, sodium and potassium fall into this category. Of trace metals, the only significant ion that is conservative is the molybdate anion MoO_4^{2-}.

Recycled trace metals have dissolved concentrations that are depleted in surface waters and enriched at depth (Figure 7.2). Recycled trace metals are also called nutrient-type trace metals (Donat and Bruland, 1995).

It is worth digressing slightly here to explain. Phytoplankton only grow in the surface waters of oceans where and when there is sufficient light to drive photosynthesis. Sunlight is rapidly absorbed on passage down the water column. Phytoplankton production ceases at about 1% of surface light levels, when the rate of photosynthesis is now exceeded by the rate of respiration of the phytoplankton. As a rough estimate, the depth of this so-called photic zone may be about 150 metres in a well-lit ocean but is reduced in less translucent coastal waters to about 30 metres or less. In the North Atlantic, there is insufficient light in the winter for phytoplankton to grow at all. When light levels are increasing in spring or decreasing in autumn, the potential for phytoplankton growth then depends on the relative depths of the photic zone and the depth of vertical mixing, taking the phytoplankton into and out of the photic zone. In the summer, there is maximum light penetration, and warming of the surface waters limits vertical mixing, even to the extent of isolating the now less dense surface water above the rest of the water column. The interface between the two water bodies is a zone of very rapid temperature change, the

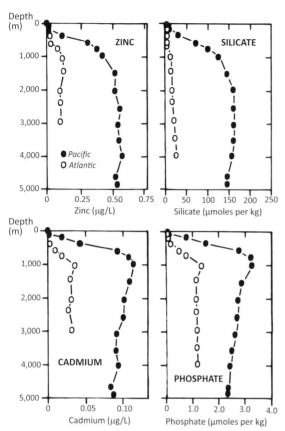

Figure 7.2 Profiles of dissolved zinc, cadmium, silicate and phosphate in the Atlantic and Pacific oceans. (After Luoma and Rainbow, 2008.)

thermocline. In the north temperate spring, increasing light levels and a decreasing depth of vertical mixing enable phytoplankton to grow, producing a spring bloom. At this time of rapid growth, the phytoplankton take up essential nutrients such as nitrate, phosphate and silicate from solution, greatly depleting dissolved nutrient concentrations in the surface waters. As the summer progresses, there is no replacement of nutrients from deeper waters because there is no vertical mixing across the thermocline. Without new nutrients, the phytoplankton cannot grow, although there is plenty of light. In fact, this is a permanent state of affairs in tropical oceans, which consequently, but initially counterintuitively, have very low levels of phytoplankton and primary production. In the North Atlantic in the autumn, cooling

surface waters start to mix again with deeper waters with high nutrient levels. The phytoplankton may, then, bloom again briefly before a decreasing depth of the photic zone interacting with an increased depth of vertical mixing causes a cessation of phytoplankton production. Another process is also taking place. Dying phytoplankton, together with the faecal pellets and bodies of the zooplankton that have grazed the phytoplankton, are sinking out of the photic zone through the thermocline to deeper water. Here they are subject to microbial decomposition of their organic matter, regenerating the inorganic nutrients into solution. Vertical mixing returns these nutrients to the surface waters as autumn turns to winter, but there is insufficient light for the phytoplankton to grow until the next spring.

The upshot of all this is that during the temperate summer, and for all the year in the subtropics and tropics, dissolved concentrations of nutrients, such as nitrate, phosphate and silicate, in oceans have vertical profiles of surface depletion and enrichment in deeper waters below the thermocline (Figure 7.2). The nutrients have been recycled. Particular trace metals such as silver, cadmium, chromium, nickel, vanadium and zinc (Figure 7.2) have the same bio-geochemical cycle as nutrients (Donat and Bruland, 1995). In a typically permanently stratified ocean, these trace metals are depleted in surface waters by uptake by phytoplankton and/or adsorption onto detritus particles derived from phytoplankton (Donat and Bruland, 1995). As dying phytoplankton and organic detritus particles sink into deeper water below the thermocline, the previously accumulated trace metals are regenerated by microbial decomposition, replenishing and maintaining higher dissolved concentrations of the trace metals in the deeper waters (Table 7.1, Figure 7.2). The vertical profile of dissolved cadmium correlates closely with that of phosphate (Figure 7.2), suggesting that cadmium is cycled with the formation and decomposition of organic tissues. The vertical distribution of dissolved zinc, on the other hand, correlates most closely with that of silicate (Figure 7.2), required particularly by diatoms to construct their valves.

Scavenged trace metals interact so strongly with particles in the ocean that they are removed rapidly from solution without recycling (Bruland, 1983; Donat and Bruland, 1995). Their dissolved concentrations are at a maximum near their sources, which may include rivers via coastal waters, atmospheric dust, bottom sediments and hydrothermal vents. Dissolved concentrations then decrease rapidly with distance away from the source ((Bruland, 1983; Donat and Bruland, 1995). Scavenged metals include aluminium, arsenic, cobalt, manganese, mercury and lead (Table 7.1). Lead enters oceans predominantly from the atmosphere, and so concentrations are high at the surface and decrease with depth (Table 7.1) (Bruland, 1983).

Some trace metals, for example copper and iron, have hybrid vertical distributions in the oceans, being influenced by both scavenging and recycling (Table 7.1) (Donat and Bruland, 1995). In areas of high primary productivity, dissolved copper and iron show a recycled vertical distribution with depletion in surface waters and regeneration at depth. Under these circumstances, dissolved copper concentrations increase with depth, but only gradually as a result of the interaction of regeneration and strong scavenging. Regenerated dissolved iron is scavenged particularly strongly throughout the water column, so there is no increase of dissolved iron concentration with depth. As well as being scavenged by particles, iron in seawater also comes out of solution by rapid removal by hydroxide colloid formation (Whitfield, 2001). In less productive oceanic waters, or in areas with high surface input via the deposition of dust, copper and iron may show maximum concentrations at the surface (Donat and Bruland, 1995).

7.3 Plankton

Prologue. Plankton represent the dominant life in the waters of coastal seas and the upper 100 metres or so of the oceans. The term 'plankton' describes the drifting organisms that are so small that they are unable to maintain their position or distribution independently of the movement of the body of water in which they are living. Plankton are divided into phytoplankton, carrying out photosynthesis, and zooplankton, consisting of small herbivorous, omnivorous and carnivorous animals. All production in the water column of seas and oceans passes through phytoplankton and zooplankton.

7.3.1 Phytoplankton

Prologue. Phytoplankton are the primary producers of the water column of coastal seas and oceans. Phytoplankton are unicellular, both eukaryotic and prokaryotic.

In British seas and adjacent oceans, diatoms are dominant members of the phytoplankton in cooler, more nutrient-rich waters, such as at the time of the spring bloom. Dinoflagellates cope better with warmer, nutrient-poor conditions during the temperate summer or in permanently stratified tropical and subtropical waters. So do the coccolithophores that we met in Chapter 4 as the major biological originators of the chalk deposits underlying so much of the British Isles. The importance in sequence of diatoms and dinoflagellates as the major primary producers in the marine waters around the British Isles and elsewhere has been subject to reinterpretation recently. It is now appreciated that the plankton nets used to collect phytoplankton were letting through photosynthetic bacteria, other prokaryotes and many very small eukaryotic members of the phytoplankton. As a result, their contribution to marine primary productivity has been grossly underestimated, particularly in warmer waters.

Whatever their systematic origin, all phytoplankton need the same resources in order to grow, namely light, carbon dioxide and a supply of essential nutrients. We have mentioned previously classic nutrients such as nitrate, phosphate and silicate. Also in the list of requirements are the essential trace metals, including iron, manganese, zinc, copper, cobalt, molybdenum and nickel in descending order of importance (Sunda, 1989). A primary essential role of trace metals is as a catalytic centre in an enzyme, as in the case of zinc in carbonic anhydrase.

7.3.1.1 Trace Metal Uptake and Accumulation

Prologue. Both essential and non-essential trace metals are taken up from solution by phytoplankton. Inorganic and organic complexation strongly affects the uptake of particular trace metals from solution by marine phytoplankton.

Dissolved trace metals are taken up by phytoplankton after binding with transporter proteins in the membrane of the cell (Sunda, 1989; Sunda and Huntsman, 1998). Each essential trace metal will bind to one or more transporters considered specific to that metal. Nevertheless, no transporter is absolutely specific to one particular trace metal. Other essential trace metals, or indeed non-essential metals, will bind to a transporter that may have evolved to transport a different essential metal (Sunda and Huntsman, 1998). Binding of different trace metals to that transporter will occur with different affinities, with the expectation (not always borne out) that the intended trace metal will bind with the strongest affinity. Such catholic binding can lead to the competitive inhibition by one trace metal on the rate of uptake of another.

The rate of trace metal uptake from solution by a phytoplankton cell will depend on the interplay of the processes in solution that deliver the trace metal to the cell membrane in the appropriate (bioavailable) chemical form for binding with the transporter, as well as the amount of these bioavailable chemical forms present (related to the dissolved concentration of the trace metal) and the number of transporter proteins in the cell membrane. The bioavailable form of a dissolved trace metal is modelled by the free metal ion, for, in addition to the free metal ion itself, bioavailable forms also include small labile inorganic complexes of these ions that dissociate to deliver the free metal ion to the transporter protein. The free metal ion is often not the most abundant chemical form of the dissolved trace metal in seawater (Table 6.3). Significant proportions of dissolved trace metals are complexed by different inorganic complexing agents, even in the absence of any dissolved organic compounds that might also form complexes with the dissolved trace metal. Thus dissolved cadmium, mercury and silver are strongly complexed by chloride ions, copper and lead by carbonate ions and iron by hydroxide ions (Table 6.3) (Sunda and Huntsman, 1998). Inorganic complexation decreases the availability of the free metal ion, restricting the rate of trace metal uptake.

Complexation of dissolved trace metals by organic complexing agents will also decrease the uptake rate of a dissolved trace metal by phytoplankton. Organic complexation in seawater is of major significance in the cases of copper and zinc (Sunda, 1989; Sunda and Huntsman, 1998). For these two trace metals, complexation by dissolved organic carbon compounds can be very strong indeed. This is particularly the case in coastal waters with raised dissolved concentrations of organic compounds, whether derived from, for example, phytoplankton or benthic macrophytic algae. More than 99% of dissolved copper in seas and oceans is organically complexed, except in deep oceanic waters (Sunda and Huntsman, 1998). Ninety-eight to 99% of dissolved zinc is also complexed by dissolved organic compounds in oceanic surface waters (Sunda and Huntsman, 1998). The exact chemical nature of much of this dissolved organic matter is still unknown.

While most trace metals occur in solution as the positive cation, a few do exist as negatively charged anions, for example arsenic as arsenate ($HAsO_4^-$), chromium as chromate (CrO_4^{2-}), molybdenum as molybdate (MoO_4^{2-}) and vanadium as vanadate (e.g., HVO_4^{2-}) (Table 6.3). Molybdate and chromate, at least, are taken up by sulphate uptake systems of phytoplankton (Sunda and Huntsman, 1998). The phosphate uptake system is another potential entry point for trace metal anions.

Further exceptions to the uptake of trace metals by binding of the cation to a membrane transporter system include the diffusion of small neutrally charged lipid-soluble compounds through the cell membrane, and the uptake of entire large trace metal complexes by specific membrane carriers (Sunda, 1989). Thus $AgCl_2^0$, $HgCl_2^0$ and methyl mercury CH_3HgCl^0 have the potential to bypass transporter proteins to diffuse across the cell membrane into the cell. Vitamin B_{12} is an organic complex of cobalt, the basis of the essential requirement for cobalt, and is transported into the phytoplankton cell by a specific membrane carrier (Sunda, 1989). Other such large organic complexes of a trace metal that can have specific membrane carriers in phytoplankton cells are the complexes of iron with siderophores (Sunda, 1989). Siderophores are large organic compounds usually secreted by prokaryotes to bind iron, mobilising it from insoluble iron hydroxide particles for uptake in this large complex form across the cell membrane.

Phytoplankton cells require more iron than any other essential trace metal (Whitfield, 2001). While iron is the most important essential trace metal for phytoplankton, it is also the trace metal that has the most complex speciation chemistry in seas and oceans. In oxygenated seawater, iron readily forms insoluble oxides and hydroxides, much reducing the proportion of total iron in the water column that is in dissolved form as the free ion Fe^{3+} (Sunda, 1989). These free iron ions will also be scavenged strongly by any suspended particles present (Whitfield, 2001). The amount of iron present as the insoluble oxide or hydroxide can be 10^{10} times greater than the amount of iron present as free Fe^{3+} ions (Sunda, 1989). And it is the quantity of free ions present that models the bioavailability of iron from solution, for the insoluble iron oxide and hydroxide particles suspended in the water are not directly available for iron uptake by phytoplankton (Sunda, 1989). Furthermore, free Fe^{3+} ions will form organic complexes with dissolved organic compounds, and such organically complexed iron probably accounts for any elevation of dissolved iron concentrations in many coastal waters (Sunda, 1989).

The low bioavailability of iron is a potential limiting factor for the growth of phytoplankton in marine systems. This is particularly the case in high-nutrient, low-chlorophyll (HNLC) regions of the world's oceans (Whitfield, 2001), for example in the Antarctic Ocean, far away from any input of iron in dust in the atmosphere derived from any nearby continent (Sunda, 1989). Most demonstrations of the alleviation of marine primary production limitation

by the addition of iron have taken place in in the equatorial upwelling system of the Pacific Ocean, the subarctic Pacific Ocean or in the Southern Ocean (Sunda, 1989; Whitfield, 2001). There is no evidence that any low bioavailability of iron is the limiting factor for the growth of phytoplankton in British coastal waters or in the North Atlantic and Arctic oceans.

Furthermore, marine phytoplankton, particularly oceanic phytoplankton, can regulate the rates of uptake of iron, manganese and zinc via an internal feedback system, in response to low-ambient free metal ion concentrations (Sunda, 1989; Sunda and Huntsman, 1998). The concentration of a trace metal accumulated in a phytoplankton cell represents a balance between the rate of metal uptake, the rate of metal excretion (if any) and the growth rate of the cells, in effect a rate of biodilution. By managing this balance, the phytoplankton cells can maintain intracellular concentrations of essential trace metals at optimal levels needed for growth and metabolism (Sunda, 1989; Sunda and Huntsman, 1998).

Any accumulated excess of an essential metal beyond immediate metabolic requirements and all accumulated non-essential metal need to be detoxified, temporarily or permanently. Accumulated cadmium, copper, mercury and zinc are detoxified in phytoplankton by binding to inducible phytochelatins (Sunda and Huntsman, 1998; Wang and Wang, 2009; Wu and Wang, 2014). Mercury may also be transformed into volatile gaseous mercury or deposited as insoluble mercury sulphide (Wu and Wang, 2014). Accumulated trace metals can be excreted from phytoplankton cells. Cadmium, for example, can be exported by the diatom *Thalassiosira weissflogii* bound to phytochelatin (Sunda and Huntsman, 1998). Both cadmium and copper can be excreted by the chlorophyte *Tetraselmis suecica* bound with organic matter (Nassiri et al., 1996), presumably based on phytochelatins. In *T. suecica*, cadmium and copper can also be extruded in vesicles during cell division (Nassiri et al., 1996), the vesicles possibly being derived from lysosomes breaking down phytochelatins.

7.3.1.2 Adaptations of Oceanic Phytoplankton

Prologue. Oceanic species of phytoplankton are adapted to grow at maximal rates at much lower free metal ion concentrations of essential metals, such as iron, manganese and zinc, than their coastal water counterparts growing in higher dissolved concentrations of these trace metals.

Different phytoplankton species have different trace metal requirements. Indeed, the primary reason for the improved performance of oceanic phytoplankton at low essential metal bioavailabilities is their substantially decreased cellular requirement for a specific essential metal (Sunda, 1989; Sunda and Huntsman, 1998). This decrease may, for example, involve the substitution of one trace metal for another in the catalytic centre of an enzyme or the substitute use of an enzyme without a trace metal component. Oceanic phytoplankton have also evolved an increased uptake ability for manganese in comparison to their coastal equivalents (Sunda, 1989; Sunda and Huntsman, 1998). Such an increased uptake ability on the part of oceanic phytoplankton is not found, however, for iron or zinc (Sunda, 1989; Sunda and Huntsman, 1998).

Oceanic phytoplankton can grow at lower cellular concentrations of manganese than can coastal species, an adaptation to low essential metal bioavailabilities seen also for iron and zinc (Sunda, 1989; Sunda and Huntsman, 1998). In addition, however, oceanic phytoplankton have an increased ability to take up manganese as a further important adaptation to the low bioavailability of manganese in the oceanic environment. The uptake of manganese by phytoplankton is carried out by a single, high affinity transporter under negative feedback regulation (Sunda, 1989; Sunda and Huntsman, 1998). The evolutionary pressure of the low oceanic bioavailability has acted to increase the affinity of the manganese transporter in oceanic phytoplankton (Sunda, 1989; Sunda and Huntsman, 1998). For example, the oceanic diatom *Thalassiosira oceanica* grows better at low manganese concentrations than the estuarine species *Thalassiosira pseudonana* (Sunda, 1989).

The manganese transporter is not absolutely manganese-specific. Copper, cadmium and zinc can all bind competitively with this transporter, potentially inhibiting the uptake of manganese and causing manganese deficiency in phytoplankton (Sunda and Huntsman, 1998). The same trace metals can also bind internally to components of the feedback system regulating manganese uptake, aggravating their ability to cause manganese deficiency (Sunda and Huntsman, 1998). In most coastal waters, however, the much higher concentrations of the free metal ion Mn^{2+} in comparison to the concentrations of the free metal ions Cu^{2+}, Cd^{2+} and Zn^{2+} are more than sufficient to compensate for the relatively strong binding of these trace metals to the manganese transporter.

Oceanic phytoplankton species also grow at lower dissolved iron concentrations than can coastal species (Sunda, 1989; Sunda and Huntsman, 1998). In the case of iron, this ability is down only to a substantially decreased cellular requirement for growth, and not to any increased uptake ability (Sunda, 1989; Sunda and Huntsman, 1998). In fact, there is probably no room for improvement of iron uptake systems, for example, by increasing the affinity of an iron transporter. The iron uptake rate of phytoplankton in general already appears to be at physical limits for the diffusion of labile inorganic iron species and free iron ions to the membrane binding sites and for the rate of exchange of iron at the transporter binding site (Sunda, 1989). Thus both the rate of delivery of dissolved bioavailable iron to the cell membrane and the rate of binding of this iron to the transporter protein are at the physical limits of each process. No increased affinity of the transporter will help.

The ability of phytoplankton to grow with low cellular concentrations of iron can be measured as the molar ratio of iron to phosphorus contents (Brand, 1991). Oceanic phytoplankton, including dinoflagellates and coccolithophores, have subsistence optimum molar ratios of iron to phosphorus of 1:10,000, while coastal species have ratios between 1:100 and 1:8,000 (Brand, 1991). Thus the oceanic dinoflagellates and coccolithophores are growing well with lower cellular concentrations of iron than their coastal counterparts. The same conclusion has been drawn for oceanic and coastal diatoms (Sunda et al., 1991). Possible mechanisms might include the replacement of ferredoxin, an iron-containing protein involved in photosynthesis, with the albeit less efficient protein flavodoxin, which lacks iron (Sunda, 1989). Cytochrome c-553, a protein containing iron and also involved in photosynthesis, is potentially replaceable by the copper-containing protein plastocyanin (Sunda, 1989).

The story for iron is repeated for zinc. Oceanic phytoplankton grow at lower dissolved zinc concentrations than can coastal species. Again, this is down to a decreased cellular requirement of the essential metal for growth, and not to any increased ability to take up zinc from solution (Sunda, 1989; Sunda and Huntsman, 1998). As for iron, the rate of uptake of zinc is also curtailed by inherent physical limits on the kinetics of the delivery of bioavailable zinc to the cell membrane and its exchange from solution to membrane transporters. Exchange kinetics are faster for zinc than for iron, so it is the diffusion rate that is more important in imposing limitations on the uptake rate of zinc by phytoplankton (Sunda and Huntsman, 1998). There are two membrane transporter systems for the uptake of zinc in eukaryotic phytoplankton such as diatoms, coccolithophores and unicellular chlorophytes. There is, firstly, a high affinity system under negative feedback control; and, secondly, a low affinity system not apparently under regulatory feedback (Sunda and Huntsman, 1998). It appears that both cadmium and cobalt will bind to the high affinity zinc transporter (Sunda and Huntsman, 1998).

So again, the primary adaptive mechanism of oceanic phytoplankton to low oceanic bioavailabilities of zinc is a reduced growth requirement for the essential metal (Sunda and Huntsman, 1992, 1998). The oceanic diatom *Thalassiosira oceanica* and the oceanic coccolithophore Emiliania huxleyi grow better at low dissolved zinc concentrations than the coastal diatoms species *Thalassiosira pseudonana* and *T. weissflogii* (Sunda and Huntsman, 1992). The two oceanic species can grow at near maximal rates at Zn^{2+} concentrations of 0.033 ng/L, whereas growth rates of the two coastal species are limited at free zinc ion concentrations below 0.65 ng/L (Sunda and Huntsman, 1992). Comparisons of molar ratios of cellular zinc to carbon show that the oceanic phytoplankton are able to grow at lower zinc bioavailabilities as a result of a reduced growth requirement for cellular zinc (Sunda and Huntsman, 1992).

The elucidation of one potential mechanism for a reduced zinc requirement in oceanic diatoms led to the exciting discovery of the use of cadmium for an essential purpose. Cadmium had, hitherto, generally been considered non-essential. In the 1990s, it became appreciated that cadmium could act as a nutrient for several oceanic species of phytoplankton, including diatoms and chlorophytes, under conditions of zinc limitation (Lee and Morel, 1995). It was then shown that the diatom *Thalassiosira weissflogii* contained a variant of the enzyme carbonic anhydrase that contained cadmium in place of the usual zinc (Lane et al., 2005). Carbonic anhydrase is a very significant enzyme in the physiology of phytoplankton, being involved with the acquisition of inorganic carbon needed for photosynthesis. The substitution of cadmium for zinc in this enzyme consequently makes an important decrease in the zinc requirements of the phytoplankton cells.

7.3.1.3 Deficiency and Toxicity

Prologue. While very low dissolved bioavailabilities of essential trace metals bring a risk of deficiency to phytoplankton, raised bioavailabilities in coastal waters may be associated with the opposite risk of ecotoxicity.

As well as having different essential trace metal requirements, different phytoplankton species have different sensitivities to high and low bioavailabilities of essential trace metals (e.g., Fe, Mn, Zn, Cu, Co, Ni), and to high bioavailabilities of non-essential metals (e.g., Ag, Hg, Pb) (Sunda, 1989). Therefore, different trace metal bioavailabilities can affect the species competition of marine phytoplankton communities.

We have discussed previously how low oceanic bioavailabilities have a direct potential to limit phytoplankton growth, as in the case of iron. Essential trace metal deficiencies may also come about by the indirect action of a high bioavailability of one trace metal inhibiting the uptake of another essential metal. Thus a high dissolved bioavailability of copper may lead to manganese deficiency in phytoplankton in upwelling systems with high copper concentrations (Sunda, 1989).

In their interaction with marine phytoplankton, most essential trace metals have the potential to follow the pattern depicted way back in Figure 1.1 – deficiency at low bioavailability and toxicity at high bioavailability.

At the high end of the dissolved concentration gradient, elevated zinc concentrations in coastal waters may inhibit photosynthesis by phytoplankton, as highlighted in British coastal waters in 1978 (Davies and Sleep, 1979). Samples of seawater and resident phytoplankton, including species of diatoms of the genus *Rhizosolenia* and dinoflagellates of the genus *Scrippsiella*, were collected 10 kilometres southwest of Plymouth. Increasing amounts of dissolved zinc were added to these seawater samples, which were considered to all have the same physicochemical characteristics that would affect the dissolved bioavailability of the zinc present. A dissolved concentration of 15 µg/L zinc, and possibly lower, inhibited photosynthesis in the phytoplankton. Dissolved zinc concentrations between 5 and 10 µg/L zinc, and sometimes exceeding 10 µg/L zinc, were not uncommon in British coastal waters at the time, suggesting that zinc inhibition of phytoplankton growth was occurring (Davies and Sleep, 1979). We saw earlier that dissolved zinc concentrations in the Bristol Channel were as high as 44 µg/L in 1971 and 12 µg/L in Liverpool Bay in the early 1970s (Abdullah et al., 1972; Abdullah and Royle, 1974). By the 1990s, dissolved zinc concentrations were about 2 µg/L in both the Bristol Channel and Liverpool Bay (Table 7.1). Nevertheless, it does seem that dissolved concentration ranges of trace metals in British coastal waters are in the same ballpark as those that might affect the growth of phytoplankton.

7.3.2 Zooplankton

Prologue. Marine zooplankton come from many different animal phyla, although they also include large protistans, such as foraminiferans or radiolarians, capable of feeding on their phytoplanktonic protistan counterparts. Zooplankton may spend their whole lifecycle in the plankton or may represent just part of the life cycle of an invertebrate otherwise living on the bottom. Many of these temporary members of the zooplankton are the larval stages of benthic adult invertebrates and may dominate coastal waters at particular times of the year. Conversely, medusae (called jellyfish when large enough) are the planktonic stages of the lifecycle of otherwise benthic cnidarians. The carnivorous medusae bear the gonads and can therefore be considered as the adult stage of the lifecycle.

7.3.2.1 Calanoid Copepods

Prologue. The zooplankton of British coastal seas and adjacent oceans are dominated by crustaceans, particularly calanoid copepods, such as *Calanus finmarchicus*. Calanoids take up and accumulate trace metals from solution and from their diet of mainly phytoplankton. Dissolved trace metals will also adsorb onto the surface of these small invertebrates with high surface area to volume ratios.

Calanoid copepods are the major herbivores of the water column of British coastal seas and the North Atlantic Ocean. They are able to filter feed on the diatoms that populate the northern temperate spring bloom. Filter feeding has been evolved secondarily, for calanoid copepods primarily feed raptorially by seizing large diatoms or small zooplanktonic animals before ingesting their contents. Filter feeding is carried out by feathery setae on the second maxillae, which sieve out diatoms from a feeding current driven by the rapid vibration of another set of head appendages, the second antennae. But this filter feeding process costs a great deal of energy, and there needs to be a sufficient level of phytoplankton in the water to make it energetically worthwhile. This is certainly the case during the spring bloom, when the calanoid copepods can cash in on the good times. During other times of the year, the copepods need to fall back on energetically less-expensive raptorial feeding to survive. Associated with the two feeding modes are different digestive strategies with different gut passage times. When phytoplankton concentrations are high, the calanoid copepods carry out superfluous feeding with rapid gut throughput. The ingested phytoplankton are in effect compressed, as if in a wine press, to release the soluble contents of their cells. The slower gut passage time associated with raptorial feeding allows time for more complete digestion.

Copepods, like other members of the zooplankton, take up trace metals from solution, both by adsorption onto the exoskeleton and by absorption through the permeable surfaces needed for respiration. They also take up trace metals trophically from ingested food. The bodies, moulted exoskeletons and faecal pellets of copepods, with their trace metal contents, sink through the water column. These are decomposed by microorganisms, usually below any thermocline present, releasing trace metals back into solution. How long these dissolved trace metals stay in solution depends on whether they are then scavenged by any suspended detritus particles present, strongly so in the case of copper. The faecal pellets will vary in their trace metal contents according to their history in the alimentary tracts of the copepods. When filter feeding on phytoplankton is taking place, gut passage times are very rapid, and only the soluble contents of the cells of phytoplankton are available for assimilation. At slower gut passage times, there is time for more complete digestion of the phytoplankton cell structures, in addition to the cytoplasm of the cells.

These two different digestive strategies have knock-on effects for the assimilation by the copepods of the trace metals accumulated by the phytoplankton. During filter feeding at high phytoplankton concentrations, the assimilation of trace metals by calanoid copepods is directly related to the cytoplasmic contents of trace metals in the filtered phytoplankton and not their total trace metal content (Reinfelder and Fisher,

1991; Rainbow et al., 2011a). During filter feeding, calanoid species of the genera *Acartia* and *Temora* assimilated cadmium, selenium, silver and zinc from the diatom *Thalassiosira pseudonana*, with efficiencies of 17%, 97%, 30%, and between 27 and 47%, respectively (Reinfelder and Fisher, 1991). Similarly, a species of the genus *Calanus*, filter feeding at a high phytoplankton concentration, had assimilation efficiencies of 35%, 56% and 30% for cadmium, selenium and zinc from the diatom *Thalassiosira weissflogii*, and assimilation efficiencies of 57%, 62% and 31% respectively when feeding on the dinoflagellate *Prorocentrum minimum* (Xu and Wang, 2001). Furthermore, the assimilation efficiencies of the trace metals in this latter study were positively related to the gut passage times of the ingested phytoplankton, confirming that digestion and subsequent alimentary absorption of the trace metals by the copepods are more efficient when the phytoplankton spend longer in the copepod gut (Xu and Wang, 2001). Nevertheless, it is apparent that there is still considerable uptake of trace metals by calanoid copepods from phytoplankton during filter feeding with associated shorter gut passage times. It is also the case that copepods can assimilate a lot of organic carbon from phytoplankton under these conditions, the assimilation efficiency of carbon reaching 84% in the first example mentioned (Reinfelder and Fisher, 1991).

Calanoid copepods can also assimilate trace metals associated with suspended detritus particles (Xu and Wang, 2002). Assimilation efficiencies are still high, being 47 to 83% for cadmium, 30 to 59% for selenium and 41 to 75% for zinc, in the case of a species of *Acartia* feeding on detritus particles derived from diatoms or a benthic macrophytic green alga (Xu and Wang, 2002).

Separately from any process of adsorption of trace metals onto the exoskeleton, calanoid copepods will also take up and accumulate trace metals from solution. Copepods are very small organisms with correspondingly high surface area to volume ratios, and their external cuticle is permeable. Trace metal uptake rates from solution can, therefore, be expected to be high in comparison with those of larger crustaceans

(Zauke et al., 1996). This is indeed the case. Accumulation rates of zinc from solution by species of *Calanus* (Zauke et al., 1996) were higher than those of a decapod, an amphipod and a barnacle (Rainbow and White, 1989). Cadmium accumulation rates from solution by the copepods exceeded those of the decapod and amphipod (Rainbow and White, 1989; Zauke et al., 1996).

The relative importance of dissolved trace metals and ingested trace metals as bioavailable sources of trace metals to calanoid copepods does inevitably depend on the respective concentrations of trace metals in solution and in the ingested diet, and on the amount of phytoplankton present to be filtered. Nevertheless, it is still possible to make some generalisations. At relatively low, but not atypical, concentrations of phytoplankton (the chlorophyte *Dunaliella viridis*), the calanoid copepod *Pseudodiaptomus coronatus* incorporated more cadmium from solution than from food (Sick and Baptist, 1979). Only at very high phytoplankton concentrations was more cadmium incorporated from the ingested phytoplankton than from solution (Sick and Baptist, 1979). Typically 60 to 80% of cobalt and silver, and 50 to 60% of cadmium accumulated by a variety of calanoid copepods were taken up from solution (Wang and Fisher, 1998; Fisher et al., 2000). On the other hand, about 80% of zinc and more than 98% of selenium accumulated by the copepods were derived from the diet (Wang and Fisher, 1998; Fisher et al., 2000). Similarly, food is the dominant (>75%) source of copper to coastal calanoid copepods (species of *Acartia* and *Temora*) in uncontaminated coastal waters (Chang and Reinfelder, 2002).

Much of the trace metal accumulated from solution by small zooplanktonic crustaceans is adsorbed onto the exoskeleton (Fowler, 1977; Fisher et al., 2000), without being absorbed and subject to physiological processing. Correspondingly, moulted exoskeletons have high concentrations of adsorbed trace metals, accentuated by further trace metals being adsorbed from solution onto the newly available surfaces on a fresh moult (Fowler, 1977; Fisher et al., 2000). Trace metals absorbed from solution into the body of a copepod, or assimilated from the diet, do undergo physiological processing, including excretion. In fact, the excretion of accumulated trace metals by calanoid copepods is remarkably fast. Calanoid copepods feeding on diatoms and dinoflagellates turned over 1 to 3% of accumulated cadmium and selenium per hour (32 to 72% per day), and between 2 and 5% of accumulated zinc per hour (50 to 115% per day) (Xu and Wang, 2001).

This excretion of accumulated trace metals returns trace metals back into solution in the photic zone, while the decomposition of egested faecal pellets, moults or dead bodies regenerates trace metals over a longer time scale, typically in deeper waters in oceans or on the bottom in coastal waters. Copepod faecal pellets have sinking rates of about 100 metres per day (Angel, 1984), and so will leave the photic zone in a day or so. Microbial regeneration of trace metals from phytoplankton cells and detritus, on the other hand, is slow, of the order of days to weeks (Hutchins and Bruland, 1994). Microbial release of trace metals from copepod faecal pellets is also slow, half-times of zinc release ranging between 2 and 14 days (Lee and Fisher, 1992). Thus trace metal regeneration from copepod faecal pellets will also occur in deeper waters below the photic zone of oceans (Lee and Fisher, 1992; Hutchins et al., 1995) or on the bottom in shallow seas. For example, about 80% of the iron in copepod faecal pellets is retained during descent out of the photic zone, being remineralised in subsurface oceanic waters (Hutchins et al., 1995).

Table 7.2 lists accumulated concentrations of trace metals in calanoid copepods from British coastal waters and the North Atlantic Ocean.

Accumulated zinc concentrations in calanoid copepods vary from about 70 µg/g to 500 µg/g (Table 7.2) or higher (Zauke et al., 1996). Surface area effects seem to be playing a strong role here, increasing the proportion of surface adsorbed zinc in the total accumulated body concentration. Species of *Calanus* are relatively large calanoid copepods. The addition of smaller copepods of the genus *Paracalanus* to a sample of *Calanus* species increased the average zinc concentrations of the copepods (Table 7.2). Similarly, the smaller species of the genera *Acartia* and *Temora* had higher zinc concentrations than did species of *Calanus* (Table 7.2). The lower concentrations of zinc (about 70 µg/g) found in the

Table 7.2 **Concentrations (mean or range, μg/g dry weight) of trace metals in calanoid copepods from British coastal waters and the North Atlantic Ocean.**

Taxon	Region	Cd	Cu	Pb	Zn
Calanus finmarchicus	North Atlantic Ocean[1]	2.4	5.9	0.4	97
Calanus spp.	Northern North Sea/Atlantic[1]	11	7.0	1.0	70
Calanus spp.	Central North Sea[1]	3.2	6.6	1.0	123
Calanus spp.	Southern North Sea[1]	1.8	7.1	1.0	129
Calanus/Paracalanus spp.	Firth of Clyde[2]	0.8	13	8.0	200
Acartia spp.	Central North Sea[1]	1.7	15	2.0	491
Acartia spp.	German Bight[1]	2.5	9.7	0.7	225
Temora spp.	German Bight[1]	1.0	14		380–498

Note: Calanus spp. represent a mixture of *C. finmarchicus* and *C. helgolandicus.*
Sources: Zauke et al. (1996);[1] Steele et al. (1973).[2]

larger copepods with a lower proportion of surface adsorbed zinc (Table 7.2) are of the same order as the theoretical calculation of the zinc requirement (35 μg/g) of enzymes for zinc in crustaceans (Table 3.5) and a theoretical total metabolic zinc requirement of 50 μg/g (Rainbow and Luoma, 2011b, c). Given the rapid turnover of accumulated zinc by calanoids, between 2 and 5% of accumulated zinc per hour, it would seem that much of the absorbed zinc that has entered the body of the copepod is not stored in a detoxified form but remains metabolically available. Higher accumulated concentrations would not imply storage detoxification in the copepods, but an increased contribution of surface adsorbed zinc.

Body concentrations of copper in calanoid copepods seem less affected by size, typically falling in the approximate range of 6 to 15 μg/g (Table 7.2). This body copper concentration is again of the same order as the theoretical calculation (15 μg/g) of the copper requirement of enzymes in the metabolising tissue of crustaceans (Table 3.5). Copepods lack haemocyanin,

the presence of which would otherwise raise the background level of copper in a crustacean (Table 3.6) (Rainbow and Luoma, 2011b).

The raised concentrations of cadmium in species of *Calanus* where the North Sea meets the North Atlantic (Table 7.2) are probably caused by local increases of dissolved cadmium concentrations caused by upwelling along fronts between the two bodies of water (Zauke et al., 1996).

7.4 Pelagic Animals

Prologue. Marine animals of different phyla occupy the epipelagic, mesopelagic and bathypelagic zones of the water column of oceans, with many undergoing diel vertical migrations up from the mesopelagic. These pelagic animals including krill, decapod crustaceans, hyperiid amphipod crustaceans and squid and show different patterns of trace metal accumulation, some with strong cadmium accumulation in the absence of cadmium contamination. Mesopelagic deep sea decapods show evidence of copper deficiency. Ocean striders living on the surface of the ocean also have very high accumulated cadmium concentrations. Predaceous arrow worms strengthen their teeth with zinc.

The water column of an ocean is conceptually divided into three zones: the epipelagic zone from the surface to 200 metres, the mesopelagic zone from 200 to 1,000 metres and the bathypelagic zone between 1,000 and 6,000 metres (Herring, 2002). The epipelagic zone marks the depth limit of the photic zone below which the intensity of daylight is not sufficient for photosynthesis. The mesopelagic zone is the so-called twilight zone, still potentially receiving some very dim sunlight from above. The bathypelagic zone is below the influence of any daylight whatsoever, any light here being only of biological origin, known as bioluminescence. The arbitrary 6,000 metre depth limit of the bathypelagic zone takes us down to the bottom of most of the world's oceans, which typically consists of abyssal plains (Herring, 2002). The bathypelagic zone, as defined here, excludes the waters of deep oceanic trenches (down to about 11,000 metres), but these make up less than 2% of the sea floor (Herring, 2002). Not all the bottom of the oceans is flat, for there are also widespread midocean ridges, the sites of new seafloor formation. Trenches, on the other hand, are sites where one tectonic plate is descending below another.

Animals of many phyla occupy the pelagic zone of our seas and oceans. We are now moving from the zooplankton to the larger animals that can determine their position in the water column independently of the movement of the body of water in which they are living. Pelagic animals, arguably still dominated by crustaceans but now including fish, ultimately depend on the primary production of phytoplankton in the epipelagic zone as their source of energy. In spite of their mixed systematic origin, pelagic animals typically undergo a migration each day from a daytime depth often in the mesopelagic up to, or towards, the epipelagic zone at night. This is called a diel vertical migration (Herring, 2002). Incidentally, the deep scattering layer, shown up particularly by the use of acoustics in the Second World War to map the bottom and search for submarines, ascends and descends on such a daily cycle. This layer probably consists of crustaceans, in addition to other invertebrates (siphonophore cnidarians) with sound-reflecting gas spaces and mesopelagic fish with swim bladders (Herring, 2002). While diel vertical migration is strong and typical in temperate and tropical waters, it is much reduced or absent in polar waters. Vertical distances travelled may be between 50 and 250 metres for a large copepod, and between 500 and 700 metres for a mesopelagic decapod crustacean (Herring, 2002). Diel vertical migration consumes energy, and would not be carried out without some net benefit. The upper regions of the oceans contain more concentrated food for pelagic animals, and there is a nutritional advantage to feed high in the water column (Herring, 2002). But why not stay there? There is a downside to being up in the better lit waters of the ocean by day – the risk of predation by visual predators. So it would appear that the energetic cost of diel vertical migration is offset by a combination of increased food availability and decreased predation pressure (Herring, 2002). From the point of view of trace metal biology, the downward migration of a pelagic animal that has fed by night in the epipelagic zone, will also remove trace metals from the photic zone to deeper waters for ultimate regeneration, be it from excretion or microbial decay of faecal pellets or bodies.

7.4.1 Krill

Prologue. The first examples of pelagic animals to be considered here are euphausiid crustaceans, commonly called krill.

Widespread in British coastal and adjacent oceanic waters in the North Atlantic region is *Meganyctiphanes norvegica*, a euphausiid that can reach up to 4.5 centimetres long (Mauchline, 1984).

Krill primarily feed raptorially, seizing relatively large prey with their thoracic limbs. Like calanoid copepods, euphausiids have also secondarily evolved the ability to filter feed on the diatoms prevalent in the spring bloom of temperate waters or in upwelling regions elsewhere in the world. The first six of the eight pairs of thoracic limbs of euphausiids form a thoracic basket bordered by a fine mesh of setae on these limbs. Many euphausiids are omnivorous, acting as herbivores on diatoms at times of high primary productivity, but able to live as carnivores when diatom populations are low.

Table 7.3 presents accumulated trace metal concentrations in *Meganyctiphanes norvegica* collected in the Firth of Clyde, Scotland and the Northeast Atlantic Ocean.

Euphausiids are malacostracan crustaceans, and, therefore, like decapods, they use the copper-bearing protein haemocyanin to carry the oxygen needed for respiration. The euphausiids from both coastal and oceanic waters have body copper concentrations (Table 7.3) matching the theoretical estimates needed by a malacostracan crustacean to meet the essential needs of both copper-bearing enzymes and haemocyanin (about 57 µg/g, Table 3.6).

Body zinc concentrations of the two euphausiid samples (Table 7.3) are of the same order as the theoretical estimate of the total amount of zinc (50 µg/g) needed for metabolism (Rainbow and Luoma, 2011c). Higher body concentrations of zinc in the euphausiids (Table 7.3) may be attributed to the contribution of surface-adsorbed zinc on the exoskeleton. The higher body concentration of zinc in the Atlantic Ocean euphausiids than in their coastal counterparts may be indicative of a higher dissolved bioavailability of zinc in the oceanic environment. At first sight this is a counterintuitive suggestion. But the oceanic euphausiids do carry out diel vertical migration. Therefore, they will be exposed by day to the raised dissolved zinc concentrations of the mesopelagic zone. Moreover, these higher regenerated dissolved concentrations in deeper water will not be associated with high dissolved concentrations of organic matter that would otherwise complex the dissolved zinc and reduce zinc bioavailability in oceanic surface waters and especially in coastal waters. So it is indeed possible that the Atlantic krill are exposed by day to higher dissolved zinc bioavailabilities than are the krill in the Firth of Clyde, with consequential effects on their accumulated body zinc concentrations.

What about cadmium? In this case, it is the Firth of Clyde euphausiids that have the higher accumulated body concentrations (Table 7.3), indicating higher bioavailabilities of cadmium in the coastal waters. Yet, dissolved cadmium concentrations, like zinc,

Table 7.3 **Concentrations of trace metals (µg/g dry weight) in the euphausiid *Meganyctiphanes norvegica* collected from the Firth of Clyde and the Northeast Atlantic Ocean in 1984–1985, expressed as the mean or the estimated concentration of a 0.05 g euphausiid* (when a size effect is present) with 95% confidence limits.**

	Firth of Clyde	NE Atlantic
Cadmium	1.06*	0.66*
	(0.47–1.80)	(0.40–1.08)
Copper	36*	58*
	(31–41)	(40–83)
Iron	39	32
	(35–43)	(12–52)
Zinc	43	102
	(40–46)	(81–123)

Source: Rainbow (1989).

have recycled (nutrient-type) distribution profiles. So the krill would be exposed by day to raised dissolved cadmium concentrations and bioavailabilities in deeper waters. The key difference here may lie in the fact that dissolved cadmium is not complexed by dissolved organic matter. Therefore, unlike the situation for zinc, the raised dissolved cadmium concentrations expected in coastal waters are not organically complexed and directly represent high dissolved bioavailabilities. Thus the coastal krill have the higher accumulated cadmium body concentrations (Table 7.3). Much of the cadmium accumulated in the body of *M. norvegica*, as opposed to that adsorbed on the exoskeleton, is bound to the detoxificatory protein metallothionein (Poirier and Cossa, 1981).

Given their abundance in marine pelagic systems, krill, like copepods, enhance the biogeochemical

Table 7.4 **Concentrations of trace metals (μg/g dry weight) in the euphausiid** *Meganyctiphanes norvegica*, **its faecal pellets and moulted exoskeletons and its microplankton food source (phytoplankton, copepods, arrow worms and detritus particles >76 μm), collected in the Mediterranean Sea off Monaco. Also shown are the percentages of trace metal body burden contained in the moult.**

	Ag	Cd	Co	Cr	Cu	Fe	Hg	Mn	Ni	Pb	Zn
Faecal pellets	2.1	9.6	3.5	38	226	24,000	0.34	243	20	34	950
Moult	2.9	2.1	0.80	5.3	35	232	0.17	12	6.7	22	146
(% body burden)	(31)	(22)	(34)	(48)	(6)	(28)	(4)	(21)	(78)	(*ca* 150)	(18)
Euphausiid body	0.71	0.74	0.18	0.85	48	64	0.35	4.2	0.66	1.1	62
Microplankton	0.67	2.1	0.87	4.9	39	570	0.05	18	8.1	11	483

Source: Fowler (1977).

cycling of trace metals in seas and oceans through the release of particulate matter such as faecal pellets and moulted exoskeletons containing trace metals (Fowler, 1977). These serve as agents for the rapid transport of trace metals out of the surface zones to deeper waters, where microbial decomposition and the regeneration of dissolved trace metals can occur. This loss of trace metals from surface waters is enhanced by the diel vertical migrations of the krill, with ingestion of food in surface waters, and, at least, a proportion of, defaecation in the mesopelagic zone. Krill such as *M. norvegica* filter phytoplankton and suspended detritus particles from the water, but also feed omnivorously on zooplankton including copepods (Fowler, 1977). Table 7.4 summarises trace metal concentrations in the bodies of *M. norvegica*, in their faecal pellets and cast moults and in the microplankton on which they feed, in this case in the Mediterranean Sea off Monaco. Typically the faecal pellets of the krill contained the highest trace metal concentrations (Table 7.4). Trace metal concentrations of the moults were also often higher than those in the bodies of the krill (Table 7.4). The moulted

exoskeletons can contain high percentages of the body burdens of trace metals in the krill (Table 7.4). While much of this trace metal component in the cast moult will have originated from the trace metal originally absorbed by the euphausiid, additional trace metal would be adsorbed from solution onto the newly exposed surfaces of the moult. This will particularly apply in the case of lead, in which case the moult now contains more lead than the body of the euphausiid from which it originated (Table 7.4). It is clear then that the particulate products of euphausiids, particularly their faecal pellets, are significant contributors to the biogenic vertical transport of trace metals out of the surface waters of seas and oceans (Fowler, 1977).

7.4.2 Deep Sea Decapods and Copper Deficiency

Prologue. Several mesopelagic decapod crustacean species may be suffering from copper deficiency when young, limiting the synthesis of the copper-containing respiratory protein haemocyanin.

Many of the deep sea invertebrates migrating daily from the mesopelagic up to the epipelagic of the Atlantic and other oceans are decapod crustaceans. These crustaceans come from each of the large systematic subdivisions of the decapods, namely the *Dendrobranchiata* containing the penaeid genera *Gennadas* and *Sergia*, and the *Pleocyemata* with the caridean genera *Acanthephyra* and *Systellaspis*.

Table 7.5 presents data on the concentrations of trace metals in mesopelagic decapod crustaceans from the Northeast Atlantic Ocean. A size effect on the accumulated race metal concentration was apparent in several cases (White and Rainbow, 1987; Ridout et al., 1989). For example, in one collection of the caridean *Systellaspis debilis* (Plate 8a), there was a negative effect of increasing body weight on the accumulated body concentrations of cadmium, iron, manganese and zinc, probably in reflection of the higher contribution of surface adsorbed metal on the exoskeleton to the total body concentration in smaller individuals (White and Rainbow, 1987). On the other hand, there

was an increase in the body concentration of copper in *S. debilis* with increasing body weight (Figure 7.3) (White and Rainbow, 1987). There was a similar increase in copper body concentrations, with increasing size in another mesopelagic decapod, in this case in the penaeid *Gennadas valens* (Ridout et al., 1989).

The increase of body copper concentration of these mesopelagic decapods with increasing size, and therefore age, means that adult individuals have much higher copper concentrations than their juveniles (Table 7.6, Figure 7.3). The copper concentrations of these decapods of different size and age (Table 7.6) can be compared with theoretical estimates of the minimum amount of copper needed to meet the total essential copper requirements of decapod crustaceans (58 µg/g; Table 3.6). It would appear, then, that juvenile decapods only have sufficient copper to meet enzyme requirements, with little copper to spare for haemocyanin manufacture. This hypothesis was tested on a subsequently collected population of *S. debilis* from the Northeast Atlantic

Table 7.5 **Concentrations of trace metals (µg/g dw) in the bodies of mesopelagic decapod crustaceans collected from the Northeast Atlantic Ocean. Concentrations are means or, in the case of a significant size effect, the estimated concentrations in decapods of 0.1 g dw.***

	Cd	Cu	Fe	Mn	Zn
Dendrobranchiata					
Penaeidae					
Bentheogennema intermedia[1]	5.2	51	35	6.4	58
Gennadas valens[1]	3.0*	34*	22*	1.9	65
Sergia robustus[1]	2.0*	21*	15*	2.2	62*
Pleocyemata					
Caridea					
Acanthephyra purpurea[1]	2.1	43	81*	2.6*	57
Systellaspis debilis[1]	8.7	36*	26	3.1	49*
Systellaspis debilis[2]	14*	34*	42*	2.6*	52*

Sources: Ridout et al. (1989);[1] White and Rainbow (1987).[2]

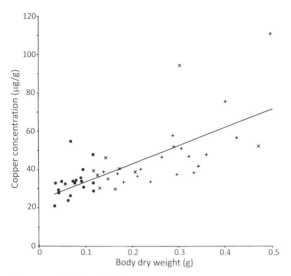

Figure 7.3 Relationship between body copper concentration (μg/g dry weight) and individual body dry weight (g) of combined juveniles (•), females (+) and males (x) of the caridean decapod *Systellaspis debilis* collected from the Northeast Atlantic Ocean. (After White and Rainbow, 1987; © 1987 Inter-Research, with permission.)

Table 7.6 **Concentrations of copper (μg/g dw) in the bodies of mesopelagic decapod crustaceans collected from the Northeast Atlantic Ocean, as estimated for individuals of 0.1, 0.2, 0.3 and 0.5 g dw.**

	0.1 g	0.2 g	0.3 g	0.5 g
Dendrobranchiata				
Penaeidae				
Gennadas valens[1]	34	64	91	144
Pleocyemata				
Caridea				
Systellaspis debilis[2]	34	43	53	72

Sources: Ridout et al. (1989);[1] White and Rainbow (1987).[2]

(Rainbow and Abdennour, 1989). Indeed, the juvenile *S. debilis* contained very little haemocyanin, while the adults contained a more typical caridean decapod quotient. In this population, a decapod of 0.10 g dw contained 9 μg/g copper bound to haemocyanin out of a total body copper concentration of 25 μg/g. A bigger individual of 0.15 g contained 58 μg/g copper as part of haemocyanin, out of a larger total body concentration of 107 μg/g copper. Furthermore, juvenile *S. debilis* undergo less distinct diel vertical migrations than adults (Roe, 1984). Perhaps this behavioural difference is related to a lack of sufficient copper for haemocyanin manufacture and consequently decreased respiratory performance that would otherwise assist vertical migration. The implication here is that these mesopelagic decapod crustaceans are approaching a state of copper deficiency in their oceanic habitat, particularly when young. Older individuals have had more time to accumulate more copper.

The body zinc concentrations of mesopelagic decapod crustaceans (Table 7.5) are consistently about the theoretical minimum (50 μg/g) needed for essential metabolic purposes (Rainbow and Luoma, 2011c). A body zinc concentration of about 50 to 65 μg/g may, therefore, represent a baseline zinc concentration for penaeid and caridean decapods. The body concentrations of iron and manganese in mesopelagic decapods (Table 7.5) may also be close to baseline concentrations. Theoretical estimates have been made of the enzyme requirements for these two trace metals, coming out as 37 μg/g iron and 3.9 μg/g manganese (White and Rainbow, 1987). Again, these theoretical estimates are similar to the minimum concentrations of iron and manganese measured in mesopelagic decapods (Table 7.5).

Cadmium concentrations in mesopelagic decapods, especially in *Systellaspis debilis* (Table 7.5), are surprisingly high in comparison to those in their coastal caridean decapod counterparts, including those from contaminated coastal waters (Table 6.27). This comparison is in direct contrast to that seen for euphausiids (Table 7.3), for which it was concluded that cadmium bioavailabilities were higher in coastal waters. Some other factor or factors seem to be

Table 7.7 **Concentrations of trace metals (µg/g dw) in the bodies of hyperiid amphipods from British coastal waters and the Atlantic Ocean. Concentrations are means or, in the case of a significant size effect, the estimated concentrations in amphipods of 0.01 g dw.***

	Location	Cd	Cu	Zn
Themisto compressa[1]	NE Atlantic	70	39	76*
Unidentified hyperiids[2]	Northern North Sea/Atlantic	51	26	72
Unidentified hyperiids[2]	Central North Sea/German Bight	7.4	14	128
Unidentified hyperiids[2]	German Bight	0.4	96	59

Sources: Rainbow (1989);[1] Zauke et al. (1996).[2]

involved here. One such factor may be a comparatively slow growth rate in the mesopelagic decapods, reducing any growth dilution of accumulated cadmium concentrations, in the absence of significant excretion. Also likely is a dietary difference between the decapods under comparison. As we shall see later in this chapter, some pelagic animals do have high accumulated cadmium concentrations. It is possible that such prey items are particularly strongly represented in the diet of *S. debilis*.

7.4.3 Cadmium and Oceanic Invertebrates

Prologue. Far from any source of anthropogenic pollution, some oceanic invertebrates have unexpectedly high body concentrations of cadmium. Such high cadmium concentrations in coastal invertebrates would raise questions of cadmium contamination, but, in fact, they appear to be perfectly natural.

7.4.3.1 Hyperiid Amphipods
Prologue. Hyperiid amphipod crustaceans are ecologically important in cold water pelagic ocean systems, forming large swarms with a very high total biomass (Raymont, 1983). Hyperiids accumulate very high concentrations of cadmium.

It is a feature of oceanic hyperiid amphipods that, in comparison to most pelagic invertebrates, including

zooplankton, they have remarkably high accumulated body concentrations of cadmium. Oceanic hyperiid amphipods, such as *Themisto compressa* in the Atlantic Ocean, can contain between 60 and 130 µg/g of cadmium (Table 7.7). In estuaries or other coastal waters, such accumulated cadmium concentrations in an amphipod, or indeed in any crustacean, would be indicative of extremely high cadmium contamination (Table 3.4). This is clearly not a case of anthropogenic contamination, for the cadmium concentrations of hyperiids in more coastal waters, including the German Bight adjacent to the output of River Elbe by Hamburg, are much lower, falling to a more expected crustacean cadmium concentration of 0.4 µg/g (Zauke et al., 1996).

So, a very high accumulated concentration of cadmium in oceanic hyperiid amphipods is a natural feature. Furthermore, the closely related hyperiid *Themisto gaudichaudii* in the Antarctic Ocean also have similarly high accumulated cadmium concentrations, up to 53 µg/g (Rainbow, 1989). The causes of this high cadmium concentration in oceanic hyperiids are unknown. Deep oceanic waters contain higher dissolved concentrations of cadmium than surface waters, at times matching dissolved concentrations in coastal waters (Table 7.1). While ocean deep water dissolved cadmium bioavailabilities may be high enough to cause some increase in accumulated cadmium concentrations in some mesopelagic crustaceans (Table 7.5), it is difficult to believe that they are so high as to cause the huge accumulated cadmium concentrations observed in the oceanic hyperiids (Table 7.7). The answer may lie in species differences between oceanic and coastal water hyperiids, and differences in their specific diets causing different trophic bioavailabilities of cadmium to these hyperiids.

Many hyperiids are associated with jellyfish or other gelatinous pelagic invertebrates, including siphonophores and pelagic tunicates, on which they feed (Raymont, 1983). Unlike this majority, species of *Themisto* are voracious generalist carnivores, feeding on the likes of copepods, krill, larval decapods, arrow worms and fish larvae (Raymont, 1983). Perhaps the rate of feeding and the presence of specific cadmium-rich prey items in the diet contribute to a high trophic bioavailability of cadmium to these oceanic hyperiids.

In turn, oceanic species of *Themisto* represent a significant part of the diet of many seabirds, particularly the smaller petrels, offering a significant trophic source of cadmium to these birds (Croxall and Prince, 1980; Rainbow, 1989).

7.4.3.2 Squid

Prologue. Amongst the pelagic predators of coastal seas and oceans are cephalopod molluscs, particularly squid. Some pelagic squid accumulate very high concentrations of cadmium.

Squid species found in British coastal waters include two species of *Loligo*. *L. forbesii*, variously referred to as the veined squid or the long-finned squid, is widespread. *L. vulgaris*, the common squid, occurs from the North Sea to the Mediterranean and West Africa. Also present are the European common squid *Alloteuthis subulata*, an otherwise eastern Atlantic species, and the European flying squid *Todarodes sagittatus*. Another pelagic cephalopod common in British coastal waters is the cuttlefish, *Sepia officinalis*, well known for its internal shell (the cuttlebone), washed up on our shores to be placed in the cages of budgerigars. In the Atlantic Ocean, there are many further species of deep sea squid.

Table 7.8 **Concentrations of cadmium (µg/g dw) in the bodies of pelagic cephalopods, with individual dry weights (g).**

	Cadmium concentration	Individual dry weight
Sepia officinalis (cuttlefish)		
English Channel	1.4	124
Loligo forbesii (long-finned squid)		
English Channel	0.45	263
West Irish Shelf	0.50	20
Loligo vulgaris (common squid)		
English Channel	0.40	14
Alloteuthis subulata (European common squid)		
English Channel	1.7	0.8
Todarodes sagittatus (European flying squid)		
West Irish Shelf	35	646
Faroe Islands	14	55
Bay of Biscay	1.2	16

Source: Bustamente et al. (1998).

Trace metals accumulated by cephalopods are typically detoxified and stored in the digestive gland. Thus 89% of the body content of cadmium in the cuttlefish *S. officinalis* is found in this organ, where it is detoxified by binding with metallothionein (Miramand and Bentley, 1992; Bustamante et al., 2002).

As Table 7.8 shows, the European flying squid *T. sagittatus* can contain a remarkably high body concentration of cadmium. Squid of this species from the West Irish Shelf, and to a lesser extent from the Faroe Islands, had much higher body cadmium concentrations than those from the Bay of Biscay

(Bustamente et al., 1998). Within the species, this apparently geographical difference may well be attributable to a size effect. Larger, older individuals of *T. sagittatus* had the highest accumulated cadmium concentrations (Table 7.8). A size effect does not, however, explain interspecific differences in accumulated cadmium concentrations. Large specimens of *Loligo fobesii* from the English Channel still had much lower cadmium concentrations than specimens of *T. sagittatus* of equivalent size (Table 7.8). Thus, *Todarodes sagittatus* does accumulate anomalously high cadmium concentrations over time, possibly from some unknown specific food source amongst its predominantly piscivorous diet.

Pelagic cephalopods are important marine predators in their own right, but in turn, they are also significant food sources for seabirds and marine mammals (Clarke, 1966; Croxall and Prince, 1996; Klages, 1996).

7.4.3.3 Ocean Striders

Prologue. Ocean striders are wingless insects living on the surface films of oceans. They also accumulate very high concentrations of cadmium from their oceanic diet.

The surface of the ocean is home to animals that are adapted to life at this interface. These animals are often divided into two categories: the pleuston and the neuston.

Members of the pleuston remain permanently at the water surface by their own buoyancy, positioned partly in the water and partly in the air. An example is the Portuguese Man o' War *Physalia physalis*, a siphonophore, with a float containing carbon monoxide.

Members of the neuston live on or under the surface film of the ocean. Included amongst the neuston are the wingless ocean striders. These insects are closely related to the more familiar pond skaters, and are sometimes called sea skaters. Ocean striders belong to the genus *Halobates*, the species occurring in the Atlantic Ocean being *Halobates micans* (Cheng, 1974). Ocean striders are often found associated with siphonophores such as *P. physalis*, and with other cnidarians of the pleuston such as the by-the-wind sailor *Velella velella*. Ocean striders, however, do not

apparently feed on these cnidarians, but rather prey on pontellid copepods, hyperiid amphipods, euphausiids and myctophid fish larvae trapped in the surface film (Cheng, 1974).

Ocean striders have remarkably high accumulated concentrations of cadmium. Specimens of *Halobates sobrinus* from off the Baja California Coast had cadmium concentrations ranging from 99 to 208 µg/g, with an average of 152 µg/g (Cheng et al., 1976). *H. micans* from the Atlantic Ocean also have high accumulated cadmium concentrations, if not an average quite so exorbitantly high. *H. micans* from the tropical Atlantic Ocean averaged 23 µg/g cadmium, although individual samples reached 309 µg/g (Bull et al., 1977). Similarly, other samples of this species from nine sites across the Atlantic Ocean had site means varying from 17 to 62 µg/g cadmium (Schulz-Baldes, 1989). Given the surface water depletion of dissolved cadmium concentrations in oceans (Figure 7.2), it is likely that the diet is a significant source of the cadmium accumulated by ocean striders. Perhaps hyperiids trapped in the surface film?

7.4.4 Zinc-Tipped Arrow Worms

Prologue. The teeth and the grasping spines around the mouth of arrow worms are impregnated with zinc for hardening.

Arrow worms belong to a small phylum of mostly planktonic, transparent marine worms called chaetognaths. They are typically abundant carnivores in the world of marine zooplankton and range from 2 to 120 millimetres long. A common genus is *Sagitta*, named after the Latin word for arrow.

The word 'chaetognath' means bristle jaw, reflecting the fact the head of an arrow worm is typically flanked by a number of grasping spines used to capture their prey. In addition to these spines, arrow worms are also provided with sets of rasping or biting teeth on either side of the mouth (Bone et al., 1983). Both spines and teeth are formed from chitin, but they are also impregnated with zinc (Bone et al., 1983). It appears that the zinc serves to harden the spines and jaws, a feature that we have already seen to have evolved independently in certain herbivorous insects and in nereidid polychaete worms.

7.5 Horn-Headed Amphipods and a Dietary Iron Challenge

Prologue. Stegocephalid amphipods characteristically have large crystals of ferritin containing iron in the epithelium of the ventral caeca of the alimentary tract. These crystals are detoxified stores of iron derived from the high levels of bioavailable iron in their cnidarian prey and are excreted with the faeces. A possible source of this trophically bioavailable iron is the iron contained in porphyrins (for example, haem groups) responsible for the red colouration in their food source.

7.5.1 Stegocephaloides christianiensis

Prologue. This stegocephalid amphipod detoxifies iron taken up from its diet in ferritin crystals subsequently lost in the faeces. A probable source of this trophically bioavailable iron is the red-coloured sea pen *Pennatula phosphorea*.

In 1974, Geoff Moore was investigating the gut contents of several amphipod crustaceans in Scottish coastal waters, in an attempt to shed light on their comparative feeding biologies (Moore, 1979). Amongst these was an amphipod rejoicing under the Latin name of *Stegocephaloides christianiensis* (Figure 7.4). This amphipod belongs to a family called the stegocephalids, literally 'horn-headed' amphipods, the biology of which had been relatively unstudied and was enigmatic. A remarkable feature of *S. christianiensis* was the presence of large octahedral crystals lying in longitudinal rows in the epithelium of the ventral caeca of the alimentary tract (Figure 7.5) (Moore, 1979). These crystals stained positively for the presence of iron, but, at that time, their functional significance could not be explained.

It was left to a follow-up study to confirm that these crystals consisted of ferritin (Moore and Rainbow, 1984). Ferritin is an ubiquitous iron-binding

Figure 7.4 *Stegocephaloides christianiensis*. The amphipod is 1.7 centimetres long. (After Moore and Rainbow, 1984, with permission.)

(a)

(b)

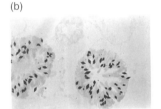

Figure 7.5 *Stegocephaloides christianiensis*: octahedral crystals (about 24 μm long), stained for iron in cells of the ventral caeca. (After Moore, 1979, with permission.)

protein in organisms that stores iron in a detoxified form. It usually exists in soluble form, and its deposition in crystalline form in cells is very rare. As a specific example, crystals of ferritin are used by limpets and chitons to store safely the iron needed for the mineralisation of the radula (Towe et al., 1963). It transpired that the ferritin crystals in the ventral caeca of *S. christianiensis* passed through the cell cycle of the epithelium before expulsion from the alimentary tract in the faeces (Moore and Rainbow, 1984). The ferritin crystals were part of a physiological system to cope with a high dietary source of iron, involving temporary storage of iron in concentrated but detoxified form before excretion from the body (Moore and Rainbow, 1984).

The question remained as to what food of the stegocephalid might offer such a strong challenge of trophically available iron. Analysis of fresh stomach contents of *S. christianiensis* revealed the presence of cnidarian stinging cells, including those of the sea pen *Pennatula phosphorea* (Moore and Rainbow, 1984). In a staged encounter in the laboratory, *S. christianiensis* fed on the tissues of the sea pen. It remained to show that the sea pen contained a high concentration of iron. Indeed it did, with iron concentrations as high as 3,160 µg/g, considerably above those in many other local benthic cnidarians, such as sea anemones (Moore and Rainbow, 1984).

This same study showed that ferritin crystals were present in the ventral caeca of nearly all specimens of other stegocephalid amphipods examined (Moore and Rainbow, 1984). Nothing in effect had been known on the feeding biology of any of these stegocephalids up to this time. It could now be concluded that it is a feature of the biology of stegocephalids that they typically contain large crystals of ferritin in the ventral caeca of the alimentary tract (Moore and Rainbow, 1984). These ferritin crystals are part of a system to expel, in detoxified form, excess iron taken up from a diet with a high trophic bioavailability of iron.

7.5.2 Parandania boecki

Prologue. *Parandania boecki* is a mesopelagic stegocephalid amphipod, containing the characteristic crystals of ferritin in the ventral caeca as detoxified temporary stores of iron derived from its diet. A possible prey item is the mesopelagic medusa *Atolla parva*, coloured red from its high concentration of iron-containing porphyrins.

Museum specimens of *Parandania boecki* had been included in the general survey of stegocephalids described previously, confirming that it too contained the characteristic crystals of ferritin (Moore and Rainbow, 1984). Nevertheless, as with so many other stegocephalids, nothing was known of its feeding biology.

(a)

(b)

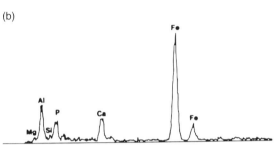

Figure 7.6 *Parandania boecki*: crystal of ferritin (about 30 µm long) in a cell of a ventral caecum of the alimentary tract, and an X-ray microanalysis spectrum confirming the presence of iron. (After Moore and Rainbow, 1989, with permission.)

Fresh material collected from the lower mesopelagic zone of the Atlantic Ocean on research cruises of the *R.R.S. Discovery* in 1985 and 1987 subsequently enabled investigation of the feeding biology of this hitherto unexamined stegocephalid amphipod (Moore and Rainbow, 1989). As now expected, ferritin crystals were prominent in the ventral caeca (Figure 7.6), with evidence also of their later expulsion into the alimentary tract and faeces (Moore and Rainbow, 1989). Furthermore, the gut of *P. boecki* contained what were identified as stinging cells of a variety of deep sea medusae, with those of the small, red, deep sea medusa *Atolla parva* being particularly prominent (Moore and Rainbow, 1989). The amphipod gut contents also tested positive for organic compounds called protoporphyrins that were specifically characteristic of *A. parva*. Protoporphyrins will bind with iron to form porphyrins such as the haem group in haemoglobin, a probable source of the red colour of the medusa. *Atolla parva* is a relatively common medusa in the emptiness of the deep sea and is probably, therefore, the prominent prey item of *P. boecki*. In another staged encounter, this stegocephalid fed readily on *A. parva*, gripping tightly the edge of the bell of the medusa as it fed (Plate 8b).

The question remained as to whether *Atolla parva* represented a particularly high dietary source of trophically available iron, reflecting a subsequent advantage for the predator of a physiological system to detoxify and/or expel the excess iron absorbed in the ventral caeca of the gut. This may well be the case. Specimens of *A. parva* had an average iron concentration of 254 µg/g (Moore and Rainbow, 1989), perhaps in correlation with a high concentration of haem responsible for its intense red colouration. In contrast, other deep sea medusae present had lower average iron concentrations, namely 14 µg/g in *Atolla wyvillei* and 19 µg/g in *Periphylla periphylla* (Moore and Rainbow, 1989). The body iron concentration of *P. boecki* itself averaged 90 µg/g. This body concentration of iron is not high compared to those of many inshore amphipods. This, however, is not surprising given the lower dissolved concentrations of iron in the deep sea than coastal waters (Table 7.1) and the fact that the very presence of a well-developed iron excretion mechanism would lower any accumulated iron concentration in *P. boecki* (Moore and Rainbow, 1989).

In hindsight, is it possible that the sea pen *Pennatula phosphorea* also contains pigmented protoporphyrins and porphyrins containing iron, as in a haem group, and that iron in this form is particularly trophically bioavailable? *P. phosphorea* certainly has a pronounced red colouration. Furthermore, red sea pens, including species of *Pennatula*, are indeed mainly pigmented by protoporphyrins (Bullock, 1970). And porphyrins are water soluble, suggesting that they would be relatively easily digested to release any constituent iron in a form readily assimilated and therefore trophically bioavailable. Moreover, iron ingested in the form of the haem group in haemoglobin in the blood of the prey of sanguivorous leeches needs detoxification in insoluble granules before expulsion from the intestinal epithelium of the leech (Jennings, 1968). This does suggest that iron does have a high trophic bioavailability when ingested as part of a porphyrin such as a haem group.

7.6 Sea Squirts and Vanadium

Prologue. Some, but not all, sea squirts contain extremely high concentrations of vanadium in specific blood cells, while other sea squirts have blood cells with high concentrations of iron. The function of the vanadium in sea squirts is not confirmed but may be associated with a role for vanadium in the formation of the tunic, perhaps as an evolutionary improvement of a previous role for iron. High concentrations of vanadium in the tunic of sea squirts may also have significance for the deterrence of would-be fish predators.

Sea squirts are common members of the marine fauna growing on the sea bottom, on rocks and other hard substrata. They will also grow on man-made structures such as piers and ships' bottoms and are important members of this fouling community. Sea squirts (ascidians) may be solitary, such as the common *Ciona intestinalis*, or may be compound, as in the case of *Botryllus schlosseri*, both of which may be found at the bottom of rocky shores on low spring tides in the British Isles.

Solitary sea squirts have a baglike body with two orifices through which seawater flows, driven by the beating cilia of a large perforated pharynx. This is a feeding current, and food suspended in this current is filtered by sheets of mucus moving across the walls of the pharynx (the branchial basket). This feeding mechanism is actually characteristic of early chordates, the phylum to which the vertebrates belong. Sea squirts belong to the tunicates, a subphylum of the chordates, for, unlikely as it might first seem, tunicates are amongst our closest evolutionary relatives. Sea squirts have a planktonic larva (the ascidian tadpole) with the further chordate characteristics of a dorsal nerve cord and a notochord. The bodies of tunicates are covered by a tunic of a cellulose-like polysaccharide, giving them their name.

While some species can withstand a short period of emersion at low tide, most sea squirts are sublittoral. Other British sea squirts include species of closely related genera in the family *Ascidiidae*, namely *Ascidia* (*A. mentula*), *Ascidiella* (*A. aspersa* and

A. scabra) and *Phallusia* (*P. mammillata*). *Ciona
intestinalis* belongs to the family *Cionidae*. It must be
remembered, though, that the assignments of species to
genera, and of genera to families, are systematic
hypotheses, likely to change between practising sys-
tematists. Both the *Ascidiidae* and the *Cionidae* are in
the order *Phlebobranchia*. The significance of this systematic relationship we shall see later in this section.

Botryllus schlosseri belongs to the family *Styelidae*,
together with species of the fouling genus *Styela*.
Relatively new to the British marine fauna is the
invasive Pacific stalked or leathery sea squirt *Styela
clava*, now common on piers, buoys, boats and fishing
gear, around the south of Britain from the Humber in
the east to the Firth of Clyde in the west. The family
Styelidae is in the order *Stolidobranchia*. Other British
ascidians in the order *Stolidobranchia* are species of
Pyura (e.g., *P. microcosmus*) in the family *Pyuridae*,
and *Molgula* (e.g., *M. occulta* and *M. manhattensis*) in
the family *Molgulidae*.

There is a third order of ascidians, the *Aplousobranchia*. Amongst the families of the *Aplousobranchia* are the *Didemnidae*, the *Polycitoridae* and the
Clavelinidae. Each family has representative species in
the British ascidian fauna. For example, in the
Didemnidae are species of *Didemnum* (e.g.,
D. maculosum) and *Trididemnum* (e.g., *T. tenerum*).
In the *Polycitoridae* are species of *Distaplia*
(*D. magnilarva*), and in the closely related *Clavelinidae* are species of *Clavelina* (e.g., *C. lepadiformis*, the
light bulb sea squirt).

What in the context of this book is the interest in
sea squirts and their systematic classification? In
1911, a German physiological chemist, Martin Henze,
reported extraordinarily high levels of vanadium in
blood cells of the sea squirt *Ascidia mentula* (Henze,
1911), well above vanadium levels ever previously
reported in any other organism (Michibata et al.,
2002). Henze showed that the vanadium is restricted
to blood cells, not free in the plasma of the blood
(Henze, 1911; Carlisle, 1968). Could vanadium possibly be playing an oxygen-binding role in a respiratory protein, as seen for iron in haemoglobin or
copper in haemocyanin (Michibata et al., 2002)? The
role of vanadium in the blood cells became even more

intriguing when it was discovered that these high
vanadium concentrations were characteristic of only
a few ascidian genera (Webb, 1939; Carlisle, 1968).
Species in the family *Ascidiidae* of the order
Phlebobranchia are the sea squirts which possess
vanadium in abundance, particularly species of
Ascidia and *Ascidiella* (Carlisle, 1968; Pirie and Bell,
1984). Sea squirt species in the closely related family,
the *Perophoridae*, also have high vanadium levels in
blood cells (Carlisle, 1968).

There are many types of cells in the blood of ascidians (Webb, 1939; Michibata et al., 2002). The high
vanadium levels, however, are restricted to blood cells
previously known as signet ring cells but now called
vanadocytes. The vanadium is present in one or more
large cell vacuoles containing material visible under
the electron microscope and rich in vanadium and
sulphur (Webb, 1939; Michibata et al., 2002). Amaz-
ingly, the sulphur is in the form of sulphuric acid, for
the vacuoles have an extremely low pH, down to 1.9,
made possible by the use of energy to pump hydrogen
ions across the vacuole membrane (Michibata et al.,
2002). While some of the vanadium in the acid-rich
vacuoles may be bound to low molecular weight proteins, much is in the form of the free V^{3+} ion (Michibata et al., 2002). Clearly the vanadium is not part of a
respiratory protein equivalent to haemoglobin or
haemocyanin. The small proteins binding vanadium
in the vanodocytes have a high specific affinity for
vanadium (Ueki et al., 2003). Now called vanabins,
these vanadium-binding proteins are rich in cysteine
residues and bind on average 16 vanadium atoms per
molecule of protein of between 12,000 and 16,000
molecular weight (Ueki et al., 2003).

Let us turn to the systematic distribution of high
vanadium concentrations amongst sea squirts.
Table 7.9 lists concentrations of vanadium and iron in
selected tissues and whole bodies of ascidians from
the three orders. It is the blood cells of ascidiid sea
squirts of the *Phlebobranchia* that have the high
vanadium contents, far above those, for example, of
sea squirts in the order *Stolidobranchia* (Table 7.9)
(Michibata et al., 1986). Ascidiid (and closely related
perophorid) sea squirts correspondingly have high
total body concentrations of vanadium (Table 7.9).

Table 7.9 **Accumulated concentrations (µg/g dw) of vanadium and iron in species of ascidians from families in the three different orders.**

Order	Vanadium	Iron
Family		
Species		
Tissue		
Phlebobranchia		
Ascidiidae		
Ascidia ahodori[1]		
Tunic	1,230	357
Blood cells	21,100	1,700
Ascidia sydneiensis[1]		
Tunic	31	1,430
Blood cells	4,680	905
Ascidia interrupta[2]		
Whole body	337	2,630
Phallusia mammillata[1]		
Tunic	15	118
Blood cells	9,860	413
Phallusia nigra[2]		
Whole body	1,980	1,210
Perophoridae		
Ecteinascidia turbinata[2]		
Whole body	1,130	555
Cionidae		
Ciona intestinalis[1]		
Tunic	2	370
Blood cells	331	424
Stolidobranchia		
Styelidae		
Styela plicata[1]		
Tunic	3	289
Blood cells	4	1,100
Botryllus planus[2]		
Whole body	<10	476
Pyuridae		
Pyura sacciformis[1]		
Tunic	1	319
Blood cells	1	451
Pyura vittata[2]		
Whole body	<10	583
Molgulidae		
Molgula manhattensis[1]		
Tunic	18	7,590
Blood cells	3	125
Aplousobranchia		
Didemnidae		
Didemnum spp.[2]		
Whole body	<2	311
Trididemnum sp.[2]		
Whole body	<5	115
Polycitoridae		
Distaplia bermudensis[2]		
Whole body	3,210	499
Clavelinidae		
Clavelina oblonga[2]		
Whole body	114	2,500

Sources: Michibata et al. (1986);[1] Stoecker (1980a).[2]

That said, high total body concentrations of vanadium are also found in some polycitorid (and to a lesser extent clavelinid) sea squirts of the third order, the *Aplousobranchia* (Table 7.9). Polycitorid sea squirts, however, lack typical vanadocytes (Stoecker, 1980a). While sea squirts in the order *Stolidobranchia* have low vanadium concentrations in their tissues, they have iron tissue concentrations that match those of sea squirts in the other two orders (Table 7.9).

The disjoint systematic distribution of high vanadium concentrations in the blood cells of sea squirts is intriguing.

In the sea squirt *Pyura stolonifera* in the order Stolidobranchia, there are iron-containing blood cells (ferrocytes) of similar morphology to the vanadocytes found in sea squirts in the order *Phlebobobranchia* (Endean, 1955; Carlisle, 1968). Furthermore, these cells are concerned with laying down fibrous elements of the outer tunic. The fibrous elements are synthesised in the iron-rich vacuoles of the blood cells before extrusion to form the tunic, consisting of a form of cellulose (Endean, 1955; Carlisle, 1968). It was subsequently shown that the same process of tunic formation occurs in the phlebobranchiate *Phallusia mammillata*, albeit in this case involving the vanadium-rich contents of the vacuoles in vanadocytes (Endean, 1961; Carlisle, 1968). Could, therefore, the primary role of both vanadocytes in phlebobranchiates and ferrocytes in stolidobranchiates be the synthesis of tunic material? Tunic formation appears to require the very acidic and powerful reducing conditions supplied by the contents of the vacuoles of these blood cells (Carlisle, 1968).

A case could be made that, over the course of evolution, vanadium has replaced iron in this process of tunic formation in particular sea squirt families. In such a scenario, this replacement would need to be associated with some selective advantage. Such an advantage might be increased efficiency of tunic formation and/or some benefit gained from having a raised vanadium content in the tunic and/or blood. Or is it the low pH of the vanadocyte vacuoles that is of functional significance (Stoecker, 1980a)? Another possibility is that under the low oxygen conditions that will affect emersed ascidians at low tide, the vanadium in the reducing conditions in vanadocyte vacuoles may be able to act as a carrier of molecular oxygen, albeit not in the guise of a respiratory protein (Carlisle, 1968).

One popular proposal for the functional significance of a high vanadium concentration in the tunic of a sea squirt is that it will act as a chemical deterrent to fouling organisms, or indeed predators or microorganisms (Stoecker, 1980a, b). The highly acidic nature of the body fluids of tunicates bearing vanadocytes may be an associated deterrent. In experiments carried out in Bermuda, however, there was no positive relationship between the vanadium concentration of a sea squirt and a lack of fouling epifauna on the tunics (Stoecker, 1980a). There was, though, a significant association between tunic acidity and a lack of such epifauna (Stoecker, 1980a). Predator deterrence may be more likely. High vanadium concentrations (above 100 µg/g wet weight) can reduce the palatability of prey to marine fish and crustaceans (Stoecker, 1980b). A wet weight concentration of 100 µg/g would translate to a dry weight concentration of 500 µg/g, assuming a 20% dry matter content. Vanadium concentrations in the bodies and tunics of some sea squirts do exceed this 500 µg/g threshold (Table 7.9).

The full functional significance of the high vanadium concentrations in the blood cells and bodies of specific sea squirts still awaits our comprehension. These remarkable accumulated vanadium concentrations remain as further indications of the variety of roles played by trace metals in the biology of the many different members of our fauna.

7.7 Polychaetes and Predator Deterrence

Prologue. The preceding section has raised again the possibility that atypically high accumulated concentrations of trace metals can act as chemical deterrents to potential predators, a concept met earlier in the case of the remarkable concentrations of arsenic accumulated by the polychaete *Aphelochaeta marioni*. In this section, we return to polychaete worms in a quest to explain other unexpectedly high concentrations of trace metals observed.

7.7.1 Fan Worms

Prologue. Sabellid polychaetes, commonly called fan worms, include species in which the crown of tentacles contains very high accumulated concentrations of either arsenic or vanadium. It has been proposed that these high concentrations act as a deterrent to potential predators.

Fan worms are suspension feeders, filtering suspended food particles from the surrounding water using a bilobed fan or crown of feathery tentacles lined with cilia. Sabellid polychaetes of the genus *Sabella* typically sit at the upper end of smooth erect tubes emerging from the sediment, with the fan extended into a crown. Fan worms are very sensitive to disturbance and will withdraw quickly into the tube. Otherwise, they are sitting pretty to be eaten by a passing fish.

There are several fan worms in British coastal waters. The common sabellid polychaete all around the British Isles is the peacock worm, *Sabella pavonina*, found in sand and mud sublittorally and in silty sand at the bottom of sheltered shores. A close relative is *Sabella spallanzanii*, the feather duster worm, which lives in sand or amongst rocks and stones. The distribution of the feather duster worm extends from the North Sea through to the eastern end of the Mediterranean, but it has also spread as an invasive species across the world to the coastal waters of South America, Africa and Australia. The tubes of the large-eyed feather duster worm *Pseudopotamilla reniformis* emerge from hard substrates such as rocks, old shells or calcareous algae into which the worm has bored. *Perkinsiana rubra*, a southern British species, similarly bores into limestone rock or calcareous algae. The twin fan worm *Bispira volutacornis* resides in rock crevices or under rocky overhangs, down to depths of 30 metres or more, although it can occur in intertidal rock pools. Other British fan worms include *Branchiomma bombyx* and *Megalomma vesiculosum*, both living in coarse sand or muddy gravel, and *Myxicola infundibulum*, which lives in a thick mucilaginous tube almost completely buried in soft sediment.

The crowns of tentacles of several sabellid fan worms are unpalatable to predators, particularly fish (Kicklighter and Hay, 2007). The presence and strength of this chemical deterrence do, however, vary between species, and even geographically between different populations of one species. Furthermore,

there appears to be a trade-off between the strategies of chemical deterrence and rapid escape down the tube when disturbed (Kicklighter and Hay, 2007). Fans and bodies of eight species of fan worms from the Caribbean and western Atlantic were presented to co-occurring fish and crabs in tests of their palatability to potential consumers (Kicklighter and Hay, 2007). Amongst these eight species were species in genera of fan worms found around Britain, namely *Sabella*, *Bispira*, *Branchiomma* and *Megalomma*. The predators either immediately rejected or rapidly regurgitated the crowns or bodies of the fan worms that they found unpalatable. The crowns of all fan worms were found to be unpalatable, as were the bodies of two *Bispira* species from North Carolina. Unpalatability increased with softness of tube and its weak resistance to tearing by a predator. The chemical deterrent causing unpalatability was not identified.

The strength of the tube and the unpalatability of the crown of tentacles are not the only defensive strategies employed by fan worms. Other mechanisms involve the toughness of the tentacles themselves and the release of mucus (Giangrande et al., 2013). Furthermore, the mucus released by sabellids does differ interspecifically in the degree of activity of lysozyme-like enzymes, taken to be an antimicrobial defence against bacterial attack (Giangrande et al., 2013). In the case of two British fan worm species, both *Sabella spallanzanii* and *Myxicola infundibulum* show high production of mucus with high lysozyme-like activity. Furthermore *S. spallanzanii* had the strongest tube of six sabellids investigated, including species of *Branchiomma* (Giangrande et al., 2013).

So where do trace metals fit in this discussion of the defensive strategies of fan worms? The answer lies in the proposal that the chemical deterrence linked to unpalatability of the crown of tentacles of several sabellids is down to the hyperaccumulation of arsenic or of vanadium.

7.7.1.2 Arsenic

Prologue. Do high accumulated concentrations of arsenic in the crown of tentacles of certain fan worms act as a chemical deterrent?

When presented with the fan worm *Sabella spallanzanii*, a British species, the white seabream, *Diplodus sargus sargus*, ate without hesitation the body of the worm but rejected the crown of tentacles as unpalatable (Fattorini and Regoli, 2004).

A clear chemical difference between these different parts of the body of the fan worm lay in their accumulated concentrations of arsenic. Concentrations of arsenic in the crown of tentacles averaged 1,040 µg/g, compared to an average 48 µg/g in the rest of the body (Fattorini and Regoli, 2004). In a survey of accumulated arsenic concentrations in polychaete worms, these crown concentrations of *S. spallanzanii* were second only to the atypically high accumulated arsenic concentrations, exceeding 2,000 µg/g, in the cirratulid polychaete *Aphelochaeta marioni*, discussed in Chapter 6 (Fattorini and Regoli, 2004). A proposed functional significance of these arsenic concentrations in *A. marioni* was also predator deterrence. Most marine organisms accumulate arsenic in organic forms (arsenobetaine, arsenocholine and arsenosugars) that are non-toxic to consumers (Neff, 1997). In the crown of *S. spallanzanii*, however, 80% of the accumulated arsenic was in the toxic form of dimethyl-arsinate, only the remaining 20% existing in the expected non-toxic organic forms (Fattorini and Regoli, 2004). The occurrence of arsenic in this toxic form strengthens the proposal that the hyperaccumulation of arsenic in the crown of *S. spallanzanii* acts as an antipredator chemical deterrent (Fattorini and Regoli, 2004).

This arsenic proposal has been explored further (Fattorini et al., 2005). The fan worm *Branchiomma luctuosum* has spread from the Indo-Pacific into the Mediterranean. The crown of tentacles of this species also has extremely high accumulated arsenic concentrations (1,590 µg/g) compared to the rest of the body (124 µg/g) (Fattorini et al., 2005). Yet, on the other hand, in an Antarctic fan worm, a species of *Perkinsiana*, later identified as *P. littoralis*, there was no hyperaccumulation of arsenic in the crown (Fattorini et al., 2005, 2010). In this case, the crown and the body both had low accumulated arsenic concentrations of 22 and 24 µg/g respectively (Fattorini et al., 2005).

It is apparent, therefore, that not all fan worms use hyperaccumulation of arsenic in the crown of tentacles as a chemical deterrent for predators. This variation is to be expected given the differential adoption of alternative defence strategies discussed earlier.

Thus, amongst polychaetes, only particular sabellid polychaetes even approach the accumulated arsenic concentrations measured in *Aphelochaeta marioni* (Fattorini et al., 2005; Waring and Maher, 2005). As is the case for the vast majority of marine organisms, including polychaetes, accumulated arsenic is usually present in non-toxic organic form (Waring and Maher, 2005). Intriguingly, this is not the case for the lugworm *Arenicola marina*. Although the lugworm typically has relatively low total accumulated arsenic concentrations (about 50 µg/g), much of this accumulated arsenic is in the particularly toxic form of arsenite, as well as in another inorganic form, arsenate (Waring and Maher, 2005). Could this be another example of antipredator chemical deterrence?

7.7.1.3 Vanadium

Prologue. The high accumulated concentrations of another trace metal, vanadium, in other sabellids may also serve the function of predator deterrence.

Arsenic is not the only trace metal contender as a source of chemical unpalatability in the crowns of tentacles of particular fan worms. Another contender is vanadium, as proposed previously for some sea squirts. As for arsenic, the concentrations of vanadium accumulated in the crowns of some fan worms are very high in some species but not in others (Fattorini and Regoli, 2012). These concentrations exceed all bioaccumulated vanadium concentrations, other than those of the particular sea squirts discussed earlier in this chapter. As in the case of arsenic, it has been proposed that this selective hyperaccumulation of vanadium by some fan worms is an antipredator device (Fattorini and Regoli, 2012).

In the Japanese fan worm *Pseudopotamilla occelata*, a close relative of the British large-eyed feather duster worm *P. reniformis*, the crown of tentacles has accumulated vanadium concentrations of between 3,000 and 7,000 µg/g, in contrast to the body concentration of up to 50 µg/g (Fattorini and Regoli, 2012). In the Antarctic and sub-Antarctic fan worm, *Perkinsiana littoralis*, a congeneric relative of the British *P. rubra*, the crown of tentacles has an average vanadium concentration of 10,500 µg/g, compared to a body vanadium concentration of 357 µg/g (Fattorini and Regoli, 2012). It was suggested earlier that a vanadium concentration above 500 µg/g dry weight (dw) might reduce the palatability of a prey item to fish and crustaceans (Stoecker, 1980b). Clearly, the crown concentrations of vanadium in at least these two fan worm species greatly exceed this proposed threshold. Indeed, it has been shown that the Antarctic emerald rock cod *Trematomus bernacchii* vigorously rejected the crowns of *P. littoralis* as a food source after tasting them, while it did not reject their bodies with the lower accumulated vanadium concentrations (Ishii et al., 1994; Fattorini and Regoli, 2012).

In *Pseudopotamilla occelata*, 90% of the total body content of vanadium in the fan worm was present in the crown of tentacles, although this represented only 7% of the total body weight (Fattorini and Regoli, 2012). As in sea squirts, the vanadium was stored as ionic V^{3+} ions in cell vacuoles, this time in the epithelial cells of the tentacles, with vanadium-binding proteins (vanabins) also present (Fattorini and Regoli, 2012). The concentrations of vanadium in these cell vacuoles of *P. occelata* are equivalent to as high as 30,000 µg/g dw. The vanadium in the crown of tentacles of *Perkinsiana littoralis* similarly constitutes more than 80% of total body vanadium, and is again in the epithelial cells partly bound to proteins (Fattorini and Regoli, 2012).

It does appear then that some fan worms use the hyperaccumulation of vanadium in the crown of tentacles as an antipredator deterrent. This vanadium hyperaccumulation appears to be an alternative to that of arsenic hyperaccumulation. The tentacles of *Sabella spallanzanii*, for example, are high in arsenic (about 1,000 µg/g), but very low in vanadium (below 0.1 µg/g) (Fattorini and Regoli, 2004, 2012). Similarly, the crown of tentacles of *Branchiomma luctuosum* has

vanadium concentrations below 1 µg/g, while its arsenic concentrations exceed 1,500 µg/g (Fattorini and Regoli, 2004, 2012). Correspondingly, the tentacle crown of *Perkinsiana littoralis* had an accumulated arsenic concentration of only 22 µg/g, in contrast to a vanadium concentration in this species of 10,500 µg/g (Fattorini and Regoli, 2005, 2012).

7.7.2 Copper in *Melinna palmata*

Prologue. Another trace metal, in this case copper, also appears to be playing a predator deterrence role in another polychaete, the ampharetid *Melinna palmata*.

Melinna palmata lives in fragile tubes in sandy muds, both intertidally and sublittorally, in southern and western British coastal waters. It is a deposit feeder, gathering food particles from the surface of the sediment with extensile feeding tentacles projecting from the head (Gibbs et al., 1981). Also present at the front of the body behind the head are eight threadlike gills. During feeding, both the head with tentacles and the upper part of the body of the worm, including the gills, protrude from the fragile tube in which the worm resides. These parts of the body are, therefore, at this time vulnerable to predation by fish, such as juvenile flounder *Platichthys flesus*, two-spotted gobies *Gobiusculus flavescens* and sand gobies *Pomatoschistus minutus* (Gibbs et al., 1981).

Whole body concentrations of copper in *Melinna palmata* from uncontaminated sediments are very high in comparison with those of other polychaetes, including another ampharetid, *Ampharete acutifrons* (Gibbs et al., 1981). Thus *M. palmata* from uncontaminated sites in the Plymouth area had body concentrations of copper between 500 and 1,100 µg/g, in comparison with 74 to 100 µg/g copper in *A. acutifrons*, and 12 µg/g in *Hediste diversicolor* from the nearby uncontaminated Yealm Estuary. The closely related species *Melinna cristata* also had a high body concentration of copper, in this case 650 µg/g in worms collected off the Northumberland coast. Much of the copper in *M. palmata* is concentrated in the gills, with copper concentrations between

3,830 and 12,300 µg/g. The gills contain 30 to 40% of the body copper, while accounting for only 3 to 4% of the body weight. The feeding tentacles also have very high copper concentrations. The gills of *M. cristata* too contain a high copper concentration (11,600 µg/g), and hold 21% of the body copper content. No other trace metals measured had atypically raised concentrations in the bodies or gills of *M. palmata*.

Feeding experiments confirmed that the gills and feeding tentacles of *M. palmata* are distasteful to the three previously discussed fish species, while the gills of the other ampharetid *A. acutifrons* with a low copper concentration (76 µg/g) were readily accepted as food (Gibbs et al., 1981). It does appear, then, that *Melinna palmata*, and presumably its congeneric species *M. cristata*, are making use of the hyperaccumulation of copper in exposed anterior parts of the body, especially the gills, to deter potential predators.

Polychaetes, therefore, have employed high accumulated concentrations of at least three different trace metals (arsenic, vanadium and copper) for the same functional purpose of the chemical deterrence of would-be predators.

7.8 Sediments

Prologue. Sediments act as sinks for trace metals, but trace metals associated with marine sediments may be bioavailable and interact with the local biota. Trace metal concentrations in the sediments of British coastal waters vary in time and space, often as a result of anthropogenic activities. One such anthropogenic activity has been the historical dumping of trace metal-contaminated sewage sludge and the ongoing dumping of dredged material in coastal waters around the British Isles.

Trace metals bound to sediments are bioavailable to different degrees to the infauna and epifauna of the sediments. They are trophically bioavailable to invertebrates that ingest sediment particles. Trace metals may also desorb from sediments into pore water and into the overlying water column, if the physico-chemical conditions are appropriate. For example,

manganese may be released from sediments to dissolve as the reduced Mn^{2+} ion in pore water in low concentrations of dissolved oxygen. According to the contribution of any oxygen-deficient pore water to the water bathing an infaunal invertebrate driving an irrigation current through its burrow, sediment-associated manganese may then become bioavailable in dissolved form. Whether taken up in dissolved form or from ingested sediment in the gut of a deposit-feeding animal, sediment-associated trace metals can be bioavailable. After bioaccumulation, these trace metals can enter coastal marine food chains, often leading to bottom-feeding fish such as flatfish, including plaice *Pleuronectes platessa*, sole *Solea solea* and dab *Limanda limanda* in British coastal waters.

7.8.1 Sediment Metal Concentrations

Prologue. As a result of considerable differences in mineralogy, grain size, organic matter content and local sources of anthropogenic inputs, coastal marine sediments show large variations in their concentrations of trace metals (Fowler, 1990). As an example, in the North Sea, trace metal concentrations in sediments may vary tenfold and often more than 100-fold, greatly affected by anthropogenic sources (North Sea Task Force, 1993). The southern North Sea, for example, receives metal-contaminated input from the estuaries of the Seine, Thames, Scheldt, Rhine, Weser, Elbe and the Humber, amongst others.

Table 7.10 lists concentrations of trace metals in the sediments of British coastal waters.

7.8.2 Dump Sites

Prologue. British coastal waters are affected by point sources of anthropogenic input of trace metals in the guise of historical and ongoing dump sites, with consequent increases in the trace metal concentrations of local coastal sediments.

The United Kingdom used to dump sewage sludge at sites in its coastal waters, a practice phased out in

Table 7.10 Means and ranges of typical concentrations (μg/g dw) of trace metals in the oxic surface sediments of British coastal waters, including sediments considered uncontaminated by trace metals and sediments from sewage sludge and dredged material dump sites.

	<μm	Digest	Ag	As	Cd	Cr	Cu	Hg	Mn	Ni	Pb	Zn
Uncontaminated												
Cawsand Bay[1]	100	1M HCl	0.02				13.7		159			55
Firth of Clyde[2,3]	204	HNO₃/HClO₄	<0.2	8 / 7–10	2 / 1–3	33 / 10–65	16 / 9–20	0.1–0.6	1,120 / 500–4,000	50 / 19–62	42 / 24–67	85 / 60–130
Liverpool Bay[4,5]	90	HNO₃/H₂O₂			<0.2	40–50	30–40	0.3–1.0		40–70	<100	180–200
Bristol Channel[6]	90	HNO₃/H₂O₂				13	24	0.1		33	43	99
Outer Thames[7]	90	HNO₃/H₂O₂			<0.2	2–10	15–20	<0.02–0.1		10–20	25–50	40–80
Sewage sludge sites												
Firth of Clyde[2,3]	204	HNO₃/HClO₄	1.6 / 0.2–5	21 / 15–24	6.4 / 3–8	164 / 48–308	269 / 250–300	1.9	911 / 500–1,000	50 / 34–70	361 / 269–403	631 / 437–681
Liverpool Bay[4,5]	90	HNO₃/H₂O₂			0.75–4.5	100–150	80–140	2.0–4.0		80–140	300–1,100	400–1,000
Bristol Channel[6]	90	HNO₃/H₂O₂				20–60	29–53	0.3–2.8		54–77	61–130	150–430
Outer Thames[7]	90	HNO₃/H₂O₂			0.5–4.5	40–98	20–110	0.3–0.7		40–75	50–300	80–340
Dredged material sites												
Liverpool Bay[4]	90	HNO₃/H₂O₂			1.5–3.4	50–110	40–150	1.5–3.9		40–80	100–200	200–500
Bristol Channel[6]	90	HNO₃/H₂O₂				23	38	0.5		53	94	180
North Sea												
North Sea[8]	2000	Total		11 / <1.2–33	0.05 / 0.01–0.38		14 / 0.1–87	<0.03–0.5		23 / 1.5–113	21 / 1.7–288	39 / 3–510
North Sea[9]	63	HNO₃/HCl/HF			0.2–6		10–100				30–120	100–1,000

Note: Summary data from the North Sea include uncontaminated sites and sites affected by anthropogenic contamination.[8,9]

Sources: Bryan and Gibbs (1987);[1] Mackay et al. (1972);[2] Thornton (1975b);[3] Norton et al. (1984a);[4] Norton et al. (1984b);[5] Murray et al. (1980);[6] Norton et al. (1981);[7] North Sea Task Force (1993);[8] Everaarts and Fischer (1992).[9]

1998. Between 1987 and 1990, the United Kingdom dumped between 5.1 and 5.7 million wet tonnes annually of sewage sludge into the North Sea alone (North Sea Task Force, 1993). In 1990, the 5.4 million wet tonnes of sewage sludge dumped into the North Sea contained 160 tonnes of zinc, as well as 77 tonnes of lead, 76 tonnes of copper, 21 tonnes of chromium, 11 tonnes of nickel, 1.2 tonnes of cadmium, 0.7 tonne of mercury and 0.1 tonne of arsenic (North Sea Task Force, 1993).

While the dumping of sewage sludge into British coastal waters has now ceased, the dumping of dredged material continues (Bolam et al., 2006; Bolam, 2012). Sediments need to be dredged in harbour areas, estuaries and navigation channels, and this dredged material can be deposited at coastal dump sites further offshore. The amount of dredged material dumped annually is considerable, and some of the dredged material, particularly from enclosed harbours and estuaries, is contaminated with trace metals. In 1990, the United Kingdom dumped nearly 22 million wet tonnes of dredged material into the North Sea, containing 2,235 tonnes of zinc, 882 tonnes of lead, 800 tonnes of chromium, 610 tonnes of copper, 288 tonnes of nickel, 200 tonnes of arsenic and just more than 4 tonnes each of cadmium and mercury (North Sea Task Force, 1993). There are more than 150 sites designated for the disposal of dredged material around England and Wales, receiving a total of about 40 million wet tonnes annually (Bolam et al., 2006). Disposal is localised over small portions of the seabed and regulated under licenses issued by the Centre for Environment, Fisheries and Aquaculture Science (CEFAS), as an executive agency of the Department for Environment, Food and Rural Affairs (DEFRA) (Bolam et al., 2006; Bolam, 2012). Other waste material also dumped at sea has historically included colliery waste and pulverised fly ash from coal-fired power stations, as off the coast of Northumberland.

Included in Table 7.10 are sediment concentrations at dump sites that received sewage sludge prior to 1998, and dump sites still receiving dredged material. With its high percentage content of organic matter, sewage sludge has a strong capacity to bind trace metals. Much of the sewage sludge requiring disposal came from large industrialised cities, such as Glasgow and London, and these sludges inevitably contained high concentrations of trace metals of domestic and industrial origin (Table 7.11). Correspondingly, sediments at sewage sludge dump sites show greatly elevated concentrations of trace metals (Table 7.10). Much more dredged material is dumped in British coastal waters than previous amounts of sewage sludge, contributing high total amounts of trace metals. Trace metals concentrations in dredged material are, however, usually lower than those in sewage sludge, depending on the source of the material. Thus trace metal concentrations in the sediments at dredged material dump sites are usually below those at sewage sludge dump sites, although often still raised above background (Table 7.10).

Unless trace metals are part of the mineral composition of a sediment, a possibility in the case of dredged material or colliery waste, the raised concentrations of trace metals in the sediments derived from dumped sludge and other material will be bioavailable to some degree to the local benthic fauna. Particular polychaete worms are good potential biomonitors of sediment trace metal bioavailabilities in and around dump sites (Bryan and Gibbs, 1987). An ideal polychaete biomonitor of trace metals in sediments affected by dumping would be a sedentary deposit-feeder, preferably one able to withstand shipboard washing and sieving after collection in a dredge or grab. The polychaete should be hardy, and common in a wide range of sediments of different trace metal bioavailabilities. It should not be too big, but of a total length suitable for easy analysis. A relatively long life span of three years or more would give time for greater accumulation of trace metals. A deposit-feeding polychaete with potential as a biomonitor is *Owenia fusiformis*, common at dump sites for sewage sludge and dredged material (Bryan and Gibbs, 1987). Unlike some other polychaetes, including *H. diversicolor* to a degree, *O. fusiformis* does not regulate its body concentrations of zinc across a range of zinc bioavailabilities. *O. fusiformis* can, therefore, be recommended as a biomonitor for

Table 7.11 **Concentrations (µg/g dw) of selected trace metals in sewage sludge dumped in British coastal waters.**

Dump site		Cd	Cr	Cu	Hg	Ni	Pb	Zn
Origin of sewage sludge	Date							
Firth of Clyde								
Glasgow[1]	ca. 1970	6–19	137–4,240	352–3,270	–	33–87	307–850	579–7,000
Outer Thames								
Beckton/Crossness[2]	ca. 1970	30–70	200–1,070	505–2,500	–	150–350	–	2,390–5,860
Beckton/Crossness[3]	1974–78	43–60	470–570	770–820	12–56	160–250	760–1,190	3,200–3,890

Sources: Halcrow et al. (1973);[1] Shelton (1971);[2] Norton et al. (1981).[3]

zinc, and also for copper, in sublittoral sediments (Bryan and Gibbs, 1987).

Carnivorous polychaetes can also act as biomonitors of the sediment bioavailabilities of trace metals, taken up and accumulated indirectly via local food chains. Thus *Nephtys hombergii*, another polychaete common at dump sites, can be used as a biomonitor for cadmium, cobalt, copper, iron, lead and silver. Another carnivorous polychaete, *Glycera tridactyla*, can be used as a biomonitor for copper, although it is recommended that the jaws, which contain copper anyway, be removed before body analysis (Bryan and Gibbs, 1987).

The resulting sediment at a sewage site dump site is characterised by abnormally high concentrations of trace metals and correspondingly raised trace metal bioavailabilities to the local marine fauna. Before seeking evidence for any subsequent ecotoxicological effects of trace metals on local benthic community structure, it is necessary to appreciate that dumped sewage sludge has other, probably more dominant, characteristics likely to cause biological changes to the local benthic community.

Firstly, there is a physical smothering effect of the sheer volume of dumped material. This material is also of very high organic content. High organic enrichment causes clear detrimental changes to a local benthic community, not least as a result of the establishment of anoxic conditions as microbes break down the organic carbon, using up any oxygen available before resorting to anaerobic respiration (Pearson and Rosenberg, 1978). Typically, in very high organic conditions, as at the centre of a dump site, the number of benthic invertebrates is very low. Nevertheless, the biomass and abundance of the few species that can survive are high (Pearson and Rosenberg, 1978). Such species are typically nematodes, together with certain oligochaete and polychaete worms, often with high haemoglobin contents (Pearson, 1987). Away from the centre of the dump site, species numbers increase, but biomass and abundance fall with decline of the high numbers of the species from the centre. A secondary biomass peak may then occur as more species are able to profit from the less severe organic enrichment, providing detrital food without associated oxygen deficiency. Beyond a dump site, benthic faunal numbers, biomass and abundance all decline to the typically lower background levels of the surrounding area (Pearson and Rosenberg, 1978; Pearson, 1987).

Such community changes driven by organic enrichment usually dominate the ecotoxicological effects of dumped sewage sludge at coastal dump sites, but not to the total exclusion of the effects of contaminants in the sludge. In addition to trace metals, dumped sewage sludge is also usually high in ecotoxic organic compounds such as petroleum hydrocarbons and organochlorines. Trace metal and organic contaminants have every potential to be taken up and accumulated by local benthic invertebrates, and ultimately be transferred via food chains to bottom-feeding fish (Galay Burgos and Rainbow, 1998, 2001).

7.8.2.1 Firth of Clyde

Prologue. The dumping of Glasgow sewage sludge at two closely adjacent sites off Garroch Head in the Firth of Clyde took place from 1904 to 1998. The resulting organically rich sediments at the sites have raised trace metal concentrations. The biodiversity of local sediment infauna and epifauna has been greatly affected, mostly as a result of smothering and organic enrichment, but with a suggestion of some interactive effect of raised cadmium bioavailability.

In 1904, dumping began of sewage sludge from the city of Glasgow into a small area about 2.6 kilometres off Garroch Head, at the southern end of the Isle of Bute in the Firth of Clyde (Figure 7.7) (Mackay et al., 1972; Rodger and Davies, 1992). The Firth of Clyde is atypically deep for British inshore coastal waters as a result of glacial action, and the dump site lay at a depth of about 70 metres. In 1974, sludge dumping moved to an area 4 kilometres further south (Figure 7.7) before ceasing altogether by 1998. By 1974, dumping had reached more than 1 million tonnes of wet sludge annually, and it is estimated that a total of about 50 million wet tonnes had by then been dumped off Garroch Head (Mackay et al., 1972; Rodger and Davies, 1992). In the 1980s, the operating authority was licensed to discharge up to 1.55 million wet tonnes (70,000 dry tonnes) of sewage sludge per year into an area of 6 square kilometres at a depth of 70 to 80 metres (Rodger and Davies, 1992).

The Firth of Clyde sites are said to be of the accumulating type (Pearson, 1987). The sludge settles out of the water column rapidly, and it is not widely distributed by any currents. Off Garroch Head,

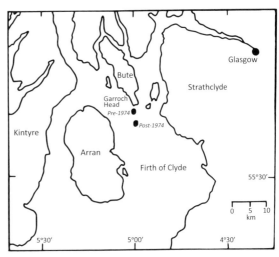

Figure 7.7 Sewage sludge dumping off Garroch Head in the Firth of Clyde, before and after 1974.

gradients of raised sediment concentrations of trace metals (Table 7.10) are evident across an area about 4 kilometres wide, approximately centring on the middle of the dump site (Mackay et al., 1972; Thornton, 1975b).

Local epifaunal invertebrates in and around the original dump site included the mobile caridean decapods *Crangon allmani* (a deeper-water relative of the common brown shrimp *C. crangon*) and *Pandalus montagui* (commonly called the pink shrimp) and the common whelk *Buccinum undatum*, a predatory and scavenging gastropod mollusc (Halcrow et al., 1973). Differences between accumulated trace metal concentrations of epifaunal invertebrates on and beyond the dump site were typically small and erratic and were unrelated to sediment trace metal concentrations (Halcrow et al., 1973; Thornton, 1975b). One exception, however, was the whelk *B. undatum*, which accumulated 1,500 µg/g zinc, 3 µg/g cadmium, 13 µg/g lead and 11 µg/g nickel in the dump site area, compared to 250 µg/g Zn, 1 µg/g Cd, 7 µg/g Pb and 3 µg/g Ni in the surrounding control area (Halcrow et al., 1973; Thornton, 1975b). Like most of the epifaunal invertebrates, local demersal fish, such as plaice *Pleuronectes platessa* and cod *Gadus morhua*, from the area of the dump site had trace metal contents not significantly different from those in fish collected elsewhere in British coastal waters (Halcrow et al., 1973).

It would appear, then that the raised trace metal concentrations in the sediments at the Garroch Head dump sites are not associated with biovailabilities high enough to be of ecotoxicological significance. Nevertheless, there have certainly been significant deleterious effects on the local benthic community as a result of the dumping of sewage sludge. The main outcomes do comply with the expected effects of organic enrichment on a macrobenthic marine community. The centre of the second dump site shows the presence of a low number of species, particularly a species of the nematode genus *Pontonema*, the oligochaete *Tubificoides benedii* and the polychaetes *Capitella capitata* and *Malacoceros* (previously *Scolelepis*) *fuliginosa* (Pearson, 1987). In high densities at the edge of the central area of the dump site are the polychaetes *Mediomastus* sp., *Notomastus latericeus* and *Chaetozone setosa* and the bivalve *Thyasira flexuosa*. Moving further out from the centre of the dump site, these species decline in number, to be replaced by the macroinvertebrate community characteristic of the surrounding unaffected area (Pearson, 1987).

And yet, on close inspection, there may be subtle effects of high trace metal bioavailabilities on the benthic macroinvertebrate community structure at the second Garroch Head sewage sludge dump site. The multivariate statistical technique of non-metric multidimensional scaling (MDS) was applied to benthic macroinvertebrate data collected along a transect of sampling sites across the dump site in 1983 (Clarke and Ainsworth, 1993). As expected, sediment organic content was identified as the major driver of change in community composition across the site. But the statistical analysis identified an additional factor separate from the percentage organic content of the sediment. This factor was represented by a combination of sediment cadmium concentration and the percentage carbon content of the sediment. The substitution, or the addition, of other trace metal concentrations for that of cadmium reduced the effect of this factor, implying that cadmium alone is the relevant trace metal. It cannot, however, be concluded with confidence that the sediment bioavailability of cadmium is directly causal in shaping the macroinvertebrate benthic community, for the sediment cadmium concentration may simply be correlated with some other unmeasured characteristic of the sediment. Nevertheless, an ecotoxicological effect that had previously been regarded as an example of simple organic enrichment may also have been affected by local raised cadmium bioavailabilities at the dump site.

7.8.2.2 Liverpool Bay

Prologue. Sewage sludge, dredged material and some industrial waste have been dumped in Liverpool Bay since the start of the twentieth century, with sewage sludge dumping ceasing at the end of that century. Dump sites are in relatively shallow water, and sediments with any dumped material are regularly reworked and dispersed widely by tidal and storm action. Consequently, the area of raised organic content and elevated trace metal concentrations in the sediment can extend up to 25 kilometres from the sewage sludge dump site. As a result of this dispersion and given the natural heterogeneity of benthic macroinvertebrate communities in Liverpool Bay, it is difficult to isolate any ecotoxicological effects clearly resulting from sludge and other dumping in Liverpool Bay.

Sewage sludge was dumped into Liverpool Bay (Figure 7.8) throughout almost all of the twentieth century. Dumping started in about 1900, when sewage treatment plants were constructed near Manchester (Norton et al., 1984a). The total amount of sewage sludge dumped annually into Liverpool Bay varied between 0.5 and 1 million wet tonnes from 1900 to 1972, rising to 1.8 million tonnes per year by 1980 (Table 7.12). This sewage sludge was associated with considerable loadings of trace metals, particularly zinc and copper (Table 7.12). In the late 1960s, there began the additional dumping of industrial waste from a number of industries within 80 kilometres of Liverpool Bay. These wastes, too, contained trace metals, but, being small in volume, contributed less than 1% of the input of trace metals from the

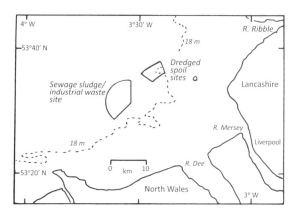

Figure 7.8 Dump sites for sewage sludge, industrial waste and dredged spoil in Liverpool Bay. (After Norton et al., 1984b, with permission.)

sewage sludge (Table 7.12) (Norton et al., 1984a, b). Dredged spoil is also now dumped into Liverpool Bay, usually at sites slightly removed from the site receiving sewage sludge and industrial waste (Figure 7.8). Dredged material is derived from the maintenance dredging of the channel approaches to the River Mersey, the River Mersey itself, docks and harbours along the Mersey and from the Manchester Ship Canal (Norton et al., 1984a). Given their sheer volume (5.7 million tonnes wet weight in 1977), dumped dredged spoils actually inputted more trace metals into Liverpool Bay than dumped sewage sludge in the 1970s (Table 7.12), even though their trace metal concentrations were typically lower. Table 7.12 also includes comparative information on the large, but more diffuse, sources of metal input from river discharge (the Rivers Mersey, Dee and Ribble) and direct sewage or industrial discharges into Liverpool Bay in 1976. While significant, these anthropogenic sources do not provide a concentrated point source of trace metals into the sediments of Liverpool Bay, as is the case for the dump sites (Figure 7.8).

Unlike the situation in the deep Firth of Clyde, the Liverpool Bay dump sites are in shallow water, at about 20 to 30 metres deep. During neap tides with limited tidal current action, there may be times when dumped sludge particles can settle out and accumulate on the bottom. At spring tides and under storm conditions, however, currents will drive considerable

reworking and dispersion of sand and mud deposits, including sludge particles (Norton et al., 1984b). As a consequence, the area of raised organic content and elevated trace metal concentrations in the sediment (Table 7.10) is more widespread than in the Firth of Clyde and can extend up to 25 kilometres from the sewage sludge dump site (Norton et al., 1984a, b).

Given this dispersion, it is difficult to isolate any ecotoxicological effects clearly resulting from sludge and other dumping in Liverpool Bay (Norton et al., 1984b). Furthermore, both sediments and associated benthic macroinvertebrates were very heterogeneous in the area including and around the dump sites, both within a single year and between years (Norton et al., 1984a). This spatial and temporal variability, resulting from essentially natural fluctuations, adds to the difficulty in detecting any effects on the benthic fauna from the dumping of sewage sludge and industrial waste. It can be concluded, though, that no gross adverse changes attributable to dumping have occurred in the benthic macroinvertebrate community in Liverpool Bay (Norton et al., 1984a).

Concentrations of mercury, cadmium, lead and zinc were measured in the 1970s in fish and shellfish in Liverpool Bay, by the then Ministry of Agriculture, Fisheries and Food (MAFF), incorporated into DEFRA in 2001 (Norton et al., 1984a; Franklin, 1987). Concentrations of mercury in the muscle tissue of fish (a combination of whiting, cod, plaice and sole) halved between 1970 and 1980, probably mainly as a result of a fall in direct industrial discharge of mercury in this period (Norton et al., 1984a). The input of mercury in sewage sludge into Liverpool Bay did also fall from 2.7 tonnes in 1976 to 0.8 tonnes in 1980, although the amount of sewage sludge dumped rose over this period (Norton et al., 1984a). The general decrease in the industrial discharge of mercury at the time, therefore, also had a knock-on effect on the concentrations of mercury in the local sewage sludge. There was no evidence for any widespread elevations of accumulated concentrations of cadmium, lead and zinc by fish and shellfish in Liverpool Bay at the time (Norton and Murray, 1983; Norton et al., 1984a). There was, however, some evidence for localised elevated accumulation of trace metals by biota close to

Table 7.12 **Quantities (tonnes) of sewage sludge, industrial wastes and dredged material dumped into Liverpool Bay in 1976, together with their trace metal loadings (tonnes). Also shown are the quantities (tonnes) of trace metals discharged from rivers and sewage and industrial sources into Liverpool Bay in 1976.**

	Wet weight (Dry weight)	Cd	Cr	Cu	Hg	Ni	Pb	Zn
Sewage sludge	1,587,000 (70,400)	2.2	93	110	2.7	13	55	265
Industrial waste	212,100 (5,470)	0	0.1	0.7	0	0.1	0.4	0.9
Dredged material	4,990,000 (2,070)	1.3	145	110	4.4	105	330	872
Discharges								
Rivers		34	80	75	0.3	98	–	155
Sewage/industrial		16	152	53	2.8	74	47	1,490

Sources: Norton et al. (1984a, 1984b).

the sludge dumping ground (Norton and Murray, 1983; Norton et al., 1984a). Some brown shrimp (*Crangon crangon*) and queen scallops (*Aequipecten opercularis*) contained cadmium and lead at about twice the concentrations found in other UK waters (Norton and Murray, 1983). Nevertheless, none of the raised accumulated concentrations appears to have been significant enough to indicate a local trace metal bioavailability with ecotoxicological potential, nor to raise any concern from a public health viewpoint (Norton and Murray, 1983).

7.8.2.3 Outer Thames

Prologue. Dumping of sewage sludge, dredged material and industrial waste at several sites in the Outer Thames Estuary started in 1887, and (with the exception of sewage sludge) continues today. Dump

sites are in relatively shallow water with associated dispersion of dumped material and receiving sediments. The benthic macroinvertebrate fauna of the Outer Thames Estuary is naturally relatively poor as a result of the mobility of the sediments under strong current scour, but there are some effects of sludge disposal apparent, particularly in the Black Deep as a result of organic enrichment. There is no evidence in the Outer Thames of trace metal ecotoxicity resulting from dumping.

Dumping of sewage sludge from the main London sewage works at Beckton and Crossness began in 1887 and continued through the twentieth century (Shelton, 1971; Norton et al., 1981). By the 1970s, more than 5 million wet tonnes (about 200,000 dry tonnes) of sewage sludge were being dumped annually into the Outer Thames Estuary (Figure 7.9)

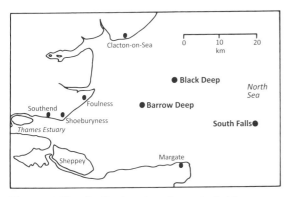

Figure 7.9 Sites for the dumping of sewage sludge, industrial waste and dredged material in the Outer Thames Estuary.

(Shelton, 1971). From 1915 to 1967, the sludge was dumped into the Black Deep, at a depth of 13 to 20 metres (Figure 7.9). Before 1915, and after 1967, the sludge was dumped into the Barrow Deep (13 to 18 metres deep) (Figure 7.9) (Shelton, 1971). Between 1915 and 1920, the Black Deep also received dredged material from other parts of the Thames Estuary (Shelton, 1971). Both the Black Deep and the Barrow Deep are swept by strong tidal currents, dispersing dumped sewage sludge beyond the immediate dump sites. A surprising but effective indicator of dumped sewage sludge in coastal sediments is the presence of tomato pips. These are resistant to digestion in the human gut and to degradation in the processing of sewage sludge and in the sea. They are also numerous enough to show how far dumped sewage sludge is moved by currents across the bottom (Shelton, 1971). Another dump site further out, the South Falls site, 110 kilometres east of Gravesend (Figure 7.9), is the dump site for dredged material and industrial waste. In 1985, the South Falls dump site received about 50,000 wet tonnes of dredged material, 55,000 wet tonnes of industrial waste and 90,000 wet tonnes of sewage sludge (Harper, 1988).

Sewage sludge from Beckton and Crossness contained very high concentrations of trace metals in the 1970s (Table 7.11), with lower but still significant concentrations today, after the effect of industrial regulations introduced towards the end of the twentieth century. In the late 1970s, the Barrow Deep was receiving annually

350 to 510 tonnes of zinc, 86 to 155 tonnes of lead, 88 to 110 tonnes of copper, 51 to 72 tonnes of chromium, 20 to 29 tonnes of nickel, 5 to 7 tonnes of cadmium and 1 to 6 tonnes of mercury in dumped sewage sludge (Norton et al., 1981). By 1990, these trace metal loads had been reduced. In 1993, a sample of sewage sludge from Beckton contained 18 μg/g cadmium and 2,280 μg/g zinc (Galay Burgos and Rainbow, 1998), concentrations below the ranges quoted for Beckton and Crossness combined in the 1970s (Table 7.11).

Trace metal concentrations in the sediments at the Outer Thames sludge dump sites are, as expected, elevated against background (Table 7.10), in association with raised organic carbon levels (Norton et al., 1981).

The benthic macroinvertebrate fauna of the Outer Thames Estuary is naturally relatively poor in comparison with those of other British coastal sediments, as a result of the mobility of the sediments under strong current scour (Shelton, 1971). The macroinvertebrate benthos of the Black Deep and Barrow Deep themselves is relatively normal for an outer estuary with strong tidal scour, but there are some effects apparent of sludge disposal, particularly in the Black Deep. The relatively high concentrations of organic matter in the sediments at the dump sites are associated with an increase in the numbers and species diversity of the polychaete fauna (Shelton, 1971). Such changes are akin to those seen away from the centre of an accumulating dump site such as at Garroch Head (Pearson, 1987). The region at the centre of the Garroch Head dump site, with very low oxygen levels and associated adapted worm fauna, is not represented in either the Black Deep or Barrow Deep, given the strong oxygenated currents present. There is no evidence in the Outer Thames of trace metal ecotoxicity resulting from dumping.

7.8.2.4 Other Dump Sites

Prologue. There are many other dump sites in British coastal waters that have been used historically for sewage sludge and presently for dredged material and some industrial waste. There is no strong evidence at any dump site for any ecotoxicological effects of trace metals on macroinvertebrate benthic communities, even when raised local trace metal bioavailabilities have been identified.

Similar stories are apparent at other dump sites around the British Isles. Usually, dumping has resulted in changes in the structure of benthic macroinvertebrate communities at the site, primarily as a result of organic enrichment of the local sediment. Even these effects may be small when dump sites are located in waters of strong tidal currents causing an already impoverished benthic fauna to be resident in the relatively unstable sediments. The Bristol Channel dump sites are similar to those of Liverpool Bay and the Outer Thames Estuary in this respect (Murray et al., 1980).

While sewage sludge dumping has now ceased in British coastal waters, the dumping of dredged materials and, to a much smaller extent, industrial waste does continue. Some generalisations can be made across the more than 150 sites designated for the disposal of dredged material around England and Wales. It is generally the case that benthic macrofaunal communities at disposal sites are faunistically impoverished to varying degrees (Bolam et al., 2006; Bolam, 2012). Abundance and biomass contribute to measures of production, and secondary production is similarly reduced at dredged material dump sites (Bolam et al., 2006; Bolam, 2012). The decrease in species diversity of benthic communities at dredged material dump sites can be further broken down into the effects of two contrasting impacts (Somerfield et al., 2006). The first impact, that of organic enrichment, leads to communities dominated by nematodes, oligochaetes and polychaetes, while many other invertebrates are absent (Pearson and Rosenberg, 1978; Somerfield et al., 2006). The other impact is that of intense physical disturbance, promoting the presence and survival of large motile or armoured forms such as epifaunal crustaceans and molluscs (Somerfield et al., 2006). These two effects may act in opposition to each other, but the general outcome is usually a reduction in species diversity (Somerfield et al., 2006).

7.8.3 Hypoxia, Manganese and Scampi

Prologue. Usually insoluble manganese can dissolve in oxygen-deficient pore waters of organically rich sediments as the reduced ion Mn^{2+}. The Norway lobster *Nephrops norvegicus* may encounter manganese in this form in sediment pore water, with consequent neurotoxicological effects on its behaviour.

The addition of organic carbon to a coastal sediment, be it from the dumping of sewage sludge or from another form of eutrophication, can result in hypoxia, a deficiency of oxygen in the environment. This hypoxia is caused by the microbial consumption of oxygen during the respiration of the high organic load. Hypoxic conditions promote an enhanced dissolved flux of manganese in the form of Mn^{2+} ions from the ample store of manganese associated with most sediments (Table 7.10) (Baden et al., 1995; Krång and Rosenqvist, 2006). Dissolved manganese concentrations in such hypoxic bottom waters can reach up to more than 1,000 times the levels in normal oxygenated conditions which promote the presence of manganese 4 and its loss from solution by precipitation and adsorption onto sediment particles. The dissolved manganese concentration in hypoxic conditions may, therefore, reach 2,000 µg/L or more (Oweson et al., 2006), compared to typical coastal water dissolved concentrations of about 1 µg/L (Table 7.1).

In excess, manganese acts as a neurotoxin and can inhibit vital functions such as muscle contraction (Baden and Neil, 1998; Krång and Rosenqvist, 2006). A possible target organism living in coastal sediments is the decapod crustacean *Nephrops norvegicus*, known by a variety of common names including the Norway lobster, the Dublin Bay prawn and scampi. *N. norvegicus* is a predator and scavenger, living in 20 to 30 centimetre deep burrows in coastal muddy sediments at depths of 20 metres or more (Krång and Rosenqvist, 2006). While the decapod will emerge from its burrow to forage, its burrowing lifestyle does mean that it is exposed to pore water and bottom water that becomes hypoxic when the sediment is enriched with organic carbon. When *N. norvegicus* were exposed to high, but environmentally realistic, dissolved concentrations of manganese under hypoxic conditions, they showed a significant reduction in the force produced by rapid extension of the abdomen in an escape response (Baden and Neil, 1998). Under these manganese exposure conditions, the decapods also showed a more than doubled reaction time to food odour stimuli in the water (Kråg and Rosenqvist, 2006). Thus, even in the absence of anthropogenic manganese contamination, hypoxic conditions in coastal bottom waters can, through the toxic agency of manganese, reduce both the escape success of *N. norvegicus* and its ability to detect and find food.

7.9 Fish

Prologue. Studies on coastal fish in British waters have concentrated on plaice, sole, dab, cod and whiting.

Moving out from the coast, the effects of anthropogenic contamination by trace metals become much less significant than in the estuaries that receive and then export the trace metals into coastal waters. Point sources such as coastal dump sites may interrupt any linear gradients of trace metal contamination out to sea. Nevertheless, local enhancement of trace metal bioavailabilities at dump sites is probably not of any ecotoxicological significance, particularly in the case of local coastal fish. Particular coastal water bodies can still, however, show raised bioavailabilities of trace metals to local fish as a result of anthropogenic contamination. The case of Liverpool Bay in the 1970s comes to mind, but any such generalised contamination of British coastal water bodies has been greatly reduced as a result of the regulation of industrial discharges in the last decades of the twentieth century.

Chapter 6 discussed the biology of trace metals in three estuarine fish: flounders, eelpout and eels. Flounders, *Platichthys flesus*, are still common beyond estuaries in deeper coastal waters, where they are joined by several of their flatfish relatives. The first of these flatfish, the plaice *Pleuronectes platessa*, lives buried in sand on the bottom. Its distribution extends from estuaries down to 100 metres, but plaice are most common from 10 to 50 metres (Henderson, 2014). Plaice tend to feed on the bottom at night. When young, they take small worms and crustaceans, but with increasing size they move onto thin-shelled

bivalves, often nipping off the ends of bivalve siphons protruding above the sediment (Muus and Dahlström, 1964; Henderson, 2014). Another flatfish, the sole *Solea solea*, also called the Dover sole, is abundant in British inshore coastal waters. It is common on sandy and muddy bottoms at depths between 5 and 100 metres. Sole can also penetrate into estuaries, and they spawn within the Thames Estuary (Henderson, 2014). Like plaice, sole will bury themselves by day, to emerge at night to feed on polychaete worms, crustaceans and bivalves.

A third flatfish is the dab *Limanda limanda*. The dab lives on sandy bottoms down to about 100 metres, and it is particularly abundant in the North Sea (Muus and Dahlström, 1964). It is a generalist predator of invertebrates, including echinoderms, crustaceans and molluscs, and will take small fish such as sand eels and gobies (Muus and Dahlström, 1964; Henderson, 2014). In the last decades of the twentieth century and in the first decade of the twenty-first century, dab were often the fish of choice in fish biomarker programmes seeking evidence of the ecotoxicological effects of anthropogenic contaminants in British coastal waters (CEFAS, 2003). While some of the biomarkers investigated were of more relevance to the ecotoxicology of organic contaminants, others did have relevance to toxic metal exposure. These included biomarkers of dab pathology and disease, such as histological changes in livers, the presence of tumours and the occurrence of DNA adducts in liver cells (CEFAS, 2003). While dab in British coastal waters at the end of the twentieth century did show the presence of these biomarkers, including liver tumours, this appears to be a reflection of the atypical susceptibility of dab to tumour formation, and not a cause for ecotoxicological concern (CEFAS, 2003). There is no biomarker evidence that high trace metal bioavailabilities are causing ecotoxicological effects in fish in British coastal waters today.

Amongst the other fish harvested in British coastal waters are two related members of the cod family, the cod itself (*Gadus morhua*) and the whiting (*Merlangius merlangus*). Cod are widely distributed as different races in British coastal waters and beyond, down to depths of 600 metres (Muus and Dahlström, 1964;

Henderson, 2014). Cod are omnivorous, feeding on worms, crustaceans and molluscs, and, with increasing age, on other fish. Cod grow very quickly and usually reach about 70 centimetres in length in coastal waters (Henderson, 2014). Whiting are also widely distributed in all British coastal waters. They are also general predators, feeding on polychaetes, shrimps and prawns, crabs, squid and small fish, becoming more piscivorous as they grow (Henderson, 2014).

7.9.1 Trace Metals in Fish

Prologue. Marine fish take up and accumulate trace metals. Trace metal concentrations in the muscles of fish from British coastal waters have been monitored for decades. These data indicate that bioavailabilities of cadmium and mercury were atypically raised in Liverpool Bay in the early 1980s, but have since fallen. Large oceanic predatory fish such as tuna have high muscle concentrations of mercury, accumulated via the food chain biomagnification of methyl mercury.

Programmes to monitor the concentrations of accumulated trace metals in British coastal fish have come to concentrate on their potential health risk to human consumers, as opposed to their use to biomonitor potentially raised bioavailabilities of trace metals in different coastal water bodies. When we eat fish and chips, the fish part of the meal consists of muscle tissue filleted from along the back of the fish. So it is the muscle tissue that is analysed as the source of trace metals in the human diet. The muscles of fish generally accumulate lower concentrations of trace metals than other tissues, such as the livers and kidneys. These latter organs, not muscles, would be the organs of choice if the major aim of the monitoring programme had been to detect differences in trace metal bioavailabilities between different coastal water bodies.

Table 7.13 presents concentrations of cadmium, copper, mercury, lead and zinc in muscle tissue of plaice, sole, dab, cod and whiting from British coastal waters from the late 1970s to the twenty-first century. Concentrations are expressed in terms of wet weight, in order to allow comparison against recommended

limits of toxic metals for human consumption. In the case of fish muscles, dry weight concentrations are about five times wet weight concentrations.

Muscle concentrations of copper and zinc typically fall into a narrow range, within and across all species, irrespective of date of collection or coastal region of origin (Table 7.13), indicative of the presence of regulation. Copper concentrations lie in the range 0.1 to 0.4 µg/g wet weight (ww), equivalent to 0.5 to 2 µg/g dw. Zinc concentrations are in the range 2 to 8 µg/g ww, 10 to 40 µg/g dw.

Cadmium concentrations in the muscles of the coastal fish typically lie below the detection limit of 0.2 µg/g ww (1 µg/g dw) applicable at the time of analysis, and, where measurable, cadmium concentrations were usually below 0.01 µg/g ww (0.05 µg/g dw) (Brown and Balls, 1997; Julshamn et al., 2013). On the other hand, plaice in Liverpool Bay in the early 1980s had a mean cadmium concentration in their muscles of 0.4 µg/g ww (2 µg/g dw) (Franklin, 1987), indicating a high local anthropogenic source of bioavailable cadmium at that time.

Lead concentrations in fish muscles are similarly often below the detection limit of 0.2 µg/g ww (1 µg/g dw) (Table 7.13). With improved analytical sensitivity, measured lead concentrations averaged 0.05 µg/g ww (0.25 µg/g dw) in plaice and 0.01 µg/g ww (0.05 µg/g dw) in whiting, from coastal waters off Northeast Scotland in the 1990s (Table 7.13). Maximum lead concentrations in the muscles of cod from the North Sea in 2010 were similarly low, reaching only 0.04 µg/g ww (0.2 µg/g dw) at most (Table 7.13). There is no evidence here to suggest any raised bioavailabilities of lead to any of these coastal fish in British waters. In comparison, we saw in Chapter 6 that in 1975 flounders from Oldbury-on-Severn, affected by the Avonmouth smelter, had average accumulated lead concentrations of 28 µg/g dw in their muscles (Hardisty et al., 1974).

A trace metal that is concentrated in the muscle tissue of fish is mercury. The analysis of mercury concentrations in the muscle tissue of coastal fish is, therefore, of biomonitoring significance as well as casting light on any ecotoxicological threat to the human consumer. It is clear from Table 7.13 that the

Table 7.13 **Mean concentrations with ranges of trace metals (μg/g wet weight) in muscles of plaice** *Pleuronectes platessa*, sole *Solea solea*, dab *Limanda limanda*, cod *Gadus morhua* and whiting *Merlangius merlangus* **from coastal waters around** *Britain*.

	Cd	Cu	Hg	Pb	Zn
Plaice					
Liverpool Bay					
1982–83[1]	0.4	0.3 0.2–0.3	0.27 0.10–0.60	<0.4	4.8 4.6–5.0
Morecambe Bay					
1983[1]	–	0.3 0.3–0.3	0.17 0.09–0.55	–	5.6 5.5–5.7
Scotland					
NE Scotland 1993–95[2]	<0.001	0.2 0.1–0.4	0.04 0.02–0.09	0.05 0.02–0.12	5.1 <0.1–7.6
English Channel East					
Hastings 1981[1]	–	0.3 0.3–0.3	0.07 0.06–0.08	–	4.6 3.2–6.5
Sole					
Liverpool Bay					
1982–83[1]	<0.2	0.3 0.3–0.3	0.25 0.07–0.51	<0.4	4.4 4.2–4.6
Morecambe Bay					
1983[1]	–	0.2 0.2–0.2	0.17 0.06–0.34	–	4.8 4.7–4.8
English Channel East					
Rye 1980[1]	<0.2	0.2 0.1–0.4	0.15 0.04–0.45	<0.2	3.6 2.8–4.7
Dab					
Liverpool Bay					
1982–83[1]	<0.2	0.3 0.2–0.3	0.21 0.09–0.64	<0.4	4.5 4.0–4.9
Morecambe Bay					
1983[1]	–	0.3 0.3–0.3	0.33 0.10–0.73	–	5.2 5.1–5.2
English Channel East					
Rye 1980[1]	<0.2	0.3 <0.2–0.6	0.12 0.05–0.41	<0.2	5.4 4.1–6.8

Table 7.13 (**cont.**)

	Cd	Cu	Hg	Pb	Zn
Cod					
Liverpool Bay					
1982–83[1]	<0.2	0.2 0.2–0.2	0.35 0.15–0.84	<0.4	3.5 3.5–3.5
Morecambe Bay					
1979[1]	<0.2	0.2 0.2–0.2	0.22 0.17–0.31	<0.2	2.4 1.9–3.2
Scotland					
NE Scotland 1993–95[2]	<0.001	0.12 0.05–0.23	0.17 0.06–0.27	<0.02	3.8 3.2–5.4
North Sea					
Northern 2010–11[3]	<0.002–0.10	–	0.12 0.01–0.54	<0.01–0.02	–
Central 2010–11[3]	<0.002–0.01	–	0.06 0.03–0.10	<0.01–0.04	–
English Channel East					
Rye 1980[1]	<0.2	0.3 <0.2–0.6	0.11 0.06–0.16	<0.2	4.3 3.0–5.0
Whiting					
Liverpool Bay					
1982–83[1]	<0.2	0.2 0.2–0.2	0.32 0.13–0.69	<0.4	3.3 3.2–3.4
Morecambe Bay					
1983[110]	–	0.2 0.2–0.2	0.60 0.32–0.98	–	3.4 3.2–3.6
Scotland					
NE Scotland 1993–95[2]	0.001 <0.001–0.004	0.2 0.1–0.2	0.04 0.02–0.09	0.01 <0.01–0.08	3.5 2.2–4.1
English Channel East					
Hastings 1977[1]	<0.2	0.3 <0.2–0.4	0.18 0.08–0.40	<0.2	4.6 3.2–6.5

Note: Dry weight concentrations are about five times wet weight concentrations.
Sources: Franklin (1987);[1] Brown and Balls (1997);[2] Julshamn et al. (2013).[3]

Table 7.14 **Average concentrations of mercury (µg/g ww) between 1970 and 1998 in muscle tissues of plaice** *Pleuronectes platessa*, **sole** *Solea solea*, **cod** *Gadus morhua* **and whiting** *Merlangius merlangus* **from four British coastal areas considered to be contaminated to above average levels in the 1970s and 1980s.**

	1970–72	1977–78	1979–80	1982	1984	1985	1994	1998
Outer Thames								
Plaice	0.35	0.19	0.09	0.08	0.09	0.04	–	–
Sole	0.24	0.15	0.19	0.09	0.11	–	–	–
Cod	0.36	0.13	0.15	0.10	0.12	0.08	–	–
Whiting	0.42	0.18	0.18	0.18	0.16	–	–	–
Swansea Bay								
Plaice	0.09	0.13	0.14	0.04	0.10	0.10	–	–
Whiting	0.13	0.23	–	0.06	0.12	–	–	–
Liverpool Bay								
Plaice	0.51	0.28	0.25	0.26	0.26	0.15	0.13	0.14
Sole	0.29	0.20	0.26	0.31	0.37	0.27	0.14	0.16
Cod	0.49	0.29	0.26	0.26	0.28	0.27	0.10	0.10
Whiting	0.65	0.41	0.35	0.29	0.41	0.32	0.27	0.30
Morecambe Bay								
Plaice	0.57	0.34	0.43	0.16	0.16	0.16	0.09	0.09
Sole	0.61	0.27	0.28	0.20	0.20	0.11	0.17	0.15
Whiting	0.59	0.26	0.41	0.34	0.44	0.27	0.27	0.29

Sources: Franklin (1987); CEFAS (1997, 2003).

muscle tissue of all five coastal fish species collected in Liverpool Bay (and to a lesser extent in Morecambe Bay to the north of Liverpool Bay) in the early 1980s had mercury concentrations much raised above those in other coastal regions at the time (for example, the eastern English Channel) or in British coastal waters generally by the 1990s and by 2010. These coastal species are reflecting the situation seen in the more estuarine flounder in Liverpool Bay in the 1970s and early 1980s. Liverpool Bay, and by extension Morecambe Bay, were then receiving considerable mercury input, not only from direct industrial and sewage sources but also from the dumping of sewage sludge and dredged material (Table 7.12). Increased local control of mercury emissions from the 1970s has brought about a welcome fall in mercury concentrations of local coastal fish (Table 7.14), just as for flounder (Table 6.29).

Fish take up mercury in inorganic form from both solution and food, and as methyl mercury from food (Luoma and Rainbow, 2008). The assimilation of methyl mercury in the alimentary tract of fish is much more efficient than that of inorganic mercury (Luoma and Rainbow, 2008). Either way, mercury taken up by fish accumulates particularly in the muscle tissue, predominantly as methyl mercury (Wang, 2012). The excretion of methyl mercury by fish is extremely slow, and methyl mercury has a half-time in large fish of one year or more (Dang and Wang, 2012; Wang and Wang, 2015). The concentration of methyl mercury in the muscle tissue of fish therefore increases with the age and therefore typically with the size of fish (Dang and Wang, 2012). Given its very limited excretion by fish and its efficient trophic uptake, methyl mercury bio-magnifies up food chains containing fish, particularly large piscivorous fish (Dang and Wang, 2012; Wang, 2012). Typically, the higher the trophic level of a fish in a food chain, the higher the methyl mercury (and therefore total mercury) concentration accumulated in the muscle tissue of that fish.

It follows then that large, long-lived, piscivorous fish have high muscle concentrations of mercury. Prime examples of such fish are species of tuna (e.g., *Thunnus*) and swordfish (*Xiphias gladius*) (Peterson et al., 1973; Wang, 2012). Table 7.15 lists concentrations of mercury in the muscle tissues of such oceanic species. These data are from the early 1970s to allow comparison with the earlier data for coastal fish, as presented in Table 7.14. The mercury concentrations in the muscles of tuna and swordfish typically match, and usually exceed, the concentrations found in fish from Liverpool Bay and Morecambe Bay at the start of the 1970s. And these particular muscle concentrations in the coastal fish had been atypically raised by local anthropogenic mercury pollution. The data for the oceanic species, however, are from ocean regions apparently remote from sources of mercury pollution. These mercury levels have been accumulated via the biomagnification of mercury along oceanic food chains leading to long-lived piscivorous fish. High concentrations of mercury would be accu-mulated naturally by tuna and swordfish without any anthropogenic input. Nevertheless, mercury is very volatile and considerable amounts are emitted into the atmosphere by such processes as coal burning and

Table 7.15 **Average concentrations with ranges of mercury (µg/g ww) in muscle tissue of tuna and swordfish species.**

	Mercury	
	Average	Range
Atlantic bluefin tuna *Thunnus thynnus*		
Gulf of Cadiz	0.68	0.46–0.91
Yellowfin tuna *Thunnus albacares*		
Atlantic (Africa)	0.49	0.29–0.77
Skipjack tuna *Katsuwonus pelamis*		
Hawaii	0.38	0.27–0.52
Swordfish *Xiphias gladius*		
Gibraltar Strait	1.36	0.99–2.01
Western Atlantic	1.15	0.05–4.90

Source: Peterson et al. (1973).

artisanal gold mining, to be transported across the globe before deposition onto the surface of oceans far away. Artisanal miners soak the ore in mercury to form an amalgam which, on subsequent heating, releases the gold but also mercury vapour into the air.

While tuna are typically fish of the open oceans, they can penetrate coastal waters. In the 1930s, for example, there was a thriving North Sea fishery for bluefin tuna, *Thunnus thynnus*, based on Scarborough in Yorkshire.

7.10 Seafood

Prologue. So, does the trace metal content of seafood pose any toxic risk to the human consumer?

One regulatory approach to this potential risk is to set maximum recommended limits for the concentrations of selected trace metals in seafood. Or, rather, in different sorts of seafood. As we have seen repeatedly throughout this book, the trophic bioavailability of a trace metal in a dietary item does depend on its particular chemical form in that food. A trace metal detoxified in a pyrophosphate granule in a consumed prey item will be less bioaccessible and trophically bioavailable to any consumer, including humans, than will a trace metal consumed bound to a detoxificatory protein such as a metallothionein (Rainbow et al., 2011a). The potential toxicity of a trace metal in food may also depend on its chemical nature in the consumed item for reasons other than ease of digestion by the consumer. For example, most arsenic accumulated by most marine invertebrates and fish is in the form of organic compounds of arsenic, particularly arsenobetaine (Neff, 1997). While organic arsenic compounds are taken up in the gut of human, and other mammalian, consumers, the arsenic is rapidly excreted again (Neff, 1997). As a result, arsenobetaine is not toxic to mammals, including ourselves. On the other hand, arsenic in a dietary item that is in inorganic form such as arsenate or, more especially, the more toxic form arsenite is potentially very toxic to the human consumer. Thus the maximum concentration of arsenic in seafood that would pose a potentially toxic risk to human consumers would be very different if the arsenic were to be in inorganic form as opposed to the organic form arsenobetaine. This chemical form will vary with the item of seafood under consideration.

A final point on the suitability of recommended maximum concentrations of potentially toxic metals in seafood. A concentration may be a useful pragmatic guideline, but ultimately whether a trace metal is toxic or not will depend on how much metal is ingested and over what time period, and on the size of the consumer. An adult can clearly cope with a larger ingested load than a small infant of much lower body weight. Furthermore, the state of development of the consumer may be relevant beyond simply a question of body weight. Mercury and lead act on the nervous system and are more dangerous to the developing child (or foetus) than to a fully developed adult. Correspondingly, a pregnant or nursing mother should ingest lower amounts of these trace metals than otherwise, to reduce any metal transfer to the embryo in the womb, or to the breast-feeding baby in mother's milk. Methyl mercury is fat soluble, and, therefore, likely to be particularly concentrated in the fat-rich milk provided by a nursing mother. The upshot is that it is not enough to be aware of simple maximum recommended concentrations of trace metals in seafood, but to limit the number of meals per week of particular types of seafood according to status – be it adult or child, male or female, pregnant or not, nursing or not, potentially childbearing or not.

So, what are the recommended maximum concentrations of different trace metals in different types of seafood? Table 7.16 lists the maximum concentrations of cadmium, lead and mercury in different seafood items under EU Commission regulations issued in 2006, 2008 and 2015.

There are no such EU maximum concentrations for other trace metals. There are, however, historical UK statutory limits for concentrations of copper, zinc and arsenic in food, emanating from Food Standards Committee Reports to the UK government in the 1950s. Thus levels of copper in food should not exceed 20 µg/g ww, but 'higher levels in shellfish are permitted if copper is of natural occurrence' (MAFF Food Standards Committee, 1956). Similarly, levels of zinc in food should not exceed 50 µg/g ww, but 'higher levels are permitted in food which naturally contain more than 50 µg/g ww such as herring and shellfish' (Ministry of Food, Food Standards Committee, 1953). Furthermore, 1959 legislation on arsenic in food decreed that no food should contain an arsenic concentration above 1 µg/g ww, except in the case of fish or edible seaweed where arsenic at a concentration above 1 µg/g 'is naturally present'. These apparently peculiar and vague exceptions go back to a day when the committees were faced with the problem of understanding why different trace metal concentrations in different seafood items showed different potential toxicities to a consumer. Moreover, some seafood products, particularly some bivalves and crustaceans, had been consumed by the British public over long historical time without apparent toxic effect, although they contained what were clearly very high accumulated concentrations of copper and zinc.

7.10.1 Fish

Prologue. The one trace metal in fish muscle that is of potential ecotoxicological concern for the human consumer is mercury in long-lived oceanic fish such as tuna.

Before we turn to a consideration of the more justifiable approach of defining tolerable weekly inputs of trace metals in consumers of seafood, let us briefly compare measured concentrations of trace metals in the muscles of marine fish (Tables 7.13 to 7.15) with the preceding recommendations (e.g., Table 7.16).

Every measured concentration of copper or zinc in the muscles of coastal fish from British waters listed in Table 7.13 is well below the recommended concentration of 20 µg/g Cu or 50 µg/g Zn previously

quoted. Concentrations of cadmium and lead in the muscles of these coastal fish in the twenty-first century (Table 7.13) are also usually well below the EU regulated concentrations of 0.05 µg/g Cd and 0.3 µg/g Pb (Table 7.16). Muscle concentrations of cadmium in plaice from some polluted coastal waters such as Liverpool Bay in the 1970s and 1980s (Table 7.13) may have exceeded the regulated concentration (Table 7.16), but those days have passed with the cleaning up of industrial effluents in the second half of the twentieth century.

The only trace metal that is really of any concern in the muscle tissue of fish is mercury. The mercury concentrations in muscles of fish species from British coastal regions affected by industrial effluent, including Liverpool Bay, Morecambe Bay and the

Table 7.16 **Maximum concentrations (µg/g ww) of cadmium, lead and mercury in different seafood items under EU Commission Regulations Nos 1881/2006, 629/2008 and 1005/2015.**

Seafood	Cadmium	Lead	Mercury
Muscle meat of fish		0.3	
Muscle meat of fish, excl. lists (a), (b) and (c)	0.05		
Muscle meat of fish, excl. list (d)			0.5
Muscle meat of list (a)	0.1		
Muscle meat of list (b)	0.2		
Muscle meat of list (c)	0.3		
Muscle meat of list (d)			1.0
Crustaceans - muscle (white meat)	0.5	0.5	0.5
Bivalve molluscs	1.0	1.5	0.5
Cephalopod molluscs (not viscera)	1.0	0.3	0.5
Dried seaweed products	3.0		

List (a): tuna, bonito, seabream, eel, grey mullet, horse mackerel, mackerel, luvar, sardine, sardinops, wedge sole.
List (b): bullet tuna
List (c): anchovy, swordfish
List (d): tuna (species of *Thunnus, Euthynnis* and *Katsuwonis pelamis*), swordfish (*Xiphias gladius*), marlin (species of *Makaira*), sail fish (*Istiophorus platypterus*), bonito, anglerfish, Atlantic catfish, eel, orange roughy, grenadier, halibut, megrim, mullet, pike, poor cod, rays, redfish, scabbard fish, seabream, shark, butterfish, sturgeon.

Outer Thames in the 1970s and 1980s (Tables 6.29 and 7.14), often exceeded the later EU-regulated concentration of 0.5 µg/g (Table 7.16). These average mercury concentrations, however, fell appreciably over time, so that by the turn of the millennium they were below this threshold. So indeed were all average concentrations (and usually all maximum individual concentrations) of mercury in muscle tissue of all British coastal fish in this century (Table 7.13).

On the other hand, mercury concentrations in the muscles of long-lived oceanic fish such as tuna and swordfish (Table 7.15) do generally approach or exceed the EU-regulated concentration of 0.5 µg/g designated for most marine fish (Table 7.16). Tuna and swordfish in fact have a higher EU-permitted level of 1 µg/g mercury (Table 7.16). This higher limit usually accommodates mercury concentrations in the muscle tissue of most tuna, but not by a wide margin. The muscle tissues of larger tuna, such as bluefin or yellowfin tuna, as expected, usually have higher mercury concentrations than those of smaller skipjack tuna, but the latter are still high (Table 7.15). And these fish do enter the British market, typically as tinned fish. So, what should be done?

Let us return to the concept of the provisional tolerable weekly input (PTWI) introduced in Chapter 4. This approach does not rely on a simple threshold concentration in food but takes into account the size of the consumer and the number of meals consumed in a particular time period. As shown in Table 7.17, PTWI are calculated in terms of total amount of trace metal to be consumed per week per unit kilogram body weight of consumer over an extended time period. Also shown in Table 7.17 are the PTWI for a 60 kilogram adult (about 9 stones, 6 pounds in old money) and for a 12 stone adult. The trace metals of concern are indeed cadmium, lead and mercury, and also arsenic, but only arsenic in inorganic form (Table 7.17). The PTWI for mercury assumes that all mercury is in the form of methyl mercury and so is appropriate for a fish diet.

Let us make some worse scenario calculations, in this case for a fish and chip supper consisting of a fillet of North Sea cod weighing 225 grams. Concentrations of trace metals in the muscle of North Sea cod

Table 7.17 **Provisional tolerable weekly inputs (PTWI, µg per week per kilogram body weight, or per 60 kilogram and 12 stone adults) of toxic metals considered of human health concern, as recommended by the Joint Food and Agriculture Organization of the United Nations (FAO) and World Health Organisation (WHO) Expert Committee on Food Additives (JECFA).**

	PTWI		
	µg per week per kg body weight	µg per week per 60 kg adult	µg per week per 12 stone adult
Inorganic arsenic	15	900	1,140
Cadmium	7	420	530
Lead	25	1,500	1,900
Mercury	1.6	96	122

Source: Morais et al. (2012).

might reach 0.01 µg/g cadmium, 0.02 µg/g lead and 0.06 µg/g mercury (Table 7.13). Thus this fish dinner would contain 2.25 µg of cadmium, 4.5 µg of lead and 13.5 µg of mercury. A 12 stone adult would have to eat nearly 250 such meals per week to reach the PTWI for cadmium, and 420 meals per week to reach the PTWI for lead (Table 7.17). Forget it.

But what about mercury? Remembering that this reflects a probably high scenario for mercury concentrations in North Sea cod, it would theoretically take nine meals a week for a 12 stone adult to reach the PTWI for mercury (Table 7.17). In practice, PTWI leaves elbow room for safety and applies to a very long period of consumption, of the order of years. So it would be difficult today to eat enough cod and chips each week for a long enough time to raise questions over mercury safety. We cannot be so confident, however, about the consumption of tinned tuna. A meal of tuna weighing 225 grams with mercury concentrations between 0.2 and 0.7 µg/g will provide between 45 and 158 µg of mercury. And the PTWI for mercury is only 122 µg per week for a 12 stone adult, and much less for children (Table 7.17). So what? One meal of tuna per week, or three meals per month, for an adult would be fine. I hesitate to make any recommendation for children, pregnant women or breast-feeding mothers. One meal of tuna per month for even young children is likely to be free of significant risk from mercury, and tuna does supply vital nutrients such as omega-3 fatty acids. Websites are available to address more specific questions, many of them based on US Environmental Protection Agency Guidelines for mercury.

Table 7.17 also lists PTWI for inorganic arsenic. There are no EU limits for concentrations of arsenic in food. There is, however, the 1959 UK statutory limit of 1 µg/g ww for arsenic in most foods, but not in seafood. The saving grace here is that almost all arsenic in almost all marine animals, including commercial fish and shellfish, is in the form of the nontoxic organic compound arsenobetaine. The muscles of North Sea and Northeast Arctic cod contain average arsenic concentrations of 4.7 and 9.3 µg/g respectively, with a maximum concentration of 20 µg/g arsenic in the North Sea cod (Julshamn et al., 2013). In our theoretical fish and chip supper of North Sea cod, the average arsenic concentration would translate to a dietary input of 1,060 µg of arsenic per meal. If this arsenic were in inorganic form, a single meal would match the PTWI for arsenic (Table 7.17). Fortunately, the arsenic exists as arsenobetaine in the fish meal, and there is no danger of arsenic toxicity.

7.10.2 Crabs, Lobsters and Shrimps

Prologue. So much for fish. What about other seafood? We start here with crustaceans – specifically the decapod crustaceans, which go under an assortment of common names, including crabs, lobsters, scampi, shrimps and prawns.

Decapod crustaceans typically regulate the concentrations of copper and zinc in their muscle tissues, and in some cases their whole bodies. An organ that does show raised concentrations of these two essential metals is the hepatopancreas. Copper concentrations in the hepatopancreas vary with the moult cycle and with associated changes in the body and blood concentrations of the copper-bearing respiratory protein haemocyanin. The hepatopancreas is also typically the organ in decapod crustaceans that is the site of detoxified storage of accumulated nonessential trace metals such as cadmium and lead. The concentrations of these non-essential trace metals may vary in muscle tissue, but these muscle concentrations of cadmium and lead are still low in comparison to the hepatopancreas.

In the commercial world of seafood, crabmeat comes in two forms. White meat is essentially muscle tissue, extracted, for example, from the claws and legs of the crab, usually the edible crab *Cancer pagurus*. Brown meat, on the other hand, consists mostly of the hepatopancreas, perhaps infiltrated at some time of the year by gonad tissue. The white meat and the brown meat of crabs, therefore, can be expected to contain different concentrations of many trace metals – higher in the brown meat. It is the muscle tissue that is the usual meat consumed in the case of lobsters, for example the European or common lobster *Homarus gammarus*. Abdominal (tail) muscles of *Nephrops norvegicus*, the Norway lobster or Dublin Bay prawn, supply the meat known as scampi. British shrimps and prawns such as the brown shrimp *Crangon crangon* and the pink shrimp *Pandalus montagui* may be eaten whole, usually after shelling (the removal of the exoskeleton), although it is often only the muscles of the abdomen (tail) that are consumed.

Table 7.18 presents concentrations of trace metals in those parts of these decapod crustaceans that are commercially available for consumption after collection in British coastal waters, both historically and in this century.

In the case of crabs, there are clear differences between the concentrations of cadmium, lead and

Table 7.18 **Mean concentrations of trace metals (μg/g ww) in the edible crab *Cancer pagurus* (white meat, brown meat, hepatopancreas, gonad), the lobster *Homarus gammarus* (claw muscle, tail meat), scampi *Nephrops norvegicus* (tail meat), whole pink shrimps *Pandalus montagui* and whole brown shrimps *Crangon crangon* from coastal waters around Britain.**

	Cd	Cu	Hg	Pb	Zn
Edible Crab *Cancer pagurus*					
Cromer, Norfolk					
1974–78 white meat[1,2]	<0.2	20	0.18	<0.2	79
1978 brown meat[1]	6.8	–	–	–	–
2013 white meat[3]	0.13	–	–	–	–
2013 brown meat[3]	1.2	–	–	–	–
Whitby, Yorkshire					
1974–77 white meat[1,2]	<0.2	19	0.26	<0.2	74
1977 brown meat[1]	2.9	–	–	–	–
Tyne					
1974 white meat[2]	<0.2	12	0.26	<0.3	67
Scotland					
Scotland 2007–8 white meat[4]	<0.02–0.03	7–10	0.16–0.61	<0.02–0.03	55–74
Scotland 2007–8 hepatopancreas[4]	6–28	18–100	0.15–0.51	0.02–0.10	17–27
Scotland 2007–8 gonad[4]	0.04–4.3	4–14	0.06–0.17	<0.02–0.04	18–90
Liverpool Bay					
1970–76 white meat[5]	<0.2	17	0.41	1.7	64
English Channel West					
Plymouth 1967–80 white meat[1,6]	<0.2	8	–	–	64
Plymouth 1967–80 hepatopancreas[1,6]	12	137	–	–	45
Lobster *Homarus gammarus*					
Whitby, Yorkshire					
1974 claw muscle[2]	<0.2	30	0.20	<0.2	51
1974 tail meat[2]	<0.2	17	0.40	<0.2	15
St Andrews					
1986 tail muscle[7]	–	–	0.09–0.22	–	–
Scampi *Nephrops norvegicus*					
Tyne					

Table 7.18 **(cont.)**

	Cd	Cu	Hg	Pb	Zn
1978 tail meat[1]	<0.4	20	0.28	<0.2	16
Scotland					
Firth of Clyde 1991 tail meat[8]	0.30	5.5	0.10	–	13
SW Scotland 1993–95 tail meat[9]	–	4.0	0.08	<0.02	13
Pink shrimp *Pandalus montagui*					
Outer Thames					
1977 whole[1]	<0.2	12	0.11	<0.4	36
Tees					
1974 whole[2]	<0.2	15	0.08	0.5	21
Tyne					
1974–77 whole[1,2]	<0.2	16	0.10–0.14	<0.2–0.5	19–20
Brown shrimp *Crangon crangon*					
Outer Thames					
1981 whole[1]	<0.4	18	0.08	<0.4	23
Tees					
1974 whole[2]	<0.2	18	0.11	0.6	18
Tyne					
1974 whole[2]	<0.2	23	0.08	<0.2	15
Liverpool Bay					
1970–76 whole[5]	0.7	25	0.17	4.1	24
1980 whole[1]	<0.4	16	0.11	<0.4	20
Morecambe Bay					
Lytham 1974 whole[2]	0.3	24	0.18	14	40
1982 whole[1]	<0.1	11	0.20	1.1	13
Bristol Channel					
1980 whole[1]	2.9	14	0.09	<0.2	20
1999 whole[10]	0.4–1.4	14–28	–	–	12–32

Sources: Franklin (1987);[1] Murray (1979);[2] Clark, Smith and Rainbow unpublished;[3] Barrento et al. (2009);[4] Norton and Murray (1983);[5] Bryan (1968);[6] Brown et al. (1988);[7] Canli and Furness (1993);[8] Brown and Balls (1997);[9] Culshaw et al. (2002).[10]

copper in the white meat and brown meat (Table 7.18), to be expected from the different roles of muscles and hepatopancreas in the physiology of these trace metals in decapods. Cadmium concentrations in white meat are always well below the maximum concentration (0.5 µg/g) under EU regulations (Table 7.16). Brown meat concentrations of cadmium, on the other hand, always exceed 0.5 µg/g, at times greatly so, as a result of cadmium accumulation in the hepatopancreas (Table 7.18), even in the absence of anthropogenic cadmium contamination. These high cadmium concentrations in brown meat do not actually exceed any EU regulations, for crab brown meat is not covered by any such regulations (Table 7.16). This is not an error of accidental omission. It is a pragmatic response to the large range of accumulated cadmium concentrations observed in the brown meat of crabs in European waters, set against the use of PTWI as a measure of risk from cadmium toxicity from the diet. Any harmful effects of cadmium would result only from prolonged exposure to high levels of cadmium. In practice, the UK Food Standards Agency is relying on the strong likelihood that the excess consumption of brown crabmeat with high cadmium will be infrequent. An occasional exceedance of the PTWI for cadmium (Table 7.17) would not cause concern.

Lead concentrations in crab white meat are now also very low, well below the limit of 0.5 µg/g (Table 7.16), after a period of lead contamination in Liverpool Bay in the 1970s (Table 7.18). The occasionally higher concentration of lead in the brown meat of crabs (Table 7.18) is subject to the arguments made above for raised cadmium concentrations, and are of no concern.

Concentrations of copper and zinc in the white meat and brown meat of crabs come under the exceptions specified above to the UK statutory limits of 20 µg/g copper and 50 µg/g zinc in food. Thus a copper concentration above 20 µg/g is permitted if the accumulated copper is 'of natural occurrence', and a zinc concentration above 50 µg/g is permissible in food such as herring and shellfish 'which naturally contain more than 50 µg/g'. Copper concentrations in crab white meat usually do not exceed 20 µg/g, but those in the hepatopancreas, constituting brown meat,

often do (Table 7.18). The zinc concentrations in the claw and leg muscles (the white meat) of crabs at least match, and often exceed, those in the hepatopancreas (the brown meat), and often exceed 50 µg/g, but not greatly so (Table 7.18). In practice, copper and zinc concentrations in both white meat and brown meat of edible crabs in British waters are not of any toxic concern.

The muscles of edible crabs can contain high concentrations of mercury (Table 7.18), as we have seen for fish. Historically, as we have come to expect, these concentrations were raised in Liverpool Bay in the 1970s, and to a lesser extent in the North Sea off Whitby and the Tyne at the same time (Table 7.18). Today, most accumulated mercury concentrations in the white meat and brown meat of edible crabs (Table 7.18) are below the EU maximum limit of 0.5 µg/g in crustacean muscle (Table 7.16). The occasional exceedance is catered for in the use of PTWI to assess any risk of mercury toxicity.

In the case of lobsters and scampi, measurements of trace metal concentrations in tail meat (Table 7.18) do not raise toxicity concerns today. There is in fact an interesting difference in the zinc concentrations of different muscles in the lobster *Homarus gammarus* (Bryan, 1967). The abdomen (tail) of the lobster contains flexor muscles with fine cross-striation that can contract very quickly, as in an escape response (Bryan, 1967). These tail muscles have a low zinc concentration of about 15 µg/g zinc (Table 7.18). Muscles in the claws and legs, on the other hand, have a coarse cross-striation and contract more slowly (Bryan, 1967). These slow muscles have a higher zinc concentration of about 60 µg/g (Table 7.18) (Bryan, 1967). The zinc-related differences in contraction properties of the two types of muscles may be related to different concentrations of zinc-containing enzymes (Bryan, 1967). One such enzyme might be arginine kinase, which plays a role in the maintenance of energy (ATP) levels in cells.

Concentrations of trace metals in different shrimps are usually quoted on a whole body basis (Table 7.18). In both the pink shrimp *Pandalus montagui* and the brown shrimp *Crangon crangon*, there was some evidence for raised trace metal concentrations historically in specific coastal areas in the 1970s (Table 7.18). Thus brown shrimps had raised mercury and lead concentrations in Liverpool Bay and Morecambe Bay at that time, and cadmium concentrations were still high then in brown shrimps from the Bristol Channel (Table 7.18). There are no concerns today about trace metal concentrations in shrimps from British coastal waters.

7.10.3 Molluscs

Prologue. The term 'shellfish' typically covers both crustaceans and molluscs. Amongst the molluscs harvested as seafood from British coastal waters are gastropods, in the form of whelks, and bivalves, such as scallops, cockles, mussels and oysters.

7.10.3.1 Whelks

Prologue. The whelk fished commercially in British coastal waters is the large common whelk *Buccinum undatum*. Trace metal concentrations in the whole bodies or the large muscular foot of whelks do not raise any toxic concerns for human consumers.

Small whelks may be consumed whole, but it is usually the muscular foot that is eaten in the case of adult whelks. In the east end of Victorian London, small whelks were a popular street food, sold in large numbers from barrows. Today, most of the whelks landed in British ports are exported to more appreciative markets in the Far East, especially South Korea.

Small common whelks are found inshore, sometimes on the lower shore, even in the outer regions of estuaries. The larger adults live sublittorally offshore on soft muddy sand bottoms, where they are fished using baited traps. As we saw in Table 6.26, common whelks have high body concentrations of zinc in particular, and also of copper, attributable partly to the presence of haemocyanin. Table 7.19 lists body concentrations of trace metals in common whelks expressed in terms of wet weight. Both zinc and copper concentrations in the bodies of *B. undatum*

(Table 7.19) exceed the general UK statutory limits of 50 µg/g and 20 µg/g respectively. Both, however, are covered by the specific exceptions permitted.

The body tissues of common whelks are dominated by the large, muscular foot, and this muscle tissue can be expected to be a site for the accumulation of mercury. Historically, the bodies of whelks collected in the mercury contaminated waters of Morecambe Bay and the Outer Thames Estuary in the 1970s did have mercury concentrations (Table 7.19) above 0.5 µg/g, the EU-permitted limit for crustacean muscle and for bivalve and cephalopod molluscs, no limit being defined for gastropods (Table 7.16). Mercury concentrations in common whelks were below 0.5 µg/g elsewhere in British coastal waters and in the Outer Thames by 1985 (Table 7.19).

A raised permitted body concentration of 1 µg/g cadmium has been designated by the EU for bivalve and cephalopod molluscs (Table 7.16), gastropods again remaining unspecified. Concentrations of cadmium in the bodies of common whelks exceeded this level in contaminated British coastal waters in the 1970s but are lower thereafter (Table 7.19).

7.10.3.2 Scallops

Prologue. It is the adductor muscle and gonad (coral) of scallops that are eaten. Trace metal concentrations in these tissues offer no cause for concern.

In the case of most bivalves, it is the whole soft tissue that is ingested. Not so with scallops. Two species of scallop are fished: the great scallop *Pecten maximus* and the queen scallop *Aequipecten opercularis*. In both these scallops, two subportions of the soft tissues are usually prepared for consumption. The first is the large cylindrical adductor muscle that brings the two valves of the shell together. The second organ is the gonad, often called the coral, orange in females but pinker in males.

Table 7.19 presents trace metal concentrations in these two organs of great scallops and queen scallops. The total soft tissues of scallops are naturally very high in cadmium and silver in comparison with other bivalves (Tables 6.16 and 6.18). These two trace

Table 7.19 **Mean concentrations of trace metals (μg/g ww) in the bodies of the common whelk** *Buccinum undatum* **and in the adductor muscle and gonad (coral) of great scallops** *Pecten maximus* **and queen scallops** *Aequipecten opercularis* **from coastal waters around Britain.**

	Cd	Cu	Hg	Pb	Zn
Common whelk *Buccinum undatum*					
Outer Thames					
Whitstable 1974[1]	5.0	44	0.22	0.2	220
1977[2]	4.2	80	0.60	<0.2	180
1985[2]	–	–	0.14	–	–
North Sea Southern Bight					
1978[2]	1.8	42	0.16	<0.2	180
Humber					
1974[1]	0.2–1.7	25–32	0.05–0.15	<0.2–0.3	44–130
1977[2]	1.4	27	0.12	<0.2	140
Tyne					
1974[1]	<0.2–3.0	20–190	0.04–0.41	<0.2–0.4	75–190
Liverpool Bay					
1980[2]	2.7	185	0.28	1.0	200
Morecambe Bay					
1974[1]	0.8	66	0.63	0.8	180
Solent					
Pre-1971[3]	0.5	40	–	1.2	138
Great scallop *Pecten maximus*					
Liverpool Bay					
1981 muscle[2]	0.8	0.3	0.02	<0.4	23
Plymouth					
1967–69 muscle[4]	–	0.2	–	0.05	16
1967–69 gonad[4]	–	3.1	–	0.3	35
English Channel					
Newhaven 1978 muscle[2]	4.0	0.9	0.04	1.2	36
Lyme Bay 1978 muscle[2]	2.8	1.2	0.04	1.3	40
Queen scallop *Aequipecten opercularis*					
Liverpool Bay					

Table 7.19 **(cont.)**

	Cd	Cu	Hg	Pb	Zn
1973–76 muscle[5]	<0.2–0.2	3.4–12	0.09–0.13	0.4–1.5	35–170
1973–76 gonad[5]	0.6–0.7	4.6–7.5	0.05–0.07	0.4–2.9	57–90
1980 muscle[2]	0.8	5.3	0.08	<0.2	54
Morecambe Bay					
1974 muscle[1]	0.3	3.4	0.09	0.4	50
Plymouth					
1967–69 muscle[4]	–	0.3	–	0.1	14
1967–69 gonad[4]	–	1.9	–	0.5	22
Salcombe, Devon					
1974 muscle[1]	0.9	2.3	0.04	0.6	36
1974 gonad[1]	0.4	2.0	0.05	1.2	49

Sources: Murray (1979);[1] Franklin (1987);[2] Segar et al. (1971);[3] Bryan (1973);[4] Norton and Murray (1983).[5]

metals are strongly accumulated in the digestive gland of scallops (Table 6.17), providing, therefore, the high cadmium and silver concentrations in the total soft tissues (Tables 6.16 and 6.17). In the case of cadmium, the muscle and gonad tissues of scallops have lower concentrations than the digestive gland and total soft tissues (Table 6.17) (Bustamente and Miramand, 2005). Thus cadmium concentrations in the adductor muscle and gonad (the coral) of queen scallops (Table 7.19) are below the EU limit for bivalves of 1 µg/g (Table 7.16). Historical (1970s) concentrations of cadmium in the muscle of great scallops from the English Channel (Table 7.19) are higher, for reasons unknown.

Copper concentrations in the total soft tissues of scallops are typically low in comparison with other bivalves (Table 6.18). Copper concentrations in the muscles and gonads of great scallops and queen scallops are similarly low (Table 7.19). Typical zinc concentrations in the total soft tissues of scallops exceed those of cockles and mussels but are below those of oysters (Table 6.18). Zinc concentrations in muscles and gonads of queen scallops can exceed 50

µg/g (Table 7.19) but are permitted by the stated exceptions to UK statutory limits.

Mercury concentrations in the muscles and gonad of scallops (Table 7.19) are well below the EU maximum permitted concentration of 0.5 µg/g.

7.10.3.3 Cockles and Mussels

Prologue. The soft tissues of cockles and mussels are generally consumed whole. Unless these bivalves are fished from metal-contaminated estuaries, there is no risk of trace metal toxicity to human consumers.

The distributions of cockles and mussels extend onto the shore and up into estuaries. The accumulation of trace metals by these two bivalves has, therefore, been discussed at some length in Chapter 6. It is the whole soft tissues that are consumed. Tables 6.14 and 6.11 list accumulated trace metal concentrations in cockles and mussels respectively, expressed on a dry weight basis. In the context of food safety, these dry weight concentrations need to be expressed on a wet weight basis, simply done by dividing by an arbitrary factor of 6.

Cockles (*Cerastoderma edule*) are fished in shallow water, often in outer estuaries, usually by dredging. Many thousands of tonnes of cockles are landed in British ports each year. Many cockles may then be exported, as in the case of cockles from the Wash to Spain via Holland. From the typical concentration ranges of trace metals in cockles from British estuaries and coastal waters, it can be concluded that there is no risk of any trace metal toxicity, unless the cockles are collected from within estuaries with a history of contamination from mining or industrial activity (Table 6.14).

Mussels (*Mytilus edulis*) are harvested from natural mussel beds, or commonly these days from rafts in mussel farms. In 2012, the United Kingdom produced 26,000 tonnes of mussels, exporting most. There is a home market for mussels, and UK consumption in 2013 was about 4,000 tonnes. Mussels are not strong accumulators of trace metals (Table 6.11). Accumulated concentrations of cadmium, lead and mercury in mussels usually fall comfortably below the EU-permitted concentrations of 1 µg/g Cd, 1.5 µg/g Pb and 0.5 µg/g Hg (all wet weight) for bivalve molluscs (Table 7.16). Similarly, accumulated copper and zinc concentrations (Table 6.11) are usually well below the UK general limits of 20 µg/g Cu and 50 µg/g Zn. Mussels growing in coastal waters known to be contaminated by trace metals from mining or industry can reach accumulated metal concentrations of particular concern, as in the case of cadmium in the Bristol Channel in the 1970s (Table 6.11). Nevertheless, as in the case of cockles, trace metal contamination of mussels as seafood is not a significant problem today in British coastal waters.

7.10.3.4 Oysters

Prologue. The soft tissues of oysters do typically contain very high accumulated concentrations of zinc and copper but are still considered safe for human consumption as allowed exceptions to defined regulatory limits. These exceptions are based on historical experience, supported by the relatively low bioaccessibility of the detoxified storage forms of many trace metals in oysters.

The two oysters of commercial interest in British waters are the native common oyster, *Ostrea edulis*, and the introduced Pacific oyster, *Crassostrea gigas*. Oysters can be collected from intertidal or sublittoral beds on the bottom by hand with rakes or tongs or by means of dredges towed by boats. Oysters are also farmed, perhaps on the bottom, but also in bags suspended from rafts in oyster farms.

In comparison with other bivalves, oysters typically have very high accumulated concentrations of zinc, and to a lesser extent of copper (Table 6.12). Table 6.12 listed accumulated soft tissue concentrations expressed on a dry weight basis of both oyster species. Expectedly, accumulated concentrations of cadmium, copper, mercury, lead and zinc in the soft tissues of oysters rise with increased bioavailabilities of these trace metals in metal-contaminated estuaries (Table 6.12).

Let us consider the accumulated concentrations of these trace metals that occur typically in the soft tissues of the Pacific oyster *C. gigas* growing in uncontaminated British coastal waters. For oysters, an arbitrary factor of 5 is often used to transform dry weight concentrations to wet weight concentrations.

Typical cadmium concentrations in the soft tissues of *C. gigas* fall into the range 0.2 to 1 µg/g ww (from Table 6.12). The top end of this range matches the maximum EU cadmium concentration in bivalves (Table 7.16). For mercury, the typical concentration range in the soft tissues is 0.02 to 0.1 µg/g ww (Table 6.12), below the EU limit of 0.5 µg/g (Table 7.16). Similarly, for lead, the typical concentration range is 0.1 to 1 µg/g ww (Table 6.12), within the EU limit for bivalves of 1.5 µg/g (Table 7.16). Typical copper concentrations in the soft tissues of Pacific oysters can reach 100 µg/g ww, above the 20 µg/g UK general limit but covered by the exception whereby 'higher levels in shellfish are permitted if copper is of natural occurrence'. Typical zinc concentrations in soft tissues of *C. gigas* greatly exceed the 50 µg/g ww UK general limit, reaching 900 µg/g ww without being indicative of an atypically raised zinc bioavailability (Table 6.12). Yet again, such exceedances are permitted by the exception whereby 'higher levels are permitted in food which naturally contain more than 50 µg/g such as herring and shellfish'.

A PTWI approach for cadmium can be applied to these oysters. A 50 gram portion of oysters would typically deliver between 10 and 50 µg of cadmium. A 12 stone adult would need to consume between 10 and 50 of these oysters per week over an extended period to match the PTWI for cadmium (Table 7.17).

There are no World Health Organisation PTWI for copper and zinc. Although the copper and zinc concentrations in the oyster tissues greatly exceed the general UK statutory limits quoted, history tells us that consumption of uncontaminated oysters does not result in dietary copper or zinc toxicity. This may not be the case for oysters collected in metal-contaminated estuaries. It needs to be remembered that comparison against arbitrary permitted concentration limits is simply a first stage in any risk assessment of the potential metal toxicity of any particular food, including oysters.

Another factor to be borne in mind is that the trace metal concentrations quoted here are total concentrations. The trophic bioavailability of a trace metal, including that to humans, will partly depend on the chemical form of the trace metal in the dietary item. Not all the trace metal in a food item will necessarily be trophically bioavailable to a consumer. The first stage in the assimilation of a trace metal in the human alimentary tract is the digestion of ingested tissue to release trace metals into the gut lumen before uptake. This stage can be assessed relatively simply by in vitro simulation of the human digestive processes to give a measure of bioaccessibility. Because of their high concentrations of trace metals, oyster tissues are commonly subjected to bioaccessibility assessments. Given the high proportion of storage detoxification in insoluble form of accumulated trace metals in oysters (see Chapter 6), it is unsurprising that not all their trace metal contents are bioaccessible (Bragigand et al., 2004; Gao and Wang, 2014). Depending on the total accumulated concentrations in the oysters, between 25 and 56% of the cadmium, between 20 and 74% of the copper and between 20 and 50% of the zinc in oysters is not bioaccessible (Bragigand et al., 2004). Low bioaccessibilities reduce trophic uptake and, thereby, toxic risk.

I am afraid that it is beyond the remit of this book to draw any conclusions on the aphrodisiac qualities of the high zinc concentrations in oyster tissues. These conclusions must be left to the personal experiences of the reader.

7.11 Seabirds

Prologue. The list of British seabirds is long and includes representatives of several different orders and families, with a variety of feeding methods and, therefore, prey types.

Seabirds are important predators in many marine food chains, particularly oceanic food chains. They may forage far and wide from land-based roosting and nesting sites, often on the cliffs of offshore islands. Many seabird species are associated with British coastal waters, even if these waters simply lie below an aerial passage to the adjacent waters of the North Atlantic and Arctic Oceans to feed. Seabirds take up and accumulate trace metals from their large assortment of marine prey.

Those seabirds most closely associated with land are the gulls (order *Charadriiformes*, family *Laridae*). Gulls are opportunist feeders, able to exploit new food sources, be it trash from a trawler, soil invertebrates turned over by a tractor ploughing a field or edible waste in urban habitats and rubbish tips. The black-headed gull *Larus ridibundus*, in particular, will feed far from the sea. Herring gulls *L. argentatus*, and their close relatives, lesser black-backed gulls *L. fuscus*, also feed inland, but will scavenge on the seashore and can prey on shellfish by dropping them from a height. Common gulls (*L. canus*) and great black-backed gulls (*L. marinus*) too are scavengers on the coast and will rob other gulls of their intended meals. In addition great black-backed gulls are active predators, for example of chicks in seabird colonies. Kittiwakes *Rissa tridactyla* also belong to the gull family and are more faithful to the sea for their food. Their diet consists mainly of fish but also invertebrates near the surface of the sea and, in modern times, of fish offal discarded from fishing boats.

A second family (*Sternidae*) in the same order contains the terns. Terns are typically long distance migrants. Those breeding on British coasts include the sandwich tern *Thalasseus* (previously *Sterna*) *sandvicensis*, the common tern *Sterna hirundo*, the Arctic tern *Sterna paradisaea*, the roseate tern *Sterna dougallii* and the little tern *Sternula albifrons*. Most terns hunt fish by diving, often hovering before plunging into the water. Occasionally, pelagic shrimps, krill and amphipods are also taken by terns.

A third family (*Alcidae*) in the order houses the auks. The flightless great auk *Pinguinus impennis* is now extinct, but its relatives carry on. Like other auks, the little auk *Alle alle* forages for food by swimming underwater, using the wings as paddles. In the summer, their main prey items are large calanoid copepods. In the winter, little auks switch to krill, hyperiid amphipods and young capelin *Mallotus villosus*. A second member of the auk family, the razorbill *Alca torda*, also feeds while swimming, usually to about 25 metres depth, but they can go deeper. Their diet generally consists of fish such as capelin, sand eels (e.g., *Hyperoplus lanceolatus* and *Ammodytes tobianus*) and juvenile cod (*Gadus morhua*), sprats (*Sprattus sprattus*) and herring *(Clupea harengus)* but can include pelagic crustaceans. A third auk is the guillemot *Uria aalge*. They also feed at sea by swimming underwater, using the wings for propulsion and the feet for steering. Guillemots mainly eat small fish, including sand eels, sprat, juvenile cod, herring and capelin. Guillemots will also prey on pelagic crustaceans, such as hyperiid amphipods and decapods, and on squid. The final auk considered here is the puffin *Fratercula arctica*. Like the other auks, puffins feed primarily by pursuit diving, rapidly beating their wings underwater for propulsion. Puffins feed on pelagic crustaceans and, characteristically, small fish such as sand eels, young herring and young capelin. Puffins can live for more than thirty years.

The last charadriiform family to be considered here is the *Stercorariidae*, containing the skuas. British examples are the Arctic skua *Stercorarius parasiticus*

and the great skua *Catharacta skua*. Skuas will steal food from other seabirds, often causing them to disgorge fish already eaten, and they will prey on chicks in seabird colonies. Skuas also scavenge on carrion and offal. In addition, great skuas will kill and eat adult seabirds. Great skuas can live to an age of 25 years or more.

A second order of seabirds is the *Procellariiformes*. The classic birds in this order are the albatrosses, residents of the southern hemisphere and, therefore, disregarded here. Of concern in this account are two families: the *Procellariidae*, containing fulmars and shearwaters, and the *Hydrobatidae*, comprised of storm petrels.

The fulmar *Fulmarus gacialis* feeds on fish, squid and decapod crustaceans in the open ocean. Fulmars are exceptionally long-lived, living for up to 40 years. Shearwaters are medium-sized, long-winged seabirds, flying close to the tops of waves in the ocean. Manx shearwaters *Puffinus puffinus* breeding on British coasts in the summer undertake very long annual migrations to the South Atlantic. Shearwaters are also long-lived. One Manx shearwater is known from ringing to have lived for more than 55 years. Shearwaters typically feed by pursuit diving to catch small fish and squid. Storm petrels are amongst the smallest of seabirds and feed on zooplanktonic crustaceans, squid and small fish picked from the surface of the ocean while hovering or pattering. The storm petrel *Hydrobates pelagicus* (Mother Carey's Chicken) can detect oily items of prey by smell. Leach's storm petrel *Oceanodroma leucorhoa* feeds on copepods, krill and hyperiid amphipods but also on myctophid lantern fish migrating to the surface at night from the mesopelagic zone of the ocean.

The third and final seabird order, the *Suliformes*, is represented by the gannet *Morus bassanus*, the largest seabird in the North Atlantic. Gannets hunt by plunge diving at speed from a height as much as 30 metres above the ocean to pursue their fish prey underwater. Any fish up to 30 centimetres long and shoaling near the surface are potential prey items, as are pelagic squid.

7.11.1 Trace Metal Accumulation in Seabirds

Prologue. Seabirds take up and accumulate trace metals from their diet. Interspecific variation in diets and their trace metal contents, in trace metal accumulation patterns and in longevity between seabird species, causes great variation in the accumulated trace metal concentrations in different tissues of different seabirds.

There are inevitably differences between seabird species in the amounts of trace metals bioavailable in their different prey items. Furthermore, there is interspecific variation in the trace metal accumulation patterns of different seabirds for different trace metals, further qualified by the different longevities of the different bird species. Seabird livers and kidneys are potential sites for the storage accumulation of detoxified trace metals. Feathers are also sites for the accumulation of certain trace metals, particularly mercury. So the moult cycle will affect the amount of such a trace metal present in the body of a seabird at any one time. Mercury, particularly methyl mercury, can be passed by a female bird to developing eggs, lowering her own mercury content but potentially exposing the developing embryo to an ecotoxicological risk under particular circumstances.

Two further factors come into play when considering the biology of trace metals in seabirds. Firstly, this chapter has already highlighted the very high concentrations of particular trace metals such as cadmium in specific oceanic animals making up the diets of several seabirds. Secondly, some pelagic seabirds are particularly long lived, as in the case of fulmars and shearwaters. Trace metals that are not excreted but stored permanently in particular organs, therefore, have the time to build up to very high tissue concentrations, necessarily in detoxified form.

Trace metals taken up from the diet in the alimentary tract of birds are passed into the blood. These trace metals are then transferred to other organs for excretion or for detoxified storage accumulation, temporary or permanent. Excretion may occur through the urine or, in some seabirds, through secretions of nasal salt glands. Organs carrying out detoxified storage of accumulated trace metals include the liver, kidney and/or bone (particularly for lead). Certain trace metals, especially mercury, will also be transferred from temporary stores in such

internal organs to the feathers. This transfer awaits the time of the next moult when the feathers are newly growing with an active blood supply. The amount of trace metal transferred into the feathers reflects the amount accumulated by the bird since the last moult. While it is typically non-essential metals (especially mercury) that are transferred seasonally to the feathers, the proportion of the total body metal content that is transferred varies with the trace metal concerned. Cadmium, for example, remains permanently stored in the body, particularly in the kidney. The accumulated concentration of cadmium in the kidney, therefore, has the potential to increase with the age of the bird. Female birds will also transfer trace metals to developing eggs, thereby offering a route of elimination from the body.

Table 7.20 lists zinc and copper concentrations in selected tissues of particular seabirds from colonies in the British Isles.

Zinc concentrations in muscle tissue such as the pectoral muscle in seabirds fall into a relatively narrow range (Table 7.20), reflecting regulation to meet essential zinc requirements. Zinc concentrations in the livers and kidneys of gulls and guillemots are similarly of the order expected to meet essential zinc requirements (Table 7.20). In the case of the longer-lived seabirds such as puffins, great skuas, Manx shearwaters and, particularly, fulmars, accumulated zinc concentrations in the livers and kidneys are raised above 100 $\mu g/g$, reaching more than 400 $\mu g/g$ in the kidneys of some fulmars (Table 7.20). These somewhat raised zinc concentrations indicate some storage detoxification of accumulated zinc, presumably before some zinc is transferred to the feathers (Table 7.20).

Copper concentrations measured in the tissues of British seabirds are all low (Table 7.20), approximately of the order necessary to meet metabolic requirements. It would appear, then, that excess copper taken up is excreted without significant storage in detoxified form in tissues.

To turn to non-essential metals, accumulated lead concentrations in the tissues of pelagic seabirds do not appear to be atypically raised in comparison with other birds (Franson and Pain, 2011). Lead is accumulated in bone tissues, and more coastal species

Table 7.20 **Zinc and copper concentrations (means and ranges, μg/g dw) in tissues of seabirds from colonies in the British Isles.**

Order	Location	Liver	Kidney	Muscle	Feather
Family	*Date*	*Mean*	*Mean*	*Mean*	*Mean*
Species		*(range)*	*(range)*	*(range)*	*(range)*
ZINC					
Charadriiformes					
Laridae					
Herring gull	Isle of May, Fife[1]	92	98	70	80
Larus argentatus	*1976*	(56–136)	(79–148)	(47–128)	(60–100)
Lesser black-backed gull	Lancashire[2]	79	88	–	–
Lara fuscus	*1991*				
Alcidae					
Guillemot	NW Scotland[3]	66	73	24	–
Uria aalge	*1988*				
Puffin	St Kilda[4]	118	164	47	108
Fratercula arctica	*1977*	(98–151)	(122–202)	(42–53)	(95–123)
Stercorariidae					
Great skua	Foula, Shetland[1]	139	164	70	96
Catharacta skua	*1976*	(61–497)	(122–213)	(38–189)	(74–124)
	Foula, Shetland[2]	103	174	–	–
	1988				
Procellariiformes					
Procellariidae					
Fulmar	St Kilda[4]	364	310	58	97
Fulmarus gacialis	*1977*	(225–688)	(185–408)	(49–66)	(89–105)
Manx shearwater	St Kilda[4]	141	176	46	87
Puffinus puffinus	*1977*	(117–170)	(157–212)	(42–50)	(87–92)

Table 7.20 **(cont.)**

Order	Location	Liver	Kidney	Muscle	Feather
Family	*Date*	*Mean*	*Mean*	*Mean*	*Mean*
Species		*(range)*	*(range)*	*(range)*	*(range)*
COPPER					
Charadriiformes					
Laridae					
Lesser black-backed gull	Lancashire[2]	20	13	–	–
Lara fuscus	*1991*				
Alcidae					
Guillemot	NW Scotland[3]	16	14	12	–
Uria aalge	*1988*				
Stercorariidae					
Great skua	Foula, Shetland[2]	19	20	–	–
Catharacta skua	*1988*				

Sources: Hutton (1981);[1] Stewart and Furness (1998);[2] Stewart et al. (1994);[3] Osborn et al. (1979).[4]

such as herring gulls have higher accumulated lead concentrations than the oceanic great skua (Hutton, 1981).

Two other non-essential trace metals are, however, accumulated to remarkable concentrations in the tissues of pelagic seabirds and are worthy of further discussion here. These two trace metals are cadmium and mercury.

7.11.1.1 Cadmium

Prologue. Accumulated cadmium concentrations are very high in the kidneys of long-lived seabirds.

Concentrations of cadmium in the tissues of adult seabirds are listed in Table 7.21. It is apparent that cadmium concentrations in the kidneys of long-lived seabirds are extremely high, matching those in earthworms (Table 4.25), the ventral caeca of wood-lice (Table 4.27) and the kidneys and livers of shrews (Table 4.31) from metal-contaminated sites. The average concentration of cadmium in the kidneys of

fulmars from St Kilda is 177 µg/g, with some individuals attaining 480 µg/g cadmium (Bull et al., 1977; Osborn et al., 1979). Concentrations of cadmium are also somewhat raised in the livers of long-lived seabirds, but to nowhere near the same extent as in the kidneys (Table 7.21). Concentrations of cadmium are extremely low to the point of non-detectability in the feathers of pelagic seabirds (Osborn et al., 1979). Thus, cadmium taken up from food is transferred for permanent detoxified storage in the kidney, with no subsequent transfer to the feathers before each moult.

Correspondingly, in great skuas sampled from Foula in Shetland in 1976, there was a significant positive relationship between cadmium concentrations in the kidney and age, as there was for the lower cadmium concentrations in the liver (Furness and Hutton, 1979). Similarly, accumulated cadmium concentrations are higher in the kidneys and livers of adult guillemots than in juveniles (Stewart et al., 1994).

Table 7.21 **Cadmium concentrations (means and ranges, μg/g dw) in livers and kidneys of seabirds from colonies in the British Isles.**

Order Family Species	Location *Date*	Liver *Mean* *(range)*	Kidney *Mean* *(range)*
Charadriiformes			
Laridae			
Herring gull	Isle of May, Fife[1]	2.0	13
Larus argentatus	*1976*	(0.8–3.3)	(9–20)
Lesser black-backed gull	Lancashire[2]	5.6	25
Lara fuscus	*1991*		
Alcidae			
Razorbill	St Kilda[3]	1.9	16
Alca torda	*1976*	(1.4–2.4)	(15–18)
Guillemot	NW Scotland[4]	1.9	9.0
Uria aalge	*1988*		
Puffin	St Kilda[3,5]	20	109
Fratercula arctica	*1976/77*	(7–33)	(61–162)
Stercorariidae			
Great skua	Foula, Shetland[1]	7.5	81
Catharacta skua	*1976*	(1.0–31)	(14–336)
	Foula, Shetland[2]	5.3	41
	1988		
Procellariiformes			
Procellariidae			
Fulmar	St Kilda[3,5]	39	177
Fulmarus gacialis	*1976/77*	(8–107)	(33–480)

Table 7.21 **(cont.)**

Order	Location	Liver	Kidney
Family	*Date*	*Mean*	*Mean*
Species		*(range)*	*(range)*
Manx shearwater	St Kilda[3,5]	20	115
Puffinus puffinus	*1976/77*	(11–40)	(60–231)
Hydrobatidae			
Storm petrel	St Kilda[3]	18	39
Hydrobates pelagicus	*1976*	(9–27)	(30–53)
Leach's storm petrel	St Kilda[3]	33	92
Oceanodroma leucorhoa	*1976*	(21–57)	(69–128)

Sources: Hutton (1981);[1] Stewart and Furness (1998);[2] Bull et al. (1977);[3] Stewart et al. (1994);[4] Osborn et al. (1979).[5]

A factor controlling accumulated cadmium concentrations in the tissues of pelagic seabirds is the concentration of cadmium in their prey. In the case of fulmars, there are two potential sources of high cadmium in their diet, namely decapod crustaceans and squid. Cadmium concentrations in mesopelagic decapods migrating to near the ocean surface at dawn and dusk are very high compared to those in coastal shrimps and prawns (Table 7.5). Some oceanic squid may also have extremely high body concentrations of accumulated cadmium (Table 7.8). Manx shearwaters too will catch oceanic squid. In the case of Leach's storm petrel, it is oceanic hyperiid amphipods that offer a high dietary source of cadmium (Table 7.7).

Cadmium accumulated in the kidneys of birds is detoxified by binding with metallothionein (Wayland and Scheuhammer, 2011), as exemplified amongst pelagic seabirds in great skuas and fulmars (Osborn, 1978; Hutton, 1981).

7.11.1.2 Mercury

Prologue. Methyl mercury accumulates in the feathers of seabirds. Feathers can be used as biomonitoring tools for the bioavailability of mercury to seabirds, highlighting changes with time and place.

As for other trace metals, mercury is taken up from the diet in the alimentary tracts of seabirds, before passing into the blood for transport elsewhere in the body. Mercury is taken up in proportion to the amount in the food, and effectively all absorbed mercury in the body tissues is in the form of methyl mercury (Furness, 2015).

Between moults, the mercury that has been taken up will accumulate in increasing concentrations in the liver and muscle tissues (Furness, 2015). At each moult, however, mercury is transferred in the blood to the newly developing feathers, and accumulated mercury concentrations in the liver and muscle fall again. Methyl mercury has a strong affinity for sulphur-containing amino acids. These amino acids are present in large amounts in keratin, the protein that makes up the greater part of a feather. Methyl mercury therefore accumulates in the new feathers, the total mercury content in these feathers reflecting the amount of mercury taken up and accumulated by the bird since the previous moult. Correspondingly, the moulted feathers take their mercury loads with them, an effective route of excretion for the mercury accumulated by the bird between the two previous moults (Furness, 2015). It follows then that seabirds do not show increasing tissue levels of mercury with age. A feature of the transfer of methyl mercury into newly developing feathers is that there is variability between feathers as to the amount of mercury transferred for accumulation (Lodenius and Solonen, 2013; Furness, 2015). The first feathers to grow have the highest mercury concentrations, with mercury concentrations decreasing in order in feathers produced later in the moult (Lodenius and Solonen, 2013; Furness, 2015).

Female seabirds will also transfer methyl mercury to the developing eggs, where it will accumulate, particularly by binding to albumins, the proteins in egg white (Becker, 1992; Furness, 2015). This is another excretion route for mercury, specific to female birds. This transfer of mercury to the egg appears to reflect the amount of mercury ingested immediately prior to egg laying (Lewis et al., 1993). In contrast, the transfer of mercury to the feathers additionally depends on the amount of mercury taken up earlier in the intermoult period and stored in the soft tissues (Lewis et al., 1993). This extra excretion route may cause females to have lower accumulated tissue concentrations of mercury than their male counterparts (Becker, 1992). Nevertheless, this maternal transfer of methyl mercury to the developing eggs is usually much less than that to the feathers at the moult. Consequently, differences in accumulated mercury concentrations between male and female birds may be insignificant (Furness, 2015). As in the case of feathers, there may be variability between eggs in a clutch as to the amount of mercury transferred and accumulated (Becker, 1992). The first egg laid by herring gulls *Larus argentatus* and common terns *Sterna hirundo* may have up to 39% more mercury than the second or third egg in the same clutch (Becker, 1992).

Concentrations of mercury in the tissues of seabirds are listed in Table 7.22. Regular transfer of mercury to the feathers for excretion at each moult means that, in contrast to cadmium, there is not one long-term storage organ accumulating mercury over the lifetime of the seabirds. Also in contrast to cadmium, the

feathers contain significant concentrations of mercury. The livers of seabirds such as the great skua, the fulmar and the Manx shearwater tend to have higher mercury concentrations than the kidney (Table 7.22), reflecting their role as temporary stores of mercury between moults. Concentrations of mercury in the muscles are lower (Table 7.22), but the greater amount of muscle tissue present means that the muscle tissues are also a significant store of mercury between moults. Fish represent a significant dietary source of mercury to seabirds, and exclusively piscivorous seabirds have higher accumulated tissue concentrations of mercury than other seabirds (Parslow and Jeffries, 1977; Hutton, 1981).

The uptake and accumulation of mercury by seabirds occur in proportion to the mercury contents of their diets (Furness, 2015). Thus there are differences in the accumulated concentrations of mercury in different populations of the same species of seabird feeding on the same prey species with different accumulated mercury concentrations. Seabirds can, therefore, be used as biomonitors of changes in mercury bioavailability, both over space and over time (Furness, 2015).

The combustion of fossil fuels and incineration of waste are particularly important anthropogenic sources of mercury released into the atmosphere. And atmospheric mercury can be transported far across the planet before deposition on dust particles and in rainfall. Thus the oceans will receive anthropogenic input of mercury with the potential to enter marine food chains, particularly after methylation to form methyl mercury in marine sediments or in the deep sea when low oxygen conditions occur (Furness, 2015). In the northern hemisphere, the anthropogenic emission of mercury may have increased fivefold in the last 150 years (Furness, 2015). Most of this mercury has entered marine systems, often far from the emission source. Mercury is a potentially very toxic trace element. It is relevant then to explore differences in the bioavailability of mercury, both geographically at any one time and historically over long periods of time.

And particularly useful as biomonitors are the feathers of seabirds (Furness, 2015). The accumulated concentrations of mercury in feathers are

Table 7.22 **Mercury concentrations (means and ranges, µg/g dw) in livers, kidneys, pectoral muscles and primary feathers of seabirds from colonies in the British Isles.**

Order	Location	Liver	Kidney	Pectoral Muscle	Primary Feather
Family	Date	Mean	Mean	Mean	Mean
Species		(range)	(range)	(range)	(range)
Charadriiformes					
Laridae					
Herring gull	Isle of May, Fife[1]	4.1	3.9	1.3	2.8
Larus argentatus	1976	(0.5–11)	(1.6–8.6)	(0.7–2.6)	(0.6–5.4)
Alcidae					
Guillemot	NW Scotland[2]	2.4	2.4	1.0	–
Uria aalge	1988				
Puffin	St Kilda[3]	4.5	5.0	<1.4	7.9
Fratercula arctica	1977	(2.9–7.1)	(2.9–7.4)	(<1.4–2.6)	(4.6–13)
Stercorariidae					
Great skua	Foula, Shetland[1]	10	8.7	1.9	6.4
Catharacta skua	1976	(3.2–30)	(3.4–20)	(1.0–3.2)	(2.0–15)
Procellariiformes					
Procellariidae					
Fulmar	St Kilda[3]	29	13	1.6	3.3
Fulmarus gacialis	1977	(20–45)	(6–24)	(0.4–2.6)	(0.9–4.7)
Manx shearwater	St Kilda[3]	10	4.7	0.9	1.2
Puffinus puffinus	1977	(5.7–14)	(3.2–62)	(0.4–1.6)	(1.0–1.3)

Sources: Hutton (1981);[1] Stewart et al. (1994);[2] Osborn et al. (1979).[3]

proportional to the amount of mercury taken up and accumulated by the seabird since the previous moult. There is, therefore, a defined time frame reflected in the measured concentrations of mercury in the feathers. This time frame is typically one year, reflecting an annual moult after the breeding season. Other time frames may apply but will be known from a knowledge of the moulting pattern of the bird species concerned. A caveat in the use of feathers for biomonitoring concerns variability between feathers. This can be accounted for by a well-designed sampling protocol. For example, the pooling of about six to ten small body feathers will provide an average measure reducing the effects of interfeather variability (Thompson et al., 1992a; Furness, 2015). In the case of feathers, it is often difficult to obtain consistent dry weights. Biomonitoring data for feathers are, therefore, often quoted in terms of fresh weight (Thompson et al., 1992a).

A biomonitoring study involving the body feathers of several species of pelagic seabirds investigated geographical differences in mercury bioavailability in the Northeast Atlantic at the end of the 1980s (Table 7.23). There was a general southwest to northeast trend of decreasing concentrations of mercury in the body feathers of seabirds from colonies in Iceland and St Kilda to those collected from Shetland, and then from Northwest and Northeast Norway (Table 7.23). There was little general evidence to suggest that differences in the bioavailability of mercury to the seabirds from different colonies were down to intraspecific differences in the diet of birds from different colonies (Thompson et al., 1992a). It is likely then that there were differences in the mercury concentrations of the same prey species of the particular seabirds between the geographical areas. The colonies lie approximately in sequence along the North Atlantic drift current. It is possible that input of anthropogenic mercury into the ocean was stronger in the southwest near North America with subsequent progressive depletion of dissolved mercury by biological activity along the North Atlantic drift current (Thompson et al., 1992a).

Kittiwakes, guillemots and fulmars from the Firth of Forth had higher mercury concentrations than those in Shetland and Norway (Thompson et al., 1992a). Perhaps this was a legacy of the mercury pollution observed upstream in the Forth Estuary in the early 1980s (Elliott and Griffiths, 1986). There were interspecific differences in the mercury concentrations of body feathers of seabirds from the same site (Table 7.23). Kittiwakes and puffins generally had relatively high feather concentrations of mercury, while guillemots generally had lower concentrations (except in the Firth of Forth) (Table 7.23).

Table 7.23 **Mercury concentrations (means and ranges, µg/g fresh weight) in body feathers of adult seabirds from colonies in the Northeast Atlantic between 1986 and 1991.**

Order	Location	Mean
Family		*(range)*
Species		
Charadriiformes		
Laridae		
Kittiwake	Iceland	5.5
Rissa tridactyla		*(1.6–10)*
	Firth of Forth	3.8
		(0.7–9.5)
	Foula, Shetland	2.9
		(1.6–5.2)
	NW Norway	4.2
		(2.9–9.4)
	NE Norway	3.1
		(1.4–6.8)
Alcidae		
Razorbill	Iceland	2.7
Alca torda		*(0.8–5.8)*
	St Kilda	2.1
		(1.0–3.3)
	Firth of Forth	2.2
		(1.2–4.6)
	Foula, Shetland	1.9
		(0.7–5.3)
	NE Norway	1.7
		(0.8–2.9)
Guillemot	Iceland	1.6
Uria aalge		*(0.7–3.7)*
	Firth of Forth	3.8
		(1.1–15)

Table 7.23 (**cont.**)

Order Family Species	Location	Mean (range)
	Foula, Shetland	1.2 (0.5–2.2)
	NE Norway	1.2 (0.5–2.2)
Puffin *Fratercula arctica*	Iceland	4.8 (2.3–7.9)
	St Kilda	5.1 (2.7–8.3)
	Firth of Forth	3.2 (0.2–7.6)
	Foula, Shetland	3.7 (1.5–11)
	NW Norway	3.0 (2.2–4.2)
	NE Norway	1.0 (0.2–2.3)
Procellariiformes		
Procellariidae		
Fulmar *Fulmarus gacialis*	Iceland	3.8 (0.8–8.0)
	St Kilda	3.3 (1.2–12)
	Firth of Forth	2.3 (1.0–3.4)
	Foula, Shetland	1.6 (0.9–3.9)

Source: Thompson et al. (1992a).

Many historical taxidermy specimens of seabirds exist in museum and other collections. With good details of provenance (date and site of collection), feathers from these specimens offer an excellent record of the bioavailability of mercury to the seabird at the time of collection. A point to be borne in mind, however, is that taxidermists would often apply inorganic mercuric chloride to specimens as a preservative, thereby affecting the total mercury concentrations of any feathers. Fortunately, there is an answer to this potential problem. As stressed earlier, mercury in feathers is in the form of organic methyl mercury. So long as the analysis of feathers extracts and measures only the methyl mercury concentration of a feather, any inorganic mercury added as a preservative is ignored (Furness, 2015). Such an organic mercury extraction procedure is now a standard step in the analysis of historical feathers for mercury content (Thompson et al., 1992b).

One such historical mercury biomonitoring study involved body feathers sampled from puffins, great skuas, fulmars, Manx shearwaters and gannets, collected from mostly British coastal sites from the middle of the nineteenth century to the 1980s (Thompson et al., 1992b). Drawing partly on this study, Figure 7.10 plots mercury concentrations in the body feathers of puffins from Scottish coastal sites between 1890 and the first decade of the twenty-first century (Furness, 2015; Thompson et al., 1992b). It is the concentration of methyl mercury that has been measured in order to eliminate the effects of any added mercuric chloride preservative, but the methyl mercury concentrations are synonymous with total mercury absorbed by the feathers of the living bird. There has been approximately a fourfold increase in the mercury concentrations of the feathers of the puffin *Fratercula arctica* over the last 150 years or so (Figure 7.10). This increase is almost exactly that predicted by models of mercury emissions and atmospheric transport in the northern hemisphere during the industrial period from 1900 (Furness, 2015; Thompson et al., 1992b). Comparisons of pre-1930 and post-1980 median concentrations of mercury in

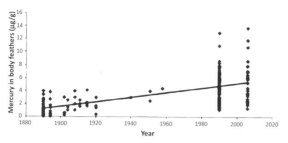

Figure 7.10 Mercury concentrations in pooled body feathers of puffins *Fratercula arctica* from Scottish coastal sites. (After Furness, 2015, with permission.)

the body feathers of puffins, great skuas, Manx shearwaters and gannets all showed significant increases over the period (Table 7.24). Manx shearwaters and puffins showed the greatest percentage increases. The increases were most pronounced in the colonies of these two seabirds to the west of Britain and Ireland (Table 7.24). Fulmars, on the other hand, showed a decrease in the mercury concentrations of body feathers over the time period (Table 7.24). It appears that this decrease resulted from a change in diet of the fulmars over the period, while the diets of the other seabirds remained unchanged, albeit with increased mercury contents. Up to about 1919, fulmars at St Kilda and in the Shetlands and Orkneys fed extensively on whale offal discarded by the whaling industry (Thompson et al., 1992b). With the cessation in 1919 of whaling in British waters, the fulmars switched to invertebrate prey at St Kilda and to lesser sand eels *Ammodytes marinus* in Shetland. Mercury concentrations in whale offal are considerably higher than in either of these other prey items, leading to a fall in the mercury intake of the fulmars against a background of a general increase in mercury concentrations of north Atlantic pelagic organisms (Thompson et al., 1992b).

Historical changes in mercury bioavailabilities can similarly be followed in the feathers of seabirds occupying coastal waters (Thompson et al., 1993). Mercury concentrations in body feathers of adult herring gulls *Larus argentatus* from the German coast of the North Sea were significantly higher

after 1940 than in earlier years. The median mercury concentration of the feathers rose from a median of 4.2 μg/g fresh weight (fw) prior to 1940, to a median of 7.3 μg/g fw post-1940 (Thompson et al., 1993). During the 1940s, these concentrations peaked at 12 μg/g fw, presumed to be an effect of high discharges of mercury into the nearby rivers Elbe and Rhine during the Second World War. Mercury concentrations in the herring gull feathers fell back in the 1950s, rose to a second lower peak (10 μg/g) in the 1970s, before finally subsiding again by the late 1980s. Mercury concentrations of body feathers of common terns *Sterna hirundo* also collected from the German coast of the North Sea similarly increased after 1941. The median mercury concentration of body feathers of adult common terns was 0.6 μg/g fw prior to 1941 and 3.0 μg/g post-1941, an increase of nearly 380% (Thompson et al., 1993).

Feathers are not the only biomonitoring tools that can be used to assess the bioavailability of mercury to seabirds. Sampling eggs can be easier than sampling feathers, and the mercury concentrations of eggs reflect the mercury contamination of food items in the local area where the seabirds have foraged in the immediate prelaying period (Furness, 2015). When sampling eggs, it is necessary to be aware of the laying order of particular eggs in a clutch, given the decrease in mercury contents transferred from the female between the first and later eggs in the same clutch. It may also be feasible to sample the blood of seabirds. Mercury concentrations in the blood reflect mercury bioavailabilities in the diet over a very short time period immediately prior to sampling.

7.11.1.3 Ecotoxicology

Prologue. Do raised bioavailabilities of mercury have ecotoxicological consequences for seabirds? Are the high accumulated concentrations of cadmium in the kidneys of long-lived seabirds associated with any toxicological consequences?

Table 7.24 **Changes in mercury concentrations (medians and ranges, μg/g fresh weight) in body feathers of North and Northeast Atlantic seabirds collected pre-1930 (museum specimens) and post-1980.**

Order Family Species	Location	Pre-1930 Mean (range)	Post-1980 Mean (range)	Change μg/g	Change %age
Charadriiformes					
Puffin	SW Britain/Ireland	1.8	4.0	+ 2.2	+ 120%
Fratercula arctica		(0.2–3.7)	(1.9–11)		
Stercorariidae					
Great skua	Iceland/Shetland/	3.7	5.6	+ 1.9	+ 53%
Catharacta skua	Orkney	(2.4–11)	(1.2–32)		
Procellariiformes					
Procellariidae					
Fulmar	Shetland/Orkney	4.0	1.4	− 2.6	− 65%
Fulmarus gacialis		(1.3–11)	(0.9–3.9)		
	St Kilda	4.5	2.9	− 1.6	− 35%
		(1.9–6.8)	(1.2–12)		
Manx shearwater	NW Britain/Ireland	1.6	4.4	+ 2.8	+ 175%
Puffinus puffinus		(0.2–3.2)	(1.4–10)		
	SW Britain/Ireland	1.3	3.3	+ 2.0	+ 152%
		(0.1–3.2)	(1.0–5.1)		
Suliformes					
Sulidae					
Gannet	Bass Rock,	6.0	7.2	+ 1.2	+ 20%
Morus bassanus	Firth of Forth	(1.8–11)	(2.6–18)		

Source: Thompson et al. (1992b).

We have seen in this chapter changes over time as well as geographical differences in the bioavailability of mercury to seabirds. We have also noted the astonishingly high concentrations of cadmium accumulated naturally in the kidneys of long-lived pelagic seabirds. The 'So what?' question naturally follows. Are there ecotoxicological effects of high bioavailabilities of trace metals to seabirds, causing high uptake rates that cannot be matched by a combination of excretion and detoxification? When considerable storage detoxification occurs, as in the case of cadmium accumulated in the kidneys of fulmars, Manx shearwaters and puffins (Table 7.21), are there consequences for the efficient functioning of these kidneys?

What about the potential effects of increased mercury bioavailabilities on the reproductive capacity of British pelagic seabirds, particularly given the raised concentrations of methyl mercury in the first eggs of a clutch? Mercury contamination correspondingly may cause increased mortality of the first chick (Wolfe et al., 1998). A study in the mid-1970s showed that great skuas in Shetland produced twice as many addled eggs and showed eight times as many deaths of chicks at hatching than other seabird species (Furness and Hutton, 1980). Such symptoms are compatible with mercury poisoning. Mercury pollution, therefore, could not be ruled out as a cause of the ecotoxicological effects seen, although PCB pollution was another candidate factor (Furness and Hutton, 1980). A later 1988 to 1989 study of great skuas from Foula in Shetland, however, failed to show any effect of mercury on their reproduction (Thompson et al., 1991). There appears, therefore, to be no unequivocal evidence for any ecotoxicological effects of mercury on populations of British pelagic seabirds.

Let us consider the kidneys of puffins, fulmars and Manx shearwaters from St Kilda (Nicholson et al., 1983; Nicholson and Osborn, 1983). Average cadmium concentrations in the kidneys of these three species of seabirds at St Kilda have been measured as 109, 177 and 115 µg/g dw respectively, with individual concentrations reaching 162, 480 and 231 µg/g (Table 7.21). Mercury does not accumulate long term in the kidneys of seabirds, most accumulation between moults taking place in the liver and muscles. Nevertheless, average concentrations of mercury in the kidneys of St Kilda puffins, fulmars and Manx shearwaters have been measured as 5.0, 13 and 4.7 µg/g dw respectively, individual concentrations peaking at 7.4, 24 and 62 µg/g (Table 7.22). Electron microscope examination of the kidneys of these St Kilda seabirds showed up features that were considered to be nephrotoxic lesions, equivalent to lesions observed in starlings experimentally exposed to high cadmium or mercury doses (Nicholson et al., 1983; Nicholson and Osborn, 1983). Pathological features included necrosis and degeneration of the epithelium of the proximal tubule and obstruction of the distal parts of the nephron with necrotic cellular debris. Abnormalities were also observed in the glomeruli of the kidneys (Nicholson et al., 1983; Nicholson and Osborn, 1983). The most widespread kidney damage occurred in the Manx shearwater, closely followed by the fulmar, while puffin kidneys showed fewer pathological features (Nicholson and Osborn, 1983). What can be concluded? The birds examined were outwardly healthy. It seems to be the case, then, that while large parts of the kidneys of these seabirds appear to be pathologically damaged, the regenerative capacity and/or the functional reserve of the kidneys are sufficiently substantial to support a healthy life (Nicholson et al., 1983). The kidneys may have such a large spare capacity that not all nephrons in the kidney need to be continuously functional (Nicholson and Osborn, 1983). While cadmium detoxification by metallothionein in the kidneys may have a metabolic energy cost, the populations of these seabirds can still grow and prosper.

7.12 Marine Mammals

Prologue. The marine mammals encountered around the British Isles consist of cetaceans and seals. Cetaceans are further divided into toothed whales (including porpoises and dolphins) and baleen whales.

Cetaceans consist of porpoises, dolphins and whales, including oceanic species which may strand on British coasts (Law et al., 1999, 2001). Cetaceans are divided into two major taxonomic groups: the odontocetes or toothed whales; and the mysticetes or baleen whales, which are rarer in British waters. The presence of teeth in the toothed whales allows them to feed on fish and squid. Mysticetes use their baleen plates to filter the likes of krill (particularly *Meganyctiphanes norvegicus* in the North Atlantic), but also, in some cases, fish such as sardines and herring.

The harbour porpoise *Phocoena phocoena*, also called the common porpoise, is the smallest, and the most abundant, cetacean in British coastal waters. It feeds on fish such as whiting and herring (Harrison Matthews, 1960).

Several dolphin species can be found in British waters (Table 7.25). The bottlenose dolphin *Tursiops truncatus* is a large dolphin that can reach 4 metres long. It is mainly piscivorous, but coastal populations will take benthic crustaceans, and offshore populations will take squid. A large pod of bottlenose dolphins lives in Cardigan Bay off west Wales, and bottlenose dolphins are also often seen in the Moray Firth in Northeast Scotland. The white-sided dolphin *Lagenorhynchus acutus* is a northern species, more commonly found in the northern North Sea, around Orkney and Shetland, and off the Outer Hebrides (Harrison Matthews, 1960). White-sided dolphins feed predominantly on small shoaling fish such as herring, mackerel, cod and sand eels but will also prey on shrimp and squid. In British waters, white-beaked dolphins *Lagenorhynchus albirostris* are chiefly inhabitants of the North Sea (Harrison Matthews, 1960). These dolphins feed on fish of the cod family, particularly cod itself, whiting and haddock. The common dolphin has been the recent subject of taxonomic reassessment, being split into two species, the short-beaked common dolphin *Delphinus delphis* and the long-beaked common dolphin *Delphinus capensis*. So data prior to the 1990s may not refer to a single species. Both common dolphins prefer warmer temperate waters, the short-beaked common dolphin being more common along the edges of continental shelves. Both species are considered to feed on many species of fish and squid, both epipelagic and meso-pelagic. There was a mass stranding of more than 50 short-beaked common dolphins in Falmouth Bay in Cornwall in June 2008 (Jepson et al., 2013). The striped dolphin *Stenella coeruleoalba* is a widespread species, often occurring in very large groups. Striped

dolphins have a catholic diet covering cephalopods, crustaceans and fish, the latter (particularly cod) being preferred in the Northeast Atlantic. Risso's dolphin *Grampus griseus* is also very widespread, choosing to live just off the continental slope where the water depth increases rapidly. In British waters, it is generally to be found off Southwest England and southern and western Ireland, but can stray further north (Harrison Matthews, 1960). The main prey of Risso's dolphin appears to be cephalopods, both pelagic squid and benthic octopus.

Another British member of the dolphin family is the pilot whale, strictly the long-finned pilot whale, *Globicephala melas* (Table 7.25). The pilot whale is sometimes called the blackfish, although this common name may cover several species. Pilot whales have long pectoral fins and a rounded head with only a very short beak, otherwise expected in a dolphin. The pilot whale may be found around the British Isles, but it is generally a northern species, abundant around Orkney and Shetland (Harrison Matthews, 1960). A pod of 21 pilot whales of all ages stranded in Fife in Scotland in September 2012 (Gajdosechova et al., 2016). Pilot whales feed on fish and cephalopods.

Moving away from the dolphin family, other toothed whales occurring in British waters, or found stranded, include sperm whales, pygmy sperm whales and beaked whales (Table 7.25).

Sperm whales *Physeter macrocephalus* are not common in British coastal waters, but occasional stragglers from further south do occur. During the winter of 1994 to 1995, 21 sperm whales stranded around the North Sea, on the coasts of England, Germany, Belgium and the Netherlands (Law et al., 1996; Holsbeek et al., 1999). Sperm whales occur more regularly in the Northeast Atlantic, for example to the west of St Kilda. Sperm whales feed on fish, but especially on oceanic squid, some of which are very large indeed (Harrison Matthews, 1960). Pygmy sperm whales *Kogia breviceps* at 3.5 metres long are not much bigger than most dolphins. Their most common prey is squid.

Beaked whales of the family *Ziphiidae* have a long, narrow snout or beak. While beaked whales are odontocetes (Table 7.25), they typically have very few teeth, and in some cases the few teeth that are present

Table 7.25 **Concentrations of trace metals (means or ranges, μg/g dw) in livers and kidneys of marine mammals in British coastal waters and the adjacent Northeast Atlantic Ocean.**

	Tissue	Location	Cadmium	Copper	Lead	Mercury	Selenium	Zinc
Cetaceans								
Odontocetes								
Delphinoidea								
Phocoenidae								
Harbour porpoise	Liver	E Scotland[1]*	0.5	22	–	12	–	130
Phocoena phocoena		S North Sea[2]	0.4	38	–	30	24	193
		England, Wales[3]*	0.6	40	0.4	47	26	173
		NE Atlantic[4]	–	31	–	60	–	129
	Kidney	E Scotland[1]	4.6	11	–	3.5	–	71
		S North Sea[2]	2.3	16	–	–	11	94
		NE Atlantic[4]	–	14	–	7.8	–	88
Delphinidae								
Bottlenose dolphin	Liver	NE Atlantic[4]	–	17	–	129	–	171
Tursiops truncatus		England, Wales[5]*	–	–	14	max 1,770	–	–
	Kidney	NE Atlantic[4]	–	15	–	36	–	84
White-sided dolphin	Liver	S Wales[6]	18	33	0.5	129	72	183
Lagenorhynchus acutus	Kidney	Faroes[7]*	23-31	–	–	1.4-2.5	–	–
White-beaked dolphin	Liver[169]	NE England[6]	<0.1	25	1.2	40	63	113
Lagenorhynchus albirostris								

Table 7.25 (**cont.**)

	Tissue	Location	Cadmium	Copper	Lead	Mercury	Selenium	Zinc
Common dolphin	Liver	W Wales[6]	1.6	39	0.3	415	171	236
Delphinus delphis		NE Atlantic[4]	–	18	–	44	–	144
	Kidney	NE Atlantic[4]	–	12	–	8.3	–	84
Striped dolphin	Liver	W Wales[6]	3.7	20	<0.2	550	211	139
Stenella coeruleoalba		NE Atlantic[4]	–	25	–	100	–	180
	Kidney	NE Atlantic[4]	–	15	–	14	–	108
Risso's dolphin	Liver	SW England[6]	0.7	17	0.6	9	15	124
Grampus griseus		W Scotland[8]*	–	–	–	3.6	–	–
	Kidney	W Scotland[8]*	–	–	–	1.8	–	–
Long-finned pilot whale	Liver	NE England[6]	0.4	18	<0.2	2.8	7.9	170
Globicephala melas		E Scotland[9]*	72	–	–	195	95	263
	Kidney	E Scotland[9]*	221	–	–	11	25	170
Physeteroidea								
Physeteridae								
Sperm whale	Liver	S North Sea[10]	52–175	5–12	0.9–3.2	9–120	6–43	90–125
Physeter macrocephalus	Kidney	S North Sea[10]	133–426	13–44	<1.0–3.6	4–20	6–14	44–140

	Tissue	Location						
Kogiidae								
Pygmy sperm whale	Liver	W Wales[6]	10	30	0.3	45	54	67
Kogia breviceps								
Ziphiidae								
Sowerby's beaked whale	Liver	E England[6]	65	62	0.4	1,040	431	269
Mesoplodon bidens								
Blainville's beaked whale	Liver	W Wales[6]	18	17	0.2	734	290	121
Mesoplodon densirostris								
Northern bottlenose whale	Liver	S North Sea[11]*	17	8	0.5	1.2	–	69
Hyperoodon ampullatus								
Mysticetes								
Balaenopteridae								
Fin whale	Liver	E England[6]	12	12	<0.5	6.0	22	175
Balaenoptera physalus								
Minke whale	Liver	SE England[6]	171	17	1.4	922	438	178
Balaenoptera acutorostrata								
Pinnipeds								
Phocidae								
Common seal	Liver	S North Sea[12]*	0.2–0.9	–	–	771–978	138–402	75–102

Table 7.25 (cont.)

Tissue	Location	Cadmium	Copper	Lead	Mercury	Selenium	Zinc
Phoca vitulina							
	S North Sea[13]	–	–	–	–	327	–
	North Sea[14]*	<0.3–0.6	8–51	0.3–1.7	5–480	–	81–168
	E England[15]	–	–	–	16	–	–
Kidney	North Sea[14]*	<0.3–1.2	7–12	0.4–1.7	5–38	–	49–98
	S North Sea[13]	–	–	–	–	21	–
	E England[15]	–	–	–	4.8	–	–
Liver	E Scotland[16]*	2.4	–	–	83	–	–
Grey seal	NE England[17]*	–	87	–	–	–	166
Halichoerus grypus	Great Britain[18]	–	–	–	–	114	–
	Dee, England[15]	–	–	–	2,080	–	–
Kidney	E Scotland[16]*	6.3	–	–	8	–	–
	NE England[17]*	–	9	–	–	–	85
	Dee, England[15]	–	–	–	66	–	–

Note: * = wet weight concentrations X3.

Sources: Falconer et al. (1983;[1] Mahfouz et al. (2014);[2] Bennett et al. (2001);[3] Aubail et al. (2013);[4] Law et al. (2012);[5] Law et al. (2001);[6] Gallien et al. (2001);[7] Zonfrillo et al. (1988;[8] Gajdosecheva et al. (2016);[9] Holsbeek et al. (1999);[10] Harms et al. (1978);[11] Koeman et al. (1973);[12] Reijnders (1980);[13] Drescher et al. (1977);[14] Simmonds et al. (1993);[15] McKie et al. (1980);[16] Caines (1978);[17] Van de Ven et al. (1979).[18]

never erupt through the skin of the jaw. Sowerby's beaked whale *Mesoplodon bidens* is a deep water whale living beyond the continental shelf across the North Atlantic. Nevertheless, it occasionally strands in the British Isles. Sowerby's beaked whale feeds on squid and also fish, such as cod. Blainville's beaked whale *Mesoplodon densirostris* is very far ranging, living in deep water across the world from the tropics to high latitudes. It feeds on deep water squid and fish. A third beaked whale found stranded on British coasts is the Northern bottlenose whale *Hyperoodon ampullatus*. The Northern bottlenose whale is an Atlantic species, the southern migration of which during the autumn and winter brings it through British coastal waters (Harrison Matthews, 1960). This whale feeds mainly on oceanic squid and fish, diving deep to feed near the ocean bottom.

Baleen whales do occasionally occur in British waters, but are usually to be found well out to the west in the Atlantic Ocean. Two families of baleen whales may be represented in records of whale strandings on British coasts, the *Balaenidae* and the *Balaenopteridae*.

Some balaenid whales are historically called right whales, being the right whales for hunting. They are relatively docile, swim slowly at the surface when feeding and their high blubber content means that they float when killed. Such a whale encountered around the British Isles is the North Atlantic right whale *Eubalaena glacialis*. The baleen plates of *E. glacialis* enable it to filter out krill and other zooplankton such as copepods occurring in patches just below the ocean surface.

Balaenopterid whales, also known as rorquals, include the huge blue whale *Balaenoptera musculus* and also the fin whale *Balaenoptera physalus*. Fin whales have stranded on British coasts, as has another rorqual, the much smaller minke whale *Balaenoptera acutorostrata* (Law et al., 2001). The name 'rorqual' derives from a Norwegian word meaning furrow whale, a reference to the series of longitudinal folds of skin on the front ventral surface of the whale. These folds allow the volume of mouth to increase greatly when feeding. Huge gulps of water with associated food can be taken into the mouth, to be subsequently pushed by the tongue through the baleen plates, so

filtering out food such as krill and large copepods but also small fish such as sardines.

Moving away from cetaceans, two species of seals are common in British coastal waters: the common seal *Phoca vitulina*, also called the harbour seal; and the grey seal *Halichoerus grypus*. The common seal prefers estuaries and coasts where offshore sandbanks are emersed at low tide (Harrison Matthews, 1960). The grey seal prefers rocky coasts with cliffs and caves, and paradoxically may be more common than the common seal (Harrison Matthews, 1960). Common seals have a varied diet. They usually prey on many different types of fish, but occasionally feed on crabs, squid and other molluscs. Grey seals also feed on a wide variety of fish, particularly sand eels, but they are opportunistic and will also prey on crabs and lobsters.

7.12.1 Trace Metals and Marine Mammals

Prologue. Marine mammals take up trace metals from the diet, with subsequent accumulation depending on the relative contributions of the physiological processes of excretion and storage detoxification, differing between trace metals and between mammals.

Trophic inputs of trace metals will vary interspecifically, according to the nature and relative importance of different prey types in the diet. The mercury content of fish is generally in the form of methyl mercury. Therefore, piscivorous marine mammals receive a relatively high input of methyl mercury. Furthermore, toothed whales occupy a higher average trophic level in a marine food chain than do baleen whales, and methyl mercury biomagnifies up food chains. On the other hand, squid tend to contain high cadmium concentrations (Bustamente et al., 1998). Squid-eating toothed whales may, therefore, receive relatively high dietary inputs of cadmium, with the subsequent potential to accumulate high cadmium concentrations.

Patterns of trace metal accumulation in marine mammals vary between trace metals. Table 7.25 presents accumulated concentrations of particular trace metals in the livers and kidneys of marine mammals in British waters and the adjacent Northeast Atlantic Ocean.

7.12.1.1 Copper and Zinc

Prologue. The regulation of tissue concentrations of essential metals such as copper and zinc is widespread among vertebrates, and marine mammals appear to be no exception.

The copper concentrations in the livers and kidneys of marine mammals as listed in Table 7.25 bear this out. It has been suggested that the normal range of regulated copper concentrations in the livers of marine mammals is about 3 to 30 µg/g ww (Law, 1996). This range does approximately cover the liver copper concentrations listed in Table 7.25, assuming a wet weight to dry weight ratio of 3. Copper concentrations in the kidneys of marine mammals appear lower than those in the liver (Table 7.25). Thus, there is little evidence of detoxified storage of copper in either organ in marine mammals, suggesting that these copper concentrations represent copper in metabolically available form to meet essential needs.

It has similarly been suggested that zinc concentrations in the livers of marine mammals are also regulated to a relatively narrow range, in this case approximately 20 to 100 µg/g ww (Law, 1996). It appears from Table 7.25 that zinc concentrations in the livers of marine mammals are indeed regulated to this range, again assuming a factor of 3 between wet and dry weights. It may well be the case that at least a proportion of the zinc accumulated in the livers of marine mammals is stored, however temporarily, bound to metallothionein. The remaining zinc is presumably metabolically available to serve essential needs, for example in zinc-based enzymes. As for copper, kidney concentrations of zinc are somewhat lower than those of livers (Table 7.25). Some of the zinc in the kidneys of marine mammals is also bound to metallothionein, between 25 and 52% in the case of sperm whales stranded on southern North Sea coasts (Holsbeek et al., 1999).

7.12.1.2 Cadmium

Prologue. Non-essential metals such as cadmium, on the other hand, often show net accumulation, being

stored in detoxified form somewhere in the body. Cadmium is predominantly accumulated in the kidneys of marine mammals, with some further accumulation in the livers.

As a generalisation, marine mammals with a high proportion of squid in their diet tend to have relatively high accumulated concentrations of cadmium, especially in the kidneys (Table 7.25). Sperm whales in particular have high kidney concentrations of cadmium (Holsbeek et al., 1999), as do pilot whales (Caurant et al., 1994). In the absence of kidney concentration data, the livers of Sowerby's beaked whales similarly reflect a high accumulation of cadmium from their squid diet (Table 7.25). The livers of minke whales too have high cadmium concentrations (Table 7.25). Minke whales filter pelagic crustaceans with their baleen plates. While oceanic krill do not individually contain particularly high cadmium concentrations, pelagic hyperiid amphipods do. It is possible that it is this component of their filtered diet of zooplankton that is supplying minke whales with a high dietary input of cadmium.

Another factor controlling the concentration of accumulated cadmium in the livers and kidneys of marine mammals is increased accumulation with age. For example, cadmium concentrations in both the livers and kidneys of harbour porpoises increase with body length (Falconer et al., 1983). Similarly, the cadmium concentrations of the livers and kidneys of pilot whales also increase significantly with age (Caurant et al., 1994; Gajdosechova et al., 2016). Cadmium accumulated in an organ needs to be detoxified, and cadmium is detoxified in marine mammals by binding with metallothionein. Thus, in the case of sperm whales stranded in the southern North Sea, between 5 and 92% of the cadmium accumulated in the livers, and between 21 and 66% of the cadmium accumulated in the kidneys was bound to metallothionein (Holsbeek et al., 1999).

7.12.1.3 Lead

Prologue. Another non-essential trace metal, lead, also typically shows net accumulation, requiring

Table 7.26 **Concentrations of lead (mean ± 1 standard deviation, µg/g dw) in bones (fifth ribs) and teeth of harbour porpoises and common dolphins stranded along Northwest European coasts between 2001 and 2002.**

	Tissue	Lead concentration
Harbour porpoise *Phocoena phocoena*	Bone	0.37 ± 0.31
	Teeth	0.75 ± 0.68
Common dolphin *Delphinus delphis*	Bone	0.66 ± 0.51
	Teeth	1.03 ± 0.91

Source: Caurant et al. (2006).

storage detoxification. The concentrations of lead accumulated in the livers and kidneys of cetaceans and seals around the British Isles are, however, unremarkable (Table 7.25).

Lead, like calcium, will bind strongly with phosphate. Thus, lead has an affinity for calcified tissues based on calcium phosphate. Hard tissues such as the bones or teeth of marine mammals, therefore, can be more appropriate than any soft tissues for studying their bioaccumulation of lead (Caurant et al., 2006). Table 7.26 presents concentrations of lead in bones and teeth of harbour porpoises and common dolphins stranded on coasts in Ireland, France and northern Spain between 1991 and 1992. Lead concentrations in bone and teeth were positively correlated, and lead concentrations in teeth were higher (about twice so; Table 7.26). Lead concentrations in both bone and teeth were also positively correlated with age, indicating ongoing accumulation of lead (Caurant et al., 2006). The analysis of different stable isotopes of lead in these cetacean tissues allowed identification of the different ultimate sources of the lead to the cetaceans. There are different lead isotope ratios, for example, between

pre-industrial background lead in the marine environment and anthropogenic lead, which used to be added to petrol as tetra-ethyl lead as an antiknocking agent. In fact, the stable isotope ratios of these cetaceans of different ages indicated the decreasing contribution of alkyl lead into European waters over the previous time period, associated with the decreasing use of leaded petrol (Caurant et al., 2006).

7.12.1.4 Mercury

Prologue. High accumulated concentrations of mercury in marine mammals can be demethylated and detoxified by binding with selenium to form metabolically inert granules.

Most mercury taken up by marine mammals, particularly piscivorous marine mammals, is in the form of methyl mercury. In a marine mammal, the accumulated mercury may remain in methyl mercury form, but it also has the potential to be detoxified into a metabolically unavailable inorganic form of mercury for storage. Methyl mercury accumulated by most mammals can be passed to growing hair and be subsequently excreted at the next moult, as in the parallel case of feathers in seabirds. Marine mammals, however, lack a pelt of hair, eliminating this route of methyl mercury excretion. There is still, however, a potential for females to lose accumulated methyl mercury, both by placental transfer to the developing embryo and in milk supplied to calves or pups. Nevertheless, mercury will be accumulated by marine mammals and will need detoxification, particularly in the cases of piscivores.

The detoxification of methyl mercury in marine mammals involves the demethylation of methyl mercury and the binding of the mercury to another trace metal, selenium, to form granules of mercuric selenide (HgSe, tiemannite) (Martoja and Viale, 1977; Martoja and Berry, 1980). Mercuric selenide granules are metabolically inert, as befits a detoxified form of a trace metal (Law et al., 2012). This detoxification of methyl mercury takes place predominantly in the liver but also occurs in other tissues including the kidney and the brain. It was the one-to-one molar correlation between mercury and

(a)

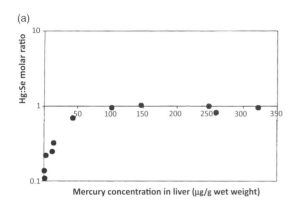

(b)

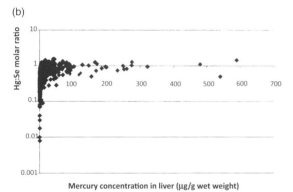

Figure 7.11 The variation in the Hg:Se molar ratio with mercury concentration (μg/g ww) in the livers of (a) harbour porpoises, and (b) a combination of cetaceans and seal species. (After Law et al., 2001, 2012, with permission.)

selenium concentrations in the livers of marine mammals that drew attention to the possible direct chemical interaction of mercury and selenium (Koeman et al., 1973). In fact, this perfect one-to-one molar ratio is not present at low accumulated mercury concentrations in marine mammal livers (Figure 7.11). Below about 100 μg/g ww mercury, the molar ratio Hg:Se is less than 1 (Figure 7.11). Mercury at this tissue concentration is not detoxified as mercuric selenide, presumably existing as methyl mercury. With increased mercury accumulation, however, the mercury is detoxified by binding with selenium, and the one-to-one molar ratio is established (Law et al., 2001, 2012).

Not surprisingly, at high rates of mercury uptake and subsequent accumulation requiring mercury detoxification, there is increased accumulation of mercury in the liver with the age of the marine mammal. In fact, increased accumulation of mercury with age will begin even at lower mercury accumulation rates, as in the cases of both the liver and the kidney of pilot whales (Caurant et al., 1994).

Table 7.25 highlights the high accumulated mercury concentrations in the livers of piscivorous odontocetes such as bottlenose dolphins, common dolphins, striped dolphins and some beaked whales. Both common seals and grey seals also have high liver mercury concentrations, expected from their fish-based diet. Where accumulated selenium concentrations have been measured in the same livers, these too are expectedly high when accumulated mercury concentrations are high (Table 7.25).

The contamination of Liverpool Bay with mercury through the 1970s and 1980s, as reflected in the mercury content of local coastal fish (Table 7.14), can also be seen in the mercury concentrations of livers and kidneys of grey seals collected at the time (Table 7.25). Grey seals from the region of the Dee Estuary in 1988/1989 had the very high mean mercury concentration of 2,080 μg/g dw in the liver (Table 7.25). Other grey seals from Liverpool Bay at about that time also showed a very high mercury concentration in the liver, in this case 430 μg/g dw (Law et al., 1992).

7.12.1.5 Organotin

Prologue. Methyl mercury is not the only organometal compound of interest in marine mammals. What of the butyl tin compounds derived from the use of tributyl tin as an antifouling agent?

Tributyl tin (TBT) will eventually degrade in the environment to form dibutyl tin (DBT) and monobutyl tin (MBT). TBT is lipophilic (as are DBT and MBT) and is taken up relatively quickly by organisms, in the case of marine mammals from the diet. As organometals, butyl tins also have the potential to be biomagnified up food chains. After uptake, animals have a varying capacity to metabolise TBT to DBT, and then MBT, by means of the cytochrome P450 enzyme system. So the TBT content of a marine mammal depends on both the concentration of TBT in its prey, and then on its ability to metabolise TBT to DBT and

MBT. As top predators, cetaceans, particularly piscivorous odontocetes feeding on fish with high lipid contents, and similarly piscivorous seals may be exposed to high butyl tin contents in their prey. Furthermore, cetaceans may have a relatively limited ability to metabolise TBT in comparison to some other animals (Law et al., 1998).

So, with the increasing use of TBT as an antifouling agent in the 1970s and thereafter, the question arose in the 1990s as to whether marine mammals living in British coastal waters were being contaminated by butyl tins. Table 7.27 includes data on the TBT, DBT, MBT and total organotin concentrations in the livers of harbour porpoises and grey seals around England and Wales between 1992 and 1996. Summed concentrations of butyl tins ranged from 22 to 640 µg/g ww (Table 7.27). DBT was the major form of butyl tin in the livers of the porpoises, indicating that some metabolic breakdown of TBT does occur there (Law et al., 1998). There were lower concentrations of butyl tins in the livers of the grey seals than the harbour porpoises, indicating either a lower intake of butyl tins from their diet, and/or a higher metabolic capacity in the seals to break down accumulated TBT (Law et al., 1998). It could be concluded that in the early 1990s, butyl tin contamination in marine mammals was widespread in coastal waters around England and Wales.

Corresponding international data from that time suggested that there was similar butyl tin contamination of marine mammals in coastal waters across the world (Law et al., 1998, 1999). A follow-up study through the 1990s was therefore carried out to ascertain whether butyl tin contamination extended to pelagic marine mammal species feeding over the outer areas of the continental shelf and the continental slope and in deep ocean waters (Law et al., 1999). These data, too, are presented in Table 7.27. Butyl tins were detected in all 16 of the liver samples analysed, with total butyl tin concentrations ranging from 19 to 312 µg/g ww (Table 7.27). Again, it was DBT that was the major form of butyl tin present in the livers of all the pelagic marine mammal species, averaging 77% of the total butyl tin concentration (Table 7.27). So it could be concluded that, in the 1990s, butyl tin contamination of marine mammals was not restricted to coastal waters but extended to deep offshore waters. Furthermore, butyl tin compounds were present in the oceanic food chains of both odontocete and mysticete cetaceans (Law et al., 1999).

We saw in Chapter 6 the endocrine disrupting characteristics of butyl tins, particularly in neogastropod molluscs. Butyl tins also have an immunosuppressive capacity in marine mammals (Kannan and Tanabe, 1997), but typically at higher accumulated concentrations than seen here (Law et al., 1999). Nevertheless, butyl tins do have the capacity to add to the immunosuppressive effects of other contaminants, not least organochlorines including PCBs, that may also be present. Fortunately, many nations banned the use of TBT on small vessels in the 1980s, and in 2008 a convention banning its use on all vessels became internationally binding to a large number of nations representing at least 25% of the tonnage of world shipping (Dafforn et al., 2011).

7.12.1.6 Ecotoxicology

Prologue. Trace metals accumulated by marine mammals found in the coastal waters of the British Isles do not appear to offer any ecotoxicological risk to these mammals.

We have discussed in earlier chapters the simplistic attraction of a threshold total trace metal concentration in an organ as indicative of the onset of toxic effects. The principle stressed in this book is that it is the rate of accumulation of a trace metal that needs to be below a threshold value for the joint processes of excretion and detoxification to match the rate of uptake of a trace metal into that organ. A total metal concentration in an organ may only have any ecotoxicological relevance in the absence of any detoxification, the total metal concentration then representing the concentration of metabolically available metal in the organ. It is the concentration of metabolically available metal in an organ that must stay below a critical threshold. We have yet to perfect analytical ways to define, isolate and measure a physical component of the accumulated metal concentration in an organ that is equivalent to the (at present theoretical) metabolically available metal

Table 7.27 **Concentrations (μg/g ww) of tributyl tin (TBT), dibutyl tin (DBT), monobutyl tin (MBT) and summated butyl tins in the livers of marine mammals stranded on the coasts of England and Wales in the 1990s.**

	Location	Date	TBT	DBT	MBT	Total butyl tins
Cetaceans						
Odontocetes						
Delphinoidea						
Phocoenidae						
Harbour porpoise	England and Wales	1992–96	<10–180	14–460	<6–110	22–640
Phocoena phocoena						
Delphinidae						
White-sided dolphin	South Wales	1994	<3	29	<3	29
Lagenorhynchus acutus						
White-beaked dolphin	East England	1995–98	36–38	101–134	<3–<4	139–170
Lagenorhynchus albirostris						
Common dolphin	West Wales	1998	53–68	132–195	<4	185–263
Delphinus delphis						
Striped dolphin	England and Wales	1996	77–82	84–230	<3–<4	161–312
Stenella coeruleoalba						
Risso's dolphin	England and Wales	1992–94	19–33	26–62	<3–7	66–81
Grampus griseus						
Long-finned pilot whale	Northeast England	1997	<3	19	3	22
Globicephala melas						

Table 7.27 (cont.)

	Location	Date	TBT	DBT	MBT	Total butyl tins
Physeteroidea						
Kogiidae						
Pygmy sperm whale	West Wales	1997	<4	50	35	85
Kogia breviceps						
Ziphiidae						
Sowerby's beaked whale	East England	1998	24	34	<4	58
Mesoplodon bidens						
Blainville's beaked whale	West Wales	1993	<6	33	<5	33
Mesoplodon densirostris						
Northern bottlenose whale	Northwest England	1998	<3	28	<3	28
Hyperoodon ampullatus						
Mysticetes						
Balaenopteridae						
Fin whale	East England	1992	<3	19	<3	19
Balaenoptera physalus						
Minke whale	East England	1996	<4	56	<4	56
Balaenoptera acutorostrata						
Pinnipeds						
Phocidae						
Grey seal	England and Wales	1993–95	<4–6	<4–11	<3–14	<12–22
Halichoerus grypus						

Sources: Law et al. (1998, 1999).

concentration (Rainbow and Luoma, 2011c; Rainbow et al., 2015).

In the case of marine mammals, we have seen that trace metals, including zinc and cadmium, are detoxified by binding with metallothionein; lead by binding with phosphate; and mercury by binding with selenium. We should not expect, therefore, that there is a critical total concentration of any such trace metal in the organ of a marine mammal that is indicative of the onset of toxic effects caused by that trace metal.

It is the rate of uptake of a toxic metal per day per unit body weight of the mammal that is crucial in investigating the onset of toxicity. In what might be considered today to be an unethical experiment, harp seals *Pagophilus groenlandicus* were fed on either 0.25 milligram or 25 milligrams of mercury as methyl mercury per kilogram body weight per day for up to 90 days (Ronald et al., 1977; Law, 1996). While the two low dosage seals showed reduced appetite and weight loss, they did survive. The two high dosage seals died on days 20 and 26, with indications of toxic hepatitis and renal failure. The failures in liver and kidney functions were associated with high accumulated total mercury concentrations in these two organs, the liver concentration of total mercury reaching 134 and 142 μg/g ww in the seals (Ronald et al., 1977; Law, 1996). Interestingly, these liver concentrations of total mercury are well below the field concentrations of 978 and 2,080 μg/g dw measured in some British common seals and grey seals respectively, even after any allowance for a wet weight to dry weight ratio (Table 7.25). This comparison again highlights the absence of any critical total mercury concentration in the livers of marine mammals. The equivalent methyl mercury concentrations of the two high dosage seals were 127 and 125 μg/g ww, indicating a lack of significant detoxification of the accumulated methyl mercury to inert inorganic mercuric selenide over the time of the experiment. In the field, the mercury in the livers of marine mammals with total mercury liver concentrations above 100 μg/g ww is all bound to selenium (Figure 7.11). Therefore, the rate of accumulation of mercury (predominantly as methyl mercury) in the two high dosage seals has been too high to be matched to any significant extent by the rate of detoxification of this mercury by demethylation and binding to selenium. It is relevant that there may indeed be a critical threshold concentration of methyl mercury in the liver of marine mammals, but certainly not of total mercury concentrations in field animals given sufficient time to detoxify the accumulated mercury.

These experimental dietary intake rates of mercury can be used to consider the fates of seals feeding in the wild on fish with high mercury contents, taken to be all in the form of methyl mercury (Law, 1996). Fish in British coastal waters today have total (and therefore methyl) mercury concentrations in muscles of about 0.1 μg/g ww (Tables 7.13 and 7.14). Higher mercury concentrations in muscles were present in fish in Liverpool Bay in the 1970s and 1980s, for example up to nearly 1 μg/g ww in whiting (Tables 7.13 and 7.14). Let us make the assumption that most of the fish body consumed by a seal consists of muscle tissue, and that the muscle concentration of mercury is a good estimate of the total body mercury concentration. A common seal *Phoca vitulina* weighing 100 kilograms and eating 10 kilograms per day of fish with a methyl mercury concentration of 0.1 μg/g ww will intake 0.01 milligram of mercury as methyl mercury per day per kg body weight. A grey seal *Halichoerus grypus* weighing 250 kilograms and eating 20 kilograms per day of fish with 0.1 μg/g ww methyl mercury will intake less than 0.01 milligrams of mercury as methyl mercury per day per kilogram body weight. The same two seals feeding on fish with 1970s Liverpool Bay methyl mercury concentrations of 1 μg/g methyl mercury would intake 0.1 and 0.08 milligram of mercury as methyl mercury per day per kilogram body weight. All these intake rates are well below even the low dosage intake rates of the experimental harp seals discussed earlier (Ronald et al., 1977; Law, 1996). There, therefore, appears to be no risk of mercury toxicity from their fish diet to either common seals or grey seals in British coastal waters today. Nor probably was there any such risk in Liverpool Bay in the 1970s and 1980s. Grey seals in Liverpool Bay in the 1980s were certainly taking in atypically high dietary loads of methyl mercury (Simmonds et al., 1993) but not such a high load as to be ecotoxicologically significant. The presence of very high accumulated loads of mercury in the livers of Dee Estuary grey seals in the 1980s (Table 7.25) indicates that the rate of accumulation of uptake of methyl mercury in these seals was high, but not so high as to exceed the rate of detoxification of the methyl mercury in the liver to mercuric selenide (Figure 7.11).

There are incidents of marine mammals suffering from infectious disease in British coastal waters. Causative links have been suggested to high local bioavailabilities, and subsequent increased uptake and accumulation of anthropogenic contaminants, including trace metals, acting via immunosuppression (Bennett et al., 2001; Law et al., 2012). In practice, however, it has proved very difficult to separate out the ecotoxicological effects of trace metals from those of, often associated, organochlorine contaminants, especially PCBs.

Postmortem investigations were carried out on 86 harbour porpoises *Phocoena phocoena* stranded dead on coasts of England and Wales between 1990 and 1994 (Bennett et al., 2001). These porpoises could be divided into two groups. Forty-nine were healthy when they died as a result of physical trauma, such as entrapment in fishing gear or bottlenose dolphin attack. On the other hand, the remaining 37 had died of infectious diseases caused by parasitic, bacterial, fungal or viral pathogens. Liver concentrations of cadmium, chromium, copper and lead did not differ between the two groups. But, liver concentrations of mercury, selenium and zinc were higher in the infected porpoises, as was the Hg:Se molar ratio in these livers.. In fact, mercury liver concentrations were not really that high, averaging 20 µg/g ww in the infected porpoises, in comparison to an average of 12 µg/g ww in the other porpoises. Furthermore, a comparison of liver Hg:Se molar ratios between infected porpoises (1:1.26) and the remainder (1:1.57) suggests that there was ample selenium in both sets of livers to detoxify the mercury present if needed. It seems unlikely, then, that there was a significant excess of methyl mercury present, even in the livers of the infected porpoises, to cause an ecotoxicological effect such as immunosuppression. The association

between a slightly raised zinc concentration in the livers of infected porpoises (77 against 43 µg/g, but see Table 7.25) and disease status may result from a physiological redistribution of zinc in the body in response to infection (Bennett et al., 2001).

A more probable cause of any immunosuppression leading to a higher likelihood of infectious disease in harbour porpoises is the associated presence of high accumulated concentrations of organochlorines such as PCBs (Jepson et al., 1999; Law et al., 2012). In a parallel study to the one described earlier (Bennett et al., 2001), a comparison was made between the summated concentrations of 25 PCB congeners in the blubber of 34 harbour porpoises which had died of physical trauma and 33 porpoises which had died as a result of infectious disease, around England and Wales in the early 1990s (Jepson et al., 1999). The average summated concentration of the 25 PCBs in the blubber of the disease-infected group (31 µg/g lipid weight) was significantly higher than that of the physical trauma group (14 µg/g lipid weight). This result is consistent with the hypothesis that chronic exposure of marine mammals to PCBs predisposes them to infectious diseases as a result of immunosuppression. A later larger comparison came to the same conclusion (Jepson et al., 2005). This follow-up study considered 175 harbour porpoises which died from physical trauma and 82 porpoises dead from infectious disease which had stranded in the United Kingdom. Again, the infected porpoises had a higher average summated concentration of 25 PCBs in the blubber (28 µg/g lipid weight) than the physical trauma porpoises (14 µg/g lipid weight).

Between these two studies, in 2000, there had been a proposal for a threshold blubber concentration of total PCBs in marine mammals that could be considered immunotoxic – namely 17 µg/g lipid weight (Kannan et al., 2000). There is no theoretical objection to such a threshold toxic concentration for organochlorine contaminants such as PCBs, for there is no equivalent storage of these contaminants in detoxified form, as mostly seen for trace metals. There is considerable justification for this proposed threshold concentration of 17 µg/g lipid weight for total PCBs in the blubber of marine mammals (Kannan et al., 2000;

Law et al., 2012). Indeed, the last study tested this proposal by dividing the disease-infected porpoises into two groups, those with summated blubber concentrations of 25 PCBs below 17 μg/g lipid weight and those with summated blubber concentrations of these PCBs above 17 μg/g lipid weight (Kannan et al., 2000). There was no significant difference between the blubber concentrations of the disease-infected group with summated PCB concentrations below 17 μg/g lipid weight and those of the physical trauma porpoises. There was, however, a significant difference between the blubber concentrations of summated PCBs of the disease infected group with summated PCB concentrations above 17 μg/g lipid weight and those of the physical trauma porpoises. These findings are consistent with the proposed threshold of 17 μg/g lipid weight for total PCBs in the blubber of marine mammals for the onset of PCB-induced ecotoxic effects, including immunosuppression (Kannan et al., 2000; Law et al., 2012).

While the average blubber concentrations in UK marine mammals of some organic contaminants such as organochlorine pesticides and brominated flame retardants fell between 1995 and 2008, those of total PCBs plateaued after an initial decline (Law et al., 2012). Thus, between 1995 and 2008, the proportion of UK stranded harbour porpoises with blubber concentrations of total PCBs above 17 μg/g lipid weight remained fairly constant at between 40 and 60% per year (Law et al., 2012). Total PCB concentrations as high as 319 μg/g lipid weight have been recorded in individual harbour porpoises between 1995 and 2008, clearly indicative of immunosuppression. The equivalent range of total PCB concentrations in the blubber concentrations of another marine mammal stranded on UK coasts in the period 1995 to 2008 was 13 to 899 μg/g lipid weight for bottlenose dolphins (Law et al., 2012). Clearly, marine mammals stranded on UK coasts are still at risk from the immunosuppressive effects of accumulated PCBs. Further reductions in total PCB concentrations in UK marine mammals are likely to take decades (Law et al., 2012).

In contrast to the situation of PCB contamination, it cannot be concluded that accumulated trace metals, including mercury accumulated as methyl mercury, offer any significant ecotoxicological threat to British marine mammals.

8 Epilogue

Box 8.1 Definitions

phytoextraction The use of plants for the removal or reduction of metals in mine waste, requiring the accumulation or hyperaccumulation of metals in plant tissues above ground. The plants are harvested and usually incinerated before disposal of the metal-rich ash as hazardous waste or its processing for recovery of the metals (a process known as *phytomining*).

phytomining The production of a metal by growing high-biomass plants that accumulate high metal concentrations from metal-contaminated soils. The plants are harvested and incinerated to produce metal-rich ash processed for recovery of the metals.

phytoremediation The use of plants (and their associated microorganisms) to stabilise or reduce contamination by, for example, a toxic metal in soil, thereby mitigating an environmental problem without the need to excavate the soil for disposal elsewhere.

phytostabilisation The formation over mine waste of a vegetative cap where uptake and accumulation processes immobilise metals within the plant rhizosphere, reducing toxic metal bioavailability and hence wildlife, livestock and human exposure.

Prologue. Trace metals have had effects on the natural history of the British Isles for more than a thousand years. Typically, these effects have been deleterious, albeit with distributions patchy in space and time. Anthropogenic activities such as mining have brought about raised local bioavailabilities of particular trace metals, thereby causing ecotoxicological effects in affected terrestrial and aquatic habitats. On the other hand, an atypically low bioavailability of an essential trace metal, a feature of certain soil types, may bring about deficiency symptoms in local vegetation and their associated fauna. Natural history in turn has an effect on the varied interactions of trace metals and the organisms that constitute the biodiversity of the British Isles, including humankind and their domesticated plants and animals. So, where are we today?

8.1 Biology of Trace Metals

Prologue. This book has been based on an exploration of the significance of accumulated trace metal concentrations in organisms. How did the large variation in bioaccumulated concentrations come about? What do they signify for the biology of the organism concerned, and, on the other hand, to researchers attempting to understand their ecotoxicological meaning? And mainly this book has been about the need to understand the interaction of the natural history of organisms and the essential and non-essential trace metals that they inevitably encounter.

Several major principles relating to the biology of trace metals have dominated the preceding chapters. In spite of their low environmental concentrations, trace metals are all toxic above a threshold bioavailability. Yet many are essential to metabolism at lower doses. The balance between essentiality and toxicity can be precarious, with only a small relative difference between the bioavailabilities of a trace metal bringing about one or the other. Yet the toxicity of high trace metal bioavailabilities is not a new feature associated with the emergence of humans. Life has evolved in the presence of trace metals that are potentially toxic because of the chemical characteristics that cause them to have an affinity to bind to many of the molecules in cells that are the very bases of metabolism. Yet it is these same chemical characteristics that have made trace metals such valuable biochemical resources in the evolution of life and its constituent metabolic processes. So many enzymes make use of trace metals in their catalytic centres, facilitating metabolism. Haemoglobin and haemocyanin are respiratory proteins reversibly binding oxygen needed for respiration. Haemoglobin relies on iron, and haemocyanin on copper. The interaction of trace metals and life processes has been in place since the year dot. So have the biochemical safety procedures needed to prevent trace metals from binding in the wrong place at the wrong time to cause toxic effects. Again, this book has made much reference to trace metal detoxification, involving the likes of metallothioneins and metal-rich granules. The upshot is that when humankind arrived to interrupt the biogeochemical cycles transporting trace metals through the different physical systems of the Earth's ecosystems, organisms were ready to cope.

Up to a point. As we have also seen, the history of the British Isles abounds with examples of trace metal ecotoxicology caused by relatively uncontrolled, excess anthropogenic activity in the fields of mining and, starting later, industry with associated metal-rich waste effluents. Such examples extend from the terrestrial habitats of the sites of mines for the likes of lead, zinc, silver, copper and arsenic, via drainage streams, rivers and estuaries to coastal waters. Cornwall has recently changed from a desolate landscape blighted by the mining industry in the nineteenth century to the holiday county of today. Beneath the yachts cruising on the serene waters of Restronguet Creek lie some of the most metal-contaminated estuarine sediments in the world. Yet we have seen how life survives even there.

We have explored the biological principles controlling the uptake of trace metals by organisms from the environment, introducing the term 'bioavailability'. Uptake typically leads to accumulation, of both essential and non-essential trace metals. Any accumulation of a non-essential trace metal requires detoxification. So does any accumulation of an essential trace metal above a low threshold concentration which reflects a minimum concentration of the essential metal in metabolically available form to meet metabolic requirements. This book has taken the brave (or foolhardy?) approach of attempting to quantify such essential requirements, particularly for copper and zinc. Please see these as hypotheses rather than statements of fact.

Excess uptake of any trace metal leads to toxicity. Please note. This sentence does not state that excess accumulation of any trace metal leads to toxicity. The latter may be true if none (or very little) of the accumulated trace metal is detoxified, but the vast majority of organisms accumulate trace metals and detoxify all or most of this accumulated load, particularly under field conditions. There are exceptions, but these are few and far between. Perhaps new zinc accumulated by the decapod crustacean *Palaemon elegans* under conditions of high zinc exposure might not be detoxified, so that the total accumulated zinc concentration equates approximately to the concentration of metabolically available zinc. Selenium in many organisms is also apparently not present in detoxified form.

Furthermore, laboratory exposures of test organisms to acute dissolved trace metal concentrations in laboratory tests might cause an influx of trace metal too rapid to allow detoxification of the newly accumulated metal. The concentration of metabolically available metal might then be nearly equivalent to the total concentration, suggesting the presence of a critical body concentration of a metal causing toxicity. Please don't go there. Toxicity occurs when the total rate of uptake of a trace metal (by all available routes summated) exceeds the combined rates of excretion (if present) and detoxification of the newly accumulating trace metal. Perhaps some readers would claim that this statement is a hypothesis. Here it is presented as a principle underlying the mechanism of trace metal toxicity.

The presence of efficient rates of storage detoxification that match rates of uptake of trace metals, even under conditions of raised bioavailability, can lead to bioaccumulated concentrations that are remarkably high in comparison to typical accumulated concentrations in other populations of that species or in other species. Zinc concentrations in the bodies of barnacles spring to mind. Pyrophosphate granules bind the zinc in detoxified form, out of harm's way. There is presumably some energetic cost in binding the zinc to these pyrophosphate granules, but the barnacles live on. High accumulated concentrations of cadmium in the kidneys of long-lived oceanic seabirds, such as fulmars and puffins, do not appear to threaten the livelihood of these birds, although the visual appearance of the kidneys would, at first sight, suggest a pathological condition.

Not all organisms do the same thing when it comes to the uptake and accumulation of trace metals. Different organisms show different patterns of accumulation of different trace metals. The accumulation of zinc by a decapod crustacean is different from that of an amphipod crustacean. In turn, both are different from that of another crustacean, a barnacle. Mussels and oysters, both suspension-feeding bivalves, have very different patterns of accumulation of zinc, with greatly contrasting background accumulated zinc concentrations. Different organisms depend to different extents on the different routes of uptake of trace metals – from solution or from the diet in the case of animals. Even the importance of ingested sediment as a source of trace metals to different deposit-feeding invertebrates will vary with the invertebrate. In short, the biology of trace metals in different organisms is not controlled only by the marine chemistry of metals. The biology of an organism, or more specifically its natural history, crucially affects how it takes up and accumulates a trace metal. It is only by appreciating the natural history of an organism that we can fully understand the interaction of that organism with a particular trace metal, including its potential ecotoxicity.

This book has stressed that trace metals have been part of the evolution of the metabolic processes of life from the start. Thus, the chemical characteristics of different trace metals (the essential metals) are key to the functioning of the metabolism of every organism on Earth, not least when metals are incorporated into enzymes. As we have seen, trace metals have also been used to other advantage in the biology of particular species, whether, for example, to increase the hardness of the biting edge of a jaw, or perhaps to act as a predator deterrent. However used, trace metals have represented a remarkable resource in evolution.

8.2 Effects of Trace Metals and the Natural History of the British Isles

Prologue. But what have been the effects of trace metals released into the environment on the natural history of the British Isles?

Discussion in this book of the effects of trace metals released into the environment on the natural history of the British Isles has inevitably centred on the ecotoxicology of raised bioavailabilities of the trace metals. Potential ecotoxicological effects have been somewhat arbitrarily divided here into effects on the organisms and ultimately ecological communities living in affected habitats, and the risks to human consumers presented by high trace metal concentrations in particular foods.

8.2.1 Past

Prologue. The economic history of the British Isles has long involved mining the inherent natural resources for trace metals of both cultural and industrial value, not least as a result of the industrial revolution. There has been an associated ecotoxicological fallout affecting British biota and habitats, on land, in streams, rivers and estuaries and in coastal waters. The worst of those days is now passed.

The British Isles have a history of mining for trace metals extending back into the Bronze Age, only now beginning to fizzle out in the face of exhaustion of economically accessible ores and international competition based on price. Nevertheless, the mining of metals in these isles has held a prominent world position over centuries, and indeed millennia. Different metals have been important at different times in our history, and British mines have produced tin, copper, arsenic, silver, lead and zinc in amounts of world significance at appropriate times.

With mining came ecological damage to local terrestrial and freshwater habitats, ranging from devastation to subtler ecotoxicological effects further afield. Ores suitable for exploitation have a widely scattered distribution over the uplands of Britain (Figure 2.2). Nevertheless, particular regions stand out as major mining centres, with associated negative effects on their local biodiversity, notably Cornwall and Devon, mid-Wales, the Lake District, the Northern Pennines and the Derbyshire Peak District. To misquote Andy Warhol, smaller specific mine sites have also had their moment in history – for example, Parys Mountain on Anglesey, the Snailbeach lead mine in Shropshire and the zinc mines of Shipham in Somerset (see Chapter 2).

Chapter 4 has described the effects of mining on the terrestrial natural history of many of these mining areas, with legacies still visible today. Local freshwater systems have also been severely affected at some time or other, not only by raised bioavailabilities of toxic metals but also by other harmful effects of high sediment loadings and acid mine drainage (see Chapter 5). The short rivers of Cornwall, such as the Carnon and the Hayle, are clear examples, as are the longer Ystwyth and Rheidol Rivers in mid-Wales. The metals that appear to have been the major drivers of ecotoxicological effects across British habitats affected by mining have been copper, lead and zinc, with a possible significant contribution too from arsenic. Conspicuous by its absence from this list is tin. A lot of tin has been mined in Cornwall, but this metal is rarely identified as an ecotoxicological scourge of the local natural history. Tin mining historically caused severe habitat damage in Cornwall, but mostly as a result of physical effects, such high sediment loadings in Cornish streams, rivers and even estuaries, caused by tin streaming. The major ore of tin in Cornwall is cassiterite, a mineral in which tin is tightly bound and relatively resistant to release in a form bioavailable to local resident organisms. Tin in cassiterite, therefore, has little scope to cause direct ecotoxicological effects.

The eighteenth century brought the industrial revolution. The harnessing of the power of coal over the next century led to increased activities of industrial processes associated, for example, with iron manufacture, smelting, foundries, electroplating and galvanisation, amongst others. Metal-rich effluents were now being emitted away from the historical centres of mining. Major rivers such as the Thames, the Humber, the Medway, the Tyne and the Tees faced increasing challenges from metal pollution, as did their estuaries.

It could not last. The second half of the twentieth century brought increased environmental regulation of industrial effluents and a consequential diminution of levels of bioavailable toxic metals in the rivers and estuaries associated with our major industrialised conurbations. Part of the cleanup process involved the building of improved sewage treatment systems. Initially, this partly transferred problems associated with trace metal pollution via sewage sludge to the coastal environment where the sludge was dumped. This transfer has now also been addressed, with sludge dumping at sea banned at the end of the twentieth century. The closure of the huge smelting works at Avonmouth led to a fall in the cadmium pollution of the lower Severn Estuary and adjacent Bristol Channel. Regulation and reduction of the emission of mercury into the River Mersey have similarly led to reduced mercury pollution in the Mersey and in Liverpool Bay.

8.2.2 Present

Prologue. There is no doubt. Over the last fifty years or so, the worst cases of trace metal pollution in the British Isles have been addressed. Many of the most affected mining and other industrialised sites have been cleaned up and rehabilitated, as have associated freshwater systems receiving their drainage. The battle is not yet won everywhere, however, but huge progress has been made.

The dumping of contaminated sewage sludge in our coastal waters is now banned. On the other hand, the marine dumping of larger amounts of, albeit less contaminated, dredged spoil material continues, as does that of a small amount of contaminated industrial solid waste. Care still needs to be taken to prevent the transfer of a trace metal pollution problem from one habitat to another. Much sewage sludge is now incinerated, and necessary measures are taken to limit the resulting airborne emission of their constituent toxic metals to be deposited on the surrounding countryside. Similarly, regulations are in place to control the application of sewage sludge with high metal loadings onto agricultural land as fertiliser.

On 1 January 1973, the United Kingdom became a member of the so-called Common Market (the European Community, or EC), the forerunner of the present European Union (EU). Since the 1970s, therefore, the United Kingdom has taken a Europe-based approach to the regulation of environmental pollutants. We have seen in this book how successful the United Kingdom has been in the last four decades or so in cleaning up many of the toxic metal-rich legacies of historical mining and industrial activities. Similarly, there is now much improved control of any

ongoing emission of trace metals in solid, dissolved or atmospheric form from industry, in comparison with the bad old days of the mid-twentieth century and before. The improvement of the ecotoxicological relevance of measures to be used in the environmental regulation of trace metal pollution is a continuing process. Early measures were relatively crude, but effective given the sorry state of our polluted habitats at the time. Refinement of such regulatory measures has sought, and is continuing to seek, increased scientific justification for their application in the real world. Thus, we are moving from a regulatory world of measured trace metal concentrations in water bodies, sediments and soils to one that understands the distinction between concentrations and bioavailabilities. This new regulatory world also appreciates that there is considerable variability in the effects of trace metal exposure on different organisms as biology intervenes. Measures to be compared against defined concentration limits are no longer straightforward and in the hands of analytical chemists. The chemists, of course, continue to have a role but are now joined by ecophysiologists and ecotoxicologists, introducing biological criteria to join their chemical counterparts.

We must not forget, however, that regulatory measures do still need to be pragmatic and achievable. We do still need simple environmental quality standards, for example drinking water standards defining maximum allowable concentrations of dissolved trace metals (Table 4.55). Critical soil concentrations of trace metals also have a role to play – not as one-size-fits-all maximum allowable limits, but as guidelines calculated for a specifically defined purpose. Such guidelines may relate, for example, to concerns about toxic metal inputs to cattle feeding on grass growing in contaminated soils of different physicochemical characteristics that would affect the uptake of the trace metals from the soil (Table 4.52). Similarly, a soil guideline value might be concerned with the risk to human health from long-term exposure to metal-contaminated soil under different land use situations, as in an old mining community (Table 4.56). Such soil guidelines are part of a risk assessment process, being used to highlight a need for further in-depth analysis.

The critical concentration concept has often been extended to attempts to define critical accumulated concentrations of toxic metals in different tissues of different organisms that can be considered as threshold concentrations for the onset of toxic effects in that organism (Tables 4.37, 4.49, 4.51 and 5.17). While clearly attractive for pragmatic simplicity, this critical tissue concentration approach has been criticised in this book as biologically unjustifiable when storage detoxification of an accumulated trace metal is taking place. And it usually is. A more ecotoxicologically justifiable approach is to consider rates of trace metal uptake into a specified organism, translated, for example, into an acceptable daily intake (ADI). Even this justified ADI approach does, however, need to be careful about using the concept of a critical tissue concentration to calculate an ADI (Tables 4.49 and 4.51). We shall return shortly to this valid ADI approach, when considering uptake of toxic metals from foods in the human diet, in the form now of a provisional tolerable weekly intake (PTWI) (Table 7.17).

The environmental health of aquatic environments in the British Isles has recently been safeguarded by the Water Framework Directive (WFD) of the European Union, introduced in 2000. The WFD commits EU member states to achieving good quantitative and qualitative status of all water bodies, including rivers, lakes, estuaries, coastal waters and groundwater. Both chemical status and ecological status need to be at least good to achieve what is termed Good Water Status. To establish good chemical status, the WFD uses ambient environmental standards to classify water bodies. Thus, the WFD establishes maximum acceptable dissolved concentrations of priority hazardous substances, which include the trace metals cadmium, lead, mercury, nickel and tin in the form of tributyl tin (TBT) (Tables 5.18 and 6.30). The WFD also requires that the current status of water bodies be determined using a formal classification scheme that characterises the ecological status of each water body based on biological quality elements. Ecological objectives are designed to protect and, where necessary, restore the structure and function of aquatic ecosystems, and thereby safeguard the sustainable use of water resources. The WFD currently focuses on community-level biological impacts on macroinvertebrates, phytoplankton, macrophytic plants and benthic algae, and fish, depending on the type of water body. The umbilical cord back to the pre-eminent role of direct chemical measures to define environmental health status has yet to be broken as biologically based criteria are under continuing development. A water body will still fail to be classified as good if the average concentration of any metal listed as a priority substance exceeds the relevant environmental quality standard for that specific metal.

Of course, it is simplistic to think only in terms of toxic metals when addressing problems of industrial pollution. Contaminant organic compounds are inevitably also going to be present. Not just obvious pollutants such as the hydrocarbons from fuel oils, but more subtle persistent pollutants and many newly emerging pollutants now being manufactured by man. Immunosuppression of exposed animals, for example seals and dolphins, is associated with organochlorines such as PCBs and dioxins, as well as with the organometal compounds methyl mercury and organotins. Endocrine disruptors mimic natural hormones, with unnatural effects on sex and reproduction in invertebrates as well as vertebrates. Endocrine disruption is similarly effected by organochlorines, as well as by TBT. Do not be fooled by exotic names. Organic endocrine disruptors may be present in familiar products, including plastic bottles, toys, cosmetics as well as in pesticides and flame retardants. As one genre of environmental problem is addressed, a new problem emerges.

8.2.2.1 TBT and Antifouling

Prologue. The banning of the antifouling agent TBT has clearly been a success in limiting significant ecotoxicological damage in our coastal environment. But?

The direct ecotoxicity of TBT, and its metabolic derivatives DBT and MBT, is now reducing in British coastal waters, even if antifouling paints based on this organotin might still be being applied to marine craft somewhere in the world beyond the influence of the International Maritime Organisation. And yet, the ecotoxicological effects of these organotins originating from antifouling paints have still not been completely flushed out of British marine systems. Seals and cetaceans living in or passing through British coastal waters still carry butyl tins in their livers and blubber – a legacy of the TBT years. And butyl tins have immunosuppression properties. These may not be significant in themselves but will make an additive contribution to the same effects caused by accumulated organochlorines such as PCBs and dioxins in our marine mammals.

Have we jumped out of the frying pan into the fire (Evans et al., 2000)? I think not, but the banning of TBT in antifouling paints applied to ships may yet have some unexpected side effects. TBT-based antifouling paints inevitably needed replacement by effective substitutes. In practice, these were copper-based products with added organic booster biocides to improve their ecotoxic efficiency (Voulvoulis et al., 2000). These organic biocides included Irgarol (an algicide), diuron (a terrestrial herbicide) and chlorothalonil and dichlofluanid (both fungicides). Chlorothalonil is also highly toxic to fish and aquatic invertebrates (Voulvoulis et al., 2000). The point here is that we are still learning about the ecotoxicity of these organic booster biocides in the marine environment, not least in coastal marinas housing large numbers of pleasure craft. Let us hope that we are not in for any unpleasant surprise.

8.2.2.2 Mercury in Tuna

Prologue. Trace metals in seafood harvested from the coastal waters of the British Isles do not today pose any significant ecotoxicological risk to human consumers. This cannot be said, however, of the methyl mercury lurking in tins of tuna destined for the dinner plate.

The concentrations of toxic metals in foods destined for human consumption are obvious starting points

in any assessment of the risk offered by trace metal-contaminated foods to the human consumer. While it is accepted that accumulated trace metal concentrations in fungi, plant or animal tissues in the human diet are not necessarily directly related to their trophic bioavailabilities to the consumer, and that different chemical forms of the accumulated metals may have different subsequent toxicities, pragmatism must rule. Thus, the United Kingdom stipulates maximum allowable concentrations of trace metals in foodstuffs in the human diet (Tables 4.54 and 7.16). While the United Kingdom did have some historical statutory limits for concentrations of particular trace metals in food before joining the EC, maximum concentrations in foods are presently defined under EU Commission Regulations for the three trace metals of any significant concern, namely cadmium, lead and mercury (Tables 4.54 and 7.16). The variety of different concentrations listed in Tables 4.54 and 7.16 represents a pragmatic response in the real world to the points made in this book concerning variation between foodstuffs in the bioavailability and toxicity of the accumulated metal, not least related to its chemical form. A further factor involved in the setting of these maximum concentrations is also the estimated frequency of ingestion of particular foodstuffs by specific parts of the population considered to be at risk. This approach is further formalised in the calculation of proposed PTWI (Table 7.17), as discussed later in this section.

Mercury has in the past been a trace metal of concern in fish harvested from British coastal waters, particularly Liverpool Bay, and destined for the dinner plate. The mercury concentrations in muscles of fish from British coastal regions such as Liverpool Bay, Morecambe Bay and the Outer Thames in the 1970s and 1980s (Tables 6.29 and 7.14) often exceeded the present EU maximum concentration of 0.5 µg/g wet weight (Table 7.16). Any threat, fortunately, has now passed with the diminution in industrial mercury emissions in the last decades of the twentieth century. Increased local control of mercury emissions has brought about a welcome fall in mercury concentrations of local coastal fish since the 1970s, so that by the turn of the millennium these were below the EU threshold.

We are still left, however, with the elephant in the room: mercury concentrations in tinned tuna consumed in the British Isles. Mercury concentrations in the muscles of tuna (Table 7.15) do generally approach or exceed the EU-regulated concentration of 0.5 µg/g designated for most marine fish (Table 7.16). Tuna actually have a higher EU-permitted level of 1 µg/g mercury (Table 7.16), which accommodates mercury concentrations in the muscle tissue of most tuna, but not by a wide margin. Theoretical calculations in Chapter 7 showed that a meal of tuna weighing 225 grams with mercury concentrations up to 0.7 µg/g will provide up to 158 µg of mercury. And the PTWI for mercury is only 122 µg per week for a 12 stone adult (Table 7.17) and much lower for children. I concluded earlier that one meal of tuna per week would be fine for most adults. Remember that PTWI calculations assume protracted intakes over long periods, providing a considerable safety margin. The situation, however, is different for children, and for pregnant women or breast-feeding mothers who will pass mercury on to the foetus or nursing baby respectively. Remember that the mercury is in the form of methyl mercury, the most toxic form of mercury which affects neural development. Tuna do provide valuable omega-3 fatty acids, but then so do many other oily fish without high mercury loads. To reassure yourself as to how much tuna can be safely eaten by a particular consumer, I do recommend consultation of relevant websites that are based on US Environmental Protection Agency Guidelines for mercury. As a middle-aged man, I certainly enjoy an occasional meal of tuna. As a grandfather, I am happier that my very young grandchildren gain their omega-3 fatty acids from sardines.

Tuna pick up their mercury far from land, as mercury is transported in the atmosphere to be deposited in the surface waters of oceans. While concentrations of mercury in long-lived oceanic fish would be high even in the absence of any anthropogenic activity, mercury is emitted into the atmosphere today in large amounts, particularly from artisanal gold mining and coal burning in Asia, Africa and Latin America (Qiu, 2013). About 6,500 tonnes of mercury were emitted into the atmosphere in 2010, with approximately 30% coming from human activities (Qiu, 2013). Mercury concentrations in the top 100 metres of the oceans have doubled in the past century (Qiu, 2013). Mercury levels in top marine predators are higher than in pre-industrial times, even in regions such as the Arctic, once considered to be pollution-free. Clearly, we have moved away from British marine waters to a situation of mercury contamination that may only be improved by international cooperation. Nevertheless, part of the fallout remains sitting in tins on our supermarket shelves.

8.2.3 Future

Prologue. What should be done with abandoned mining sites that have no or only limited associated biodiversity? How will the acidification of seawater caused by climate change affect the bioavailabilities of trace metals? The weight of evidence approach to the environmental regulation of trace metals is favoured going forward, but what is the future of environmental regulation in the United Kingdom after Brexit?

8.2.3.1 Rehabilitation and Restoration

Prologue. Mining sites can be rehabilitated but to what state?

We have reached a point where we can appreciate much of the history of mining for metals in the British Isles and its subsequent effects on particular locations and their associated natural history. Unless we do something about it, the sites of abandoned mines and their spoil heaps are biodiversity deserts, with no vegetation because of high bioavailabilities of toxic metals, exacerbated by poor soil structure, acidification and a lack of nutrients (Batty, 2005). Moving away from a mine site (or indeed a site of a metal smelter or refinery), there will be a gradient over which biodiversity is still negatively affected, but decreasingly so, until we are far enough away that the toxic metals have no significant ecotoxicological effect on the local natural history.

What can be done to rehabilitate a mine site to a state that restores it (probably inevitably imperfectly) to its pre-disturbance condition? One option may be the physical removal of metal-contaminated mine waste, for use as infill or as aggregate in road, embankment, dam and building construction (Lottermoser, 2011). A typical aim of rehabilitation at a mine site is the restoration of vegetation as a starting point to ecological recovery. This needs the importation of soil, usually after capping of the mine waste by the installation of a break layer to isolate the metal-rich waste from the imported soil (Downing et al., 1998; Lottermoser, 2011). The break layer is designed to prevent the upward migration of contaminant metals by capillary action into the upper soil layers. Various materials have been used to form such a break layer, including waste rock such as limestone, clays and organic wastes, or a man-made geocomposite, as in the reclamation of the Snailbeach lead mine in Shropshire (Downing et al., 1998). Installation of a suitable drainage system is important, often with a settlement pond to settle out any metal-contaminated silt prior to discharge into local water courses. The next stage in the rehabilitation process is the establishment of vegetation, but what vegetation?

Restoration means the recreation of both the structure and the function of an ecosystem, the resetting of an 'ecological clock' (Cairns, 1991). However, restoration cannot be perfect, and any pre-disturbance condition, however desirable, may not be a viable option (Cairns, 1991). So, how should the clock be set? To a time just before the disturbance occurred, a close approximation of the ecological condition before mining started? Or to an approximation of the condition that natural succession would have produced had no damage occurred, a landscape-compatible option? Given our typical lack of knowledge of pre-disturbance conditions, restoration efforts are usually exercises in approximations, reconstructing naturalistic rather than natural assemblages of organisms and their pre-disturbance physical environments (Cairns, 1991).

Crucial to any restoration programme is an explicit statement beforehand of the objectives, and spatial and temporal scales of the programme, so that its success can be monitored (Cairns, 1991). It is not enough simply to remove the stress from an ecosystem and let nature take its course.

Phytoremediation   Prologue. The aim of phytostabilisation is the formation of a vegetative cap, where uptake and accumulation processes immobilise metals within the plant rhizosphere, reducing metal bioavailability and hence wildlife, livestock and human exposure. Phytoextraction involves the use of plants for the removal or reduction of metals in mine waste, requiring the accumulation of metals in plant tissues aboveground. The plants are harvested and usually incinerated before disposal of the metal-rich ash as hazardous waste or processing of the ash for recovery of the metals (phytomining).

Phytoremediation is the use of plants (and their associated microorganisms) to stabilise or reduce contamination of, in this case, a toxic metal in soil, thereby mitigating an environmental problem without the need to excavate the soil for disposal elsewhere. Implicit in the use of phytoremediation is the principle that it will be cheaper than conventional physical or chemical remediation. There are two common approaches to the phytoremediation of mine waste: phytostabilisation and phytoextraction (Mendez and Maier, 2008). Both approaches require the establishment and growth of plants on mine waste, and so typically require the addition of compost and nutrients. Examples of added organic amendments include composted green waste or manure, biosolids, sewage sludge and woodchips which improve soil structure, water-holding capacity and nutrient contents. In addition, such organic amendments decrease the bioavailabilities of metals in the mine waste, by binding with free metal ions in the soil solution. Lime may be added to reduce acidification, and inorganic fertilisers may also be applied. It is not surprising that metallophytes are popular choices to be grown in phytoremediation schemes, depending on the specific objectives of the programme.

The aim of phytostabilisation is the formation of a vegetative cap, where uptake and accumulation processes immobilise metals within the plant rhizosphere, reducing metal bioavailability and hence wildlife, livestock and human exposure (Mendez and Maier, 2008). The phytostabilisation of mine tailings requires the use of metal-hardy native plants with root systems sufficiently extensive to create a stable vegetative cap that does not accumulate toxic metals into plant tissues aboveground. Furthermore, successful establishment of plants should create an abundant and diverse microbial community, promoting plant growth and soil structure development.

In the British Isles, prime candidates for use in phytostabilisation programmes are metallophytic grasses, particularly metal-tolerant grasses of the genera *Agrostis* (bentgrass) and *Festuca* (fescue) (Table 4.19) (Smith and Bradshaw, 1972 and 1979). Grasses are particularly suitable because of their rapid growth and extensive root systems. The use of metal-tolerant grasses, together with fertiliser application, does indeed enable a vegetation cover to be established even on mining sites heavily contaminated with metals, the grasses proving persistent for many years. Choice of grass species does, however, need to take into account the calcium status of the contaminated soil. *Agrostis stolonifera* (creeping bentgrass) and *Festuca rubra* (red fescue) are dominant on calcareous material, *A. tenuis* (common bentgrass) and *F. ovina* (sheep's fescue) on acidic waste (Smith and Bradshaw, 1972 and 1979). Metal-tolerant strains that have been recommended for use in the reclamation of mine wastes are *A. tenuis* from the Goginan mine near Aberystwyth for acid lead/zinc mine waste, *F. rubra* (cultivar Merlin) for calcareous lead/zinc mine waste and *A. tenuis* from Parys Mountain, Anglesey, for copper mine waste (Smith and Bradshaw, 1979).

Phytoextraction involves the use of plants for the removal or reduction of metals in mine waste. Unlike phytostabilisation, phytoextraction does require the accumulation of metals in plant tissues aboveground. The plants are harvested and usually incinerated before disposal of the metal-rich ash as

hazardous waste or processing of the ash for recovery of the metals (phytomining). By definition, there needs to be a higher concentration of metal in the harvested plant tissue than in the soil for phytoextraction to be effective. When considering phytoextraction, it is necessary to take three aspects into account: the concentration of metal in the harvestable plant material, the biomass of the crop and the specificity of the plant for the metal or metals of interest.

Inevitably, then, it is to hyperaccumulators that we turn (Chapter 4). Unfortunately, high biomass hyperaccumulators do not feature in the British flora (Macnair, 2003). The British flora does, though, have plants showing specific hyperaccumulation of particular metals, mainly nickel, cadmium and zinc. (Macnair, 2003). While high nickel bioavailability is a feature of naturally occurring serpentine soils, anthropogenic nickel contamination of soils from mining or smelting, or even sewage sludge application, is not a significant problem in Britain, in contrast, for example, to many regions of Canada. The potential of a hyperaccumulator for more than one metal is attractive, but hyperaccumulators in the British Isles are usually specific to one of the three aforementioned metals. An exception is to be found in the case of populations of alpine pennycress *Thlaspi caerulescens* with multiple hyperaccumulation of nickel, cadmium and zinc (Macnair, 2003). In continental Europe, a species of rockcress, *Arabidopsis halleri*, hyperaccumulates both cadmium and zinc (Macnair, 2003). Populations of both these species have been used in phytoextraction experiments to clean up zinc from agricultural soils contaminated with zinc from the long-term application of sewage sludge (Baker et al., 1994; Macnair, 2003). These hyperaccumulators were able to reduce the zinc loading in the contaminated soil from 444 µg/g Zn dry weight to the acceptable level of 300 µg/g after 14 (*T. caerulescens*) and 37 croppings (*A. halleri*). Comparative non-hyperaccumulating plant species took between 800 and more than 2,000 croppings to achieve the same result (Baker et al., 1994). Another example confirms the phytoextraction potential of alpine pennycress for zinc and cadmium, in this case from mine waste. *T. caerulescens* accumulated between 50 and 160 µg/g Cd dry weight and 13,000 to 19,000 µg/g Zn dry weight, from mine tailings and mine spoil at British mining sites, with higher concentrations in the plants than in the soils (Knight et al., 1997; Mendez and Maier, 2008). In a greenhouse experiment, alpine pennycress accumulated 250 µg/g Cd and 8,000 µg/g Zn from a mine spoil containing 58 µg/g Cd and 3,300 µg/g Zn (cf. Table 4.2). While this sounds promising, alpine pennycress does have a low biomass. Between 100 and 1,200 croppings would, therefore, be needed to remove the cadmium and between 200 and 600 croppings to remove the zinc from the mine tailings at the British sites examined (Knight et al., 1997; Mendez and Maier, 2008).

Attention has also been paid to the use of fast-growing trees in phytoremediation projects. Although not qualifying as hyperaccumulators, poplars and willows, in the family Salicaceae, are known to accumulate relatively high concentrations of cadmium and zinc in their leaves (Migeon et al., 2009; Remon et al., 2013). A hybrid poplar *Populus tremula x Populus tremuloides* growing in a phytostabilisation revegetation project at the site of the Metaleurop Nord zinc and lead production plant in northern France achieved leaf concentrations of 950 µg/g Zn and 44 µg/g Cd dry weight (Migeon et al., 2009). Similarly, another willow, in this case the osier *Salix viminalis*, may have some potential as a candidate for the phytoextraction of cadmium and zinc from moderately polluted calcareous urban soils (2.5 µg/g Cd and 400 µg/g Zn), extracting 0.13% of total Cd and 0.29% of total Zn per year (Jensen et al., 2009). Even though the osier has presumably extracted the most bioavailable, and therefore potentially the most ecotoxicologically significant, fractions of these metals in the soil, such phytoextraction rates are hardly impressive. It would appear, therefore, that poplars and willows are good candidates for phytostabilisation programmes as opposed to phytoextraction schemes.

Thus the use of phytoextraction to remove toxic metals from contaminated soils would have limited potential, except in cases of mild contamination with a metal that can be hyperaccumulated. In the British Isles, this means nickel, cadmium and zinc, with more of an environmental call for cadmium and zinc than for nickel. If metal contamination is high, as at mine and smelter sites (Tables 4.2 and 4.3), simply too many croppings of the likes of alpine pennycress would be required to make any significant change to soil metal concentrations. The use of phytoextraction to clean up agricultural soils contaminated, for example, by sewage sludge application to concentrations less extreme than at mine and smelter sites (Table 4.1), is more feasible (Mendez and Maier, 2008).

We should not, however, forget the potential of genetic modification techniques to develop transgenic hyperaccumulators that are more effective in phytoextraction (Pilon-Smits and Freeman, 2006; Mendez and Maier, 2008). It is potentially possible to develop hyperaccumulators with selected multiple specificities and preferably large body mass. For example, willows and poplars already show a great deal of genetic variation, so offering a genetic resource that can be developed for potential phytoextraction, as well as phytostabilisation (Migeon et al., 2009).

Phytoextraction can lead to the possibility of phytomining. Hyperaccumulating plants are grown on soils rich in the trace metal of interest. After harvesting and incineration, the resulting ash is processed in a smelter or refinery to recover the metal in question in commercially viable quantities. The potential of phytomining is high for nickel but potentially also for cobalt, thallium and gold, but not in the British Isles. Much attention has been paid to the hyperaccumulation qualities of species of *Alyssum*, a genus in the family Brassicaceae with high species diversity in the region of the Mediterranean. Yellowtuft (*Alyssum murale*) from the Balkans and Turkey is a hyperaccumulator of nickel and cobalt. It grows well in well-drained, rocky soils, including serpentine soils. Given its hyperaccumulation properties, yellowtuft offers potential not only for phytoextraction but also for phytomining of nickel. It is also a plant that can reach a metre in height, offering considerable biomass. Indeed, phytomining trials of *A. murale* and another related imported species, *Alyssum corsicum* from Corsica, have been carried out on the serpentine soils of Illinois, United States but proved not economically viable. This example highlights a potential ecological problem with phytomining, if preferred plants are imported into regions in which they are non-native. Yellowtuft is a prolific seeder. After its importation into the Illinois valley for phytomining trials, it has escaped to thrive in such areas as the serpentine soils of southern Oregon, as an undesirable alien outcompeting local plants, including endangered endemics. Furthermore, yellowtuft is poisonous to livestock. If phytomining has a future on British metal-contaminated soils, in the absence of endemic hyperaccumulators of any size, it is vital that the dangers of introducing potentially undesirable alien plants are thoroughly considered.

We must not ignore the fact that plant life is dependent on interactions with soil microorganisms, including rhizobacteria and arbuscular mycorrhizal fungi (AMF). AMF adapted to metal-rich soils can promote the growth of diverse plants in such soils (Bothe, 2011). The yellow zinc violet *Viola calaminaria* lives on zinc-rich soils in the area between Aachen in Germany and Liège in Belgium. A mycorrhizal fungus *Glomus intraradices* Br 1 isolated from the roots of the yellow zinc violet consistently outcompeted a conventional AMF isolate in the growth promotion of plants in metal-rich soils (Bothe, 2011). Such AMF could be exploited for phytoremediation purposes. It is ironic, however, that members of the plant family Brassicaceae, the family to which the metallophyte alpine pennycress belongs, are generally AMF-negative (Bothe, 2011). The ability of AMF to confer on plants a resistance to metal exposure is not uncommon and therefore of phytoremediation relevance (Meier et al., 2011, 2012). Mycorrhizal plants can improve phytostabilisation because metals are confined to the fungal hyphae and roots, without translocation to aerial parts (Meier et al., 2012). The metals, therefore, remain in the soil and do not become bioavailable to local terrestrial food webs.

8.2.3.2 Conservation

Prologue. Some metal-contaminated terrestrial habitats host rare endemic species and are worthy of conservation.

An aspect of the natural history of metal-contaminated terrestrial habitats, such as abandoned mine sites, needs important consideration. Abandoned mine sites often support rare and threatened endemic species, worthy of conservation, as well exemplified by the zinc violet *Viola calaminaria*, endemic to a small region in eastern Belgium, southern Netherlands and western Germany (Bizoux et al., 2004).

In the British Isles, several abandoned mine sites in the Northern Pennines have been designated as Sites of Special Scientific Interest (SSSI) because of their distinctive metallophyte floras, not just of green plants but also of lichens (Cooke and Morrey, 1981). Haggs Bank in the valley of the River Nent in Cumbria, with its calaminarian grassland community, has also been designated as an SSSI. This community has been recognised for its scarcity in Europe, and it is listed under Annex 1 of the European Communities Directive 92/43/EEC, on the Conservation of Natural Habitats and of Wild Fauna and Flora – the Habitats Directive. In the Peak District, ancient lead deposits similarly host distinctive plant communities with the facultative metallophytes spring sandwort *Minuartia verna* (called lead sandwort in parts of northern England), alpine pennycress *Thlaspi caerulescens* and the more widespread mountain pansy *Viola lutea* and Pyrenean scurvygrass *Cochlearia pyrenaica* (Batty, 2005). These sites are also important for metallophytic lichens, and many have been included in the Peak District Biodiversity Action Plan (Batty, 2005).

Amongst the lichens worthy of conservation is *Lecidea inops*, known in the United Kingdom only from copper-rich sites in the Lake District and now protected under Schedule 8 of the UK Wildlife and Countryside Act (1981) (Purvis, 1996). The first and only UK record of the lichen *Gyalidea roseola* is from acid lead– and zinc-rich mine spoil at the Feish Dhomhnuill mine in Strontian, Scotland (Gilbert,

2000). The bryophyte *Scopelophila cataractae*, tongue-leaf copper moss, is also very rare, found only on very moist, very toxic, zinc-rich mine spoil (Atherton et al., 2010).

It is easy to forget that biodiversity includes microbial biodiversity. Putting aside for a moment the familiar arguments for the conservation of endangered species which still apply to microorganisms, the argument to conserve the microbial diversity of the metal-rich soils of mining sites additionally includes the opportunity that the microbial flora represents for biotechnological exploitation, not least in the bioremediation of metal-contaminated sites. Reference has been made briefly earlier to genetic engineering, and the genes likely to be present in the microbial flora of abandoned mine waste could prove to be of inestimable value in the future.

Are the remediation and conservation of old mining sites mutually exclusive? Not at all necessarily. Remediation may be absolutely required, for example, if metal-rich runoff from a site is deleteriously affecting local water courses and freshwater communities. Nevertheless, it should still be possible alongside any remediation scheme to keep an area of sufficient size as a designated and deliberate conservation site.

8.2.3.3 Climate Change and Seawater Acidification

Prologue. Acidification will change the speciation of dissolved trace metals in seawater, with consequent effects on equilibrium concentrations of free metal ions, and hence dissolved bioavailabilities.

The concentration of carbon dioxide in the Earth's atmosphere is rising as a result of the burning of fossil fuels and land use changes. If current trends continue, this concentration will have more than doubled by 2100. Atmospheric carbon dioxide equilibrates with the carbon dioxide dissolved in the ocean. When more carbon dioxide dissolves in seawater, chemical equilibria cause an increase in concentrations of hydrogen (H^+) and bicarbonate (HCO_3^-) ions and

decreases in concentrations of hydroxide (OH^-) and carbonate (CO_3^{2-}) ions (Millero et al., 2009). With the increase in hydrogen ion concentration, the pH will drop. Over the next 150 years or so, the pH of the oceans will decrease from about 8.1 to 7.4 (Millero et al., 2009).

While decreases in the concentrations of carbonate ions will decrease the production of calcium carbonate shells by the likes of molluscs, we are concerned here with the potential effects of a decrease in pH on the speciation of dissolved trace metals. The chemical speciation of dissolved trace metals controls the concentration of their free metal ions. And free metal ions best model the bioavailable form of a dissolved trace metal for uptake from solution by organisms. Crudely, any increase in free metal ion concentration will increase the rate of uptake of the trace metal from solution, even in the absence of any increase in total dissolved concentration, and vice versa. We have also seen that trace metals dissolved in seawater are usually complexed by inorganic ions such as chloride (Cl^-), carbonate (CO_3^{2-}) and hydroxide (OH^-), with proportions varying greatly between trace metals (Table 3.1).

What will be the predicted effects on the proportions of different dissolved trace metals in seawater present as the free metal ion, when concentrations of hydroxide and carbonate drop at the lower pH previously predicted? Carbonate and hydroxide concentrations will decrease by 77% and 82% in this scenario (Millero et al., 2009). Dissolved trace metals that form complexes with the chloride ion, or are mainly present as the free metal ion, will not be strongly affected by the decrease in pH. There will be little change in the speciation of dissolved silver (Ag^+), cadmium (Cd^{2+}) or mercury (Hg^{2+}) (Millero et al., 2009). The proportion of the free metal ion present will increase by only a few percent in the cases of cobalt (Co^{2+}), manganese (Mn^{2+}) and zinc (Zn^{2+}) (Millero et al., 2009). Dissolved trace metals which form strong complexes with carbonate and hydroxide will have a higher proportion of total dissolved metal present as the free metal ion. One trace metal ion that forms a strong complex with carbonate in seawater is the copper (Cu^{2+}) ion. The decrease in pH will cause a 30% increase in the concentration of the free copper ion (Millero et al., 2009), with an associated increase in the bioavailability of dissolved copper. Some dissolved nickel is complexed by carbonate, and the proportion of dissolved nickel present as the free metal ion will increase by 13% (Millero et al., 2009). Dissolved lead forms significant complexes with both carbonate and chloride. As the pH decreases, there will be an approximate 10% increase in the concentration of the free lead ion, and a large increase in chloride complexation of the remaining dissolved lead (Millero et al., 2009).

The case of dissolved iron is not so straightforward. At the current pH (8.1) of oceanic seawater, the oxidised form of iron (iron 3) is at its minimum solubility, with very little present as the Fe^{3+} ion. The total concentration of dissolved iron 3 in seawater is maintained by the strong complexation of dissolved iron with dissolved organic matter (DOM) (Millero et al., 2009; Shi et al., 2010). A decrease in pH would theoretically cause a decrease in inorganic complexation of dissolved iron 3 by hydroxide to promote the level of the free Fe^{3+} ion (Millero et al., 2009). This change, however, is insignificant compared to the effects of a decrease in pH on the binding of dissolved iron 3 to DOM. The pH decrease causes increased binding to DOM, decreasing the concentration of the free Fe^{3+} ion and decreasing the bioavailability of dissolved iron to phytoplankton such as diatoms and coccolithophores by between 10 and 20% (Shi et al., 2010). We have seen in Chapter 7 that the already low bioavailability of iron to phytoplankton is a limiting factor to oceanic primary production in some regions of the world's oceans.

Acidification of the seas and oceans will therefore increase the proportion of particular dissolved trace metals present as the free metal ion, with an associated increase in their dissolved bioavailability. Copper in particular will show a predicted increase of 30% in this proportion and will have the largest increase in dissolved bioavailability. Is this of any ecotoxicological concern? Probably not, given the low bioavailabilities of dissolved trace metals in the oceans, and the abilities of marine organisms to cope with manyfold increases in dissolved trace metal uptake rates in contaminated coastal environments. The decrease in the bioavailability of iron to phytoplankton in selected areas of the world's oceans is likely to be of more ecological significance, adding to the role of iron as a limiting factor of oceanic primary production.

8.2.3.4 Ecotoxicology

Prologue. The weight of evidence (WOE) approach to ecological risk assessment is that favoured going forward.

The WOE approach combines information from multiple lines of evidence (LOE) to determine the ecotoxicological status of a metal-contaminated habitat. These LOE will consist of both chemical and biological measurements, including field evidence. Amongst the chemical LOE will be Environmental Quality Standards (EQS), such as Water Quality Standards (Tables 5.18 and 6.30). Exceedance of an EQS will highlight the need for further in-depth biological assessment.

The use of biological criteria as LOE is particularly important as we seek to increase the relevance of ecotoxicological measures in the real world. The choice of biological criteria as LOE in different habitats is the subject of ongoing research, as discussed in the chapters of this book. In this context, the WFD presently uses community-level biological impacts of environmental contaminants, for example on benthic macroinvertebrates and phytoplankton. These biological criteria need expansion to include biomarkers at different levels of biological organisation, from the molecular to the community. Biomarkers at the lowest level of biological organisation are the most sensitive to the toxic action of trace metals and are the most easily measured. Biological responses at the population and community levels are the most ecotoxicologically relevant, but are more difficult to assess unless completely devastating. Great progress is now being made in linking the responses of biomarkers at different levels. The day will come when an easily measured biomarker at a low level of biological organisation will provide information on linked ecotoxicological effects at the community level.

Beyond biomarkers, biotic indices are under development and refinement as relevant measures of ecotoxicological effects of contaminants in different habitats. Concentrations of accumulated trace metals in chosen biomonitors have the potential to replace concentrations of metals in physical media such as waters or soils as indicative of local trace metal bioavailabilities. High accumulated trace metal concentrations in biomonitors indicate the presence of atypically high bioavailabilities of those metals, strictly to that biomonitor, but often by extension to other organisms in the habitat. The high accumulated concentrations do not themselves indicate that these high metal bioavailabilities are having a toxicological effect on the biomonitor itself. At least, they will not until the day that we can identify and measure that component of the accumulated trace metal content of an organism that has not been detoxified and is metabolically available. And yet, if particular accumulated metal concentrations in selected biomonitors can be correlated with independently measured ecotoxicological effects on a local community, then they can be used as bioaccumulated metal guidelines indicative of these ecotoxicological effects. For example, accumulated copper concentrations in larvae of species of the caddisfly *Hydropsyche* have been validated as suitable bioaccumulated metal guidelines, correlating with copper-induced changes in the local mayfly community assemblages of streams affected by mining. Such use of bioaccumulated metal guidelines can be extended to terrestrial and coastal environments, using, for example, earthworms or woodlice in soils and leaf litter, and ragworms, mussels or oysters in estuaries and beyond. If accumulated metal concentrations in such biomonitors are calibrated against negative changes in associated biological communities, then a bioaccumulated concentration in the biomonitor becomes a surrogate measure of a significant ecotoxicological effect at the community level.

Implicit in any weight of evidence approach to the assessment of ecological risk created by the trace metal contamination of a particular habitat is the collection of several lines of evidence. These all LOE need to be assessed and integrated into a final conclusion as to whether trace metal contamination is of ecotoxicological significance in the habitat under consideration.

The environmental regulation of contaminants such as trace metals in the United Kingdom since 1973 has been in line with the rest of the EU. Great progress has been made. As Brexit looms, we need to be wary of simplistic and ill-informed statements that the introduction of EU legislation has simply increased bureaucracy and red tape to the hindrance of British industry. Environmental legislation should not just be abandoned because of its European origin, a genesis in which the United Kingdom played a significant role. Of course, it is cheaper for industry not to have restrictions on the emission of toxic metals into the environment. But such restrictions have been vital in the restoration of many British terrestrial, freshwater, estuarine and marine habitats previously suffering from toxic metal pollution to the detriment of local biodiversity. We also know now of the importance of a thriving environment for good human mental and physical health. In any fallout from Brexit, it is important that the United Kingdom is committed to ensuring an at least equivalent level of environmental protection as in the European Union.

8.3 Postscript

This book represents a personal viewpoint on the natural history of metals in the British Isles. It is perhaps inevitable that particular tenets that I consider self-evident have not always prevailed as the accepted norm, particularly amongst ecotoxicologists carrying out routine acute toxicity testing. I have supplied a host of supporting references from the research literature. I hope that these have not daunted the non-expert reader.

I strongly believe that it is important to have a knowledge of the natural history of an organism in order to understand the interaction between its biology and trace metals. The countryside, towns, streams, rivers, estuaries and coastal waters of the British Isles are littered with examples of the effects of metals, mostly via historical mining or subsequent industrial development. I have attempted to give an account that encompasses history, economics, geography, geology, chemistry, biochemistry, physiology, ecology, ecotoxicology and above all natural history. Examples abound of interactions between organisms and trace metals in the terrestrial, freshwater, estuarine, coastal and oceanic environments. Many of these interactions have nothing to do with metal pollution. All organisms are affected from bacteria, plants and invertebrates to charismatic species such as seabirds, seals, dolphins and whales. All have a tale to tell.

REFERENCES

Abdullah, A. M. and Ireland, M. P. (1986). Cadmium Content, Accumulation and Toxicity in Dog Whelks Collected around the Welsh Coastline. *Marine Pollution Bulletin*, **17**, 557–561.

Abdullah, M. I. and Royle, L. G. (1972). Heavy Metal Content of Some Rivers and Lakes in Wales. *Nature*, **238**, 329–330.

  (1974). A Study of the Dissolved and Particulate Trace Elements in the Bristol Channel. *Journal of the Marine Biological Association of the United Kingdom*, **54**, 581–597.

Abdullah, M. I., Royle, L. G. and Morris, A. W. (1972). Heavy Metal Concentrations in Coastal Waters. *Nature*, **235**, 158–160.

Abel, P. D. and Green, D. W. J. (1981). Ecological and Toxicological Studies on Invertebrate Fauna of Two Rivers in the Northern Pennine Orefield. In *Heavy Metals in Northern England: Environmental and Biological Aspects*, ed. P. J. Say and B. A. Whitton. Durham: Department of Botany, University of Durham, pp. 109–122.

Abrahams, P. W. and Thornton, I. (1987). Distribution and Extent of Land Contamination by Arsenic and Associated Metals in Mining Regions of Southwest England. *Transactions of the Institution of Mining and Metallurgy*, **96**, B1–B8.

Ackroyd, D. R., Bale, A. J., Howland, R. J. M., et al. (1986). Distributions and Behaviour of Dissolved Cu, Zn and Mn in the Tamar Estuary. *Estuarine, Coastal and Shelf Science*, **23**, 621–640.

Adams, W. J., Blust, R., Borgmann, U., et al. (2010). Utility of Tissue Residues for Predicting Effects of Metals on Aquatic Organisms. *Integrated Environmental Assessment and Management*, **7**, 75–98.

Alexander, W. and Street, A. (1951). *Metals in the Service of Man*, 4th edn. Harmondsworth: Penguin.

Allen, J. I. and Moore, M. N. (2004). Environmental Prognostics: Is the Current Use of Biomarkers Appropriate for Environmental Risk Evaluation? *Marine Environmental Research*, **58**, 227–232.

Alloway, B. J. and Davies, B. E. (1971). Trace Element Content of Soils Affected by Base Metal Mining in Wales. *Geoderma*, **5**, 197–208.

Alzieu, C. (1986). TBT Detrimental Effects on Oyster Culture in France – Evolution since Antifouling Paint Regulation. Institute of Electrical and Electronics Engineers, New York, *Proceedings of Oceans 86 Conference Record*. Organotin Symposium 4, 1130–1134.

  (2000). Impact of Tributyl Tin in Invertebrates. *Ecotoxicology*, **9**, 71–76.

Amiard, J. C. and Amiard-Triquet, C. (2013). Molecular and Histocytological Biomarkers. In *Ecological Biomarkers: Indicators of Ecotoxicological Effects*, ed. C. Amiard-Triquet, J. C. Amiard and P. S. Rainbow, Boca Raton, FL: CRC Press, pp. 75–105.

Amiard, J. C., Amiard-Triquet, C., Barka, S., Pellerin, J. and Rainbow, P. S. (2006). Metallothioneins in Aquatic Invertebrates: Their Role in Metal Detoxification and Their Use as Biomarkers. *Aquatic Toxicology*, **76**, 160–202.

Amiard, J. C., Amiard-Triquet, C., Berthet, B. and Métayer, C. (1987). Comparative Study of the Patterns of Bioaccumulation of Essential (Cu, Zn) and Non-Essential (Cd, Pb) Trace Metals in Various Estuarine and Coastal Organisms. *Journal of Experimental Marine Biology and Ecology*, **106**, 73–89.

Amiard, J. C., Amiard-Triquet, C. and Métayer, C. (1985). Experimental Study of Bioaccumulation, Toxicity and Regulation of Some Trace Metals in Various Estuarine and Coastal Organisms. *Symposia Biologica Hungarica*, **29**, 313–323.

Amiard-Triquet, C. (2009). Behavioral Disturbances: The Missing Link between Sub-Organismal and Supra-Organismal Responses to Stress? Prospects Based on Aquatic Research. *Human and Ecological Risk Assessment*, **15**, 87–110.

Amiard-Triquet, C. and Amiard, J. C. (2013). Behavioral Ecotoxicology. In *Ecological Biomarkers: Indicators of Ecotoxicological Effects*, ed. C. Amiard-Triquet, J. C. Amiard and P. S. Rainbow. Boca Raton, FL: CRC Press, pp. 253–277.

Amiard-Triquet, C. and Rainbow, P. S. (Eds) (2009). *Environmental Assessment of Estuarine Ecosystems*. Boca Raton, FL: CRC Press.

Amiard-Triquet, C., Amiard, J. C. and Rainbow, P. S. (Eds) (2013). *Ecological Biomarkers: Indicators of Ecotoxicological Effects.* Boca Raton, FL: CRC Press.

Amiard-Triquet, C., Rainbow, P. S. and Roméo, M. (Eds) (2011). *Tolerance to Environmental Contaminants.* Boca Raton, FL: CRC Press.

Ancion, P.-Y., Lear, G., Dopheide, A. and Lewis, G. D. (2013). Metal Concentrations in Stream Biofilm and Sediments and Their Potential to Explain Biofilm Microbial Community Structure. *Environmental Pollution*, **173**, 117–124.

Andrews, M. J. (1984). Thames Estuary: Pollution and Recovery. In *Effects of Pollution at the Ecosystem Level*, ed. P. J. Sheehan, D. R. Miller, G. C. Butler and P. Bourdeau. Chichester: SCOPE, John Wiley and Sons, pp. 195–217.

Andrews, M. J. and Rickard, D. G. (1980). Rehabilitation of the Inner Thames Estuary. *Marine Pollution Bulletin*, **11**, 327–332.

Andrews, S. M., Johnson, M. S. and Cooke, J. A. (1984). Cadmium in Small Mammals from Grassland Established on Metalliferous Mine Waste. *Environmental Pollution A*, **33**, 153–162.

(1989a). Distribution of Trace Element Pollutants in a Contaminated Grassland Ecosystem Established on Metalliferous Fluorspar Tailings. 1: Lead. *Environmental Pollution*, **58**, 73–85.

(1989b). Distribution of Trace Element Pollutants in A Contaminated Grassland Ecosystem Established on Metalliferous Fluorspar Tailings. 2: Zinc. *Environmental Pollution*, **59**, 241–252.

Angel, M. V. (1984). Detrital Organic Fluxes through Pelagic Ecosystems. In *Flows of Energy and Materials in Marine Ecosystems*, ed. M. J. R. Fasham. New York, NY: Plenum Press, pp. 475–516.

Anon (1998) Summary of Data for Selected Sites along the Thames Estuary. In *A Rehabilitated Estuarine Ecosystem*, ed. M. J. Attrill. London: Kluwer Academic Publishers, appendix A.

Antonovics, J., Bradshaw, A. M. and Turner, R. G. (1971). Heavy Metal Tolerance in Plants. *Advances in Ecological Research*, **7**, 1–85.

ANZECC and ARMCANZ [Australian and New Zealand Environment and Conservation Council and Agricultural and Resource Management Council of Australia and New Zealand] (2000). *National Water Quality Management Strategy, Vol. 1, The Guidelines. Paper No. 4.* Australia: Department of Agriculture and Resources. www.agriculture.gov.au/SiteCollectionDocuments/water/nwqms-guidelines-4-vol1.pdf

Appleton, J. D., Cave, M. R., Palumbo-Roe, B. and Wragg, J. (2013). Lead Bioaccessibility in Topsoils from Lead Mineralisation and Urban Domains, UK. *Environmental Pollution*, **178**, 278–287.

Appleton, J. D., Cave, M. R. and Wragg, J. (2012). Modelling Lead Bioaccessibility in Urban Topsoils Based on Data from Glasgow, London, Northampton and Swansea, UK. *Environmental Pollution*, **171**, 265–272.

Apte, S. C., Gardner, M. J., Gunn, A. M., Ravenscroft, J. E. and Vale, J. (1990). Trace Metals in the Severn Estuary: A Reappraisal. *Marine Pollution Bulletin*, **8**, 393–396.

Archer, F. C. and Hodgson, I. H. (1987). Total and Extractable Trace Element Contents of Soils in England and Wales. *Journal of Soil Science*, **38**, 421–431.

Ardestani, M. M., van Straalen, N. M. and van Gestel, C. A. M. (2014). Uptake and Elimination Kinetics of Metals in Soil Invertebrates: A Review. *Environmental Pollution*, **193**, 277–295.

Armitage, P. D. (1980). The Effects of Mine Drainage and Organic Enrichment on Benthos in the River Nent System, Northern Pennines. *Hydrobiologia*, **74**, 119–128.

Armitage, P. D. and Blackburn, J. H. (1985). Chironomidae in a Pennine Stream System Receiving Mine Drainage and Organic Enrichment. *Hydrobiologia*, **121**, 165–172.

Arnold, R. E., Hodson, M. E. and Langdon, C. J. (2008). A Cu Tolerant Population of the Earthworm *Dendrodrilus rubidus* (Savigny, 1862) at Coniston Copper Mines, Cumbria, UK. *Environmental Pollution*, **152**, 713–722.

Aston, S. R., Thornton. I., Webb, J. S., Purves, J. B. and Milford, B. L. (1974). Stream Sediment Composition: An Aid to Water Quality Assessment. *Water, Air, and Soil Pollution*, **3**, 321–325.

Atherton, I., Bosanquet, S. and Lawley, M. (Eds) (2010). *Mosses and Liverworts of Britain and Ireland: A Field Guide.* Plymouth: British Bryological Society, Latimer Trend & Co. Ltd.

Atkinson, R. L. (1985). *Tin and Tin Mining.* Princes Risborough: Shire Publications Ltd.

(1987). *Copper and Copper Mining.* Princes Risborough: Shire Publications Ltd.

Attrill, M. J. and Thomas, R. M. (1995). Heavy Metal Concentrations in Sediment from the Thames Estuary, UK. *Marine Pollution Bulletin*, 11, 742–744.

Au, D. W. T. (2004). The Application of Histo-Cytopathological Biomarkers in Marine Pollution Monitoring: A Review. *Marine Pollution Bulletin*, 48, 817–834.

Aubail, A., Méndez-Fernandez, P., Bustamente, P., Churlaud, C., Ferreira, M. and Vingada, J. V. (2013). Use of Skin and Blubber Tissues of Small Cetaceans to Assess the Trace Element Content of Internal Organs. *Marine Pollution Bulletin*, 76, 158–169.

Austen, M. C. and Somerfield, P. J. (1997). A Community-Level Sediment Bioassay Applied to an Estuarine Heavy Metal Gradient. *Marine Environmental Research*, 43, 315–328.

Awrahman, Z. A., Rainbow, P. S., Smith, B. D., Khan, F. R., Bury, N. R. and Fialkowski, W. (2015). Bioaccumulation of Arsenic and Silver by the Caddisfly Larvae *Hydropsyche siltalai* and *H. pellucidula*: A Biodynamic Modeling Approach. *Aquatic Toxicology*, 161, 196–207.

Awrahman, Z. A., Rainbow, P. S., Smith, B. D., Khan, F. R. and Fialkowski, W. (2016). Caddisflies *Hydropsyche* spp. as biomonitors of Trace Metal Bioavailability Thresholds Causing Disturbance in Freshwater Stream Benthic Communities. *Environmental Pollution*, 216, 793–805.

Baden, S. P. and Neil, D. M. (1998). Accumulation of Manganese in the Haemolymph, Nerve and Muscle Tissue of *Nephrops norvegicus* (L.) and Its Effect on Neuromuscular Performance. *Comparative Biochemistry and Physiology*, 119A, 351–359.

Baden, S. P., Eriksson, S. P. and Weeks, J. M. (1995). Uptake, Accumulation and Regulation of Manganese during Experimental Hypoxia and Normoxia by the Decapod *Nephrops norvegicus* (L.). *Marine Pollution Bulletin*, 31, 93–102.

Badsha, K. S. and Goldspink, C. R. (1982). Preliminary Observations on the Heavy Metal Content of Four Species of Freshwater Fish in NW England. *Journal of Fish Biology*, 21, 251–267.

Baker, A. J. M. (1981). Accumulators and Excluders: Strategies in the Response of Plants to Heavy Metals. *Journal of Plant Nutrition*, 3, 643–654.

(1987). Metal Tolerance. *New Phytologist*, 106 (Suppl.), 93–111.

Baker, A. J. M., Ernst, W. H. O., Van der Ent, A., Malaisse, F. and Ginocchio, R. (2010). Metallophytes: The Unique Biological Resource, Its Ecology and Conservational Status in Europe, Central Africa and Latin America. In *Ecology of Industrial Pollution*, ed. L. C. Batty and K. B. Hallberg. Cambridge: British Ecological Society, Cambridge University Press, pp. 7–40.

Baker, A. J. M., McGrath, S. P., Sidoli, C. M. D. and Reeves, R. D. (1994). The Possibility of in Situ Heavy Metal Decontamination of Polluted Soils Using Crops Of Metal-Accumulating Plants. *Resources, Conservation and Recycling*, 11, 41–49.

Balassone, G., Rossi, M., Boni, M., Stanley, G. and McDermott, P. (2008). Mineralogical and Geochemical Characterization of Nonsulfide Zn-Pb at Silvermines and Galmoy (Irish Midlands). *Ore Geology Reviews*, 33, 168–186.

Ballan-Dufrançais, C. (2002). Localization of Metals in Cells of Pterygote Insects. *Microscopy Research and Technique*, 56, 403–420.

Balls, P. W. and Topping, G. (1987). The Influence of Inputs to the Firth of Forth on the Concentrations of Trace Metals in Coastal Waters. *Environmental Pollution*, 45, 159–172.

Barak, N. A.-E. and Mason, C. F. (1990). A Survey of Heavy Metal Levels in Eels *(Anguilla anguilla)* from Some Rivers in East Anglia, England: The Use of Eels as Pollution Indicators. *Internationale Revue der gesamten Hydrobiologie und Hydrographie*, 75, 827–833.

Barkay, T., Tripp, S. C. and Olson, B. H. (1985). Effect of Metal-Rich Sewage Sludge Application on the Bacterial Communities of Grasslands. *Applied and Environmental Microbiology*, 49, 333–337.

Barltrop, D., Strehlow, C. D., Thornton, I. and Webb, J. S. (1975). Absorption of Lead from Dust and Soil. *Postgraduate Medical Journal*, 51, 801–804.

Barnes, H. (1948). The Leaching Rate and Behaviour of Antifouling Compositions. *Journal of the Oil and Colour Chemists' Association*, 31, 455–461.

Barnes, R. D. (1980). *Invertebrate Zoology*, 4th edn. Philadelphia, PA: Saunders College.

Barnes, R. S. K. (1974). *Estuarine Biology.* London: Edward Arnold (Publishers) Ltd.

Barnett, B. E. and Ashcroft, C. R. (1985). Heavy Metals in *Fucus vesiculosus* in the Humber Estuary. *Environmental Pollution*, 9, 193–213.

Barrento, S., Marques, A., Teixeira, B., Carvalho, M. L., Vaz-Pires, P. and Nunes, M. L. (2009). Influence of Season and Sex on the Contents of Minerals and Trace Elements in Brown Crab (*Cancer pagurus*, Linnaeus, 1758). *Journal of Agricultural and Food Chemistry*, 57, 3253–3260.

Barrington, E. J. W. (1967). *Invertebrate Structure and Function*. London: Nelson.

Barton, D. B. (1971a). *Essays in Cornish Mining History*, vol. 2. Truro: D. Bradford Barton Ltd.

  (1971b). Arsenic Production in West Cornwall. In *Essays in Cornish Mining History*, vol. 2, ed. D. B. Barton. Truro: D. Bradford Barton Ltd., pp. 101–125.

Basu, M., Pande, M., Bhadoria, P. B. S. and Mahapatra, S. C. (2009). Potential Fly-Ash Utilization in Agriculture: A Global Review. *Progress in Natural Science*, 19, 1173–1186.

Bat, L. (1998). Influence of Sediment on Heavy Metal Uptake by the Polychaete *Arenicola marina*. *Turkish Journal of Zoology*, 22, 341–350.

Batty, L. C. (2005). The Potential Importance of Mine Sites for Biodiversity. *Mine Water and the Environment*, 24, 101–103.

Bayley, M., Baatrup, E. and Bjerregaard, P. (1997). Woodlouse Behavior in the Assessment of Clean and Contaminated Field Sites. *Environmental Toxicology and Chemistry*, 16, 2309–2314.

Beane, S. J., Comber, S. D. W,. Rieuwerts, J. and Long, P. (2016). Abandoned Metal Mines and Their Impact on Receiving Waters: A Case Study from Southwest England. *Chemosphere*, 153, 294–306.

Bebianno, M. J. and Langston, W. J. (1993). Turnover Rate of Metallothionein and Cadmium in *Mytilus edulis*. *BioMetals*, 6, 239–244.

  (1995). Induction of Metallothionein Synthesis in the Gill and Kidney of *Littorina littorea* Exposed to Cadmium. *Journal of the Marine Biological Association of the United Kingdom*, 75, 173–186.

Becker, P. H. (1992). Egg Mercury Levels Decline with Laying Sequence in Charadriiformes. *Bulletin of Environmental Contamination and Toxicology*, 48, 762–767.

Beernaert, J., Scheirs, J., Van den Brande, G., et al. (2008). Do Wood Mice (*Apodemus sylvaticus* L.) Use

Food Selection as a Means to Reduce Heavy Metal Intake? *Environmental Pollution*, 151, 599–607.

Bellinger, D. C. (2004). Lead. *Pediatrics*, 113 (Suppl. 4), 1016–1022.

Beltman, D. J., Clements, W. H., Lipton, J. and Cacela, D. (1999). Benthic Invertebrate Metals Exposure, Accumulation, and Community-Level Effects Downstream from a Hard-Rock Site. *Environmental Toxicology and Chemistry*, 18, 299–307.

Benedetti, M., Ciaprini, F., Piva, F., et al. (2011). A Multidisciplinary Weight of Evidence Approach for Classifying Polluted Sediments: Integrating Sediment Chemistry, Bioavailability, Biomarkers Responses and Bioassays. *Environment International*, 38, 17–28.

Bengtsson, G. and Tranvik, L. (1989). Critical Metal Concentrations for Forest Soil Invertebrates. *Water, Air and Soil Pollution*, 47, 381–417.

Bengtsson, G., Ek, H. and Rundgren, S. (1992). Evolutionary Response of Earthworms to Long-Term Metal Exposure. *Oikos*, 63, 289–297.

Bennett, P. M., Jepson, P. D., Law, R. J., et al. (2001). Exposure to Heavy Metals and Infectious Disease Mortality in Harbour Porpoises from England and Wales. *Environmental Pollution*, 112, 33–40.

Benson, L. M., Porter, E. K. and Peterson, P. J. (1981). Arsenic Accumulation, Tolerance and Genotypic Variation in Plants on Arsenical Mine Wastes in S.W. England. *Journal of Plant Nutrition*, 3, 655–666.

Bérard, A., Mazzia, C., Sappin-Didier, V. Capowiez, L. and Capowiez, Y. (2014). Use of the MicroResp™ Method to Assess Pollution-Induced Community Tolerance in the Context of Metal Soil Contamination. *Ecological Indicators*, 40, 27–33.

Berrow, M. L. and Burridge, J. C. (1980). Trace Element Levels in Soils: Effects of Sewage Sludge. In *Inorganic Pollution and Agriculture*. MAFF/ADAS Reference Book 326. London: HMSO, pp. 159–184.

Berry, R. J. (1977). *Inheritance and Natural History*. London: New Naturalist, Collins.

Berthet, B., Mouneyrac, C., Amiard, J.-C., et al. (2003). Accumulation and Soluble Binding of Cadmium, Copper, and Zinc in the Polychaete *Hediste diversicolor* from Coastal Sites with Different Trace Metal Bioavailabilities. *Archives of Contamination and Toxicology*, 45, 468–478.

Bervoets, L., De Jonge, M. and Blust, R. (2016). Identification of Threshold Body Burdens of Metals for the Protection of the Aquatic Ecological Status Using Two Benthic Invertebrates. *Environmental Pollution*, **210**, 76–84.

Bervoets, L., Int Panis, L. and Verheyen, R. (1994). Trace Metal Levels in Water, Sediments and *Chironomus gr. thumni*, from Different Water Courses in Flanders (Belgium). *Chemosphere*, **29**, 1591–1601.

Bervoets, L., Knapen, D., De Jonge, M., Van Campenhout, K. and Blust, R. (2013). Differential Hepatic Metal and Metallothionein Levels in Three Feral Fish Species along a Metal Pollution Gradient. *PLoS ONE*, **8**(3), doi:10.1371/journal.pone.0060805.

Bhattacharya, P., Welch, A. H., Stollenwerk, K. G., McLaughlin, M. J., Bundschuh, J. and Panaullah, G. (2007). Arsenic in the Environment: Biology and Chemistry. *Science of the Total Environment*, **379**, 109–120.

Bick, D. (1982). *The Old Copper Mines of Snowdonia*. Newent, Gloucestershire: Pound House.

Bierkens, J., Buekers, J., Van Holderbeke, M. and Torfs, R. (2012). Health Impact Assessment and Monetary Valuation of IQ Loss in Pre-School Children Due to Lead Exposure through Locally Produced Food. *Science of the Total Environment*, **414**, 90–97.

Bizoux, J. P., Brevers, F., Meerts, P., Graitson, E. and Mahy, G. (2004). Ecology and Conservation of Belgian Populations of *Viola calaminaria*, a Metallophyte with a Restricted Geographic Distribution. *Belgian Journal of Botany*, **137**, 91–104.

Blaber, S. J. M. (1970). The Occurrence of a Penis-Like Outgrowth behind the Right Tentacle in Spent Females of *Nucella lapillus* (L.). *Proceedings of the Malacological Society of London*, **39**, 231–233.

Blanck, H. (2002). A Critical Review of Procedures and Approaches Used for Assessing Pollution-Induced Community Tolerance (PICT) in Biotic Communities. *Human and Ecological Risk Assessment*, **8**, 1003–1034.

Blanck, H. and Wängberg, S. A. (1988). Induced Community Tolerance in Marine Periphyton Established under Arsenate Stress. *Canadian Journal of Fisheries and Aquatic Sciences*, **45**, 1816–1819.

Blanck, H., Wängberg, S. A. and Molander, S. (1988). Pollution-Induced Community Tolerance: A New

Ecotoxicological Tool. In *Functional Testing of Aquatic Biota for Estimating Hazards of Chemicals ASTM 988*, ed. J. Cairns and J. R. Pratt. Philadelphia, PA: American Society for Testing and Materials, pp. 219–230.

Bland, S., Ackroyd, D. R., Marsh, J. G. and Millward, G. E. (1982). Heavy Metal Content of Oysters from the Lynher Estuary, U.K. *Science of the Total Environment*, **22**, 235–241.

Boisson, F., Goudard, F., Durand, J. P., et al. (2003). Comparative Radiotracer Study of Cadmium Uptake, Storage, Detoxification and Depuration in the Oyster *Crassostrea gigas*: Potential Adaptive Mechanisms. *Marine Ecology Progress Series*, **254**, 177–186.

Bolam, S. G. (2012). Impacts of Dredged Material Disposal on Macrobenthic Invertebrate Communities: A Comparison of Structural and Functional (Secondary Production) Changes at Disposal Sites around England and Wales. *Marine Pollution Bulletin*, **64**, 2199–2210.

Bolam, S. G., Rees, H. L., Somerfield, P., et al. (2006). Ecological Consequences of Dredged Material Disposal in the Marine Environment: A Holistic Assessment of Activities around the England and Wales Coastline. *Marine Pollution Bulletin*, **52**, 415–426.

Boldina-Cosqueric, I., Amiard, J.-C., Amiard-Triquet, C., et al. (2010). Biochemical, Physiological and Behavioural Biomarkers in the Endobenthic Bivalve *Scrobicularia plana* as Tools for the Assessment of Estuarine Sediment Quality. *Ecotoxicology and Environmental Safety*, **73**, 1733–1741.

Bone, Q., Ryan, K. P. and Pulsford, A. L. (1983). The Structure and Composition of the Teeth and Grasping Spines of Chaetognaths. *Journal of the Marine Biological Association of the United Kingdom*, **63**, 929–939.

Bonneris, E., Perceval, O., Masson, S., Hare, L. and Campbell, P. G. C. (2005). Sub-Cellular Partitioning of Cd, Cu and Zn in Tissues of Indigenous Unionid Bivalves Living along a Metal Exposure Gradient and Links to Metal-Induced Effects. *Environmental Pollution*, **135**, 195–208.

Bortolotti, G. R. (2010). Flaws and Pitfalls in the Chemical Analysis of Feathers: Bad News–Good News for Avian Chemoecology and Toxicology. *Ecological Applications*, **20**, 1766–1774.

Boscher, A., Gobert, S., Guignard, C., et al. (2010). Chemical Contaminants in Fish Species from Rivers in the North of Luxembourg: Potential Impact on the Eurasian Otter (*Lutra lutra*). *Chemosphere*, **78**, 785–792.

Bothe, H. (2011). Plants in Heavy Metal Soils. In *Soil Biology, Vol. 30: Detoxification of Heavy Metals*, ed. I. Sherameti and A. Varma. Berlin–Heidelberg: Springer-Verlag, pp. 35–57.

Boult, S., Collins, D. N., White, K. N. and Curtis, C. D. (1994). Metal Transport in a Stream Polluted by Acid Mine Drainage: The Afon Goch, Anglesey, UK. *Environmental Pollution*, **84**, 279–284.

Bowen, G. G., Dussek, C. and Hamilton, R. M. (1998). Pollution Resulting from the Abandonment and Subsequent Flooding of Wheal Jane Mine in Cornwall, UK. *Geological Society London, Special Publications*, **128**, 93–99.

Bowen, H. J. M. (1979). *Environmental Chemistry of the Elements*. London and New York, NY: Academic Press.

Boyd, R. S. (2004). Ecology of Hyperaccumulation. *New Phytologist*, **162**, 563–567.

Boyd, R. S. and Wall, M. A. (2001). Responses of Generalist Predators Fed High-Ni *Melanotrichus boydi* (Heteroptera: Miridae): Elemental Defense against the Third Trophic Level. *American Midland Naturalist*, **146**, 186–198.

Boyden, C. R. (1975). Distribution of Some Trace Metals in Poole Harbour, Dorset. *Marine Pollution Bulletin*, **6**, 180–187.

(1977). Effect of Size upon Metal Content of Shellfish. *Journal of the Marine Biological Association of the United Kingdom*, **57**, 675–714.

Boyden, C. R. and Romeril, M. G. (1974). A Trace Metal Problem in Pond Oyster Culture. *Marine Pollution Bulletin*, **5**, 74–78.

Boyden, C. R., Aston, S. R. and Thornton, I. (1979). Tidal and Seasonal Variations of Trace Elements in Two Cornish Estuaries. *Estuarine and Coastal Marine Science*, **9**, 303–317.

Boyden, C. R., Watling, H. and Thornton, I. (1975). Effect of Zinc on the Settlement of the Oyster *Crassostrea gigas*. *Marine Biology*, **31**, 227–234.

Bradshaw, A. D. (1952). Populations of *Agrostis tenuis* Resistant to Lead and Zinc Poisoning. *Nature*, **169**, 1098.

Bragigand, V., Berthet, B., Amiard, J. C. and Rainbow, P. S. (2004). Estimates of Trace Metal Bioavailability to Humans Ingesting Contaminated Oysters. *Food and Chemical Toxicology*, **42**, 1893–1902.

Brand, L. E. (1991). Minimum Iron Requirements of Marine Phytoplankton and the Implications for the Biogeochemical Control of New Production. *Limnology and Oceanography*, **36**, 1756–1771.

Brittain, J. E. (1982). Biology of Mayflies. *Annual Review of Entomology*, **27**, 119–147.

Broadway, A., Cave, M. R., Wragg, J., et al. (2010). Determination of the Bioaccessibility of Chromium in Glasgow Soil and the Implications for Human Health Risk Assessment. *Science of the Total Environment*, **409**, 267–277.

Brook, F. and Allbutt, M. (1973). *The Shropshire Lead Mines*. Cheddleton, Leek, Staffordshire: Moorland Publishing Company.

Brookes, P. C. (1995). The Use of Microbial Parameters in Monitoring Soil Pollution by Heavy Metals. *Biology and Fertility of Soils*, **19**, 269–279.

Brooks, R. R. (1987). *Serpentine and Its Vegetation*. London and Sydney: Croom Helm.

Brown, B. E. (1976). Observations on the Tolerance of the Isopod *Asellus meridianus* Rac. to Copper and Lead. *Water Research*, **10**, 555–559.

(1977a). Effects of Mine Drainage on the River Hayle, Cornwall. A) Factors Affecting Concentrations of Copper, Zinc and Iron in Water, Sediments and Dominant Invertebrate Fauna. *Hydrobiologia*, **52**, 221–233.

(1977b). Uptake of Copper and Lead by a Metal-Tolerant Isopod *Asellus meridianus* Rac. *Freshwater Biology*, **7**, 235–244.

(1978). Lead Detoxification by a Copper-Tolerant Isopod. *Nature*, **276**, 388–390.

Brown, F. M. J. and Balls, P. W. (1997). Trace Metals in Fish and Shellfish from Scottish Waters. *Scottish Fisheries Research Report*, **60**, 1–36.

Brown, J. H., Buchanan, J. S. and Whitley, J. E. (1988). Uptake and Excretion of Inorganic Mercury in the Lobster *Homarus gammarus* (L.) White 1847: Long-Term Effects of Exposure to Low Levels of the Metal. *Ecotoxicology and Environmental Safety*, **15**, 125–141.

Bruland, K. W. (1983). Trace Elements in Sea-Water. *Chemical Oceanography*, **8**, 157–220.

Bryan, G. W. (1967). Zinc Concentrations of Fast and Slow Contracting Muscles in the Lobster. *Nature*, 213, 1043–1044.

(1968). Concentrations of Zinc and Copper in the Tissues of Decapod Crustaceans. *Journal of the Marine Biological Association of the United Kingdom*, 48, 303–321.

(1973). The Occurrence and Seasonal Variation of Trace Metals in the Scallops *Pecten maximus* (L.) and *Chlamys opercularis* (L.). *Journal of the Marine Biological Association of the United Kingdom*, 53, 145–166.

(1976). Some Aspects of Heavy Metal Tolerance in Aquatic Organisms. In *Effects of Pollutants on Aquatic Organisms*, ed. A. P. M. Lockwood. Cambridge: Cambridge University Press, pp. 7–34.

(1984). Pollution Due to Heavy Metals and Their Compounds. In *Marine Ecology, Part 3*, ed. O. Kinne. London: John Wiley & Sons, pp. 1289–1430.

(1985). Bioavailability and Effects of Heavy Metals in Marine Deposits. In *Wastes in the Ocean, Vol. 6: Nearshore Waste Disposal*, ed. B. H. Ketchum, J. M. Capuzzo, W. V. Burt, I. W. Duedall, P. K. Park and D. R. Kester. New York, NY: John Wiley and Sons, pp. 42–79.

Bryan, G. W. and Gibbs, P. E. (1979). Zinc: A Major Inorganic Component of Nereid Polychaete Jaws. *Journal of the Marine Biological Association of the United Kingdom*, 59, 969–973.

(1980). Metals in Nereid Polychaetes: The Contribution of Metals in the Jaws to the Total Body Burden. *Journal of the Marine Biological Association of the United Kingdom*, 60, 641–654.

(1983). Heavy Metals in the Fal Estuary, Cornwall: A Study of Long-Term Contamination by Mining Waste and Its Effects on Estuarine Organisms. *Occasional Publications of the Marine Biological Association of the United Kingdom*, 2, 1–112.

(1987). Polychaetes as Indicators of Heavy-Metal Availability in Marine Deposits. In *Oceanic Processes in Marine Pollution, Vol. 1: Biological Processes and Wastes in the Ocean*, ed. J. M. Capuzzo and D. R. Kester. Malabar, FL: Robert E. Krieger Publishing Company, pp. 37–49.

Bryan, G. W. and Hummerstone, L. G. (1971). Adaptation of the Polychaete *Nereis diversicolor* to Estuarine Sediments Containing High Concentrations of Heavy Metals. 1. General Observations and Adaptations to Copper. *Journal of the Marine Biological Association of the United Kingdom*, 51, 845–863.

(1973a). Brown Seaweed as an Indicator of Heavy Metals in Estuaries in South-West England. *Journal of the Marine Biological Association of the United Kingdom*, 53, 705–720.

(1973b). Adaptation of the Polychaete *Nereis diversicolor* to Estuarine Sediments Containing High Concentrations of Zinc and Cadmium. *Journal of the Marine Biological Association of the United Kingdom*, 53, 839–857.

(1977). Indicators of Heavy-Metal Contamination in the Looe Estuary (Cornwall) with Particular Regard to Silver and Lead. *Journal of the Marine Biological Association of the United Kingdom*, 57, 75–92.

(1978). Heavy Metals in the Burrowing Bivalve *Scrobicularia plana* from Contaminated and Uncontaminated Estuaries. *Journal of the Marine Biological Association of the United Kingdom*, 58, 401–419.

Bryan, G. W. and Langston, W. J. (1992). Bioavailability, Accumulation and Effects of Heavy Metals in Sediments with Special Reference to United Kingdom Estuaries: A Review. *Environmental Pollution*, 76, 89–131.

Bryan, G. W. and Uysal. H. (1978). Heavy Metals in the Burrowing Bivalve *Scribicularia plana* from the Tamar Estuary in Relation to Environmental Levels. *Journal of the Marine Biological Association of the United Kingdom*, 58, 89–108.

Bryan, G. W., Gibbs, P. E., Burt, G. R. and Hummerstone, L. G. (1987b). The Effects of Tributyltin (TBT) Accumulation on Adult Dog-Whelks, *Nucella lapillus*: Long-Term Field and Laboratory Experiments. *Journal of the Marine Biological Association of the United Kingdom*, 67, 525–544.

Bryan, G. W., Gibbs, P. E., Hummerstone, L. G. and Burt, G. R. (1986). The Decline of the Gastropod *Nucella lapillus* around South-West England: Evidence for the Effect of Tributyl Tin from Antifouling Paints. *Journal of the Marine Biological Association of the United Kingdom*, 66, 611–640.

(1987a). Copper, Zinc, and Organotin as Long-Term Factors Governing the Distribution of Organisms in the Fal Estuary in Southwest England. *Estuaries*, 10, 208–219.

Bryan, G. W., Langston, W. J. and Hummerstone, L. G. (1980). The Use of Biological Indicators of Heavy Metal Contamination in Estuaries. *Occasional Publications of the Marine Biological Association of the United Kingdom*, 1, 1–73.

Bryan, G. W., Langston, W. J., Hummerstone, L. G. and Burt, G. R. (1985). A Guide to the Assessment of Heavy-Metal Contamination in Estuaries. *Occasional Publications of the Marine Biological Association of the United Kingdom*, 4, 1–92.

Bryan, G. W., Langston, W. J., Hummerstone, L. G., Burt, G. R. and Ho, Y. B. (1983). An Assessment of the Gastropod, *Littorina littorea*, as an Indicator of Heavy-Metal Contamination in United Kingdom Estuaries. *Journal of the Marine Biological Association of the United Kingdom*, 63, 327–345.

Brydie, J. R. and Polya, D. A. (2003). Metal Dispersion in Sediments and Waters of the River Conwy Draining the Llanrwst Mining Field, North Wales. *Mineralogical Magazine*, 67, 289–304.

Buckley, J. A. (1992). *The Cornish Mining Industry: A Brief History*. Redruth, Cornwall: Tor Mark Press.

Bull, K. R., Every, W. J., Freestone, P., Hall, J. R. and Osborn, D. (1983). Alkyl Lead Pollution and Bird Mortalities on the Mersey Estuary, UK, 1979–1981. *Environmental Pollution A*, 31, 239–259.

Bull, K. R., Murton, R. K., Osborn, D., Ward, P. and Cheng, L. (1977). High Levels of Cadmium in Atlantic Seabirds and Sea-Skaters. *Nature*, 269, 507–509.

Bullock, E. (1970). Occurrence of Free Porphyrins in Certain Coelenterates. *Comparative Biochemistry and Physiology*, 33, 711–712.

Bundy, J. G., Kille, P., Liebeke, M. and Spurgeon, D. J. (2014). Metallothioneins May Not Be Enough: The Role of Phytochelatins in Invertebrate Metal Detoxification. *Environmental Science and Technology*, 48, 885–886.

Burger, J. (1993). Metals in Avian Feathers: Bioindicators of Environmental Pollution. *Reviews in Environmental Toxicology*, 5, 197–306.

Burrows, I. G. and Whitton, B. A. (1983). Heavy Metals in Water, Sediments and Invertebrates from a Metal-Contaminated River Free of Organic Pollution. *Hydrobiologia*, 106, 263–273.

Burt, R., Waite, P. and Burnley, R. (1987). *Cornish Mines: Metalliferous and Associated Minerals 1845–1913*. Exeter: University of Exeter in Association with the Northern Mine Research Society NMRA. Wheaton and Co. Ltd.

Burton, M. A. S. and Peterson, P. J. (1979). Metal Accumulation by Aquatic Bryophytes from Polluted Mine Streams. *Environmental Pollution*, 19, 39–46.

Burton, S. M., Rundle, S. D. and Jones, M. B. (2001). The Relationship between Trace Metal Contamination and Stream Meiofauna. *Environmental Pollution*, 111, 159–167.

Bury, N. R. and Handy, R. D. (2010). Copper and Iron Uptake in Teleost Fish. In *Surface Chemistry, Bioavailability and Metal Homeostasis in Aquatic Organisms: An Integrated Approach. Essential Reviews in Experimental Biology*, vol. 2. Ed. N. R. Bury and R. D. Handy. London: Society for Experimental Biology, pp. 107–127.

Bury, N. R. and Wood, C. M. (1999). Mechanism of Branchial Apical Silver Uptake by Rainbow Trout Is via the Proton-Coupled Na^+ Channel. *American Journal of Physiology*, 277, R1385–R1391.

Bustamente, P. and Miramand, P. (2005). Subcellular and Body Distributions of 17 Trace Elements in the Variegated Scallop *Chlamys varia* from the French Coast of the Bay of Biscay. *Science of the Total Environment*, 337, 59–73.

Bustamente, P., Caurant, F., Fowler, S.W. and Miramand, P. (1998). Cephalopods as a Vector for the Transfer of Cadmium to Top Marine Predators in the North-East Atlantic Ocean. *Science of the Total Environment*, 220, 71–80.

Bustamente, P., Cosson, R. P., Gallien, I., Caurant, F. and Miramand, P. (2002). Cadmium Detoxification Processes in the Digestive Gland of Cephalopods in Relation to Accumulated Cadmium Concentrations. *Marine Environmental Research*, 53, 227–241.

Butterworth, J., Lester, P. and Nickless, G. (1972). Distribution of Heavy Metals in the Severn Estuary. *Marine Pollution Bulletin*, 3, 72–74.

Cain, B. W., Sileo, L., Franson, J. C. and Moore, J. (1983). Effects of Dietary Cadmium on Mallard Ducklings. *Environmental Research*, 32, 286–297.

Caines, L. A. (1978). Heavy Metal Residues in Grey Seals (*Halichoerus grypus*) from the Farne Islands. *International Council for the Exploration of the Seas ICES, 1980/E*, 40, 1–13.

Cairns, J. (1991). The Status of the Theoretical and Applied Status of Restoration Ecology. *Environmental Professional*, 13, 186–194.

Caldow, R., McGrorty, S., West, A., Durell, S. E. A. le V. dit, Stillman, R. and Anderson, S. (2005), Macro-Invertebrate Fauna in the Intertidal Mudflats. In *The Ecology of Poole Harbour*, ed. J. Humphreys and V. May. Amsterdam: Elsevier, pp. 91–108.

Callahan, D. L., Baker, A. J. M., Kolev, S. D. and Wedd, A. G. (2006). Metal Ion Ligands in Hyperaccumulating Plants. *Journal of Biological Inorganic Chemistry*, 11, 2–12.

Campbell, J. I. A., Jacobsen, C. S. and Sørensen, J. (1995). Species Variation and Plasmid Incidence among Fluorescent *Peudomonas* Strains Isolated from Agricultural and Industrial Soils. *FEMS Microbiology Ecology*, 18, 51–62.

Campbell, P. G. C. (1995). Interaction between Trace Metals and Aquatic Organisms: A Critique of the Free-Ion Activity Model. In *Metal Speciation and Aquatic Systems*, ed. A. Tessier and D. R. Turner. New York, NY: Wiley, pp. 45–102.

Camusso, M., Polesello, S., Valsecchi, S. and Vignati, D. A. L. (2012). Importance of Dietary Uptake of Trace Elements in the Benthic Deposit-Feeding *Lumbriculus variegatus*. *Trends in Analytical Chemistry*, 36, 103–112.

Canadian Council of Ministers of the Environment (CCME) (1999a). *Canadian Water Quality Guidelines for the Protection of Aquatic Life*. Winnipeg: Canadian Council of Ministers of the Environment.

(1999b). *Canadian Sediment Quality Guidelines for the Protection of Aquatic Life: Summary Tables. Canadian Environmental Quality Guidelines, 1999*. Winnipeg: Canadian Council of Ministers of the Environment.

Canli, M. and Furness, R. W. (1993). Heavy Metals in Tissues of the Norway Lobster *Nephrops norvegicus*: Effects of Sex, Size and Season. *Chemistry and Ecology*, 8, 19–32.

Carlisle, D. B. (1968). Vanadium and Other Metals in Ascidians. *Proceedings of the Royal Society London B*, 171, 31–42.

Carpenter, K. E. (1924). A Study of the Fauna of Rivers Polluted by Lead Mining in the Aberystwyth District of Cardiganshire. *Annals of Applied Biology*, 11, 1–23.

Casado-Martinez, M. C., Smith, B. D., DelValls, T. A. and Rainbow, P. S. (2009a). Pathways of Trace Metal Uptake in the Lugworm *Arenicola marina*. *Aquatic Toxicology*, 92, 9–17.

Casado-Martinez, M. C., Smith, B. D., DelValls, T. A., Luoma, S. N. and Rainbow, P. S. (2009b). Biodynamic Modelling and the Prediction of Accumulated Trace Metal Concentrations in the Polychaete *Arenicola marina*. *Environmental Pollution*, 157, 2743–2750.

Casado-Martinez, M. C., Smith, B. D., Luoma, S. N. and Rainbow, P. S. (2010a). Metal Toxicity in a Sediment-Dwelling Polychaete: Threshold Body Concentrations or Overwhelming Accumulation Rates? *Environmental Pollution*, 158, 3071–3076.

(2010b). Bioaccumulation of Arsenic from Water and Sediment by a Deposit-Feeding Polychaete (*Arenicola marina*): A Biodynamic Modelling Approach. *Aquatic Toxicology*, 98, 34–43.

Caurant, F., Amiard, J. C., Amiard-Triquet, C. and Sauriau, P. G. (1994). Ecological and Biological Factors Controlling the Concentrations of Trace Elements (As, Cd, Cu, Hg, Se, Zn) in Delphinids *Globicephala melas* from the North Atlantic Ocean. *Marine Ecology Progress Series*, 103, 207–219.

Caurant, F., Aubail, A., Lahaye, V., et al. (2006). Lead Contamination of Small Cetaceans in European Waters: The Use of Stable Isotopes for Identifying the Sources of Lead Exposure. *Marine Environmental Research*, 62, 131–148.

Cave, R. R., Ledoux, L., Turner, K., Jickells, T. Andrews, J. E. and Davies, H. (2003). The Humber Catchment and Its Coastal Area: From UK to European Perspectives. *Science of the Total Environment*, 314–316, 31–52.

CEFAS (Centre for Environment, Fisheries and Aquaculture Science) (1997). Monitoring and Surveillance of Non-Radioactive Contaminants in the Aquatic Environment and Activities Regulating the Disposal of Wastes at Sea, 1994. *Aquatic Environment Monitoring Report*, 47, 1–59.

(2001). Monitoring and Surveillance of Non-Radioactive Contaminants in the Aquatic Environment and Activities Regulating the Disposal of Wastes at Sea, 1998. *Aquatic Environment Monitoring Report*, 53, 1–75.

(2003). Monitoring of the Quality of the Marine Environment, 1999–2000. *Aquatic Environment Monitoring Report*, 54, 1–98.

Chadwick, E. A., Simpson, V. R., Nicholls, A. E. L. and Slater, F. M. (2011). Lead levels in Eurasian Otters Decline with Time and Reveal Interactions between

Sources, Prevailing Weather and Stream Chemistry. *Environmental Science and Technology*, **45**, 1911–1916.

Chan, H. M and Rainbow, P. S. (1993). The Accumulation of Dissolved Zinc by the Shore Crab *Carcinus maenas* (L.). *Ophelia*, **38**, 13–30.

Chan, H. M., Bjerregaard, P., Rainbow, P. S. and Depledge, M. H. (1992). Uptake of Zinc and Cadmium by Two Populations of Shore Crabs *Carcinus maenas* at Different Salinities. *Marine Ecology Progress Series*, **86**, 91–97.

Chang, S. I. and Reinfelder, J. R. (2002). Relative Importance of Dissolved versus Trophic Bioaccumulation of Copper in Marine Copepods. *Marine Ecology Progress Series*, **231**, 179–186.

Chapman, P. M. (2007). Determining When Contamination Is Pollution: Weight of Evidence Determinations for Sediments and Effluents. *Environment International*, **33**, 492–501.

Chapman, P. M. and Hollert, H. (2006). Should the Sediment Quality Triad Become a Tetrad, a Pentad, or Possibly Even a Hexad? *Journal of Soils and Sediments*, **6**, 4–8.

Chapman, P. M. and Long, E. R. (1983). The Use of Bioassays as a Part of a Comprehensive Approach to Marine Pollution Assessment. *Marine Pollution Bulletin*, **14**, 81–84.

Chapman, P. M., Swartz, R. C., Roddie, B., Phelps, H. L., van den Hurk, P. and Butler, R. (1992). An International Comparison of Sediment Toxicity Tests in the North Sea. *Marine Ecology Progress Series*, **91**, 253–264.

Chapman, P. M., Wang, F., Janssen, C., Persoone, G. and Allen, H. E. (1998). Ecotoxicology of Metals in Aquatic Sediments: Binding and Release, Bioavailability, Risk Assessment and Remediation. *Canadian Journal of Fisheries and Aquatic Sciences*, **55**, 2221–2243.

Chassard-Bouchaud, C., Boutin, J. F., Hallegot, P. and Galle, P. (1989). Chromium Uptake, Distribution and Loss in the Mussel *Mytilus edulis*: A Structural, Ultrastructural and Microanalytical Study. *Diseases of Aquatic Organisms*, **7**, 117–136.

Chatelain, M., Gasparini, J., Jacquin, L. and Frantz, A. (2014). The Adaptive Function of Melanin-Based Plumage Coloration to Trace Metals. *Biology Letters*, **10**, 20140164. dx.doi.org/10.1098/rsbl.2014.0164.

Cheng, L. (1974). Notes on the Ecology of the Oceanic Insect *Halobates. Marine Fisheries Review*, **36**, 1–7.

Cheng, L., Alexander, G. V. and Franco, P. J. (1976). Cadmium and Other Heavy Metals in Sea-Skaters (Gerridae: *Halobates, Rheumobates*). *Water, Air and Soil Pollution*, **6**, 33–38.

Cheung, M. S. and Wang, W.-X. (2005). Influences of Subcellular Compartmentalization in Different Prey on the Transfer of Metals to a Predatory Gastropod from Different Prey. *Marine Ecology Progress Series*, **286**, 155–166.

Chisholm, J. E., Jones, G. C. and Purvis, O. W. (1987). Hydrated Copper Oxalate, Moolooite, in Lichens. *Mineralogical Magazine*, **51**, 715–718.

Christian, G. (1967). *Ashdown Forest*. Lewes, Sussex: Farncombe & Co. (1928) Ltd.

Chubb, C. J., Dale, R. P. and Stoner, J. H. (1980). Inputs into Swansea Bay. In *Industrialised Embayments and Their Environmental Problems*, ed. M. B. Collins, F. T. Banner, P. A. Tyler, S. J. Wakefield and A. E. James. Oxford: Pergamon Press, pp. 307–327.

Clark, H. F., Hausladen, D. M. and Brabander, D. J. (2008). Urban Gardens: Lead Exposure, Recontamination Mechanisms, and Implications for Remediation Design. *Environmental Research*, **107**, 312–319.

Clarke, K. R. and Ainsworth, M. (1993). A Method of Linking Multivariate Community Structure to Environmental Variables. *Marine Ecology Progress Series*, **92**, 205–219.

Clarke, M. R. (1996). Cephalopods as Prey. III. Cetaceans. *Philosophical Transactions of the Royal Society London B*, **351**, 1053–1065.

Clemens, S. (2006). Toxic Metal Accumulation, Responses to Exposure and Mechanisms of Tolerance in Plants. *Biochimie*, **88**, 1707–1719.

Clements, W. H. (2000). Integrating Effects of Contaminants across Levels of Biological Organization: An Overview. *Journal of Aquatic Ecosystem Stress and Recovery*, **7**, 113–116.

(2004). Small-Scale Experiments Support Causal Relationships between Metal Contamination and Macroinvertebrate Community Responses. *Ecological Applications*, **14**, 954–957.

Clements, W. H. and Newman, M. C. (2002). *Community Ecotoxicology*. Chichester: John Wiley & Sons, Ltd.

Clements, W. H. and Rohr, J. R. (2009). Community Responses to Contaminants: Using Basic Ecological

Principles to Predict Ecotoxicological Effects. *Environmental Toxicology and Chemistry*, **28**, 1789–1800.

Clements, W. H., Carlisle, D. M., Lazorchak, J. M. and Johnson, P. C. (2000). Heavy Metals Structure Benthic Communities in Colorado Mountain Streams. *Ecological Applications*, **10**, 626–638.

Clyde River Purification Board (1984). Concentrations of Trace Metals in Water and Brown Trout from the Glengonnar Water. Leadhills. Clyde River Purification Board Technical Report No. 82.

Cobbett, C. and Goldsbrough, P. (2002). Phytochelatins and Metallothioneins: Roles in Heavy Metal Detoxification and Homeostasis. *Annual Review of Plant Biology*, **53**, 159–182.

Colbourn, P. and Thornton, I. (1978). Lead Pollution in Agricultural Soils. *Journal of Soil Science*, **29**, 513–526.

Colbourn, P., Alloway, B. J. and Thornton, I. (1975). Arsenic and Heavy Metals in Soils Associated with Regional Geochemical Anomalies in South-West England. *Science of the Total Environment*, **4**, 359–363.

Cole, S., Codling, I. D., Parr, W. and Zabel, T. (1999). *Guidelines for Managing Water Quality Impacts within UK European Marine Sites*. Swindon: Water Research Council (WRC).

Condry, W. M. (1981). *The Natural History of Wales*. London: New Naturalist, Collins.

Cooke, J. A. (2011). Cadmium in Small Mammals. In *Environmental Contaminants in Biota: Interpreting Tissue Concentrations*, 2nd edn., ed. W. N. Beyer and J. P. Meador. Boca Raton, FL: Taylor and Francis Books, pp. 627–642.

Cooke, J. A. and Morrey, D. R. (1981). Heavy Metals and Fluoride in Soils and Plants Associated with Metaliferous Mine Wastes in the Northern Pennines. In *Heavy Metals in Northern England: Environmental and Biological Aspects*, ed. P. J. Say and B. A. Whitton. Durham: Department of Botany, University of Durham, pp. 153–164.

Cooper, S., Hare, L. and Campbell, P. G. C. (2010). Subcellular Partitioning of Cadmium in the Freshwater Bivalve, *Pyganodon grandis*, after Separate Short-Term Exposures to Waterborne or Diet-Borne Metal. *Aquatic Toxicology*, **100**, 303–312.

Corp, N. and Morgan, A. J. (1991). Accumulation of Heavy Metals from Polluted Soils by the Earthworm,

Lumbricus rubellus: Can Laboratory Exposure of 'Control' Worms Reduce Biomonitoring Problems? *Environmental Pollution*, **74**, 39–52.

Cotter-Howells, J. and Thornton, I. (1991). Sources and Pathways of Environmental Lead to Children in a Derbyshire Mining Village. *Environmental Geochemistry and Health*, **13**, 127–135.

Coughtrey, P. J. and Martin, M. H. (1976a). The Distribution of Pb, Zn, Cd and Cu within the Pulmonate Mollusc *Helix aspersa* Müller. *Oecologia*, **23**, 315–322.

(1976b). Comparisons between the Levels of Lead, Zinc and Cadmium within a Contaminated Environment. *Chemosphere*, **1**, 15–20.

(1977). Cadmium Tolerance of *Holcus lanatus* from a Site Contaminated by Aerial Outfall. *New Phytologist*, **79**, 273–280.

Coughtrey, P. J., Jones, C. H., Martin, M. H. and Shales, S. W. (1979). Litter Accumulation in Woodlands Contaminated by Pb, Zn, Cd and Cu. *Oecologia*, **39**, 51–60.

Coyle, G. (2010). *The Riches Beneath Our Feet*. Oxford: Oxford University Press.

Creamer, R. E., Rimmer, D. L. and Black, H. I. J. (2008). Do Elevated Soil Concentrations of Metals Affect the Diversity and Activity of Soil Invertebrates in the Long-Term? *Soil Use and Management*, **24**, 37–46.

Croteau, M.-N. and Luoma, S. N. (2008). A Biodynamic Understanding of Dietborne Metal Uptake by a Freshwater Invertebrate. *Environmental Science and Technology*, **42**, 1801–1806.

(2009). Predicting Dietborne Metal Toxicity from Metal Influxes. *Environmental Science and Technology*, **43**, 4915–4921.

Crothers, J. H. (1967). The Biology of the Shore Crab *Carcinus maenas* (L.). *Field Studies*, **2**, 407–434.

Crowder, A. (1991). Acidification, Metals and Macrophytes. *Environmental Pollution*, **71**, 171–203.

Croxall, J. P. and Prince, P. A. (1980). Food, Feeding Ecology and Ecological Segregation of Seabirds at South Georgia. *Biological Journal of the Linnean Society*, **14**, 103–131.

(1996). Cephalopods as Prey. I. Seabirds. *Philosophical Transactions of the Royal Society London B*, **351**, 1023–1043.

Culbard, E. B., Thornton, I., Watt, J., Wheatley, M., Moorcroft, S. and Thompson, M. (1988). Metal

Contamination in British Urban Dusts and Soils. *Journal of Environmental Quality*, **17**, 226–234.

Culshaw, C., Newton, L. C. Weir, I. and Bird, D. J. (2002). Concentrations of Cd, Zn and Cu in Sediments and Brown Shrimp (*Crangon crangon*) from the Severn Estuary and Bristol Channel, UK. *Marine Environmental Research*, **54**, 331–334.

Dafforn, K. A., Lewis, J. A. and Johnston, E. L. (2011). Antifouling Strategies: History and Regulation, Ecological Impacts and Mitigation. *Marine Pollution Bulletin*, **62**, 453–465.

Daka, E. R. and Hawkins, S. J. (2004). Tolerance to Heavy Metals in *Littorina saxatilis* from a Metal Contaminated Estuary in the Isle of Man. *Journal of the Marine Biological Association of the United Kingdom*, **84**, 393– 400.

Dallinger, R., Berger, B., Hunziker, P. E. and Kägi, J. H. R. (1997). Metallothionein in Snail Cd and Cu Metabolism. *Nature*, **388**, 237–238.

Dallinger, R., Chabicovsky, M., Hödl, E., Prem, C., Hunziker, P. and Manzl, C. (2005). Copper in *Helix pomatia* (Gastropoda) Is Regulated by One Single Cell Type: Differently Responsive Metal Pools in Rhogocytes. *American Journal of Physiology – Regulatory, Integrative and Comparative Physiology*, **289**, 1185–1195.

Dang, F. and Wang, W.-X. (2012). Why Mercury Concentration Increases with Fish Size? Biokinetic Explanation. *Environmental Pollution*, **163**, 192–198.

Darlington, S. T. and Gower, A. M. (1990). Location of Copper in Larvae of *Plectrocnemia conspersa* (Curtis) (Trichoptera) Exposed to Elevated Metal Concentrations in a Mine Drainage Stream. *Hydrobiologia*, **196**, 91–100.

Darlington, S. T., Gower, A. M. and Ebdon, L. (1987). Studies on *Plectrocnemia conspersa* (Curtis) in Copper Contaminated Streams in South West England. In *Series Entomologica, Vol. 39: Proceedings of the 5th International Symposium on Trichoptera, Lyon, France, July 1986*, ed. M. Bournaud and H. Trachet. Netherlands: Springer, pp. 353–357.

Dauwe, T., Bervoets, L., Blust, R., Pinxten, R. and Eens, M. (1999). Are Eggshells and Egg Contents of Great and Blue Tits Suitable as Indicators of Heavy Metal Pollution? *Belgian Journal of Zoology*, **129**, 439–447.

Dauwe, T., Janssens, E., Bervoets, L., Blust, R. and Eens, M. (2005a). Heavy-Metal Concentrations in Female Laying Great Tits (*Parus major*) and Their Clutches. *Archives of Environmental Contamination and Toxicology*, **49**, 249–256.

Dauwe, T., Janssens, E., Pinxten, R. and Eens, M. (2005b). The Reproductive Success and Quality of Blue Tits (*Parus caeruleus*) in a Heavy Metal Pollution Gradient. *Environmental Pollution*, **136**, 243–251.

Davies, A. G. and Sleep, J. A. (1979). Photosynthesis in Some British Coastal Waters May Be Inhibited by Zinc Pollution. *Nature*, **277**, 292–293.

Davies, B. E. (1971). Trace Metal Content of Soils Affected by Base Metal Mining in the West of England. *Oikos*, **22**, 366–372.

(1987). Consequences of Environmental Contamination by Lead Mining in Wales. *Hydrobiologia*, **149**, 213–220.

Davies, B. E. and Ballinger, R. C. (1990). Heavy Metals in Soils in North Somerset, England, with Special Reference to Contamination from Base Metal Mining in the Mendips. *Environmental Geochemistry and Health*, **12**, 291–300.

Davies, B. E. and Ginnever, R. C. (1979). Trace Metal Contamination of Soils and Vegetables in Shipham, Somerset. *Journal of Agricultural Science*, **93**, 753–756.

Davies, B. E. and Roberts, L. J. (1978). The Distribution of Heavy Metal Contaminated Soils in Northeast Clwyd, Wales. *Water, Air, and Soil Pollution*, **9**, 507–518.

DEFRA and Environment Agency (2002). *The Contaminated Land Exposure Assessment Model (CLEA): Technical Basis and Algorithms. R & D Publication CLR 10. ISBN 1 857 05749 X*. Bristol: Environment Agency.

De Jonge, M., Belpaire, C., Geerarts, C., De Cooman, W., Blust, R. and Bervoets, L. (2012). Ecological Impact Assessment of Sediment Remediation in a Metal-Contaminated Lowland River Using Translocated Zebra Mussels and Resident Macroinvertebrates. *Environmental Pollution*, **171**, 99–108.

De Jonge, M., Blust, R. and Bervoets, L. (2010). The Relation between Acid Volatile Sulphides (AVS) and Metal Accumulation in Aquatic Invertebrates: Implications of Feeding Behaviour and Ecology. *Environmental Pollution*, **158**, 1381–1391.

De Jonge, M., Dreesen, F., De Paepe, J., Blust, R. and Bervoets, L. (2009). Do Acid Volatile Sulphides (AVS) Influence the Accumulation of Sediment-Bound Metals to Benthic Invertebrates under Natural Field Conditions? *Environmental Science and Technology*, 43, 4510–4516.

De Jonge, M., Lofts, S., Bervoets, L. and Blust, R. (2014). Relating Metal Exposure and Chemical Speciation to Trace Metal Accumulation in Aquatic Insects under Natural Field Conditions. *Science of the Total Environment*, 496, 11–21.

De Jonge, M., Tipping, E., Lofts, S., Bervoets, L. and Blust, R. (2013). The Use of Invertebrate Body Burdens to Predict Ecological Effects of Metal Mixtures in Mining-Impacted Waters. *Aquatic Toxicology*, 142–143, 294–302.

De Jonge, M., Van de Vijver, B., Blust, R. and Bervoets, L. (2008). Responses of Aquatic Organisms to Metal Pollution in a Lowland River in Flanders: A Comparison of Diatoms and Macroinvertebrates. *Science of the Total Environment*, 407, 615–629.

Del Val, C., Barea, J. M. and Azcón-Aguilar, C. (1999). Diversity of Arbuscular Mycorrhizal Fungus Populations in Heavy-Metal-Contaminated Soils. *Applied and Environmental Microbiology*, 65, 718–723.

Department of Environment (1989). *Code of Practice for Agricultural Use of Sewage Sludge*. London: HMSO. http://adlib.everysite.co.uk/resources/000/247/164/sludge-report.pdf

De Pauw, N. and Heylen, S. (2001). Biotic Index for Sediment Quality Assessment of Watercourses in Flanders, Belgium. *Aquatic Ecology*, 35, 121–133.

de Vries, W., Römkens, F. A. M. and Schütze, G. (2007). Critical Soil Concentrations of Cadmium, Lead and Mercury in View of Health Effects on Humans and Animals. *Reviews of Environmental Contamination and Toxicology*, 191, 91–130.

Dines, H. G. (1969). *The Metalliferous Mining Region of South-West England, Vols. I and II.* 2nd Impression with Amendments. London: Her Majesty's Stationery Office.

Di Toro, D. M., Mahony, J. D., Hansen, D. J., Scott, K. J., Carlson, A. R. and Ankley, G. T. (1992). Acid Volatile Sulfide Predicts the Acute Toxicity of Cadmium and Nickel in Sediments. *Environmental Science and Technology*, 26, 96–101.

Di Toro, D. M., Zarba, C. S., Hansen, D. J., et al. (1991). Technical Basis for Establishing Sediment Quality Criteria for Nonionic Organic Chemicals Using Equilibrium Partitioning. *Environmental Toxicology and Chemistry*, 10, 1541–1583.

Di Veroli, A., Santoro, F., Pallatoni, M., et al. (2014). Deformities of Chironomid Larvae and Heavy Metal Pollution: From Laboratory to Field Studies. *Chemosphere*, 112, 9–17.

Dmowski, K. (1999). Birds as Bioindicators of Heavy Metal Pollution: Review and Examples Concerning European Species. *Acta Ornithologica*, 34, 1–25.

Donat, J. R. and Bruland, K. W. (1995). Trace Elements in the Oceans. In *Trace Elements in Natural Waters*, ed. B. Salbu and E. Steinnes. Boca Raton, FL: CRC Press, pp. 47–281.

Dorrington, V. H. and Pyatt, F. B. (1982). History, Ecology and Physical and Chemical Nature of Two Metalliferous Spoil Tips in South-West England. *International Journal of Environmental Studies*, 18, 177–185.

Downing, M. G., Staff, M. G. and Sheldrake, P. J. (1998). The Legacy of Metalliferous Mining in Great Britain. In *GREEN 2. Contaminated and Derelict Land*, ed. R. W. Sarsby. London: Thomas Telford, pp. 52–59.

Downs, S. G., Macleod, C. L., Jarvis, K., Birkett, J. W. and Lester, J. N. (1999). Comparison of Mercury Bioaccumulation in Eel (*Anguilla anguilla*) and Roach (*Rutilus rutilus*) from River Systems in East Anglia, UK: 1. Concentrations in Fish Tissue. *Environmental Technology*, 20, 1189–1200.

Dowson, P. H., Bubb, J. M. and Lester, J. N. (1992). Organotin Distribution in Sediments and Waters of Selected East Coast Estuaries in the UK. *Marine Pollution Bulletin*, 24, 492–498.

(1993). Temporal Distributions of Organotins in the Aquatic Environment: Five Years after the 1987 UK Retail Ban on TBT Based Antifouling Paints. *Marine Pollution Bulletin*, 26, 487–494.

Drescher, H. E., Harms, U. and Huschenbeth, E. (1977). Organochlorines and Heavy Metals in the Harbour Seal *Phoca vitulina* from the German North Sea Coast. *Marine Biology*, 41, 99–106.

Drobne, D. and Hopkin, S. P. (1995). The Toxicity of Zinc to Terrestrial Isopods in a "Standard" Laboratory Test. *Ecotoxicology and Environmental Safety*, 31, 1–6.

Duquesne, S., Newton, L. C., Giusti, L., Marriott, S. B., Stärk, H.-J. and Bird, D. J. (2006). Evidence for Declining Levels of Heavy Metals in the Severn Estuary and Bristol Channel, U.K. and Their Spatial Distribution in Sediments. *Environmental Pollution*, **143**, 187–196.

Durou, C., Poirier, L., Amiard, J.-C., Budzinski, H., Gnassia-Barelli, M., Lemenech, K., Peluhet, L., Mouneyrac, C., Roméo, M. and Amiard-Triquet, C. (2007b). Biomonitoring in a Clean and a Multi-Contaminated Estuary Based on Biomarkers and Chemical Analyses in the Endobenthic Worm *Nereis diversicolor*. *Environmental Pollution*, **148**, 445–458.

Durou, C., Smith, B. D., Roméo, M., Rainbow, P. S., Mouneyrac, C., Mouloud, M., Gnassia-Barelli, M., Gillet, P., Deutch, B. and Amiard-Triquet, C. (2007a). From Biomarkers to Population Responses in *Nereis diversicolor:* Assessment of Stress in Estuarine Ecosystems. *Ecotoxicology and Environmental Safety*, **66**, 402–411.

Durrant, C. J., Stevens, J. R., Hogstrand, C. and Bury, N. R. (2011). The Effect of Metal Pollution on the Population Genetic Structure of Brown Trout (*Salmo trutta* L.) Residing in the River Hayle, Cornwall, UK. *Environmental Pollution*, **159**, 3595–3603.

Edington, J. M. and Hildrew, A. G. (1981). *A Key to the Caseless Caddis Larvae of the British Isles with Notes on Their Ecology*. Freshwater Biological Association Scientific Publication No. 43. Kendal: Titus Wilson & Son Ltd.

Edwards, K. C., Swinnerton, H. H. and Hall, R. H. (1962). *The Peak District*. London: Collins, New Naturalist.

Eisler, R. (1981). *Trace Metal Concentrations in Marine Organisms*. New York, NY: Pergamon Press.

Elderfield, H., Hepworth, A., Edwards, P. N. and Holliday, L. M. (1979). Zinc in the Conwy River and Estuary. *Estuarine and Coastal Marine Science*, **9**, 403–422.

Elderfield, H., Thornton, L. and Webb, J. S. (1971). Heavy Metals and Oyster Culture in Wales. *Marine Pollution Bulletin*, **2**, 44–47.

Elliott, M. and Griffiths, A. H. (1986). Mercury Contamination in Components of an Estuarine Ecosystem. *Water, Science and Technology*, **18**, 161–170.

Emmerson, R. H. C., O'Reilly-Wiese, S. B., Macleod, C. L. and Lester, J. N. (1997). A Multivariate Assessment of Metal Distribution in Inter-Tidal Sediments of the Blackwater Estuary, UK. *Marine Pollution Bulletin*, **34**, 960–968.

Endean, R. (1955). Studies on the Blood and Tests of Some Australian Ascidians. III. The Formation of the Test of *Pyura stolonifera* (Heller). *Australian Journal of Marine and Freshwater Research*, **6**, 157–164.

(1961). The Test of the Ascidian, *Phallusia mammillata*. *Quarterly Journal of Microscopical Science*, **102**, 107–117.

Environment Agency (2008a). *Environmental Quality Standards for Trace Metals in the Aquatic Environment*. Environment Agency Science Report SC030194. Bristol: Environment Agency.

(2008b). *Assessment of Metal-Mining Contaminated River Sediments in England and Wales*. Environment Agency Science Project Report SC030136/SR4. Bristol: Environment Agency.

(2009a). *Ecological Indicators for Abandoned Mines, Phase 1: Review of the Literature*. Environment Agency Science Project Report SC030136/R49. Bristol: Environment Agency.

(2009b). *Using Soil Guideline Values*. Science Report SC050021/SGV Introduction. Bristol, UK: Environment Agency. www.gov.uk/government/uploads/system/uploads/attachment_data/file/297676/scho0309bpqm-e-e.pdf

(2009c). *Updated Technical Background to the CLEA Model*. Science Report SC050021/SR3. Bristol: Environment Agency. www.gov.uk/government/uploads/system/uploads/attachment_data/file/297676/scho0309bpqm-e-e.pdf

Eraly, D., Hendrickx, F., Backeljau, T., Bervoets, L. and Lens, L. (2011). Direct and Indirect Effects of Metal Stress on Physiology and Life History Variation in Field Populations of a Lycosid Spider. *Ecotoxicology and Environmental Safety*, **74**, 1489–1497.

Eraly, D., Hendrickx, F., Bervoets, L. and Lens, L. (2010). Experimental Exposure to Cadmium Affects Metallothionein-Like Protein Levels but Not Survival and Growth in Wolf Spiders from Polluted and Reference Populations. *Environmental Pollution*, **158**, 2124–2131.

Erry, B. V., Macnair, M. R., Meharg, A. A. and Shore, R. F. (2000). Arsenic Contamination in Wood Mice (*Apodemus sylvaticus*) and Bank Voles (*Clethrionomys glareolus*) on Abandoned Mine Sites in Southwest Britain. *Environmental Pollution*, **110**, 179–187.

Erry, B. V., Macnair, M. R., Meharg, A. A., Shore, R. F. and Newton, I. (1999). Arsenic Residues in Predatory Birds from an Area of Britain with Naturally and Anthropogenically Elevated Arsenic Levels. *Environmental Pollution*, **106**, 91–95.

European Food Safety Authority (2008). Scientific Opinion of the Panel on Plant Protection Products and Their Residues on a Request from the EFSA PRAPeR Unit on Risk Assessment for Birds and Mammals. *European Food Safety Authority Journal*, **734**, 1–181.

Evans, P. R. and Moon, S. J. (1981). Heavy Metals in Shorebirds and Their Prey in North-East England. In *Heavy Metals in Northern England: Environmental and Biological Aspects*, ed. P. J. Say and B. A. Whitton. Durham: Department of Botany, University of Durham, pp. 181–190.

Evans, P. R., Uttley, J. D., Davidson, N. C. and Ward, P. (1987). Shorebirds (S.Os Charadrii and Scolopaci) as Agents of Transfer of Heavy Metals within and between Estuarine Ecosystems. In *Pollutant Transport and Fate in Ecosystems, Special Publication Number 6 of the British Ecological Society*, ed. P. J. Coughtrey, M. H. Martin and M. H. Unsworth. Oxford: Blackwell Scientific Publications, pp. 337–352.

Evans, S. M., Birchenough, A. C. and Brancato, M. S. (2000). The TBT Ban: Out of the Frying Pan into the Fire. *Marine Pollution Bulletin*, **40**, 204–211.

Evans, S. M., Evans, P. M. and Leksono, T. (1996). Widespread Recovery of Dogwhelks, *Nucella lapillus* (L.), from Tributyltin Contamination in the North Sea and Clyde Sea. *Marine Pollution Bulletin*, **32**, 263–269.

Evans, S. M., Hutton, A., Kendall, M. and Samosir, A. M. (1991). Recovery in Populations of Dogwhelks *Nucella lapillus* (L.) Suffering from Imposex. *Marine Pollution Bulletin*, **22**, 331–333.

Everaarts, J. M. and Fischer, C. V. (1992). The Distribution of Heavy Metals (Cu, Zn, Cd, Pb) in the Fine Fraction of Surface Sediments of the North Sea. *Netherlands Journal of Sea Research*, **29**, 232–331.

Everard, M. and Denny, P. (1984). The Transfer of Lead by Freshwater Snails in Ullswater, Cumbria. *Environmental Pollution A*, **35**, 299–314.

Everett, J. L., Day, C. L. and Reynolds, D. (1967). Comparative Survey of Lead at Selected Sites in the British Isles in Relation to Air Pollution. *Food and Cosmetics Toxicology*, **5**, 29–35.

Falasco, E., Bona, F., Badino, G., Hoffmann, L. and Ector, L. (2009). Diatom Teratological Forms and Environmental Alterations: A Review. *Hydrobiologia*, **623**, 1–35.

Falconer, C. R., Davies, I. M. and Topping, G. (1983). Trace Metals in the Common Porpoise, *Phocoena phocoena*. *Marine Environmental Research*, **8**, 119–127.

Faria, M. S., Lopes, R. J., Malcato, J., Nogueira, A. J. A. and Soares, A. M. V. M. (2008). In Situ Bioassays with *Chironomus riparius* Larvae to Biomonitor Metal Pollution in Rivers and to Evaluate the Efficiency of Restoration Measures in Mine Areas. *Environmental Pollution*, **151**, 213–221.

Fattorini, D. and Regoli, F. (2004). Arsenic Speciation in Tissues of the Mediterranean Polychaete *Sabella spallanzanii*. *Environmental Toxicology and Chemistry*, **23**, 1881–1887.

(2012). Hyper-Accumulation of Vanadium in Polychaetes. In *Vanadium: Biochemical and Molecular Biological Approaches*, ed. H. Michibata. Dordrecht: Springer Science, pp. 73–92.

Fattorini, D., Notti, A., Halt, M. N., Gambi, M. N. and Regoli, F. (2005). Levels and Chemical Speciation of Arsenic in Polychaetes: A Review. *Marine Ecology*, **26**, 255–264.

Fattorini, D., Notti, A., Nigro, M. and Regoli, F. (2010). Hyperaccumulation of Vanadium in the Antarctic Polychaete *Perkinsiana littoralis* as a Natural Chemical Defense against Predation. *Environmental Science and Pollution Research*, **17**, 220–228.

Fialkowski, W. and Rainbow, P. S. (2006). The Discriminatory Power of Two Biomonitors of Trace Metal Bioavailabilities in Freshwater Streams. *Water Research*, **40**, 1805–1810.

Fialkowski, W., Calosi, P., Dahlke, S., et al. (2009). The Sandhopper *Talitrus saltator* (Crustacea: Amphipoda) as a Biomonitor of Trace Metal Bioavailabilities in European Coastal Waters. *Marine Pollution Bulletin*, **58**, 39–44.

Fialkowski, W. F., Klonowska-Olejnik, M., Smith, B. D. and Rainbow, P. S. (2003). Mayfly Larvae (*Baetis rhodani* and *B. vernus*) as Biomonitors of Trace Metal Pollution in Streams of a Catchment Draining a Zinc and Lead Mining Area of Upper Silesis, Poland. *Environmental Pollution*, **121**, 253–267.

Ficetola, G. F., Miaud, C., Pompanon, F. and Taberlet, P. (2008). Species Detection Using Environmental DNA from Water Samples. *Biology Letters*, **4**, 423–425.

Fisher, I. J., Pain, D. J. and Thomas, V. G. (2006). A Review of Lead Poisoning from Ammunition Sources in Terrestrial Birds. *Biological Conservation*, **131**, 421–432.

Fisher, N. S., Stupakoff, I., Sañudo-Wilhelmy, S., Wang, W.-X., Teyssié, J.-L., Fowler, S. W. and Crusius, J. (2000). Trace Metals in Marine Copepods: A Field Test of a Bioaccumulation Model Coupled to Laboratory Uptake Kinetics Data. *Marine Ecology Progress Series*, **194**, 211–218.

Florence, T. M. (1977). Trace Metal Species in Fresh Waters. *Water Research*, **11**, 681–687.

Ford, G. C., Harrison, P. M., Rice, D. W., et al. (1984). Ferritin: Design and Formation of an Iron-Storage Molecule. *Philosophical Transactions of the Royal Society London B*, **304**, 551–565.

Fordyce, F. (2007). Selenium Geochemistry and Health. *Ambio*, **36**, 94–97.

Förstner, U. and Wittmann, G. T. W. (1983). *Metal Pollution in the Aquatic Environment*, 2nd edn. Berlin: Springer-Verlag.

Fossi Tankoua, O., Buffet, P. E., Amiard, J.-C., et al. (2012). Intersite Variation of a Battery of Biomarkers at Different Levels of Biological Organisation in the Estuarine Endobenthic Worm *Nereis diversicolor* (Polychaeta, Nereididae). *Aquatic Toxicology*, **114–115**, 96–103.

Foster, P. (1976). Concentrations and Concentration Factors of Heavy Metals in Brown Algae. *Environmental Pollution*, **10**, 45–53.

Foster, P., Hunt, D. T. E. and Morris, A. W. (1978). Metals in an Acid Mine Stream and Estuary. *Science of the Total Environment*, **9**, 75–86.

Foster, P. L. (1977). Copper Exclusion as a Mechanism of Heavy Metal Tolerance in a Green Alga. *Nature*, **269**, 322–323.

(1982a). Species Associations and Metal Contents of Algae from Rivers Polluted by Heavy Metals. *Freshwater Biology*, **12**, 17–39.

(1982b). Metal Resistances of Chlorophyta from Rivers Polluted by Heavy Metals. *Freshwater Biology*, **12**, 41–61.

Foulds, S. A., Brewer, P. A., Macklin, M. G., Haresign, W., Betson, R. E. and Rassner, S. M. E. (2014). Flood-Related Contamination in Catchments Affected by Historical Metal Mining: An Unexpected and Emerging Hazard of Climate Change. *Science of the Total Environment*, **476–477**, 165–180.

Fowler, S. W. (1977). Trace Elements in Zooplankton Particulate Products. *Nature*, **269**, 51–53.

(1990). Critical Review of Selected Heavy Metal and Chlorinated Hydrocarbon Concentrations in the Marine Environment. *Marine Environmental Research*, **29**, 1–64.

Frankland, E. and Morton, J. C. (1873). *Fifth Report of the Rivers Pollution Commission*. London: HMSO.

Franklin, A. (1987). The Concentrations of Metals, Organochlorine Pesticide and PCB Residues in Marine Fish and Shellfish: Results from MAFF Fish and Shellfish Monitoring Programmes, 1977–84. *MAFF Aquatic Monitoring Report*, **16**, 1–38.

Franson, J. C. and Pain, D. J. (2011). Lead in Birds. In *Environmental Contaminants in Biota: Interpreting Tissue Concentrations*, 2nd edn., ed. W. N. Beyer and J. P. Meador. Boca Raton, FL: Taylor and Francis Books, pp. 563–593.

Fraústo da Silva, J. J. R. and Wiliams, R. J. P. (1993). *The Biological Chemistry of the Elements*. Oxford: Clarendon Press.

Furness, R. W. (2015). Birds as Monitors of Mercury Pollution. In *Fundamentals of Ecotoxicology*, 4th edn., ed. M. C. Newman. Boca Raton, FL: CRC Press, pp. 160–165.

Furness, R. and Hutton, M. (1979). Pollutant Levels in the Great Skua *Catharacta skua*. *Environmental Pollution*, **19**, 261–268.

(1980). Pollutants and Impaired Breeding of Great Skuas *Catharacta skua* in Britain. *Ibis*, **122**, 88–94.

Gabriels, W., Lock, K., De Pauw, N. and Goethals, P. L. M. (2010). Multimetric Macroinvertebrate Index Flanders (MMIF) for Biological Assessment of Rivers and Lakes in Flanders (Belgium). *Limnologica*, **40**, 199–207.

Gajdosechova, Z., Brownlow, A., Cottin, N. T., et al. (2016). Possible Link between Hg and Cd Accumulation in the Brain of Long-Finned Pilot Whales (*Globicephala melas*). *Science of the Total Environment*, **545–546**, 407–413.

Galay Burgos, M. and Rainbow, P. S. (1998). Uptake, Accumulation and Excretion by *Corophium volutator* (Crustacea: Amphipoda) of Zinc, Cadmium and Cobalt Added to Sewage Sludge. *Estuarine, Coastal and Shelf Science*, **47**, 603–620.

(2001). Availability of Cadmium and Zinc from Sewage Sludge to the Flounder, *Platichthys flesus*, via a Marine Food Chain. *Marine Environmental Research*, 51, 417–439.

Gallien, I., Caurant, F., Bordes, M., et al. (2001). Cadmium-Containing Granules in Kidney Tissue of the Atlantic White-Sided Dolphin (*Lagenorhynchus acutus*) off the Faroe Islands. *Comparative Biochemistry and Physiology C*, 130, 389–395.

Gao, S. and Wang, W.-X. (2014). Oral Bioaccessibility of Toxic Metals in Contaminated Oysters and Relationships with Metal Internal Sequestration. *Ecotoxicology and Environmental Safety*, 110, 261–268.

Gao, Y. and Bradshaw, A. D. (1995). The Containment of Toxic Wastes: II. Metal Movement in Leachate and Drainage at Parc Lead–Zinc Mine, North Wales. *Environmental Pollution*, 90, 379–382.

Garty, J. (2001). Biomonitoring Atmospheric Heavy Metals with Lichens: Theory and Application. *Critical Reviews in Plant Sciences*, 20, 309–371.

Gault, N. F. S., Tolland, E. L. C. and Parker, J. G. (1983). Spatial and Temporal Trends in Heavy Metal Concentrations in Mussels from Northern Ireland Coastal Waters. *Marine Biology*, 77, 307–316.

Geens, A., Dauwe, T., Bervoets, L., Blust, R. and Evans, M. (2010). Haematological Status of Wintering Great Tits (*Parus major*) along a Metal Pollution Gradient. *Science of the Total Environment*, 408, 1174–1179.

Geffard, A., Amiard, J.-C. and Amiard-Triquet, C. (2002). Use of Metallothionein in Gills from Oysters (*Crassostrea gigas*) as a Biomarker: Seasonal and Intersite Variations. *Biomarkers*, 7, 123–137.

Geffard, A., Amiard-Triquet, C., Amiard, J.-C. and Mouneyrac, C. (2001). Temporal Variations of Metallothionein and Metal Concentrations in the Digestive Gland of Oysters (*Crassostrea gigas*) from a Clean and a Metal-Rich Site. *Biomarkers*, 6, 91–107.

Geffard, A., Smith, B. D., Amiard-Triquet, C., Jeantet, A. Y. and Rainbow, P. S. (2005). Kinetics of Trace Metal Accumulation and Excretion in the Polychaete *Nereis diversicolor*. *Marine Biology*, 147, 1291–1304.

Geiszinger, A. E., Goessler, W. and Francesconi, K. (2002). Biotransformation of Arsenate to the Tetramethylarsonium Ion in the Marine Polychaetes *Nereis diversicolor* and *Nereis virens*. *Environmental Science and Technology*, 36, 2905–2910.

Gensemer, R. W. and Playle, R. C. (1999). The Bioavailability and Toxicity of Aluminum in Aquatic Environments. *Critical Reviews in Environmental Science and Technology*, 29, 315–450.

George, S. G. (1983). Heavy Metal Detoxication in the Mussel *Mytilus edulis* – Composition of Cd-Containing Kidney Granules (Tertiary Lysosomes). *Comparative Biochemistry and Physiology*, 76C, 53–57.

George, S. G. and Pirie, B. J. S. (1980). Metabolism of Zinc in the Mussel *Mytilus edulis* (L.): A Combined Ultrastructural and Biochemical Study. *Journal of the Marine Biological Association of the United Kingdom*, 60, 575–590.

George, S. G., Carpene, E., Coombs, T. L., Overnell, J. and Youngson, A. (1979). Characterisation of Cadmium-Binding Proteins from Mussels, *Mytilus edulis* (L.) Exposed to Cadmium. *Biochimica et Biophysica Acta*, 580, 225–233.

George, S. G., Pirie, B. J. S., Cheyne, A. R., Coombs, T. L. and Grant, P. T. (1978). Detoxication of Metals by Marine Bivalves: An Ultrastructural Study of the Compartmentation of Copper and Zinc in the Oyster *Ostrea edulis*. *Marine Biology*, 45, 147–156.

George, S. G., Pirie, B. J. S. and Coombs, T. L. (1980). Isolation and Elemental Analysis of Metal-Rich Granules from the Kidney of the Scallop, *Pecten maximus* (L.). *Journal of Experimental Marine Biology and Ecology*, 42, 143–156.

Giangrande, A., Licciano, M., Schirosi, R., Musco, L. and Stabili, L. (2013). Chemical and Structural Defensive External Strategies in Six Sabellid Worms (Annelida). *Marine Ecology*, 35, 36–45.

Gibb, J. O. T., Svendsen, C., Weeks, J. M. and Nicholson, J. K. (1997). ^{1}H NMR Spectroscopic Investigations of Tissue Metabolite Biomarker Response to Cu(II) Exposure in Terrestrial Invertebrates: Identification of Free Histidine as a Novel Biomarker of Exposure to Copper in Earthworms. *Biomarkers*, 2, 295–302.

Gibbs, P. E. and Bryan, G. W. (1980). Copper: The Major Component of Glycerid Polychaete Jaws. *Journal of the Marine Biological Association of the United Kingdom*, 60, 205–214.

(1984). Calcium Phosphate Granules in Muscle Cells of *Nephtys* (Annelida, Polychaeta): A Novel Skeleton? *Nature*, 310, 494–495.

(1986). Reproductive Failure in Populations of the Dog-Whelk, *Nucella lapillus*, Caused by Imposex Induced by Tributyltin from Antifouling Paints. *Journal of the Marine Biological Association of the United Kingdom*, **66**, 767–777.

Gibbs, P. E., Bryan, G. W., Pascoe, P. L. and Burt, G. R. (1987). The Use of the Dog-Whelk, *Nucella lapillus*, as an Indicator of Tributyltin (TBT) Contamination. *Journal of the Marine Biological Association of the United Kingdom*, **67**, 507–523.

Gibbs, P. E., Bryan, G. W. and Ryan, K. P. (1981). Copper Accumulation by the Polychaete *Melinna palmata*: An Antipredation Mechanism? *Journal of the Marine Biological Association of the United Kingdom*, **61**, 707–722.

Gibbs, P. E., Langston, W. J., Burt, G. R. and Pascoe, P. L. (1983). *Tharyx marioni* (Polychaeta): A Remarkable Accumulator of Arsenic. *Journal of the Marine Biological Association of the United Kingdom*, **63**, 313–325.

Gibbs, P. E., Pascoe, P. L. and Burt, G. R. (1988). Sex Change in the Female Dog-Whelk, *Nucella lapillus*, Induced by Tributyltin from Antifouling Paints. *Journal of the Marine Biological Association of the United Kingdom*, **68**, 715–731.

Giesy, J. P., Verbrugge, D. A., Othout, R. A., et al. (1994). Contaminants in Fishes from Great Lakes–influenced Sections and above Dams of Three Michigan Rivers. II. Implications for Health of Mink. *Archives of Environmental Contamination and Toxicology*, **27**, 213–223.

Gilbert, O. (2000). *Lichens*. London: New Naturalist, Collins.

Giller, K. E., Witter, E. and McGrath, S. P. (1998). Toxicity of Heavy Metals to Microorganisms and Microbial Processes in Agricultural Soils: A Review. *Soil Biology and Biochemistry*, **30**, 1389–1414.

Gillet, P., Mouloud, M., Durou, C. and Deutsch, B. (2008). Response of *Nereis diversicolor* Population (Polychaeta, Nereididae) to the Pollution Impact. *Estuarine, Coastal and Shelf Science*, **76**, 201–210.

Gimbert, F., Vijver, M. G., Coeurdassier, M., et al. (2008). How Subcellular Partitioning Can Help to Understand Heavy Metal Accumulation and Elimination Kinetics in Snails. *Environmental Toxicology and Chemistry*, **27**, 1284–1292.

Giska, I., van Gestel, C. A. M., Skip, B. and Laskowski, R. (2014). Toxicokinetics of Metals in the Earthworm *Lumbricus rubellus* Exposed to Natural Polluted Soils: Relevance of Laboratory Tests to the Field Situation. *Environmental Pollution*, **190**, 123–132.

Goede, A. A. (1985). Mercury, Selenium, Arsenic and Zinc in Waders from the Dutch Wadden Sea. *Environmental Pollution A*, **37**, 287–309.

Gonçalves, M. T., Gonçalves, S. C., Portugal, A., Silva, S., Sousa, J. P. and Freitas, H. (2007). Effects of Nickel Hyperaccumulation in *Alyssum pintodasilvae* on Model Arthropods Representatives of Two Trophic Levels. *Plant and Soil*, **293**, 177–188.

Goodman, G. T. and Roberts, T. M. (1971). Plants and Soils as Indicators of Metals in the Air. *Nature*, **231**, 287–292.

Gore, F., Fawell, J. and Bartram, J. (2010). Too Much or Too Little? A Review of the Conundrum of Selenium. *Journal of Water and Health*, **8**, 405–416.

Gower, A. M. and Darlington, S. T. (1990). Relationships between Copper Concentrations in Larvae of *Plectrocnemia conspersa* (Curtis) (Trichoptera) and in Mine Drainage Streams. *Environmental Pollution*, **65**, 155–168.

Gower, A. M., Myers, G., Kent, M. and Foulkes, M. E. (1994). Relationships between Macroinvertebrate Communities and Environmental Variables in Metal-Contaminated Streams in South-West England. *Freshwater Biology*, **32**, 199–221.

Grant, A. and Middleton, R. (1990). An Assessment of Metal Contamination of Sediments in the Humber Estuary, U.K. Estuarine, *Coastal and Shelf Science*, **31**, 71–85.

Grant, A. and Millward, R. N. (1997). Detecting Community Responses to Pollution. In *Responses of Marine Organisms to Their Environment. Proceedings of the 30th European Marine Biology Symposium*, ed. L. E. Hawkins, S. Hutchinson, A. C. Jensen, M. Sheader and J. A. Williams. Southampton: Southampton Oceanography Centre, University of Southampton, pp. 201–209.

Grant, A., Hateley, J. G. and Jones, N. V. (1989). Mapping the Ecological Impact of Heavy Metals in the Estuarine Polychaete *Nereis diversicolor* Using Inherited Metal Tolerance. *Marine Pollution Bulletin*, **20**, 235–238.

Gray, N. F. (1996). The Use of an Objective Index for the Assessment of the Contamination of Surface Water and Groundwater by Acid Mine Drainage. *Journal of the Chartered Institution of Water and Environmental Management*, **10**, 332–340.

(1997). Environmental Impact and Remediation of Acid Mine Drainage: A Management Problem. *Environmental Geology*, **30**, 62–71.

Gray, N. F. and Delaney, E. (2008). Comparison of Benthic Macroinvertebrate Indices for the Assessment of the Impact of Acid Mine Drainage on an Irish River below an Abandoned Cu-S Mine. *Environmental Pollution*, **155**, 31–40.

Green, J. (1968). *The Biology of Estuarine Animals*. London: Sidgwick and Jackson.

Gregory, R. P. G. and Bradshaw, A. D. (1965). Heavy Metal Tolerance in Populations of *Agrostis tenuis* Sibth. and Other Grasses. *New Phytologist*, **64**, 131–135.

Griffith, J. J. (1919). Influence of Mines upon Land and Livestock in Cardiganshire. *Journal of Agricultural Science*, **9**, 366–395.

Grosell, M. and Brix, K. V. (2009). High Net Calcium Uptake Explains the Hypersensitivity of the Freshwater Snail, *Lymnaea stagnalis*, to Chronic Lead Exposure. *Aquatic Toxicology*, **91**, 302–311.

Halcrow, W., Mackay, D. W. and Thornton, I. (1973). The Distribution of Trace Metals and Fauna in the Firth of Clyde in Relation to the Disposal of Sewage Sludge. *Journal of the Marine Biological Association of the United Kingdom*, **53**, 721–739.

Harding, J. P. C. and Whitton, B. A. (1976). Resistance to Zinc of *Stigeoclonium tenue* in the Field and in the Laboratory. *British Phycological Journal*, **11**, 417–426.

Hardisty, M. W., Kartar, S. and Sainsbury, M. (1974). Dietary Habits and Heavy Metal Concentrations in Fish from the Severn Estuary and Bristol Channel. *Marine Pollution Bulletin*, **5**, 61–63.

Hargreaves, J. W., Lloyd, E. J. H. and Whitton, B. A. (1975). Chemistry and Vegetation of Highly Acidic Streams. *Freshwater Biology*, **5**, 563–576.

Harley, M. B. (1950). Occurrence of a Filter Feeding Mechanism in the Polychaete *Nereis diversicolor*. *Nature*, **165**, 734–735.

Harmens, H., Ilyin, I., Mills, G., et al. (2012). Country-Specific Correlations across Europe between Modelled Atmospheric Cadmium and Lead Deposition and Concentrations in Mosses. *Environmental Pollution*, **166**, 1–9.

Harms, U., Drescher, H. E. and Huschenbeth, E. (1978). Further Data on Heavy Metals and Organochlorines in Marine Mammals from German Coastal Waters. *Meeresforsch*, **26**, 153–161.

Harper, D. J. (1988). Dissolved Cadmium and Lead in the Thames Estuary. *Marine Pollution Bulletin*, **19**, 535–538.

Harris, J. R. (1952). *An Angler's Entomology*. London: New Naturalist, Collins.

Harrison Matthews, L. (1960). *British Mammals*. London: New Naturalist, Collins.

Harrison, P. M. and Hoare, R. J. (1980). *Metals in Biochemistry*. London: Chapman & Hall.

Hartley, W., Dickinson, N. M., Clemente, R., et al. (2009). Arsenic Stability and Mobilization in Soil at an Amenity Grassland Overlying Chemical Waste (St. Helens, UK). *Environmental Pollution*, **157**, 847–856.

Hateley, J. G., Grant, A. and Jones, N. V. (1989). Heavy Metal Tolerance in Estuarine Populations of *Nereis diversicolor*. In *Reproduction, Genetics and Distribution of Marine Organisms, Proceedings 23rd European Marine Biological Symposium*, ed. J. S. Ryland and P. A. Tyler. Fredensborg: Olsen and Olsen, pp. 379–385.

Hawkins, S. J., Gibbs, P. E., Pope, N. D., et al. (2002). Recovery of Polluted Ecosystems: The Case for Long-Term Studies. *Marine Environmental Research*, **54**, 215–222.

Hayward, P. J. (1995). Introduction. In *Handbook of the Marine Fauna of North-West Europe*, ed. P. J. Hayward and J. S. Ryland. Oxford: Oxford University Press, pp. 1–18.

(2004). *A Natural History of the Seashore*. London: New Naturalist, Collins.

(2016). *Shallow Seas of Northwest Europe*. London: New Naturalist, Collins.

Heinz, G. H. (1976). Methylmercury: 2nd-Year Feeding Effects on Mallard Reproduction and Duckling Behaviour. *Journal of Wildlife Management*, **40**, 82–90.

Henderson, P. A. (2014). *Identification Guide to the Inshore Fish of the British Isles*. Pennington, Hampshire: Pisces Conservation Ltd.

Hendrickx, F., Maelfait, J.-P. and Langenbick, F. (2003). Absence of Cadmium Excretion and High Assimilation Result in Cadmium Biomagnification in a Wolf Spider. *Ecotoxicology and Environmental Safety*, **55**, 287–292.

Henze, M. (1911). Untersuchungen über das Blut der Ascidien. I. Mitteilung Die Vanadiumverbindung der Blutkörperchen. *Hoppe-Seyler's Zeitschrift fur physiologische Chemie*, **72**, 215–228.

Herring, P. (2002). *The Biology of the Deep Ocean*. Oxford: Oxford University Press.

Herrington, R., Stanley, C. and Symes, R. (1999). *Gold*. London: Natural History Museum.

Hildrew, A. G. (2009). Sustained Research on Stream Communities: A Model System and the Comparative Approach. *Advances in Ecological Research*, **41**, 176–312.

Hill, I. G., Worden, R. H.and Meighan, I. G. (2000). Geochemical Evolution of a Palaeolaterite: The Interbasaltic Formation, Northern Ireland. *Chemical Geology*, **166**, 65–84.

Hillerton, J. E. and Vincent, J. F. V. (1982). The Specific Location of Zinc in Insect Mandibles. *Journal of Experimental Biology*, **101**, 333–336.

Hinzmann, M. F., Lopes-Lima, M., Bobos, I., Ferreira, J., Domingua, B. and Machado, J. (2014). Morphological and Chemical Characterization of Mineral Concretions in the Freshwater Bivalve *Anodonta cygnaea* (Unionidae). *Journal of Morphology*, **276**, 65–76.

Hirst, H., Jüttner, I. and Ormerod, S. J. (2002). Comparing the Responses of Diatoms and Macroinvertebrates to Metals in Upland Streams of Wales and Cornwall. *Freshwater Biology*, **47**, 1752–1765.

Hobbelen, P. H. F., Koolhaas, J. E. and van Gestel, C. A. M. (2006). Bioaccumulation of Heavy Metals in the Earthworms *Lumbricus rubellus* and *Aporrectodea calaginosa* in Relation to Total and Available Metal Concentrations in Field Soils. *Environmental Pollution*, **144**, 639–646.

Hödl, E., Felder, E., Chabicovsky, M. and Dallinger, R. (2010). Cadmium Stress Stimulates Tissue Turnover in *Helix pomatia*: Increasing Cell Proliferation from Metal Tolerance to Exhaustion in Molluscan Midgut Gland. *Cell and Tissue Research*, **341**, 159–171.

Hoffmann, C., Schubert, G. and Calvignac-Spencer, S. (2016). Aquatic Biodiversity Assessment for the Lazy. *Molecular Ecology*, **25**, 846–848.

Hogstrand, C. and Wood, C. M. (1996). The Physiology and Toxicology of Zinc in Fish. In *Toxicology of Aquatic Pollution. Physiological, Cellular and Molecular Approaches*, ed. E. W. Taylor. Cambridge: Cambridge University Press, pp. 61–84.

Hollamby, S., Afema-Azikuru, J., Waigo, S., Cameron, K., Gandolf, A. R., Norris, A. and Sikarskie, J. G. (2006). Suggested Guidelines for Use of Avian Species as Biomonitors. *Environmental Monitoring and Assessment*, **118**, 13–20.

Holliday, L. M. and Liss, P. S. (1976). The Behaviour of Dissolved Iron, Manganese and Zinc in the Beaulieu Estuary, S. England. *Estuarine and Coastal Marine Science*, **4**, 349–353.

Holsbeek, L., Joiris, C. R., Debacker, V., Ali, I. B., Roose, P., Nellissen, J.-P., Gobert, S., Bouquegneau, J.-M. and Bossicart, M. (1999). Heavy Metals, Organochlorines and Polycyclic Aromatic Hydrocarbons in Sperm Whales Stranded in the Southern North Sea during the 1994/1995 Winter. *Marine Pollution Bulletin*, **38**, 304–313.

Hopkin, S. P. (1989). *Ecophysiology of Metals in Terrestrial Invertebrates*. Barking: Elsevier Applied Science.

(1990a). Species-Specific Differences in the Net Assimilation of Zinc, Cadmium, Lead, Copper and Iron by the Terrestrial Isopods *Oniscus asellus* and *Porcellio scaber*. *Journal of Applied Ecology*, **27**, 460–474.

(1990b). Critical Concentrations, Pathways of Detoxification and Cellular Ecotoxicology of Metals in Terrestrial Arthropods. *Functional Ecology*, **4**, 321–327.

(1991). A Key to the Woodlice of Britain and Ireland. *Field Studies*, **7**, 599–650.

(1993). Ecological Implications of 95% Protection Levels for Metals in Soil. *Oikos*, **66**, 137–141.

Hopkin, S. P. and Hames, C. A. C. (1994). Zinc, among a 'Cocktail' of Metal Pollutants, Is Responsible for the Absence of the Terrestrial Isopod *Porcellio scaber* from the Vicinity of a Primary Smelting Works. *Ecotoxicology*, **2**, 68–78.

Hopkin, S. P. and Martin, M. H. (1983). Heavy Metals in the Centipede *Lithobius variegatus* (Chilopoda). *Environmental Pollution B*, **6**, 309–318.

(1985). Assimilation of Zinc, Cadmium, Lead, Copper, and Iron by the Spider *Dysdera crocata*, a Predator of Woodlice. *Bulletin of Environmental Contamination and Toxicology*, **34**, 183–187.

Hopkin, S. P. and Nott, J. A. (1979). Some Observations on Concentrically Structured Intracellular Granules in the Hepatopancreas of the Shore Crab *Carcinus maenas* (L.). *Journal of the Marine*

Biological Association of the United Kingdom, **59**, 867–877.

Hopkin, S. P., Hames, C. A. C. and Bragg, S. (1989a). Terrestrial Isopods as Biological Indicators of Zinc Pollution in the Reading Area, South East England. *Monitore Zoologico Italiano (NS)*, **4**, 477–488.

Hopkin, S. P., Hardisty, G. N. and Martin, M. H. (1986). The Woodlouse *Porcellio scaber* as a 'Biological Indicator' of Zinc, Cadmium, Lead and Copper Pollution. *Environmental Pollution*, **11B**, 271–290.

Howard, A. G. and Nickless, G. (1977). Heavy Metal Complexation in Polluted Molluscs. 1. Limpets (*Patella vulgata* and *Patella intermedia*). *Chemico-Biological Interactions*, **16**, 107–114.

Howard, A. G., Arbab-Zavar, M. H. and Apte, S. (1984). The Behaviour of Dissolved Arsenic in the Estuary of the River Beaulieu. *Estuarine, Coastal and Shelf Science*, **19**, 493–504.

Hudson-Edwards, K. A., Macklin, M. G., Curtis, C. D. and Vaughan, D. J. (1996). Processes of Formation and Distribution of Pb-, Zn-, Cd-, and Cu-Bearing Minerals in the Tyne Basin, Northeast England: Implications for Metal-Contaminated River Systems. *Environmental Science and Technology*, **30**, 72–80.

Hummel, H., Modderman, R., Amiard-Triquet, C. et al. (1997). A Comparative Study on the Relation between Copper and Condition Index in Marine Bivalves and the Relation with Copper in the Sediment. *Aquatic Toxicology*, **38**, 165–181.

Humphreys, J., Caldow, R. W. G., McGrorty, S., West, A. D. and Jensen, A. C. (2007). Population Dynamics of Naturalised Manila Clams *Ruditapes philippinarum* in British Coastal Waters. *Marine Biology*, **151**, 2255–2270.

Hunt, L. E. and Howard, A. G. (1994). Arsenic Speciation and Distribution in the Carnon Estuary Following the Acute Discharge of Contaminated Water from a Disused Mine. *Marine Pollution Bulletin*, **28**, 33–38.

Hunt, R. F. R. S. (2011). *British Mining: A Treatise on the History and Future Prospect of Metalliferous Mines in the United Kingdom. With Illustrations. Book III.* London: British Library, Historical Print Editions.

Hunter, B. A., Johnson, M. S. and Thompson, D. J. (1987a). Ecotoxicology of Copper and Cadmium in a Contaminated Grassland Ecosystem. I. Soil and Vegetation Contamination. *Journal of Applied Ecology*, **24**, 573–586.

(1987b). Ecotoxicology of Copper and Cadmium in a Contaminated Grassland Ecosystem. II. Invertebrates. *Journal of Applied Ecology*, **24**, 587–599.

(1987c). Ecotoxicology of Copper and Cadmium in a Contaminated Grassland Ecosystem. III. Small Mammals. *Journal of Applied Ecology*, **24**, 601–614.

(1989). Ecotoxicology of Copper and Cadmium in a Contaminated Grassland Ecosystem. IV. Tissue Distribution and Age Accumulation in Small Mammals. *Journal of Applied Ecology*, **26**, 89–99.

Hutchins, D. A. and Bruland, K. W. (1994). Grazer-Mediated Regeneration and Assimilation of Fe, Zn and Mn from Planktonic Prey. *Marine Ecology Progress Series*, **110**, 259–269.

Hutchins, D. A., Wang, W.-X. and Fisher, N. S. (1995). Copepod Grazing and the Biogeochemical Fate of Diatom Iron. *Limnology and Oceanography*, **40**, 989–994.

Hutton, M. (1980). Metal Contamination of Feral Pigeons *Columba livia* from the London area. Part 2. Biological Effects of Lead Exposure. *Environmental Pollution*, **22A**, 281–293.

(1981). Accumulation of Heavy Metals and Selenium in Three Seabird Species from the United Kingdom. *Environmental Pollution*, **26A**, 129–145.

(1984). Impact of Airborne Metal Contamination on a Deciduous Woodland System. In *Effects of Pollution at the Ecosystem Level*, ed. P. J. Sheehan, D. R. Miller, G. C. Miller and P. Bourdeau. Chichester: SCOPE, John Wiley & Sons, Ltd., pp. 365–375.

Hutton, M. and Goodman, G. T. (1980). Metal Contamination of Feral Pigeons *Columba livia* from the London Area: Part 1. Tissue Accumulation of Lead, Cadmium and Zinc. *Environmental Pollution*, **22A**, 207–217.

Hutton, M. and Symon, C. (1986). The Quantities of Cadmium, Lead, Mercury and Arsenic Entering the U.K. Environment from Human Activities. *Science of the Total Environment*, **57**, 129–150.

Icely, J. D. and Nott, J. A. (1980). Accumulation of Copper within the 'Hepatopancreatic' Caeca of *Corophium volutator* (Crustacea: Amphipoda). *Marine Biology*, **57**, 193–199.

Ireland, M. P. (1973). Result of Fluvial Zinc Pollution on the Zinc Content of Littoral and Sub-Littoral Organisms in Cardigan Bay, Wales. *Environmental Pollution*, **4**, 27–35.

(1974). Variations in the Zinc, Copper, Manganese and Lead Content of *Balanus balanoides* in Cardigan Bay, Wales. *Environmental Pollution*, 7, 65–75.

(1979a). Distribution of Metals in the Digestive Gland-Gonad Complex of the Marine Gastropod *Nucella lapillus*. *Journal of Molluscan Studies*, 45, 322–327.

(1979b). Distribution of Essential and Toxic Metals in the Terrestrial Gastropod *Arion ater*. *Environmental Pollution*, 20, 271–278.

Ireland, M. P. and Richards, K. S. (1977). The Occurrence and Localisation of Heavy Metals and Glycogen in the Earthworms *Lumbricus rubellus* and *Dendrobaena rubidus* from a Heavy Metal Site. *Histochemistry*, 51, 153–166.

Ireland, M. P. and Wootton, R. J. (1977). Distribution of Lead, Zinc, Copper and Manganese in the Marine Gastropods, *Thais lapillus* and *Littorina littorea*, around the Coast of Wales. *Environmental Pollution*, 12, 27–41.

Ishii, T., Otake, T., Okoshi, K., Nakahara, M. and Nakamura, R. (1994). Intracellular Localization of Vanadium in the Fan Worm *Pseudopotamilla occelata*. *Marine Biology*, 121, 143–151.

Izagirre, U. and Marigómez, I. (2009). Lysosomal Enlargement and Lysosomal Membrane Destabilisation in Mussel Digestive Cells Measured by an Integrative Index. *Environmental Pollution*, 157, 1544–1553.

Jackson, A. P. and Alloway, B. J. (1992). The Transfer of Cadmium from Agricultural Soils to the Human Food Chain. In *Biogeochemistry of Trace Metals*, ed. D. C. Adriano. Boca Raton, FL: Lewis Publishers, pp. 109–158.

Jain, S. K. and Bradshaw, A. D. (1966). Evolutionary Divergence among Adjacent Plant Populations. I. The Evidence and Its Theoretical Analysis. *Heredity*, 21, 407–441.

Janssen, M. P. M., Bruins, A., De Vries, T. H. and van Straalen, N. M. (1991). Comparison of Cadmium Kinetics in Four Soil Arthropod Species. *Archives of Environmental Contamination and Toxicology*, 20, 305–312.

Janssens, E., Dauwe, T., Bervoets, L. and Eens, M. (2001). Heavy Metals and Selenium in Feathers of Great Tits (*Parus major*) along a Metal Pollution Gradient. *Environmental Toxicology and Chemistry*, 20, 2815–2820.

Janssens, E., Dauwe, T., Pinxten, R. and Eens, M. (2003). Breeding Performance of Great Tits (*Parus major*) along a Gradient of Heavy Metal Pollution. *Environmental Toxicology and Chemistry*, 22, 1140–1145.

Janssens de Bisthoven, I. G., Timmermans, K. R. and Ollevier, F. (1992). The Concentration of Cadmium, Lead, Copper and Zinc in *Chironomus* gr. *thummi* Larvae (Diptera, Chironomidae) with Deformed *versus* Normal Menta. *Hydrobiologia*, 239, 141–149.

Janz, D. M., DeForest, D. K., Brooks, M. L., et al. (2010). Selenium Toxicity to Aquatic Organisms. In *Ecological Assessment of Selenium in the Aquatic Environment*, ed. P. M. Chapman, W. J. Adams, M. L. Brooks, et al. Boca Raton, FL: CRC Press, pp. 141–231.

Järup, L. and Åkesson, A. (2009). Current Status of Cadmium as an Environmental Health Problem. *Toxicology and Applied Pharmacology*, 238, 201–208.

Jarvis, A. P. and Younger, P. L. (1997). Dominating Chemical Factors in Mine Water Induced Impoverishment of the Invertebrate Fauna of Two Streams in the Durham Coalfield, UK. *Chemistry and Ecology*, 13, 249–270.

(2000). Broadening the Scope of Mine Water Environmental Impact Assessment. *Environmental Impact Assessment Review*, 20, 85–96.

Jeffries, M. and Mills, D. (1990). *Freshwater Ecology: Principles and Applications*. Chichester: John Wiley & Sons.

Jennings, J. B. (1968). Nutrition and Digestion. In *Chemical Zoology*, vol. 2, ed. M. Florkin and B. T. Scheer. New York, NY: Academic Press, pp. 303–326.

Jensen, A., Carrier, I. and Richardson, N. (2005b). Marine Fisheries of Poole Harbour. In *The Ecology of Poole Harbour*, ed. J. Humphreys and V. May. Amsterdam: Elsevier, pp. 195–203.

Jensen, A., Humphreys, J., Caldow, R. and Cesar, C. (2005a). The Manila Clam in Poole Harbour. In *The Ecology of Poole Harbour*, ed. J. Humphreys and V. May. Amsterdam: Elsevier, pp. 163–173.

Jensen, J. K., Holm, P. E., Nejrup, J., Larsen, M. B. and Borggaard, O. K. (2009). The Potential of Willow for Remediation of Heavy Metal Polluted Calcareous

Urban Soils. *Environmental Pollution*, **157**, 931–937.

Jepson, P. D., Bennett, P. M., Allchin, C. R., et al. (1999). Investigating Potential Associations between Chronic Exposure to Polychlorinated Biphenyls and Infectious Disease Mortality in Harbour Porpoises from England and Wales. *Science of the Total Environment*, **243–244**, 339–348.

Jepson, P. D., Bennett, P. M., Deaville, R., Allchin, C. R., Baker, J. R. and Law, R. J. (2005). Relationships between PCBs and Health Status in Harbour Porpoises (*Phocoena phocoena*) Stranded in the United Kingdom. *Environmental Toxicology and Chemistry*, **24**, 238–248.

Jepson, P. D., Deaville, R., Acevedo-Whitehouse, K., et al. (2013). What Caused the UK's Largest Common Dolphin (*Delphinus delphis*) Mass Stranding Event? *PLoS ONE*, **8**, e60953.

Johnson, M. S. and Eaton, J. W. (1980). Environmental Contamination through Residual Trace Metal Dispersal from a Derelict Lead–Zinc Mine. *Journal of Environmental Quality*, **9**, 175–179.

Johnson, M. S., Roberts, R. D., Hutton, M. and Inskip, M. J. (1978). Distribution of Lead, Zinc and Cadmium in Small Mammals from Polluted Environments. *Oikos*, **30**, 153–159.

Johnston, E. L. and Keough, M. J. (2002). Direct and Indirect Effects of Repeated Pollution Events on Marine Hard-Substrate Assemblages. *Ecological Applications*, **12**, 1212–1228.

(2003). Competition Modifies the Response of Organisms to Toxic Disturbance. *Marine Ecology Progress Series*, **251**, 15–26.

Johnston, E. L., Marzinelli, E. M., Wood, C. A., Speranza, D. and Bishop, J. D. D. (2011). Bearing the Burden of Boat Harbours: Heavy Contaminant and Fouling Loads in a Native Habitat-Forming Alga. *Marine Pollution Bulletin*, **62**, 2137–2144.

Johnstone, K. M., Rainbow, P. S., Clark, P. F., Smith, B. D. and Morritt, D. (2016). Trace Metal Bioavailabilities in the Thames Estuary: Continuing Decline in the 21st Century. *Journal of the Marine Biological Association of the United Kingdom*, **96**, 205–216.

Jonas, P. J. C. and Millward, G. E. (2010). Metals and Nutrients in the Severn Estuary and Bristol Channel: Contemporary Inputs and Distribution. *Marine Pollution Bulletin*, **61**, 52–67.

Jones, D. T. and Hopkin, S. P. (1998). Reduced Survival and Body Size in the Terrestrial Isopod *Porcellio scaber* from a Metal-Polluted Environment. *Environmental Pollution*, **99**, 215–223.

Jones, J. I., Davy-Bowker, J., Murphy, J. F. and Pretty, J. L. (2010). Ecological Monitoring and Assessment of Pollution in Rivers. In *Ecology of Industrial Pollution*, ed. L. C. Batty and K. B. Hallberg. Cambridge: British Ecological Society, Cambridge University Press, pp. 126–146.

Jones, J. I., Spencer, K., Rainbow, P. S., et al. (2016). *The Ecological Impacts of Contaminated Sediment from Abandoned. Metal Mines*. Final Report, WT0970 Characterisation and Targeting of Measures for (Non-Coal) Polluted Mine Waters: Impacts of Contaminated Sediment on Ecological Recovery. London: DEFRA, pp. 1–352.

Jones, J. R. E. (1940). A Study of the Zinc-Polluted River Ystwyth in North Cardiganshire, Wales. *Annals of Applied Biology*, **27**, 368–378.

(1949). An Ecological Study of the River Rheidol, North Cardiganshire, Wales. *Journal of Animal Ecology*, **18**, 67–88.

(1958). A Further Study of the Zinc-Polluted River Ystwyth. *Journal of Animal Ecology*, **27**, 1–14.

Jones, K. C., Peterson, P. J. and Davies, B. E. (1985). Silver and Other Metals in Some Aquatic Bryophytes from Streams in the Lead Mining District of Mid-Wales, Great Britain. *Water, Air and Soil Pollution*, **24**, 329–338.

Julshamn, K., Duinker, A., Nilsen, B. M., Ndreaas, K. and Maage, A. (2013). A Baseline of Metals in Cod (*Gadus morhua*) from the North Sea and Coastal Norwegian Waters, with Focus on Mercury, Arsenic, Cadmium and Lead. *Marine Pollution Bulletin*, **72**, 264–273.

Jürgens, M. D., Johnson, A. C., Jones, K. C., Hughes, D. and Lawlor, A. J. (2013). The Presence of EU Priority Substances Mercury, Hexachlorobenzene, Hexachlorobutadiene and PBDEs in Wild Fish from Four English Rivers. *Science of the Total Environment*, **461–462**, 441–452.

Kalman, J., Bonnail-Miguel, E., Smith, B. D., Bury, N. R. and Rainbow, P. S. (2015). Toxicity and the Fractional Distribution of Trace Metals Accumulated from Contaminated Sediments by the Clam *Scrobicularia plana* Exposed in the Laboratory and the Field. *Science of the Total Environment*, **506–507**, 109–117.

Kalman, J., Smith, B. D., Bury, N. R. and Rainbow, P. S. (2014). Biodynamic Modelling of the Bioaccumulation of Trace Metals by an Infaunal Estuarine Invertebrate, the Clam *Scrobicularia plana*. *Aquatic Toxicology*, **154**, 121–130.

Kammenga, J. E., Dallinger, R., Donker, M. H., Köhler, H.-R., Simonsen, V., Triebskorn, R. and Weeks, J. M. (2000). Biomarkers in Terrestrial Invertebrates for Ecotoxicological Soil Risk Assessment. *Reviews of Environmental Contamination and Toxicology*, **164**, 93–147.

Kannan, K. and Tanabe, S. (1997). Response to Comment on 'Elevated Accumulation of Tributyltin and Its Breakdown Products in Bottlenose Dolphins (*Tursiops truncatus*) Found Stranded along the US Atlantic and Gulf Coasts'. *Environmental Science and Technology*, **31**, 3035–3036.

Kannan, K., Blakenship, A. L., Jones, P. D. and Giesy, J. P. (2000). Toxicity Reference Values for the Toxic Effects of Polychlorinated Biphenyls to Aquatic Mammals. *Human and Ecological Risk Assessment*, **6**, 181–201.

Kelly, M. (1988). *Mining and the Freshwater Environment*. London and New York, NY: Elsevier Applied Science.

Kelly, M. G., Girton, C. and Whitton, B. A. (1987). Use of Moss-Bags for Monitoring Heavy Metals in Rivers. *Water Research*, **21**, 1429–1435.

Kelly, M. G., Juggins, S., Guthrie, R., et al. (2008). Assessment of Ecological Status in UK Rivers Using Diatoms. *Freshwater Biology*, **53**, 403–422.

Khan, F. R., Irving, J. R., Bury, N. R. and Hogstrand, C. (2011). Differential Tolerance of Two *Gammarus pulex* Populations Transplanted from Different Metallogenic Regions to a Polymetal Gradient. *Aquatic Toxicology*, **102**, 95–103.

Kicklighter, C. E. and Hay, M. E. (2007). To Avoid or Deter: Interactions among Defensive and Escape Strategies in Sabellid Worms. *Oecologia*, **151**, 161–173.

Kilkenny, B. and Good, J. A. (1998). Rehabilitation of Abandoned Metalliferous Mine Spoil Using Composted Sewage Sludge at Avoca Mines, County Wicklow, (Ireland). In *GREEN 2. Contaminated and Derelict Land*, ed. R. W. Sarsby. London: Thomas Telford, pp. 476–482.

Kirby, J., Delany, S. and Quinn, J. (1994). Mute swans in Great Britain: A Review, Current Status and Long-Term Trends. *Hydrobiologia*, **280**, 467–482.

Klages, N. T. W. (1996). Cephalopods as Prey. II. Seals. *Philosophical Transactions of the Royal Society London B*, **351**, 1045–1052.

Klumpp, D. W. and Peterson, P. J. (1979). Arsenic and Other Trace Elements in the Waters and Organisms of an Estuary in SW England. *Environmental Pollution*, **19**, 11–20.

Knight, B., Zhao, F. J., McGrath, S. P. and Shen, Z. G. (1997). Zinc and Cadmium Uptake by the Hyperaccumulator *Thlaspi caerulescens* in Contaminated Soils and Its Effects on the Concentration and Speciation of Metals in Soil Solution. *Plant and Soil*, **197**, 71–78.

Knight, M. and Parke, M. (1950). A Biological Study of *Fucus vesiculosus* L. and *F. serratus* L. *Journal of the Marine Biological Association of the United Kingdom*, **29**, 439–514.

Kochian, L. V. (1995). Cellular Mechanisms of Aluminium Toxicity and Resistance in Plants. *Annual Review of Plant Physiology and Plant Molecular Biology*, **46**, 237–260.

Koeman, J. H., Peeters, W. H. M., Koudstaal-Hol, C. H. M, Tjioe, P. S. and de Goeij, J. J. M. (1973). Mercury-Selenium Correlations in Marine Mammals. *Nature*, **245**, 385–386.

Köhler, H.-R. (2002). Localization of Metals in Cells of Saprophagous Soil Arthropods (Isopoda, Diplopoda, Collembola). *Microscopy Research and Technique*, **56**, 393–401.

Köhler, H.-R., Körtje, K.-H. and Albert, G. (1995). Content, Absorption Quantities and Intracellular Storage Site of Heavy Metals in Diplopoda. *BioMetals*, **8**, 37–46.

Kraak, M. H. S., Scholten, M. C. T., Peeters, W. H. M. and de Kock, W. C. (1991). Biomonitoring of Heavy Metals in the Western European Rivers Rhine and Meuse Using the Freshwater Mussel *Dreissena polymorpha*. *Environmental Pollution*, **74**, 101–114.

Kramarz, P. (1999). Dynamics of Accumulation and Decontamination of Cadmium and Zinc in Carnivorous Invertebrates. 1. The Ground Beetle, *Poecilus cupreus* L. *Bulletin of Environmental Contamination and Toxicology*, **63**, 531–537.

Krämer, U., Cotter-Howells, J. D., Charnock, J. M., Baker, A. J. M. and Smith, J. A. (1996). Free Histidine as

A Metal Chelator in Plants that Accumulate Nickel. *Nature*, **379**, 635–638.

Krång, A.-S. and Rosenqvist, G. (2006). Effects of Manganese on Chemically Induced Food Search Behaviour of the Norway Lobster, *Nephrops norvegicus* (L.). *Aquatic Toxicology*, **78**, 284–291.

Lancaster, J. and B. J. Downes (2013). *Aquatic Entomology*. Oxford: Oxford University Press.

Lane, T. W., Saito, M. A., George, G. N., Pickering, I. J., Prince, R. C. and Morel, F. M. M. (2005). A Cadmium Enzyme from a Marine Diatom. *Nature*, **435**, 42.

Langdon, C. J., Piearce, T. G., Black, S. and Semple, K. T. (1999). Resistance to Arsenic-Toxicity in a Population of the Earthworm *Lumbricus rubellus*. *Soil Biology and Biochemistry*, **31**, 1963–1967.

Langdon, C. J., Piearce, T. G., Meharg, A. A. and Semple, K. T. (2001). Resistance to Copper Toxicity in Populations of the Earthworms *Lumbricus rubellus* and *Dendrodrilus rubidus* from Contaminated Mine Wastes. *Environmental Toxicology and Chemistry*, **20**, 2336–2341.

Langston, W. J. (1980). Arsenic in U.K. Estuarine Sediments and Its Availability to Benthic Organisms. *Journal of the Marine Biological Association of the United Kingdom*, **60**, 869–881.

  (1982). The Distribution of Mercury in British Estuarine Sediments and Its Availability to Deposit-Feeding Bivalves. *Journal of the Marine Biological Association of the United Kingdom*, **62**, 667–684.

  (1983). The Behaviour of Arsenic in Selected United Kingdom Estuaries. *Canadian Journal of Fisheries and Aquatic Sciences*, **40** (Suppl. 2), 143–150.

  (1984). Availability of Arsenic to Estuarine and Marine Organisms: A Field and Laboratory Evaluation. *Marine Biology*, **80**, 143–154.

  (1986). Metals in Sediments and Benthic Organisms in the Mersey Estuary. *Estuarine, Coastal and Shelf Science*, **23**, 239–261.

  (1995). Tributyl Tin in the Marine Environment: A Review of Past and Present Risks. *Pesticide Outlook*, Dec, 18–24.

Langston, W. J. and Zhou, M. (1987). Cadmium Accumulation, Distribution and Metabolism in the Gastropod *Littorina littorea*: The Role of Metal-Binding Proteins. *Journal of the Marine Biological Association of the United Kingdom*, **67**, 585–601.

Langston, W. J., Bebianno, M. J. and Burt, G. R. (1998). Metal Handling Strategies in Molluscs. In *Metal Metabolism in Aquatic Environments*, ed. W. J. Langston and M. Bebianno. London: Chapman and Hall, pp. 219–283.

Langston, W. J., Chesman, B. S. and Burt, G. R. (2006). Characterisation of European Marine Sites. *Mersey Estuary SPA. Occasional Publications of the Marine Biological Association of the United Kingdom*, **18**, 1–185.

Langston, W. J., Chesman, B. S., Burt, G. R., Pope, N. D. and McEvoy, J. (2002). Metallothionein in Liver of Eels *Anguilla anguilla* from the Thames Estuary: An Indicator of Environmental Quality? *Marine Environmental Research*, **53**, 263–293.

Langston, W. J., Pope, N. D., Chesman, B. S. and Burt, G. R. (2004). Bioaccumulation of Metals in the Thames Estuary: 2001. *Thames Estuary Environmental Quality Series*, **10**, 1–139.

Langston, W. J., Pope, N. D., Davey, M., et al. (2015). Recovery from TBT Pollution in English Channel Environments: A Problem Solved? *Marine Pollution Bulletin*, **95**, 551–564.

Langston, W. J., Pope, N. D., Jonas, P. J. C., et al. (2010). Contaminants in Fine Sediments and Their Consequences for Biota of the Severn Estuary. *Marine Pollution Bulletin*, **61**, 68–82.

Laslett, R. E. (1995). Concentrations of Dissolved and Suspended Particulate Cd, Cu, Mn, Ni, Pb and Zn in Surface Waters around the Coasts of England and Wales and in Adjacent Seas. *Estuarine, Coastal and Shelf Science*, **40**, 67–85.

Law, R. J. (1996). Metals in Marine Mammals. In *Environmental Contaminants in Wildlife: Interpreting Tissue Concentrations*, ed. W. N. Beyer, G. H. Heinz and A. W. Redmon-Norwood. Boca Raton, FL: Lewis Publishers, pp. 357–376.

Law, R. J., Barry, J., Barber, J. L., et al. (2012). Contaminants in Cetaceans from UK Waters: Status as Assessed within the Cetacean Strandings Investigation Programme from 1990 to 2008. *Marine Pollution Bulletin*, **64**, 1485–1494.

Law, R. J., Bennett, M. E., Blake, S. J., Allchin, C. R., Jones, B. R. and Spurrier, C. J. H. (2001). Metals and Organochlorines in Pelagic Cetaceans Stranded on the Coasts of England and Wales. *Marine Pollution Bulletin*, **42**, 522–526.

Law, R. J., Blake, S. J., Jones, B. R. and Rogan, E. (1998). Organotin Compounds in Liver Tissue of Harbour Porpoises (*Phocoena phocoena*) and Grey Seals

(*Halichoerus grypus*) from the Coastal Waters of England and Wales. *Marine Pollution Bulletin*, **36**, 241–247.

Law, R. J., Blake, S. J. and Spurrier, C. J. H. (1999). Butyltin Compounds in Liver Tissues of Pelagic Cetaceans Stranded on the Coasts of England and Wales. *Marine Pollution Bulletin*, **38**, 1258–1261.

Law, R. J., Jones, B. R., Baker, J. R., Kennedy, S., Milne, R. and Morris, R. J. (1992). Trace Metals in Livers of Marine Mammals from the Welsh Coast and the Irish Sea. *Marine Pollution Bulletin*, **24**, 296–304.

Law, R. J., Stringer, R. L., Allchin, C. R. and Jones, B. R. (1996). Metals and Organochlorines in Sperm Whales (*Physeter macrocephalus*) Stranded around the North Sea during the 1994/1995 Winter. *Marine Pollution Bulletin*, **32**, 72–77.

Law, R. J., Waldock, M. J., Allchin, C. R., Laslett, R. E. and Bailey, K. J. (1994). Contaminants in Seawater around England and Wales: Results from Monitoring Surveys, 1990–1992. *Marine Pollution Bulletin*, **28**, 668–675.

Lee, B.-G. and Fisher, N.S. (1992). Decomposition and Release of Elements from Zooplankton Debris. *Marine Ecology Progress Series*, **88**, 117–128.

Lee, S. V. and Cundy, A. B. (2001). Heavy Metal Contamination and Mixing Processes in Sediments from the Humber Estuary, Eastern England. *Estuarine, Coastal and Shelf Science*, **53**, 619–636.

Lee, T. W and Morel, F. M. M. (1995). Replacement of Zinc by Cadmium in Marine Phytoplankton. *Marine Ecology Progress Series*, **127**, 305–309.

Leffler, P. E. and Nyholm, N. E. (1996). Nephrotoxic Effects in Free-Living Bank Voles in a Heavy Metal Polluted Environment. *Ambio*, **25**, 417–420.

Lemly, A. D. (2004). Aquatic Selenium Pollution Is a Global Environmental Safety Issue. *Ecotoxicology and Environmental Safety*, **59**, 44–56.

Lenz, M. and Lens, P. N. L. (2009). The Essential Toxin: The Changing Perception of Selenium in Environmental Sciences. *Science of the Total Environment*, **407**, 3620–3633.

Lepp, N. W. (1992). Uptake and Accumulation of Metals in Bacteria and Fungi. In *Biogeochemistry of Trace Metals*, ed. D. C. Adriano. Boca Raton, FL: Lewis Publishers, pp. 277–298.

Lewis, S. A., Becker, P. H. and Furness, R. W. (1993). Mercury Levels in Eggs, Tissues, and Feathers of Herring Gulls *Larus argentatus* from the German Wadden Sea Coast. *Environmental Pollution*, **80**, 293–299.

Li, X. and Thornton, I. (1993a). Multi-Element Contamination of Soils and Plants in Old Mining Areas, U.K. *Applied Geochemistry*, **Suppl. Issue 2**, 51–56.

(1993b). Arsenic, Antimony and Bismuth in Soil and Pasture Herbage in Some Old Metalliferous Mining Areas in England. *Environmental Geochemistry and Health*, **15**, 135–144.

Lichtenegger, H. C., Schöberl, T., Bartl, M. H., Waite, J. H. and Stucky, G. D. (2002). High Abrasion Resistance with Sparse Mineralization: Copper Biomineral in Worm Jaws. *Science*, **298**, 389–392.

Lichtenegger, H. C., Schöberl, T., Ruokolainen, J. T., et al. (2003). Zinc and Mechanical Prowess in the Jaws of *Nereis*, a Marine Worm. *Proceedings of the National Academy of Sciences*, **100**, 9144–9149.

Lincoln, R. J. (1979). *British Marine Amphipoda: Gammaridea*. London: British Museum (Natural History).

Lindberg, P. and Odsjö, T. (1983). Mercury Levels in Feathers of Peregrine Falcon *Falco peregrinus* Compared with Total Mercury Content in Some of Its Prey Species in Sweden. *Environmental Pollution B*, **5**, 297–318.

Lisk, D. J. (1988). Environmental Implications of Incineration of Municipal Solid Waste and Ash Disposal. *Science of the Total Environment*, **74**, 39–66.

Little, D. I. and Smith, J. (1994). Appraisal of Contaminants in Sediments of the Inner Bristol Channel and Severn Estuary. *Biological Journal of the Linnean Society*, **51**, 55–69.

Little, P. and Martin, M. H. (1972). A Survey of Zinc, Lead and Cadmium in Soil and Natural Vegetation around a Smelting Complex. *Environmental Pollution*, **3**, 241–254.

Liu, F. and Wang, W.-W. (2011). Metallothionein-Like Proteins Turnover, Cd and Zn Biokinetics in the Dietary Cd-Exposed Scallop *Chlamys nobilis*. *Aquatic Toxicology*, **105**, 361–368.

(2012). Proteome Pattern in Oysters as a Diagnostic Tool for Metal Pollution. *Journal of Hazardous Materials*, **239–240**, 241–248.

Ljung, K., Selinus, O., Ottabbong, E. and Berglund, M. (2006). Metal and Arsenic Distribution in Soil

Particle Size Relevant to Soil Ingestion by Children. *Applied Geochemistry*, 21, 1613–1624.

Lobel, P. B. (1987a). Short-Term and Long-Term Uptake of Zinc by the Mussel, *Mytilus edulis*: A Study in Individual Variability. *Archives of Environmental Contamination and Toxicology*, 16, 723–732.

(1987b). Intersite, Intrasite and Inherent Variability of the Whole Soft Tissue Zinc Concentrations of Individual Mussels *Mytilus edulis*: The Importance of the Kidney. *Marine Environmental Research*, 21, 59–71.

Lobel, P. B. and Wright, D. A. (1983). Frequency Distribution of Zinc Concentrations in the Common Mussel, *Mytilus edulis* (L.). *Estuaries*, 6, 154–159.

Lobel, P. B., Mogie, P., Wright, D. A. and Wu, B. L. (1982). Metal Accumulation in Four Molluscs. *Marine Pollution Bulletin*, 13, 170–174.

Lodenius, M. and Solonen, T. (2013). The Use of Feathers of Birds of Prey as Indicators of Metal Pollution. *Ecotoxicology*, 22, 1319–1334.

Lofts, S., Spurgeon, D. J., Svendsen, C. and Tipping, E. (2004). Deriving Soil Critical Limits for Cu, Zn, Cd, and Pb: A Method Based on Free Ion Concentrations. *Environmental Science and Technology*, 38, 3623–3631.

Long, E. R. and Chapman, P. M. (1985). A Sediment Quality Triad: Measures of Sediment Contamination, Toxicity and Infaunal Community Composition in Puget Sound. *Marine Pollution Bulletin*, 16, 405–415.

Lottermoser, B. G. (2011). Recycling, Reuse and Rehabilitation of Mine Wastes. *Elements*, 7, 405–410.

Lousley, J. E. (1950). *Wild Flowers of Chalk and Limestone*. London: New Naturalist, Collins.

Lunn, A. (2004). *Northumberland*. London: New Naturalist, Collins.

Luoma, S. N. (1977). Detection of Trace Contaminant Effects in Aquatic Ecosystems. *Journal of the Fisheries Research Board of Canada*, 34, 436–439.

(1989). Can We Determine the Biological Availability of Sediment-Bound Trace Elements? *Hydrobiologia*, 176/177, 379–396.

Luoma, S. N. and Bryan, G. W. (1978). Factors Controlling the Availability of Sediment-Bound Lead to the Estuarine Bivalve *Scrobicularia plana*. *Journal of the Marine Biological Association of the United Kingdom*, 58, 793–802.

(1982). A Statistical Study of Environmental Factors Controlling Concentrations of Heavy Metals in the Burrowing Bivalve *Scrobicularia plana* and the Polychaete *Nereis diversicolor*. *Estuarine, Coastal and Shelf Science*, 15, 95–108.

Luoma, S. N. and Rainbow, P. S. (2005). Why Is Metal Bioaccumulation So Variable? Biodynamics as a Unifying Concept. *Environmental Science and Technology*, 39, 1921–1931.

(2008). *Metal Contamination in Aquatic Environments: Science and Lateral Management*. Cambridge: Cambridge University Press.

Luoma, S. N., Cain, D. J. and Rainbow, P. S. (2010). Calibrating Biomonitors to Ecological Disturbance: A New Technique for Deciphering Metal Effects in Natural Waters. *Integrated Environmental Assessment and Management*, 6, 199–209.

Ma, W. C. (1987). Heavy Metal Accumulation in the Mole, *Talpa europea*, and Earthworms as an Indicator of Metal Bioavailability in Terrestrial Environments. *Bulletin of Environmental Contamination and Toxicology*, 39, 933–938.

(1989). Effect of Soil Pollution with Metallic Lead Pellets on Lead Bioaccumulation and Organ/Body Weight Alterations in Small Mammals. *Archives of Environmental Contamination and Toxicology*, 18, 617–622.

(2011). Lead in Mammals. In *Environmental Contaminants in Biota: Interpreting Tissue Concentrations*, 2nd edn., ed. W. N. Beador and J. P. Meador. Boca Raton, FL: Taylor and Francis Books, pp. 595–607.

MacGregor, N. (2010). *A History of the World in 100 Objects. 19 Mold Gold Cape*. London: Allen Lane, Penguin Group.

Mackay, D. W., Halcrow, W. and Thornton, I. (1972). Sludge Dumping in the Firth of Clyde. *Marine Pollution Bulletin*, 3, 7–11.

Macnair, M. R. (2003). The Hyperaccumulation of Metals by Plants. *Advances in Botanical Research*, 40, 63–105.

Mahfouz, C., Henry, F., Courcot, L., et al. (2014). Harbour Porpoises (*Phocoena phocoena*) Stranded along the Southern North Sea: An Assessment through Metallic Contamination. *Environmental Research*, 133, 266–273.

Maitland, P. S. and Campbell, R. N. (1992). *Freshwater Fishes*. London: New Naturalist, Collins.

Maltby, L. and Naylor, C. (1990). Preliminary Observations on the Ecological Relevance of the *Gammarus* 'Scope for Growth' Assay: Effect of Zinc on Reproduction. *Functional Ecology*, 4, 393–397.

Maltby, L., Clayton, S. A, Wood, R. M. and McLoughlin, N. (2002). Evaluation of the *Gammarus pulex in situ* Feeding Assay as a Biomonitor of Water Quality: Robustness, Responsiveness, and Relevance. *Environmental Toxicology and Chemistry*, 21, 361–368.

Maltby, L., Naylor, C. and Calow, P. (1990a). Field Deployment of a Scope for Growth Assay Involving *Gammarus pulex*, a Freshwater Benthic Detritivore. *Ecotoxicology and Environmental Safety*, 19, 292–300.

(1990b). Effect of Stress on a Freshwater Benthic Detritivore: Scope for Growth in *Gammarus pulex*. *Ecotoxicology and Environmental Safety*, 19, 285–291.

Manly, R. and George, W. O. (1977). The Occurrence of Some Heavy Metals in Populations of the Freshwater Mussel *Anodonta anatina* (L.) from the River Thames. *Environmental Pollution*, 14, 139–154.

Mantel, L. H. and Farmer, L. L. (1983). Osmotic and Ionic Regulation. In *The Biology of Crustacea, Vol. 5: Internal Anatomy and Physiological Regulation*, ed. L. H. Mantel. New York, NY: Academic Press, pp. 54–161.

Mantoura, R. F. C., Dickson, A. and Riley, J. P. (1978). The Complexation of Metals with Humic Materials in Natural Waters. *Estuarine and Coastal Marine Science*, 6, 387–408.

Marchant, B. P., Tye, A. M. and Rawlins, B. G. (2011). The Assessment of Point-Source and Diffuse Soil Metal Pollution Using Robust Geostatistical Methods: A Case Study in Swansea (Wales, UK). *European Journal of Soil Science*, 62, 346–358.

Marigómez, I., Soto, M., Carajaville, M. P., Angulo, E. and Giamberini, L. (2002). Cellular and Subcellular Distribution of Metals in Molluscs. *Microscopy Research and Technique*, 56, 358–392.

Martin, D. J. and Rainbow, P. S. (1998). The Kinetics of Zinc and Cadmium in the Haemolymph of the Shore Crab *Carcinus maenas*. *Aquatic Toxicology*, 40, 203–231.

Martin, M. H., Coughtrey, P. J. and Young, E. W. (1976). Observations on the Availability of Lead, Zinc, Cadmium and Copper in Woodland Litter and the Uptake of Lead, Zinc and Cadmium by the Woodlouse, *Oniscus asellus*. *Chemosphere*, 5, 313–318.

Martin, M. H., Duncan, E. M. and Coughtrey, P. J. (1982). The Distribution of Heavy Metals in a Contaminated Woodland Ecosystem. *Environmental Pollution*, 3, 147–157.

Martín-Díaz, M. L., Blasco, J., Sales, D. and DelValls, T. A. (2009). The Use of a Kinetic Biomarker Approach for *in Situ* Monitoring of Littoral Sediments Using the Crab *Carcinus maenas*. *Marine Environmental Research*, 68, 82–88.

Martinez, E. A., Moore, B. C., Schaumloffel, J. and Dasgupta, N. (2003). Morphological Abnormalities in *Chironomus tentans* Exposed to Cadmium- and Copper-Spiked Sediments. *Ecotoxicology and Environmental Safety*, 55, 204–212.

Martoja, R. and Berry, J.-P. (1980). Identification of Tiemannite as a Probable Product of Demethylation of Mercury by Selenium in Cetaceans: A Complement to the Scheme of the Biological Cycle of Mercury. *Vie et Milieu*, 30, 7–10.

Martoja, R. and Viale, D. (1977). Accumulation de granules de séléniure mercurique dans le foie d'Odontocètes (Mammifères, Cétacés): un mécanisme possible de détoxication du méthyl-mercure par le sélénium. *Comptes Rendus de l'Académie des Sciences D*, 285, 109–112.

Martoja, M., Tue, V. T. and Elkaïm, B. (1980). Bioaccumulation du cuivre chez *Littorina littorea* (L.) (Gastéropode Prosobranche): signification physiologique et écologique. *Journal of Experimental Marine Biology and Ecology*, 43, 251–270.

Mason, A. Z. and Jenkins, K. D. (1995). Metal Detoxification in Aquatic Organisms. In *Metal Speciation and Aquatic Systems*, ed. A. Tessier and D. R. Turner. New York, NY:.Wiley, pp. 479–608.

Mason, A.Z. and Nott, J. A. (1981). The Role of Intracellular Biomineralized Granules in the Regulation and Detoxification of Metals in Gastropods with Special Reference to the Marine Prosobranch *Littorina littorea*. *Aquatic Toxicology*, 1, 239–256.

Mason, A. Z. and Simkiss, K. (1983). Interactions between Metals and Their Distribution in Tissues of *Littorina littorea* (L.) Collected from Clean and Polluted Sites. *Journal of the Marine Biological Association of the United Kingdom*, **63**, 661–672.

Mason, A. Z., Simkiss, K. and Ryan, K. P. (1984). The Ultrastructural Localization of Metals in Specimens of *Littorina littorea* Collected from Clean and Polluted Sites. *Journal of the Marine Biological Association of the United Kingdom*, **64**, 699–720.

Mason, C. F. (1987). A Survey of Mercury, Lead and Cadmium in Muscle of British Freshwater Fish. *Chemosphere*, **16**, 901–906.

Matsumoto, H. (2000). Cell Biology of Aluminium Toxicity and Tolerance in Higher Plants. *International Review of Cytology*, **200**, 1–46.

Matthews, H. and Thornton, I. (1982). Seasonal and Species Variation in the Content of Cadmium and Associated Metals in Pasture Plants at Shipham. *Plant and Soil*, **66**, 181–193.

Matthiessen, P., Bifield, S., Jarrett, F., et al. (1998). An Assessment of Sediment Toxicity in the River Tyne Estuary, UK by Means of Bioassays. *Marine Environmental Research*, **45**, 1–15.

Mauchline, J. (1984). *Euphausiid, Stomatopod and Leptostracan Crustaceans. Synopses of the British Fauna (New Series) No. 30.* Leiden: Linnean Society of London and the Estuarine and Brackish-Water Sciences Association, E. J. Brill/Dr W. Backhuys.

Maund, S. J., Taylor, E. J. and Pascoe, D. (1992). Population Responses of the Freshwater Amphipod Crustacean *Gammarus pulex* (L.) to Copper. *Freshwater Biology*, **28**, 29–36.

McEvoy, J., Langston, W. J., Burt, G. R. and Pope, N. D. (2000). Bioaccumulation of Metals in the Thames Estuary: 1997. *Thames Estuary Environmental Quality Series*, **2**, 1–116.

McGeer, J. C., Brix, K. V., Skeaff, J. M., et al. (2003). Inverse Relationship between Bioconcentration Factor and Exposure Concentration for Metals: Implications for Hazard Assessment of Metals in the Aquatic Environment. *Environmental Toxicology and Chemistry*, **22**, 1017–1037.

McKenzie, L. A., Brooks, R. and Johnston, E. L. (2011). Heritable Pollution Tolerance in a Marine Invader. *Environmental Research*, **111**, 926–932.

(2012a). A Widespread Contaminant Enhances Invasion Success of a Marine Invader. *Journal of Applied Ecology*, **49**, 767–773.

(2012b). Using Clones and Copper to Resolve the Genetic Architecture of Metal Tolerance in a Marine Invader. *Ecology and Evolution*, **2**, 1319–1329.

McKie, J. C., Davies, I. M. and Topping, G. (1980). Heavy Metals in Grey Seals *(Halichoerus grypus)* from the East Coast of Scotland. *International Council for the Exploration of the Seas ICES*, *1980/E*, **41**, 1–13.

McLean, R. O. and Jones, A. K. (1975). Studies of Tolerance to Heavy Metals in the Flora of the Rivers Ystwyth and Clarach, Wales. *Freshwater Biology*, **5**, 431–444.

Meador, J. P. (2000). Predicting the Fate and Effects of Tributyltin in Marine Systems. *Reviews in Environmental Contamination and Toxicology*, **166**, 1–48.

Meharg, A. A. and Macnair, M. R. (1991). Uptake, Accumulation and Translocation of Arsenate in Arsenate-Tolerant and Non-Tolerant *Holcus lanatus* L. *New Phytologist*, **117**, 225–231.

Meier, S., Azcón, R., Cartes, P., Borie, F. and Cornejo, P. (2011). Alleviation of Cu Toxicity in *Oenothera picensis* by Copper-Adapted Arbuscular Mycorrhizal Fungi and Treated Agrowaste Residue. *Applied Soil Ecology*, **48**, 117–124.

Meier, S., Borie, F., Bolan, N. and Cornejo, P. (2012). Phytoremediation of Metal-Polluted Soils by Arbuscular Mycorrhizal Fungi. *Critical Reviews in Environmental Science and Technology*, **42**, 741–775.

Mendez, M. O. and Maier, R. M. (2008). Phytoremediation of Mine Tailings in Temperate and Arid Environments. *Reviews in Environmental Science and Biotechnology*, **7**, 47–59.

Mersch, J., Jeanjean, A., Spor, H. and Pihan, J.-C. (1992). The Freshwater Mussel *Dreissena polymorpha* as a Bioindicator for Trace Metals, Organochlorines and Radionuclides. In *The Zebra Mussel Dreissena polymorpha: Ecology, Biological Monitoring and First Applications in Water Quality Management*, ed. D. Neumann and H. A. Jenner. Stuttgart, Jena and New York, NY: Gustav Fischer Verlag, pp. 227–244.

Metcalfe, J. L. (1989). Biological Water Quality Assessment of Running Waters Based on Macroinvertebrate Communities: History and Present Status in Europe. *Environmental Pollution*, **60**, 101–139.

Michailova, P., Ilkova, J., Kerr, R. and White, K. (2009). Chromosome Variability in *Chironomus acidophilus* Keyl, 1960 from the Afon Goch, UK: A River Subject to Long-Term Trace Metal Pollution. *Aquatic Insects*, **31**, 213–225.

Michibata, H., Terada, T., Anada, N., Yamakawa, K. and Numakunai, T. (1986). The Accumulation and Distribution of Vanadium, Iron, and Manganese in Some Solitary Ascidians. *Biological Bulletin*, **171**, 672–681.

Michibata, H., Uyama, T., Ueki, T. and Kanamori, K. (2002). Vanadocytes, Cells Hold the Key to Resolving the Highly Selective Accumulation and Reduction of Vanadium in Ascidians. *Microscopy Research and Technique*, **56**, 421–434.

Migeon, A., Richaud, P., Guinet, F., Chalot, M. and Blaudez, D. (2009). Metal Accumulation by Woody Species on Contaminated Sites in the North of France. *Water, Air and Soil Pollution*, **204**, 89–101.

Miller, B. S. (1986). Trace Metals in the Common Mussel *Mytilus edulis* (L.) in the Clyde Estuary. *Proceedings of the Royal Society of Edinburgh*, **90B**, 377–391.

Millero, F. J., Woosley, R., Ditrolio, B. and Waters, J. (2009). Effect of Ocean Acidification on the Speciation of Metals in Seawater. *Oceanography*, **22**, 72–85.

Millward, R. N. (1996). Intracellular Inclusions in the Nematode *Tripyloides marinus* from Metal-Enriched and Cleaner Estuaries in Cornwall, South-West England. *Journal of the Marine Biological Association of the United Kingdom*, **76**, 885–895.

Millward, R. N. and Grant, A. (1995). Assessing the Impact of Copper on Nematode Communities from a Chronically Metal-Enriched Estuary Using Pollution-Induced Community Tolerance. *Marine Pollution Bulletin*, **30**, 701–706.

(2000). Pollution-Induced Tolerance to Copper of Nematode Communities in the Severely Contaminated Restronguet Creek and Adjacent Estuaries, Cornwall, United Kingdom. *Environmental Toxicology and Chemistry*, **19**, 454–461.

Milton, A., Cooke, J. A. and Johnston, M. S. (2003). Accumulation of Lead, Zinc, and Cadmium in a Wild Population of *Clethrionomys glareolus* from an Abandoned Lead Mine. *Archives of Environmental Contamination and Toxicology*, **44**, 405–411.

Ministry of Agriculture, Fisheries and Food (MAFF), Food Standards Committee (1956). *Report on Copper: Revised Recommendations for Limits for Copper Content of Foods*. London: Her Majesty's Stationery Office.

Ministry of Food, Food Standards Committee (1953). *Report on Zinc*. London: Her Majesty's Stationery Office.

Miramand, P. and Bentley, D. (1992). Concentration and Distribution of Heavy Metals in Tissues of Two Cephalopods, *Eledone cirrhosa* and *Sepia officinalis*, from the French Coast of the English Channel. *Marine Biology*, **114**, 407–414.

Mitchell, P. and Barr, D. (1995). The Nature and Significance of Public Exposure to Arsenic: A Review of Its Relevance to South West England. *Environmental Geochemistry and Health*, **17**, 57–82.

Moffat, W. E. (1989). Blood Lead Determinants of a Population Living in a Former Lead Mining Area in Southern Scotland. *Environmental Geochemistry and Health*, **11**, 3–9.

Moore, M. N., Allen, J. I. and McVeigh, A. (2006). Environmental Prognostics: An Integrated Model Supporting Lysosomal Stress Responses as Predictive Biomarkers of Animal Health Status. *Marine Environmental Research*, **61**, 278–304.

Moore, M. N., Viarengo, A. G., Somerfield, P. J. and Sforzini, S. (2013). Linking Lysosomal Biomarkers and Ecotoxicological Effects at Higher Biological Levels. In *Ecological Biomarkers: Indicators of Ecotoxicological Effects*, ed. C. Amiard-Triquet, J. C. Amiard and P. S. Rainbow. Boca Raton, FL: CRC Press, pp. 107–130.

Moore, P. G. (1979). Crystalline Structures in the Gut Caeca of the Amphipod *Stegocephaloides christianiensis* Boeck. *Journal of Experimental Marine Biology and Ecology*, **39**, 223–229.

Moore, P. G. and Rainbow, P. S. (1984). Ferritin Crystals in the Gut Caeca of *Stegocephaloides christianiensis* Boeck and Other Stegocephalidae (Amphipoda : Gammaridea): A Functional Interpretation. *Philosophical Transactions of the Royal Society London B*, **306**, 219–245.

(1987). Copper and Zinc in an Ecological Series of Talitroidean Amphipoda (Crustacea). *Oecologia*, **73**, 120–126.

(1989). Feeding of the Mesopelagic Gammaridean Amphipod *Parandania boecki* (Stebbing, 1888) (Crustacea : Amphipoda: Stegocephalidae) from the Atlantic Ocean. *Ophelia*, **30**, 1–19.

Moore, P. G., Rainbow, P. S. and Hayes, E. (1991). The Beach-Hopper *Orchestia gammarellus* (Crustacea: Amphipoda) as a Biomonitor for Copper and Zinc: North Sea Trials. *Science of the Total Environment*, **106**, 221–23

Morais, S., Garcia e Costa, F. and de Lourdes Pereira, M. (2012). Heavy Metals and Human Health. In *Environmental Health: Emerging Issues and Practice*, ed. J. Oosthuizen. Den Haag: InTech. DOI:10.5772/1519.

Morgan, A. J. and Morris, B. (1982). The Accumulation and Intracellular Compartmentation of Cadmium, Lead, Zinc and Calcium in Two Earthworm Species (*Dendrobaena rubida* and *Lumbricus rubellus*) Living in Highly Contaminated Soils. *Histochemistry*, **75**, 269 –287.

Morgan, J. E. and Morgan, A. J. (1988). Earthworms as Biological Monitors of Cadmium, Copper, Lead and Zinc in Metalliferous Soils. *Environmental Pollution*, **54**, 123–138.

(1989a). Zinc Sequestration by Earthworm (Annelida: Oligochaeta) Chloragocytes. *Histochemistry*, **90**, 405–411.

(1989b). The Effect of Lead Incorporation on the Elemental Composition of Earthworm (Annelida, Oligochaeta) Chloragosome Granules. *Histochemistry*, **92**, 237–241.

Moriarty, F., Bull, K. R., Hanson, H. M. and Freestone, P. (1982). The Distribution of Lead, Zinc and Cadmium in Sediments of an Ore-Enriched Lotic Ecosystem, the River Ecclesbourne, Derbyshire. *Environmental Pollution B*, **4**, 45–68.

Moriarty, F., Hanson, H. M. and Freestone, P. (1984). Limitations of Body Burden as an Index of Environmental Contamination: Heavy Metals in Fish *Cottus gobio* L. from the River Ecclesbourne, Derbyshire. *Environmental Pollution A*, **34**, 297–320.

Morillo, J. and Usero, J. (2008). Trace Metal Bioavailability in the Waters of Two Different Habitats in Spain: Huelva Estuary and Algeciras Bay. *Ecotoxicology and Environmental Safety*, **71**, 851–859.

Morris, A. W. (1984). The Chemistry of the Severn Estuary and the Bristol Channel. *Marine Pollution Bulletin*, **15**, 57–61.

(1986). Removal of Trace Metals in the Very Low Salinity Region of the Tamar Estuary, England. *Science of the Total Environment*, **49**, 297–304.

Morrison, L., Bennion, M., McGrory, E., Hurley, W. and Johnson, M. P. (2017). *Talitrus saltator* as a Biomonitor: An Assessment of Trace Element Contamination on an Urban Coastline Gradient. *Marine Pollution Bulletin*, **120**, 232–238.

Morrissey, C. A., Bendell-Young, L. I. and Elliott, J. E. (2005). Assessing Trace-Metal Exposure to American Dippers in Mountain Streams of Southwestern British Columbia, Canada. *Environmental Toxicology and Chemistry*, **24**, 836–845.

Mouneyrac, C., Amiard, J.-C. and Amiard-Triquet, C. (1998). Effects of Natural Factors (Salinity and Body Weight) on Cadmium, Copper, Zinc and Metallothionein-Like Protein Levels in Resident Populations of Oysters *Crassostrea gigas* from a Polluted Estuary. *Marine Ecology Progress Series*, **162**, 125–135.

Mouneyrac, C., Mastain, O., Amiard, J.-C., et al. (2003). Trace-Metal Detoxification and Tolerance of the Estuarine Worm *Hediste diversicolor* Chronically Exposed in Their Environment. *Marine Biology*, **143**, 731–744.

Mouneyrac, C., Perrein-Ettajani, H. and Amiard-Triquet, C. (2010). Influence on Anthropogenic Stress on Fitness and Behaviour of a Key-Species of Estuarine Ecosystems, the Ragworm *Nereis diversicolor*. *Environmental Pollution*, **158**, 121–128.

Mudge, G. P. (1983). The Incidence and Significance of Ingested Lead Pellet Poisoning in British Wildfowl. *Biological Conservation*, **27**, 333–372.

Murphy, B. L., Toole, A. P. and Bergstrom, P. D. (1989). Health Risk Assessment for Arsenic Contaminated Soil. *Environmental Geochemistry and Health*, **11**, 163–170.

Murray, A. J. (1979). Metals, Organochlorine Pesticides and PCB Residue Levels in Fish and Shellfish Landed in England and Wales during 1974. *MAFF Directorate of Fisheries Research. Aquatic Environment Monitoring Report*, **2**, 1–52.

Murray, L. A., Norton, M. G., Nunny, R. S. and Rolfe. M. S. (1980). The Field Assessment of Effects of Dumping Wastes at Sea: 7. Sewage Sludge and Industrial Waste Disposal in the Bristol Channel. *MAFF Directorate of Fisheries Research. Fisheries Research Technical Report*, **59**, 1–40.

Muskett, C. J., Roberts, L. H. and Page, B. J. (1979). Cadmium and Lead Pollution from Secondary Metal Refinery Operations. *Science of the Total Environment*, 11, 73–87.

Muus, B. J. and Dahlstrøm, P. (1964). *Collins Guide to the Sea Fishes of Britain and North-Western Europe.* London: Collins.

Nahmani, J., Hodson, M. E. and Black, S. (2007). A Review of Studies Performed to Assess Metal Uptake by Earthworms. *Environmental Pollution*, 145, 402–424.

Nakatsu, C. and Hutchinson, T. C. (1988). Extreme Metal and Acid Tolerance of *Euglena mutabilis* and an Associated Yeast from Smoking Hills, Northwest Territories, and Their Apparent Mutualism. *Microbial Ecology*, 16, 213–231.

NAMHO (National Association of Mining History Organisations) (2013). Mining and Quarrying Assessments 5: Iron and Ironstone. Advanced draft. 16/06/2014. www.namho.org/research/SECTION_5_Iron_20131209.pdf

Nasrolahi, A., Smith, B. D., Ehsanpour, M., Afkhami, M. and Rainbow, P. S. (2014). Biomonitoring of Trace Metal Bioavailabilities to the Barnacle *Amphibalanus amphitrite* along the Iranian Coast of the Persian Gulf. *Marine Environmental Research*, 101, 215–224.

Nassiri, Y., Ginsburger-Vogel, T., Mansot, J. L. and Wéry, J. (1996). Effects of Heavy Metals on *Tetraselmis suecica*: Ultrastructural and Energy-Dispersive X-ray Spectroscopic Studies. *Biology of the Cell*, 86, 151–160.

Nassiri, Y., Rainbow, P. S., Amiard-Triquet, C., Rainglet, F. and Smith, B. D. (2000). Trace Metal Detoxification in the Ventral Caeca of *Orchestia gammarellus* (Crustacea: Amphipoda). *Marine Biology*, 136, 477–484.

Neal, C., Whitehead, P. G., Jeffery, H. and Neal, M. (2005). The Water Quality of the River Carnon, West Cornwall, November 1992 to March 1994: The Impacts of Wheal Jane Discharges. *Science of the Total Environment*, 338, 23–39.

Neff, J. M. (1997). Ecotoxicology of Arsenic in the Marine Environment. *Environmental Toxicology and Chemistry*, 16, 917–927.

Neira, C., Levin, L. A., Mendoza, G. and Zirino, A. (2013). Alteration of Benthic Communities Associated with Copper Contamination Linked to Boat Moorings. *Marine Ecology*, 35, 46–66.

Newton, I., Bogan, J. A. and Haas, M. B. (1989). Organochlorines and Mercury in the Eggs of British Peregrines *Falco peregrinus*. *Ibis*, 131, 355–376.

Newton, I., Wyllie, I. and Asher, A. (1993). Long-Term Trends in Organochlorine and Mercury Residues in Some Predatory Birds in Britain. *Environmental Pollution*, 79, 143–151.

Newton, L. (1944). Pollution of the Rivers of West Wales by Lead and Zinc Mine Effluent. *Annals of Applied Biology*, 31, 1–11.

Ng, T. Y.-T., Pais, N. M. and Wood, C. M. (2011). Mechanisms of Waterborne Cu Toxicity to the Pond Snail *Lymnaea stagnalis*: Physiology and Cu Bioavailability. *Ecotoxicology and Environmental Safety*, 74, 1471–1479.

Nicholson, J. K. and Osborn, D. (1983). Kidney Lesions in Pelagic Seabirds with High Tissue Levels of Cadmium and Mercury. *Journal of Zoology*, 200, 99–118.

Nicholson, J. K., Kendall, M. D. and Osborn, D. (1983). Cadmium and Mercury Nephrotoxicity. *Nature*, 304, 633–635.

Nickless, G., Stenner, R. and Terrille, N. (1972). Distribution of Cadmium, Lead and Zinc in the Bristol Channel. *Marine Pollution Bulletin*, 3, 188–190.

Noël-Lambot, F., Bouquegneau, J. M., Frankenne, F. and Disteche, A. (1980). Cadmium, Zinc and Copper Accumulation in Limpets *(Patella vulgata)* from the Bristol Channel with Special Reference to Metallothioneins. *Marine Ecology Progress Series*, 2, 81–89.

North Sea Task Force (1993). *North Sea Quality Status Report 1993. Oslo and Paris Commissions, London.* Fredensborg: Olsen & Olsen.

Norton, M. G. and Murray, A. J. (1983). The Metal Content of Fish and Shellfish in Liverpool Bay. *Chemistry in Ecology*, 1, 159–171.

Norton, M. G., Eagle, R. A., Nunny, R. S., Rolfe, M. S., Hardiman, P A. and Hampson, B. L. (1981). The Field Assessment of Effects of Dumping Wastes at Sea: 8. Sewage Sludge Dumping in the Outer Thames Estuary. *MAFF Directorate of Fisheries Research. Fisheries Research Technical Report*, 62, 1–62.

Norton, M. G., Franklin, A., Rowlatt, S. M., Nunny, R. S. and Rolfe, M. S. (1984a). The Field Assessment of Effects of Dumping Wastes at Sea: 12. The Disposal of Sewage Sludge, Industrial Wastes and Dredged

Spoils in Liverpool Bay. *MAFF Directorate of Fisheries Research. Fisheries Research Technical Report*, **76**, 1–50.

Norton, M. G., Rowlatt, S. M. and Nunny, R. S. (1984b). Sewage Sludge Dumping and Contamination of Liverpool Bay Sediments. *Estuarine, Coastal and Shelf Science*, **19**, 69–87.

Nørum, U., Bondgaard, M., Pedersen, T. V. and Bjerregaard, P. (2005). In Vivo and in Vitro Cadmium Accumulation during the Moult Cycle of the Male Shore Crab *Carcinus maenas:* Interaction with Calcium Metabolism. *Aquatic Toxicology*, **72**, 29–44.

Nott, J. A. and Nicolaidou, A. (1989). The Cytology of Heavy Metal Accumulations in the Digestive Glands of Three Marine Gastropods. *Proceedings of the Royal Society London B*, **237**, 347–362.

(1990). Transfer of Metal Detoxification along Marine Food Chains. *Journal of the Marine Biological Association of the United Kingdom*, **70**, 905–912.

(1994). Variable Transfer of Detoxified Metals from Snails to Hermit Crabs in Marine Food Chains. *Marine Biology*, **120**, 369–377.

Nriagu, J. O. (1989). A Global Assessment of Natural Sources of Atmospheric Trace Metals. *Nature*, **338**, 47–49.

Nriagu, J. O. and Pacyna, J. M. (1988). Quantitative Assessment of Worldwide Contamination of Air, Water and Soils by Trace Metals. *Nature*, **333**, 134–139.

Nugegoda, D. and Rainbow, P. S. (1988). Zinc Uptake and Regulation by the Sublittoral Prawn *Pandalus montagui* (Crustacea: Decapoda). *Estuarine, Coastal and Shelf Science*, **26**, 619–632.

(1989a). Effects of Salinity Changes on Zinc Uptake and Regulation by the Decapod Crustaceans *Palaemon elegans* and *Palaemonetes varians*. *Marine Ecology Progress Series*, **51**, 57–75.

(1989b). Salinity, Osmolality and Zinc Uptake in *Palaemon elegans* (Crustacea: Decapoda). *Marine Ecology Progress Series*, **55**, 149–157.

Nuttall, C. A. and Younger, P. L. (1999). Reconnaissance Hydrogeochemical Evaluation of an Abandoned Pb–Zn Orefield, Nent Valley, Cumbria, UK. *Proceedings of the Yorkshire Geological Society*, **52**, 395–405.

O'Brien, A. L. and Keough, M. J. (2013). Detecting Benthic Community Responses to Pollution in Estuaries: A Field Mesocosm Approach. *Environmental Pollution*, **175**, 45–55.

O'Connor, D.J. and Nielsen, S.W. (1981). Environmental Survey of Methylmercury Levels in Wild Mink (*Mustela vison*) and Otter (*Lutra canadensis*) from the Northeastern United States and Experimental Pathology of Methylmercurialism in the Otter. In *Proceedings Worldwide Furbearer Conference Frostburg, MD, USA, August 1980*, ed. J. A. Chapman and D. Pursley. Vancouver: R. R. Donnelly, pp. 1728–1745.

O'Donohoe, J., Chalkley, S., Richmond, J. and Barltrop, D. (1998). Blood Lead in U.K. Children: Time for a Lower Action Level? *Clinical Science*, **95**, 219–223.

Ogilvie, L. and Grant, A. (2008). Linking Pollution Induced Community Tolerance (PICT) and Microbial Community Structure in Chronically Metal Polluted Estuarine Sediments. *Marine Environmental Research*, **65**, 187–198.

O'Halloran, J., Myers, A. A. and Duggan, P. F. (1988). Lead Poisoning in Swans and Sources of Contamination in Ireland. *Journal of Zoology*, **216**, 211–223.

(1991). Lead Poisoning in Mute Swans *Cygnus olor* in Ireland: A Review. *Wildfowl*, **Suppl.** 1, 389–395.

Olson, B. H. and Thornton, I. (1982). The Resistance Patterns to Metals of Bacterial Populations in Contaminated Land. *Journal of Soil Science*, **33**, 271–277.

Oomen, A. G., Hack, A., Minekus, M., et al. (2002). Comparison of Five in Vitro Digestion Models to Study the Bioaccessibility of Soil Contaminants. *Environmental Science and Technology*, **36**, 3326–3334.

Orton, J. H. (1923). An Account of Investigations into the Cause or Causes of the Unusual Mortality among Oysters in English Oyster Beds during 1920 and 1921. *Ministry of Agriculture and Fisheries Fishery Investigations Series II*, **7**, 1–199.

Osborn, D. (1978). A Naturally Occurring Cadmium and Zinc Binding Protein from the Liver and Kidney of *Fulmarus glacialis*, a Pelagic North Atlantic Seabird. *Biochemical Pharmacology*, **27**, 822–824.

Osborn, D., Harris, M. P. and Nicholson, J. K. (1979). Comparative Tissue Distribution of mercury, Cadmium and Zinc in Three Species of Pelagic Seabirds. *Comparative Biochemistry and Physiology*, **64C**, 61–67.

O'Shaughnessy, A. W. E. (1866). On Green Oysters. *Annals and Magazine of Natural History*, 18, 221–228.

OSPAR Commission (2007). *Background Document on Biological Effects Monitoring Techniques*. OSPAR Publication 333/2007 Assessment and Monitoring Series. London: OSPAR.

Owens, M. (1984). Severn Estuary: An Appraisal of Water Quality. *Marine Pollution Bulletin*, 15, 41–47.

Oweson, C. A. M., Baden, S. P. and Hernroth, B. E. (2006). Manganese Induced Apoptosis in Haematopoietic Cells of *Nephrops norvegicus* (L.). *Aquatic Toxicology*, 77, 322–328.

Packer, D. M., Ireland, M. P. and Wootton, R. J. (1980). Cadmium, Copper, Lead, Zinc and Manganese in the Polychaete *Arenicola marina* from Sediments around the Coast of Wales. *Environmental Pollution A*, 22, 309–321.

Pain, D. J., Cromie, R. and Green, R. E. (2015). Poisoning of Birds and Other Wildlife from Ammunition-Derived Lead in the UK. In *Lead Ammunition: Understanding and Minimising the Risks to Human and Environmental Health. Proceedings of the Oxford Lead Symposium, Edward Grey Institute, University of Oxford, December 2014*, ed. R. J. Delahey and C. J. Spray. Oxford: Edward Grey Institute, pp. 58–84.

Pain, D. J., Sears, J. and Newton, I. (1995). Lead Concentrations in Birds of Prey in Britain. *Environmental Pollution*, 87, 173–180.

Palacios, O., Pagani, A., Pérez-Rafael, S., et al. (2011). Shaping Mechanisms of Metal Specificity in a Family of Metazoan Metallothioneins: Evolutionary Differentiation of Mollusc Metallothioneins. *BMC Biology*, 9, 4. http://biomedcentral.com/1741-7007/9/4.

Parkman, R. H., Curtis, C. D. and Vaughan, D. J. (1996). Metal Fixation and Mobilisation in the Sediments of the Afon Goch Estuary: Dulas Bay, Anglesey. *Applied Geochemistry*, 11, 203–210.

Paris, J. R., King, R. A. and Stevens, J. R. (2015). Human Mining Activity across the Ages Determines the Genetic Structure of Modern Brown Trout (*Salmo trutta* L.) Populations. *Evolutionary Applications*, 8, 573–585.

Parslow, J. L. F. (1973). Mercury in Waders from the Wash. *Environmental Pollution*, 5, 295–304.

Parslow, J. L. F. and Jefferies, D. J. (1977). Gannets and Toxic Chemicals. *British Birds*, 70, 366–372.

Patterson, G. and Whitton, B. A. (1981). Chemistry of Water, Sediments and Algal Filaments in Groundwater Draining an Old Lead–Zinc Mine. In *Heavy Metals in Northern England: Environmental and Biological Aspects*, ed. P. J. Say and B. A. Whitton. Durham: Department of Botany, University of Durham, pp. 65–72.

Pearson, T. H. (1987). Benthic Ecology in an Accumulating Sludge-Disposal Site. In *Oceanic Processes in Marine Pollution, Vol. 1: Biological Processes and Wastes in the Ocean*, ed. J. M. Capuzzo and D. R. Kester. Malabar, FL: Robert E. Krieger Publishing Company, pp. 195–200.

Pearson, T. H. and Rosenberg, R. (1978). Macrobenthic Succession in Relation to Organic Enrichment and Pollution of the Marine Environment. *Oceanography and Marine Biology Annual Review*, 16, 229–311.

Peden, J. D., Crothers, J. H., Waterfall, C. E. and Beasley, J. (1973). Heavy Metals in Somerset Marine Organisms. *Marine Pollution Bulletin*, 4, 7–9.

Pedersen, K. L., Bach, L. T. and Bjerregaard, P. (2014). Amount and Metal Composition of Midgut Gland Metallothionein in Shore Crabs (*Carcinus maenas*) after Exposure to Cadmium in Food. *Aquatic Toxicology*, 150, 182–188.

Pedersen, K. L., Pedersen, S. N., Højrup, P., et al. (1994). Purification and Characterization of a Cadmium-Induced Metallothionein from the Shore Crab *Carcinus maenas* (L.). *Biochemical Journal*, 297, 609–614.

Peltier, G. L., Meyer, J. L., Jagoe, C. H and Hopkins, W. A. (2008). Using Trace Element Concentrations in *Corbicula fluminea* to Identify Potential Sources of Contamination in an Urban River. *Environmental Pollution*, 154, 283–290.

Perales-Vela, H. V., Peña-Castro, J. M. and Cañizares-Villanueva, R. O. (2006). Heavy Metal Detoxification in Eukaryotic Microalgae. *Chemosphere*, 64, 1–10.

Peralta-Videa, J. R., Lopez, M. L., Narayan, M., Saupe, G. and Gardea-Torresdey, J. (2009). The Biochemistry of Environmental Heavy Metal Uptake by Plants: Implications for the Food Chain. *International Journal of Biochemistry and Cell Biology*, 41, 1665–1677.

Perkins, J. W. (1972). *Geology Explained: Dartmoor and the Tamar Valley*. Newton Abbot: David and Charles.

Peterson, C. L., Klawe, W. L. and Sharp, G. D. (1973). Mercury in Tunas: A Review. *Fishery Bulletin*, **71**, 603–613.

Peterson, L. R., Trivett, V., Baker, A. J. M., Aguiar, C. and Pollard, A. J. (2003). Spread of Metals through an Invertebrate Food Chain as Influenced by a Plant that Hyperaccumulates Nickel. *Chemoecology*, **13**, 103–108.

Phillips, D. J. H. (1994). Macrophytes as Biomonitors of Trace Metals. In *Biomonitoring of Coastal Waters and Estuaries*, ed. K. J. M. Kramer. Boca Raton, FL: CRC Press, pp. 85–106.

Phillips, D. J. H. and Rainbow, P. S. (1988). Barnacles and Mussels as Biomonitors of Trace Elements: A Comparative Study. *Marine Ecology Progress Series*, **49**, 83–93.

(1994). *Biomonitoring of Trace Aquatic Contaminants*, 2nd edn. London: Chapman and Hall.

Pilon-Smits, E. A. H. and Freeman, J. L. (2006). Environmental Cleanup Using Plants: Biotechnological Advances and Ecological Considerations. *Frontiers in Ecology and the Environment*, **4**, 203–210.

Piola, R. F. and Johnston, E. L. (2008). Pollution Reduces Native Diversity and Increases Invader Dominance in Marine Hard-Substrate Communities. *Diversity and Distributions*, **14**, 329–342.

Pirie, B. J. S. and Bell, M. V. (1984). The Localization of Inorganic Elements, Particularly Vanadium and Sulphur, in Haemolymph from the Ascidians *Ascidia mentula* (Müller) and *Ascidiella aspersa* (Müller). *Journal of Experimental Marine Biology and Ecology*, **74**, 187–194.

Pirie, B. J. S., George, S. G., Lytton, D. G. and Thomson, J. D. (1984). Metal-Containing Blood Cells of Oysters: Ultrastructure, Histochemistry and X-ray Microanalysis. *Journal of the Marine Biological Association of the United Kingdom*, **64**, 115–123.

Pirrie, D., Beer, A. J. and Camm, G. S. (1999). Early Diagenetic Sulphide Minerals in the Hayle Estuary, Cornwall. *Geoscience in South-West England*, **9**, 325–332.

Poirier, L. and Cossa, D. (1981). Distribution tissulaire du cadmium chez *Meganyctiphanes norvegica* (Euphausiacée): état naturel et accumulation expérimentale de formes solubles. *Canadian Journal of Fisheries and Aquatic Sciences*, **38**, 1449–1453.

Pollard, A. J. and Baker, A. J. M. (1997). Deterrence of Herbivory by Zinc Hyperaccumulation in *Thlaspi caerulescens*. *New Phytologist*, **135**, 655–658.

Pook, C., Lewis, C. and Galloway, T. (2009). The Metabolic and Fitness Costs Associated with Metal Resistance in *Nereis diversicolor*. *Marine Pollution Bulletin*, **58**, 1063–1071.

Pope, N. D. and Langston, W. J. (2011). Sources, Distribution and Temporal Variability of Trace Metals in the Thames Estuary. *Hydrobiologia*, **672**, 49–68.

Porter, E. K. and Peterson, P. J. (1977). Arsenic Tolerance in Grasses Growing on Mine Waste. *Environmental Pollution*, **14**, 255–265.

Portmann, J. E. (1979). Chemical Monitoring of Residue Levels in Fish and Shellfish Landed in England and Wales during 1970–73. *Aquatic Environment Monitoring Report. MAFF Directorate of Fisheries Research*, **1**, 1–70.

Posthuma, L. (1990). Genetic Differentiation between Populations of *Orchesella cincta* (Collembola) from Heavy Metal Contaminated Sites. *Journal of Applied Ecology*, **27**, 609–622.

Posthuma, L., Hogervorst, R. F., Joosse, E. N. G. and van Straalen, N. M. (1993). Genetic Variation and Covariation for Characteristics Associated with Cadmium Tolerance in Natural Populations of the Springtail *Orchesella cincta* (L.). *Evolution*, **47**, 619–631.

Posthuma, L., Hogervorst, R. F. and van Straalen, N. M. (1992). Adaptation to Soil Pollution by Cadmium Excretion in Natural Populations of *Orchesella cincta* (L.) (Collembola). *Archives of Environmental Contamination and Toxicology*, **22**, 145–156.

Power, M., Attrill, M. J. and Thomas, R. M. (1999). Heavy Metal Concentration Trends in the Thames Estuary. *Water Research*, **33**, 1672–1680.

Proctor, J. (1971a). The Plant Ecology of Serpentine. II. Plant Response to Serpentine Soils. *Journal of Ecology*, **59**, 397–410.

(1971b). The Plant Ecology of Serpentine. III. The Influence of a High Magnesium/Calcium Ratio and High Nickel and Chromium Levels in some British and Swedish Serpentine Soils. *Journal of Ecology*, **59**, 827–842.

Proctor, J. and Woodell, S. R. J. (1971). The Plant Ecology of Serpentine. I. Serpentine Vegetation of England and Scotland. *Journal of Ecology*, **59**, 375–395.

Pulford, I. D., MacKenzie, A. B., Donatello, S. and Hastings, L. (2009). Source Term Characterisation Using Concentration Trends and Geochemical Associations of Pb and Zn in River Sediments in the Vicinity of a Disused Mine Site: Implications for Contaminant Metal Dispersion Processes. *Environmental Pollution*, **157**, 1649–1656.

Pullen J. S. H. and Rainbow, P. S. (1991). The Composition of Pyrophosphate Heavy Metal Detoxification Granules in Barnacles. *Journal of Experimental Biology and Ecology*, **150**, 249–266.

Purvis, O. W. (1966). Interactions of Lichens with Metals. *Science Progress*, **79**, 283–309.

(2010a). Lichens and Industrial Pollution. In *Ecology of Industrial Pollution*, ed. L. C. Batty and K. B. Hallberg. Cambridge: British Ecological Society, Cambridge University Press, pp. 41–69.

(2010b). *Lichens*. London: Natural History Museum.

Purvis, O. W. and Halls, C. (1996). A Review of Lichens in Metal-Enriched Environments. *Lichenologist*, **28**, 571–601.

Purvis, O. W. and James, P. W. (1985). Lichens of the Coniston Copper Mines. *Lichenologist*, **17**, 221–237.

Purvis, O. W., Elix, J. A., Broomhead, J. A. and Jones, G. C. (1987). The Occurrence of Copper-Norstictic Acid in Lichens from Cupriferous Substrata. *Lichenologist*, **19**, 193–203.

Pynnönen, K., Holwerda, D. A. and Zandee, D. I. (1987) Occurrence of Calcium Concretions in Various Tissues of Freshwater Mussels, and Their Capacity for Cadmium Sequestration. *Aquatic Toxicology*, **10**, 101–114.

Qiu, J. (2013). Tough Talk over Mercury Treaty. *Nature*, **493**, 144–145.

Quicke, D. L. J., Wyeth, P., Fawke, J. D., Basibuyuk, H. H. and Vincent, J. F. V. (1998). Manganese and Zinc in the Ovipositors and Mandibles of Hymenopterous Insects. *Zoological Journal of the Linnean Society*, **124**, 387–396.

Radford, P. J., Uncles, R. J. and Morris, A. W. (1981). Simulating the Impact of Technological Change on Dissolved Cadmium Distribution in the Severn Estuary. *Water Research*, **15**, 1045–1052.

Rainbow, P. S. (1985). Accumulation of Zn, Cu and Cd by Crabs and Barnacles. *Estuarine, Coastal and Shelf Science*, **21**, 669–686.

(1987). Heavy Metals in Barnacles. In *Barnacle Biology: Crustacean Issues 5*, ed. A. J. Southward. Rotterdam: A. A. Balkema, pp. 405–417.

(1989). Copper, Cadmium and Zinc Concentrations in Oceanic Amphipod and Euphausiid Crustaceans, as a Source of Heavy Metals to Pelagic Seabirds. *Marine Biology*, **103**, 513–518.

(1997). Trace Metal Accumulation in Marine Invertebrates: Marine Biology or Marine Chemistry? *Journal of the Marine Biological Association of the United Kingdom*, **77**, 195–210.

(1998). Phylogeny of Trace Metal Accumulation in Crustaceans. In *Metal Metabolism in Aquatic Environments*, ed. W. J. Langston and M. Bebianno. London: Chapman and Hall, pp. 285–319.

(2002). Trace Metal Concentrations in Aquatic Invertebrates: Why and So What? *Environmental Pollution*, **120**, 497–507.

(2007). Trace Metal Bioaccumulation: Models, Metabolic Availability and Toxicity. *Environment International*, **33**, 576–582.

(2011). Charles Darwin and Marine Biology. *Marine Ecology*, **32** (Suppl. 1), 130–134.

Rainbow, P. S. and Abdennour, C. (1989). Copper and Haemocyanin in the Mesopelagic Decapod Crustacean *Systellaspis debilis*. *Oceanologica Acta*, **12**, 91–94.

Rainbow, P. S. and Kwan, M. K. H. (1995). Physiological Responses and the Uptake of Cadmium and Zinc by the Amphipod Crustacean Orchestia gammarellus. *Marine Ecology Progress Series*, **127**, 87–102.

Rainbow, P. S. and Luoma, S. N. (2005). Lessons from History: A Cornish Tale. Learned Discourse. *SETAC Globe*, **6**, 25–26.

(2011a). Biodynamic Parameters of the Accumulation of Toxic Metals, Detoxification, and the Acquisition of Metal Tolerance. In *Tolerance to Environmental Contaminants*, ed. C. Amiard-Triquet, P. S. Rainbow and M. Roméo. Boca Raton, FL: CRC Press, pp. 127–151.

(2011b). Trace Metals in Aquatic Invertebrates. In *Environmental Contaminants in Biota: Interpreting Tissue Concentrations*, 2nd edn., ed. W. N. Beyer and J. P. Meador. Boca Raton, FL: Taylor and Francis Books, pp. 231–252.

(2011c). Metal Toxicity, Uptake and Bioaccumulation in Aquatic Invertebrates: Modelling Zinc in Crustaceans. *Aquatic Toxicology*, **105**, 455–465.

(2015). Bioavailability of Metals to Aquatic Biota. In *Fundamentals of Ecotoxicology*, 4th edn., ed. M. C. Newman. Boca Raton, FL: CRC Press, pp. 140–143.

Rainbow, P. S. and Moore, P. G. (1986). Comparative Metal Analyses in Amphipod Crustaceans. *Hydrobiologia*, 14, 273–289.

(1990). Seasonal Variation in Copper and Zinc Concentrations in Three Talitrid Amphipods (Crustacea). *Hydrobiologia*, 196, 65–72.

Rainbow, P. S. and Phillips, D. J. H. (1993). Cosmopolitan Biomonitors of Trace Metals. *Marine Pollution Bulletin*, 26, 593–601.

Rainbow, P. S. and Smith, B. D. (2010). Trophic Transfer of Trace Metals: Subcellular Compartmentalisation in Bivalve Prey and Comparative Assimilation Efficiencies of Two Invertebrate Predators. *Journal of Experimental Marine Biology and Ecology*, 390, 143–148.

(2013). Accumulation and Detoxification of Copper and Zinc by the Decapod Crustacean *Palaemonetes varians* from Diets of Field-Contaminated Polychaetes *Nereis diversicolor*. *Journal of Experimental Marine Biology and Ecology*, 449, 312–320.

Rainbow, P. S. and Wang, W.-X. (2001). Comparative Assimilation of Cr, Cr, Se, and Zn by the Barnacle *Elminius modestus* from Phytoplankton and Zooplankton Diets. *Marine Ecology Progress Series*, 218, 239–248.

Rainbow, P. S. and White, S. L. (1989). Comparative Strategies of Heavy Metal Accumulation by Crustaceans: Zinc, Copper and Cadmium in a Decapod, an Amphipod and a Barnacle. *Hydrobiologia*, 174, 245–262.

Rainbow, P. S., Amiard, J. C., Amiard-Triquet, C., et al. (2007). Trophic Transfer of Trace Metals: Subcellular Compartmentalization in Bivalve Prey, Assimilation by a Gastropod Predator and *in Vitro* Digestion Simulations. *Marine Ecology Progress Series*, 348, 125–138.

Rainbow, P. S., Amiard-Triquet, C., Amiard, J. C., et al. (1999). Trace Metal Uptake Rates in Crustaceans (Amphipods and Crabs) from Coastal Sites in NW Europe Differentially Enriched with Trace Metals. *Marine Ecology Progress Series*, 183, 189–203.

Rainbow, P. S., Geffard, A., Jeantet, A.-Y., Smith, B. D., Amiard, J. C. and Amiard-Triquet, C. (2004). Enhanced Food Chain Transfer of Copper from a Diet of Copper-Tolerant Estuarine Worms. *Marine Ecology Progress Series*, 271, 183–191.

Rainbow, P. S., Hildrew, A. G., Smith, B. D., Geatches, T. and Luoma S. N. (2012). Caddisflies as Biomonitors Identifying Thresholds of Toxic Metal Bioavailability that Affect the Stream Benthos. *Environmental Pollution*, 166, 196–207.

Rainbow, P. S., Kriefman, S., Smith, B. D. and Luoma, S. N. (2011c). Have the Bioavailabilities of Trace Metals to a Suite of Biomonitors Changed over Three Decades in SW England Estuaries Historically Affected by Mining? *Science of the Total Environment*, 409, 1589–1602.

Rainbow, P. S., Liu, F. and Wang, W.-X. (2015). Metal Accumulation and Toxicity: The Critical Accumulated Concentration of Metabolically Available Zinc in an Oyster Model. *Aquatic Toxicology*, 162, 102–108.

Rainbow, P. S., Luoma, S. N. and Wang, W.-X. (2011a). Trophically Available Metal: A Variable Feast. *Environmental Pollution*, 159, 2347–2349.

Rainbow, P. S., Malik, I. and O'Brien, P. (1993). Physico-Chemical and Physiological Effects on the Uptake of Dissolved Zinc and Cadmium by the Amphipod Crustacean *Orchestia gammarellus*. *Aquatic Toxicology*, 25, 15–30.

Rainbow, P. S., Moore, P. G. and Watson, D. (1989). Talitrid Amphipods as Biomonitors for Copper and Zinc. *Estuarine, Coastal and Shelf Science*, 28, 567–582.

Rainbow, P. S, Poirier, L., Smith, B. D., Brix, K. V. and Luoma, S. N. (2006a). Trophic Transfer of Trace Metals: Subcellular Compartmentalization in a Polychaete and Assimilation by a Decapod Crustacean. *Marine Ecology Progress Series*, 308, 91–100.

(2006b). Trophic Transfer of Trace Metals from the Polychaete Worm *Nereis diversicolor* to the Polychaete *Nereis virens* and the Decapod Crustacean *Palaemonetes varians*. *Marine Ecology Progress Series*, 321, 167–181.

Rainbow, P. S., Scott, A. G., Wiggins, E. A. and Jackson, R. W. (1980). Effect of Chelating Agents on the Accumulation of Cadmium by the Barnacle *Semibalanus balanoides*, and Complexation of Soluble Cd, Zn and Cu. *Marine Ecology Progress Series*, 2, 143–152.

Rainbow, P. S., Smith, B. D. and Casado-Martinez, M. C. (2011b). Biodynamic Modelling of the Bioaccumulation of Arsenic by the Polychaete *Nereis diversicolor. Environmental Chemistry*, **8**, 1–8.

Rainbow, P. S., Smith, B. D. and Lau, S. S. S. (2002). Biomonitoring of Trace Metal Availabilities in the Thames Estuary Using a Suite of Littoral Biomonitors. *Journal of the Marine Biological Association of the United Kingdom*, **82**, 793–799.

Rainbow, P. S., Smith, B. D. and Luoma, S. N. (2009a). Differences in Trace Metal Bioaccumulation Kinetics among Populations of the Polychaete *Nereis diversicolor* from Metal-Contaminated Estuaries. *Marine Ecology Progress Series*, **376**, 173–184.

(2009b). Biodynamic Modelling and the Prediction of Ag, Cd and Zn Accumulation from Solution and Sediment by the Polychaete *Nereis diversicolor. Marine Ecology Progress Series*, **390**, 145–155.

Raistrick, A. (1973). *Lead Mining in the Mid-Pennines.* Truro: D. Bradford Barton Ltd.

(1989). *Dynasty of Iron Founders*, 2nd edn. York: William Sessions Limited, Ebor Press.

Raistrick, A. and Jennings, B. (1989). *A History of Lead Mining in the Pennines. Newcastle upon Tyne, UK: Davis Books Ltd and Littleborough*, Lancashire: George Kelsall Publishing.

Raymont, J. E. G. (1972). Some Aspects of Pollution in Southampton Water. *Proceedings of the Royal Society of London B*, **180**, 451–468.

(1983). *Plankton and Productivity in the Oceans, Volume 2: Zooplankton*, 2nd edn. Oxford: Pergamon Press.

Read, H. J., Martin, M. H. and Rayner, J. M. V. (1998). Invertebrates in Woodland Polluted by Heavy Metals: An Evaluation Using Canonical Correspondence Analysis. *Water, Air, and Soil Pollution*, **106**, 17–42.

Reed, R. H. and Moffat, L. (1983). Copper Toxicity and Copper Tolerance in *Enteromorpha compressa* (L.) Grev. *Journal of Experimental Marine Biology and Ecology*, **69**, 85–103.

Reese, M. J. (1937). The Microflora of the Non-Calcareous Streams Rheidol and Melindwr with Special Reference to Water Pollution from Lead Mines in Cardiganshire. *Journal of Ecology*, **25**, 385–407.

Reeves, P. D. (2006). Hyperaccumulation of Trace Elements by Plants. In *Earth and Environmental Sciences, Vol. 68, Phytoremediation of Metal-Contaminated Soils*. NATO Science Series IV, ed. J.-L. Morel, G. Echevarria and N. Goncharova, Dordrecht: Springer, pp. 25–52.

Regoli, F. (2000). Total Oxyradical Scavenging Capacity (TOSC) in Polluted and Translocated Mussels: A Predictive Biomarker of Oxidative Stress. *Aquatic Toxicology*, **50**, 351–361.

Reijnders, P. J. H. (1980). Organochlorine and Heavy Metal Residues in Harbour Seals from the Wadden Sea and Their Possible Effects on Reproduction. *Netherlands Journal of Sea Research*, **14**, 30–65.

Reinfelder, J. R. and Fisher, N. S. (1991). The Assimilation of Elements Ingested by Marine Copepods. *Science*, **251**, 794–796.

Remon, E., Bouchardon, J.-L., Le Guédard, M., Bessoule, J.-J., Conord, C. and Faure, O. (2013). Are Plants Useful as Accumulation Indicators of Metal Bioavailability? *Environmental Pollution*, **175**, 1–7.

Ridout, P. S., Rainbow, P. S., Roe, H. S. J. and Jones, H. R. (1989). Concentrations of V, Cr, Mn, Fe, Ni, Co, Cu, Zn, As and Cd in Mesopelagic Crustaceans from the North East Atlantic Ocean. *Marine Biology*, **100**, 465–471.

Rieuwerts, J. S., Mighanetara, K., Braungardt, C. B., Rollinson, G. K., Pirie, D. and Azizi, F. (2014). Geochemistry and Mineralogy of Arsenic in Mine Wastes and Stream Sediments in a Historic Metal Mining Area in the UK. *Science of the Total Environment*, **472**, 226–234.

Rieuwerts, J. S., Searle, P. and Buck, R. (2006). Bioaccessible Arsenic in the Home Environment in Southwest England. *Science of the Total Environment*, **371**, 89–98.

RNO (2006). *Surveillance du Milieu Marin. Travaux du RNO. Edition 2006.* Paris: Ifremer et Ministère de l'Ecologie et du Dévelopement Durable.

Roberts, D. F., Elliott, M. and Read, P. A. (1986). Cadmium Contamination, Accumulation and Some Effects in Mussels from a Polluted Marine Environment. *Marine Environmental Research*, **18**, 165–183.

Roberts, R. D., Johnson, M. S. and Hutton, M. (1978). Lead Contamination of Small Mammals from Abandoned Metalliferous Mines. *Environmental Pollution*, **15**, 61–69.

Rodger, G. K. and Davies, I. M. (1992). The Recovery of a Sewage Sludge Dumping Ground: Trace Metal Geochemistry in the Sediment. *Science of the Total Environment*, 119, 57–75.

Rodrigues, E. T. and Pardal, M. A. (2014). The Crab *Carcinus maenas* as a Suitable Experimental Model in Ecotoxicology. *Environment International*, 70, 158–182.

Roe, H. S. J. (1984). The Diel Migration and Distribution within a Mesopelagic Community in the North-East Atlantic: 2 – Vertical Migrations and Feeding of Mysids and Decapod Crustacea. *Progress in Oceanography*, 13, 269–318.

Rollinson, G. K., Pirrie, D., Power, M. R., Cundy, A. and Camm, G. S. (2007). Geochemical and Mineralogical Record of Historical Mining, Hayle Estuary, Cornwall, UK. *Geoscience in South-West England*, 11, 326–337.

Roméo, M., Poirier, L. and Berthet, B. (2009). Biomarkers Based upon Biochemical Responses. In *Environmental Assessment of Estuarine Ecosystems*, ed. C. Amiard-Triquet and P. S. Rainbow. Boca Raton, FL: CRC Press, pp. 59–81.

Romeril, M. G. (1974). Trace Metals in Sediments and Bivalve Mollusca in Southampton Water and the Solent. *Revue Internationale d'Océanographie Medicale*, 23, 31–47.

Romero-Ruiz, A., Alhama, J., Blasco, J., Gómez-Ariza, J. L. and López-Barea, J. (2008). New Metallothionein Assay in *Scrobicularia plana*: Heating Effect and Correlation with Other Biomarkers. *Environmental Pollution*, 156, 1340–1347.

Ronald, K., Tessaro, S. V., Uthe, J. F., Freeman, H. C. and Frank, R. (1977). Methylmercury Poisoning in the Harp Seal (*Pagophilus groenlandicus*). *Science of the Total Environment*, 38, 153–166.

Rout, G. R., Samantaray, S. and Das, P. (2001). Aluminium Toxicity in Plants: A Review. *Agronomie*, 21, 3–21.

Rowan, J. S., Barnes, S. J. A., Hetherington, S. L., Lambers, B. and Parsons, F. (1995). Geomorphology and Pollution: The Environmental Impacts of Lead Mining, Leadhills, Scotland. *Journal of Geochemical Exploration*, 52, 57–65.

Rowlands, J. (1966). *Copper Mountain*. Denbigh: Anglesey Antiquarian Society, Gee and Son.

Ruiz, J. M. (2004). Oil Spills versus Shifting Baselines. *Marine Ecology Progress Series*, 282, 307–309.

Runham, N. W. (1961). The Histochemistry of the Radula of *Patella vulgata*. *Quarterly Journal of Microscopical Science*, 102, 371–380.

Runham, N. W., Thornton, P. R., Shaw, D. A. and Wayte, R. C. (1969). The Mineralization and Hardness of the Radular Teeth of the Limpet *Patella vulgata* L. *Zeitschrift für Zellforschung und Mikroskopische Anatomie*, 99, 608–626.

Russell, G. and Morris, O. P. (1970). Copper Tolerance in the Marine Fouling Alga *Ectocarpus siliculosus*. *Nature*, 228, 288–289.

Rygg, B. (1985), Effect of Sediment Copper on Benthic Fauna. *Marine Ecology Progress Series*, 25, 83–89.

Salice, C. J. and Miler, T. J. (2003). Population-Level Responses to Long-Term Cadmium Exposure in Two Strains of the Freshwater Gastropod *Biomphalaria glabrata*: Results from a Life-Table Experiment. *Environmental Toxicology and Chemistry*, 22, 678–688.

Sauvé, S., Dumestre, A., McBride, M. and Hendershot, W. (1998). Derivation of Soil Quality Criteria Using Predicted Chemical Speciation of Pb^{2+} and Cu^{2+}. *Environmental Toxicology and Chemistry*, 17, 1481–1489.

Savari, A., Lockwood, A. P. M. and Sheader, M. (1991). Effects of Season and Size (Age) on Heavy Metal Concentrations of the Common Cockle (*Cerastoderma edule* (L.)) from Southampton Water. *Journal of Molluscan Studies*, 57, 45–57.

Say, P. J. and Giani, N. (1981). The Riou Mort, a Tributary to the River Lot Polluted by Heavy Metals. II. Accumulation of Zinc by Oligochaetes and Chironomids. *Acta Oecologica*, 2, 339–355.

Say, P. J. and Whitton, B. A. (1980). Changes in Flora down a Stream Showing a Zinc Gradient. *Hydrobiologia*, 76, 255–262.

(1981). Chemistry and Ecology of Zinc-Rich Streams in the Northern Pennines. In *Heavy Metals in Northern England: Environmental and Biological Aspects*, ed. P. J. Say and B. A. Whitton. Durham: Department of Botany, University of Durham, pp. 55–63.

Say, P. J., Burrows, I. G. and Whitton, B. A. (1990). *Enteromorpha* as a Monitor of Heavy Metals in Estuaries. *Hydrobiologia*, 195, 119–126.

Say, P. J., Diaz, B. M. and Whitton, B. A. (1977). Influence of Zinc on Lotic Plants. I. Tolerance of

Hormidium Species to Zinc. *Freshwater Biology*, **7**, 357–376.

Say, P. J., Harding, P. C. and Whitton, B. A. (1981). Aquatic Mosses as Monitors of Heavy Metal Contamination in the River Etherow, Great Britain. *Environmental Pollution B*, **2**, 295–307.

Scaps, P. (2002). A Review of the Biology, Ecology and Potential Use of the Common Ragworm *Hediste diversicolor* (O.F. Müller) (Annelida: Polychaeta). *Hydrobiologia*, **470**, 203–218.

Scheifler, R., Coeurdassier, M., Morilhat, C., et al. (2006). Lead Concentrations in Feathers and Blood of Common Blackbirds (*Turdus merula*) and in Earthworms Inhabiting Unpolluted and Moderately Polluted Urban Areas. *Science of the Total Environment*, **371**, 197–205.

Scheinberg, H. (1991). Copper. In *Metals and Their Compounds in the Environment*, ed. E. Merian. Weinheim: VCH, pp. 893–908.

Scheuhammer, A. M. (1987). The Chronic Toxicity of Aluminium, Cadmium, Mercury, and Lead in Birds: A Review. *Environmental Pollution*, **46**, 263–295.

(1989). Monitoring Wild Bird Populations for Lead Exposure. *Journal of Wildlife Management*, **53**, 759–764.

(1991). Effects of Acidification on the Availability of Toxic Metals and Calcium to Wild Birds and Mammals. *Environmental Pollution*, **71**, 329–375.

Scheuhammer, A. M. and Norris, S. L. (1996). The Ecotoxicology of Lead Shot and Lead Fishing Weights. *Ecotoxicology*, **5**, 279–295.

Schill, R. O. and Köhler, H.-R. (2004). Energy Reserves and Metal-Storage Granules in the Hepatopancreas of *Oniscus asellus* and *Porcellio scaber* (Isopoda) from a Metal Gradient at Avonmouth, UK. *Ecotoxicology*, **13**, 787–796.

Schmidt, T. S., Kraus, J. M., Walters, D. M. and Wanty, R. B. (2013). Emergence Flux Declines Disproportionately to Larval Density along a Stream Metals Gradient. *Environmental Science & Technology*, **47**, 8784–8792.

Schofield, R. and Lefevre, H. (1989). High Concentrations of Zinc in the Fangs and Manganese in the Teeth of Spiders. *Journal of Experimental Biology*, **144**, 577–581.

Schofield, R., Lefevre, H. and Shaffer, M. (1989). Complementary Microanalysis of Zn, Mn and Fe in the Chelicera of Spiders and Scorpions Using Scanning MeV-ion and Electron Microprobes. *Nuclear Instruments and Methods in Physics Research*, **B40/41**, 698–701.

Schulz-Baldes, M. (1978). Lead Transport in the Common Mussel *Mytilus edulis*. In *Physiology and Behaviour of Marine Organisms*, ed. D. S. McLusky and A. J. Berry. Oxford: Pergamon Press, pp. 211–218.

(1989). The Sea-Skater *Halobates micans*: An Open Ocean Bioindicator for Cadmium Distribution in Atlantic Surface Waters. *Marine Biology*, **102**, 211–215.

Scott-Fordsmand, J. J. and Depledge, M. H. (1997). Changes in Tissue Concentrations and Contents of Calcium, Copper and Zinc in the Shore Crab *Carcinus maenas* (L.) (Crustacea: Decapoda) during the Moult Cycle and Following Copper Exposure during Ecdysis. *Marine Environmental Research*, **44**, 397–414.

Scottish Environment Protection Agency (2011). *Review of Metal Concentrations Data Held for Glengonnar Water and Wanlock Water, South Central Scotland*. Stirling: Scottish Environment Protection Agency.

Scullion, J. and Edwards, R. W. (1980a). The Effect of Coal Industry Pollutants on the Macro-Invertebrate Fauna of a Small River in the South Wales Coalfield. *Freshwater Biology*, **10**, 141–162.

(1980b). The Effect of Pollutants from the Coal Industry on the Fish Fauna of a Small River in the South Wales Coalfield. *Environmental Pollution A*, **21**, 141–153.

Segar, D. A., Collins, J. D. and Riley, J. P. (1971). The Distribution of the Major and Some Minor Elements in Marine Animals. Part II. Molluscs. *Journal of the Marine Biological Association of the United Kingdom*, **51**, 131–136.

Shacklette, H. T. and Boerngen, J. G. (1984). *Element Concentrations in Soils and Other Surficial Materials of the Conterminous United States*. US Geological Survey Professional Paper 1270. Alexandria, VA: US Geological Survey.

Shaw, W. T. (1975). *Mining in the Lake Counties*. Clapham, Lancaster: Dalesman Publishing Company Ltd.

Shelton, R. G. J. (1971). Sludge Dumping in the Thames Estuary. *Marine Pollution Bulletin*, **2**, 24–27.

Sherlock, E. (2012). *Key to the Earthworms of the UK and Ireland*. London: Field Studies Council and Natural History Museum.

Shi, D., Xu, Y., Hopkinson, M. and Morel, F. M. M. (2010). Effect of Ocean Acidification on Iron Availability to Marine Phytoplankton. *Science*, **327**, 676–679.

Shore, R., Carney, G. and Stygall, T. (1975). Cadmium Levels and Carbohydrate Metabolism in Limpets. *Marine Pollution Bulletin*, **6**, 187–189.

Shore, R. F. and Douben, P. E. T. (1994a). The Ecotoxicological Significance of Cadmium Intake and Residues in Terrestrial Small Mammals. *Ecotoxicology and Environmental Safety*, **29**, 101–112.

(1994b). Predicting Ecotoxicological Impacts of Environmental Contaminants on Terrestrial Small Mammals. *Reviews of Environmental Contamination and Toxicology*, **134**, 48–89.

Shore, R. F., Pereira, M. G., Walker, L. A. and Thompson, D. R. (2011). Mercury in Nonmarine Birds and Mammals. In *Environmental Contaminants in Biota: Interpreting Tissue Concentrations*, 2nd edn., ed. W. N. Beyer and J. P. Meador. Boca Raton, FL: Taylor and Francis Books, pp. 609–624.

Sick, L. V. and Baptist, G. J. (1979). Cadmium Incorporation by the Marine Copepod *Pseudodiaptomus coronatus*. *Limnology and Oceanography*, **24**, 453–462.

Simkiss, K. and Mason, A. Z. (1984). Cellular Responses of Molluscan Tissues to Environmental Metals. *Marine Environmental Research*, **14**, 103–118.

Simkiss, K., Taylor, M. G. and Greaves, G. N. (1990). Form of the Anion in the Intracellular Granules of the Crab. *Journal of Inorganic Biochemistry*, **39**, 17–23.

Simmonds, M. P., Johnston, P. A. and French, M. C. (1993). Organochlorine and Mercury Contamination in United Kingdom Seals. *Veterinary Record*, **132**, 291–295.

Simon, E. (1977). Cadmium Tolerance in Populations of *Agrostis tenuis* and *Festuca ovina*. *Nature*, **265**, 328–330.

Skorupa, J. P. (1998). Selenium Poisoning of Fish and Wildlife in Nature: Lessons from Twelve Real-World Examples. In *Environmental Chemistry of Selenium*, ed. W. T. Frankenberger Jr. and R. A. Engberg. New York, NY: Marcel Dekker Inc., pp. 315–354.

Slingsby, D. R. and Brown, D. H. (1977). Nickel in British Serpentine Soils. *Journal of Ecology*, **65**, 597–618.

Slingsby, D. R., Hopkins, J., Carter, S., Dalrymple, S. and Slingsby, A. (2010). Change and Stability: Monitoring the Keen of Hamar 1978–2006. Scottish Natural Heritage Report. www.snh.org.uk/pdfs/publications/nnr/Keen_of_Hamar_NNR_Change_and_Stability.pdf

Sloane, P. I. W. and Norris, R. H. (2003). Relationships of AUSRIVAS-Based Macroinvertebrate Predictive Model Outputs to a Metal Pollution Gradient. *Journal of the North American Benthological Society*, **22**, 457–471.

Smaldon, G., Holthuis, L. B. and Fransen, C. H. J. M. (1993). *Coastal Shrimps and Prawns. Synopses of the British Fauna (New Series) No. 15*, 2nd edn. Dorchester: Linnean Society of London and the Estuarine and Coastal Sciences Association, Henry Ling Ltd, Dorset Press.

Smith, B. S. (1980). The Estuarine Mud Snail, *Nassarius obsoletus*: Abnormalities in the Reproductive System. *Journal of Molluscan Studies*, **46**, 247–256.

(1981). Male Characteristics on Female Mud Snails Caused by Antifouling Bottom Paints. *Journal of Applied Toxicology*, **1**, 22–25.

Smith, C. W., Aptroot, A., Coppins, B. J., et al. (Eds) (2009). *The Lichens of Great Britain and Ireland*. Bodmin and King's Lynn: British Lichen Society, MPG Books Group.

Smith, R. A. H. and Bradshaw, A. D. (1972). Stabilisation of Toxic Mine Wastes by the Use of Tolerant Plant Populations. *Transactions of the Institute of Mining and Metallurgy, Section A: Mining Technology*, **81**, A230–A237.

(1979). The Use of Metal Tolerant Plant Populations for the Reclamation of Metalliferous Wastes. *Journal of Applied Ecology*, **16**, 595–612.

SNIFFER (2011) *River Fish Classification Tool: Science Work*. Phase 3 Report FINAL. Project WFD68c. Edinburgh: Scotland and Northern Ireland Forum for Environmental Research (SNIFFER).

Sohal, R. S., Peters, P. D. and Hall, T. A. (1977). Origin, Structure, Composition and Age-Dependence of Mineralized Dense Bodies (Concretions) in the Midgut Epithelium of the Adult Housefly, *Musca domestica*. *Tissue and Cell*, **9**, 87–102.

Somerfield, P. J., Atkins, M., Bolam, S. G., et al. (2006). Relative Impacts at Sites of Dredged-Material Relocation in the Coastal Environment: A Phylum-Level Meta-Analysis Approach. *Marine Biology*, **148**, 1231–1240.

Somerfield, P. J., Gee, J. M. and Warwick, R. M. (1994). Soft Sediment Meiofaunal Community Structure in Relation to a Long-Term Heavy Metal Gradient in the Fal Estuary System. *Marine Ecology Progress Series*, 105, 79–88.

Southgate, T., Slinn, D. J. and Eastham, J. F. (1983). Mine-Derived Metal Pollution in the Isle of Man. *Marine Pollution Bulletin*, 14, 137–140.

Southward, A. J. (2008). *Barnacles: Synopses of the British Fauna (New Series) No. 57*. Dorchester: Linnean Society of London and Estuarine and Coastal Sciences Association, Henry Ling, Ltd, Dorset Press.

Sparks, T. (Ed) (2000). *Statistics in Ecotoxicology*. Chichester: John Wiley & Sons Ltd.

Spence, D. H. N. (1970). Scottish Serpentine Vegetation. *Oikos*, 21, 22–31.

Spence, S. K., Bryan, G. W., Gibbs, P. E., Masters, D., Morris, L. and Hawkins, S. J. (1990). Effect of TBT Contamination on *Nucella* Populations. *Functional Ecology*, 4, 425–432.

Spurgeon, D. J. and Hopkin, S. P. (1995). Extrapolation of the Laboratory Based OECD Earthworm Toxicity Test to Metal Contaminated Field Sites. *Ecotoxicology*, 4, 190–205.

(1996a). The Effects of Metal Contamination on Earthworm Populations around a Smelting Works: Quantifying Species Effects. *Applied Soil Ecology*, 4, 147–160.

(1996b). Risk Assessment of the Threat of Secondary Poisoning by Metals to Predators of Earthworms in the Vicinity of a Primary Smelting Works. *Science of the Total Environment*, 187, 167–183.

(1999a). Tolerance to Zinc in Populations of the Earthworm *Lumbricus rubellus* from Uncontaminated and Metal-Contaminated Ecosystems. *Archives of Environmental Contamination and Toxicology*, 37, 332–337.

(1999b). Comparisons of Metal Accumulation and Excretion Kinetics in Earthworms (*Eisenia fetida*) Exposed to Contaminated Field and Laboratory Soils. *Applied Soil Ecology*, 11, 227–243.

Spurgeon, D. J., Hopkin, S. P. and Jones, D. T. (1994). Effects of Cadmium, Copper, Lead and Zinc on Growth, Reproduction and Survival of the Earthworm *Eisenia fetida* (Savigny): Assessing the Environmental Impact of Point-Source Metal Contamination in Terrestrial Ecosystems. *Environmental Pollution*, 84, 123–130.

Spurgeon, D. J., Rowland, P., Ainsworth, G., Rothery, P., Long, S. and Black, H. I. J. (2008). Geographical and Pedological Drivers of Distribution and Risks to Soil Fauna of Seven Metals (Cd, Cu, Cr, Ni, Pb, V and Zn) in British Soils. *Environmental Pollution*, 153, 273–283.

Spurgeon, D. J., Weeks, J. M. and van Gestel, C. A. M. (2003). A Summary of Eleven Years Progress in Earthworm Ecotoxicology. *Pedobiologia*, 47, 588–606.

Statham, P. J., Auger, Y., Burton, J. D., et al. (1993). Fluxes of Cd, Co, Cu, Fe, Mn, Ni, Pb, and Zn through the Strait of Dover into the Southern North Sea. *Oceanologica Acta*, 16, 541–552.

Steele, J. H., McIntyre, A. D., Johnston, R., Baxter, I. G., Topping, G. and Dooley, H. D. (1973). Pollution in the Clyde Sea Area. *Marine Pollution Bulletin*, 4, 153–157.

Sterenborg, I. and Roelofs, D. (2003). Field-Selected Cadmium Tolerance in the Springtail *Orchesella cincta* Is Correlated with Increased Metallothionein mRNA Expression. *Insect Biochemistry and Molecular Biology*, 33, 741–747.

Stewart, F. M. and Furness, R. W. (1998). The Influence of Age on Cadmium Concentrations in Seabirds. *Environmental Monitoring and Assessment*, 50, 159–171.

Stewart, F. M., Thompson, D. R., Furness, R. W. and Harrison, N. (1994). Seasonal Variation in Heavy Metal Levels in Tissues of Common Guillemots, *Uria aalge* from Northwest Scotland. *Archives of Environmental Contamination and Toxicology*, 27, 168–175.

Stoecker, D. (1980a). Relationships between Chemical Defense and Ecology in Benthic Ascidians. *Marine Ecology Progress Series*, 3, 257–265.

(1980b). Chemical Defenses of Ascidians against Predators. *Ecology*, 61, 1327–1334.

Stürzenbaum, S. R., Kille, P. and Morgan, A. J. (1998). The Identification, Cloning and Characterization of Earthworm Metallothionein. *FEBS Letters*, 431, 437–442.

Sunda, W. G. (1989). Trace Metal Interactions with Phytoplankton. *Biological Oceanography*, 6, 41–442.

Sunda, W. G. and Huntsman, S. A. (1992). Feedback Interactions between Zinc and Phytoplankton in Seawater. *Limnology and Oceanography*, 37, 25–40.

(1998). Processes Regulating Cellular Metal Accumulation and Physiological Effects: Phytoplankton as Model Systems. *Science of the Total Environment*, **219**, 165–181.

Sunda, W. G., Swift, D. G. and Huntsman, S. A. (1991). Low Iron Requirement for Growth in Oceanic Phytoplankton. *Nature*, **351**, 55–57.

Taberlet, P., Coissac, E., Hajibabaei, M. and Rieseberg, L. H. (2012). Environmental DNA. *Molecular Ecology*, **21**, 1789–1793.

Tappin, A. D., Barriada, J. L., Braungardt, C. B., Evans, E. H., Patey, M. D. and Achterberg, E. P. (2010). Dissolved Silver in European Estuarine and Coastal Waters. *Water Research*, **44**, 4204–4216.

Tappin, A. D., Millward, G. E., Statham, P. J., Burton, J. D. and Morris, A. W. (1995). Trace Metals in the Central and Southern North Sea. *Estuarine, Coastal and Shelf Science*, **41**, 275–323.

Taylor, A. C. and Spicer, J. I. (1986). Oxygen-Transporting Properties of the Blood of Two Semi-Terrestrial Amphipods, *Orchestia gammarellus* (Pallas) and *O. mediterranea* (Costa). *Journal of Experimental Marine Biology and Ecology*, **97**, 135–150.

Taylor, D. (1982). Distribution of Heavy Metals in the Water of a Major Industrialised Estuary. *Environmental Technology Letters*, **3**, 137–144.

Taylor, M. D. (1997). Accumulation of Cadmium Derived from Fertilisers in New Zealand Soils. *Science of the Total Environment*, **208**, 123–126.

Tessier, A., Campbell, P. G. C. and Bisson, M. (1979). Sequential Extraction Procedure for the Speciation of Particulate Trace Metals. *Analytical Chemistry*, **51**, 844–854.

Tête, N., Afonso, E., Crini, N., Drouhot, S., Prudent, A.-S. and Scheifler, R. (2014a). Hair as a Noninvasive Tool for Risk Assessment: Do the Concentrations of Cadmium and Lead in the Hair of Woodmice (*Apodemus sylvaticus*) Reflect Internal Concentrations? *Ecotoxicology and Environmental Safety*, **108**, 233–241.

Tête, N., Durfort, M., Rieffel, D., Scheifler, R. and Sánchez-Chardi, A. (2014b). Histopathology Related to Cadmium and Lead Bioaccumulation in Chronically Exposed Wood Mice, *Apodemus sylvaticus*, around a Former Smelter. *Science of the Total Environment*, **481**, 167–177.

Thain, J. E. and Waldock, M. J. (1986). The Impact of Tributyl Tin (TBT) Antifouling Paints on Molluscan Fisheries. *Water Science and Technology*, **18**, 193–202.

Thompson, D. R. (1996). Mercury in Birds and Terrestrial Mammals. In *Environmental Contaminants in Wildlife: Interpreting Tissue Concentrations*, ed. W. N. Beyer, G. H. Heinz and A. W. Redmon-Norwood. Boca Raton, FL: Lewis Publishers, pp. 341–356.

Thompson, D. R., Becker, P. H. and Furness, R. W. (1993). Long-Term Changes in Mercury Concentrations in Herring Gulls *Larus argentatus* and Common Terns *Sterna hirundo* from the German North Sea Coast. *Journal of Applied Ecology*, **30**, 316–320.

Thompson, D. R., Furness, R. W. and Barrett, R. T. (1992a). Mercury Concentrations in Seabirds from Colonies in the Northeast Atlantic. *Archives of Environmental Contamination and Toxicology*, **23**, 383–389.

Thompson, D. R., Furness, R. W. and Walsh, P. M. (1992b). Historical Changes in the Marine Ecosystem of the North and North-East Atlantic Ocean as Indicated by Seabird Feathers. *Journal of Applied Ecology*, **29**, 79–84.

Thompson, D. R., Hamer, K. C. and Furness, R. W. (1991). Mercury Accumulation in Great Skuas *Catharacta skua* of Known Age and Sex, and Its Effects upon Breeding and Survival. *Journal of Applied Ecology*, **28**, 672–684.

Thornton, I. (1975a). Some Aspects of Environmental Geochemistry in Britain. In *Symposium Proceedings of International Conference on Heavy Metals in the Environment, Toronto, Ontario, Canada, October 27–31, 1975*, vol. 2. Toronto: Institute for Environmental Studies, University of Toronto, pp. 17–38.

(1975b). Geochemical Parameters in the Assessment of Estuarine Pollution. In *The Ecology of Resource Degradation and Renewal. The 15th Symposium of the British Ecological Society, July 1973*, ed. M. J. Chadwick and G. T. Goodman. Oxford: Blackwell Scientific Publications, pp. 157–169.

(1993). Environmental Geochemistry and Health in the 1990s: A Global Perspective. *Applied Geochemistry*, Suppl. Issue 2, 203–210.

Thornton, I., Watling, H. and Darracott, A. (1975). Geochemical Studies in Several Rivers and Estuaries

Used for Oyster Rearing. *Science of the Total Environment*, 4, 325–345.

Thornton, I. and Webb. J. S. (1979). Geochemistry and Health in the United Kingdom. *Philosophical Transactions of the Royal Society London B*, 288, 151–168.

Thornton, I., Culbard, E., Moorcroft, S., Watt, J., Wheatley, M. and Thompson, M. (1985). Metals in Urban Dusts and Soils. *Environmental Technology Letters*, 6, 137–144.

Tipping, E. (1994). WHAM: A Chemical Equilibrium Model and Computer Code for Waters, Sediments, and Soils Incorporating a Discrete Site/Electrostatic Model of Ion-Binding by Humic Substances. *Computers and Geoscience*, 20, 973–1023.

Tipping, E., Lofts, S. and Lawlor, A. J. (1998). Modelling the Chemical Speciation of Trace Metals in the Surface Waters of the Humber System. *Science of the Total Environment*, 210, 63–77.

Tipping, E., Lofts, S. and Sonke, J. E. (2011a). Humic Ion-Binding Model VII: A Revised Parameterisation of Cation-Binding by Humic Substances. *Environmental Chemistry*, 8, 225–235.

Tipping, E., Poskitt, J. M., Lawlor, A. J., Wadsworth, R. A., Norris, D. A. and Hall, J. R. (2011b). Mercury in United Kingdom Topsoils: Concentrations, Pools, and Critical Limit Exceedances. *Environmental Pollution*, 159, 3721–3729.

Tlili, A. and Montuelle, B. (2011). Microbial Pollution-Induced Community Tolerance. In *Tolerance to Environmental Contaminants*, ed. C. Amiard-Triquet, P. S. Rainbow and M. Roméo. Boca Raton, FL: CRC Press, pp. 85–108.

Topping G. (1973). Heavy Metals in Shellfish from Scottish Waters. *Aquaculture*, 1, 379–384.

Towe, K. M. and Lowenstam, H. A. (1967). Ultrastructure and Development of Iron Mineralization in the Radular Teeth of *Cryptochiton stelleri* (Mollusca). *Journal of Ultrastructure Research*, 17, 1–13.

Towe, K. M., Lowenstam, H. A. and Nesson, M. H. (1963). Invertebrate Ferritin: Occurrence in Mollusca. *Science*, 142, 63–64.

Townsend, C. R., Hildrew, A. G. and Francis, J. (1983). Community Structure in Some Southern English Streams: The Influence of Physicochemical Factors. *Freshwater Biology*, 13, 521–544.

Truchet, M., Martoja, R. and Berthet, B. (1990). Conséquences histologiques de la pollution métalliques d'un estuaire sur deux mollusques, *Littorina littorea* L. et *Scrobicularia plana* da Costa. *Comptes Rendus de l'Académie des Sciences, Série III*, 311, 261–268.

Tschan, M., Robinson, B. H. and Schulin, R. (2009). Antimony in the Soil-Plant System: A Review. *Environmental Chemistry*, 6, 106–115.

Turnbull, L. (2006). *The History of Lead Mining in the North East of England*. Hexham: Ergo Press.

Turner, A. (2010). Marine Pollution from Antifouling Paint Particles. *Marine Pollution Bulletin*, 60, 159–171.

(2011). Oral Bioaccessibility of Trace Metals in Household Dust: A Review. *Environmental Geochemistry and Health*, 33, 331–341.

Turner, D. R., Whitfield, M. and Dickson, A. G. (1981). The Equilibrium Speciation of Dissolved Components in Freshwater and Seawater at 25°C and 1 atm Pressure. *Geochimica et Cosmochimica Acta*, 45, 855–881.

Tyler, G. (1990). Bryophytes and Heavy Metals: A Literature Review. *Botanical Journal of the Linnean Society*, 104, 231–253.

Ueki, T., Adachi, T., Kawano, S., Aoshima, M., Yamaguchi, N., Kanamori, K. and Michibata, H. (2003). Vanadium-Binding Proteins (Vanabins) from a Vanadium-Rich Ascidian *Ascidia sydneiensis samea*. *Biochimica et Biophysica Acta*, 1626, 43–50.

Underhill-Day, J. and Dyrynda, P. (2005). Non-Native Species in and around Poole Harbour. In *The Ecology of Poole Harbour*, ed. J. Humphreys and V. May. Amsterdam: Elsevier, pp. 159–162.

Underwood, E. J. (1962). *Trace Elements in Human and Animal Nutrition*, 2nd edn. New York, NY, and London: Academic Press.

Valentini, A., Taberlet, P., Miaud, C., et al. (2016). Next-Generation Monitoring of Aquatic Biodiversity Using Environmental DNA Metabarcoding. *Molecular Ecology*, 24, 929–942.

Van de Ven, W. S. M., Koeman, J. H. and Svenson, A. (1979). Mercury and Selenium in Wild and Experimental Seals. *Chemosphere*, 8, 539–555.

Van der Wal, P. (1989). Structural and Material Design of Mature Mineralized Radula Teeth of *Patella vulgata* (Gastropoda). *Journal of Ultrastructural and Molecular Structure Research*, 102, 147–161.

van Gestel, C. A. M. (2008). Physico-Chemical and Biological Parameters Determine Metal Bioavailability in Soils. *Science of the Total Environment*, **406**, 385–395.

van Straalen, N. M. (1993). Soil and Sediment Quality Criteria Derived from Invertebrate Toxicity Data. In *Ecotoxicology of Metals in Invertebrates*, ed. R. Dallinger and P. S. Rainbow. Boca Raton, FL: Lewis Publishers, pp. 427–441.

(1998). Evaluation of Bioindicator Systems Derived from Soil Arthropod Communities. *Applied Soil Ecology*, **9**, 429–437.

van Straalen, N. M. and Denneman, C. A. J. (1989). Ecotoxicological Evaluation of Soil Quality Criteria. *Ecotoxicology and Environmental Safety*, **18**, 241–251.

van Straalen, N. M. and Roelofs, D. (2005). Cadmium Tolerance in a Soil Arthropod: A Model of Real-Time Microevolution. *Entomologische Berichten*, **65**, 105–110.

van Straalen, N. M., Burghouts, T. B. A., Doornhof, M. J., et al. (1987). Efficiency of Lead and Cadmium Excretion in Populations of *Orchesella cincta* (Collembola) from Various Contaminated Forest Soils. *Journal of Applied Ecology*, **24**, 953–968.

van Straalen, N. M., Donker, M. H., Vijver, M. G. and van Gestel, C. A. M. (2005). Bioavailability of Contaminants Estimated from Uptake Rates into Soil Invertebrates. *Environmental Pollution*, **136**, 409–417.

Viarengo, A., Pertica, M., Mancinelli, G., et al. (1984). Possible Role of Lysosomes in the Detoxication of Copper in the Digestive Gland Cells of Metal-Exposed Mussels. *Marine Environmental Research*, **14**, 469–470.

Viarengo, A., Zanicchi, G., Moore, M. N. and Orunesu, M. (1981). Accumulation and Detoxication of Copper by the Mussel, *Mytilus galloprovincialis* Lam.: A Study of the Subcellular Distribution in the Digestive Gland Cells. *Aquatic Toxicology*, **1**, 147–157.

Vijver, M. G., Vink, J. P. M., Miermans, C. J. H. and van Gestel, C. A. M. (2003). Oral Sealing Using Glue: A New Method to Distinguish between Intestinal and Dermal Uptake of Metals in Earthworms. *Soil Biology and Biochemistry*, **35**, 125–132.

Vilas Boas, L., Gonçalves, S. C., Portugal, A., Freitas, H. and Gonçalves, M. T. (2014). A Ni Hyperaccumulator and a Congeneric Non-Accumulator Reveal Equally Effective Defences against Herbivory. *Science of the Total Environment*, **466–467**, 11–15.

Vinceti, M., Crespi, C. M., Bonvicini, F., et al. (2013). The Need for Reassessment of the Safe Upper Limit of Selenium in Drinking Water. *Science of the Total Environment*, **443**, 633–642.

Virsek, M. K., Hubad, B. and Lapanje, A. (2013). Mercury Induced Community Tolerance in Microbial Films Is Related to Pollution Gradients in a Long-Term Polluted River. *Aquatic Toxicology*, **144–145**, 208–217.

Vivian, C. M. G. (1980). Trace Metal Studies in the River Tawe and Swansea Bay. In *Industrialised Embayments and Their Environmental Problems*, ed. M. B. Collins, F. T. Banner, P. A. Tyler, S. J. Wakefield and A. E. James. Oxford: Pergamon Press, pp. 329–341.

Vivian, C. M. G. and Massie, K. S. (1977). Trace Metals in Waters and Sediments of the River Tawe, South Wales, in Relation to Local Sources. *Environmental Pollution*, **14**, 47–61.

Voulvoulis, N., Scrimshaw, M. D. and Lester, J. N. (2000). Occurrence of Four Biocides Utilized in Antifouling Paints, as Alternatives to Organotin Compounds, in Waters and Sediments of a Commercial Estuary in the UK. *Marine Pollution Bulletin*, **40**, 938–946.

Wadge, A., Hutton, M. and Peterson, P. J. (1986). The Concentrations and Particle Size Relationships of Selected Trace Elements in Fly Ashes from U.K. Coal-Fired Power Plants and a Refuse Incinerator. *Science of the Total Environment*, **54**, 13–27.

Walker, G. (1977). 'Copper' Granules in the Barnacle *Balanus balanoides*. *Marine Biology*, **39**, 343–349.

(1977b). Personal communications.

Walker, G., Rainbow, P. S., Foster, P. and Holland, D. L. (1975). Zinc Phosphate Granules in Tissue Surrounding the Midgut of the Barnacle *Balanus balanoides*. *Marine Biology*, **33**, 161–166.

Walker, L. A., Simpson, V. R., Rockett, L., Wienburg, C. L. and Shore, R. F. (2007). Heavy Metal Contamination in Bats in Britain. *Environmental Pollution*, **148**, 483–490.

Wallace, W. G. and Lopez, G. R. (1996). Relationship between the Subcellular Cadmium Distribution in Prey and Cadmium Transfer to a Predator. *Estuaries*, **19**, 923–930.

(1997). Bioavailability of Biologically Sequestered Cadmium and the Implications of Metal Detoxification. *Marine Ecology Progress Series*, **147**, 149–157.

Wallace, W. G. and Luoma, S. N. (2003). Subcellular Compartmentalization of Cd and Zn in Two Bivalves. II. The Significance of Trophically Available Metal (TAM). *Marine Ecology Progress Series*, **257**, 125–137.

Wallace, W. G., Lee, B. G. and Luoma, S. N. (2003). Subcellular Compartmentalization of Cd and Zn in Two Bivalves. I. Significance of Metal-Sensitive Fractions (MSF) and Biologically Detoxified Metal (BDM). *Marine Ecology Progress Series*, **249**, 183–197.

Wallace, W. G., Lopez, G. R. and Levinton, J. S. (1998). Cadmium Resistance in an Oligochaete and Its Effect on Cadmium Trophic Transfer to an Omnivorous Shrimp. *Marine Ecology Progress Series*, **172**, 225–237.

Walley, K. A., Khan, M. S. I. and Bradshaw, A. D. (1974). The Potential for Evolution of Heavy Metal Tolerance in Plants. I. Copper and Zinc Tolerance in *Agrostis tenuis*. *Heredity*, **32**, 309–319.

Walsh, A. R. and O'Halloran, J. (1998). Accumulation of Chromium by a Population of Mussels (*Mytilus edulis* (L.)) Exposed to Leather Tannery Effluent. *Environmental Toxicology and Chemistry*, **17**, 1429–1438.

Wang, M.-J. and Wang, W.-X. (2009). Cadmium in Three Marine Phytoplankton: Accumulation, Subcellular Fate and Thiol Induction. *Aquatic Toxicology*, **95**, 99–107.

Wang, W.-X. (2002). Interactions of Trace Metals and Different Marine Food Chains. *Marine Ecology Progress Series*, **243**, 295–309.

(2012). Biodynamic Understanding of Mercury Accumulation in Marine and Freshwater Fish. *Advances in Environmental Research*, **1**, 15–35.

Wang, W.-X. and Fisher, N. S. (1998). Accumulation of Trace Elements in a Marine Copepod. *Limnology and Oceanography*, **43**, 273–283.

Wang, W.-X. and Rainbow, P. S. (2006). Subcellular Partitioning and the Prediction of Cadmium Toxicity to Aquatic Organisms. *Environmental Chemistry*, **3**, 395–399.

Wang, W.-X., Fisher, N. S. and Luoma, S. N. (1995). Assimilation of Trace Elements Ingested by the Mussel *Mytilus edulis*: Effects of Algal Food Abundance. *Marine Ecology Progress Series*, **129**, 165–176.

(1996). Kinetic Determinations of Trace Element Bioaccumulation in the Mussel *Mytilus edulis*. *Marine Ecology Progress Series*, **140**, 91–113.

Wang, W.-X., Qiu, J.-W. and Qian, P. Y. (1999). Significance of High Trophic Transfer in Predicting the High Concentration of Zinc in Barnacles. *Environmental Science and Technology*, **33**, 2905–2909.

Wang, X. and Wang, W.-X. (2015). Physiologically Based Pharmacokinetic Model for Inorganic and Methylmercury in a Marine Fish. *Environmental Science and Technology*, **49**, 10173–10181.

Wardlaw, J. (2005). Water Quality and Pollution Monitoring in Poole Harbour. In *The Ecology of Poole Harbour*, ed. J. Humphreys and V. May. Amsterdam: Elsevier, pp. 219–222.

Waring, J. and Maher, W. (2005). Arsenic Bioaccumulation and Species in Marine Polychaeta. *Applied Organometallic Chemistry*, **19**, 917–929.

Warwick, R. M. (2001). Evidence for the Effects of Metal Contamination on the Intertidal Macrobenthic Assemblages of the Fal Estuary. *Marine Pollution Bulletin*, **42**, 145–148.

Warwick, W. F. (1988). Morphological Deformities in Chironomidae (Diptera) Larvae as Biological Indicators of Toxic Stress. In *Toxic Contaminants and Ecosystem Health: A Great Lakes Focus*, ed. M. S. Evans. New York, NY: J Wiley & Sons, Inc., pp. 281–320.

Wayland, M. and Scheuhammer, A. M. (2011). Cadmium in Birds. In *Environmental Contaminants in Biota: Interpreting Tissue Concentrations*, 2nd edn., ed. W. N. Beyer and J. P. Meador. Boca Raton, FL: Taylor and Francis Books, pp. 645–666.

Webb, D. A. (1939). Observations on the Blood of Certain Ascidians, with Special Reference to the Biochemistry of Vanadium. *Journal of Experimental Biology*, **16**, 499–523.

Webb, J. S., Thornton, I., Thompson, M., Howarth, R. J. and Lowenstein, P. L. (1978). *The Wolfson Geochemical Atlas of England and Wales*. Oxford: Oxford University Press.

Weeks, J. M. (1992a). The Use of the Terrestrial Amphipod *Arcitalitrus dorrieni* (Crustacea: Amphipoda: Talitridae) as a Potential Biomonitor of Ambient Zinc and Copper Availabilities in Leaf-Litter. *Chemosphere*, **24**, 1505–1522.

(1992b). The Talitrid Amphipod (Crustacea) *Platorchestia platensis* as a Biomonitor of Trace Metals (Cu and Zn) in Danish Waters. In *Physiological and Biochemical Strategies in Baltic Organisms: New Approaches in Ecotoxicological Research. Proceedings of the 12th Baltic Marine Biologists Symposium, Helsingør, Denmark, 25–30 August 1991*, ed. E. Bjørnestad, L. Hagerman and K. Jensen. Fredensborg: Olsen & Olsen, pp. 173–178.

Weeks, J. M and Rainbow, P. S. (1993). The Relative Importance of Food and Seawater as Sources of Copper and zinc to Talitrid Amphipods (Crustacea; Amphipoda; Talitridae). *Journal of Applied Ecology*, **30**, 722–735.

Weeks, J. M., Rainbow, P. S. and Moore, P. G. (1992). The Loss, Uptake and Tissue Distribution of Copper and Zinc during the Moult Cycle in an Ecological Series of Talitrid Amphipods (Crustacea: Amphipoda). *Hydrobiologia*, **245**, 15–25.

WFD-UKTAG (2008a). *UKTAG River Assessment Methods. Benthic Invertebrate Fauna. River Invertebrate Classification Tool (RICT)*. Edinburgh: Water Framework Directive–United Kingdom Advisory Group, SNIFFER.

(2008b). *Proposals for Environment Quality Standards for Annex VIII Substances*. Final Report SR1–2007. Edinburgh: Water Framework Directive–United Kingdom Advisory Group.

(2009). *UKTAG River Assessment Methods. Macrophytes and Phytobenthos. Macrophytes (River LEAFPACS)*. Edinburgh: Water Framework Directive-United Kingdom Advisory Group, SNIFFER.

(2013). *Updated Recommendations on Environmental Standards. River Basin Management (2015–21)*. Final Report November 2013 (Minor amendments January 2014). Report 2004112013. Edinburgh: Water Framework Directive–United Kingdom Advisory Group.

Wheeler, A. C. (1979). *The Tidal Thames. The History of a River and Its Fishes*. London: Routledge and Kegan Paul.

White, D. H., Finley, M. T. and Ferrell, J. F. (1978). Histopathological Effects of Dietary Cadmium on Kidneys and Testes of Mallard Ducks. *Journal of Toxicology and Environmental Health*, **4**, 551–558.

White, S. L. and Rainbow, P. S. (1982). Regulation and Accumulation of Copper, Zinc and Cadmium by the Shrimp *Palaemon elegans*. *Marine Ecology Progress Series*, **8**, 95–101.

(1984). Regulation of Zinc Concentration by *Palaemon elegans* (Crustacea: Decapoda): Zinc Flux and Effects of Temperature, Zinc Concentration and Moulting. *Marine Ecology Progress Series*, **16**, 135–147.

(1987). Heavy Metal Concentrations and Size Effects in the Mesopelagic Decapod Crustacean *Systellaspis debilis*. *Marine Ecology Progress Series*, **37**, 147–151.

Whitehead, P. G. and Neal, C. (2005). The Wheal Jane Remediation Study: Some General Conclusions. *Science of the Total Environment*, **338**, 155–157.

Whitehead, P. G. and Prior, H. (2005). Bioremediation of Acid Mine Drainage: An Introduction to the Wheal Jane Wetlands Project. *Science of the Total Environment*, **338**, 15–21.

Whiteley, J. D. and Pearce, N. J. G. (2003). Metal Distribution during Diagenesis in the Contaminated Sediments of Dulas Bay, Anglesey, N. Wales, UK. *Applied Geochemistry*, **18**, 901–913.

Whitfield, M. (2001). Interactions between Phytoplankton and Trace Metals in the Ocean. *Advances in Marine Biology*, **41**, 1–128.

Whiting, S. N., Neumann, P. M. and Baker, A. J. M. (2003). Nickel and Zinc Hyperaccumulation by *Alyssum murale* and *Thlaspi caerulescens* (Brassicaceae) Do Not Enhance Survival and Whole-Plant Growth under Drought Stress. *Plant, Cell and Environment*, **26**, 351–360.

Whitton, B. A. (1970). Toxicity of Heavy Metals to Freshwater Algae: A Review. *Phykos*, **9**, 116–125.

Whitton, B. A., Say, P. J. and Wehr, J. D. (1981). Use of Plants to Monitor Heavy Metals in Rivers. In *Heavy Metals in Northern England: Environmental and Biological Aspects*, ed. P. J. Say and B. A. Whitton. Durham: Department of Botany, University of Durham, pp. 135–145.

Widdows, J., Donkin, P., Brinsley, M. D., Evans, S. V., Salkeld, P. N., Franklin, A., Law, R. J. and Waldock, M. J. (1995). Scope for Growth and Contaminant Levels in North Sea Mussels. *Marine Ecology Progress Series*, **127**, 131–148.

Widdows, J., Donkin, P., Staff, F. J., et al. (2002). Measurement of Stress Effects (Scope for Growth) and Contaminant Levels in Mussels (*Mytilus edulis*) Collected from the Irish Sea. *Marine Environmental Research*, **53**, 327–356.

Wildish, D. J. (1987). Estuarine Species of *Orchestia* (Crustacea: Amphipoda: Talitroidea) from Britain. *Journal of the Marine Biological Association of the United Kingdom*, **67**, 571–583.

Wilkes, R., Bennion, M., McQuaid, N., et al. (2017). Intertidal seagrass in Ireland: Pressures, WFD Status and an Assessment of Trace Metal Contamination in Intertidal Habitats Using *Zostera noltei*. *Ecological Indicators*, **82**, 117–130.

Wilkins, P. (1977). Observations on Ecology of *Mielichhoferia elongata* and Other 'Copper Mosses' in the British Isles. *Bryologist*, **80**, 175–181.

Williams, P. R., Attrill, M. J. and Nimmo, M. (1998). Heavy Metal Concentrations and Bioaccumulation within the Fal Estuary, UK: A Reappraisal. *Marine Pollution Bulletin*, **36**, 643–645.

Williams, R. J. P. and Fraústo da Silva, J. J. R. (1996). *The Natural Selection of the Elements*. Oxford: Clarendon Press.

(2003). Evolution Was Chemically Constrained. *Journal of Theoretical Biology*, **220**, 323–343.

Williamson, P. and Evans, P. R. (1973). A Preliminary Study of the Effects of High Levels of Inorganic Lead on Soil Fauna. *Pedobiologia*, **13**, 16–21.

Willies, L. (1982). *Lead and Leadmining*. Princes Risborough: Shire Publications Ltd.

Wilson, G. V. (1921). The Lead, Zinc, Copper and Nickel Ores of Scotland. *Special Reports on the Mineral Resources of Great Britain*, **XVII**, 1–159.

Wilson, J. G. (1982). Heavy Metals in *Littorina rudis* along a Copper Contamination Gradient. *Journal of Life Sciences Royal Dublin Society*, **4**, 27–35.

Wilson, K. W., Head, P. C. and Jones, P. D. (1986). Mersey Estuary (U.K.) Bird Mortalities: Causes, Consequences and Correctives. *Water Science and Technology*, **18**, 171–180.

Wolfe, M. F., Schwarzbach, S. and Sulaiman, R. A. (1998). Effects of Mercury on Wildlife: A Comprehensive Review. *Environmental Toxicology and Chemistry*, **17**, 146–160.

Wong, M. H. (1982). Metal Cotolerance to Copper, Lead and Zinc in *Festuca rubra*. *Environmental Research*, **29**, 42–47.

Woods, R. (1988). Further Neglected Habitats. *British Lichen Society Bulletin*, **65**, 15–16.

Woods Hole Oceanographic Institution (1952). *Marine Fouling and Its Prevention. United States Naval Institute, Annapolis, Maryland*. Menasha, WI: George Banta Publishing Company.

Wragg, J., Cave, M. R., Basta, N., et al. (2011). An Inter-Laboratory Trial of the Unified BARGE Bioaccessibility Method for Arsenic, Cadmium and Lead in the Soil. *Science of the Total Environment*, **409**, 4016–4030.

Wright, D. A. (1976). Heavy Metals in Animals from the North East Coast. *Marine Pollution Bulletin*, **7**, 36–38.

Wright, G., Misstear, B., Gallagher, V., Suilleabhain, D. O. and O'Connor, P. (1999). Avoca Mines: Uncontrolled Acid Mine Drainage in Ireland. In *Mine, Water and Environment for the 21st Century, Vol. II: Proceedings of the 1999 International Mine Water Association IMWA International Congress, Seville, Spain: Mine/Quarry Waste Disposal and Closure*, ed. R. F. Fernández Rubio. Wendelstein: IMWA, pp. 551–556.

Wright, J. F., Moss, D., Armitage, P. D. and Furse, M. T. (1984). A Preliminary Classification of Running Water Sites in Great Britain Based on Macroinvertebrate Species and the Prediction of Community Type Using Environmental Data. *Freshwater Biology*, **14**, 221–256.

Wu, L. and Bradshaw, A. D. (1972). Aerial Pollution and the Rapid Evolution of Copper Tolerance. *Nature*, **238**, 167–169.

Wu, Y. and Wang, W.-X. (2014). Intracellular Speciation and Transformation of Inorganic Mercury in Marine Phytoplankton. *Aquatic Toxicology*, **148**, 122–129.

Xu, J. and Thornton, I. (1985). Arsenic in Garden Soils and Vegetable Crops in Cornwall, England: Implications for Human Health. *Environmental Geochemistry and Health*, **7**, 131–133.

Xu, Y. and Wang, W.-X. (2001). Individual Responses of Trace-Element Assimilation and Physiological Turnover by the Marine Copepod *Calanus sinicus* to Changes in Food Quantity. *Marine Ecology Progress Series*, **218**, 227–238.

(2002). The Assimilation of Detritus-Bound Metals by the Marine Copepod *Acartia spinicauda*. *Limnology and Oceanography*, **47**, 604–610.

Yamaguchi, N., Gazzard, D., Scholey, G. and Macdonald, D. W. (2003). Concentrations and Hazard Assessment of PCBs, Organochlorine Pesticides and Mercury in Fish Species from the Upper Thames: River Pollution and Its Potential Effects on Top Predators. *Chemosphere*, **50**, 265–273.

Yon, J.-N. and Lead, J. R. (2008). Manufactured Nanoparticles: An Overview of Their Chemistry, Interactions and Potential Environmental Implications. *Science of the Total Environment*, **400**, 396–414.

Yonge, C. M. (1960). *Oysters*. London: New Naturalist, Collins.

Yonge, C. M. and Thompson, T. E. (1976). *Living Marine Molluscs*. London: Collins.

Young, M. L. (1977). The Roles of Food and Direct Uptake from Water in the Accumulation of Zinc and Iron in the Tissues of the Dogwhelk, *Nucella lapillus* (L.). *Journal of Experimental Marine Biology and Ecology*, **30**, 315–325.

Younger, P. L. (1998). Adit Hydrology in the Long-Term: Observations from the Pb-Zn Mines of Northern England. In *Mine Water and Environmental Impacts 2. Proceedings of the International Mine Water Association Symposium, Johannesburg, South Africa*, ed. P. J. L. Nel. Sydney: IMWA, Cape Breton University, pp. 321–330.

Zauke, G.-P., Krause, M. and Weber, A. (1996). Trace Metals in Mesoplankton of the North Sea: Concentrations in Different Taxa and Preliminary Results on Bioaccumulation in Copepod Collectives. *Internationale Revue der gesamten Hydrobiologie und Hydrographie*, **81**, 141–160.

Zhang, G., Fang, X. Guo, X., et al. (2012). The Oyster Genome Reveals Stress Adaptation and Complexity of Shell Formation. *Nature*, **490**, 49–54.

Zhou, J. L., Liu, Y. P. and Abrahams, P. W. (2003). Trace Metal Behaviour in the Conwy Estuary, North Wales. *Chemosphere*, **51**, 429–440.

Zonfrillo, B., Sutcliffe, R., Furness, R. W. and Thompson, D. R. (1988). Notes on a Risso's Dolphin from Argyll with Analyses of Its Stomach Contents and Mercury Levels. *Glasgow Naturalist*, **21**, 297–303.

INDEX